AUTOCAD 2010
A PROBLEM-SOLVING APPROACH

Sham Tickoo

D1088279

DELMAR
CENGAGE Learning

Australia • Brazil • Japan • Korea • Mexico • Singapore • Spain • United Kingdom • United States

AutoCAD 2010:
A Problem-Solving Approach
Sham Tickoo

Vice President, Career and Professional Editorial: Dave Garza

Director of Learning Solutions: Sandy Clark

Acquisitions Editor: Stacy Masucci

Managing Editor: Larry Main

Senior Product Manager: John Fisher

Senior Editorial Assistant: Dawn Daugherty

Vice President Marketing, Career and Professional: Jennifer Baker

Executive Marketing Manager: Deborah S. Yarnell

Marketing Manager: Jimmy Stephens

Associate Marketing Manager: Mark Pierro

Production Director: Wendy Troeger

Production Manager: Mark Bernard

Art Director: David Arsenault

Content Project Manager: Angela Iula

Technology Project Manager: Joe Pliss

Production Technology Analyst: Tom Stover

Technical Editors:
  D. Saravanan
  Swapna D.
  Vivek Singh
  (CADCIM Technologies)

Copy Editors:
  Anju Jethwani
  Sudam C. Mishra
  Rajendar Badola
  (CADCIM Technologies)

Library of Congress Control Number: 2009905132
ISBN-13: 978-1-4390-5567-0
ISBN-10: 1-4390-5567-X

Delmar Cengage Learning
5 Maxwell Drive
Clifton Park, NY 12065-2919
USA

Cengage Learning is a leading provider of customized learning solutions with office locations around the globe, including Singapore, the United Kingdom, Australia, Mexico, Brazil, and Japan. Locate your local office at:
international.cengage.com/region

Cengage Learning products are represented in Canada by Nelson Education, Ltd.

To learn more about Delmar, visit **www.cengage.com/delmar**

Purchase any of our products at your local college store or at our preferred online store **www.ichapters.com**

Notice to the Reader
Publisher does not warrant or guarantee any of the products described herein or perform any independent analysis in connection with any of the product information contained herein. Publisher does not assume, and expressly disclaims, any obligation to obtain and include information other than that provided to it by the manufacturer. The reader is expressly warned to consider and adopt all safety precautions that might be indicated by the activities described herein and to avoid all potential hazards. By following the instructions contained herein, the reader willingly assumes all risks in connection with such instructions. The publisher makes no representations or warranties of any kind, including but not limited to, the warranties of fitness for particular purpose or merchantability, nor are any such representations implied with respect to the material set forth herein, and the publisher takes no responsibility with respect to such material. The publisher shall not be liable for any special, consequential, or exemplary damages resulting, in whole or part, from the readers' use of, or reliance upon, this material.

Printed in the United States of America
1 2 3 4 5 6 7 13 12 11 10 09

DEDICATION

*To teachers, who make it possible to disseminate knowledge
to enlighten the young and curious minds
of our future generations*

*To students, who are dedicated to learning new technologies
and making the world a better place to live*

THANKS

*To the faculty and students of the MET Department of
Purdue University Calumet for their cooperation*

*To engineers of CADCIM Technologies and
Santosh Tickoo for their valuable help*

Table of Contents

Preface xxx

AutoCAD Part I

Chapter 1   Introduction to AutoCAD

Starting AutoCAD* 1-2
AutoCAD Screen Components* 1-5
    Drawing Area 1-5
    Command Window 1-5
    Application Status Bar 1-6
    Drawing Status Bar 1-9
    Status Bar Tray Options 1-9
Invoking Commands in AutoCAD* 1-11
    Keyboard 1-12
    Ribbon 1-12
    Application Menu 1-13
    Tool Palettes 1-13
    Menu Bar 1-14
    Toolbar 1-15
    Shortcut Menu 1-15
AutoCAD Dialog Boxes 1-16
Starting a New Drawing 1-18
    Open a Drawing 1-19
    Start from Scratch 1-19
    Use a Template 1-19
    Use a Wizard 1-20
Saving Your Work 1-26
    Save Drawing As Dialog Box 1-26
    Views List 1-28
    Create New Folder Button 1-29
    Up one level Button 1-29
    Search the Web 1-29
    Tools List 1-29
Automatic Timed Save 1-29
Creation of Backup Files 1-30
    Changing Automatic Timed Saved and Backup Files into AutoCAD Format.................1-30
    Using the Drawing Recovery Manager to Recover Files 1-30
Closing a Drawing 1-31
Opening an Existing Drawing 1-31
    Opening an Existing Drawing Using the Select File Dialog Box.................1-32
    Opening an Existing Drawing Using the Startup Dialog Box 1-35
    Opening an Existing Drawing Using the Drag and Drop Method................1-35
Quitting AutoCAD 1-35
Dynamic Input Mode 1-36
    Enable Pointer Input 1-36
    Enable Dimension Input Where Possible 1-37
    Show Command Prompting and Command Input Near the Crosshairs.........1-38

Drafting Tooltip Appearance 1-38
Understanding the Concept of Sheet Sets 1-39
    Creating a Sheet Set 1-40
    Adding a Subset to a Sheet Set 1-44
    Adding Sheets to a Sheet Set or a Subset 1-45
    Archiving a Sheet Set 1-45
    Resaving All the Sheets in a Sheet Set 1-46
Creating and Managing Workspaces 1-46
    Creating a New Workspace 1-48
    Modifying the Workspace Settings 1-48
AutoCAD'S HELP 1-49
    Help Menu 1-49
    InfoCenter Bar 1-52
Additional Help Resources 1-52

Chapter 2   Getting Started with AutoCAD

Drawing Lines in AutoCAD 2-2
    The Continue Option 2-4
    The Close Option 2-5
    The Undo Option 2-5
Coordinate Systems 2-6
    Absolute Coordinate System 2-7
    Relative Coordinate System 2-9
    Relative Polar Coordinates 2-11
    Direct Distance Entry 2-13
Erasing Objects 2-16
Canceling and Undoing a Command 2-16
Object Selection Methods 2-17
    The Window Option 2-17
    The Crossing Option 2-18
Drawing Circles 2-19
    The Center and Radius Option 2-20
    The Center and Diameter Option 2-20
    The Two-Point Option 2-21
    The Three-Point Option 2-21
    The Tangent Tangent Radius Option 2-22
    The Tangent, Tangent, Tangent Option 2-22
Basic Display Commands 2-24
    Zooming Drawings 2-24
    Panning in Realtime 2-26
Setting Units 2-26
    Setting Units Using the Drawing Units Dialog Box 2-26
Setting Limits of the Drawing 2-32
    Setting Limits 2-33
    Limits for Architectural Drawings 2-34
    Limits for Metric Drawings 2-35
Introduction to Plotting Drawings 2-36
    Basic Plotting 2-37
Modifying AutoCAD Settings Using the Options Dialog Box 2-41
    Files 2-41
    Display 2-41
    Open and Save 2-41
    Plot and Publish 2-42
    System 2-43

User Preferences 2-43
Drafting 2-43
3D Modeling 2-43
Selection 2-43
Profiles 2-43

Chapter 3   Starting with the Advanced Sketching

Drawing Arcs 3-2
    The 3-Point Option 3-2
    The Start, Center, End Option 3-3
    The Start, Center, Angle Option 3-4
    The Start, Center, Length Option 3-5
    The Start, End, Angle Option 3-6
    The Start, End, Direction Option 3-6
    The Start, End, Radius Option 3-8
    The Center, Start, End Option 3-8
    The Center, Start, Angle Option 3-9
    The Center, Start, Length Option 3-9
    Continue Option 3-10
    Continue (LineCont:) Option 3-11
Drawing Rectangles 3-11
    Drawing Rectangles Using Two Opposite Corners 3-12
    Drawing Rectangles by Specifying the Area and One Side 3-13
    Drawing Rectangles by Specifying their Dimensions 3-13
    Drawing Rectangle at an Angle 3-14
    Chamfer 3-14
    Fillet 3-15
    Width 3-15
    Thickness 3-15
    Elevation 3-16
Drawing Ellipses 3-17
    Drawing an Ellipse Using the Axis and Endpoint Option 3-17
    Drawing Ellipse Using the Center Option 3-19
    Drawing Elliptical Arcs 3-20
Drawing Regular Polygons 3-23
    The Center of Polygon Option 3-23
    The Edge Option 3-25
Drawing Polylines 3-26
    Next Point of Line 3-26
    Width 3-27
    Halfwidth 3-28
    Length 3-28
    Undo 3-28
    Close 3-28
    Arc 3-29
Drawing Donuts 3-31
Placing Points 3-34
    Changing the Point Type 3-34
    Changing the Point Size 3-35
Drawing Infinite Lines 3-36
    Drawing XLINE 3-36
    Drawing RAY 3-39
Writing A Single Line Text 3-39

Chapter 4   Working with Drawing Aids

Understanding the Concept and use of Layers 4-2
     Advantages of Layers 4-2
Working with Layers 4-3
     Creating New Layers 4-4
     Making a Layer Current... 4-5
     Controlling the Display of Layers 4-6
     Deleting Layers 4-11
     Managing the Display of Columns... 4-11
     Selective Display of Layers 4-11
     Layer States 4-13
     Reconciling New Layers 4-15
     Isolating and Unisolating Layers 4-15
     Controlling the Layer Settings 4-16
Changing the Linetype 4-21
LWEIGHT Command... 4-22
Object Properties 4-23
     Properties Panel... 4-23
     Properties Palette 4-24
     Quick Properties Palette 4-25
Global and Current Linetype Scaling 4-26
LTSCALE Factor for Plotting 4-27
     Changing Linetype Scale Using the PROPERTIES Command........................... 4-28
Working with the DesignCenter 4-28
Drafting Settings dialog box 4-28
     Setting Grid.. 4-29
     Setting Snap 4-31
     Snap Type... 4-32
Drawing Straight Lines Using the Ortho Mode 4-33
Working with Object Snaps.. 4-34
     AutoSnap 4-35
     Endpoint... 4-36
     Midpoint... 4-36
     Nearest.. 4-37
     Center... 4-37
     Tangent... 4-38
     Quadrant... 4-38
     Intersection... 4-39
     Apparent Intersection 4-40
     Perpendicular... 4-40
     Node 4-41
     Insert 4-42
     None 4-43
     Parallel.. 4-43
     Extension.. 4-44
     From 4-44
     Midpoint Between 2 Points 4-45
     Temporary Tracking Point 4-45
     Combining Object Snap Modes 4-46
Running Object Snap Mode 4-46
     Overriding the Running Snap 4-48
     Cycling through Snaps 4-48
     Setting the Priority for Coordinate Entry 4-49

Using AutoTracking 4-50
    Object Snap Tracking 4-50
    Polar Tracking 4-51
    AutoTrack Settings 4-51
Function and Control Keys 4-52

Chapter 5   Editing Sketched Objects-I

Creating a Selection Set 5-2
    Last 5-2
    Previous 5-3
    WPolygon 5-3
    CPolygon 5-4
    Remove 5-4
    Add 5-5
    ALL 5-5
    Fence 5-6
    Group 5-6
    BOX 5-6
    AUto 5-7
    Multiple 5-7
    Undo 5-7
    SIngle 5-7
    SUbobject 5-7
    Object 5-8
Editing Sketches 5-8
Moving Sketched Objects 5-8
Copying Sketched Objects 5-9
    Creating Multiple Copies 5-9
    Creating a Single Copy 5-10
Copying Objects Using the Base Point 5-10
Pasting Contents from the Clipboard 5-11
Pasting Contents Using Original Coordinates 5-11
Offsetting Sketched Objects 5-11
    Through Option 5-12
    Erase Option 5-13
    Layer Option 5-13
Rotating Sketched Objects 5-13
Scaling Sketched Objects 5-16
Filleting Sketches 5-17
    Creating Fillets Using the Radius Option 5-18
    Creating Fillets Using the Select First Object Option 5-18
    Creating Fillets Using the Trim Option 5-19
    Creating Fillets Using the Polyline Option 5-19
    Creating Fillets Using the Multiple Option 5-20
    Filleting Objects with a Different UCS 5-20
    Setting the TRIMMODE System Variable 5-21
Chamfering Sketches 5-21
    Creating Chamfer Using the Distance Option 5-21
    Creating Chamfer Using the Select First Line Option 5-22
    Creating Chamfer Using the Polyline Option 5-22
    Creating Chamfer Using the Angle Option 5-22
    Creating Chamfer Using the Trim Option 5-22
    Creating Chamfer Using the Method Option 5-23

Creating Chamfer Using the Multiple Option...5-23
Setting the Chamfering System Variables 5-24
Trimming Sketched Objects...5-24
Select object to trim Option 5-25
Shift-select to extend Option 5-25
Edge Option 5-25
Project Option 5-25
Fence Option 5-26
Crossing Option 5-26
Erase Option 5-27
Undo Option 5-27
Extending Sketched Objects 5-27
Select object to extend Option 5-28
Shift-select to trim Option...5-28
Project Option 5-28
Edge Option 5-28
Undo Option 5-29
Trimming and Extending with Text, Region, or Spline.....................................5-29
Stretching Sketched Objects 5-30
Lengthening Sketched Objects 5-31
Select an object Option 5-31
DElta Option 5-31
Percent Option 5-32
Total Option 5-32
DYnamic 5-33
Arraying Sketched Objects...5-33
Rectangular Array 5-33
Polar Array 5-35
Mirroring Sketched Objects..5-38
Text Mirroring 5-39
Breaking the Sketched Objects 5-40
1 Point Option 5-40
2 Points Option..5-41
2 Points Select Option 5-41
Measuring Sketched Objects...5-42
Dividing Sketched Objects 5-43
Joining Sketched Objects 5-44
Joining Collinear Lines 5-44
Joining Arcs 5-45
Joining Elliptical Arcs 5-45
Joining Splines 5-45
Joining Polylines..5-45

Chapter 6   Editing Sketched Objects-II

Editing with Grips...6-2
Types of Grips 6-2
Adjusting Grip Settings...6-3
Grip Size Area..6-3
Grips Area..6-4
Editing Objects with Grips...6-5
Stretching Objects with Grips (Stretch Mode)..6-5
Moving Objects with Grips (Move Mode) 6-8
Rotating Objects with Grips (Rotate Mode) 6-8
Scaling Objects with Grips (Scale Mode)..6-10
Mirroring Objects with Grips (Mirror Mode) 6-12

Loading Hyperlinks 6-13
Editing Gripped Objects 6-13
Changing Properties Using PropertiesPalette 6-14
Changing Properties Using Grips...6-14
Matching Properties of Sketched Objects 6-15
Quick Selection of Sketched Objects...6-16
    Apply to 6-16
    Object type 6-16
    Properties 6-16
    Operator 6-17
    Value 6-17
    How to apply Area 6-17
    Append to current selection set 6-17
Managing Contents Using the DesignCenter...6-18
    Folders Tab 6-19
    Open Drawings Tab 6-21
    History Tab 6-21
    Autodesk seek design content Link* 6-21
Making Inquiries About Objects and Drawings...6-23
    Measuring Area of Objects 6-24
    Measuring the Distance between Two Points 6-27
    Identifying the Location of a Point on the Screen 6-28
    Listing Information About Objects 6-29
    Listing Information about All Objects in a Drawing 6-30
    Checking Time-Related Information 6-30
    Obtaining Drawing Status Information 6-30
    Displaying Drawing Properties 6-31
Basic Display Options 6-32
Redrawing the Screen 6-32
Regenerating Drawings 6-33
Zooming Drawings 6-33
    Realtime Zooming 6-34
    All Option 6-34
    Center Option 6-35
    Extents Option 6-36
    Dynamic Option 6-36
    Previous Option 6-38
    Window Option 6-38
    Scale Option 6-39
    Object Option 6-41
    Zoom In and Out...6-41
Panning Drawings 6-42
    Panning in Realtime 6-42
Creating Views...6-43
    View Manager Dialog Box...6-43
Placing Views on a Sheet of a Sheet Set 6-47
Aerial View 6-47
    Menus 6-48

Chapter 7   Creating Text and Tables

Annotative Objects 7-2
Annotation Scale 7-2
    Assigning Annotative Property and Annotation Scales...7-2
    Customizing Annotation Scale 7-3

Multiple Annotation Scales..7-3
    Assigning Multiple Annotation Scales to Annotative Objects...................................7-4
Controlling the Display of Annotative Objects...7-5
Creating Text..7-6
    Writing Single Line Text..7-6
Entering Special Characters...7-11
Creating Multiline Text..7-11
    Text Window...7-13
    Text Editor Tab..7-13
    More..7-23
Editing Text..7-28
    Editing Text Using the DDEDIT Command...7-29
    Editing Text Using the Properties Palette..7-29
    Modifying the Scale of the Text..7-29
    Modifying the Justification of the Text...7-30
Inserting Table in the Drawing...7-30
    Table Style Area..7-31
    Insert Options Area...7-32
    Insertion Behavior Area...7-34
    Column and Row Settings Area...7-35
    Set Cell Styles Area..7-35
Creating a New Table Style...7-36
    Starting table Area...7-37
    General Area..7-37
    Cell Styles Area..7-38
Setting a Table Style As Current...7-40
Modifying a Table Style..7-40
Modifying Tables...7-40
    Modifying Rows..7-40
    Modifying Columns..7-41
    Merge Cells...7-41
    Match Cells...7-41
    Cell Styles...7-41
    Cell Borders..7-41
    Text Alignment...7-41
    Locking..7-41
    Data Format...7-42
    Insert Block...7-42
    Field..7-43
    Formula...7-43
    Manage Cell Content...7-45
    Match Cells...7-46
    Cell Styles...7-46
    Link Cell...7-46
    Download from source...7-46
Substituting Fonts...7-47
    Specifying an Alternate Default Font..7-48
Creating Text Styles..7-48
Determining Text Height..7-50
Creating Annotative Text..7-51
Checking Spelling...7-51
Text Quality and Text Fill...7-53
Finding and Replacing Text*..7-54
    Find Text...7-54
    Replacing Text...7-54
Creating Title Sheet Table in a Sheet Set...7-54

Chapter 8   Basic Dimensioning, Geometric Dimensioning, and Tolerancing

Need for Dimensioning 8-2
Dimensioning in AutoCAD 8-2
Fundamental Dimensioning Terms 8-3
    Dimension Line 8-3
    Dimension Text 8-3
    Arrowheads 8-3
    Extension Lines 8-4
    Leader 8-4
    Center Mark and Centerlines 8-5
    Alternate Units 8-5
    Tolerances 8-5
    Limits 8-5
Associative Dimensions 8-6
Definition Points 8-7
Annotative Dimensions 8-8
Selecting Dimensioning Commands 8-8
    Using the Ribbon and Toolbar 8-8
    Using the Command Line 8-8
Dimensioning a Number of Objects Together 8-9
Creating Linear Dimensions 8-10
    DIMLINEAR Command Options 8-11
Creating Aligned Dimensions 8-14
Creating Arc Length Dimensions 8-15
Creating Rotated Dimensions 8-16
Creating Baseline Dimensions 8-17
Creating Continued Dimensions 8-18
Creating Angular Dimensions 8-20
    Dimensioning the Angle between Two Nonparallel Lines 8-21
    Dimensioning the Angle of an Arc 8-21
    Angular Dimensioning of Circles 8-22
    Angular Dimensioning based on Three Points 8-22
Creating Diameter Dimensions 8-23
Creating Jogged Dimensions 8-24
Creating Radius Dimensions 8-25
Creating Jogged Linear Dimensions 8-25
Generating Center Marks and Centerlines 8-26
Creating Ordinate Dimensions 8-26
Maintaining Equal Spacing Between Dimensions 8-28
Creating Dimension Breaks 8-29
    Multiple 8-30
    Auto 8-30
    Remove 8-30
    Manual 8-31
Creating Inspection dimensions 8-31
    Inspection Label 8-31
    Dimension Value 8-32
    Inspection Rate 8-32
Working with True Associative Dimensions 8-33
    Removing the Dimension Associativity 8-33
    Converting a Dimension into a True Associative Dimension 8-33
Drawing Leaders 8-34
    Annotation Tab 8-34

Leader Line & Arrow Tab 8-36
Attachment Tab 8-37
Multileaders 8-38
Drawing Multileaders..8-39
leader Landing First 8-39
Content First...8-39
Options...8-40
Adding leaders to the Existing Multileader...8-41
Removing Leaders From the Existing Multileader..8-41
Aligning Multileaders 8-42
Distribute...8-42
Make Leader Segments Parallel 8-43
Specify Spacing..8-44
Use Current Spacing 8-44
Collecting Multiple Notes to Single Landing..8-45
Geometric Dimensioning and Tolerancing 8-46
Geometric Characteristics and Symbols 8-47
Adding Geometric Tolerance..8-48
Feature Control Frame 8-48
Geometric Characteristics Symbol 8-49
Tolerance Value and Tolerance Zone Descriptor 8-49
Material Condition Modifier...8-50
Datum...8-50
Complex Feature Control Frames...8-50
Combining Geometric Characteristics...8-50
Composite Position Tolerancing 8-51
Using Feature Control Frames with Leaders 8-52
Projected Tolerance Zone 8-53
Creating Annotative Dimensions, Tolerances, Leaders, and Multileaders 8-56

Chapter 9   Editing Dimensions

Editing Dimensions Using Editing Tools 9-2
Editing Dimensions by Stretching..9-2
Editing Dimensions by Trimming and Extending................................9-3
Flipping Dimension Arrow..9-5
Modifying the Dimensions..9-5
New..9-5
Rotate...9-5
Home..9-6
Oblique..9-6
Editing Dimension Text..9-6
Left 9-7
Right...9-7
Center...9-7
Home..9-7
Angle 9-7
Updating Dimensions..9-7
Editing Dimensions with Grips 9-8
Editing Dimensions Using the Properties Palette 9-8
Properties Palette (Dimension)..9-8
Properties Palette (Multileader) 9-10
Model Space and Paper Space Dimensioning 9-11

Chapter 10 Dimension Styles, Multileader Styles, and System Variables

Using Styles and Variables to Control Dimensions..10-2
Creating and Restoring Dimension Styles 10-2
New Dimension Style Dialog Box 10-3
    Lines Tab 10-4
    Symbols and Arrows Tab 10-7
Controlling the Dimension Text Format..10-11
    Text Tab 10-11
Fitting Dimension Text and Arrowheads 10-16
    Fit Tab 10-16
Formatting Primary Dimension Units 10-19
    Primary Units Tab 10-19
Formatting Alternate Dimension Units 10-23
    Alternate Units Tab 10-23
Formatting the Tolerances 10-25
    Tolerances Tab..10-25
Dimension Style Families 10-29
Using Dimension Style Overrides...10-32
Comparing and Listing Dimension Styles 10-34
Using Externally Referenced Dimension Styles 10-34
Creating and Restoring Multileader Styles 10-35
Modify Multileader Style Dialog Box 10-36
    Leader Format...10-36
    Leader Structure Tab..10-38
    Content Tab...10-40

Chapter 11 Adding Constraints to Sketches

Introduction 11-2
Adding Geometric Constraints* 11-2
    Applying the Horizontal Constraint 11-2
    Applying the Vertical Constraint...11-3
    Applying the Coincident Constraint..11-4
    Applying the Fix Constraint 11-5
    Applying the Collinear Constraint...11-5
    Applying the Perpendicular Constraint 11-6
    Applying the Parallel Constraint..11-6
    Applying the Concentric Constraint..11-7
    Applying the Tangent Constraint...11-7
    Applying the Symmetric Constraint...11-7
    Applying the Equal Constraint...11-8
    Applying the Smooth Constraint 11-8
Controlling the Display of Constraints* 11-8
Applying Constraints Automatically* 11-16
Applying Dimensional Constraints* 11-17
Converting a Dimensional Constraint into an Annotational Constraint*..................................11-18
Concept of a Fully-Defined Sketch* 11-19
    User-Defined 11-19
    Fully-Defined 11-19
    Over-Defined 11-19
Controlling the Display of the Dimensional Constraint* 11-20
Working with Equations* 11-25
    Adding Equations while Applying Dimensional Constraints..11-25
    Adding Equations Using Parameters Manager 11-26

Chapter 12 Model Space Viewports, Paper Space Viewports, and Layouts

Model Space and Paper Space/Layouts 12-2
Model Space Viewports (Tiled Viewports) 12-3
 Creating Tiled Viewports 12-3
Making a Viewport Current 12-5
Joining Two Adjacent Viewports 12-6
Paper Space Viewports (Floating Viewports) 12-7
 Creating Floating Viewports 12-8
 Creating Polygonal Viewports 12-9
 Converting an Existing Closed Object into a Viewport 12-10
Temporary Model Space 12-10
Editing Viewports 12-12
 Controlling the Display of Objects in Viewports 12-13
 Locking the Display of Objects in Viewports 12-13
 Controlling the Display of Hidden Lines in Viewports 12-13
 Clipping Existing Viewports 12-14
 Maximizing Viewports 12-15
Manipulating the Visibility of Viewport Layers 12-15
 ? Option 12-16
 Color Option 12-16
 Ltype Option 12-16
 LWeight Option 12-16
 PStyle Option 12-16
 Freeze Option 12-16
 Thaw Option 12-16
 Reset Option 12-17
 Newfrz (New Freeze) Option 12-17
 Vpvisdflt (Viewport Visibility Default) Option 12-17
Controlling the Layers in Viewports Using the Layer Properties Manager Dialog Box 12-17
 VP Freeze 12-18
 VP Color 12-18
 VP Linetype 12-18
 VP Lineweight 12-18
 VP Plot Style 12-18
 Removing Viewport Overrides 12-19
Paper Space Linetype Scaling (PSLTSCALE System Variable) 12-19
Inserting Layouts 12-21
 New Option 12-21
 Copy Option 12-21
 Delete Option 12-22
 Template Option 12-22
 Rename Option 12-23
 SAveas Option 12-23
 Set Option 12-23
 ? Option 12-23
Importing Layouts to Sheet Sets 12-23
Inserting a Layout Using the Wizard 12-24
Defining Page Settings 12-24
Working with the MVSETUP Command 12-26
 Using the MVSETUP Command on the Model Tab 12-26
 Using the MVSETUP Command on the Layout Tab 12-27
Converting the Distance Between Model Space and Paper Space 12-30
Controlling the Display of Annotative Objects in Viewports 12-30

Chapter 13   Template Drawings

Creating Template Drawings 13-2
Standard Template Drawings.. 13-2
Loading a Template Drawing 13-8
Customizing Drawings with Layers and Dimensioning Specifications.......... 13-9
Customizing a Drawing with Layout.. 13-14
Customizing Drawings with Viewports.. 13-18
Customizing Drawings According to Plot Size and Drawing Scale............... 13-21

Chapter 14   Plotting Drawings

Plotting Drawings in AutoCAD... 14-2
Plotting Drawings Using the Plot Dialog Box 14-2
    Page Setup Area... 14-3
    Printer/Plotter Area 14-3
    Paper size Area 14-4
    Number of Copies Area... 14-4
    Plot Area Area 14-4
    Plot Offset (Origin Set to Printable area) Area 14-6
    Plot Scale Area 14-6
    Plot Style Table (Pen Assignments) Area 14-7
    Shaded Viewport Options Area 14-7
    Plot Options Area 14-8
    Drawing orientation Area 14-9
    Preview.. 14-10
Adding Plotters 14-11
    The PLOTTERMANAGER Command.. 14-11
Editing the Plotter Configuration... 14-13
    General Tab 14-13
    Ports Tab 14-13
    Device and Document Settings Tab.. 14-14
Importing PCP/PC2 Configuration Files 14-15
Setting Plot Parameters.. 14-15
    Working with Page Setups 14-15
Using Plot Styles 14-19
    Adding a Plot Style 14-20
    Plot Style Table Editor 14-22
    Applying Plot Styles... 14-26
    Setting the Current Plot Style 14-27
Plotting Sheets in a Sheet Set 14-29

Chapter 15   Hatching Drawings

Hatching 15-2
Hatch Patterns... 15-2
    Hatch Boundary 15-3
Hatching Drawings Using the Hatch and Gradient Dialog Box................... 15-3
Hatch and Gradient Dialog Box Options.. 15-5
    Hatch Tab 15-5
    User defined Hatch Patterns... 15-8
    Custom Hatch Patterns.. 15-9
    Selecting Hatch Boundary 15-10
    Options Area 15-14
    Inherit Properties 15-15

Preview 15-15
Hatch origin Area 15-16
More Options 15-17
Islands Area 15-17
Inherit options Area 15-20
The HATCH Shortcut Menu 15-20
Gradient Tab 15-21
Creating Annotative Hatch 15-23
Hatching the Drawing Using the Tool Palettes 15-23
Drag and Drop Method 15-24
Select and Place Method 15-24
Modifying the Properties of the Predefined Patterns Available in the Tool Palettes 15-24
Hatching Around Text, Dimensions, and Attributes 15-26
Editing Hatch Patterns 15-26
Using the HATCHEDIT Command 15-26
Ignoring Hatch Pattern Entities While Snapping 15-28
Using the Properties Command 15-28
Editing Hatch Boundary 15-30
Using Grips 15-30
Trimming the Hatch Patterns 15-31
Using AutoCAD Editing Commands 15-31
Hatching Blocks and Xref Drawings 15-32
Creating a Boundary Using Closed Loops 15-32
Other Features of Hatching 15-34

Chapter 16   Working with Blocks

The Concept of Blocks 16-2
Advantages of Using Blocks 16-2
Formation of Blocks 16-3
Drawing Objects for Blocks 16-3
Checking Layers 16-4
Converting Entities into a Block 16-5
Inserting Blocks 16-8
Name 16-8
Insertion point Area 16-10
Scale Area 16-10
Rotation Area 16-11
Block Unit Area 16-11
Explode Check Box 16-12
Basepoint 16-13
Scale 16-13
X, Y, Z 16-13
Rotate 16-14
Creating and Inserting Annotative Blocks 16-15
Block Editor 16-18
Dynamic Blocks 16-19
Block Editor Tab 16-20
Adding Parameters and Assigning Actions to
        Dynamic Blocks (Block Authoring Palettes) 16-22
Adding Parameter and Action Simultaneously Using Parameter Sets 16-36
Inserting Blocks Using the DesignCenter 16-37
Using Tool Palettes to Insert Blocks 16-38
Inserting Blocks in the Drawing 16-38
Modifying Properties of the Blocks in the Tool Palettes 16-39

Adding Blocks in Tool Palettes 16-40
        Drag and Drop Method 16-40
        Shortcut Menu 16-40
Modifying Existing Blocks in the Tool Palettes 16-41
Layers, Colors, Linetypes, and Lineweights for Blocks 16-42
Nesting of Blocks 16-43
Inserting Multiple Blocks 16-45
Creating Drawing Files Using the Write Block Dialog Box 16-47
Defining the Insertion Base Point 16-49
Editing Blocks 16-49
        Editing Blocks in Place 16-50
        Exploding Blocks Using the XPLODE Command 16-53
Renaming Blocks 16-55
Deleting Unused Blocks 16-56
Applying Constraints to Blocks* 16-57

AutoCAD Part II

Chapter 17   Defining Block Attributes

Understanding Attributes 17-2
Defining Attributes 17-2
        Mode Area 17-2
        Attribute Area 17-4
        Insertion Point Area 17-5
        Text Settings Area 17-5
Editing Attribute Definition 17-8
        Using the Properties Palette 17-9
Inserting Blocks with Attributes 17-9
Managing Attributes 17-11
        Select block 17-13
        Block Drop-down List 17-13
        Sync 17-13
        Move Up 17-14
        Move Down 17-14
        Edit 17-14
        Settings 17-17
        Remove 17-18
Extracting Attributes 17-18
        Begin 17-18
        Define Data Source 17-19
        Select Objects 17-22
        Select Properties 17-23
        Refine Data 17-23
        Choose Output 17-30
        Table Style 17-31
        Finish 17-32
        The ATTEXT Command for Attribute Extraction 17-32
Controlling Attribute Visibility 17-38
Editing Block Attributes 17-39
        Editing Attributes Using the Enhanced Attribute Editor 17-39
        Editing Attributes Using the Edit Attributes Dialog Box 17-40
        Global Editing of Attributes 17-43

Redefining a Block with Attributes 17-49
      In-place Editing of Blocks with Attributes 17-49
Inserting Text Files in the Drawing 17-50

Chapter 18   Understanding External References

External References 18-2
Dependent Symbols 18-2
Managing External References in a Drawing 18-4
      List View 18-4
      Attaching an Xref Drawing (Attach Option) 18-6
      Points to Remember about Xref 18-8
      Detaching an Xref Drawing (Detach Option) 18-9
      Updating an Xref Drawing (Reload Option) 18-10
      Unloading an Xref Drawing (Unload Option) 18-10
      Adding an Xref Drawing (Bind Option) 18-10
      Editing an XREF's Path 18-11
The Overlay Option 18-12
Working with the ATTACH Command* 18-16
Working with Underlays* 18-17
      Editing an Underlay 18-18
Opening an Xreffed Object in a Separate Window 18-19
Using the DesignCenter to Attach a Drawing as an Xref 18-19
Adding xref Dependent Named Objects 18-20
Clipping External References* 18-22
Displaying Clipping Frame* 18-24
Demand Loading 18-25
      Spatial and Layer Indexes 18-26
Editing References In-Place* 18-26

Chapter 19   Working with Advanced Drawing Options

Understanding the Use of Multilines 19-2
Defining the Multiline Style 19-2
      Styles List Box 19-2
      Set Current 19-2
      New 19-2
      Load 19-6
      Save 19-6
      Rename 19-6
      Preview of Area 19-6
Drawing Multilines 19-7
      Justification Option 19-7
      Scale Option 19-7
      STyle Option 19-8
Editing Multilines Using Grips 19-8
Editing Multilines Using the Dialog Box 19-9
      Cross Intersection (CC/OC/MC) 19-9
      Tee Intersection (CT/OT/MT) 19-10
      Corner Joint (CJ) 19-11
      Adding and Deleting Vertices (AV/DV) 19-11
      Cutting and Welding Multilines (CS/CA/WA) 19-12
Creating Revision Clouds 19-14

Creating Wipeouts..19-15
Creating NURBS..19-16
     Options for Creating Splines...19-17
Editing Splines...19-18
     Options for Editing Splines...19-18

Chapter 20   Grouping and Advanced Editing of Sketched Objects

Grouping Sketched Objects Using the Object Grouping Dialog Box 20-2
     Group Name List Box 20-2
     Group Identification Area 20-3
     Create Group Area 20-4
     Change Group Area 20-4
Selecting Groups 20-7
Changing Properties of an Object 20-8
     Using the Properties Palette 20-8
Exploding Compound Objects 20-14
Editing Polylines 20-15
     Editing Single Polyline 20-16
     Editing Multiple Polylines 20-30
Undoing Previous Commands...20-32
     Undoing Previous Commands Using the Quick Access Toolbar 20-32
     Undoing Previous Commands Using the Command Line 20-32
     Number (N) Option 20-32
     Control (C) Option 20-33
     Auto (A) Option...20-34
     BEgin (BE) and End (E) Options 20-35
     Mark (M) and Back Options...20-36
Reversing the Undo Operation 20-37
Renaming Named Objects..20-37
Removing Unused Named Objects...20-39
     View Items You Can Purge 20-39
     View Items You Cannot Purge 20-40
Setting Selection Modes Using the Options Dialog Box...20-41
     Noun/Verb Selection..20-41
     Use Shift to Add to Selection Option 20-43
     Press and Drag...20-43
     Implied Windowing...20-43
     Object Grouping..20-44
     Associative Hatch...20-44
     Pickbox Size 20-44

Chapter 21   Working with Data Exchange and Object Linking and Embedding

Understanding the Concept of Data Exchange in AutoCAD...21-2
Creating Data Interchange (DXF) Files 21-2
     Creating a Data Interchange File 21-2
     Information in a DXF File...21-3
     Converting DXF Files into Drawing Files...21-3
OTHER DATA EXCHANGE FORMATS 21-4
     DXB File Format..21-4
     Creating and Using an ACIS File 21-4

Importing 3D Studio Files..21-5
Creating and Using a Windows Metafile...21-5
Creating and Using a V8 DGN File..21-5
Creating a BMP File 21-6
Raster Images..21-6
Attaching Raster Images 21-7
Managing Raster Images...21-9
Editing Raster Image Files 21-10
Clipping Raster Images 21-10
Adjusting Raster Images...21-12
Modifying the Image Quality 21-13
Modifying the Transparency of an Image 21-13
Controlling the Display of Image Frames 21-14
Changing the Display Order...21-14
Other Editing Operations 21-14
Scaling Raster Images 21-14
Working With PostScript Files 21-15
Creating the PostScript Files..21-15
Object Linking and Embedding (OLE) 21-17
Clipboard..21-18
Object Embedding 21-18
Linking Objects 21-21
Linking Information into AutoCAD...21-23
Linking Objects into AutoCAD 21-23
Embedding Objects into AutoCAD 21-24
Working with OLE Objects..21-25

Chapter 22   Technical Drawing with AutoCAD

Multiview Drawings...22-2
Understanding X, Y, and Z Axes...22-2
Orthographic Projections...22-2
Positioning Orthographic Views 22-7
Dimensioning...22-11
Dimensioning Components 22-12
Basic Dimensioning Rules 22-13
Sectional Views..22-23
Full Section 22-23
Half Section 22-24
Broken Section...22-25
Revolved Section 22-25
Removed Section 22-25
Offset Section 22-27
Aligned Section 22-27
Cutting Plane Lines 22-28
Spacing for Hatch Lines 22-29
Direction of Hatch Lines 22-30
Points to Remember 22-30
Auxiliary Views..22-33
How to Draw Auxiliary Views 22-34
Detail Drawing, Assembly Drawing, and Bill of Materials...22-37
Detail Drawing 22-37
Assembly Drawing 22-38
Bill of Materials 22-38

Chapter 23   Isometric Drawings

Isometric Drawings 23-2
Isometric Projections..23-2
Isometric Axes and Planes 23-3
Setting the Isometric Grid and Snap 23-3
Drawing Isometric Circles..23-8
      Creating Fillets in the Isometric Drawings...23-9
Dimensioning Isometric Objects...23-9
Isometric Text 23-11

Chapter 24   The User Coordinate System

The World Coordinate System (WCS) 24-2
Controlling the Visibility of the UCS Icon...24-2
      ON 24-3
      OFF...24-3
      All   24-3
      Noorigin 24-3
      ORigin 24-3
      Properties 24-3
Defining the New UCS 24-5
      W (World) Option 24-6
      F (Face) Option 24-6
      OB (OBject) Option 24-6
      V (View) Option 24-8
      X/Y/Z Options...24-8
      ZA (ZAxis) Option 24-10
      P (Previous) Option 24-11
      NA (NAmed) Option 24-11
      D (Delete) Option 24-12
      A (Apply) Option 24-13
      O (Origin) Option 24-13
      3 (3-Point) Option 24-14
Managing the UCS Through the Dialog Box..24-17
      Named UCSs Tab 24-17
      Setting the UCS to Preset Orthographic UCSs Using the Orthographic UCSs Tab.......24-18
      Settings Tab 24-20
System Variables...24-21

Chapter 25   Getting Started with 3D

Starting Three Dimensional (3D) Modeling In AutoCAD..25-2
Use of Three-dimensional drawing 25-2
Types of 3D models..25-3
      Wireframe Models 25-3
      Surface Models 25-3
      Solid Models 25-4
Conventions in AutoCAD...25-5
Changing the Viewpoint to View 3D Models 25-6
      Changing the Viewpoint Using the Viewpoint Presets Dialog Box.................25-7
      Changing the Viewpoint Using the Command Line 25-9
      Changing the Viewpoint Using the Toolbar or the Ribbon 25-12
      Changing the Viewpoint Using the ViewCube 25-12

3D Coordinate Systems 25-15
    Absolute Coordinate System 25-15
    Relative Coordinate System 25-18
Trim, Extend, and Fillet Commands in 3D 25-21
Setting Thickness and Elevation for the New Objects 25-22
    ELEV Command 25-22
Suppressing the Hidden Edges 25-24
Creating 3D Polylines 25-24
Converting Wireframe Models Into Surface Models 25-25
    Creating 3D Faces 25-25
    Creating Polyface Meshes 25-26
    Controlling the Visibility of the 3D Face Edges 25-28
Drawing Standard Surface Primitives 25-29
    Creating a Box-Shaped Surface Mesh 25-29
    Creating a Cone-Shaped Surface Mesh 25-29
    Creating a Surface Mesh Dish 25-31
    Creating a Surface Mesh Dome 25-31
    Creating a Planar Mesh 25-32
    Creating a Pyramid Surface Mesh 25-32
    Creating a Spherical Surface Mesh 25-35
    Creating a Surface Mesh Torus 25-36
    Creating a Surface Mesh Wedge 25-36
Creating Planar Surfaces 25-37
The 3DMESH Command 25-37
Editing the Surface Mesh 25-38
    The PEDIT Command 25-38
Dynamic Viewing of 3D Objects 25-41
    Using the SteeringWheels 25-41
    Dynamically Rotating the View of the Model 25-45
    Clipping the View of the Model Dynamically 25-52

Chapter 26   Creating Solid Models

What is Solid Modeling? 26-2
Creating Predefined Solid Primitives 26-2
    Creating a Solid Box 26-2
    Creating a Solid Cone 26-5
    Creating a Solid Cylinder 26-7
    Creating a Solid Sphere 26-10
    Creating a Solid Torus 26-11
    Creating a Solid Wedge 26-13
    Creating a Pyramid 26-13
    Creating a Polysolid 26-14
    Creating a Helix 26-17
Modifying the Visual Styles of Solids 26-20
    Available Visual Styles in Drawing 26-20
    2D Wireframe options 26-21
    2D Hide - Obscured Lines 26-22
    2D - Intersection Edges 26-23
    2D Hide - Miscellaneous 26-24
    Halo gap % 26-24
    Controlling the Display of Edges 26-24
    Controlling the Display of Faces 26-26
    Controlling Backgrounds and Shadows 26-27

Creating Complex Solid Models...26-28
    Creating Regions...26-28
    Creating Complex Solid Models by Applying Boolean Operations 26-28
    Combining Solid Models..26-29
    Subtracting One Solid From the Other 26-29
    Intersecting Solid Models...26-29
    Checking Interference in Solids..26-30
Dynamic UCS..26-35
Defining the New UCS Using the ViewCube and the Ribbon...............................26-36
Creating Extruded Solids...26-37
    Extruding along the Z Axis 26-37
    Extruding along a Direction...26-38
    Extruding along a Path 26-38
Creating Revolved Solids 26-39
    Specifying the Start Point for the Axis of Revolution.....................................26-39
    Object 26-40
    X 26-40
    Y...26-40
    Z...26-40
Creating Swept Solids 26-41
    Alignment...26-42
    Base Point 26-42
    Scale...26-43
    Twist...26-43
Creating Lofted Solids 26-44
    Guide..26-45
    Path..26-45
    Cross Sections Only 26-46
Creating Presspull Solids 26-50

Chapter 27   Modifying 3D Objects

Filleting Solid Models 27-2
    Undo...27-3
    Chain..27-3
    Radius...27-3
Chamfering Solid Models...27-3
    Undo...27-4
    Loop 27-4
Rotating Solid Models in 3D Space 27-4
    2points Option 27-5
    Object Option...27-5
    Last Option..27-6
    View Option...27-6
    Xaxis Option...27-6
    Yaxis Option...27-7
    Zaxis Option...27-7
Rotating Solid Models About an Axis 27-7
Mirroring Solid Models in 3D Space 27-8
    3points Option 27-8
    Object 27-9
    Zaxis...27-10
    View Option...27-10
    XY/YZ/ZX 27-11
Moving Models in 3d Space...27-14

Creating Arrays in 3D Space..27-15
    3D Rectangular Array..27-15
    3D Polar Array..27-16
Aligning Solid Models..27-16
    Aligning the Objects Using One Pair of Points..27-17
    Aligning the Objects Using Two Pairs of Points..27-17
    Aligning the Objects Using Three Pairs of Points..27-18
Aligning Solids by Defining an Alignment Plane 27-19
Extracting Edges of a Solid Model 27-20
Converting Objects to Surfaces...27-21
Converting Objects to Solids...27-22
Converting Surfaces to Solids..27-22
Slicing Solid Models..27-23
    3points..27-23
    Object 27-24
    Surface 27-24
    Zaxis 27-25
    View 27-25
    XY, YZ, ZX 27-26
Creating the Cross-Sections of the Solids 27-26
    3points..27-27
    Object 27-27
    View 27-27
    XY, YZ, ZX 27-27

Chapter 28   Editing 3D Objects

Editing Solid Models...28-2
    Face Option 28-2
    Edge Option 28-7
    Body Option 28-8
    Undo Option 28-10
    Exit Option 28-10
Generating a Section by Defining a Section Plane 28-11
    Defining a Section Plane by Selecting a Face 28-11
    Defining a Section Plane by Specifying Two Points..28-13
    Defining a Section Plane by Drawing the Section Line 28-14
    Defining a Section Plane by Selecting a Preset Orthographic Plane 28-15
    Generating 2D and 3D Sections...28-15
    Live Section Settings 28-18
Generating Drawing Views of the Solid Model..28-19
    SOLVIEW Command...28-19
    SOLDRAW Command..28-22
    SOLPROF Command...28-22
    FLATSHOT Command...28-22
Calculating the Mass Properties of the Solid Models 28-26
Recording Your Drawing Steps Using Action Recorder 28-28
Using ShowMotion for Presentation Purposes 28-30
    Playing the Animation..28-31

Chapter 29   Mesh Modeling

IntroducTion...29-2
Creating Mesh Primitives*..29-2
    Creating a Mesh Box...29-2

Creating Surface Meshes..29-5
    Creating Revolved Surface Meshes 29-5
    Creating Edge Surface Meshes..29-6
    Creating Ruled Surface Meshes 29-7
    Creating Tabulated Surface Meshes 29-8
Modifying Mesh Objects*..29-10
    Adding Smoothness to Meshes 29-10
    Refining the Meshes 29-14
    Adding Crease to Meshes 29-15
Editing mesh faces*...29-16
    Splitting the Mesh Faces...29-17
    Extruding the Mesh Faces 29-17
Converting Mesh Objects*...29-20
    Converting Mesh Objects into Solids...29-20
    Converting Mesh Objects into Surfaces 29-22
Working with Gizmos 29-24
    Move Gizmo 29-24
    Rotate Gizmo...29-24
    Scale Gizmo 29-25

Chapter 30　Rendering and Animating Designs

Understanding the Concept of Rendering 30-2
Selecting and Attaching Materials 30-2
    Selecting a Material for a Design 30-2
    Attaching Selected Materials to Objects..30-4
    Attaching Material by Layers 30-7
Understanding Elementary Rendering 30-8
Adding Lights to the Design...30-10
    Default Light 30-11
    Point Light...30-12
    Spotlight 30-20
    Distant Light..30-22
    Sun Light 30-23
Converting Pre-AutoCAD 2010 Lights to AutoCAD 2010 Format 30-30
Modifying Lights..30-30
Creating New Materials 30-31
Converting Pre-AutoCAD 2010 Materials to AutoCAD 2010 Format 30-36
Modifying Material 30-37
    Rendering Quality...30-37
    General Rollout 30-38
    Ray Tracing Rollout..30-40
    Indirect Illumination Rollout 30-41
    Diagnostic Rollout 30-42
    Processing Rollout 30-43
    Managing Render Presets...30-44
Controlling the Rendering Environment 30-45
    Rendering with a Background..30-46
    Rendering with Fog 30-46
    Adjusting the Lighting Exposure to Rendered Image 30-47
Mapping Materials on Objects...30-48
    Box 30-49
    Planar 30-49
    Spherical..30-50
    Cylindrical 30-50

copY mapping to 30-50
Reset mapping 30-50
Adjusting Bitmap Placement on a Model 30-50
Rendering a Model with Different Render Settings 30-51
Render Again 30-52
Save 30-52
Save Copy 30-52
Make Render Settings Current 30-52
Remove From the List 30-52
Delete Output File 30-53
Obtaining Rendering Information 30-54
Saving a Rendering 30-55
Saving the Rendered Image to a File 30-55
Saving the Viewport Rendering 30-55
Saving the Rendered Image from Render Window 30-56
Plotting Rendered Images 30-56
Unloading AutoCAD Render 30-57
Creating Animations 30-57
Creating Animation of 3D Navigations 30-57
Creating Animation by Defining the Path of Camera Movement 30-59

Student Projects 1

Index 1

> **Chapters available for free download**

The following chapters are available on the author's and publisher's Web site for free download. To download these free chapters, log on to:

www.cadcim.com/students/students.htm or
www.cengage.com/cad/autodeskpress (Follow the **Online Companions** link)

Chapter 31 Accessing External Database
Chapter For Free Download

Understanding Database 31-2
AutoCAD Database Connectivity 31-3
Database Configuration 31-4
      Dbconnect Manager 31-4
Viewing and Editing Table Data from AutoCAD 31-7
Creating Links with Graphical Objects 31-9
Creating Labels 31-13
AutoCAD SQL Environment (ASE) 31-17
AutoCAD Query Editor 31-17
Forming Selection Sets Using the Link Select 31-22

Chapter 32 AutoCAD on the Internet
Chapter For Free Download

Introduction 32-2
Changed Internet Commands 32-3
Understanding URLS 32-4
Drawings on the Internet 32-6
Inserting a Block from the Internet 32-12
Accessing Other Files on the Internet 32-12
i-Drop 32-13
Saving the Drawing on the Internet 32-13
Online Resources 32-14
Using Hyperlinks with AutoCAD 32-15
Pasting as Hyperlink 32-22
Editing Hyperlinks 32-23
Removing Hyperlinks from Objects 32-23
The Drawing Web Format 32-23
Creating a DWF File 32-24

Chapter 33 Script Files and Slide Shows
Chapter For Free Download

What are Script Files? 33-2
Running Script Files 33-4
Repeating Script Files 33-10
Introducing Time Delay in Script Files 33-11

Resuming Script Files..33-12
Command Line Switches 33-12
Running a Script File while Loading AutoCAD...33-12
What is a Slide Show?...33-22
What are Slides?...33-22
Creating Slides...33-22
Viewing Slides 33-24
Preloading Slides..33-26
Slide Libraries..33-28

Chapter 34   Creating Linetypes and Hatch Patterns   *Chapter For Free Download*

Standard Linetypes 34-2
Linetype Definitions 34-2
Elements of Linetype Specification 34-3
Creating Linetypes..34-3
Alignment Specification..34-9
LTSCALE Command 34-10
LTSCALE Factor for Plotting..34-12
Current Linetype Scaling (CELTSCALE) 34-13
Alternate Linetypes..34-14
Modifying Linetypes 34-15
Complex Linetypes 34-18
Creating a String Complex Linetype...34-18
Hatch Pattern Definition..34-26
How Hatch Works 34-29
Simple Hatch Pattern 34-29
Effect of Angle and Scale Factor on Hatch 34-30
Hatch Pattern with Dashes and Dots 34-31
Hatch with Multiple Descriptors..34-34
Saving Hatch Patterns in a Separate File 34-39
Custom Hatch Pattern File...34-39

Chapter 35   Customizing the ACAD.PGP File   *Chapter For Free Download*

What is the ACAD.PGP File?...35-2
Sections of the ACAD.PGP File..35-8
Reinitializing the ACAD.PGP File..35-12

Preface

AutoCAD, developed by Autodesk Inc., is the most popular PC-CAD system available in the market. Today, over 7 million people use AutoCAD and other AutoCAD based design products. 100% of the Fortune 100 firms are Autodesk customers and 98% of the Fortune 500 firms are Autodesk customers. AutoCAD's open architecture has allowed third-party developers to write an application software that has significantly added to its popularity. For example, the author of this book has developed a software package "**SMLayout**" for sheet metal products that generates a flat layout of various geometrical shapes such as transitions, intersections, cones, elbows, and tank heads. Several companies in Canada and the United States are using this software package with AutoCAD to design and manufacture various products. AutoCAD also facilitates customization that enables the users to increase their efficiency and improve their productivity.

This book contains a detailed explanation of AutoCAD 2010 commands and their applications to solve drafting and design problems. Every AutoCAD command is thoroughly explained with the help of examples and illustrations. This makes it easy for the users to understand its function and application in the drawing. After reading this textbook, you will be able to use AutoCAD commands to make a drawing, create text, create and insert symbols, dimension a drawing, apply constraints to sketches, create blocks and dynamic blocks, create 3D objects, mesh objects and solid models, and render a scene.

The book also covers basic drafting and design concepts that provide you with the essential drafting skills to solve the drawing problems in AutoCAD. These include orthographic projections, dimensioning principles, sectioning, auxiliary views, and assembly drawings. While going through this textbook, you will discover some new unique applications of AutoCAD that will have a significant effect on your drawings. In addition, you will be able to understand why AutoCAD has become such a popular software package and an international standard in PC-CAD.

Please refer to the following table for conventions used in this text.

Convention

- Command names are capitalized and bold.

- A key icon appears when you have to respond by pressing the ENTER or the RETURN key.

Example

The **MOVE** command

Convention

- AutoCAD 2010 features in the text are designated by an asterisk symbol at the end of the feature.

- All the exercises that are designated by a double asterisk symbol at the end of the question are courtesy of CADCIM Technologies.

- Command sequences are indented. The responses are indicated by a boldface. The directions are indicated by italics and the comments are enclosed in parentheses.

- The command selection from the **Ribbon**, **Menu Bar**, **Quick Access toolbar**, **Tool Palettes**, **Application menu**, toolbars, Status Bar, and Command prompt are enclosed in a shaded box.

Example

Mesh Box*, **Refine Mesh***

Draw a detail drawing whose top, side, and section views are given in Figure 15-53. Then, hatch the section view. **

Command: **MOVE**
Select object: **G**
Enter group name: *Enter a group name (the group name is group1)*

Ribbon:	Draw > Line
Menu Bar:	Draw > Line
Tool Palettes:	Draw > Line
Toolbars:	Draw > Line
Command:	LINE or L

e-resource

This is an educational resource that creates a truly electronic classroom. It is a CD-ROM containing tools and instructional resources that enrich your classroom and reduce your preparation time. The features of **e-resource** link directly to the text and tie together to provide a unified instructional system. With **e-resource**, you can spend quality time teaching, rather than preparing to teach.

Features contained in an e-resource include:

Syllabus: Lesson plans created for each chapter. You have the option of using these lesson plans with your own course information.

Chapter Hints: Objectives and teaching hints that provide the basis for a lecture outline and help you present concepts and material.

Answers to Review Questions: These solutions enable you to grade and evaluate the end-of-chapter tests.

PowerPoint® Presentations: These slides provide the basis for a lecture outline and help you present concepts and material. The key points and concepts can be graphically highlighted for student retention.

Exam View Test Bank: Over 800 questions of varying levels of difficulty are provided in true/false and multiple-choice formats, so that you can assess the knowledge of students.

AVI Files: AVI files, listed by topic, enable you to view a quick video that illustrates and explains key concepts.

Drawing Files
The drawing files have been used in examples and exercises.

Author's Web sites
For Faculty: Please contact the author at **stickoo@calumet.purdue.edu** to access the Web site that contains the following:

1. PowerPoint presentations, program listings, and drawings used in this textbook.
2. Syllabus, chapter objectives and hints, and questions with answers of every chapter.

For Students: You can download drawings, exercises, tutorials, programs, and free chapters by accessing the author's Web site **www.cadcim.com/students/students.htm**. On this Web page, scroll to the AutoCAD section and click on the AutoCAD 2010 textbook link.

Publisher's Web site
You can also get learning materials from the publisher's website:
www.cengage.com/cad/autodeskpress

AutoCAD
Part I

Author's Web sites

For Faculty: Please contact the author at **stickoo@calumet.purdue.edu** to access the Web site that contains the following:

1. PowerPoint presentations, program listings, and drawings used in this textbook.
2. Syllabus, chapter objectives and hints, and questions with answers for every chapter.

For Students: You can download drawings, exercises, tutorials, programs, and free chapters by accessing the author's Web site **www.cadcim.com/students/students.htm** or use the link **www.cengage.com/cad/autodeskpress**. On this Web page, choose the Online Companions option and click on the AutoCAD 2010: A Problem-Solving Approach link.

Chapter 1

Introduction to AutoCAD

Learning Objectives

After completing this chapter, you will be able to:

• *Start AutoCAD and start a drawing.*
• *Understand components of the initial AutoCAD screen.*
• *Invoke AutoCAD commands from the keyboard, menu, toolbar, shortcut menu, **Tool Palettes,** and **Ribbon**.*
• *Understand the functioning of dialog boxes in AutoCAD.*
• *Start a new drawing using the **QNEW** command and the **Startup** dialog box.*
• *Save the work using various file-saving commands.*
• *Close a drawing.*
• *Open an existing drawing.*
• *Understand the concept of Multiple Document Environment.*
• *Exit AutoCAD.*
• *Use various options of AutoCAD's help.*
• *Understand the use of Active Assistance, Learning Assistance, and other interactive help topics.*

STARTING AutoCAD*

When you turn on your computer, the operating systems such as Microsoft® Windows Vista™, Windows® XP Home, and Professional (SP2), Windows® 2000 (SP4), Windows® XP Professional x64 Edition, and Windows Vista® 64-bit will be automatically loaded on it and the Windows screen will be displayed with a number of application icons. You can start AutoCAD by double-clicking on the AutoCAD 2009 icon on the desktop of your computer. You can also load AutoCAD from the Windows taskbar by choosing the **Start** button at the bottom left-corner of the screen (default position) to display the menu. From this menu choose **All Programs** to display the program folders. Now, choose **Autodesk > AutoCAD 2010** folder to display AutoCAD programs and then choose **AutoCAD 2010-English** to start AutoCAD, see Figure 1-1.

*Figure 1-1 Starting AutoCAD 2010 from the **Start** menu*

When you start AutoCAD for the first time, the **Welcome to AutoCAD 2010** page of the **Initial Setup** wizard will be displayed prompting you to specify the industry that closely describes your work to start the default drawing environment, see Figure 1-2. Select the industry that describes your work and choose the **Next** button; the **Optimize your Default Workspace** page will be displayed, prompting you to optimize the default workspace, see Figure 1-3. Select the options according to your requirement and choose the **Next** button; the **Specify a Drawing Template File** page will be displayed prompting you to specify the default drawing template, see Figure 1-4. If you have an existing drawing file template, select the **Use my existing drawing template file** radio button and specify the location of the

*Figure 1-2 The **Welcome to AutoCAD 2010** page of the **Initial Setup** wizard*

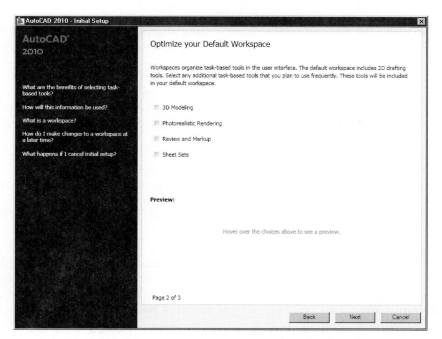

*Figure 1-3 The **Optimize your Default Workspace** page*

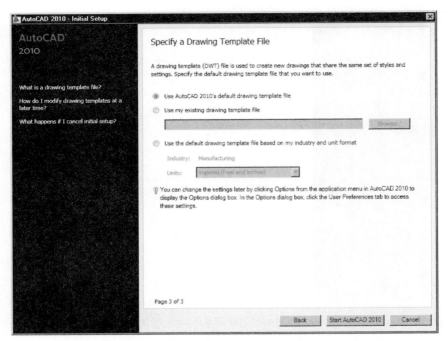

*Figure 1-4 The **Specify a Drawing Template File** page*

template file using the **Browse** button. After setting all options, choose the **Start AutoCAD 2010** button. From now onward, AutoCAD will open according to the specified settings.

 Note
*You can change these initial setup settings according to your requirement at anytime. To do so, choose the **Options** button from the **Application Menu**; the **Options** dialog box will be displayed. Choose the **User Preferences** tab and choose the **Initial Setup** button to change the initial setup parameters.*

In this textbook, following parameters are set in the initial setup:
*Drawing environment: **Manufacturing**, Tools in default workspace: All check boxes selected, Drawing template: **AutoCAD 2010's default drawing template***
So, the default settings for the display of some tool palettes, Ribbon, or tool bars will be based on these parameters.

When you start AutoCAD 2010, the **New Features Workshop** window will be displayed, see Figure 1-5. This window enables you to preview the recent enhancements in AutoCAD. To preview the enhancements, select the **Yes** radio button and then choose the **OK** button from the **New Features Workshop** window; the **New Features Workshop** window will be displayed. This window lists the recent enhancements in AutoCAD in the left pane. When you click on a topic, its description is displayed in the right pane. After previewing the enhancements, close the modified **New Features Workshop** window to start AutoCAD. If you do not want to preview the enhancements, select the **Maybe later** or **No, don't show me this again** radio button from the **New Features Workshop** window.

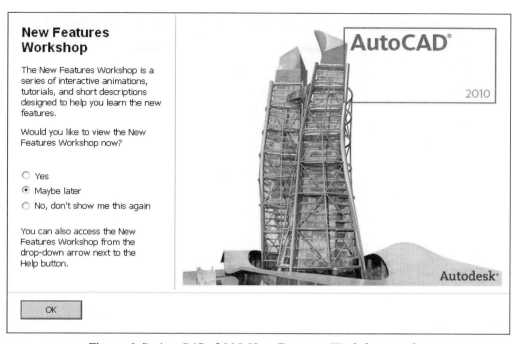

*Figure 1-5 AutoCAD 2010 **New Features Workshop** window*

AUTOCAD SCREEN COMPONENTS*

Various components of the initial AutoCAD screen are drawing area, command window, menu bar, several toolbars, model and layout tabs, and status bar (Figure 1-6). A title bar that has AutoCAD symbol and the current drawing name is displayed on top of the screen.

Drawing Area

The drawing area covers the major portion of the screen. Here, you can draw the objects and use the commands. To draw the objects, you need to define the coordinate points, which can be selected by using your pointing device. The position of the pointing device is represented on the screen by the cursor. There is a coordinate system icon at the lower left corner of the drawing area. The window also has the standard Windows buttons such as close, minimize, scroll bar, and so on, on the top right corner. These buttons have the same functions as for any other standard window.

Command Window

The command window at the bottom of the drawing area has the Command prompt where you can enter the commands. It also displays the subsequent prompt sequences and the messages. You can change the size of the window by placing the cursor on the top edge (double line bar known as the grab bar) and then dragging it. This way you can increase its size to see all the previous commands you have used. By default, the command window displays only three lines. You can also press the F2 key to display **AutoCAD Text window**, which displays the previous commands and prompts.

*Figure 1-6 AutoCAD screen components in AutoCAD **Initial Setup Workspace***

Tip
*You can hide all toolbars displayed on the screen by pressing the CTRL+0 keys or by choosing
View > Clean Screen from the menu bar. To turn on the display of the toolbars again, press
the CTRL+0 keys. Note that the 0 key on the numeric keypad of the keyboard cannot be used
for the **Clean Screen** option. You can also choose the **Clean Screen** button in the Status Bar
to hide all toolbars.*

Application Status Bar

The Status Bar is displayed at the bottom of the screen, as shown in Figure 1-7. It contains
some useful information and buttons that will make it easy to change the status of some
AutoCAD functions. You can toggle between on and off states of most of these functions by
choosing them.

Drawing Coordinates

The information about the coordinates is displayed on the left-corner of the Status Bar. You
can select this coordinate button to toggle between the on and off states. The **COORDS**
system variable controls the type of display of the coordinates. If the value of the **COORDS**
variable is set to 0, the coordinate display is static, that is, the coordinate values displayed in
the Status Bar change only when you specify a point. If the value of the **COORDS** variable is
set to 1 or 2, the coordinate display is dynamic. When the variable is set to 1, AutoCAD
constantly displays the absolute coordinates of the graphics cursor with respect to the UCS

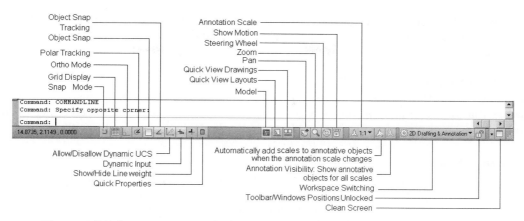

Figure 1-7 *The Status Bar displayed in the* **2D Drafting & Annotation** *workspace*

origin. The polar coordinates (length<angle) are displayed if you are in an AutoCAD command and the **COORDS** variable is set to 2. You can also click on the **Drawing Coordinates** area to change the coordinate status from on to off and vice versa.

Snap Mode
If the **Snap Mode** button is chosen in the Status Bar, the snap mode is on. So, you can move the cursor in fixed increments. The F9 key acts as a toggle key to turn the snap off or on.

Grid Display
The grid lines are used as reference lines to draw objects in AutoCAD. If the **Grid Display** button is chosen in the Status Bar, the grid display is on and the grid lines are displayed on the screen. The F7 function key can be used to turn the grid display on or off.

Ortho Mode
If the **Ortho Mode** button is chosen in the Status Bar, you can draw lines at right angles only. You can use the F8 function key to turn ortho on or off.

Polar Tracking
If you turn the polar tracking on, the movement of the cursor is restricted along a path based on the angle set as the polar angle. Choose the **Polar Tracking** button in the Status Bar to turn the polar tracking on. You can also use the F10 function key to turn on this option. Note that turning the polar tracking on, automatically turns off the ortho mode.

Object Snap
When the **Object Snap** button is chosen in the Status Bar, you can use the running object snaps to snap on to a point. You can also use the F3 function key to turn the object snap on or off. The status of **OSNAP** (off or on) does not prevent you from using the immediate mode object snaps.

Object Snap Tracking
When you choose this button in the Status Bar, the inferencing lines will be displayed.

Inferencing lines are dashed lines that are displayed automatically when you select a sketching tool and track a particular keypoint on the screen. Choosing this button turns the object snap tracking on or off.

Allow/Disallow Dynamic UCS

Choosing this button allows or disallows the use of dynamic UCS. Allowing the dynamic UCS ensures that the XY plane of the UCS got dynamically aligned with the selected face of the model. You can also use the F6 function key to turn the **DUCS** button on or off.

Dynamic Input

The **Dynamic Input** button is used to turn the **Dynamic Input** on or off. Turning it on facilitates the heads-up design approach because all commands, prompts, and dimensional inputs will now be displayed in the drawing area and you do not need to look at the Command prompt all the time. This saves the design time and also increases the efficiency of the user. If the **Dynamic Input** mode is turned on, you will be allowed to enter the commands through the **Pointer Input** boxes, and the numerical values through the **Dimensional Input** boxes. You will also be allowed to select the command options through the **Dynamic Prompt** options in the graphics window. To turn the **Dynamic Input** on or off, use the CTRL+D keys.

Show/Hide Lineweight

Choosing this button in the Status Bar allows you to turn on or off the display of lineweights in the drawing. If this button is not chosen, the display of Lineweight will be turned off.

Quick Properties

If you select a sketched entity when this button is chosen in the Status Bar, the properties of the selected entity will be displayed in a panel.

Model

The **Model** button is chosen by default because you are working in the model space to create drawings. You will learn more about the model space in later chapters.

Layout1

Choose this button to shift the drawing to the layouts (paper space) for creating drawing views.

Quick View Layouts

Choose this button to display a flyout from which you can choose the layout you need to invoke.

Quick View Drawings

Choose this button to display a flyout from which you can choose the drawings you need to invoke.

Pan

This button allows you to view the portion of the drawing that is outside the current display area. To do so, choose this button in the Status Bar, press and hold the left mouse button and then drag the drawing area. Press ESC to exit this command.

Zoom

Choose this button to enlarge the view of the drawing on the screen without affecting the actual size of the objects. You will learn more about zoom in later chapters.

SteeringWheels

The SteeringWheels has a set of navigation tools such as pan, zoom, and so on. You will learn more about the SteeringWheel in the later chapters.

ShowMotion

Choose this button to capture different views in a sequence and animate them when required.

Annotation Scale

The annotation scale controls the size and display of the annotative objects in the model space. The **Annotation Scale** button has a drop-down list that displays all the annotation scales available for the current drawing.

Annotation Visibility

This button is used to control the visibility of the annotative objects that do not support the current annotation scale in the drawing area.

Automatically Add Scale

This button, if chosen, automatically adds all the annotation scales that are set current to all the annotative objects present in the drawing.

Drawing Status Bar

The **Drawing Status Bar** is displayed in between the drawing area and the command window. Choose the **Application Status Bar Menu** arrow and select the **Drawing Status Bar** option from the flyout; the **Drawing Status Bar** will be displayed, as shown in Figure 1-8. Turn on the **Drawing Status Bar**; the **Annotation Scale**, **Annotation Visibility**, and **Automatically Add Scale** buttons will move automatically to the **Drawing Status Bar**. If you turn off the **Drawing Status Bar**, these buttons will move back to the application status bar.

Tip
*In AutoCAD 2010, you can control the display of the options in the Status Bar, Drawing Status Bar, and Text window by choosing the corresponding button from the **Windows** panel of the **View** tab in the **Ribbon**.*

Status Bar Tray Options

The status bar tray options are displayed at the lower-right corner of the screen. These options are used to access the frequently used commands in AutoCAD.

Toolbar/Window Positions Unlocked

The **Toolbar/Window Positions Unlocked** icon is used to lock and unlock the positions of

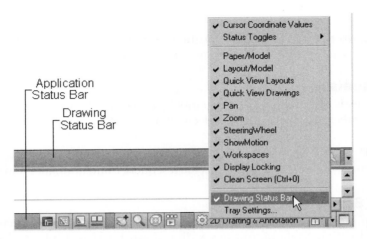

*Figure 1-8 The **Drawing Status Bar***

toolbars and windows. When you click on this icon, a shortcut menu is displayed. Choosing the **Floating Toolbars/Panels** option allows you to lock the current position of the floating toolbars. A checkmark will be displayed in the shortcut menu on the type of toolbars that are currently locked. Choosing the **Docked Toolbars/Panels** option from the shortcut menu allows you to lock the current position of all the docked toolbars. Similarly, you can lock or unlock the position of floating and docked windows, such as the **Properties** window or the **Tool Palettes**. If you move the cursor on the **All** option, a cascading menu is displayed that provides the option to lock and unlock all the toolbars and windows.

Note
The LOCKUI system variable is responsible for the locking and unlocking of the toolbars and windows. The following are the values of the system variable:

Lockui<0> No toolbar or window locked
Lockui<1> Locks all docked toolbars
Lockui<2> Locks all docked windows
Lockui<4> Locks all floating toolbars
Lockui<8> Locks all floating windows

Status Bar Menu
Click on the **Status Bar Menu** arrow to display the drop-down list showing options of the Status Bar. Select the tool that you want to be displayed or clear the check mark near the tool to remove it from display.

Clean Screen

The **Clean Screen** button is on the lower right corner of the screen. This button, when chosen, displays an expanded view of the drawing area by hiding all the toolbars except the command window, Status Bar, and menu bar. The expanded view of the drawing area can also be displayed by choosing **View > Clean Screen** from the menu bar or by using the CTRL+0 keys. Choose the **Clean Screen** button again to restore the previous display state.

Plot/Publish Details Report Available

This icon is displayed when some plotting or a publishing activity was performed in the background. When you click on this icon, the **Plot and Publish Details** dialog box, which provides the details about the plotting and publishing activity, will be displayed. You can copy this report to the clipboard by choosing the **Copy to Clipboard** button from the dialog box.

Manage Xrefs

The **Manage Xrefs** icon is displayed whenever an external reference drawing is attached to the selected drawing. This icon displays a message and an alert whenever the Xreffed drawing needs to be reloaded. The detailed information regarding the status of each Xref in the drawing and the relation between the various Xrefs, click on the **Manage Xrefs** icon; the **External References Palette** will be displayed. The Xrefs are discussed in detail in Chapter 18, Understanding External References.

CAD Standards

The **CAD Standards** icon is displayed whenever a standard drawing is associated with the selected drawing to compare the standards. This icon displays a message and an alert whenever a standard violation occurs in the drawing. The drawing can be checked for the standard violation and edited using the **Check Standards** dialog box, which is invoked by clicking on the **CAD Standards** icon.

Unreconciled New Layers

The **Unreconciled New Layers** icon is displayed whenever a new layer is inserted or an external reference is attached to the current drawing. This icon displays a message when a new layer is added. This helps in preventing any new layer being added to the current drawing without the users information.

Validate Digital Signatures

The **Validate Digital Signatures** icon is displayed whenever an AutoCAD drawing has a valid digital signature. You can validate the digital signature by clicking on this icon.

INVOKING COMMANDS IN AutoCAD*

On starting AutoCAD, when you are in the drawing area, you need to invoke AutoCAD commands to perform any operation. For example, to draw a line, first you need to invoke the **LINE** command and then define the start point and the endpoint of the line. Similarly, if you want to erase objects, you must enter the **ERASE** command and then select the objects for erasing. AutoCAD has provided the following methods to invoke the commands:

Keyboard	Ribbon	Application Menu	Tool Palettes
Tool Palettes	Menu Bar	Shortcut menu	Toolbar

Keyboard

You can invoke any AutoCAD command from the keyboard by typing the command name and then pressing the ENTER key. If the **Dynamic Input** is on and the cursor is in the drawing area, by default, the command will be entered through the **Pointer Input** box. The **Pointer Input** box is a small box displayed on the right of the cursor, as shown in Figure 1-9. However, if the cursor is currently placed on any toolbar or menu bar, or if the **Dynamic Input** is turned off, the command will be entered through the Command prompt. Before you enter a command, the Command

*Figure 1-9 The **Pointer Input** box displayed when the **Dynamic Input** is on*

prompt is displayed as the last line in the command window area. If it is not displayed, you must cancel the existing command by pressing the ESC (Escape) key. The following example shows how to invoke the **LINE** command using the keyboard:

> Command: **LINE** or **L** [Enter]   (L is command alias)

Ribbon

This is a new method of invoking commands while working with annotation and 3D modeling. Most of the commands for creating, modifying, and annotating a 2D & 3D design are available in the **Ribbon** instead of being spread out in the entire drawing area in different toolbars and menus, see Figure 1-10.

*Figure 1-10 The **Ribbon** for the **2D Drafting & Annotation** workspace*

When you start AutoCAD session for the first time, the **Ribbon** is displayed horizontally below the **Quick Access** toolbar, by default. The **Ribbon** consists of various tabs. The tabs have different panels, which in turn, have buttons arranged in rows. Some of the panels and buttons have small black down arrow. This indicates that the corresponding panels and buttons have some more buttons in the form of flyout. Click on this down arrow to access hidden buttons. If you choose a button from the flyout, the corresponding command will be invoked and the button that you have chosen will be displayed in the panel. For example, to draw a circle using the **2-Point** option, click on the down arrow next to the **Center, Radius** button in the **Draw** panel of the **Home** tab; a flyout will be displayed. Choose the **2-Point** button from the flyout and draw the circle. You will observe that the **2-Point** button is displayed in place of the **Center, Radius** button. In this textbook, this command selection sequence will be referenced as, choose **Home > Draw > Center, Radius > 2-Point** from the **Ribbon**.

Choose the down arrow to expand the panel. You will notice that a push pin is at the left end of the panel. Click on the push pin to keep the panel in the expanded state. Also, some of the panels have an inclined arrow at the lower-right corner. When you left click on an inclined arrow, a dialog box is displayed. You can define the setting of the corresponding panel in the dialog box.

You can reorder the panels in the tab. To do so, press and hold the left mouse button on the panel to be moved and drag it to the required position. To undock the **Ribbon**, right-click on the blank space in the **Ribbon** and choose the **Undock** option. You can move, resize, anchor, and auto-hide the **Ribbon** using the shortcut menu that will be displayed when you right-click on the heading strip. To anchor the floating **Ribbon** to the left or right of the drawing area in the vertical position, right-click on the heading strip of the floating **Ribbon**; the shortcut menu is displayed. Choose the corresponding option from this shortcut menu. The **Auto-hide** option will hide the **Ribbon** into the heading strip and will display it only when you move the cursor over this strip.

You can customize the display of tabs and panels in the **Ribbon**. To customize the **Ribbon**, right-click on any one of the buttons in it; a shortcut menu will be displayed. On moving the cursor over one of the options, a flyout will be displayed with a tick mark before all options and the corresponding tab or panel will be displayed in the **Ribbon**. Select/clear appropriate option to display/hide a particular tab or panel.

Application Menu

The **Application Menu** is available at the top-left of the AutoCAD window. It contains some of the tools that are available in the **Standard** toolbar. Click the down arrow on the **Application Menu** to display the tools. Alternatively, press ALT+F to display the tools in the **Application Menu**, as shown in Figure 1-11. You can search a command using the search field on the top of the **Application Menu**. To search a command, enter the complete or partial name of the command in the search field; the possible command will be listed. If you click on a command from the list, the corresponding command will get activated.

By default, the **Recent Document** button is chosen in the **Application Menu**. Therefore, the recently opened drawings will be listed. If you have opened multiple drawing files, choose the **Open Documents** button; the documents that are opened will be listed in the **Application Menu**. To set the preferences of the file, choose the **Options** button available at the bottom-right of the **Application Menu**. To exit AutoCAD, choose the **Exit** button next to the **Options** button.

Tool Palettes

AutoCAD has provided **Tool Palettes** as an easy and convenient way of placing and sharing hatch patterns and blocks in the current drawing. By default, the **Tool Palettes** are not displayed. Choose **View > Palettes > Tool Palettes Windows** from the **Ribbon** or choose the CTRL+3 keys to display the **Tool Palettes** as a window on the right of the drawing area. You can resize the **Tool Palettes** using the resizing cursor that is displayed when you place the cursor on the top or bottom extremity of the **Tool Palettes.** The **Tool Palettes** are discussed in detail in Chapter 15, *Hatching Drawings*.

*Figure 1-11 The **Application Menu***

Menu Bar

You can also select commands from the menu bar. Menu Bar is not displayed by default. To display the menu bar, choose the down arrow in the **Quick Access** toolbar; a flyout is displayed. Choose the **Show Menu Bar** option from it; the menu bar will be displayed. As you move the cursor over the menu bar, different titles are highlighted. You can choose the desired item by pressing the pick button of your pointing device. Once the item is selected, the corresponding sub menu is displayed directly under the title. You can invoke a command from the menu by pressing the pick button of your pointing device. Some of the menu items display an arrow on the right side, which indicates that they have a cascading menu. The cascading menu provides various options to execute the same AutoCAD command. You can display the cascading menu by choosing the menu item or by moving the arrow pointer to the right of that item. You can then choose any item from the cascading menu by highlighting the item or command and pressing the pick button of your pointing device. For example, to draw an ellipse using the **Center** option, choose the **Draw** menu and then choose the **Ellipse** option; a cascading menu will be displayed. From the cascading menu, choose the **Center** option. In this text, this command selection sequence will be referenced as choosing **Draw > Ellipse > Center** from the menu bar.

Toolbar

Toolbars are not displayed by default. To display a toolbar, first display the menu bar and choose **Tools > Toolbars > AutoCAD** from the menu bar; the list of toolbars will be displayed. Select the required toolbar. In a toolbar a set of buttons representing various AutoCAD commands are grouped together. When you move the cursor over the button of a toolbar, the button gets lifted and a three dimensional (3D) box encloses it. The tooltip (name of the button) is also displayed below the button. Once you locate the desired button, the command associated with it can be invoked by choosing it.

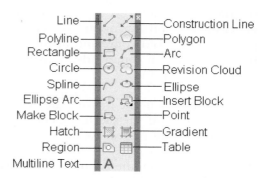

*Figure 1-12 The **Draw** toolbar*

For example, you can invoke the **LINE** command by choosing the **Line** button from the **Draw** toolbar, see Figure 1-12.

Some of the buttons in a toolbar have a small triangular arrow at the lower right corner. This indicates that the button has a flyout attached to it. If you hold the triangular arrow button down, a flyout that contains the options for the command will be displayed. Choose a command from the toolbar; the Command prompts are displayed in the command window.

Moving and Resizing Toolbars

The toolbars can be moved anywhere on the screen by placing the cursor on the strip and then dragging it to the desired location. You must hold the pick button down while dragging. While moving the toolbars, you can dock them to the top or sides of the screen by dropping them in the docking area. You may also prevent docking by holding the CTRL key when moving the toolbar to a desired location. You can also change the size of the toolbars by placing the cursor anywhere on the border of the toolbar where it takes the shape of a double arrow (Figure 1-13), and then pulling it in the desired direction (Figure 1-14). You can also customize toolbars to meet your requirements.

*Figure 1-13 Reshaping the **Draw** toolbar*

*Figure 1-14 The **Draw** toolbar reshaped*

Shortcut Menu

AutoCAD has provided shortcut menus as an easy and convenient way of invoking the recently used commands. These shortcut menus are context-sensitive which means that the commands present in them are dependent on the place/ object for which they are displayed. This shortcut menu is invoked by right-clicking and is displayed at the cursor location. You can right-click anywhere in the drawing area to display the general shortcut menu. It generally contains an option to select the previously invoked command again, apart from the common commands for Windows, refer to Figure 1-15.

Figure 1-15 *Shortcut menu with the recently used commands*

If you right-click in the drawing area while a command is active, the shortcut menu is displayed. It contains the options of that particular command. Figure 1-16 shows the shortcut menu when the **POLYLINE** command is active.

If you right-click on the **Layout** tabs, the shortcut menu displayed contains the options for layouts (Figure 1-17).

You can also right-click on the command window to display the shortcut menu. This menu displays the six most recently used commands and some of the window options like **Copy** and **Paste** (Figure 1-18). The commands and their prompt entries are displayed in the History window (previous command lines not visible) and can be selected, copied, and pasted in the command line using the shortcut menu. As you press the up arrow key, the previously entered commands are displayed in the command window. Once the desired command is displayed at the Command prompt, you can execute it by simply pressing the ENTER key. You can also copy and edit any previously invoked command by locating it in the History window and then selecting the lines. Right-click in the command window to display the shortcut menu (Figure 1-19); select copy, and then paste the selected lines in the command line. Once the lines are pasted, you can edit them.

You can right-click on the coordinate display area of the Status Bar to display the shortcut menu. This menu contains the options to modify the display of coordinates, as shown in Figure 1-19. You can also right-click on any of the toolbars to display the shortcut menu from where you can choose any toolbar to be displayed.

Tip
A shortcut menu is available for any situation while working in AutoCAD. You should try to make use of it frequently by right-clicking at various positions.

AutoCAD DIALOG BOXES

There are certain commands, which when invoked, display a dialog box. A dialog box is a convenient method of a user interface. When you choose an item in the menu bar with the ellipses [**...**], it displays the dialog box. For example, **Options** in the **Tools** menu displays the **Options** dialog box. A dialog box contains a number of parts like the dialog label, radio buttons, text or edit boxes, check boxes, slider bars, image boxes, and command buttons.

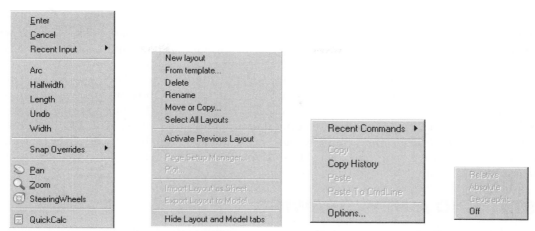

Figure 1-16 Shortcut menu with the **POLYLINE** command active

Figure 1-17 Shortcut menu for the **Layout** tab

Figure 1-18 Command line window shortcut menu

Figure 1-19 The Status Bar shortcut menu

These components are also referred to as tiles. Some of the components of a dialog box are shown in Figure 1-20.

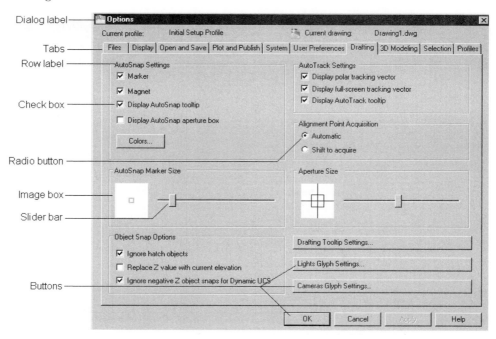

Figure 1-20 The components of a dialog box

You can select the desired tile using the pointing device, which is represented by an arrow when a dialog box is invoked. The title bar displays the name of the dialog box. The **tabs** specify the various sections with a group of related options under them. The **check boxes** are toggle buttons for making the particular option available or unavailable. The **drop-down list** displays an item and an arrow on the right which when selected displays a list of items to choose from. You can make a selection in the **radio buttons**. Only one can be selected at a time. The **image box** displays the preview image of the item selected. The **text box** is an area where you can enter a text like a file name. It is also called an **edit box**, because you can make any change to the text entered. In some dialog boxes, there is the **[...]** button, which displays another related dialog box. There are certain **command buttons** (OK, Cancel, Help) at the bottom of the dialog box. The name implies their functions. The button with a dark border is the default button. The dialog box has a **Help** button for getting help on the various features of the dialog box.

STARTING A NEW DRAWING

Application Menu:	New > Drawing	**Command:**	NEW or QNEW
Toolbar:	Quick Access > QNew	**Menu Bar:**	New > Drawing

You can open a new drawing using the **QNEW** command. When you invoke the **QNew** command, by default AutoCAD will display the **Select template** dialog box, as shown in Figure 1-21. This dialog box displays a list of default templates available in AutoCAD 2009. The default template is *acad.dwt*, which starts the 2D drawing environment.

Figure 1-21   *The **Select template** dialog box*

You can select the *acad3D.dwt* template to start the 3D modeling environment. Alternatively, you can select any other template to start a new drawing that will use the settings of the selected template. You can also open any drawing without using any template either in metric or imperial system. To do so, choose the down arrow on the right of the **Open** button and select either the **Open with no Template-Metric** or the **Open with no Template-Imperial** option from the flyout.

You can also open a new drawing using the **Use a Wizard** and **Start from Scratch** options from the **Create New Drawing** dialog box. By default, this dialog box is not invoked. To turn on the display of the **Create New Drawing** dialog box, enter **STARTUP** in the command window and then enter **1** as the new value for this system variable. After setting 1 as the new value for the system variable, whenever you invoke the **QNEW** command, the **Create New Drawing** dialog box will be displayed, as shown in Figure 1-22. The options in this dialog box are discussed next.

*Figure 1-22 The **Create New Drawing** dialog box*

Open a Drawing
By default, this option is not available.

Start from Scratch
When you choose the **Start from Scratch** button (Figure 1-23), AutoCAD provides you with options to start a new drawing that contains the default AutoCAD setup for Imperial (*Acad.dwt*) or Metric drawings (*Acadiso.dwt*). If you select the Imperial default setting, the limits are 12X9, text height is 0.20, and dimensions and linetype scale factors are 1.

Use a Template
When you choose the **Use a Template** button in the **Create New Drawing** dialog box, AutoCAD displays a list of templates, see Figure 1-24. The default template file is *acad.dwt* or *acadiso.dwt*, depending on the installation. You can directly start a new file in the 2D sketching

Figure 1-23 *Options that are displayed to start a new drawing when you choose the **Start from Scratch** button*

environment by selecting the *acad.dwt* or *acadiso.dwt* template. If you use a template file, the new drawing will have the same settings as specified in the template file. All the drawing parameters of the new drawing such as units, limits, and other settings are already set according to the template file used. The preview of the template file selected is displayed in the dialog box. You can also define your own template files that are customized to your requirements (see Chapter 13, *Template Drawings*). To differentiate the template files from the drawing files, the template files have a *.dwt* extension whereas the drawing files have a *.dwg* extension. Any drawing file can be saved as a template file. You can use the **Browse** button to select other template files. When you choose the **Browse** button, the **Select a template file** dialog box is displayed with the **Template** folder open, displaying all the template files.

Figure 1-24 *The default templates that are displayed when you choose the **Use a Template** button*

Use a Wizard

The **Use a Wizard** option allows you to set the initial drawing settings before actually starting a new drawing. When you choose the **Use a Wizard** button, AutoCAD provides you with the option of using the **Quick Setup** or **Advanced Setup**, see Figure 1-25. In the **Quick Setup**, you can specify the units and the limits of the work area. In the **Advanced Setup**, you can set the units, limits, and the different types of settings for a drawing.

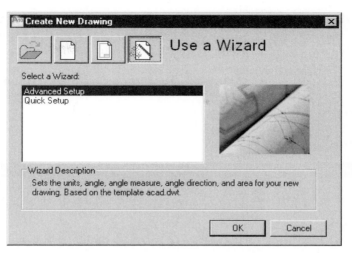

Figure 1-25 *The wizard options that are displayed when you choose the **Use a Wizard** button*

Advanced Setup

This option allows you to preselect the parameters of a new drawing such as the units of linear and angular measurements, type and direction of angular measurements, approximate area desired for the drawing, precision for displaying the units after decimal, and so on. When you select the **Advanced Setup** wizard option from the **Create New Drawing** dialog box and choose the **OK** button, the **Advanced Setup** dialog box is displayed. The **Units** page is displayed by default, as shown in Figure 1-26.

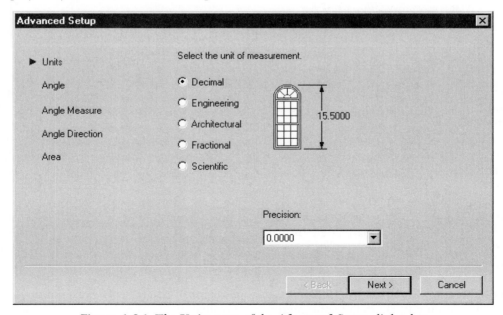

Figure 1-26 *The **Units** page of the **Advanced Setup** dialog box*

This page is used to set the units for measurement in the current drawing. You can select the required unit of measurement by selecting the respective radio button. You will notice that the preview image is modified accordingly. The different units of measurement you can choose from are Decimal, Engineering, Architectural, Fractional, and Scientific. You can also set the precision for the measurement units by selecting it from the **Precision** drop-down list.

Choose the **Next** button to open the **Angle** page, as shown in Figure 1-27. You will notice that an arrow appears on the left of **Angle** in the **Advanced Setup** dialog box. This suggests that this page is current.

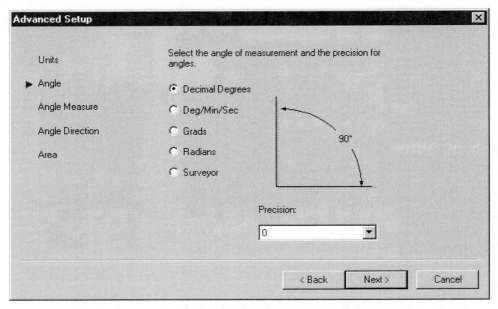

*Figure 1-27 The **Angle** page of the **Advanced Setup** dialog box*

This page is used to set the units for angular measurements and its precision. The units for angle measurement are Decimal Degrees, Deg/Min/Sec, Grads, Radians, and Surveyor. The units for angle measurement can be set by selecting any one of these radio buttons as required. The preview of the selected angular unit is displayed on the right of the radio buttons. The precision format changes automatically in the **Precision** drop-down list depending on the angle measuring system selected. You can then select the precision from the drop-down list.

The next page is the **Angle Measure** page, as shown in Figure 1-28. This page is used to select the direction of the base angle from which the angles will be measured. You can also set your own direction by selecting the **Other** radio button and then entering the value in its edit box. This edit box is available when you select the **Other** radio button.

Choose **Next** to display the **Angle Direction** page (Figure 1-29) to set the orientation for the angle measurement. By default the angles are positive, if measured in a counterclockwise direction.

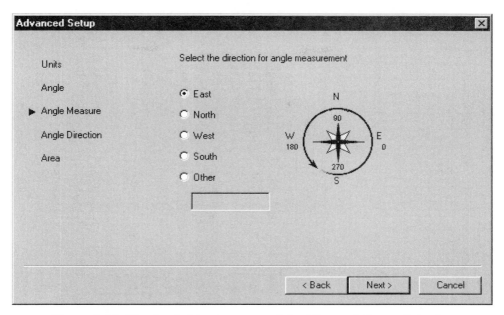

Figure 1-28 *The* **Angle Measure** *page of the* **Advanced Setup** *dialog box*

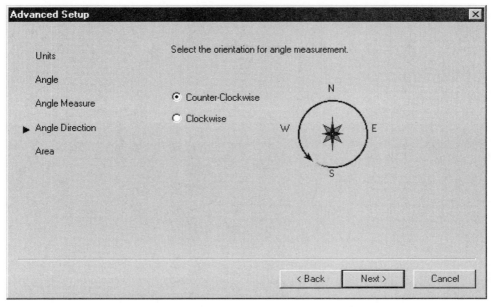

Figure 1-29 *The* **Angle Direction** *page of the* **Advanced Setup** *dialog box*

This is because the **Counter-Clockwise** radio button is selected. If you select the **Clockwise** radio button, the angles will be considered positive when measured in the clockwise direction. To set the limits of the drawing, choose the **Next** button. The **Area** page is displayed, as shown in Figure 1-30. You can enter the width and length of the drawing area in the respective edit boxes.

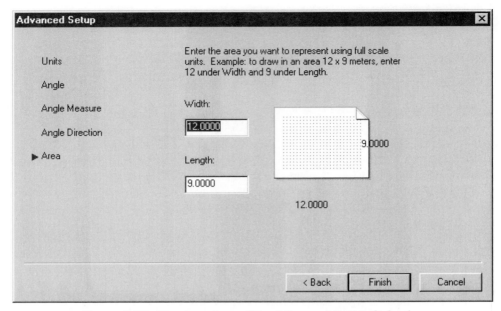

*Figure 1-30 The **Area** page of the **Advanced Setup** dialog box*

Note

*Even after you increase the limits of the drawing, the drawing display area is not increased. You need to invoke the **ZOOM** command and then invoke the **All** option to increase the drawing display area.*

Quick Setup

When you select **Quick Setup** and choose the **OK** button, the **Quick Setup** dialog box is displayed. This dialog box has just two pages: **Units** and **Area**. The **Units** page is opened by default, as shown in Figure 1-31. The options in the **Units** page are similar to those in the **Units** page of the **Advanced Setup** dialog box. The only difference is that you cannot set the precision for the units in this dialog box.

Choose **Next** to display the **Area** page, as shown in Figure 1-32. The **Area** page is similar to that of the **Advanced Setup** dialog box where you can set the drawing limits.

Tip

*By default, when you open an AutoCAD session, a drawing opens automatically. But you can open a new drawing using options such as **Start from Scratch** and **Wizards** before entering into AutoCAD environment using the **Startup** dialog box. As mentioned earlier, the display of the **Startup** dialog box is turned off by default. Refer to the section of **Starting a New Drawing** to know how to turn on the display of this dialog box.*

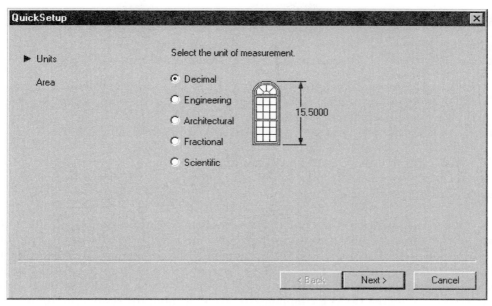

Figure 1-31 *The **Units** page of the **QuickSetup** dialog box*

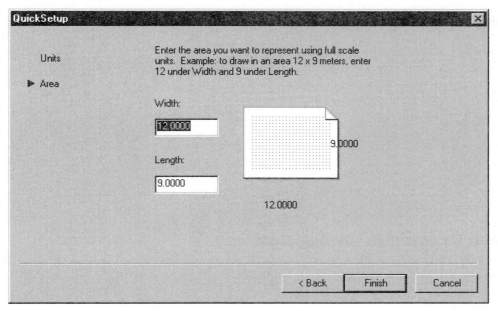

Figure 1-32 *The **Area** page of the **QuickSetup** dialog box*

SAVING YOUR WORK

Application Menu:	SAVEAS, SAVE	Command:	QSAVE, SAVEAS, SAVE
Toolbar:	Quick Access > Save	Menu Bar:	Save or Save As

In AutoCAD or any computer system, you must save your work before you exit from the drawing editor or turn the system off. Also, it is recommended that you save your drawings after regular time intervals, so that in the event of a power failure or an editing error, all work saved before the problem started will be retained.

AutoCAD has provided the **QSAVE**, **SAVEAS**, and **SAVE** commands that allow you to save your work. These commands allow you to save your drawing by writing it to a permanent storage device, such as a hard drive or in any removable drive.

When you choose **Save** from the **Quick Access toolbar** or **Application Menu**, the **QSAVE** command is invoked. If the current drawing is unnamed and you save the drawing for the first time in the present session, the **QSAVE** command will prompt you to enter the file name in the **Save Drawing As** dialog box, as shown in Figure 1-30. You can enter the name for the drawing and then choose the **Save** button. If you have saved a drawing file once and then edited it, you can use the **QSAVE** command to save it, without the system prompting you to enter a file name. This allows you to do a quick save.

When you choose the **SAVEAS** option from the **Application Menu**, the **Save Drawing As** dialog box is displayed, similar to that shown in Figure 1-33. Even if the drawing has been saved with a file name, this command gives you an option to save it with a different file name. In addition to saving the drawing, it sets the name of the current drawing to the file name you specify, which is displayed in the title bar. This command is used when you want to save a previously saved drawing under a different file name. You can also use this command when you make certain changes to a template and want to save the changed template drawing but leave the original template unchanged.

The **SAVE** command is the most rarely used command and can be invoked only from the command line by entering **SAVE** at the Command prompt. It is similar to the **SAVEAS** command and displays the **Save Drawing As** dialog box. With this command, you can save a previously saved drawing under a different file name, but this command does not set it as the current drawing.

Save Drawing As Dialog Box

The **Save Drawing As** dialog box displays the information related to the drawing files on your system. The various components of the dialog box are described next.

Places List

A column of icons is displayed on the left side of the dialog box. These icons contain the shortcuts to the folders that are frequently used. You can quickly save your drawings in one of these folders. The **History** folder displays the list of the most recently saved drawings. You

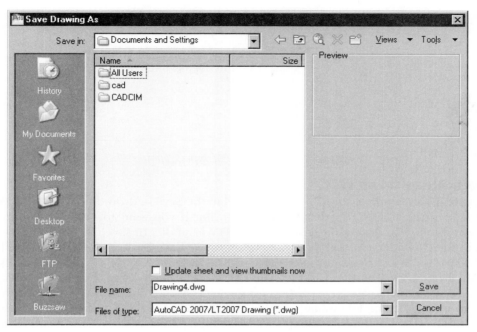

Figure 1-33 *The* **Save Drawing As** *dialog box*

can save your personal drawings in the **My Documents** or the **Favorites** folder. The **FTP** folder displays the list of the various FTP sites that are available for saving the drawing. By default, no FTP sites are shown in the dialog box. To add a FTP site to the dialog box, choose the **Tools** button on the upper-right corner of the dialog box to display a shortcut menu and select **Add/Modify FTP Locations**. The **Desktop** folder displays the list of contents on the desktop. The **Buzzsaw** icons connect you to their respective pages on the Web. You can add a new folder in this list for an easy access by simply dragging the folder on to the **Places** list area. You can rearrange all these folders by dragging them and then placing them at the desired locations. It is also possible to remove the folders, which are not in frequent use. Right-click on the particular folder and then select **Remove** from the shortcut menu.

File name Edit Box

To save your work, enter the name of the drawing in the **File name** edit box by typing the file name or selecting it from the drop-down list. If you select the file name, it automatically appears in the **File name** edit box. If you have already assigned a name to the drawing, the current drawing name is taken as the default name. If the drawing is unnamed, the default name *Drawing1* is displayed in the **File Name** edit box. You can also choose the down arrow at the right of the edit box to display the names of the previously saved drawings and choose a name here.

Files of type Drop-Down List

The **Files of type** drop-down list (Figure 1-34) is used to specify the drawing format in which you want to save the file. For example, to save the file as an AutoCAD 2004 drawing file, select **AutoCAD 2004/LT 2004 Drawing (*.dwg)** from the drop-down list.

*Figure 1-34 The **Files of type** drop-down list*

Save in Drop-Down List

The current drive and path information is listed in the **Save in** drop-down list. AutoCAD will initially save the drawing in the default folder, but if you want to save the drawing in a different folder, you have to specify the path. For example, to save the present drawing as *house* in the *C1* folder, choose the arrow button in the **Save in** drop-down list to display the drop-down list. Select **C**: from the drop-down list; all folders in the C drive will be listed in the **File** list box. Double-click on the **C1** folder, if it is already listed there or create a folder C1 by choosing the **Create New Folder** button. Select *house* from the list, if it is already listed there, or enter it in the **File name** edit box and then choose the **Save** button. Your drawing (*house*) will be saved in the *C1* folder (*C:\C1\house.dwg*). Similarly, to save the drawing in the D drive, select **D:** in the **Save in** drop-down list.

Tip
The file name you enter to save a drawing should match its contents. This helps you to remember the drawing details and makes it easier to refer to them later. Also, the file name can be 255 characters long and can contain spaces and punctuation marks.

Views List

The **Views** drop-down list has options for the type of listing of files and displaying the preview images (Figure 1-35).

List, Details, Thumbnails, and Preview Options

*Figure 1-35 The **Views** list*

If you choose the **Details** option, it will display the detailed information about the files (size, type, date, and time of modification) in the **Files** list box. In the detailed information, if you click on the **Name** label, the files are listed with the names in alphabetical order. If you double-click on the **Name** label, the files will be listed in reverse order. Similarly, if you click on the **Size** label, the files are listed according to their size in ascending order. Double-clicking on the **Size** label will list the files in descending order of size. Similarly, you can click on the **Type** label or the **Modified** label to list the files accordingly. If you choose the **List** option, all files present in the current folder will be listed in the **File** list box. If you select the **Preview** option, the list box displays the **Preview** image box wherein the bitmap image of the file chosen is displayed. If cleared, the **Preview** image box is not displayed. If you select the **Thumbnails** option, the list box displays the preview of all the drawings, along with their names displayed at the bottom of the drawing preview.

Create New Folder Button

 If you choose the **Create New Folder** button, AutoCAD creates a new folder under the name **New Folder**. The new folder is displayed in the **File** list box. You can accept the name or change it to your requirement.

Up one level Button

 The **Up one level** button displays the folders that are up by one level. For example, if you are in the *Sample* subfolder of the *AutoCAD 2010* folder, then choosing the **Up one level** button will take you to the *AutoCAD 2010* folder.

Search the Web

 It displays the **Browse the Web** dialog box that enables you to access and store AutoCAD files on the Internet. You can also use the ATL+3 keys to browse the Web when this dialog box is available on the screen.

Tools List

The **Tools** drop-down list (Figure 1-36) has an option for adding or modifying the FTP sites. These sites can then be browsed from the FTP shortcut in the **Places** list. The **Add Current Folder to Places** and **Add to Favorites** options add the folder displayed in the **Save in** edit box to the **Places** list or to the Favorites folder. The **Options** button displays the **Saveas Options** dialog box where you can save the proxy images of the custom objects. It has the **DWG Options** and **DXF Options** tabs. The **Security Options** button displays the **Security Options** dialog box, which is used to configure the security options of the drawing.

*Figure 1-36 The **Tools** list*

AUTOMATIC TIMED SAVE

AutoCAD allows you to save your work automatically at specific intervals. To change the time intervals, you can enter the intervals duration in minutes in the **Minutes between saves** text box in the **File Safety Precautions** area in the **Options** dialog box (**Open and Save** tab). This dialog box can be invoked by choosing the **Options** button from the **Application Menu**. Depending on the power supply, hardware, and type of drawings, you should decide on an appropriate time and assign it to this variable. AutoCAD saves the drawing with the file extension *.sv$*. You can also change the time interval by using the **SAVETIME** system variable.

 Tip
Although the automatic save saves your drawing, after a certain time interval, you should not completely depend on it because the procedure for converting the ac$ file into a drawing file is cumbersome. Therefore, it is recommended that you save your files regularly, using the **QSAVE** *or* **SAVEAS** *commands.*

CREATION OF BACKUP FILES

If the drawing file already exists and you use **SAVE** or **SAVEAS** commands to update the current drawing, AutoCAD creates a backup file. AutoCAD takes the previous copy of the drawing and changes it from a file type *.dwg* to *.bak*, and the updated drawing is saved as a drawing file with the *.dwg* extension. For example, if the name of the drawing is *myproj.dwg*, AutoCAD will change it to *myproj.bak* and save the current drawing as *myproj.dwg*.

Changing Automatic Timed Saved and Backup Files into AutoCAD Format

Sometimes, you may need to change the automatic timed saved and backup files into AutoCAD format. To change the backup file into an AutoCAD format, open the folder, in which you have saved the backup or the automatic timed saved drawing using **My Computer** or **Windows Explorer**. Choose the **Tools > Folder Options** from the menu bar to invoke the **Folder Option**s dialog box. Choose the **View** tab and under the **Advanced settings** area, and clear the **Hide extensions for known file types** text box, if selected. Exit the dialog box. Rename the automatic saved drawing or the backup file with a different name and also change the extension of the drawing from *.sv$* or *.bak* to *.dwg*. After you rename the drawing, you will notice that the icon of the automatic saved drawing or the backup file is replaced by the AutoCAD icon. This indicates that the automatic saved drawing or the backup file is changed to an AutoCAD drawing.

Using the Drawing Recovery Manager to Recover Files

The automatically saved files can also be retrieved using the **Drawing Recovery Manager**. If the automatic save operation is performed in the drawing and the system crashes accidentally, the next time you run AutoCAD, the **Drawing Recovery** dialog box will be displayed, as shown in Figure 1-37.

The dialog box informs you that the program unexpectedly failed and you can open the most suitable among the backup files created by AutoCAD. Choose the **Close** button from the **Drawing Recovery** dialog box; the **Drawing Recovery Manager** is displayed on the left of the drawing area, as shown in Figure 1-38. The **Backup Files** rollout lists the original files, backup files, and the automatically saved files. Select the file; its preview will be displayed in the **Preview** rollout. Also, the information corresponding to the selected file is displayed in the **Details** rollout. To open the backup file, double-click on its name in the **Backup Files** rollout. Alternatively, right-click on the file name and choose **Open** from the shortcut menu. It is recommended that you save the backup file at the desired location before you start working on it.

Tip
*You can open the **Drawing Recovery Manager** again by choosing **Drawing Utilities > Open the Drawing Recovery Manager** from the **Application Menu** or by entering **DRAWINGRECOVERY** at the Command prompt.*

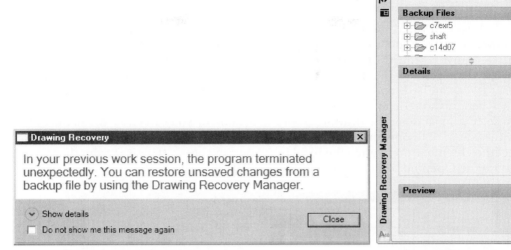

Figure 1-37 The **Drawing Recovery** dialog box

Figure 1-38 The **Drawing Recovery Manager**

CLOSING A DRAWING

You can use the **CLOSE** command to close the current drawing file without actually quitting AutoCAD. If you choose **Close** from the **Application Menu** or enter **CLOSE** at the Command prompt, the current drawing file will be closed. If multiple drawing files are opened, choose **Close > All Drawings** from the **Application Menu**. If you have not saved the drawing after making the last changes to it and you invoke the **CLOSE** command, AutoCAD displays a dialog box that allows you to save the drawing before closing. This box gives you an option to discard the current drawing or the changes made to it. It also gives you an option to cancel the command. After closing the drawing, you are still in AutoCAD from where you can open a new or an already saved drawing file. You can also use the close button (**X**) of the drawing area to close the drawing.

Note
You can close a drawing even if a command is active.

OPENING AN EXISTING DRAWING

Application Menu: Open > Drawing	**Toolbar:**	Quick Access > Open
Menu Bar:	File > Open	**Command:**	OPEN

You can open an existing drawing file that has been saved previously. There are three methods that can be used to open a drawing file, by using the **Select File** dialog box, by using the **Startup** dialog box, and by **Dragging and Dropping**.

Opening an Existing Drawing Using the Select File Dialog Box

If you are already in the drawing editor, and you want to open a drawing file, choose the **Open** button from the **Quick Access** toolbar; the **Select File** dialog box will be displayed. Alternatively, invoke the **OPEN** command to display the **Select File** dialog box, see Figure 1-39. You can select the drawing to be opened using this dialog box. This dialog box is similar to the standard dialog boxes. You can choose the file you want to open from the folder in which it is stored. You can change the folder from the **Look in** drop-down list. You can then select the name of the drawing from the list box or you can enter the name of the drawing file you want to open in the **File name** edit box. After selecting the drawing file, you can select the **Open** button to open the file. Here, you can choose *Drawing1* from the list and then choose the **Open** button to open the drawing.

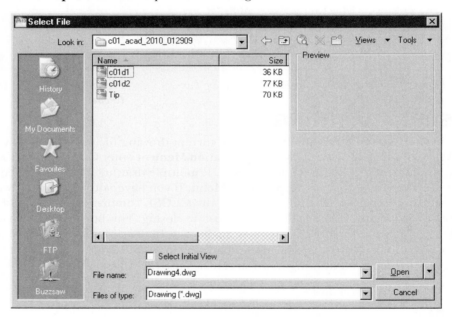

*Figure 1-39 The **Select File** dialog box*

When you select a file name, its image is displayed in the **Preview** box. If you are not sure about the file name of a particular drawing but know the contents, you can select the file names and look for the particular drawing in the **Preview** box. You can also change the file type by selecting it in the **Files of type** drop-down list. Apart from the *dwg* files, you can open the *dwt* (template) files or the *dxf* files. You have all the standard icons in the **Places** list, which can be used to open the drawing files from different locations. The **Open** button has a drop-down list, as shown in Figure 1-40. You can choose a method for opening the file using this drop-down list. These methods are discussed next.

*Figure 1-40   The **Open** drop-down list*

Open Read-Only

To view a drawing without altering it, you must select the **Open Read-Only** option from the drop-down list. In other words, read only protects the drawing file from changes. AutoCAD does not prevent you from editing the drawing, but if you try to save the opened drawing with the original file name, AutoCAD warns you that the drawing file is **write protected**. However, you can save the edited drawing to a file with a different file name using the **SAVEAS** command. This way you can preserve your drawing.

Partial Open

The **Partial Open** option enables you to open only a selected view or a selected layer of a selected drawing. This option can be used to edit small portions of a complicated drawing and then save it with the complete drawing. When you choose the **Partial Open** option from the **Open** drop-down list, the **Partial Open** dialog box (Figure 1-41) is displayed, which contains the different views and layers of the selected drawing. When you select a check box for a layer and then choose the **Open** button, only the objects drawn in that particular layer for the drawing are displayed in the new drawing window. You can make the changes and then save it. For example, in the *C/Program Files/AutoCAD 2010* folder, double-click on the **Sample** folder and then choose *Blocks and Tables - Metric.dwg* from the list. Now, choose the down arrow on the right of the **Open** button to display the drop-down list and choose **Partial Open**. All the views and layers of this drawing are displayed in the **Partial Open** dialog box. Select the check box on the right of the layer that you want to open. When you choose the **Open** button, after selecting the layers, only the selected layers of the drawing will be opened.

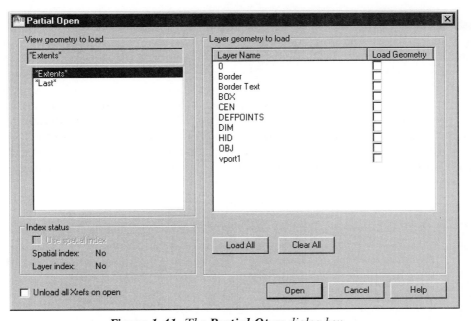

Figure 1-41 *The **Partial Open** dialog box*

Note
The concept of layers is discussed in Chapter 4, Working with Drawing Aids.

Loading Additional Objects to Partially Opened Drawing

Once you have opened a part of a drawing and made the necessary changes, you may want to load additional objects or layers on the existing ones. This can be done by using the **PARTIALOAD** command, which can be invoked by choosing **File > Partial Load** from the menu bar or by entering **PARTIALOAD** at the Command prompt. This command displays the **Partial Load** dialog box, which is similar to the **Partial Open** dialog box. You can choose another layer and the objects drawn in it will be added to the partially loaded drawing.

Note

*The **Partial Load** option is not enabled in the **File** menu unless a drawing is partially opened.*

Loading a drawing partially is a good practice when you are working with objects on a specific layer in a large complicated drawing.

*In the **Select File** dialog box, the preview of a drawing which was partially opened and then saved is not displayed.*

Tip

If a drawing was partially opened and saved previously, it is possible to open it again with the same layers and views. AutoCAD remembers the settings so that while opening a previously partially opened drawing, a dialog box is displayed asking for an option to fully open it or restore the partially opened drawing.

Select Initial View

A view is defined as the way you look at an object. Select the **Select Initial View** check box if you want to load a specific view initially when AutoCAD loads the drawing. This option will work, if the drawing has saved views. This is generally used while working on a large complicated drawing, in which you want to work on a particular portion of the drawing. You can save that particular portion as a view and then select it to open the drawing next time. You can save a desired view, by using AutoCAD's **VIEW** command (see "**Creating Views**", Chapter 6). If the drawing has no saved views, selecting this option will load the last view. If you select the **Select Initial View** check box and then the **OK** button, AutoCAD will display the **Select Initial View** dialog box. You can select the view name from this dialog box, and AutoCAD will load the drawing with the selected view displayed.

Tip

*Apart from opening a drawing from the **Startup** dialog box or the **Select File** dialog box, you can also open a drawing from the **Application Menu**. By default, the **Recent Documents** option is chosen in the **Application Menu**, so the most recently opened drawings will be displayed and you can open the required file from it.*

It is possible to open an AutoCAD 2000 drawing in AutoCAD 2010. When you save this drawing, it is automatically converted and saved as an AutoCAD 2010 drawing file.

Opening an Existing Drawing Using the Startup Dialog Box

If you have configured the settings to show the **Startup** dialog box by setting the **STARTUP** system variable value as **1**, **Startup** dialog box will be displayed every time you start a new AutoCAD session. The first button in this dialog box is the **Open a Drawing** button. When you choose this button, a list of the most recently opened drawings will be displayed for you to select from, see Figure 1-42. The **Browse** button displays the **Select File** dialog box, which allows you to browse to another file.

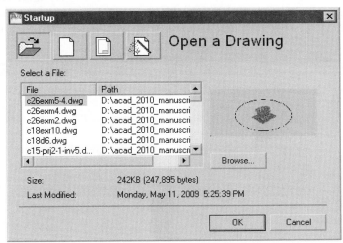

Figure 1-42 *List of the recently opened drawings*

Note

The display of the dialog boxes related to opening and saving drawings is disabled, if the STARTUP system variable is set to 0 and the FILEDIA system variable is set to 0. The initial value of these variable is 1.

Opening an Existing Drawing Using the Drag and Drop Method

You can also open an existing drawing in AutoCAD by dragging it from the Window Explorer and dropping it into AutoCAD. If you drop the selected drawing in the drawing area, the drawing will be inserted as a block and as a result you cannot modify it. But, if you drag the drawing from the Window Explorer and drop it anywhere other than the drawing area, AutoCAD opens the selected drawing.

QUITTING AUTOCAD

You can exit the AutoCAD program by using the **EXIT** or **QUIT** commands. Even if you have an active command, you can choose **Exit AutoCAD** from the **Application Menu** to quit the AutoCAD program. In case the drawing has not been saved, it allows you to save the work first through a dialog box. Note that if you choose **No** in this dialog box, all the changes made in the current list till the last save will be lost. You can also use the close button (**X**) of the main AutoCAD window (present in the title bar) to end the AutoCAD session.

DYNAMIC INPUT MODE

As mentioned earlier, turning the **Dynamic Input** mode on allows you to enter the commands through the pointer input and the dimensions using the dimensional input. When this mode is turned on, all the prompts are available at the tooltip as dynamic prompts and you can select the command options through the dynamic prompt. The settings for the **Dynamic Input** mode are done through the **Dynamic Input** tab of the **Drafting Settings** dialog box. To invoke the **Drafting Settings** dialog box, right-click on the **Dynamic Input** button in the Status Bar; a shortcut menu will be displayed. Choose the **Settings** option from shortcut menu; the **Drafting Settings** dialog box will be displayed, as shown in Figure 1-43. The options in this tab are discussed next.

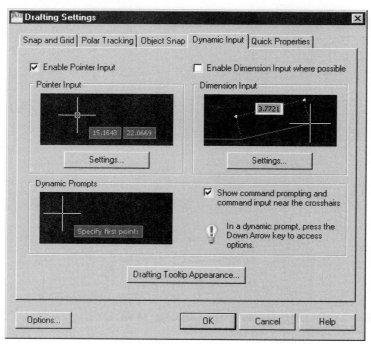

*Figure 1-43 The **Dynamic Input** tab of the **Drafting Settings** dialog box*

Enable Pointer Input

If the **Enable Pointer Input** check box selected, you can enter the commands through the pointer input. Figure 1-44 shows the **CIRCLE** command entered through the pointer input. If this check box is cleared, the **Dynamic Input** will be turned off and commands have to be entered through the Command prompt, in a way similar to the old releases of AutoCAD.

Figure 1-44 Entering a command using the pointer input

Choosing the **Settings** button from the **Pointer Input** area displays the **Pointer Input Settings** dialog box, as shown in Figure 1-45. The radio buttons in the **Format** area of this dialog box are used to set the default settings for specifying the other points, after specifying the first point. By default, the **Polar format** and **Relative coordinates** radio buttons are selected. As a result, the coordinates will be specified in the polar form and with respect to the relative coordinates system. You can select the **Cartesian format** radio button to enter the coordinates in cartesian form. Likewise, if you select the **Absolute coordinates** radio button, the numerical entries will be measured with respect to the absolute coordinate system.

The **Visibility** area in the **Pointer Input Settings** dialog box is used to set the visibility of the coordinates tool tips. By default, the **When a command asks for a point** radio button is selected. You can select the other radio buttons to modify this display.

Figure 1-45 The *Pointer Input Settings* dialog box

Enable Dimension Input where possible

This check box is selected by default. As a result, the dimension input field is displayed in the graphics area showing the preview of that dimension. Figure 1-46 displays the dimension input field. The dotted lines shows the geometric parameters like length, radius, or diameter corresponding to that dimension. Figure 1-44 shows a line being drawn using the **Pline** command. The two dimension inputs that are shown are for the length of the line and the angle with a positive direction of the X axis.

Using the TAB key, you can toggle between the dimension input fields. As soon as you have specified one dimension and moved to the other, the previous dimension is locked. If the **Enable Dimensional Input where possible** check box is cleared, the preview of dimensions will not be displayed. You can only enter the dimensions in the dimension input field below the cursor, as shown in Figure 1-47. Choose the **Settings** button from the **Dimension Input** area to display the **Dimension Input Settings** dialog box, as shown in Figure 1-48.

By default, the **Show 2 dimension input field at a time** radio button is selected. As a result, two dimension input fields will be displayed in the drawing area while stretching a sketched entity. The two input fields will depend on the entity that is being stretched. For example, if you stretch a line using one of its endpoints, the input field will show the total length of the line and the change in its length. Similarly, while stretching a circle using a grip on its circumference, the input fields will show the total radius and the change in the radius. You can set the priority to display only one input field or various input fields, simultaneously, by selecting their respective check boxes.

Figure 1-46 *Input fields displayed when the* **Enable Dimension Input where possible** *check box is selected*

Figure 1-47 *Input fields displayed when the* **Enable Dimension Input where possible** *check box is cleared*

Tip
If multiple dimension input fields are available, use the TAB key to switch between the dimension input fields

Show command prompting and command input near the crosshairs

If this check box is selected, the prompt sequences will be dynamically displayed near the crosshairs. Whenever a blue arrow appears at the pointer input, it suggests that the access options are available. To access these options, press the down arrow key to see the dynamic prompt listing all options. In the dynamic prompt, you can use the cursor or the down arrow key to jog through the options. A black dot will appear before the option that is currently active. In Figure 1-47, the **Length** option is currently active; press ENTER to confirm the polyline creation with the **Length** option.

Figure 1-48 *The* **Dimension Input Settings** *dialog box*

Drafting Tooltip Appearance

When you choose the **Drafting Tooltip Appearance** button, the **Tooltip Appearance** dialog box will be displayed, as shown in Figure 1-49. This dialog box contains the options to customize the tooltip appearance. The **Colors** button is chosen to change the color of the tooltip in the model space or layouts.

*Figure 1-49 The **Tooltip Appearance** dialog box*

The edit box in the **Size** area is used to specify the size of the tooltip. You can also use the slider to control the size of the tool tip. The preview is displayed in the **Model Preview** area and the **Layout Preview** area, as soon as the value is changed in the **Size** edit box. Likewise, the transparency of the tooltip can be controlled using the edit box or the slider in the **Transparency** area.

Selecting the **Override OS settings for all drafting tooltips** radio button in the **Apply to** area ensures that changes made in the **Tooltip Appearance** dialog box will be applied to all drafting tooltips. If you select the **Use settings only for Dynamic Input tooltips** radio button, the changes will be applied only to the **Dynamic Input** tooltips. For example, if you change any of the parameters using the **Tooltip Appearance** dialog box and select the **Use settings only for Dynamic Input tooltips** radio button, the tooltips for the dynamic input will be modified, but for the polar tracking it will consider the original values. On the other hand, if you select the **Override OS settings for all drafting tooltips** radio button, the tooltips displayed for the polar tracking will also be modified based on the values in the **Tooltip Appearance** dialog box.

UNDERSTANDING THE CONCEPT OF SHEET SETS

The sheet sets feature allows you to logically organize a set of multiple drawings as a single unit, called the sheet set. For example, consider a setup, in which there are a number of drawings in different folders in the hard drive of a computer. Organizing or archiving these drawings is tedious and time consuming. However, this can be easily and efficiently done by creating sheet sets. In a sheet set, you can import the layouts from an existing drawing or create a new sheet with a new layout and place the views in the new sheet. You can easily plot and publish all the drawings in the sheet set. You can also manage and create sheet sets using the **Sheet Set Manager**. To do so, choose the **Sheet Set Manager** button from the **Palettes** panel in the **View** tab of the **Ribbon** or press the CTRL+4 keys.

Creating a Sheet Set

AutoCAD allows you to create two different types of sheet sets. The first one is an example sheet set that uses a well organized structure of settings. The second one is used to organize existing drawings. The procedure for creating both these types of sheet sets are discussed next.

Creating an Example Sheet Set

To create an example sheet set, select **New Sheet Set** from the **Open** drop-down list in the **Sheet Set Manager** or choose **New > Sheet Set** from the **Application Menu**. You can also enter **NEWSHEETSET** at the Command prompt. When you invoke this command, the **Create Sheet Set** wizard will be displayed with the **Begin** page, as shown in Figure 1-50. From this page, select the **An example sheet set** radio button, if it is not selected by default. Choose the **Next** button. The **Sheet Set Example** page will be displayed, as shown in Figure 1-51.

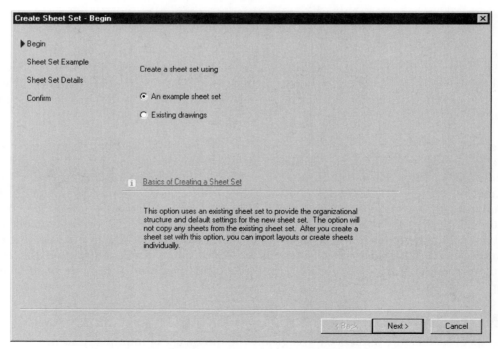

*Figure 1-50 The **Begin** page of the **Create Sheet Set** wizard*

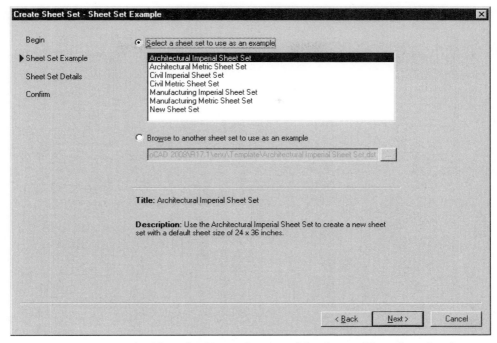

Figure 1-51 The **Sheet Set Example** *page of the* **Create Sheet Set** *wizard*

By default, the **Select a sheet set to use as an example** radio button is selected in this page. The list box below this radio button displays the list of sheet sets that you can use as an example. Each of these sheet sets has structurally organized settings for the sheets. You can select the required sheet set from this list box. The title and the description related to the selected sheet set is displayed in the lower portion of the dialog box.

You can also select the **Browse to another sheet set to use as an example** radio button to select another sheet set located in a different location. You can enter the location of the sheet set in the edit box below this radio button or choose the [**...**] button to display the **Browse for Sheet Set** dialog box. Using this dialog box, you can locate the sheet set file, which is saved with the *.dst* extension.

After selecting the sheet set to use as an example, choose the **Next** button. The **Sheet Set Details** page will be displayed, as shown in Figure 1-52.

Enter the name of the new sheet set in the **Name of new sheet set** edit box. By default, some description is added in the **Description (optional)** area. You can enter additional description in this area. The **Store sheet set data file (.dst) here** edit box displays the default location in which the sheet set data file will be stored. You can modify this location by entering the new location or by selecting the folder using the **Browse for Sheet Set Folder** dialog box, which is displayed by choosing the [**...**] button.

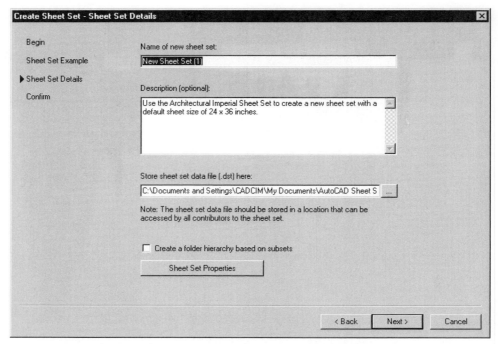

*Figure 1-52 The **Sheet Set Details** page of the **Create Sheet Set** wizard*

You can modify the sheet set properties such as name, storage location, template, description, and so on by choosing the **Sheet Set Properties** button.

Once all the parameters on this page are configured, choose the **Next** button. The **Confirm** page is displayed, as shown in Figure 1-53. This page shows the detailed structure of the sheet set and also lists its parameters and properties.

After checking all parameters and properties, choose the **Finish** button; the **Sheet Set Manager** displays the sheet structure in the **Sheets** area and the details of that sheet set in the **Details** area, as shown in Figure 1-54. If the **Details** area is not displayed, right-click in the **Sheet Set Manager** and select the **Preview/Details Pane** option.

Creating a Sheet Set Using Existing Drawings

As mentioned earlier, this sheet set is used to organize and archive an existing set of drawings. To create this type of sheet set, select the **Existing drawings** radio button from the **Begin** page of the **Create Sheet Set** wizard and choose **Next**. The **Sheet Set Details** page is displayed, which is similar to the one shown in Figure 1-49. Enter the name of the sheet set and the description on this page. Note that by default, there will be no description given about the new sheet set. After setting the parameters on this page, choose the **Next** button. The **Choose Layouts** page is displayed, as shown in Figure 1-55.

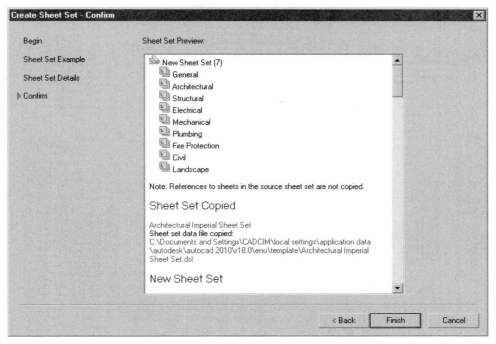

*Figure 1-53 The **Confirm** page of the **Create Sheet Set** dialog box*

Choose the **Browse** button from this page and browse for the folder in which the files to be included in the sheet set are saved. All the drawing files, along with their initialized layouts, are displayed in the list box available below the **Browse** button. You can select as many folders as you want by choosing the **Browse** button.

Tip
You can remove the folders from the list box in the **Choose Layouts** *page by selecting them and pressing the DELETE key.*

When you select a folder, all the drawings in it and all the initialized layouts in the drawings have a check mark on their left. This suggests that all these drawings and layouts will be included in the sheet set. You can clear the check box of the folder to clear all the check boxes and then select the check boxes of only the required drawings and layouts. You can modify the import options by using the **Import Options** dialog box, which is displayed by choosing the **Import Options** button.

*Figure 1-54 The **Sheet Set Manager***

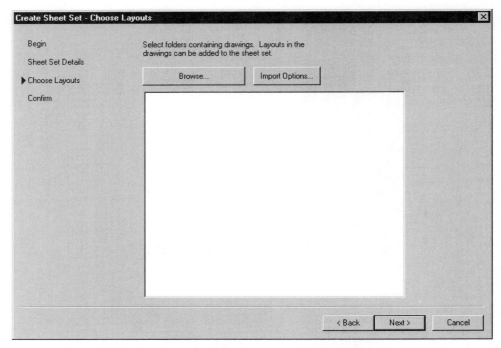

*Figure 1-55 The **Choose Layouts** page of the **Create Sheet Set** dialog box*

After selecting the layouts to be included, choose the **Next** button to display the **Confirm** page similar to that shown in Figure 1-53. This page lists all the layouts that will be included in the sheet set. Choose **Finish** to complete the process of creating the sheet set.

Adding a Subset to a Sheet Set

For a better and more efficient organization of a sheet set, it is recommended that you add subsets to the sheet set. For example, consider a case where you have created a sheet set called Mechanical Drawings, in which you want to store all the mechanical drawings. In this sheet set, you can create subsets such as Bolts, Nuts, Washers, and so on and place the sheets of bolts, nuts, and washers for a more logical organization of the sheet set.

To add a subset to a sheet set, right-click on the sheet set or subset and choose **New Subset** from the shortcut menu. The **Subset Properties** dialog box is displayed, see Figure 1-56.

Enter the name of the subset in the **Subset name** field. Also, specify the location for saving the DWG file and the template for creating the sheets using this dialog box. If you specify **Yes** in the **Create Folder Hierarchy** field, a new folder will be created under the sheet set. If you specify Yes in the **Prompt for template** check box, which will prompt you to select the template for the drawings. Choose **OK** after configuring all the parameters. A new subset will be added to the sheet set or the subset that you selected.

Figure 1-56 *The* **Subset Properties** *dialog box*

Adding Sheets to a Sheet Set or a Subset

To add a new sheet to a sheet set or a subset, right-click on it in the **Sheet Set Manager** window and choose **New Sheet** from the shortcut menu. The **New Sheet** dialog box is displayed. In this dialog box, enter the number and the title of the sheet, along with the file name. You can also set the path of the folder and the sheet template to be used, using this dialog box. Choose **OK** after configuring all the parameters. A new sheet will be added to the sheet set or the subset that you selected.

Archiving a Sheet Set

AutoCAD allows you to archive the sheet set as a zip file, a self-extracting executable file (exe), or a file folder. All the files related to the sheet set are automatically included in the zip file. To archive a sheet set, right-click on its name in the **Sheet Set Manager** and choose **Archive** from the shortcut menu. After AutoCAD gathers the archive information, the **Archive a Sheet Set** dialog box is displayed, as shown in Figure 1-57.

Before archiving the sheet set, you can modify the archiving options by using the **Modify Archive Setup** dialog box, as shown in Figure 1-58. This dialog box is displayed when you choose the **Modify Archive Setup** button. Using the **Archive package type** drop-down list, you can specify whether the archived file is a zip file, a self-extracting executable file (exe), or a file folder. You can also specify the format, in which you want to save the files. You can select the current release format, AutoCAD 2007/LT 2007, AutoCAD 2004/LT 2004, or AutoCAD 2000/LT 2000 formats for archiving the files.

*Figure 1-57 The **Archive a Sheet Set** dialog box*

After setting the parameters, choose the **OK** button from the **Archive a Sheet Set** dialog box. The standard save dialog box will be displayed, which can be used to specify the name and the location of the resultant file.

Resaving All the Sheets in a Sheet Set

The **Sheet Set Manager** allows you to easily resave all the sheets in a sheet set. To save all the sheets in a sheet set again, right-click on the name of the sheet set in the **Sheet Set Manager** and choose **Resave All Sheets** from the shortcut menu, as shown in Figure 1-59. All the sheets in the sheet set will be resaved once again.

CREATING AND MANAGING WORKSPACES

A workspace is defined as a customized arrangement of Ribbon, toolbars, menus, and window palettes in the AutoCAD environment. Workspaces are required when you need to customize the AutoCAD environment for a specific use, in which you need only a certain set of toolbars and menus. For such requirements, you can create your own workspaces, in which only specified toolbars, menus, and palettes are available. By default, the **Initial Setup Workspace** is current when you start AutoCAD. You can choose any other predefined workspace from the shortcut menu that will be displayed on choosing the **Workspace Switching** button on the Status Bar, see Figure 1-60, or by choosing **Tools > Workspaces > Name of the Workspace** from the menu bar. You can also choose the workspace using the **Workspaces** toolbar.

Figure 1-58 The *Modify Archive Setup* dialog box

Figure 1-59 Resaving all the sheets in a
sheet set using the **Sheet Set Manager**

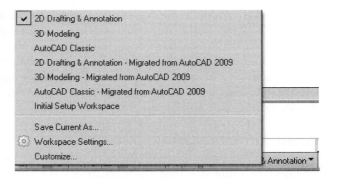

Figure 1-60 The predefined workspaces

Creating a New Workspace

To create a new workspace, invoke the toolbars and the window palettes that you want to be displayed in the new workspace. Next, select the **Save Current As** option from the drop-down list in the **Workspaces** toolbar; the **Save Workspace** dialog box will be displayed, as shown in Figure 1-61. Enter the name of the new workspace in the **Name** edit box and choose **Save** button.

Figure 1-61 The Save Workspace dialog box

The new workspace is now current in the shortcut menu that is displayed on choosing the **Workspaces** button on the Status Bar. Likewise, you can create workspaces as per your requirement and can switch from one workspace to the other by selecting the name from the drop-down list in the **Workspaces** toolbar.

Modifying the Workspace Settings

AutoCAD allows you to modify the workspace settings. To do so, choose the **Workspace Settings** option in the shortcut menu that is displayed on choosing the **Workspaces** button on the Status Bar; the **Workspace Settings** dialog box will be displayed, as shown in Figure 1-62. All workspaces that are created are listed in the **My Workspace** drop-down list. You can make any of the workspaces as My Workspace by selecting it in the **My Workspace** drop-down list. You can also choose the **My Workspace** button from the **Workspaces** toolbar to change the current workspace to the one that was set as My Workspace in the **Workspace Settings** dialog box. The other options in this toolbar are discussed next.

Menu Display and Order Area

The options in this area are used to set the order of display of workspaces in the drop-down list of the **Workspaces** toolbar. By default, workspaces are listed in the sequence of their creation. To change the order, select the workspace and choose the **Move Up** or **Move Down** button. You can also add a separator between workspaces by choosing the **Add Separator** button. A separator is a line that is placed between two workspaces in the shortcut menu that is displayed on choosing the **Workspaces** button on the Status Bar or by choosing **Tools > Workspaces > Name of the Workspace** in the **Application Menu**. Figure 1-63 shows separators added between all the workspaces in the shortcut menu that is displayed on choosing the **Workspaces** button on the Status Bar.

When Switching Workspaces Area

By default, the **Do not save changes to workspace** radio button is selected in this area. This ensures that while switching the workspaces, the changes made in the current workspace will

Figure 1-62 The **Workspace Settings** dialog box

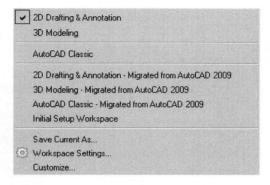

Figure 1-63 *The shortcut menu after adding separators*

not be saved. If you select the **Automatically save workspace changes** radio button, the changes made in the current workspace will be automatically saved when you switch to the other workspace.

AutoCAD'S HELP

| **Titlebar:** | ? > Help | **Shortcut Key:** | F1 | **Command:** | HELP or ? |

You can get the on-line help and documentation about the working of AutoCAD 2010 commands from the **Help** menu in the title bar, see Figure 1-64. You can also access the **Help** menu by pressing the F1 function key. From the previous release of AutoCAD, a new **InfoCenter** bar has been added in the menu bar that will help you search information by using certain keywords, see Figure 1-65. The options in the **Help** menu and the **InfoCenter** bar are discussed next.

Help Menu
Help

Choosing the **Help** button displays the **AutoCAD 2010 Help** dialog box, as shown in Figure 1-66.

You can use this dialog box to access help on different topics and commands. It has three tabs: **Contents**, **Index**, and **Search**, that display the corresponding help topics. If you are in the middle of a command and require help regarding it, choosing the **Help** button will display information about that particular command in the dialog box.

Figure 1-64 The **Help** menu

Figure 1-65 The **InfoCenter** bar

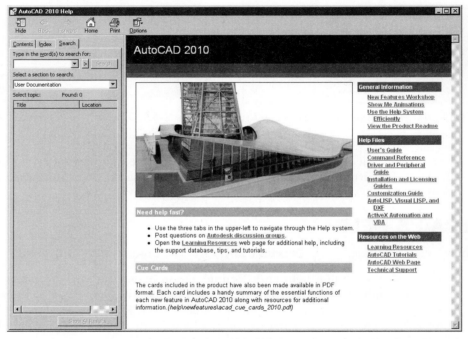

Figure 1-66 The **AutoCAD 2010 Help** dialog box (**Contents** tab)

Contents. This tab displays the help topics, which are organized by categories, pertaining to different sections of AutoCAD such as the User's Guide, Command Reference, and so on. To select a category, double-click on the corresponding book icon or choose the plus sign on the left. The icon becomes an open book with a minus (-) sign and a list of headings associated with that category is displayed. Use the plus sign (+) to further open the headings until you reach the help topic, which has a question mark (?) displayed with it. Choose the topic to display information about the selected topic or command in the window present on the right side of the dialog box.

Index. This tab displays the complete index (search keywords) in an alphabetical order. To display information about an item or a command, type the item (word) or command name in the edit box. With each letter entered, the listing keeps on changing in the list area, displaying the possible topics. When you enter the word and if AutoCAD finds it, the word is automatically highlighted in the list area. Choose the **Display** button to display the information about it.

Search. This tab creates a word list based on all the keywords present in the on-line help files. When you type any word and then choose the **Search** button, a list of matching words is displayed in a window below to narrow down your search. This search is dependent on the option you have selected in the **select a section to search** drop-down list. Use the scroll bar to scroll through the list, double-click on the topic to display the related help in the right pane.

New Features Workshop

This option gives you an interactive list of all the new features in AutoCAD 2010. You can choose this option from the **Help** menu, which displays the **New Features Workshop** window with a list of various topics in the right pane (Figure 1-67). When you choose a topic, the description of the feature improvements will be displayed in the right pane.

Figure 1-67 The New Features Workshop window

Additional Resources

This utility connects you to the **Product Support**, **Training**, Customization, **Autodesk User Group International** Web pages and Web sites through the Microsoft Internet Explorer. The **Developer Help** option provides a detailed help on customizing AutoCAD. You can click on any link on the right of the window that is displayed when you choose this option.

Customer Involvement Program

This option is used to share information about your system configuration and uses of Autodesk products with Autodesk. The collected information is used by Autodesk for the improvement of Autodesk software.

About

This option gives you information about the Release, Serial number, Licensed to, and also the legal description about AutoCAD.

InfoCenter Bar

InfoCenter

InfoCenter is an easy way to get the desired help documentation. Choose the **InfoCenter** button from the **InfoCenter** bar and enter the keywords or questions to be searched in the textbox; the result will be displayed as a link in the drop-down panel, as shown in Figure 1-68. To display the required information in the **AutoCAD 2010 User's Guide** window, click on any one of the search topics from the drop-down panel.

Communication Center

Choose the **Communication Center** button to display the drop-down panel that contains the information regarding software updates and other products of Autodesk. You can also access the information about subscriptions, articles, tips, and patches of the software using the **Communication Center** button.

*Figure 1-68 The display of **InfoCenter** search results*

Favorites

Choose this button from the **InfoCenter** bar to display the drop-down panel showing the list of favorite links. You can mark any link displayed on the **InfoCenter** and the **Communication Center** as favorite. To do so, click on the favorite symbol displayed on the right of each search result in the **InfoCenter** or **Communication Center** drop-down panel. To remove the entry from the favorites, click on the red cross displayed on the right of each favorite entry in the favorite drop-down panel.

ADDITIONAL HELP RESOURCES

1. You can get help for a command while working by pressing the F1 key. The **Help** dialog box containing information about the command is displayed. You can exit the dialog box and continue with the command.

2. You can get help about a dialog box by choosing the **Help** button in that dialog box.

3. Some of the dialog boxes have a **question mark** (?) button at the top right corner just adjacent to the **close** button. When you choose this button the ? gets attached to the cursor. You can then drop it on any item in the dialog box to display information about that particular item.

4. Autodesk has provided several resources that you can use to get assistance with your AutoCAD questions. The following is a list of some of the resources:

    a. Autodesk Web site *http://www.autodesk.com*
    b. AutoCAD Technical Assistance Web site *http://www.autodesk.com/support*
    c. AutoCAD Discussion Groups Web site *http://discussion.autodesk.com/index.jspa*
    d. Autodesk Press Web Site *http://www.cengage.com/cad/autodeskpress*

5. You can also get help by contacting the author, Sham Tickoo, at *stickoo@calumet.purdue.edu*.

6. You can download AutoCAD drawings, programs, and special topics by accessing the author's Web site at *http://ww.cadcim.com/students/students.htm*.

Self-Evaluation Test

Answer the following questions and then compare them to those given at the end of this chapter:

1. You can press the F3 key to display the **AutoCAD** text window, which displays the previous commands and prompts. (T/F)

2. You cannot create a new sheet set using the **Sheet Set Manager**. (T/F)

3. If a drawing was partially opened and saved previously, it is not possible to open it again with the same layers and views. (T/F)

4. If the current drawing is unnamed and you save the drawing for the first time in the present session, the **QSAVE** command will prompt you to enter the file name in the **Save Drawing As** dialog box. (T/F)

5. You can archive a sheet set by right-clicking on the name of the sheet set in the **Sheet Set Manager** and choosing _____ from the shortcut menu.

6. The _____displays a message and an alert whenever Autodesk provides the latest information regarding software updates and their other products.

7. If you want to work on a drawing without altering the original, you must select the _____ option from the **Open** drop-down list in the **Select File** dialog box.

8. The _____ enables you to open only a selected view or a selected layer of a selected drawing.

9. You can use the _____ command to close the current drawing file without actually quitting AutoCAD.

10. The _____ system variable can be used to change the time interval for automatic save.

Review Questions

Answer the following questions:

1. The shortcut menu invoked by right-clicking in the command window displays the most recently used commands and some of the window options such as **Copy**, **Paste**, and so on. (T/F)

2. It is possible to open an AutoCAD 2002 drawing in AutoCAD 2009. (T/F)

3. The file name you enter to save a drawing in the **Save Drawing As** dialog box file name can be 255 characters long, but cannot contain spaces and punctuation marks. (T/F)

4. You can close a drawing in AutoCAD 2010 even if a command is active. (T/F)

5. Which one of the following combination of keys should be pressed to hide all toolbars displayed on the screen?

    (a) CTRL+3   (b) CTRL+0
    (c) CTRL+5   (d) CTRL+2

6. Which one of the following combination of keys should be pressed to turn on or off the display of the **Tool Palettes** window?

    (a) CTRL+3   (b) CTRL+0
    (c) CTRL+5   (d) CTRL+2

7. Which of the following commands is used to exit from the AutoCAD program?

    (a) **QUIT**   (b) **END**
    (c) **CLOSE**   (d) None

8. Which of the following options in the **Startup** dialog box is used to set the initial drawing settings before actually starting a new drawing?

    (a) **Start from Scratch**   (b) **Use a Template**
    (c) **Use a Wizard**   (d) None

9. When you choose **Save** from the **File** menu or choose the **Save** button in the **Quick Access** toolbar, which of the following commands is invoked?

    (a) **SAVE**   (b) **LSAVE**
    (c) **QSAVE**   (d) **SAVEAS**

10. AutoCAD has provided _____ as an easy and convenient way of placing and sharing hatch patterns and blocks in the current drawing.

11. By default, the angles are positive if measured in the _____ direction.

12. You can change the size of toolbars by placing the cursor anywhere on the _____ of the toolbar where it takes the shape of a double-sided arrow.

13. To differentiate the template files from the drawing files, the template files have the _____ extension, whereas the drawing files have the _____ extension.

14. You can also use _____ and _____ instead of dragging and dropping the objects from one drawing to another while multiple drawings are opened.

15. The _____ tab of the **AutoCAD 2010 Help** window displays the help topics that are organized by categories pertaining to different sections of AutoCAD.

Answers to Self-Evaluation Test
1. F, 2. F, 3. F, 4. T, 5. Archive, 6. Communication Center, 7. Open Read-Only, 8. Partial Open, 9. CLOSE, 10. SAVETIME

Chapter 2

Getting Started with AutoCAD

Learning Objectives

After completing this chapter, you will be able to:

- Draw lines using the **LINE** command and its options.
- Understand various coordinate systems used in AutoCAD.
- Use the **ERASE** command to clear the drawing area.
- Understand the two basic object selection methods: Window and Crossing.
- Draw circles using the options of the **CIRCLE** command.
- Use the **ZOOM** and **PAN** display commands.
- Set up units using the **UNITS** command.
- Set up and determine limits for a given drawing.
- Plot drawings using the basic plotting options.
- Use the **Options** dialog box to specify settings.

DRAWING LINES IN AutoCAD

Ribbon:	Home > Draw > Line	**Toolbar:**	Draw > Line
Menu Bar:	Draw > Line	**Tool Palettes:**	Draw > Line
Command:	LINE or L		

The most fundamental object in a drawing is line. A line can be drawn between any two points by using the **LINE** command. You can invoke the **LINE** command by choosing the **Line** button from the **Draw** panel of the **Home** tab in the **Ribbon**, as shown in Figure 2-1. Besides this, you can choose the **Line** button from the **Draw** tab of the **Tool Palettes**. To invoke the **Tool Palettes**, choose **View > Palettes > Tool Palettes** from the **Ribbon**, as shown in Figure 2-2. Alternatively, you can invoke the **LINE** command from the **Draw** toolbar, as shown in Figure 2-3. However, the **Draw** toolbar is not displayed by default. To invoke this toolbar, choose **Tools > Toolbars > AutoCAD > Draw** from the menu bar.

*Figure 2-1 Invoking the **LINE** command from the **Ribbon***

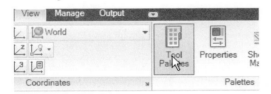

*Figure 2-2 Invoking the **Tool Palettes** from the **Ribbon***

*Figure 2-3 Invoking the **LINE** command from the **Draw** toolbar*

You can also invoke the **LINE** command by entering **LINE** or **L** (L is the command alias of the LINE) at the Command prompt. Once you have invoked the **LINE** command, the next prompt, the **Specify first point**, requires you to specify the starting point of the line. You can either select a point using the pointing device or you can enter its coordinates. After the first point is selected, AutoCAD will prompt you to enter the second point at the **Specify next point or [Undo]** prompt. At this point, you may continue to select points or terminate the **LINE** command by pressing ENTER, ESC, or SPACEBAR. You can also right-click to display the shortcut menu and choose the **Enter** or **Cancel** options to exit from the **LINE** command. After terminating the **LINE** command, AutoCAD will again display the Command prompt. The prompt sequence for the drawing shown in Figure 2-4 is given next.

Start a new file with the *acad.dwt* template in the **2D Drafting & Annotation** or **AutoCAD Classic** workshop.

Command: **LINE** ⏎
Specify first point: *Move the cursor (mouse) and left-click to specify the first point.*
Specify next point or [Undo]: *Move the cursor and left-click to specify the second point.*

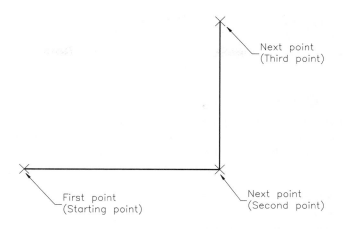

Figure 2-4 *Drawing lines using the **LINE** command*

Specify next point or [Undo]: *Specify the third point.*
Specify next point or [Close/Undo]: ⎡Enter⎤ *(Press ENTER to exit the **LINE** command.)*

The **LINE** command has the following three options:

Continue   **Close**   **Undo**

Tip
When you select the start point of the line by pressing the left mouse button, a rubber-band line appears that stretches between the selected point and the current position of the cursor. This line is sensitive to the movement of the cursor and helps you to select the direction and the placement of the next point for the line.

Note
*To clear the drawing area to work out the new exercises and examples, choose the **Erase** button from the **Modify** panel, which is present in the **Home** tab of the **Ribbon** or type **ERASE** at the Command prompt and press ENTER. The screen crosshairs will change into a box called pick box and AutoCAD will prompt you to select objects. You can select an object by positioning the pick box anywhere on it and then by pressing the left mouse button. Once you have finished selecting the objects, press ENTER to terminate the **ERASE** command and the selected objects will be erased. If you enter **All** at the **Select objects** prompt, AutoCAD will erase all objects from the screen. (See "Erasing Objects" discussed later in this chapter.) You can use the **U** (undo) command to undo the last command. Alternatively, choose the **Undo** button from the **Quick Access** toolbar to undo the last command.*

Command: **ERASE** or **E** ⎡Enter⎤   *(**E** is the command alias of the **ERASE** command.)*
Select objects: *Select objects.*   *(Select objects using the pick box.)*
Select objects: ⎡Enter⎤

Command: **ERASE** ⏎
Select objects: **ALL** ⏎
Select objects: ⏎
Command: **U** ⏎   *(The **U** command will undo the last command.)*

The Continue Option

After exiting the **LINE** command, you may want to draw another line starting from the point where the previous line ended. In such cases, you can use the **Continue** option. This option enables you to grab the endpoint of the previous line and continue drawing the line from that point, as shown in Figure-2-5. The prompt sequence for the **Continue** option is given next.

Command: **LINE** or **L** ⏎   *(**L** is the command alias of the **LINE** command.)*
Specify first point: *Pick first point of the line.*
Specify next point or [Undo]: *Pick second point.*
Specify next point or [Undo]: ⏎

Command: **LINE** ⏎   *(Or choose **Repeat LINE** from the shortcut menu.)*
Specify first point: ⏎   *(Press ENTER or right-click to continue the line from the last line.)*
Specify next point or [Undo]: *Pick second point of second line (third point in Figure 2-5).*
Specify next point or [Undo]: ⏎

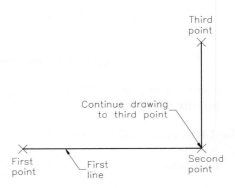

Figure 2-5 *Using the **Continue** option with the **LINE** command*

Tip
*You can also type the @ symbol to start the line from the last point. For example, if you draw a circle and then immediately start the **LINE** command, the @ will snap to the center point of the circle.*

*The **Continue** option snaps to the endpoint of the last line or arc even if other points have been defined to draw entities such as circles, ellipses, and so on, after the line was drawn.*

Command: **LINE** ⏎
_line Specify first point: *Pick first point of the line.*
Specify next point or [Undo]: *Pick second point.*
Specify next point or [Undo]: ⏎

Command: **LINE** or **L** ⏎   *(**L** is the command alias of the **LINE** command.)*
_line Specify first point: **@** ⏎   *(Continues drawing the line from the last point.)*
Specify next point or [Undo]: *Pick second point of the second line.*
Specify next point or [Undo]: ⏎

The Close Option

The **Close** option can be used to join the current point with the initial point of the first line when two or more lines are drawn in continuation. For example, this option can be used when an open figure needs one more line to close it and make a polygon (a polygon is a closed figure with at least three sides, for example, a triangle or rectangle). The following is the prompt sequence for the **Close** option, refer to Figure 2-6.

> Command: **LINE** [Enter]
> _line Specify first point: *Pick first point.*
> Specify next point or [Undo]: *Pick second point.*
> Specify next point or [Undo]: *Pick third point.*
> Specify next point or [Close/Undo]: *Pick fourth point.*
> Specify next point or [Close/Undo]: **C** [Enter]
> *(Joins the fourth point with the first point)*

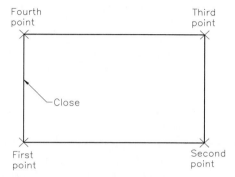

You can also choose the **Close** option from the shortcut menu, which appears when you right-click in the drawing area.

*Figure 2-6 Using the **Close** option with the **LINE** command*

The Undo Option

If you draw a line, and then realize that you made an error, you can remove the line using the **Undo** option of the **LINE** command. If you need to remove more than one line, you can use this option multiple times and go as far back as you want. In this option, you can type **Undo** (or just **U**) at the **Specify next point or [Undo]** prompt. You can also right-click to display the shortcut menu, which gives you the **Undo** option. The following example illustrates the use of the **Undo** option, refer to Figure 2-7.

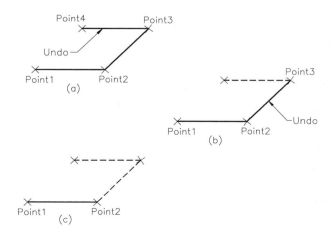

*Figure 2-7 Removing lines using the **Undo** option of the **LINE** command*

Command: **LINE** or **L** ⏎   (**L** *is the command alias of the **LINE** command.*)
_line Specify first point: *Pick first point (Point 1 in Figure 2-7).*
Specify next point or [Undo]: *Pick second point (Point 2).*
Specify next point or [Undo]: *Pick third point.*
Specify next point or [Close/Undo]: *Pick fourth point.*
Specify next point or [Close/Undo]: **U** ⏎   (*Removes last line from Point 3 to Point 4.*)
Specify next point or [Close/Undo]: **U** ⏎   (*Removes next line from Point 2 to Point 3.*)
Specify next point or [Close/Undo]: ⏎

Tip
*AutoCAD allows you to enter the command aliases in place of the complete command name. For example, you can enter **L** instead of **LINE** at the Command prompt to invoke the **LINE** command.*

Note
*By default, whenever you open a new drawing, you need to modify the drawing display area. To modify the display area, type **ZOOM** at the Command prompt and press ENTER. In the command sequence, type **ALL** and press ENTER; the drawing display is modified. You will learn more about the **ZOOM** command later in this chapter.*

COORDINATE SYSTEMS

To specify a point in a plane, take two mutually perpendicular lines as references. The horizontal line is called the X axis, and the vertical line is called the Y axis. The point of intersection of these two axes is called the origin. The X and Y axes divide the XY plane into four parts, generally known as quadrants. The X coordinate measures the horizontal distance from the origin (how far left or right) on the X axis. The Y coordinate measures the vertical distance from the origin (how far up or down) on the Y axis. The origin has the coordinate values of X = 0, Y = 0. The origin is taken as the reference for locating any point on the XY plane. The X coordinate is positive if measured to the right of the origin, and negative if measured to the left of the origin. The Y coordinate is positive if measured above the origin, and negative if measured below the origin. This method of specifying points is called the Cartesian coordinate system, see Figure 2-8. In AutoCAD, the default origin is located at the lower left corner of the graphics area of the screen. AutoCAD uses the following coordinate systems to locate a point in an XY plane.

1. Absolute coordinates

2. Relative coordinates
    a. Relative rectangular coordinates
    b. Relative polar coordinates

3. Direct distance entry

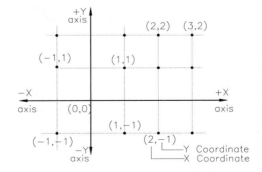

Figure 2-8 *Cartesian coordinate system*

Absolute Coordinate System

In the absolute coordinate system, the points are located with respect to the origin (0,0). For example, a point with X = 4 and Y = 3 is measured 4 units horizontally (displacement along the *X* axis) and 3 units vertically (displacement along the *Y* axis) from the origin, as shown in Figure 2-9. In AutoCAD, the absolute coordinates are specified at the Command prompt by entering *X* and *Y* coordinates, separated by a comma. However, remember that if you are specifying the coordinates using the **Dynamic Input** mode, you need to add # as the suffix to the X coordinate value. For example, enter #1,1 in the dynamic input boxes to use absolute coordinate system. The following example illustrates the use of absolute coordinates at the Command prompt to draw the rectangle shown in Figure 2-10.

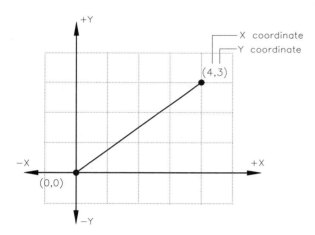

Figure 2-9 *Absolute coordinate system*

Figure 2-10 *Lines created using the absolute coordinates*

Command: **LINE** [Enter]
_line Specify first point: **1,1** [Enter]
                    (X = 1 and Y = 1.)
Specify next point or [Undo]: **4,1** [Enter]
                    (X = 4 and Y = 1.)
Specify next point or [Undo]: **4,3** [Enter]
Specify next point or [Close /Undo]: **1,3** [Enter]
Specify next point or [Close/Undo]: **C** [Enter]

Example 1   *General*

For Figure 2-11, enter the absolute coordinates of the points given in the following table. Then, draw the figure using the absolute coordinates. Save the drawing under the name *Exam1.dwg*.

Point	Coordinates	Point	Coordinates
1	3,1	5	5,2
2	3,6	6	6,3
3	4,6	7	7,3
4	4,2	8	7,1

Start a new file with the *acad.dwt* template in the **2D Drafting & Annotation** or **AutoCAD Classic** workspace. Once the coordinates of the points are known, you can draw the figure by using the **LINE** command. But before you proceed with drawing the object, you need to turn off the dynamic input mode, if it is on by default, so that you can view the command you are entering in the command window. The prompt sequence is given next.

Figure 2-11 *Drawing a figure using the absolute coordinates*

   Choose the **Dynamic Input** button from the Status Bar, if it is already chosen.
   Command: **ZOOM** [Enter]
   Specify corner of window, enter a scale factor (nX or nXP), or
   [All/Center/Dynamic/Extents/Previous/Scale/Window/Object] <real time>: **All** [Enter]
   Command: **LINE** [Enter]
   _line Specify first point: **3,1** [Enter]   (*Start point.*)
   Specify next point or [Undo]: **3,6** [Enter]
   Specify next point or [Undo]: **4,6** [Enter]
   Specify next point or [Close/Undo]: **4,2** [Enter]
   Specify next point or [Close/Undo]: **5,2** [Enter]
   Specify next point or [Close/Undo]: **6,3** [Enter]
   Specify next point or [Close/Undo]: **7,3** [Enter]
   Specify next point or [Close/Undo]: **7,1** [Enter]
   Specify next point or [Close/Undo]: **C** [Enter]

Choose the **Save** button from the **Quick Access** toolbar to display the **Save Drawing As** dialog box. Enter the name *Exam1* in the **File name** edit box to replace *Drawing1.dwg* and then choose the **Save** button. The drawing will be saved with the specified name in the default *My Documents* folder.

Exercise 1   *General*

For Figure 2-12, enter the absolute coordinates of the points given in the following table. Then, use these coordinates to draw the same figure. The distance between the dotted lines is 1 unit.

Point	Coordinates	Point	Coordinates
1	2, 1	6	_____
2	_____	7	_____
3	_____	8	_____
4	_____	9	_____
5	_____		

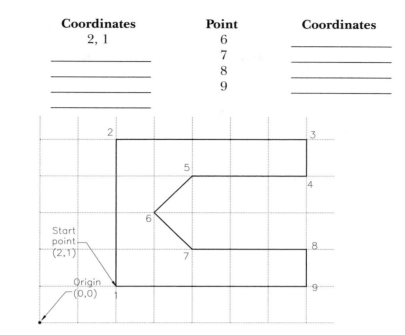

Figure 2-12 *Drawing for Exercise 1*

Relative Coordinate System

There are two types of relative coordinates: relative rectangular and relative polar.

Relative Rectangular Coordinates

In the relative rectangular coordinate system, the displacements along the *X* and *Y* axes (DX and DY) are measured with reference to the previous point rather than to the origin. If the dynamic input is turned on, by default the coordinates are entered using relative rectangular method, You can simply enter the X coordinate then type comma (,) and then enter the Y coordinate. However, if the dynamic input is turned off, the relative coordinate system is designated by the symbol @ and it should precede any relative entry. The following prompt sequence illustrates the use of the relative rectangular coordinate system to draw a rectangle with the lower left corner at the point (1,1). The length of the rectangle is 4 units and the width is 3 units, as shown in Figure 2-13.

Figure 2-13 *Drawing lines using the relative rectangular coordinates*

Command: **LINE** Enter
_line Specify first point: **1,1** Enter   *(Start point)*
Specify next point or [Undo]: **@4,0** Enter   *(Second point DX = 4, DY = 0.)*

Specify next point or [Undo]: **@0,3** Enter   *(Third point DX = 0, DY = 3.)*
Specify next point or [Close/Undo]:
@-4,0 Enter   *(Fourth point DX = -4, DY = 0.)*
Specify next point or [Close/Undo]: **@0,-3** Enter *(Start point DX = 0, DY = -3.)*
Specify next point or [Close/Undo]: Enter

Remember that if **Dynamic Input** is on, you need to input a comma (,) after entering the first value in the dynamic input boxes. Else, AutoCAD will take coordinates in relative polar form.

Sign Convention. As just mentioned, in the relative rectangular coordinate system the displacements along the X and Y axes are measured with respect to the previous point. Imagine a horizontal line and a vertical line passing through the previous point so that you get four quadrants. If the new point is located in the first quadrant, the displacements DX and DY are both positive. If the new point is located in the third quadrant, the displacements DX and DY are both negative. In other words, up or right are positive and down or left are negative.

Example 2   *General*

Draw Figure 2-14 using the relative rectangular coordinates of the points given in the table that follows.

Point	Coordinates	Point	Coordinates
1	3,1	8	@-1,-1
2	@4,0	9	@-1,1
3	@0,1	10	@-1,0
4	@-1,0	11	@0,-2
5	@1,1	12	@1,-1
6	@0,2	13	@-1,0
7	@-1,0	14	@0,-1

To draw the sketch of this example, start a new file with the *acad.dwt* template in the **2D Drafting & Annotation** or **AutoCAD Classic** workspace. Before you proceed, you need to make sure the dynamic input is turned on.

Command: **ZOOM** Enter
Specify corner of window, enter a scale factor (nX or nXP), or
[All/Center/Dynamic/Extents/Previous/Scale/Window/Object] <real time>: **All** Enter
Command: **LINE** Enter
_line Specify first point: **Type 3,1 in the dynamic input boxes and press** Enter
(Start point)

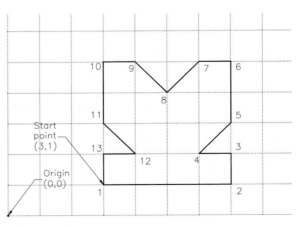

Figure 2-14 *Using relative rectangular coordinates with the **LINE** command*

Specify next point or [Undo]: **Type 4,0 in the dynamic input boxes and press** [Enter]
Specify next point or [Undo]: **Type 0,1 in the dynamic input boxes and press** [Enter]
Specify next point or [Close/Undo]: **Type -1,0 in the dynamic input boxes and press** [Enter]
Specify next point or [Close/Undo]: **Type 1,1 in the dynamic input boxes and press** [Enter]
Specify next point or [Close/Undo]: **Type 0,2 in the dynamic input boxes and press** [Enter]
Specify next point or [Close/Undo]: **Type -1,0 and press** [Enter]
Specify next point or [Close/Undo]: **Type -1,-1 and press** [Enter]
Specify next point or [Close/Undo]: **Type -1,1 and press** [Enter]
Specify next point or [Close/Undo]: **Type -1,0 and press** [Enter]
Specify next point or [Close/Undo]: **Type 0,-2 and press** [Enter]
Specify next point or [Close/Undo]: **Type 1,-1 and press** [Enter]
Specify next point or [Close/Undo]: **Type -1,0 and press** [Enter]
Specify next point or [Close/Undo]: **Type 0,-1 and press** [Enter]
Specify next point or [Close/Undo]: [Enter]

Exercise 2   *General*

For Figure 2-15, enter the relative rectangular coordinates of the points given in the following table. Next, use these coordinates to draw the figure. The distance between the dotted lines is 1 unit.

Point	Coordinates	Point	Coordinates
1	2, 1	17	_____
2	_____	18	_____
3	_____	19	_____
4	_____	20	_____
5	_____	21	_____
6	_____	22	_____
7	_____		
8	_____		
9	_____		
10	_____		
11	_____		
12	_____		
13	_____		
14	_____		
15	_____		
16	_____		

Relative Polar Coordinates

In the relative polar coordinate system, a point is located by defining both the distance of the point from the current point and the angle that the line between the two points makes with the positive X axis. The prompt sequence to draw a line from a point at 1,1 to a point at

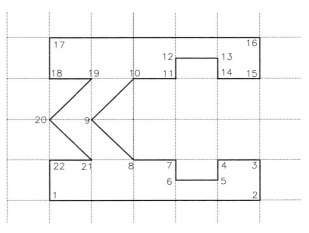

Figure 2-15 *Drawing for Exercise 2*

a distance of 5 units from the point (1,1), and at an
angle of 30-degree to the *X* axis, refer to Figure 2-16.

> Command: **LINE** [Enter]
> Specify first point: **1,1** [Enter]
> Specify next point or [Undo]: **@5<30** [Enter]

If the **Dynamic Input** mode is on, the relative polar
coordinate mode is activated and the second input box
shows the angle value, preceded by the < symbol.
Therefore, you do not need to input the @ symbol.
You can simply enter the distance value and then press
the TAB key to shift to the second input box for
specifying the angle value.

*Figure 2-16 Drawing a line using
relative polar coordinates*

Sign Convention. By default, in the relative polar coordinate system, the angle is measured
from the horizontal axis (3 'O' clock) as the zero degree baseline. Also, the angle is positive
if measured in a counterclockwise direction and negative if measured in a clockwise direction.
Here, it is assumed that the default setup of the angle measurement has not been changed.

 Note
*You can modify the default settings of the angle measurement direction using the **UNITS**
command, which is discussed later.*

Example 3   *General*

For Figure 2-17, enter the relative polar coordinates of each point given in the table, and
then draw the sketch. Use absolute coordinates for the start point (1.5, 1.75). The dimensions
are shown in the drawing. Also, save this drawing as *Exam3.dwg*.

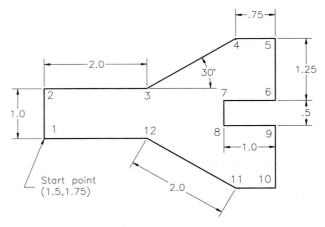

Figure 2-17 Drawing for Example 3

Point	Coordinates	Point	Coordinates
1	1.5,1.75	7	@1.0<180
2	@1.0<90	8	@0.5<270
3	@2.0<0	9	@1.0<0
4	@2.0<30	10	@1.25<270
5	@0.75<0	11	@0.75<180
6	@1.25<-90 (or <270)	12	@2.0<150

Start a new file with the *acad.dwt* template in the **2D Drafting & Annotation** or **AutoCAD Classic** workspace. Now, you need to modify the drawing display area. To do so, choose the **ZOOM** button from the Status Bar, enter **All** at the Command prompt and then press ENTER. Next, turn off the dynamic input by choosing the **Dynamic Input** button from the Status Bar.

Command: **LINE** [Enter]
_line Specify first point: **1.5,1.75** [Enter]   *(Start point)*
Specify next point or [Undo]: **@1<90** [Enter]
Specify next point or [Undo]: **@2.0<0** [Enter]
Specify next point or [Close/Undo]: **@2<30** [Enter]
Specify next point or [Close/Undo]: **@0.75<0** [Enter]
Specify next point or [Close/Undo]: **@1.25<-90** [Enter]
Specify next point or [Close/Undo]: **@1.0<180** [Enter]
Specify next point or [Close/Undo]: **@0.5<270** [Enter]
Specify next point or [Close/Undo]: **@1.0<0** [Enter]
Specify next point or [Close/Undo]: **@1.25<270** [Enter]
Specify next point or [Close/Undo]: **@0.75<180** [Enter]
Specify next point or [Close/Undo]: **@2.0<150** [Enter]
Specify next point or [Close/Undo]: **C** [Enter]   *(Joins the last point with the first point.)*

Save this drawing by choosing the **Save** button from the **Quick Access** toolbar. The **Save Drawing As** dialog box will be displayed. Enter the name *Exam3* in the **File name** edit box and then choose the **Save** button. The drawing will be saved with the specified name in the *My Documents* folder.

Exercise 3   *General*

Draw the object shown in Figure 2-18. Locate the points using the absolute, relative rectangular, and relative polar coordinate systems. Do not draw the dimensions; they are for reference only.

Direct Distance Entry

You can draw a line by specifying the length of the line and its direction using the Direct Distance Entry, as shown in Figure 2-19. The direction is determined by the position of the cursor and the length of the line is entered from the keyboard. If the **Ortho** mode is on, you can draw lines along the *X* or *Y* axis by specifying the length of the line and positioning the cursor along the ortho. You can also use it with the other draw commands like the **RECTANGLE** command. You can also use Direct Distance Entry with polar tracking and **SNAPANG**. For example, if **SNAPANG** is 45-degree and **Ortho** mode is on, you can draw a

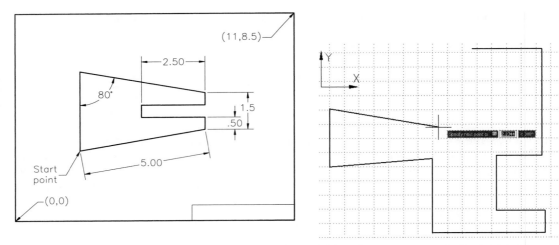

Figure 2-18 *Drawing for Exercise 3*   **Figure 2-19** *Using the* **Direct Distance Entry** *method to draw lines*

line at 45 or 135-degree direction by positioning the cursor and entering the distance from the keyboard. Similarly, if the polar tracking is on, you can position the cursor at the predefined angles and then enter the length of the line from the keyboard.

> Command: **LINE** Enter
> _line Specify first point: *Start point.*
> Specify next point or [Undo]: *Position the cursor and then enter distance.*
> Specify next point or [Undo]: *Position the cursor and then enter distance.*

Example 4   *General*

In this example, you will draw the drawing, as shown in Figure 2-20, using Direct Distance Entry method. The starting point is 2,2.

Figure 2-20 *Drawing for Example 4*

In this example, you will use the polar tracking option to draw lines. The polar tracking option allows you to track lines that are drawn at the specified angles. The default angle that is specified for polar tracking is 90-degree. As a result, you can use the polar tracking to draw lines at an angle that is divisible by 90, such as 90, 180, 270, and 360. This is the reason you first need to add another angle of 45-degree to the polar tracking that will allow you to track the lines drawn at an angle divisible by 45, such as 45, 90, 135, and so on.

Start a new file with the **acad.dwt** template in the **2D Drafting & Annotation** or **AutoCAD Classic** workspace. Now, to add a 45-degree angle to polar tracking, right-click on the **Polar Tracking** button on the Status Bar and choose **45** from the shortcut menu. Again choose the **Polar Tracking** button in the Status Bar to turn the polar tracking on. As you move the cursor to draw lines, a tooltip that shows the length of the line and the angle will be displayed.

Note

*To add the angular values other than those in the shortcut menu, choose **Settings** from the shortcut menu. Next, select the **Additional angles** check box in the **Polar Angle Settings** area and choose the **New** button. Enter a new value in the field that appears and then press ENTER. Choose **OK** to close the dialog box.*

The following is the Command prompt sequence for drawing the sketch in Figure 2-20. It is presumed that you have modified the drawing display area.

Command: **LINE** [Enter]
Specify first point: **2,2**
Specify next point or [Close/Undo]: *Move the cursor horizontally toward the right and when the tooltip appears with the polar angle, enter **2**.*
Specify next point or [Close/Undo]: *Move the cursor at an angle close to 45-degree and when the tooltip appears with the polar angle, enter **0.7071**.*
Specify next point or [Close/Undo]: *Move the cursor vertically upward and when the tooltip appears with the polar angle, enter **1**.*
Specify next point or [Close/Undo]: *Move the cursor horizontally toward the left and when the tooltip appears with the polar angle, enter **3**.*
Specify next point or [Close/Undo]: *Move the cursor vertically downward and when the tooltip appears with the polar angle, enter **1**.*
Specify next point or [Close/Undo]: **C**

Note

You will learn more about polar tracking in Chapter 4, Working with Drawing Aids.

Exercise 4   *General*

Use the direct distance entry method to draw a parallelogram. The base of the parallelogram equals 4 units, the side equals 2.25 units, and the angle equals 45-degree. Draw the same parallelogram using the absolute, relative, and polar coordinates. Note the differences and the advantages of using Direct Distance Entry.

ERASING OBJECTS

Ribbon:	Home > Modify > Erase	**Toolbar:**	Modify > Erase
Menu Bar:	Modify > Erase	**Tool Palettes:**	Modify > Erase
Command:	ERASE or E		

After drawing some objects, you may need to erase some of them. To erase, choose the **Erase** button from the **Modify** panel of the **Ribbon**, see Figure 2-21. This command is used exactly the same way as an eraser is used in manual drafting to delete the unwanted information. You can also invoke the **ERASE** command from the **Modify** toolbar, as shown in Figure 2-22. To invoke the **Modify** toolbar, choose **Tools > Toolbars > AutoCAD > Modify** from the menu bar. When you invoke the **ERASE** command, a small box, known as pick box, replaces the screen cursor. To erase an object, move the pick box so that it touches the object. Select the object by pressing the left mouse button, as shown in Figure 2-23. AutoCAD confirms the selection by changing the selected objects into dashed lines and the **Select objects** prompt returns. You can either continue selecting the objects or press ENTER to terminate the object selection process and erase the selected objects. To enter the command from the keyboard, type **E** or **ERASE**. The prompt sequence is given next.

Figure 2-21 Invoking the Erase command from the Ribbon

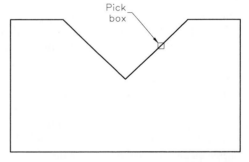

Figure 2-22 Invoking the Erase command from the Modify toolbar

Command: **ERASE** [Enter]
Select objects: *Select first object.*
Select objects: *Select second object.*
Select objects: [Enter]

If you enter **All** at the **Select objects** prompt, AutoCAD will erase all objects in the drawing, even if the objects are outside the screen display area.

Command: **ERASE** [Enter]
Select objects: **All**
Select objects: [Enter]

Figure 2-23 Selecting object by positioning the pick box at the top of the object and then pressing the pick button on the pointing device

You can also first select the objects to be erased from the drawing and then right-click in the drawing area to display the shortcut menu. From this menu, you can choose the **Erase** option.

CANCELING AND UNDOING A COMMAND

If you are in a command and you want to cancel or exit, press the ESC (Escape) key on the keyboard.

Command: **ERASE** 〔Enter〕
Select objects: *Press ESC (Escape) to cancel the command.*

Similarly, sometimes you unintentionally erase some object. When you discover such an error, you can correct it by restoring the erased object by means of the **OOPS** command. The **OOPS** command restores objects that have been accidentally erased by the previous **ERASE** command, refer to Figure 2-24. You can also use the **U** (Undo) command to undo the last command.

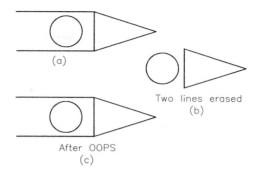

Figure 2-24 Use of the OOPS command

Command: **OOPS** 〔Enter〕   *(Restores erased objects.)*
Command: **U** 〔Enter〕   *(Undoes the last command.)*

OBJECT SELECTION METHODS

One of the ways to select objects is to select them individually, which can be time-consuming, if you have a number of objects to select. This problem can be solved by creating a selection set that enables you to select several objects at a time. The selection set options can be used with those commands that require object selection, such as **ERASE** and **MOVE**. There are many object selection methods, such as **All**, **Last**, **Add**, and so on. At this point, you will learn two options: **Window** and **Crossing**. The remaining options are discussed in Chapter 5.

The Window Option

This option is used to select an object or group of objects by enclosing them in a box or window. The objects to be selected should be completely enclosed within the window; those objects that lie partially inside the boundaries of the window are not selected. You can select the **Window** option by entering **W** at the **Select objects** prompt. You are prompted to select the two opposite corners of the window. After selecting the first corner, you can select the other corner by moving the cursor to the desired position and specifying the particular point. As you move the cursor, a violet color window is displayed that changes in size as you move the cursor. The objects selected by the **Window** option are displayed as dashed objects, as shown in Figure 2-25.

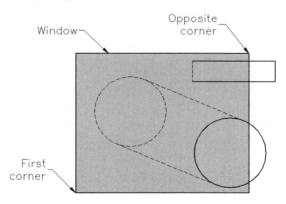

Figure 2-25 Selecting objects using the Window option

The prompt sequence for using the **Window** option with the **ERASE** command is given next.

Command: **ERASE** [Enter]
Select objects: **W** [Enter]
Specify first corner: *Select the first corner.*
Specify opposite corner: *Select the second corner.*
Select objects: [Enter]

You can also invoke the **Window** option by selecting a blank point on the screen at the **Select objects** prompt. This is automatically taken as the first corner of the window. Moving the cursor to the right will display a blue shaded window. After getting all the objects to be selected inside this window, you can specify the other corner with your pointing device. The objects that are completely enclosed within the window will be selected and highlighted. The following is the prompt sequence for automatic window selection with the **ERASE** command.

Command: **ERASE** [Enter]
Select objects: *Select a blank point as the first corner of the window.*
Specify opposite corner: *Drag the cursor to the right to select the other corner of the window.*
Select objects: [Enter]

The Crossing Option

This option is used to select an object or group of objects by creating a box or window around them. The objects to be selected should be touching the window boundaries or completely enclosed within it. You can invoke the **Crossing** option by entering **C** at the **Select objects** prompt. After you choose the **Crossing** option, AutoCAD prompts you to select the first corner at the **Specify first corner** prompt. Once you have selected the first corner, a shaded dashed box or window of light green color is drawn. By moving the cursor, you can change the size of the crossing box, hence putting the objects to be selected within (or touching) the box. Here, you can select the other corner. The objects selected by the **Crossing** option are highlighted by displaying them as dashed objects, as shown in Figure 2-26. The following prompt sequence illustrates the use of the **Crossing** option when you choose the **Erase** button.

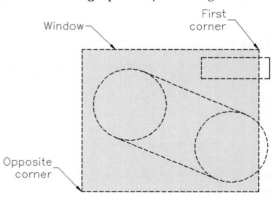

*Figure 2-26 Selecting objects using the **Crossing** option*

Select objects: **C** [Enter]
Specify first corner: *Select the first corner of the crossing window.*
Specify opposite corner: *Select the other corner of the crossing window.*
Select objects: [Enter]

You can also select the **Crossing** option automatically by selecting a blank point on the screen at the **Select objects** prompt and moving the cursor to the left. The blank point you selected becomes the first corner of the crossing window and AutoCAD will then prompt you to select the other corner. As you move the cursor, a green shaded box or window drawn with dashed lines is displayed. The objects that are touching or completely enclosed within the window will be selected. The objects selected by the **Crossing** option are highlighted by being displayed as dashed objects. The prompt sequence for automatic crossing selection when you choose the **Erase** button is given next.

Select objects: *Select a blank point as the first corner of the crossing window.*
Specify opposite corner: *Drag the cursor to the left to select the other corner of the crossing window.*
Select objects: Enter

Note
*In AutoCAD, the entities are highlighted when you move the cursor over them. This feature is known as the selection preview. To set the selection preview, choose the **Visual Effects Settings** button in the **Selection Preview** area of the **Selection** tab of the **Options** dialog box; the **Visual Effect Settings** dialog box will be displayed. This dialog box is used to set the type of appearance during highlighting. The preview will be a dashed line if the **Dash** radio button is selected. The preview will be thicken line, if the **Thicken** line radio button is selected. If the **Both** radio button is selected, the preview will be dashed thicken line. The selection preview of the different lines is shown in Figure 2-27.*

Figure 2-27 *The selection preview with the* ***Both****,* ***Thicken****, and* ***Dash*** *radio buttons selected*

DRAWING CIRCLES

Ribbon:	Home > Draw > Circle	**Toolbar:**	Draw > Circle
Menu Bar:	Draw > Circle	**Tool Palettes:**	Draw > Circle
Command:	CIRCLE or C		

To draw a circle, you can use the **CIRCLE** command. You can invoke the **CIRCLE** command from the **Draw** panel of the **Ribbon**, as shown in Figure 2-28. You can also invoke **CIRCLE** command from the **Draw** toolbar or the **Tool Palettes**. The prompt sequence for the **CIRCLE** command is given next.

Command: **CIRCLE** Enter
Specify center point for circle or [3P/2P/Ttr (tan tan radius)]:

*Figure 2-28 Invoking the **CIRCLE** command from the **Ribbon***

The options of the **CIRCLE** command are explained in the following sections:

The Center and Radius Option

Using this option, you can draw a circle by defining the center and the radius of the circle, as shown in Figure 2-29. After entering the **CIRCLE** command, AutoCAD will prompt you to enter the center of the circle, which can be selected by specifying a point on the screen or by entering the coordinates of the center point. Next, you will be prompted to enter the radius of the circle. Here you can accept the default value, enter a new value, or select a point on the circumference of the circle to specify the radius.

The following is the prompt sequence for drawing a circle with a center at 3,2 and a radius of 1 unit.

> Command: **CIRCLE** `Enter`
> Specify center point for circle or [3P/2P/Ttr (tan tan radius): **3,2** `Enter`
> Specify radius of circle or [Diameter]<current>: **1** `Enter`

Note
*You can also set the radius by assigning a value to the **CIRCLERAD** system variable. The value you assign becomes the default value of the radius.*

The Center and Diameter Option

You can use this option to draw a circle by defining its center and diameter. After invoking the **CIRCLE** command, AutoCAD prompts you to enter the center of the circle, which can be selected by specifying a point on the screen or by entering the coordinates of the center point. Next, you will be prompted to enter the radius of the circle. At this prompt, enter **D**. After this, you will be prompted to enter the diameter of the circle. For entering the diameter, you can accept the default value, enter a new value, or drag the circle to the desired diameter and select a point. If you use a menu option to select the **CIRCLE** command with the **Diameter** option, the menu automatically enters the **Diameter** option and prompts for the diameter after you specify the center. The following is the prompt sequence for drawing a circle with the center at (2,3) and a diameter of 2 units, as shown in Figure 2-30.

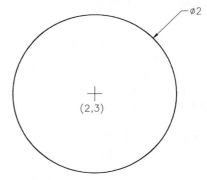

Figure 2-29 *A circle drawn by specifying its center and radius*

Figure 2-30 *A circle drawn by specifying its center and diameter*

Command: **CIRCLE** Enter
Specify center point for circle or [3P/2P/Ttr(tan tan radius): **2, 3** Enter
Specify radius of circle or [Diameter]<current>: **D** Enter
Specify diameter of circle <current>: **2** Enter

The Two-Point Option

You can also draw a circle using the **Two-Point** option. In this option, AutoCAD lets you draw the circle by specifying the two endpoints of the circle's diameter. For example, if you want to draw a circle that passes through the points (1,1) and (2,1), you can use the **CIRCLE** command with 2P option, as shown in the following example, refer to Figure 2-31.

Command: **CIRCLE** Enter
Specify center point for circle or [3P/2P/Ttr (tan tan radius)]: **2P** Enter
Specify first end point of circle's diameter: **1,1** Enter
Specify second end point of circle's diameter: **2,1** Enter   *(You can also use the relative coordinates.)*

The Three-Point Option

You can use this option to draw a circle by defining three points on its circumference. The three points may be entered in any order. To draw a circle that passes through the points (3,3), (3,1), and (4,2), refer to Figure 2-32. The prompt sequence for drawing a circle with three points option is given next.

Command: **CIRCLE** Enter
Specify center point for circle or [3P/2P/Ttr(tan tan radius)]: **3P** Enter
Specify first point on circle: **3,3** Enter
Specify second point on circle: **3,1** Enter
Specify third point on circle: **4,2** Enter

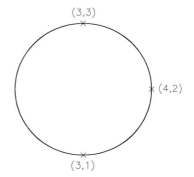

Figure 2-31 *A circle drawn using the*
Two-Point *option*

Figure 2-32 *A circle drawn using the*
Three-Point *option*

You can also use relative rectangular coordinates to define the points.

Command: **CIRCLE** [Enter]
Specify center point for circle or [3P/2P/Ttr(tan tan radius)]: **3P** [Enter]
Specify first point on circle: **3,3** [Enter]
Specify second point on circle: **@0,-2** [Enter]
Specify third point on circle: **@1,1** [Enter]

The Tangent Tangent Radius Option

A tangent is an object (line, circle, or arc) that contacts the circumference of a circle at only one point. In this option, AutoCAD uses the Tangent object snap to locate two tangent points on the selected objects that are to be tangents to the circle. Then you have to specify the radius of the circle. The prompt sequence for drawing a circle using the **Ttr** option is given next.

Command: **CIRCLE** [Enter]
Specify center point for circle or [3P/2P/Ttr(tan tan radius)]: **T** [Enter]
Specify point on object for first tangent of circle: *Select first line, circle, or arc.*
Specify point on object for second tangent of circle: *Select second line, circle, or arc.*
Specify radius of circle <current>: **0.75** [Enter]

In Figures 2-33 through 2-36, the dotted circles represent the circles that are drawn by using the **Ttr** option. The circle actually drawn depends on how you select the objects that are to be tangent to the new circle. The figures show the effect of selecting different points on the objects. If you specify too small or large radius, you may get unexpected results or the "**Circle does not exist**" prompt.

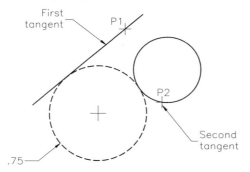

Figure 2-33 *Drawing a circle using the* **Tangent Tangent Radius** *option*

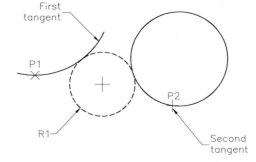

Figure 2-34 *Drawing a circle using the* **Tangent Tangent Radius** *option*

The Tangent, Tangent, Tangent Option

To invoke this option, click on the down arrow on the right of the **Center, Radius** tool in the **Draw** panel of the **Ribbon**; a flyout will be displayed. Choose the **Tan , Tan, Tan** option from it. In this option, AutoCAD uses the Tangent Object Snap to locate three points on the three selected objects to which the circle is drawn tangent. The following is the prompt sequence to draw a circle, as shown in Figure 2-37, using the **Tan, Tan, Tan** option.

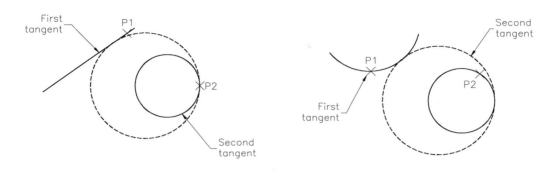

Figure 2-35 *Drawing a circle using the* **Tangent Tangent Radius** *option*

Figure 2-36 *Drawing a circle using the* **Tangent Tangent Radius** *option*

Command: *Choose* **Center, Radius > Tan, Tan, Tan** *from the* **Draw** *panel of the* **Ribbon**.
_circle Specify center point for circle or [3P/2P/Ttr (tan tan radius)]: _3p Specify first point on circle: _tan to *Select the first object.*
Specify second point on circle: _tan to *Select the second object.*
Specify third point on circle: _tan to *Select the third object.*

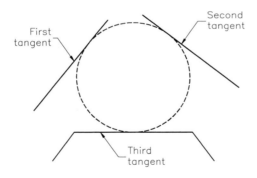

Figure 2-37 *Circle drawn using the* **Tan, Tan, Tan** *option*

Exercise 5   *Mechanical*

Draw Figure 2-38 using various options of the **LINE** and **CIRCLE** commands. Use absolute, relative rectangular, or relative polar coordinates for drawing the triangle. The vertices of the triangle will be used as the center of the circles. The circles can be drawn using the **Center and Radius**, **Center and Diameter**, or **Tan, Tan, Tan** options. (Height of triangle = 4.5 X sin 60 = 3.897.) Do not draw the dimensions; they are for reference only.

Chapter 2

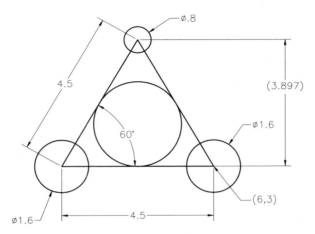

Figure 2-38 Drawing for Exercise 5

BASIC DISPLAY COMMANDS

Drawing in AutoCAD is much simpler than manual drafting in many ways. Sometimes while drawing, it is very difficult to see and alter minute details. In AutoCAD, you can overcome this problem by viewing only a specific portion of the drawing. This is done using the **ZOOM** command. This command lets you enlarge or reduce the size of the drawing displayed on the screen. Some of the drawing display commands such as **ZOOM** and **PAN** will be introduced here. A detailed explanation of these commands appears in Chapter 7.

Zooming Drawings

Ribbon:	View > Navigate > Zoom	Status Bar:	Zoom
Toolbar:	Standard > Zoom or Zoom toolbar	Command:	ZOOM or Z

The **Zoom** command is used to enlarge or reduce the view of a drawing on the screen, without affecting the actual size of entities. Some **ZOOM** commands are available in the **Standard** toolbar, as shown in Figure 2-39. The **ZOOM** command can be invoked by choosing the **Zoom** button from the Status Bar. After invoking the **ZOOM** command, you can use various zooming options to obtain the desired display. To invoke these options directly, press the down arrow next to the **Extents** button in the **Navigate** panel of the **View** tab in the **Ribbon**; all options of the **ZOOM** command will be displayed in

Figure 2-39 The **ZOOM** commands in the *Standard* toolbar

the flyout, refer to Figure 2-40. You can also invoke the **ZOOM** command by choosing **View > Zoom** from the menu bar. To display the menu bar, left-click on the arrow in the **Quick Access** toolbar and then select **Show Menu Bar**.

Following is the prompt sequence of the **ZOOM** command:

Command: **ZOOM** or **Z** Enter
Specify corner of window, enter a scale factor (nX or nXP), or
[All/Center/Dynamic/Extents/Previous/Scale/Window/Object]
:

Realtime Zooming

You can use the **Realtime Zoom** to zoom in and zoom out interactively. To zoom in, invoke the command, and then hold the pick button down and move the cursor up. If you want to zoom in further, bring the cursor down, specify a point, and move the cursor up. Similarly, to zoom out, hold the pick button down and move the cursor down. Realtime zoom is the default setting for the **ZOOM** command. At the Command prompt, pressing ENTER after invoking the **ZOOM** command automatically invokes the realtime zoom. To exit the Realtime Zoom, right-click to display the shortcut menu and choose **Exit**. You can also press ESC or the ENTER key to exit the command.

Window Option

This is the most commonly used option of the **ZOOM** command. It lets you specify the area you want to zoom in on by letting you specify two opposite corners of a rectangular window. The center of the specified window becomes the center of the new display screen. The area inside the window is magnified in size to fill the drawing area as completely as possible. The points can be specified by selecting them with the help of the pointing device or by entering their coordinates.

Figure 2-40 *Invoking the **ZOOM** command from the **Navigate** panel*

Previous Option

While working on a complex drawing, you may need to zoom in on a portion of the drawing to edit some minute details. When you have completed the editing you may want to return to the previous view. This can be done using the **Previous** option. AutoCAD remembers the last ten views that can be restored using the **Previous** option.

All Option

This option zooms to the drawing limits or the extents, whichever is greater. Whenever you increase the limits (**Quick Setup** in **Use a Wizard** dialog box) the current display is not affected and hence does not show. You need to use the **Zoom All** option to display the limits of the drawing. Sometimes it is possible that the objects are drawn beyond the limits. In such a case, the **Zoom All** option zooms to fill the drawn objects in the drawing area, irrespective of its limits.

Note

If the default view of the acad3d.dwt file is current, the ZOOM command will show only the Extents, Window, Previous, and Realtime options when invoked from the Command prompt.

You will learn about the other options of the ZOOM command in detail in Chapter 7.

Panning in Realtime

You can use **Pan Realtime** to pan the drawing interactively, by sliding the drawing and placing it at the required position. To pan a drawing, invoke the command from the Status Bar and then hold the pick button down and move the cursor in any direction. When you select realtime pan, AutoCAD displays an image of a hand, indicating that you are in the PAN mode. You can drag the hand anywhere on the screen to move the drawing. To exit realtime pan, right-click to display the shortcut menu and choose **Exit**. You can also press ESC or the ENTER key to exit the command.

Note

The Realtime Zoom, Window Zoom, Zoom Extents, and Pan Realtime options can also be invoked from the Status Bar.

Tip

You can right-click to display a shortcut menu while the realtime zoom or realtime pan options are active. The Realtime Zoom, Realtime Pan, Exit, and other ZOOM command options are available in this shortcut menu.

SETTING UNITS

| **Application Menu:** | Drawing utilities > Units | **Command:** | UNITS |

In Chapter 1, you have already learned to set units while starting a drawing from the **Startup** dialog box using the **Wizards** option. If you want to change the units while you are already working on a drawing, the **UNITS** command can be used.

Setting Units Using the Drawing Units Dialog Box

The **UNITS** command is used to select a format for the units of distance and angle measurement. To invoke this command, enter **UNITS** at the Command prompt; the **Drawing Units** dialog box will be displayed, as shown in Figure 2-41. You can now specify the precision for the length and angle in the respective **Precision** drop-down lists in this dialog box, refer to Figure 2-42. You can also invoke the **Drawing Units** dialog box by choosing **Format > Units** from the menu bar.

Specifying Units

In the **Drawing Units** dialog box, you can select a desired format of units from the drop-down list displayed when you choose the down arrow to the right of the **Type** edit box. You can select one of the following five formats.

Figure 2-41 *The **Drawing Units** dialog box*   **Figure 2-42** *Specifying the precision from the **Precision** drop-down list*

1. Architectural (0'-01/16")   2. Decimal (0.00)   3. Engineering (0'-0.00")
4. Fractional (0 1/16)   5. Scientific (0.00E+01)

If you select the scientific, decimal, or fractional format, you can enter the distances or coordinates in any of these three formats, but not in engineering or architectural units. In the following example, the units are set as decimal, scientific, fractional, and decimal and fractional to enter the coordinates of different points.

Command: **LINE** `Enter`
Specify first point: **1.75,0.75** `Enter`   (*Decimal*)
Specify next point or [Undo]: **1.75E+01,3.5E+00** `Enter`   (*Scientific*)
Specify next point or [Undo]: **10-3/8,8-3/4** `Enter`   (*Fractional*)
Specify next point or [Close/Undo]: **0.5,17/4** `Enter`   (*Decimal and fractional*)

If you choose the engineering or architectural format, you can enter the distances or coordinates in any of the five formats. In the following example, the units are set as architectural; hence, different formats are used to enter the coordinates of points.

Command: **LINE** `Enter`
Specify first point: **1-3/4,3/4** `Enter`   (*Fractional.*)
Specify next point or [Undo]: **1'1-3/4",3-1/4** `Enter`   (*Architectural.*)
Specify next point or [Undo]: **0'10.375,0'8.75** `Enter`   (*Engineering.*)
Specify next point or [Close/Undo]: **0.5,4-1/4"** `Enter`   (*Decimal and engineering.*)

Note

The inch symbol (") is optional. For example, 1'1-3/4" is the same as 1'1-3/4, and 3/4" is the same as 3/4.

You cannot use the feet (') or inch (") symbols if you have selected scientific, decimal, or fractional unit formats.

Specifying Angle

You can select one of the following five angle measuring systems.

    1. Decimal Degrees (0.00)   2. Deg/min/sec (0d00'00")
    3. Grads (0.00g)   4. Radians (0.00r)
    5. Surveyor's Units (N 0d00'00" E)

If you select any of the first four measuring systems, you can enter the angle in the Decimal, Degrees/minutes/seconds, Grads, or Radians system, but you cannot enter the angle in Surveyor's units. However, if you select Surveyor's units, you can enter the angles in any of the five systems. If you enter an angle value without any indication of a measuring system, it is taken in the current system. To enter the value in another system, use the appropriate suffixes and symbols, such as r (Radians), d (Degrees), g (Grads), or the others shown in the following examples. In the following example, the system of angle measure is Surveyor's units and different systems of angle measure are used to define the angle of the line.

Command: **LINE** [Enter]
Specify first point: **3,3** [Enter]
Specify next point or [Undo]: **@3<45.5** [Enter]   *(Decimal degrees)*
Specify next point or [Undo]: **@3<90d30'45"** [Enter]   *(Degrees/min/sec)*
Specify next point or [Close/Undo]: **@3<75g** [Enter]   *(Grads)*
Specify next point or [Close/Undo]: **@3<N45d30'E** [Enter] *(Surveyor's units)*

In Surveyor's units, you must specify the bearing angle that the line makes with the north-south direction, as shown in Figure 2-43. For example, if you want to define an angle of 60-degree with north, in the Surveyor's units the angle will be specified as N60dE. Similarly, you can specify angles such as S50dE, S50dW, and N75dW, refer to Figure 2-43. You cannot specify an angle that exceeds 90-degree (N120E). The angles can also be specified in radians or grads, for example, 180-degree is equal to □ (3.14159) radians. You can

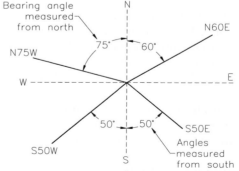

Figure 2-43 *Specifying the angle in Surveyor's units*

convert degrees into radians or radians into degrees using the equations given next.

    radians = degrees X 3.14159/180; degrees = radians X 180/3.14159

Grads are generally used in land surveys. There are 400 grads or 360-degree in a circle. A 90-degree angle is equal to 100 grads.

Tip

*An example corresponding to the type of unit and angle selected from the **Length** or **Angle** area of the dialog box can be seen in the **Sample Output** area of the dialog box.*

You can also specify the units to be used for the measurement of intensity of the photometric lights. The photometric lights are used for the rendering of the drawings. Select the desired unit system from the **Units for specifying the intensity of lighting** drop-down list in the **Lighting** area of the **Drawing Units** dialog box. While using the photometric lights in rendering, you cannot use the **Generic** unit for the measurement of the intensity of light. The concept of rendering and lights is discussed in Chapter 30.

In AutoCAD, by default the angles are positive, if measured in the counterclockwise direction, and negative if measured in the clockwise direction. Also, the angles are measured from the positive X axis, see Figures 2-44 and 2-45. If you want the angles to be measured as positive in the clockwise direction, select the **Clockwise** check box from the **Angle** area. Now, the angles will be positive if measured in the clockwise direction, and the negative if measured in the counterclockwise direction.

Figure 2-44 *North, South, East, and West directions*

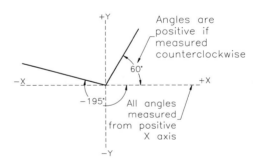

Figure 2-45 *Measuring angles counterclockwise from the positive X axis (default)*

When you choose the **Direction** button in the **Drawing Units** dialog box, the **Direction Control** dialog box appears, which gives you an option of selecting the setting for direction of the base angle, see Figure 2-46.

If you select the **Other** option, you can set your own direction for the base angle by entering a value in the **Angle** edit box or by choosing the **Pick an Angle** button to pick two points on the screen to specify the angle. After selecting an angle, you can choose the **OK** button to apply the settings. This will redisplay the **Drawing Units** dialog box.

You can also set the units of measure while inserting a block or a drawing. In the **Drawing Units** dialog box, select any measuring unit from the **Units to scale inserted content** drop-down list, see Figure 2-47. Now, while inserting a block or a drawing from the **DesignCenter**, AutoCAD inserts the block with the specified unit. Even if the block was created using a different measuring unit, AutoCAD scales it and inserts it using the specified measuring unit. If you want to insert the block with the original units, then choose **Unitless** from the drop-down list.

Figure 2-46 *Setting direction from the* *Direction Control* *dialog box*

Figure 2-47 *Selecting measuring units for inserting* *drawings and blocks using the drag-and-drop method*

Note
The insertion of blocks into a drawing is discussed in detail in Chapter 16.

The **Sample Output** area in the **Drawing Units** dialog box shows an example of the current format of the units and angles. When you change the type of length and angle measure in the **Length** and **Angle** areas of the **Drawing Units** dialog box, the corresponding example is displayed in the **Sample Output** area.

Example 5   *General*

In this example, you will set the units of a drawing according to the given specifications and then draw Figure 2-48. You also need to do the following.

a.   Set the units of length to fractional, with the denominator of the smallest fraction equal to 32.

b.   Set the angular measurement to surveyor's units, with the number of fractional places for display of angles equal to zero.

c.   Set the direction to 90-degree (north) and the direction of measurement of angles to clockwise (angles measured positive in clockwise direction), refer to Figure 2-48.

Figure 2-48 *Drawing for Example 5*

The procedure of completing this example is given in the following steps.

1. Start a new file with the *acad.dwt* template in the **2D Drafting & Annotation** or the **AutoCAD Classic** workspace and invoke the **Drawing Units** dialog box by choosing the **Drawing Utilities > Units** from the **Application Menu**. You can also invoke this dialog box by entering **UNITS** at the Command prompt.

2. In the **Length** area of the dialog box, select **Fractional** from the **Type** drop-down list. From the **Precision** drop-down list, select **0 1/32**, see Figure 2-49.

3. In the **Angle** area of the dialog box, select **Surveyor's Units** from the **Type** drop-down list. From the **Precision** drop-down list, select **N 0d E**, if it is not already selected. Also, select the **Clockwise** check box to set the clockwise angle measurement as positive.

Figure 2-49 *Setting units for Example 5 in the* **Drawing Units** *dialog box*

4. Choose the **Direction** button to display the **Direction Control** dialog box. Next, select the **North** radio button. Choose the **OK** button to exit the **Direction Control** dialog box.

5. Choose the **OK** button to exit the **Drawing Units** dialog box.

6. With the units set, draw Figure 2-48 using the relative polar coordinates. Here, the units are fractional and the angles are measured from north (90-degree axis). Also, the angles are measured as positive in the clockwise direction and negative in the counterclockwise direction, see Figure 2-50.

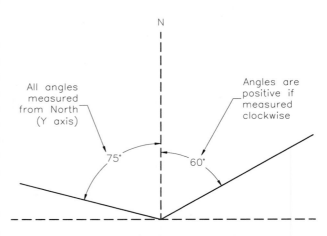

Figure 2-50 Angles measured from north (Y axis)

7. To modify the drawing display area, choose the **ZOOM** button from the Status Bar, enter **All** at the Command prompt, and press ENTER.

8. Turn off the dynamic input. The prompt sequence to complete the sketch is as follows:

Command: **LINE** Enter
Specify first point: **2,2** Enter
Specify next point or [Undo]: **@2.0<0** Enter
Specify next point or [Undo]: **@2.0<60** Enter
Specify next point or [Close/Undo]: **@1<180** Enter
Specify next point or [Close/Undo]: **@1<90** Enter
Specify next point or [Close/Undo]: **@1<180** Enter
Specify next point or [Close/Undo]: **@2.0<60** Enter
Specify next point or [Close/Undo]: **@0.5<90** Enter
Specify next point or [Close/Undo]: **@2.0<180** Enter
Specify next point or [Close/Undo]: **C** Enter

SETTING LIMITS OF THE DRAWING

Command:	LIMITS

In AutoCAD, the drawings must be drawn full scale and, therefore, the limits are needed to size up a drawing area. The limits of the drawing area are usually determined by the following factors.

1. The actual size of the drawing.
2. The space needed for putting down the dimensions, notes, bill of materials, and other necessary details.
3. The space between various views so that the drawing does not look cluttered.
4. The space for the border and title block, if any.

In Chapter 1, you have already learned to set the limits while starting a drawing from the **Startup** dialog box using the **Wizards** option. If you want to change the limits while you are already working in a drawing, the **LIMITS** command can be used. When you start AutoCAD using the **Imperial** file, the default limits are 12,9. The following is the prompt sequence of the **LIMITS** command for setting the limits of 24,18.

> Command: **LIMITS** `Enter`
> Reset Model space limits:
> Specify lower left corner or [ON/OFF]<current>: **0,0** `Enter`
> Specify upper right corner <current>: **24,18** `Enter`

At the preceding two prompts you are required to specify the lower left corner and the upper right corner of the sheet. Normally you choose (0,0) as the lower left corner, but you can enter any other point. If the sheet size is 24 X 18, enter (24,18) as the coordinates of the upper right corner.

Tip
*Whenever you increase the drawing limits, the display area does not change. You need to use the **All** option of the **ZOOM** command to display the complete area inside the drawing area.*

Setting Limits

To get a good idea of how to set up limits, it is always better to draw a rough sketch of the drawing to help calculate the area needed. For example, if an object has a front view size of 5 X 5, a side view size of 3 X 5, and a top view size of 5 X 3, the limits should be set so that they can accommodate the drawing and everything associated with it. In Figure 2-51, the space between the front and side views is 4 units and between the front and top views is 3 units. Also, the space between the border and the drawing is 5 units on the left, 5 units on the right, 3 units

Figure 2-51 *Setting limits in a drawing*

at the bottom, and 2 units at the top. (The space between the views and between the borderline and the drawing depends on the drawing.)

After you know the sizes of various views and have determined the space required between views, between the border and the drawing, and between the borderline and the edges of the paper, you can calculate the space you need as follows.

> Space along (X axis) = 1 + 5 + 5 + 4 + 3 + 5 + 1 = 24
> Space along (Y axis) = 1 + 3 + 5 + 3 + 3 + 2 + 1 = 18

Thus, the space or work area you need for the drawing is 24 X 18. Once you have determined the space, select the sheet size that can accommodate your drawing. In the case just explained, you will select a D size (34 X 22) sheet. Therefore, the actual drawing limits are 34,22.

Limits for Architectural Drawings

Most architectural drawings are drawn at a scale of 1/4" = 1', 1/8" = 1', or 1/16" = 1'. You must set the limits accordingly. The following example illustrates how to calculate the limits in architectural drawings.

Given
Sheet size = 24 X 18
Scale is 1/4" = 1'

Calculate limits
Scale is 1/4" = 1'
      or 1/4" = 12"
       or 1" = 48"
     X limit = 24 X 48
           = 1152" or 1152 Units
           = 96'
    Y limit = 18 X 48
           = 864" or 864 Units
           = 72'

Thus, the scale factor is 48 and the limits are 1152",864", or 96',72'.

Example 6   *General*

In this example, you will calculate the limits and determine an appropriate drawing scale factor for Figure 2-52. The drawing is to be plotted on a 12" X 9" sheet. Assume the missing dimensions.

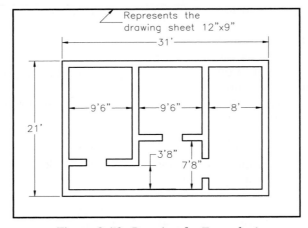

Figure 2-52   *Drawing for Example 6*

The calculation for the scale factor is given next.

Given or known
Overall length of the drawing = 31'
Length of the sheet = 12"
Approximate space between the drawing and the edges of the paper = 2"

Calculate scale factor
To calculate the scale factor, you have to try various scales until you find one that satisfies the given conditions. After some experience, you will find this fairly easy to do. For this example, assume a scale factor of 1/4" = 1'.

Scale factor 1/4" = 1'   or 1" = 4'
Thus, a line 31' long will be = 31'/4' = 7.75" on paper. Similarly, a line 21' long = 21'/4' = 5.25".

Approximate space between the drawing and the edges of paper = 2"
Therefore, total length of the sheet = 7.75 + 2 + 2 = 11.75"
Similarly, total width of the sheet = 5.25 + 2 + 2 = 9.25"

Because you selected the scale 1/4" = 1', the drawing will definitely fit on the given sheet of paper (12" x 9"). Therefore, the scale for this drawing is 1/4" = 1'.

Calculate limits
Scale factor = 1" = 48" or 1" = 4'
The length of the sheet is 12"
Therefore, X limit = 12 X 4' = 48'
Also, Y limit = 9 X 4' = 36'

Limits for Metric Drawings

When the drawing units are metric, you must use **standard metric size sheets** or calculate the limits in millimeters (mm). For example, if the sheet size you decide to use is 24 X 18, the limits, after conversion to the metric system, will be 609.6,457.2 (multiply length and width by 25.4). You can round these numbers to the nearest whole numbers 610,457. Note that metric drawings do not require any special setup, except for the limits. Metric drawings are like any other drawings that use decimal units. As with architectural drawings, you can draw metric drawings to a scale. For example, if the scale is 1:20, you must calculate the limits accordingly. The following example illustrates how to calculate the limits for metric drawings.

Given
Sheet size = 24" X 18"
Scale = 1:20

Calculate limits
Scale is 1:20
Therefore, scale factor = 20

X limit = 24 X 25.4 X 20 = 12192 units
Y limit = 18 X 25.4 X 20 = 9144 units

Exercise 6   *General*

Set the units of the drawing according to the following specifications and then make the drawing shown in Figure 2-53 (leave a space of 3 to 5 units around the drawing for dimensioning and title block). The space between the dotted lines is 1 unit.

1. Set **UNITS** to decimal units, with two digits to the right of the decimal point.

2. Set the angular measurement to decimal degrees, with the number of fractional places for display of angles equal to 1.

3. Set the direction to 0-degree (east) and the direction of measurement of angles to counterclockwise (angles measured positive in a counterclockwise direction).

4. Set the limits leaving a space of 3 to 5 units around the drawing for dimensioning and title block.

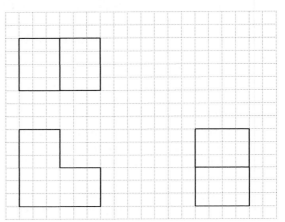

Figure 2-53 *Drawing for Exercise 6*

INTRODUCTION TO PLOTTING DRAWINGS

Ribbon:	Output > Plot > Plot	Toolbar:	Quick Access > Plot
Application Menu:	Print > Plot	Command:	PLOT or PRINT

Once you have created a drawing in the current session of AutoCAD, you may need to have its hard copy for your reference or for sending it to the client. This hard copy is very useful in the industry and can be created by plotting and printing it on a sheet of paper. Suppose, you have drawn an architectural plan on the computer, you can print it and send its hard copy to a site for implementation. Similarly, if you have created a mechanical component, you can print it and send its hard copy to a shop floor for manufacturing. Drawings can be plotted using the **PLOT** command. If you invoke this command, the **Plot** dialog box will be displayed. If the dialog box is not expanded by default, choose the **More Options** button at the lower right corner of the dialog box to expand it, as shown in Figure 2-54.

The values in this dialog box are the ones that were set during the configuration of AutoCAD. If the displayed values conform to your requirements, you can start plotting without making any changes. If necessary, you can make changes in the default values.

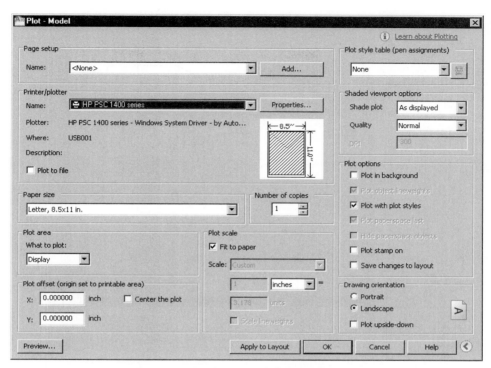

Figure 2-54 *The expanded* **Plot** *dialog box*

Basic Plotting

In this section, you will learn to set up the basic plotting parameters. Later, you will learn about the advance options that allow you to plot according to your plot drawing specifications. Basic plotting involves selecting the correct output device (plotter), specifying the area to plot, selecting paper size, specifying the plot origin, orientation, and the plot scale.

Example 7   *General*

You will plot the drawing shown in Figure 2-55 using the **Window** option to select the area to plot. The drawing was drawn in Example 3 of this chapter and here it is assumed to be open on the screen. Assume that AutoCAD is configured for two output devices: Default System Printer and **Linotronic 300v52.3**.

1. Invoke the **Plot** dialog box by choosing the **Plot** button in the **Quick Access** toolbar or the **Plot** panel of the **Output** tab in the **Ribbon**, or by entering **PLOT** at the Command prompt.

2. The name of the default system printer is displayed in the **Name** drop-down list in the **Printer/plotter** area. In this example, it is **Linotronic 300v52.3**, see Figure 2-56. You can use any other printer by selecting the name of the device from the **Name** drop-down list.

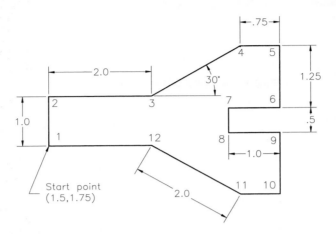

Figure 2-55 *Drawing created in Example 3*

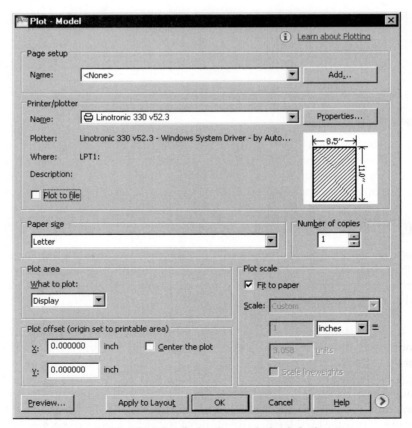

Figure 2-56 *The **Plot** dialog box with the default printer*

Note
The default system printer varies from system to system. It is based on the printer chosen as the default printer while configuring your system for the output device.

3. Select the **Window** option from the **What to plot** drop-down list in the **Plot area** area. The dialog box is temporarily closed and the drawing area will appear. Now, select the two opposite corners to define a window that specifies the plot area (the area you want to plot). Note that the complete drawing, along with the dimensions should be enclosed in the window. Once you have defined the two corners, the **Plot** dialog box will reappear.

4. To set the size of the plot, you can select a size from the drop-down list in the **Paper size** area, which lists all the plotting sizes that the present plotter can support. You can select any one of the sizes listed in the dialog box or specify a size (width and height) of your own through the **Plotter Manager**. (This option is discussed later in Chapter 14, Plotting Drawings.) Once you select a size, you can also select the orientation of the paper. However, to set the orientation, you need to expand the **Plot** dialog box by choosing the **More Options** button at the lower right corner of the dialog box. The expanded form of this dialog box is shown in Figure 2-57. To set the orientation, select the **Landscape** or **Portrait** radio button from the **Drawing orientation** area. The sections in the **Plot**

*Figure 2-57 The expanded form of the **Plot** dialog box*

dialog box related to paper size and orientation are automatically revised to reflect the new paper size and orientation. In this example, you will specify **A4** Paper size and **Portrait** orientation.

5.   You can also modify values for the plot offset from the **Plot offset** area; the default values for X and Y are 0. For this example, you can select the **Center the plot** check box to get the drawing in the center of the paper.

6.   In AutoCAD, you can enter values for the plot scale from the **Plot scale** area. Clear the **Fit to paper** check box, if selected and then open the **Scale** drop-down list in the **Plot scale** area to display various scale factors. From this list, you can select a scale factor that you want to use. For example if you select the scale factor **1/4" = 1'-0"**, the edit boxes below the drop-down list will show 0.25 inches = 12 units. If you want the drawing to be plotted so that it fits on the specified sheet of paper, select the **Fit to paper** check box. When you select this check box, AutoCAD will determine the scale factor and display it in the edit boxes. In this example, you will plot the drawing so that it scales to fit the paper. Therefore, select the **Fit to paper** check box and notice the change in the edit boxes. You can also enter your own values in the edit boxes.

7.   You can preview the plot on the specified paper size before actually plotting it. This way you can save time and stationery. To preview a plot, choose the **Preview** button. Once regeneration is complete, the preview image is displayed on the screen, see Figure 2-57. Here, in place of the cursor, a realtime zoom icon is displayed. You can hold the pick button of your pointing device and then move it up to zoom into the preview image and move the cursor down to zoom out of the preview image.

8.   If the plot preview is satisfactory, you can directly plot your drawing by choosing **Plot** from the shortcut menu, as shown in Figure 2-58. If you want to make some changes in the settings, choose **Exit** in the shortcut menu or press the ESC or the ENTER key to get back to the dialog box. Choose the **OK** button in the dialog box to plot the drawing.

Figure 2-58 *Plot preview with the shortcut menu*

MODIFYING AutoCAD SETTINGS USING THE OPTIONS DIALOG BOX

| **Application Menu:** | Options | **Command:** | OPTIONS |

You can use the **Options** dialog box to change the default settings that affect the drawing environment or the AutoCAD interface and customize them to your requirements using the **Options** dialog box. For example, you can use this dialog box to turn off the settings to display the shortcut menu by right-clicking or specify the support directories that contain the files you need. The most convenient way of invoking this dialog box is by right-clicking in the command window or in the drawing area when no command is active or no object is selected and choosing **Options** from the shortcut menu. The **Options** dialog box is shown in Figure 2-59. The dialog box contains ten tabs that display the sections to change the various environmental aspects. The current profile and current drawing names are displayed on the top, above the tabs. You can save a set of custom settings in a profile to be used later for other drawings. If you do not specify a profile, the current settings are stored with the name **Unnamed Profile**. The various tabs in the **Options** dialog box are discussed next.

Files

This tab stores the directories in which AutoCAD looks for the driver, support, menu, project, template, and other files. It uses three icons: folder, paper stack, and file cabinet. The folder icon is for a search path, the paper stack icon is for files, and the file cabinet icon is for a specific folder. Suppose you want to know the path of the font mapping file. You can select the **Text Editor, Dictionary, and Font File Names** folder and then select the **Font Mapping Files** icon to display the path, see Figure 2-59. Similarly, you can define a custom hatch pattern file and then add its search path. This way AutoCAD can locate the custom hatch pattern.

Display

This tab controls the drawing and window settings like screen menu display and scroll bar. For example, if you want to display the screen menu, select the **Display screen menu** check box in the **Window Elements** area. You can also change the color of the graphics window background, layout window background, command line background, and also the color of the command line text using the **Drawing Window Colors** dialog box that is displayed by choosing the **Colors** button. This tab also allows you to modify the display resolution and display performance. You can also set the smoothness and resolutions of certain objects such as the circle, arc, rendered object, and polyline curve. Here you can toggle on and off the various layout elements such as the layout tabs on the screen, margins, paper background, and so on. You can also toggle on and off the display performance such as the pan and zoom with raster images, apply the solid fills, and so on.

Open and Save

This tab controls the parameters related to opening and saving of files in AutoCAD. You can specify the file type for saving while using the SAVEAS command. The various formats are

Chapter 2

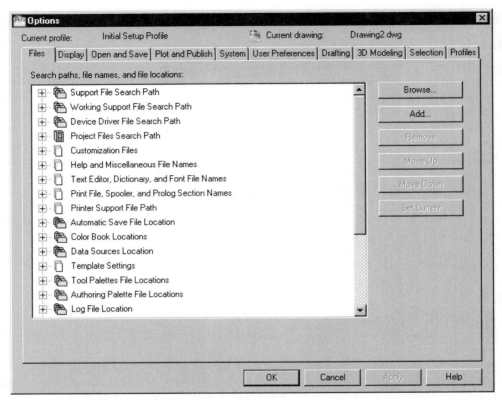

*Figure 2-59 The **Options** dialog box with the **Files** tab chosen*

AutoCAD 2007 Drawing (*.dwg), AutoCAD 2004/LT 2004 Drawing (*.dwg), AutoCAD 2000/ LT2000 Drawing (*.dwg), AutoCAD R14/LT98/LT97 Drawing (*.dwg), AutoCAD Drawing Template(*.dwt), AutoCAD 2007 DXF (*.dxf), AutoCAD 2004/LT2004 DXF (*.dxf), AutoCAD 2000/LT2000 DXF(*.dxf), AutoCAD R12/LT2 DXF (*.dxf), and so on. You can also set the various file safety precautions such as the Automatic Save feature, or the creation of a backup copy. You can add a password and digital signatures to your drawing while saving using the **Security Options** button in the **File Safety Precautions** area. You can control the display of the digital signature information when a file with a valid digital signature is opened with the help of the **Display digital signature information** check box. You can change the number of recently saved files to be displayed in the **File open** area for opening. You can also set the various parameters for external references and the ObjectARX applications.

Plot and Publish

The **Plot and Publish** tab controls the parameters related to the plotting and publishing of the drawings in AutoCAD. You can set the default output device and also add a new plotter. You can set the general parameters such as the layout or plot device paper size and the background processing options while plotting or publishing. It is possible to select the spool alert for the system printer and also the OLE plot quality. You can also set the parameters for the plot style such as using the color-dependent plot styles or the named plot styles.

System

This tab contains AutoCAD system settings options such as the 3D graphics display and pointing device settings options where you can choose the pointing device driver. Here you can also set the system parameters such as the single drawing mode instead of MDE, the display of the **Startup** option while opening a new session of AutoCAD and the **OLE Properties** dialog box, and beep for wrong user input. You also have options to set the parameters for database connectivity.

User Preferences

This tab controls settings that depend on the way the user prefers working on AutoCAD, such as the right-click customization where you can change the shortcut menus. You can set the units parameters for the **DesignCenter** as well as the priorities for various data entry methods. Here it is possible to set the order of object sorting methods and also set the lineweight options.

Drafting

This tab controls settings such as the autosnap settings and the aperture size. Here you can also set the toggles on and off for the various autotracking settings. Using this tab, you can also set the tool tip appearance in the **Model** tab and layouts for **Dynamic Input** mode.

3D Modeling

This tab controls settings related to 3D modeling, such as the display of cursor in the 3D modeling environment, visual styles, 3D navigation, and so on.

Selection

This tab controls settings related to the methods of object selection such as the grips, which enable you to change various grip colors and the grip size. You can also set the toggles on or off for the various selection modes.

Profiles

This tab saves and restores the system settings. To set the profile in the **Options** dialog box, choose the **Profiles** tab and then choose the **Add to List** button; the **Add Profile** dialog box is displayed. Enter the profile name and description. Then choose the **Apply & Close** button. Next, make the new profile current and then specify the settings. The settings are saved in the new profile and can be restored anytime by making the profile current.

Note

*The options in various tabs of the **Options** dialog box have been discussed throughout the book wherever applicable.*

Tip

*Some options in various tabs of the **Options** dialog box have a drawing file icon in front of them. For example, the options in the **Display resolution** area of the **Display** tab have drawing file icons. This specifies that these parameters are saved with the current drawing* *only and therefore affect it. The options without the drawing file icon are saved with the current profile and affect all the drawings present in that AutoCAD session or future sessions.*

Example 8   *General*

In the following example, you will use the **Options** dialog box to create a profile that contains the specified settings.

1. Choose **Options** from the **Application Menu** to invoke the **Options** dialog box. Alternatively, right-click in the drawing area to display the shortcut menu and choose **Options** to invoke this dialog box.

2. Choose the **Profiles** tab and then choose the **Add to List** button to display the **Add Profile** dialog box. Enter the name of the new profile as **Myprofile1** and a description of the new profile, and then choose the **Apply & Close** button to exit.

3. Select **Myprofile1** profile and then choose the **Set Current** button to make **Myprofile1** current. You will notice that the **Current Profile** name above the tabs changes from **<<Unnamed Profile>>** to **Myprofile1**.

4. Choose the **Display** tab and then choose the **Colors** button; the **Drawing Window Colors** dialog box is displayed. Select the **2D model space** in the **Context** area and **Uniform background** in the **Interface element** area. Select **White** from the **Color** drop-down list, see Figure 2-60. This changes the color of the model tab background to white. Choose the **Apply & Close** button to return to the **Options** dialog box. Notice that the screen color changes to white.

5. Choose the **Drafting** tab and change the **AutoSnap Marker Size** to the maximum using the slider bar. Choose the **Apply** button and then the **OK** button to exit the dialog box.

6. Draw a line and then choose the **OSNAP** button from the Status Bar. Again, invoke the **LINE** command and move the cursor on the previously drawn line; a marker is displayed at the endpoint. Notice the size of the marker now.

7. Invoke the **Options** dialog box again and choose the **Profiles** tab. Double-click on the default profile (**<<Unnamed Profile>>**) to reload the default settings. The screen settings will change as specified in the default profile.

Self-Evaluation Test

Answer the following questions and then compare them to those given at the end of this chapter:

1. You can draw a line by specifying the length of the line and its direction, using **Direct Distance Entry**. (T/F)

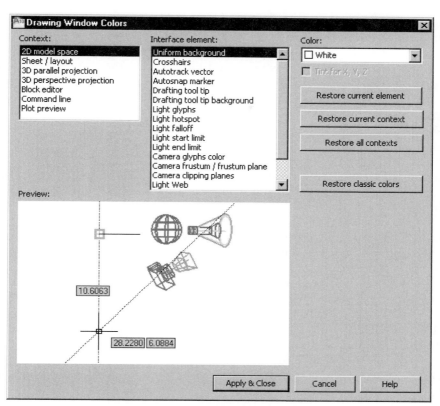

*Figure 2-60 The **Drawing Window Colors** dialog box*

2. Using the **Crossing** method of object selection, only those objects that are completely enclosed within the boundaries of the crossing box are selected. (T/F)

3. The **Three-Point** option of the **CIRCLE** command lets you draw a circle by specifying the two endpoints of the circle's diameter. (T/F)

4. If you choose the engineering or architectural format for units in the **Drawing Units** dialog box, you can enter the distances or coordinates in any of the five formats. (T/F)

5. You can erase a previously drawn line using the _____ option of the **LINE** command.

6. The _____ option of the **CIRCLE** command can be used to draw a circle, if you want the circle to be tangent to two previously drawn objects.

7. The _____ command enlarges or reduces the view of the drawing on the screen, but it does not affect the actual size of the entities.

8. After increasing the drawing limits, you need to use the _____ option of the **ZOOM** command to display the complete area inside the drawing area.

9. In _____ units, you must specify the bearing angle that the line makes with the north-south direction.

10. You can preview the plot before the actual plotting using the _____ button in the **Plot** dialog box.

Review Questions

Answer the following questions:

1. In the **Relative rectangular** coordinate system, the displacements along the X and Y axes (DX and DY) are measured with reference to the previous point rather than to the origin. (T/F)

2. In AutoCAD, by default the angles are positive if measured in the counterclockwise direction and the angles are measured from the positive X axis. (T/F)

3. You can also invoke the **PLOT** command by choosing **Plot** from the shortcut menu, which is displayed by right-clicking on the Command window. (T/F)

4. The **Files** tab of the **Options** dialog box stores the directories, in which AutoCAD looks for the driver, support, menu, project, template, and other files. (T/F)

5. Which of the following keys on the keyboard is used to terminate the **LINE** command at the **Specify next point or [Close/Undo]:** prompt?

    (a) SPACEBAR   (b) BACKSPACE
    (c) ENTER   (d) ESC

6. Which of the following options of the **ZOOM** command zooms to the drawing limits or the extents, whichever is greater?

    (a) **Previous**   (b) **Window**
    (c) **All**   (d) **Realtime**

7. How many formats of units can you choose from the **Drawing Units** dialog box?

    (a) Three   (b) Five
    (c) Six   (d) Seven

8. Which of the following input methods cannot be used to invoke the **OPTIONS** command that displays the **Options** dialog box?

    (a) Menu   (b) Toolbar
    (c) Shortcut menu   (d) Command prompt

9. When you define the direction by specifying the angle, the output of the angle does not depend on which one of the following factors?

    (a) Angular units   (b) Angle value
    (c) Angle direction   (d) Angle base

10. The _____ option of the **LINE** command can be used to join the current point with the initial point of the first line when two or more lines are drawn in continuation.

11. The _____ option of drawing the circle cannot be invoked by entering the command at the Command prompt.

12. When you select any type of unit and angle in the **Length** or **Angle** area of the **Drawing Units** dialog box, the corresponding example is displayed in the _____ area of the dialog box.

13. If you want the drawing to be plotted so that it fits on the specified sheet of paper, select the _____ option in the **Plot** dialog box.

14. The _____ tab in the **Options** dialog box stores the details of all profiles available in the current drawing.

15. You can use the _____ command to change the settings that affect the drawing environment or the AutoCAD interface.

Exercises

Exercise 7 *General*

Use the following relative rectangular and absolute coordinate values in the **LINE** command to draw an object.

Point	Coordinates	Point	Coordinates
1	3.0, 3.0	5	@3.0,5.0
2	@3,0	6	@3,0
3	@-1.5,3.0	7	@-1.5,-3
4	@-1.5,-3.0	8	@-1.5,3

Exercise 8 *General*

For the drawing shown in Figure 2-61, enter the relative rectangular and relative polar coordinates of the points in the following table, and then use these coordinates to draw the figure. The distance between the dotted lines is 1 unit. Save this drawing as *Exer8.dwg*.

Point	Coordinates	Point	Coordinates
1	3.0, 1.0	9	_____

Chapter 2

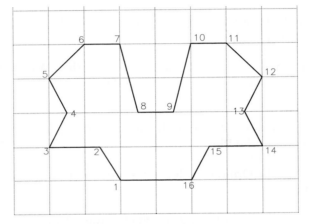

Figure 2-61 Drawing for Exercise 8

Exercise 9   *Mechanical*

For the drawing shown in Figure 2-62, enter the relative polar coordinates of the points in the following table. Then use these coordinates to draw the figure. Do not draw the dimensions.

Point	Coordinates	Point	Coordinates
1	1.0, 1.0	6	_____
2	_____	7	_____
3	_____	8	_____
4	_____	9	_____
5	_____		

Exercise 10   *Mechanical*

Draw the sketch shown in Figure 2-63, using the **LINE** and **CIRCLE** commands. The distance between the dotted lines is 1.0 units.

Exercise 11   *Mechanical*

Draw the sketch shown in Figure 2-64 using the **LINE** command and the **Ttr** option of the **CIRCLE** command.

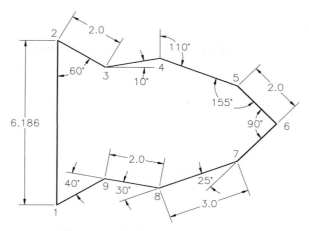

Figure 2-62 *Drawing for Exercise 9*

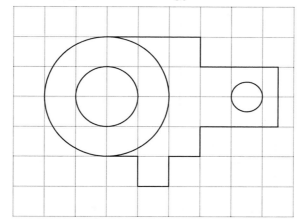

Figure 2-63 *Drawing for Exercise 10*

Figure 2-64 *Drawing for Exercise 11*

Chapter 2

Exercise 12 *Mechanical*

Set the units for a drawing according to the following specifications.
1. Set the **UNITS** to architectural, with the denominator of the smallest fraction equal to 16.
2. Set the angular measurement to degrees/minutes/seconds, with the number of fractional places for display of angles equal to 0d00'.
3. Set the direction to 0-degree (east) and the direction of measurement of angles to counterclockwise (angles measured positive in a counterclockwise direction).

Based on Figure 2-65, determine and set the limits of the drawing. The scale for this drawing is 1/4" = 1'. Leave enough space around the drawing for dimensioning and title block. (HINT: Scale factor = 48; sheet size required is 12 x 9; therefore, the limits are 12 X 48, 9 X 48 = 576, 432. Use the **ZOOM** command and then select the **All** option to display the new limits.)

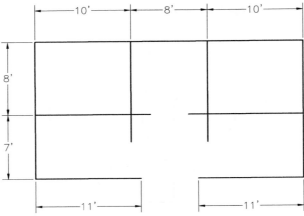

Figure 2-65 *Drawing for Exercise 12*

Exercise 13 *Mechanical*

Draw the sketch shown in Figure 2-66. The distance between the dotted lines is 10 feet. Determine the limits for this drawing and use the Architectural units with 0'-01/32" precision.

Exercise 14 *General*

Draw the object shown in Figure 2-67. The distance between the dotted lines is 5 inches. Determine the limits for this drawing and use the Fractional units with 1 1/16 precision.

Exercise 15 *General*

Draw the object shown in Figure 2-68. The distance between the dotted lines is 1 unit. Determine the limits for this drawing and use the Decimal units with 0.00 precision.

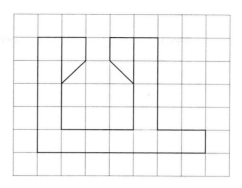

Figure 2-66 *Drawing for Exercise 13*

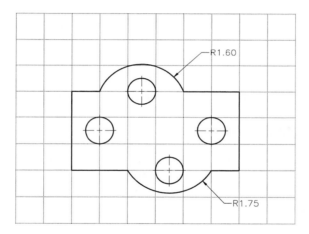

Figure 2-67 *Drawing for Exercise 14*

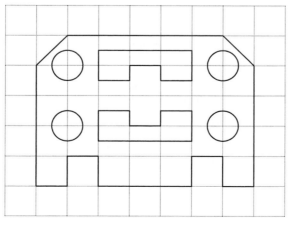

Figure 2-68 *Drawing for Exercise 15*

Chapter 2

Exercise 16   *General*

Draw the object shown in Figure 2-69. The distance between the dotted lines is 10 feet. Determine the limits for this drawing and use the Engineering units with 0'0.00" precision.

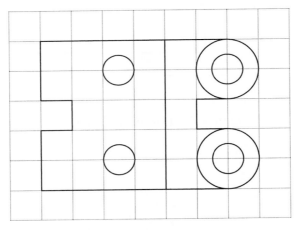

Figure 2-69 *Drawing for Exercise 16*

Problem-Solving Exercise 1   *Mechanical*

Draw the object shown in Figure 2-70, using the **LINE** and **CIRCLE** commands. In this exercise only the diameters of the circles are given. To draw the lines and small circles (Dia 0.6), you need to find the coordinate points for the lines and the center points of the circles. For example, if the center of concentric circles is at 5,3.5, then the *X* coordinate of the lower left corner of the rectangle is 5.0 - 2.4 = 2.6.

Figure 2-70 *Drawing for Problem-Solving Exercise 1*

Problem-Solving Exercise 2

Mechanical

Draw the object shown in Figure 2-71 using various options of the **CIRCLE** and **LINE** commands. In this exercise, you have to find the coordinate points for drawing the lines and circles. Also, you need to determine the best and easiest method to draw the 0.85 diameter circles along the outermost circle.

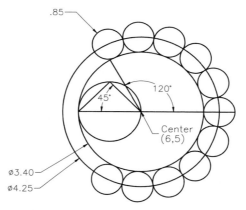

Figure 2-71 *Drawing for Problem-Solving Exercise 2*

Problem-Solving Exercise 3

Mechanical

Draw the object shown in Figure 2-72 using the absolute, relative rectangular, or relative polar coordinate system. Draw according to the dimensions shown in the figure, but do not dimension the sketch. The dimensions are only for your reference.

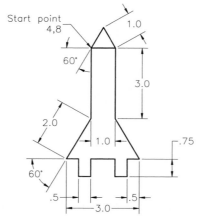

Figure 2-72 *Drawing for Problem-Solving Exercise 3*

Chapter 2

Problem-Solving Exercise 4   *Mechanical*

Draw the object shown in Figure 2-73 according to the given dimensions. Also, draw a centerline as a continuous line in the object. Note that the dimensions are given only for your reference. **

Figure 2-73 *Drawing for the Problem-Solving Exercise 4*

Problem-Solving Exercise 5   *Mechanical*

Draw the object shown in Figure 2-74 according to the given dimensions. Also, draw a centerline as a continuous line in the object. Note that the dimensions are given only for your reference. **

Figure 2-74 *Drawing for the Problem-Solving Exercise 5*

Answers to Self-Evaluation Test
1. T, **2.** F, **3.** F, **4.** T, **5. Undo**, **6. Tan, Tan, Radius**, **7. ZOOM**, **8. All**, **9.** Surveyor's, **10. Preview**

Chapter **3**

Starting with the Advanced Sketching

DRAWING ARCS

Ribbon:	Home > Draw > 3-Point	Toolbar:	Draw > Arc
Menu Bar:	Draw > Arc	Command:	ARC or A

An arc is defined as a part of a circle. In AutoCAD, it can be drawn using the **ARC** command. You can invoke the **ARC** command from the **Draw** toolbar, as shown in Figure 3-1. AutoCAD provides eleven different options to draw an arc. To view these options, click on the down arrow on the **3-Point** button in the **Draw** panel of the **Home** tab in the **Ribbon**; a flyout will be displayed, as shown in Figure 3-2. The default option for drawing an arc is the **3-Point** option. After invoking the **ARC** command, other options can be invoked by entering the appropriate letter in the command window or by right-clicking and choosing the appropriate option from the flyout. The option that was used last while creating the arc will be displayed in the **Draw** panel.

*Figure 3-1 Invoking the **ARC** command from the **Draw** toolbar*

The 3-Point Option

When you choose the **Arc** button from the **Draw** toolbar, or enter **ARC** at the Command prompt, the **3-Point** option automatically gets activated. You can also choose the **3-Point** button from the **Draw** panel of the **Home** tab in the **Ribbon**. On choosing this button, you will be prompted to specify the start point, second point, and endpoint of the arc, see Figure 3-3. The following is the prompt sequence to draw an arc with a start point at (2,2), second point at (3,3), and an endpoint at (3,4). (You can also specify the points by moving the cursor and then specifying them on the screen.)

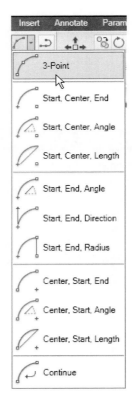

Command: *Turn off the dynamic input and then choose the **3-Point** button from the **Draw** panel in the **Ribbon** (**3-Point** is the default option).*
_arc Specify start point of arc or [Center]: **2,2** `Enter`
Specify second point of arc or [Center/End]: **3,3** `Enter`
Specify end point of arc: **3,4** `Enter`

*Figure 3-2 The options in the **Ribbon** for drawing an arc*

Exercise 1   *General*

Draw several arcs using the **3-Point** option. The points can be selected by entering coordinates or by specifying points on the screen. Also, try to create a circle by drawing two separate arcs and a single arc. Notice the limitations of the **ARC** command.

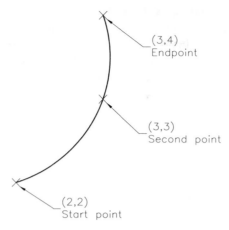

Figure 3-3 *Drawing an arc using the **3-Point** option*

The Start, Center, End Option

This option is slightly different from the **3-Point** option. In this option, instead of entering the second point, you need to enter the center of the arc. If you know the start point, endpoint, and center point of the arc, choose this option. The arc is drawn in a counterclockwise direction from the start point to the endpoint around the specified center. The endpoint specified need not be on the arc and is used only to calculate the angle at which the arc ends. Thse radius of the arc is determined by the distance between the center point and start point. The prompt sequence for drawing an arc, shown in Figure 3-4, with a start point at (3,2), center point at (2,2), and endpoint at (2,3.5) is given next.

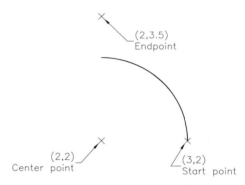

Figure 3-4 *Drawing an arc using the **Start, Center, End** option*

> Command: *Turn off the dynamic input and then enter **ARC***
> _arc Specify start point of arc or [Center]: **3,2** Enter
> Specify second point of arc or [Center/End]: **C** Enter
> Specify center point of arc: **2,2** Enter
> Specify end point of the arc or [Angle/chord Length]: **2,3.5** Enter

Note

*If you choose **3-Point > Start, Center, End** from the **Draw** panel of the **Ribbon**, the **Center** option is automatically invoked. This way you bypass the prompt **Specify second point of arc or [Center/End]**. Therefore, you need to specify only the center point.*

The Start, Center, Angle Option

This option is the best choice, if you know the included angle of the arc. The included angle is the angle formed by the start and endpoint of the arc with the specified center. This option draws an arc in the counterclockwise direction with the specified center and start point spanning the indicated angle, see Figure 3-5. If the specified angle is negative, the arc is drawn in the clockwise direction, see Figure 3-6.

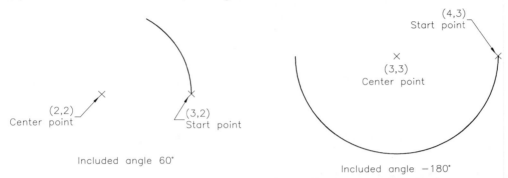

Figure 3-5 *Drawing an arc using the* ***Start,*** ***Center, Angle*** *option*

Figure 3-6 *Drawing an arc using a negative angle in the* ***Start, Center, Angle*** *option*

The prompt sequence for drawing an arc with center at (2,2), start point at (3,2), and an included angle of 60-degree, as shown in Figure 3-5, is given next.

Command: *Turn off the dynamic input and then enter* ***ARC***.
Specify start point of arc or [Center]: **3,2** Enter
Specify second point of arc or [Center/End]: **C** Enter
Specify center point of arc: **2,2** Enter
Specify end point of the arc or [Angle/chord Length]: **A** Enter
Specify included angle: **60** Enter

You can draw arcs with negative angle values in the **Start, Center, Included Angle** (St,C,Ang) option by entering "-" (negative sign), followed by the required angle at the **Specify included angle** prompt (Figure 3-6). The following prompt sequence is displayed, when you invoke the **ARC** command:

_arc Specify start point of arc or [Center]: **4,3** Enter
Specify second point of arc or [Center/End]: **C** Enter
Specify center point of arc: **3,3** Enter
Specify end point of the arc or [Angle/ chord Length]: **A** Enter
Specify included angle: **-180** Enter

Note
If you choose ***3-Point > Start, Center, Angle*** *from the* ***Draw*** *panel of the* ***Ribbon***, *you need to specify the start point, the center point, and the angle.*

Exercise 2

Mechanical

a. Draw an arc using the **St, C, Ang** option. The start point is (6,3), the center point is (3,3), and the angle is 240-degree.

b. Make the drawing shown in Figure 3-7. The distance between the dotted lines is 1.0 unit. Create the radii by using the arc command options indicated in the drawing.

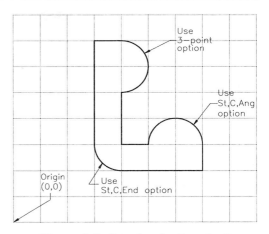

Figure 3-7 Drawing for Exercise 2

The Start, Center, Length Option

In this option, you are required to specify the start point, center point, and length of the chord. A chord is defined as a straight line connecting the start point and endpoint of an arc. The chord length needs to be specified so that AutoCAD can calculate the ending angle. Identical start, center, and chord length specifications can be used to define four different arcs. AutoCAD always draws this type of arc counterclockwise from the start point. Therefore, a positive chord length gives the smallest possible arc with that length. This arc is known as the minor arc. The minor arc is less than 180-degree. A negative value for the chord length results in the largest possible arc, also known as the major arc. The chord length can be determined, by using the standard chord length tables or the mathematical relation $(L = 2*Sqrt [h(2r-h)])$. For example, an arc of radius 1 unit, with an included angle of 30-degree, has a chord length of 0.51764 units. The prompt sequence for drawing an arc, shown in Figure 3-8 that has a start point (3,1), center (2,2), and the chord length of 2 units is given next.

Command: *Turn off the dynamic input and choose* **3-Point > Start, Center, Length** *from the* **Draw** *panel of the* **Ribbon**.
_arc Specify start point of arc or [Center]: **3,1** `Enter`
Specify second point of arc or [Center/End]: _c Specify center point of arc: **2,2** `Enter`
Specify end point of the arc or [Angle/ chord Length]: _l Specify length of chord: **2** `Enter`

You can draw the major arc by defining the length of the chord as negative, see Figure 3-9. In this case, the arc with a start point at (3,1), center point at (2,2), and a negative chord length of -2, is drawn with the following prompt sequence.

Command: *Choose* **3-Point > Start, Center, Length** *from the* **Draw** *panel of the* **Ribbon**
_arc Specify start point of arc or [Center]: **3,1** `Enter`
Specify second point of arc or [Center/End]: _c Specify center point of arc: **2,2** `Enter`
Specify end point of the arc or [Angle/ chord Length]: _l Specify length of chord: **-2** `Enter`

Chapter 3

Figure 3-8 *Drawing an arc using the **Start, Center, Length** option*

Figure 3-9 *Drawing an arc using a negative chord length in the **Start, Center, Length** option*

Note
*The points that you have specified for drawing the arcs using different options are quite close to each other and therefore the arcs created might overlap each other. You can use the **ERASE** command to erase some of the previously drawn arcs and gain space to draw new arcs.*

Exercise 3   *General*

Draw a minor arc with the center point at (3,4), start point at (4,2), and chord length of 4 units.

The Start, End, Angle Option

With this option, you can draw an arc by specifying the start point of the arc, endpoint, and the included angle. A positive included angle value draws an arc in the counterclockwise direction from the start point to the endpoint, spanning the included angle; a negative included angle value draws the arc in the clockwise direction, as shown in Figure 3-10. The prompt sequence for drawing an arc with a start point at (3,2), endpoint at (2,4), and included angle of 120-degree is given next.

Command: *Turn off the dynamic input and choose* **3-Point > Start, End, Angle** *from the **Draw** panel of the **Ribbon**.*
_arc Specify start point of arc or [Center]: **3,2** Enter
Specify second point of arc or [Center/End]: _e Specify end point of arc: **2,4** Enter
Specify center point of arc or [Angle/Direction/Radius]: _a Specify included angle: **120** Enter

Figure 3-10 *Drawing an arc using the **Start, End, Angle** option*

The Start, End, Direction Option

In this option, you can draw an arc by specifying the start point, endpoint, and starting direction of the arc, in degrees. In other words, the arc starts in the direction you specify (the start of the arc is established tangent to the direction you specify). This option can be used to draw a major or minor arc, in a clockwise or counterclockwise direction, whose size and position are determined by the distance between the start point and endpoint and the direction specified. To illustrate the positive direction option shown in Figure 3-11, the prompt sequence for an arc with a start point at (4,3), endpoint at (3,5), and direction of 90-degree is given next.

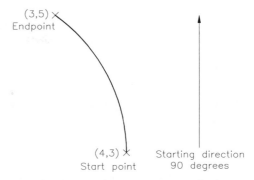

Figure 3-11 *Drawing an arc using the **Start, End, Direction** option*

> Command: *Turn off the dynamic input and choose **3-Point > Start, End, Direction** from the **Draw** panel of the **Ribbon**.*
> _arc Specify start point of arc or [Center]: **4,3** [Enter]
> Specify second point of arc or [Center/End]: _e Specify end point of arc: **3,5** [Enter]
> Specify center point of arc or [Angle/Direction/Radius]: _d Specify tangent direction for the start point of arc: **90** [Enter]

To illustrate the option of using a negative direction degree specification, shown in Figure 3-12, the prompt sequence for an arc with a start point at (4,3), endpoint at (3,4), and direction of -90-degree, is given next.

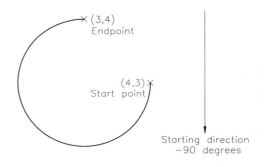

Figure 3-12 *Drawing an arc in the negative direction using the **Start, End, Direction** option*

> Command: *Choose **3-Point > Start, End, Direction** from the **Draw** panel of the **Ribbon**.*
> _arc Specify start point of arc or [Center]: **4,3** [Enter]
> Specify second point of arc or [Center/End]: e_e Specify end point of arc: **3,4** [Enter]
> Specify center point of arc or [Angle/Direction/Radius]: _d Specify tangent direction for the start point of arc: **-90** [Enter]

Note

*With the **Start, End, Direction** option, if you do not specify a start point but press ENTER at the **Specify start point of arc or [Center]** prompt, the start point and direction of the arc will be taken from the **endpoint and ending direction** of the previous line or arc drawn on the current screen. You are then required to specify only the endpoint of the arc.*

Exercise 4   *Mechanical*

a. Specify the directions and coordinates of two arcs in such a way that they form a circular figure.

b. Make the drawing shown in Figure 3-13. Create the curves using the **ARC** command. The distance between the dotted lines is 1.0 unit and the diameter of the circles is 1 unit.

The Start, End, Radius Option

This option is used when you know the start point, endpoint, and radius of the arc. The same values for the three variables (start point, endpoint, and radius) can result in four different arcs. AutoCAD resolves this by always drawing this type of arc in a counterclockwise direction from the start point. Therefore, a negative radius value results in a **major arc** (the largest arc between two endpoints), Figure 3-14(a), while a positive radius value results in a **minor arc** (smallest arc between the start point and endpoint), Figure 3-14(b). The prompt sequence to draw a major arc with a start point at (3,3), endpoint at (2,5), and radius of -2, Figure 3-14(a), is given next.

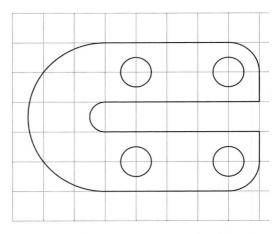

Figure 3-13 *Drawing for Exercise 4*

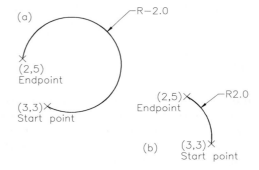

Figure 3-14 *Drawing an arc using the **Start, End, Radius** option*

> Command: *Turn off the dynamic input and choose **3-Point > Start, End, Radius** from the **Draw** panel of the **Ribbon**.*
> _arc Specify start point of arc or [Center]: **3,3** Enter
> Specify second point of arc or [Center/End]: _e Specify end point of arc: **2,5** Enter

Specify center point of arc or [Angle/Direction/Radius]: _r Specify radius of arc: **-2** Enter

The prompt sequence to draw a minor arc with its start point at (3,3), endpoint at (2,5), and radius as 2, as shown in Figure 3-14(b) is given next.

> Command: *Choose **3-Point > Start, End, Radius** from the **Draw** panel of the **Ribbon**.*
> _arc Specify start point of arc or [Center]: **3,3** Enter
> Specify second point of arc or [Center/End]: _e Specify end point of arc: **2,5** Enter
> Specify center point of arc or [Angle/Direction/Radius]: _r Specify radius of arc: **2** Enter

The Center, Start, End Option

The **Center, Start, End** option is a modification of the **Start, Center, End** option. Use this option, whenever it is easier to start drawing an arc by establishing the center first. Here the arc is always drawn in a counterclockwise direction from the start point to the endpoint, around the specified center. The prompt sequence for drawing the arc, shown in Figure 3-15, that has a center point at (3,3), start point at (5,3), and endpoint at (3,5) is given next.

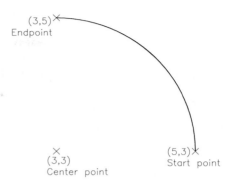

Figure 3-15 *Drawing an arc using the* **Center, Start, End** *option*

Command: *Choose* **3-Point > Center, Start, End** *from the* **Draw** *panel of the* **Ribbon**.
_arc Specify start point of arc or [Center]: _**c** Specify center point of arc: **3,3** [Enter]
Specify start point of arc: **5,3** [Enter]
Specify end point of arc or [Angle/chord Length]: **3,5** [Enter]

The Center, Start, Angle Option

This option is a variation of the **Start, Center, Angle** option. Use this option, whenever it is easier to draw an arc by giving the center first. The prompt sequence for drawing the arc shown in Figure 3-16, that has a center point at (4,5), start point at (5,4), and included angle of 120-degree is given next.

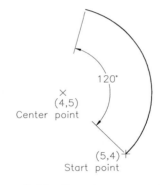

Figure 3-16 *Drawing an arc using the* **Center, Start, Angle** *option*

Command: *Turn off the dynamic input and choose* **3-Point > Center, Start, Angle** *from the* **Draw** *panel of the* **Ribbon**.
_arc Specify start point of arc or [Center]: _c
Specify center point of arc: **4,5** [Enter]
Specify start point of arc: **5,4** [Enter]
Specify end point of arc or [Angle/chord Length]: _a Specify included angle: **120** [Enter]

The Center, Start, Length Option

The **Center, Start, Length** option is a modification of the **Start, Center, Length** option. This option is used, whenever it is easier to draw an arc by establishing the center first. The prompt sequence for drawing the arc shown in Figure 3-17, that has a center point at (2,2), start point at (4,3), and length of chord as 3 is given next.

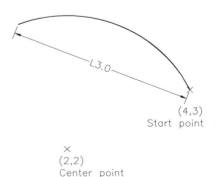

Figure 3-17 *Drawing an arc using the* **Center, Start, Length** *option*

Command: *Turn off the dynamic input and choose* **3-Point > Center, Start, Length** *from the* **Draw** *panel of the* **Ribbon**.
_arc Specify start point of arc or [Center]: **2,2** Enter
Specify start point of arc: **4,3** Enter
Specify end point of arc or [Angle/chord Length]: _l Specify length of chord: **3** Enter

Continue Option

With this option, you can continue drawing an arc from a previously drawn arc or line. When you choose **3-Point > Continue** from the **Draw** panel of the **Ribbon**, the start point and direction of the arc will be taken from the endpoint and ending direction of the previous line or arc. When this option is used to draw arcs, each successive arc will be tangent to the previous one. Most often, this option is used to draw arcs tangent to a previously drawn line. The prompt sequence to draw the arc which is tangent to an earlier drawn line, as shown in Figure 3-18, using the **Continue** option is given next.

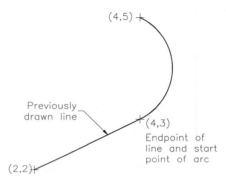

Figure 3-18 *Drawing an arc using the* **Continue** *option*

Command: **LINE** Enter
Specify first point: **2,2** Enter
Specify next point or [Undo]: **4,3** Enter
Specify next point or [Undo]: Enter

Command: *Choose* **3-Point > Continue** *option from the* **Draw** *panel of the* **Ribbon**.
Command: _arc Specify start point of arc or [Center]:
Specify endpoint of arc: **4,5** Enter

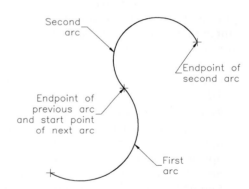

Figure 3-19 *Drawing an arc using the* **Continue** *option*

The prompt sequence to draw an arc continued from a previously drawn arc, as shown in Figure 3-19, is given next.

Command: **ARC** Enter
Specify start point of arc or [Center]: **2,2** Enter
Specify second point of arc or [Center/End]: **E** Enter
Specify endpoint of arc : **3,4** Enter
Specify center point of arc or [Angle/Direction/Radius]: **R** Enter
Specify radius of arc: **2** Enter

Command: *Choose* **3-Point > Continue** *option from the* **Draw** *panel of the* **Ribbon**.
Command: _arc Specify start point of arc or [Center]:
Specify end point of arc: **5,4** Enter

Tip

*You can also invoke the **Continue** option automatically. To do so, first draw a line or an arc and choose any option of the **ARC** command from the **Draw** panel of the **Ribbon**. Now, press ENTER at the **Specify start point of arc or [Center]** prompt; the **Continue** option will be invoked automatically. The endpoint of the line or the arc drawn previously will be selected as the start point of the arc. Also, you will be prompted to specify the endpoint to complete the arc.*

Continue (LineCont:) Option

This option is used when you want to continue drawing a line from the endpoint of a previously drawn arc. When you use this option, the start point and direction of the line will be taken from the endpoint and ending direction of the previous arc. In other words, the line will be tangent to the arc drawn on the current screen. This option is invoked when you press ENTER at the **Specify first point** prompt of the **LINE** command. The prompt sequence to draw a line, tangent to an earlier drawn arc, as shown in Figure 3-20, is given next.

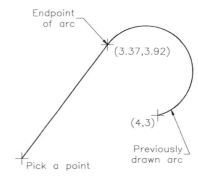

Figure 3-20 Drawing a line from the endpoint of an arc

    Command: **ARC** Enter
    Specify start point of arc or [Center]: **4,3** Enter
    Specify second point of arc or [Center/End]: **E** Enter
    Specify end point of arc: **3.37,3.92** Enter
    Specify center point of arc or [Angle/Direction/Radius]: *Specify the center point.*

    Command: **LINE** Enter
    Specify first point: Enter
    Length of line: *Enter a value or pick a point.*

Exercise 5   *General*

a.   Use the **Center, Start, Angle** and **Continue** options to draw the figures shown in Figure 3-21.

b.   Make the drawing shown in Figure 3-22. The distance between the dotted lines is 1.0 units. Create the radii as indicated in the drawing by using the **ARC** command options.

DRAWING RECTANGLES

Ribbon:	Home > Draw > Rectangle	**Toolbar:**	Draw > Rectangle
Tool Palettes:	Draw > Rectangle	**Command:**	RECTANG

Figure 3-21 *Drawing for Exercise 5(a)*   **Figure 3-22** *Drawing for Exercise 5(b)*

A rectangle can be drawn by choosing the **Rectangle** button from the **Draw** panel in the **Home** tab of the **Ribbon**, as shown in Figure 3-23, or by choosing the **Rectangle** button from the **Draw** toolbar. You can also enter **RECTANG** at the Command prompt. In AutoCAD, you can draw rectangles by specifying two opposite corners of the rectangle, by specifying the area and the size of one of the sides, or by specifying the dimensions of the rectangle. All these methods of drawing rectangles are discussed next.

Figure 3-23 *Invoking the* **RECTANGLE** *command from the* **Ribbon**

Drawing Rectangles Using Two Opposite Corners

On invoking the rectangle command, you are prompted to specify the first corner of the rectangle. Here, you can enter the coordinates of the first corner or specify the desired point with the pointing device. The first corner can be any one of the four corners. Next, you are prompted to specify the other corner. This corner is taken as the corner diagonally opposite to the first corner. You can specify the coordinates for the other corner or simply move the cursor to specify it, refer to Figure 3-24. The prompt sequence for drawing a rectangle with (3,3) as its lower left corner coordinate and (6,5) as its upper right corner (Figure 3-24), is given next. It is presumed that the dynamic input is turned off.

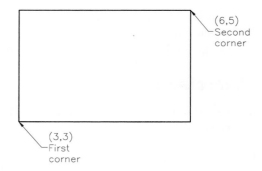

Figure 3-24 *Drawing a rectangle by specifying two opposite corners*

Command: **RECTANG** ⌨Enter
Specify first corner point or [Chamfer/Elevation/Fillet/Thickness/Width]: **3,3** ⌨Enter *(Lower left corner location.)*
Specify other corner point or [Area/Dimensions/Rotation]: **6,5** ⌨Enter *(Upper right corner location)*

Drawing Rectangles by Specifying the Area and One Side

This option allows you to draw a rectangle by specifying its area and the value of one of the sides. This can be done by choosing the **Area** option from the shortcut menu at the **Specify other corner point or [Area/Dimensions/Rotation]** prompt. The following is the prompt sequence to use this option:

Command: **RECTANG** ⌨Enter
Specify first corner point or [Chamfer/Elevation/Fillet/Thickness/Width]: **3,3** ⌨Enter
Specify other corner point or [Area/Dimensions/Rotation]: **A** ⌨Enter
Enter area of rectangle in current units <100.000>: **15** ⌨Enter
Calculate rectangle dimensions based on [Length/Width] <Length>: **L** ⌨Enter
Enter rectangle length <10.0000>: **5** ⌨Enter

In the above case, the area and length of the rectangle were entered. The system automatically calculates its width using the following formula:

Area of rectangle = Length X Width
Width = Area of rectangle/Length
Width =15/5
Width =3 units

Drawing Rectangles by Specifying their Dimensions

You can also draw a rectangle by specifying its dimensions. This can be done by choosing the **Dimensions** option from the shortcut menu at the **Specify other corner point or [Area/ Dimensions/Rotation]** prompt, which then allows you to enter the length and width of the rectangle. The prompt sequence for drawing a rectangle with a length of **5** units and width of **3** units is given next.

Command: **RECTANG** ⌨Enter
Specify first corner point or [Chamfer/Elevation/Fillet/Thickness/Width]: **3,3** ⌨Enter
Specify other corner point or [Area/Dimensions/Rotation]: **D** ⌨Enter
Specify length for rectangles <0.0000>: **5** ⌨Enter
Specify width for rectangles <0.0000>: **3** ⌨Enter
Specify other corner point or [Area/Dimensions/Rotation]: *Click on the screen to specify the orientation of rectangle.*

Here, you are allowed to choose any one of the four locations for placing the rectangle. You can move the cursor to see the four locations. Depending on the location of the cursor, the corner point which is specified first holds the position of either the lower left corner, the

Chapter 3

lower right corner, the upper right corner, or the upper left corner. After deciding the position, you can then click to place the rectangle.

Drawing Rectangle at an Angle

You can also draw a rectangle which is inclined at an angle. This can be done by choosing the **Rotate** option from the shortcut menu, at the **Specify other corner point or [Area/Dimensions/Rotation]** prompt, which then allows you to enter the rotation angle. After entering the rotation angle, you can continue sizing the rectangle using any of the methods. Once you have specified the rotation angle, next time whenever you draw the rectangle, it will be drawn at an angle. Set the rotation angle to Zero, if you do not want to draw the rectangle at an angle. The prompt sequence for drawing the rectangle at a rotation of 45 degree is:

Command: **RECTANG** Enter
Specify first corner point or [Chamfer/Elevation/Fillet/Thickness/Width]: *Select a point as lower left corner location.*
Specify other corner point or [Area/Dimensions/Rotation]: R Enter
Specify rotation angle or [Pick points] <current>: 45 Enter
Specify other corner point or [Area/Dimensions/Rotation]: *Select a diagonally opposite point.*

While specifying the other corner point, you can place the rectangle in any of the four quadrants. Move the cursor in different quadrants and select a point in the quadrant, in which you need to draw the rectangle. Figure 3-25 shows a rectangle drawn at an angle of 45-degree.

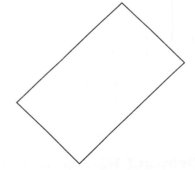

Figure 3-25 Rectangle drawn at an angle

The remaining options of the **RECTANG** command are discussed next.

Chamfer

The **Chamfer** option creates a chamfer, which is an angled corner, by specifying the chamfer distances, see Figure 3-26. The chamfer is created at all the four corners. You can give two different chamfer values to create an unequal chamfer.

Command: **RECTANG** Enter
Specify first corner point or [Chamfer/Elevation/Fillet/Thickness/Width]: C Enter
Specify first chamfer distance for rectangles <0.0000>: *Enter a value.*
Specify second chamfer distance for rectangles <0.0000>: *Enter a value.*
Specify first corner point or [Chamfer/Elevation/Fillet/Thickness/Width]: *Select a point as lower left corner location.*
Specify other corner point or [Area/Dimensions/Rotation]: *Select a point as upper right corner location.*

Note
The first corner point that you specify need not be the location of the lower left corner. While selecting the other corner, you can select a location such that the first corner point becomes the lower right corner, the upper left corner, or the upper right corner.

Fillet

The **Fillet** option allows you to create a filleted rectangle, see Figure 3-27. You can specify the required fillet radius. The following is the prompt sequence for specifying the fillet.

Specify first corner point or [Chamfer/Elevation/Fillet/Thickness/Width]: **F** [Enter]
Specify fillet radius for rectangles <0.0000>: *Enter a value.*

Note that the rectangle will be filleted only if the length and width of the rectangle are equal to or greater than twice the value of the specified fillet. Otherwise, AutoCAD will draw a rectangle without fillets.

Figure 3-26 Drawing a rectangle with chamfers

Figure 3-27 Drawing a rectangle with fillets

Note
*You can draw a rectangle either with chamfers or with fillets. If you specify the chamfer distances first and then specify the fillet radius in the same **RECTANG** command, the rectangle will be drawn with fillets only.*

Width

The **Width** option allows you to create a rectangle whose line segments have some specified width, as shown in Figure 3-28.

Specify first corner point or [Chamfer/Elevation/Fillet/Thickness/Width]: **W** [Enter]
Specify line width for rectangles <0.0000>: *Enter a value.*

Thickness

The **Thickness** option allows you to draw a rectangle that is extruded in the Z direction by a specified value of thickness. For example, if you draw a rectangle with thickness of 2 units, you will get a rectangular box whose height is 2 units, see Figure 3-29. To view the box, choose **View > 3D Views > SE Isometric** from the **Ribbon**. Next, you can restore the view back to the plan view by choosing **View > 3D Views > Top** from the **Ribbon**.

Specify first corner point or [Chamfer/Elevation/Fillet/Thickness/Width]: **T** Enter
Specify thickness for rectangles <0.0000>: *Enter a value.*

Figure 3-28 *Drawing a rectangle with a specified width*

Figure 3-29 *Drawing rectangles with thickness and elevation specified*

Elevation

The **Elevation** option allows you to draw a rectangle at a specified distance from the *XY* plane, along the *Z* axis. For example, if the elevation is 2 units, the rectangle will be drawn two units above the *XY* plane. If the thickness of the rectangle is 1 unit, you will get a rectangular box of 1 unit height, located 2 units above the *XY* plane, see Figure 3-29.

Chamfer/Elevation/Fillet/Thickness/Width/<First corner>: **E** Enter
Specify elevation for rectangles <0.0000>: *Enter a value.*

To view the drawn rectangle in 3D space, change the viewpoint by choosing **View > 3D Views > SE Isometric** from the **Ribbon**. Next, to restore the top view, choose **View > 3D Views > Top** from the **Ribbon**.

Note
*The value you enter for fillet, width, elevation, and thickness becomes the current value for the subsequent **RECTANG** command. Therefore, you must reset the values, if they are different from the current values. The thickness of a rectangle is always controlled by its thickness settings.*

*The rectangle generated on the screen is treated as a single object. Therefore, the individual sides can be edited only after the rectangle has been exploded using the **EXPLODE** command.*

Tip
*You can combine different options in one **RECTANG** command and then draw the rectangle with the specified characteristics. When you invoke the **RECTANG** command again, the previously set options and their values are displayed before the first prompt. This allows you to change the settings according to the new specifications.*

Exercise 6   *General*

Draw a rectangle 4 unit long, 3 unit wide, and with its first corner at (1,1). Draw another rectangle of length 2 unit and width 1 unit, with its first corner at 1.5,1.5, and at an angle of 65-degree.

DRAWING ELLIPSES

Ribbon:	Home > Draw >Ellipse	Toolbar:	Draw >Ellipse
Tool Palettes:	Draw > Ellipse	Command:	ELLIPSE

If a circle is observed from an angle, the shape seen is called an ellipse, which can be created in AutoCAD using the **ELLIPSE** command. An ellipse can be created using various options and all these options are grouped together in the **Draw** panel of the **Ribbon**, refer to Figure 3-30. AutoCAD creates a true ellipse, also known as a NURBS-based (Non-Uniform Rational Bezier Spline) ellipse. The true ellipse has a center and quadrant points. If you select it, the grips (small blue squares) will be displayed at the center and quadrant points of the ellipse. If you move one of the grips located on the perimeter of the ellipse, the major or minor axis will change, which changes the size of the ellipse.

*Figure 3-30 Invoking the **ELLIPSE** command from the **Ribbon***

Once you invoke the **ELLIPSE** command, you will be prompted to **Specify axis endpoint of ellipse or [Arc/Center]** or **Specify axis endpoint of ellipse or [Arc/Center/Isocircle]** (if isometric snap is on). The response to this prompt depends on the option you choose. The various options are explained next.

Note

*The **Isocircle** option is not available by default in the **ELLIPSE** command. To display this option, you have to select the **Isometric snap** radio button in the **Snap and Grid** tab of the **Drafting Settings** dialog box. The **Isocircle** option will be discussed in Chapter 23.*

Drawing an Ellipse Using the Axis and Endpoint Option

In this option, you draw an ellipse by specifying one of its axes and the endpoint of the other axis. To use this option, acknowledge the **Specify axis endpoint of ellipse or [Arc/Center]** prompt by specifying a point, either by using a pointing device or by entering its coordinates. This is the first endpoint of one axis of the ellipse. AutoCAD will then respond with the prompt **Specify other endpoint of axis**. Here, specify the other endpoint of the axis. The angle at which the ellipse is drawn depends on the angle made by these two axis endpoints. Your response to the next prompt determines whether the axis is the **major axis** or the **minor axis**.

The next prompt is **Specify distance to other axis or [Rotation]**. If you specify a distance, it is presumed as half the length of the second axis. You can also specify a point. The distance from this point to the midpoint of the first axis is again taken as half the length of this axis. The ellipse will pass through the selected point only if it is perpendicular to the midpoint of the first axis. To visually analyze the distance between the selected point and the midpoint of the first axis, AutoCAD appends an elastic line to the crosshairs, with one end fixed at the midpoint of the first axis. You can also drag the point, dynamically specifying half of the other axis distance. This helps you to visualize the ellipse, refer to Figure 3-31. The prompt sequence for drawing an ellipse with one axis endpoint located at (3,3), the other at (6,3), and the distance of the other axis being 1 is given next.

> Command: **ELLIPSE** Enter
> Specify axis endpoint of ellipse or [Arc/Center]: **3,3** Enter
> Specify other endpoint of axis: **6,3** Enter
> Specify distance to other axis or [Rotation]: **1** Enter

Another example for drawing an ellipse using this option is shown in Figure 3-32 and illustrated by the following prompt sequence.

> Command: **ELLIPSE** Enter
> Specify axis endpoint of ellipse or [Arc/Center]: **3,3** Enter
> Specify other endpoint of axis: **4,2** Enter
> Specify distance to other axis or [Rotation]: **2** Enter

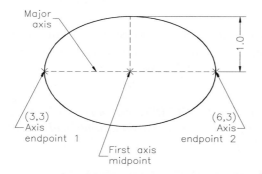

Figure 3-31 Drawing an ellipse using the **Axis and Endpoint** option

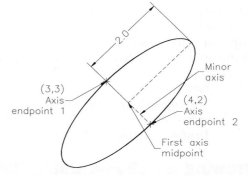

Figure 3-32 Drawing an ellipse using the **Axis and Endpoint** option

If you enter **Rotation** or **R** at the **Specify distance to other axis or [Rotation]** prompt, the first axis specified is automatically taken as the major axis of the ellipse. The next prompt is **Specify rotation around major axis**. The major axis is taken as the diameter line of the circle, and the rotation takes place around this diameter line into the third dimension. The ellipse is formed, when AutoCAD projects this rotated circle into the drawing plane. You can enter the rotation angle value in the range of 0 to 89.4-degree only, because an angle value greater than 89.4-degree changes the circle into a line. Instead of entering a definite angle value at the **Specify rotation around major axis** prompt, you can specify a point relative to the midpoint of the major axis. This point can be dragged to specify the ellipse dynamically. The following is the prompt sequence for a rotation of 0-degree around the major axis, as shown in Figure 3-33(a):

Command: **ELLIPSE** [Enter]
Specify axis endpoint of ellipse or [Arc/Center]: *Select point (P1).*
Specify other endpoint of axis: *Select another point (P2).*
Specify distance to other axis or [Rotation]: **R** [Enter]
Specify rotation around major axis: **0** [Enter]

Figure 3-33 also shows rotations of 45-degree, 60-degree, and 89.4-degree.

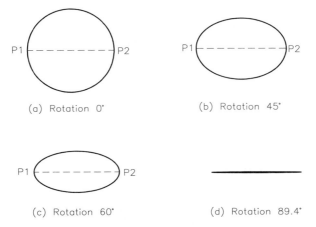

Figure 3-33 *Rotation about the major axis*

Exercise 7   *General*

Draw an ellipse whose major axis is 4 units and the rotation around this axis is 60-degree. Draw another ellipse, whose rotation around the major axis is 15-degree.

Drawing Ellipse Using the Center Option

In this option, you can construct an ellipse by specifying the center point, endpoint of one axis, and length of the other axis. The only difference between this method and the ellipse by axis and endpoint method is that instead of specifying the second endpoint of the first axis, the center of the ellipse is specified. The center of an ellipse is defined as the point of intersection of the major and minor axes. In this option, the first axis need not be the major axis. For example, draw an ellipse with center at (4,4), axis endpoint at (6,4), and length of the other axis as 1 unit, as shown in Figure 3-34.

Chapter 3

The prompt sequence is given next.

Command: **ELLIPSE** Enter
Specify axis endpoint of ellipse or
[Arc/Center]: **C** Enter
Specify center of ellipse: **4,4** Enter
Specify endpoint of axis: **6,4** Enter
Specify distance to other axis or [Rotation]:
1 Enter

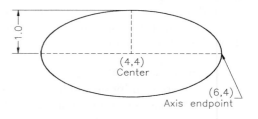

Instead of entering the distance, you can enter
Rotation or **R** at the **Specify distance to other
axis [Rotation]** prompt. This takes the first axis

*Figure 3-34   Drawing an ellipse using the
Center option*

specified as the major axis. The next prompt, **Specify rotation around major axis**, prompts
you to enter the rotation angle value. The rotation takes place around the major axis, which
is taken as the diameter line of the circle. The rotation angle values should range from 0 to
89.4-degree.

Drawing Elliptical Arcs

Ribbon:	Home > Draw > Center > Elliptical Arc	**Toolbar:**	Draw > Ellipse Arc
Tool Palettes:	Draw > Ellipse Arc	**Command:**	ELLIPSE > Arc

AutoCAD allows you to draw elliptical arcs
using the **Ellipse Arc** button in the **Draw** toolbar or by
choosing **Center > Elliptical Arc** from the **Draw** panel
of the **Ribbon,** as shown in Figure 3-35. When you
choose this button, the **ELLIPSE** command is invoked
with the **Arc** option selected and you will be prompted
to enter information about the geometry of the ellipse
and the arc limits. You can define the arc limits by using
the following options:

1. Start and End angle of the arc.
2. Start and Included angle of the arc.
3. Start and End parameters.

*Figure 3-35   Invoking **Elliptical Arc**
from the **Ribbon***

The angles are measured from the first point and in counterclockwise direction, if AutoCAD's
default setup is not changed. The next example illustrates the use of these three options.

Example 1   *General*

Draw the following elliptical arcs, as shown in Figures 3-36 and 3-37.

a.   Start angle = -45, end angle = 135

b. Start angle = -45, included angle = 225
c. Start parameter = @1,0, end parameter = @1<225

Specifying Start and End Angle of the Elliptical Arc [Figure 3-36(a)]

Command: *Choose* **Center** > **Elliptical Arc** *from the* **Draw** *panel of the* **Ribbon.**
Specify axis endpoint of ellipse or [Arc/Center]: **A**
Specify axis endpoint of elliptical arc or [Center]: *Select the first endpoint.*
Switch on the Ortho mode by pressing F8, if it is not already chosen
Specify other endpoint of axis: *Select the second point to the left of the first point.*
Specify distance to other axis or [Rotation]: *Select a point or enter a distance.*
Specify start angle or [Parameter]: **-45** Enter
Specify end angle or [Parameter/Included angle]: **135** Enter *(Angle where arc ends.)*

Specifying Start and Included Angle of the Elliptical Arc [Figure 3-36(b)]

Command: *Choose* **Center** > **Elliptical Arc** *from the* **Draw** *panel of the* **Ribbon.**
Specify axis endpoint of ellipse or [Arc/Center]: **A**
Specify axis endpoint of elliptical arc or [Center]: *Select the first endpoint.*
Specify other endpoint of axis: *Select the second point.*
Specify distance to other axis or [Rotation]: *Select a point or enter a distance.*
Specify start angle or [Parameter]: **-45** Enter
Specify end angle or [Parameter/Included angle]: **I** Enter
Specify included angle for arc<current>: **225** Enter *(Included angle.)*

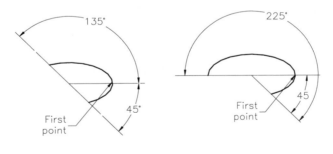

(a) Start angle=-45°
    End angle =135°

(b) Start angle=-45°
    Included angle=225°

Figure 3-36 *Drawing elliptical arcs*

Specifying Start and End Parameters (Figure 3-37)

Command: *Choose* **Center** > **Elliptical Arc** *from the* **Draw** *panel of the* **Ribbon.**
Specify axis endpoint of ellipse or [Arc/Center]: **A**
Specify axis endpoint of elliptical arc or [Center]: *Select the first endpoint.*

Specify other endpoint of axis: *Select the second endpoint.*
Specify distance to other axis or [Rotation]: *Select a point or enter a distance.*
Specify start angle or [Parameter]: **P**
Specify start parameter or [Angle]: **@1,0**
Specify end parameter or [Angle/ Included angle]: **@1<225**

Figure 3-37 *Drawing an elliptical arc by specifying the start and end parameters*

Exercise 8   *General*

a. Construct an ellipse with center at (2,3), axis endpoint at (4,6), and the other axis endpoint at a distance of 0.75 units from the midpoint of the first axis.

b. Make the drawing, as shown in Figure 3-38. The distance between the dotted lines is 1.0 unit. Create the elliptical arcs using the **ELLIPSE** command options.

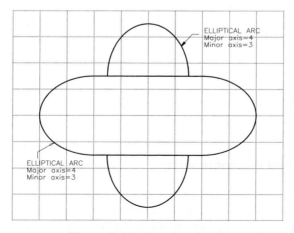

Figure 3-38 *Drawing for Exercise 8*

Note

From now on, all command sequences in this book are given with presumption that the dynamic input is turned off on your computer.

DRAWING REGULAR POLYGONS

Ribbon:	Home > Draw > Polygon	**Toolbar:**	Draw > Polygon
Tool Palettes:	Draw > Polygon	**Command:**	POLYGON

A regular polygon is a closed geometric entity with equal sides. The number of sides varies from **3** to **1024**. For example, a triangle is a three-sided polygon and a pentagon is a five-sided polygon. In AutoCAD, the **POLYGON** command is used to draw regular 2D polygons. Choose the **Polygon** button from the **Draw** panel in the **Home** tab of the **Ribbon** to invoke this command, refer to Figure 3-39. The characteristics of a polygon drawn in AutoCAD are those of a closed polyline having a 0 width. You can change the width of the polyline forming the polygon. The prompt sequence is given next.

*Figure 3-39 Invoking the **POLYGON** command from the Ribbon*

Command: **POLYGON** [Enter]
Enter number of sides <4>:

Once you invoke the **POLYGON** command, it prompts you to enter the number of sides. The number of sides determine the type of polygon (for example, six sides define a hexagon). The default value for the number of sides is **4**. You can change the number of sides to your requirement and then the new value becomes the default value. You can also set a default value for the number of sides, by using the **POLYSIDES** system variable.

The Center of Polygon Option

After you specify the number of sides, the next prompt is **Specify center of polygon or [Edge]**, where the default option prompts you to select a point as the center point of the polygon. If you specify the center point, then the **Enter an option [Inscribed in circle/Circumscribed about circle]<I> prompt** is displayed at the Command prompt. A polygon is said to be **inscribed** when it is drawn inside an imaginary circle and its vertices (corners) touch the circle, see Figure 3-40. Whereas, a polygon is **circumscribed** when it is drawn outside the imaginary circle and the sides of the polygon are tangent to the circle (midpoint of each side of the polygon will lie on the circle), see Figure 3-41. If you want to have an inscribed polygon, choose **Inscribed in circle** option from the shortcut menu. The next prompt issued is **Specify radius of circle**. Here, you are required to specify the radius of the circle on which all the vertices of the polygon will lie. Once you specify it, a polygon will be generated. If you want to select the circumscribed option, choose the **Circumscribed about circle** option from the shortcut menu at the **Enter an option [Inscribed in circle/Circumscribed about circle]<I>** prompt. You can also select these options from the dynamic preview. After this, enter the radius of the circle. Alternatively, specify the radius of the circle dynamically by moving the mouse and clicking. The inscribed or circumscribed

Chapter 3

circle will not be displayed. The prompt sequence for drawing an inscribed octagon shown in Figure 3-40, with the center at (4,4) and the radius of 1.5 units, is given next.

Command: **POLYGON** Enter
Enter number of sides<4>: **8** Enter
Specify center of polygon or [Edge]: **4,4** Enter
Enter an option[Inscribed in circle/Circumscribed about circle]<I>: **I** Enter
Specify radius of circle: **1.5** Enter

The prompt sequence for drawing a circumscribed pentagon shown in Figure 3-41 with center at (4,4) and a radius of 1.5 units, is given next.

Command: **POLYGON** Enter
Enter number of sides<4>: **5** Enter
Specify center of polygon or [Edge]: **4,4** Enter
Enter an option[Inscribed in circle/Circumscribed about circle]<I>: **C** Enter
Specify radius of circle: **1.5** Enter

Inscribed octagon

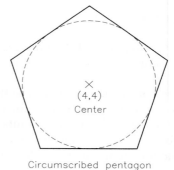

Circumscribed pentagon

Figure 3-40 *Drawing an inscribed polygon using the **Center of Polygon** option*

Figure 3-41 *Drawing a circumscribed polygon using the **Center of Polygon** option*

Note
If you select a point to specify the radius of an inscribed polygon, one of the vertices is positioned on the selected point. In case of circumscribed polygons, the midpoint of an edge is placed on the point you have specified. In this manner, you can specify the size and rotation of the polygon.

In case of the numerical specification of the radius, the bottom edge of the polygon is rotated by the prevalent snap rotation angle.

Exercise 9   *General*

Draw a circumscribed polygon of eight sides. The polygon should be drawn by the **Center of Polygon** method.

The Edge Option

The other method for drawing a polygon is to select the **Edge** option. This can be done by entering E at the **Specify center of polygon or [Edge]** prompt. The next two prompts issued are **Specify first endpoint of edge** and **Specify second endpoint of edge**. Here, you need to specify the two endpoints of an edge of the polygon. The polygon is drawn in a counterclockwise direction, with the two entered points defining its first edge. To draw a hexagon (six-sided polygon), as shown in Figure 3-42, using the **Edge** option, with the first endpoint of the edge at (2,4) and the second endpoint of the edge at (2,2.5), the prompt sequence is given next.

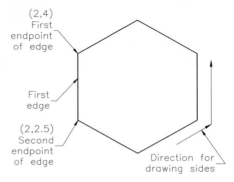

Figure 3-42 *Drawing a polygon (hexagon) using the* ***Edge*** *option*

        Command: **POLYGON** [Enter]
        _polygon Enter number of sides<4>: **6** [Enter]
        Specify center of polygon or [Edge]: **E** [Enter]
        Specify first endpoint of edge: **2,4** [Enter]
        Specify second endpoint of edge: **2,2.5** [Enter]

Exercise 10   *General*

Draw a polygon with ten sides using the **Edge** option and an elliptical arc, as shown in Figure 3-43. Let the first endpoint of the edge be at (7,1) and the second endpoint be at (8,2).

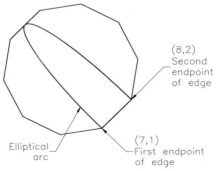

Figure 3-43 *Polygon and elliptical arc for Exercise 10*

DRAWING POLYLINES

Ribbon:	Home > Draw > Polyline	**Toolbar:**	Draw > Polyline
Tool Palettes:	Draw > Polyline	**Command:**	PLINE (or PL)

The term POLYLINE can be broken into two parts: POLY and LINE. POLY means "many". This signifies that a polyline can have many lines. Invoke the **PLINE** command by choosing the **Polyline** button from the **Draw** panel in the **Home** tab of the **Ribbon**, see Figure 3-44. A polyline is a line that has different features. Some of the features are listed next.

Figure 3-44 Invoking the ***POLYLINE*** *command from the* ***Ribbon***

1. Polylines can be thick lines with a desired width. They are very flexible and can be used to draw any shape, such as a filled circle or a doughnut.
2. Polylines can be used to draw objects in any linetype (for example, hidden linetype).
3. Advanced editing commands can be used to edit them (for example, the **PEDIT** command).
4. A single polyline object can be formed by joining polylines and polyarcs of different thicknesses.
5. It is easy to determine the area or perimeter of a polyline feature. Also, it is easy to offset when drawing walls.

The **PLINE** command functions fundamentally like the **LINE** command, except that additional options are provided and all the segments of the polyline form a single object. After invoking the **PLINE** command, the following prompt is displayed:

Specify start point: *Specify the starting point or enter its coordinates.*
Current line width is nn.nnnn

Current line width is nn.nnnn is displayed automatically, which indicates that the polyline drawn will have nn.nnnn width. If you want a different width, invoke the **Width** option at the next prompt and set it. Next, the following prompt is displayed.

Specify next point or [Arc/Halfwidth/Length/Undo/Width]: *Specify the next point or enter an option.*

Depending on your requirements, the options that can be invoked at this prompt are as follows.

Next Point of Line

This option is maintained as the default option and is used to specify the next point of the current polyline segment. If additional polyline segments are added to the first polyline, AutoCAD automatically makes the endpoint of the previous polyline segment the start point of the next polyline segment. The prompt sequence is given next.

Command: **PLINE** `Enter`
Specify start point: *Specify the starting point of the polyline.*
Current line width is 0.0000.
Specify next point or [Arc/Halfwidth/Length/Undo/Width]: *Specify the endpoint of the first polyline segment.*
Specify next point or [Arc/Close/Halfwidth/Length/Undo/Width]: *Specify the endpoint of the second polyline segment, or press ENTER to exit the command.*

Width

You can change the current polyline width by entering **W** (width option) at the last prompt. You can also right-click and choose the **Width** option from the shortcut menu. Next, you are prompted to specify the starting and ending width of the polyline.

Specify next point or [Arc/Halfwidth/Length/Undo/Width]: **W** `Enter`
Specify starting width <current>: *Specify the starting width.*
Specify ending width <starting width>: *Specify the ending width.*

The starting width value is taken as the default value for the ending width. Therefore, to have a uniform polyline, you need to press ENTER at the **Specify ending width < >** prompt. To draw a uniform polyline, shown in Figure 3-45, with a width of 0.25 units, start point at (4,5), endpoint at (5,5), and the next endpoint at (3,3), use the following prompt sequence:

Command: **PLINE** `Enter`
Specify start point: **4,5** `Enter`
Current line-width is 0.0000
Specify next point or [Arc/Halfwidth/Length/Undo/Width]: **W** `Enter`
Specify starting width <current>: **0.25** `Enter`
Specify ending width <0.25>: `Enter`
Specify next point or [Arc/Halfwidth/Length/Undo/Width]: **5,5** `Enter`
Specify next point or [Arc/Close/Halfwidth/Length/Undo/Width]: **3,3** `Enter`
Specify next point or [Arc/Close/Halfwidth/Length/Undo/Width]: `Enter`

You can get a tapered polyline, by entering two different values at the starting width and the ending width prompts. To draw a tapered polyline, shown in Figure 3-46, with a starting width of 0.5 units and an ending width of 0.15 units, a start point at (2,4), and an endpoint at (5,4), use the prompt sequence given next.

Command: **PLINE** `Enter`
Specify start point: **2,4** `Enter`
Current line-width is 0.0000
Specify next point or [Arc/Halfwidth/Length/Undo/Width]: **W** `Enter`
Specify starting width <0.0000>: **0.50** `Enter`
Specify ending width <0.50>: **0.15** `Enter`
Specify next point or [Arc/Halfwidth/Length/Undo/Width]: **5,4** `Enter`
Specify next point or [Arc/Close/Halfwidth/Length/Undo/Width]: `Enter`

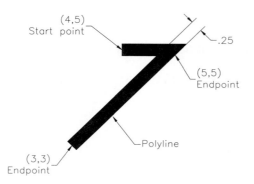

Figure 3-45 *Drawing a uniform polyline using the **PLINE** command*

Figure 3-46 *Drawing a tapered polyline using the **PLINE** command*

Halfwidth

With this option, you can specify the starting and ending halfwidth of a polyline. This halfwidth distance is equal to half of the actual width of the polyline. This option can be invoked by entering **H** or choosing **Halfwidth** from the shortcut menu at the following prompt:

> Specify next point or [Arc/Halfwidth/Length/Undo/Width]: **H** `Enter`
> Specify starting half-width <0.0000>: **0.12** `Enter` *(Specify desired starting halfwidth)*
> Specify ending half-width <0.1200>: **0.05** `Enter` *(Specify desired ending halfwidth)*

Length

This option prompts you to enter the length of a new polyline segment. The new polyline segment will be the length you have entered. It will be drawn at the same angle as the last polyline segment or tangent to the previous polyarc segment. This option can be invoked by entering **L** at the following prompt or by choosing **Length** from the shortcut menu.

> Specify next point or [Arc/Close/Halfwidth/Length/Undo/Width]: **L** `Enter`
> Specify length of line: *Specify the desired length of the Pline.*

Undo

This option erases the most recently drawn polyline segment. It can be invoked by entering **U** at the following prompt.

> Specify next point or [Arc/Close/Halfwidth/Length/Undo/Width]: **U** `Enter`

You can use this option repeatedly until you reach the start point of the first polyline segment. Further use of **Undo** option evokes the message **All segments already undone**.

Close

This option is available when at least one segment of the polyline is drawn. It closes the polyline by drawing a polyline segment from the most recent endpoint to the initial start point. At the same time, it exits from the **PLINE** command. The width of the closing segment can be changed by using the **Width/Halfwidth** option, before invoking the **Close** option.

Arc

This option is used to switch from drawing polylines to drawing polyarcs, and provides you the options associated with drawing polyarcs. The prompt sequence is given next.

> Specify next point or [Arc/Close/Halfwidth/Length/Undo/Width]: **A** Enter
> Specify endpoint of arc or [Angle/CEnter/CLose/Direction/Halfwidth/Line/Radius/Second pt/Undo/Width]: *Enter an option.*

By default, the arc segment is drawn tangent to the previous segment of the polyline. The direction of the previous line, arc, or polyline segment is the default direction for the polyarc. The preceding prompt contains options associated with the PLINE Arc. The detailed explanation of each of these options is as follows.

Angle

This option prompts you to enter the included angle for the arc. If you enter a positive angle, the arc is drawn in a counterclockwise direction from the start point to the endpoint. If the angle specified is negative, the arc is drawn in a clockwise direction. The prompts are given next.

> Specify included angle: *Specify the included angle.*
> Specify endpoint of arc or [Center/Radius]: *Specify the endpoint of arc or choose an option.*

Center refers to the center of the arc segment, Radius refers to the radius of the arc, and Endpoint draws the arc.

CEnter

This option prompts you to specify the center of the arc to be drawn. As mentioned before, usually the arc segment is drawn such that it is tangent to the previous polyline segment; in such cases AutoCAD determines the center of the arc automatically. Therefore, the **CEnter** option provides the freedom to choose the center of the arc segment. This option can be invoked by entering **CE** at the **Specify end point of arc or [Angle/CEnter/CLose/Direction/ Halfwidth/Line/Radius/Second pt/Undo/Width]** prompt. Once you specify the center point, AutoCAD issues the following prompt.

> Specify endpoint of arc or [Angle/Length]: *Specify the endpoint of arc or choose an option.*

Angle refers to the included angle, Length refers to the length of the chord, and Endpoint refers to the endpoint of the arc.

CLose

This option is used to join the last point of a polyline with its start point in the form of an arc. This option is available only when you specify two or more than two points for creating the polyline segments.

Chapter 3

Direction

Usually, the arc drawn, with the **PLINE** command, is tangent to the previous polyline segment. In other words, the starting direction of the arc is the ending direction of the previous segment. The **Direction** option allows you to specify the tangent direction of your choice for the arc segment to be drawn. You can specify the direction by specifying a point. The prompts are given next.

Specify tangent direction for the start point of arc: *Specify the direction.*
Specify endpoint of arc: *Specify the endpoint of arc.*

Halfwidth

This option is the same as for the **PLine** and prompts you to specify the starting and ending halfwidth of the arc segment.

Line

This option takes you back to the **Line** mode. You can draw polylines only in this mode.

Radius

This option prompts you to specify the radius of the arc segment. The prompt sequence is given next.

Specify radius of arc: *Specify the radius of the arc segment.*
Specify endpoint of arc or [Angle]: *Specify the endpoint of arc or choose an option.*

If you specify a point, the arc segment is drawn. If you enter an angle, you will have to specify the angle and the direction of the chord at the **Specify included angle** and **Specify direction of chord for arc<current>** prompts, respectively.

Second pt

This option selects the second point of an arc in the three-point arc option. The prompt sequence is given next.

Specify second point of arc: *Specify the second point on the arc.*
Specify endpoint of arc: *Specify the third point on the arc.*

Undo

This option reverses the changes made in the previously drawn segment.

Width

This option prompts you to enter the width of the arc segment. To draw a tapered arc segment, you can enter different values at the starting width and ending width prompts. The prompt sequence is identical to that of the polyline. Also, a specified point on a polyline refers to the midpoint on its width.

Endpoint of Arc

This option is maintained as the default and prompts you to specify the endpoint of the current arc segment. The following is the prompt sequence for drawing an arc, shown in Figure 3-47, with start point at (3,3), endpoint at (3,5), starting width of 0.50 units, and ending width of 0.15 units.

Command: **PLINE** Enter
Specify start point: **3,3** Enter
Current line-width is 0.0000
Specify next point or [Arc/Halfwidth/Length/Undo/Width]: **A** Enter
Specify endpoint of arc or [Angle/CEnter/CLose/Direction/Halfwidth/Line/Radius/Second pt/Undo/Width]: **W** Enter
Specify starting width <current>: **0.50** Enter
Specify ending width <0.50>: **0.15** Enter
Specify endpoint of arc or [Angle/CEnter/CLose/Direction/Halfwidth/Line/Radius/Second pt/Undo/Width]: **3,5** Enter
Specify endpoint of arc or [Angle/CEnter/CLose/Direction/Halfwidth/Line/Radius/Second pt/Undo/Width]: Enter

Figure 3-47 Drawing a polyarc

Tip
*After invoking the **PLINE** command and specifying the start point, you can right-click to display the shortcut menu. You can choose any of the options under the **PLINE** command, directly from the shortcut menu, instead of entering the appropriate letters at the Command prompt. Similarly, after invoking the **ARC** option of the **PLINE** command, you can right-click to display the shortcut menu and choose any polyarc option.*

Note
*If **FILL** is on or if **FILLMODE** is 1, the polylines drawn are filled. If you change **FILL** to off or **FILLMODE** to 0, only the outlines are drawn for the new plines and previously drawn plines are also changed from filled to no-fill. However, note that the change is effective only on regeneration. Similarly, it works in reverse also.*

*Also, the **PLINEGEN** system variable controls the linetype pattern between the vertex points of a 2D polyline. A value of 0 centers the linetype for each polyline segment and 1 makes it continuous.*

Exercise 11   *General*

Draw the objects shown in Figures 3-48 and 3-49. Assume the width of different polylines.

DRAWING DONUTS

| **Ribbon:** | Home > Draw > Donut | **Command:** | DONUT |

In AutoCAD, the **DONUT** or **DOUGHNUT** command is issued to draw an object that looks like a filled circle ring called a donut. Actually, AutoCAD's donuts are made of two semicircular

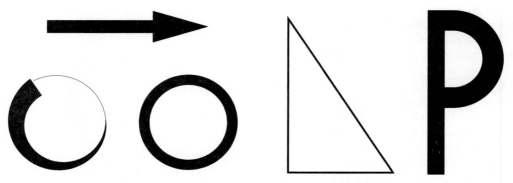

Figure 3-48 *Drawing for Exercise 11*   **Figure 3-49** *Drawing for Exercise 11*

polyarcs with a certain width. Therefore, the **DONUT** command allows you to draw a thick circle. The donuts can have any inside and outside diameters. If **FILLMODE** is off, the donuts look like circles (if the inside diameter is zero) or concentric circles (if the inside diameter is not zero). After specifying the two diameters, the donut gets attached to the crosshairs. You can select a point for the center of the donut anywhere on the screen with the help of a pointing device, and then place the donut. You can place it by clicking your pointing device. The prompt sequence for drawing donuts is given next.

> Command: **DONUT** ⌷Enter⌷
> Specify inside diameter of donut <current>: *Specify the inner diameter of the donut.*
> Specify outside diameter of donut <current>: *Specify the outer diameter of the donut.*
> Specify center of donut or <exit>: *Specify the center of the donut.*
> Specify center of donut or <exit>: *Specify the center of the donut to draw more donuts of previous specifications or give a null response to exit.*

The defaults for the inside and outside diameters are the respective diameters of the most recent donut drawn. The values for the inside and outside diameters are saved in the **DONUTID** and **DONUTOD** system variables. A solid-filled circle is drawn by specifying the inside diameter as zero (**FILLMODE** is on). Once the diameter specification is completed, the donuts are formed at the crosshairs and can be placed anywhere on the screen. For the location, you can enter the coordinates of the point or specify the point by dragging the center point. Once you have specified the center of the donut, AutoCAD repeats the **Specify center of donut or <exit>** prompt. As you go on specifying the locations for the center point, donuts with the specified diameters are drawn at specified locations. To end the **DONUT** command, give a null response to this prompt by pressing ENTER. Since donuts are circular polylines, the donut can be edited with the **PEDIT** command or any other editing command that can be used to edit polylines.

Example 2   *General*

You will draw an unfilled donut shown in Figure 3-50 with an inside diameter of 0.75 units, an outside diameter of 2.0 units, and centered at (2,2). You will also draw a filled donut and a solid-filled donut with the given specifications.

The following is the prompt sequence to draw an unfilled donut shown in Figure 3-51.

> Command: **FILLMODE** Enter
> New value for FILLMODE <1>: **0** Enter
> Command: **DONUT** Enter
> Specify inside diameter of donut<0.5000>: **0.75** Enter
> Specify outside diameter of donut <1.000>: **2** Enter
> Specify center of donut or <exit>: **2,2** Enter
> Specify center of donut or <exit>: Enter

The prompt sequence for drawing a filled donut, shown in Figure 3-51, is given next, with an inside diameter of 0.5 units, outside diameter of 2.0 units, centered at a specified point.

> Command: **FILLMODE** Enter
> Enter new value for FILLMODE <0>: **1** Enter
> Command: **DONUT** Enter
> Specify inside diameter of donut<0.5000>: **0.50** Enter
> Specify outside diameter of donut <1.000>: **2** Enter
> Specify center of donut or <exit>: *Specify a point.*
> Specify center of donut or <exit>: Enter

FILLMODE=0

Figure 3-50 *Drawing an unfilled donut using the **DONUT** command*

FILLMODE=1

Figure 3-51 *Drawing a filled donut using the **DONUT** command*

To draw a solid-filled donut, shown in Figure 3-52, with an outside diameter of 2.0 units, use the following prompt sequence.

> Command: **DONUT** Enter
> Specify inside diameter of donut <0.50>: **0** Enter
> Specify outside diameter of donut <1.0>: **2** Enter
> Specify center of donut or <exit>: *Specify a point.*
> Specify center of donut or <exit>: Enter

Donut with inside diameter zero

Figure 3-52 *Solid-filled donut*

PLACING POINTS

Ribbon:	Home > Draw > Point	**Toolbar:**	Draw > Point
Tool Palettes:	Draw > Point	**Command:**	POINT

The point is the basic drawing object. Points are invaluable in building a drawing file. To place a point anywhere on the screen, AutoCAD provides the **POINT** command.

> Command: **POINT** [Enter]
> Current point modes: PDMODE=n PDSIZE=n.nnnn
> Specify a point: *Specify the location where you want to place the point.*

You can invoke the **POINT** command from the **Draw** toolbar or from the **Ribbon** (**Multiple Point** option), as shown in Figure 3-53. You can place as many points as you desire in a single command. Then, press ESC to exit from the **POINT** command. If you invoke this command by entering **POINT** at the Command prompt or use the **Point > Single Point** option from the menu bar, you can draw only one point in a single point command.

Figure 3-53 Invoking the
POINT command from the
Ribbon

Note
*It is possible to have a temporary construction marker for the point known as blip. A mark appears on the screen where you place the point. Then this blip mark can be cleared, once the screen is redrawn using the **REDRAW** command, and the point is left on the screen. The visibility of a blip can be controlled using the **BLIPMODE** system variable.*

Changing the Point Type

Ribbon:	Utilities > Point Style	**Command:**	DDPTYPE

The point type can be set from the **Point Style** dialog box, as shown in Figure 3-54. To invoke this dialog box, choose **Utilities > Point Style** from the **Ribbon**; the **Point Style** dialog box will be displayed. There are twenty combinations of point types in this dialog box. You can choose a point style in this dialog box, which is indicated by highlighting that particular point style. Next, choose the **OK** button. Now, all the points will be drawn in the selected style, until you change it to a new style. The type of point drawn is stored in the **PDMODE** (Point Display MODE) system variable. You can also change the point style by entering a numeric value in the **PDMODE** variable.

Figure 3-55 shows the **PDMODE** values for different point type areas.

Figure 3-54 *The* **Point Style** *dialog box*

Pdmode Value	Point Style	Pdmode Value	Point Style
0		64+0=64	□
1		64+1=65	□
2	+	64+2=66	⊞
3	×	64+3=67	⊠
4	'	64+4=68	⊡
32+0=32	○	96+0=96	⊡
32+1=33	○	96+1=97	⊡
32+2=34	⊕	96+2=98	⊕
32+3=35	⊗	96+3=99	⊠
32+4=36	⊙	96+4=100	⊡

Figure 3-55 *Different point styles for* **PDMODE** *values*

Exercise 12   *General*

Check what types of points are drawn for each value of the **PDMODE** variable.

Changing the Point Size

Ribbon: Utilities > Format > Point Style	**Command:** DDPTYPE

The size of a point can be set from the **Point Style** dialog box by entering the desired point size in the **Point Size** edit box, refer to Figure 3-54. You can generate the point at a specified percentage of the graphics area height or define an absolute size for the point. An absolute size for the point can be specified by selecting the **Set Size in Absolute Units** radio button in the **Point Style** dialog box and then entering a value in the **Point Size** edit box. The point size can also be set by changing the value of **PDSIZE** (Figure 3-56). The variable **PDSIZE** governs the size of the point (except for the **PDMODE** values of 0 and 1). You can set the size in absolute units by specifying a positive value for the **PDSIZE** variable. If the **Set Size Relative to** **Screen** radio button is selected in the **Point Style** dialog box, the size is taken as a percentage of the viewport size. This can also be set by entering a negative value for the **PDSIZE** variable. For example, a setting of 5 makes the point 5 units high; a setting of -5 makes the point 5 percent of the current drawing area.

PDSIZE=2   PDSIZE=4

Figure 3-56 *Changing the point size using the* **PDSIZE** *variable*

Chapter 3

Exercise 13   *General*

a.   Try various combinations of the **PDMODE** and **PDSIZE** variables.

b.   Check the difference between the points generated from negative values of **PDSIZE** and the points generated from positive values of **PDSIZE**.

DRAWING INFINITE LINES

The **XLINE** and **RAY** commands can be used to draw construction or projection lines. These are the lines that aid in construction or projection and are drawn very lightly, when drafting manually. An **xline** (construction line) is a 3D line that extends to infinity at both ends. As the line is infinite in length, it does not have any endpoints. A **ray** is a 3D line that extends to infinity at only one end. The other end of the ray has a finite endpoint. The xlines and rays have zero extents. This means that the extents of the drawing will not change, if you use the commands that change the drawing extents, such as the **ZOOM** command with the **All** option. Most of the object snap modes work with both xlines and rays, with some limitations: You cannot use the **Endpoint** object snap with the xline because, by definition an xline does not have any endpoints. However, for rays you can use the **Endpoint** snap on one end only. Also, xlines and rays take the properties of the layer, in which they are drawn.

Tip
Xlines and rays plot like any other objects in a drawing may create confusion. Therefore, it is a good idea to create the construction lines in a different layer altogether, such that you can recognize them easily. You will learn about layers in later chapters.

Drawing XLINE

Ribbon:	Home > Draw > Construction Line	**Toolbar:**	Draw > Construction Line
Tool Palettes:	Draw > Construction Line	**Command:** XLINE	

You can choose the **Construction Line** button from the **Draw** panel in the **Home** tab of the **Ribbon** to invoke the **XLINE** command, refer to Figure 3-57. The prompt sequence on invoking the **XLINE** command is as follows:

Command: **XLINE** ⏎
Specify a point or [Hor/Ver/Ang/Bsect/Offset]: *Specify an option or select a point through which the xline will pass.*

The various options of the command are discussed next.

Figure 3-57 Choosing the Construction Line button from the Ribbon

Point

If you use the default option, AutoCAD will prompt you to select two points through which the xline shall pass at the **Specify a point** and the **Specify through point** prompts. After you select the first point, AutoCAD will dynamically rotate the xline through the specified point, as you move the cursor. When you select the second point, an xline will be created that passes through the first and second points (Figure 3-58).

> Command: **XLINE** Enter
> Specify a point or [Hor/Ver/Ang/Bisect/Offset]: *Specify a point.*
> Specify through point: *Specify the second point.*

You can continue to select more points to create more xlines. All these xlines will pass through the first point you had selected at the **Specify a point** prompt. This point is also called the root point. Right-click or press ENTER to end the command.

Horizontal

This option will create horizontal xlines of infinite length that pass through the selected points. The xlines will be parallel to the *X* axis of the current UCS, see Figure 3-59. As you invoke this option, the horizontal xline gets attached to the cursor. You are prompted to select only one point through which the horizontal xline passes. You can continue selecting points to draw horizontal xlines and right-click or press ENTER to end the command.

Vertical

This option will create vertical xlines of infinite length that pass through the selected points. The xlines will be parallel to the *Y* axis of the current UCS, see Figure 3-59. As you invoke this option, the vertical xline gets attached to the cursor. You are prompted to select only one point through which the vertical xline passes. You can continue selecting points to draw vertical xlines and right-click or press ENTER to end the command.

Figure 3-58 Drawing xlines

Figure 3-59 Horizontal and vertical xlines

Angular

This option will create xlines of infinite length that pass through the selected point at a specified angle (in Figure 3-60, the angle specified is 38-degree). The angle can be specified by entering a value at the keyboard. You can also use the reference option by selecting an object and then specifying an angle relative to it. The **Reference** option is useful, when the actual angle is not known but the angle relative to an existing object can be specified.

Command: **XLINE** Enter
_xline Specify a point or [Hor/Ver/Ang/Bisect/Offset]: **A** Enter
Enter angle of xline (0) or [Reference]: **R** Enter   *(Here you use the **Reference** method for specifying the angle)*
Select a line object: *Select a line.*
Enter angle of xline <0>: *Enter angle (the angle will be measured counterclockwise with respect to the selected line)*
Specify through point: *Specify the second point.*

Bisect

This option will create an xline that passes through the angle vertex and bisects the angle you specify by selecting two points. The xline created using this option will lie in the plane defined by the selected points. You can use the object snaps to select the points on the existing objects. The following is the prompt sequence for this option, refer to Figure 3-61.

Command: **XLINE** Enter
_xline Specify a point or [Hor/Ver/Ang/Bisect/Offset]: **B** Enter
Specify angle vertex point: *Enter a point (P1).*
Specify angle start point: *Enter a point (P2).*
Specify angle end point: *Enter a point (P3).*
Specify angle end point: *Select more points or press ENTER or right-click to end the command.*

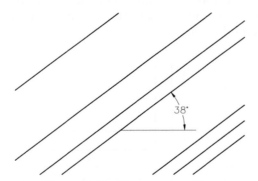

Figure 3-60 *The angular xlines*

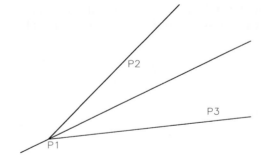

Figure 3-61 *Using the **Bisect** option to draw xlines*

Offset

The **Offset** option creates xlines that are parallel to the selected line/xline at a specified offset distance. You can specify the offset distance by entering a numerical value or by selecting two points on the screen. If you select the **Through** option, the offset line will pass through the selected point. This option works like the **OFFSET** editing command. The prompts at the command line are as follows.

Command: **XLINE** Enter
_xline Specify a point or [Hor/Ver/Ang/Bisect/Offset]: **O** Enter
Specify offset distance or [Through] <Through>: *Press ENTER to accept the **Through***

option or specify a distance from the selected line object at which the xline shall be drawn.
Select a line object: *Select the object to which the xline is drawn parallel at a specified distance.*
Specify through point: *Select a point through which the xline should pass.*

If you specify the offset distance, and after you have selected a line object, you are prompted to specify the direction in which the xline is to be offset. You can continue drawing xlines or right-click or press ENTER to end the command.

Drawing RAY

Ribbon:	Home > Draw > Ray	Command:	RAY

A ray is a 3D line similar to the xline construction line, with the difference being that it extends to infinity only in one direction. It starts from a point you specify and extends to infinity through the specified point. The prompt sequence is given next.

Command: **RAY** `Enter`
Specify start point: *Select the starting point for the ray.*
Specify through point: *Specify the second point.*

Press ENTER or right-click to exit the command.

Note
When you trim an xline, it gets converted into a ray, and when a ray is trimmed at the end that is infinite, it gets converted into a line object.

WRITING A SINGLE LINE TEXT

Ribbon:	Annotation > Multiline Text > Single Line Text	Command:	TEXT

 The **TEXT** command lets you write a single line text in the drawing. Although you can write more than one line of text using this command, but each line will be a separate text entity. After invoking this command, you need to specify the start point for the text. Then you need to specify the text height and also the rotation angle. The characters appear on the screen, as you enter them. When you press ENTER, after typing a line, the cursor automatically places itself at the start of the next line and repeats the prompt for entering another line. You can end the command by pressing the ENTER key. You can use the BACKSPACE key to edit the text on the screen while you are writing it. The prompt sequence is given next.

Command: **TEXT** `Enter`
Current text style: "Standard" Text height: 0.2000 Annotative: No
Specify start point of text or [Justify/Style]: *Specify the starting point of the text.*
Specify height<current>: *Enter the text height.*
Specify rotation angle of text <0>: `Enter`
Enter the first line of text in the text box displayed in the drawing window.
Enter the second line of text in the text box displayed in the drawing window.
Press ENTER.

Note

The other commands to enter text are discussed in detail in Chapter 7.

To move the objects, use the **MOVE** *command and then select the objects, specify the base point, and the second point of displacement. The* **MOVE** *command is discussed in detail in Chapter 5.*

Self-Evaluation Test

Answer the following questions and then compare them to those given at the end of this chapter:

1. A negative value for a chord length in the **Start, Center, Length** option of the **ARC** command results in the largest possible arc, also known as the major arc. (T/F)

2. In the **ARC** command, if you do not specify a start point but just press ENTER or choose the **Continue** option, the start point and direction of the arc is taken from the endpoint and ending direction of the previous line or arc drawn on the current screen. (T/F)

3. If the **PELLIPSE** is set to 1, AutoCAD creates a true ellipse, also known as NURBS-based (Non-Uniform Rational Bezier Spline) ellipse. (T/F)

4. The start and end parameters of an elliptical arc are determined by specifying a point on the circle whose diameter is equal to the minor diameter of the ellipse. (T/F)

5. The _____ option of the **RECTANG** command allows you to draw a rectangle at a specified distance from the *XY* plane along the *Z* axis.

6. If the **FILLMODE** is set to _____, only the outlines are drawn for the new polyline.

7. You can get a _____ polyline by entering two different values at the starting width and the ending width prompts.

8. In case of a(n) _____ polyline, the vertices are not stored as separate entities, but as a single object with an array of information.

9. If you invoke the _____ command from the **Draw** panel of the **Ribbon**, you can draw as many points as you desire in a single command.

10. The size of a point is taken as a percentage of the viewport size, if you enter a _____ value for the **PDSIZE** variable.

Review Questions

Answer the following questions:

1. Using the **Start, End, Angle** option of the **ARC** command, a negative included angle value draws the arc in a clockwise direction. (T/F)

2. When the **Continue** option of the **ARC** command is used to draw arcs, each successive arc is perpendicular to the previous one. (T/F)

3. If you specify the chamfer distances first and then specify the fillet radius in the same **RECTANG** command, the rectangle will be drawn with chamfers only. (T/F)

4. Using the **RECTANG** command, the rectangle drawn is treated as a combination of different objects; therefore, the individual sides can be edited independently. (T/F)

5. Using the **Start, Center, Length** option of the **ARC** command, a positive chord length generates the smallest possible arc (minor arc) with this length, and the arc is always less than

   (a) 90-degree   (b) 180-degree
   (c) 270-degree   (d) 360-degree

6. Which one of the following options of the **RECTANG** command allows you to draw a rectangle that is extruded in the Z direction by the specified value?

   (a) **Elevation**   (b) **Thickness**
   (c) **Extrude**   (d) **Width**

7. Which of the following commands draws a line anywhere in 3D space, which starts from a point that you specify and the other end extending to infinity?

   (a) **PLINE**   (b) **RAY**
   (c) **XLINE**   (d) **MLINE**

8. If the old 2D polylines should not get converted to lightweight polylines, when opening the drawings in AutoCAD 2010, and also the new polylines drawn should be 2D polylines, the **PLINETYPE** variable should be set to which of the following values?

   (a) 0   (b) 1
   (c) 2   (d) 3

9. Which of the following values should be assigned to the **PDMODE** variable such that a cross mark (X) is generated through the specified point?

   (a) 1   (b) 3
   (c) 5   (d) 7

Chapter 3

10. A polygon is said to be _____, when it is drawn inside an imaginary circle and its vertices (corners) touch the circle.

11. If additional polyline segments are added to the first polyline, AutoCAD automatically makes the _____ of the first polyline segment the start point of the next polyline segment.

12. To create a solid filled circle by the **DONUT** command, the value of the inside diameter of the circle should be _____.

13. With the **DONUT** command, you can draw a solid-filled circle by specifying the inside diameter as _____ and keeping the **FILLMODE** on.

14. The visibility of blips can be controlled using the _____ system variable.

15. An absolute size for the point can be specified by entering a _____ value for the **PDSIZE** variable.

16. The _____ option of the **XLINE** command creates xlines of infinite length that are parallel to the *Y* axis of the current UCS.

17. You can use the _____ key to edit the text on the screen, while you are writing it using the **TEXT** command.

Exercises

Exercise 14   *Mechanical*

Draw the sketch shown in Figure 3-62. The distance between the dotted lines is 1.0 unit. Create the radii, using appropriate **ARC** command options.

Figure 3-62 Drawing for Exercise 14

Exercise 15   *Graphics*

Draw the sketch shown in Figure 3-63. The distance between the dotted lines is 1.0 unit. Create the arcs, using appropriate **ARC** command options.

Figure 3-63 Drawing for Exercise 15

Exercise 16   *Mechanical*

Draw the sketch shown in Figure 3-64. The distance between the dotted lines is 0.5 unit. Create the ellipses, using the **ELLIPSE** command.

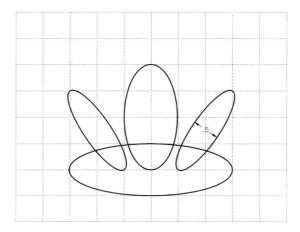

Figure 3-64 Drawing for Exercise 16

Exercise 17   *Mechanical*

Draw the sketch shown in Figure 3-65 using the **LINE, CIRCLE**, and **ARC** commands. The distance between the dotted lines is 1.0 unit and the diameter of the circles is 1.0 units.

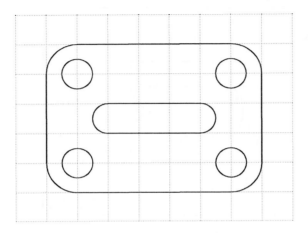

Figure 3-65 *Drawing for Exercise 17*

Exercise 18   *Mechanical*

Draw the sketch shown in Figure 3-66 using the **LINE, CIRCLE**, and **ARC** commands or their options. The distance between the grid lines is 1.0 unit and the diameter of the circle is 1.0 unit.

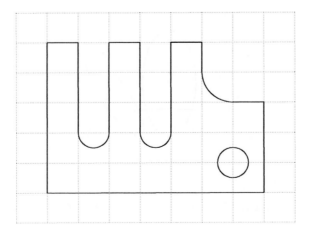

Figure 3-66 *Drawing for Exercise 18*

Problem-Solving Exercise 1   *Mechanical*

Draw the sketch shown in Figure 3-67. Create the radii by using the arc command options indicated in the drawing. (Use the @ symbol to snap to the previous point. Example: Specify start point of arc or [Center]: @)

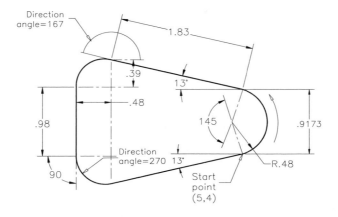

Figure 3-67 *Drawing for Problem-Solving Exercise 1*

Problem-Solving Exercise 2   *Mechanical*

Draw the sketch shown in Figure 3-68. Create the radii by using the arc command options. The distance between the dotted lines is 0.5 units.

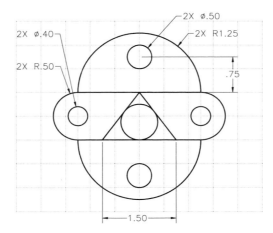

Figure 3-68 *Drawing for Problem-Solving Exercise 2*

Problem-Solving Exercise 3   *Mechanical*

Draw the sketch shown in Figure 3-69. Create the radii by using the **ARC** command options. The distance between the dotted lines is 1.0 unit.

Figure 3-69 *Drawing for Problem-Solving Exercise 3*

Problem-Solving Exercise 4   *Mechanical*

Draw the sketch shown in Figure 3-70 using the **POLYGON**, **CIRCLE**, and **LINE** commands.

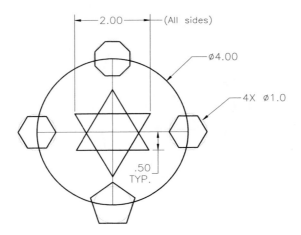

Figure 3-70 *Drawing for Problem-Solving Exercise 4*

Problem-Solving Exercise 5   *Mechanical*

Draw the sketch shown in Figure 3-71 using the draw commands. Note, Sin30=0.5, Sin60=0.866. The distance between the dotted lines is 1 unit.

Figure 3-71 Drawing for Problem-Solving Exercise 5

Chapter 3

Problem-Solving Exercise 6   *Mechanical*

Draw the sketch shown in Figure 3-72 using the draw commands. Also, draw the hidden lines and centerlines as continuous line. Note that the dimensions are given only for your reference. **

Figure 3-72 *Drawing for the Problem-Solving Exercise 6*

Answers to Self-Evaluation Test
1. T, **2.** T, **3.** F, **4.** F, **5. Elevation**, **6.** 0, **7.** tapered, **8.** optimized, **9.** Multiple Point, **10.** negative

Chapter 4

Working with Drawing Aids

Learning Objectives

After completing this chapter, you will be able to:

- *Set up layers and assign colors and line types to them.*
- *Use the **Properties** toolbar to directly change the general object properties.*
- *Change the properties of objects using the **PROPERTIES** command.*
- *Determine current, global line type scaling, and **LTSCALE** factor for plotting.*
- *Set up Grid, Snap, and Ortho modes on the basis of drawing requirements.*
- *Use Object Snaps and understand their applications.*
- *Combine Object Snap modes and set up running Object Snap modes.*
- *Use AutoTracking to locate keypoints in a drawing.*

In this chapter, you will learn about the drawing setup and the factors that affect the quality and accuracy of a drawing. This chapter contains a detailed description of how to set up layers. You will also learn about some other drawing aids, such as Grid, Snap, and Ortho. These aids will help you to create drawings accurately and quickly.

UNDERSTANDING THE CONCEPT AND USE OF LAYERS

The concept of layers can be best explained by using the concept of overlays in manual drafting. In manual drafting, different details of the drawing can be drawn on different sheets of paper or overlays. Each overlay is perfectly aligned with the others. Once all of them are placed on top of each other, you can reproduce the entire drawing. As shown in Figure 4-1, the object lines are drawn in the first overlay and the dimensions in the second overlay. You can place these overlays on top of each other and get a combined look at the drawing.

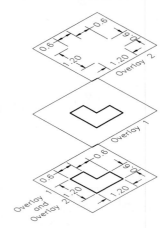

Instead of using overlays in AutoCAD, you can use layers. Each layer is assigned a name. You can also assign a color and line type to these layers. For example, in Figure 4-2, the object lines are drawn in the OBJECT layer and the dimensions are drawn in the DIM layer. The object lines will be red because red has been assigned to the OBJECT layer. Similarly, the dimension lines will be green because the DIM layer has been assigned the green color. You can display the layers together, individually, or in any combination.

Figure 4-1 *Drawing lines and dimensions in different overlays*

Advantages of Layers

1. Each layer can be assigned a different color. Assigning a particular color to a group of objects is very important for plotting. For example, if all object lines are red, at the time of plotting you can assign the red color to a slot (pen) that has the desired tip width (e.g., medium). Similarly, if the dimensions are green, you can assign the green color to another slot (pen) that has a thin tip. By assigning different colors to different layers you can control the width of the lines when the drawing is plotted. You can also make a layer plottable or nonplottable.

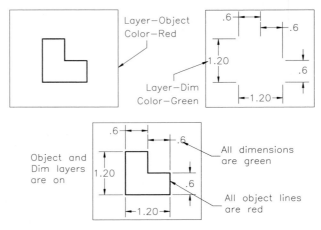

Figure 4-2 *Drawing lines and dimensions in different layers*

2. The layers are useful for some editing operations. For example, to erase all dimensions in a drawing, you can freeze or lock all layers except the dimension layer and then erase all dimensions by using the **Crossing** option to select objects.
3. You can turn a layer off or freeze a layer that you do not want to be displayed or plotted.
4. You can lock a layer that will prevent the user from accidentally editing the objects in it.
5. The colors also help you distinguish different groups of objects. For example, in architectural drafting, the plans for foundation, floors, plumbing, electrical, and heating systems may all be made in different layers. In electronic drafting and in PCB (printed circuit board), the design of each level of a multilevel circuit board can be drawn on a separate layer. Similarly, in mechanical engineering, the main components of an assembly can be made in one layer, other components such as nuts, bolts, keys, and washers can be made in another layer, and the annotations such as datum symbols and identifiers, texture symbols, Balloons, and Bill of Materials can be made in yet another layer.

WORKING WITH LAYERS

| **Ribbon:** | Layers > Layer Properties | **Command:** | LAYER or LA |
| **Toolbar:** | Layers > Layer Properties Manager | | |

You can choose the **Layer Properties** button from the **Layers** panel in the **Home** tab of the **Ribbon**, refer to Figure 4-3 or **Layers** toolbar, see Figure 4-4 to invoke the **Layer Properties Manager** dialog box. You can perform the functions associated with the layers, using this dialog box. For example, you can create new layers, assign them colors

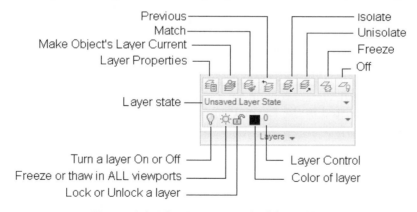

*Figure 4-3 The **Layers** panel of the **Ribbon***

*Figure 4-4 The **Layers** toolbar*

and linetypes, or perform any operation that is shown in the dialog box. You can also perform some layer functions, such as freeze, thaw, lock, unlock, and so on directly from the **Layers** panel in the **Ribbon**. When you invoke the **Layer Properties Manager** dialog box, a default

Chapter 4

layer with the name **0** is displayed. It is the current layer and any object you draw is created in it. There are certain features associated with each layer such as color, linetype, and lineweight. Layer 0 has a default color white, linetype continuous, and lineweight default.

Creating New Layers

To create new layers, choose the **New Layer** button in the **Layer Properties Manager** dialog box. Alternatively, press ALT+N, after the **Layer Properties Manager** dialog box is displayed; a new layer with the name Layer1 and the properties of the 0 layer is created and listed just below the 0 layer in the dialog box, see Figure 4-5.

Figure 4-5 *Partial view of the* **Layer Properties Manager** *dialog box with a new layer created*

If you have more layers, in addition to layer 0, the new layer will have the properties of the layer that is selected in the **Layer Properties Manager** dialog box. You can change or edit the name by selecting it and then entering a new name. If more than one layer is selected, the new layer is placed at the end of the layers list and has the properties of the layer selected last. Right-clicking anywhere in the **Layers** list area of the **Layer Properties Manager** dialog box displays a shortcut menu that also gives you an option to create a new layer. You can right-click on the layer whose properties you want to use in the new layer and then select **New Layer** from the shortcut menu.

You can also create a layer that is automatically frozen in all viewports. Choose the **New Layer VP Frozen in All Viewports** button; a new layer that is frozen in all viewports will be created and it will be listed in the **Layer Properties Manager** dialog box.

Note
You will learn more about viewports in Chapter 12.

Layer names

1. A layer name can be up to 255 characters long, including letters (a-z), numbers (0-9), special characters ($ _ -), and spaces. Any combination of lower and uppercase letters can be used while naming a layer. However, characters such as <>;:,'?"=, and so on are not valid characters while naming a layer.

2. The layers should be named to help the user identify the contents of the layer. For example, if the layer name is HATCH, a user can easily recognize it and also its contents. On the other hand, if the layer name is X261, it is hard to identify its contents.

3. Layer names should be short, but should also convey the meaning.

Note

*The length of the layer name is controlled by the **EXTNAMES** system variable that has a default value 1. If you change it to 0, the layer name is allowed to be up to thirty-one characters long but it cannot include special characters and spaces.*

Tip

If you exchange drawings with or provide drawings to consultants or others, it is very important that you standardize and coordinate layer names and other layer settings.

Making a Layer Current

To draw an object in a particular layer, you need to make it the current layer. Only one layer can be made current, in which new objects will be drawn. To make a layer current, double-click on it in the list box; the selected layer is made current. Alternatively, select the name of the desired layer and then choose the **Set Current** button in the dialog box. AutoCAD will display a green color check mark in the **Status** column of that row in the **Layer Properties Manager** dialog box. Also, the name of the current layer will be displayed at the upper-left corner of the dialog box. Choose the **Close** button to exit the dialog box.

Right-clicking on a layer in the layer list box displays a shortcut menu that gives you an option (**Set current**) to make the selected layer current, see Figure 4-6.

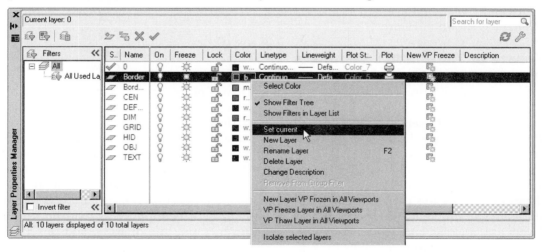

Figure 4-6 *The **Layer Properties Manager** dialog box with the shortcut menu*

The name and properties of the current layer are displayed in the **Layers** toolbar. You can also make a layer current by selecting it from the **Layer** drop-down list in the **Layers** toolbar or from the **Layers** panel of the **Ribbon**. To make the layer current, you can also use the **CLAYER** system variable from the command window. Choose the **Make Object's Layer Current** button from the **Layers** panel of the **Ribbon**; you will be prompted to select the object whose layer you want to make current. After selecting the object, the layer associated with that object will be made current.

Note

*When you select more than one layer at a time using the SHIFT or CTRL key, the **Make Current** option is not displayed in the shortcut menu in the **Layer Properties Manager** dialog box. This is because only one layer can be made current at a time.*

Controlling the Display of Layers

You can control the display of the layers by selecting the **Turn a layer On or Off**, **Freeze or thaw in ALL viewports,** and **Lock or Unlock a layer** toggle buttons in the list box of any particular layer.

Turn a Layer On or Off

With the **Turn a layer On or Off** toggle icon (light bulb), you can turn the layers on or off. The layers that are turned on are displayed and can be plotted while the layers that are turned off are not displayed and cannot be plotted. You can perform all the operations such as drawing and editing in the layer that has been turned off. You can turn the current layer off, but AutoCAD will display a message box informing you that the current drawing layer has been turned off. You can also turn the layer on or off by clicking on the **On/Off** toggle icon from the **Layer** drop-down list in the **Layers** toolbar or the **Layers** panel of the **Ribbon**, as shown in Figure 4-7.

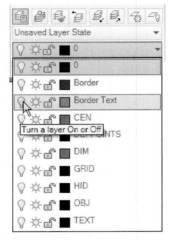

*Figure 4-7 Turning layer Border Text off in the **Ribbon***

Freeze or Thaw in ALL Viewports

While working on a drawing, if you do not want to see certain layers, you can also use the **Freeze or thaw in ALL viewports** toggle icon (sun/snowflakes) to freeze them. You can use the **Layers** toolbar, **Layers** panel of the **Ribbon**, or the **Layer Properties Manager** dialog box to freeze or thaw a layer. No modifications can be done in the frozen layer. For example, while editing a drawing, you may not want the dimensions to be changed and displayed on the screen. To avoid this, you can freeze the **Dim** layer, in which you are dimensioning the objects. The frozen layers are invisible and cannot be plotted. The **Thaw** option negates the effect of the **Freeze** option, and the frozen layers are restored to normal. The difference between the **Off** option and the **Freeze** option is that the frozen layers are not calculated by the computer while regenerating the drawing, and this saves time. The current layer cannot be frozen.

New VP Freeze

If you want to freeze some layers in the new floating viewports, select the **New VP Freeze** toggle icon; AutoCAD will freeze the selected layers in the subsequently created new viewports, without affecting the viewports that already exist.

VP Freeze

When you select a layout in the **Model/Layout** tab (by clicking on Layout1), or you set the **TILEMODE** variable to 0 (see Chapter 12, Model Space Viewports, Paper Space Viewports, and Layouts), you can freeze or thaw the selected layers in the active floating viewport by selecting the **VP Freeze** icon for the selected layers. The selected layer can be frozen in the current viewport if it is thawed, but a layer that is frozen in the model space cannot be thawed in the current viewport. Once you are in a floating viewport, the **VP Freeze** icon is added in the **Layer Properties Manager** dialog box on the right, refer Figure 4-8.

*Figure 4-8 Partial view of the **Layer Properties Manager** dialog box with the **VP Freeze** icons*

If the icon is not visible, you can move the scroll bar at the bottom of the layer list box to display it. Also, the **Freeze or thaw in current viewport** icon in the **Layers** toolbar becomes available once you have viewports. Selecting this icon makes the selected layers invisible in the active floating viewport only. The frozen layers will still be visible in other viewports. (Paper space is discussed in Chapter 12) Also, check the **VPLAYER** command for selectively freezing layers in viewports.

Tip
*The width of the column headings in the **Layer Properties Manager** dialog box can be decreased or increased by positioning the cursor between the column headings on the separator (the cursor turns into a two-sided arrow). Now, hold down the pick button of your pointing device and then drag the cursor to the right or left. This way you can vary the width of the column headings.*

Lock or Unlock a Layer

While working on a drawing, if you do not want to accidentally edit some objects on a particular layer but still need to have them visible, you can use the **Lock/Unlock** toggle icon to lock the layers. When a layer is locked, you can still use the objects in the locked layer for Object Snaps and inquiry commands such as **LIST**. You can also make the locked layer the current layer and draw objects on it. Note that you can also plot the locked layer. The **Unlock** option negates the **Lock** option and allows you to edit objects on the layers previously locked.

Make a Layer Plottable or Nonplottable

If you do not want to plot a particular layer, for example, construction lines, you can use the **Plot** toggle icon (printer) to make the layer plottable or nonplottable. This icon is available in the **Layer Properties Manager** dialog box. The construction lines will not be plotted if its layer is made nonplottable.

Chapter 4

Tip
*It is faster and convenient to use the **Layer** drop-down list in the **Layers** panel of the **Ribbon** to make a layer current and control the display features of the layer (On/Off, Freeze/Thaw, Lock/Unlock).*

Assigning Linetype to Layer

By default, layers are assigned continuous linetypes and white color, if no layer is selected at the time of creating a new layer. Otherwise, the new layer takes the properties of the selected layer. To assign a new linetype to a layer, click on the field under the **Linetype** column of that layer in the **Layer Properties Manager** dialog box. AutoCAD will display the **Select Linetype** dialog box. This dialog box displays the linetypes that are defined and loaded on your computer. Select the new linetype and then choose the **OK** button. The linetype that you select is assigned to the layer you selected initially.

If you have not loaded the linetypes and are opening the **Select Linetype** dialog box for the first time, only the **Continuous** linetype is displayed, as shown in Figure 4-9.

*Figure 4-9 The **Select Linetype** dialog box*

You need to load the linetypes that you want and then assign them to the layers. To load the linetypes, choose the **Load** button in the **Select Linetype** dialog box. This displays the **Load or Reload Linetypes** dialog box, see Figure 4-10. This dialog box displays all linetypes in the *acad.lin* file. In this dialog box, you can select a single linetype or a number of linetypes by pressing and holding the SHIFT or CTRL key and then selecting the linetypes. If you right-click, AutoCAD displays the shortcut menu, which you can use to select all linetypes. Then choose the **OK** button; the selected linetypes are loaded and displayed in the **Select Linetype** dialog box. Now, select the desired linetype and choose **OK**. The selected linetype is assigned to the selected layer.

Note
*By default, the linetypes in the acad.lin file are displayed in the **Load or Reload Linetypes***

*dialog box. You can select the linetypes in the acadiso.lin file by choosing the **File** button in the dialog box and then opening the acadiso.lin file from the **Select Linetype Files** dialog box.*

Tip
*You can also create your own linetypes. This is discussed in detail in the **Customizing AutoCAD 2010** textbook by Prof. Sham Tickoo.*

Figure 4-10 *The **Load or Reload Linetypes** dialog box*

Assigning Color to Layer

To assign a color, select the color swatch in a particular layer in the **Layer Properties Manager** dialog box; AutoCAD will display the **Select Color** dialog box. Select the desired color and then choose the **OK** button. The color you selected will be assigned to the selected layer. The number of colors is determined by your graphics card and monitor. Most color systems support eight or more colors. If your computer allows it, you may choose a color number between 0 and 255 (256 colors). The following are the first seven standard colors:

Color number	Color name	Color number	Color name
1	Red	5	Blue
2	Yellow	6	Magenta
3	Green	7	White
4	Cyan		

Note
The use of nonstandard colors may cause compatibility problems if the drawings are used on other systems with different colors. On a light background, white color appears black. Certain colors are hard to see on light backgrounds and others are hard to see on dark backgrounds. So, you may have to use different colors than those specified in some examples and exercises.

Chapter 4

Assigning Lineweight to Layer

Lineweight is used to give thickness to the objects in a layer. This thickness is displayed on the screen if the display of the lineweight is on. The lineweight assigned to the objects can also be plotted. If you make a sectional plan at a certain height, you can assign a layer with a larger value of lineweight to create the objects through which the section is made. Another layer with a lesser lineweight can be used to show the objects through which the section does not pass. To assign a lineweight to a layer, select the layer and then click on the lineweight associated with it; the **Lineweight** dialog box is displayed, see Figure 4-11. Select a lineweight from the **Lineweights** list. Choose **OK** to return to the **Layer Properties Manager** dialog box.

*Figure 4-11 The **Lineweight** dialog box*

Note
*The **LWEIGHT** command for editing and displaying the lineweight is discussed in detail later in this chapter.*

Assigning Plot Style to Layer

Plot style is a group of property settings such as color, linetype, and lineweight that can be assigned to a layer. The applied plot style affects the drawings while plotting only. The drawing, in which you are working, should be in a named plot style mode (*.stb*) to make the plot style available in the **Layer Properties Manager** dialog box. If the **Plot Style** icon in the dialog box is not available, then you are in a color-dependent mode (*.ctb*). To make it available, choose **Options** button from the **Application Menu**; the **Options** dialog box will be displayed. Choose the **Plot and Publish** tab from this dialog box. Next, choose the **Plot Style Table Settings** button to invoke the **Plot Style Table Settings** dialog box. In this dialog box, select the **Use named plot style** radio button from the **Default plot style behavior for new drawings** area and then exit both the dialog boxes. After changing the plot style to named plot style dependent, you have to start a new AutoCAD session to apply this setting. Start a new AutoCAD session with templates like *acad -Named Plot Styles* and invoke the **Layer Properties Manager**. You will notice that the default plot style is **Normal,** in which the color, linetype, and lineweight are BYLAYER. To assign the plot style to a layer, select the layer and then click on its plot style; the **Select Plot Style** dialog box will be displayed, as shown in Figure 4-12. In this dialog box, you can select a specific plot style from the **Plot styles** list of available plot styles. Plot styles need to be created before you can use them (see Chapter 14, *Plotting Drawings*). Choose **OK** to return to the **Layer Properties Manager** dialog box.

Note
When you are in any viewport, you can override the color, linetype, lineweight, and plot style settings assigned to a selected layer for the particular viewport (current viewport). These settings for all other viewports and model space will remain unaffected. All these overrides are highlighted

*in different background color. Also, the icon of the selected layer under the **Status** column will change to inform you that some of the properties of the selected layer have been overridden by the new one (see Chapter 12, Model Space Viewports, Paper Space Viewports, and Layouts).*

Tip

*You can also change the plot style mode from the command line by using the **PSTYLEPOLICY** system variable. A value of **0** sets it to the color-dependent mode and a value of **1** sets it to the named mode.*

Deleting Layers

You can delete a layer by selecting and then choosing the **Delete Layer** button in the **Layer Properties Manager** dialog box. Remember that to delete a layer, it is necessary that it should not contain any objects. You cannot delete layers 0, Defpoints (created while dimensioning), and Ashade (created while rendering), current layer, and an Xref-dependent layer.

Figure 4-12 The Select Plot Style dialog box

Managing the Display of Columns

AutoCAD allows you to change the display order of the columns in the list. To do so, drag and drop the column head that you want to move at the desired location. You can also change the display order by using the shortcut menu that is invoked when you right-click on any column head, as shown in Figure 4-13. Choose the **Customize** option from the shortcut menu to invoke the **Customize Layer Columns** dialog box, as shown in Figure 4-14. Select the desired column from the list and choose the **Move Up** or **Move Down** button to change the display order.

You can also control the column heads to be displayed in the **Layer Properties Manager** dialog box by using the shortcut menu, as shown in Figure 4-13. The tick marked entries in the shortcut menu will be displayed in the list. Choose the entry that you want to display in the **Layer Properties Manager** dialog box from the shortcut menu.

Selective Display of Layers

If the drawing has a limited number of layers it is easy to scan through them. However, if the drawing has a large number of layers, it is sometimes difficult to search through the layers. To solve this problem, you can use layer filters. By defining filters, you can specify the properties and only the layers that match those properties will be displayed in the **Layer Properties Manager** dialog box. By default, **All** and **All Used Layers** filters are created. The **All** filter is selected by default, which ensures that all the layers are displayed. If you select the **All Used**

Chapter 4

Figure 4-13 *The shortcut menu to manage column heads*

Layers filter, only those layers will be displayed that are used in the drawing. Rest of the layers are not displayed.

AutoCAD allows you to create a layer property filter or a layer group filter. A layer group filter can have additional layer property filters. To create a filter, choose the **New Property Filter** button, which is the first button on the top left corner of the dialog box. When you choose this button, a new property filter is added to the list and the **Layer Filter Properties** dialog box is displayed, as shown in Figure 4-15.

The current name of the filter is displayed in the **Filter name** edit box. You can enter any name for the filter in this edit box. Using this dialog box, you can create filters based on any property column in the **Filter definition** area. The layers that will be actually displayed in the **Layer Properties Manager** dialog box, based on the filter that you create, are shown in the **Filter preview** area. For example, to list only those layers that are red in color, click on

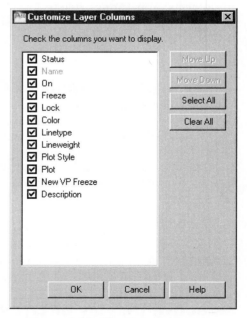

Figure 4-14 *The **Customize Layer Columns** dialog box*

the field under the **Color** column. A swatch **[...]** button is displayed in this field. Choose this button to display the **Select Color** dialog box. Select red from this dialog box and then exit it. You will notice that the filter row color is changed to red and the display of layers in the **Filter preview** is modified such that only the red layers are displayed.

Figure 4-15 *The **Layer Filter Properties** dialog box*

After creating the filter, exit the **Layer Filter Properties** dialog box. The layer filter is selected automatically in the **Layer Properties Manager** dialog box and only the layers that satisfy the filter properties are displayed. You can restore the display of all the layers again by clicking on the **All** filter.

Choose the **New Group Filter** button (second button on the top left corner of the dialog box) to group all the filters under a common group name. You can create groups and subgroups of filters that helps you to categorize the filters with similar criteria.

Note

*To modify a layer filter, double-click on it; the **Layer Filter Properties** dialog box is displayed. Modify the filter and then exit the dialog box.*

In the **Layer Properties Manager** dialog box, when you select the **Invert filter** check box, you invert the filter that you have selected. For example, if you have selected the filter to show all the layers, none of the layers will be displayed. Note that in the **Layer Properties Manager** dialog box, the current layer is not displayed in the list box if it is not among the filtered layers.

Layer States

You can save and then restore the properties of all layers in a drawing using the **Layer States Manager** dialog box, which is invoked using the **Layer States Manager** button in the **Layer Properties Manager** dialog box or in the **Layers** panel of the **Ribbon**. While working on a drawing at any point in time, you can save all the layers with their present properties settings under one name and then restore it anytime later. Invoke the **Layer States Manager** dialog box, as shown in Figure 4-16. Choose the **New** button to invoke the **New Layer State to Save** dialog box. Specify the name and description related to the state of a layer in the **New layer state name** and **Description** edit boxes respectively, and then choose the **OK** button. You can specify the different states and properties of the layers that you want to save in the **Layers properties to restore** area by selecting the corresponding check boxes. Choosing the **OK** button saves the checked states and properties of the layers. However, this state is saved only for the current file. If you want to use the current layer state in other files also, choose the **Export** button to invoke the **Export layer state** dialog box. Enter the name for the layer state. The layer state is exported with the *.las* extension.

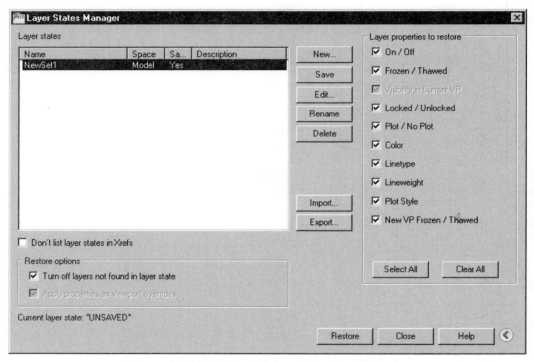

Figure 4-16 The **Layer States Manager** dialog box

You can import the layer state later in any other file using the **Layer States Manager** dialog box. In this dialog box, you are allowed to edit the states and properties of the saved state. You can also rename and delete a state.

Select the **Don't list layer states in Xrefs** check box; the layer states that are defined in an externally referenced drawing will be referenced from the **Layer states** list box. Select the **Turn off layers not found in layer state** check box to turn off the newly created layers (that were created after defining the layer state) so that the drawing looks the same way as it was when the layer state was saved. The **Apply properties as viewport overrides** check box applies the layer property overrides to the current viewport. This check box is selected by default and is available only when a viewport is active.

Tip
*The linetypes will be restored with the layer state in a new drawing file only if those linetypes are already loaded using the **Select Linetype** dialog box.*

Reconciling New Layers

When you save or plot a drawing for the first time, AutoCAD compiles a list of existing layers. These layers are known as reconciled layers. When new layers are inserted or an external reference is attached to the current drawing, AutoCAD compares these new layers (unreconciled layers) with the previously created list. This helps in preventing any new layers being added to the current drawing without users information.

If you need AutoCAD to notify the unreconciled layer, choose the **Settings** button from the **Layer Properties Manager** dialog box; the **Layer Settings** dialog box will be displayed. Select the **Evaluate all new layers** and **Notify when new layers are present** check boxes from the **Layer Settings** dialog box. You also need to select the options under these check boxes based on your requirement. When a new unreconciled layer is externally referenced with the current drawing after selecting the **Notify when new layers are present** check box and the corresponding options, AutoCAD, by default, displays the **Unreconciled New Layers** information bubble whenever new unreconciled layers are found in the current drawing or in the externally referenced drawings, see Figure 4-17. As soon as the unreconciled layers are found in the drawing, an **Unreconciled New Layers** filter is automatically created in the **Layer Properties Manager** dialog box. Click on the blue link in the information bubble or open the **Layer Properties Manager** dialog box to view the unreconciled layers. To reconcile the layers, select all the unreconciled layers in the **Layer Properties Manager** dialog box, right-click on them and choose the **Reconcile Layer** option from the shortcut menu.

Figure 4-17 The information bubble for unreconciled layers

Isolating and Unisolating Layers

Ribbon:	Layers > Isolate/Unisolate	**Command:**	LAYISO/LAYUNISO
Toolbar:	Layers II > Layer Isolate/Layer Unisolate		

This option is used to lock or turn off all layers except for those of the selected objects. Choose the **Isolate** button from the **Layers** panel of the **Ribbon**; you will be prompted to select the objects. Select one or more objects and press the ENTER key. All layers containing the selected objects will get isolated and unlocked. All other layers will be turned off or frozen in the current viewport or locked depending on the current settings. The unisolated layers (locked) will be displayed as faded. This will help you to differentiate between the isolated and unisolated layers. You can change the settings by choosing the

Chapter 4

Settings option from the shortcut menu at the **Select objects on the layer(s) to be isolated or [Settings]** prompt. The options at the **Settings** prompt are discussed next.

Off. This option turns off or freezes all layers except the selected layers. Choose the **Off** suboption to turn off all layers except the selected one in all viewports, and choose the **Vpfreeze** suboption to freeze all layers except the selected one only in the current floating viewport.

Lock and fade. This option locks all layers except the selected one and prompts you to enter the fade value to specify the fading intensity. You can also control the fading intensity by dragging the **Locked layer fading** sliding bar in the **Layers** panel of the **Ribbon**. You can also turn off or on the fading effect by choosing the **Locked Layer Fading** button from the **Layers** panel of the **Ribbon**

 To unisolate the previously isolated layer, choose the **Layer Unisolate** button from the **Layers** panel of the **Ribbon**.

Controlling the Layer Settings

The **Settings** button in the **Layer Properties Manager** dialog box is used to control the new layer notification settings. You can also change the background color used to highlight the layers properties that are overridden in the current viewport. Choose the **Settings** button in the **Layer Properties Manager** dialog box; the **Layer Settings** dialog box will be displayed (see Figure 4-18). The options in the **Layer Settings** dialog box are discussed next.

New Layer Notification Settings Area

By default, the **Evaluate new layers added to drawing** check box is selected. If this check box is cleared, AutoCAD will not evaluate the new layers. Also, the user will not be informed about the unreconciled layers. All other options in the **New Layer Notification Settings** area will be available only when this check box is selected. The **Evaluate new xref layers only** radio button is selected by default and allows

*Figure 4-18 The **Layer Settings** dialog box*

AutoCAD to evaluate the new layers added only to the externally referenced drawings. Select the **Evaluate all new layers** radio button to evaluate all the layers in the drawing, including the unreconciled layers of the externally referenced drawings. This setting can also be controlled by **LAYEREVAL** system variable by setting its value as 2.

The **Notify when new layers are present** check box is selected by default and is used to specify the operations after which AutoCAD will evaluate the unreconciled files. These operations include open, save, attach/reload xrefs, insert, restore layer states, and plot. Select the respective check boxes to specify your preferences. These settings can also be controlled by the **LAYERNOTIFY** system variable. The values for this system variable are 0, 1, 2, 4, 8, 16, and 32. The respective operations after which an information bubble will be displayed are Off, Plot, Open, Load/Reload/Attach for xrefs, Restore layer state, Save, and Insert.

Isolate Layer Settings Area

The options in this area are used to set the display state of the unisolated layers. If you need to display the locked layer in a faded mode, select the **Lock and fade** radio button and set the intensity of fading using the slider. Similarly, you can make the unisolated layers turned off or VP Freeze in the viewport by selecting the **Off** radio button and the corresponding option.

Dialog Settings Area

You can apply the current layer filter to the **Filter applied** drop-down list in the **Layers** toolbar. To do so, invoke the **Layer Filter Properties** dialog box by choosing the **New Property Filter** button from the **Layer Properties Manager** dialog box. Select the layer state or property from the **Filter definition** list box and choose the **OK** button. Only the filtered layers will be displayed in the **Filter applied** drop-down list of the **Layers** toolbar.

Example 1   *Mechanical*

Set up three layers with the following linetypes and colors. Then create the drawing shown in Figure 4-19 (without dimensions).

Figure 4-19 *Drawing for Example 1*

Chapter 4

Layer name	Color	Linetype	Lineweight
Obj	Red	Continuous	0.012"
Hid	Yellow	Hidden	0.008"
Cen	Green	Center	0.006"

In this example, assume that the limits and units are already set. Before drawing the lines, you need to create layers and assign colors, linetypes, and lineweights to them. Also, depending on the objects that you want to draw, you need to set that layer as current. In this example, you will create the layers using the **Layer Properties Manager** dialog box. You will use the **Layers** panel of the **Ribbon** to set the layers current and then draw the figure.

1. As the lineweights specified are in inches, first change the units for the lineweight, if they are in millimeters. Invoke the **Lineweight Settings** dialog box by right-clicking on the **Show/Hide Lineweight** button on the Status Bar and choose the **Settings** option from the shortcut menu. Select the **Inches [in]** radio button in the **Units for Listing** area of the dialog box and then choose the **OK** button.

2. Choose the **Layer Properties** button from the **Layers** panel in the **Home** panel of the **Ribbon** to display the **Layer Properties Manager** dialog box. The layer **0** with default properties is displayed in the list box.

3. Choose the **New Layer** button; AutoCAD automatically creates a new layer (Layer1) having the default properties and displays it in the list box. Change its name by entering **Obj** in place of Layer1.

4. Now choose the color of this layer to display the **Select Color** dialog box. Select the **Red** color and then choose **OK**. Red color is assigned to the **Obj** layer.

5. Choose the lineweight of the layer to display the **Lineweight** dialog box. Select **0.012"** and then choose **OK**. A lineweight of 0.012" is assigned to the **Obj** layer.

6. Again, choose the **New Layer** button; a new layer (Layer1) having the properties of layer **Obj** is created. Change its name by entering **Hid** in place of Layer1.

7. Choose the color of this layer to display the **Select Color** dialog box. Select the **Yellow** color and then choose **OK**.

8. Choose the linetype of the layer to display the **Select Linetype** dialog box. If the linetype **HIDDEN** is not displayed in the dialog box, choose the **Load** button to display the **Load and Reload Linetypes** dialog box. Select **HIDDEN** from the list and choose the **OK** button. Select **HIDDEN** in the **Select Linetype** dialog box and then choose **OK**.

9. Choose the lineweight of the layer to display the **Lineweight** dialog box. Select **0.008"** and then choose **O**K. A lineweight of 0.008" is assigned to the **Hid** layer.

10. Similarly create the new layer **Cen** and assign color **Green**, linetype **CENTER**, and lineweight **0.006"** to it.

11. Select the **Obj** layer and then choose the **Set Current** button to make the **Obj** layer current, see Figure 4-20. Choose the **Close** button to exit the dialog box.

Figure 4-20 Layers created for Example 1

12. Choose the **Show/Hide Lineweight** button from the Status Bar to turn on the display of the lines to be drawn. Also, choose **Dynamic Input** to turn off the dynamic input. Using the **LINE** command, draw the object lines in the **Obj** layer. The prompt sequence is given next.

Command: **LINE** Enter
Specify first point: **9,1** Enter
Specify next point or [Undo]: **9,9** Enter
Specify next point or [Undo]: **11,9** Enter
Specify next point or [Close/Undo]: **11,1** Enter
Specify next point or [Close/Undo]: **9,1** Enter
Specify next point or [Close/Undo]: **1,3** Enter
Specify next point or [Close/Undo]: **1,7** Enter
Specify next point or [Close/Undo]: **9,9** Enter
Specify next point or [Close/Undo]: Enter

Command: Enter *(Invokes the LINE command.)*
Specify first point: **1,3** Enter
Specify next point or [Undo]: **9,3** Enter
Specify next point or [Undo]: Enter

Command: Enter *(Invokes the LINE command.)*
Specify first point: **1,7** Enter
Specify next point or [Undo]: **9,7** Enter
Specify next point or [Undo]: Enter

Chapter 4

13. Select the down arrow in the **Layer** drop-down list from the **Layers** panel of the **Ribbon** to display the list of layers and then select the **Hid** layer from the list to make it current, as shown in Figure 4-21.

14. Now draw the following lines, using the **LINE** command.

> Command: **LINE** [Enter]
> Specify first point: **1,4** [Enter]
> Specify next point or [Undo]: **11,4** [Enter]
> Specify next point or [Undo]:   [Enter]
>
> Command: [Enter]   *(Invokes the LINE command.)*
> Specify first point: **1,6**   [Enter]
> Specify next point or [Undo]: **11,6** [Enter]
> Specify next point or [Undo]:   [Enter]

*Figure 4-21 Selecting the **Hid** layer from the **Layers** panel of **Ribbon** to set it as the current layer*

15. Again, select the down arrow in the **Layer** drop-down list from the **Layers** panel of the **Ribbon** and select the **Cen** layer from the list to make it current.

16. Now draw the following line using the **LINE** command.

> Command: **LINE**   [Enter]
> Specify first point: **0,5** [Enter]
> Specify next point or [Undo]: **12,5** [Enter]
> Specify next point or [Undo]: [Enter]

Exercise 1   *Mechanical*

Set up layers with the following linetypes and colors. Then make the drawing (without dimensions) as shown in Figure 4-22. The distance between the dotted lines is 1 unit.

Layer name	Color	Linetype
Object	Red	Continuous
Hidden	Yellow	Hidden
Center	Green	Center
Dimension	Blue	Continuous

Tip
*Remember that the **Linetype Control**, **Lineweight Control**, **Color Control**, and **Plot Style Control** list boxes in the **Properties** panel of the **Ribbon** should display **ByLayer** as the current property of objects. This is to ensure that when you draw an object, it takes the properties assigned to the layer in which the object is drawn.*

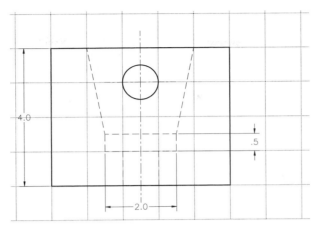

Figure 4-22 Drawing for Exercise 1

CHANGING THE LINETYPE

The **LINETYPE** command can be used to load, delete, or make a linetype as current, which can be used to draw an object. To invoke this command, enter **LINETYPE** at the Command prompt or select the **Other** option from the **Linetype** drop-down list in the **Properties** panel in the **Home** tab of the **Ribbon**; the **Linetype Manager** dialog box will be displayed, as shown in Figure 4-23. By default, the linetypes displayed are **ByLayer**, **ByBlock**, and

Figure 4-23 The Linetype Manager dialog box

Continuous. Choosing the **Load** button displays the **Load or Reload Linetypes** dialog box, which is also displayed from the **Layer Properties Manager** dialog box. The linetypes that you load here will be displayed in the **Select Linetypes** dialog box also. Similarly, if you have loaded some linetypes using the **Layer Properties Manager** dialog box, they will be displayed in the **Linetype Manager** dialog box. You can make a linetype current by using the **Current** button. All the new objects will be drawn with the current linetype. If you have chosen the current linetype as **ByLayer**, then the object will be drawn with the linetype of the layer. You can also make any linetype current other than ByLayer, and the object will be drawn with the particular linetype. The current linetype is also displayed in the **Linetype Control** drop-down list of the **Properties** toolbar.

LWEIGHT COMMAND

The **LWEIGHT** command is used to assign a lineweight to an object. To invoke this command, enter **LWEIGHT** at the Command prompt or select the **Lineweight Settings** option from the **Lineweight** drop-down list in the **Properties** panel of the **Home** tab in the **Ribbon**; the **Lineweight Settings** dialog box will be displayed, as shown in Figure 4-24. You can also choose the **Settings** option from the shortcut menu that is displayed on right-clicking on the **Hide/Show Lineweight** button on the Status Bar, to invoke this dialog box. In this dialog box, you can choose the current lineweight for the objects. You can change the units for the lineweight and also the display of lineweights for the current drawing. The lineweights are displayed in pixel widths and depending on the lineweight value chosen, the lineweights are displayed if the **Display Lineweight** check box is selected. For a large drawing, this increases the regeneration time and should be cleared. As mentioned earlier, you can also turn the display of lineweights to on or off directly from the Status Bar by choosing the **Hide/Show Lineweight** button. The slider bar for **Adjust Display Scale** affects the regeneration time as well. You can keep the slider bar at **Max** for getting a good display of different lineweights on the screen in the Model space; otherwise, keep it at **Min** for a faster regeneration.

*Figure 4-24 The **Lineweight Settings** dialog box*

Tip
For the drawings that have a large number of entities, you should keep the display scale at the minimum value for increasing the performance of AutoCAD. It is also recommended that you turn off the display of the lineweight from the Status Bar to reduce the regeneration time.

OBJECT PROPERTIES

An object, when created, has certain properties such as color, linetype, lineweight, plot style, and layer associated with it. These properties of an object can be changed and made current using the **Properties** palette or the **Properties** panel in the **Home** tab of the **Ribbon**.

Properties Panel

You can use the **Properties** panel of the **Ribbon** to directly change the general properties of the selected objects and also set those properties current.

Color

Select the object whose color you want to change. The current color of the object is displayed in the **Object Color** drop-down list. To display the list of colors, select the down arrow, as shown in Figure 4-25, and select a new color. The color of the selected object is changed to the new color. If you want to set a different color as current, choose a color from the drop-down list without selecting any object. By default, the color of an object is **ByLayer** and hence, it takes the color assigned to the layer in which it is created. From the **Object Color** drop-down list, you can select any other color to assign the color explicitly to the objects drawn. This color is set as current and all the new objects will be drawn in this color. If you want to assign a color that is not displayed in the list, select the **Select Colors** option to display the **Select Color** dialog box. You can select the desired color in this dialog box, and then choose the **OK** button.

*Figure 4-25 Different colors in the **Object Color** drop-down list*

Linetype

You can select the new linetype and assign it to the selected object from the **Linetype** drop-down list in the **Properties** panel of the **Ribbon**, as shown in Figure 4-26. You can also make a linetype current by selecting it from the list. By default, the linetype of an object is **ByLayer**. The linetypes that are loaded from the **LAYER** or the **LINETYPE** command will be listed. To assign a linetype that is not displayed in the list, choose the **Other** option to display the **Linetype Manager** dialog box. You can use the **Load** button to load other linetypes.

*Figure 4-26 The **Linetype** drop-down list*

Lineweight

Similarly, to assign the lineweight to the selected object, you can select a different lineweight value from the **Lineweight** drop-down

Chapter 4

list in the **Properties** panel of the **Ribbon**. Also, you can set a new lineweight current by selecting it from the list.

Plot Style

Expand the **Properties** panel by selecting the down arrow. You can select a different plot style for the selected object from the **Plot style** drop-down list. Also, you can set a new plot style current by selecting it from the list.

Note

*When you set a linetype, lineweight, or color current all objects drawn thereafter will have the current linetype, lineweight, and color. The properties of the current layer are not considered. You need to set current those properties to **ByLayer** again such that the objects assume the linetype, lineweight, color, or plot style of the layer, in which they are created.*

Properties Palette

Ribbon:	Home > View > Palettes > Properties	Toolbar:	Standard > Properties
Command:	PROPERTIES, CH, MO		

You can use the **Properties** palette to change the general properties of the selected objects and also set those properties current. The **Properties** palette can be displayed by invoking the **Properties** command. It can also be invoked by selecting an object and then right-clicking to display a shortcut menu. From the shortcut menu, choose **Properties**. The **Properties** palette is displayed, as shown in Figure 4-27, and you can change the different properties of the selected object. When you select the objects whose properties you want to change, the **Properties** palette displays the properties of the selected object. Depending on the object selected, the properties differ. You can also make the different properties current by selecting no object and then selecting the particular properties in the window. Right-clicking in the **Properties** palette displays a shortcut menu from where you can choose to **Allow Docking** or **Hide** the palette.

*Figure 4-27 The **Properties** palette*

When you select **Color** from the **General** list, the color of the selected object is displayed, along with a drop-down arrow. From this list, you can select any color to assign to the selected object. If you want to assign a color that is not displayed in the list, choose the **Select Color** option to display the **Select Color** dialog box. You can select the desired color in this dialog box, and then choose the **OK** button.

Similarly, you can set current the linetype, lineweight, linetype scale, and other properties for the other objects individually. You can also assign a hyperlink to an object. The **Hyperlink** field in this window displays the name and description of the hyperlink, if any, assigned to the object. If there is no hyperlink attached to the object, this field is blank. To add a hyperlink to the selected object, click on this field; the **[...]** button is displayed in this field. Choose the **[...]** button; the **Insert Hyperlink** dialog box is displayed where you can enter the path of the URL or file you want to link to the selected object.

Tip
You can keep the **Properties** *palette open while working in a drawing. When no object is selected, the current settings of the drawing are listed. Whenever you need to change a certain property of an object, select the object so that its properties are listed in the* **Properties** *palette and make the required changes.*

Note
You can use the **CHPROP** *command to change the properties of an object from the command line. You can also change the general properties of an object by using the* **Properties** *option of the* **CHANGE** *command. Both these commands, along with the* **Geometry** *portion of the* **Properties** *palette to edit the objects, have been discussed in Chapter 20.*

The system variables **CECOLOR**, **CELTYPE**, **CELWEIGHT** *control the current color, linetype, and lineweight of the object. You can set the properties of the objects current using these variables from the command line.*

When you set the property of the object current, it does not consider the property of the layer, in which the object will be drawn.

Tip
It is easier and faster to set the properties of the object current from the **Properties** *toolbar.*

It is better to leave the different properties of objects at **ByLayer** *to avoid confusion.*

Exercise 2   *General*

Draw a hexagon on the **Obj** layer in red. Let the linetype be hidden. Now, use the **Properties** palette to change the layer to some other existing layer, the color to yellow, and the linetype to continuous.

Quick Properties Palette

Status Bar: Quick Properties

In AutoCAD, you can view some of the properties of the selected entity in the **Quick Properties** palette or in the **Properties** palette. To view the properties, keep the **Quick Properties** button chosen in the Status Bar and then select an entity; some of the properties of the selected entity will be displayed in the **Quick Properties** palette. You can change the properties that are displayed in the **Quick Properties** palette. Figure 4-28 shows the **Quick Properties** palette that is displayed on selecting a line.

If you select multiple objects of different types, then in the drop-down list available in the

Quick Properties palette, the name of the objects will be defined as **All** and the number of selected entities will be displayed in the parenthesis. After selecting multiple entities, you can also select a particular type of object from the drop-down list to view the properties of the particular object type. If you move the cursor over the **Quick Properties** palette, it will expand if any of the properties is hidden.

*Figure 4-28 The **Quick Properties** palette*

To set the properties displayed in the **Quick Properties** palette, right-click on the **Quick Properties** button in the Status Bar; the **Draft Settings** dialog box will be displayed with the **Quick Properties** tab chosen. You can specify the display of the object type, position of the palette on invoking, and the height of the palette by selecting suitable options in this dialog box.

You can customize the **Quick Properties** palette such that it displays all the required properties of the selected entity. To do so, choose the **Customize** button available next to the drop-down list; the **Customize User Interface** dialog box will be displayed. In the dialog box, different object types and their corresponding properties will be displayed on the right. If you select an object type, you will notice that the check boxes of some of the properties are selected. The properties, whose check boxes are selected, will be listed in the **Quick Properties** palette. You can also select the check boxes of the properties that you want to be displayed in the **Quick Properties** palette.

If you place the cursor over an entity and pause, a tooltip, called the rollover tooltip, will be displayed. You can synchronize the properties displayed in the rollover tooltips and the **Quick Properties** panel. To do so, invoke the **Customize User Interface** dialog box, right-click on the **Rollover Tooltips** subnode in it and choose the **Synchronize with Quick Properties** option.

GLOBAL AND CURRENT LINETYPE SCALING

The **LTSCALE** system variable controls the global scale factor of the lines in a drawing. For example, if **LTSCALE** is set to 2, all lines in the drawing will be affected by a factor of 2. Like **LTSCALE**, the **CELTSCALE** system variable controls the linetype scaling. The difference is that **CELTSCALE** determines the current linetype scaling. For example, if you set **CELTSCALE** to 0.5, all lines drawn after setting the new value for **CELTSCALE** will have the linetype scaling factor of 0.5. The value is retained in the **CELTSCALE** system variable. Line (a) in Figure 4-29 is drawn with a **CELTSCALE** factor of 1; line (b) is drawn with a **CELTSCALE** factor of 0.5. The length of the

*Figure 4-29 Using **CELTSCALE** to control current linetype scaling*

dash is reduced by a factor of 0.5 when **CELTSCALE** is 0.5. The net scale factor is equal to the product of **CELTSCALE** and **LTSCALE**. Figure 4-29(c) shows a line that is drawn with **LTSCALE** of 2 and **CELTSCALE** of 0.25. The net scale factor = **LTSCALE** X **CELTSCALE** = 2 X 0.25 = 0.5. You can also change the global and current scale factors by entering a desired value in the **Linetype Manager** dialog box. If you choose the **Show details** button, the properties associated with the selected linetype are displayed, as shown in Figure 4-30. You can change the values according to your drawing requirements.

*Figure 4-30 The **Details** area of the **Linetype Manager** dialog box*

LTSCALE FACTOR FOR PLOTTING

The **LTSCALE** factor for plotting depends on the size of the sheet you use to plot the drawing. For example, if the limits are 48 by 36, the drawing scale is 1:1, and to plot the drawing on a 48" by 36" size sheet, then the **LTSCALE** factor is 1. If you check the specification of the Hidden linetype in the *acad.lin* file, the length of each dash is 0.25. Hence, when you plot a drawing with 1:1 scale, the length of each dash in a hidden line is 0.25.

However, if the drawing scale is 1/8" = 1' and you want to plot the drawing on 48" by 36" paper, the **LTSCALE** factor must 8 X 12 = 96. The length of each dash in the hidden line will increase by a factor of 96, because the **LTSCALE** factor is 96. Therefore, the length of each dash will be (0.25 X 96 = 24) units. At the time of plotting, the scale factor must be 1:96 to plot the 384' by 288' drawing on 48" by 36" paper. Each dash of the hidden line that was 24" long on the drawing will be 24/96 = 0.25" long when plotted. Similarly, if the desired text size on the paper is 1/8", the text height in the drawing must be 1/8 X 96 = 12".

LTSCALE factor for PLOTTING = Drawing Scale

Sometimes your plotter may not be able to plot a 48" by 36" drawing, or you might like to decrease the size of the plot so that the drawing fits within a specified area. To get the correct dash lengths for hidden, center, or other lines, you must adjust the **LTSCALE** factor. For example, if you want to plot the previously mentioned drawing in a 45" by 34" area, the correction factor is:

Correction factor	= 48/45
	= 1.0666
New **LTSCALE** factor	= **LTSCALE** factor x Correction factor

$$= 96 \times 1.0666$$
$$= 102.4$$

New **LTSCALE** factor for PLOTTING = Drawing Scale x Correction Factor

Note
*If you change the **LTSCALE** factor, all lines in the drawing are affected by the new ratio.*

Changing Linetype Scale Using the PROPERTIES Command

You can also change the linetype scale of an object by using the **PROPERTIES** command. When you invoke this command, AutoCAD will display the **Properties** palette. All the properties of the selected objects are displayed in the **Properties** palette. To change the current linetype scale, select the objects and then invoke the **Properties** command to display the **Properties** palette. In this window, locate the **Linetype scale** edit box in the **General** list and enter the new linetype scale in the edit box. The linetype scale of the selected objects is changed to the new value you entered.

WORKING WITH THE DESIGNCENTER

Ribbon:	View > Palettes > DesignCenter	**Command:**	ADCENTER
Toolbar:	Standard > DesignCenter		

The **DesignCenter** allows you to reuse and share contents in different drawings. You can use this window to locate the drawing data with the help of search tools and then use it in your drawing. You can insert layers, linetypes, blocks, layouts, external references, and other drawing content in any number of drawings. Hence, if a layer is created once it can be repeatedly used any number of times.

Open the **DesignCenter** palette and choose the **Tree View Toggle** button to display the **Tree pane** and the **Palette** side by side (if they are not already displayed). Open the folder, in which the drawing is saved containing the layers and linetypes you want to insert. You can use the **Load** and the **Up** buttons to open the folder you want. For example, you want to insert some layers and linetypes from the drawing *c4d1* stored in the *D:\AutoCAD 2010\c04_acad_2010* folder. Browse to the folder and open *c4d1* to display its contents. Select **Layers** and all the layers created in *c4d1* are displayed in the Palette. Press and hold the CTRL key down and select the **Border** and **CEN** layers. Right-click to display the shortcut menu and choose **Add Layer[s]**, as shown in Figure 4-31. The two layers are added to the current drawing. You can also drag and drop the desired layers into the current drawing. If you open the drop-down list in the **Layers** toolbar, you will notice that the two layers are listed there. Similarly, you can insert the other elements such as linetypes, blocks, dimension styles, and so on. You will learn more about **DesignCenter** in Chapter 6.

DRAFTING SETTINGS DIALOG BOX

Command:	DSETTINGS

You can use the **Drafting Settings** dialog box to set the drawing modes such as Grid, Snap,

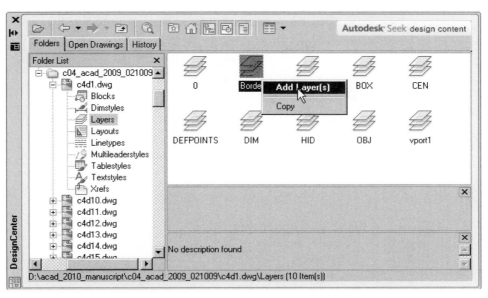

Figure 4-31 *The **DesignCenter** with the shortcut menu*

Object Snap, Polar, Object Snap tracking, and Dynamic Input. All these aids help you to draw accurately and also increase the drawing speed. You can right-click on the **Snap Mode**, **Grid Display**, **Ortho Mode**, **Polar Tracking**, **Object Snap**, **Object Snap Tracking**, **Allow/Disallow Dynamic UCS**, **Dynamic Input**, or **Quick Properties** buttons on the Status Bar to display a shortcut menu, as shown in Figure 4-32. From this shortcut menu, choose **Settings** to display the **Drafting Settings** dialog box, as shown in Figure 4-33. This dialog box operates under five main tabs: **Snap and Grid**X, **Object Snap**, **Polar Tracking**, **Dynamic Input**, and **Quick Properties**. On starting AutoCAD, you are provided with the default settings of these aids. You can change them according to your requirements by using the **Drafting Settings** dialog box, shown in Figure 4-33.

Figure 4-32 Settings in the shortcut menu

Setting Grid

The grid lines are dotted lines on the screen at predefined spacing, see Figure 4-34. These dotted lines act as a graph that can be used as reference lines in a drawing. You can change the distance between the grid dots as per your requirement. The grid pattern appears within the drawing limits, which helps to define the working area. The grid also gives you a sense of the size of the drawing objects.

Grid On (F7): Turning the Grid On or Off

You can turn the grid on/off by using the **Grid On** check box in the **Drafting Settings** dialog box. You can also turn the grid on or off by choosing the **Grid Display** button in the Status Bar or using the **On** or **Off** option in the shortcut menu, when you right-click on the **Grid Display** button in the Status Bar, or using the **GRID** command. The function key F7 acts as

*Figure 4-33 The **Snap and Grid** tab of the **Drafting Settings** dialog box*

a toggle key for turning the grid on or off. When the grid is turned on after it has been off, the grid is set to the previous grid spacing.

Grid X Spacing and Grid Y Spacing

The **Grid X spacing** and **Grid Y spacing** edit boxes in the **Drafting Settings** dialog box are used to define a desired grid spacing along the *X* and *Y* axes. For example, to set the grid spacing to 0.5 units, enter 0.5 in the **Grid X spacing** and **Grid Y spacing** edit boxes, see Figure 4-35. You can also enter

Figure 4-34 Grid lines

different values for horizontal and vertical grid spacing, see Figure 4-36. If you enter only the grid X spacing and then choose the **OK** button in the dialog box, the corresponding Y spacing value is automatically set to match the X spacing value. Therefore, if you want different X and Y spacing values, you need to set the X spacing first, and then the Y spacing.

Note
Grids are especially effective in drawing when the objects in the drawing are placed at regular intervals.

You can also use the **GRID** command to set the grid from the command line. At the prompt sequence for the **GRID** command, you can use the **Aspect** option to assign a different value to the horizontal and vertical grid spacings.

Tip
The grid and the snap grid (discussed in the next section) are independent of each other. However, you can automatically display the grid lines at the same resolution as that of the snap grid by using the **Snap** option of the **GRID** command. When you use this option, AutoCAD will automatically change the grid spacing to zero and display the grid lines at the same resolution as set for Snap. Therefore, in the **Drafting Settings** dialog box if the grid spacing is specified as zero, it automatically adjusts to equal the Snap resolution.

In the **GRID** command, to specify the grid spacing as a multiple of the Snap spacing, enter X after the value (2X).

Figure 4-35 Controlling grid spacing

Figure 4-36 Creating unequal grid spacing

Setting Snap

The snap is used to set increments for the cursor movement. While moving the cursor, it is sometimes difficult to position a point accurately. The **SNAP** command allows you to set up an invisible grid, see Figure 4-37, that allows the cursor to move in fixed increments from one snap point to another. The snap points are the points where the invisible snap lines intersect. The snap spacing is independent of the grid spacing, and so the two can have equal or different values. You generally set snap to an increment of the grid setting, for example, Snap=2 and Grid=10.

Figure 4-37 Invisible Snap grid

Snap On

You can turn the snap on/off and set the snap spacing from the **Drafting Settings** dialog box. As in the case of GRID, the Snap On/Off selection turns the invisible snap grid on or off and moves the cursor by increments. If the SNAP is off, the cursor will not move by increments. It will move freely as you move the pointing device. When you turn the snap off, AutoCAD remembers the value of the snap, and this value is restored when you turn the snap back on. You can also turn the snap on or off by choosing the **Snap** button in the Status Bar, or from the shortcut menu displayed by right-clicking on the Snap button in the Status Bar, or by using the function key F9 as a toggle key for turning the snap on or off.

Snap Type

There are two Snap types, **PolarSnap** and **Grid snap**. **Grid snap** snaps along the grid and is either of **Rectangular snap** or **Isometric snap**. Rectangular is the default standard snap.

Isometric Snap/Grid

You can select the **Isometric snap** radio button in the **Snap type** area, shown in Figure 4-38, of the **Drafting Settings** dialog box to set the snap grid to isometric mode. The default is off (standard). The isometric mode is used to make isometric drawings. In isometric drawings, the isometric axes are at angles of 30, 90, and 150-degree. The Isometric snap/grid enables you to display the grid lines along these axes, see Figure 4-39. Once you select the **Isometric snap** radio button and choose **OK** in the dialog box, AutoCAD automatically changes the cursor to align with the isometric axis. You can adjust the cursor orientation when you are working on the left, top, or right plane of the drawing by using the F5 key or holding down the CTRL key and then pressing the E key to cycle the cursor through different isometric planes (left, right, top). You can change the vertical snap and grid spacing by entering values in the **Snap Y spacing** and **Grid Y spacing** edit boxes. You will notice that the X spacing is not available for this option.

Figure 4-38 Selecting *Isometric snap* in the *Snap type* area

Figure 4-39 Isometric snap grid

PolarSnap

You can use the polar snap with polar tracking. Here the cursor snaps to points at a specified distance along the Polar alignment angles. You can select the **PolarSnap** radio button in the **Snap type** area to set the snap grid to the polar mode. When you choose this option, the normal snap options are not available. You can enter a desired distance in the **Polar distance** edit box. If this value is zero, it assumes the same value as the Snap X spacing. The cursor snaps along an imaginary path on the basis of Polar tracking angles, relative to the last point selected or acquired. The angles can be set in the **Polar Tracking** tab of the **Drafting Settings** dialog box (discussed later in the chapter).

For example, select the **PolarSnap** radio button, enter 0.5 in the **Polar distance** edit box, and then choose the **OK** button. Choose the **Polar Tracking** button in the Status Bar for polar tracking. Invoke the **LINE** command and select the starting point anywhere on the screen. Now move the cursor; an imaginary path will be displayed at 0, 90, 180, and 270 angles (polar tracking) with small cross marks displayed on this imaginary line at a distance of 0.5 (polar snap).

Note

*The polar snap works along with polar and object tracking. Hence, it is used when the polar tracking is on by choosing the **Polar Tracking** button on the Status Bar or by selecting the **Polar Tracking On** check box in the **Polar Tracking** tab of the **Drafting Settings** dialog box.*

*The snap type is also controlled by the **SNAPTYPE** system variable.*

DRAWING STRAIGHT LINES USING THE ORTHO MODE

You can turn the Ortho mode on or off by choosing the **Ortho Mode** button in the Status Bar, or by using the function key F8, or using the **ORTHO** command. The Ortho mode allows you to draw lines at right angles only. Whenever you use the pointing device to specify the next point, the movement of the rubber-band line connected to the cursor is either horizontal (parallel to the *X* axis) or vertical (parallel to the *Y* axis). To draw a line in the Ortho mode, specify the starting point at the **Specify first point** prompt. To specify the second point, move the cursor with the pointing device and specify a desired point. The line drawn will be either vertical or horizontal, depending on the direction, in which you moved the cursor, see Figures 4-40 and 4-41.

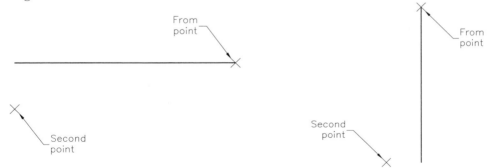

Figure 4-40 Drawing a horizontal line using the ORTHO mode

Figure 4-41 Drawing a vertical line using the ORTHO mode

Tip

You can use the Status Bar at the bottom of the graphics area to easily and conveniently toggle between on or off for the different drafting functions like the Snap, Grid, and Ortho.

Chapter 4

WORKING WITH OBJECT SNAPS

Toolbar: Object Snap

Object snaps are one of the most useful features of AutoCAD. They improve your performance and the accuracy of your drawing and make drafting much simpler than it normally would be. The term object snap refers to the cursor's ability to snap exactly to a geometric point on an object. The advantage of using object snaps is that you do not have to specify an exact point. For example, to place a point at the midpoint of a line, you may not be able to specify the exact point. Using the **MIDpoint** object snap, all you do is to move the cursor somewhere on the object. You will notice a marker (in the form of a geometric shape, a triangle for Midpoint) is automatically displayed at the middle point (snap point). You can click to place a point at the position of the **marker**. You also have a **tooltip** with the object snap marker. When you place the cursor on the marker, a tooltip with the name of the object snap will be displayed. The object snaps recognize only the objects that are visible on the screen, which include the objects on locked layers. The objects on the layers that are turned off or frozen are not visible, and so they cannot be used for object snaps. Object snaps can be invoked from the shortcut menu that is displayed when you press and hold the SHIFT key down and right-click, see Figure 4-42 or from the **Object Snap** toolbar shown in Figure 4-43.

*Figure 4-42   The **Object snap** modes shortcut menu*

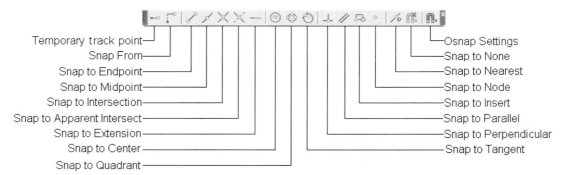

*Figure 4-43   The **Object Snap** toolbar*

You can invoke the shortcut menu to access the Object Snaps by holding the SHIFT key and then right-clicking in the drawing window. You can also invoke this shortcut menu when you are inside any other sketching command. For example, invoke the **CIRCLE** command and then right-click to invoke the shortcut menu. From the shortcut menu, choose the **Snap Overrides** option to display the **Object Snaps** shortcut menu. The following are the object snap modes in AutoCAD.

ENDpoint	CENter	PERpendicular	NEArest
MIDpoint	QUAdrant	PARallel	NONe
INTersection	TANgent	INSert	From
APParent Intersection	EXTension	NODe	Midpoint Between two points

AutoSnap

The AutoSnap feature controls the various characteristics for the object snap. As you move the target box over the object, AutoCAD displays the geometric marker corresponding to the shapes shown in the **Object Snap** tab of the **Drafting Settings** dialog box. You can change the different AutoSnap settings such as attaching a target box to the cursor when you invoke any object snap, or change the size and color of the marker. These settings can be changed from the **Options** dialog box (**Drafting** tab). You can invoke the **Options** dialog box from the **Application Menu**, or from the shortcut menu when you right-click when no command is active, or by entering **OPTIONS** at the Command prompt. You can also invoke it by selecting the **Options** button in the **Drafting Settings** dialog box. When you choose the **Drafting** tab in the **Options** dialog box, the **AutoSnap Settings** are displayed, see Figure 4-44. You can select the **Marker** check box to toggle the display of the marker. You can use the **Magnet** check box to toggle the magnet that snaps the crosshairs to the particular point of the object for that object snap. You can use the **Display AutoSnap tooltip** and **Display AutoSnap aperture**

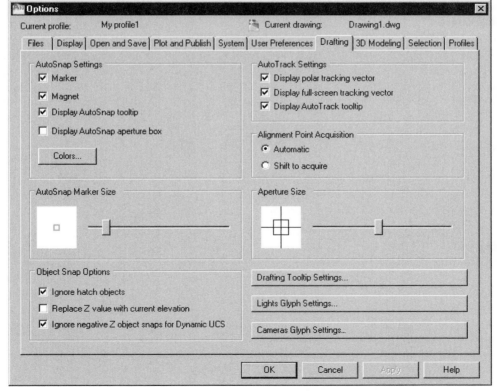

Figure 4-44 The **Drafting** tab of the **Options** dialog box

Chapter 4

box check boxes to toggle the display of the tooltip and the aperture box. Selecting the **Display AutoSnap tooltip** check box shows a flag that gives the name of the object snap that AutoCAD has detected. You can change the size of the marker and the aperture box by moving the **AutoSnap Marker Size** and **Aperture Size** slider bars, respectively. You can also change the color of the markers through the **Colors** button. The size of the aperture is measured in pixels, short for picture elements. Picture elements are dots that make up the screen picture. The aperture size can also be changed using the **APERTURE** command. In AutoCAD, the default value for the aperture size is 10 pixels. The display of the marker and the tooltip is controlled by the **AUTOSNAP** system variable. The following are the bit values for **AUTOSNAP**.

Bit Values	Functions
0	Turns off the Marker, AutoSnap Tooltip, and Magnet
1	Turns on the Marker
2	Turns on the AutoSnapTooltip
4	Turns on the Magnet
47	Turns on the Marker, AutoSnap Tooltip, and Magnet

The basic functionality of Object Snap modes is discussed next.

Endpoint

The **ENDpoint** Object Snap mode snaps to the closest endpoint of a line or an arc. To use this Object Snap mode, select the Endpoint button, and move the cursor (crosshairs) anywhere close to the endpoint of the object. The marker will be displayed at the endpoint; click to specify that point. AutoCAD will grab the endpoint of the object. If there are several objects near the cursor crosshairs, AutoCAD will grab the endpoint of the object that is closest to the crosshairs, or if the Magnet is on, you can move to grab the desired endpoint. For Figure 4-45, invoke the **LINE** command from the **Draw** toolbar. The following is the prompt sequence:

Specify first point: *Select the **Snap to Endpoint** button from the **Object Snap** toolbar.*
_endp of *Move the crosshair and select the arc.*
Specify next point or [Undo]: *Select the endpoint of the line.*

Midpoint

The **MIDpoint** Object Snap mode snaps to the midpoint of a line or an arc. To use this Object Snap mode, select Midpoint osnap and select the object anywhere. AutoCAD will grab the midpoint of the object. For Figure 4-46, invoke the **LINE** command from the **Draw** toolbar. The following is the prompt sequence:

Specify first point: *Select the starting point of the line.*
Specify next point or [Undo]: *Choose the **Snap to Midpoint** button from the **Object Snap** toolbar.*
_mid of *Move the cursor and select the original line.*

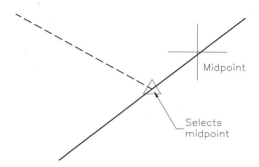

Figure 4-45 *The **ENDpoint** Object Snap mode*   **Figure 4-46** *The **MIDpoint** Object Snap mode*

Nearest

The **NEArest** Object Snap mode selects a point on an object (line, arc, circle, or ellipse) that is visually closest to the graphics cursor (crosshairs). To use this mode, enter the command, and then choose the Nearest object snap. Move the crosshairs near the intended point on the object so as to display the marker at the desired point and then select the object. AutoCAD will grab a point on the line where the marker was displayed. For Figure 4-47, invoke the **LINE** command from the **Draw** toolbar. The following is the prompt sequence.

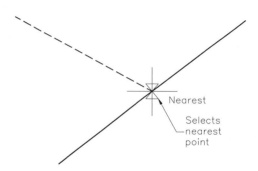

Figure 4-47 *The **NEArest** Object Snap mode*

> Specify first point: *Choose the **Snap to Nearest** button from the **Object Snap** toolbar.*
> _nea to *Select a point near an existing object.*
> Specify next point or [Undo]: *Select endpoint of the line.*

Center

The **CENter** Object Snap mode allows you to snap to the center point of an ellipse, circle, or arc. After selecting this option, you must point to the visible part of the circumference of a circle or arc. For Figure 4-48, invoke the **LINE** command from the **Draw** toolbar. The following is the prompt sequence.

> Specify first point: *Choose the **Snap to Center** button from the **Object Snap** toolbar.*
> _cen of *Move the cursor and select the circle.*
> Specify next point or [Undo]: *Select the endpoint of the line.*

Tangent

The **TANgent** Object Snap allows you to draw a tangent to or from an existing ellipse, circle, or arc. To use this object snap, place the cursor on the circumference of the circle or arc to select it. For Figure 4-49, invoke the **LINE** command from the **Draw** toolbar. The prompt sequence is given next.

Specify first point: *Select the starting point of the line.*
Specify next point or [Undo]: *Choose the **Snap to Tangent** button from the **Object Snap** toolbar.*
_tan to *Move the cursor and select the circle.*
Specify next point or [Undo]: *Select the endpoint of the line (tangent of the circle).*

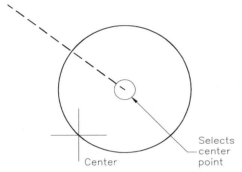

Figure 4-48 *The **CENter** Object Snap mode*

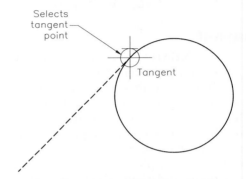

Figure 4-49 *The **TANgent** Object Snap mode*

Note
If the start point of a line is defined using the Tangent Object Snap, the tip shows Deferred Tangent. However, if you end the line using this Object Snap, the tip shows Tangent.

Figure 4-50 shows the use of **NEArest**, **ENDpoint**, **MIDpoint**, and **TANgent** Object Snap modes.

Quadrant

The **QUAdrant** Object Snap mode is used when you need to snap to a quadrant point of an ellipse, arc, or a circle. A circle has four quadrants, and each quadrant subtends an angle of 90-degree. The

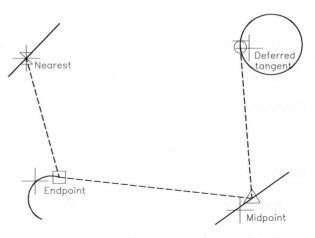

Figure 4-50 *Using the **NEArest**, **ENDpoint**, **MIDpoint**, and **TANgent** Object Snap modes*

quadrant points are located at 0, 90, 180, and 270-degree positions. If the circle is inserted as a block (see Chapter 16), that is rotated, the quadrant points are also rotated by the same amount, see Figures 4-51 and 4-52.

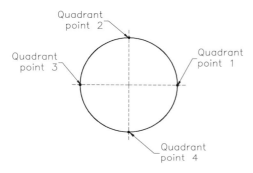

Figure 4-51 *Location of the circle quadrants*

Figure 4-52 *Quadrants in a rotated circle*

To use this object snap, position the cursor on the circle or arc closest to the desired quadrant. The prompt sequence for drawing a line from the third quadrant of a circle, as shown in Figure 4-53, is given next.

Specify first point: *Choose the **Snap to Quadrant** button from the **Object Snap** toolbar.*
_qua of *Move the cursor close to the third quadrant of the circle and select it.*
Specify next point or [Undo]: *Select the endpoint of the line.*

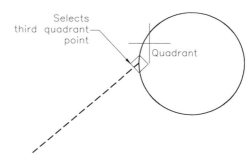

Figure 4-53 *The **QUAdrant** Object Snap mode*

Intersection

The **INTersection** Object Snap mode is used to snap to a point where two or more lines, circles, ellipses, or arcs intersect. To use this object snap, move the cursor close to the desired intersection so that the intersection is within the target box, and then specify that point. For Figure 4-54, invoke the **LINE** command. The prompt sequence is given next.

Specify first point: *Choose the **Snap to Intersection** button from the **Object Snap** toolbar.*
_ int of *Position the cursor near the intersection and select it.*
Specify next point or [Undo]: *Select the endpoint of the line.*

After selecting the **Intersection** Object Snap, if your cursor is close to an object and not close to an actual intersection, the tooltip displays **Extended Intersection**. If you select this object now, AutoCAD prompts **and**, for the selection of another object. If your cursor is close to another object, AutoCAD marks the extended intersection point between these two objects. This mode selects extended or visual intersections of lines, arcs, circles, or ellipses (Figure 4-55).

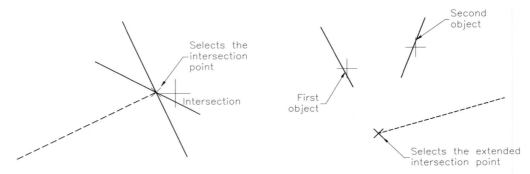

Figure 4-54 *The **INTersection** Object Snap mode*   ***Figure 4-55*** *The **Extended Intersection** Object Snap mode*

The extended intersections are the intersections that do not exist at present, but are imaginary and formed if the line or arc is extended.

Apparent Intersection

The **APParent Intersection** Object Snap mode selects projected or visual intersections of two objects in 3D space. Sometimes, two objects appear to intersect one another in the current view, but in 3D space the two objects do not actually intersect. The **Apparent Intersection** snap mode selects such visual intersections. This mode works on wireframes in 3D space (See Chapter 25 for wireframe models). If you try to use this object snap mode in 2D, it works like an extended intersection.

Perpendicular

The **PERpendicular** Object Snap mode is used to draw a line perpendicular to or from another line, or normal to or from an arc or circle, or to an ellipse. When you use this mode and select an object, AutoCAD calculates the point on the selected object so that the previously selected point is perpendicular to the line. The object can be selected by positioning the cursor anywhere on the line. First invoke the **LINE** command; the prompt sequence to draw a line perpendicular to a given line (Figure 4-56) is given next.

Specify first point: *Select the starting point of the line.*
Specify next point or [Undo]: *Choose the **Snap to Perpendicular** button from the **Object Snap** toolbar.*
_per to   *Select the line on which you want to draw perpendicular.*

When you select the line first, the rubber-band feature of the line is disabled. The line will appear only after the second point is selected. Invoke the **LINE** command. The prompt sequence for drawing a line perpendicular from a given line (Figure 4-57) is given next.

Specify first point: *Choose the **Snap to Perpendicular** button from the **Object Snap** toolbar.*
_per to   *Select the line on which you want to draw perpendicular.*
Specify next point or [Undo]: *Select the endpoint of the line.*

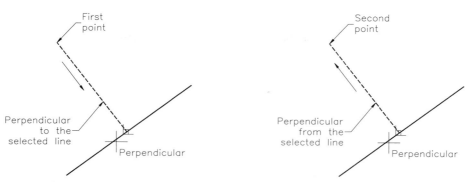

Figure 4-56 *Selecting the start point and then the perpendicular snap*

Figure 4-57 *Selecting the perpendicular snap first*

Figure 4-58 shows the use of the various object snap modes.

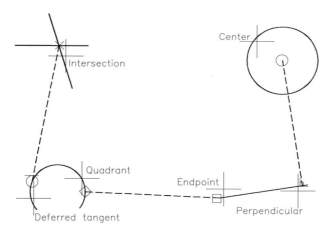

Figure 4-58 *Using various object snap modes to locate points*

Exercise 3   *General*

Draw the sketch shown in Figure 4-59. P1 and P2 are the center points of the top and bottom arcs. The space between the dotted lines is 1 unit.

Node

You can use the **NODe** Object Snap to snap to a point object drawn using the **POINT** command, or placed using the **DIVIDE** or **MEASURE** commands. In Figure 4-60, three points have been drawn using the AutoCAD **POINT** command. You can snap to these points by using the NODe snap mode. Invoke the **LINE** command and the prompt sequence is as follows (Figure 4-61).

Chapter 4

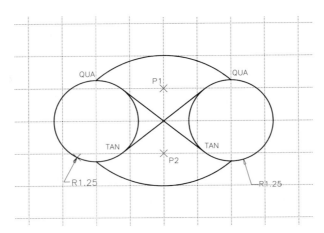

Figure 4-59 *Drawing for Exercise 3*

Specify first point: *Choose the* **Snap to Node** *button from the* **Object Snap** *toolbar.*
_nod of *Select point P1.*
Specify next point or [Undo]: *Choose the* **Snap to Node** *button from the* **Object Snap** *toolbar.*
_nod of *Select point P2.*
Specify next point or [Undo]: *Choose the* **Snap to Node** *button from the* **Object Snap** *toolbar.*
_nod of *Select point P3.*

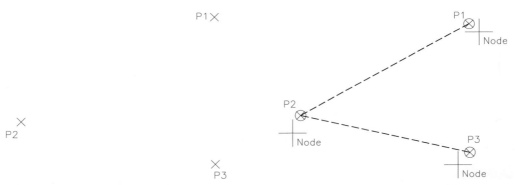

Figure 4-60 *Three Point objects*   **Figure 4-61** *Using* **NODe** *object snap*

Insert

The **INSert** Object Snap mode is used to snap to the insertion point of a text, shape, block, attribute, or attribute definition. In Figure 4-62, the text **WELCOME** is left-justified and the text **AutoCAD** is center-justified. The point, with respect to which the text is justified, is the insertion point of that text string. If you want to snap to these insertion points or the insertion point of a block, you must use the **INSert** Object Snap mode. Invoke the **LINE** command and following is the prompt sequence.

Specify first point: *Choose the **Snap to Insert** button from the **Object Snap** toolbar.*
_ins of   *Select WELCOME text.*
Specify next point or [Undo]: *Choose the **Snap to Insert** button from the **Object Snap** toolbar.*
_ins of   *Select the block.*
Specify next point or [Undo]: *Choose the **Snap to Insert** button from the **Object Snap** toolbar.*
_ins of   *Select AutoCAD text.*

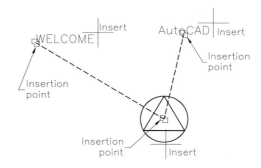

*Figure 4-62 The **INSert** Object Snap mode*

None

The **NONe** Object Snap mode turns off any running object snap (see the section "Running Object Snap Mode" that follows) for one point only. The following example illustrates the use of this Object Snap mode.

Invoke the **Drafting Settings** dialog box and select the **Object Snap** tab. Select the **Midpoint** and **Center** check boxes. This sets the object snaps to Mid and Cen. Now, suppose you want to draw a line whose starting point is a point closer to the endpoint of another line. After you invoke the **LINE** command and move the cursor to the desired position on the previous line, it automatically snaps to its midpoint. You can disable this using the **NONe** object snap by choosing it from the **Object Snap** toolbar. You can select the desired point on the line. Once you have selected a point, it continues the command with the specified object snaps.

Parallel

When you need to draw a line parallel to a line or polyline on the screen, you can use the **PARallel** Object Snap (Figure 4-63). For example, when you are in the middle of the **LINE** command, and you have to draw a line parallel to the one already on the screen, you can use the **PARallel** object snap as follows.

Command: *Choose **Line** from the **Draw** toolbar.*
Specify first point: *Select a point on the screen.*
Specify next point or [Undo]: *Choose the **Parallel** button from the **Object Snap** toolbar.*
_par to *Specify object to which parallel is to be drawn.*

When you specify the reference object, a parallel sign is displayed. You should briefly pause on that line so that a small plus sign appears on it to indicate that it has been selected. Now, on moving the cursor close to an angle parallel to the line, an imaginary parallel line (construction line) appears, on which you can select the next point. A tooltip which has the relative polar coordinates is displayed with the cursor as you move it on the construction line. This helps you to select the next point. This line which has been drawn is parallel to the selected object.

Extension

The **EXTension** Object Snap gives you an option to locate a point on the extension path of a line or an arc (Figure 4-64). It can also be used with Intersection to determine the point of the extended intersection. To use the extension, choose the **Line** button from the **Draw** toolbar and then choose the **Snap to Extension** button from the **Object Snap** toolbar. Briefly pause at the end of the line or arc you want to use. A small plus sign (+) appears at the end of the line or arc, indicating that it has been selected. If you move the cursor along the extension path, a temporary extension path is displayed and the tooltip displays relative polar coordinates from the end of the line. Select a point or enter a distance to begin a line and then select another point to finish the line.

*Figure 4-63 Using the **PARallel** Object Snap mode*   *Figure 4-64 Using the **EXTension** Object Snap mode*

Note
*While using the **PARallel** object snap, you can choose more than one line as reference lines that are indicated by a plus sign. Depending upon the direction in which you move the cursor, AutoCAD will choose any one of these reference lines. The line that is chosen gets marked by the parallel symbol in place of the plus sign.*

*Similarly, while using the **EXTension** object snap, you can choose more than one endpoint as the reference.*

From

The **From** Object Snap mode can be used to locate a point relative to a given point (Figure 4-65). For example, to locate a point that is 2.5 units up and 1.5 units right from the endpoint of a given line, you can use the **From** object snap as follows.

Command: **LINE**
Specify first point: *Choose the **Snap From** button from the **Object Snap** toolbar.*

*Figure 4-65 Using the **From** Object Snap mode to locate a point*

_from Base point: *Choose the **Snap to Endpoint** button from the **Object Snap** toolbar.*
_endp of *Specify the endpoint of the given line*
<Offset>: **@1.5,2.5**

Note
*The **From** Object Snap cannot be used as the running object snap.*

Midpoint Between 2 Points

This Object Snap mode allows you to select the midpoint of an imaginary line drawn between two selected points. Note that this Object Snap mode can only be invoked from the shortcut menu. To understand the working of this Object Snap mode, refer to the sketch shown in Figure 4-66.

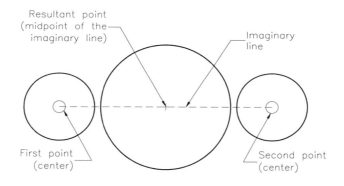

*Figure 4-66 Using the **Midpoint Between 2 Points** Object Snap
mode to locate a point*

In this sketch, there are two circles and you need to draw another circle with the center point at the midpoint of an imaginary line drawn between the center points of the two existing circles. The following is the prompt sequence.

Command: **CIRCLE** Enter
Specify center point for circle or [3P/2P/Ttr (tan tan radius)]: *Right-click and choose **Snap Overrides > Midpoint Between 2 Points** from the shortcut menu.*
_m2p First point of mid: *Select the left circle.*
Second point of mid: *Select the right circle.*
Specify radius of circle or [Diameter] <current>: *Specify the radius of the new circle.*

Temporary Tracking Point

The Temporary Tracking Point can be used to locate a point with respect to two different points. When you select a point that is indicated by a plus sign, depending on the direction in which you move the cursor, an orthogonal imaginary line is displayed either horizontally or vertically. Then you can select another point to display another orthogonal imaginary line in the other direction. The desired point will be located where these two imaginary lines intersect. You can use this option along with the other Object Snap modes to locate a point. For Figure 4-67, invoke the **LINE** command from the **Draw** toolbar and then use the temporary tracking as follows.

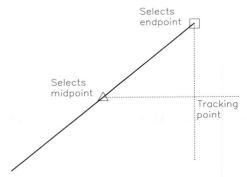

Figure 4-67 Using temporary tracking point

Specify first point: *Choose the **Temporary Tracking Point** button from the **Object Snap** toolbar.*
Specify temporary OTRACK point: *Choose the **Snap to Midpoint** button from the **Object Snap** toolbar.*
Select the midpoint of the line and move the cursor horizontally toward the right.
Specify first point: *Choose the **Temporary Tracking Point** button from the **Object Snap** toolbar.*
Specify temporary OTRACK point: *Choose the **Snap to Endpoint** button from the **Object Snap** toolbar.*
Select the upper endpoint of the line and move the cursor vertically down.

As you move your cursor down, both the horizontal and vertical imaginary lines are displayed. Select the point at their intersection point and then draw the line.

Combining Object Snap Modes

You can also combine the snaps from the command line by separating the snap modes with a comma. AutoCAD will search for the specified modes and grab the point on the object that is closest to the point where the object is selected. The prompt sequence for using the Midpoint and Endpoint object snaps is given next.

Command: **LINE** [Enter]
Specify first point: **MID, END** [Enter]   *(MIDpoint or ENDpoint object snap.)*
Select the object.

Note
In the discussion of object snaps, "line" generally includes xlines, rays, and polyline segments, and "arc" generally includes polyarc segments.

Exercise 4	*General*

Draw the sketch shown in Figure 4-68. The space between the dotted lines is 1 unit.

RUNNING OBJECT SNAP MODE

Toolbar:	Object Snap > Osnap Settings	**Command:**	OSNAP

In the previous sections, you have learned to use the object snaps to snap to different

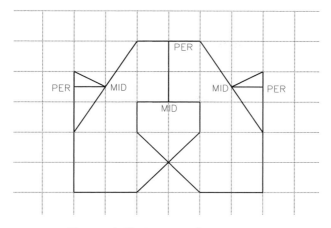

Figure 4-68 *Drawing for Exercise 4*

points of an object. One of the drawbacks of these object snaps is that you have to select them every time you use them, even if it is the same snap mode. This can be solved by using **running object snaps**. The Running Osnap can be invoked from the **Object Snap** tab of the **Drafting Settings** dialog box (Figure 4-69). If you choose the **Object Snap Settings** button from the toolbar, or enter **OSNAP** at the Command line, or choose **Settings** from the shortcut menu displayed when you right-click on the **OSNAP** button in the Status Bar, the **Object Snap** tab is displayed in the **Drafting Settings** dialog box. In this tab, you can set the running object snap modes by selecting the check boxes next to the snap modes. For example, to set

Figure 4-69 *The **Drafting Settings** dialog box (**Object Snap** tab)*

Endpoint as the running Object Snap mode, select the **Endpoint** check box and then the **OK** button in the dialog box.

Once you set the running Object Snap mode, you are automatically in that mode and the marker is displayed when you move the crosshairs over the snap points. If you had selected a combination of modes, AutoCAD selects the mode that is closest to the screen crosshairs. For example, you have selected the **Endpoint**, **Midpoint**, and **Center** check boxes in the dialog box. Now for selection of a point when you move the cursor over a previously drawn line and whenever your cursor is closer to the midpoint of the line, the marker is displayed there. You can also move the cursor to the end of the line to invoke the Endpoint object snap. If you place the cursor over the circumference of a previously drawn circle, the Center osnap is invoked. The running Object Snap mode can be turned on or off, without losing the object snap settings, by choosing the **OSNAP** button in the Status Bar. You can also accomplish this by pressing the function key F3 or CTRL+F keys.

Note
When you use a temporary object snap and fail to specify a suitable point, AutoCAD issues an Invalid point prompt and reissues the pending command's prompt. However, with a running Object Snap mode, if you fail to specify a suitable point, AutoCAD accepts the point specified without snapping to an object.

*The running osnap works only when it is turned on by using the **Osnap** button in the Status Bar.*

Overriding the Running Snap
When you select the running object snaps, all other Object Snap modes are ignored unless you select another Object Snap mode. Once you select a different osnap mode, the running OSNAP mode is temporarily overruled. After the operation has been performed, the running OSNAP mode goes into effect again. If you want to discontinue the current running Object Snap modes totally, choose the **Clear all** button in the **Drafting Settings** dialog box. If you want to temporarily disable the running object snap, choose the **OSNAP** button (off position) in the Status Bar.

If you override the running object snap modes for a point selection and no point is found to satisfy the override Object Snap mode, AutoCAD displays a message to this effect. For example, if you specify an override Object Snap mode of Center and no circle, ellipse, or arc is found at that location, AutoCAD will display the message "**No center found for specified point. Point or option keyword required**".

Cycling through Snaps
AutoCAD displays the geometric marker corresponding to the shapes shown in the **Object snap settings** tab of the **Drafting Settings** dialog box. You can use the TAB key to cycle through the snaps. For example, if you have a circle with an intersecting rectangle as shown in Figure 4-70 and you want to snap to one of the geometric points on the circle, you can use the TAB key to cycle through the geometric points. The geometric points for a circle are the center point, quadrant points, and the intersecting points with the rectangle. To snap to

one of these points, the first thing you need to do is to set the running object snaps (center, quadrant, and intersection object snaps) in the **Drafting Settings** dialog box. After entering a command, when you drag the cursor over the objects, AutoSnap displays a marker and a SnapTip. You can cycle through the snap points for an object by pressing the TAB key. For example, if you press the TAB key while the aperture box is on the circle and the rectangle (near the lower left intersection point), AutoSnap will display the intersection, center, and quadrant points one by one. When a certain osnap is displayed, it's pertaining object is highlighted by making it dashed. For example, when the center or quadrant snaps are displayed the circle becomes

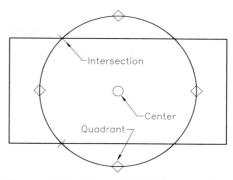

Figure 4-70 *Using the TAB key to cycle through the snaps*

dashed, and when the intersection snap is displayed, both the circle and the rectangle become dashed. By selecting a point, you can snap to one of these points.

Setting the Priority for Coordinate Entry

Sometimes you may want the keyboard entry to take precedence over the running Object Snap modes. This is useful when you want to locate a point that is close to the running osnap. By default, when you specify the coordinates (by using the keyboard) of a point located near a running osnap, AutoCAD ignores the point and snaps to the running osnap. For example, you select the **Intersection** object snap in the **Object Snap** tab of the **Drafting Settings** dialog box. Now, if you enter the coordinates of the endpoint of a line very close to the intersection point (Figure 4-71a), the line will snap to the intersection point and ignore the keyboard entry point. You can set the priority between the keyboard entry and object snap through the **User Preferences** tab of the

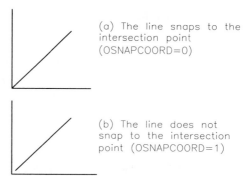

(a) The line snaps to the intersection point (OSNAPCOORD=0)

(b) The line does not snap to the intersection point (OSNAPCOORD=1)

Figure 4-71 *Setting the coordinate entry priority*

Options dialog box, see Figure 4-72. Select the **Keyboard entry** radio button and then choose **OK**. Now, if you enter the coordinates of the endpoint of a line very close to the intersection point (Figure 4-71b), the starting point will snap to the coordinates specified and the running osnap (intersection) is ignored. This setting is stored in the **OSNAPCORD** the system variable and you can also set the priority through this variable. A value **0** gives priority to running osnap, value **1** gives priority to the keyboard entry, and value **2** gives priority to the keyboard entry except scripts.

Tip
*If you do not want the running osnap to take precedence over the keyboard entry, you can simply disable the running osnap temporarily by choosing the **OSNAP** button on the Status*

Chapter 4

*Bar to the off position. This lets you specify a point close to a running osnap. This way you do not have to change the settings for coordinate entry in the **Options** dialog box.*

Figure 4-72 *Setting the priority for coordinate data entry*

USING AUTOTRACKING

When using AutoTracking, the cursor moves along temporary paths to locate key points in a drawing. It can be used to locate points with respect to other points or objects in the drawing. There are two types of AutoTracking options: **Object Snap Tracking** and **Polar Tracking**.

Object Snap Tracking

Object Snap Tracking tracks the movement of the cursor along the alignment paths based on the Object Snap points (running osnaps) that are selected in the **Object Snap** tab of the **Drafting Settings** dialog box. Selecting the **Object Snap Tracking On (F11)** check box in the **Object Snap** tab of the **Drafting Settings** dialog box, choosing the **Object Snap Tracking** button on the Status Bar, or using the function key F11 sets the object snap tracking on.

The direction of the path is determined by the motion of the cursor or the point you select on an object. For example, if you want to draw a circle whose center is located in the line with the center of two existing circles (Figure 4-73), you can use AutoTracking. First select **Center** in the **Object Snap** tab of the **Drafting Settings** dialog box to set **Center** as the running osnap. Also choose the Osnap button on the Status Bar to the On position. Now, you can activate the object tracking by choosing the **Object Snap Tracking** button on the Status Bar. Choose the **Circle** button from the **Draw** toolbar. At the next prompt **Specify center point for circle or [3P/2P/Ttr (tan, tan, radius)]**, pause the cursor at the center of the first circle to attain the point (plus sign) and then move it slightly to get the imaginary horizontal line. Move the cursor toward the second circle; the horizontal path disappears. Now, place the cursor briefly at the center of the second circle and move it slightly to get the imaginary vertical line. Move the cursor up along the vertical alignment path and when you are nearly in line with the other circle, the horizontal path is also displayed. Select the intersection of the two alignment paths and it becomes the center of the circle. At the next prompt, **Specify radius of circle or [Diameter] <current>**, enter radius or specify a point.

Similarly, you can use AutoTracking in combination with the **midpoint** object snap to locate the center of the rectangle and then draw a circle (Figure 4-74).

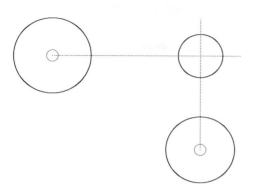

Figure 4-73 *Using **AutoTracking** to locate a point (center of circle)*

Figure 4-74 *Using **AutoTracking** to locate a point (midpoint of rectangle)*

Note
*Object tracking works only when **OSNAP** is on and some running object snaps have been set.*

Polar Tracking

Polar Tracking is used to locate points on an angular alignment path. Polar tracking can be selected by choosing the **Polar Tracking** button on the Status Bar, by using the function key F10, or by selecting the **Polar Tracking On (F10)** check box in the **Polar Tracking** tab of the **Drafting Settings** dialog box (Figure 4-75). Polar Tracking constrains the movement of the cursor along a path that is based on the polar angle settings. For example, if the **Increment angle** list box value is set to 15-degree in the **Polar Angle Settings** area, the cursor will move along the alignment paths that are multiples of 15-degree (0, 15, 30, 45, 60, and so on) and a tooltip will display a distance and angle. Selecting the **Additional angles** check box and choosing the **New** button allows you to add an additional angle value. The imaginary path will also be displayed at these new angles, apart from the increments of the increment angle selected. For example, if the increment angle is set at **15** and you add an additional angle of **22**, the imaginary path will be displayed at 0, 15, 22, 30, 45, and the increments of 15. Polar tracking is on only when the Ortho mode is off.

In the **Drafting Settings** dialog box (**Polar Tracking** tab), you can set the polar tracking to absolute or relative to the last segment. If you select the **Absolute** radio button, which is the default, the base angle is taken from 0. If you select the **Relative to last segment** radio button, the base angle for the increments is set to the last segment drawn. You can also use Polar tracking together with Object tracking (Otrack). You can select the **Track using all polar angle settings** radio button in the dialog box.

AutoTrack Settings

You have different settings while working with autotracking. These can be set in the **Drafting** tab of the **Options** dialog box (Figure 4-76). If you choose the **Options** button in the **Polar Tracking** tab of the **Drafting Settings** dialog box, the **Options** dialog box with the **Drafting** tab open is displayed. You can use the **Display polar tracking vector** check box to toggle the display of the angle alignment path for Polar tracking. You can also use the **Display full-screen**

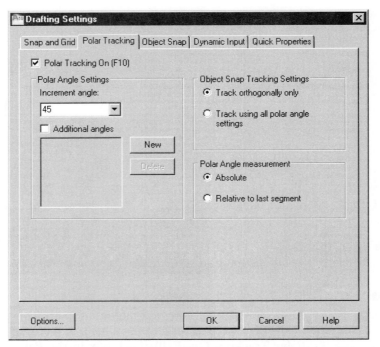

Figure 4-75 *The **Polar Tracking** tab of the **Drafting Settings** dialog box*

tracking vector check box to toggle the display of a full-screen construction line for Otrack. You can use the **Display AutoTrack tooltip** check box to toggle the display of tooltips with the paths. You can also use the **TRACKPATH** system variable to set the path display settings.

Figure 4-76 *The **AutoTrack Settings** area in the **Options** dialog box (**Drafting** tab)*

Tip
*You need the **Options** dialog box quite frequently to change the different drafting settings. When you choose the **Options** button from the **Drafting Settings** dialog box, it directly opens the required tab. After making the changes in the dialog box, choose the **OK** button to get back to the **Drafting Settings** dialog box. The **Options** button is available in all the three tabs of the **Drafting Settings** dialog box.*

FUNCTION AND CONTROL KEYS

You can also use the function and control keys to change the status of the coordinate display, Snap, Ortho, Osnap, tablet, screen, isometric planes, running Object Snap, Grid, Polar, and Object tracking. The following is a list of function and control keys.

F1	Help	F7	Grid On/Off (CTRL+G)
F2	Graphics Screen/AutoCAD Text Window	F8	Ortho On/Off (CTRL+L)
F3	Osnap On/Off (CTRL+F)	F9	Snap On/Off (CTRL+B)

F4 Tablet mode On/Off (CTRL+T)   F10 Polar tracking On/Off
F5 Isoplane top/right/left (CTRL+E)   F11 Object Snap tracking
F6 Coordinate display On/Off (CTRL+D)   F12 Dynamic Input On/Off

Self-Evaluation Test

Answer the following questions and then compare them to those given at the end of this chapter:

1. The layers that are turned off are displayed on the screen but cannot be plotted. (T/F)

2. The drawing, in which you are working, should be in the named plot style mode (*.stb*) to make the plot style available in the **Layer Properties Manager** dialog box. (T/F)

3. The grid pattern appears within the drawing limits, which helps define the working area. (T/F)

4. If the circle is inserted as a rotated block, the quadrant points are not rotated by the same amount. (T/F)

5. You can also change the plot style mode from the command line by using the _____ system variable.

6. The _____ command enables you to set up an invisible grid that allows the cursor to move in fixed increments from one snap point to another.

7. The _____ snap works along with polar and object tracking only.

8. The _____ Object Snap mode selects the projected or visual intersections of two objects in 3D space.

9. The _____ can be used to locate a point with respect to two different points.

10. You can set the priority between the keyboard entry and the object snap through the _____ tab of the **Options** dialog box.

Review Questions

Answer the following questions:

1. You cannot enter different values for horizontal and vertical grid spacing. (T/F)

2   You can lock a layer, which will prevent the user from accidentally editing the objects in that layer. (T/F)

3. When a layer is locked, you cannot use the objects in the locked layer for Osnaps. (T/F)

Chapter 4

4. The thickness given to the objects in a layer, using the **Lineweight** option, is displayed on the screen and is plotted. (T/F)

5. Which of the following options is not displayed in the **Layer** shortcut menu in the **Layer Properties Manager** dialog box when you select more than one layer using the SHIFT key?

     (a) **New Layer**   (b) **Select All**
     (c) **Make Current**   (d) **Clear All**

6. Which of the following function keys acts as a toggle key for turning the grid on or off?

     (a) F5   (b) F6
     (c) F7   (d) F8

7. Which one of the following object snap modes turns off any running object snap for one point only?

     (a) **NODe**   (b) **NONe**
     (c) **From**   (d) **NEArest**

8. Which one of the following object snaps cannot be used as the running object snap?

     (a) **EXTension**   (b) **PARallel**
     (c) **From**   (d) **NODe**

9. Which of the following keys can be used to cycle through the different running object snaps?

     (a) ENTER   (b) SHIFT
     (c) CTRL   (d) TAB

10. While working on a drawing, at any point of time you can save all layers with their present properties' settings under one name and then restore them anytime later using the _____ button in the **Layer Properties Manager** dialog box.

11. You can use the _____ window to locate the drawing data with the help of the search tools and then use it in your drawing.

12. The grid pattern appears within the drawing _____, which helps define the working area.

13. The difference between the **Off** option and the **Freeze** option is that the frozen layers are not _____ by the computer while regenerating the drawing.

14. The size of the aperture is measured in _____ , short for picture elements.

15. Using the **Extension** object snap, moving the cursor along the path displays a temporary extension path and the tooltip displays _____ coordinates from the end of the line.

Exercises

Exercise 5   *Mechanical*

Set up layers with the following linetypes and colors. Then make the drawing, as shown in Figure 4-77. The distance between the dotted lines is 1.0 unit.

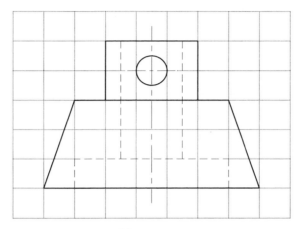

Figure 4-77 *Drawing for Exercise 5*

Layer name	Color	Linetype
Object	Red	Continuous
Hidden	Yellow	Hidden
Center	Green	Center

Exercise 6   *Mechanical*

Set up layers, linetypes, and colors, as given in Exercise 5. Then make the drawing shown in Figure 4-78. The distance between the dotted lines is 1.0 unit.

Exercise 7   *Mechanical*

Set up layers, linetypes, and colors and then make the drawing shown in Figure 4-79. The distance between the dotted lines is 1.0 unit.

Exercise 8   *Mechanical*

Set up layers, linetypes, and colors and then make the drawing shown in Figure 4-80. Use the object snaps as indicated.

Chapter 4

Figure 4-78 *Drawing for Exercise 6*

Figure 4-79 *Drawing for Exercise 7*

Figure 4-80 *Drawing for Exercise 8*

Problem-Solving Exercise 1   *Mechanical*

Draw the object shown in Figure 4-81. First draw the lines and then draw the arcs using appropriate **ARC** command options.

Figure 4-81   *Drawing for Problem-Solving Exercise 1*

Problem-Solving Exercise 2   *Mechanical*

Draw the object shown in Figure 4-82. First draw the front view (bottom left) and then the side and top views. Assume the missing dimensions.

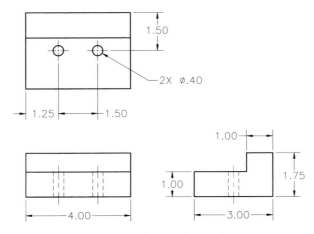

Figure 4-82   *Drawing for Problem-Solving Exercise 2*

Chapter 4

Problem-Solving Exercise 3   *Mechanical*

Draw the object shown in Figure 4-83. First, draw the front view (bottom left) and then the
side and top views. Assume the missing dimensions. **

Figure 4-83 *Drawing for the Problem-Solving Exercise 3*

Answers to Self-Evaluation Test
1. F, **2.** T, **3.** T, **4.** F, **5. PSTYLEPOLICY**, **6.** SNAP, **7.** Polar , **8. Apparent Intersection**,
9. Temporary Tracking, **10. User Preferences**

Chapter 5

Editing Sketched Objects-I

CREATING A SELECTION SET

In Chapter 2, only two options of the selection set were discussed (Window and Crossing). In this chapter, you will learn additional selection set options that can be used to select objects, when certain commands are invoked. The following options are explained here.

Last	CPolygon	Add	Undo	Previous	SUbobject
Fence	BOX	SIngle	ALL	Group	AUto
WPolygon	Remove	Multiple	Object		

Note
The default object selection method for most of the commands is to use the pick box to select one entity at a time. If you click in a blank area using a pick box, the window or the crossing option is invoked.

Last

This option is used to select the most recently drawn object that is partially or fully visible in the current display of the screen. This is the convenient option to select the most recently drawn object that is visible on the screen. Keep in mind that if the last drawn object is not in the current display, the object in the current display that was drawn last will be selected. Although a selection set is being formed using the **Last** option, only one object is selected. However, you can use the Last option a number of times. You can use the Last selection option with any command that requires selection of objects (e.g., **COPY**, **MOVE**, and **ERASE**). After invoking the particular command, enter LAST or L at the **Select objects** prompt to use this object selection method.

Exercise 1 *General*

Using the **LINE** command, draw Figure 5-1(a). Then use the **ERASE** command with the **Last** option to erase the three most recently drawn lines to obtain Figure 5-1(d).

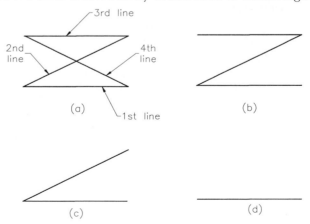

Figure 5-1 *Erasing objects using the **Last** selection option*

Previous

The **Previous** option automatically selects the objects in the most recently created selection set. To invoke this option, you can enter **P** at the **Select objects** prompt. AutoCAD saves the previous selection set and lets you select it again by using this option. In other words, with the help of the **Previous** option, you can edit the previous set without reselecting its objects individually. Another advantage of the **Previous** option is that you need not remember the objects if more than one editing operation has to be carried out on the same set of objects. For example, if you want to copy a number of objects and then move them, you can use the **Previous** option to select the same group of objects with the **MOVE** command. The prompt sequence will be as follows.

> Command: **COPY** [Enter]
> Select objects: *Select the objects.*
> Select objects: [Enter]
> Specify base point or [displacement] <Displacement>: *Specify the base point.*
> Specify second point of displacement or <use first point as displacement>: *Specify the point for displacement.*
>
> Command: **MOVE** [Enter]
> Select objects: **P** [Enter]
> # found

The previous selection set is cleared by various deletion operations and the commands associated with them, like **UNDO**. You cannot select the objects in the model space and then use the same selection set in paper space, or vice versa. This is because AutoCAD keeps the record of the space (paper space or model space), in which the individual selection set is created.

WPolygon

This option is similar to the **Window** option, except that in this option you can define a window that consists of an irregular polygon. You can specify the selection area by specifying points around the object you want to select (Figure 5-2). Similar to the window method, all the objects to be selected using this method should be completely enclosed within the polygon. The polygon is formed as you specify the points and can take any shape except the one that is self-intersecting. The last segment of the polygon is automatically drawn to close the polygon. The polygon can be created by specifying the coordinates of the points or by specifying the points with the help of a

*Figure 5-2 Selecting objects using the **WPolygon** option*

pointing device. With the **Undo** option, the most recently specified WPolygon point can be undone. To use the **WPolygon** option with object selection commands (like **ERASE, MOVE, COPY**), invoke the command, and then enter **WP** at the **Select objects** prompt. The prompt sequence for selecting the objects using the **WPolygon** option is given next.

> Command: *Invoke any command such as **Erase, Move, Copy**, and so on.*
> Select objects: **WP** [Enter]
> First polygon point: *Specify the first point.*
> Specify endpoint of line or [Undo]: *Specify the second point.*
> Specify endpoint of line or [Undo]: *Specify the third point.*
> Specify endpoint of line or [Undo]: *Specify the fourth point.*
> Specify endpoint of line or [Undo]: *Press ENTER after specifying the last point of polygon.*

Exercise 2   *General*

Draw a number of objects on the screen, and then erase some of them using the **WPolygon** option to select the objects you want to erase.

CPolygon

This method of selection is similar to the WPolygon method except that just like a crossing, a **CPolygon** also selects those objects that are not completely enclosed within the polygon, but touch the polygon boundaries. In other words, an object lying partially inside the polygon or even touching it is also selected, in addition to those objects that are completely enclosed within the polygon (Figure 5-3). CPolygon is formed as you specify the points. The points can be specified at the Command line or by specifying points with the pointing device. Just as in the **WPolygon** option, the crossing polygon can take any shape except the one in which it intersects itself. Also, the last segment of the polygon is drawn

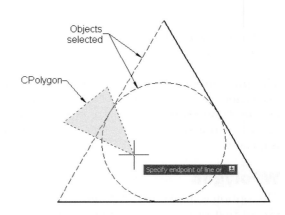

*Figure 5-3 Selecting objects using the **CPolygon** option*

automatically, and so the CPolygon is closed at all times. The prompt sequence for the **CPolygon** option is given next.

> Select objects: **CP** [Enter]
> First polygon point: *Specify the first point.*
> Specify endpoint of line or [Undo]: *Specify the second point.*
> Specify endpoint of line or [Undo]: *Specify the third point.*
> Specify endpoint of line or [Undo]: *Press ENTER after specifying the last point of the polygon.*

Remove

The **Remove** option is used to remove the objects from the selection set (but not from the drawing). After selecting many objects from a drawing by any selection method, there may be a need for removing some of the objects from the selection set. The following prompt sequence displays the use of the **Remove** option for removing the objects from the selection set shown in Figure 5-4.

> Command: *Choose the **Erase** button.*
> Select objects: *Select objects (a), (b), and (c).*
> Select objects: **R**
> Remove objects: *Select object (a).*
> 1 found, 1 removed, 2 total.
> Remove objects: [Enter]

Figure 5-4 Removing objects using the **Remove** *option*

Tip

*The objects can also be removed from the selection set using the SHIFT key. For example, pressing the SHIFT key and selecting object (a) with the pointing device will remove it from the selection set. AutoCAD will display the following message; **1 found, 1 removed, 2 total**.*

Add

You can use the **Add** option to add objects to the selection set. When you begin creating a selection set, you are in the **Add** mode. After you create a selection set by any selection method, you can add more objects by simply selecting them with the pointing device, when the system variable **PICKADD** is set to 1 (default). When it is set to 0, to add objects to the selection set you will have to press the SHIFT key and then select the objects.

ALL

The **ALL** selection option is used to select all the objects in the current working environment of the current drawing. Note that the objects in the "OFF" layers are also selected with the **ALL** option. However, the objects that are in the frozen layers are not selected using this method. You can use this selection option with any command that requires object selection. After invoking the command, the **ALL** option can be entered at the **Select objects** prompt. Once you enter this option, all the objects drawn on the screen will be highlighted (dashed). For example, if there are four objects on the screen and you want to erase all of them, the prompt sequence is given next.

> Command: *Choose the **Erase** button.*
> Select objects: **ALL** [Enter]
> 4 found
> Select objects: [Enter]

You can use this option in combination with the other selection options. For example, consider that there are five objects on the drawing screen and you want to erase three of them. After invoking the **ERASE** command, enter **ALL** at the **Select objects** prompt. Then press the SHIFT key and select the two objects you want to remove from the selection set; the remaining three objects are erased.

Chapter 5

Fence

In the **Fence** option, a selection set is created by drawing an open polyline fence through the objects to be selected. Any object touched by the fence polyline is selected (Figure 5-5). The selection fence can be created by entering the coordinates at the Command line or by specifying the points with the pointing device. With this option, more flexibility for selection is provided because the fence can intersect itself. The **Undo** option can be used to undo the most recently selected fence point. Like the other selection options, this option is also used with the commands that need an object selection. The prompt sequence for using the **Fence** option is given next.

Figure 5-5 *Erasing objects using the **Fence** option*

> Select objects: **F** Enter
> First fence point: *Specify the first point.*
> Specify endpoint of line or [Undo]: *Specify the second point.*
> Specify endpoint of line or [Undo]: *Specify the third point.*
> Specify endpoint of line or [Undo]: *Specify the fourth point.*
> Specify endpoint of line or [Undo]: Enter

Group

The **Group** option enables you to select a group of objects by their group name. You can create a group and assign a name to it with the help of the **GROUP** command (see Chapter 20). Once a group has been created, you can select it using the **Group** option for editing purposes. This makes the object selection process easier and faster, since a set of objects is selected by entering just the group name. The prompt sequence is given next.

> Command: **MOVE** Enter
> Select objects: **G** Enter
> Enter group name: *Enter the name of the predefined group you want to select.*
> 4 found
> Select objects: Enter

Exercise 3   *General*

Draw six circles and select all of them to erase using the **ALL** option of the **ERASE** command. Now change the selection set contents by removing alternate circles from the selection set by using the fence so that alternate circles are erased.

BOX

When the system variable **PICKAUTO** is set to 1 (default), the **BOX** selection option is used to select the objects inside a rectangle. After you enter **BOX** at the **Select objects** prompt, you are required to specify the two corners of a rectangle at the **Specify first corner** and the **Specify opposite corner** prompts. If you define the box from the right to left, it is equivalent to the **CPolygon** selection option. Therefore, it also selects the objects that touch the rectangle boundaries, in addition to the ones that are completely enclosed within the rectangle. If you define the box from the left to right, this option is equivalent to the Window option and selects only those objects that are completely enclosed within the rectangle. The prompt sequence is given next.

Select objects: **BOX** [Enter]
Specify first corner: *Specify a point.*
Specify opposite corner: *Specify the opposite corner point of the box.*
Select objects: [Enter]

AUto

The **AUto** option is used to establish an automatic selection. You can select a single object or a number of objects, by creating a window or a crossing. If you select a single object, it is selected; if you specify a point in the blank area, you are automatically in the **BOX** selection option and the point you have specified becomes the first corner of the box. **Auto** and **Add** are the default selections.

Multiple

On entering **M** (Multiple) at the **Select objects** prompt, you can select multiple objects at a single **Select objects** prompt. The objects will not be highlighted. Once you give a null response to the **Select objects** prompt, all the selected objects are highlighted together.

Undo

This option removes the most recently selected object from the selection set.

SIngle

When you enter **SI** (SIngle) at the **Select objects** prompt, the selection takes place in the SIngle selection mode. Once you select an object or a number of objects using a Window or Crossing option, the **Select objects** prompt is not repeated. AutoCAD proceeds with the command for which the selection is made.

Command: *Choose the **Erase** button.*
Select objects: **SI** [Enter]
Select objects: *Select the object for erasing; the selected object is erased.*

SUbobject

This option is used to select sub-entities such as edges and faces of 3D solids. This option can also be used for selecting individual constituting members of a composite solid created by applying Boolean operations such as union, intersection, and so on. You can use this option on solid models with modify operations such as erase, move, rotate, and so on.

Chapter 5

Tip
Subobjects can also be selected by pressing and holding the CTRL key down when you are prompted to select objects for the selection set.

Object

This option is used to exit the **SUbobject** option and it enables you to select objects as a whole again.

You can also create a selection set using the **SELECT** command. The prompt sequence to do so is given next.

> Command: **SELECT** [Enter]
> Select objects: *Use any selection method.*

EDITING SKETCHES

To use AutoCAD effectively, you need to know the editing commands and how to use them. In this section, you will learn about the editing commands. These commands can be invoked from the Ribbon, toolbar, or entered at the Command prompt. Some of the editing commands such as **ERASE** and **OOPS** have been discussed in Chapter 2 (Getting Started with AutoCAD). The rest of them will be discussed in this section.

MOVING SKETCHED OBJECTS

Ribbon:	Home > Modify > Move	Toolbar:	Modify > Move
Command:	MOVE		

Sometimes the objects are not located at the position where they actually should be. In these situations, you can use the **MOVE** command (Figure 5-6). This command allows you to move one or more objects from their current location to a new location. This change in the location of the objects does not change their size or orientation. On invoking this command, you will be prompted to select the objects to be moved. You can use any of the object selection techniques for selecting one or more objects. Once you have selected the objects, you will be prompted to specify the base point. This base point is the reference point with which the object will be picked and moved. It is

*Figure 5-6 Invoking the **MOVE** command from the **Modify** panel*

advisable to select the base point on the object selected to be moved. Next, you will be prompted to specify the second point of displacement. This is the new location point where you want to move the object. When you specify this point, the selected objects will be moved to this point. Figure 5-7 shows an object moved using the **MOVE** command. The prompt sequence that will be followed when you choose the **Move** button from the **Modify** toolbar is given next.

> Select objects: *Select the objects to be moved.*
> Select objects: [Enter]

Specify base point or [Displacement] <Displacement>: *Specify the base point for moving the selected object(s).*
Specify second point or <use first point as displacement>: *Specify second point or press ENTER to use the first point.*

If you press ENTER at the **Specify second point of displacement or <use first point as displacement>** prompt, AutoCAD interprets the first point as the relative value of the displacement in the X axis and Y axis directions. This value will be added in the X

Base point   Original object after selecting

Second point of displ.   Object after moving to new location

Figure 5-7 *Moving the objects to a new location*

and Y axis coordinates and the object will be automatically moved to the resultant location. For example, draw a circle with its center at (3,3) and then select the center point of the circle as the base point. Now, at the **Specify second point of displacement or <use first point as displacement>** prompt, press ENTER. You will notice that the circle is moved such that its center is now placed at 6,6. This is because 3 units (initial coordinates) are added along both the X and Y directions.

COPYING SKETCHED OBJECTS

Ribbon:	Home > Modify > Copy	Toolbar:	Modify > Copy
Command:	COPY		

The **COPY** command is used to copy an existing object. This command is similar to the **MOVE** command in the sense that it makes copies of the selected objects and places them at specified locations, but the originals are left intact. In this command, you need to select the objects and then specify the base point. Next, you are required to specify the second point. This point is where you want the copied object to be placed. Figure 5-8 shows the objects copied using this command. The prompt sequence that will be followed when you choose **Home > Modify > Copy** from the **Ribbon** is given next.

Select objects: *Select the objects to copy.*
Select objects: Enter
Specify base point or [Displacement/mOde]<Displacement>: *Specify the base point.*
Specify second point or <use first point as displacement>: *Specify a new position on the screen using the pointing device or entering coordinates.*
Specify second point or [Exit/Undo] <Exit>: Enter

Creating Multiple Copies

You can use the **COPY** command to make multiple copies of the same object (Figure 5-9). When you select the second point of displacement, a copy is placed at this point and the prompt is automatically repeated until you press ENTER to exit the **COPY** command. You can continue to specify the points for placing multiple copies of the selected entities. You can use **U (Undo)** to Undo the last copied instance at any stage of the **COPY** command. After

Chapter 5

entering U, you can again specify the position of the last instance. The prompt sequence for
the **COPY** command for multiple copies is given next.

Specify base point or displacement: *Specify the base point.*
Specify second point or <use first point as displacement>: *Specify a point for placement.*
Specify second point or <Exit/Undo>: *Specify another point for placement.*
Specify second point or <Exit/Undo>: *Specify another point for placement.*
Specify second point or <Exit/Undo>: Enter

Figure 5-8 *Copying the object using the COPY command*

Figure 5-9 *Making multiple copies*

Creating a Single Copy

By default, AutoCAD creates multiple copies of the selected
objects. But, you can also create a single copy of the selected
object. To create a single copy of the selected object, choose
the **mOde** option from the shortcut menu at the **Specify base
point or [Displacement/mOde]** <current> prompt. Next,
choose the **Single** option from the shortcut menu at the **Enter
a copy mode option [Single/Multiple]** <current> prompt.
After specifying the second point of displacement, a copy
will be placed at that point and you can now exit the **COPY**
command.

COPYING OBJECTS USING THE BASE POINT

Command:	COPYBASE

The command **COPYBASE** is used to specify the base point
of the objects to be copied. This command can also be invoked
from the shortcut menu that is displayed upon right-clicking.
Choose **Copy with Base Point** to invoke this command, see
Figure 5-10. This command can be very useful while pasting

Figure 5-10 *Invoking the COPYBASE command from the shortcut menu*

an object very precisely in the same diagram or into another diagram. When you invoke this command from the shortcut menu, you will be prompted to specify the base point for copying the objects, and then select the objects. Unlike the **COPY** command, where the objects are dragged and placed at the desired location, this command copies the selected object on to the Clipboard. Then from the Clipboard, the objects are placed at the specified location. The Clipboard is defined as a medium for storing the data while transferring the data from one place to the other. Note that the contents of the Clipboard are not visible. The objects copied using this command can be copied from one drawing file to the other or from one working environment to the other. After copying objects with the **COPYBASE** command, you can paste the objects at the desired place with a greater accuracy.

Note
You will learn more about the working environments and the Clipboard in the following chapters.

PASTING CONTENTS FROM THE CLIPBOARD

Ribbon: Clipboard > Paste > Paste as Block   **Command:** PASTEBLOCK

The **PASTEBLOCK** command is used to paste the contents of the Clipboard into a new drawing or in the same drawing at a new location. You can also invoke the **PASTEBLOCK** command from the shortcut menu by right-clicking in the drawing area and choosing **Paste as Block**.

PASTING CONTENTS USING ORIGINAL COORDINATES

Ribbon: Clipboard > Paste > Paste to Original Coordinates   **Command:** PASTEORIG

The **PASTEORIG** command is used to paste the contents of the Clipboard into a new drawing using the coordinates from the original drawing. You can also invoke the **PASTEORIG** command from the shortcut menu by right-clicking in the drawing area and choosing **Paste to Original Coordinates**. The **PASTEORIG** command is available only when the Clipboard contains AutoCAD data from a drawing other than the current drawing.

Exercise 4   *Mechanical*

In this exercise, you will draw the object shown in Figure 5-11. Use the **COPY** command for creating the drawing.

OFFSETTING SKETCHED OBJECTS

Ribbon:	Home > Modify > Offset	**Toolbar:**	Modify > Offset
Command:	OFFSET		

 To draw parallel lines, polylines, concentric circles, arcs, curves, and so on, you can use the **OFFSET** command (Figure 5-12). This command creates another object that is similar to the selected one. Remember that you are allowed to select only one entity at a time to be offset. When offsetting an object, you can specify the offset distance and

the side to offset, or you can specify a
distance through which you want to
offset the selected object. Depending
on the side to offset, you can create
smaller or larger circles, ellipses, and
arcs. If the offset side is toward the
inner side of the perimeter, the arc,
ellipse, or circle will be smaller than
the original. The prompt sequence
that will follow when you choose
Modify > Offset from the **Home** tab
of the **Ribbon** menu is given next.

Figure 5-11 *Drawing for Exercise 4*

Current settings: Erase
source=No Layer=Source OFFSETGAPTYPE=0
Specify offset distance or [Through/Erase/Layer] <Through>: *Specify the offset distance*
Select object to offset or [Exit/Undo]<Exit>: *Select the object to offset*
Specify point on side to offset or <Exit/Multiple/Undo>: *Specify a point on side to offset*
Select object to offset or [Exit/Undo]<Exit>*Select another object to offset or press* Enter.

Through Option

The offset distance can be specified by entering a value or by specifying two points with the
pointing device. AutoCAD will measure the distance between these two points and use it as
the offset distance. The **Through** option is generally used for creating orthographic views. In
this case, you do not have to specify a distance; you simply specify an offset point, see
Figure 5-13. The offset distance is stored in the **OFFSETDIST** system variable. A negative
value indicates that the offset value is set to the **Through** option. You can offset lines, arcs, 2D
polylines, xlines, circles, ellipses, elliptical arcs, rays, and planar splines. If you try to offset
objects other than these, the message **Cannot offset that object** is displayed.

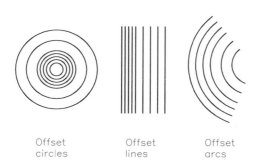

Figure 5-12 *Using the **OFFSET** command multiple
times to create multiple offset entities*

Figure 5-13 *Using the **Through** option*

Erase Option

The **Erase** option is used to specify whether the source object has to be deleted or not. Enter **Yes** at the **Erase source object after offsetting** prompt; the source object will be deleted after being offset. The prompt sequence that will follow when you choose the **Erase** option is given next.

    Current settings: Erase source=No Layer=Source OFFSETGAPTYPE=0
    Specify offset distance or [Through/Erase/Layer] <current>: **E**
    Erase source object after offsetting? [Yes/No] <No>: **Y**
    Specify offset distance or [Through/Erase/Layer] <current>:

Layer Option

Use the **Layer** option to specify whether the offset entity will be placed in the current layer or the layer of the source object. The prompt sequence for offsetting the entity in the Source layer is given next.

    Current settings: Erase source=Yes Layer=Current OFFSETGAPTYPE=0
    Specify offset distance or [Through/Erase/Layer] <current>: L
    Enter layer option for offset objects [Current/Source] <Current>: S
    Specify offset distance or [Through/Erase/Layer] <current>:

Exercise 5	*General*

Use the **OFFSET** edit command to draw Figures 5-14 and 5-15.

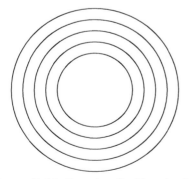

Figure 5-14 *Drawing for Exercise 5*

Figure 5-15 *Drawing for Exercise 5*

ROTATING THE SKETCHED OBJECTS

Ribbon:	Home > Modify > Rotate	**Toolbar:**	Modify > Rotate
Command:	ROTATE		

While creating designs there are many occasions when you have to rotate an object or a group of objects. You can accomplish this by using the **ROTATE** command. When you invoke this command, AutoCAD will prompt you to select the objects and the base point about which the selected objects will be rotated. You should be careful when selecting the base point as it is easy to get confused, if the base point is not located on a known object. After you specify the base point, you are required to enter a rotation angle. By default, positive angles produce a counterclockwise rotation and negative angles produce a clockwise rotation (Figure 5-16). The **ROTATE** command can also be invoked from the shortcut menu by selecting the object and clicking the right mouse button in the drawing area and choosing **Rotate**. The prompt sequence that will follow when you choose the **Rotate** button is given next.

Current positive angle in UCS: ANGDIR=*current* ANGBASE=*current*
Select objects: *Select the objects for rotation.*
Select objects: Enter
Specify base point: *Specify a base point about which the selected objects will be rotated.*
Specify rotation angle or [Copy/Reference]<current>: *Enter a positive or negative rotation angle, or specify a point.*

If you need to rotate objects with respect to a known angle, you can do this in two different ways using the **Reference** option. The first way is to specify the known angle as the reference angle, followed by the proposed angle to which the objects will be rotated (Figure 5-17). Here the object is first rotated clockwise from the X axis, through the reference angle. Then the object is rotated through the new angle from this reference position in a counterclockwise direction. The prompt sequence is given next.

Current positive angle in UCS: ANGDIR=*current* ANGBASE=*current*
Select objects: *Select the objects for rotation.*
Select objects: Enter
Specify base point: *Specify the base point.*
Specify rotation angle or [Copy/Reference]<current>: **R** Enter
Specify the reference angle <0>: *Enter reference angle.*
Specify the new angle or [Points]: *Enter new angle or enter* **P** *to select two points for specifying the angle value.*

Figure 5-16 Rotation of objects with different rotation angles

*Figure 5-17 Rotation using the **Reference** option*

The other method is used when the reference angle and the new angle are not known. In this case, you can use the edges of the original object and the reference object to specify the original object and reference angle, respectively. Figure 5-18 shows a model created at an unknown angle and a line also at an unknown angle. In this case, this line will be used as a reference object for rotating the object. In such cases, remember that the base point should be taken on the reference object. This is because you cannot define two points for specifying the new angle. You have to directly enter the angle value or specify only one point. Therefore, the base point will be taken as the first point and the second point can be defined for the new angle. Figure 5-19 shows the model after rotating it with reference to the line such that the line and the model are inclined at similar angles. The prompt sequence to rotate the model in Figure 5-18 is given next.

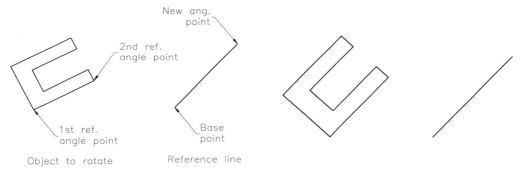

Figure 5-18 *Rotating the objects using a reference line*

Figure 5-19 *The model after rotating with reference to the line*

Current positive angle in UCS: ANGDIR=*current* ANGBASE=*current*
Select objects: *Select the object for rotation.*
Select objects: Enter
Specify base point: *Specify the base point as the lower endpoint of the line, see Figure 5-18.*
Specify rotation angle or [Copy/Reference]<current>: **R** Enter
Specify the reference angle <0>: *Specify the first point on the edge of the model, see Figure 5-18.*
Specify second point: *Specify the second point on the same edge of the model, see Figure 5-18.*
Specify the new angle or [Points]<0>: *Select the other endpoint of the reference line, see Figure 5-18.*

You can also specify the new angle by entering the numerical value at the **Specify the new angle** prompt.

If you want to retain the original object and create a copy while rotating, you can use the **Copy** option. The source entity is retained in its original orientation and a new instance is created and rotated through the specified angle. The prompt sequence with **Copy** option is given next.

Current positive angle in UCS: ANGDIR=counterclockwise ANGBASE=0
Select objects: *Select the object for rotation.*
Select objects: Enter

Specify base point: *Specify a base point about which the selected objects will be rotated.*
Specify rotation angle or [Copy/Reference] <current>: **C** [Enter]
Rotating a copy of the selected objects.
Specify rotation angle or [Copy/Reference] <current>: *Enter a positive or negative rotation angle, or specify a point.*

SCALING SKETCHED OBJECTS

Ribbon:	Home > Modify > Scale	Toolbar:	Modify > Scale
Command:	SCALE		

Many times you will need to change the size of objects in a drawing. You can do this with the **SCALE** command. This command dynamically enlarges or shrinks the selected object about a base point, keeping the aspect ratio of the object constant. This means that the size of the object will be increased or reduced equally in the X, Y, and Z directions. The dynamic scaling property allows you to view the object as it is being scaled. This can be viewed by moving the pointing device in the drawing area after selecting the base point. Application of the identical scale factor to the X, Y, and Z dimensions ensures that the shape of the objects being scaled do not change. This is a useful and timesaving editing command because instead of redrawing objects to the required size, you can scale the objects with a single **SCALE** command. Another advantage of this command is that if you have already put the dimensions on the drawing, they will also change accordingly. You can also invoke the **SCALE** command from the shortcut menu by right-clicking in the drawing area and choosing **Scale**. The prompt sequence that will follow when you choose the **Scale** button is given next.

Select objects: *Select objects to be scaled.*
Select objects: [Enter]
Specify base point: *Specify the base point, preferably a known point.*
Specify scale factor or [Copy/Reference]<current>: *Specify the scale factor.*

The base point will not be moved from its position and the selected object(s) will be scaled around the base point, as shown in Figures 5-20 and 5-21. To reduce the size of an object, the scale factor should be less than 1 and to increase its size, the scale factor should be greater than 1. You can enter a scale factor or select two points to specify a distance as a factor. When you select two points to specify a distance as a factor, the first point should be on the referenced object.

Sometimes it is time-consuming to calculate the relative scale factor. In such cases, you can scale the object by specifying a desired size in relation to the existing size (a known dimension). In other words, you can use a **reference length**. This can be done by entering **R** at the **Specify scale factor or [Copy/Reference]** prompt. Then, you either specify two points to specify the length or enter a length. At the next prompt, enter the length relative to the reference length. For example, if a line is **2.5** units long and you want its length to be **1.00** units, instead of calculating the relative scale factor, you can use the **Reference** option. The prompt sequence for using the **Reference** option is given next.

Figure 5-20 *Original object before scaling*   **Figure 5-21** *After scaling to 0.5 of the actual size*

Select objects: *Select the object to scale.*
Select objects: Enter
Specify base point: *Specify the base point.*
Specify scale factor or [Copy/Reference]<current>: **R** Enter
Specify reference length <1>: *Specify the reference length.*
Specify new length [Point]: *Specify the new length or enter **P** to select two points for specifying the angle value.*

Similar to the **ROTATE** command, here also you can scale one object using the reference of another object. Again, the base point has to be taken on the reference object as you can define only one point for the new length.

You can use the **Copy** option to retain the source object and scale the copied instance of the source object. The command prompt for using the **Copy** option is given next.

Select objects: *Select the object to scale.*
Select objects: Enter
Specify base point: *Specify the base point.*
Specify scale factor or [Copy/Reference] <current>: **C** Enter
Scaling a copy of the selected objects.
Specify scale factor or [Copy/Reference] <current>: *Specify the scale factor.*

Tip
*If you have not used the required drawing units for a drawing, you can use the **Reference** option of the **SCALE** command to correct the error. Select the entire drawing with the help of the **All** selection option. Specify the **Reference** option, and then select the endpoints of the object whose desired length you know. Specify the new length, and all objects in the drawing will be scaled automatically to the desired size.*

FILLETING SKETCHES

Ribbon:	Home > Modify > Fillet	**Toolbar:**	Modify > Fillet
Command:	FILLET		

Chapter 5

The edges in the design are generally filleted to reduce the area of stress concentration. The **FILLET** command helps you form round corners between any two entities by allowing you to define two entities that form a sharp vertex. The result is that a smooth round arc is created that connects the two objects. A fillet can also be created between two intersecting or parallel lines as well as nonintersecting and nonparallel lines, arcs, polylines, xlines, rays, splines, circles, and true ellipses. The fillet arc created will be tangent to both the selected entities. The default fillet radius is 0.0000. As a result, when you invoke this command, you first need to specify the radius value. The prompt sequence that will follow when you choose the **Fillet** button is given next.

Current Settings: Mode= TRIM, Radius= 0.0000
Select first object or [Undo/Polyline/Radius/Trim/Multiple]:

Creating Fillets Using the Radius Option

The fillet you create depends on the radius distance you specify. The default radius is 0.0000. You can enter a distance or two points. The new radius you enter becomes the default radius and remains in effect until changed. The prompt sequence is given next.

Select first object or [Undo/Polyline/Radius/Trim/Multiple]: **R** Enter
Specify fillet radius <current>: *Enter a fillet radius or press ENTER to accept the current value.*

Note
*The **FILLETRAD** system variable controls and stores the current fillet radius. The default value of this variable is 0.0000.*
You can also fillet 3D objects in different planes. This is discussed in later chapters.

Tip
A fillet with a zero radius creates sharp corners and is used to clean up lines at corners if they overlap or have a gap.

Creating Fillets Using the Select First Object Option

This is the default method to fillet two objects. As the name implies, it prompts for the first object required for filleting. The prompt sequence to use this option is given next.

Current Settings: Mode= TRIM, Radius= modified value
Select first object or [Undo/Polyline/Radius/Trim/Multiple]: *Specify first object.*
Select second object or shift-select to apply corner: *Select second object or press down the SHIFT key while selecting the object to create sharp corner.*

The **FILLET** command can also be used to cap the ends of two parallel as well as non parallel lines, see Figure 5-22. The cap is a semicircle whose radius is equal to half the distance between the two parallel lines. The cap distance is calculated automatically when you select the two parallel lines for filleting. You can select lines with the Window, Crossing, or Last option, but to avoid unexpected results, select the objects by picking objects individually. Also, selection by picking objects is necessary in the case of arcs and circles that have the possibility of more than one fillet. They are filleted closest to the select points, as shown in Figure 5-23.

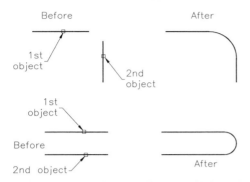

Figure 5-22 *Filleting the parallel and non parallel lines*

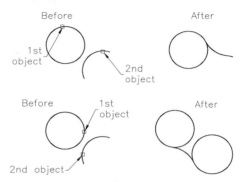

Figure 5-23 *Using the **FILLET** command on circles and arcs*

Creating Fillets Using the Trim Option

When you create a fillet, an arc is created and the selected objects are either trimmed or extended at the fillet endpoint. This is because the **Trim** mode is set to **Trim**. If it is set to **No Trim** they are left intact. Figure 5-24 shows a model filleted with the **Trim** mode set to **Trim** and **No Trim**. The prompt sequence is given next.

Select first object or [Undo/Polyline/Radius/ Trim/Multiple]: ***T*** Enter

Specify Trim mode option [Trim/No trim] <current>: *Enter **T** to trim the edges, **N** to leave them intact. You can also choose the required option from the dynamic preview.*

Figure 5-24 *Filleting the lines with the **Trim** mode set to **Trim** and **No Trim***

Creating Fillets Using the Polyline Option

Using the **Polyline** option, you can fillet a number of entities that comprise a single polyline (Figure 5-25). The polylines can be created using the **POLYLINE** or the **RECTANG** command. If the object is created using the **POLYLINE** command, it must be closed using the **Close** option. Otherwise, the last corner of the polyline will not be filleted. When you select this option, AutoCAD prompts you to select a polyline. All the vertices of the polyline are filleted when you select the polyline. The fillet radius for all the vertices will be the same. If the selected polyline is not closed, then the last corner is not filleted. The prompt sequence for using this option is given next.

Current Settings: Mode= *current*, Radius= *current*

Select first object or [Undo/Polyline/
Radius/Trim/Multiple]: **P** [Enter]
Select 2D polyline: *Select the polyline.*

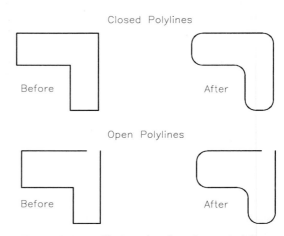

Figure 5-25 Filleting closed and open polylines

Creating Fillets Using the Multiple Option

When you invoke the **Fillet** command, by
default the fillet is created between a single
set of entities only. But with the help of the
Multiple option, you can add fillets to more
than one set of entities. When you select this
option, AutoCAD prompts you to select the
first object and then the second object. A
fillet will be created between the two entities.
Next, you are again prompted to select the
first object and then the second object. This
prompt continues until you press ENTER
to terminate the **Fillet** command. The prompt sequence to use this option is given next.

Current Settings: Mode= TRIM, Radius= current
Select first object or [Undo/Polyline/Radius/Trim/Multiple]: **M** [Enter]
Select first object or [Undo/Polyline/Radius/Trim/Multiple]: *Specify the first object of one
set.*
Select second object or shift-select to apply corner: *Select the second object or press down
the SHIFT key while selecting the object to create sharp corner.*
Select first object or [Undo/Polyline/Radius/Trim/Multiple]: *Specify the first object of the
other set.*
Select second object or shift-select to apply corner: *Select the second object or press down
the SHIFT key while selecting the object to create sharp corner.*
Select first object or [Undo/Polyline/Radius/Trim/Multiple]: *Specify the first object of the
other set or press* [Enter]

Note
*Use the **Undo** option to go back to the previous fillet when you fillet entities using the **Multiple**
option.*

Filleting Objects with a Different UCS

The fillet command will also fillet objects that are not in the current UCS plane. To create a
fillet for these objects, AutoCAD will automatically change the UCS transparently so that it
can generate a fillet between the selected objects.

Note
*Filleting objects in different planes is discussed in detail in Chapter 25 (Getting Started with
3D).*

Setting the TRIMMODE System Variable

The **TRIMMODE** system variable eliminates any size restriction on the **FILLET** command. By setting **TRIMMODE** to **0**, you can create a fillet of any size without actually cutting the existing geometry. Also, there is no restriction on the fillet radius. This means that the fillet radius can be larger than one or both objects that are being filleted. The default value of this variable is 1.

Note
TRIMMODE = 0   *Fillet or chamfer without cutting the existing geometry.*
TRIMMODE = 1   *Extend or trim the geometry.*

CHAMFERING SKETCHES

Ribbon:	Home > Modify > Fillet > Chamfer	**Toolbar:**	Modify > Chamfer
Command:	CHAMFER		

Chamfering the sharp corners is another method of reducing the areas of stress concentration in the design. Chamfering is defined as the process, in which the sharp edges or corners are beveled. In simple words, it is defined as the taper provided on a surface. A beveled line connects two separate objects to create a chamfer. The size of a chamfer depends on its distance from the corner. If a chamfer is equidistant from the corner in both directions, it is a 45-degree chamfer. A chamfer can be drawn between two lines that may or may not intersect. However, remember that the lines are not parallel because parallel lines cannot be chamfered. This command also works on a single polyline. In AutoCAD, the chamfers can be created using two methods: by defining two distances, or by defining one distance and the chamfer angle. The prompt sequence that will follow, when you choose **Modify > Fillet > Chamfer** from the **Ribbon** is given next.

(TRIM mode) Current chamfer Dist1 = 0.0000, Dist2 = 0.0000
Select first line or [Undo/Polyline/Distance/Angle/Trim/mEthod/Multiple]:

Creating Chamfer Using the Distance Option

This option is used to enter the chamfer distance. The default values of the distances are 0.0000 and 0.0000. This option can be invoked by entering **D** at the **Select first line or [Polyline/Distance/Angle/ Trim/Method/mUltiple]** prompt. Next, enter the first and second chamfer distances. The first distance is the distance of the corner calculated along the edge selected first. Similarly, the second distance is calculated along the edge that is selected last. The new chamfer distances remain in effect until you change them. Instead of entering the distance values, you can specify two points to indicate each distance, see Figure 5-26. The prompt sequence is given next.

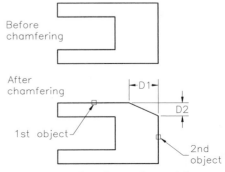

Figure 5-26 *Chamfering the model using the **Distance** option*

Select first line or [Undo/Polyline/Distance/Angle/Trim/mEthod/Multiple]: **D** Enter
Specify first chamfer distance <0.0000>: *Enter a distance value or specify two points.*
Specify second chamfer distance <0.0000>: *Enter a distance value or specify two points.*

The first and second chamfer distances are stored in the **CHAMFERA** and **CHAMFERB** system variables. The default value of these variables is 0.0000.

Tip
*If Dist 1 and Dist 2 are set to zero, the **CHAMFER** command will extend or trim the selected lines so that they end at the same point.*

Creating Chamfer Using the Select First Line Option

In this option, you need to select two nonparallel objects so that they are joined with a beveled line. The size of the chamfer depends upon the values of the two distances. The prompt sequence is given next.

(TRIM mode) Current chamfer Dist1 = current, Dist2 = current
Select first line or [Undo/Polyline/Distance/Angle/Trim/mEthod/Multiple]: *Specify the first line.*
Select second object or shift-select to apply corner: *Select the second object or press down the SHIFT key while selecting the object to create sharp corner.*

Creating Chamfer Using the Polyline Option

You can use the **CHAMFER** command to chamfer all corners of a closed or open polyline, as shown in Figure 5-27. With a closed polyline, all the corners of the polyline are chamfered to the set distance values. Sometimes the polyline may appear closed. But, if the last segment was not created using the **Close** option, it may not actually be closed. In this case, the last corner is not chamfered. The prompt sequence is given next.

(TRIM mode) Current chamfer Dist1 = current, Dist2 = current
Select first line or [Undo/Polyline/Distance/Angle/Trim/mEthod/Multiple]: **P** Enter
Select 2D polyline: *Select the polyline.*

Creating Chamfer Using the Angle Option

The other method of creating chamfer is by specifying the distance and the chamfer angle, as shown in Figure 5-28. The chamfer angle can be defined using this option. The prompt sequence is given next.

Select first line or [Undo/Polyline/Distance/Angle/Trim/mEthod/Multiple]: **A** Enter
Specify chamfer length on the first line <current>: *Specify a length.*
Specify chamfer angle from the first line <current>: *Specify an angle.*

Creating Chamfer Using the Trim Option

Depending on this option, the selected objects are either trimmed or extended to the endpoints of the chamfer line or left intact. The prompt sequence is given next.

Figure 5-27 *Chamfering a polyline*   **Figure 5-28** *Chamfering using the **Angle** option*

Select first line or [Undo/Polyline/Distance/Angle/Trim/mEthod/Multiple]: **T** Enter
Enter Trim mode option [Trim/No Trim] <current>:

Creating Chamfer Using the Method Option
This option is used to toggle between the **Distance** method and the **Angle** method for creating the chamfer. The current settings of the selected method will be used for creating the chamfer. The prompt sequence is given next.

Select first line or [Undo/Polyline/Distance/Angle/Trim/mEthod/Multiple]: **E** Enter
Enter trim method [Distance/Angle] <current>: *Enter **D** for Distance option, **A** for Angle option.*

Note
*If you set the value of the **TRIMMODE** system variable to **1** (default value), the objects will be trimmed or extended after they are chamfered and filleted. If **TRIMMODE** is set to zero, the objects are left untrimmed.*

Creating Chamfer Using the Multiple Option
When you invoke the **Chamfer** command, by default the chamfer is created between a single set of entities only. But with the help of the **Multiple** option, you can add chamfers to the multiple sets of entities. When you select this option, AutoCAD prompts you to select the first line and then the second line. The chamfer is added to the two selected lines. Next, you are again prompted to select the first line and then the second line. This prompt continues until you press ENTER to terminate the **Chamfer** command. The prompt sequence that follows to create multiple chamfers is given next.

(TRIM mode) Current chamfer Dist1 = current, Dist2= current
Select first line or [Undo/Polyline/Distance/Angle/Trim/mEthod/Multiple]: **M** Enter
Select first line or [Undo/Polyline/Distance/Angle/Trim/mEthod/Multiple]: *Specify the first line of one set.*
Select second object or shift-select to apply corner: *Select the second object or press down the SHIFT key while selecting the object to create sharp corner.*

Chapter 5

Select first line or [Undo/Polyline/Distance/Angle/Trim/mEthod/Multiple]: *Specify the first line of the other set.*
Select second object or shift-select to apply corner: *Select the second object or press down the SHIFT key while selecting the object to create sharp corner.*
Select first line or [Undo/Polyline/Distance/Angle/Trim/mEthod/Multiple]: *Specify the first line of the other set or press* Enter *to terminate the* **Chamfer** *command.*

Note
Use the **Undo** *option to go back to the previous chamfer when you are chamfering entities using the* **Multiple** *option.*

Setting the Chamfering System Variables

The chamfer modes, distances, length, and angle can also be set using the following variables:

CHAMMODE = 0   Distance/Distance (default)
CHAMMODE = 1   Length/Angle
CHAMFERA   Sets first chamfer distance on the first selected line (default = 0.0000)
CHAMFERB   Sets second chamfer distance on the second selected line (default = 0.0000)
CHAMFERC   Sets the chamfer length (default = 0.0000)
CHAMFERD   Sets the chamfer angle from the first line (default = 0)

TRIMMING THE SKETCHED OBJECTS

Ribbon:	Modify > Trim	Toolbar:	Modify > Trim
Command:	TRIM		

When creating a design, there are a number of places where you have to remove the unwanted and extending edges. Breaking individual objects takes time if you are working on a complex design with many objects. In such cases, you can use the **TRIM** command. This command trims the objects that extend beyond a required point of intersection. When you invoke this command, you will be prompted to select the cutting edges or boundaries. These edges can be lines, polylines, circles, arcs, ellipses, xlines, rays, splines, text, blocks, or even viewports. There can be more than one cutting edge and you can use any selection method to select them. After the cutting edge or edges are selected, you must select each object to be trimmed. An object can be both a cutting edge and an object to be trimmed. You can trim lines, circles, arcs, polylines, splines, ellipses, xlines, and rays. The prompt sequence is given next.

Current settings:Projection=UCS Edge=None
Select cutting edges...
Select objects or <select all>: *Select the cutting edges.*
Select objects: Enter
Select object to trim or shift-select to extend or [Fence/Crossing/Project/Edge/eRase/Undo]:

Select object to trim Option

Here you have to specify the objects you want to trim and the side from which the object will be trimmed. This prompt is repeated until you press ENTER. This way you can trim several objects with a single **TRIM** command (Figure 5-29). The prompt sequence is given next.

> Current settings: Projection= UCS Edge= None
> Select cutting edges...
> Select objects or <select all>: *Select the first cutting edge.*
> Select objects: *Select the second cutting edge.*
> Select objects: Enter
> Select object to trim or shift-select to extend or [Fence/Crossing/Project/Edge/eRase/Undo]: *Select the first object.*
> Select object to trim or shift-select to extend or [Fence/Crossing/Project/Edge/eRase/Undo]: *Select the second object.*
> Select object to trim or shift-select to extend or [Fence/Crossing/Project/Edge/eRase/Undo]: Enter

Shift-select to extend Option

This option is used to switch to the extend mode. It is used to extend the object instead of trimming. In case the object to extend does not intersect with the cutting edge, you can press the SHIFT key and then select the object to be extended. The selected edge will be extended taking the cutting edge as the boundary for extension.

Edge Option

This option is used whenever you want to trim those objects that do not intersect the cutting edges, but would intersect if the cutting edges were extended, see Figure 5-30. The prompt sequence is given next.

> Command: **TRIM** Enter
> Current settings: Projection= UCS Edge= None
> Select cutting edges...
> Select objects: *Select the cutting edge.*
> Select objects: Enter
> Select object to trim or shift-select to extend or [Fence/Crossing/Project/Edge/eRase/Undo]: **E** Enter
> Enter an implied edge extension mode [Extend/No extend] <current>: **E** Enter
> Select object to trim or shift-select to extend or [Fence/Crossing/Project/Edge/eRase/Undo]: *Select object to trim.*

Project Option

In this option, you can use the Project mode while trimming the objects. The prompt sequence is given next.

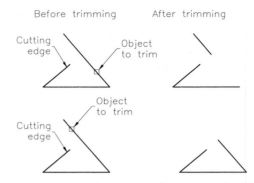

Figure 5-29 *Using the* **TRIM** *command*

Figure 5-30 *Trimming an object using the* **Edge** *option (* **Extend** *)*

> Select object to trim or shift-select to extend or [Fence/Crossing/Project/Edge/eRase/
> Undo]: **P** Enter
> Enter a projection option [None/Ucs/View] <current>:

The **None** option is used whenever the objects to trim intersect the cutting edges in 3D space. If you want to trim those objects that do not intersect the cutting edges in 3D space, but do visually appear to intersect in a particular UCS or the current view, use the **UCS** or **View** options. The UCS option projects the objects to the *XY* plane of the current UCS, while the **View** option projects the objects to the current view direction (trims to their apparent visual intersections).

Fence Option

As discussed earlier, the **Fence** option is used for the selection purpose. Using the **Fence** option, all the objects crossing the selection fence are selected. The prompt sequence for the **Fence** option is given next.

> Select object to trim or shift-select to extend or [Fence/Crossing/Project/Edge/eRase/
> Undo]: **F** Enter
> Specify first fence point: Specify the first point of the fence
> Specify next fence point or [Undo]: Specify the second point of the fence
> Specify next fence point or [Undo]: Specify the third point of the fence
> Specify next fence point or [Undo]: Enter

Crossing Option

The **Crossing** option is used to select the entities using a crossing window. The objects touching the window boundaries or completely enclosing it are selected. The prompt sequence is given next.

> Select object to trim or shift-select to extend or [Fence/Crossing/Project/Edge/eRase/
> Undo]: C Enter
> Specify first corner: *Select the first corner of the crossing window.*
> Specify opposite corner: *Select the first corner of the crossing window.*

Erase Option

The **Erase** option in the **Trim** command is used to erase the entities without canceling the **Trim** command.

> Select object to trim or shift-select to extend or [Fence/Crossing/Project/Edge/eRase/Undo]: **R** [Enter]
> Select objects to erase or <exit>: *Select object to erase.*
> Select objects to erase: *Select object to erase or* [Enter]
> Select objects to erase: [Enter]

Undo Option

If you want to remove the previous change created by the **TRIM** command, enter **U** at the **Select object to trim or** [Fence/Crossing/Project/Edge/eRase/Undo] prompt.

Exercise 6   *Mechanical*

Draw the top illustration in Figure 5-31 and then use the **FILLET**, **CHAMFER**, and **TRIM** commands to obtain the following figure. (Assume the missing dimensions.)

Figure 5-31 *Drawing for Exercise 6*

EXTENDING THE SKETCHED OBJECTS

| **Ribbon:** | Home > Modify > Extend | **Toolbar:** | Modify > Extend |
| **Command:** | EXTEND | | |

The **EXTEND** command may be considered the opposite of the **TRIM** command. In the **TRIM** command you trim objects; in the **EXTEND** command you can extend lines, polylines, rays, and arcs to meet other objects. This command does not extend closed loops. The command format is similar to that of the **TRIM** command. You are required to select the boundary edges first. The boundary edges are those objects that the selected lines or arcs extend to meet. These edges can be lines, polylines, circles, arcs, ellipses, xlines, rays, splines, text, blocks, or even viewports. The prompt sequence that will follow when you choose the **Extend** button is given next.

　　Current settings: Projection= UCS, Edge=None
　　Select boundary edges...
　　Select objects or <select all>: *Select boundary edges.*
　　Select objects: `Enter`
　　Select object to extend or shift-select to trim or [Fence/Crossing/Project/Edge/Undo]:

Select object to extend Option

Here you have to specify the object you want to extend to the selected boundary (Figure 5-32). This prompt is repeated until you press ENTER. Now you can select a number of objects in a single **EXTEND** command.

Shift-select to trim Option

This option is used to switch to the trim mode in the **EXTEND** command, if two entities are intersecting. You can press the SHIFT key and then select the object to be trimmed. In this case, the boundary edges will be taken as the cutting edges.

Project Option

In this option, you can use the projection mode while trimming objects. The prompt sequence that will follow when you choose the **Extend** button is given next.

　　Current settings: Projection= UCS Edge= None
　　Select boundary edges...
　　Select objects or <select all>: *Select boundary edges.*
　　Select objects: `Enter`
　　Select object to extend or shift-select to trim or [Fence/Crossing/Project/Edge/Undo]: **P**
　　`Enter`
　　Enter a projection option [None/UCS/View] <current>:

The **None** option is used whenever the objects to be extended intersect with the boundary edge in 3D space. If you want to extend those objects that do not intersect the boundary edge in 3D space, use the **UCS** or **View** option. The **UCS** option projects the objects to the *XY* plane of the current UCS, while the **View** option projects the objects to the current view.

Edge Option

You can use this option whenever you want to extend objects that do not actually intersect the boundary edge, but would intersect its edge if the boundary edges were extended (Figure 5-33). If you enter **E** at the prompt, the selected object is extended to the implied boundary edge. If you enter **N** at the prompt, only those objects that would actually intersect the real boundary edge are extended (the default). The prompt sequence is given next.

　　Current settings: Projection= UCS Edge= None
　　Select boundary edges...

Select objects or <select all>: *Select the boundary edge.*
Select objects: [Enter]
Select object to extend or shift-select to trim or [Fence/Crossing/Project/Edge/Undo]: **E** [Enter]

Enter an implied extension mode [Extend/No extend] <current>: **E** [Enter]
Select object to extend or shift-select to trim or [Project/Edge/Undo]: *Select the line to extend.*

Note
*Working of the **Fence** and **Crossing** options in the **EXTEND** command is the same as that in **TRIM** command.*

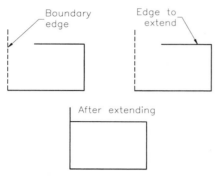

Figure 5-32 Extending an edge

*Figure 5-33 Extending an edge using the **Edge** option (**Extend**)*

Undo Option

If you want to remove the previous change created by the **EXTEND** command, enter **U** at the **Select object to extend or** [Fence/Crossing/Project/Edge/Undo]: prompt.

Trimming and Extending with Text, Region, or Spline

The **TRIM** and **EXTEND** commands can be used with text, regions, or splines as edges (Figure 5-34). This makes the **TRIM** and **EXTEND** commands two of the most useful editing commands in AutoCAD. The **TRIM** and **EXTEND** commands can also be used with arcs, elliptical arcs, splines, ellipses, 3D Pline, rays, and lines. See Figure 5-35. The system variables **PROJMODE** and **EDGEMODE** determine how the **TRIM** and **EXTEND** commands are executed. The values that can be assigned to these variables are discussed next.

Value	PROJMODE	EDGEMODE
0	True 3D mode	Use regular edge without extension (default)
1	Project to current Extend the edge to the natural boundary UCS *XY* plane (default)	
2	Project to current view plane	

Chapter 5

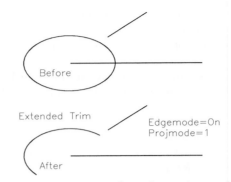

Figure 5-34 *Using the* **TRIM** *and* **EXTEND** *commands with text, spline, and region*

Figure 5-35 *Using the* **Edge** *option to do an implied trim*

STRETCHING THE SKETCHED OBJECTS

Ribbon:	Home > Modify > Stretch	**Toolbar:**	Modify > Stretch
Command:	STRETCH		

This command can be used to stretch objects, altering the selected portions of the objects. With this command, you can lengthen objects, shorten them, and alter their shapes, see Figure 5-36. You must use a Crossing, or CPolygon selection to specify the objects to be stretched. The prompt sequence that will follow when you choose the **Stretch** button is given next.

> Select objects to stretch by crossing-window or crossing-polygon...
> Select objects: *Select objects using crossing window or polygon.*
> Select objects: Enter

Figure 5-36 *Stretching the entities*

After selecting the objects, you have to specify the point of displacement. You should select only that portion of the object that needs stretching.

> Specify base point or displacement <displacement>: *Specify the base point.*
> Specify second point of displacement or <use first point as displacement>: *Specify the displacement point or press ENTER to use the first point as the displacement.*

You normally use a Crossing or CPolygon selection with **STRETCH** because if you use a Window, those objects that cross the window are not selected, and the objects selected because they are fully within the window are moved, not stretched. The object selection and stretch specification process of **STRETCH** is a little unusual. You are really specifying two things: first, you are selecting objects. Second, you are specifying the portions of those selected objects to be stretched. You can use a Crossing or CPolygon selection to simultaneously specify both, or you can select objects by any method, and then use any window or crossing specification to specify what parts of those objects to stretch. Objects or portions of selected objects completely within the window or crossing specification are moved. If selected objects cross the window or crossing specification, their defining points within the window or crossing specification are moved, their defining points outside the window or crossing specification remain fixed, and the parts crossing the window or crossing specification are stretched. Only the last window or crossing specification made determines what is stretched or moved. Figure 5-36 illustrates using a crossing selection to simultaneously select the two angled lines and specify that their right ends will be stretched. Alternatively, you can select the lines by any method and then use a Crossing selection (which will not actually select anything) to specify that their right ends will be stretched.

Note
The regions and solids cannot be stretched. When you select them, they will be moved instead of stretching.

LENGTHENING THE SKETCHED OBJECTS

Ribbon:	Modify > Lengthen	**Command:**	LENGTHEN

 Like the **TRIM** and **EXTEND** commands, the **LENGTHEN** command can be used to extend or shorten lines, polylines, elliptical arcs, and arcs. The **LENGTHEN** command has several options that allow you to change the length of objects by dynamically dragging the object endpoint, entering the delta value, entering the percentage value, or entering the total length of the object. This command also allows the repeated selection of the objects for editing. This command has no effect on closed objects such as circles. The selected entity will be increased or decreased from the endpoint closest to the selection point. The prompt sequence that will follow when you invoke the **Lengthen** command is given next.

Select an object or [DElta/Percent/Total/DYnamic]:

Select an object Option

This is the default option that returns the current length or the included angle of the selected object. If the object is a line, AutoCAD returns only the length. However, if the selected object is an arc, AutoCAD returns the length and the angle. The same prompt sequence will be displayed, after you select the object.

DElta Option

The **DElta** option is used to increase or decrease the length or angle of an object by defining the distance or angle by which the object will be extended. The delta value can be entered by entering a numerical value or by specifying two points. A positive value will increase (Extend) the length of the selected object and a negative value will decrease it (Trim), see Figure 5-37. The prompt sequence to use this option is given next.

Select an object or [DElta/Percent/Total/DYnamic]: **DE** Enter
Enter delta length or [Angle] <current>: *Specify the length or enter A for angle.*
Enter delta angle <current>: *Specify the delta angle.*
Select object to change or [Undo]: *Select the object to be extended.*
Select object to change or [Undo]: Enter

Percent Option

The **Percent** option is used to extend or trim an object by defining the change as a percentage of the original length or the angle, Figure 5-37. The current length of the line is taken as 100 percent. If you enter a value more than 100, the length will increase by that amount. Similarly, if you enter a value less than 100, then the length will decrease by that amount. For example, a positive number of 150 will increase the length by 50 percent and a positive number of 75 will decrease the length by 25 percent of the original value (negative values are not allowed).

Total Option

The **Total** option is used to extend or trim an object by defining the new total length or angle. For example, if you enter a total length of 1.25, AutoCAD will automatically increase or decrease the length of the object so that the new length is 1.25. The value can be entered by entering a numerical value or by specifying two points. The object is shortened or lengthened with respect to the endpoint that is closest to the selection point. The selection point is determined by where the object was selected. The prompt sequence for using this option is given next.

Specify total length or [Angle] <current>: *Specify the length or enter* **A** *for angle.*

If you enter a numeric value in the previous prompt, the length of the selected objects is changed accordingly, see Figure 5-38. You can change the angle of an arc by entering **A** and giving the new value.

Figure 5-37 *Using the **DElta** and **Percent** options*   **Figure 5-38** *Using the **Total** option*

Note

*By default, the **Lengthen** button is not available in the **Modify** toolbar. To add the **Lengthen** button in the **Modify** toolbar, right-click on any toolbar to display the shortcut menu. Choose **Customize** from the shortcut menu to display the **Customize User Interface** dialog box. Select the **Modify** option in the **Filter the command list by category** drop-down list; all the tools that can be added or are currently available in the **Modify** toolbar are displayed. Drag the **Lengthen** button from the **Command list** area to the **Modify** toolbar in the drawing area. Choose the **OK** button from the **Customize User Interface** dialog box.*

DYnamic

The **DYnamic** option allows you to dynamically change the length or angle of an object by specifying one of the endpoints and dragging it to a new location. The other end of the object stays fixed and is not affected by dragging. The angle of lines, radius of arcs, and shape of elliptical arcs are unaffected.

ARRAYING THE SKETCHED OBJECTS

Ribbon:	Home > Modify > Array	Toolbar:	Modify > Array
Command:	ARRAY		

An array is defined as the method of creating multiple copies of the selected object and arranging them in a rectangular or circular fashion. In some drawings, you may need to create an object multiple times in a rectangular or circular arrangement. For example, suppose you have to draw six chairs around a table. This job can be accomplished by drawing each chair separately or by using the **COPY** command to make multiple copies of the chair. But it is a very tedious process and also the alignment of the chairs will have to be adjusted. This can be easily done using the **ARRAY** command. All you have to do is to create just one chair and the remaining five will be created and automatically arranged around the table by the **ARRAY** command. This method is more efficient and less time-consuming. This command allows you to make multiple copies of the selected objects in a rectangular or polar fashion. Each resulting element of the array can be controlled separately. When you choose the **Array** button, the **Array** dialog box is displayed. You can use this dialog box for creating a rectangular array or a polar array.

Rectangular Array

A rectangular array is formed by making copies of the selected object along the X and Y directions of an imaginary rectangle (along rows and columns). The rectangular array can be created by selecting the **Rectangular Array** radio button in the **Array** dialog box, see Figure 5-39.

Rows

This edit box is used to specify the number of rows in the rectangular array, which are arranged along the X axis of the current UCS.

*Figure 5-39 The **Rectangular Array** option in the **Array** dialog box*

Columns

This edit box is used to specify the number of columns in the rectangular array, which are arranged along the *Y* axis of the current UCS.

Offset distance and direction Area

The options under this area are used to define the distance between the rows and the columns and the angle of the array.

Row offset. This edit box is used to specify the distance between the rows, see Figure 5-40. You can either enter the distance value in this edit box or choose the **Pick Row Offset** button to define the distance. This button is provided on the right of the **Row offset** edit box. When you choose this button, the **Array** dialog box will be temporary closed and you will be prompted to specify two points on the screen to define the row offset distance. A positive value of the row offset will create the rows along the positive *X* axis direction. A negative value of the row offset will create the rows along the negative *X* axis direction, see Figure 5-41.

Column offset. This edit box is used to specify the distance between the columns, see Figure 5-40. You can either enter the distance value in this edit box or choose the **Pick Column Offset** button to define it. This button is provided on the right of the **Column offset** edit box. When you choose this button, the **Array** dialog box will be temporarily closed and you will be prompted to specify two points on the screen to define the column offset distance. A positive value of the row offset will create the rows along the positive *Y* axis direction. A negative value of row offset will create the rows along the negative *Y* axis direction, see Figure 5-41.

You can define both the row offset distance and the column offset distance simultaneously by

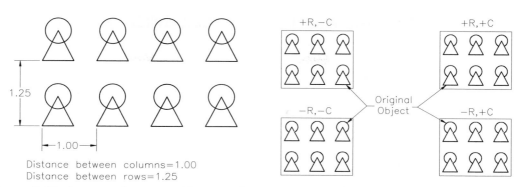

Figure 5-40 *A rectangular array with row and column distance*

Figure 5-41 *Specifying the direction of arrays*

choosing the **Pick Both Offsets** button. When you choose this button, the **Array** dialog box will be temporarily closed and you will be prompted to define a unit cell by specifying two opposite corners. The distance along the *X* axis of the unit cell defines the distance between the column and the distance along the *Y* axis of the unit cell defines the distance between the rows.

Angle of array. This edit box is used to define the angle of the array. This is the value by which the rows and the columns will be rotated, as shown in Figure 5-42. A positive value will rotate them in the counterclockwise direction and a negative value will rotate them in the clockwise direction. You can also define the angle by choosing the **Pick Angle of Array** button and specifying two points on the screen.

The object for an array can be selected by choosing the **Select objects** button. You can preview the creation of an array by choosing the **Preview** button. When you choose this button, the preview of the array will be displayed. If you are satisfied with the array, right-click to accept the array. If you are not satisfied

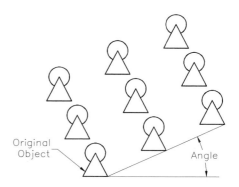

Figure 5-42 *Rotated rectangular array*

with the array, left-click on the screen or press ESC; the **Array** dialog box will be redisplayed on the screen and you can make the necessary changes.

Polar Array

A polar array is an arrangement of the objects around a point in a circular fashion. This kind of array is created by selecting the **Polar Array** radio button in the **Array** dialog box, Figure 5-43.

Center point

The center point of the array is defined as the point around which the selected items will be arranged. It is considered as the center point of the imaginary circle on whose circumference

Chapter 5

*Figure 5-43 The **Polar Array** option in the **Array** dialog box*

the items will be placed. The coordinates of the center of the array can be specified in the **X** and **Y** edit boxes. You can either enter the values in these edit boxes or select the center point of the array from the screen. To select the center point of the array from the screen, choose the **Pick Center Point** button provided on the right side of the **Y** edit box. When you choose this button, the **Array** dialog box will be temporarily closed and you will be prompted to select the center of the array. The **Array** dialog box will be redisplayed when you select the center of the array. Figure 5-44 shows a polar array displaying the center of the imaginary circle about which the objects are rotated.

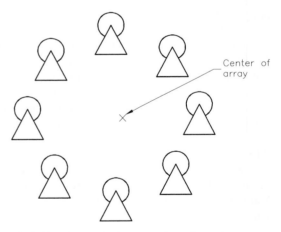

Figure 5-44 Center point of an array

Method and values Area

The options under this area are used to set the parameters related to the method that will be employed to create the polar array. The other options in this area will be available based on the method you select for creating the polar array.

Method. This drop-down list provides you with three methods for creating the polar array. These three methods are discussed next.

Total number of items & Angle to fill. This method is used to create a polar array by specifying the total number of items in the array and the total included angle between the first and the last item of the array, see Figure 5-45. The number of items and the angle to be filled can be specified in the **Total number of items** and **Angle to fill** edit boxes, respectively. You can also specify the angle to fill on the screen. This is done by choosing the **Pick Angle to Fill** button provided on the right side of the **Angle to fill** edit box. The **Array** dialog box will be temporarily closed and you will be prompted to specify the angle to fill.

Total number of items & Angle between items. This method is used when you want to create a polar array by specifying the total number of items in the array and the included angle between two adjacent items, see Figure 5-46. The angle between the items is also called the incremental angle. The number of items and the angle between the items can be specified in the **Total number of items** and **Angle between items** edit boxes, respectively. You can also specify the angle between the items on the screen. This is done by choosing the **Pick Angle Between Items** button on the right side of the **Angle between items** edit box. The **Array** dialog box will be temporarily closed and you will be prompted to specify the angle between the items.

Figure 5-45 *Array created by specifying the number of items and the angle to fill*

Figure 5-46 *Array created by specifying the number of items and the angle between items*

Angle to fill & Angle between items. This method is used when you want to specify the angle between the items and total angle to fill, as shown in Figure 5-47. In this case, the number of items is not specified, but is automatically calculated using the total angle to fill in the array and the angle between the items. The angle to fill and the angle between the items can be entered in the respective edit boxes or can be specified on the screen using the respective buttons.

Figure 5-47 *Array created by specifying the angle to fill and the angle between items*

Rotate items as copied

This check box is selected to rotate the objects as they are copied around the center point. If this check box is cleared, the objects are not rotated as they are copied. This means that the replicated objects remain in the same orientation as the original object. You can have the items rotated around the center point as they are copied by selecting this check box. Here, the same face of each object points toward the pivot point. Figure 5-48 shows a polar array with items rotated as copied and Figure 5-49 shows a polar array with items not rotated as copied.

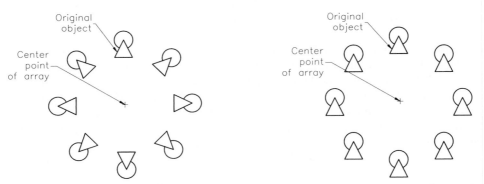

Figure 5-48 Items rotated as copied   *Figure 5-49* Items not rotated as copied

More

When this button is chosen, the **Array** dialog box expands providing the options of defining the base point of the objects, see Figure 5-50. The base point will maintain a constant distance from the center of the array when the items are copied in a polar fashion.

Figure 5-50 More options of the **Array** dialog box

You can use the object's default base points by selecting the **Set to object's default** check box. You can also specify the user-defined points by clearing this check box. Specify the *X* and *Y* coordinate values of the center of the array in the **X** and **Y** edit boxes in this area. You can also specify the base point on the screen by choosing the **Pick Base Point** button.

You can preview the array using the **Preview** button. Right-click if you are satisfied with the array or press ESC or left-click to make any modifications.

Tip
*By default, in a polar array, the objects are arrayed in the counterclockwise direction. But if you change the angle measuring direction from counterclockwise to clockwise using the **Units** dialog box, then the objects will be arrayed in the clockwise direction and not in the counterclockwise direction.*

MIRRORING SKETCHED OBJECTS

Ribbon:	Home > Modify > Mirror	**Toolbar:**	Modify > Mirror
Command:	MIRROR		

The **MIRROR** command creates a mirror copy of the selected objects. The objects can be mirrored at any angle. This command is helpful in drawing symmetrical figures.

When you invoke this command, AutoCAD will prompt you to select the objects and then the mirror line.

After you select the objects to be mirrored, AutoCAD prompts you to enter the first point of the mirror line and the second point of the mirror line. A mirror line is an imaginary line about which objects are reflected. You can specify the endpoints of the mirror line by specifying the points on the screen or by entering their coordinates. The mirror line can be specified at any angle. After the first point on the mirror line has been selected, AutoCAD displays the selected objects as they would appear after mirroring. Next, you need to specify the second endpoint of the mirror line. Once this is accomplished, AutoCAD prompts you to specify whether you want to retain the original figure (Figure 5-51) or delete it and just keep the mirror image of the figure (Figure 5-52). The prompt sequence when you choose this button is given next.

Select objects: *Select the objects to be mirrored.*
Select objects: Enter
Specify first point of mirror line: *Specify the first endpoint.*
Specify second point of mirror line: *Specify the second endpoint.*
Delete source objects? [Yes/No] <N>: *Enter Y for deletion, N for retaining previous objects.*

To mirror the objects at some angle, define the mirror line accordingly. For example, to mirror an object such that the mirrored object is placed at an angle of 90-degree from the original object, define the mirror line at an angle of 45-degree, see Figure 5-53.

Figure 5-51 *Creating a mirror image of an object using the **MIRROR** command*

Figure 5-52 *Retaining and deleting old objects after mirroring*

Text Mirroring

By default, the **MIRROR** command reverses all objects, except texts. But, if you want the text to be mirrored, (written backward), you need to modify the value of the **MIRRTEXT** system variable. This variable has two values that are given next (Figure 5-54).

1 = Text is reversed in relation to the original object.
0 = Restricts the text from being reversed with respect to the original object. This is the default value.

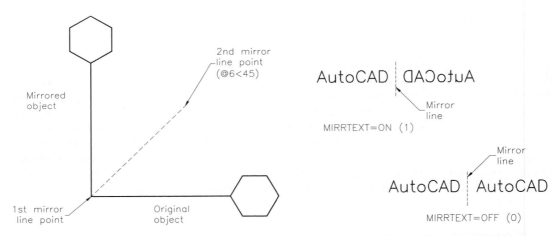

Figure 5-53 *Mirroring the object at an angle*   **Figure 5-54** *Using the **MIRRTEXT** system variable for mirroring the text*

Therefore, if you want the existing object to be mirrored, but at the same time you want the text to be readable and not reversed, set the value of the **MIRRTEXT** variable to **0**, and then mirror the objects.

BREAKING THE SKETCHED OBJECTS

Ribbon:	Home > Modify > Break at Point, Break	**Command:**	BREAK
Toolbar:	Modify > Break at Point, Break		

The **BREAK** command breaks an existing object into two or erases portions of the objects. This command can be used to remove a part of the selected objects or to break objects such as lines, arcs, circles, ellipses, xlines, rays, splines, and polylines. You can break the objects using the methods discussed next.

1 Point Option

This method of breaking the existing entities can be directly invoked by choosing the **Break at Point** button from the **Modify** panel of the **Ribbon**. Choose this button to break the object into two parts by specifying a breakpoint. When you choose this button, you will be prompted to select the object to be broken. Once you select the object to be broken, you will be prompted to specify the first break point. You will not be prompted to specify the second break point and the object will be automatically broken at the first point. The prompt sequence that will follow when you choose this button is given next.

Select object: *Select the object to be broken.*
Specify second break point or [First point]: _f
Specify first break point: *Specify the point at which the object should be broken.*
Specify second break point: @

2 Points Option

This method allows you to break an object between two selected points. The portion of the object between the two selected points is removed. The point at which you select the object becomes the first break point and then you are prompted to enter the second break point. The prompt sequence that will follow when you choose this button is given next.

> Select object: *Select the object to be broken.*
> Specify second break point or [First point]: *Specify the second break point on the object.*

The object is broken between these two points and the in-between portion of the object is removed, see Figure 5-55.

Figure 5-55 *Using the 2 Point option for breaking the line*

2 Points Select Option

This method is similar to the **2 Points** option; the only difference is that instead of making the selection point as the first break point, you are allowed to specify a new first point, see Figure 5-56. The prompt sequence that will follow when you choose the **Break** button is given next.

> Select object: *Select the object to be broken.*
> Enter second break point or [First point]: **F** ⏎
> Specify first break point: *Specify a new break point.*
> Specify second break point: *Specify the second break point on the object.*

If you need to work on arcs or circles, make sure that you work in the counterclockwise direction, or you may end up cutting the wrong part. In this case, the second point should be selected in the counterclockwise direction with respect to the first one (Figure 5-57).

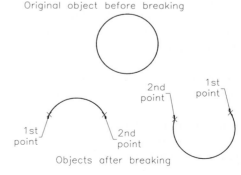

Figure 5-56 *Re-specifying the first break point for breaking the line*

Figure 5-57 *Breaking a circle*

You can use the **2 Points** and **2 Points Select** method to break an object into two without removing a portion in between. This can be achieved by specifying the same point on the object as the first and the second break points. If you specify the first break point on the line and the second break point beyond the end of the line, one complete end starting from the first break point will be removed. The extrusion direction of the selected object need not be parallel to the *Z* axis of the UCS.

Exercise 7   *General*

Break a line at five different places and then erase the alternate segments. Next, draw a circle and break it into four equal parts.

MEASURING SKETCHED OBJECTS

| **Ribbon:** | Home > Draw > Multiple Points > Measure   **Command:** MEASURE |

While drawing, you may need to segment an object at fixed distances without actually dividing it. You can use the **MEASURE** command (Figure 5-58) to accomplish it. This command places points or blocks (nodes) on the given object at a specified distance. The shape of these points is determined by the **PDMODE** system variable and the size is determined by the **PDSIZE** variable. The **MEASURE** command starts measuring the object from the endpoint closest to where the object is selected. When a circle is to be measured, an angle from the center is formed that is equal to the Snap rotation angle. This angle becomes the starting point of measurement, and the markers are placed at equal intervals in a counterclockwise direction.

Figure 5-58 *Invoking the* **MEASURE** *command from the* **Draw** *panel*

This command goes on placing markers at equal intervals of the specified distance without considering whether the last segment is at the same distance or not. Instead of entering a value, you can also select two points that will be taken as the distance. In Figure 5-59, a line and a circle is measured. The Snap rotation angle is 0-degree. The **PDMODE** variable is set to 3 so that X marks are placed as markers. The prompt sequence that will follow when you invoke this option from the **Draw** menu is given next.

Select object to measure: *Select the object to be measured.*
Specify length of segment or [Block]: *Specify the length for measuring the object.*

You can also place blocks as markers (Figure 5-60), but the block must already be defined within the drawing. You can align these blocks with the object to be measured. The prompt sequence is given next.

Select object to measure: *Select the object to be measured.*

Specify length of segment or [Block]: **B** `Enter`
Enter name of block to insert: *Enter the name of the block.*
Align block with object? [Yes/No] <Y>: *Enter **Y** to align, **N** to not align.*
Specify length of segment: *Enter the measuring distance.*

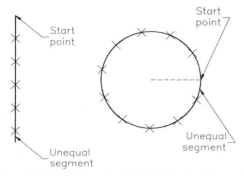

Figure 5-59 *Using the **MEASURE** command*

Figure 5-60 *Using blocks with the **MEASURE** command*

Note
You will learn more about creating and inserting blocks in Chapter 16 (Working with Blocks).

DIVIDING THE SKETCHED OBJECTS

| **Ribbon:** | Home > Draw > Multiple Points > Divide | **Command:** | MEASURE |

The **DIVIDE** command is used to divide an object into a specified number of equal length segments without actually breaking it. This command is similar to the **MEASURE** command except that here you do not have to specify the distance. The **DIVIDE** command calculates the full length of the object and places markers at equal intervals. This makes the last interval equal to the rest of the intervals. If you want a line to be divided, invoke this command and select the object to be divided. After this, you enter the number of divisions or segments. The number of divisions entered can range from 2 to 32,767. The prompt sequence that will follow when you invoke this command is given next.

Select object to divide: *Select the object you want to divide.*
Enter number of segments or [Block]: *Specify the number of segments.*

You can also place blocks as markers, but the block must be defined within the drawing. You can align these blocks with the object to be measured. The prompt sequence is given next.

Select object to divide: *Select the object to be measured.*
Enter number of segments or [Block]: **B** `Enter`
Enter name of block to insert: *Enter the name of the block.*
Align block with object? [Yes/No] <Y>: *Enter **Y** to align, **N** to not align.*
Enter number of segments: *Enter the number of segments.*

Chapter 5

Figure 5-61 shows a line and a circle divided using this command and Figure 5-62 shows the use of blocks for dividing the selected segment.

Dividing into five equal parts

Blocks not aligned with object

Blocks aligned with object

Figure 5-61 *Using the* **DIVIDE** *command*

Figure 5-62 *Using blocks with the* **DIVIDE** *command*

Note
The size and shape of the points placed by the **DIVIDE** *and* **MEASURE** *commands are controlled by the* **PDSIZE** *and* **PDMODE** *system variables.*

JOINING THE SKETCHED OBJECTS

| **Ribbon:** | Home > Modify > Join | **Command:** | Join |

The **JOIN** command is used to join two or more collinear lines or arcs lying on the same imaginary circle. The lines or arcs can have gap between them. This command can also be used to join two or more polylines or splines, but they should be on the same plane and should not have any gap between them. Note that this command can also be used to join the two end points of an arc to convert it into a circle. As you invoke this command, you will be prompted to select the source object. The source object is the one with which the other objects will be joined. The prompt sequence depends on the type of source object selected. The following prompt sequence is displayed when you invoke this command:

Select source object: *Select a sketched entity (line, polyline, arc, elliptical arc or spline).*

Joining Collinear Lines

If you have selected line as the source object, you will be prompted to select lines to join to source. Remember that the lines will be joined only if they are collinear. If there is any gap between the endpoint of the lines, it will be filled. The prompt sequence for joining two lines with a source line is given next.

Select lines to join to source: *Select the first line to join to the source.*
Select lines to join to source: *Select the second line to join to the source.*
Select lines to join to source: `Enter`
2 lines joined to source

Joining Arcs

Arcs can be joined using the **JOIN** command only if they lie on the same imaginary circle on which the source arc lie. The prompt sequence that follows when you select an arc as the source object is given next.

> Select arcs to join to source or [cLose]: *Select the first arc to join to the source.*
> Select arcs to join to source or: *Select the second arc to join to the source or* Enter

Enter **L** at the **Select arcs to join to source or [cLose]** prompt to close the arc and convert it into a circle.

Joining Elliptical Arcs

Elliptical arcs can be joined using the **JOIN** command only if they lie on the same imaginary ellipse on which the source elliptical arc lie. The prompt sequence that follows when you select an elliptical arc as the source object is given next.

> Select elliptical arcs to join to source or [cLose]: *Select the first elliptical arc to join to the source.*
> Select elliptical arcs to join to source or: *Select the second arc to join to the source or* Enter

Enter **L** at the **Select arcs to join to source or [cLose]** prompt to close the arc and convert it into an ellipse.

Joining Splines

Select a spline as the source object. You are now prompted to select splines to join to the source. The splines will be joined only if they are coplanar and are connected at their ends. The prompt sequence that follows when you when you select **Spline** as the source object is given next.

> Select splines to join to source: *Select the first spline to join to the source.*
> Select splines to join to source: *Select the second spline to join to the source or* Enter

Joining Polylines

Only the end-connected polyline can be joined using this command. The prompt sequence when you select a polyline as the source object is given next.

> Select objects to join to source: *Select a polyline/arc/line to join to the source.*
> Select objects to join to source: *Select a polyline/arc/line to join to the source or* Enter

Self-Evaluation Test

Answer the following questions and then compare them to those given at the end of this chapter:

1. When you shift a group of objects using the **MOVE** command, the size and orientation of these objects is changed. (T/F)

2. The **COPY** command makes the copies of the selected object, leaving the original object intact. (T/F)

3. A fillet cannot be created between two parallel and nonintersecting lines. (T/F)

4. With the **BREAK** command, when you select an object, the selection point becomes the first break point. (T/F)

5. Depending on the side to be offset, you can create smaller or larger circles, ellipses, and arcs with the _____ command.

6. The _____ command prunes the objects that extend beyond a required point of intersection.

7. The offset distance is stored in the _____ system variable.

8. Instead of specifying the scale factor, you can use the _____ option to scale an object with the reference of another object.

9. If the _____ system variable is set to a value of 1, the mirrored text is not reversed with respect to the original object. (T/F)

10. There are two types of arrays: _____ and _____.

Review Questions

Answer the following questions:

1. In the case of the **Through** option of the **OFFSET** command, you do not have to specify a distance; you simply have to specify an offset point. (T/F)

2. With the **BREAK** command, you cannot break an object in two parts without removing a portion in between. (T/F)

3. With the **FILLET** command, the extrusion direction of the selected object must be parallel to the *Z* axis of the UCS. (T/F)

4. A rectangular array can be rotated. (T/F)

5. Which object selection allows you to create a polygon that selects all the objects, touching or lying inside it?

   (a) **Last**   (b) **WPolygon**
   (c) **CPolygon**   (d) **Fence**

6. Which command is used to copy an existing object using a base point to another drawing?

   (a) **COPY**   (b) **COPYBASE**
   (c) **MOVE**   (d) **None**

7. Which command is used to change the size of an existing object with respect to an existing entity?

   (a) **ROTATE**   (b) **SCALE**
   (c) **MOVE**   (d) **None**

8. Which option of the **LENGTHEN** command modifies the length of the selected entity such that irrespective of the original length, the entity acquires the specified length?

   (a) **DElta**   (b) **DYnamic**
   (c) **Percent**   (d) **Total**

9. Which selection option is used to select the most recently drawn entity?

   (a) **Last**   (b) **Previous**
   (c) **Add**   (d) **None**

10. If an entire selected object is within a window or crossing specification, the _____ command works like the **MOVE** command.

11. The _____ option of the **EXTEND** command is used to extend objects to the implied boundary.

12. With the **FILLET** command, using the **Polyline** option, if the selected polyline is not closed, the _____ corner is not filleted.

13. When the chamfer distance is zero, the chamfer created is in the form of a _____.

14. AutoCAD saves the previous selection set and lets you select it again by using the _____ selection option.

15. The _____ selection option is used to select objects by creating a section line touching the objects to be selected.

Chapter 5

Exercises

Exercise 8   *General*

Draw the object shown in Figure 5-63. Use the **DIVIDE** command to divide the circle and use the **NODE** object snap to select the points. Assume the dimensions of the drawing.

Exercise 9   *Mechanical*

Draw the object shown in Figure 5-64 and save the drawing. Assume the missing dimensions.

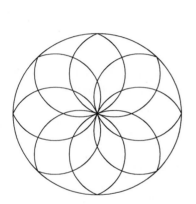

Figure 5-63 *Drawing for Exercise 8*

Figure 5-64 *Drawing for Exercise 9*

Exercise 10   *Mechanical*

Draw the drawing shown in Figure 5-65 and save the drawing.

Figure 5-65 *Drawing for Exercise 10*

Exercise 11   *Mechanical*

Draw the object shown in Figure 5-66 and save the drawing.

Figure 5-66 *Drawing for Exercise 11*

Exercise 12   *Mechanical*

Draw the object shown in Figure 5-67 and save the drawing. Assume the missing dimensions.

Figure 5-67 *Drawing for Exercise 12*

Problem-Solving Exercise 1   *Mechanical*

Draw the object shown in Figure 5-68 and save the drawing. Assume the missing dimensions.

Problem-Solving Exercise 2   *Mechanical*

Draw the object shown in Figure 5-69 and save the drawing. Assume the missing dimensions.

Chapter 5

Fillets and Rounds=R.1

Figure 5-68 *Drawing for Problem Solving Exercise 1*

Figure 5-69 *Drawing for Problem Solving Exercise 2*

Problem-Solving Exercise 3   *Architectural*

Draw the dining table with chairs, as shown in Figure 5-70, and save the drawing. Assume the missing dimensions.

Problem-Solving Exercise 4   *Architectural*

Draw the reception table with chairs, as shown in Figure 5-71, and save the drawing. The dimensions of the chairs are the same as those in Problem-Solving Exercise 3.

Problem-Solving Exercise 5   *Architectural*

Draw the center table with chairs, as shown in Figure 5-72, and save the drawing. The dimensions of the chairs are the same as those in Problem-Solving Exercise 3.

Figure 5-70 *Drawing for Problem-Solving Exercise 3*

Figure 5-71 *Drawing for Problem-Solving Exercise 4*

Figure 5-72 *Drawing for Problem-Solving Exercise 5*

Problem-Solving Exercise 6   *Architectural*

Draw the object, as shown in Figure 5-73, and save the drawing. Refer to the note mentioned in the drawing to create an arc of radius 30.

Figure 5-73 *Drawing for Problem-Solving Exercise 6*

Problem-Solving Exercise 7   *Mechanical*

Draw the object, as shown in Figure 5-74, and save the drawing.

Figure 5-74 *Drawing for Problem-Solving Exercise 7*

Problem-Solving Exercise 8   *Mechanical*

Draw the object shown in Figure 5-75, and save the drawing.

Figure 5-75 *Drawing for Problem-Solving Exercise 8*

Problem-Solving Exercise 9   *Mechanical*

Draw the object shown in Figure 5-76. The dimensions given in the figure are only for your reference. **

Figure 5-76 *Drawing for Problem-Solving Exercise 9*

Problem-Solving Exercise 10　　　　*Mechanical*

Draw the object shown in Figure 5-77. The dimensions given in the figure are only for your reference. **

Figure 5-77　Drawing for Problem-Solving Exercise 10

Answers to Self-Evaluation Test

1. F, **2.** T, **3.** F, **4.** T, **5. OFFSET**, **6. TRIM**, **7. OFFSETDIST**, **8. Reference**, **9. MIRRTEXT**, **10. Rectangular, Polar**

Chapter 6

Editing Sketched Objects-II

Learning Objectives

After completing this chapter, you will be able to:
- *Understand the concept of grips and adjust grip settings.*
- *Stretch, move, rotate, scale, and mirror objects with grips.*
- *Use the **MATCHPROP** command to match the properties of the selected object.*
- *Use the **Properties** palette for editing objects.*
- *Use the **QSELECT** command to select objects.*
- *Manage contents using the **DesignCenter**.*
- *Use the Inquiry commands.*
- *Use the **REDRAW** and **REGEN** commands.*
- *Use the **ZOOM** command and its options.*
- *Understand the **PAN** and **VIEW** commands.*
- *Understand the use of the **Aerial View** window.*

EDITING WITH GRIPS

Grips provide a convenient and quick means of editing objects. With grips you can stretch, move, rotate, scale, and mirror objects, change properties, and load the Web browser. Grips are small squares that are displayed on an object at its definition points when it is selected. The number of grips depends on the selected object. For example, a line has three grip points, a polyline segment has two, and an arc has three triangular grips along with the small square grips. These triangular grips resemble arrowheads. These arrowheads point towards the direction, in which the arc can be edited dynamically. Similarly, a circle has five grip points and a dimension (vertical) has five. When you select the **Enable grips** and **Noun/verb selection** check boxes in the **Selection** tab of the **Options** dialog box, a small square (aperture box) at the intersection of crosshairs is displayed (Figure 6-1). The grip location of some of the objects is shown in Figure 6-2.

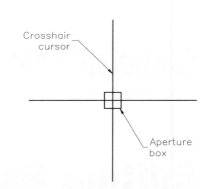

Figure 6-1 *Aperture box at the intersection of crosshairs*

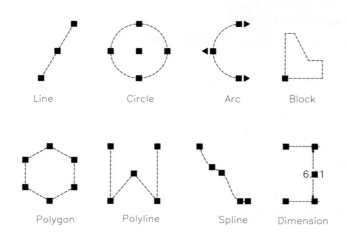

Figure 6-2 *Grip location of various objects*

TYPES OF GRIPS

Grips can be classified into three types: unselected grips, hover grips, and selected grips. The selected grips are also called hot grips. When you select an object, the grips are displayed at the definition points of the object, that is highlighted as a dashed line. These grips are called unselected grips (blue). Now, if you move the cursor over an unselected grip, and pause for a second, the grid will be displayed in orange. These grips are called hover grips. Dimensions corresponding to a hover grip are displayed when you place the cursor on the grip, as shown in Figure 6-3. Next, if you select a grip on this object, it becomes a hot grip (filled red square). Once the grip is hot, the object can be edited. To cancel the grip, press ESC. If you press ESC once, the hot grip changes to an unselected grip. You can also snap to the unselected grip.

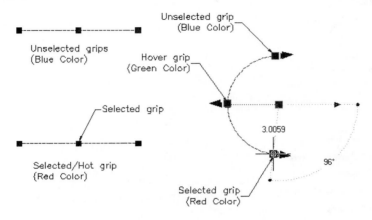

Figure 6-3 *Hover grip dimensions*

ADJUSTING GRIP SETTINGS

Application Menu:	Options
Command:	OPTIONS

The grip settings can be adjusted using the options under the **Selection** tab of the **Options** dialog box. This dialog box can also be invoked by choosing **Options** from the shortcut menu, see Figure 6-4. The shortcut menu is displayed upon right-clicking in the drawing area.

The options in the **Selection** tab of the **Options** dialog box (Figure 6-5) are discussed next.

Grip Size Area

The **Grip Size** area of the **Selection** tab of the **Options** dialog box consists of a slider bar and a rectangular box that displays the size of the grip. To adjust the size, move the slider box left or right. The size can also be adjusted by using the **GRIPSIZE** system variable. The **GRIPSIZE** variable is defined in pixels, and its value can range from 1 to 255 pixels.

Figure 6-4 *Invoking the **Options** dialog box from the shortcut menu*

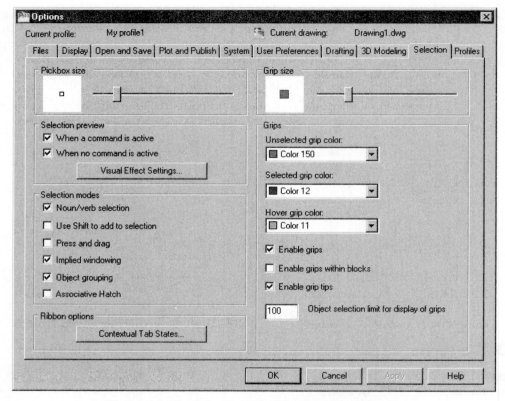

*Figure 6-5 The **Selection** tab of the **Options** dialog box*

Grips Area

The **Grips** area is used to control the display and the color of the grips.

Unselected grip color

To set the color of the unselected grip, select a color from this drop-down list. Alternatively, select the **Select Color** option to display the **Select Color** dialog box and select the color from the dialog box. This color can also be set using the **GRIPCOLOR** system variable.

Selected grip color

To set the color of the selected grip, select a color from this drop-down list. Alternatively, select the **Select Color** option to display the **Select Color** dialog box and select the color from the dialog box This color can also be set using the **GRIPHOT** system variable.

Hover grip color

To set the color of the hover grip, select a color from this drop-down list. Alternatively, select the **Select Color** option to display the **Select Color** dialog box and select the color from the dialog box This color can also be set using the **GRIPHOVER** system variable.

The **Grips** area has three check boxes; **Enable grips**, **Enable grips within blocks**, and **Enable grip tips**. The grips can be enabled by selecting the **Enable Grips** check box. They can also be enabled by setting the **GRIPS** system variable to 1. The second check box, **Enable grips within blocks**, enables the grips within a block. If you select this box, AutoCAD will display grips for every object in the block. If you disable the display of grips within a block, the block will have only one grip at its insertion point. You can also enable the grips within a block by setting the value of the **GRIPBLOCK** system variable to 1 (On). If **GRIPBLOCK** is set to 0 (Off), AutoCAD will display only one grip for a block at its insertion point (Figure 6-6). The third check box, **Enable grip tips**, enables

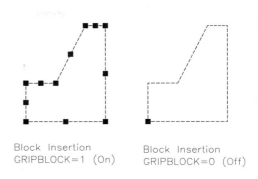

Block Insertion
GRIPBLOCK=1 (On)   Block Insertion
                        GRIPBLOCK=0 (Off)

*Figure 6-6 Block insertion with **GRIPBLOCK** set to 1 and 0*

you to display the grip tips when the cursor moves over the custom object that supports grip tips. If you disable this check box, the grip tips are not displayed when the cursor moves over the custom object. You can also enable the grip tips by setting the value of the **GRIPTIPS** system variable to 0 (Off). If **GRIPTIPS** is set to 1 (On), AutoCAD will display the grip tips for the custom object.

Note
*If the block has a large number of objects, and if **GRIPBLOCK** is set to 1 (On), AutoCAD will display grips for every object in the block. Therefore, it is recommended that you set the system variable **GRIPBLOCK** to 0 or clear the **Enable grips within blocks** check box in the **Selection** tab of the **Options** dialog box.*

Object selection limit for display of grips
This text box is used to specify the maximum number of objects that can be selected at a single attempt for the display of grips. If you select objects more than that specified in the text box using a single selection method, the grips will not be displayed. Note that this limit is set only for those objects that are selected at a single attempt using any of the **Crossing**, **Window**, **Fence**, or **All** options.

EDITING OBJECTS WITH GRIPS
As mentioned earlier, you can perform different kinds of editing operations using the selected grip. The editing operations are discussed next.

Stretching Objects with Grips (Stretch Mode)
If you select an object, AutoCAD displays the unselected grips at the definition points of the object. When you select a grip for editing, you are automatically in the **Stretch** mode. The **Stretch** mode has a function similar to the **STRETCH** command. When you select a grip, it acts as a base point and is called a base grip. You can also select several grips by holding the

SHIFT key down and then selecting the grips. Now, release the SHIFT key and select one of the hot grips to stretch them simultaneously. The geometry between the selected base grips is not altered. You can also make copies of the selected objects or define a new base point. When selecting grips on text objects, blocks, midpoints of lines, centers of circles and ellipses, and point objects in the stretch mode, the selected objects are moved to a new location. The following example illustrates the use of the **Stretch** mode.

1. Use the **PLINE** command to draw a W-shaped figure shown in Figure 6-7(a).

2. Select the object that you want to stretch [Figure 6-7(a)]. When you select the object, grips will be displayed at the endpoints of each object. A polyline has two grip points. If you use the **LINE** command to draw the object, AutoCAD will display three grips for each object.

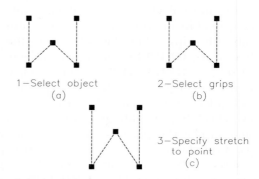

3. Hold the SHIFT key down, and select the grips that you want to stretch [grips on the lower endpoints of the two vertical lines in Figure 6-7(b)]. The selected grips will become hot grips and the color will change from blue to red.

Figure 6-7 *Using the* *Stretch* *mode to stretch the lines*

Note
You have to make sure that you hold the SHIFT key before selecting even the first grip. You cannot hold the SHIFT key and select more grips if the first grip is selected without holding it.

4. Select one of the selected (hot grip) grips, and specify a point to which you want to stretch the line [Figure 6-7(c)]. When you select a grip, the following prompt is displayed in the Command prompt area.

    STRETCH
    Specify stretch point or [Base point/Copy/Undo/eXit]:

    The Stretch mode has several options: **Base point**, **Copy**, **Undo**, and **eXit**. You can use the **Base point** option to define the base point and the **Copy** option to make copies.

5. Select the grip where the two lines intersect. Right-click to display the shortcut menu (Figure 6-8) and choose the **Copy** option. Select the points, as shown in Figure 6-9(a). Each time you select a point, AutoCAD will make a copy Figure 6-9(b). You can also specify stretch points by entering the numerical value in the dimensional input boxes in the drawing area. If you press the CTRL key when specifying the point to which the object is to be stretched, without selecting the copy option, then also AutoCAD allows you to make multiple copies of the selected object. Also, if you press the CTRL key again

when specifying the next point, the cursor snaps to a point whose location is based on the distance between the first two points, that is, the distance between the selected object and the location of the copy of the selected object.

6. Make a copy of the drawing, as shown in Figure 6-9(c). Select the object, and then select the grip where the two lines intersect. When AutoCAD displays the **STRETCH** prompt, choose the **Base Point** option from the shortcut menu or enter B at the Command prompt. Select the bottom left grip as the base point, and then give the displacement point, as shown in Figure 6-9(d).

7. To terminate the grip editing mode, right-click when the grip is hot to display the shortcut menu and then select **eXit**. You can also enter X at the Command prompt or press ESC to exit.

Figure 6-8 Shortcut menu with editing options

Note
*You can select an option (**Copy** or **Base Point**) from the shortcut menu that can be invoked by right-clicking your pointing device after selecting a grip. The different modes can also be selected from the shortcut menu. You can also cycle through all the different modes by selecting a grip and pressing the ENTER key or the SPACEBAR.*

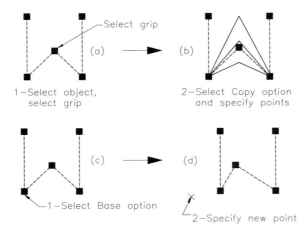

*Figure 6-9 Using the **Copy** and **Base point** options of the **Stretch** mode*

Moving Objects with Grips (Move Mode)

The **Move** mode lets you move the selected objects to a new location. When you move objects, their size and angles do not change. You can also use this mode to make the copies of the selected objects or to redefine the base point. The following example illustrates the use of the **Move** mode.

1. Use the **LINE** command to draw the shape, as shown in Figure 6-10(a). When you select the objects, grips will be displayed at the definition points and the object will be highlighted.

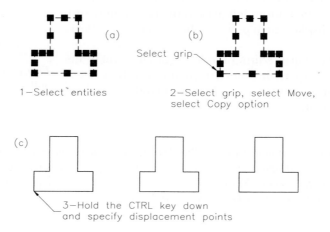

Figure 6-10 *Using the **Move** mode to move and make copies of the selected objects*

2. Select the grip located at the lower left corner and then choose **Move** from the shortcut menu. You can also invoke the **Move** mode by giving a null response by pressing the SPACEBAR or ENTER key. AutoCAD will display the following prompt in the Command prompt area:

    MOVE
    Specify move point or [Base point/Copy/Undo/eXit]:

3. Hold down the CTRL key, and then enter the first displacement point. The distance between the first and the second object defines the snap offset for subsequent copies. While holding down the CTRL key, move the screen crosshairs to the next snap point and select the point. AutoCAD will make a copy of the object at this location. If you release the CTRL key, you can specify any point where you want to place a copy of the object. You can also enter coordinates to specify the displacement.

Rotating Objects with Grips (Rotate Mode)

The **Rotate** mode allows you to rotate objects around the base point without changing their size. The options of the **Rotate** mode can be used to redefine the base point, specify a reference angle, or make multiple copies that are rotated about the specified base point. You can access the **Rotate** mode by selecting the grip and then selecting **Rotate** from the shortcut menu, or by giving a null response twice by pressing the SPACEBAR or the ENTER key. The following example illustrates the use of the **Rotate** mode.

1. Use the **LINE** command to draw the shape, as shown in Figure 6-11(a). When you select the objects, grips will be displayed at the definition points and the shape will be highlighted.

2. Select the grip located at the lower left corner and then invoke the **Rotate** mode. AutoCAD will display the following prompt.

   ROTATE
   Specify a rotation angle or [Base point/Copy/Undo/Reference/eXit]:

3. At this prompt, enter the rotation angle. By default the rotation angle will be entered through the dimensional input below the cursor in the drawing area. AutoCAD will rotate the selected objects by the specified angle [Figure 6-11(b)].

4. Make a copy of the original drawing, as shown in Figure 6-11(c). Select the objects, and then select the grip located at its lower left corner. Invoke the **Rotate** mode and then select the **Copy** option from the shortcut menu or enter **C** (Copy) at the Command prompt or the **Copy** option from the dynamic preview. Enter the rotation angle. AutoCAD will rotate the copy of the object through the specified angle [Figure 6-11(d)].

Figure 6-11 *Using the* **Rotate** *mode to rotate and make the copies of the selected objects*

5. Make another copy of the object, as shown in Figure 6-12(a). Select it, and then select the grip at point (P0). Access the **Rotate** mode and the copy option as described earlier. Select the **Reference** option from the shortcut menu, enter **R** at the prompt sequence, or select the **Reference** option from the dynamic preview. The prompt sequence is given next.

ROTATE (multiple) **
Specify rotation angle or [Base point/Copy/Undo/Reference/eXit]: **R**
Specify reference angle <0>: *Select the grip at (P1).*
Specify second point: *Select the grip at (P2).*
Specify new angle or [Base point/Copy/Undo/Reference/eXit]: **45**

In response to the **Specify reference angle <0>:** prompt, select the grips at points (P1) and (P2) to define the reference angle. When you enter the new angle, AutoCAD will rotate and insert a copy at the specified angle [Figure 6-12(c)]. For example, if the new angle is 45-degree, the selected objects will be rotated about the base point (P0) so that the line P1P2 makes a 45-degree angle with respect to the positive *X* axis.

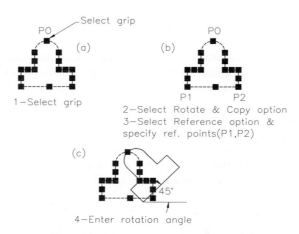

Figure 6-12 *Using the **Rotate** mode to rotate by giving a reference angle*

Scaling Objects with Grips (Scale Mode)

The **Scale** mode allows you to scale objects with respect to the base point without changing their orientation. The options of **Scale** mode can be used to redefine the base point, specify a reference length, or make multiple copies that are scaled with respect to the specified base point. You can access the **Scale** mode by selecting the grip and then choosing **Scale** from the shortcut menu, or giving a null response three times by pressing the SPACEBAR or the ENTER key. The following example illustrates the use of the **Scale** mode.

1. Use the **PLINE** command to draw the shape shown in Figure 6-13(a). When you select the objects, they will be highlighted and the grips will be displayed at the definition points.

2. Select the grip located at the lower left corner as the base grip, and then invoke the **Scale** mode. AutoCAD will display the following prompt in the Command prompt area.

SCALE
Specify scale factor or [Base point/Copy/Undo/Reference/eXit]:

3. At this prompt, enter the scale factor or move the cursor and select a point to specify a new size. AutoCAD will scale the selected objects by the specified scale factor [Figure 6-13(b)]. If the scale factor is less than 1 (<1), the objects will be scaled down by the specified factor. If the scale factor is greater than 1 (>1), the objects will be scaled up.

4. Make a copy of the original drawing, as shown in Figure 6-13(c). Select the objects, and then select the grip located at their lower left corner. Invoke the **Scale** mode. At the following prompt, enter C (Copy), and then enter B for the base point.

    SCALE (multiple) **
    Specify scale factor or [Base point/Copy/Undo/Reference/eXit]: **B**

5. At the **Specify base point** prompt, select the point (P0) as the new base point, and then enter **R** at the following prompt.

    SCALE (multiple) **
    Specify scale factor or [Base point/Copy/Undo/Reference/eXit]: **R**
    Specify reference length <1.000>: *Select grips at (P1) and (P2).*

After specifying the reference length at the **Specify new length or [Base point/Copy/ Reference/eXit]** prompt, enter the actual length of the line. AutoCAD will scale the objects so that the length of the bottom edge is equal to the specified value [Figure 6-13(c)].

4—Select base point, Select Reference option
Select ref. points (P1,P2), Specify actual length

Figure 6-13 *Using the **Scale** mode to scale and make the copies of the selected objects*

Mirroring Objects with Grips (Mirror Mode)

The **Mirror** mode allows you to mirror the objects across the mirror axis without changing their size. The mirror axis is defined by specifying two points. The first point is the base point, and the second point is the point that you select when AutoCAD prompts for the second point. The options of the **Mirror** mode can be used to redefine the base point and make a mirror copy of the objects. You can access the **Mirror** mode by selecting a grip and then choosing **Mirror** from the shortcut menu, or giving a null response four times by pressing the SPACEBAR or the ENTER key. The following is the example for the **Mirror** mode.

1. Use the **PLINE** command to draw the shape shown in Figure 6-14(a). When you select the object, grips will be displayed at the definition points and the object will be highlighted.

2. Select the grip located at the lower right corner (P1), and then invoke the **Mirror** mode. The following prompt is displayed.

    MIRROR
    Specify second point or [Base point/Copy/Undo/eXit]:

3. At this prompt, enter the second point (P2). AutoCAD will mirror the selected objects with line P1P2 as the mirror axis, as shown in Figure 6-14(b).

4. Make a copy of the original figure, as shown in Figure 6-14(c). Select the object, and then select the grip located at its lower right corner (P1). Invoke the **Mirror** mode and then choose the **Copy** option to make a mirror image while retaining the original object. Alternatively, you can also hold down the SHIFT key and make several mirror copies by specifying the second point.

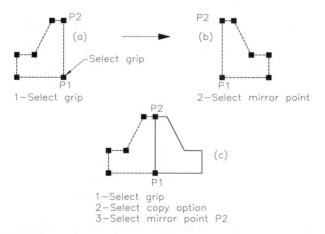

Figure 6-14 *Using the* **Mirror** *mode to create a mirror image of the selected objects*

5. Select point (P2) in response to the prompt **Specify second point or [Base point/Copy/ Undo/eXit]**. AutoCAD will create a mirror image, and the original object will be retained.

Note
*You can use some editing commands such as **ERASE, MOVE, ROTATE, SCALE, MIRROR**, and **COPY** on an object with unselected grips. However, this is possible only if the **PICKFIRST** system variable is set to 1 (On).*

You cannot select an object when the grip is hot.

To remove an object from the selection set displaying grips, press the SHIFT key and then select the particular object. This object, which is removed from the selection set, will no longer be highlighted.

LOADING HYPERLINKS

If you have already added a hyperlink to the object, you can also use the grips to open a file associated with it. For example, the hyperlink could start a word processor, or activate the Web browser and load a Web page that is embedded in the selected object. To launch the Web browser that provides hyperlinks to other Web pages, select the URL-embedded object and then right-click to display the shortcut menu. In the shortcut menu, choose the **Hyperlink > Open** option and AutoCAD will automatically load the Web browser. When you move the cursor over or near the object that contains a hyperlink, AutoCAD displays the hyperlink information with the cursor.

Note
Working with hyperlinks is discussed in later chapters.

EDITING GRIPPED OBJECTS

You can also edit the properties of the gripped objects by using the **Properties** panel in the **Home** tab of the **Ribbon**, see Figure 6-15. When you select objects without invoking a command, the grips (rectangular boxes) are displayed on the objects. The gripped objects are highlighted and will display grips (rectangular boxes) at their grip points. For example,

*Figure 6-15 The **Properties** panel*

to change the color of the gripped objects, select the **Object Color** drop-down list in the **Properties** panel and then select a color from it. The color of the gripped objects will change to the selected color. Similarly, to change the layer, lineweight, or linetype of the gripped objects, select the required linetype, lineweight, or layer from the corresponding drop-down lists. If the gripped objects have different colors, linetypes, or lineweights, the **Object Color**, **Linetype**, and **Lineweight** drop-down lists will appear blank. You can also change the plot style of the selected objects using this panel.

CHANGING PROPERTIES USING PROPERTIES PALETTE

Ribbon:	View > Palettes > Properties	**Command:** PROPERTIES
Application Menu:	Drawing Utilities > Properties	
Toolbar:	Quick Access > Properties (*Customize to Add*)	

 As mentioned earlier, each object has a number of properties associated to it such as the color, layer, linetype, line weight, and so on. You can modify the properties of an object by using the **PROPERTIES** command. When you invoke this command, AutoCAD will display the **Properties** palette, see Figure 6-16. The **Properties** palette can also be displayed when you double-click on the object to be edited. The contents of the **Properties** palette change according to the objects selected. For example, if you select a text entity, the related properties such as its height, justification, style, rotation angle, obliquing factor, and so on, will be displayed.

The **Properties** palette can also be invoked from the shortcut menu. To do so, select an entity and right-click in the drawing area. Choose the **Properties** option from the shortcut menu. If you select more than one entity, the common properties of the selected entities will be displayed in the **Properties** palette. To change the properties of the selected entities, click in the cell next to the name of the property and change the values manually. Alternatively, you can choose from the available options in the drop-down list, if one is available. You can cycle through the options by double-clicking in the property cell.

Figure 6-16 *The* ***Properties*** *palette for editing the properties of an entity*

> **Note**
> *Some of the options of the* ***Properties*** *palette have been explained in Chapter 4. Other options are explained in detail in later chapters.*

CHANGING PROPERTIES USING GRIPS

You can also use the grips to change the properties of a single or multiple objects. To change the properties, select the object to display the grips and then right-click to display the shortcut menu. Choose the **Properties** option to display the **Properties** palette. If you select a circle, AutoCAD will display **Circle** in the **No selection** drop-down list on the upper left corner of the **Properties** palette. Similarly, if you select text, **Text** is displayed in the drop-down list. If you select several objects, AutoCAD will display all the objects in the selection drop-down list of the **Properties** palette. You can use this palette to change the properties (color, layer, linetype, linetypes scale, lineweight, thickness, and so on) of the gripped objects.

MATCHING PROPERTIES OF SKETCHED OBJECTS

Ribbon:	Home > Clipboard > Match Properties	**Command:**	MATCHPROP
Toolbar:	Quick Access > Match Properties (*Customize to Add*)		
Menu Bar:	Modify > Match Properties		

The **MATCHPROP** command can be used to change some properties like color, layer, linetype, and linetype scale of the selected objects. However, in this case, you need a source object whose properties will be forced on the destination objects. When you invoke this command, AutoCAD will prompt you to select the source object and then the destination objects. The properties of the destination objects will be changed to that of the source object. This command is a transparent command and can be used inside another command. The prompt sequence that will follow when you choose the **Match Properties** button from the **Properties** panel of the **Home** tab in the **Ribbon** is given next.

Select source object: *Select the source object.*
Current active settings: Color Layer Ltype Ltscale Lineweight Thickness PlotStyle Dim Text Hatch Polyline Viewport Table Material Shadow display Multileader
Select destination object(s) or [Settings]:

If you select the destination object in the **Select destination object(s) or [Settings]** prompt, the properties of the source object will be forced on it. If you select the **Settings** option, AutoCAD displays the **Property Settings** dialog box (Figure 6-17). The properties displayed are those of the source object. You can use this dialog box to edit the properties that are copied from the source to destination objects.

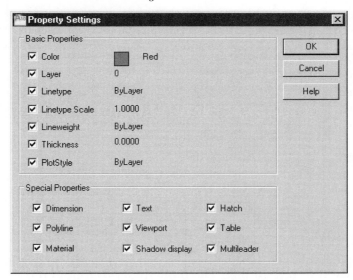

*Figure 6-17 The **Property Settings** dialog box*

QUICK SELECTION OF SKETCHED OBJECTS

Ribbon:	Home > Utilities > Quick Select	Command:	QSELECT

The **QSELECT** command creates a new selection set that will either include or exclude all objects that match the specified object type and property criteria. The **QSELECT** command can be applied to the entire drawing or existing selection set. If a drawing is partially opened, **QSELECT** does not consider the objects that are not loaded. The **QSELECT** command can be invoked by choosing the **Quick Select** button in the **Utilities** panel of the **Ribbon**. In the shortcut menu, the **QSELECT** command can be invoked by choosing **Quick Select**. When you invoke this command, the **Quick Select** dialog box will be displayed, see Figure 6-18. The **Quick Select** dialog box specifies the object filtering criteria and creates a selection set from it.

Figure 6-18 The Quick Select dialog box

Apply to

The **Apply to** drop-down list specifies whether to apply the filtering criteria to the entire drawing or to the current selection set. If there is an existing selection set, the **Current selection** is the default value. Otherwise, the entire drawing is the default value. You can select the objects to create a selection set by choosing the **Select objects** button on the right side of this drop-down list. The **Quick Select** dialog box is temporarily closed when you choose this button and you will be prompted to select the objects. The dialog box will be redisplayed once a selection set is made.

Object type

This drop-down list specifies the type of object to be filtered. It lists all the available object types and if some objects are selected, it lists all the selected object types. **Multiple** is the default setting.

Properties

This list box displays the properties to be filtered. All the properties related to the object type will be displayed in this list box. The property selected from this list box will define the options that will be available in the **Operator** and **Value** drop-down list.

Operator

This drop-down list specifies the range of the filter for the chosen property. The filters that are available are given next.

- Equals   =   • Not Equal   < >
- Greater than   >   • Less than   <
- Select All
- Wildcard Match (Available only for text objects that can be edited and if you choose **Hyperlink** option from the **Properties** list.)

Note
*The **Value** drop-down list will not be available when you select **Select All** from the **Operator** drop-down list.*

Value

This drop-down list specifies the property value of the filter. If the values are known, it becomes a list of the available values from which you can select a value. Otherwise, you can enter a value.

How to apply Area

The options under this area are used to specify whether the filtered entities will be included or excluded from the new selection set. This area provides the following two radio buttons.

Include in new selection set

If this radio button is selected, the filtered entities will be included in the new selection set. If selected, this radio button creates a new selection set composed only of those objects that conform to the filtering criteria.

Exclude from new selection set

If this radio button is selected, the filtered entities will be excluded from the new selection set. This radio button creates a new selection set of objects that do not conform to the filtering criteria.

Append to current selection set

This creates a cumulative selection set by using multiple uses of Quick Select. It specifies whether the objects selected using the **QSELECT** command replace the current selection set or append the current selection set.

Tip
*Quick Select supports objects and their properties that are created by some other applications. If the object have properties other than AutoCAD, then the source application of the object should be running for the properties to be available in the **QSELECT**.*

MANAGING CONTENTS USING THE DESIGNCENTER

Ribbon:	View > Palettes > DesignCenter	Command:	ADCENTER
Menu Bar:	Tools > Palettes > DesignCenter		

The **DesignCenter** window is used to locate and organize drawing data, and to insert blocks, layers, external references, and other customized drawing content. These contents can be selected from either your own files, local drives, a network, or the Internet. You can even access and use the contents between the files or from the Internet. You can use the **DesignCenter** to conveniently drag and drop any information that has been previously created into the current drawing. This powerful tool reduces the repetitive tasks of creating information that already exists. To invoke the **DesignCenter** window, choose the **DesignCenter** button from the **Palettes** panel in the **View** tab of the **Ribbon**; the **DesignCenter** window will be displayed, see Figure 6-19.

*Figure 6-19 The **DesignCenter** window*

This window can be moved to any location on the screen by picking and dragging it with the grab bar located on the left of the window. You can also resize it by clicking the borders and dragging them to the right or left. Right-clicking on the title bar of the window displays a shortcut menu that gives the options to move, resize, close, dock, and hide the **DesignCenter** window. The **Auto-Hide** button on the grab bar acts as a toggle for hiding and displaying the **DesignCenter**. Also, double-clicking on the title bar of the window docks the **DesignCenter** window if it is undocked and vice versa. To use this option, make sure that the **Allow Docking** option is selected from the shortcut menu that is displayed by right-clicking on the grab bar.

Note

*The **DesignCenter** can be turned on and off by pressing the CTRL+2 keys.*

Figure 6-20 shows the **DesignCenter** toolbar buttons. When you choose the **Tree View Toggle** button on the **DesignCenter** toolbar, it displays the **Tree View** (Left Pane) with a tree view of the contents of the drives. If the tree view is not displayed, you can also right-click in the window and choose **Tree** from the shortcut menu that is displayed. Now, the window is divided into two parts, the **Tree View** (left pane) and the **Palette** (right pane). The Palette displays folders, files, objects in a drawing, images, Web-based content, and custom content. You can also resize both the **Tree View** and the **Palette** by clicking and dragging the bar between them to the right or the left.

*Figure 6-20 The **DesignCenter** toolbar buttons*

The **DesignCenter** has four tabs provided below the **DesignCenter** toolbar buttons. They are **Folders**, **Open drawings**, **History**, and **DC Online**. The description of these tabs is given next.

Folders Tab

The **Folders** tab lists all the folders and files in the local and network drives. When this tab is selected, the **Tree View** displays the tree view of the contents of the drives and the palette displays the various folders, and files in a drawing, images, and the Web-based content in the selected drive.

In the **Tree View**, you can browse the contents of any folder by clicking on the plus sign (+) adjacent to it to expand the view. Further, expanding the contents of a file, displays the categories such as **Blocks**, **Dimstyles**, **Layers**, **Layouts**, **Linetypes**, **Textstyles**, and **Xrefs**. Clicking on any one of these categories in the **Tree View** displays the listing under the selected category in the Palette (Figure 6-21). Alternately, right-clicking a particular folder, file, or category of the file contents displays a shortcut menu.

The **Explore** option in this shortcut menu also further expands the selected folder, file, or category of contents to display the listing of the contents respectively. Choosing the **Preview** button from the toolbar displays an image of the selected object or file in a **Preview pane** below the Palette. Choosing the **Description** button displays a brief text description of the selected item, if it has one in the **Description** box. When you click on a specific block name in the palette, its preview image and description that was defined earlier when creating the block are displayed in the **Preview pane** and the **Description box**, respectively.

*Figure 6-21 The **DesignCenter** displaying **Tree pane**, **Palette**, **Preview pane**, and the **Description box***

You can drag and drop any of the contents into the current drawing, or add them by double-clicking on them. These are then reused as part of the current drawing. When you double-click on specific Xrefs and blocks, AutoCAD displays the **External Reference** dialog box and the **Insert** dialog box, respectively, to help in attaching the external reference and inserting the block, respectively. Right-clicking a block displays the options of **Insert Block**, **Copy**, or **Create Tool Palette** and right-clicking an Xref displays the options of **Attach Xref** or **Copy** in the shortcut menus. Similarly, when you double-click a layer, text style, dimstyle, layout, or linetype style, they also get added to the current drawing. If any of these named objects already exist in the current drawing, duplicate definition is ignored and it is not added again. When you right-click on a specific linetype, layer, textstyle, layout, or dimstyle in the palette, a shortcut menu is displayed that gives you an option to **Add** or **Copy**. The **Add** option directly adds the selected named object to the current drawing. The **Copy** option copies the specific named object to the clipboard from where you can paste it into a particular drawing.

 Note
You will learn more about inserting blocks in Chapter 16, Working with Blocks.

Right-clicking a particular folder or file in the **Tree View** displays a shortcut menu. The various options in the shortcut menu, besides those discussed earlier, are **Add to Favorites**, **Organize Favorites**, **Create Tool Palette**, and **Set as Home**. **Add to Favorites** adds the selected file or folder to the **Favorites** folder, which contains the most often accessed files and folders. **Organize Favorites** allows you to reorganize the contents of the **Favorites** folder. When you select **Organize Favorites** from the shortcut menu, the **Autodesk** folder is opened in a window. **Create Tool Palette** adds the blocks of the selected file or folder to the **Tool Palettes** window, which contains the predefined blocks. **Set as Home** sets the selected file or folder as the **Home** folder. You will notice that when the **Design Center** command is invoked the next time, the file that was last set as the **Home** folder is displayed selected in the **DesignCenter**.

Open Drawings Tab

The **Open Drawings** tab lists all the drawings that are open, including the current drawing which is being worked on. When you select this tab, the **Tree View** (left pane) displays the tree view of all the drawings that are currently open and the **Palette** (right pane) displays the various contents in the selected drawing.

History Tab

The **History** tab lists the most recent locations accessed through the **DesignCenter**. When you select this tab, the **Tree View** (left pane) and the **Palette** (right pane) are replaced by a list box. Right-clicking a particular file displays a shortcut menu. The various options in the shortcut menu are **Explore, Folders, Open Drawings, Delete**, and **Search**. The **Explore** option invokes the **Folders** tab of the **DesignCenter** with the file selected in the **Tree View** and the contents in the selected file displayed in the **Palette View**. The **Folders** option invokes the **Folders** tab of the **DesignCenter**. The **Open Drawings** option invokes the **Open Drawings** tab of the **DesignCenter**. The **Delete** option deletes the selected drawing from the History list. The **Search** option allows you to search for drawings or named objects such as blocks, textstyles, dimstyles, layers, layouts, external references, or linetypes.

Autodesk seek design content Link*

The Autodesk seek design content link allows you to download the *.dwg*, *.dwf*, *.pdf*, *.dgn* files of various products from the Autodesk online source for product specification and design files. To access the online source, click on the **Autodesk seek design content** in the **DesignCenter**; a new web page will be displayed. In this page, you can search for the required file type using the **Search** option. Once you get the required files, you can download them to your system. You can also drag and drop the drawing files of a particular design. To do so, left-click on the thumbnail of the drawing file when the i-drop indicator is displayed. Next, drag and drop the file in the current drawing.

Choosing the **Back** button in the **DesignCenter** toolbar displays the last item selected in the **DesignCenter**. If you pick the down arrow on the **Back** button, a list of the recently visited items is displayed. You can view the desired item in the **DesignCenter** by selecting it from the list. The **Forward** button is available only if you have chosen the **Back** button once. This button displays the same page as the current page before you choose the **Back** button. The **Up** button moves one level up in the tree structure from the current location. Choosing the **Favorites** button displays shortcuts to files and folders that are accessed frequently by you and are stored in the **Favorites** folder. This reduces the time you take to access these files or folders from their normal location. Choosing the **Tree View Toggle** button in the **DesignCenter** toolbar displays or hides the tree pane with the tree view of the contents in a hierarchical form. Choosing the **Load** button displays the **Load** dialog box, whose options are similar to those of the standard **Select file** dialog box. When you select a file here and choose the **Open** button, AutoCAD displays the selected file and its contents in the **DesignCenter**.

The **Views** button gives four display format options for the contents of the palette: **Large icons**, **Small icons**, **List**, and **Details**. The **List** option lists the contents in the palette, while the **Details** option gives a detailed list of the contents in the palette with the name, file size, and type.

Right-clicking in the palette displays a shortcut menu with all the options provided in the **DesignCenter** in addition to the **Add to Favorites**, **Organize favorites**, **Refresh**, and **Create Tool Palette of Blocks** options. The **Refresh** option refreshes the palette display if you have made any changes to it. The **Create Tool Palette of Blocks** option adds the drawings of the selected file or folder to the **Tool Palettes**, which contains the predefined blocks The following example will illustrate how to use the **DesignCenter** to locate a drawing and then use its contents into a current drawing.

Example 1   *Architectural*

Use the **DesignCenter** to locate and view the contents of the drawing *Kitchens.dwg*. Also, use the **DesignCenter** to insert a block from this drawing and import a layer and a textstyle from the *Blocks and Tables - Imperial.dwg* file located in the **Sample** folder. Use these to make a drawing of a Kitchen plan (*MyKitchen.dwg*) and then add text to it, as shown in Figure 6-22.

1. Open a new drawing using the **Start from Scratch** option. Make sure to select the **Imperial (feet and inches)** option in the **Create New Drawing** dialog box.

2. Change the units to **Architectural** using the **Drawing Units** dialog box. Increase the limits to 10',10'. Invoke the **ALL** option of the **ZOOM** command to increase the drawing display area.

3. Choose the **DesignCenter** button from the **Palettes** panel in the **View** tab of the **Ribbon**; the **DesignCenter** window is displayed at its default location.

Figure 6-22 Drawing for Example 1

4. In the **DesignCenter** toolbar, choose the **Tree View Toggle** button to display the **Tree View** and the **Palette** (if not already displayed). Also, choose the **Preview** button, if preview is not displayed already. You can resize the window, if need be, to view both the **Tree View** and the **Palette**, conveniently.

5. Choose the **Search** button in the **DesignCenter** to display the **Search** dialog box. Here, select **Drawings** from the **Look for** drop-down list and **C:** (or the drive in which AutoCAD 2010 is installed) from the **In** drop-down list. Select the **Search subfolders** check box. In the **Drawings** tab, type **Kitchens** in the **Search for the word(s)** edit box and select **File Name** from the **In the field(s)** drop-down list. Now, choose the **Search Now** button to commence the search. After the drawing has been located, its details and path are displayed in a list box at the bottom of the dialog box.

6. Now, right-click on *Kitchens.dwg* in the list box of the **Search** dialog box and choose **Load into Content Area** from the shortcut menu. You will notice that the drawing and its contents are displayed in the Tree view.

7. Close the **Search** dialog box, if it is still open.

8. Double-click on *Kitchens.dwg* in the Tree View to expand the tree view and display its contents, in case they are not displayed. You can also expand the contents by clicking on the + sign located on the left of the file name in the Tree view.

9. Select **Blocks** in the Tree View to display the list of blocks in the drawing in the **Palette**. Using the left mouse button, drag and drop the block **Kitchen Layout-7x8 ft** in the current drawing.

10. Now, double-click on the *AutoCAD Textstyles* and *Linetypes.dwg* file located in the **DesignCenter** folder in the same directory to display its contents in the **Palette**.

11. Double-click on **Textstyles** to display the list of text styles in the **Palette**. Select **Dutch Bold Italic** in the Palette and drag and drop it in the current drawing. You can use this textstyle for adding text to the current drawing.

12. Use the imported textstyle to add text to the current drawing and complete it refer to Figure 6-22.

 Note
1. To create the text, refer to Figure 6-22, you need to change the text height.
2. You can import Blocks, Dimstyles, Layers, and so on to the current drawing from any existing drawing.

13. Save the current drawing as *MyKitchen.dwg*.

MAKING INQUIRIES ABOUT OBJECTS AND DRAWINGS

When you create a drawing or examine an existing one, you often need some information about it. In manual drafting, you can inquire about the drawing by performing measurements and calculations manually. Similarly, when drawing in an AutoCAD environment, you will need to make inquiries about the data pertaining to your drawing. The inquiries can be about the distance from one location on the drawing to another, the area of an object like a polygon or circle, coordinates of a location on the drawing, and so on. AutoCAD keeps track of all these details. Since inquiry commands are used to obtain information about the selected objects, they do not affect the drawings in any way. The following is the list of Inquiry commands:

AREA	DIST	ID	LIST	DBLIST
STATUS	TIME	DWGPROPS	MASSPROP	

Note
*The **MASSPROP** command will be discussed in Chapter 28.*

For most of the Inquiry commands, you are prompted to select objects; once the selection is complete, AutoCAD switches from the graphics mode to the text mode, and all the relevant information about the selected objects is displayed. For some commands, information is displayed in the AutoCAD Text Window. The display of the text screen can be tailored to your requirements using a pointing device. Therefore, by moving the text screen to one side, you can view the drawing screen and the text screen, simultaneously. If you choose the **minimize** button or the **Close** button, you will return to the graphics screen. You can also return to the graphics screen by entering the **GRAPHSCR** command at the Command prompt. Similarly, you can return to the AutoCAD Text Window by entering **TEXTSCR** at the Command prompt.

Measuring Area of Objects

		Command:	AREA
Ribbon:	Home > Utilities >Measure > Area		
Menu Bar:	Tools > Inquiry > Area		

Finding the area of a shape or an object manually is time consuming. In AutoCAD, the **AREA** command is used to automatically calculate the area of an object in square units. This command saves time when calculating the area of shapes, especially when the shapes are complicated or irregular.

You can use the default option of the **AREA** command to calculate the area and perimeter or circumference of the space enclosed by the sequence of specified points. For example, to find the area of an object (one which is not formed of a single object) you have created with the help of the **LINE** command (Figure 6-23), you need to select all the vertices of that object. By selecting the points, you can define the shape of the object whose area is to be found. This is the default method for determining the area of an object. The only restriction is that all the points you specify should be in a plane parallel to the *XY* plane of the current UCS. You can

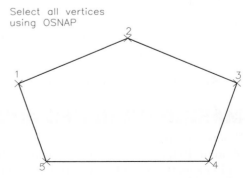

Figure 6-23 *Using the **AREA** command*

make the best possible use of the object snaps such as the ENDpoint, INTersect, and TANgent, or even use running Osnaps, to help you select the vertices quickly and accurately. For AutoCAD, to find the area of a shape, the shapes need not have been drawn with polylines; nor do the lines need to be closed. However, curves must be approximated with short straight segments. In such cases, AutoCAD computes the area by assuming that the first point and the last point are joined. The prompt sequence that will follow when you choose **Utilities > Measure > Area** in the **Home** tab of the **Ribbon** is given next.

> Specify first corner point or [Object/Add area/Subtract area] <current>: *Specify first point.*
> Specify next corner point or press ENTER for total: *Specify the second point.*
> Specify next corner point or press ENTER for total: *Continue selecting until all points enclosing the area have been selected.*
> Specify next corner point or press ENTER for total: Enter
> Area = X, Perimeter = Y

Here, X represents the numerical value of the area and Y represents the circumference/perimeter. It is not possible to accurately determine the area of a curved object, such as an arc, with the default (Point) option. However, the approximate area under an arc can be calculated by specifying several points on the given arc. If the object whose area you want to find is not closed (formed of independent segments) and has curved lines, you should use the following steps to determine the accurate area of such an object.

1. Convert all the segments in that object into polylines using the **PEDIT** command.
2. Join all the individual polylines into a single polyline. Once you have performed these operations, the object becomes closed and you can then use the **Object** option of the **AREA** command to determine the area.

If you specify two points on the screen, the **AREA** command will display the value of the area as 0.00; the perimeter value is the distance between the two points.

Object Option

You can use the **Object** option to find the area of objects such as polygons, circles, polylines, regions, solids, and splines. If the selected object is a polyline or polygon, AutoCAD displays the area and perimeter of the polyline. In case of open polylines, the area is calculated assuming that the last point is joined to the first point but the length of this segment is not added to the polyline length, unlike the default option. If the selected object is a circle, ellipse, or planar closed spline curve, AutoCAD will provide information about its area and circumference. For a solid, the surface area is displayed. For a 3D polyline, all vertices must lie in a plane parallel to the *XY* plane of the current UCS. The extrusion direction of a 2D polyline whose area you want to determine should be parallel to the *Z* axis of the current UCS. In case of polylines which have a width, the area and length of the polyline are calculated using the centerline. If any of these conditions is violated, an error message is displayed on the screen. The following prompt sequence appears when you choose the **Area** button.

Specify first corner point or [Object/Add/Subtract]: **O** Enter
Select objects: *Select an object.* Enter
Area = (X), Perimeter = (Y)

X represents the numerical value of the area, and Y represents the circumference/perimeter.

Tip
*In many cases, the easiest and most accurate way to find the area of a region enclosed by
multiple objects is to use the **BOUNDARY** command to create a polyline, and then use the
AREA Object option.*

Add Option

Sometimes you want to add areas of different objects to determine a total area. For example,
in the plan of a house, you need to add the areas of all the rooms to get the total floor area.
In such cases, you can use the **Add** option. Once you invoke this option, AutoCAD activates
the **Add** mode. By using the **First corner point** option at the **Specify first corner point or
[Object/Subtract]** prompt, you can calculate the area and perimeter by selecting points on
the screen. Pressing ENTER, after you have selected the points defining the area that is to be
added, calculates the total area, since the **Add** mode is on. The Command prompt is as follows:

Specify next corner point or press ENTER for total [ADD mode]:

If the polygon whose area is to be added is not closed, the area and perimeter are calculated
assuming that a line that connects the first point to the last point is added to close the
polygon. The length of this area is added in the perimeter. The **Object** option adds the areas
and perimeters of selected objects. While using this option, if you select an open polyline,
the area is calculated considering the last point is joined to the first point but the perimeter
does not consider the length of this assumed segment, unlike the **First corner point** option.
When you select an object, the area of the selected object is displayed on the screen. At this
time the total area is equal to the area of the selected object. When you select another object,
AutoCAD displays the area of the selected object as well as the combined area (total area) of
the previous object and the currently selected object. In this manner, you can add areas of
different objects. Until the **Add** mode is active, the string **ADD mode** is displayed along with
all subsequent object selection prompts to remind you that the **Add** mode is active. When the
AREA command is invoked, the total area is initialized to zero.

Subtract Option

The action of the **Subtract** option is the reverse of that of the **Add** option. Once you invoke
this option, AutoCAD activates the **Subtract** mode. The **First corner point** and **Object** options
work similar to the way they work in the **ADD** mode. When you select an object, the area of
the selected object is displayed on the screen. At this time, the total area is equal to the area
of the selected object. When you select another object, AutoCAD displays the area of the
selected object, as well as the area obtained by subtracting the area of the currently selected
object from the area of the previous object. In this manner, you can subtract areas of objects
from the total area. Until the **Subtract** mode is active, the string **Subtract** mode is displayed,
along with all subsequent object selection prompts, to remind you that the **Subtract** mode

is active. To exit the **AREA** command, press ENTER (null response) at the **Specify first corner point or [Object/Add/Subtract]** prompt. The prompt sequence for these two modes, as shown in Figure 6-24, is given next.

Specify first corner point or [Object/Add/ Subtract]: **A** Enter
Specify first corner point or [Object/ Subtract]: **O** Enter
(ADD mode) Select objects: *Select the polyline.*
Area = 2.4438, Perimeter = 6.4999
Total area = 2.4438
(ADD mode) Select objects: Enter
Specify first corner point or [Object/Subtract]: **S** Enter
Specify first corner point or [Object/Add]: **O** Enter
(SUBTRACT mode) Select object: *Select the circle.*
Area = 0.0495, Circumference = 0.7890
Total area = 2.3943
(SUBTRACT mode) Select objects: *Select the second circle.*
Area = 0.0495, Circumference = 0.7890
Total area = 2.3448
(SUBTRACT mode) Select object: Enter
Specify first corner point or [Object/Add]: Enter

*Figure 6-24 Measuring the area of a sketch using the **Add** and **Subtract** options*

The **AREA** and **PERIMETER** system variables hold the area and perimeter (or circumference in the case of circles) of the previously selected polyline (or circle). Whenever you use the **AREA** command, the **AREA** variable is reset to zero.

Tip
*If an architect wants to calculate the area of flooring and skirting in a room, the **Area** command provides you with its area and perimeter. You can use these parameters to calculate the skirting.*

Measuring the Distance between Two Points

Ribbon:	Home > Utilities > Measure > Distance	Command:	DIST
Menu Bar:	Tools > Inquiry > Distance		

The **DIST** command is used to measure the distance between two selected points, as shown in Figure 6-25. The angles that the selected points make with the *X* axis and the *XY* plane are also displayed. The measurements are displayed in the current units. Delta X (horizontal displacement), delta Y (vertical displacement), and delta Z are also displayed. The distance computed by the **DIST** command is saved in the **DISTANCE** variable.

The prompt sequence that will follow when you choose the **Distance** button is given next.

Specify first point: *Specify a point.*
Specify second point: *Specify a point.*

A=XY location
B=XY location
L=Length
M=Delta X
P=Delta Y
Q=Angle

AutoCAD returns the following information.

Distance = *Calculated distance between
the two points.*
Angle in XY plane = *Angle between the
two points in the XY plane.*
Angle from XY plane = *Angle the
specified points make with the XY plane.*
Delta X = *Change in X,* Delta Y =
Change in Y, Delta Z = *Change in Z.*

Figure 6-25 Using the DIST command

If you enter a single number or fraction at
the **Specify first point** prompt, AutoCAD will convert it into the current unit of measurement and display it in the command line.

Command: **DIST**
First point: 3-3/4 *(Enter a number or a fraction.)*
Distance = 3.7500

Note
*The Z coordinate is used in 3D distances. If you do not specify the Z coordinates of the two
points between which you want to know the distance, AutoCAD takes the current elevation as
the Z coordinate value.*

Identifying the Location of a Point on the Screen

Ribbon:	Home > Utilities > ID Point	**Command:**	ID
Menu Bar:	Tools > Inquiry > ID Point		

The **ID** command is used to identify the position of a point you specify by displaying
the *X, Y,* and *Z* coordinates of the point. The prompt sequence that will follow, when
you choose the **Locate Point** button from the **Inquiry** toolbar is given next.

Specify point: *Specify the point to be identified.*
X = X coordinate   Y = Y coordinate   Z = Z coordinate

AutoCAD takes the current elevation as the Z coordinate value. If an **Osnap** mode is used to
snap to a 3D object in response to the **Specify point** prompt, the Z coordinate displayed will
be that of the selected feature of the 3D object. You can also use the **ID** command to identify
the location on the screen. This can be realized by entering the coordinate values you want to
locate on the screen. AutoCAD identifies the point by drawing a blip mark at that location, if
the **BLIPMODE** system variable is on. For example, the following is the prompt sequence to
find where the position X = 2.345, Y = 3.674, and Z = 1.0000 is located on the screen.

Chapter 6

Specify point: 2.345,3.674,1.00 [Enter]
X = 2.345 Y = 3.674 Z = 1.0000

The coordinates of the point specified in the **ID** command are saved in the **LASTPOINT** system variable. You can locate a point with respect to the **ID** point by using the relative or polar coordinate system. You can also snap to this point by typing @ when AutoCAD prompts for a point.

Listing Information About Objects

| **Ribbon:** | Home > Properties > List | **Command:** | LIST |
| **Menu Bar:** | Tools > Inquiry > List | | |

 The **LIST** command displays all the information pertaining to the selected objects. The information is displayed in the AutoCAD Text Window. The prompt sequence that follows, when you choose the **List** button from the **Inquiry** toolbar is given next.

Select objects: *Select objects whose data you want to list.*
Select objects: [Enter]

Once you select the objects to be listed, AutoCAD shifts you from the graphics screen to the AutoCAD Text Window. The information displayed (listed) varies from object to object. The information on an object's type, its coordinate position with respect to the current UCS (user coordinate system), the name of the layer on which it is drawn, and whether the object is in model space or paper space is listed for all types of objects. If the color, lineweight, and the linetype are not BYLAYER, they are also listed. Also, if the thickness of the object is greater than 0, that is also displayed. The elevation value is displayed in the form of a Z coordinate (in the case of 3D objects). If an object has an extrusion direction different from the Z axis of the current UCS, the object's extrusion direction is also provided.

More information based on the objects in the drawing is also provided. For example, for a line, the following information is displayed.

1. The coordinates of the endpoints of the line.
2. Its length (in 3D).
3. The angle made by the line with respect to the X axis of the current UCS.
4. The angle made by the line with respect to the XY plane of the current UCS.
5. Delta X, Delta Y, and Delta Z: this is the change in each of the three coordinates from the start point to the endpoint.
6. The name of the layer in which the line was created.
7. Whether the line is drawn in Paper space or Model space.

The center point, radius, true area, and circumference of circles are displayed. For polylines, this command displays the coordinates. In addition, for a closed polyline, its true area and perimeter are also given. If the polyline is open, AutoCAD lists its length and also calculates the area by assuming a segment connecting the start point and endpoint of the polyline. In the case of wide polylines, all computation is done based on the centerlines of the wide segments. For a selected viewport, the **LIST** command displays whether the viewport is on and active, on and inactive, or off. Information is also displayed about the status of Hideplot and the scale relative to paper space. If you use the **LIST** command on a polygon mesh, the size of the mesh (in terms of M, X, N), the coordinate values of all the vertices in the mesh, and whether the mesh is closed or open in M and N directions, are all displayed. As mentioned before, if all the information does not fit on a single screen, AutoCAD pauses to allow you to press ENTER to continue the listing.

Listing Information about All Objects in a Drawing

Command: DBLIST

The **DBLIST** command displays information pertaining to all the objects in the drawing. Once you invoke this command, information is displayed in the Command prompt. If you want to display the drawing information in the AutoCAD Text Window, press the F2 key on the keyboard. If the information does not fit on a single screen, AutoCAD pauses to allow you to press ENTER to continue the listing. To terminate the command, press ESC. To return to the graphics screen, close the AutoCAD Text Window. This command can be invoked by entering **DBLIST** at the Command prompt.

Checking Time-Related Information

Menu Bar: Tools > Inquiry > Time   **Command:**   TIME

The time and date maintained by your system are used by AutoCAD to provide information about several time factors related to the drawings. Hence, you should be careful about setting the current date and time in your computer. The **TIME** command can be used to display information pertaining to time related to a drawing and the drawing session. The display obtained by invoking the **TIME** command is similar to the following:

Command: **TIME**
Current time: Tuesday, March 05, 2009 at 6:59:41:157 PM
Times for this drawing:
Created: Tuesday, March 05, 2009 at 3:51:19:396 PM
Last updated: Tuesday, March 05, 2009 at 3:51:19:396 PM
Total editing time: 0 days 03:08:22.522
Elapsed timer (on):   0 days 03:08:21.961
Next automatic save in: 0 days 00:37:11.432

Enter option [Display/ON/OFF/Reset]: *Enter the required option.*

Obtaining Drawing Status Information

Ribbon: Tools > Drawing Utilities > Status   **Command:**   STATUS

The **STATUS** command displays information about the prevalent settings of various drawing parameters, such as snap spacing, grid spacing, limits, current space, current layer, current color, and various memory parameters. Once you enter this command, AutoCAD displays information similar to the following.

106 objects in Drawing.dwg

Model space limits are	X: 0.0000	Y: 0.0000(On)	
	X: 6.0000	Y: 4.4000	
Model space uses	X:0.6335	Y:-0.2459	**Over
	X:8.0497	Y: 4.9710	**Over
Display shows	X: 0.0000	Y:-0.2459	
	X: 8.0088	Y: 5.5266	
Insertion base is	X: 0.0000	Y: 0.0000	Z: 0.0000
Snap resolution is	X: 0.2500	Y: 0.2500	
Grid spacing is	X: 0.2500	Y: 0.2500	

Current space: Model space
Current layout: Model
Current layer: OBJ
Current color: BYLAYER .... 7 (white)
Current linetype: BYLAYER .... CONTINUOUS
Current lineweight: BYLAYER
Current elevation: 0.0000   thickness: 0.0000
Fill on Grid on Ortho off Qtext off Snap off Tablet on
Object snap modes: None
Free dwg disk space: 2047.7 MBytes
Free temp disk: 2047.7 MBytes
Free physical memory: 15.0 MBytes
Free swap file space: 1987.5 MBytes

All the values (coordinates and distances) on this screen are given in the format declared in the **UNITS** command. You will also notice **Over in the **Model space uses** or **Paper space uses** line. This signifies that the drawing is not confined within the drawing limits. The amount of memory on the disk is given in the **Free dwg disk space** line. Information on the name of the current layer, current color, current space, current linetype, current lineweight, current elevation, snap spacing (snap resolution), grid spacing, various tools that are on or off (such as Ortho, Snap, Fill, Tablet, Qtext), and which object snap modes are active is also provided by the display obtained by invoking the **STATUS** command.

Displaying Drawing Properties

Application Menu: Drawing Utilities > Drawing Properties   **Command:** DWGPROPS

The **DWGPROPS** command displays information about drawing properties. Choose **Drawing Utilities > Drawing Properties** from the **Application Menu**; the **Drawing Properties** dialog box will be displayed, as shown in Figure 6-26. This dialog box has four tabs under which information about the drawing is displayed. This information helps you look for the drawing more easily. These tabs are discussed next.

General

This tab displays general properties about the drawing like the **Type**, **Size**, and **Location**.

Summary

The **Summary** tab displays predefined properties like the Author, title, and subject. Also this tab is displayed at the last that allows you to specify the information related to the drawing.

Statistics

This tab stores and displays data such as the file size and data such as the dates when the drawing was last saved or modified.

Custom

This tab displays custom file properties including values assigned by you.

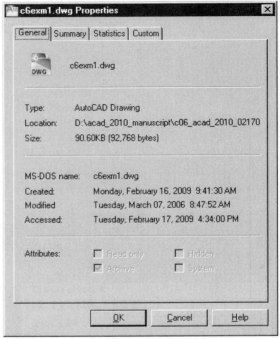

BASIC DISPLAY OPTIONS

*Figure 6-26 The **Drawing Properties** dialog box*

Drawing in AutoCAD is much simpler than manual drafting in many ways. Sometimes while drawing, it is very difficult to see and alter minute details. In AutoCAD, you can overcome this problem by viewing only a specific portion of the drawing. For example, if you want to display a part of the drawing on a larger area, you can use the **ZOOM** command, which lets you enlarge or reduce the size of the drawing displayed on the screen. Similarly, you can use the **REGEN** command to regenerate the drawing and **REDRAW** to refresh the screen. In this chapter, you will learn some of the drawing display commands, such as **REDRAW**, **REGEN**, **PAN**, **ZOOM**, and **VIEW**. These commands can also be used in the transparent mode. Transparent commands are commands that can be used while another command is in progress. Once you have completed the process involved with a transparent command, AutoCAD automatically returns you to the command with which you were working before you invoked the transparent command.

REDRAWING THE SCREEN

Menu Bar:	View > Redraw	**Command:**	REDRAW

The **REDRAW** command redraws the screen and is used to remove the small cross marks (blips) that appear when a point is specified on the screen when BLIPMODE is set to on. The blip mark is not treated as an element of the drawing. The **REDRAW** command also redraws the objects that do not display on the screen as a result of editing some other object. In AutoCAD, several commands redraw the screen automatically (for example, when a grid is turned off), but it is sometimes useful to redraw the screen explicitly. In AutoCAD, the

REDRAW command can also be used in the transparent mode. This can be done by typing an apostrophe in front of the command. The apostrophe appended to a command indicates that the command is to be used as a transparent command (Command: **REDRAW**). Use of the **REDRAW** command does not involve a prompt sequence; instead, the redrawing process takes place without any prompting for information.

The **REDRAW** command affects only the current viewport. If you have more than one viewport you can use the **REDRAWALL** command to redraw all the viewports. **Redraw** in the **View** menu is the **REDRAWALL** command.

Tip
*While working on complex drawings, it may be better to set the **BLIPMODE** variable as **Off** (default), instead of using the **REDRAW** command to clear blips.*

REGENERATING DRAWINGS

Menu Bar:	View > Regen	**Command:**	REGEN

The **REGEN** command makes AutoCAD regenerate the entire drawing to update it. The need for regeneration usually occurs when you change certain aspects of the drawing. All objects in the drawing are recalculated and redrawn in the current viewport. One of the advantages of this command is that the drawing is refined by smoothing out circles and arcs. To use this command, enter **REGEN** at the Command prompt. AutoCAD displays the message **Regenerating model** while it regenerates the drawing. The **REGEN** command affects only the current viewport. If you have more than one viewport, you can use the **REGENALL** command to regenerate all of them. The **REGEN** command can be aborted by pressing ESC. This saves time if you are going to use another command that causes automatic regeneration.

Tip
*Under certain conditions, the **ZOOM** and **PAN** commands automatically regenerate the drawing. Some other commands also perform regenerations under certain conditions.*

ZOOMING DRAWINGS

Ribbon:	View > Navigate > Extents	**Command:**	ZOOM
Menu Bar:	View > Zoom		

Creating drawings on the screen would not be of much use if you can not magnify the drawing view to work on the minute details. Getting close to or away from the drawing is the function of the **ZOOM** command. In other words, this command enlarges or reduces the view of the drawing on the screen, but it does not affect the actual size of the objects. In this way, the **ZOOM** command functions like the zoom lens on a camera. When you magnify the apparent size of a section of the drawing, you see that area in greater detail. On the other hand, if you reduce the apparent size of the drawing, you see a larger area. With this release of AutoCAD, zooming of the entities is animated. You can see the transition of entities from the initial state to the zoomed state.

The ability to zoom in, or magnify, has been helpful in creating the minuscule circuits used in the electronics and computer industries. This is one of the most frequently used commands. Also, this command can be used transparently, which means that it can be used while working in other commands. The **ZOOM** command can be also invoked from the **Zoom** flyout, as shown in Figure 6-27, or by right-clicking in the drawing area and choosing **Zoom** from the menu. This option is also available in the shortcut menu that is displayed even when you are inside some other command, as shown in Figure 6-28.

This command has several options and can be used in a number of ways. The following is the prompt sequence that is displayed when you invoke this command.

> Command: **ZOOM** [Enter]
> Specify corner of window, enter a scale factor (nX or nXP), or [All/Center/Dynamic/Extents/Previous/Scale/Window/Object] <real time>:

Figure 6-27 The **Zoom** *flyout*

Realtime Zooming

You can invoke the **Realtime Zoom** command by choosing **Navigate > Extents > Realtime** from the **View** tab of the **Ribbon** to zoom in and zoom out interactively. To zoom in, invoke the command, then hold the pick button down and move the cursor up. If you want to zoom in further, release the pick button and bring the cursor down. Specify a point and move the cursor up again. Similarly, to zoom out, hold the pick button down and move the cursor down. If you move the cursor vertically up from the midpoint of the screen to the top of the window, the drawing is magnified by 100% (zoom in 2x magnification). Similarly, if you move the cursor vertically down from the midpoint of the screen to the bottom of the window, the drawing display is reduced 100% (zoom out 0.5x magnification). Realtime zoom is the default setting for the **ZOOM** command. Pressing ENTER after entering the **ZOOM** command automatically invokes the realtime zoom.

Figure 6-28 *Invoking the* **ZOOM** *command from the* **LINE** *command shortcut menu*

When you use the realtime zoom, the cursor becomes a magnifying glass and displays a plus sign (+) and a minus sign (–). When you reach the zoom out limit, AutoCAD does not display the minus sign (–) while dragging the cursor. Similarly, when you reach the zoom in limit, AutoCAD does not display the plus sign (+) while dragging the cursor. To exit the realtime zoom, press ENTER or ESC, or select **Exit** from the shortcut menu.

All Option

This option of the **ZOOM** command adjusts the display area on the basis of the drawing limits (Figure 6-29) or extents of the object, whichever is greater. Even if the objects are not within the limits, they are still included in the display. Hence, with the help of the **All** option, you can view the entire drawing in the current viewport (Figure 6-30).

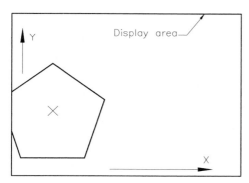

Figure 6-29 *Drawing showing the limits*

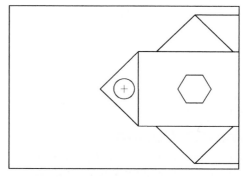

Figure 6-30 *Using the **Zoom All** option*

Center Option

This option lets you define a new display window by specifying its center point (Figures 6-31 and 6-32) and the magnification or height. Here, you are required to enter the center and the height of the subsequent screen display. If you press ENTER instead of entering a new center point, the center of the view will remain unchanged. Instead of entering a height, you can enter the **magnification factor** by typing a number. If you press ENTER at the height prompt, or if the height you enter is the same as the current height, magnification does not take place. For example, if the current height is 2.7645 and you press ENTER at the **magnification or height <current>** prompt, magnification will not take place. The smaller the value, the greater is   the enlargement of the image. You can also enter a number followed by **X**. This indicates the change in magnification, not as an absolute value, but as a value relative to the current screen.

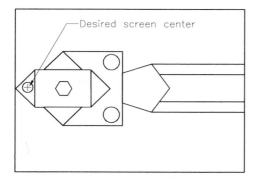

Figure 6-31 *Drawing before using the **ZOOM** Center option*

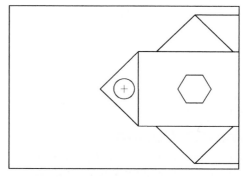

Figure 6-32 *Drawing after using the **ZOOM** Center option*

The prompt sequence is given next.

> Command : **ZOOM** Enter
> Specify corner of window, enter a scale factor (nX or nXP), or
> [All/Center/Dynamic/Extents/Previous/Scale/Window/Object] <real time>: **C** Enter
> Specify center point: *Specify a center point.*
> Enter magnification or height <current>: **5X** Enter

In Figure 6-32, the current magnification height is 5X, which magnifies the display five times. If you enter the value **2**, the size (height and width) of the zoom area changes to 2 X 2 around the specified center. In Figure 6-33, if you enter .12 as the height after specifying the center point as the circle's center, the circle will zoom to fit in the display area since its diameter is 0.12. The prompt sequence is given next.

> Command: **ZOOM** Enter
> Specify corner of window, enter a scale factor (nX or nXP), or
> [All/Center/Dynamic/Extents/Previous/Scale/Window/Object]<real time>: **C** Enter
> Center point: *Select the center of the circle.*
> Magnification or Height <5.0>: **0.12** Enter

Extents Option

As the name indicates, this option lets you zoom to the extents of the biggest object in the drawing. The extents of the drawing comprise the area that has the drawings in it. The rest of the empty area is neglected. With this option, all objects in the drawing are magnified to the largest possible display, see Figure 6-34.

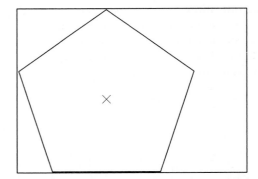

Figure 6-33 *Drawing after using the* **ZOOM** *Center option*

Figure 6-34 *Drawing after using the* **ZOOM** *Extents option*

Dynamic Option

This option displays the portion of the drawing that you have already specified. The prompt sequence for using this option of the **ZOOM** command is given next.

Command: **ZOOM** Enter

Specify corner of window, enter a scale factor (nX or nXP), or
[All/Center/Dynamic/Extents/Previous/Scale/Window/Object]<real time>: **D** Enter

You can then specify the area you want to be displayed by manipulating a view box representing your viewport. This option lets you enlarge or shrink the view box and move it around. When you have the view box in the proper position and size, the current viewport is cleared by AutoCAD and a special view selection screen is displayed. This special screen comprises information regarding the current view as well as the available views. In a color display, the different viewing windows are very easy to distinguish because of their different colors, but in a monochrome monitor, they can be distinguished by their shape.

Blue dashed box representing drawing extents

Drawing extents are represented by a dashed blue box (Figure 6-35), which constitutes the drawing limits or the actual area occupied by the drawing.

Green dashed box representing the current view

A green dashed box is formed to represent the area that the current viewport comprises when the **Dynamic** option of the **ZOOM** command is invoked (Figure 6-36).

Figure 6-35 *Box representing drawing extents*   *Figure 6-36* *Representation of the current view*

Panning view box (X in the center)

A view box initially of the same size as the current view box is displayed with an X in the center (Figure 6-37). You can move this box with the help of your pointing device. This box, known as the panning view box, helps you to find the center point of the zoomed display you want. When you have found the center, press the pick button to display the zooming view box appear.

Zooming view box (arrow on the right side)

After you press the pick button at the center of the panning view box, the X at the center of the view box is replaced by an arrow pointing to the right edge of the box. This zooming view box (Figure 6-38) indicates the ZOOM mode. You can now increase or decrease the area of this box according to the area you want to zoom into. To shrink the box, move the pointer to the left; to increase it, move the pointer to the right. The top, right, and bottom sides of the zooming view box move as you move the pointer, but the left side remains fixed, with the zoom base point at the midpoint of the left side. You can slide it up or down along the left side. When you have the zooming view box in the desired size for your zoom display, press ENTER to complete the command and zoom into the desired area of the drawing. Before pressing ENTER, if you want to change the position of the zooming view box, click the pick button of your pointing device to redisplay the panning view box. After repositioning, press ENTER.

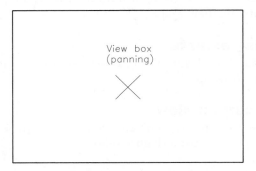

Figure 6-37 *The panning view box*

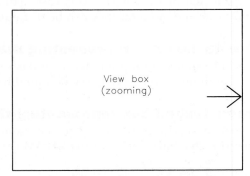

Figure 6-38 *The zooming view box*

Previous Option

While working on a complex drawing, you may need to zoom in on a portion of the drawing to edit some minute details. Once the editing is over you may want to return to the previous view. This can be done by choosing **Extents > Previous** from the **Utilities** tab of the **Ribbon**. Without this command, it would be very tedious to zoom back to the previous views. AutoCAD saves the view specification of the current viewport whenever it is being altered by any of the ZOOM options or by the PAN, VIEW Restore, **DVIEW**, or **PLAN** commands (which are discussed later). Up to ten views are saved for each viewport. The prompt sequence for this option is given next.

Command: **ZOOM** Enter
Specify corner of window, enter a scale factor (nX or nXP), or
[All/Center/Dynamic/Extents/Previous/Scale/Window/Object]<real time>: **P** Enter

Successive **ZOOM > P** commands can restore up to ten previous views. **VIEW** here refers to the area of the drawing defined by its display extents. If you erase some objects and then issue a **ZOOM > Previous** command, the previous view is restored, but the erased objects are not.

Window Option

This is the most commonly used option of the **ZOOM** command. It lets you specify the area you want to zoom in on, by letting you specify two opposite corners of a rectangular window. The center of the specified window becomes the center of the new display screen. The area inside the window is magnified or reduced in size to fill the display as completely as possible. The points can be specified either by selecting them with the help of the pointing device or by entering their coordinates. The prompt sequence is given next.

> Command: **ZOOM** Enter
> Specify corner of window, enter a scale factor (nX or nXP), or
> [All/Center/Dynamic/Extents/Previous/Scale/Window/Object]<real time>: *Specify a point.*
> Specify opposite corner: *Specify another point.*

Whenever the **ZOOM** command is invoked, the window method is one of two default options. This is illustrated by the previous prompt sequence where you can specify the two corner points of the window without invoking any option of the **ZOOM** command. The **Window** option can also be used by entering **W**. In this case, the prompt sequence is given next.

> Command: **ZOOM** Enter
> Specify corner of window, enter a scale factor (nX or nXP), or
> [All/Center/Dynamic/Extents/Previous/Scale/Window/Object]<real time>: **W** Enter
> Specify first corner: *Specify a point.*
> Specify opposite corner: *Specify another point.*

Scale Option

 The **Scale** option of the **ZOOM** command is a very versatile option. It can be used in the following ways:

Scale: Relative to full view

This option of the **ZOOM** command lets you magnify or reduce the size of a drawing according to a scale factor (Figure 6-39). A scale factor equal to 1 displays an area equal in size to the area defined by the established limits. This may not display the entire drawing if the previous view was not centered on the limits or if you have drawn outside the limits. To get a magnification relative to the full view, you can enter any other number. For example, you can type 4 if you want the displayed image to be enlarged four times. If you want to decrease the magnification relative to the full view, you need to enter a number that is less than 1. In Figure 6-40, the image size is decreased because the scale factor is less than 1. In other words, the image size is half of the full view because the scale factor is 0.5.

The prompt sequence is given next.

> Command: **ZOOM** Enter
> Specify corner of window, enter a scale factor (nX or nXP), or
> [All/Center/Dynamic/Extents/Previous/Scale/Window/Object]<real time>: **S** Enter
> Enter a scale factor (nX or nXP): **0.5** Enter

Scale: Relative to current view

The second way to scale is with respect to the current view (Figure 6-41). In this case, instead

Figure 6-39 *Drawing before the ZOOM Scale option selected*

Figure 6-40 *Drawing after the ZOOM Scale option selected*

of entering only a number, enter a number followed by an **X**. The scale is calculated with reference to the current view. For example, if you enter **0.25X**, each object in the drawing will be displayed at one-fourth (¼) of its current size. The following example increases the display magnification by a factor of 2 relative to its current value (Figure 6-42).

Command: **ZOOM** [Enter]
Specify corner of window, enter a scale factor (nX or nXP), or
[All/Center/Dynamic/Extents/Previous/Scale/Window/Object]<real time>: **2X** [Enter]

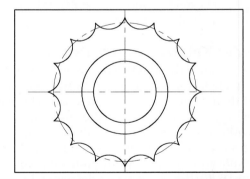

Figure 6-41 *Drawing before the ZOOM Scale (X) option selected*

Figure 6-42 *Drawing after the ZOOM Scale (X) option selected*

Scale: Relative to paper space units

The third method of scaling is with respect to paper space. You can use paper space in a variety of ways and for various reasons. For example, you can array and plot various views of your model in the paper space. To scale each view relative to paper space units, you can use the **ZOOM XP** option. Each view can have an individual scale. The drawing view can be at any scale of your choice in a model space viewport. For example, to display a model space at one-fourth (¼) the size of the paper space units, the prompt sequence is given next.

Command: **ZOOM** Enter
Specify corner of window, enter a scale factor (nX or nXP), or
[All/Center/Dynamic/Extents/Previous/Scale/Window/Object]<real time>: **1/4XP** Enter

Note
For a better understanding of this topic, refer to "Model Space Viewports, Paper Space Viewports and Layouts" in Chapter 12.

Object Option

The **Object** option of the **ZOOM** command is used to select one or more than one objects and display them at the center of the screen in the largest possible size.

Zoom In and Out

You can also zoom into the drawing using the **In** option, which doubles the image size.

Similarly, you can use the **Out** option to decrease the size of the image by half. To invoke these options from the command line, enter **ZOOM 2X** for the **In** option or **ZOOM .5X** for the **Out** option at the Command prompt. The center of the screen is taken as the reference point for enlarging or reducing the view of the drawing.

Exercise 1   *Mechanical*

Draw Figure 6-43 according to the given dimensions. Use the **ZOOM** command to get a bigger view of the drawing. Do not dimension the drawing.

Figure 6-43 *Drawing for Exercise 1*

Chapter 6

PANNING DRAWINGS

Ribbon:	View > Navigate > Pan	**Command:**	PAN
Menu Bar:	View > Pan > Pan Realtime		

You may want to view or draw on a particular area outside the current display. You can do this using the **PAN** command. If done manually, this would be like holding one corner of the drawing and dragging it across the screen. The **PAN** command allows you to bring into view the portion of the drawing that is outside the current display area. This is done without changing the magnification of the drawing. The effect of this command can be illustrated by imagining that you are looking at a big drawing through a window (**display window**) that allows you to slide the drawing right, left, up, and down to bring the part you want to view inside this window. You can invoke the **PAN** command from the shortcut menu also.

Panning in Realtime

You can use the **Realtime Pan** to pan the drawing interactively. To pan a drawing, invoke the command and then hold the pick button down and move the cursor in any direction. When you select the realtime pan, AutoCAD displays an image of a hand indicating that you are in the **PAN** mode. **Realtime pan** is the default setting for the **PAN** command. Choosing the **Pan Realtime** button in the toolbar and entering **PAN** at the Command prompt automatically invokes the realtime pan. To exit the realtime pan, press ENTER or ESC, or choose **Exit** from the shortcut menu.

For the **PAN** command, there are various options that can be used to pan the drawing in a particular direction. These items can be invoked only from the menu bar, as shown in Figure 6-44.

*Figure 6-44 Invoking the **PAN** command options from the **View** menu*

Point

In this option, you are required to specify the displacement. To do this, you need to specify in what direction to move the drawing and by what distance. You can give the displacement either by entering the coordinates of the points or by specifying the coordinates by using a pointing device. The coordinates can be entered in two ways. One way is to specify a single coordinate pair. In this case, AutoCAD takes it as a relative displacement of the drawing with respect to the screen. For example, in the following case, the **PAN** command would shift the displayed portion of the drawing 2 units to the right and 2 units up.

Command: *Select the **Point** option from the **View** menu.*
Specify base point or displacement: **2,2** Enter
Specify second point: Enter

In the second case, you can specify two coordinate pairs. AutoCAD computes the displacement from the first point to the second. Here, the displacement is calculated between point (3,3) and point (5,5).

Command: *Specify the Point item from the View menu.*
Specify base point or displacement: **3,3** `Enter`　*(Or specify a point.)*
Specify second point: **5,5** `Enter`　　　　*(Or specify a point.)*

Left

Moves the drawing left so that some of the right portion of the drawing is brought into view.

Right

Moves the drawing right so that some of the left portion of the drawing is brought into view.

Up

Moves the drawing up so that some of the bottom portion of the drawing is brought into view.

Down

Moves the drawing down so that some of the top portion of the drawing is brought into view.

 Tip
*You can use the scroll bars to pan the drawing vertically or horizontally. The scroll bars are located at the right side and the bottom of the drawing area. You can control the display of the scroll bars in the **Display** tab of the **Options** dialog box.*

CREATING VIEWS

| Ribbon : | View > Views > Named Views | Command: | VIEW |
| Menu Bar: | View > Named Views | | |

While working on a drawing, you may frequently be working with the **ZOOM** and **PAN** commands, and you may need to work on a particular drawing view (some portion of the drawing) more often than others. Instead of wasting time by recalling your zooms and pans and selecting the same area from the screen over and over again, you can store the view under a name and restore it using the name you have given. Choose the **Named Views** button from the **View** panel (Figure 6-45) to invoke the **View Manager** dialog box. This dialog box is used to save the current view under a name so that you can restore (display) it later. It does not save any drawing object data, only the view parameters needed to redisplay that portion of the drawing.

Figure 6-45 *Cursor choosing the **Named Views** button from the **View** panel*

View Manager Dialog Box

You can save and restore the views from the **View Manager** dialog box, as shown in Figure 6-46.

Figure 6-46 *The **View Manager** dialog box*

This dialog box is very useful when you are saving and restoring many view names. With this dialog box, you can name the current view or restore some other view. The **Views** area lists the current view and all the existing named views in the model space and layouts. This area also lists all the preset views such as orthographic and isometric views. As soon as you select any view from the list box, the information about it will be displayed in the **Information** area provided on the right of the list box.

The following are the options in the **Views** area of the **View Manager** dialog box.

Set Current

The **Set Current** button allows you to replace the current view by the view you select from the list box in the **Views** area. You can select any predefined view or any preset view to be made current. AutoCAD uses the center point and magnification of each saved view and executes a **ZOOM Center** with this information when a view is restored.

New

The **New** button allows you to create and save a new view by giving it a name. When you choose the **New** button, the **New View / Shot Properties** dialog box is displayed, as shown in Figure 6-47. The options of this dialog box are discussed next.

View name. You can enter the name for the view in the **View name** edit box.

View category. You can specify the category of the view from the **View category** edit box. The categories include the front view, top view, and so on. If there are some existing categories, you can select them from this drop-down list also.

Figure 6-47 *The New View / Shot Properties dialog box*

View type. You can specify the type of view in this drop-down list.

Boundary Area. This area is used to specify the boundary of the view. To save the current display as the view, select the **Current display** radio button. To define a window that will specify the new view (without first zooming in on that area), select the **Define window** radio button. As soon as you select this radio button, the dialog boxes will be temporarily closed and you will be prompted to define two corners of the window. You can modify the window by choosing the **Define view window** button. You can also enter the X and Y coordinates in the **Specify first corner** and **Specify other corner** in the command lines.

Settings Area. This area allows you to save the layer visibility, UCS, live section, and view style in which you want to visualize the object with the new view. To save the settings of the visibility of the current layers with the view, select the **Save layer snapshot with view** check box. You can select the WCS or the UCS to be saved with the view from the **UCS** drop-down list. If you have created a live section in model space for 3D objects, you can select it from the **Live section** drop-down list, in order to save it with the view. You can also select the visual style to view the object from the **Visual style** drop-down list.

Note

The concept of WCS, UCS, live section, and visual style will be discussed in later chapters.

Background Area. This area allows you to save a view in different background types. To do so, select the type of background from the drop-down list in the **Background** area; the **Background** dialog box will be displayed. You can select the background type from the **Type** drop-down list. Alternatively, you can select these options directly from the drop-down list in the **Background** area of the **New View** dialog box. On selecting the **Solid** type background, you can select any one color to be displayed in the background. To change the color, click on the **Color** area and select the type of color from the **Select Color** dialog box. On selecting the **Gradient** type background, you can apply two or three colors to the background. On selecting the **Image** type background, you can insert an image in the background. To select an image file, choose the **Browse** button from the **Image options** area of the **Background** dialog box. You can also set sun and sky in the background of a view. To do so, select the **Sun & Sky** option from the drop-down list in the **Background** area of the **New View / Shot Properties** dialog box; the **Adjust Sun & Sky Background** dialog box will be displayed. Specify the settings of the sun and sky background in this dialog box and choose the **OK** button. To save the specified sun and sky properties with the view, select the **Save sun properties with view** check box from the **New View / Shot Properties** dialog box.

Note
The effect of background can be visualized only if you have saved the view in the model space with a visual style other than the 2D Wireframe.

*The options in the **Adjust Sun & Sky Background** dialog box and the **Adjust Image** button in the **Image options** area of the **Background** dialog box will be explained in detail in Chapter 30 (Rendering and Animating Designs) of this book.*

*The options in the **Shot Properties** tab are used when you create Show Motion animations. This is discussed in detail in Chapter 28.*

Update Layers

The **Update Layers** button is chosen to update the layer information saved with an existing view.

Edit Boundaries

The **Edit Boundaries** button is chosen to edit the boundary that was defined using the **New View Manager** dialog box while creating the new view.

Tip
You can use the shortcut menu to rename or delete any named view in the dialog box. You can also update the layer information or edit the boundary of a view using this shortcut menu.

*The options of the **VIEW** command can be used transparently by entering '**-VIEW** at the Command prompt.*

PLACING VIEWS ON A SHEET OF A SHEET SET

As mentioned in Chapter 1, you can place views in the sheets of a sheet set. To place a view on the sheet, open the sheet by double-clicking on it in the **SHEET SET MANAGER**. Next, you need to locate the file from which you want to copy the view. To locate this file, choose the **Model Views** tab. If the location of the file is not already listed in this tab, double-click on **Add New Location** to display the **Browse for Folder** dialog box. Using this dialog box, select the folder in which you have saved the file. All the files of that folder are listed. Click on the plus sign (+) located on the left of the file to display all the views created in that file.

*Figure 6-48 The **Model Views** tab of the **Sheet Set Manager** displaying the views in a selected drawing*

Now, right-click on the view and choose **Place on Sheet** from the shortcut menu, as shown in Figure 6-48. The preview of the view will be attached to the cursor and you will be prompted to specify the insertion point. By default, the view will be placed with 1:1 scale. But you can change the view scale before placing the view. To change the scale, right-click to display the shortcut menu that has various preset view scales. Choose the desired view scale from this shortcut menu and then place the view. The view is placed inside the paper space viewport in the layout. Also, the name of the view is displayed below the view in the layout where you placed the view. You can view the list of all the views in the current sheet set by choosing the **View List** tab.

AERIAL VIEW

Menu Bar:	View > Aerial View	**Command:**	DSVIEWER

As a navigational tool, AutoCAD provides you with the option of opening another drawing display window along with the graphics screen window you are working on. This window, called the **Aerial View** window, can be used to view the entire drawing and select those portions that you want to quickly zoom or pan. The AutoCAD graphics screen window resets itself to display the portion of the drawing you have selected for Zoom or Pan in the **Aerial View** window. You can keep the **Aerial View** window open as you work on the graphics screen, or minimize it so that it stays on the screen as a button which can be restored when required (Figure 6-49). As with all windows, you can resize or move it. The **Aerial View** window can also be invoked when you are in the midst of any command other than the **DVIEW** command. The **Aerial View** window has two menus, **View** and **Options**. It also has a toolbar containing zoom in, zoom out, and zoom global options. The following is a description of the available options in the **Aerial View** window.

Figure 6-49 *AutoCAD graphics screen with the* ***Aerial View*** *window*

The **Aerial View** window has three buttons: **Zoom In**, **Zoom Out**, and **Global** (Figure 6-50).

Zoom In

This option leads to the magnification by a factor of **2** centered on the current view box.

Zoom Out

This option leads to the reduction by **half** centered on the current view box.

Global

The **Global** option displays the complete drawing in the **Aerial View** window.

Menus

The menus in the menu title bar are **View** and **Options**. The **View** menu contains **Zoom In**, **Zoom Out**, and **Global** options, which are the same as their respective buttons in the toolbar. The options in the **Options** menu are discussed next.

Auto Viewport

When you are working with viewports, you may need to change the view in the **Aerial View** window to display the current viewport view. You can achieve this by selecting the **Auto Viewport** option from the **Options** menu. If you are working in multiple viewports and you zoom or pan in the **Aerial View** window, only the view in the current viewport is affected.

Dynamic Update

When you make any changes in the current drawing, the view box is updated simultaneously in the **Aerial View** window if you have selected the **Dynamic Update** option. If **Dynamic Update** is not selected, the drawing is not updated simultaneously.

*Figure 6-50 The **Aerial View** window*

The drawing in the **Aerial View** window is updated only when you move the cursor in the **Aerial View** window or invoke any of the menus or the toolbar buttons.

Realtime Zoom

This option dynamically updates the view in the drawing area when you are zooming in the **Aerial View** window in realtime.

Self-Evaluation Test

Answer the following questions and then compare them to those given at the end of this chapter:

1. The number of grips depends on the selected object. (T/F)

2. You can use the **Options** dialog box to modify grip parameters. (T/F)

3. You need at least one source object while using the **MATCHPROP** command. (T/F)

4. You cannot drag and drop entities from the **DESIGNCENTER** window. (T/F)

5. The **ALL** option of the **ZOOM** command displays drawing limits or extents, whichever is greater. (T/F)

6. While using the **ZOOM scale** option, with respect to the current view, you need to enter a number followed by an **X**. (T/F)

7. You can enable grips within a block by setting the system variable _____ to 1 (On).

8. The color of the unselected grips can also be changed by using the _____ system variable.

9. You can access the **Mirror** mode by selecting a grip and then entering _____ or _____ from the keyboard or giving a null response by pressing the SPACEBAR four times.

10. The _____ drop-down list will not be available if you select **Select All** from the **Operator** drop-down list in the **Quick Select** dialog box.

Review Questions

Answer the following questions:

1. Which system variable is used to modify the color of the selected grip?

   (a) **GRIPCOLOR**   (b) **GRIPHOT**
   (c) **GRIPCOLD**   (d) **GRIPBLOCK**

2. Which system variable is used to enable the display of grips inside blocks?

   (a) **GRIPCOLOR**   (b) **GRIPHOT**
   (c) **GRIPCOLD**   (d) **GRIPBLOCK**

3. Which system variable is used to modify the size of grips?

   (a) **GRIPCOLOR**   (b) **GRIPSIZE**
   (c) **GRIPCOLD**   (d) **GRIPBLOCK**

4. How many views are saved with the **Previous** option of the **ZOOM** command?

   (a) 6   (b) 8
   (c) 10   (d) 12

5. Which command recalculates all objects in a drawing and redraws the current viewport only?

   (a) **REDRAW**   (b) **REDRAWALL**
   (c) **REGEN**   (d) **REGENALL**

6. If you select a grip of an object, the grip becomes a hot grip. (T/F)

7. If the **BLIPMODE** variable is set to on, blip marks do not appear on the screen. (T/F)

8. The **Rotate** mode allows you to rotate objects around the base point without changing their size. (T/F)

9. If you have already added a hyperlink to an object, you can also use the grips to open a file associated with it. (T/F)

10. You can view the entire drawing (even if it is beyond limits) with the help of the _____ option.

11. The **GRIPSIZE** is defined in pixels, and its value can range from _____ to _____ pixels.

12. The **VIEW** command does not save any drawing object data. Only the _____ parameters required to redisplay that portion of the drawing are saved.

13. The _____ mode lets you move the selected objects to a new location.

14. The _____ mode allows you to scale objects with respect to the base point without changing their orientation.

15. The **Mirror** mode allows you to mirror objects across the _____ without changing the size of the objects.

16. With the **ZOOM** command, the actual size of an object changes. (T/F)

17. The **REDRAW** command can be used as a transparent command. (T/F)

Exercises

Exercise 2   *General*

1. Use the **LINE** command to draw the shape shown in Figure 6-51(a).

2. Use grips (**Stretch** mode) to get the shape shown in Figure 6-51(b).

3. Use the **Rotate** and **Stretch** modes to get the copies shown in Figure 6-51(c).

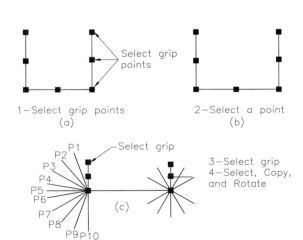

Figure 6-51 *Drawing for Exercise 2*

Exercise 3   *Mechanical*

Use the draw and editing commands to create the sketch shown in Figure 6-52. Use the display commands to facilitate the process.

Figure 6-52 *Drawing for Exercise 3*

Exercise 4   *Mechanical*

Use the draw and editing command to create the sketch shown in Figure 6-53. Use the display commands to facilitate the process.

Figure 6-53 *Drawing for Exercise 4*

Problem-Solving Exercise 1

Mechanical

Use the draw and editing commands to create the drawing shown in Figure 6-54. Use the display commands to facilitate the process.

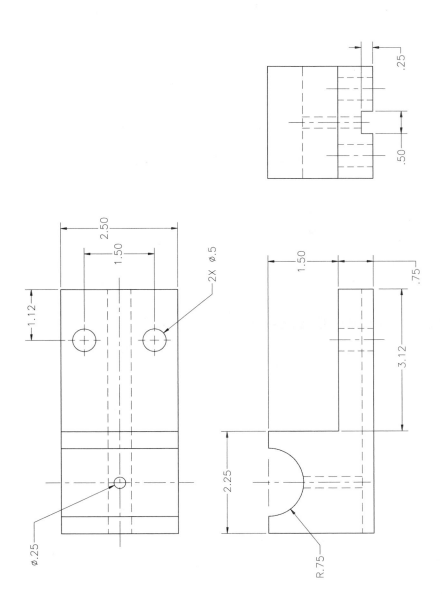

Figure 6-54 Drawing for Problem-Solving Exercise 1

Problem-Solving Exercise 2   *Mechanical*

Draw the sketch shown in Figure 6-55 using draw and edit commands. Use the **MIRROR** command to mirror the shape 9 units across the *Y* axis so that the distance between two center points is 9 units. Mirror the shape across the *X* axis and then reduce the mirrored shape by 75 percent. Join the two ends to complete the shape of the open end spanner. Save the file. Assume the missing dimensions. Note that this is not a standard size spanner.

Figure 6-55 *Drawing for Problem-Solving Exercise 2*

Problem-Solving Exercise 3   *Mechanical*

Use the draw and editing commands to create the drawing shown in Figure 6-56. Use the display commands to facilitate the process. Assume the missing dimensions.

Figure 6-56 *Drawing for Problem-Solving Exercise 3*

Problem-Solving Exercise 4

Architectural

Draw the reception desk shown in Figure 6-57. To get the dimensions of the chairs, refer to Problem-Solving Exercise 3 of Chapter 5.

Figure 6-57 *Drawing for Problem-Solving Exercise 4*

Problem-Solving Exercise 5   *Mechanical*

Draw the views of Vice Body shown in Figure 6-58. The dimensions given in the figure are only for your reference. **

Figure 6-58 *Views and dimensions of the Vice Body*

Answers to Self-Evaluation Test
1. T, 2. T, 3. T, 4. F, 5. T, 6. T, 7. **GRIPBLOCK**, 8. **GRIPCOLOR**, 9. **MIRROR, MI**, 10. **Value**

Chapter 7

Creating Text and Tables

* Understand the concept of annotative objects.
* Draw text using the **TEXT** command.
* Create paragraph text using the **MTEXT** command.
* Edit text using the **DDEDIT** command.
* Draw tables using the **TABLE** command.
* Create and modify the table style.
* Modify the tables.
* Use the **Properties** palette to change properties of the text.
* Substitute fonts and specify alternate default fonts.
* Create text styles using the **STYLE** command.
* Determine text height.
* Draw annotative text.
* Check spellings and find and replace text.

ANNOTATIVE OBJECTS

One of the recent enhancements in AutoCAD is the improved functionality of the drawing annotation that enables you to annotate your drawing easily and efficiently. Annotations are the notes and objects that are inserted into a drawing to add important information to it. The list of annotative objects is given next.

Text	Mtext	Dimensions	Leaders	Multileaders
Hatches	Tolerances	Blocks	Attributes	

All the above mentioned annotative objects have a new annotative property. This annotative property allows you to automate the process of display of the drawings at a correct desired scale on the paper.

ANNOTATION SCALE

Annotation Scale is a feature of AutoCAD that allows you to control the size of the annotative objects in model space. In the earlier releases of AutoCAD, you had to specify the height of the annotative object in model space, keeping in mind the extent upto which the factor would be scaled down while printing the drawing. But, the **Annotation scale** option allows you to set the height of the printed text while annotating, and then allows you to control the height of the object in model space.

Generally, the formula used to calculate the height of the annotative objects in model space is:

Height of the annotative object in model space = Annotation scale X Height of the annotative object in paper space.

For example, if you need to create an annotative object with a height of 1/4" on paper, and assume that annotation scale is set to 1/2" = 1' (implies 1" = 2', or 1" = 24"). Then, the height of the object in model space will be 1/4" x 24"= 6" and the object will be displayed in model space with its height equal to 6".

Note
The above calculation is performed automatically by AutoCAD and has been explained to make the concept clear.

Assigning Annotative Property and Annotation Scales

You can apply the annotative properties to objects in many ways. The objects (Mtext, Hatches, Blocks, and Attributes) that are drawn with the help of dialog boxes have the **Annotative** check boxes in their respective dialog box. You can also create the annotative styles, and all the objects created using that style will have the annotative property automatically. The objects that are created using the command window can be converted into annotative after creating them through the **Properties** palette. You can also override the annotative property of an individual object through the **Properties** palette, see Figure 7-1.

All annotative objects are indicated with an annotative symbol attached to the cursor, see Figure 7-2. When you move the cursor over an annotative object, this annotative symbol starts rolling on. All the annotative styles are indicated with the same annotative symbol displayed next to the cursor.

Figure 7-1 *Assigning annotative property through the* **Properties** *palette*

Figure 7-2 *Annotation symbol attached to the cursor*

To assign the annotation scale while working in model space, choose the **Annotation Scale** button displayed on the right of the status bar; a flyout will be displayed. Select the desired annotation scale from the flyout. This flyout displays 33 commonly used imperial and metric scales, and this list can also be customized. Among these scales, 1:1 annotation scale is selected by default. To change the default annotation scale, enter **CANNOSCALE** in the command window and specify the new annotation scale. To assign the annotation scale while working in viewport, select a viewport first and then assign the annotation scale.

Customizing Annotation Scale

In any drawing, only few annotation scales are used. So, you can delete the unwanted scales from the list according to your requirement. To customize the **Annotation Scale** list, choose the **Custom** option from the flyout that is displayed when you click on the **Annotation Scale** button; the **Edit Scale List** dialog box will be displayed, as shown in Figure 7-3. With this dialog box, you can add, modify, delete, change the sequence of scales displayed, and restore the default list of scales.

Note
You cannot delete the annotation scales that have been used to draw any annotative object in the drawing.

Figure 7-3 *The* **Edit Scale List** *dialog box*

MULTIPLE ANNOTATION SCALES

Ribbon: Annotate > Annotation Scaling > Add/Delete Scales   **Command:** OBJECTSCALE
Menu Bar: Modify > Annotative Object Scale > Add/Delete Scales

You can assign more than one annotation scale to an annotative object. This enables you to display or print the same annotative object in different sizes. The annotative object is displayed or printed based on the current annotation scale. Assigning multiple annotation scales to annotative objects saves a considerable time that is lost while creating a set of objects with different scales in different layers.

Assigning Multiple Annotation Scales to Annotative Objects

To assign multiple annotation scales to objects, enter the **OBJECTSCALE** command in the Command window; you will be prompted to select an annotative object to which you want to add another annotation scale. Select one or more annotation objects and then press ENTER; the **Annotation Object Scale** dialog box will be displayed, see Figure 7-4. Alternatively, select an annotative object and right click to display the shortcut menu. Now, choose **Annotative Object Scale > Add / Delete Scales** from the shortcut menu to display the **Annotation Object Scale** dialog box. Using this dialog box, you can add annotation scale to an annotative object or delete the

*Figure 7-4 The **Annotation Object Scale** dialog box*

annotation scale assigned to the selected annotative object. However, you cannot delete the current annotation scale using this dialog box. When you place the cursor over an annotative object, all annotative objects with multiple annotation scales will be indicated by a double-annotation symbol attached to the cursor, see Figure 7-5. You can also control the assignment of multiple annotation scales to annotative objects using the **Annotation Scaling** panel in the **Annotate** tab of the **Ribbon**, see Figure 7-6.

Figure 7-5 Multiple annotation scale symbol attached to the cursor

*Figure 7-6 The **Annotation Scaling** panel of the **Ribbon***

The annotation scales that have been set as current can be assigned to annotative objects without invoking the **Annotation Object Scale** dialog box. To do so, choose the **Automatically add scales to annotative objects when the annotation scale changes** button located at the right corner of the Status Bar to activate it. Now onward, all annotation scales that are set as current will be automatically added to the annotation objects.

By default, the annotation scale will be applied to all objects irrespective of their layers. However, you can control the assignment of annotation scale depending upon the layer of the object by using the **ANNOAUTOSCALE** system variable. This system variable can have values ranging from -4 to 4 except 0. The description of these values is given next.

1 = This value assigns the current annotation scale to all annotative objects, except the annotative objects that are in the locked, turned off, frozen, and viewport frozen layers.
2 = This value assigns the current annotation scale to all annotative objects, except the annotative objects that are in the turned off, frozen, and viewport frozen layers.
3 = This value assigns the current annotation scale to all annotative objects, except the ones that are in the locked layers.
4 = This value assigns the current annotation scale to all objects.
-1 = On assigning this value, the **ANNOAUTOSCALE** gets turned off, but when turned on again, it regains the value 1.
-2 = On assigning this value, the **ANNOAUTOSCALE** gets turned off, but when turned on again, it regains the value 2.
-3 = On assigning this value, the **ANNOAUTOSCALE** gets turned off, but when turned on again, it regains the value 3.
-4 = On assigning this value, the **ANNOAUTOSCALE** gets turned off, but when turned on again, it regains the value 4.

CONTROLLING THE DISPLAY OF ANNOTATIVE OBJECTS

 After assigning annotation scale to objects, the paper height display scale (in case of plotting) or the viewport scale (in case of viewports) is automatically applied to annotative objects and they are displayed accordingly in the model space or viewport. The annotative objects that do not support the current annotation scale will not be scaled and will be displayed according to the previous annotation scale applied to them.

You can control the display of annotative objects by using the **Annotation Visibility** button on the right of the **Annotation Scale** flyout in the Status Bar. If this button is turned on, all annotative objects will be displayed in the model space or viewport regardless of the current annotation scale. If this button is turned off, only the annotative objects that have annotation scale equal to that of the current annotation scale will be displayed. Note that the **Annotation Visibility** buttons in the model and layout tabs are independent of each other.

When you select an annotative object, you can observe that the selected annotative object is displayed in all annotative scales assigned to it. The annotation scale that is equal to the current scale is displayed in dark dashed lines, whereas the other scales of the object are displayed in faded dashed lines. You can position the annotation object of different scales at different locations with the help of grips. However, if you need to synchronize the start point of annotative objects of different scale factors with current annotative objects, select the annotative object and then right-click in the drawing area; a shortcut menu will be displayed. Choose **Annotative Object Scale > Synchronize Multiple-Scale positions** from the shortcut menu; the start point of annotative objects will be synchronized. Alternatively, you can also use the **ANNORESET** command to synchronize the start point.

SELECTIONANNODISPLAY system variable is used to control the display of the supported scale representations. If you assign 0 to the system variable, it will stop displaying all the supported scale representations selected; only the representations with the current annotation scale will be displayed. If you assign 1 to this system variable, it will display all the supported scale representations selected.

Note

Whenever you create a drawing that involves annotative objects in it, use the newly introduced **2D Drafting & Annotation** *workspace. This workspace displays the* **Ribbon** *with drafting and annotation panels visible by default and hides many of the tools. This also makes the drawing and annotating work much more easy. For all examples and exercises given onward and involving annotative objects, you can use this workspace.*

CREATING TEXT

In manual drafting, lettering is accomplished by hand using a lettering device, pen, or pencil. This is a very time-consuming and tedious job. Computer-aided drafting has made this process extremely easy. Engineering drawings invoke certain standards to be followed in connection with the placement of a text in a drawing. In this section, you will learn how a text can be added in a drawing by using the **TEXT** and **MTEXT** commands.

Writing Single Line Text

Ribbon:	Home > Annotation > Multiline text > Single Line Or Annotate > Text > Multiline text > Single Line **Command:** TEXT or DTEXT
Menu Bar:	Draw > Text > Single Line Text **Toolbar:** Text > Single Line Text

 The **TEXT** command lets you write text on a drawing and allows you to delete what has been typed by using the BACKSPACE key. You can enter multiple lines in one command. The **TEXT** command displays a line in the drawing area that identifies the start point, the height of the text, and the rotation angle. The characters appear on the screen as you enter them. When you press ENTER after typing a line, the cursor automatically places itself at the start of the next line and you can enter the next line of the text. You can end the command by pressing the ESC key or clicking outside the text box.

The screen crosshairs can be moved irrespective of the cursor line for the text. If you specify a point, this command will complete the current line of the text and move the cursor line to the point you selected. This cursor line can be moved and placed anywhere on the screen; therefore, multiple lines of text can be entered at any desired location in the drawing area with a single **TEXT** command. By pressing BACKSPACE, you can delete one character to the left of the current position of the cursor box. Even if you have entered several lines of text, you can use BACKSPACE and continue deleting, until you reach the start point of the first line.

This command can be used with most of the text alignment modes, although it is most useful in the case of left-justified texts. In case of aligned texts, this command assigns a height appropriate for the width of the first line to every line of the text. Even if you select the

Justify option, the text is first left-aligned at the selected point. The prompt sequence that follows when you choose this command is given next.

> Specify start point of text or [Justify/Style]: *Specify the start point.*
> Specify height <0.2000>: **0.15** ⌷Enter⌷
> Specify rotation angle of text <0>: ⌷Enter⌷

Start Point Option

This is the default and the most commonly used option in the **TEXT** command. By specifying a start point, the text is left-justified along its baseline starting from the location of the starting point. Before AutoCAD draws the text, it needs the text height, the rotation angle for the baseline, and the text string to be drawn. It prompts you for all this information. The prompt sequence is given next.

> Specify height <default>: *Specify the height of the text.*
> Specify rotation angle of text <default>: *Specify the angle to which the text must be rotated.*

The **Specify height** prompt determines the distance by which the text extends above the baseline, measured by the capital letters. This distance is specified in drawing units. You can specify the text height by specifying two points or entering a value. In the case of a null response, the default height, that is, the height used for the previous text drawn in the same style, will be used.

The **Specify rotation angle of text** prompt determines the angle at which the text line will be drawn. The default value of the rotation angle is **0-degree** (3 o'clock, or east), and in this case, the text is drawn horizontally from the specified start point. The rotation angle is measured in a counterclockwise direction. The last angle specified becomes the current rotation angle, and if you give a null response, the last angle specified will be used as default. You can also specify the rotation angle by specifying two points. The text is drawn upside down if a point is specified at a location to the left of the start point.

You can now enter the text in the **Text Editor**. Spaces are allowed between words. After entering the text, click outside the **Text Editor**.

Justify Option

AutoCAD offers various options to align the text. **Alignment** refers to the layout of the text. The main text alignment modes are **left**, **center**, and **right**. You can align a text using a combination of modes; for example, top/middle/bottom and left/center/right (Figure 7-7). **Top** refers to the line along which lie the top points of the capital letters; **Baseline** refers to the line along which lie their bases. Letters with descenders (such as p, g, y) dip below the baseline to the bottom. When the **Justify** option is invoked, the user can place the text in one of the fourteen various alignment types by selecting the desired alignment option using the Command prompt or the selection preview (Figure 7-8). The orientation of the text style determines the command interaction for Text Justify. (Text styles and fonts are discussed later in this chapter). For now, assume that the text style orientation is horizontal. The prompt sequence that will follow when you choose the **Single Line Text** button to use this option is given next.

> Specify start point of text or [Justify/Style]: **J** ⏎
> Enter an option [Align/Fit/Center/Middle/Right/TL/TC/TR/ML/MC/MR/BL/BC/ BR]: *Select any of these options.*

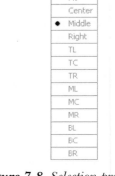

Figure 7-7 *Text alignment positions*

Figure 7-8 *Selection preview for the **Justify** option*

If the text style is vertically oriented (refer to the "Creating Text Styles" section later in this chapter), only four alignment options are available. The prompt sequence is given next.

> Specify start point of text or [Justify/Style]: **J** ⏎
> Enter an option [Align/Center/Middle/Right]: *You can select the desired alignment option.*

If you know what justification you want, you can enter it directly at the **Specify start point of text or [Justify/Style]** prompt instead of first entering **J** to display the justification prompt. If you need to specify a style as well as a justification, you must specify the style first. The various alignment options are as follows.

Align Option

In this option, the text string is written between two points (Figure 7-9). You must specify the two points that act as the endpoints of the baseline. The two points may be specified horizontally or at an angle. AutoCAD adjusts the text width (compresses or expands) so that it fits between the two points. The text height is also changed, depending on the distance between points and the number of letters.

> Specify start point of text or [Justify/Style]: **J** ⏎
> Enter an option [Align/Fit/Center/Middle/Right/TL/TC/TR/ML/MC/MR/BL/BC/BR]: **A** ⏎

Specify first endpoint of text baseline: *Specify a point.*
Specify second endpoint of text baseline: *Specify a point.*

Fit Option

This option is very similar to the previous one. The only difference is that in this case, you select the text height, and it does not vary according to the distance between the two points. AutoCAD adjusts the letter width to fit the text between the two given points, but the height remains constant (Figure 7-9). The **Fit** option is not accessible for vertically oriented text styles. If you try the **Fit** option on the vertical text style, you will notice that the text string does not appear in the prompt. The prompt sequence is given next.

Enter an option [Align/Fit/Center/Middle/Right/TL/TC/TR/ML/MC/MR/BL/BC/BR]: **F**
Specify first endpoint of text baseline: *Specify a point.*
Specify second endpoint of text baseline: *Specify a point.*
Specify height<current>: *Enter the height.*

Note
*You do not need to select the **Justify** option (J) for selecting the text justification. You can enter the text justification by directly entering justification when AutoCAD prompts "Specify start point of text or [Justify/Style]:"*

Center Option

You can use this option to select the midpoint of the baseline for the text. This option can be invoked by entering **Justify** and then **Center** or **C**. After you select or specify the center point, you must enter the letter height and the rotation angle (Figure 7-9).

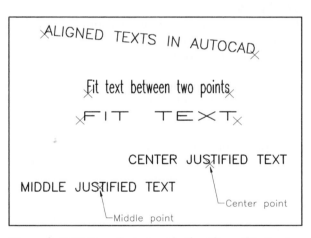

*Figure 7-9 Writing the text using the **Align**, **Fit**, **Center**, and **Middle** options*

Specify start point of text or [Justify/Style]: **C** Enter
Specify center point of text: *Specify a point.*
Specify height<current>: **0.15** Enter
Specify rotation angle of text<0>: Enter

Middle Option

Using this option, you can center text not only horizontally, as with the previous option, but also vertically. In other words, you can specify the middle point of the text string (Figure 7-9). You can alter the text height and the angle of rotation to meet your requirement. The prompt sequence that follows when you choose the **Single Line Text** button from the **Annotate** panel in the **Home** tab of the **Ribbon** is given next.

Specify start point of text or [Justify/Style]: **M** `Enter`
Specify middle point of text: *Specify a point.*
Specify height<current>: **0.15** `Enter`
Specify rotation angle of text<0>: `Enter`

Right Option

This option is similar to the default left-justified start point option. The only difference is that the text string is aligned to the lower right corner (the endpoint you specify), that is, the text is **right-justified** (Figure 7-9). The prompt sequence that will follow when you choose this button is given next.

Specify start point of text or [Justify/Style]: **R** `Enter`
Specify right endpoint of text baseline: *Specify a point.*
Specify height<current>: **0.15** `Enter`
Specify rotation angle of text<0>: `Enter`

TL Option

In this option, the text string is justified from top left (Figure 7-10). The prompt sequence is given next.

Specify start point of text or [Justify/Style]: **TL** `Enter`
Specify top-left point of text: *Specify a point.*
Specify height<current>: **0.15** `Enter`
Specify rotation angle of text <0>: `Enter`

Note
The rest of the text alignment options are similar to those already discussed, and you can try them on your own. The prompt sequence is almost the same as those given for the previous examples.

Figure 7-10 *Writing text using the **Right**, **Top-Left**, **Top-Center**, and **Top-Right** options*

Style Option

With this option, you can specify another existing text style. Different text styles can have different text fonts, heights, obliquing angles, and other features. This option can be invoked by entering **TEXT** and then **S** at the next prompt. The prompt sequence that will follow when you choose the **Single Line Text** button for using this option is given next.

Specify start point of text or [Justify/Style]: **S** `Enter`
Enter style name or [?] <current>: *Specify the desired style or enter **?** to list all the styles.*

If you want to work in the previous text style, just press ENTER at the last prompt. If you want to activate another text style, enter the name of the style at the last prompt. You can also choose from a list of available text styles, which can be displayed by entering **?**. After you enter **?**, the next prompt is given next.

Enter text style(s) to list **<*>**: *Specify the text styles to list or enter * to list all the styles.*

Press ENTER to display the available text style names and the details of the current styles and commands in the **AutoCAD Text Window**.

Note
*With the help of the **Style** option of the **TEXT** command, you can select a text style from an existing list. If you want to create a new style, use the **STYLE** command, which is explained later in this chapter.*

ENTERING SPECIAL CHARACTERS

In almost all drafting applications, you need to draw **special characters** (symbols) in the normal text and in the dimension text. For example, you may want to draw the degree symbol (°) or the diameter symbol (ø), or you may want to underscore or overscore some text. This can be achieved with the appropriate sequence of control characters (control code). For each symbol, the control sequence starts with a percent sign written twice (%%). The character immediately following the double percent sign depicts the symbol. The control sequences for some of the symbols are given next.

Control sequence	Special character
%%c	Diameter symbol (ø)
%%d	Degree symbol (°)
%%p	Plus/minus tolerance symbol (±)

For example, if you want to draw 25° Celsius, you need to enter **25%%dCelsius**. If you enter **43.0%%c**, you get 43.0ø on the drawing screen.

In addition to the control sequences shown earlier, you can use the %%nnn control sequence to draw special characters. The nnn can take a value in the range of 1 to 126. For example, to draw the & symbol, enter the text string **%%038**.

To insert an euro symbol, press and hold the ALT key and enter 0128 from the numeric keypad and then release the ALT key. Before entering the numbers, make sure that the NUM LOCK is on.

CREATING MULTILINE TEXT

Ribbon:	Home > Annotate > Multiline Text	**Tool Palette:**	Draw > MText
	Or Annotate > Text > Multiline Text	**Command:**	MTEXT
Menu Bar: Draw > Text > Multiline Text		**Toolbar:**	Draw, Text > Multiline Text

You can use the **MTEXT** command (Figure 7-11) to write a multiline text whose width can be specified by defining two corners of the text boundary or by entering a width, using the coordinate entry. The text created by the **MTEXT** command is a single object regardless of the number of lines it contains.

Note

*When you invoke the **MTEXT** command, a sample text is attached to the cursor. By default, the text "abc" is attached to the cursor. The **MTJIGSTRING** system variable stores the default contents of the sample text. You can specify a string of ten alphanumeric characters as the default sample text.*

*Figure 7-11   Choosing the **Multiline Text** command from the **Home** tab*

After specifying the width, you have to enter the text in the **In-Place Text Editor**. The prompt sequence is given next.

Command: **MTEXT** [Enter]
Current text style: "Standard". Text height: 0.2000 Annotative: No
Specify first corner: *Select a point to specify first corner.*
Specify opposite corner or [Height/Justify/Line spacing/Rotation/Style/Width/Columns]: *Select an option or select a point to specify other corner.*

After selecting the first corner, you can move the pointing device so that a box that shows the location and size of the paragraph text is formed. An arrow is displayed within the boundary, which indicates the direction of the text flow. When you define the text boundary, it does not mean that the text paragraph will fit within the defined boundary. AutoCAD only uses the width of the defined boundary as the width of the text paragraph. The height of the text boundary has no effect on the text paragraph. Once you have defined the boundary of the paragraph text, AutoCAD displays the **Text Editor** tab along with a **Text Window**, as shown in Figure 7-12, and are discussed next.

*Figure 7-12   The **Text Editor** tab and the text window*

Note

Although the box boundary that you specify controls the width of the paragraph, a single word is not broken to adjust inside the boundary limits. This means that if you write a single word whose width is more than the box boundary specified, AutoCAD will write the word irrespective of the box width and therefore, will exceed the boundary limits.

Text Window

The **Text window** is used to enter the multiline text. The width of the active text area is determined by the width of the window that you specify when you invoke the **MTEXT** command. You can increase or decrease the size of the text window by dragging the double-headed arrow provided at its top-right and bottom left corner of the text window. You can also use the scroll bar to move up or down to display the text.

The ruler on the top of the text window is used to specify the indentation of the current paragraph. The top slider of the ruler specifies the indentation of the first line of the paragraph while the bottom slider specifies the indentation of the other lines of the paragraph.

Text Editor Tab

The options in this tab are as discussed next.

Text Style

The **Text Style** button is located in the **Style** panel of the **Text Editor** tab in the **Ribbon**. If you choose this button, a flyout containing text styles created in the current drawing will be displayed. You can select the desired text style from this flyout. You can also create a new text style using the **STYLE** command, which is explained later in this chapter.

Annotative

The **Annotative** button is used to write the annotative text. The annotative text scales itself according to the current annotative setting. The annotative text also keeps on scaling according to the change in the annotative scale. The annotative texts are defined in the drawing area in terms of the paper height and the current annotation scale setting.

 Note
*While editing the text, the **Annotative** button is used to convert the non-annotative multiline text into annotative multiline text and vice versa.*

*You can also change the non-annotative text to annotative by setting the text's **Annotative** property to **Yes** in the **Properties PALETTES**. This applies to both the single line and the multiline text.*

Text Height

The **Text Height** drop-down list is used to specify the text height of the multiline text. The default value in this drop-down list is 0.2000. Once you modify the height, AutoCAD retains that value unless you change it. Remember that the multiline text height does not affect the height specified for the **TEXT** command.

Bold, Italic, Underline, Overline

You can use the appropriate buttons in the **Formatting** panel to make the selected text bold, italics, underlined, or create overlined text. Bold and italics are not supported by SHX fonts and hence, they will not be available for the particular fonts. These four buttons toggle between on and off.

Font

The **Font** drop-down list in the **Formatting** panel displays all fonts in AutoCAD. You can select the desired font from this drop-down list, see Figure 7-13.

Note
Irrespective of the font assigned to a text style, you can assign a different font to that style for the current multiline text using the Font drop-down list.

Figure 7-13 The Font drop-down list

Uppercase

When you select text in the text editor and choose this button, the alphabets written in the lowercase are converted to the uppercase.

Lowercase

Choosing this button converts the alphabets written in the uppercase to the lowercase.

Color

The **Color** drop-down list is used to set the color for the multiline text. You can also select the color from the **Select Color** dialog box that is displayed by selecting **Select Colors** in the drop-down list.

Background Mask

This option is used to define a background color for the multiline text. When you choose this option, the **Background Mask** dialog box is displayed, as shown in Figure 7-14. To add a background mask, select the **Use background mask** check box and select the required color from the drop-down list in the **Fill Color** area. You can set the size of the colored background behind the text using the **Border offset factor** edit box. The box that defines the background color will be offset

Figure 7-14 The Background Mask dialog box

from the text by the value you define in this edit box. The value in this edit box is based on the height of the text. If you enter **1** as the value, the height of the colored background will be equal to the height of the text and will extend through the length of the window defined to write the multiline text. Similarly, if you enter **2** as the value, the height of the colored box will be twice the height of the text and will be equally offset above and below the text. However, the length of the colored box will still be equal to the length of the window defined to write the multiline text. You can select the **Use drawing background color** check box in the **Fill Color** area to use the color of the background of the drawing area to add the background mask.

Oblique Angle
Expand the **Formatting** panel to view this option. This option is used to specify the slant angle for the text. Enter the angle in the **Oblique Angle** edit box or use spinner to specify the slant angle, which is measured from the positive direction of X-axis in the clockwise direction.

Tracking
Expand the **Formatting** panel to view this option. This option is used to control the spacing between the selected characters. Select the characters and enter the value of the spacing in the **Tracking** edit box. You can also use the **Tracking** spinner to specify this value. The default tracking value is 1.000. You can specify values ranging from 0.7500 to 4.0000.

Width Factor
Expand the **Formatting** panel to view this option. This option is used to control the width of characters. By default, the value is set to 1.000. Select the characters and use the **Width Factor** edit box or the **Width Factor** spinner to specify the width of characters.

Justification
In large complicated technical drawings, the **Justification** option is used to fit the text matter with a specified justification and alignment. For example, if the text justification is bottom-right (BR), the text paragraph will spill to the left and above the insertion point, regardless of how you define the width of the paragraph. When you choose the **Justification** button in the **Paragraph** panel of the **Text Editor** in the **Ribbon**, a cascading menu appears that displays the predefined text justifications. By default, the text is **Top Left** justified. You can choose the new justification from the cascading menu. The various justifications are TL, ML, BL, TC, MC, BC, TR, MR, and BR. Figure 7-15 shows various text justifications for multiline text.

Paragraph
This button is used to control the setting for the paragraph such as tab, indent, alignment, paragraph spacing, and line spacing, Choose this button to display the **Paragraph** dialog box on the screen, as shown in Figure 7-16. The options in the **Paragraph** dialog box are discussed next.

Tab. This area is used to specify the tab type by selecting the required radio button. This area is also used to add or remove the additional tab positions up to which the cursor is moved while pressing the TAB key once at the start of a new paragraph for the left, center, right, and decimal justification. This option can also be set using the mouse by clicking on the ruler of the **In-Place Text** editor.

Figure 7-15 *Text justifications for multiline text (P1 is the text insertion point)*

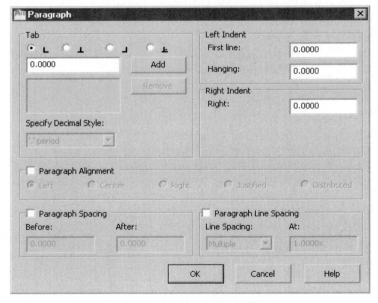

Figure 7-16 *The **Paragraph** dialog box*

Left Indent. In this area, you can set the indent of the first line and the following lines of multiline text by entering the values in the **First Line** and the **Hanging** edit boxes.

Right Indent. In this area, you can set the right indent of the paragraph of the multiline text by entering the value in the **Right** edit box.

Paragraph Alignment. Select the **Paragraph Alignment** check box to set the alignment for the current or the selected paragraph.

Paragraph Spacing. Select the **Paragraph Spacing** check box to set the gap before the start of the paragraph and after the end of the paragraph. According to the gap set by you, AutoCAD calculates the gap between the two consecutive paragraphs. This gap is equal to the sum of after paragraph gap for the upper paragraph and the before paragraph gap for the lower paragraph.

Paragraph Line Spacing. Select the **Paragraph Line Spacing** check box to set the gap between the two consecutive lines. Select the **Exactly** option from the **Line Spacing** drop-down list to maintain the gap exactly specified by you in the **At** edit box, irrespective of the text height. On selecting the **At least** option, both the user specified arbitrary value as well as the text height will be considered to determine the gap. If the text height is smaller than the specified value, the line space will be determined by the user specified value. If the text height is larger, the line spacing will be equal to the text height value. By selecting the **Multiple** option, you can specify the spacing according to the text height. When the text height is not equal in a line, the line space will be determined by the largest text height value of that line.

Left

This radio button is used to left align the text written in the **Text Window**.

Center

This radio button is used to center align the text written in the **Text Window**.

Right

This radio button is used to right align the text written in the **Text Window**.

Justify

This radio button is used to adjust horizontal spacing so that the text is aligned evenly in between the left and right margins. Justifying the text creates a smooth edge on both sides.

Distribute

This button is used to adjust the horizontal spacing so that the text is evenly distributed throughout the width of the column in the **In-Place Text Editor**.

 Note
The spaces entered at the end of the line are also considered while justifying the text.

Line Spacing

The line spacing is the distance between the two consecutive lines in a multiline text. Choose **Paragraph > Line Spacing** from the **Ribbon** to display a flyout containing options to select some predefined line spacing or opening the **Paragraph** dialog box, see Figure 7-17. The **1.0x**, **1.5x**, **2x**, **2.5x** options set the line spacing in the multiples of the largest text height value in the same line. The **More** option is chosen to open the **Paragraph** dialog box that has been discussed earlier. The **Clear Line Space** option removes the space setting for the lines of the selected or the current paragraph. As a result, the line spacing will retrieve the default mtext line space setting.

Bullets and Numbering

This option is used to control the settings to display the bullets and numbering in the multiline text. Choose the **Numbering** button to display a flyout, as shown in Figure 7-18. The options in this flyout are discussed next.

*Figure 7-17 Options in the **Line Spacing** flyout*

Off. Select this option to remove letters, numbers, and bullets from the selected text.

Lettered. This option inserts the letters for list formatting. You can choose the uppercase letters and lower case letters to be used for the list formatting.

Note
If the number of items exceed the number of alphabets, the numbering continues again from the start, but by using the double alphabet.

Numbered. This option inserts numbers for the list formatting.

*Figure 7-18 Options in the **Numbering** flyout*

Bulleted. This option inserts bullets for the list formatting.

Start. This option restarts a new letter or a numbering sequence for the list formatting.

Continue. This option adds the selected paragraphs to the last list and continues the sequence.

Allow Auto-list. This option applies the list formatting automatically when you enter the text. The following characters can be used as punctuation after letters and numbers: period (.), comma (,), close angle bracket (>), close square bracket (]), close parenthesis ()), and close curly bracket (}). Auto-listing will start automatically with the following operations:

1. Start a line of test by entering numbers.
2. Provide space by pressing the TAB key.
3. Press the ENTER key to move to the next line.
4. Press the SHIFT+ENTER key to add a plain paragraph.

To end the Auto-listing, press the ENTER key twice.

Use Tab Delimiter Only. If this option is checked, list formatting will be applied to the text only when the space after the letter and number is created by the TAB key.

Allow Bullets and Lists. If this option is checked, the list formatting will be applied to the whole text lines in the multiline text area. If you clear this option, any list formatting in the multiline text object will be removed and the items will be converted into the plain text. And, all the **Bullets and Lists** options, including the **Allow Auto-list** available on right-clicking in the text window will become unavailable. To make these options available, again right-click in the text window and choose the **Allow Bullets and Lists** sub-option from the **Bullets and Lists** option.

Combine Paragraphs

This option is used to combine the selected paragraphs into a single paragraph. AutoCAD replaces the returns between all the paragraphs by a space. As a result, the lines in the resultant paragraph are in continuation.

Columns

This button is chosen to create columns in the multiline text. On choosing this button from the **Insert** panel, a flyout will be displayed that has the options to control the number, height, and gap between the columns. These options are shown in Figure 7-19, and are discussed next.

Figure 7-19 *The options in the **Column** flyout*

No Columns. This option does not create any column. The whole of the multiline text is written in a single column. When you start a multiline text, this option is selected by default.

Dynamic Columns. The dynamic columns are the text driven columns that change according to the quantity of the text. According to the quantity of text written, the columns are added or removed automatically. There are two options to control the height of the dynamic columns, **Auto height** and **Manual height**. The **Auto height** option automatically starts a new column when the text exceeds the range that the preset column can accommodate. The default height of the first column is decided by the height of the window that you create while starting the multiline text. All the other new columns generated are equal in height to the first column. If you change the height of the first column, the height of the remaining columns changes accordingly, in order to be in level with the first one. The **Manual height** option, on the other hand, provides the flexibility to control the height of each column individually. A new column will not start if the text exceeds the range that the present column can accommodate. The text will keep on adding to the same column by increasing the height of that column. Later on, you can add extra columns by adjusting the height of the present column. When you decrease the height of the column in which you have entered the text that does not fit the height of the new column, a new column will start automatically. Similarly, you can add more columns by adjusting the height of the last column.

Static Columns. The **Static Columns** option allows you to specify the number of columns before writing the text. All the generated columns are of the same height and width. You can also change the number of columns after writing the text.

Insert Column Break Alt + Enter. This option is used to start writing the text in a new column manually. When you insert a column break, the cursor automatically moves to the start of the next column. You can also press the ALT+ENTER key to insert a column break.

Column Settings. This option controls the setting to create a column. Choose this option to invoke the **Column Settings** dialog box, see Figure 7-20. The **Column Settings** dialog box has four areas: **Column Type**, **Column Number**, **Height**, and **Width**. The **Column Type** area specifies the type of column you want to create. All options in this area are similar to the options discussed above. In the **Column Number** area, you can specify the number of columns to be generated when you create the manual type of columns. In the **Height** area, you can specify the height of the columns to be generated for **Auto Height Dynamic Columns**, and **Static Column** type. The **Width** area controls the width of the columns and the gap between two consecutive columns.

*Figure 7-20 The **Column Settings** dialog box*

Symbol

This option is used to insert the special characters in the text. When you choose the **Symbol** button from the **Insert** panel of the **Text Editor** tab in the **Ribbon**, a flyout will be displayed with some predefined special characters, see Figure 7-21. You can also choose **Other** from the flyout to display the **Character Map** dialog box. This dialog box has a number of other special characters that you can insert in the multiline text. To insert characters from the dialog box, select the character that you want to copy and then choose the **Select** button. Once you have selected all the required special characters, choose the **Copy** button and then close the dialog box. Now in the text window, position the cursor where you want to insert the special characters and right-click to display the shortcut menu. Choose **Paste** to insert the selected special character in the **Text Editor**.

Field

AutoCAD allows you to insert a field in the multiline text. The field contains data that is associated with the property that defines the field. For example, you can insert a field having the author's name of the current drawing. If you have already defined the author of the current drawing in the **Drawing Properties** dialog box, it will automatically be displayed in the field. If you modify the author and update the field, the changes will automatically be made in the text. When you choose this option, the **Field** dialog box is displayed, as shown in Figure 7-22. You can select the field to be added in the **Field names** list box and define the format of the field using the **Format** list box. Choose **OK** after selecting the field and format. If the data in the selected field is already defined, it will be displayed in the **Text window**. If not, the field will display dashes (----).

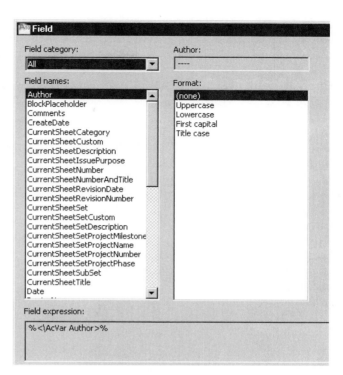

Figure 7-21 Flyout displayed when you choose the Symbol button from the Insert panel

Figure 7-22 The Field dialog box

Note

*To update a field after the text of that field is modified, double-click on the multiline text to display the **In-Place Text Editor**. Click on the field in the **Text window** to select it and then right-click to display the shortcut menu. Choose **Update Field** from this menu; the field will be updated. You can also edit the field or convert it into text using the same shortcut menu.*

Spell Check*

In AutoCAD 2010, you can also check spelling errors. Choose this button; the misspelt words will appear with a red underline. You can specify settings to check spelling errors as per your requirement. To do so, left-click on the inclined arrow in the **Spell Check** panel; the **Check Spelling Settings** dialog box will be displayed. This dialog box is used to specify what to include for spell check such as the dimension text, block attributes, suboption, and text in external references. You can also specify what not to include like capitalized words, mixed cases, uppercase, and words with numbers or punctuation.

Edit Dictionaries

This option is used to match the word with the words in dictionaries while a text is run on spell check. On choosing this option, the **Dictionaries** dialog box will be displayed. This dialog box has two areas, **Main dictionary** and **Custom dictionary**. In the **Main dictionary** area where you can select the required dictionary from a list of different language dictionaries. And in the **Custom dictionary** area where you can add the commonly used words to a .cus file. You can also import the words from other dictionaries to the current customed .cus file.

Check Spelling Settings. Choose this sub-option to display the **Check Spelling Settings** dialog box.

Find and Replace

When you choose the **Find & Replace** button from the **Tools** panel, the **Find and Replace** dialog box is displayed, as shown in Figure 7-23. You can also choose the options available in this dialog box are discussed next.

Find what. This option is used to the text string that you need to find in this edit box and locate a part of a word or a complete word available in this text box.

Figure 7-23 The Find and Replace dialog box

Replace with. If you want some text to be replaced, enter new text for replacement in this text box.

Match case. If this check box is selected, AutoCAD will find out the word only if the case of all characters in the word are identical to the word mentioned in the **Find what** edit box.

Find whole words only. If this check box is selected, AutoCAD will match the word in the text only if it is a single word, identical to the one mentioned in the **Find what** text box. For example, if you enter **and** in the **Find what** edit box and select the **Find whole words only** check box, AutoCAD will not match the string **and** in words like sand, land, band, and so on.

Use wildcards. This check box is selected to use the wildcard characters such as * , ?, and so on in the **Find what** edit box and then replace the selected text.

Match diacritics. This check box is selected to match the Latin characters, diacritical marks, or assent in your search.

Match half/full width forms (East Asian languages). If this check box is selected, you can match half or full width characters in your search.

Find Next. Choose this button to continue the search for the text entered in the **Find what** box.

Replace. Choose this button to replace the highlighted text with the text entered in the **Replace with** text box.

Replace All. If you choose this button, all the words in the current multiline text that match the word specified in the **Find what** text box will be replaced with the word entered in the **Replace with** box.

Import Text

Expand the **Tools** panel to view this option. When you choose this option, AutoCAD displays the **Select File** dialog box. In this dialog box, select any text file that you want to import as the multiline text; the imported text will be displayed in the text area. Note that only the ASCII or RTF file is interpreted properly.

AutoCAPS

Expand the **Tools** panel to view this option. If you choose this option, the case of all the text written or imported after choosing this option will be changed to uppercase. However, the case of the text written before choosing this option is not changed.

More

To edit text, select it and then choose the **More** button from the **Options** panel in the **Text Editor** tab of the **Ribbon**; a flyout will be displayed, as shown in Figure 7-24. The options in the **Options** flyout are discussed next.

Character Set. This option is used to define the character set for the current font. You can select the desired character set from the cascading menu that is displayed when you choose this option.

Figure 7-24 The sub-options of Editor Settings option

Remove Formatting. This option is used to remove the formatting such as bold, italics, or underline from the selected text. To use this option, select the text whose formatting you need to change and then right-click to display the shortcut menu. In the menu, choose the **Remove Formatting** option. The formatting of the selected text will be removed.

Editor Settings. The **Editor Settings** option has sub-options to control the display of the **Text Formatting** toolbar. These sub-options are shown in Figure 7-24 and are discussed next.

> **Always Display as WYSIWYG (What you see is what you get).** This option controls the display size and orientation of the text in the **Text Editor**. If you choose this sub-option, the text in the **Text Editor** will be displayed in the **In-Place Text Editor** with its original size, position, and orientation. But, if the text is very small, or very large or rotated, then you can deselect this sub-option, so that the displayed text is oriented horizontally and is displayed in the readable size. Thus making it easy to read and edit.

> **Show Toolbar**. This sub-option is used to show or hide the **Text Formatting** toolbar.

> **Opaque Background**. Choose this sub-option to make the background of the **Text**

Window opaque. By default, the background is transparent. Invoke the text window over the existing objects/entities in the drawing window to see the effect of the opaque background sub-option.

Ruler

The **Ruler** button in the **Options** panel is used to turn on or off the display of ruler in the **Text Window**.

Undo

The **Undo** button allows you to undo the actions in the **In-Place Text Editor**. You can also press the CTRL+Z keys to undo the previous actions.

Redo

The **Redo** button allows you to redo the actions in the **In-Place Text Editor**. You can also press the CTRL+Y keys to redo the previous actions.

The options discussed next are not available in the **Text Editor** tab. However, in the **Text Formatting** toolbar, these options will be displayed on choosing **More > Editor Settings > Show Toolbar** from the **Text Editor** tab.

Stack

To create a fraction text, you must use the stack option with the special characters /, ^, and #. In AutoCAD 2010, this option is not available in the **Text Editor** tab of the **Ribbon**. However, if you enter two numbers separated by / or ^ and then press the ENTER key or SPACEBAR, the **AutoStack Properties** dialog box will be displayed, as shown in Figure 7-25. You can use this dialog box to control stacking properties. The character / stacks the text vertically with a line, and the character ^ stacks the text vertically without a line (tolerance stack). The character # stacks the text with a diagonal line. After you enter the text with the required special character between them, select the text, and then choose the **Stack/Unstack** button. The stacked text is displayed equal to 70 percent of the actual height.

Stack

Note that this option is available only if you select stack characters from the text. The **Stack** option is used to stack the selected text if there are any stack characters (characters separated by /, #, or ^) available in the multiline text.

Unstack

This option is also available only if there are some stacked characters in the text. The **Unstack** option is used to unstack the selected stacked text.

Stack Properties

This option is available only when you select a stacked text. When you choose this option, the **Stack Properties** dialog box will be displayed, as shown in Figure 7-26. This dialog box is used to edit the text and the appearance of the selected stacked text. The options in this dialog box are discussed next.

Figure 7-25 *The **AutoStack Properties** dialog box*   **Figure 7-26** *The **Stack Properties** dialog box*

Text Area. You can change the upper and the lower values of the stacked text by entering their values in the **Upper** and **Lower** text boxes, respectively.

Appearance Area. You can change the style, position, and size of the stacked text by entering their values in the **Style**, **Position**, and **Text size** text boxes, respectively.

Defaults. This option allows you to restore the default values or save the new settings as the default settings for the selected stacked text.

AutoStack. When you choose this button, the **AutoStack Properties** dialog box is displayed, in the same way as shown earlier in Figure 7-25. The options in this dialog box were discussed under the **Stack** heading.

Show Ruler. The **Show Ruler** sub-option is used to show or hide the **Ruler** in the **Text Window**.

Tip
*The options in the **Options** flyout are also available in the shortcut menu that will be displayed when you select the text in the **Text window** and right-click. The shortcut menu is shown in Figure 7-27.*

Note
*The **Cut** and **Copy** sub-options of the **Edit** option can be used to move or copy the text from the text editor to any other application. Similarly, using the **Paste** sub-option, you can paste text from any windows text-based application to the **In-Place Text Editor**.*

Figure 7-27 *The shortcut menu of the **Text Editor***

Example 1   *General*

In this example, you will use the **In-Place Text Editor** dialog box to write the following text in the drawing window.

For long, complex entries, create multiline text using the MTEXT option. The angle is 10-degree. Dia = 1/2" and length = 32 1/2".

The font of the text is **Swis721 BT**, text height is 0.20, color is red, and written at an angle of 10-degree with Middle-Left justification. Make the word "multiline" bold, underline the text "multiline text", and make the word "angle" italic. The line spacing type and line spacing between the lines are **At least** and **1.5x**, respectively. Use the symbol for degrees. After writing the text in the **Text window**, replace the word "option" with "command".

1. Choose the **Multiline Text** button from the **Annotation** panel in the **Draw** tab of the **Ribbon** in the **2D Drafting & Annotation** workspace. After invoking this command, specify the first corner on the screen to define the first corner of the paragraph text boundary. You need to specify the rotation angle of the text before specifying the second corner of the paragraph text boundary. The prompt sequence is given next.

    Current text style: STANDARD. Text height: 0.2000. Annotative: No
    Specify first corner: *Select a point to specify the first corner.*
    Specify opposite corner or [Height/Justify/Line spacing/Rotation/Style/Width/Columns]: **R** Enter
    Specify rotation angle <0>: 10.
    Specify opposite corner or [Height/Justify/Line spacing/Rotation/Style/Width/Columns]: **L** Enter
    Enter line spacing type [At least/Exactly] <At least>: Enter
    Enter line spacing factor or distance <1x>: **1.5x**
    Specify opposite corner or [Height/Justify/Line spacing/Rotation/Style/Width/Columns]: *Select another point to specify the other corner.*

    The **In-Place Text Editor** and the **Text Editor** tab is displayed.

2. Select the **Swis721 BT** true type font from the **Font** drop-down list of the **Formatting** panel in the **Text Editor** tab of the **Ribbon**.

3. Enter **0.20** in the **Text height** edit box of the **Style** panel in the **Ribbon**, if the value in this edit box is not 0.2.

4. Select **Red** from the **Color** drop-down list in the **Formatting** panel of the **Ribbon**.

5. Now, enter the text in the **In-Place Text Editor** dialog box, as shown in Figure 7-28. To add the degree symbol, choose the **Symbol** button from the **Insert** panel of the **Text Editor** tab in the **Ribbon**; a flyout is displayed. Choose **Degrees** from the flyout. When you type 1/2 after Dia = and then press the " key, AutoCAD displays the **AutoStack**

Properties dialog box. Select the **Convert it to a diagonal fraction** radio button, if it is not already selected. Also, make sure the **Enable AutoStacking** check box is selected. Now, close the dialog box.

Similarly, when you type 1/2 after length = 32 1/2 and then press the " key, AutoCAD displays the **AutoStack Properties** dialog box. Select the **Convert it to a diagonal fraction** radio button and then close the dialog box.

*Figure 7-28 The **In-Place Text Editor** dialog box*

6. Double-click on the word "multiline" to select it (or pick and drag to select the text) and then choose the **Bold** button to make it boldface and the **Underline** button to underline it.

7. Highlight the words "multiline text" and choose the **Underline** button to underline it.

8. Highlight the word "angle" by double-clicking on it and then choose the **Italic** button.

9. Choose **Paragraph > Justification > Middle Left** from the **Ribbon**.

10. Choose the **Find & Replace** button from the **Tools** panel in the **Text Editor** tab of the **Ribbon**. Alternatively, right-click on the Text Window and select **Find and Replace** from the shortcut menu. Alternatively, you can use CTRL+R keys. On doing so, the **Find and Replace** dialog box is displayed.

11. In the **Find what** edit box, enter **option** and in the **Replace with** edit box, enter **command**.

12. Choose the **Find Next** button. AutoCAD finds the word "option" and highlights it. Choose the **Replace** button to replace **option** by **command**. The **AutoCAD** information box is displayed informing you that AutoCAD has finished searching for the word. Choose **OK** to close the information box.

Chapter 7

Note
To set the width of the multiline text objects, hold and drag the arrowhead on the right side of the ruler. In this example, the text has been accommodated into two lines.

13. Now, choose **Cancel** to close the **Find and Replace** dialog box and return to the **In-Place Text Editor**. To exit the editor mode, click outside the **Text Editor**; the text is displayed on the screen, as shown in Figure 7-29.

For long, complex entires, create **multiline text** using MTEXT command. The *angle* is 10°. Dia =½" and length = 32 ½"

Figure 7-29 Multiline text for Example 1

Tip
*The text in the **In-Place Text Editor** can be selected by double-clicking on the word, by holding the left mouse button of the pointing device and then dragging the cursor, or by triple-clicking on the text to select the entire line or paragraph.*

Exercise 1   *General*

Write the text on the screen, as shown in Figures 7-30 and 7-31. Use the special characters and text justification options shown in the drawing, using the **TEXT** and **MTEXT** commands. The text height is 0.1 and 0.15 for Figures 7-30 and 7-31, respectively.

87.3%
42°Fahr.
3.50∅ is the diameter of this arc
25.0±0.25 is the tolerance
UNDERSCORED TEXT IN AUTOCAD
OVERSCORED TEXT IN AUTOCAD
OVERSCORED AND UNDERSCORED TEXT

Figure 7-30 Drawing with special characters   *Figure 7-31 Drawing for Exercise 1*

EDITING TEXT

The contents of **MTEXT** and **TEXT** object can be edited by using the **DDEDIT** and **Properties** commands. You can also use the AutoCAD editing commands, such as **MOVE**, **ERASE**, **ROTATE**, **COPY**, **MIRROR**, and **GRIPS** with any text object.

In addition to editing, you can also modify the text in AutoCAD. The modification that you can perform on the text include changing its scale and justification. The various editing and modifying operations are discussed next.

Editing Text Using the DDEDIT Command

Menu Bar:	Modify > Object > Text > Edit	Toolbar:	Text > Edit
Command:	DDEDIT		

You can use the **DDEDIT** command to edit the text. The most convenient way of invoking this command is by double-clicking on the text. If you double-click on the single line text written using the **TEXT** command, AutoCAD creates an edit box around the text and highlights it. You can modify the text string in this edit box. The size of the bounding box increases or decreases as you add more text or remove the existing text. Note that for the text object, you cannot modify any of its properties in the bounding box. However, if you double-click on the multiline text written using the **MTEXT** command, AutoCAD displays the text in the **In-Place Text Editor** for multiline text. Alternatively, choose **Annotate > Text > Edit** from the **Ribbon** to edit the multiline text. You are prompted to select the multiline text. The **In-Place Text Editor** is displayed after selecting the text. You can make changes using various options in the editor. Apart from changing the text string, you are also allowed to change the properties of the paragraph text.

You can also invoke the **Edit Text** dialog box and the **In-Place Text Editor** from the shortcut menu. Select the text for editing and then right-click in the drawing area. A shortcut menu is displayed and depending on the text object you have selected, the **Edit** or **Mtext Edit** options are available in the menu. Selecting these options displays the respective dialog boxes.

Editing Text Using the Properties Palette

Using the **DDEDIT** command with the text object, you can only change the text string, and not its properties such as height, angle, and so on. In this case, you can use the **Properties** palette for changing the properties. Select the text, right-click to invoke the shortcut menu, and choose the **Properties** option from the shortcut menu. AutoCAD displays the **Properties** palette with all properties of the selected text, as shown in Figure 7-32. Here, you can change any value and also the text string. To edit a single line text, you can change the text string in the window. But for a multiline text, you must choose the **Full editor** button in the **Contents** edit box. AutoCAD automatically switches to the **In-Place Text Editor**, where you can make changes to paragraph text.

Figure 7-32 The **Properties** *palette*

Modifying the Scale of the Text

Ribbon:	Annotate > Text > Scale		
Menu Bar:	Modify > Object > Text > Scale		
Toolbar:	Text > Scale	**Command:**	SCALETEXT

You can modify the scale factor of the text using the **Scale** button in the extended options of the **Text** panel of the **Annotate** tab in the **Ribbon**. AutoCAD uses one of the justification options as the base point to scale the text. The prompt sequence that follows when you invoke this command is given next.

Select objects: *Select the text object to scale.*
Select objects: Enter
Enter a base point option for scaling
[Existing/Left/Center/Middle/Right/TL/TC/TR/ML/MC/MR/BL/BC/BR] <Existing>: *Specify an option that will be used as the base point to scale the text.*
Specify new model height or [Paper height/Match object/Scale factor] <0.2000>: *Enter the new height or select an option.*

Paper Height

It scales the text height depending on the annotative properties of the drawing. The **Paper height** option can only be applied to the annotative objects.

Match object

You can use the **Match object** option to select an existing text whose height will be used to scale the selected text.

Scale factor

You can use the **Scale factor** option to specify a scale factor to scale the text. You can also use the **Reference** option to specify the scale factor for the text.

Modifying the Justification of the Text

Ribbon:	Annotate > Text > Justify	**Toolbar:**	Text > Justify
Menu bar:	Modify > Object > Text > Justify	**Command:**	JUSTIFYTEXT

You can modify the justification of the text by choosing the **Justify** button in the extended options of the **Text** panel in the **Annotate** tab of the **Ribbon**. Note that even after modifying the justification using this command, the location of the text is not changed. The prompt sequence that follows when you invoke this tool is given next.

Select objects: *Select the text object whose justification needs to be changed.*
Select objects: Enter
Enter a justification option
[Left/Align/Fit/Center/Middle/Right/TL/TC/TR/ML/MC/MR/BL/BC/BR] <Left>: *Specify the new justification for the text.*

INSERTING TABLE IN THE DRAWING

Ribbon:	Home > Annotation > Table	**Command:**	TABLE
	Or Annotate > Tables > Table	**Tool Palette:**	Draw > Table
Menu Bar:	Draw > Table	**Toolbar:**	Draw > Table

A number of mechanical, architectural, electric, or civil drawings require a table in which some information about the drawing is displayed. For example, a drawing of an assembly needs the Bill of Material, which is a table providing details such as the number of parts in the drawing, their names, their material, and so on. To enter this information, AutoCAD allows you to create tables using the **TABLE** command, see Figure 7-33. When you invoke this command, the **Insert Table** dialog box is displayed, as shown in Figure 7-34. The options in this dialog box are discussed next.

*Figure 7-33 The **Tables** panel in the Annotate tab of the **Ribbon***

*Figure 7-34 The **Insert Table** dialog box*

Table style Area

The drop-down list in this area displays the names of the various table styles in the current drawing. By default, the **Standard** table style is displayed in the drop-down list. Choose the **Table Style** button on the right of this drop-down list to display the **Table Style** dialog box. This dialog box can be used to create a new table style, modify, or delete an existing table style. You can also set the selected table style as the current table style. You will learn more about creating a new table style in the next section.

Insert options Area

The options in this area enable you to create tables with different types of data. You can add static data, externally-linked data, or object data to a table. The options in this area are discussed next.

Start from empty table

This option enables you to enter the data manually in the table. An empty table is created in which you have to enter the values manually. This type of table data is known as the static data. This is the default option selected for creating a table.

Note

*You can copy an existing excel spreadsheet and paste it on the existing drawing as a table with the static data. To do so, use the **PASTESPEC** command and paste the copied data as AutoCAD entity. The resulting table will be similar to the table created using the **Start from empty table** option.*

From a data link

This option enables you to create a table automatically in AutoCAD from an excel sheet. This excel sheet remains linked with the drawing and any changes made within the excel sheet can be easily updated in the table. From the drop-down list, you can select an already linked excel sheet or you can also attach a new excel sheet to the drawing. To link a new excel file, choose the **Data Link Manager** button; the **Select a Data Link** dialog box will be displayed, see Figure 7-35. Select the **Create a new Excel Data Link** option; the **Enter Data Link Name** dialog box will be invoked. Enter a name for the new data link that you are creating and choose the **OK** button. Next, the **New Excel Data Link** dialog box will be invoked, see Figure 7-36. Choose the **Browse [...]** button; the **Save As** dialog box will be invoked. Specify the location of the excel file to be linked with the current drawing and choose the **Open** button from the **Save As** dialog box. After choosing the **Open** button, more options are added in the **New Excel Data Link** dialog box, see Figure 7-37. The options in the modified **New Excel Data Link** dialog box are discussed next.

Choose an Excel file. This drop-down list displays the excel file attached to the current drawing. Choose the **Browse [...]** button to change the excel file to be linked with the current drawing.

Link options Area. In this area, you can specify the part of the excel file to be linked with the drawing. Select the **Link entire sheet** radio button to link the entire worksheet in the form of a table to the drawing. Select the **Link to a named range** radio button to link the already defined name ranges from the excel sheet to the drawing. Select the **Link to range** radio button to enter the range of cells to be included in the data link. Choose the **Preview** button; the preview of the range of cells included in the data link will be displayed. The preview of the attached data from the excel sheet will be displayed in the preview area. Clear the **Preview** check box to disable the preview of the table.

Cell contents Area. The **Keep data formats and formulas** radio button in this area allows you to link the excel file data along with the supported formulas. The **Keep data formats,**

Figure 7-35 *The **Select a Data Link*** *dialog box*

Figure 7-36 *The **New Excel Data Link*** *dialog box*

solve formulas in Excel radio button allows you to import data from the excel file, but the calculations made using the formulas are done in Excel. The **Convert data formats to text, solve formulas in Excel** radio button in this area is selected by default and lets you import the data of the excel file as text, and the calculations are done in Excel. The **Allow writing to source file** check box is selected by default. This check box enables you to choose the **Download from Source** button from the **Tables** panel of the **Annotate** tab in the **Ribbon** to update the changes made in the drawing table that is linked to the excel sheet.

Cell formatting Area. Select the **Use Excel formatting** check box to import any formatting settings if specified in the excel sheet. Other options in this area will be available only when the **Use Excel formatting** check box is selected. The **Keep table updated to Excel formatting** radio button allows you to update the drawing table according to the changes made in the excel sheet. Choose **Annotate > Tables > Upload to Source** button from the **Ribbon**; the changes made in the excel sheet linked to the drawing table will be updated. If you select the **Start with Excel formatting, do not update** radio button, any formatting changes specified in the excel sheet will be imported to the table, but the changes made in the excel sheet after importing will not be updated in the drawing table. Choose the **OK** button twice to display the **Insert Table** dialog box.

From object data in the drawing (Data Extraction)

This option enables you to create a table by automatically extracting data related to the objects created in the present drawing file. Select the **From object data in the drawing (Data Extraction)** radio button and choose the **OK** button; AutoCAD will start a **Data**

*Figure 7-37 The modified **New Excel Data Link** dialog box*

Extraction wizard that will guide you step by step to insert the desired data into the current drawing.

Note
*For more information regarding the **Data Extraction** wizard, refer to Chapter-17 (Defining Block Attributes).*

Insertion behavior Area

The options in this area are used to specify the method of placing the table in the drawing. These options are discussed next.

Specify insertion point

This radio button is selected to place the table using the upper left corner of the table. If this radio button is selected and you choose **OK** from the **Insert Table** dialog box, you will be prompted to select the insertion point, which is by default the upper left corner of the table. By creating a different table style, you can change the point using which the table is inserted.

Specify window

If this radio button is selected and you choose **OK** from the **Insert Table** dialog box, you will be prompted to specify two corners for placing the table. The number of rows and columns in the table will depend on the size of the window you define.

Column and row settings Area

The options in this area are used to specify the number and size of rows and columns. The availability of these options depend on the option selected from the **Insert options** and **Insertion behavior** area. These options are discussed next.

Columns

This spinner is used to specify the number of columns in the table.

Column width

This spinner is used to specify the width of columns in the table.

Data rows

This spinner is used to specify the number of rows in the table.

Row height

This spinner is used to specify the height of rows in the table. The height is defined in terms of lines and the minimum value is one line.

Set cell styles Area

The options in this area are used to assign different cell styles for the rows in the new table. You can assign different cell styles to all the rows of a table. The options in this area are discussed next.

First row cell style

This drop-down list is used to specify a cell style for the first row of the table. By default, the **Title** cell style is selected for the first row of the table.

Second row cell style

This drop-down list is used to specify a cell style for the second row of the table. By default, the **Header** cell style is selected for the second row of the table.

All other row cell styles

This drop-down list is used to specify a cell style for all the other rows of the table. By default, the **Data** cell style is selected for the remaining rows of the table.

After setting the parameters in the **Insert Table** dialog box, choose the **OK** button. Depending on the type of insertion behavior selected, you will be prompted to insert the table. As soon as you complete the insertion procedure, the **In-Place Text editor** is displayed and you are allowed to enter the parameters in the first row of the table. By default, the first row is the

Chapter 7

title of the table. After entering the data, press ENTER. The first field of the first column is highlighted, which is the column head, and you are allowed to enter the data in it.

AutoCAD allows you to use the arrow keys on the keyboard to move to the other cells in the table. You can enter the data in the field and then press the arrow key to move to the other cells in the table. After entering the data in all the fields, press ENTER to exit the **Text Formatting** toolbar.

Tip
*You can also right-click while entering the data in the table to display the shortcut menu. This shortcut menu is similar to that shown in the **In-Place Text Editor** and can be used to insert field, symbols, text, and so on.*

CREATING A NEW TABLE STYLE

Ribbon:	Home > Annotation > Table Style	**Command:**	TABLESTYLE
Menu Bar:	Format > Table Style	**Toolbar:**	Styles > Table Style

To create a new table style, choose **Table Style** from the extended options of the **Annotation** panel in the **Home** tab of the **Ribbon**; the **Table Style** dialog box will be displayed, as shown in Figure 7-38. You can also invoke this dialog box by choosing the inclined arrow of the **Tables** panel in the **Annotate** panel.

*Figure 7-38   The **Table Style** dialog box*

To create a new table style, choose the **New** button from the **Table Style** dialog box; the **Create New Table Style** dialog box will be displayed, as shown in Figure 7-39.

Enter the name of the table style in the **New Style Name** edit box. Select the style on which you want to base the new style from the **Start With** drop-down list. By default, this drop-down list shows only **Standard** table style. After specifying the settings, choose the **Continue** button from the **Create New Table Style** dialog box; the **New Table Style** dialog box will be displayed, see Figure 7-40. The options in this dialog box are discussed next.

Starting table Area

The options in this area enable you to specify a table in your drawing to be used as reference for formatting the current table style. Once the table is selected, the table style associated with the

Figure 7-39 The **Create New Table Style** *dialog box*

corresponding table gets copied to the current table style and then you can modify it according to your requirement. The **Remove Table** button allows you to remove the initial table style from the current table style. This button is highlighted only when a table style is attached to the current one.

Figure 7-40 The **New Table Style** *dialog box*

General Area

The **Table direction** drop-down list from the **General** area is used to specify the direction of the table. By default, this direction is down. As a result, the title and headers will be at the top and the data fields will be below them. If you select **Up** from the **Table direction** drop-down list, the title and headers will be at the bottom of this table and the data fields will be on the top of the table.

Cell styles Area

This area has the options to define a new cell style or to modify the existing ones. The **Cell style** drop-down list displays the existing cell styles within the table. It also displays the options to create a new cell style or to manage the existing ones. These options are discussed next.

Create new cell style

To create a new cell style, select this option from the **Cell styles** drop-down list. Alternatively, you can choose the **Create a new cell style** button on the right of the **Cell styles** drop-down list. On choosing this button, the **Create New Cell Style** dialog box will be invoked. Next, enter a style name for the new cell style. Select the existing cell style from the **Start With** drop-down list; the settings from the existing style will be used as a reference for the new one to be created. Next, choose the **Continue** button; the new cell style will get added to the **Cell style** drop-down list. Next, you can modify this new cell style by modifying the options in the **General**, **Text**, and **Borders** tab of the **New Table Style** dialog box.

General Tab

The options under this tab are used to control the general appearance, alignment, and formatting of the table cells. These options are shown in Figure 7-41 and are discussed next.

Fill color. This drop-down list is used to specify the fill color for the cells.

Alignment. This drop-down list is used to specify the alignment of the text entered in the cells. The default alignment is top center.

Format. If you choose the **Browse** [...] button available on the right of **Format**, the **Table Cell Format** dialog box will be displayed. This dialog box

*Figure 7-41 The **General** tab of the **New Table Style** dialog box*

is used to specify the data type and format of the data type to be entered in the table. While creating the table, once the data type and format are specified, you cannot enter any other type of data or format without changing or modifying the table style. The default data type is **General**. In this data type, you can enter any alphanumeric characters.

Type. This drop-down list is used to specify the cell style either as a **Label** or **Data**.

Horizontal. This edit box is used to specify the minimum spacing between the data entered in the cells and the left and right border lines of the cells.

Vertical. This edit box is used to specify the minimum spacing between the data entered in the cells and the top and bottom border lines of the cells.

Merge cells on row/column creation. This check box is selected to merge all the new rows and columns created by using this cell style into one cell.

Text Tab

The options under this tab are used to control the display of the text to be written in the cells. These options are shown in Figure 7-42 and are discussed next.

Text style. This drop-down list is used to select the text style that is used for entering the text in the cells. By default, it shows only **Standard**, which is the default text style. You will learn to create more text styles later in this chapter.

Text height. This edit box is used to specify the height of the text to be entered in the cells.

*Figure 7-42 The **Text** tab of the **New Table Style** dialog box*

Text color. This drop-down list is used to specify the color of the text that will be entered in the cells. If you select the **Select Color** option, the **Select Color** dialog box will be displayed, which can be used to select from index color, true color, or from the color book.

Text angle. This edit box is used to specify the slant angle of the text to be entered in the cell.

Borders Tab

The options in this area are used to set the properties of the border of the table, see Figure 7-43. The lineweight and color settings that you specify using this area will be applied to all borders, outside borders, inside borders, bottom border, left border, top border, right border, or without border depending on which button is chosen from this area. Select the **Double line** check box to display the borders with double lines. You can also control the spacing between the double lines by entering the gap value in the **Spacing** edit box.

*Figure 7-43 The **Borders** tab of the **New Table Style** dialog box*

Manage Cell Styles

To create a new cell style, select this option from the **Cell style** drop-down list. Alternatively, you can choose the **Manage Cell Style dialog** button on the right of the **Cell style** drop-down list. On choosing this button, the **Manage Cell Styles** dialog box will be invoked, see Figure 7-44. This dialog box displays all the cell styles in the current table. You can also create a new cell style, delete, or rename the existing one. Note that the cell styles **Title**, **Header**, and **Data** are the default table styles provided in AutoCAD, and they cannot be deleted or renamed.

SETTING A TABLE STYLE AS CURRENT

To set a table style as the current style for creating all the new tables, invoke the **Table Style** dialog box by choosing **Table Style** from the extended options of the **Annotation** panel in the **Home** tab of the **Ribbon**. Next, select the table style from the **Styles** list box in the **Table Style** dialog box and choose the **Set Current** button. You can also set a table style current by selecting it from the **Selects a table style** drop-down list in the **Tables** panel of the **Ribbon**. This is a convenient method of setting a table style current.

*Figure 7-44 The **Manage Cell Styles** dialog box*

MODIFYING A TABLE STYLE

To modify a table style, invoke the **Table Style** dialog box by choosing **Table Style** in the extended options of the **Annotation** panel in the **Home** tab of the **Ribbon**. Select the table style from the **Styles** list box in the **Table Style** dialog box and choose the **Modify** button; the **Modify Table Style** dialog box is displayed. This dialog box is similar to the **New Table Style** dialog box. Modify the options in the various tabs and areas of this dialog box and then choose **OK** to exit from the **Modify Table Style** dialog box.

MODIFYING TABLES

Select any cell from the table; the **Table Cell** tab will be added to the **Ribbon**, as shown in Figure 7-45. The options in the **Table Cell** tab are used to modify the table, insert block, add formulas, and perform other operations.

*Figure 7-45 The **Table Cell** tab of the **Ribbon***

Modifying Rows

To insert a row above a cell, select a cell and choose **Rows > Insert Above** from the **Table Cell** tab. To insert a row below a cell, select a cell and choose **Rows > Insert Below** from the **Table Cell** tab. To delete the selected row, choose **Rows > Delete Row(s)** from the **Table Cell** tab. You can also add more than one row by selecting more than one row in the table.

Modifying Columns

This option is used to modify columns. To add a column to the left of a cell, select the cell and choose **Columns > Insert Left** from the **Table Cell** tab. To add a column to the right of a cell, select a cell and choose **Columns > Insert Right** from the **Table Cell** tab. To delete the selected column, choose **Columns >Delete Column(s)** from the **Ribbon**. You can also add more than one column by selecting more than one row in the table.

Merge Cells

This button is used to merge cells. Choose this button; the **Merge Cells** flyout is displayed. There are three options available in the flyout. Select multiple cells using the **SHIFT** key and then choose **Merge All** from the flyout to merge all the selected cells. To merge all the cells in the row of the selected cell, choose the **Merge By Row** option from the flyout. Similarly, to merge all cells in the column of the selected cell, select the **Merge By Column** option from the flyout. You can also divide the cells by choosing the **Unmerge Cells** button from the **Merge** panel.

Match Cells

This option is used to inherit the properties of one cell into the other. For example, if you have specified **Top Left Cell Alignment** in the source cell, then using the **Match Cells** option, you can inherit this property to the destination cell. This option is useful if you have assigned a number of properties to one cell, and you want to inherit these properties in some specified number of cells. Choose **Cell Styles > Match Cell** from the **Table Cell** tab; the cursor is changed to the match properties cursor and you are prompted to choose the destination cell. Choose the cells to which you want the properties to be inherited and then press ENTER.

Cell Styles

This drop-down list displays the preexisting cell styles or options to modify the existing ones. Select the desired cell style to be assigned to the selected cell. The **Cell Styles** drop-down list also has the options to create a new cell style or manage the existing ones. These options have been discussed earlier in the **Creating a New Table Style** topic.

Cell Borders

Choose **Cell Styles > Cell Borders** from the **Table Cell** tab; the **Cell Border Properties** dialog box is displayed, as shown in Figure 7-46. The options in this dialog box are similar to those in the **Border** tab of the **New Table Style** dialog box.

Text Alignment

The down arrow on the right of the **Middle-Center** button of the **Cell Styles** panel in the **Ribbon** is used to align the text written in cells with respect to the cell boundary. Choose this button; the **Text Alignment** flyout will be displayed. Select the desired text alignment from this flyout; the text of the selected cell will get aligned accordingly.

Locking

This button is used to lock the cells so that they cannot be edited by accident. Select a cell and choose **Cell Format > Cell Locking** from the **Table Cell** tab; the **Cell Locking** flyout will be displayed. Four options are available in the flyout. The **Unlocked** option is chosen by default. Choose the **Content Locked** option to prevent the modification in the content of the text, but you can modify the formatting of the text. Choose the **Format Locked** option to prevent the modification in the formatting of the text, but in this case, you can modify the content of the text. Select the **Content and Format Locked** option to prevent the modification of both formatting and content of the text.

Data Format

The display of the text in the cell depends on the format type selected. Choose **Cell Format** to change the format of the text in the cell. On doing so, the **Data Format** flyout is displayed. Choose the required format from it. You can also select the **Custom Table Cell** option and choose the required format from the **Data type** list box in the **Table Cell Format** dialog box and then choose the **OK** button.

Insert Block

This button is used to insert a block in the selected cell. Choose **Insert > Block** from the **Ribbon**; the **Insert a Block in a Table Cell** dialog box is displayed, as shown in Figure 7-47. Enter the name of the block in the **Name** edit box or choose the **Browse** button to locate the destination file of the block. If you have browsed the file path, it will be displayed in the **Path** area. The options available in the **Insert a Block in a Table Cell** dialog box are discussed next.

Properties Area

The options in this area are discussed next.

Scale. This drop-down list is used to specify the scale of the block. By default, this edit

Figure 7-46 The **Cell Border Properties** *dialog box*

Figure 7-47 The **Insert a Block in a Table Cell** *dialog box*

box is not available because the **AutoFit** check box is selected below this drop-down list. Selecting the **AutoFit** check box ensures that the block is scaled such that it fits in the selected cell.

Rotation angle. The **Rotation angle** edit box is used to specify the angle by which the block will be rotated before being placed in the cell.

Overall cell alignment. This drop-down list is used to define the block alignment in the selected cell.

Field

You can also insert a field in the cell. The field contains the data that is associative to the property that defines the field. For example, you can insert a field that has the name of the author of the current drawing. If you have already defined the author of the current drawing in the **Drawing Properties** dialog box, it will automatically be displayed in the field. If you modify the author name and update the field, the changes will automatically be made in the text. When you choose the **Field** button from the **Insert** panel of the **Ribbon**, the **Field** dialog box will be displayed. You can select the field to be added from the **Field names** list box and select the format of the field from the **Format** list box. Choose **OK** after selecting the field and format. If the data in the selected field is already defined, it will be displayed in the **Text window**. If not, the field will display dashes (----).

Formula

Choose **Insert > Formula** from the **Ribbon**; a flyout is displayed. This flyout contains the formulas that can be applied to a given cell. The formula calculates the values for that cell using the values of other cell. In a table, the columns are named with letters (like A, B, C, ...) and rows are named with numbers (like 1, 2, 3, ...). The **TABLEINDICATOR** system variable controls the display of column letters and row numbers. By default, the **TABLEINDICATOR** system variable is set to 1, which means the row numbers and column letters will be displayed when **In-Place Text Editor** is invoked. Set the system variable to zero to turn off the visibility of row numbers and column letters. The nomenclature of cells is done using the column letters and row numbers. For example, the cell corresponding to column A and row 2 is A2. For a better understanding, some of the cells have been labeled accordingly in Figure 7-48.

*Figure 7-48 The **Table** showing nomenclature for **Columns**, **Rows**, and **Cells***

Chapter 7

Formulas are defined by the range of cells. The range of cells is specified by specifying the name of first and the last cell of the range, separated by a colon (:). The range takes all the cells falling between specified cells. For example, if you write A2 : C3, this means all the cells falling in 2nd and 3rd rows, Column A and B will be taken into account. To insert a formula, double click on the cell; **In-Place Text Editor** is invoked. You can now write the syntax of the formula in the cell. The syntax for different formulas are discussed later while explaining different formulas. Formulas can also be inserted by using the **Formula** flyout. Different formulas available in the **Formula** flyout are discussed next.

Sum

The **Sum** option gives output for a given cell as the sum of the numerical values entered in a specified range of cells. Choose **Insert > Formula > Sum** from the **Table Cell** tab; you will be prompted to select the first corner of the table cell range and then the second corner. The sum of values of all the cells that fall between the selected range will be displayed as the output. As soon as you specify the second corner, the **In-Place Text Editor** is displayed and also the formula is displayed in the cell. In addition to the formula, you can also write multiline text in the cell. Choose **Text Editor > Close > Close Text Editor** from the **Ribbon**, to exit the editor. When you exit the text editor, the formula is replaced by a hash (#). Now, if you enter numerical values in the cells included in the range, the hash (#) is replaced according to the addition of those numerical values. The prompt sequence, when you select the **Sum** option, is given next.

Select first corner of table cell range: *Specify a point in the first cell of the cell range.*
Select second corner of table cell range: *Specify a point in the last cell of the cell range.*

 Note
*The syntax for the **Sum** option is: =Sum{Number of the first cell of cell range (for example: A2): Number of the last cell of the cell range (for example: C5)}*

Average

This option is used to insert a formula that calculates the average of values of the cells falling in the cell range. Prompt sequence is the same as for the **Sum** option.

 Note
*The syntax for the **Average** option is: =Average{Number of the first cell of the cell range (for example: A2): Number of the last cell of the cell range (for example: C5)}*

Count

This option is used to insert a formula that calculates the number of cells falling under the cell range. The prompt sequence is the same as for the **Sum** option.

 Note
*The syntax for the **Count** option is: =Count{Number of the first cell of cell range (for example: A2): Number of the last cell of cell range (for example: C5)}*

Cell

This option equates the current cell with the selected cell. Whenever there is a change in the value of the selected cell, the change is automatically updated in the other cell. To do so, choose **Insert > Formula > Cell** from the **Table Cell** tab; you are prompted to select a table cell. Select the cell with which you want to equate the current cell. The prompt sequence for the **Cell** option is given next.

Select table cell: *Select a cell to equate with the current cell.*

Note
*The syntax for the **Cell** option is: =Number of the cell.*

Equation

Using this option, you can manually write equations. The syntax for writing the equations should be the same as explained earlier.

Manage Cell Content

The **Manage Cell Content** button is used to control the sequence of display of the blocks in the cell, if there are more than one blocks in a cell. Choose **Insert > Manage Cell Content** from the **Table Cell** tab; the **Manage Cell Content** dialog box will be displayed, see Figure 7-49. The options in the **Manage Cell Content** dialog box are discussed next.

Cell content Area

This area lists all the blocks entered in the cell according to the order of their insertion sequence.

Move Up

This button is used to move the selected block to one level up in the display order.

Move Down

This button is used to move the selected block to one level down in the display order.

*Figure 7-49 The **Manage Cell Content** dialog box*

Delete

This button is used to delete the selected block from the table cell.

Options Area

The options in this area are used to control the direction in which the inserted block is placed in the cell. If you select the **Flow** radio button, the direction of placement of the blocks in the cell will depend on the width of the cell. Select the **Stacked horizontal** radio button to place the inserted blocks horizontally. Similarly, select the **Stacked vertical** radio button to place the inserted blocks vertically. You can also specify the gap to be maintained between the two consecutive blocks by entering the desired gap value in the **Content spacing** edit box.

Match Cells

This option is used to inherit the properties of one cell into the other. For example, if you have specified **Top Left Cell Alignment** to the source cell, then using the **Match Cells** option, you can inherit this property to the destination cell. This option is useful if you have assigned a number of properties to one cell and you want to inherit these properties in some specified number of cells. Choose the **Match Cell** option from the shortcut menu; the cursor is changed to the match properties cursor and you are prompted to select the destination cell. Select the cells, in which you want the properties to be inherited and then press ENTER.

Cell Styles

This drop-down list displays the preexisting cell styles or options to modify the existing ones. Select the desired cell style to be assigned to the selected cell. The **Cell Styles** drop-down list also has the options to create a new cell style or to manage the existing ones. These options have been discussed earlier in the **Creating a New Table Style** topic.

Link Cell

This button is used to insert data from an excel spreadsheet into the selected table. Choose **Data > Link Cell** from the **Table Cell** tab; the **Select a Data Link** dialog box will be displayed. The options in this dialog box are similar to the **Select a Data Link** dialog box that has been discussed earlier in the "**Inserting Table in the Drawing**" topic of this chapter.

Download from source

If the contents of the attached excel spreadsheet are changed after linking it to a cell, you can update these changes in the table by choosing this button. AutoCAD will inform you about the changes in the content of the attached excel sheet by displaying an information bubble at the lower-right corner of the screen.

Figure 7-50 The shortcut menu of the ***Table***

Tip
*The options in the **Table** toolbar are also available in the shortcut menu that is displayed when you select any cell in the table and then right-click on it. The shortcut menu is shown in Figure 7-50.*

Note
*The **Cut** and **Copy** options can be used to move or copy the content from one cell to another. Using the **Paste** option, you can paste the content that you have cut or copied from one cell to another. Place the cursor on the **Recent Input** option; a flyout containing the recent command inputs will be displayed. Choose the **Remove All Property Overrides** option to restore all the default properties of the table.*

SUBSTITUTING FONTS

AutoCAD provides you the facility to designate the fonts that you want to substitute for the other fonts used in the drawing. The information about font mapping is specified in the font mapping file (*acad.fmp*). The font mapping has the following advantages.

1. You can specify a font to be used when AutoCAD cannot find a font used in the drawing.

2. You can enforce the use of a particular font in the drawings. If you load a drawing that uses different fonts, you can use font mapping to substitute the desired font for the fonts used in the drawing.

3. You can use *.shx* fonts while creating or editing a drawing. When you are done and ready to plot the drawing, you can substitute other fonts for *.shx* fonts.

The font mapping file is an ASCII file with *.fmp* extension containing one font mapping per line. The format on the line is given next.

Base name of the font file; **Name of the substitute font with extension** (ttf, shx, etc.)

For example, if you want to substitute the ROMANC font for *SWISS.TTF*; the entry is given next.

SWISS;ROMANC.SHX

You can enter this line in the *acad.fmp* file or create a new file. To create a new font mapping file, you need to specify this new file. You can use the **Options** dialog box to specify the new font mapping file. Choose the **Options** button available in the **Application Menu**; the **Options** dialog box will be displayed. Choose the **Files** tab and click on the **plus** sign next to **Text Editor, Dictionary,** and **Font File Names**. Now, click on the plus sign next to **Font Mapping File** to display the path and the name of the font mapping file. Double-click on the file; **Select a file** dialog box will be displayed. Next, select the new font mapping file and exit the **Options** dialog box. At the Command prompt, enter **REGEN** to convert the existing text font to the font as specified in the new font mapping file. You can also use the **FONTMAP** system variable to specify the new font map file.

Command: **FONTMAP** [Enter]
Enter new value for FONTMAP, or . for none <"path and name of the current font mapping file">: *Enter the name of the new font mapping file.*

The following file is a partial listing of the *acad.fmp* file with the new font mapping line added (swiss;romanc.shx).

swiss;romanc.shx	cibt;CITYB___.TTF	cobt;COUNB___.TTF
eur;EURR____.TTF	euro;EURRO___.TTF	par;PANROMAN.TTF
rom;ROMANTIC.TTF	romb;ROMAB___.TTF	romi;ROMAI___.TTF
sas;SANSS___.TTF	sasb;SANSSB__.TTF	sasbo;SANSSBO_.TTF
saso;SANSSO__.TTF		

Note

The text styles that were created using the PostScript fonts are substituted with an equivalent TrueType font and plotted using the substituted font.

Specifying an Alternate Default Font

When you open a drawing file that specifies a font file that is not on your system or is not specified in the font mapping file, AutoCAD, by default, substitutes the *simplex.shx* font file. You can specify a different font file in the **Options** dialog box or do so by changing the **FONTALT** system variable.

       Command: **FONTALT** [Enter]
       Enter new value for FONTALT, or . for none <"simplex.shx">: *Enter the font file name.*

CREATING TEXT STYLES

Ribbon:	Annotate > Text > Text Style	**Menu Bar:**	Format > Text Style
Toolbar:	Text > Text Style or Styles > Text Style	**Command:**	STYLE

By default, the text in AutoCAD is written using the default text style which is called **Standard**. This text style is assigned a default text font (*txt.shx*). Another default text style that is available is **Annotative**. This text style is also assigned a default text font (*txt.shx*). If you need to write a text using some other fonts and other parameters, you need to use the **In-Place Text Editor**. This is because you can change the formatting and font of the text only by using this command.

However, it is a tedious job to use the **In-Place Text Editor** every time to write the text and change its properties. That is why, AutoCAD provides you with an option for modifying the default text style or creating a new text style. After creating a new text, you can make it current. All the texts written after making the new style current will use this style.

To create a new text style or modify the default style, left click on the inclined arrow in the **Text** panel of the **Annotate** tab of the **Ribbon**; the **Text Style** dialog box will be displayed, as shown in Figure 7-51.

The **Styles** area displays the styles present in the drawing along with the current style highlighted in blue. An annotative symbol is displayed in front of the annotative text styles. The **Style List Filter** drop-down list below the **Styles** area is used to specify whether all the

*Figure 7-51 The **Text Style** dialog box*

styles will be displayed or only the styles that have been used in the drawing will be displayed. To create a new style, choose the **New** button from the **Text Style** dialog box; the **New Text Style** dialog box will be displayed, as shown in Figure 7-52, and then choose the **OK** button from the **Style Name** edit box.

*Figure 7-52 The **New Text Style** dialog box*

A new style having the entered name and the properties present in the **Text Style** dialog box will be created. To modify this style, select the style name from the list box and then change the different settings by entering new values in the appropriate boxes. You can change the font by selecting a new font from the **Font Name** drop-down list. Similarly, you can change the text height, width, and oblique angle.

Remember that if you have already specified the height of the text in the **Text Style** dialog box, AutoCAD will not prompt you to enter the text height while writing the text using the **TEXT** command. The text will be created using the height specified in the text style. If you want AutoCAD to prompt you for the text height, specify 0 text height in the dialog box. Select the **Annotative** check box to automate the process of scaling the text height. Annotative texts are defined according to the height of the text to be displayed on the paper. According to the annotation scale set for the spaces, the text will be displayed in the viewports and the model space. Select the **Match text orientation to layout** check box to match the orientation of the text in the paper space viewport with the orientation of the layout.

For **Width Factor,** 1 is the default value. If you want the letters expanded, enter a width factor greater than 1. For compressed letters, enter a width factor less than 1. Similarly, for the **Oblique Angle**, 0 is the default. If you want the slant of the letters toward the right, the value should be greater than 0; to slant the letters toward the left, the value should be less than 0. You can also force the text to be written upside down, backwards, and vertically by checking their respective check boxes. As you make the changes, you can see their effect in the **Preview** box. After making the desired changes, choose the **Apply** button and then the **Close** button to exit the dialog box. Figure 7-53 shows the text objects with all these settings.

Figure 7-53 *Specifying different features to text style files*

DETERMINING TEXT HEIGHT

The actual text height is equal to the product of the **scale factor** and the **plotted text height**. Therefore, scale factors are important numbers for plotting the text at the correct height. This factor is a reciprocal of the drawing plot scale. For example, if you plot a drawing at a scale of ¼ = 1, you calculate the scale factor for text height given next.

   ¼" = 1"   (i.e., the scale factor is 4)

The scale factor for an architectural drawing that is to be plotted at a scale of ¼" = 1'0" is calculated as given next

   ¼" = 1'0", or ¼" = 12", or 1 = 48
   Therefore, in this case, the scale factor is 48.

For a civil engineering drawing with a scale 1"= 50', the scale factor is shown next.

   1" = 50', or 1" = 50X12", or 1 = 600

   Therefore, the scale factor is 600.

Next, calculate the height of the AutoCAD text. If it is a full-scale drawing (1=1) and the text is to be plotted at 1/8" (0.125), it should be drawn at that height. However, in a civil engineering drawing, a text drawn 1/8" high will look like a dot. This is because the scale for a civil engineering drawing is 1"= 50', which means that the drawing you are working on is 600 times larger. To draw a normal text height, multiply the text height by 600. Now, the height will be as calculated below.

   0.125" x 600 = 75

Similarly, in an architectural drawing, which has a scale factor of 48, a text that is to be 1/8"
high on paper must be drawn 6 units high, as shown in the following calculation:

$0.125 \times 48 = 6.0$

It is very important to evaluate scale factors and0 text heights before you begin a drawing. It
would be even better to include the text height in your prototype drawing by assigning the
value to the **TEXTSIZE** system variable.

CREATING ANNOTATIVE TEXT

One of the recent inclusions in AutoCAD is now you do not need to calculate the text height
in advance. While creating the annotative text, if you decide the text height to be displayed
on the paper, the current annotation scale will automatically decide the display size of the
text in the model space or the paper space viewport. For example, if you want the text to be
displayed at a height of 1/4" on the paper, you can define a text style having a **Paper Text
Height** of 1/4". When you add text to a viewport having a scale of 1/4"=1'0", the current
annotation scale, which is set as the same scale as of the viewport, automatically scales the
text to display appropriately at 12". You can create annotative text by assigning the annotative
text style that has been explained earlier.

To write a single line annotative text, choose an annotative type text style from the **Text Style**
dialog box and set it as the current text style. The annotative type text style will be displayed
with an annotative symbol on its side. Next, enter the text in the drawing using the **TEXT**
command in the Command prompt. To write multiline annotative text, enter the **MTEXT**
command at the Command prompt and specify the two opposite corners of the box denoting
the width of the multiline text; the **In-Place Text Editor** will be displayed on the screen. You
can also select an existing annotative text style from **Text Style** option in the **Text Editor** tab
of the **Ribbon** or choose **Annotative** button in the **Style** panel **Text Editor** tab of the **Ribbon**
to create the annotative multiline text.

The existing non-annotative texts whether it is single line or multiline can also be converted
into the annotative text. To do so, select the text object and right-click on it to choose the
Properties option from the shortcut menu. In the **Properties** palette below the **Text** area,
click on the **Annotative** edit box and select the **Yes** option from the drop-down list.

CHECKING SPELLING

Ribbon:	Annotate > Text > Check Spelling	Menu Bar:	Tools > Spelling
Toolbar:	Text > Spell Check	Command:	SPELL

You can check the spelling of all the text in the drawing. The text that can be checked
for spelling using the **SPELL** command are: single line or multiline text, dimension
text, text in external references, and block attribute text. To check the spellings in
the given the text, choose **Annotate > Text > Check Spelling** from the **Ribbon**; the
Check Spelling dialog box will be displayed, see Figure 7-54. To start the spell check, choose

*Figure 7-54 The **Check Spelling** dialog box*

the **Start** button from the **Check Spelling** dialog box. If there is no spelling mistake in the drawing, the **AutoCAD Message** dialog box with the message '**Spelling check complete**' will be displayed. Choose the **OK** button from the **AutoCAD Message** dialog box to close it. If any spelling is incorrect in the drawing, AutoCAD will highlight the word and zoom it according to the size of the window so that it is easily visible. The options in the **Check Spelling** dialog box are discussed next.

The **Where to check** drop-down list displays three options to specify the portion of the drawing to be Spell checked during the spelling check. These options are: **Entire drawing**, **Current space/layout**, and **Selected objects**. If you select the **Selected objects** option, the **Select text objects** button will be highlighted to enable you to select the text in the drawing to be checked. The **Not in dictionary** box displays the text that is found to be misspelled. The correct spelled alternate words are listed in the **Suggestions** box. The **Main dictionary** drop-down list displays the language options in which you want to check the text.

The **Add to Dictionary** button is used to add the identified misspelled word to the dictionary. The **Ignore** button leaves the identified misspelled words intact. The **Ignore All** button leaves all the identified misspelled words intact. The **Change** button substitutes the present word with the word suggested in the **Suggestions** box. The **Change All** button substitutes all the words that are similar to the currently selected word with the word suggested in the **Suggestions** box. You can also use the MS Word dictionary or any other dictionary for checking spellings. Choose the **Dictionaries** button; the **Dictionaries** dialog box will be displayed, see Figure 7-55. Next, select the **Manage custom dictionaries** option from the **Current custom dictionary** drop-down list; the **Manage Custom Dictionaries** dialog box will be displayed, see Figure 7-56. From this dialog box, you can create a new dictionary, add an existing dictionary, or delete a dictionary from the **Custom dictionaries** list. To add the words listed in the custom dictionary, choose the **Add** button from the

Dictionaries dialog box. The **Settings** button in the **Check Spelling** dialog box is used to specify the type of text to be checked for spelling in your drawing.

Note
*The dictionaries can also be changed by specifying the name in the **DCTMAIN** or **DCTCUST** system variable.*

TEXT QUALITY AND TEXT FILL

AutoCAD supports **TrueType fonts**. You can use your own **TrueType fonts** by adding them to the Fonts directory. You can also keep your fonts in a separate directory, in that case you must specify the location of your fonts directory in the AutoCAD search path.

*Figure 7-55 The **Dictionaries** dialog box*

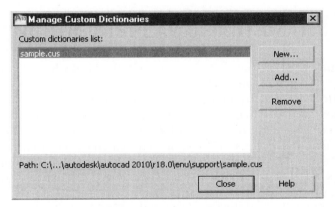

*Figure 7-56 The **Manage Custom Dictionaries** dialog box*

The resolution and text fill of the **TrueType font** text is controlled by the **TEXTFILL** and **TEXTQLTY** system variables. If **TEXTFILL** is set to 1, the text will be filled. If the value is set to 0, the text will not be filled. On the screen the text will appear filled, but when it is plotted the text will not be filled. The **TEXTQLTY** variable controls the quality of the **TrueType font** text. The value of this variable can range from 0 to 100. The default value is 50, which gives a resolution of 300 dpi (dots per inch). If the value is set to 100, the text will be drawn at 600 dpi. The higher the resolution, the more time it takes to regenerate or plot the drawing.

FINDING AND REPLACING TEXT*

Ribbon:	Annotate > Text > Find text	Menu Bar:	Edit > Find
Toolbar:	Text > Find	Command:	FIND

You can use the **FIND** command to find and replace the text. The text could be a line text created by the **TEXT** command, paragraph text created by the **MTEXT** command, dimension annotation text, block attribute value, hyperlinks, or hyperlink description. When you invoke this command, AutoCAD displays the **Find and Replace** dialog box, as shown in Figure 7-57. You can use this dialog box to perform the following functions:

*Figure 7-57 The **Find and Replace** dialog box*

Find Text

To find the text, enter the text that you want to find in the **Find what** edit box. You can search the entire drawing or confine your search to the selected text. To select the text, choose the **Select objects** button available to the right of the **Find where** drop-down list. The **Find and Replace** dialog box is temporarily closed and AutoCAD switches to the drawing window. Once you have selected the text, the **Find and Replace** dialog box is redisplayed. In the **Find where** drop-down list, you can specify if you need to search the entire drawing or the current selection. Choose the **Options** button available at the bottom left corner of the **Find and Replace** dialog box; the **Find and Replace** dialog box gets expanded, as shown in Figure 7-58. In this dialog box, you can specify whether to find the whole word and whether to match the case of the specified text. To find the text, choose the **Find** button. The text found text along with the surrounding text will be displayed in the **Text String** column of the **List results** area, provided the **List results** check box is selected. To find the next occurrence of the text, choose the **Find** button again.

Replacing Text

If you want to replace the specified text with the new text, enter the new text in the **Replace with** edit box. Now, if you choose the **Replace** button, only the found text will be replaced. If you choose the **Replace All** button, all occurrences of the specified text will be replaced with the new text.

CREATING TITLE SHEET TABLE IN A SHEET SET

While working with the sheet set, it is recommended that you create a title sheet that has the

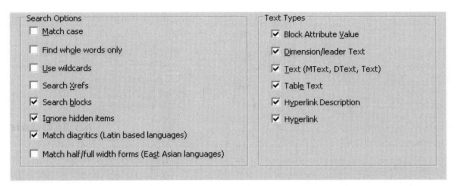

Figure 7-58 *Expanded view of the **Find and Replace** dialog box*

details about the sheets in the sheet set. You can enter the details about the sheets in the table. The advantage of using a table is that the information in it can be automatically updated if there is a change in the sheet number or name. Also, if a sheet is removed from the current sheet set, you can easily update the table to reflect the change in the sheet set.

To create a table in a sheet set, open the title sheet by double-clicking on it.

Tip
*If the title sheet is displayed at the bottom of the list in the **Sheet Set Manager**, you can drag and move it to top, below the name of the sheet set. Hold the left mouse button down on the title sheet and drag the cursor upward. Next, release the left mouse button below the name of the sheet set.*

When the title sheet is opened, right-click on the name of the sheet set in the **Sheet Set Manager** and choose the **Insert Sheet List Table** option from the shortcut menu, as shown in Figure 7-59. The **Insert Sheet List Table** dialog box will be displayed, as shown in Figure 7-60. If the **Insert Sheet List Table** option is not highlighted, either the model tab is active or the layout is not from the current sheet set.

Figure 7-59 *Inserting table in the title sheet using the **Sheet Set Manager***

Specify the table style settings using the options in the **Table Style Settings** area. You can also display the names of the subsets in the table by selecting the **Show Subheader** check box from the **Table Style Settings** area.

Chapter 7

*Figure 7-60 The **Sheet List Table** dialog box*

Enter the title of the table in the **Title Text** text box in the **Table Data Settings** area. By default, two rows are displayed for each subset in the table. This is because there are only two rows displayed in the **Column Settings** area. You can add additional rows by choosing the **Add** button. To change the data type of a row, click on the field in the **Data type** column. The field is changed into a drop-down list. Select the required data type from this drop-down list. After specifying all the parameters, choose **OK** from the **Insert Sheet List Table** dialog box.

The table will be attached to the cursor and you will be prompted to specify the insertion point. Specify the insertion point for the table. The table is inserted and based on the parameters selected, the sheets are displayed in the table. Figure 7-61 shows a sheet set table inserted in a title sheet.

If some changes are made in the sheet set numbering or any other property of the sheets in the sheet set, you can easily highlight the changes in the sheet list table by updating it. To update the sheet list table, right-click on it and choose **Update Sheet List Table** from the shortcut menu. The sheet list table is automatically updated.

Sheet List Table	
Sheet Number	Sheet Title
1	Title Sheet
Main Parts	
2	Base Part
3	Brasses
4	Cap
Fasteners	
5	Bolt
6	Nut
7	Washer

Figure 7-61 Sheet list table

Self-Evaluation Test

Answer the following questions and then compare them to those given at the end of this chapter:

1. Tables in AutoCAD are created using the **TABLET** command. (T/F)

2. An annotative text automatically gets scaled according to the viewport's scale. (T/F)

3. You can insert a block into a table cell. (T/F)

4. The **Standard** text style cannot be used for creating annotative text. (T/F)

5. You can control the height of an individual column separately by choosing the **Manual height** sub-option from **Text Editor > Insert > Columns >Dynamic Columns** of the **Ribbon**. (T/F)

6. Multiple lines of text can be entered at any desired location in the drawing area with a single _____ command.

7. With the _____ justification option of the **TEXT** command, AutoCAD adjusts the letter width to fit the text between the two given points, but the height remains constant.

8. In the **MTEXT** command, the height specified in the **In-Place Text Editor** does not affect the _____ system variable.

9. You can change the text string in the edit box using the _____ command for editing a single line text. However, for a multiline text, you must choose the **Full editor** button in the **Content** edit box of the **Properties** palette.

10. You can use the _____ system variable to specify the new font mapping file.

Review Questions

Answer the following questions:

1. You cannot insert a field using the **MTEXT** command. (T/F)

2. You can add or delete rows and columns from a table by using the **Table** toolbar. (T/F)

3. The **TEXT** command does not allow you to see the text on the screen as you type it. (T/F)

4. Which of the following buttons in the **Table** toolbar is used to inherit the properties from one cell to another?

    (a) **Cell Styles**   (b) **Match Cells**
    (c) **Link Cells**   (d) **Manage Cell Content**

5. Which of the following text styles is not present in AutoCAD by default?

    (a) **Standard**   (b) **Annotative**
    (c) **Auto text**   (d) None of the above

Chapter 7

6. If you want the text to be displayed at a height of 1/4" on the paper, then you can define a text style having a paper height of 1/4". When you add a text to a viewport having a scale of 1/4"=1'0", the current annotation scale, which is set to the same scale as the viewport, automatically scales the text to display approximately at _____.

    (a) 0.1875"   (b) 0.375"
    (c) 2.66"   (d) 4.8"

7. Which of the following characters in the **Text Editor tab** of the **Ribbon** is used to stack the text with a diagonal line without using the **Autostack Properties** dialog box?

    (a) ^   (b) /
    (c) #   (d) @

8. Which of the following commands can be used to create a new text style and modify the existing ones?

    (a) **TEXT**   (b) **MTEXT**
    (c) **STYLE**   (d) **SPELL**

9. The columns created by the _____ sub-option of the **Dynamic Columns** option of the **Columns** button in the **Insert** panel of the **Text Editor** tab are equal in size. (T/F)

10. The _____ sub-option in the **Formula** flyout of the **Insert** panel in the **Table** tab of the **Ribbon** is used to equate the current cell with the selected cell.

11. The four main text alignment modes are _____, _____, _____, and _____.

12. You can use the _____ command to write a paragraph text whose width can be specified by defining the _____ of the text boundary.

13. When the **Justify** option is invoked, the user can place the text in one of the _____ various types of alignment by choosing the desired alignment option.

14. The text created by the _____ command is a single object irrespective of the number of lines it contains.

15. Using the **MTEXT** command, the character _____ stacks the text vertically without a line (tolerance stack).

16. If you want to edit text, select it and then right-click such that various editing options _____ in the menu are available.

17. The _____ command is used to check the spelling of all the texts written in the current drawing.

Exercises

Exercise 2   *General*

Write the text, shown in Figure 7-62, on the screen. Use the text justification that will produce the text as shown in the drawing. Assume a value for text height. Use the **Properties** palette to change the text, as shown in Figure 7-63.

Figure 7-62 *Drawing for Exercise 2*

Figure 7-63 *Drawing for Exercise 2 (After changing the text)*

Exercise 3   *General*

Write the text on the screen, as shown in Figure 7-64. You must first define new text styles using the **STYLE** command with the attributes, as shown in the figure. The text height is 0.25 units.

Figure 7-64 *Drawing for Exercise 3*

Exercise 4   *Mechanical*

Draw the sketch shown in Figure 7-65 using the draw, edit, text, and display commands. Do not dimension the drawing.

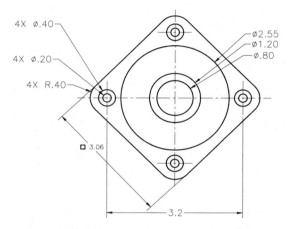

Figure 7-65 *Drawing for Exercise 4*

Exercise 5   *Mechanical*

Draw the sketch shown in Figures 7-66 and 7-67. Use the **MIRROR** command to duplicate the features that are identical. Also, add the text shown in the figures. Do not dimension the drawing.

Figure 7-66 *Drawing for Exercise 5*   **Figure 7-67** *Drawing for Exercise 5*

Exercise 6   *Mechanical*

Draw Figure 7-68 and also add the text shown in the figure. Do not dimension the drawing.

Figure 7-68 *Drawing for Exercise 6*

Problem-Solving Exercise 1   *Architectural*

Draw the sketch shown in Figure 7-69 using the draw, edit, and display commands. Also add the text to the drawing. Assume the missing dimensions. Do not dimension the drawing.

Figure 7-69 *Drawing for Problem-Solving Exercise 1*

Problem-Solving Exercise 2   *Architectural*

Draw Figure 7-70 using AutoCAD's draw, edit, and display commands. Also, add text to the drawing. Assume the missing dimensions. Do not dimension the drawing.

Figure 7-70 *Drawing for Problem Solving Exercise 2*

Answers to Self-Evaluation Test
1. F, 2. T, 3. T, 4. F, 5. T, 6. TEXT, 7. Fit, 8. TEXTSIZE, 9. Properties, 10. FONTMAP

Chapter **8**

Basic Dimensioning, Geometric Dimensioning, and Tolerancing

Learning Objectives

After completing this chapter, you will be able to:

- *Understand the need for dimensioning in drawings.*
- *Understand the fundamental dimensioning terms.*
- *Understand associative dimensioning.*
- *Understand and create annotative dimensioning.*
- *Use the **Quick Dimension** option for quick dimensioning.*
- *Create various types of dimensions in the drawing.*
- *Create center marks and centerlines.*
- *Attach leaders to objects.*
- *Attach and modify multileaders.*
- *Use geometric tolerancing, feature control frames, and characteristic symbols.*
- *Combine geometric characteristics and create composite position tolerancing.*
- *Use the projected tolerance zone.*
- *Use feature control frames with leaders.*

NEED FOR DIMENSIONING

To make designs more informative and practical, a drawing must convey more than just the graphic picture of a product. To manufacture an object, the drawing of that object must contain size descriptions such as the length, width, height, angle, radius, diameter, and location of features. These informations are added to the drawing by dimensioning. Some drawings also require information about tolerances with the size of features. The informations conveyed through dimensioning are vital and often as important as the drawing itself. With the advances in computer-aided design/drafting and computer-aided manufacturing, it has become mandatory to draw part to actual size so that dimensions reflect the actual size of features. At times, it may not be necessary to draw the object of the same size as the actual object would be when manufactured, but it is absolutely essential that the dimensions be accurate. Incorrect dimensions will lead to manufacturing errors.

By dimensioning, you not only give the size of a part, but also give a series of instructions to a machinist, an engineer, or an architect. The way the part is positioned in a machine, the sequence of machining operations, and the location of various features of the part depend on how you dimension the part. For example, the number of decimal places in a dimension (2.000) determines the type of machine that will be used to do that machining operation. The machining cost of such an operation is significantly higher than for a dimension that has only one digit after the decimal (2.0). Similarly, whether a part is to be forged or cast, the radii of the edges, and the tolerance you provide to these dimensions determine the cost of the product, the number of defective parts, and the number of parts you get from a single die.

DIMENSIONING IN AutoCAD

The objects that can be dimensioned in AutoCAD range from straight lines to arcs. The dimensioning commands provided by AutoCAD can be classified into four categories:

Dimension Drawing Commands   **Dimension Style Commands**
Dimension Editing Commands   **Dimension Utility Commands**

While dimensioning an object, AutoCAD automatically calculates the length of the object or the distance between two specified points. Also, settings such as the gap between the dimension text and the dimension line, the space between two consecutive dimension lines, arrow size, and text size are maintained and used when the dimensions are being generated for a particular drawing. The generation of arrows, lines (dimension lines, extension lines), and other objects that form a dimension is automatically performed by AutoCAD to save the user's time. This also results in uniform drawings. However, you can override the default measurements computed by AutoCAD and change the settings of various standard values. The modification of dimensioning standards can be achieved through the dimension variables.

The dimensioning functions offered by AutoCAD provide you with extreme flexibility in dimensioning by letting you dimension various objects in a variety of ways. This is of great help because different industries, such as architectural, mechanical, civil, or electrical, have different standards for the placement of dimensions.

FUNDAMENTAL DIMENSIONING TERMS

Before studying AutoCAD's dimensioning commands, it is important to know and understand various dimensioning terms that are common to linear, angular, radius, diameter, and ordinate dimensioning. Figures 8-1 and 8-2 show various dimensioning parameters.

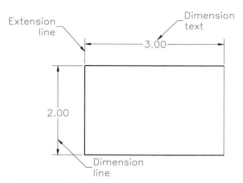

Figure 8-1 *Various dimensioning parameters*

Figure 8-2 *Various dimensioning parameters*

Dimension Line

The dimension line indicates which distance or angle is being measured. Usually, this line has arrows at both ends, and the dimension text is placed along the dimension line. By default the dimension line is drawn between the extension lines (Figure 8-1). If the dimension line does not fit inside, two short lines with arrows pointing inward are drawn outside the extension lines. The dimension line for angular dimensions (which are used to dimension angles) is an arc. You can control the positioning and various other features of the dimension lines by setting the parameters in the dimension styles. (The dimension styles are discussed in Chapter 10.)

Dimension Text

The dimension text is a text string that reflects the actual measurement (dimension value) between the selected points as calculated by AutoCAD. You can accept the value that AutoCAD returns or enter your own value. In case you use the default text, AutoCAD can be supplied with instructions to append the tolerances to it. Also, you can attach prefixes or suffixes of your choice to the dimension text.

Arrowheads

An arrowhead is a symbol used at the end of a dimension line (where dimension lines meet the extension lines). Arrowheads are also called terminators because they signify the end of the dimension line. Since drafting standards differ from company to company, AutoCAD allows you to draw arrows, tick marks, closed arrows, open arrows, dots, right angle arrows, or user-defined blocks (Figure 8-3). The user-defined blocks at the two ends of the dimension line can be customized to your requirements. The size of the arrows, tick marks, user blocks, and so on can be regulated by using the dimension variables.

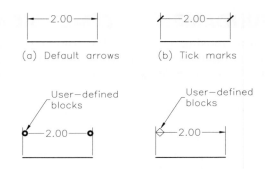

Extension Lines

Extension lines are drawn from the object measured to the dimension line (Figure 8-4). These lines are also called witness lines. Extension lines are used in linear and angular dimensioning. Generally, extension lines are drawn perpendicular to the dimension line. However, you can make extension lines inclined at an angle by

Figure 8-3 Using arrows, tick marks, and user-defined blocks

using the **DIMEDIT** command (**Oblique** option) or by choosing **Annotate > Dimensions > Oblique** from the **Ribbon**. Alternatively, you can choose **Dimension Edit** from the **Dimension** toolbar. AutoCAD also allows you to suppress either one or both extension lines in a dimension (Figure 8-5). You can insert breaks in a dimension or an extension line, in case they intersect other geometric objects or dimension entities. You can also control various other features of the extension lines by setting parameters in dimension styles. (Dimension styles are discussed in Chapter 10.)

Figure 8-4 Extension lines

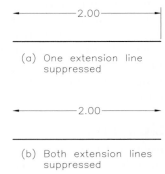

Figure 8-5 Extension line suppressed

Leader

A leader is a line that stretches from the dimension text to the object being dimensioned. Sometimes the text for dimensioning and other annotations do not adjust properly near the object. In such cases, you can use a leader and place the text at the end of the leader line. For example, the circle shown in Figure 8-6 has a keyway slot that is too small to be dimensioned. In this situation, a leader can be drawn from the

Figure 8-6 Leader used to attach annotation

text to the keyway feature. Also, a leader can be used to attach annotations such as part numbers, notes, and instructions to an object. You can also draw multileaders that are used to connect one note to different places or many notes to one place.

Center Mark and Centerlines

The center mark is a cross mark that identifies the center point of a circle or an arc. Centerlines are mutually perpendicular lines passing through the center of the circle/arc and intersecting the circumference of the circle/arc. A center mark or centerlines are automatically drawn when you dimension a circle or arc (see Figure 8-7). The length of the center mark and the extension of the centerline beyond the circumference of the circle are determined by the value assigned to the **DIMCEN** dimension variable or the value entered in the **Center marks** area in the **Symbols and Arrows** tab of the **Dimension Style Manager** dialog box. To invoke this dialog box, left-click on the inclined arrow in the **Dimensions** panel of the **Annotate** tab of the **Ribbon**. You can also choose **Annotate > Dimensions > Center Mark** from the **Ribbon** to dimension the center mark of the circle.

Alternate Units

With the help of alternate units, you can generate dimensions for two systems of measurement at the same time (Figure 8-8). For example, if the dimensions are in inches, you can use the alternate units dimensioning facility to append metric dimensions to the dimensions (controlling the alternate units through the dimension variables is discussed in Chapter 10).

Figure 8-7 *Center mark and centerlines*   **Figure 8-8** *Using alternate units for dimensioning*

Tolerances

Tolerance is the amount by which the actual dimension can vary (Figure 8-9). AutoCAD can attach the plus/minus tolerances to the dimension text (actual measurement computed by AutoCAD). This is also known as deviation tolerance. The plus and minus tolerances that you specify can be same or different. You can use the dimension variables to control the tolerance feature (these variables are discussed in Chapter 10).

Limits

Instead of appending tolerances to the dimension text, you can apply tolerances to the measurement itself (Figure 8-10). Once you define the tolerances, AutoCAD will automatically calculate the upper and lower limits of the dimension. These values are then displayed as a dimension text.

For example, if the actual dimension as computed by AutoCAD is 2.6105 units and the tolerance values are +0.025 and -0.015, the upper and lower limits are 2.6355 and 2.5955. After calculating the limits, AutoCAD will display them as dimension text, as shown in Figure 8-10. The dimension variables that control the limits are discussed in Chapter 10.

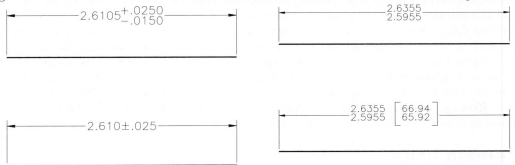

Figure 8-9 *Using tolerances with dimensions*   **Figure 8-10** *Using limits with dimensioning*

ASSOCIATIVE DIMENSIONS

The Associative dimensioning is a method of dimensioning, in which the dimension is associated with the object that is dimensioned. In other words, the dimension is influenced by the changes in the size of the object. In the previous releases of AutoCAD, the dimensions were not truly associative, but were related to the objects being dimensioned by definition points on the DEFPOINTS layer. To cause the dimension to be associatively modified, these definition points had to be adjusted along with the object being changed. If, for example, you use the **SCALE** command to change an object's size and select the object, the dimensions will not be modified. If you select the object and its defpoints (using the Crossing selection method), then the dimension will be modified. However, in AutoCAD 2002, a new concept called **true associative dimensioning** was introduced. This concept ensures that if the dimensions are associated to the object and the object changes its size, the dimensions will also change automatically. With the introduction of the true associative dimensions, there is no need to select the definition points along with the object. This eliminates the use of definition points for updating the dimensions.

The associative dimensions automatically update their values and location if the value or location of the object is modified. For example, if you edit an object using simple editing operations such as breaking, using the **BREAK** command, then the true associative dimension will be modified automatically. The dimensions can be converted into the true associative dimensions using the **DIMREASSOCIATE** command. You can also choose **Annotate > Dimensions > Reassociate** from the **Ribbon** to reassociate the dimension. The association of the dimensions with the objects can be removed using the **DIMDISASSOCIATE** command. Both these commands will be discussed later in this chapter.

The dimensioning variable **DIMASSOC** controls the associativity of dimensions. The default value of this variable is **2**, which means the dimensions are associative. When the value is **1**, the dimensions placed are non-associative. When the **DIMASSOC** is turned off, then the dimension will be placed in the exploded format. This means that the dimensions will now be placed as a combination of individual arrowheads, dimension lines, extension lines, and text. Also, note that the exploded dimensions cannot be associated to any object.

DEFINITION POINTS

Definition points are the points drawn at the positions used to generate a dimension object. The definition points are used by the dimensions to control their updating and rescaling. AutoCAD draws these points on a special layer called **DEFPOINTS**. These points are not plotted by the plotter because AutoCAD does not plot any object on the **DEFPOINTS** layer. If you explode a dimension (which is as good as turning **DIMASSOC** off), the definition points are converted to point objects on the **DEFPOINTS** layer. In Figure 8-11, the small circles indicate the definition points for different objects.

Figure 8-11 *Definition points of linear, angular, and ordinate dimensions*

The definition points for linear dimensions are the points used to specify the extension lines and the point of intersection of the first extension line and the dimension line. The definition points for the angular dimension are the endpoints of the lines used to specify the dimension and the point used to specify the dimension line arc. For example, for a three-point angular dimension, the definition points are the extension line endpoints, angle vertex, and the point used to specify the dimension line arc.

The definition points for the radius dimension are the center point of the circle or arc, and the point where the arrow touches the object. The definition points for the diameter dimension are the points where the arrows touch the circle. The definition points for the ordinate dimension are the UCS origin, feature location, and leader endpoint.

Note
In addition to the definition points just mentioned, the middle point of the dimension text serves as the definition point for all types of dimensions.

ANNOTATIVE DIMENSIONS

When all the elements of a dimension such as text, spacing, and arrows get scaled according to the specified annotation scale, it is known as Annotative Dimension. They are created in the drawing by assigning annotative dimension styles to them. You can also change the non-annotative dimensions to annotative by changing their **Annotative** property to **Yes** in the **Properties** palette.

SELECTING DIMENSIONING COMMANDS

Using the Ribbon and Toolbar

You can select the dimensioning commands from the **Dimensions** panel of the **Annotate** tab in the **Ribbon** (Figure 8-12) or the **Dimension** toolbar (Figure 8-13). The **Dimension** toolbar can be displayed by choosing **Tools > Toolbars > AutoCAD > Dimension** from the menu bar, if it is displayed.

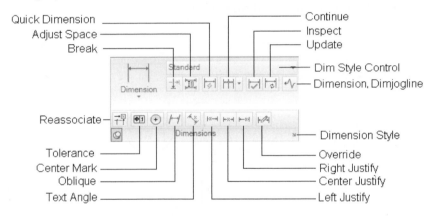

Figure 8-12 The **Dimensions** panel

Figure 8-13 The **Dimension** toolbar

Using the Command Line

You can directly enter a dimensioning command in the Command line or use the **DIM** or the **DIM1** commands to invoke the dimensioning commands. Both these methods of using the command line are discussed next.

AutoCAD has provided the following fundamental dimensioning types:

Quick dimensioning   **Linear dimensioning**   **Diameter dimensioning**
Radius dimensioning   **Angular dimensioning**   **Ordinate dimensioning**
Arc Length dimensioning

Figures 8-14 and 8-15 show various fundamental dimension types.

Note
*The **DIMDEC** variable sets the number of decimal places for the value of primary dimension and the **DIMADEC** variable for angular dimensions. For example, if **DIMDEC** is set to **3**, AutoCAD will display the decimal dimension up to three decimal places (2.037).*

Figure 8-14 Linear and angular dimensions

Figure 8-15 Radius, diameter, and ordinate dimensions

DIMENSIONING A NUMBER OF OBJECTS TOGETHER

Ribbon: Annotate > Dimensions > Quick Dimension	**Command:** QDIM
Menu bar: Dimension > Quick Dimension	**Toolbar:** Dimension > Quick Dimension

The **QDIM** command allows you to dimension a number of objects at the same time. It also allows you to quickly edit dimension arrangements already existing in the drawing and also create new dimension arrangements. It is especially useful when creating a series of baseline or continuous dimensions. It also allows you to dimension multiple arcs and circles at the same time. When you are using the **QDIM** command, you can relocate the datum base point for baseline and ordinate dimensions. The prompt sequence that will follow when you choose this button is given next.

Select geometry to dimension: *Select the objects to be dimensioned and press ENTER.*
Select geometry to dimension: [Enter]
Specify dimension line position, or [Continuous/Staggered/Baseline/Ordinate/Radius/Diameter/datumPoint/Edit/seTtings] <Continuous>: *Press ENTER to accept the default dimension arrangement and specify dimension line location or enter new dimension arrangement or edit the existing dimension arrangement.*

For example, you can dimension all circles in a drawing (Figure 8-16) by using quick dimensioning as follows:

Associative dimension priority = Endpoint.
Select geometry to dimension: *Select all circles.*

*Figure 8-16 Using the **QDIM** command to quickly dimension multiple circles*

Select geometry to dimension: `Enter`
Specify dimension line position, or
[Continuous/Staggered/Baseline/Ordinate/Radius/Diameter/datumPoint/Edit/
seTtings] <Continuous>: *Press D for diameter dimensioning and select a point where you want
to position the radial dimension.*

CREATING LINEAR DIMENSIONS

Ribbon:	Annotate > Dimensions > Linear	**Command:**	DIMLIN or DIMLINEAR
Menu Bar:	Dimension > Linear	**Toolbar:**	Dimension > Linear

Linear dimensioning applies to those dimensioning commands that measure the
shortest distance between two points. You can directly select the object to dimension
or select two points. The points can be any two points in the space, endpoints of an
arc or line, or any set of points that can be identified. To achieve accuracy, points must be
selected with the help of object snaps or by selecting an object to dimension. In case, the
object selected is aligned, then the linear dimensions will add **Horizontal** or **Vertical**
dimensions to the object. The prompt sequence that will follow when you choose this button
is given next.

Specify first extension line origin or <select object>: `Enter`
Select object to dimension: *Select the object.*
Specify dimension line location or
[Mtext/Text/Angle/Horizontal/Vertical/Rotated]: *Select a point to locate the position of the
dimension.*

Instead of selecting the object, you can also select the two endpoints of the line that you want
to dimension (Figure 8-17). Usually the points on the object are selected by using the **object
snaps** (endpoints, intersection, center, etc.). The prompt sequence is as follows.

Specify first extension line origin or <select object>: *Select a point.*

Specify second extension line origin:
Select second point.
Specify dimension line location or
[Mtext/Text/Angle/Horizontal/Vertical/
Rotated]: *Select a point to locate the position of
the dimension.*

When using the **DIMLINEAR** command, you
can obtain the horizontal or vertical dimension
by simply defining the appropriate dimension
location point. If you select a point above or
below the dimension, AutoCAD creates a
horizontal dimension. If you select a point that
is on the left or right of the dimension, AutoCAD creates a vertical dimension
through that point.

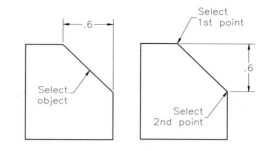

Figure 8-17 *Drawing linear dimensions*

DIMLINEAR Command Options

The options under this command are discussed next.

Mtext Option

The **Mtext** option allows you to override the default dimension text and also change the font,
height, and so on, using the **In-Place Text Editor** dialog box. When you enter **M** at the
Specify dimension line location or [Mtext/Text/Angle/Horizontal/Vertical/Rotated] prompt,
the **In-Place Text Editor** is displayed. You can change the text by entering a new text. You
can also use the various options of the text editor (explained in Chapter 7) and then choose
OK. However, if you override the default dimensions, the dimensional associativity of the
dimension text is lost. This means that if you modify the object using the definition points,
AutoCAD will not recalculate the dimension text. Even if the dimension is a true associative
dimension, the text will not be recalculated when the object is modified. The prompt
sequence to invoke this option is given next.

Specify first extension line origin or <select object>: *Specify a point.*
Specify second extension line origin: *Specify the second point.*
Specify dimension line location or
[Mtext/Text/Angle/Horizontal/Vertical/Rotated]: **M** *(Enter the dimension text in the **In-Place
Text Editor** dialog box and then click outside the text editor to accept the changes.)*
Specify dimension line location or
[Mtext/Text/Angle/Horizontal/Vertical/Rotated]: *Specify the dimension location.*

Text Option

This option also allows you to override the default dimension. However, this option will
prompt you to specify the new text value in the Command prompt itself, see Figure 8-18. The
prompt sequence to invoke this option is given next.

Specify first extension line origin or <select object>: *Select a point.*
Specify second extension line origin: *Select second point.*

Specify dimension line location or
[Mtext/Text/Angle/Horizontal/Vertical/Rotated]: **T**
Enter dimension text <Current>: *Enter new text.*
Specify dimension line location or
[Mtext/Text/Angle/Horizontal/Vertical/Rotated]: *Specify the dimension location.*

Angle Option

This option lets you change the angle of the dimension text, see Figure 8-18.

Rotated Option

This option lets you create a dimension that is rotated at a specified angle, see Figure 8-18.

Horizontal Option

This option lets you create a horizontal dimension regardless of where you specify the dimension location, see Figure 8-19.

Vertical Option

This option lets you create a vertical dimension regardless of where you specify the dimension location, see Figure 8-19.

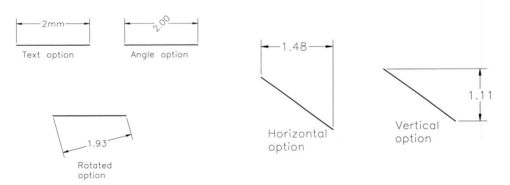

*Figure 8-18 The **Text**, **Angle**, and **Rotated** options*

*Figure 8-19 The **Horizontal** and **Vertical** options*

Note

If you override the default dimensions, the dimensional associativity of the dimension text is lost and AutoCAD will not recalculate the dimension when the object is scaled.

Example 1   *General*

In this example, you will use linear dimensioning to dimension a horizontal line of 4 units length. The dimensioning will be done first by selecting the object and later on by specifying the first and second extension line origins. Using the **Text Editor**, modify the default text such that the dimension is underlined.

Selecting the Object

1. Start a new file in the **2D Drafting & Annotation** workspace and draw a line of 4 units length.

2. Choose **Annotate > Dimensions > Linear** from the **Ribbon**. The prompt sequence is as follows:

    Specify first extension line origin or <select object>: `Enter`
    Select object to dimension: *Select the line.*
    Specify dimension line location or
    [Mtext/Text/Angle/Horizontal/Vertical/Rotated]: **M** `Enter`
    *The **Text Editor** will be displayed, as shown in Figure 8-20. Select the default dimension value and then choose the **Underline** button from the **Formatting** panel of the **Text Editor** tab of the **Ribbon** to underline the text. Click anywhere in the drawing area to exit the **Text Editor** tab.*
    Specify dimension line location or
    [Mtext/Text/Angle/Horizontal/Vertical/Rotated]: *Place the dimension.*
    Dimension text = 4.0000

*Figure 8-20 The **Text Editor***

Specifying Extension Line Origins

1. Choose **Annotate > Dimensions > Linear** from the **Ribbon**. The prompt sequence is as follows:

    Specify first extension line origin or <select object>: *Select the first endpoint of the line using the **Endpoint** object snap, see Figure 8-21.*

    Specify second extension line origin: *Select the second endpoint of the line using the **Endpoint** object snap, see Figure 8-21.*
    Specify dimension line location or
    [Mtext/Text/Angle/Horizontal/Vertical/Rotated]: **M**
    *Select the text and then choose the **Underline** button from the **Ribbon** to underline the text in the **In-Place Text Editor**.*
    Specify dimension line location or
    [Mtext/Text/Angle/Horizontal/Vertical/Rotated]: *Place the dimension.*
    Dimension text = 4.00

Chapter 8

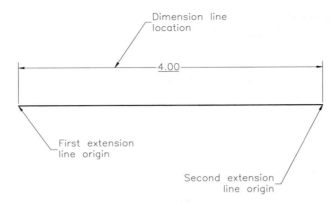

Figure 8-21 *Line for Example 1*

CREATING ALIGNED DIMENSIONS

Ribbon:	Annotate > Dimensions > Aligned	**Command:**	DIMALIGNED
Menu bar: Dimensions > Aligned		**Toolbar:**	Dimension > Aligned

Generally, the drawing consists of various objects that are neither parallel to the X axis nor to the Y axis. Dimensioning of such objects can be done using aligned dimensioning. In horizontal or vertical dimensioning, you can only measure the shortest distance from the first extension line origin to the second extension line origin along the horizontal or vertical axis, respectively, whereas, with the help of aligned dimensioning, you can measure the true aligned distance between the two points. The working of the **ALIGNED** dimension command is similar to that of the other linear dimensioning commands. The dimension created with the **ALIGNED** command is parallel to the object being dimensioned. The prompt sequence that will follow when you choose this button is given next.

Specify first extension line origin or <select object>: *Specify the first point or press ENTER to select the object.*
Specify second extension line origin: *Specify second point.*
Specify dimension line location or [Mtext/Text/Angle]: *Specify the location for the dimension line.*
Dimension text = Current

The options provided under this command are similar to those under the **DIMLINEAR** command. Figure 8-22 illustrates the aligned dimensioning.

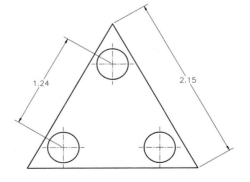

Figure 8-22 *The aligned dimensioning*

Exercise 1

Mechanical

Draw the object shown in Figure 8-23 and then use linear and aligned dimensioning to dimension the part. The distance between the dotted lines is 0.5 units. The dimensions should be up to 2 decimal places. To get dimensions up to 2 decimal places, enter DIMDEC at the Command prompt and then enter 2. (There will be more information about dimension variable in Chapter 10.)

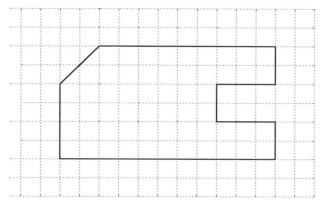

Figure 8-23 *Drawing for Exercise 1*

CREATING ARC LENGTH DIMENSIONS

Ribbon: Annotate > Dimensions > Linear > Arc Length	**Command:** DIMARC
Menu Bar: Dimension > Arc Length	**Toolbar:** Dimension > Arc

The Arc Length dimensioning is used to dimension the length of an arc or the polyline arc segment. You are required to select an arc or a Polyline arc segment and the dimension location. Figure 8-24 shows the Arc Length dimensioning of an arc. You can invoke this command by choosing **Annotate > Dimensions > Linear > Arc Length** from the **Ribbon**. The prompt sequence that will follow is given next.

Select arc or polyline arc segment: *Select arc or polyline arc segment to dimension.*
Specify arc length dimension location, or [Mtext/Text/Angle/Partial/Leader]: *Specify the location for the dimension line.*
Dimension text = *Current.*

Using the **Partial** option, you can dimension a selected portion of the arc, as shown in Figure 8-25. The prompt sequence for the **Partial** option is given next.

Select arc or polyline arc segment: *Select arc or polyline arc segment to dimension.*
Specify arc length dimension location, or [Mtext/Text/Angle/Partial]: **P** [Enter]

Specify first point for arc length dimension: *Specify the first point on arc.*
Specify second point for arc length dimension: *Specify the second point on arc.*
Specify arc length dimension location, or [Mtext/Text/Angle/Partial]: *Specify the location for the dimension line.*
Dimension text = *Current*

Using the **Leader** option, you can attach a leader to the dimension text, starting from its circumference. This leader is drawn radial to the arc, as shown in Figure 8-26.

Figure 8-24 Arc Length dimensioning   *Figure 8-25* Partial Arc Length dimensioning

 Note
*The **Leader** option is displayed only when the arc subtends an included angle greater than 90 degrees at its centre.*

CREATING ROTATED DIMENSIONS

Rotated dimensioning is used when you want to place the dimension line at an angle (if you do not want to align the dimension line with the extension line origins selected), as shown in Figure 8-27. The **ROTATED** dimension option will prompt you to specify the dimension line angle. You can invoke this command by using **DIMLINEAR** (Rotate option) or by entering **ROTATED** at the **Dim:** Command prompt. The prompt sequence is given next.

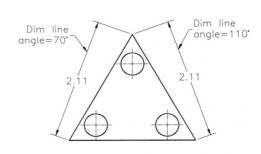

Figure 8-26 Leader Arc Length dimensioning   *Figure 8-27* Rotated dimensioning

Dim: **ROTATED**
Specify angle of dimension line <0>: **110**
Specify first extension line origin or <select object>: *Select the lower right corner of the triangle.*
Specify second extension line origin: *Select the top corner.*
Specify dimension line location or [MText/Text/Angle]: *Select the location for the dimension line.*
Enter dimension text <2.0597>: [Enter]

Tip
You can draw horizontal and vertical dimensioning by specifying the rotation angle of 0-degree for horizontal dimensioning and 90-degree for vertical dimensioning.

CREATING BASELINE DIMENSIONS

Ribbon: Annotate > Dimensions > Continue > Baseline **Toolbar:** Dimension > Baseline
Menu Bar: Dimension > Baseline   **Command:** DIMBASE or DIMBASELINE

Sometimes in manufacturing, you may want to locate different points and features of a part with reference to a fixed point (base point or reference point). This can be accomplished by using the Baseline dimensioning (Figure 8-28). To invoke the Baseline dimensions, you need to choose **Annotate > Dimensions > Continue > Baseline** from the **Ribbon**. Using this command you can continue a linear dimension from the first extension line origin of the first dimension to the dimension point. The new dimension line is automatically offset by a fixed amount to avoid overlapping of the dimension

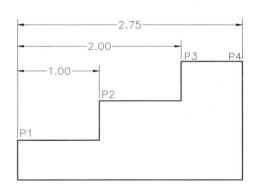

Figure 8-28 Baseline dimensioning

lines. This has to be kept in mind that there must already exist a linear, ordinate, or angular associative dimension to use the Baseline dimensions. When you choose the **Baseline** button, the last linear, ordinate, or angular dimension created, will be selected and used as the baseline. The prompt sequence that will follow when you choose this button is given next.

Specify a second extension line origin or [Undo/Select] <Select>: *Select the origin of the second extension line.*
Dimension text = current
Specify a second extension line origin or [Undo/Select] <Select>: *Select the origin of the second extension line.*
Dimension text = current
Specify a second extension line origin or [Undo/Select] <Select>: *Select the origin of the second extension line or press ENTER.*
Select base dimension: [Enter]

When you use the **DIMBASELINE** command, you cannot change the default dimension text. However, the **DIM** mode commands allow you to override the default dimension text.

Command: **DIM**
Dim: **HOR**
Specify first extension line origin or <select object>: *Select left corner (P1, Figure 8-28).* *(Use Endpoint object snap.)*
Specify second extension line origin: *Select the origin of the second extension line (P2).*
Specify dimension line location or [MText/Text/Angle]: **T**
Enter dimension text <1.0000>: **1.0**
Specify dimension line location or [MText/Text/Angle]: *Select the dimension line location.*
Dim: **BASELINE (or BAS)**
Specify a second extension line origin or [Select] <Select>: *Select the origin of the next extension line (P3).*
Enter dimension text <2.0000>: **2.0**
Dim: **BAS**
Specify a second extension line origin or [Select] <Select>: *Select the origin of the next extension line (P4).*
Enter dimension text <3.000>: **2.75**

The next dimension line is automatically spaced and drawn by AutoCAD.

CREATING CONTINUED DIMENSIONS

Ribbon: Annotate > Dimensions > Continue	**Toolbar:** Dimension > Continue	
Menu Bar: Dimension > Continue	**Command:** DIMCONT or DIMCONTINUE	

Using this command, you can continue a linear dimension from the second extension line of the previous dimension. This is also called as Chained or Incremental dimensioning. Note that there must exist linear, ordinate, or angular associative dimension to use the Continue dimensions. The prompt sequence that will follow when you choose this button is given next.

Specify a second extension line origin or [Undo/Select] <Select>: *Select the origin of the second extension line or press ENTER to select the existing dimension.*
Select continued dimension: *Select the dimension.*
Specify a second extension line origin or [Undo/Select] <Select>: *Specify the point on the origin of the second extension line.*
Dimension text = current
Specify a second extension line origin or [Undo/Select] <Select>: *Specify the point on the origin of the second extension line.*
Dimension text = current
Specify a second extension line origin or [Undo/Select] <Select>: Enter
Select continued dimension: Enter

Also, in this case, the **DIM** command should be used if you want to change the default dimension text.

Command: **DIM**
Dim: **HOR**
Specify first extension line origin or <select object>: *Select left corner (P1, see Figure 8-29). (Use Endpoint object snap.)*
Specify second extension line origin: *Select the origin of the second extension line (P2, see Figure 8-29).*
Specify dimension line location or [MText/Text/Angle]: **T**
Enter dimension text <current>: **0.75**
Specify dimension line location or [MText/Text/Angle]: *Select the dimension line location.*
Dim: **CONTINUE**

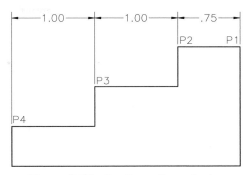

Figure 8-29 *Continue dimensioning*

Specify a second extension line origin or [Select] <Select>: *Select the origin of the next extension line (P3, see Figure 8-29).*
Enter dimension text <current>: Enter
Dim: **CONTINUE**
Specify a second extension line origin or [Select] <Select>: *Select the origin of next extension line (P4, see Figure 8-29).*
Enter dimension text <current>: Enter

The default base (first extension line) for the dimensions created with the **CONTINUE** command is the previous dimension's second extension line. You can override the default extension by pressing ENTER at the **Specify a second extension line origin or [Select] <Select>** prompt, and then specifying the other dimension. The extension line origin nearest to the selection point is used as the origin for the first extension line.

Tip
*You can use the **Select** option of the **DIMBASELINE** or the **DIMCONTINUE** command to select any other existing dimension to be used as the baseline or continuous dimension.*

Exercise 2   *Mechanical*

Draw the object shown in Figure 8-30 and then use the baseline dimensioning to dimension the top half and continue dimensioning to dimension the bottom half. The distance between the dotted lines is 0.5 unit.

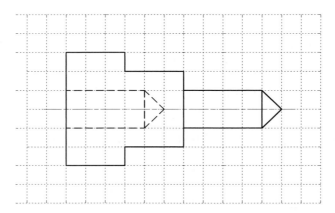

Figure 8-30 *Drawing for Exercise 2*

CREATING ANGULAR DIMENSIONS

Ribbon: Annotate > Dimensions > Linear > Angular **Toolbar:**   Dimension > Angular
Menu Bar: Dimension > Angular   **Command:**   DIMANGULAR or DIMANG

The Angular dimensioning is used when you want to dimension an angle. This command generates a dimension arc (dimension line in the shape of an arc with arrowheads at both ends) to indicate the angle between two nonparallel lines. This command can also be used to dimension the vertex and two other points, a circle with another point, or the angle of an arc. For every set of point, there exists one acute angle and one obtuse angle (inner and outer angles). If you specify the dimension arc location between the two points, you will get the acute angle; if you specify it outside the two points, you will get the obtuse angle. Figure 8-31 shows the four ways to dimension two nonparallel lines. The prompt sequence that will follow when you choose this button is given next:

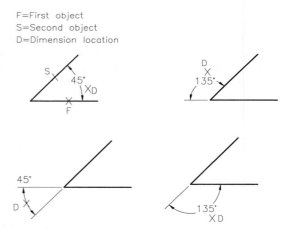

Figure 8-31 *Angular dimensioning between two nonparallel lines*

Select arc, circle, line, or <specify vertex>: *Select the object or press ENTER to select a vertex point where two segments meet.*
Select second line: *Select the second object.*
Specify dimension arc line location or [Mtext/Text/Angle/Quadrant]: *Place the dimension or select an option.*
Dimension text = current

The methods of dimensioning various entities using this command are discussed next.

Dimensioning the Angle between Two Nonparallel Lines

The angle between two nonparallel lines or two straight line segments of a polyline can be dimensioned using the **DIMANGULAR** dimensioning command. The vertex of the angle is taken as the point of intersection of the two lines.

The following example illustrates the dimensioning of two nonparallel lines using the **DIMANGULAR** command. Alternatively, the angular dimensioning can be done by choosing **Annotate > Dimensions > Linear > Angular** from the **Ribbon**.

Select arc, circle, line, or <specify vertex>: *Select the first line.*
Select second line: *Select the second line.*
Specify dimension arc line location or [Mtext/Text/Angle/Quadrant]: **M** *(Enter the new value in the **Text Editor** dialog box).*
Specify dimension arc line location or [Mtext/Text/Angle/Quadrant]: *Specify the dimension arc location or select an option.*

The location of the extension lines and the dimension arc is determined by the way you locate the dimension arc. One of the recent enhancements in AutoCAD is that now you can place the dimension text outside the quadrant in which you measure the angle by extending the dimension arc. Choose the **Quadrant** option from the shortcut menu from the **Specify dimension arc line location or [Mtext/ Text/Angle/Quadrant]** prompt. Next, you will be prompted to specify the quadrant in which you want to measure the angle. Then, specify the quadrant with mouse. If the dimension arc line is lying outside the quadrant that is being measured, the dimension arc will be extended

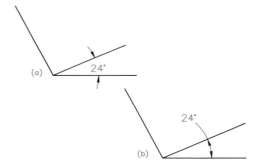

Figure 8-32 *Drawing displaying the angular dimension created with and without choosing the **Quadrant** option*

up to that location with the help of extension line. Figure 8-32(a) shows a dimension created without using the **Quadrant** option and Figure 8-32(b) shows the same dimension created by using the **Quadrant** option.

Dimensioning the Angle of an Arc

Angular dimensioning can also be used to dimension the angle of an arc. In this case, the center point of the arc is taken as the vertex and the two endpoints of the arc are used as the extension line origin points for the extension lines (Figure 8-33). The following example illustrates the dimensioning of an arc using the **DIMANGULAR** command:

Select arc, circle, line, or <specify vertex>: *Select the arc.*
Specify dimension arc line location or [Mtext/Text/Angle/Quadrant]: *Specify a location for the arc line or select an option.*

Angular Dimensioning of Circles

The angular feature associated with the circle can be dimensioned by selecting a circle object at the **Select arc, circle, line, or <specify vertex>** prompt. The center of the selected circle is used as the vertex of the angle. The first point selected (when the circle is selected for angular dimensioning) is used as the origin of the first extension line. In a similar manner, the second point selected is taken as the origin of the second extension line (Figure 8-34). The following is the prompt sequence for dimensioning a circle:

Select arc, circle, line, or <specify vertex>: *Select the circle at the point where you want the first extension line.*
Specify second angle endpoint: *Select the second point on or away from the circle.*
Specify dimension arc line location or [Mtext/Text/Angle/Quadrant]: *Select the location for the dimension line.*

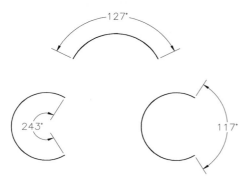

Figure 8-33 Angular dimensioning of arcs

Figure 8-34 Angular dimensioning of circles

Angular Dimensioning based on Three Points

If you press ENTER at the **Select arc, circle, line, or <specify vertex>** prompt, AutoCAD allows you to select three points to create an angular dimension. The first point is the vertex point, and the other two points are the first and second angle endpoints of the angle (Figure 8-35). The coordinate specifications of the first and the second angle endpoints must not be identical. However, the angle vertex coordinates and one of the angle endpoint coordinates can be identical. The following example illustrates angular dimensioning by defining three points:

Select arc, circle, line, or <specify vertex>: [Enter]

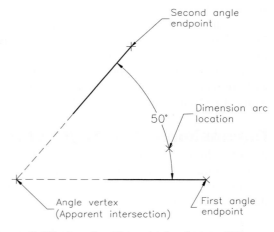

Figure 8-35 Angular dimensioning by specifying three points

Specify angle vertex: *Specify the first point, vertex. This is the point where two segments meet. If the two segments do not meet, use the **Apparent Intersection** object snap.*
Specify first angle endpoint: *Specify the second point. This point will be the origin of the first extension line.*
Specify second angle endpoint: *Specify the third point. This point will be the origin of the second extension line.*
Specify dimension arc line location or [Mtext/Text/Angle/Quadrant]: *Select the location for the dimension line.*
Dimension text = current

Note

*If you use the **DIMANGULAR** command, you cannot specify the text location. With the **DIMANGULAR** command, AutoCAD positions the dimensioning text automatically. If you want to manually define the position of the dimension text, use the **ANGULAR** option of the **DIM** command.*

Exercise 3   *Mechanical*

Make the drawing shown in Figure 8-36 and then use angular dimensioning to dimension all angles of the part. The distance between the dotted lines is 0.5 unit.

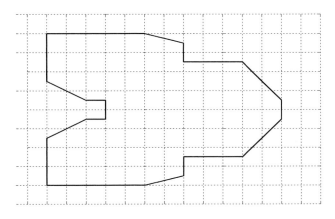

Figure 8-36 *Drawing for Exercise 3*

CREATING DIAMETER DIMENSIONS

| **Ribbon:** Annotate > Dimensions > Dimension > Diameter | **Command:** DIMDIA |
| **Toolbar:** Dimension > Diameter | **Menu Bar:** Dimension > Diameter |

Diameter dimensioning is used to dimension a circle or an arc. Here, the measurement is done between two diametrically opposite points on the circumference of the circle or arc (Figure 8-37). The dimension text generated by AutoCAD commences with the Ø symbol to indicate a diameter dimension. The prompt sequence that will follow when

you choose this button from the **Linear** flyout in
the **Dimensions** panel is given next.

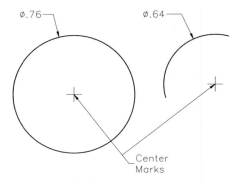

> Select arc or circle: *Select an arc or circle by*
> *selecting a point anywhere on its circumference.*
> Dimension text = Current
> Specify dimension line location or [Mtext/
> Text/Angle]: *Specify a point to position the*
> *dimension.*

Figure 8-37 *Diameter dimensioning*

If you want to override the default value of the
dimension text, use the **Mtext** or the **Text** option.
The control sequence %%C is used to obtain the
diameter symbol ø. It is followed by the dimension text that should appear in the diameter
dimension. For example, if you want to write a text that displays a value ø20, then enter
%%c20 at the text prompt.

Tip
The control sequence %%D can be used to generate the degree symbol " ° " (45°).

CREATING JOGGED DIMENSIONS

Ribbon: Annotate > Dimensions > Dimension > Jogged	**Command:** DIMJOGGED	
Menu Bar: Dimension > Jogged	**Toolbar:** Dimension > Jogged	

The necessity of the jogged dimension is due to the space constraint in certain
drawings and to avoid merging of the dimension line with other dimensions. Also,
there are instances when it is not possible to show the center of the circle in the sheet.
In such situations, the jogged dimensions are used, as shown in Figure 8-38. The jogged
dimensions can be added using the **Jogged** tool. Note that with this command, you can add
only jogged radius dimensions. To add the jogged dimensions, choose **Annotate >**
Dimensions > Dimension > Jogged from the **Ribbon** and select an arc or a circle from the
drawing area; you are now prompted to select
the center location override. The center
location override specifies the start point of the
dimension line. Next, specify the dimension
line location and the jog location. The
Command prompt when you invoke the **Jogged**
tool is given next.

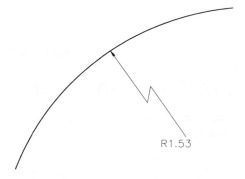

> Select arc or circle: *Select arc or circle to be*
> *dimensioned.*
> Specify center location override: *specify a*
> *point which will currently override the actual*
> *center point location.*
> Dimension text = *Current*

Figure 8-38 *Jogged dimensioning*

Specify dimension line location or [Mtext/Text/Angle]: *Specify a point to position dimension line.*
Specify jog location: *Specify a point for the positioning of jog.*

CREATING RADIUS DIMENSIONS

Ribbon:	Annotate > Dimensions > Dimension > Radius **Command:**	DIMRAD
Menu Bar:	Dimension > Radius **Toolbar:** Dimension > Radius	

 The Radius dimensioning is used to dimension a circle or an arc (Figure 8-39). Radius and diameter dimensioning are similar; the only difference is that instead of the diameter line, a radius line is drawn (half of the diameter line), which is measured from the center to any point on the circumference. The dimension text generated by AutoCAD is preceded by the letter **R** to indicate a radius dimension. If you want to use the default dimension text (dimension text generated automatically by AutoCAD), simply specify a point to position the dimension at the **Specify dimension line location or [Mtext/Text/Angle]** prompt. You can also enter a new value or specify a prefix or suffix, or suppress the entire text by entering a blank space following the **Enter dimension text <current>** prompt. A center mark for the circle/arc is drawn automatically, provided the center mark value controlled by the **DIMCEN** variable is not 0. The prompt sequence that will follow when you choose this button is given next.

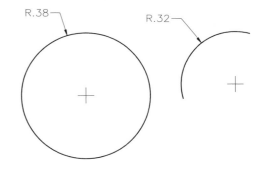

Select arc or circle: *Select the object that you want to dimension.*
Dimension text = Current
Specify dimension line location or [Mtext/Text/Angle]: *Specify the dimension location.*

If you want to override the default value of the dimension text, use the **Text** or the **Mtext** option.

Figure 8-39 *Radius dimensioning*

Note
 *In case of diametric, radial, and jogged radial dimensions, you can extend the dimension line beyond the endpoints of the arc by creating the arc extension line beyond the endpoints. These arc extension lines are similar to the other extension lines. The system variables **DIMSE1** and **DIMSE2** allow the display of the first and the second extension line, respectively. These variables should be **ON** in order to suppress the respective extension line.*

CREATING JOGGED LINEAR DIMENSIONS

Ribbon:	Annotate > Dimensions > Dimension, Dimjogline **Command:** DIMJOGLINE	
Menu Bar:	Dimension > Jogged Linear **Toolbar:** Dimension > Jogged Linear	

The **Dimension, Dimjogline** button is used to add or remove a jog or break in the preexisting dimensions. This kind of dimensioning technique is generally used to dimension the components that have a high length to width ratio, see Figure 8-40. To add a jog to a linear dimension, choose **Annotate > Dimensions > Dimension, Dimjogline** from the **Ribbon** and select the dimension to which you want to add a jog. Next, pick a point along the dimension line to specify the location of the jog placement. Alternatively, you can press the ENTER key to place the jog automatically. To remove a jog, invoke the **DIMJOGLINE** command and choose **Remove** from the shortcut menu. Next, specify the dimension line from which you want to remove the jog. You can modify the location of the jog symbol with the help of grips. Note that you can add a jog only to the linear or aligned dimensions with this command.

Figure 8-40 *Jogged linear dimensioning*

Note
*The height of the jog symbol can be changed by changing the **Jog height factor** in the **Lines & Arrows** rollout of the **Properties** palette.*

GENERATING CENTER MARKS AND CENTERLINES

| **Ribbon:** | Annotate > Dimensions > Center Mark | **Command:** | DIMCENTER |
| **Menu Bar:** | Dimension > Center Mark | **Toolbar:** | Dimension > Center Mark |

When circles or arcs are dimensioned with the **DIMRADIUS** or **DIMDIAMETER** command, a small mark known as center mark, or a line known as centerline, may be drawn at the center of the circle/arc. Sometimes, you may want to mark the center of a circle or an arc without using these dimensioning commands. This can be achieved with the help of the **DIMCENTER** command. You can invoke this command by choosing **Annotate > Dimensions > Center Mark** from the **Ribbon** or by entering **CENTER** (or **CEN**) at the **Dim:** prompt. When you invoke this command, you are prompted to select the arc or the circle. The result of this command will depend upon the value of the **DIMCEN** variable. If the value of this variable is positive, the center marks are drawn, see Figure 8-41 and if the value is negative, centerlines are drawn, see Figure 8-42.

Note
*The center marks created by **DIMCENTER** or **DIM CENTER** are lines, but not associative dimensioning objects, and they have an explicit linetype.*

CREATING ORDINATE DIMENSIONS

| **Ribbon:** | Annotate > Dimensions > Dimension > Ordinate | **Command:** | DIMORD |
| **Menu Bar:** | Dimension > Ordinate | **Toolbar:** | Dimension > Ordinate |

Ordinate dimensioning is used to dimension the *X* and *Y* coordinates of the selected point. This type of dimensioning is also known as arrowless dimensioning because no arrowheads are drawn in it. Ordinate dimensioning is also called datum

Figure 8-41 *Using a positive value for* **DIMCEN**

Figure 8-42 *Using a negative value for* **DIMCEN**

dimensioning because all dimensions are related to a common base point. The current UCS (user coordinate system) origin becomes the reference or the base point for ordinate dimensioning. With ordinate dimensioning, you can determine the X or Y displacement of a selected point from the current UCS origin.

In ordinate dimensioning, AutoCAD automatically places the dimension text (*X* or *Y* coordinate value) and the leader line along the *X* or *Y* axis (Figure 8-43). Since ordinate dimensioning pertains to either the *X* coordinate or the *Y* coordinate, you should keep ORTHO on. When ORTHO is off, the leader line is automatically given a bend when you select the second leader line point that is offset from the first point. This allows you to generate offsets and avoid overlapping text on closely spaced dimensions. In ordinate dimensioning, only one extension line (leader line) is drawn.

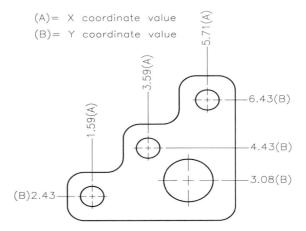

Figure 8-43 *Ordinate dimensioning*

The leader line for an *X* coordinate value will be drawn perpendicular to the *X* axis, and the leader line for a *Y* coordinate value will be drawn perpendicular to the *Y* axis. Since you cannot override this, the leader line drawn perpendicular to the *X* axis will have the dimension text aligned with the leader line. The dimension text is the X datum of the selected point. The leader line drawn perpendicular to the *Y* axis will have the dimension text, which is the Y datum of the selected point, aligned with the leader line. Any other alignment specification for the dimension text is nullified. Hence, changes in the Text Alignment in the **Dimension Style Manager** dialog box (**DIMTIH** and **DIMTOH** variables) have no effect on the alignment of the dimension text. You can specify the coordinate value you want to dimension at the **Specify leader endpoint or [Xdatum/ Ydatum/MText/Text/Angle]** prompt.

If you select or enter a point, AutoCAD checks the difference between the feature location and the leader endpoint. If the difference between the X coordinates is greater, the dimension measures the Y coordinate; otherwise, the X coordinate is measured. In this manner AutoCAD determines whether it is an X or Y type of ordinate dimension. However, if you enter Y instead of specifying a point, AutoCAD will dimension the Y coordinate of the selected feature. Similarly, if you enter X, AutoCAD will dimension the X coordinate of the selected point. The prompt sequence that follows when you choose this button is given next.

Specify feature location: *Select a point on an object.*
Specify leader endpoint or [Xdatum/Ydatum/Mtext/Text/Angle]: *Enter the endpoint of the leader.*

You can override the default text with the help of the **Mtext** or the **Text** option. If you use the **Mtext** option, the **In-Place Text Editor** will be displayed. If you use the **Text** option, you will be prompted to specify the new text in the Command line itself.

Exercise 4   *General*

Draw the model shown in Figure 8-44 and then use ordinate dimensioning to dimension the part. The distance between the dotted lines is 0.5 units.

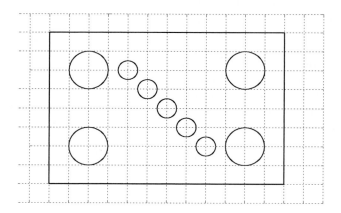

Figure 8-44 Drawing for Exercise 4

MAINTAINING EQUAL SPACING BETWEEN DIMENSIONS

Ribbon: Annotate > Dimensions > Adjust Space	**Command:** DIMSPACE
Menu Bar: Dimension > Dimension Space	**Toolbar:** Dimension > Dimension Space

 This command is used to space the overlapping or unequally spaced linear and angular dimensions equally. Maintaining the equal spacing between the dimension lines increases the clarity of the drawing display. Figure 8-45 shows a drawing with unequally spaced dimensions, whereas Figure 8-46 shows the same drawing with equally spaced dimensions after the use of the **DIMSPACE** command. Figure 8-47 shows a drawing with its angular dimensions equally spaced.

Figure 8-45 *Drawing with its linear dimensions unequally spaced*

Figure 8-46 *Drawing after its linear dimensions equally spaced*

The prompt sequence that will follow when you choose this button is given next.

    Command: **_DIMSPACE**
    Select base dimension: *Select a linear or an angular dimension with respect to which all other dimensions will get spaced.*
    Select dimensions to space: *Select the dimensions that you want to be equally spaced from the specified base dimension.*
    Select dimensions to space: *Select more similar dimensions or press* `Enter`
    Enter value or [Auto] <Auto>: *Enter a value to specify the spacing between the selected dimensions or enter **A** to calculate automatically the gap between the dimensions.*

Figure 8-47 *Drawing with its angular dimensions equally spaced*

Enter **0** in the above prompt to align all the selected dimensions in a single line. With the **Auto** option selected, AutoCAD will calculate the space value on the basis of the dimension text height. The spacing is automatically maintained to a value that is double of the dimension text height.

CREATING DIMENSION BREAKS

Ribbon: Annotate > Dimensions > Break	**Command:** DIMBREAK
Menu Bar: Dimension > Dimension Break	**Toolbar:** Dimension > Dimension Break

 This button is used to create a break in the dimension line, extension line, or multileaders at the point where they intersect with other drawing entities or annotations. Inserting dimension breaks will keep annotative objects separate from the main drawing, which increases the clarity of the drawing reading, see Figure 8-48. The prompt sequence that will follow when you choose this button is given next.

Command: **_DIMBREAK**
Select a dimension to add/remove break or [Multiple]: *Select a dimension in which you want to insert a break.*
Select object to break dimension or [Auto/Manual/Remove] <Auto>: *Select objects or dimensions that intersect the above selected dimension, enter an option or press* Enter *to choose the **Auto** option.*

The options in the **DIMBREAK** command are discussed next.

Multiple

This option allows you to select more than one dimension for adding or removing a break. The prompt sequence that will follow is given next.

Figure 8-48 Drawing with dimension breaks

Select dimensions: *Select the dimensions by any object selection method and press* Enter
Select object to break dimensions or [Auto/Remove] <Auto>: *Enter an option or press* Enter

The **Auto** option, available in the above prompt, creates breaks automatically at the intersections of the selected dimensions with the other drawing and annotation entities. Also, if the selected dimensions are self-intersecting, then the **Auto** option, if chosen, creates a break at the point of intersection. The **Remove** option removes all the dimension breaks that exist in the selected dimensions.

Auto

This option creates the break automatically at the intersection of the selected dimension with the other drawing and annotation entities.

Note
*All dimension breaks created by the **Multiple** and **Auto** options are updated automatically when you modify the broken dimension or the intersecting objects. But, the new objects created over the broken dimension will have no effect on the display of the broken dimension. To break the newly created dimension, you have to use the **DIMBREAK** command again.*

Remove

This option removes all the dimension breaks from the selected dimension.

Manual

This option is used to create a break in the dimension manually. You have to specify two points between which you want to break the dimension or the extension line. You can also create the dimension break at only one location with this option.

Note

*The dimension breaks created by the **Manual** option do not update automatically when you modify the broken dimension or the intersecting objects. For updating the dimension break, you need to remove the dimension, and then again add the dimension break.*

You cannot add dimension breaks to the straight leaders, spline leaders, and spline multileaders.

CREATING INSPECTION DIMENSIONS

Ribbon:	Annotate > Dimensions > Inspect	**Command:**	DIMINSPECT
Menu Bar:	Dimension > Inspection	**Toolbar:**	Dimension > Inspection

The inspection rate is used to specify how frequently the dimensions are needed to be checked in order to ensure the variations in the dimensions are within the range. Select a dimension from the drawing and choose **Annotate > Dimensions > Inspect** from the **Ribbon**; the **Inspection Dimension** dialog box will be displayed, as shown in Figure 8-49. Choose the **OK** button; the selected dimension will be changed to the inspection dimension that consists of three fields, which are discussed next.

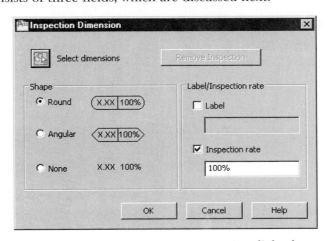

*Figure 8-49 The **Inspection Dimension** dialog box*

Inspection Label

The inspection label is located at the extreme left of the inspection dimension and is used to specify the particular inspection dimension, see Figure 8-50.

Dimension Value

The dimension value is located in the middle of the inspection dimension and is used to display the value of the dimension to be maintained, see Figure 8-50.

Inspection Rate

The inspection rate is located at the right-end of the inspection dimension and is used to specify the frequency with which the inspection dimension should be measured, see Figure 8-50. The inspection frequency is specified in term of percentage value. The value of 100% signifies that the dimension of each and every component should be measured, if the value is 50%, then each alternate component should be inspected for that dimension, and if the value is 25%, then every fourth component should be inspected for that dimension.

If you have not selected the dimension before invoking the **Inspect** command, choose the **Select dimensions** button from the **Inspection Dimension** dialog box (Refer to Figure 8-49); the dialog box will temporarily disappear from the screen, thereby enabling you to select the dimension that you want to convert into the inspection dimension. Select the dimensions and press the ENTER key; the **Inspection Dimension** dialog box will be displayed again. All the selected dimensions will display the same value of the label and inspection dimension. Next, specify the shape of the boundary around the inspection dimension from the **Shape** area. Figure 8-50 displays the shapes available in the **Inspection Dimension** dialog box. Next, enter the values for the label and inspection rate in their respective edit boxes in the **Label/Inspection rate** area and choose the **OK** button. You can also control the display of label and inspection rate in the inspection dimension by clearing or selecting the **Label** and **Inspection rate** check boxes respectively. To remove the inspection dimension of the selected dimension, choose the **Remove Inspection** button from the **Inspection Dimension** dialog box and then choose the **OK** button.

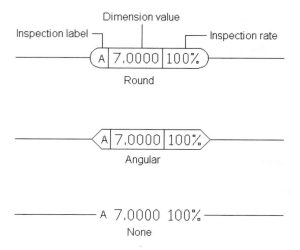

Figure 8-50 *Shapes and components of inspection dimension*

Exercise 5   *Mechanical*

Draw and dimension the object, as shown in Figure 8-51.

Figure 8-51 Drawing for Exercise 5

WORKING WITH TRUE ASSOCIATIVE DIMENSIONS

The true associative dimensions are the dimensions that are automatically modified when the objects to which they are associated are modified. By default, all dimensions in AutoCAD are true associative dimensions. If the dimension attached to the object is true associative, then it will be modified automatically when the object is modified. In this case, you do not have to select the definition points of the dimensions. Any dimension in AutoCAD can be converted into a disassociated dimension and then back to the true associative dimension. This is discussed next.

Removing the Dimension Associativity

Command:	DIMDISASSOCIATE

The **DIMDISASSOCIATE** command is used to remove the associativity of the dimensions from the object to which they are associated. When you invoke this command, you will be prompted to select the dimensions to be disassociated. The true association of the selected dimensions is automatically removed once you exit this command. The number of dimensions disassociated is displayed in the Command prompt.

Converting a Dimension into a True Associative Dimension

Ribbon:	Annotate > Dimensions > Reassociate	Command:	DIMREASSOCIATE
Menu Bar:	Dimension > Reassociate Dimensions		

The **DIMREASSOCIATE** command is used to create a true associative dimension by associating the selected dimension to the specified object. When you invoke this command, you will be prompted to select the objects. These objects are the dimensions to be associated. Once you select the dimensions to be associated, a cross is displayed and you are prompted to select the feature location. This cross implies that the dimension is not associated. You can define a new association point for the dimensions by selecting the objects or by using the object snaps. If you select a dimension that has already been associated to an object, the cross will be displayed inside a box. The prompt sequence that is displayed varies depending upon the type of dimension selected. In case of **linear**, aligned, radius, and diameter dimensions, you can directly select the object to associate the dimension. If the arcs or circles are assigned angular dimensions using three points, then also you can select these arcs or circles directly for associating the dimensions. For rest of the dimension types, you can use the object snaps to specify the point to associate the dimensions.

DRAWING LEADERS

Command:	QLEADER

The leader line is used to attach annotations to an object or when the user wants to show a dimension without using another dimensioning command. Sometimes, leaders of the dimensions of circles or arcs are so complicated that you need to construct a leader of your own. The leaders can be created using the **QLEADER** command. The leaders drawn by using this command create the arrow and the leader lines as a single object. The text is created as a separate object. This command can create multiline annotations and offer several options such as copying existing annotations and so on. You can customize the leader and annotation by selecting the **Settings** option at the **Specify first leader point, or [Settings] <Settings>** prompt. The prompt sequence that will follow when you use this command is given next.

> Specify first leader point, or [Settings] <Settings>: *Specify the start point of the leader.*
> Specify next point: *Specify the endpoint of the leader.*
> Specify next point: *Specify the next point.*
> Specify text width <current>: *Enter the text width of multiline text.*
> Enter first line of annotation text <Mtext>: *Press ENTER; AutoCAD displays the **In-Place Text Editor** dialog box. Enter text in the dialog box and then choose **OK** to exit.*

If you press ENTER at the **Specify first leader point, or [Settings] <Settings>** prompt, then the **Leader Settings** dialog box is displayed. The **Leader Settings** dialog box gives you a number of options for the leader line and the text attached to it. It has the following tabs:

Annotation Tab
This tab provides you with various options to control annotation features, see Figure 8-52.

Annotation Type Area
MText. When this radio button is selected, AutoCAD uses the **In-Place Text Editor** to create an annotation. Therefore, on selecting this radio button, the options in the **MText options** area become available.

Copy an Object. This option allows you to copy an existing annotation object (like multiline text, single line text, tolerance, or block) and attach it at the end of the leader. For example, if you have a text string in the drawing that you want to place at the end of the leader, you can use the **Copy an Object** option to place it at the end of the leader.

Tolerance. Select the **Tolerance** radio button and choose **OK** to exit from the **Leader Settings** dialog box. Then, specify the next point to complete the leader line; AutoCAD displays the **Geometric Tolerance** dialog box on the screen. Specify the tolerance in this dialog box and choose **OK** to exit. AutoCAD will place the specified geometric tolerance with the feature control frame at the end of the leader (Figure 8-53).

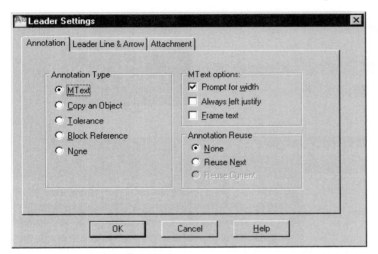

*Figure 8-52 The **Annotation** tab of the **Leader Settings** dialog box*

*Figure 8-53 Leader with **MText**, **Block**, **Tolerance**, and **MText with Frame** annotations*

Block Reference. The **Block Reference** option allows you to insert a predefined block at the end of the leader. When you select this option, AutoCAD will prompt you to enter the block name and insertion point.

None. This option creates a leader without placing any annotation at the end of the leader.

MText options Area

The options under this area will be available only if the **MText** radio button is selected from the **Annotation Type** area. This area provides you with the following options.

Prompt for width. Selecting this check box allows you to specify the width of the multiline text annotation.

Always left justify. This option left justifies the multiline text annotation in all situations. Selecting this check box makes the Prompt for the width option unavailable.

Frame text. Selecting this check box draws a box around the multiline text annotation.

Annotation Reuse Area

The options under this area allow you to reuse the annotation.

None. When selected, AutoCAD does not reuse the leader annotation.

Reuse Next. This option allows you to reuse the annotation that you are going to create next for all subsequent leaders.

Reuse Current. This option allows you to reuse the current annotation for all subsequent leaders.

Leader Line & Arrow Tab

The options under this area are related to the leader parameters, see Figure 8-54.

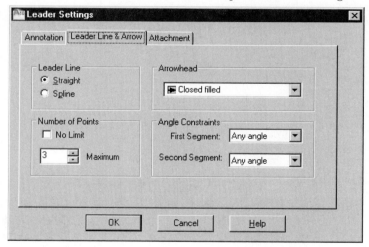

*Figure 8-54 The **Leader Line & Arrow** tab of the **Leader Settings** dialog box*

Leader Line Area

This area gives the options for the **leader line type** such as straight or spline. The **Spline** option draws a spline through the specified leader points and the **Straight** option draws straight lines. Figure 8-55 shows straight and spline leader lines.

Number of Points Area

The options provided under this area are used to specify the number of points in the leader.

No Limit. If this check box is selected, you can define as many number of points as you want in the

Figure 8-55 Straight and Splined leaders

leader line. AutoCAD will keep prompting you for the next point, until you press ENTER at this prompt.

Maximum. This spinner is used to specify the maximum number of points on the leader line. The default value is 3, which means that by default, there will be only three points in the leader line. You can specify the maximum number of points using this spinner. This has to be kept in mind that the start point of the leader is the first leader point. This spinner will be available only if the **No Limit** check box is cleared. Figure 8-55 shows a leader line with five number of points.

Arrowhead Area

The drop-down list under this area allows you to define a leader arrowhead. The arrowhead is the same as the one for dimensioning. You can also use the user-defined arrows by selecting **User Arrow** from the drop-down list.

Angle Constraints Area

The options provided under this area are used to define the angle for the segments of the leader lines.

First Segment. This drop-down list is used to specify the angle at which the first leader line segment will be drawn. You can select the predefined values from this drop-down list.

Second Segment. This drop-down list is used to specify the angle at which the second leader line segment will be drawn.

Attachment Tab

The **Attachment** tab (Figure 8-56) will be available only if you have selected **MText** from the **Annotation Type** area of the **Annotation** tab. The options provided under this tab are used

Figure 8-56 *The **Attachment** tab of the **Leader Settings** dialog box*

Chapter 8

for attaching the multiline text to the leader. It has two columns: **Text on left side** and **Text on right side**. Both these columns have five radio buttons below them. Each radio button corresponds to the option of attaching the multiline text. If you draw a leader from the right to the left, AutoCAD uses the settings under **Text on left side**. Similarly, if you draw a leader from the left to the right, AutoCAD uses the settings as specified under **Text on right side**. This area also provides you with the **Underline bottom line** check box. If this check box is selected, then the last line of the multiline text will be underlined.

Note
*You can use the Command Line to create leaders with the help of the **LEADER** command. The options under this command are similar to those under the **QLEADER** command. The only difference is that the **LEADER** command uses the Command Line.*

Exercise 6   *Mechanical*

Make the drawing shown in Figure 8-57 and then use the **QLEADER** command to dimension the part, as shown. The distance between the dotted lines is 0.5 units.

Figure 8-57 Drawing for Exercise 6

MULTILEADERS

Ribbon:	Annotate > Leaders > Multileader	**Command:**	MLEADER
Menu Bar:	Multileader > Multileader	**Toolbar:**	Multileader > Multileader

Multileaders are the enhanced leaders; wherein the leader and its content is part of the same object. A Multileader provides you enough flexibility so that you can create either the arrowhead or the tail end first. Alternatively, you can specify the content of the leader first and then draw the leader. Multileaders can have more than one leader so that a single comment can be pointed at more than one location. Besides this, multiple notes can be attached to a single leader so that more than one comment can be pointed at a single location. You can add or remove leaders from the previously created multileaders. You can also control and modify the appearance of the multileaders.

DRAWING MULTILEADERS

To draw a multileader, choose **Annotate > Leaders > Multileader** from the **Ribbon**. The prompt sequence that follows is given next.

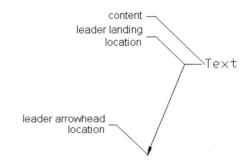

> Command: **_MLEADER**
> Specify leader arrowhead location or [leader Landing first/Content first/ Options] <Options>: *Specify the location for the arrowhead from where the leader will start.*
> Specify leader landing location: *Specify the leader landing location, see Figure 8-58.*

*AutoCAD displays the **In-Place Text Editor** dialog box. Enter the text in the dialog box and click outside the editor to accept and exit the command. If you click outside the editor without entering any text and exit the command, no text will be attached to the multileader.*

Figure 8-58 Image displaying the options to locate the leader start point

The other options in the **MLEADER** command are discussed next.

leader Landing first

This option is used to locate the leader landing line first. Choose the **leader Landing first** option from the shortcut menu at the **Specify leader arrowhead location or [leader Landing first/Content first/Options] <Options>** prompt to invoke this option. Alternatively, if the **Dynamic Input** button is turned on in the Status Bar, then you can choose this option from the dynamic input box associated with the cursor. Now, you will be prompted to specify a point where the leader landing should be placed. Next, you will be prompted to specify the leader arrowhead location. The leader landing will automatically adjust itself to the leader landing point, depending upon the direction in which you create the leader. For example, if you create the leader to the left of the leader landing location, the landing will automatically be created to the right of the leader landing location.

Content first

This option is used to specify the content (multiline text or block) of the leader first. Choose the **Content first** option from the shortcut menu from the **Specify leader arrowhead location or [leader Landing first/Content first/Options] <Options>** prompt to invoke this option. Alternatively, if the **Dynamic Input** button is turned on in the Status Bar, you can choose this option from the dynamic input box associated with the cursor. After specifying the content to be attached with the leader, a horizontal leader landing will automatically be attached to the content and then you can specify the leader arrowhead location.

Note

If you have drawn the multileader by using any one of the above-mentioned options previously, then the succeeding multileaders will be created using that option only, until otherwise specified.

Options

The suboptions within this option are used to control the leader type, content to be attached with the leader, size of the leader landing, and so on. These suboptions are discussed next.

Leader type

This option provides the suboptions for the leader line type such as straight, spline, or none. The **sPline** suboption draws a spline through the specified leader points and the **Straight** option draws the straight lines. Figure 8-55 shows the straight and spline leader lines. The **None** option enables you to draw the content of the leader without drawing the leader itself.

leader lAnding

This option provides you the suboptions to control the display of the horizontal landing line. The **No** suboption will create a leader with no horizontal landing and the **Yes** suboption will create a leader with the horizontal landing. If you choose the **Yes** suboption, then you will be prompted to specify the length of the horizontal landing. The new leader will be created with the specified landing length. Figure 8-59 shows the leaders created by using various suboptions of the **leader lAnding**.

Figure 8-59 Leaders drawn by using various types of landing

Content type

This option is used to specify the type of content to be attached with the leader. The **Mtext** suboption uses the AutoCAD **In-Place Text Editor** to create the content. The **Block** suboption allows you to insert a predefined block at the end of the leader. When you select this option, AutoCAD will prompt you to enter the block name and the specified block will be attached to the new leader. The **None** suboption creates a leader without placing any content at the end of the leader.

Maxpoints

This option is used to specify the maximum number of points on the leader line. The default value is 2, which means there will be only two points on the leader line. You can specify the maximum number of points by entering the new value. This has to be kept in mind that the start point of the leader is the first leader point.

First angle

This option is used to specify the angle that the first leader line will measure from the horizontal at the start point. You can also specify a value which is the natural number multiple of 15-degree.

Second angle

This option is used to specify the angle that the second leader line will measure from the horizontal at the second point. You can only specify a value which is the natural number multiple of 15-degree.

eXit options

This option will return you back to the main options of the multileader.

Note

All settings specified in the Options heading of the multileader will last only for that particular leader, and the next leader will be created with the default settings. You can also change the default settings of the multileader from the Multileader Style Manager dialog box that will be discussed in detail in Chapter-10.

ADDING LEADERS TO THE EXISTING MULTILEADER

| **Ribbon:** | Annotate > Leaders > Add Leader | **Command:** | MLEADEREDIT |
| **Toolbar:** | Multileader > Add Leader | | |

The **Add Leader** button is used to add leaders to the existing multileader, so that you can point a single content to more than one location. Choose **Annotate > Leaders > Add Leader** from the **Ribbon**; you will be prompted to select the multileader to which you want to attach the new leader. After selecting the desired multileader, a new leader will branch out from the landing location of the existing multileader. Next, you will be prompted to specify the location of the arrowhead of the new leader. Next, you can attach any number of leaders to the selected multileader. To finish this command, press the ENTER key.

REMOVING LEADERS FROM THE EXISTING MULTILEADER

| **Ribbon:** | Annotate > Leaders > Remove Leader | **Command:** | MLEADEREDIT |
| **Toolbar:** | Multileader > Remove Leader | | |

The **Remove Leader** button is used to remove leaders from the existing multileader. To remove a leader, choose **Annotate > Leaders > Remove Leader** from the **Ribbon**;

Chapter 8

you will be prompted to select the multileader from which you want to remove leaders. Next, you will be prompted to select the leaders to be removed from the selected multileader. Select the leaders that you want to remove and then press the ENTER key. You can also select all the leaders of the specified multileader. But, in such cases, the entire selected multileader will get deleted from the drawing area.

ALIGNING MULTILEADERS

| **Ribbon:** | Annotate > Leaders > Align | **Command:** | MLEADERALIGN |
| **Toolbar:** | Multileader > Align Multileaders | | |

This button is used to arrange the selected multileaders by aligning them about a line or making them parallel or by maintaining the spacing between the horizontal landing of the leaders. To do so, choose **Annotate > Leaders > Align** from the **Ribbon**; you will be prompted to select the multileaders to be aligned. After selecting the multileaders, press the ENTER key. Next, you will be prompted to select one of the multileaders with respect to which all other multileaders will get aligned. Also, you can choose the type of alignment from the different options available at the Command prompt. The prompt sequence for the **MLEADERALIGN** command is given next.

Command: **MLEADERALIGN**
Select multileaders: *Select the multileaders to be aligned and* Enter.
Current mode: *Use current spacing.*

All further prompt sequences will vary with the alignment type used in aligning the leaders. The alignment type used previously will become the default alignment type for the current command.

Select multileader to align to or [Options]: *Enter O to choose the type of alignment.*
Enter an option [Distribute/make leader segments Parallel/specify Spacing/Use current spacing] <Use current spacing>: *Specify an option and* Enter.

The options in the above prompt sequence are discussed next.

Distribute

This option is used to accommodate all the selected multileaders at constant spacing between the two specified points. The prompt sequence that will be followed is given next.

Enter an option [Distribute/make leader segments Parallel/specify Spacing/Use current spacing] <Distribute>: **D** Enter.
Specify first point or [Options]: *Specify the location of the first point.*
Specify second point: *Specify the location of the second point and its orientation with respect to the first point.*

The tail end of all the selected leaders landing will get aligned with equal spacing along the line joining the first and the second specified point. Figure 8-60 shows a drawing with its multileaders nonaligned. Figure 8-61 displays the same drawing with its multileaders aligned along an imaginary vertical line using the **Distribute** option.

Figure 8-60 *Drawing with multileaders not aligned*

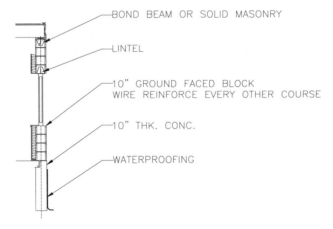

Figure 8-61 *Drawing with its multileaders aligned using the*
Distribute option

make leader segments Parallel

This option is used to make all the selected multileaders parallel to one of the leaders selected as reference. The prompt sequence that will be followed is given next.

Enter an option [Distribute/make leader segments Parallel/specify Spacing/Use current spacing] <current>: **P** [Enter].

Select multileader to align to or [Options]: *Select the leader with which you want to make all other selected multileaders parallel.*

Figure 8-62 shows the drawing with its leader aligned parallel to the lowest leader of the drawing.

Figure 8-62 Drawing with its multileaders aligned using the make leader segments Parallel option

specify Spacing

This option is used to align the selected multileaders in the specified direction, and also maintains a constant spacing between the landing lines. You need to specify one reference leader, with respect to which it will measure the specified interval and direction. The prompt sequence that will be followed is given next.

Enter an option [Distribute/make leader segments Parallel/specify Spacing/Use current spacing] <current>: **S** Enter.
Specify spacing <current>: *Enter the distance value to be maintained between the two consecutive leader landing lines.*
Select multileader to align to or [Options]: *Select one of the leaders with respect to which the specified distance will be measured.*
Specify direction: *Specify the desired direction in the drawing area for the alignment of multileaders.*

When you select a multileader to align with other multileaders, an imaginary line will be displayed in the drawing area. This line allows you to specify the direction of alignment. All the multileaders will get aligned to the selected multileader at the specified distance and direction. Figure 8-63 shows the drawing with its multileaders aligned vertical using the **specify Spacing** option.

Use current spacing

This option is used to align the selected multileaders in the specified direction and also to

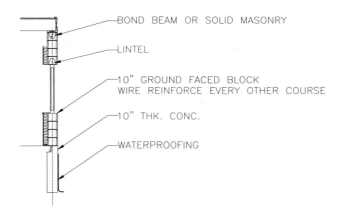

BOND BEAM OR SOLID MASONRY

LINTEL

10" GROUND FACED BLOCK
WIRE REINFORCE EVERY OTHER COURSE

10" THK. CONC.

WATERPROOFING

Figure 8-63 *Drawing with its multileaders aligned using the*
specify Spacing *option*

retain the distance between the landing lines intact. You need to specify one reference leader, with respect to which the distance will be measured in the specified direction. The prompt sequence that will be followed is given next.

Enter an option [Distribute/make leader segments Parallel/specify Spacing/Use current spacing] <specify Spacing>: **U** Enter.
Select multileader to align to or [Options]: *Select one of the leaders with respect to which the direction to be specified is maintained.*
Specify direction: *Specify the desired direction in the drawing area for the alignment of the multileaders.*

COLLECTING MULTIPLE NOTES TO SINGLE LANDING

Ribbon:	Annotate > Leaders > Collect	**Command:**	MLEADERCOLLECT
Toolbar:	Multileader > Collect Multileaders		

The **Collect** button is used to collect the content of the selected multileaders and attach them to a single leader landing. Note that only the multileaders having predefined blocks as content can be collected by this command. To collect the multileaders, choose **Annotate > Leaders > Collect** from the **Ribbon**. Select the multileaders in the order in which you want them to be collected. The content of the last selected multileader will be retained to its landing position and the content of all other multileaders will move to the last multileader. Next, specify a point in the drawing area where the upper left-corner of the collected content of the multileaders will be placed. You can collect the content horizontally or vertically by placing the content one after another, or by specifying the space between which the contents will be placed, and the content exceeding the specified distance will be sent to the next row. The prompt sequence for the **MLEADERCOLLECT** command is given next.

Chapter 8

Command: **MLEADERCOLLECT**
Select multileaders: *Select the multileaders in the order in which you want to collect the content of these multileaders and press* Enter.
Specify collected multileader location or [Vertical/Horizontal/Wrap] <current>: *Specify a point in the drawing area where the upper left-corner of the collected content will be placed or choose an option.*

If you choose **Horizontal** from the shortcut menu from the **Specify collected multileader location or [Vertical/Horizontal/Wrap] <current>** prompt sequence, the contents of the selected multileaders will get collected horizontally. Figure 8-64 shows the multileaders without being collected, Figure 8-65(a) shows the same leaders after the horizontal alignment. If you choose **Vertical** from the shortcut menu from the **Specify collected multileader location or [Vertical/Horizontal/Wrap] <current>** prompt sequence, the contents of the selected multileaders will get collected vertically, see Figure 8-65(b).

If you choose the **Wrap** option from the right-click shortcut menu at the **Specify collected multileader location or [Vertical/Horizontal/Wrap] <current>** prompt sequence, you will be prompted to enter a value for the wrap width. The contents of the selected multileaders will be placed side by side and the contents that could not be accommodated into the specified wrap width will be sent to the next row. In Figure 8-65(c), three blocks get accommodated into the specified wrap width and the remaining fourth one is sent to the next row. Alternatively, you can also enter the number of blocks to be accommodated into one row and the remaining content will automatically be sent to the next row. Choose **Number** from the right-click shortcut menu from the **Specify wrap width or [Number]** prompt sequence. Next, enter the number of blocks to be accommodated into each row. In Figure 8-65(d), only two blocks are specified to be accommodated into each row, so the remaining two blocks have been sent to the second row.

Figure 8-64 Multileaders prior to collection   *Figure 8-65 Contents of the multileaders after being collected*

GEOMETRIC DIMENSIONING AND TOLERANCING

One of the most important parts of the design process is giving the dimensions and tolerances, since every part is manufactured from the dimensions given in the drawing. Therefore, every designer must understand and have a thorough knowledge of the standard practices used in the industry to make sure that the information given on the drawing is correct and can be understood by other people. Tolerancing is equally important, especially in the assembled parts. Tolerances and fits determine how the parts will fit. Incorrect tolerances could result in a product that is not usable. In addition to dimensioning and tolerancing, the function and the relationship that exists between the mating parts is important if the part is to perform the way it was designed. This aspect of the design process is addressed by geometric dimensioning and tolerancing, generally known as GDT.

Geometric dimensioning and tolerancing is a means to design and manufacture parts with respect to the actual function and relationship that exists between different features of the same part or the features of the mating parts. Therefore, a good design is not achieved by just giving dimensions and tolerances. The designer has to go beyond dimensioning and think of the intended function of the part and how the features of the part are going to affect its function. For example, Figure 8-66 shows a part with the required dimensions and tolerances. In this drawing, there is no mention of the relationship between the pin and the plate. Is the pin perpendicular to the plate? If it is, to what degree should it be perpendicular? Also, it does not mention on which surface the perpendicularity of the pin is to be measured. A design like this is open to individual interpretation based on intuition and experience. This is where geometric dimensioning and tolerancing play an important part in the product design process.

Figure 8-67 has been dimensioned using geometric dimensioning and tolerancing. The feature symbols define the datum (reference plane) and the permissible deviation in the perpendicularity of the pin with respect to the bottom surface. From a drawing like this, chances of making a mistake are minimized.

Figure 8-66 *Traditional dimensioning and tolerancing technique*

Figure 8-67 *Geometric dimensioning and tolerancing*

GEOMETRIC CHARACTERISTICS AND SYMBOLS

Before discussing the application of AutoCAD commands in geometric dimensioning and tolerancing, you need to understand the following feature symbols and tolerancing components. Figure 8-68 shows the geometric characteristics and symbols used in geometric dimensioning and tolerancing.

Kind of feature	Type of feature	Characteristics	
Related	Location	Position	⊕
		Concentricity or Coaxiality	◎
		Symmetry	=
	Orientation	Parallelism	//
		Perpendicularity	⊥
		Angularity	∠
Individual	Form	Cylindricity	⌀
		Flatness	▱
		Circularity or Roundness	○
		Straightness	—
Individual or related	Profile	Surface Profile	⌒
		Line Profile	⌒
Related	Runout	Circular Runout	↗
		Total Runout	↗↗

Figure 8-68 *Characteristics and symbols used in GTOL*

Note
These symbols are the building blocks of geometric dimensioning and tolerancing.

ADDING GEOMETRIC TOLERANCE

Ribbon: Annotate > Dimensions > Tolerance	**Command:** TOLERANCE
Menu Bar: Dimension > Tolerance	**Toolbar:** Dimension > Tolerance

 Geometric tolerance displays the deviations of profile, orientation, form, location, and runout of a feature. In AutoCAD, geometrical tolerancing is displayed by feature control frames. The frames contain all the information about tolerances for a single dimension. To display feature control frames with various tolerancing parameters, the specifications are entered in the **Geometric Tolerance** dialog box (Figure 8-69).

Various components that constitute GTOL are shown in Figures 8-70 and 8-71.

Feature Control Frame
The **feature control frame** is a rectangular box that contains the geometric characteristics symbols and tolerance definition. The box is automatically drawn to standard specifications; you do not need to specify its size. You can copy, move, erase, rotate, and scale the feature control frame. You can also snap to them using various Object snap modes. You can edit feature control frames using the **DDEDIT** command or you can also edit them using **GRIPS**. The system variable **DIMCLRD** controls the color of the feature control frame. The system variable **DIMGAP** controls the gap between the feature control frame and the text.

Figure 8-69 *The Geometric Tolerance dialog box*

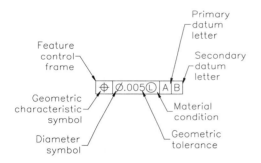

Figure 8-70 *Components of GTOL*

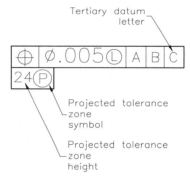

Figure 8-71 *Components of GTOL*

Geometric Characteristics Symbol

The geometric characteristics symbols indicate the characteristics of a feature like straightness, flatness, perpendicularity, and so on. You can select the required symbol from the **Symbol** dialog box (Figure 8-72). This dialog box is displayed by selecting the box provided in the **Sym** area of the **Geometric Tolerance** dialog box. To select the required symbol, just pick the symbol using the left mouse button. The symbol will now be displayed in the box under the **Sym** area.

Figure 8-72 The Symbol dialog box

Tolerance Value and Tolerance Zone Descriptor

The tolerance value specifies the tolerance on the feature as indicated by the tolerance zone descriptor. For example, a value of .003 indicates that the feature must be within a 0.003 tolerance zone. Similarly, ⬜.003 indicates that this feature must be located at a true position within a 0.003 diameter. The tolerance value can be entered in the edit box provided under the **Tolerance 1** or the **Tolerance 2** area of the **Geometric Tolerance** dialog box. The tolerance zone descriptor can be invoked by selecting the box located to the left of the edit box. The system variable **DIMCLRT** controls the color of the tolerance text, variable **DIMTXT** controls the tolerance text size, and variable **DIMTXSTY** controls the style of the tolerance text. Using the **Projected Tolerance Zone**, inserts a projected tolerance zone symbol, which is an encircled P, after the projected tolerance zone value.

Material Condition Modifier

The material condition modifier specifies the material condition when the tolerance value takes effect. For example, ▢.003(M) indicates that this feature must be located at a true position within a 0.003 diameter at maximum material condition (MMC). The material condition modifier symbol can be selected from the **Material Condition** dialog box (Figure 8-73). This dialog box can be invoked by selecting the boxes located on the right side of the edit boxes under the **Tolerance 1**, **Tolerance 2**, **Datum 1**, **Datum 2**, and **Datum 3** areas of the **Geometric Tolerance** dialog box.

*Figure 8-73 The **Material Condition** dialog box*

Datum

The datum is the origin, surface, or feature from which the measurements are made. The datum is also used to establish the geometric characteristics of a feature. The datum feature symbol consists of a reference character enclosed in a feature control frame. You can create the datum feature symbol by entering characters (like -A-) in the **Datum Identifier** edit box in the **Geometric Tolerance** dialog box and then selecting a point where you want to establish this datum.

You can also combine datum references with geometric characteristics. AutoCAD automatically positions the datum references on the right end of the feature control frame.

COMPLEX FEATURE CONTROL FRAMES
Combining Geometric Characteristics

Sometimes, it is not possible to specify all geometric characteristics in one frame. For example, Figure 8-74 shows the drawing of a plate with a hole in the center.

In this part, it is determined that surface C must be perpendicular to surfaces A and B within 0.002 and 0.004, respectively. Therefore, we need two frames to specify the geometric characteristics of surface C. The first frame specifies the allowable deviation in perpendicularity of surface C with respect to surface A. The second frame specifies the allowable deviation in perpendicularity of surface C with respect to surface B. In addition to these two frames, we need a third frame that identifies datum surface C.

All the three feature control frames can be defined in one instance of the **TOLERANCE** command.

1. Choose the **Tolerance** button to invoke the **Geometric Tolerance** dialog box. Select the box under the **Sym** area to display the **Symbol** dialog box. Select the **perpendicular** symbol. AutoCAD will display the selected symbol in the first row of the **Sym** area.

Figure 8-74 *Combining feature control frames*

2. Enter **.002** in the first row edit box under the **Tolerance 1** area and enter **A** in the first row edit box under the **Datum 1** area.

3. Select the second row box under the **Sym** area to display the **Symbol** dialog box. Select the **perpendicular** symbol. AutoCAD will display the selected symbol in the second row box of the **Sym** area.

4. Enter **.004** in the second row edit box under the **Tolerance 1** area and enter **B** in the second row edit box under the **Datum 1** area.

5. In the **Datum Identifier** edit box, enter **C**, and then choose the **OK** button to exit the dialog box.

6. In the graphics screen, select the position to place the frame.

7. Similarly, create the remaining feature control frames.

Composite Position Tolerancing

Sometimes the accuracy required within a pattern is more important than the location of the pattern with respect to the datum surfaces. To specify such a condition, composite position tolerancing may be used. For example, Figure 8-75 shows four holes (pattern) of diameter 0.15. The design allows a maximum tolerance of 0.025 with respect to datum A, B, and C at the maximum material condition (holes are smallest). The designer wants to maintain a closer positional tolerance (0.010 at MMC) between the holes within the pattern. To specify this requirement, the designer must insert the second frame. This is generally known as composite position tolerancing. AutoCAD provides the facility to create two composite position tolerance frames by means of the **Geometric Tolerance** dialog box. The composite tolerance frames can be created as follows:

Figure 8-75 *Composite position tolerancing*

1. Invoke the **TOLERANCE** command to display the **Geometric Tolerance** dialog box. Select the box under the **Sym** area to display the **Symbol** dialog box. Select the **position** symbol. AutoCAD will display the selected symbol in the first row of the **Sym** area.

2. In the first row of the **Geometric Tolerance** dialog box, enter the geometric characteristics and the datum references required for the first position tolerance frame.

3. In the second row of the **Geometric Tolerance** dialog box, enter the geometric characteristics and the datum references required for the second position tolerance frame.

4. When you have finished entering the values, choose the **OK** button in the **Geometric Tolerance** dialog box, and then select the point where you want to insert the frames. AutoCAD will create the two frames and automatically align them with the common position symbol, as shown in Figure 8-76.

USING FEATURE CONTROL FRAMES WITH LEADERS

The **Leader Settings** dialog box invoked using the **QLEADER** command has the **Tolerance** option that allows you to create the feature control frame and attach it to the end of the leader extension line, see Figure 8-76. The following is the prompt sequence for using the **QLEADER** command with the **Tolerance** option:

Specify first leader point, or [Settings] <Settings>: *Press ENTER to display the **Leader Settings** dialog box. Choose the **Annotation** tab. Now, select the **Tolerance** radio button from the **Annotation Type** area. Choose **OK**.*
Specify first leader point, or [Settings] <Settings>: *Specify the start point of the leader line.*
Specify next point: *Specify the second point.*
Specify next point: *Specify the third point to display the **Geometric Tolerance** dialog box.*

Now, specify the tolerance value and choose the **OK** button; the **Feature Control Frame** will be displayed with leader.

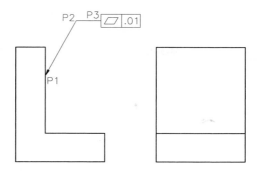

Figure 8-76 *Using the feature control frame with leader*

PROJECTED TOLERANCE ZONE

Figure 8-77 shows two parts joined with a bolt. The lower part is threaded, and the top part has a drilled hole. When these two parts are joined, the bolt that is threaded in the lower part will have the orientation error that exists in the threaded hole. In other words, the error in the threaded hole will extend beyond the part thickness, which might cause interference, and the parts may not assemble. To avoid this problem, projected tolerance is used. The projected tolerance establishes a tolerance zone that extends above the surface. In Figure 8-77, the position tolerance for the threaded hole is 0.010, which extends 0.1 above the surface (datum A). By using the projected tolerance, you can ensure that the bolt is within the tolerance zone up to the specified distance.

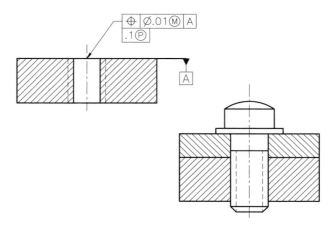

Figure 8-77 *Projected tolerance zone*

You can use the AutoCAD GDT feature to create feature control frames for the projected tolerance zone as discussed next.

Chapter 8

1. Invoke the **QLEADER** command and then press ENTER at the **Specify first leader point, or [Settings] <Settings>** prompt to display the **Leader Settings** dialog box.

2. Choose the **Annotation** tab and then select the **Tolerance** radio button from the **Annotation Type** area. Choose **OK** to return to the Command line. Specify the first, second, and the third leader point, refer to Figure 8-78. As soon as you specify the third point, the **Geometric Tolerance** dialog box will be displayed. Choose the position symbol from the **Symbol** dialog box.

3. In the first row of the **Geometric Tolerance** dialog box, enter the geometric characteristics and the datum references required for the first position tolerance frame.

4. In the **Height** edit box, enter the height of the tolerance zone (.1 for the given drawing) and select the box on the right of **Projected Tolerance Zone**. The projected tolerance zone symbol will be displayed in the box.

5. Choose the **OK** button in the **Geometric Tolerance** dialog box. AutoCAD will create two frames and automatically align them, refer to Figure 8-78.

6. Similarly, attach the datum symbol also.

Example 2   *Mechanical*

In the following example, you will create a feature control frame to define the perpendicularity specification, see Figure 8-78.

Figure 8-78 *Drawing for Example 2*

1. Choose **Annotate > Dimensions > Tolerance** from the **Ribbon** to display the **Geometric Tolerance** dialog box. Choose the upper box from the **Sym** area to display the **Symbol** dialog box. Select the **perpendicularity** symbol. This symbol will now be displayed in the **Sym** area.

2. Select the box on the left of the upper edit box under the **Tolerance 1** area. A diameter

symbol will appear to denote a cylindrical tolerance zone.

3. Enter **.005** in the upper edit box under the **Tolerance 1** area.

4. Enter **A** in the edit box under the **Datum 1** area. Choose the **OK** button to accept the changes made in the **Geometric Tolerance** dialog box.

5. The **Enter tolerance location** prompt is displayed in the Command line area and the **Feature Control Frame** is attached to the cursor at its middle left point. Select a point to insert the frame.

6. To place the datum symbol, use the **TOLERANCE** command to display the **Geometric Tolerance** dialog box. In the **Datum Identifier** edit box, enter **A**.

7. Choose the **OK** button to accept the changes to the **Geometric Tolerance** dialog box, and then select a point to insert the frame.

Example 3
Mechanical

In the following example, you will create a leader with a combination feature control frames to control runout and cylindricity, see Figure 8-79.

1. Enter the **QLEADER** command. The prompt sequence is as follows.

   Specify first leader point, or [Settings] <Settings>: *Press ENTER to display the **Leader Settings** dialog box. Choose the **Annotation** tab. Select the **Tolerance** radio button from the **Annotation Type** area. Choose **OK**.*
   Specify first leader point, or [Settings] <Settings>: *Specify the leader start point, as shown in Figure 8-79.*
   Specify next point: *Specify the second point of the leader.*
   Specify next point: *Specify the third point of the leader line to display the **Tolerance** dialog box.*

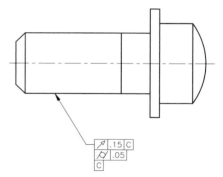

Figure 8-79 *Drawing for Example 3*

2. Choose the runout symbol from the **Symbol** dialog box. The **runout** symbol will be displayed on the first row of the **Sym** area. Enter **.15** in the first row edit box under the Tolerance 1 area.

3. Enter **C** in the edit box under the **Datum 1** area.

4. Select the edit box on the second row of the **Sym** area and select the **cylindricity** symbol. The **cylindricity** symbol will be displayed in the second row of the **Sym** area.

5. Enter **.05** in the second row edit box of the Tolerance 1 area.

6. Enter **C** in the **Datum Identifier** edit box.

7. Choose the **OK** button to accept the changes to the **Geometric Tolerance** dialog box. The control frames will be automatically attached at the end of the leader.

CREATING ANNOTATIVE DIMENSIONS, TOLERANCES, LEADERS, AND MULTILEADERS

Annotative dimensions are drawn in the drawing by assigning annotative dimension style to the current drawing. To do so, select the dimension style from the **Annotate > Dimensions > Dimension Style** in the **Ribbon** or from the **Dimension** toolbar. Note that the dimension style selected should have an annotative symbol next to it. You can also change the non-annotative dimensions to annotative by changing the dimension's **Annotative** property to **Yes** in the **Properties** palette.

For drawing the annotative geometric tolerances and leaders, you first need to add the simple tolerances to the drawing and then select all tolerances. Next, you need to modify the **Annotative** property under the **Misc** area of the **Properties** palette to **Yes**.

The annotative leaders are drawn by assigning annotative dimension style, whereas the annotative multileaders are drawn by assigning annotative multileader style. The leaders are created with two components: the leader and its content. So, drawing annotative leaders does not ensure that the content attached to the leader will also be annotative. But, the multileaders are drawn as a single component, so the multileaders drawn with annotative style will have annotative content also.

 Note
Creating and modifying dimension styles and multileader styles will be discussed in detail in Chapter 10.

Self-Evaluation Test

Answer the following questions and then compare them to those given at the end of this chapter:

1. You can specify dimension text of your own or accept the measured value computed by AutoCAD. (T/F)

2. The **DIMSPACE** command is used to maintain equal spacing between the linear dimensions or angular dimensions. (T/F)

3. The leaders drawn by using the QLEADER command create the arrow, the leader lines, and text as a single object. (T/F)

4. You cannot combine GTOL with leaders. (T/F)

5. The _____ command is used to dimension only the X coordinates of the selected object.

6. The _____ symbols are the building blocks of geometric dimensioning and tolerancing.

7. In the _____ dimensions, the dimension text is aligned by default with the object being dimensioned.

8. The _____ dimensions are automatically updated when the object to which they are assigned are changed.

9. The _____ point is taken as the vertex point of the angular dimensions while dimensioning an arc or a circle.

10. The extra leaders can be added or removed from the existing multileaders by using the _____ command.

Review Questions

Answer the following questions:

1. Only inner angles (acute angles) can be dimensioned with angular dimensioning. (T/F)

2. In addition to the most recently drawn dimension (the default base dimension), you can use any other linear dimension as the base dimension. (T/F)

3. In the continued dimensions, the base for successive continued dimensions is the base dimension's first extension line. (T/F)

4. You cannot add dimension break to straight multileaders. (T/F)

5. Using the **MLEADER** command, you can specify the content of multileader before draw-
    ing the leader itself. (T/F)

6. Which of the following commands can be used to convert dimensions into true associa-
    tive dimensions?

    (a) **DIMREASSOCIATE**   (b) **DIMASSOCIATE**
    (c) **DIMDISASSOCIATE**   (d) **None**

7. Which command can be used to dimension more than one object in a single effort?

    (a) **DIMLINEAR**   (b) **DIMANGULAR**
    (c) **QDIM**   (d) **None**

8. Which command is used to dimension different points and features of a part with reference
    to a fixed point?

    (a) **DIMBASELINE**   (b) **DIMANGULAR**
    (c) **DIMRADIUS**   (d) **DIMALIGNED**

9. Which command can be used to add geometric dimensions and tolerance to the current
    drawing?

    (a) **GTOL**   (b) **TOLERANCE**
    (c) **TEXT**   (d) **None**

10. Which of the following options of the **MLEADERALIGN** command can be used to
    maintain a constant spacing between the selected multileader landing lines?

    (a) **Distribute**   (b) **make leader segments Parallel**
    (c) **Specify Spacing**   (d) **Use current spacing**

11. Geometric dimensioning and tolerancing is generally known as _____.

12. Give three examples of geometric characteristics that indicate the characteristics of a
    feature. _____.

13. The three ways to return to the Command prompt from the **Dim:** prompt (dimensioning
    mode) are _____, _____, and _____.

14. The six fundamental dimensioning types provided by AutoCAD are _____
    _____.

15. Horizontal dimensions measure displacement along the _____.

Exercises

Exercise 7   *Mechanical*

Draw the object shown in Figure 8-80 and then dimension it. Save the drawing as a **DIMEXR7** drawing file.

Figure 8-80 *Drawing for Exercise 7*

Exercise 8   *Mechanical*

Draw and dimension the object shown in Figure 8-81. Save the drawing as **DIMEXR8**.

Figure 8-81 *Drawing for Exercise 8*

Chapter 8

Exercise 9   *Mechanical*

Draw the object shown in Figure 8-82 and then dimension it. Save the drawing as **DIMEXR9**.

Figure 8-82 *Drawing for Exercise 9*

Exercise 10   *Mechanical*

Draw the object shown in Figure 8-83 and then dimension it. Save the drawing as **DIMEXR10**.

Figure 8-83 *Drawing for Exercise 10*

Exercise 11   *Mechanical*

Draw the object shown in Figure 8-84 and then dimension it. Save the drawing as **DIMEXR11**.

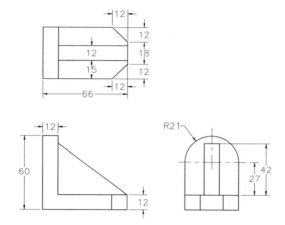

Figure 8-84 *Drawing for Exercise 11*

Exercise 12   *Mechanical*

Draw the object shown in Figure 8-85 and then dimension it. Save the drawing as **DIMEXR12**.

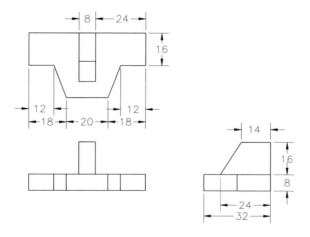

Figure 8-85 *Drawing for Exercise 12*

Chapter 8

Exercise 13   *Mechanical*

Draw the object shown in Figure 8-86 and then dimension it. Save the drawing as **DIMEXR13**.

Figure 8-86 Drawing for Exercise 13

Problem-Solving Exercise 1   *Mechanical*

Draw Figure 8-87 and dimension it, as shown in the drawing. The **FILLET** command should be used where needed. Save the drawing as **DIMPSE1**.

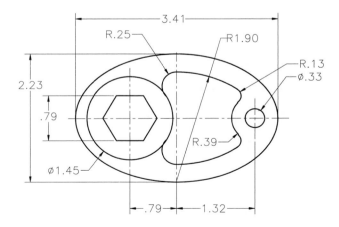

Figure 8-87 Drawing for Problem-Solving Exercise 1

Problem-Solving Exercise 2

Mechanical

Draw Figure 8-88 and then dimension it, as shown in the drawing. Save the drawing as **DIMPSE2**.

Figure 8-88 *Drawing for Problem-Solving Exercise 2*

Problem-Solving Exercise 3

Mechanical

Draw Figure 8-89 and then dimension it, as shown in the drawing. Save the drawing as **DIMPSE3**.

Figure 8-89 *Drawing for Problem-Solving Exercise 3*

Chapter 8

Problem-Solving Exercise 4   *Mechanical*

Draw the orthographic views of an object, as shown in Figure 8-90 and then dimension it, as shown in the drawing. Save the drawing as **DIMPSE4**.

Figure 8-90 *Drawing for Problem-Solving Exercise 4*

Answers to Self-Evaluation Test
1. T, **2.** T, **3.** F, **4.** F, **5.** **DIMORDINATE**, **6.** geometric characteristics, **7.** aligned, **8.** true associative dimensions, **9.** center, **10.** **MLEADEREDIT**

Chapter 9

Editing Dimensions

Learning Objectives

After completing this chapter, you will be able to:

- *Edit dimensions.*
- *Stretch, extend, and trim dimensions.*
- *Use the **Oblique** and **Text Angle** command options to edit dimensions.*
- *Update dimensions using the **Update** command.*
- *Use the **PROPERTIES** command to edit dimensions.*
- *Dimension in model space and paper space.*

EDITING DIMENSIONS USING EDITING TOOLS

For editing dimensions, AutoCAD has provided some special editing commands that work with dimensions. These editing commands can be used to define a new dimension text, return to the home text, create oblique dimensions, and rotate and update the dimension text. You can also use the **TRIM**, **STRETCH**, and **EXTEND** commands to edit the dimensions. In case the dimension assigned to the object is a true associative dimension, it will be automatically updated if the object is modified. However, if the dimension is not true associative dimension, you will have to include the dimension along with the object in the edit selection set. The properties of the dimensioned objects can also be changed using the **Properties palette** or the **Dimension Style Manager**.

Editing Dimensions by Stretching

You can edit a dimension by stretching it. However, to stretch a dimension, appropriate definition points must be included in the selection crossing or window. As the middle point of the dimension text is a definition point for all types of dimensions, you can easily stretch and move the dimension text to any location you want. When you stretch the dimension text, the gap in the dimension line gets filled automatically. While editing, the definition points of the dimension being edited must be included in the selection crossing box. The dimension is automatically calculated when you stretch the dimension.

 Note
The dimension type remains the same after stretching. For example, the vertical dimension maintains itself as a vertical dimension and measures only the vertical distance even after the line it dimensions is modified and converted into an inclined line. The following example illustrates the stretching of object lines and dimensions.

Example 1   *Mechanical*

In this example, you will stretch the objects and dimensions shown in Figure 9-1 to a new location using grips. The new location of the lines and dimensions is at a distance of 0.5 in the positive *Y* axis direction. See Figure 9-2.

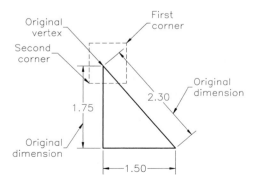

Figure 9-1 *Original location of lines and dimensions*

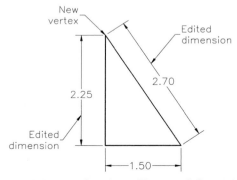

Figure 9-2 *New location of lines and dimensions*

1. Choose **Home > Modify > Stretch** from the **Ribbon**. The prompt sequence that will follow is given below:

   Select objects to stretch by crossing-window or crossing-polygon
   Select objects: Specify opposite corner: *Define a crossing window using the first and second corner, as shown in Figure 9-1.*
   Select objects: Enter
   Specify base point or [Displacement]<Displacement>: *Select original vertex using the osnaps as the base point.*
   Specify second point of displacement or <use first point as displacement>: **@0.5<90**

2. The selected entities will be stretched to the new location. The dimension that was initially 1.75 will become 2.25 and the dimension that was initially 2.30 will become 2.70, see Figure 9-2. Press ESC to remove the grip points from the objects.

Exercise 1   *General*

The two dimensions in Figure 9-3(a) are too close. Fix the drawing by stretching the dimension, as shown in Figure 9-3(b).

1. Stretch the outer dimension to the right so that there is some distance between the two dimensions.
2. Stretch the dimension text of the outer dimension so that the dimension text is staggered (lower than the first dimension).

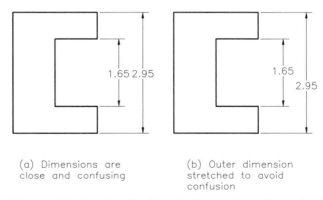

(a) Dimensions are close and confusing

(b) Outer dimension stretched to avoid confusion

Figure 9-3 Drawing for Exercise 1, stretching dimensions

Editing Dimensions by Trimming and Extending

Trimming and extending operations can be carried out with all types of linear dimensions (horizontal, vertical, aligned, rotated) and the ordinate dimension. Even if the dimensions are true associative, you can trim and extend them. AutoCAD trims or extends a linear dimension between the extension line definition points and the object used as a boundary or trimming edge. To extend or trim an ordinate dimension, AutoCAD moves the feature location

(location of the dimensioned coordinate) to the boundary edge. To retain the original ordinate value, the boundary edge to which the feature location point is moved should be orthogonal to the measured ordinate. In both cases, the imaginary line drawn between the two extension line definition points is trimmed or extended by AutoCAD, and the dimension is adjusted automatically. Figures 9-4 and 9-5 show the dimensions edited by trimming and extending.

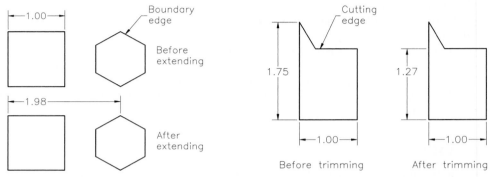

Figure 9-4 *Dimensions edited by extending*   **Figure 9-5** *Edgemode extended trimming*

Exercise 2   *Mechanical*

Use the **Edge > Extend** option of the **TRIM** command to trim the dimension given in Figure 9-6(a), so that it looks like Figure 9-6(b).

1. Make the drawing and dimension it, as shown in Figure 9-6(a). Assume the missing dimensions.

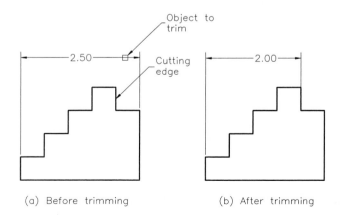

Figure 9-6 *Drawing for Exercise 2*

2. Trim the dimensions by using the **Edge > Extend** option of the **TRIM** command.

Flipping Dimension Arrow

You can flip the arrowheads individually. To flip the arrow, select the dimension. Place the cursor on the grip corresponding to the arrowhead that you want to flip. When the color of the grip turns red, invoke the shortcut menu by right-clicking and choose the **Flip Arrow** option from the shortcut menu.

MODIFYING THE DIMENSIONS

Ribbon:	Annotate > Dimensions > Oblique	**Command:**	DIMEDIT
Toolbar:	Dimension > Dimension Edit	**Menu Bar:**	Dimension > Oblique

The dimensions can be modified by choosing the **Dimension Edit** button from the **Dimension** toolbar (Figure 9-7). Alternatively, you can use the **DIMEDIT** command to modify the dimensions. This command has four options: **New**, **Rotate**, **Home**, and **Oblique**. The prompt sequence that will follow when you choose this button is given below:

Enter type of dimension editing (Home/New/Rotate/Oblique) <Home>: *Enter an option.*

└Dimension Edit

*Figure 9-7 Invoking the **DIMEDIT** command from the **Dimension** toolbar*

New

The **New** option is used to replace the existing dimension with a new text string. When you invoke this option, the **In-Place Text Editor** will be displayed. By default, 0.0000 will be displayed. Using this **In-Place Text Editor**, write the dimension or the text string with which you want to replace the existing dimension. Once you have written a new dimension in the editor and chosen **OK**, you will be prompted to select the dimension to be changed. Select the dimension; it will be replaced with the new dimension.

Rotate

The **Rotate** option is used to position the dimension text at a specified angle. With this option, you can change the orientation (angle) of the dimension text of any number of associative dimensions. The angle can be specified by entering its value at the **Specify angle for dimension text** prompt or by specifying two points at the required angle. Once you have specified the angle, you will be prompted to select the dimension text to be rotated. You will notice that the text rotates around its middle point, see Figure 9-8.

*Figure 9-8 Using the **DIMEDIT** command to edit dimensions*

Home

The **Home** option restores the text of a dimension to its original (home/default) location if the position of the text has been changed by stretching or editing, see Figure 9-8.

Oblique

In linear dimensions, extension lines are drawn perpendicular to the dimension line. The **Oblique** option bends the linear dimensions. It draws extension lines at an oblique angle (Figure 9-9). This option is particularly important to create isometric dimensions and can be used to resolve conflicting situations due to the overlapping of extension lines with other objects. Making an existing dimension oblique by specifying an angle oblique to it does not affect the generation of new linear dimensions. The oblique angle is maintained even after performing most editing operations. (See Chapter 23 for details about how to use this option.) When you invoke this option, you will be prompted to select the dimension to be edited. After selecting it, you will be prompted to specify the obliquing angle. The extensions lines will be bent at the angle specified. You can also invoke this option by choosing **Dimension > Oblique** from the **Ribbon**.

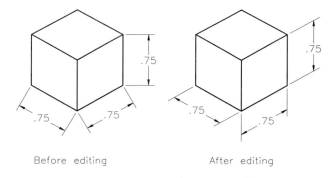

Figure 9-9 Using the **Oblique** option to edit dimensions

EDITING DIMENSION TEXT

| **Ribbon:** Annotate > Dimension > Text Angle | **Command:** DIMTEDIT |
| **Toolbar:** Dimension > Dimension Text Edit | **Menu Bar:** Dimension > Align Text > Angle |

The dimension text can be edited by using the **DIMTEDIT** command. This command is used to edit the placement and orientation of a single existing dimension. You can apply this command, for example, in cases where dimension texts of two or more dimensions are too close together, creating confusion. In such cases, the **DIMTEDIT** command is invoked to move the dimension text to some other location so that there is no confusion. You can also choose **Annotate > Dimension > Text Angle** from the **Ribbon**. The prompt sequence that will follow when you choose this button is given next.

Select dimension: *Select the dimension to modify.*
Specify new location for dimension text or [Left/Right/Center/Home/Angle]:

Left

With this option, you can left-justify the dimension text along the dimension line. The vertical placement setting determines the position of the dimension text. The horizontally aligned text is moved to the left and the vertically aligned text is moved down, see Figure 9-10. This option can be used only with the linear, diameter, or radial dimensions.

Right

With this option, you can right-justify the dimension text along the dimension line. Similar to the Left option, the vertical placement setting determines the position of the dimension text. The horizontally aligned text is moved to the right, and the vertically aligned text is moved up, see Figure 9-10. This option can be used only with linear diameter and radius dimensions.

Center

With this option, you can center-justify the dimension text for linear and aligned dimensions, see Figure 9-10. The vertical setting controls the vertical position of the dimension text.

Home

The **Home** option is used to restore (move) the dimension text of a dimension to its original (home/default) location, if the position of the text has been changed, see Figure 9-10.

Angle

With the **Angle** option, you can position the dimension text at the angle you specify, see Figure 9-10. The angle can be specified by entering its value at the **Specify angle for dimension text** prompt or by specifying two points at the required angle. You will notice that the text rotates around its middle point. If the dimension text alignment is set to Orient Text Horizontally, the dimension text is aligned with the dimension line. If information about the dimension style is available on the selected dimension, AutoCAD uses it to redraw the dimension, or the prevailing dimension variable settings are used for the redrawing process. Entering 0-degree angle changes the text to its default orientation.

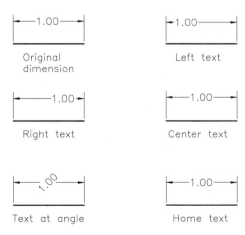

*Figure 9-10 Using the **DIMTEDIT** command to edit dimensions*

UPDATING DIMENSIONS

The **Update** option regenerates and updates the prevailing dimension entities (such as arrows heads and text height) using the current settings for the dimension variables, dimension style, text style, and units. On choosing this button, you will be prompted to select the dimensions to be updated. You can select all the dimensions or specify those that should be updated.

Chapter 9

EDITING DIMENSIONS WITH GRIPS

| **Ribbon:** | Annotate > Dimensions > Update | **Menu Bar:** Dimension > Update |
| **Toolbar:** | Dimension > Dimension Update | |

You can also edit dimensions by using the GRIP editing modes. GRIP editing is the easiest and quickest way to edit dimensions. You can perform the following operations with GRIPS.

1. Position the text anywhere along the dimension line. Note that you cannot move the text and position it above or below the dimension line.
2. Stretch a dimension to change the spacing between the dimension line and the object line.
3. Stretch the dimension along the length. When you stretch a dimension, the dimension text automatically changes.
4. Move, rotate, copy, mirror, or scale the dimensions.
5. Relocate the dimension origin.
6. Change properties such as color, layer, linetype, and linetype scale.
7. Load Web browser (if any *Universal Resource Locator* is associated with the object).

EDITING DIMENSIONS USING THE PROPERTIES PALETTE

| **Ribbon:** | View > Palettes > Properties | **Command:** | PROPERTIES |
| **Toolbar:** | Standard > Properties | **Menu Bar:** | Modify > Properties |

You can also modify a dimension or leader by using the **Properties** palette. The **Properties** palette is displayed when you choose the **View > Palettes > Properties** from the **Ribbon**. It can also be invoked by double-clicking on the dimension to be edited. All the properties of the selected object are displayed in the **Properties** palette (see Figure 9-11).

Properties Palette (Dimension)

You can use the **Properties** palette (Figure 9-11) to change the properties of a dimension, the dimension text style, or geometry, format, and annotation-related features of the selected dimension. The changes take place dynamically in the drawing. The **Properties** palette provides the following categories for the modification of dimensions.

General

In the general category, the various parameters displayed are **Color**, **Layer**, **Linetype**, **Linetype scale**, **Plot style**, **Lineweight**, **Hyperlink,** and **Associative** with their current values. To change the color of the selected object, select **Color** property

*Figure 9-11 The **Properties** palette for dimensions*

and then select the required color from the drop-down list. Similarly, layer, plot style, linetype, and lineweight can be changed from the respective drop-down lists. The linetype scale can be changed manually in the corresponding cell.

Misc

This category displays the dimension style by name (for the **DIMSTYLE** system variable, use **SETVAR**) and also specifies whether the dimension text will be annotative or not. You can change the dimension style and the annotative property of the text from the respective drop-down lists for the selected dimension.

Lines & Arrows

The various parameters of the lines and arrows in the dimension such as arrowhead size, type, arrow lineweight, and so on can be changed in this category.

Text

The various parameters that control the text in the dimension object such as text color, text height, vertical position text offset, and so on can be changed in this category.

Fit

In the fit category, various parameters are **Dim line forced**, **Dim line inside**, **Dim scale overall**, **Fit**, **Text inside**, and **Text movement**. All parameters can be changed in the drop-down list, except **Dim scale overall** (which can be changed manually).

Primary Units

In the primary units category, the parameters displayed are **Decimal separator**, **Dim prefix**, **Dim suffix**, **Dim roundoff**, **Dim scale linear**, **Dim units**, **Suppress leading zeros**, **Suppress trailing zeros**, **Suppress zero feet**, **Suppress zero inches**, and **Precision**. Among these properties, **Dim units**, **Suppress leading zeros**, **Suppress trailing zeros**, **Suppress zero feet**, **Suppress zero inches**, and **Precision** properties can be changed with the help of corresponding drop-down lists. The other parameters can be changed manually.

Alternate Units

Alternate units are required when a drawing is to be read in two different units. For example, if an architectural drawing is to be read in both metric and feet-inches, you can turn the alternate units on. The primary units can be set to metric and the alternate units to architectural. As a result, the dimensions of the drawing will be displayed in metric units as well as in engineering. In the alternate unit category, there are various parameters for the alternate units. They can be changed only if the **Alt enabled** parameter is **on**. The parameters such as **Alt format**, **Alt precision**, **Alt suppress leading zeros**, **Alt suppress trailing zeros**, **Alt suppress zero feet**, and **Alt suppress zero inches** can be changed from the respective drop-down lists and others can be changed manually.

Tolerances

The parameters of this category can be changed only if the Tolerances display parameter has

Chapter 9

some mode of the tolerance selected. The parameters listed in this rollout depend on the mode of tolerance selected.

Properties Palette (Multileader)

The **Properties** palette for **Multileader** can be invoked by selecting a Leader and then choosing **View > Palettes > Properties** from the **Ribbon**. You can also invoke the **Properties** palette (Figure 9-12) from the shortcut menu by right-clicking in the drawing area and choosing **Properties**. This palette can also be invoked by double-clicking in the leader to be edited. The various properties under the **Properties** palette (Multileader) are described next.

Figure 9-12 The **Properties** *palette for modifying multileaders*

General

The parameters in the general category are the same as those discussed in the previous section (**Properties** palette for dimensions).

Misc

This category displays the **Overall scale**, **Multileader style**, **Annotative**, and **Annotative scale** of the Multileader. You can change the style, annotative property, and annotation scale of the Multileader by using the respective drop-down lists.

Leaders

This category controls the display of certain elements such as lines, arrowheads, landing distance, and so on. The Multileader is composed of these elements.

Text

This category displays the properties that are used to control the text appearance and location of the multileader text. These properties can be changed by selecting the desired option from the respective drop-down lists.

Example 2   *Mechanical*

In this example, you will modify the dimensions given in Figure 9-13 so that they match the dimensions given in Figure 9-14. Assume the missing dimensions.

1. Choose **Annotate > Text > Text Style** from the **Ribbon** and create a style with the name **ROMANC**. Select **romanc.shx** as the font for the style.

2. Double-click on the dimension 2.25 to display the **Properties** palette.

Figure 9-13 *Drawing for Example 2*

Figure 9-14 *Drawing after editing the dimensions*

3. In the **Text** category, select the **Text style** drop-down list and then select **ROMANC** style from this drop-down list. The changes will take place dynamically.

4. Select **0.0000** from the **Precision** drop-down list in the **Primary Units** area.

5. Once all the required changes are made in the linear dimension, choose the **Select Object** button in the **Properties** palette. You will be prompted to select the object. Select the leader line and then press ENTER.

6. The **Properties** palette will not display the leader options. Select **Spline** from the **Leader type** drop-down list in the **Leaders** category; the straight line will be converted into a spline with an arrow dynamically.

7. Close the **Properties** palette.

8. Choose the **Dimension Edit** button from the **Dimension** toolbar. Enter **N** in the prompt sequence to display the **In-Place Text Editor**.

9. Enter **%%C.25 DRILL AND REAM** in the text editor and then click outside the text editor to accept the changes.

10. You will be prompted to select the object to be changed. Select the diameter dimension and then press ENTER. The diameter dimension will be modified to the new value.

MODEL SPACE AND PAPER SPACE DIMENSIONING

Dimensioning objects can be drawn in the model space or paper space. If the drawings are in model space, associative dimensions should also be created in them. If the drawings are in the model space and the associative dimensions are in the paper space, the dimensions will not change when you perform such editing operations as stretching, trimming, and extending, or such display operations as zoom and pan in the model space viewport. The definition points of a dimension are located in the space where the drawing is drawn. You can select the **Scale dimensions to layout** radio button under the **Scale for dimension features** area in the **Fit** tab. This tab is available when you choose **Modify, New,** or **Override** buttons from the

Chapter 9

Dimension Style Manager dialog box, depending on whether you want to modify the present style or you want to create a new style (see Figure 9-15). Choose **OK** and then **Close** to exit from the dialog boxes. AutoCAD calculates a scale factor that is compatible with the model space and the paper space viewports. Now, choose **Annotate > Dimensions > Update** from the **Ribbon** and select the dimension objects for updating.

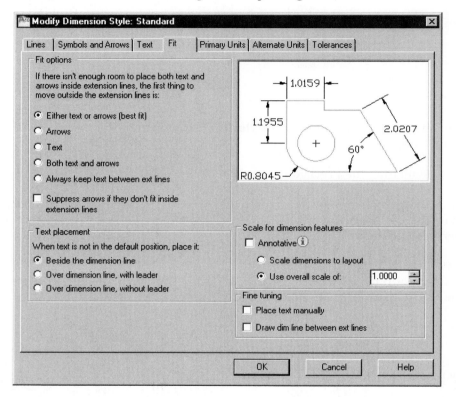

Figure 9-15 *Selecting paper space scaling in the **Modify Dimension Style** dialog box*

The drawing shown in Figure 9-16 uses paper space scaling. The main drawing and detail drawings are located in different floating viewports (paper space). The zoom scale factors for these viewports are different: 0.3XP, 1.0XP, and 0.5XP, respectively. When you use paper scaling, AutoCAD automatically calculates the scale factor for dimensioning so that the dimensions are uniform in all the floating viewports (model space viewports).

Figure 9-16 *Dimensioning in paper model space viewports using paper space scaling or setting* **DIMSCALE** *to 0*

Self-Evaluation Test

Answer the following questions and then compare them to those given at the end of this chapter:

1. In associative dimensioning, the items constituting a dimension (such as dimension lines, arrows, leaders, extension lines, and dimension text) are drawn as a single object. (T/F)
2. If the value of the variable **DIMDASSOC** is set to zero, the dimension lines, arrows, leaders, extension lines, and dimension text are drawn as independent objects. (T/F)

3. You cannot edit dimensions using grips. (T/F)

4. The true associative dimensions cannot be trimmed or extended. (T/F)

5. You can use the _____ command to break the dimensions into individual entities.

6. The _____ option of the **DIMTEDIT** command is used to justify the dimension text toward the left side.

7. The _____ option of the **DIMTEDIT** command is used to justify the dimension text to the center of the dimension.

8. The _____ option of the **DIMEDIT** command is used to create a new text string.

9. The _____ option of the **DIMEDIT** command is used to bend the extension lines through the specified angle.

10. The _____ button from the **Dimension** panel is used to update the dimensions.

Chapter 9

Review Questions

Answer the following questions:

1. The horizontal, vertical, aligned, and rotated dimensions cannot be edited using grips. (T/F)
2. Trimming and extending operations can be carried out with all types of linear (horizontal, vertical, aligned, and rotated) dimensions and with the ordinate dimension. (T/F)

3. To extend or trim an ordinate dimension, AutoCAD moves the feature location (location of the dimensioned coordinate) to the boundary edge. (T/F)

4. Once moved from the original location, the dimension text cannot be restored to its original position. (T/F)

5. With the _____ or _____ commands, you can edit the dimension text.

6. The _____ command is particularly important for creating isometric dimensions and is applicable in resolving conflicting situations due to overlapping of extension lines with other objects.

7. The _____ command is used to edit the placement and orientation of a single existing dimension.

8. The _____ command regenerates (updates) prevailing associative dimension objects (like arrows and text height) using the current settings for the dimension variables, dimension style, text style, and units.

9. Explain when to use the **EXTEND** command and how it works with dimensions.
   _____.

10. Explain the use and working of the **PROPERTIES** command for editing dimensions.
    _____.

Exercises

Exercise 3   *General*

1. Create the drawing shown in Figure 9-17. Assume the dimensions where necessary.
2. Dimension the drawing, as shown in Figure 9-17.
3. Edit the dimensions so that they match the dimensions, as shown in Figure 9-18.

Figure 9-17 *Drawing for Exercise 3,*

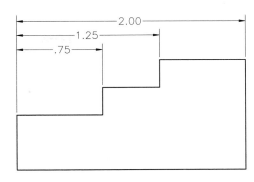

Figure 9-18 *Drawing for Exercise 3, after*

Exercise 4   *Mechanical*

1. Draw the object shown in Figure 9-19(a). Assume the dimensions where necessary.
2. Dimension the drawing, as shown in Figure 9-19(a).
3. Edit the dimensions so that they match the dimensions shown in Figure 9-19(b).

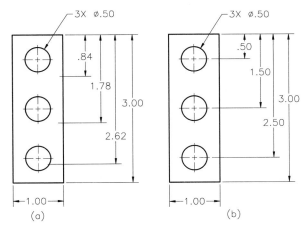

Figure 9-19 *Drawings for Exercise 4*

Exercise 5   *Architectural*

Create the drawing shown in Figure 9-20 and then dimension it. Assume the dimensions wherever necessary. After dimensioning the drawing, edit them so that they match the dimensions shown in Figure 9-20.

Chapter 9

Figure 9-20 *Drawing for Exercise 5*

Problem-Solving Exercise 1   *Mechanical*

Create the drawing shown in Figure 9-21 and then dimension it. Edit the dimensions so that they are positioned as shown in the drawing. You may change the dimension text height and arrow size to 0.08 units.

Figure 9-21 *Drawing for Problem-Solving Exercise 1*

Problem-Solving Exercise 2   *Mechanical*

Draw the front and side view of an object shown in Figure 9-22 and then dimension the two views. Edit the dimensions so that they are positioned as shown in the drawing. You may change the dimension text height and arrow size to 0.08 units.

Figure 9-22 *Drawing for Problem-Solving Exercise 2*

Problem-Solving Exercise 3   *Architectural*

Create the drawing of the floor plan shown in Figure 9-23 and then apply the dimensions as shown. Edit the dimensions, if needed, so that they are positioned as shown in the drawing.

Figure 9-23 Drawing for Problem-Solving Exercise 3

Answers to Self-Evaluation Test
1. T, 2. T, 3. F, 4. F, 5. EXPLODE, 6. Left, 7. Center, 8. New, 9. Oblique, 10. Update

Chapter 10

Dimension Styles, Multileader Styles, and System Variables

Learning Objectives

After completing this chapter, you will be able to:
- *Use styles and variables to control dimensions.*
- *Create dimensioning styles.*
- *Set dimension variables using various tabs of the **New**, **Modify**, and **Override Dimension Style** dialog boxes.*
- *Set other dimension variables that are not in the dialog boxes.*
- *Understand dimension style families and how to apply them in dimensioning.*
- *Use dimension style overrides.*
- *Compare and list dimension styles.*
- *Import externally referenced dimension styles.*
- *Create, restore, and modify multileader styles.*

USING STYLES AND VARIABLES TO CONTROL DIMENSIONS

In AutoCAD, the appearance of dimensions on the drawing screen and the manner in which they are saved in the drawing database are controlled by a set of dimension variables. The dimensioning commands use these variables as arguments. The variables that control the appearance of dimensions can be managed using dimension styles. You can use the **Dimension Style Manager** dialog box to control the dimension styles and dimension variables through a set of dialog boxes.

CREATING AND RESTORING DIMENSION STYLES

Ribbon:	Annotate > Dimensions > Dimension Style (The inclined arrow)
Toolbar:	Dimension > Dimension Style or Styles > Dimension Style
Menu Bar:	Dimension > Dimension Style or Format > Dimension Style
Command:	DIMSTYLE

The dimension styles control the appearance and positioning of dimensions and leaders in the drawing. If the default dimensioning style (Standard or Annotative) does not meet your requirements, you can select another existing dimensioning style or create a new one. The default names of the dimension style file are **Standard** and **Annotative**. Dimension styles can be created by using the **Dimension Style Manager** dialog box. Left-click on the inclined arrow in the **Dimensions** panel of the **Annotate** tab of the **Ribbon** to invoke the **Dimension Style Manager** dialog box (Figure 10-1).

*Figure 10-1 The **Dimension Style Manager** dialog box*

From the **Dimension Style Manager** dialog box, choose the **New** button to display the **Create New Dimension Style** dialog box (Figure 10-2). Enter the dimension style name in the **New Style Name** text box and then select a style on which you want to base your style from the **Start With** drop-down list. Select the **Annotative** check box to make the new dimension style annotative. Choose the **i** (information) button to get the help and information about the annotative objects. The **Use for** drop-down list allows you to select the dimension type to which you want to apply the new dimension style. For example, if you wish to use the new style for only the diameter dimension, you can select **Diameter dimensions** from the **Use**

Figure 10-2 *The* **Create New Dimension Style** *dialog box*

for drop-down list. Choose the **Continue** button to display the **New Dimension Style** dialog box. Specify the parameters of the new dimension style and choose **OK**. The parameters of the **New Dimension Style** dialog box are discussed in the next section.

In the **Dimension Style Manager** dialog box, the current dimension style name is shown in front of **Current dimension style** and is also shown highlighted in the **Styles** list box. A brief description of the current style (its differences from the default settings) is also displayed in the **Description** area. The **Dimension Style Manager** dialog box also has a **Preview of** window that displays the preview of the current dimension style. A style can be made current (restored) by selecting the name of the dimension style you want to make current from the list of defined dimension styles and choosing the **Set Current** button. You can also make a style current by double-clicking on the style name in the **Styles** list box. The drop-down list in the **Dimensions** panel under the **Annotate** tab of the **Ribbon** also displays the dimension styles. Selecting a dimension style from this list also sets it current. The list of dimension styles displayed in the **Styles** list box is dependent on the option selected from the **List** drop down-list. If you select the **Styles in use** option, only the dimension styles in use will be listed in the **Style** list box. If you right-click on a style in the **Styles** list box, a shortcut menu is displayed that provides you with the options to **Set current**, **Rename**, or **Delete** a dimension style. Selecting the **Don't list styles in Xrefs** check box does not list the names of Xref styles in the **Styles** list box. Choosing the **Modify** button displays the **Modify Dimension Style** dialog box where you can modify an existing style. Choosing the **Override** button displays the **Override Current Style** dialog box where you can define overrides to an existing style (discussed later in this chapter). Both these dialog boxes along with the **New Dimension Style** dialog box have identical properties. Choosing the **Compare** button displays the **Compare Dimension Styles** dialog box (also discussed later in this chapter) that allows you to compare two existing dimension styles.

NEW DIMENSION STYLE DIALOG BOX

The **New Dimension Style** dialog box can be used to specify the dimensioning attributes (variables) that affect the various properties of the dimensions. The various tabs provided under the **New Dimension Style** dialog box are discussed next.

Chapter 10

Lines Tab

The options in the **Lines** tab (Figure 10-3) of the **New Dimension Style** dialog box are used to specify the dimensioning attributes (variables) that affect the format of the dimension lines. For example, the appearance and behavior of the dimension lines and extension lines can be changed with this tab. If the settings of the dimension variables have not been altered in the current editing session, the settings displayed in the dialog box are the default settings.

*Figure 10-3 The **Lines** tab of the **New Dimension Style** dialog box*

Dimension lines Area

This area provides you with the options of controlling the display of the dimension lines and leader lines. These options are discussed next.

Color. This drop-down list is used to set the colors for the dimension lines and arrowheads. Its dimension arrowheads have the same color as the dimension line because arrows constitute a part of the dimension line. The color you set here will also be assigned to the leader lines and arrows. The default color for the dimension lines and arrows is ByBlock. You can specify the color of the dimension line by selecting it from the **Color** drop-down list. You can also select **Select Color** from the **Color** drop-down list to display the **Select Color** dialog box where you can choose a specific color. The color number or the special color label is stored in the **DIMCLRD** variable.

Linetype. This drop-down list is used to set the linetype for the dimension lines.

Lineweight. This drop-down list is used to specify the lineweight for the dimension line. You can select the required lineweight by selecting it from this drop-down list. This value is also stored in the **DIMLWD** variable. The default value is ByBlock. Remember that you cannot assign the lineweight to the arrowheads using this drop-down list.

Extend beyond ticks. The **Extend beyond ticks** (Oblique tick extension) spinner will be available only when you select the oblique, Architectural tick, or any such arrowhead type in the **First** and **Second** drop-down lists in the **Arrowheads** area. This spinner is used to specify the distance by which the dimension line will extend beyond the extension line. The extension value, entered in the **Extend beyond ticks** edit box, gets stored in the **DIMDLE** variable. By default, this edit box is disabled because the oblique arrowhead type is not selected.

Baseline spacing. The **Baseline spacing** (baseline increment) spinner is used to control the spacing between successive dimension lines drawn using the baseline dimensioning, see Figure 10-4. You can specify the dimension line increment to your requirement by specifying the desired value using the **Baseline spacing** spinner. The default value displayed in the **Baseline spacing** spinner is 0.38 units. This spacing value is also stored in the **DIMDLI** variable.

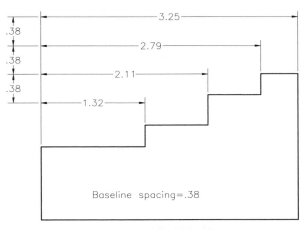

Figure 10-4 *Baseline increment*

Suppress. The **Suppress** check boxes control the display of the first and second dimension lines. By default, both dimension lines will be drawn. You can suppress one or both the dimension lines by selecting their corresponding check boxes. The values of these check boxes are stored in the **DIMSD1** and **DIMSD2** variables.

 Note
The first and second dimension lines are determined by how you select the extension line origins. If the first extension line origin is on the right, the first dimension line will also be on the right.

Extension lines Area
The options in this area are discussed next.

Color. This drop-down list is used to control the color of the extension lines. The default extension line color is ByBlock. You can assign a new color to the extension lines by selecting it from this drop-down list. The color number or the color label is saved in the **DIMCLRE** variable.

Chapter 10

Linetype ext line 1. The options in this drop-down list are used to specify the linetype of extension line 1. By default, the line type for the extension line 1 is set to ByBlock. You can change the linetype by selecting a new value of linetype from the drop-down list.

Linetype ext line 2. The options in this drop-down list are same as for the Linetype ext line 1. The options are used to specify the line type of extension line 2.

Note
*You can specify different linetypes for the dimensions and the extensions lines. Use the **Linetype** drop-down list in the **Dimension Lines** area to specify the linetype for dimensions. Use the **Linetype ext line 1** and **Linetype ext line 2** drop-down lists to specify the linetype for the extension lines.*

Lineweight. This drop-down list is used to modify the lineweight of the extension lines. The default value is ByBlock. You can change the lineweight value by selecting a new value from this drop-down list. The value for lineweight is stored in the **DIMLWE** variable.

Extend beyond dim lines. It is the distance by which the extension lines extend past the dimension lines, see Figure 10-5. You can change the extension line offset using the **Extend beyond dim lines** spinner. This value is also stored in the **DIMEXE** variable. The default value for the extension distance is 0.1800 units.

Offset from origin. It is the distance by which the extension line is offset from the point you specify as the origin of the extension line, see Figure 10-6. You may need to override this setting for specific dimensions while dimensioning curves and angled lines. You can specify an offset distance of your choice using this spinner. AutoCAD stores this value in the **DIMEXO** variable. The default value for this distance is 0.0625.

Fixed length extension lines. With the selection of the **Fixed length extension lines** check box, you can specify a fixed length (starting from the dimension line to the origin) for the extension lines in the **Length** edit box. By default, the check box is cleared and the length spinner is not available. Once the check box is selected, you can specify the length of the extension line either by entering a numerical value in the **Length** edit box or using a spinner. Figure 10-7 displays a drawing with full length extension lines and Figure 10-8 displays the same drawing with fixed length extension lines.

Suppress. The **Suppress** check boxes are used to control the display of the extension lines. By default, both extension lines will be drawn. You can suppress one or both of them by selecting the corresponding check boxes (Figure 10-9). The values of these check boxes are stored in the **DIMSE1** and **DIMSE2** variables.

Note
The first and second extension lines are determined by how you select the extension line origins. If the first extension line origin is on the right, the first extension line will also be on the right.

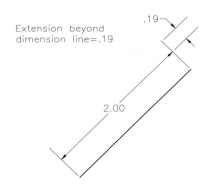

Figure 10-5 *Extension beyond dimension lines*

Figure 10-6 *The offset from origin*

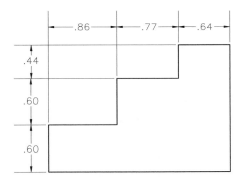

Figure 10-7 *Drawings with full length extension lines*

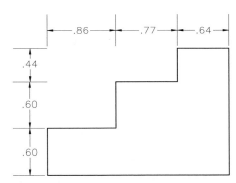

Figure 10-8 *Drawing with fixed length extension lines*

Symbols and Arrows Tab

The options in the **Symbols and Arrows** tab (Figure 10-10) of the **New Dimension Style** dialog box are used to specify the variables and attributes that affect the format of the symbols and arrows. You can change the appearance of symbols and arrows.

Arrowheads Area

The options in this area are used to specify the arrowheads and they are discussed next.

First/Second. When you create a dimension, AutoCAD draws the terminator symbols at the two ends of the dimension line. These terminator symbols, generally referred to as **arrowheads**, represent the beginning and end of a dimension. AutoCAD has provided nineteen standard termination symbols you can apply

Figure 10-9 *Suppressing the extension lines*

Chapter 10

*Figure 10-10 The **Symbols and Arrows** tab of the **New Dimension Style** dialog box*

at each end of the dimension line. In addition to these, you can create your own arrows or terminator symbols. By default, the same arrowhead type is applied at both ends of the dimension line. If you select the first arrowhead, it is automatically applied to the second end by default. However, if you want to specify a different arrowhead at the second dimension line endpoint, you must select the desired arrowhead type from the **Second** drop-down list. The first endpoint of the dimension line is the intersection point of the first extension line and the dimension line. The first extension line is determined by the first extension line origin. However, in angular dimensioning, the second endpoint is located in a counterclockwise direction from the first point, regardless of how the points were selected when creating the angular dimension. The specified arrowhead types are selected from the **First** and **Second** drop-down lists. The first arrowhead type is saved in the **DIMBLK1** system variable and the second arrowhead type is saved in the **DIMBLK2** system variable.

AutoCAD provides you with an option of specifying a user-defined arrowhead. To define a user-defined arrow, you must create one as a block. (See Chapter 16 for information regarding blocks.) Now, from the **First** or the **Second** drop-down list, select **User Arrow**; the **Select Custom Arrow Block** dialog box will be displayed, see Figure 10-11.

All the blocks in the current drawing will be available in the **Select from Drawing Blocks** drop-down list. You can select the desired block and it will become the current arrowhead.

*Figure 10-11 The **Select Custom Arrow Block** dialog box*

Creating an Arrowhead Block

1. To create a block for an arrowhead, you will use a 1 X 1 box, refer to Figure 10-12. AutoCAD automatically scales the X and Y scale factors of the block to the arrowhead size multiplied by the overall scale. You can specify the arrow size in the **Arrow size** spinner. The **DIMASZ** variable controls the length of the arrowhead. For example, if **DIMASZ** is set to 0.25, the length of the arrow will be 0.25 units. Also, if the length of the arrow is not 1 unit, it will leave a gap between the dimension line and the arrowhead block.

2. The arrowhead must be drawn as it would appear on the right side of the dimension line. Choose **Home > Block > Create** from the **Ribbon** to convert it into a block.

3. The insertion point of the arrowhead block must be the point that will coincide with the extension line, see Figure 10-12.

Leader. The **Leader** drop-down list displays the arrowhead types for the Leader arrow. Here, also, you can either select the standard arrowheads from the drop-down list or select **User Arrow** that allows you to define and use a user-defined arrowhead type.

Arrow size. This spinner is used to define the size of the arrowhead, see Figure 10-13. The default value is 0.18 units, which is stored in the **DIMASZ** system variable.

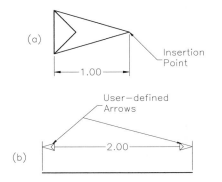

Figure 10-12 Creating user-defined arrows

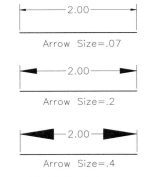

Figure 10-13 Defining arrow sizes

Center marks Area

This area deals with the options that control the appearance of the center marks and centerlines in the radius and diameter dimensioning. However, keep in mind that the center marks or the centerlines will be drawn only when dimensions are placed outside the circle.

None. If you select **None** radio button, no mark or line will be drawn at the center of the circle.

Mark. If you select the **Mark** radio button, a mark will be drawn at the center of the circle.

Line. If you select the **Line** radio button, the centerlines will be drawn at the center of the circles.

The spinner in the **Center marks** area is used to set the size of the center marks or centerlines. This value is stored in the **DIMCEN** variable. The default value of the **DIMCEN** variable is 0.09.

> **Note**
> *If you use the **DIMCEN** command, a positive value will create a center mark, whereas a negative value will create a centerline. If the value is 0, AutoCAD does not create center marks or centerlines.*

Dimension Break Area

The **Break size** spinner in this area is used to control the default break length in the dimension while applying the break dimensions. The default value of **Break size** is 0.125. Figure 10-14 shows a drawing with dimensioning breaks.

Arc length symbol Area

The radio buttons in this area are used to specify the position of arc when applying the arc length dimension.

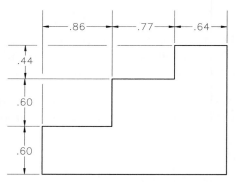

Preceding dimension text. If you select the **Preceding dimension text** radio button, the arc symbol will appear before the dimension text while applying the arc length dimension.

Figure 10-14 Drawing with dimension breaks

Above dimension text. With the **Above dimension text** radio button selected, the arc symbol in the arc length dimension will appear above the dimension text.

None. Select the **None** radio button, if you do not want the arc symbol to appear with the arc length dimension.

Radius jog dimension Area

The **Jog angle** edit box in this area is used to specify the angle of jog that appears while applying the jogged radius dimension. By default, its value is 45-degree.

Linear jog dimension Area

The option in this area is used to control the height of the jog (vertical height from the vertex of one jog angle to another) that appears while applying the jogged linear dimension. The jog height is always maintained with respect to the text height of the dimension. The **Jog height factor** spinner is used to change the factor that will be multiplied with the text height to calculate the linear jog height.

Note

*Unlike specifying a negative value for the **DIMCEN** variable, you cannot enter a negative value in the **Size** spinner. Selecting **Line** from the **Type** drop-down list automatically treats the value in the **Size** spinner as the size for the centerlines and sets **DIMCEN** to the negative of the value shown.*

Exercise 1 *Mechanical*

Draw and dimension the drawing, as shown in Figure 10-15. Assume the missing dimensions.

Figure 10-15 Drawing for Exercise 1

CONTROLLING THE DIMENSION TEXT FORMAT
Text Tab

You can control the dimension text format through the **Text** tab of the **New Dimension Style** dialog box (Figure 10-16). In the **Text** tab, you can control the parameters such as the placement, appearance, horizontal and vertical alignment of the dimension text, and so on. For example, you can force AutoCAD to align the dimension text along the dimension line. You can also force the dimension text to be displayed at the top of the dimension line. You can save the settings in a dimension style file for future use. The **New Dimension Style** dialog box has a **Preview** window that updates dynamically to display the text placement as the settings are changed. Individual items of the **Text** tab and the related dimension variables are described next.

*Figure 10-16 The **Text** tab of the **New Dimension Style** dialog box*

Text appearance Area

The options in this area are discussed next.

Text style. The **Text style** drop-down list displays the names of the predefined text styles. From this list, you can select the style name that you want to use for dimensioning. You must define the text style before you can use it in dimensioning (see "**Creating Text Styles**" in Chapter 7). Choosing the [**...**] button displays the **Text Style** dialog box that allows you to create a new or modify an existing text style. The value of this setting is stored in the **DIMTXSTY** system variable. The change in the dimension text style does not affect the text style you are using to draw the other text in the drawing.

Text color. This drop-down list is used to modify the color of the dimension text. The default color is **ByBlock**. If you choose the **Select Color** option from the **Text color** drop-down list, the **Select Color** dialog box is displayed, where you can choose a specific color. This color or color number is stored in the **DIMCLRT** variable.

Fill color. This drop-down list is used to set the fill color of the dimension text. A box of the selected color will be placed around the dimension text.

Text height. This spinner is used to modify the height of the dimension text, see Figure 10-17. You can change the dimension text height only when the current text style does not have a fixed height. In other words, the text height specified in the **STYLE** command should be zero. This is because a predefined text height (specified in the **STYLE** command) overrides any other setting for the dimension text height. This value is stored in the **DIMTXT** variable. The default text height is 0.1800 units.

Fraction height scale. This spinner is used to set the scale of the fractional units in relation to the dimension text height. This spinner will be available only when you select a format for the primary units, in which you can define the values in fractions, such as architectural or fractional. This value is stored in the **DIMTFAC** variable.

Draw frame around text. When selected, this check box draws a frame around the dimension text, see Figure 10-18. This value is stored as a negative value in the **DIMGAP** system variable.

Figure 10-17 *Changing the dimension height* ***Figure 10-18*** *Dimension text inside the frame*

Text placement Area
The options in this area are discussed next.

Vertical. The **Vertical** drop-down list displays the options that control the vertical placement of the dimension text. The current setting is highlighted. Controlling the vertical placement of the dimension text is possible only when the dimension text is drawn in its normal (default) location. This setting is stored in the **DIMTAD** system variable. The options in this drop-down list are discussed next.

> **Centered**. If this option is selected, the dimension text gets positioned on the dimension line in such a way that the dimension line is split to allow for the placement of the text, see Figure 10-19. If the **1st** or **2nd Extension Line** option is selected in the **Horizontal** drop-down list, this centered setting will position the text on the extension line, not on the dimension line.

> **Above**. If this option is selected, the dimension text is placed above the dimension line, except when the dimension line is not horizontal and the dimension text inside the extension lines is horizontal. The distance of the dimension text from the dimension line

is controlled by the **DIMGAP** variable. This results in an unbroken solid dimension line being drawn under the dimension text, see Figure 10-19.

Outside. This option places the dimension text on the side of the dimension line.

JIS. This option lets you place the dimension text to conform to the **JIS** (Japanese Industrial Standards) representation.

Note
*The horizontal and vertical placement options selected are reflected in the dimensions shown in the **Preview** window.*

Horizontal. This drop-down list is used to control the horizontal placement of the dimension text. You can select the required horizontal placement from this list. However, remember that these options will be useful only when the **Place text manually** check box in the **Fine Tuning** area of the **Fit** tab is cleared. The options in this drop-down list are discussed next.

Centered. This option places the dimension text between the extension lines. This is the default option.
At Ext Line 1. This option is selected to place the text near the first extension line along the dimension line, see Figure 10-20.

At Ext Line 2. This option is selected to place the text near the second extension line along the dimension line, see Figure 10-20.

Over Ext Line 1. This option is selected to place the text over the first extension line and also along the first extension line, see Figure 10-20.

Over Ext Line 2. This option is selected to place the text over the second extension line and also along the second extension line, see Figure 10-20.

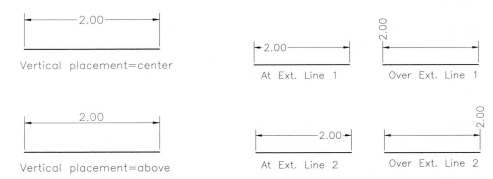

Figure 10-19 *Vertical text placement*   ***Figure 10-20*** *Horizontal text placement*

Offset from dim line. This spinner is used to specify the distance between the dimension line and the dimension text (Figure 10-21). You can set the text gap you need using this spinner. The text gap value is also used as the measure of minimum length for the segments of the dimension line and in basic tolerance. The default value specified in this box is 0.09 units. The value of this setting is stored in the **DIMGAP** system variable.

Text alignment Area
The options in this area are discussed next.

Horizontal. This is the default option and if selected, the dimension text is drawn horizontally with respect to the current UCS (user coordinate system). The alignment of the dimension line does not affect the text alignment. Selecting this radio button turns both the **DIMTIH** and **DIMTOH** system variables **on**. The text is drawn horizontally even if the dimension line is at an angle.

Aligned with dimension line. Selecting this radio button aligns the text with the dimension line (Figure 10-22) and both the system variables **DIMTIH** and **DIMTOH** are turned off.

ISO standard. If you select the **ISO Standard** radio button, the dimension text is aligned with the dimension line, only when the dimension text is inside the extension lines. Selecting this option turns the system variable **DIMTOH** on, that is, the dimension text outside the extension line is horizontal, regardless of the angle of the dimension line.

Figure 10-21 Offset from the dimension line

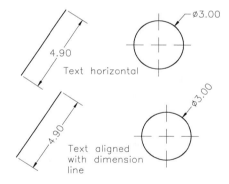

Figure 10-22 Specifying the text alignment

Exercise 2   *Mechanical*

Draw Figure 10-23 and then set the values in the **Lines**, **Symbols and Arrows**, and **Text** tabs of the **New Dimension Style** dialog box to dimension the drawing, as shown in the figure. (Baseline spacing = 0.25, Extension beyond dimension lines = 0.10, Offset from origin = 0.05, Arrow size = 0.09, Text height = 0.08.)

Figure 10-23 *Drawing for Exercise 2*

FITTING DIMENSION TEXT AND ARROWHEADS
Fit Tab

The options in this area are discussed next.

The **Fit** tab provides you with the options that control the placement of dimension lines, arrowheads, leader lines, text, and the overall dimension scale (Figure 10-24).

Fit options Area

These options are used to set the priorities for moving the text and arrowheads outside the extension lines, if the space between the extension lines is not enough to fit both of them.

Either text or arrows (best fit). This is the default option. In this option, AutoCAD places the dimension where it fits best between the extension lines.

Arrows. When you select this option, AutoCAD places the text and arrowheads inside the extension lines if there is enough space to fit both. If the space is not available, the arrows are moved outside the extension lines. If there is not enough space for text, both text and arrowheads are placed outside the extension lines.

Text. When you select this option, AutoCAD places the text and arrowheads inside the extension lines, if there is enough space to fit both. If there is enough space to fit the arrows, the arrows will be placed inside and the dimension text moves outside the extension lines. However, if there is not enough space for either the text or the arrowheads, both are placed outside the extension lines.

Both text and arrows. If you select this option, AutoCAD will place the arrows and dimension text between the extension lines, if there is enough space available to fit both. Otherwise, both the text and arrowheads are placed outside the extension lines.

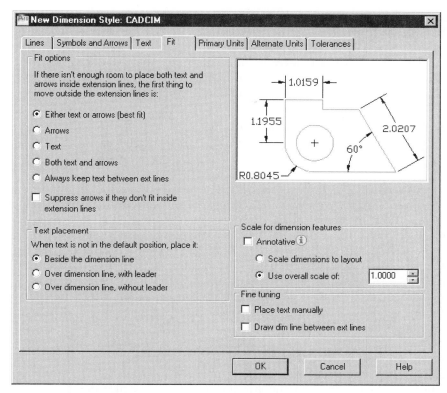

*Figure 10-24 The **Fit** tab of the **New Dimension Style** dialog box*

Always keep text between ext lines. This option always keeps the text between the extension lines even in cases where AutoCAD would not do so. Selecting this radio button does not affect the radius and diameter dimensions. The value is stored in the **DIMTIX** variable and the default value is **off**.

Suppress arrows if they don't fit inside extension lines. If you select this check box, the arrowheads are suppressed if the space between the extension lines is not enough to adjust them. The value is stored in the **DIMSOXD** variable and the default value is **off**.

Text placement Area

This area provides you with the options to position the dimension text when it is moved from the default position. The value is stored in the **DIMTMOVE** variable. The options in this area are as follows:

Beside the dimension line. This option places the dimension text beside the dimension line.

Over dimension line, with leader. Selecting this option places the dimension text away from the dimension line and a leader line is created, which connects the text to the dimension line. But, if the dimension line is too close to the text, a leader is not drawn. The Horizontal placement decides whether the text is placed to the right or left of the leader.

Over dimension line, without leader. In this option, AutoCAD does not create a leader line, if there is insufficient space to fit the dimension text between the extension lines. The dimension text can be moved freely, independent of the dimension line.

Scale for dimension features Area

The options under this area are used to set the value for the overall dimension scale or scaling to the paper space.

Annotative. This check box is used to specify that the selected dimension style is annotative. With this option, you can also convert an existing non-annotative dimension style to annotative and vice-versa. If you select the **Annotative** check box from the **Create New Dimension Style** dialog box, then the **Annotative** check box will also be selected by default.

Use overall scale of. The current general scaling factor that pertains to all of the size-related dimension variables, such as text size, center mark size, and arrowhead size, is displayed in the **Use overall scale of** spinner. You can alter the scaling factor to your requirement by entering the scaling factor of your choice in this spinner. Altering the contents of this box alters the value of the **DIMSCALE** variable, since the current scaling factor is stored in it. The overall scale (**DIMSCALE**) is not applied to the measured lengths, coordinates, angles, or tolerance. The default value for this variable is 1.0. In this condition, the dimensioning variables assume their preset values and the drawing is plotted at a full scale. The scale factor is the reciprocal of the drawing size and so the drawing is to be plotted at the half size. The overall scale factor (**DIMSCALE**) will be the reciprocal of ½, which is 2.

Note
*If you are in the middle of the dimensioning process and you change the **DIMSCALE** value and save the changed setting in a dimension style file, the dimensions with that style will be updated.*

Tip
*When you increase the limits of the drawing, you need to increase the overall scale of the drawing using the **Use overall scale of** spinner before dimensioning. This will save the time required in changing the individual scale factors of all the dimension parameters.*

Scale dimensions to layout. If you select the **Scale dimensions to layout** radio button, the scale factor between the current model space viewport and the floating viewport is computed automatically. Also, by selecting this radio button, you disable the **Use overall scale of** spinner (it is disabled in the dialog box) and the overall scale factor is set to 0. When the overall scale factor is assigned a value of 0, AutoCAD calculates an acceptable default value based on the scaling between the current model space viewport and the paper space. If you are in the paper space (**TILEMODE=0**), or are not using the **Scale dimensions to layout** feature, AutoCAD sets the overall scale factor to 1; otherwise, AutoCAD calculates a scale factor that makes it possible to plot text sizes, arrow sizes, and other scaled distances at the values, in which they have been previously set. (For further details regarding model space and layouts, refer to Chapter 12.)

Fine tuning Area

The **Fine tuning** area provides additional options governing placement of the dimension text. The options are as follows.

Place text manually. When you dimension, AutoCAD places the dimension text in the middle of the dimension line (if there is enough space). If you select the **Place text manually** check box, you can position the dimension text anywhere along the dimension line. You will also notice that when you select this check box, the **Horizontal Justification** is ignored. This setting is saved in the **DIMUPT** system variable. The default value of this variable is **off**. Selecting this check box enables you to position the dimension text anywhere along the dimension line.

Draw dim line between ext lines. This check box is selected when you want the dimension line to appear between the extension lines, even if the text and dimension lines are placed outside the extension lines. When you select this option in the radius and diameter dimensions (when default text placement is horizontal), the dimension line and arrows are drawn inside the circle or arc, and the text and leader are drawn outside. If you select the **Draw dim line between ext lines** check box, the **DIMTOFL** variable is set to on by AutoCAD. Its default setting is off.

FORMATTING PRIMARY DIMENSION UNITS
Primary Units Tab

You can use the **Primary Units** tab of the **New Dimension Style** dialog box to control the dimension text format and precision values (Figure 10-25). You can use the options under this tab to control Units, Dimension Precision, and Zero Suppression for dimension measurements. AutoCAD lets you attach a user-defined prefix or suffix to the dimension text. For example, you can define the diameter symbol as a prefix by entering %%C in the **Prefix** edit box; AutoCAD will automatically attach the diameter symbol in front of the dimension text. Similarly, you can define a unit type, such as **mm**, as a suffix; AutoCAD will then attach **mm** at the end of every dimension text. This tab also enables you to define zero suppression, precision, and dimension text format.

Linear dimensions Area

The options in this area are discussed next.

Unit format. This drop-down list provides you with the options of specifying the units for the primary dimensions. The formats include **Decimal**, **Scientific**, **Architectural**, **Engineering**, **Fraction**, and **Windows Desktop**. Remember that by selecting a dimension unit format, the drawing units (which you might have selected by using the **UNITS** command) are not affected. The unit setting for linear dimensions is stored in the **DIMLUNIT** system variable.

*Figure 10-25 The **Primary Units** tab of the **New Dimension Style** dialog box*

Precision. This drop-down list is used to control the number of decimal places for the primary units. The setting for precision (number of decimal places) is saved in the **DIMDEC** variable.

Fraction format. This drop-down list is used to set the fraction format. The options are Diagonal, Horizontal, and not stacked. This drop-down list will be available only when you select **Architectural** or **Fractional** from the **Unit format** drop-down list. The value is stored in the **DIMFRAC** variable.

Decimal separator. This drop-down list is used to select an option that will be used as the decimal separator. For example, Period [.], Comma [,] or Space [ ]. If you have selected Windows desktop units in the **Unit Format** drop-down list, AutoCAD uses the Decimal symbol settings. The value is stored in the **DIMDSEP** variable.

Round off. The **Round off** spinner sets the value for rounding off the dimension values. For example, if the **Round off** spinner is set to 0.05, a value of 0.06 will round off to 0.10. The number of decimal places of the round off value you enter in the edit box should be less than or equal to the value in the **Precision** edit box. The value is stored in the **DIMRND** variable and the default value in the **Round off** edit box is 0, see Figure 10-26.

Prefix. You can append a prefix to the dimension measurement by entering it in this edit

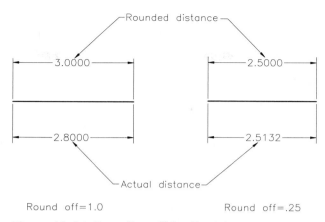

Figure 10-26 *Rounding off the dimension measurements*

box. The dimension text is converted into **Prefix<dimension measurement>** format. For example, if you enter the text "Abs" in the **Prefix** edit box, "Abs" will be placed in front of the dimension text (Figure 10-27). The prefix string is saved in the **DIMPOST** system variable.

Figure 10-27 *Adding prefix and suffix to the dimensions*

 Note
*Once you specify a prefix, default prefixes such as **R** in radius dimensioning and ø in diameter dimensioning are cancelled.*

Suffix. Just like appending a prefix, you can append a suffix to the dimension measurement by entering the desired suffix in this edit box. For example, if you enter the text **mm** in the **Suffix** edit box, the dimension text will have <dimension measurement>mm format, see Figure 10-25. AutoCAD stores the suffix string in the **DIMPOST** variable.

 Tip
*The **DIMPOST** variable is used to append both prefix and suffix to the dimension text. This*

Chapter 10

variable takes a string value as its argument. For example, if you want to have a suffix for centimeters, set **DIMPOST** *to cm. To establish a prefix to a dimension text, type the prefix text string and then "<>".*

Measurement scale Area

The options in this area are discussed next.

Scale factor. You can specify a global scale factor for the linear dimension measurements by setting the desired scale factor in the **Scale factor** spinner. All the linear distances measured by dimensions, which include radii, diameters, and coordinates, are multiplied by the existing value in this spinner. For example, if the value of the **Scale factor** spinner is set to 2, two unit segments will be dimensioned as 4 units (2 X 2). However, the angular dimensions are not affected. In this manner, the value of the linear scaling factor affects the contents of the default (original) dimension text (Figure 10-28). The default value for linear scaling is 1. With the default value, the dimension text generated is the actual measurement of the object being dimensioned. The linear scaling value is saved in the **DIMLFAC** variable.

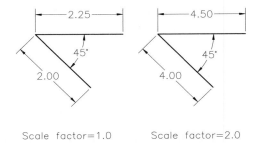

Figure 10-28 *Identical figures dimensioned using different scale factors*

Note
The linear scaling value is not exercised on rounding a value, or on plus or minus tolerance value. Therefore, changing the linear scaling factor will not affect the tolerance values.

Apply to layout dimensions only. When you select the **Apply to layout dimensions only** check box, the scale factor value is applied only to the dimensions in the layout. The value is stored as a negative value in the **DIMLFAC** variable. If you change the **DIMLFAC** variable from the **Dim:** prompt, AutoCAD displays the viewport option to calculate the **DIMLFAC** variable. First, set the **TILEMODE** to 0 (paper space), and then invoke the **MVIEW** command to get the **Viewport** option.

Zero suppression Area

The options in this area are used to suppress the leading or trailing zeros in the dimensioning. This area provides you with four check boxes. These check boxes can be selected to suppress the leading or trailing zeros or zeros in the feet and inches. The **0 feet** and the **0 inches** check boxes will be available only when you select **Engineering** or **Architectural** from the **Unit format** drop-down list. When the Architectural units are being used, the **Leading** and **Trailing** check boxes are disabled. For example, if you select the **0 feet** check box, the dimension text 0'-8 ¾" becomes 8 ¾". By default, the 0 feet and 0 inches value is suppressed. If you want to suppress the inches part of a feet-and-inches dimension when the distance in the feet portion is an integer value and the inches portion is zero, select the **0 inches** check box. For example, if you select the **0 inches** check box, the dimension text 3'-0" becomes 3'.

Similarly, if you select the **Leading** check box, the dimension that was initially 0.53 will become .53. If you select the **Trailing** check box, the dimension that was initially 2.0 will become 2.

Angular dimensions Area

This area provides you with the options to control the units format, precision, and zero suppression for the Angular units.

Units format. The **Units format** drop-down list displays a list of unit formats for the angular dimensions. The default value, in which the angular dimensions are displayed, is **Decimal Degrees**. The value governing the unit setting for the angular dimensions is stored in the **DIMAUNIT** variable.

Precision. You can select the number of decimal places for the angular dimensions from this drop-down list. This value is stored in the **DIMADEC** variable.

Zero suppression Area. Similar to the linear dimensions, you can suppress the **leading**, **trailing**, neither, or both zeros in the angular dimensions by selecting the respective check boxes in this area. The value is stored in the **DIMAZIN** variable.

FORMATTING ALTERNATE DIMENSION UNITS
Alternate Units Tab

By default, the **Alternate Units** tab of the **New Dimension Style** dialog box is disabled and the value of the **DIMALT** variable is turned off. If you want to perform alternate units dimensioning, select the **Display alternate units** check box. By doing so, AutoCAD activates various options in this area (Figure 10-29). This tab sets the format, precision, angles, placement, scale, and so on for the alternate units in use. In this tab, you can specify the values that will be applied to the alternate dimensions.

Alternate units Area

The options in this area are identical to those under the **Linear dimensions** area of the **Primary Units** tab. This area provides you with the options to set the format for all dimension types.

Unit format. You can select a unit format to apply to the alternate dimensions from this drop-down list. The options under this drop-down list include Scientific, Decimal, Engineering, Architectural stacked, Fractional stacked, Architectural, Fractional, and Windows Desktop. The value is stored in the **DIMALTU** variable. The relative size of fractions is governed by the **DIMTFAC** variable.

Precision. You can select the number of decimal places for the alternate units from the **Precision** drop-down list. The value is stored in the **DIMALTD** variable.

Multiplier for alt units. To generate a value in the alternate system of measurement, you need a factor with which all the linear dimensions will be multiplied. The value for this factor can be set using the **Multiplier for alt units** spinner. The default value of 25.4 is for

*Figure 10-29 The **Alternate Units** tab of the **New Dimension Style** dialog box*

dimensioning in inches with the alternate units in millimeters. This scaling value (contents of the **Multiplier for alt units** spinner) is stored in the **DIMALTF** variable.

Round distances to. This spinner is used to set a value to which you want all your measurements (made in alternate units) to be rounded off. This value is stored in the **DIMALTRND** system variable. For example, if you set the value of the **Round distances to** spinner to 0.25, all the alternate dimensions get rounded off to the nearest .25 unit.

Prefix/Suffix. The **Prefix** and **Suffix** edit boxes are similar to the edit boxes in the **Linear dimensions** area of the **Primary Units** tab. You can enter the text or symbols that you want to precede or follow the alternate dimension text. The value is stored in the **DIMAPOST** variable. You can also use control codes and special characters to display special symbols.

Zero suppression Area

This area allows you to suppress the leading or trailing zeros in decimal unit dimensions by selecting either, both, or none of the **Trailing** and **Leading** check boxes. Similarly, selecting the **0 feet** check box suppresses the zeros in the feet area of the dimension, when the dimension value is less than a foot. Selecting the **0 inches** check box suppresses the zeros in the inches area of the dimension. For example, 1'-0" becomes 1'. The **DIMALTZ** variable controls the suppression of zeros for alternate unit dimension values. The values that are between 0 and 3 affect the feet-and-inch dimensions only.

Placement Area

This area provides the options that control the positioning of the Alternate units. The value is stored in the **DIMAPOST** variable.

After primary value. Selecting the **After primary value** radio button places the alternate units dimension text after the primary units. This is the default option, see Figure 10-30.

Below primary value. Selecting the **Below primary value** radio button places the alternate units dimension text below the primary units, see Figure 10-30.

Figure 10-30 illustrates the result of entering information in the **Alternate Units** tab. The decimal places get saved in the **DIMALTD** variable, the scaling value (contents of the **Multiplier for alt units** spinner) in the **DIMALTF** variable, and the suffix string (contents of the **Suffix** edit box) in the **DIMAPOST** variable. Similarly, the units format for alternate units are in **DIMALTU**, and suppression of zeros for alternate units decimal values are in **DIMALTZ**.

Figure 10-30 Placement of alternate units

FORMATTING THE TOLERANCES
Tolerances Tab

The **Tolerances** tab (Figure 10-31) allows you to set the parameters for options that control the format and display of the tolerance dimension text. These include the alternate unit tolerance dimension text.

Tolerance format Area

The **Tolerance format** area of the **Tolerances** tab (Figure 10-31) lets you specify the tolerance method, tolerance value, position of tolerance text, and precision and height of the tolerance text. For example, if you do not want a dimension to deviate more than plus 0.01 and minus 0.02, you can specify this by selecting **Deviation** from the **Method** drop-down list and then specifying the plus and minus deviation in the **Upper Value** and the **Lower Value** edit boxes. When you dimension, AutoCAD will automatically append the tolerance to it. The **DIMTP** variable sets the maximum (or upper) tolerance limit for the dimension text and **DIMTM** variable sets the minimum (or lower) tolerance limit for the dimension text. Different settings and their effects on relevant dimension variables are explained in the following sections.

Figure 10-31 *The* **Tolerances** *tab of the* **New Dimension Style** *dialog box*

Method. The **Method** drop-down list lets you select the tolerance method. The tolerance methods supported by AutoCAD are **Symmetrical**, **Deviation**, **Limits**, and **Basic**. These tolerance methods are described next.

None. Selecting the **None** option sets the **DIMTOL** variable to 0 and does not add tolerance values to the dimension text, that is, the **Tolerances** tab is disabled.

Symmetrical. This option is used to specify the symmetrical tolerances. When you select this option, the **Lower Value** spinner is disabled and the value specified in the **Upper Value** spinner is applied to both plus and minus tolerances. For example, if the value specified in the **Upper Value** spinner is 0.05, the tolerance appended to the dimension text is ±0.05, see Figure 10-32. The value of **DIMTOL** is set to 1 and the value of **DIMLIM** is set to 0.

Deviation. If you select the **Deviation** tolerance method, the values in the **Upper Value** and **Lower Value** spinners will be displayed as plus and minus dimension tolerances. If you enter values for the plus and minus tolerances, AutoCAD appends a plus sign (+) to the positive values of the tolerance and a negative sign (–) to the negative values of the tolerance. For example, if the upper value of the tolerance is 0.005 and the lower value of the tolerance is 0.002, the resulting dimension text generated will have a positive tolerance of 0.005 and a negative tolerance of 0.002 (Figure 10-32). Even if one of the tolerance values is 0, a sign is appended to it. On specifying the deviation tolerance, AutoCAD sets the **DIMTOL** variable value to 1 and the **DIMLIM** variable value to 0. The values in the **Upper Value** and **Lower Value** edit boxes are saved in the **DIMTP** and **DIMTM** system variables, respectively.

Limits. If you select the **Limits** tolerance method from the **Method** drop-down list, AutoCAD adds the upper value (contents of the **Upper Value** spinner) to the dimension text (actual measurement) and subtracts the lower value (contents of the **Lower Value** spinner) from the dimension text. The resulting values are displayed as the dimension text, see Figure 10-32. Selecting the **Limits** tolerance method results in setting the **DIMLIM** variable value to 1 and the **DIMTOL** variable value to 0. The numeral values in the **Upper Value** and **Lower Value** edit boxes are saved in the **DIMTP** and **DIMTM** system variables, respectively.

Basic. A basic dimension text is a dimension text with a box drawn around it (Figure 10-32). The basic dimension is also called a reference dimension. Reference dimensions are used primarily in geometric dimensioning and tolerances. The basic dimension can be realized by selecting the basic tolerance method. The distance provided around the dimension text (distance between dimension text and the rectangular box) is stored as a negative value in the **DIMGAP** variable. The negative value signifies the basic dimension. The default setting is off, resulting in the generation of dimensions without the box around the dimension text.

Figure 10-32 Specifying tolerance using Various tolerancing methods

Precision. The **Precision** drop-down list is used to select the number of decimal places for the tolerance dimension text. The value is stored in **DIMTDEC** variable.

Upper value/Lower value. In the **Upper value** spinner, the positive upper or maximum value is specified. If the method of tolerances is symmetrical, the same value is used as the lower value also. The value is stored in the **DIMTP** variable. In the **Lower** spinner, the lower or minimum value is specified. The value is stored in the **DIMTM** variable.

Scaling for height. The **Scaling for height** spinner is used to specify the height of the dimension tolerance text relative to the dimension text height. The default value is 1, which means the height of the tolerance text is the same as the dimension text height. If you want the tolerance text to be 75 percent of the dimension height text, enter 0.75 in the **Scaling for height** edit box. The ratio of the tolerance height to the dimension text height is calculated by AutoCAD and then stored in the **DIMTFAC** variable. **DIMTFAC = Tolerance Height/Text Height.**

Vertical position. This drop-down list allows you to specify the location of the tolerance text for deviation and symmetrical methods only. The three alignments that are possible are with the **Bottom**, **Middle**, or **Top** of the main dimension text. The settings are saved in the **DIMTOLJ** system variable (Bottom=0, Middle=1, and Top=2).

Tolerance alignment Area

The options in this area are used to control the alignment of the tolerance value text when they are placed in stacked condition. These options get highlighted only when you select the **Deviation** or **Limit** option from the **Method** drop-down list. Select the **Align decimal separators** radio button to align the tolerance text vertically along the decimal point, see Figure 10-33(a). And select the **Align operational symbols** radio button to align the tolerance text vertically along the plus sign (+) and negative sign(-), see Figure 10-33(b).

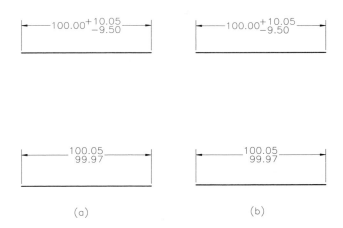

(a)   (b)

Figure 10-33 *Various tolerance alignment options*

Zero suppression Area

This area controls the zero suppression in the tolerance text depending on which one of the check boxes is selected. Selecting the **Leading** check box suppresses the leading zeros in all the decimal tolerance text. For example, 0.2000 becomes .2000. Selecting the **Trailing** check box suppresses the trailing zeros in all the decimal tolerance text. For example, 0.5000 becomes 0.5. Similarly, selecting both the boxes suppresses both the trailing and leading zeros and selecting none, suppresses none. If you select **0 feet** check box, the zeros in the feet portion of the tolerance dimension text are suppressed if the dimension value is less than one foot. Similarly, selecting the **0 inches** check box suppresses the zeros in the inches portion of the dimension text. The value is stored in the **DIMTZIN** variable.

Alternate unit tolerance Area

The options in this area define the precision and zero suppression settings for the Alternate unit tolerance values. The options under this area will be available only when you display the alternate units along with the primary units.

Precision. This drop-down list is used to set the number of decimal places to be displayed in the tolerance text of the alternate dimensions. This value is stored in the **DIMALTTD** variable.

Zero suppression Area

Selecting the respective check boxes controls the suppression of the **Leading** and **Trailing** zeros in decimal values and the suppression of zeros in the Feet and Inches portions for dimensions in the feet and inches format. The value is stored in the **DIMALTTZ** variable.

Exercise 3	*Mechanical*

Draw Figure 10-34 and then set the values in various tabs of the **New Dimension Style** dialog box to dimension it, as shown. (Baseline spacing = 0.25, Extension beyond dim lines = 0.10, Offset from origin = 0.05, Arrowhead size =0.07, Text height = 0.08.) Assume the missing dimensions.

Figure 10-34 *Drawing for Exercise 3*

DIMENSION STYLE FAMILIES

The dimension style feature of AutoCAD lets the user define a dimension style with values that are common to all dimensions. For example, the arrow size, dimension text height, or color of the dimension line are generally the same in all types of dimensioning such as linear, radial, diameter, and angular. These dimensioning types belong to the same family because they have some characteristics in common. In AutoCAD, this is called a **dimension style family,** and the values assigned to the family are called **dimension style family values.**

After you have defined the dimension style family values, you can specify variations on it for other types of dimensions such as radial and diameter. For example, if you want to limit the number of decimal places to two in radial dimensioning, you can specify that value for radial

Chapter 10

dimensioning. The other values will stay the same as the family values to which this dimension type belongs. When you use the radial dimension, AutoCAD automatically uses the style that was defined for radial dimensioning; otherwise, it creates a radial dimension with the values as defined for the family. After you have created a dimension style family, any changes in the parent style are applied to family members, if the particular property is the same. Special suffix codes are appended to the dimension style family name that correspond to different dimension types. For example, if the dimension style family name is MYSTYLE and you define a diameter type of dimension, AutoCAD will append $4 at the end of the dimension style family name. The name of the diameter type of dimension will be MYSTYLE$4. The following are the suffix codes for different types of dimensioning.

Suffix Code	Dimension Type	Suffix Code	Dimension Type
0	Linear	2	Angular
3	Radius	4	Diameter
6	Ordinate	7	Leader

Example 1   *Mechanical*

The following example illustrates the concepts of dimension style families, Figure 10-35.

1. Specify the values for the dimension style family.
2. Specify the values for the linear dimensions.
3. Specify the values for the diameter and radius dimensions.
4. After creating the dimension style, use it to dimension the given drawing.

Figure 10-35 *Drawing for Example 1*

1. Start a new file in the **2D Drafting & Annotation** workspace, and draw an object, as shown in Figure 10-35.

2. Left-click on the inclined arrow in the **Dimensions** panel of the **Annotate** tab; the **Dimension Style Manager** dialog box is displayed. The **Standard** and **Annotative** options are displayed in the **Styles** list box. Select the **Annotative** style from the **Styles** list box.

3. Choose the **New** button to display the **Create New Dimension Style** dialog box. In this dialog box, enter **MyStyle** in the **New Style Name** edit box. Select **Annotative** from the **Start With** drop-down list. Also select **All dimensions** from the **Use for** drop-down list. Now, choose the **Continue** button to display the **New Dimension Style: MyStyle** dialog box. In this dialog box, first choose the **Lines** tab and then enter the following values in their respective options. Next, choose the **Symbols and Arrows** tab and enter the following values.

Lines tab	**Symbols and Arrows** tab
Baseline Spacing: 0.15	Center Mark for circle, Size: 0.05
Offset from origin: 0.03	Arrow size: 0.09
Extension beyond dim line: 0.07	Center Mark: Line

4. Choose the **Text** tab and change the following values:

Text Height: 0.09	Offset from dimension line: 0.03

5. Choose the **Fit** tab and make sure the **Annotative** check box is selected.

6. After entering the values, choose the **OK** button to return to the **Dimension Style Manager** dialog box. This dimension style contains the values that are common to all dimension types.

7. Now, choose the **New** button again in the **Dimension Style Manager** dialog box to display the **Create New Dimension Style** dialog box. AutoCAD displays **Copy of MyStyle** in the **New Style Name** edit box. Select **MyStyle** from the **Start With** drop-down list if it is not already selected. From the **Use for** drop-down list, select **Linear dimensions**. Choose the **Continue** button to display the **New Dimension Style: MyStyle: Linear** dialog box and set the following values in the **Text** tab:

    a. Select the **Aligned with dimension line** radio button in the **Text alignment** area.
    b. In the **Text placement** area, select **Above** from the **Vertical** drop-down list.

8. In the **Primary Units** tab, set the precision to two decimal places. Select the **Leading** check box in the **Zero Suppression** area. Next, choose the **OK** button to return to the **Dimension Style Manager** dialog box.

9. Choose the **New** button again to display the **Create New Dimension Style** dialog box. Select **MyStyle** from the **Start With** drop-down list and the **Diameter dimensions** type from the **Use for** drop-down list. Next, choose the **Continue** button; the **New dimension Style: MyStyle: Diameter** dialog box is displayed.

10. Choose the **Primary Units** tab and then set the **Precision** to two decimal places. Select the **Leading** check box in the **Zero Suppression** area. Next, choose the **OK** button to return to the **Dimension Style Manager** dialog box.

Chapter 10

11. Under this dialog box, choose the **New** button to display the **Create New Dimension Style** dialog box. Select **MyStyle** from the **Start With** drop-down list and **Radius dimensions** from the **Use for** drop-down list. Choose the **Continue** button to display the **New Dimension Style: MyStyle: Radial** dialog box.

12. Choose the **Primary Units** tab and then set the precision to two decimal places. Select the **Leading** check box in the **Zero Suppression** area. Next, enter **Rad** in the **Prefix** edit box.

13. In the **Fit** tab, select the **Text** radio button in the **Fit options** area. Next, choose the **OK** button to return to the **Dimension Style Manager** dialog box.

14. Select **MyStyle** from the **Styles** list box and choose the **Set current** button. Choose the **Close** button to exit the dialog box.

15. Use the linear and baseline dimensioning to draw the linear dimensions, as shown in Figure 10-35. You will notice that when you enter any linear dimensioning, AutoCAD automatically uses the values that were defined for the linear type of dimensioning.

16. Use the diameter dimensioning to dimension the circles, refer to Figure 10-35. Again, notice that the dimensions are drawn according to the values specified for the diameter type of dimensioning.

17. Now, use the radius dimensioning to dimension the fillet, refer to Figure 10-35.

USING DIMENSION STYLE OVERRIDES

Most of the dimension characteristics are common in a production drawing. The values that are common to different dimensioning types can be defined in the dimension style family. However, at times, you might have different dimensions. For example, you may need two types of linear dimensioning: one with tolerance and one without. One way to draw these dimensions is to create two dimensioning styles. You can also use the dimension variable overrides to override the existing values. For example, you can define a dimension style (**MyStyle**) that draws dimensions without tolerance. Now, to draw a dimension with tolerance or update an existing dimension, you can override the previously defined value through the **Dimension Style Manager** dialog box or by setting the variable values at the Command prompt. The following example illustrates how to use the dimension style overrides.

Example 2   *Mechanical*

In this example, you will update the overall dimension (3.00) so that the tolerance is displayed with the dimension. You will also add linear dimensions, as shown in Figure 10-36.

This problem can be solved by dimension style overrides as well as using the **Properties** palette. However, here only the dimension style overrides method is discussed.

Figure 10-36 *Drawing for Example 2*

1. Invoke the **Dimension Style Manager** dialog box. Select **MyStyle** from the **Styles** list box and choose the **Override** button to display the **Override Current Style: MyStyle** dialog box. The options in this dialog box are identical to the **New Dimension Style** dialog box discussed earlier in this chapter.

2. Choose the **Tolerances** tab and select **Symmetrical** from the **Method** drop-down list.

3. Set the value of the **Precision** spinner to two decimal places. Set the value of the **Upper value** spinner to **0.02** and select the **Leading** check box in the **Zero suppression** area. Next, choose the **OK** button to exit the dialog box (this does not save the style). You will notice that **<style overrides>** is displayed under **MyStyle** in the **Styles** list box, indicating that the style overrides the **MyStyle** dimension style.

4. The **<style overrides>** is displayed until you save it under a new name or under the style it is displayed, or until you delete it. Select **<style overrides>** and right-click to display the shortcut menu. Choose the **Save to current style** option from the shortcut menu to save the overrides to the current style. Choosing the **Rename** option allows you to rename the style override and save it as a new style.

5. Choose **Annotate > Dimension > Update** from the **Ribbon** and select the dimension that measures **3.00**. The dimension now displays the symmetrical tolerance.

6. Draw the remaining two linear dimensions. They will automatically appear with the tolerances, see Figure 10-36.

 Tip
*You can also use the **DIMOVERRIDE** command to apply the change to the existing dimensions. Apply the changes to the **DIMTOL**, **DIMTP**, and **DIMTM** variables.*

COMPARING AND LISTING DIMENSION STYLES

Choosing the **Compare** button in the **Dimension Style Manager** dialog box displays the **Compare Dimension Styles** dialog box where you can compare the settings of two dimensions styles or list all the settings of one of them (Figure 10-37).

The **Compare** and the **With** drop-down lists display the dimension styles in the current drawing. Selecting the dimension styles from the respective lists compare the two styles. In the **With** drop-down list, if you select **<none>** or the same style as selected from the **Compare** drop-down list, all the properties of the selected style are displayed. The comparison results are displayed under four headings: **Description** of the Dimension Style property, the **System Variable** controlling a particular setting, and the **values of the variable for both the dimension styles** which differ in the two styles in comparison. The number of differences between the selected dimension styles are displayed below the **With** drop-down list. The button provided in this dialog box prints the comparison results to the Windows clipboard from where they can be pasted to other Windows applications.

Figure 10-37 The **Compare Dimension Styles** dialog box

USING EXTERNALLY REFERENCED DIMENSION STYLES

The externally referenced dimensions cannot be used directly in the current drawing. When you Xref a drawing, the drawing name is appended to the style name and the two are separated by the vertical bar (|) symbol. It uses the same syntax as the other externally dependent symbols. For example, if the drawing (FLOOR) has a dimension style called DECIMAL and you Xref this drawing in the current drawing, AutoCAD will rename the dimension style to FLOOR|DECIMAL. You cannot make this dimension style current, nor can you modify or override it. However, you can use it as a template to create a new style. To accomplish this, invoke the **Dimension Style Manager** dialog box. If the **Don't list styles in Xrefs** check box is selected, the styles in the Xref are not displayed. Clear this check box to display the Xref dimension styles and choose the **New** button. In the **New Style Name** edit box of the **New**

Dimension Style dialog box, enter the name of the dimension style. AutoCAD will create a new dimension style with the same values as those of the externally referenced dimension style (FLOOR|DECIMAL).

CREATING AND RESTORING MULTILEADER STYLES

Ribbon:	Annotate > Leaders > Multileader Style Manager (*Inclined arrow*)
Toolbar:	Multileader > Multileader Style
	or Styles > Multileader Style
Command:	MLEADERSTYLE

The multileader styles control the appearance and positioning of multileaders in the drawing. If the default multileader styles (**Standard** and **Annotative**) do not meet your requirement, you can select any other existing multileader style as per your requirement. The default multileader style file names are **Standard** and **Annotative**. Left-click on the inclined arrow in the **Multileaders** panel of the **Annotate** tab of the **Ribbon**; the **Multileader Style Manager** dialog box will be displayed, as shown in Figure 10-38.

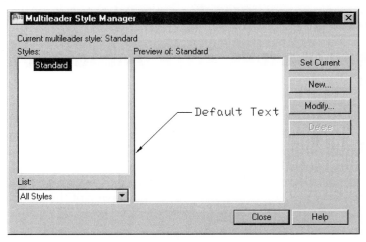

Figure 10-38 The **Multileader Style Manager** *dialog box*

From the **Multileader Style Manager** dialog box, choose the **New** button to display the **Create New Multileader Style** dialog box, see Figure 10-39. Enter the multileader style name in the **New style name** text box and then select a style from the **Start with** drop-down list on which you want to base your current style. Select the **Annotative** check box to specify that the new dimension style should be annotative. Click on the **i** (information) button to get the help and information about the annotative objects. Choose the **Continue** button to display the **Modify Multileader Style** dialog box in which you

Figure 10-39 The **Create New Multileader Style** *dialog box*

can define the new style. After defining the new style, choose the **OK** button. The parameters of the **New Dimension Style** dialog box are discussed in the next section.

In the **Multileader Style Manager** dialog box, the current multileader style name is shown in front of the **Current multileader style** option and is also highlighted in the **Styles** list box. The **Multileader Style Manager** dialog box also has a **Preview of** window that displays a preview of the current multileader style. A style can be made current (restored) by selecting the name of the multileader style that you want to make current from the list of defined multileader styles and choosing the **Set Current** button. You can also make a style current by double-clicking on the style name in the **Styles** list box. The **Multileader Style Control** drop-down list in the **Multileaders** panel of the **Ribbon** also displays the multileader styles. Select the required multileader style from this list to set it as current. The list of multileader styles displayed in the **Styles** list box depends on the option selected from the **List** drop down-list. If you select the **Styles in use** option, only the multileader styles in use will be listed in the **Styles** list box. If you right-click on a style in the **Styles** list box, a shortcut menu will be displayed to provide you with the options such as **Set current**, **Modify**, **Rename**, and **Delete**. Choose the **Modify** button to display the **Modify Multileader Style** dialog box in which you can modify an existing style. Choose the **Delete** button to delete the selected multileader style that has not been used in the drawing.

MODIFY MULTILEADER STYLE DIALOG BOX

The **Modify Multileader Style** dialog box can be used to specify the multileader attributes (variables) that affect various properties of the multileader. The various tabs provided under the **Modify Multileader Style** dialog box are discussed next.

Leader Format

The options in the **Leader Format** tab (Figure 10-40) of the **Modify Multileader Style** dialog box are used to specify the multileader attributes that affect the format of the multileader lines. For example, the appearance and behavior of the multileader lines and the arrow head can be changed with this tab.

General Area

This area provides you the options for controlling the display of the multileader lines. These options are discussed next.

Type. This option is used to specify the type of lines to be used for creating the multileaders. You can select the **Straight**, **Spline**, or **None** option from the **Type** drop-down list. Select the **None** option to create the multileader with no leader lines. You will only be able to draw the content of that multileader. The default line type for the multileaders is **Straight**.

Color. This drop-down list is used to set the colors for the leader lines and arrowheads. The multileader arrowheads have the same color as the multileader lines because the arrows are a part of the multileader lines. The default color for the multileader lines and arrows is **ByBlock**. Select the **Select Color** option from the **Color** drop-down list to display the **Select Color** dialog box from where you can choose a specific color.

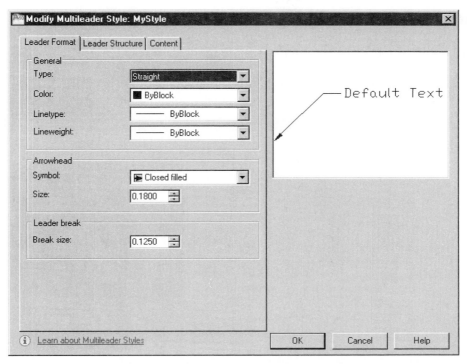

*Figure 10-40 The **Leader Format** tab of the **Modify Multileader Style** dialog box*

Linetype. This drop-down list is used to set the linetype for the multileader lines.

Lineweight. This drop-down list is used to specify the lineweight for the multileader lines. You can select the required lineweight from this drop-down list. The default value is **ByBlock**. Note that you cannot assign the lineweight to the arrowheads using this drop-down list.

Arrowhead Area
This area provides you the options for controlling the shape and size of the arrowhead. The options in this area are discussed next.

Symbol. When you create a multileader, AutoCAD draws the terminator symbol at the starting point of the multileader line. This terminator symbol is generally referred to as the arrowhead, and it represents the beginning of the multileader. AutoCAD provides you nineteen standard termination symbols that can be selected from the **Symbol** drop-down list of the **Arrowhead** area. In addition to this, you can create your own arrows or terminator symbols.

Size. This spinner is used to specify the size of the arrowhead.

Leader break Area
The **Break size** spinner in this area is used to specify the break length in the multileader while applying the dimension breaks. The default value of **Break size** is 0.125.

Leader Structure Tab

The options in the **Leader Structure** tab (Figure 10-41) of the **Modify Multileader Style** dialog box are used to specify the dimensioning attributes that affect the structure of the multileader lines. The attributes that can be controlled by using the **Leader Structure** tab are the number of lines to be drawn before adding the content, adding landing before the content, length of the landing line, multiline to be annotative or not, and so on.

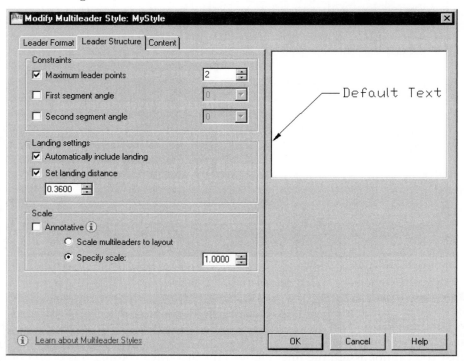

*Figure 10-41 The **Leader Structure** tab of the **Modify Multileader Style** dialog box*

Constraints Area

This area provides you the options to control the number of multileader points and the direction of the multileader.

Maximum leader points. This check box is used to specify the maximum number of points in the multileader line. The default value is **2**, which means that there will be only two points in the leader line. Select the check box to specify the maximum number of points. You can change the value of the maximum leader points in the spinner in front of this option. Note that the start point of the multileader is the first multileader point and it should also be included in the counting.

First segment angle. This check box is used to specify the angle of the first multileader line from the horizontal. Select this check box to specify a value in the drop-down list displayed in front of it.

Second segment angle. This check box is used to specify the angle of the second multileader line from the horizontal. Select this check box to specify a value in the drop-down list displayed in front of it.

Note

*The angle values for the **First segment angle** and the **Second segment angle** can be in multiples of the specified angle in the respective spinners.*

Landing settings Area

This area provides you the options to control the inclusion of landing line and its length.

Automatically include landing. Select this check box to attach a landing line to the multileader. By default, this check box is selected and is used to attach the landing line to the multileaders. **Set landing distance**. By default, this check box is selected and used to specify the length of the landing line to be attached with the multileader. You can specify the value in the spinner that is below this check box. This spinner gets highlighted only when you select the **Set landing distance** check box. The default value of the landing length is set to 0.36.

Scale Area

The options under this area are used to set the value for the overall multileader scale or the scale of the multileader in the paper space.

Annotative. This check box is used to set the selected multileader style to be annotative. With this option, you can also convert an existing non-annotative multileader style to annotative and vice-versa. If you select the **Annotative** check box from the **Create New Multileader Style** dialog box, then the **Annotative** check box in the **Modify Multileader Style** dialog box will also be selected by default.

Scale multileaders to layout. If you select the **Scale multileaders to layout** radio button, the scale factor between the current model space viewport and the floating viewport (paper space) will be computed automatically. Also, you can disable the **Specify scale** spinner (it is disabled in the dialog box) and the overall scale factor is set to 0 by selecting this radio button. When the overall scale factor is assigned a value of 0, AutoCAD calculates an acceptable default value based on the scaling between the current model space viewport and the paper space. AutoCAD sets the overall scale factor to 1 if you are in the paper space or not using the **Scale multileaders to layout** feature. Otherwise, AutoCAD calculates a scale factor that makes it possible to plot the multileaders at the values in which they have been previously set. For further details regarding the model space and layouts, refer to Chapter 12.

Specify scale. All the current multileaders are scaled with a value specified in the **Specify scale** spinner. You can alter the scaling factor as per your requirement by entering the scaling factor of your choice in this spinner. The default value for this variable is **1.0**. With this value, AutoCAD assumes its preset value and the drawing is plotted in a full scale. The scale factor is the reciprocal of the drawing size. So, for plotting the drawing at the one-fourth size, the overall scale factor will be 4.

Chapter 10

Content Tab

The options in the **Content** tab (Figure 10-42) of the **Modify Multileader Style** dialog box are used to specify the multileader attributes that affect the content and the format of the text or block to be attached with the multileader.

You can attach the multiline text or block to the multileader. You can also attach nothing as a content to the multileader. Select the desired multileader type from the **Multileader type** drop-down list; the other options in the **Content** tab of the **Modify Multileader Style** dialog box will vary with the change in the multileader type. Figure 10-42 displays the options in the **Modify Multileader Style** dialog box for the **Mtext** multileader type and these options are discussed next.

Text options Area

The options in this area are discussed next.

Default text. This option is used to specify the default text to be attached with the multileader. Choose the **[...]** button; the **In-Place Text Editor** will be displayed. Enter the text that you want to display with the multileader by default and then, click outside the text editor window; the preview window in the **Content** tab will display the multileader with the default text attached to it. While creating a new multileader, AutoCAD prompts you to specify whether you want to retain the default text specified or overwrite it with the new text.

Text style. This drop-down list is used to select the text style to be used for writing the content of the multileader. Only the predefined and default text styles are displayed in the drop-down list.

Text angle. This drop-down list is used to control the rotation of the multiline text with respect to the landing line.

Text color. This drop-down list is used to select the color to be used for writing the content of the multileader.

Text height. This spinner is used to specify the height of the multiline text to be attached with the multileader.

Always left justify. Select this check box to left justify the text attached to the multileader. On clearing the **Always left justify** check box, a multileader with the multileader style shown in Figure 10-43 (a) is displayed. Whereas, on selecting this check box, a multileader with the multileader style as shown in Figure 10-43 (b) is displayed.

Frame text. Select this check box, it will enable you to draw a rectangular frame around the multiline text to be attached with the multileader, see Figure 10-44.

Leader connection Area

The options provided in this area are used for controlling the attachment of the multiline text to the multileader.

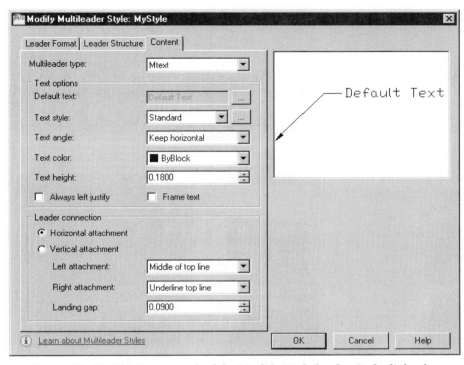

*Figure 10-42 The **Content** tab of the **Modify Multileader Style** dialog box*

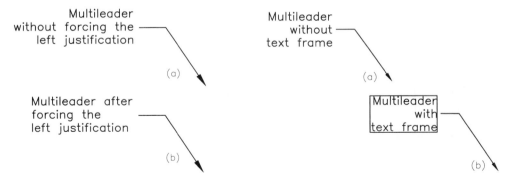

*Figure 10-43 Multileaders displaying the effect of the **Always left justify** option*

*Figure 10-44 Multileaders displaying the effect of the **Frame text** option*

Left attachment. If you draw a multileader from the right to the left, then the option selected from this drop-down list will be used to attach the multiline text with the landing line.

Right attachment. If you draw a multileader from the left to the right, then the option selected from this drop-down list will be used to attach the multiline text with the landing line.

Chapter 10

Landing gap. This spinner is used to specify the gap between the landing line and the multiline text at the point of attachment.

If you select **Block** from the **Multileader type** drop-down list, the options will change accordingly. Figure 10-45 displays the options in the **Modify Multileader Style** dialog box for the **Block** option of the **Multileader type** drop-down list and these options are discussed next.

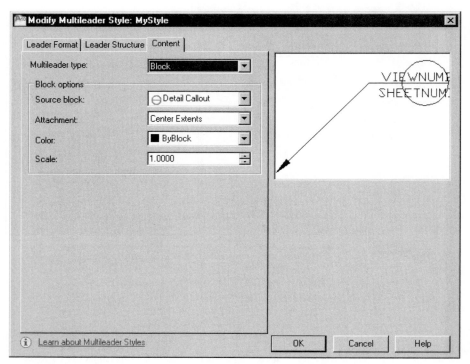

Figure 10-45 *The* ***Content*** *tab of the* ***Modify Multileader Style*** *dialog box for the* ***Block*** *option of the* ***Multileader type*** *drop-down list*

Block options Area

Source block. This drop-down list is used to select the block to be attached with the multileader as content. You can attach the standard block listed in the drop-down list or attach the user defined blocks to the multileader by selecting **User Block** from the drop-down list. Six standard shapes of block are available in this drop-down list with their previews on the side of block name. While creating the multileader, if you select the **Detail Callout** option from the **Source block** drop-down list, you will be prompted to specify the **View number** and **Sheet number** to be displayed within the block in stacked form. For the other remaining standard blocks, you will be prompted to specify the **Tag number** only to be displayed within the block. To attach the user-defined block to the multileader, select **User Block** from the **Source block** drop-down list; the **Select Custom Content Block** dialog box will be displayed with the list of all the blocks defined in the current drawing under the **Select from Drawing Blocks** drop-down list. Select the desired block and choose the **OK**

button; the selected block will be added to the **Source block** drop-down list for the current drawing. Also, it will be attached to the multileader as content.

Attachment. This drop-down list is used to specify the point of attachment of the block to the leader. Select **Center Extents** from the drop-down list to attach the leader landing line in the middle of the outer boundary of the block. Select **Insertion point** to attach the leader landing line to the insertion point specified while defining the block.

Color. This drop down list is used to specify the color of the block to be attached with the leader landing line.

Self-Evaluation Test

Answer the following questions and then compare them to those given at the end of this chapter:

1. You can invoke the **Dimension Style Manager** dialog box using both the **Annotate** tab and the **Home** tab of the **Ribbon**. (T/F)

2. The size of the arrow block is determined by the value stored in the **Arrow size** edit box. (T/F)

3. A block can be set as default content to be attached with the multileader. (T/F)

4. The size of the tolerance text with respect to dimensions can be defined. (T/F)

5. The **DIMTVP** variable is used to control the _____ position of the dimension text.

6. When you select the **Arrows** option from the **Fit** tab of the **Dimension Style** dialog box, AutoCAD places the text and arrowheads _____.

7. A basic dimension text is dimension text with a _____ drawn around it.

8. The **Suppress** check boxes in the **Dimension Lines** area control the display of _____ and _____.

9. You can specify the tolerance using _____ methods.

10. The _____ button in the **Dimension Style Manager** dialog box is used to override the current dimension style.

Review Questions

Answer the following questions:

1. You cannot replace the default arrowheads at the end of dimension lines. (T/F)

2. When the **DIMTVP** variable has a negative value, the dimension text is placed below the dimension line. (T/F)

3. The length of the multileader landing line cannot be changed. (T/F)

4. The named dimension style associated with the dimension being updated by overriding is not updated. (T/F)

5. Which of the following buttons can be used to make a dimension style active for dimensioning?

    (a) **Set Current**   (b) **New**
    (c) **Override**   (d) **Modify**

6. Which of the following tabs in the **Dimension Style Manager** dialog box is used to add a suffix **mm** to the dimensions?

    (a) **Fit**   (b) **Text**
    (c) **Primary Units**   (d) **Alternate Units**

7. Which tab of the **Dimension Style Manager** dialog box will be used, if you want to place the dimension text manually every time you create a dimension?

    (a) **Fit**   (b) **Text**
    (c) **Primary Units**   (d) **Alternate Units**

8. The size of the _____ is determined by the value stored in the **Arrow size** edit box.

9. When **DIMSCALE** is assigned a value of _____, AutoCAD calculates an acceptable default value based on the scaling between the current model space viewport and the paper space.

10. If you use the **DIMCEN** command, a positive value will create a center mark, whereas a negative value will create a _____.

11. If you select the _____ check box, you can position the dimension text anywhere along the dimension line.

12. You can append a prefix to the dimension measurement by entering the desired prefix in the **Prefix** edit box of the _____ dialog box.

13. If you select the **Limits** tolerance method from the **Method** drop-down list, AutoCAD _____ the upper value to the dimension and _____ the lower value from the dimension text.

14. You can also use the _____ command to override a dimension value.

15. What is the dimension style family, and how does it help in dimensioning? _____
_____ .

Exercises

Exercises 4 through 9 *Mechanical*

Create the drawings shown in Figures 10-46 through 10-51. You must create dimension style
and multileader style files and specify the values for the different dimension types such as
linear, radial, diameter, and ordinate. Assume missing dimensions.

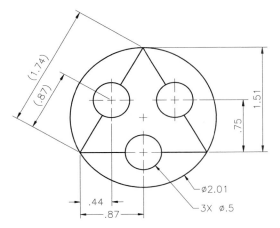

Figure 10-46 *Drawing for Exercise 4*

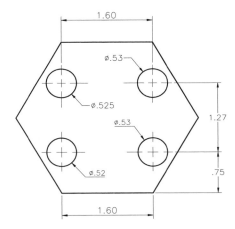

Figure 10-47 *Drawing for Exercise 5*

Figure 10-48 *Drawing for Exercise 6*

HOLE	X	Y	Z
DIA.	.25	.3	.40

Figure 10-49 *Drawing for Exercise 7*

HOLE	X1	X2	X3	Y1	Y2	Z1
DIM.	R.2	R.2	R.2	R.1	R.1	R.15
QTY.	1	1	1	1	1	1
X	.25	1.375	2.50	.75	2.0	1.375
Y	.25	.25	.25	.75	.75	1.125
Z	THRU	THRU	THRU	1.0	1.0	THRU

Figure 10-50 *Drawing for Exercise 8*

Figure 10-51 *Drawing for Exercise 9*

SPANNER NO.	A	B	C	D
S1	.85	.65	.50	.38
S2	1.00	.75	.59	.44
S3	1.15	.88	.67	.52
S4	1.25	.95	.74	.56

Exercise 10   *Mechanical*

Draw the sketch shown in Figure 10-52. You must create the dimension style and multileader style. Specify different dimensioning parameters. Also, suppress the leading and trailing zeros in the dimension style. Assume missing dimensions.

Figure 10-52 *Drawing for Exercise 10*

Exercise 11   *Mechanical*

Draw the sketch shown in Figure 10-53. You must create the dimension style and specify different dimensioning parameters in the dimension style. Assume missing dimensions.

Chapter 10

Figure 10-53 *Drawing for Exercise 11*

Exercise 12   *Mechanical*

Draw the sketch shown in Figure 10-54. You must create the dimension style and specify different dimensioning parameters in the dimension style. Assume missing dimensions.

Figure 10-54 *Drawing for Exercise 12*

Problem-Solving Exercise 1   *Mechanical*

Create the drawing shown in Figure 10-55. You must create the dimension style and specify different dimensioning parameters in the dimension style. Also, suppress the leading and trailing zeros in the dimension style. Assume missing dimensions.

Figure 10-55 *Drawing for Problem-Solving Exercise 1*

Problem-Solving Exercise 2   *Mechanical*

Draw the shaft shown in Figure 10-56. You must create the dimension style and specify the dimensioning parameters based on the given drawing. Assume missing dimensions.

Figure 10-56 Drawing for Problem-Solving Exercise 2

Problem-Solving Exercise 3   *Mechanical*

Draw the connecting rod shown in Figure 10-57. You must create the dimension style and specify the dimensioning parameters based on the given drawing. Assume missing dimensions.

Figure 10-57 Drawing for Problem-Solving Exercise 3

Problem-Solving Exercise 4   *Architectural*

Draw the elevation of the house shown in Figure 10-58. Assume missing dimensions.

Figure 10-58 *Drawing for Problem-Solving Exercise 4*

Problem-Solving Exercise 5   *Architectural*

Create the drawing shown in Figure 10-59. Assume missing dimensions.

Figure 10-59 Drawing for Problem-Solving Exercise 5

Problem-Solving Exercise 6   *Architectural*

Create the drawing shown in Figure 10-60. Assume missing dimensions.

Chapter 10

Figure 10-60 Drawing for Problem-Solving Exercise 6

Answers to Self-Evaluation Test
1. T, **2.** T, **3.** T, **4.** T, **5.** vertical, **6.** inside, **7.** frame, **8.** first, second dimension lines, **9.** four, **10.** Override

Chapter *11*

Adding Constraints to Sketches

Learning Objectives

After completing this chapter, you will be able to:

- *Add geometric constraints to sketches.*
- *Control the display of constraints.*
- *Apply constraints automatically.*
- *Apply dimensional constraints to sketches.*
- *Convert a dimensional constraint into an annotational constraint.*
- *Understand the concept of a fully-defined sketch.*
- *Control the display of dimensional constraints.*
- *Work with equations.*

INTRODUCTION

In AutoCAD 2010, a set of new tools has been introduced that helps you create parametric drawings. These tools are available in the **Parametric** tab of the **Ribbon** as well as in the **Parametric** toolbar. The parametric nature of a drawing implies that the shape and size of a geometry can be defined using standard properties or parameters. The main function of parametric property is to drive a new size or shape to the selected geometry without considering its original dimensions. Also, you can modify the shape and size of any entity of a sketch at any stage of the design process. This property makes the designing process very easy.

To make a drawing parametric, you need to apply constraints. When you apply constraints to sketches, it restricts their degrees of freedom, thereby making them stable. The stability ensures that the size, shape, and location of sketches do not change unexpectedly with respect to surroundings. You can apply two types of constraints to sketches, geometric and dimensional. Geometric constraints are applied between two entities in a sketch and they control the shape of a drawing, whereas dimensional constraints are applied to a single entity and they control the size of a drawing. Remember that dimensions applied and dimensional constraints applied are different. You will learn about the difference in this chapter.

If you need to draw a sketch of parts such as bolt, nut, valves, and so on in which the size of one entity is driven by one basic dimension, then in AutoCAD 2010, you can draw these parts quickly and easily using the **Parameters Manager**.

ADDING GEOMETRIC CONSTRAINTS*

AutoCAD provides you with twelve types of geometric constraints. Usually, these constraints are applied between two entities in a sketch and control the shape of a drawing. However, there are constraints, such as Fix, that are applied to a single entity. Whenever you apply a geometric constraint between two entities, the entity, which is selected second will be affected. For example, if you apply the parallel constraint between two lines, the line selected second will orient toward the first line. It is always recommended to first apply geometric constraints so that the shape of the sketched entity does not change on applying dimensional constraints. The different geometric constraints and the procedure to apply them are discussed next.

Applying the Horizontal Constraint

Ribbon:	Parametric > Geometric > Horizontal
Menu Bar:	Parametric > Geometric Constraints > Horizontal

The **Horizontal** constraint forces a selected line, polyline, or two points to become horizontal. To apply this constraint using the Command window, enter **GEOMCONSTRAINT** at the Command prompt. The prompt sequence to apply this constraint is given next.

Command: **GEOMCONSTRAINT**

Enter constraint type
[Horizontal/Vertical/Perpendicular/PArallel/Tangent/SMooth/Coincident/CONcentric
/COLlinear/Symmetric/Equal/Fix] <Horizontal>: [Enter]
Select an object or [2Points] <2Points>: *Select a line or polyline.*

Remember that if you apply this constraint to a line on which no other constraint is applied, then the line will orient such that the endpoint nearer to the location you clicked on will become the fixed point and the other endpoint will become horizontal, thereby making the line horizontal.

If you need to make two points horizontal, press ENTER when the **Select an object or [2Points] <2Points>** prompt is displayed. The following prompt sequence will be displayed:

Select an object or [2Points] <2Points>: [Enter]
Select first point: *Select the first point*
Select second point: *Select the second point*

You can also apply this constraint by choosing the **Horizontal** button from the **Geometric** panel of the **Parametric** tab in the **Ribbon**.

Applying the Vertical Constraint

Ribbon:	Parametric > Geometric > Vertical
Menu Bar:	Parametric > Geometric Constraints > Vertical

The **Vertical** constraint forces a selected line, polyline, or two points to become vertical. To apply this constraint using the Command window, enter **GEOMCONSTRAINT** at the Command prompt. The prompt sequence to apply this constraint is given next.

Command: **GEOMCONSTRAINT**
Enter constraint type
[Horizontal/Vertical/Perpendicular/PArallel/Tangent/SMooth/Coincident/CONcentric
/COLlinear/Symmetric/Equal/Fix] <Horizontal>: *Enter V and* [Enter]
Select an object or [2Points] <2Points>: *Select a line or polyline*

Remember that if you apply this constraint to a line on which no other constraint is applied, then the line will orient such that the endpoint nearer to the location you clicked on will become the fixed point and the other endpoint will become vertical, thereby making the line vertical.

If you need to make two points vertical, press ENTER when the **Select an object or [2Points] <2Points>** prompt is displayed. The following prompt sequence will be displayed:

Select an object or [2Points] <2Points>: [Enter]
Select first point: *Select the first point*
Select second point: *Select the second point*

You can also apply this constraint by choosing the **Vertical** button from the **Geometric** panel of the **Parametric** tab in the **Ribbon**.

Note
On applying a constraint to an entity, the corresponding symbol is displayed near the entity, see Figure 11-1. If you move the cursor over the symbol, the symbol will get highlighted and a cross mark (X) will be displayed. Click on the cross mark to hide the symbol. However, if you place the cursor on the entity on which a constraint is already applied, a glyph will be displayed stating that a constraint is applied.

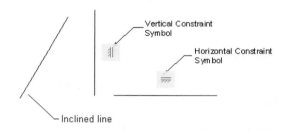

Figure 11-1 The Vertical and Horizontal Constraint symbols

Note
*You can apply a constraint by invoking a particular command using the tools in the **Parametric** tab of the **Ribbon** or by entering **GEOMCONSTRAINT** at the Command prompt and choosing the required option from the Command prompt. In this chapter, you will learn how to apply a constraint using the tools from the **Parametric** tab of the **Ribbon**. (The options in the Command prompt will be discussed, wherever necessary).*

Applying the Coincident Constraint

Ribbon:	Parametric > Geometric > Coincident
Menu Bar:	Parametric > Geometric Constraints > Coincident

The coincident constraint forces a selected point or keypoint of an entity to be coincident with the point or keypoint of another entity. The entity selected can be a line, polyline, circle, an arc or ellipse. Similarly, the selected point can be a point, or keypoints such as an endpoints or midpoint of a line or a polyline, or the centerpoint of an arc, ellipse, or a circle. To apply this constraint, choose the **Coincident** button from the **Parametric** tab of the **Ribbon**; the cursor will be replaced by a selection box along with the coincident symbol. Also, you will be prompted to select an object. Move the cursor over an entity; the keypoint that is close to the cursor will be highlighted in red. Next, click to select it; you will be prompted to select the second object to be coincident with the first object. Select the keypoint of the second entity; the keypoint of the second entity will move and coincide with that of the first entity.

Even if you have drawn the complete sketch by snapping keypoints, the coincident constraint will not be applied automatically. To apply the coincident constraints at all intersections at the same time, invoke the **Coincident** constraint tool and enter **A** at the **Select first point or [Object/Autoconstrain] <Object>:** prompt; you will be prompted to select objects. Select the objects individually or by using the selection window and press ENTER; the coincident constraint will be applied to all selected entities automatically. The Command prompt sequence to apply the **Autoconstrain** is given next.

Command: **GEOMCONSTRAINT**
Enter constraint type [Horizontal/Vertical/Perpendicular/PArallel/Tangent/SMooth/Coincident/CONcentric/COLlinear/Symmetric/Equal/Fix] <Coincident>: Enter
Select first point or [Object/Autoconstrain] <Object>: *Enter* **A** *to apply* ***Autoconstrain,*** Enter
Select objects: *Select objects individually or by using the selection window:*
Select objects: *Select more objects or* Enter

Note
The options to change the default settings for the ***Autoconstrain*** *option will be discussed later in this chapter.*

Applying the Fix Constraint

Ribbon:	Parametric > Geometric > Fix
Menu Bar:	Parametric > Geometric Constraints > Fix

The **Fix** constraint forces the selected entity to become fixed at a position. If you apply this constraint to a line or to an arc, its location will be fixed with respect to the selected keypoint. But, you can change the size of the entity by dragging its endpoints. To apply the **Fix** constraint, choose **Geometric > Fix** from the **Parametric** tab of the **Ribbon**; you will be prompted to select a point or an object. Next, move the cursor over the entity; the point closer to the cursor will be highlighted. Click to fix the highlighted keypoint. Now, you cannot move the fixed keypoint. However, you can drag and relocate other keypoints.

To fix all keypoints of an object, invoke the **Fix** constraint and press ENTER when the **Select point or [Object] <Object>** Command prompt is displayed; you will be prompted to select an object. Select the object; all its keypoints will be fixed. Now, you can drag the keypoints to resize or rotate the object, but you cannot relocate it.

Applying the Collinear Constraint

Ribbon:	Parametric > Geometric > Collinear
Menu Bar:	Parametric > Geometric Constraints > Collinear

The **Collinear** constraint forces the selected lines to lie on the same infinite line. To apply this constraint, choose **Geometric > Collinear** from the **Parametric** tab of the **Ribbon**; you will be prompted to select the first object. Select a line; you will be prompted to select the second object. Select another line; the line selected later will become collinear with the first line.

To apply the collinear constraint to multiple lines, choose the **Collinear** button; the **Select first object or [Multiple]** prompt will be displayed at the Command prompt. Enter **M** and press ENTER; you will be prompted to select the first object. Select a line; you will be prompted to select the second object. Select the lines that need to be made collinear with the line selected first and then press ENTER; the selected lines will become collinear to the line selected first.

Applying the Perpendicular Constraint

Ribbon:	Parametric > Geometric > Perpendicular
Menu Bar:	Parametric > Geometric Constraints > Perpendicular

The **Perpendicular** constraint forces the selected lines to become perpendicular to each other. To apply this constraint, choose **Geometric > Perpendicular** from the **Parametric** tab of the **Ribbon**; you will be prompted to select an object. Select a line or a polyline; you will be prompted to select the second object. Select another line or a polyline. The object selected second will become perpendicular to the object selected first. Figure 11-2 shows two lines before and after applying the perpendicular constraint.

Applying the Parallel Constraint

Ribbon:	Parametric > Geometric > Parallel
Menu Bar:	Parametric > Geometric Constraints > Parallel

The **Parallel** constraint forces the selected lines to become parallel to each other. To apply this constraint, choose **Geometric > Parallel** from the **Parametric** tab of the **Ribbon**; you will be prompted to select the first object. Select a line or a polyline; you will be prompted to select the second object. Select another line or a polyline; the object selected second will become parallel to the object selected first. Figure 11-3 shows two lines before and after applying the parallel constraint.

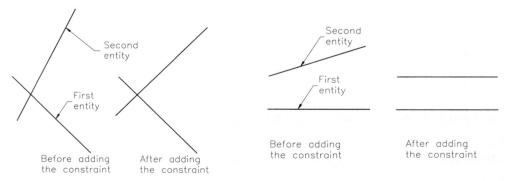

Figure 11-2 *Entities before and after applying the perpendicular constraint*

Figure 11-3 *Entities before and after applying the parallel constraint*

Applying the Concentric Constraint

Ribbon: Parametric > Geometric > Concentric
Menu Bar: Parametric > Geometric Constraints > Concentric

The concentric constraint forces the selected arc or circle to share the center point of another arc, circle, or ellipse. To apply this constraint, choose **Geometric > Concentric** from the **Parametric** tab of the **Ribbon**; you will be prompted to select the first object. Select an arc, ellipse, or a circle; you will be prompted to select the second object. Select another arc, ellipse, or circle; the object selected second will become concentric to the object selected first.

Applying the Tangent Constraint

Ribbon: Parametric > Geometric > Tangent
Menu Bar: Parametric > Geometric Constraints > Tangent

The tangent constraint forces the selected arc, circle, spline, or ellipse to become tangent to another arc, circle, spline, ellipse, or line. To apply this constraint, choose **Geometric > Tangent** from the **Parametric** tab of the **Ribbon**; you will be prompted to select an object. Select an arc, circle, spline, line, or an ellipse; you will be prompted to select the second object. Select another arc, circle, spline, ellipse, or line; the object selected second will become tangent to the object selected first. Figure 11-4 shows a line and a circle before and after applying the **Tangent** constraint. Figure 11-5 shows two arcs before and after applying the **Tangent** constraint.

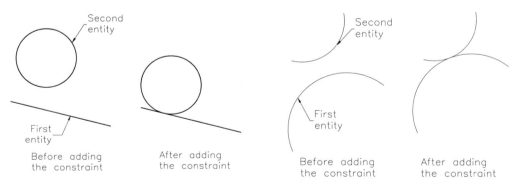

Figure 11-4 *A line and a circle before and after applying the tangent constraint*

Figure 11-5 *Two arcs before and after applying the tangent constraint*

Applying the Symmetric Constraint

Ribbon: Parametric > Geometric > Symmetric
Menu Bar: Parametric > Geometric Constraints > Symmetric

The symmetric constraint forces two selected lines, arcs, circles, or ellipses to remain equidistant from the centerline. To apply this constraint, choose **Geometric > Symmetric** from the **Parametric** tab of the **Ribbon**; you will be prompted to select the first object or two points. Select an arc, an ellipse, a line, or a circle; you will be prompted to select the second object. Select another arc, ellipse, circle, or line; you will be prompted to select the symmetry line. Select a line; the selected objects will become symmetric with respect to the selected line. You can also make two points symmetric. To do so, press ENTER at the **Select first object or (2 points)** Command prompt. Next, select two points and the symmetric line; the selected points will become symmetric to the selected line.

Applying the Equal Constraint

Ribbon:	Parametric > Geometric > Equal
Menu Bar:	Parametric > Geometric Constraints > Equal

The equal constraint forces the selected lines or polylines to have equal length, or the selected circles, arcs, or an arc and a circle to have equal radii. To apply this constraint, choose **Geometric > Equal** from the **Parametric** tab of the **Ribbon**; you will be prompted to select an object. Select the first object; you will be prompted to select the second object. Select another object; the size of the second object will become equal to that of the object selected first.

If you need to apply the equal constraint to multiple objects, choose the **Equal** button from the **Parametric** tab of the **Ribbon**; the **Select first object or [Multiple]** prompt will be displayed at the Command prompt. Enter **M** and press ENTER; you will be prompted to select the first object. Select an object; you will be prompted to select the second object. Select the other objects that need to be equal and press ENTER; the size of the objects selected later will become equal to that of the object selected first.

Applying the Smooth Constraint

Ribbon:	Parametric > Geometric > Smooth
Menu Bar:	Parametric > Geometric Constraints > Smooth

The smooth constraint is used to apply a curvature continuity between a spline and an entity connected together. The entities that can be connected to a spline include a line, spline, polyline, or an arc. To apply this constraint, choose **Geometric > Smooth** from the **Parametric** tab of the **Ribbon**; you will be prompted to select the first object. Select a spline as the first object; you will be prompted to select the second object. Select another object; the smooth constraint will be applied between the selected objects.

CONTROLLING THE DISPLAY OF CONSTRAINTS*

On applying a constraint, the symbol of that particular constraint is displayed in a constraint bar, by default. However, if you do not need the symbol of particular constraint to be displayed by default, click on the inclined arrow in the **Geometric** panel of the **Parametric** tab of the **Ribbon**; the **Constraint Settings** dialog box will be displayed, as shown in Figure 11-6. In this dialog box, the **Geometric** tab is chosen and check boxes of all constraints are selected, by default. Clear the check box of the constraints that you do not want to be displayed. You can also control the transparency of the constraint bar by setting the value in the **Constraint bar Transparency** area of this dialog box. After specifying the parameters, choose **OK**.

Figure 11-6 *The **Constraint Settings** dialog box with the* ***Geometric*** *tab chosen*

On moving the cursor over a constraint bar, the constraint symbol and entities on which the constraint was applied, will be highlighted. If a sketch has multiple constraints and all are displayed, it will be confusing. Therefore, you need to hide or display the constraint symbols. To hide a particular constraint, move the cursor over that constraint; the constraint bar will be highlighted and a cross-mark will be displayed on it. Click on the cross-mark to hide that constraint. To hide all constraints, choose the **Hide All** button in the **Geometric** panel of the **Parametric** tab of the **Ribbon**. To display all hidden constraints, choose the **Show All** button from the **Geometric** panel. If you need to display constraints of a particular entity, choose the **Show** button from the **Geometric** panel in the **Parametric** tab of the **Ribbon**; you will be prompted to select an entity. Select the entity from the drawing area and press ENTER; the constraints of that particular entity will be displayed. Remember that hiding a constraint does not delete that constraint. To delete a constraint, move the cursor over the constraint bar and press the DELETE key. If you need to delete all applied constraints, choose the **Delete Constraints** button in the **Manage** panel of the **Parametric** tab of the **Ribbon**; you will be prompted to select entities. Select the entities individually or by using the selection box, and then press ENTER; the constraints applied on the selected entities will be deleted.

Tip
If you move the cursor over a constraint bar and right-click, a shortcut menu will be displayed. Choose the corresponding options to delete or hide a constraint. Also, you can choose the ***Constraint Bar Settings*** *option to display the* ***Constraint Settings*** *dialog box.*

Example 1

Draw the sketch shown in Figure 11-7 and apply appropriate constraints to it. Do not dimension the sketch.

Figure 11-7 Sketch for Example 1

You will create the outer loop of the sketch using the **LINE** command and apply the required constraints by following the steps given below:

1. Start a new file with the *acad.dwt* template in the **2D Drafting & Annotation** workspace and draw the outer profile similar to that shown in Figure 11-7. Make sure that the endpoints are snapped so that they form a closed sketch, as shown in Figure 11-8.

2. Draw a vertical centerline that passes through the midpoint of the horizontal line at the bottom, as shown in Figure 11-9.

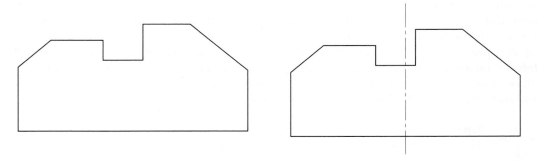

Figure 11-8 *Outer loop of the sketch*   **Figure 11-9** *Sketch after drawing the centerline*

3. Choose the **Coincident** button from the **Parametric** tab of the **Ribbon**; the cursor changes to a selection box with coincident symbol attached to it. Also, you are prompted to select the first point.

4. Move the cursor over the left vertical line such that the endpoint near the bottom horizontal line is highlighted.

5. Click to select the endpoint as the first point; you are prompted to select the second point.

6. Move the cursor over the bottom horizontal line such that its endpoint near the left vertical line is highlighted.

7. Click to select the endpoint as the second point; a blue dot is displayed at the intersection of the two lines indicating that coincident constraint has been applied. If you move the cursor near the blue dot, the coincident constraint is displayed in the constraint bar.

Applying the coincident constraint at each intersecting endpoint is a time consuming activity. Therefore, you need to apply this constraint collectively to the remaining positions.

8. Choose the **Coincident** button from the **Parametric** tab of the **Ribbon**; the cursor changes to a selection box with the symbol of this constraint attached to it. Also, you are prompted to select the first point.

9. Enter **A** at the **Select first point or [Object/Autoconstrain] <Object>:** prompt and press ENTER; you are prompted to select all objects.

10. Draw the selection box such that all entities of the sketch are selected and then press ENTER; the coincident constraint is applied at all intersections of endpoints, as shown in Figure 11-10.

11. Choose the **Perpendicular** button from the **Parametric** tab of the **Ribbon**; the cursor changes to a selection box with the symbol of this constraint attached to it. Also, you are prompted to select the first object.

12. Select the horizontal line at the bottom as the first object; you are prompted to select the second object.

13. Select the vertical line on the left as the second object; the perpendicular constraint is applied and the constraint bar is displayed.

14. Similarly, apply the perpendicular constraint between the horizontal line at the bottom and the vertical line on the right.

Chapter 11

15. Apply the perpendicular constraint between the vertical lines and the small horizontal line. The sketch after applying all perpendicular constraints is shown in Figure 11-11.

Figure 11-10 *Sketch after applying the coincident constraint*

Figure 11-11 *Sketch after applying the perpendicular constraint*

16. Choose the **Horizontal** button from the **Parametric** tab of the **Ribbon**; the cursor changes to a selection box with the symbol of the horizontal constraint attached to it. Also, you are prompted to select an object.

17. Select the horizontal line at the bottom; the horizontal constraint is applied.

18. Similarly, apply the horizontal constraint to the other three horizontal lines. The sketch after applying the horizontal constraint is shown in Figure 11-12.

    Next, you need to apply the equal constraints.

19. Choose the **Equal** button from the **Parametric** tab of the **Ribbon**; the cursor changes to a selection box with the **Equal** constraint's symbol attached to it. Also, you are prompted to select the first object.

20. Select the vertical line on the left as the first object; you are prompted to select the second object.

21. Select the vertical line on the right as the second object; the equal constraint is applied and the constraint bar is displayed.

22. Similarly, apply the equal constraint between the two inclined lines, two small vertical lines, and two small horizontal lines. The sketch after applying all constraints is shown in Figure 11-13.

    Next, you need to apply the symmetric constraint between entities on either side of the centerline.

23. Choose the **Symmetric** button from the **Parametric** tab of the **Ribbon**; the cursor changes

Figure 11-12 *Sketch after applying the horizontal constraint*

Figure 11-13 *Sketch after applying the equal constraint*

to a selection box with symmetric symbol attached to it. Also, you are prompted to select the first object.

24. Select the vertical line on the left as the first object; you are prompted to select the second object.

25. Select the vertical line on the right as the second object; you are prompted to select the symmetry line.

26. Select the centerline as the symmetry line; the vertical lines will become symmetric to the centerline and also the symmetric symbol will be displayed, as shown in Figure 11-14. Note that in this figure, other constraints are hidden for the purpose of clarity.

27. Similarly, apply the symmetric constraints between the two inclined lines and then between two small horizontal lines. The three symmetric constraints applied are shown in Figure 11-15.

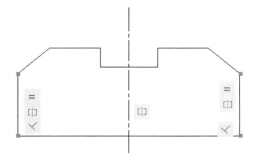

Figure 11-14 *Sketch after applying the symmetric constraint*

Figure 11-15 *Centerline with three symmetric constraints*

28. Choose the **Hide All** button from the **Parametric** tab of the **Ribbon** to hide all constraint bars. Figure 11-16 shows the sketch after hiding all constraint bars.

Next, you need to create the inner loop using the **Arc**, **Circle**, and **Line** tools at arbitrary positions. But, the profile should be similar to that given in Figure 11-17. Then, you need to apply the constraints.

1. Draw two circles of any radius inside the outer loop.

2. Draw a slot using the **Center, Start, End** option of the **ARC** tool and the **LINE** command. The sketch after drawing the circles and slot is shown in Figure 11-17. You can also draw the profile of the slot using the **PLINE** command. However, the sketch in this tutorial has been drawn using the **ARC** and **LINE** commands.

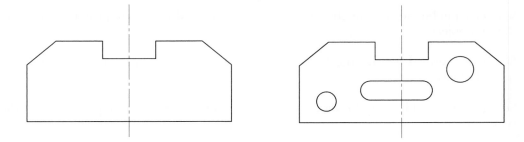

Figure 11-16 *Outer loop with all constraint bars hidden*

Figure 11-17 *Sketch after creating circles and a slot*

3. Apply the equal constraint first between two circles and then, between two arcs.

4. Apply the coincident constraint between lines and arcs in the slot.

5. Apply the symmetric constraint between the circles as well as between the arcs.

6. Choose the **Tangent** button from the **Parametric** tab of the **Ribbon**; the cursor changes to a selection box with the tangent symbol attached to it. Also, you are prompted to select the first object.

7. Select an arc; you are prompted to select the second object.

8. Select a line; the tangent constraint will be applied at one end of the line and also, the tangent symbol will be displayed, as shown in Figure 11-18.

9. Similarly, apply the tangent constraint at all other places in the slot, as shown in Figure 11-19.

Next, you need to make the centerpoint of the circles and the slot to be in a straight line.

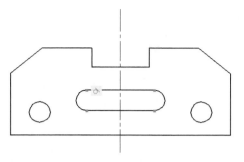

Figure 11-18 *Sketch after applying the tangent constraint*

Figure 11-19 *Sketch after applying the tangent constraint at all places of the slot*

10. Choose the **Horizontal** button from the **Parametric** tab of the **Ribbon**; the prompt sequence that follows is given next.

Command: **GEOMCONSTRAINT**
Enter constraint type
[Horizontal/Vertical/Perpendicular/PArallel/Tangent/SMooth/Coincident/CONcentric
/COLlinear/Symmetric/Equal/Fix] <Horizontal>:Horizontal
Select an object or [2Points] <2Points>: Enter
Select first point: *Select the circle on the left side* Enter
Select second point: *cen* Enter
of: *Select the center of the arc on the left side* Enter

The sketch after applying the horizontal constraint is shown in Figure 11-20.

11. Choose the **Hide All** button from the **Parametric** tab of the **Ribbon** to hide the display of all constraint bars and save the drawing. The sketch after applying the horizontal constraint is shown in Figure 11-21.

Figure 11-20 *Sketch after applying the horizontal constraint between a circle and an arc*

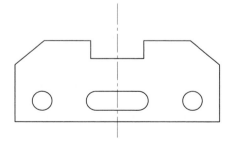

Figure 11-21 *Sketch after applying all constraints*

APPLYING CONSTRAINTS AUTOMATICALLY*

In the previous section, twelve types of constraints were discussed. Out of these, eight constraints can be applied automatically, provided you have a closed sketch. To apply constraints automatically, choose the **AutoConstrain** button in the **Geometric** panel of the **Parametric** tab of the **Ribbon**; you will be prompted to select an object. Select the entities individually or by using the selection box and then, press ENTER; all possible constraints will be applied automatically.

You can also specify the settings of the constraints that need to be applied automatically. To do so, choose the **AutoConstrain** button; the **Select objects or [Settings]** prompt will be displayed at the Command window. Enter **S** at the Command prompt and press the ENTER key; the **Constraint Settings** dialog box will be displayed, as shown in Figure 11-22.

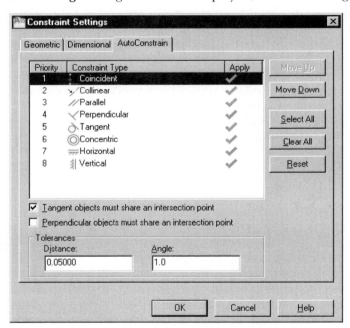

*Figure 11-22 The **Constraint Settings** dialog box with the
AutoConstrain tab chosen*

In this dialog box, the **AutoConstrain** tab is chosen by default and the eight constraints that can be applied automatically are displayed. You can rearrange these constraints according to your priorities by choosing the **Move Up** or **Move Down** button. If you do not need a particular constraint to be applied automatically to the sketch, click on the green tick mark corresponding to the constraint; the tick mark will turn white indicating that the corresponding constraints will not be applied automatically. If the **Tangent objects must share an intersection point** check box is selected, the tangent constraint will be applied, only if the two entities intersect at a point. If you clear this check box, the tangent constraint will be applied, only if the distance between the endpoints satisfies the value specified in the **Tolerances** area. Similarly, if the **Perpendicular objects must share an intersection point** check box is selected, the perpendicular constraint will be applied, only if the two entities intersect at a point. If you clear this check box, the perpendicular constraint will be applied, only if the distance between the endpoints satisfies the tolerances value specified in the **Tolerances** area.

Exercise 1

Draw the sketch shown in Figure 11-23 and apply the appropriate constraints to it. Do not dimension the sketch.

Figure 11-23 *Sketch for Exercise 1*

APPLYING DIMENSIONAL CONSTRAINTS*

The earlier versions of AutoCAD were mainly used during the documentation stage of a design. However, the introduction of dimensional constraint has made AutoCAD 2010 a complete design package. In AutoCAD 2010, you can apply dimensional constraint to an entity and change the size of the entity with ease. Therefore, now you can use AutoCAD at the initial design stage itself. Remember that the dimensions, which you apply using the tools in the **Dimensions** panel of the **Annotate** tab of the **Ribbon**, do not change the size of the entity. However, because of their associative nature, if you change the size of an entity by dragging grips, the dimension will get changed.

You can apply the horizontal, vertical, aligned, angular, radius, and diameter dimensional constraint to a sketch. The procedure to apply a dimensional constraint to an entity is similar to that of applying dimension to an entity. For example, to apply the horizontal dimensional constraint, choose **Linear > Horizontal** from the **Parametric** tab of the **Ribbon**; you will be prompted to select the first constraint point. Move the cursor near one of the keypoints; it will be highlighted. Click to select the keypoint; you will be prompted to select the second constraint point. Move the cursor near the other keypoint and click, when it gets highlighted; the dimension will be attached to the cursor and you will be prompted to place the dimension. On placing the dimension, the dimension value will be displayed in a textbox. Press ENTER to accept the value. If you need to change the value, enter a new value in the dimension box and press ENTER. A lock symbol will be displayed next to the dimension value indicating that it is a dimensional constraint. In this way, you can apply other dimensional constraints as well.

Instead of selecting keypoints to apply the dimensional constraint, you can also select the entity to be dimensioned. To do so, choose **Linear > Horizontal** from the **Parametric** tab of the **Ribbon** and press ENTER when the **Specify first constraint point or [Object] <Object>** prompt is displayed; you will be prompted to select an object. Select the object; the dimension will be attached to the cursor. Place it at the required position.

Tip
You can also enter the arithmetic functions directly in the text box that is displayed on placing a dimensional constraint. For example, if you have an arithmetic function such as (22*5.5)-30+5, which is equal to 96, you do not need to calculate this function. Just enter the function in the text box and press ENTER; AutoCAD will automatically solve the mathematical expression. However, only the mathematical expression will be displayed in the text box and not the result.

CONVERTING A DIMENSIONAL CONSTRAINT INTO AN ANNOTATIONAL CONSTRAINT*

As discussed earlier, the dimensional constraint applied and the dimensions applied are different. The dimensional constraints are applied by using the tools available in the **Dimensional** panel of the **Parametric** tab of the **Ribbon**. Whereas, dimensions are applied using the tools in the **Dimensions** panel of the **Annotate** tab of the **Ribbon**. The dimensions that are applied using the tools from the **Dimensions** panel are called annotational constraints as they are used only for annotational purpose and cannot control the size of an entity.

If you have applied a dimension (annotational constraint) to an entity, you can convert it into a dimensional constraint. To do so, choose the **Convert** button from the **Parametric** tab of the **Ribbon**; you will be prompted to select a dimension. Select the dimension; its value will be displayed in a text box with a default name. Change the dimensional value, if required, else, press ENTER; the dimension (annotational constraint) will be converted into the dimensional constraint.

You can also convert a dimensional constraint into an annotational constraint. To do so, select a dimensional constraint and invoke the **Properties** palette. Select **Annotational** from the **Constraint** field in the **Constraint** rollout; the dimensional constraint will change to the annotational constraint and the options in the **Properties** rollout will change accordingly.

You can also make a dimensional constraint as a reference constraint. A reference constraint is displayed in brackets and is driven by a dimensional or annotational constraint. To make a dimensional constraint as reference, invoke the **Properties** palette of a dimensional constraint and select **Yes** in the **Reference** field of the **Constraint** rollout. If you have changed a dimensional constraint to an annotational constraint, you can make this annotational constraint as a reference constraint using the **Properties** palette.

CONCEPT OF A FULLY-DEFINED SKETCH*

It is important for you to understand the concept of fully-defined sketches. If you have created a sketch in the design stage, you need to add the required geometrical and dimensional constraints to the sketch to constrain its degrees of freedom with respect to the surrounding. After applying the required constraints, the sketch may exist in any of the following three states:

1.   Under-defined
2.   Fully-defined
3.   Over-defined

Under-defined

An under-defined sketch is the one in which some of the geometrical constraints or dimensional constraints are not defined. As a result, the degrees of freedom of the sketch are not fully constrained, and therefore, the entities may move or change their size unexpectedly. So, you need to add some geometrical constraints or dimensional constraints to it to make it fully-defined.

Fully-defined

A fully-defined sketch is the one in which all entities and their positions are fully-defined by geometrical constraints, dimensional constraints, or both. In a fully-defined sketch, all degrees of freedom of a sketch are constrained, and therefore, the sketched entities cannot move or change their size and location unexpectedly.

Over-defined

An over-defined sketch is the one in which some of the geometrical constraints, dimensional constraints, or both are conflicting. Also, if the geometrical constraints or dimensional constraints in a sketch exceed the required number, the sketch will be over-defined. If you try to apply more geometrical constraints or dimensional constraints to a fully-defined entity, a message box will be displayed with the message that applying this constraint will over define the entity, as shown in Figures 11-24 and 11-25.

*Figure 11-24 The **Geometric Constraints** message box displayed on applying more than the required geometric constraints*

Figure 11-25 *The* **Dimensional Constraints** *message box displayed on applying more than the required dimensional constraints*

Note
If you apply a dimensional constraint to a fully-defined entity, the **Dimensional Constraints** *message box will be displayed, refer to Figure 11-25. Choose the* **Create a reference dimension** *option from the message box to convert that particular dimension into a reference dimension.*

CONTROLLING THE DISPLAY OF THE DIMENSIONAL CONSTRAINT*

On applying a dimensional constraint to an entity, a value will be displayed along with a name. If you have also applied annotational constraints to the sketch, then the sketch will be cluttered with dimensions. Therefore, you need to hide these constraints. To do so, choose the **Show Dynamic Constraints** button from the **Dimensional** panel of the **Parametric** tab in the **Ribbon**. Remember that this is a toggle button, therefore, you can choose this button again to display all dimensional constraints. To hide the annotational constraint, turn off the corresponding layer from the **Layer Properties Manager**.

When you apply a dimensional constraint to an entity, a default name and a lock symbol will be displayed along with the value. If you need to display only the value, click on the inclined arrow on the **Dimension** panel of the **Parametric** tab of the **Ribbon**; the **Constraint Settings** dialog box will be displayed with the **Dimensional** tab chosen, as shown in Figure 11-26. Specify the format of the dimensional constraint to be displayed in the **Dimension name format** drop-down list available in the **Dimensional constraint format** area of this dialog box.

If you convert a dimensional constraint into an annotational constraint, a lock symbol will be displayed. If you do not need the lock symbol to be displayed for an annotational constraint, clear the **Show lock icon for annotational constraints** check box from the **Constraint Settings** dialog box. In case, you want to view the dynamic constraints of an entity whose dimensional constraints are hidden by choosing the **Show Dynamic Constraints** button, select the **Show hidden dynamic constraints for selected objects** check box in the **Constraint Settings** dialog box. Now, if you select an entity whose dynamic constraints are hidden, the dynamic constraints will be displayed.

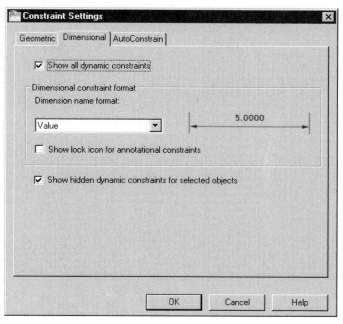

Figure 11-26 *The **Constraint Settings** dialog box with the **Dimensional** tab chosen*

Example 2

In this example, you will draw the sketch shown in Figure 11-27 and then, apply the geometric and dimensional constraints to it.

Figure 11-27 *Sketch for Example 2*

First, you will draw the profile and apply the geometrical constraints.

1. Start a new file with the *acad.dwt* template in the **2D Drafting & Annotation** workspace and draw the outer profile with dimensions close to that given in Figure 11-27. Make sure that the endpoints are snapped so that they form a closed sketch, as shown in Figure 11-28.

2. Choose the **Auto Constrain** button from the **Parametric** tab of the **Ribbon**, enter **S** at the Command prompt, and press ENTER; the **Constraint Settings** dialog box is displayed.

3. Make sure that all constraints are selected. As there is no need to apply the concentric or tangent constraints in this sketch, select the **Concentric** and **Tangent** constraints individually and move them down using the **Move Down** button.

4. Choose the **OK** button; you will be prompted to select objects. Draw a selection box, select all entities of the sketch and then, press ENTER; all possible constraints are applied to the sketch, as shown in Figure 11-29.

Note
*The constraints applied using the **AutoConstrain** command may vary depending upon the profile. In your case, if the constraints are different than that shown in Figure 11-29, apply the missing constraints.*

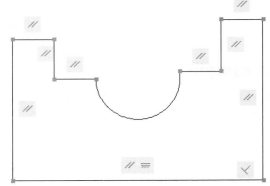

Figure 11-28 *Closed sketch drawn*   **Figure 11-29** *Constraints applied automatically*

5. Now, you need to apply the equal constraint. To do so, choose the **Equal** button from the **Geometric** panel in the **Parametric** tab and apply the equal constraint between the lines that are supposed to have equal dimensions. For dimensions, refer to Figure 11-27. The equal constraints need to be applied to the two pairs of horizontal and two pairs of vertical lines in the sketch. The sketch after applying the equal constraint is shown in Figure 11-30.

Note
If you do not want to apply the equal constraint to vertical lines, you need to apply collinear constraint to the two pairs of horizontal lines. But, after applying the equal constraint, if you apply the collinear constraint, a message box will be displayed stating that this constraint cannot be applied and conflicts the existing constraints or is over-constraining the geometry.

Next, you need to make the centerpoint of the circle and the horizontal line to be in a line.

6.   Choose the **Horizontal** button from the **Parametric** tab of the **Ribbon**; the prompt sequence that will follow is given next.

Command: **GEOMCONSTRAINT**
Enter constraint type
[Horizontal/Vertical/Perpendicular/PArallel/Tangent/SMooth/Coincident/CONcentric /COLlinear/Symmetric/Equal/Fix] <Horizontal>: **Horizontal**
Select an object or [2Points] <2Points>: Enter
Select first point: *Move the cursor over the horizontal line and click when the endpoint near the arc is highlighted* Enter
Select second point: *cen* Enter *and select the arc on the left side.*

The sketch after applying the horizontal constraint is shown in Figure 11-31.

    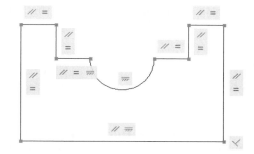

Figure 11-30 *Sketch after applying the equal constraint*

Figure 11-31 *Sketch after applying the horizontal constraint*

7.   Choose the **Hide All** button from the **Parametric** tab of the **Ribbon**; all constraint bars disappear.

Now, you need to apply dimensional constraints.

1.   Choose the **Linear** button from the **Dimensional** panel of the **Parametric** tab; the prompt sequence that will follow is given next.

Command: **DIMCONSTRAINT**
Current settings: Constraint form = Dynamic
Select associative dimension to convert or
[LInear/Horizontal/Vertical/Aligned/ANgular/Radial/Diameter/Form]
<Aligned>:_Linear
Specify first constraint point or [Object] <Object>: Enter
Select object: *Select the horizontal line at the bottom.*
Specify dimension line location: *Move the cursor down and place the horizontal dimension at the required location. Enter **60** in the text box.*

2. Press the ENTER key or choose the **Linear** button from the **Dimensional** panel of the **Parametric** tab; the prompt sequence that will follow is given next.

    Command: _DimConstraint
    Current settings: Constraint form = Dynamic
    Select associative dimension to convert or
    [LInear/Horizontal/Vertical/Aligned/ANgular/Radial/Diameter/Form]
    <Aligned>:_Linear
    Specify first constraint point or [Object] <Object>: `Enter`
    Select object: *Select the vertical line on the left*
    Specify dimension line location: *Move the cursor on the left and place the vertical dimension at the required location. Enter* **35** *in the textbox.*

3. Similarly, apply the dimension to the horizontal and vertical lines (refer to Figure 11-27) and change the value to **10**.

4. Choose the **Radial** button from the **Dimensional** panel of the **Parametric** tab; you are prompted to select the arc or circle.

5. Select the arc and enter **10** as radius.

Now, you need to apply the dimension between the center of the arc and the vertical line on the left, as shown in Figure 11-27.

6. Choose **Linear > Horizontal** from the **Dimensional** panel of the **Parametric** tab; the prompt sequence that will follow is given next.

    Command: _DimConstraint
    Current settings: Constraint form = Dynamic
    select associative dimension to convert or
    [LInear/Horizontal/Vertical/Aligned/ANgular/Radial/Diameter/Form]
    <Radial>:_Horizontal
    Specify first constraint point or [Object] <Object>: *Select the keypoint at the top of the smaller vertical line on the left.*
    Specify second constraint point: *cen* `Enter` *and select the arc.*
    Specify dimension line location: *Place the dimension at the required location and enter* **20***; the* ***Dimensional Constraints****, message box is displayed.*

7. Choose the **Create a reference dimension** option from the message box. The value **20** is displayed in parenthesis stating that it is a reference dimension. The sketch after applying all dimensional constraints is shown in Figure 11-32.

Figure 11-32 Sketch with all dimensional constraints

WORKING WITH EQUATIONS*

Equations are mathematical relations between dimensions. Consider the sketch of a bolt. All dimensions of a bolt are driven by the base diameter. In AutoCAD 2010, if you need to draw the sketch of an object whose dimensions are driven by a base dimension, you can draw it easily by using equations in which all dimensions are equated with base dimensions. If you want to relate one or two dimensions with base dimension, you can do so by applying a dimensional constraint. But if most of the dimensions need to be represented as equations, then you need to use the **Parameters Manager**. Both these methods are discussed next.

Adding Equations while Applying Dimensional Constraints

When you apply the dimensional constraint using the tools from the **Dimensional** panel of the **Ribbon**, the dimensions are displayed along with a default name. For the next dimensional constraint to be placed, you can represent it as an equation with the name of the first constraint. For example, to create a rectangle of length l units and width w units, which is half the length, you need to follow the steps given below.

1. Draw a rectangle and apply the required geometric constraints to it.
2. Choose the **Linear** button from the **Dimensional** panel of the **Ribbon** and select one of the lines representing the length and place the dimension; a text box will be displayed.
3. Enter **10** in the text box and press ENTER; a new dimension will be displayed with the default name d1=10.0000.
4. Choose the **Linear** button from the **Dimensional** panel of the **Ribbon** and select one of the lines representing the width and place the dimension; a text box will be displayed.
5. Enter **0.5*d1** in the text box and press ENTER; the rectangle will resize and a new dimension will be displayed with the default name d2=0.5*d1.
6. Now, if you change the dimension of the length, the width will also change automatically.

Note that, if you change the default name, d1, the name in the expression d2 will also change.

Adding Equations Using Parameters Manager

Adding equations to dimensions while applying constraints may look simpler, but in actual practice, when there are more dimensions, it is quite difficult. This problem can be solved by applying equations to dimensions using the **Parameters Manager**. After applying dimensional constraints, choose the **Parameters Manager** button from the **Manage** panel of the **Parametric** tab of the **Ribbon**; the **Parameters Manager** palette will be displayed with all dimensions and expressions, as shown in Figure 11-33. The names of the dimensional constraints applied will be listed under the **Name** column in the **Dimensional Constraints** node. Also, the expression of the corresponding constraints will be listed in the **Expression** column.

Now, you can change the existing expressions or add new expressions. To do so, double-click in the corresponding **Expression** field and add a new expression. If you want to add a list of new user-defined expressions, choose the **Creates a new user parameter** button from the **Parameters Manager** palette; a new field will be added under the **User Variables** node. You can add user-defined expression in the field and specify the constraints as expression in any one of the fields under the **Dimensional Constraints** node.

Figure 11-33 *The **Parameters Manager** palette*

Example 3

In this example, you will open the sketch created in Exercise 1 of this chapter. You had applied geometrical constraints to this sketch. Now, you will apply dimensional constraints and then equate all dimensions, except the dimension 20 mm, with respect to the base length, 150 mm. To the dimension 20 mm, you will apply user-defined expressions as $d150 = 20$, $d75 = 10$, $d300 = 40$, and $d100 = 13.3$.

1. Choose the **Open** button from the **Quick Access** toolbar and open the drawing created in Exercise 1 of this chapter.

2. Make sure that you have applied all required geometrical constraints, see Figure 11-34.

Note
If you have not applied all constraints to the sketch, it may get deformed while applying dimensional constraints.

3. Choose the **Linear** button from the **Dimensional** panel of the **Parametric** tab; the prompt sequence that follows is given next.

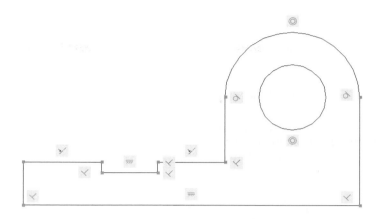

Figure 11-34 *Sketch with all constraints applied*

Command: _DimConstraint
Current settings: Constraint form = Dynamic
Select associative dimension to convert or
[LInear/Horizontal/Vertical/Aligned/ANgular/Radial/Diameter/Form]
<Aligned>:_Linear
Specify first constraint point or [Object] <Object>: Enter
Select object: *Select the horizontal line at the bottom.*
Specify dimension line location: *Move the cursor down and place the horizontal dimension at the required location. Enter* ***150*** *in the text box and press ENTER.*

4. Press ENTER or choose the **Linear** button from the **Dimensional** panel of the **Parametric** tab; the prompt sequence that follows is given next.

Command: _DimConstraint
Current settings: Constraint form = Dynamic
Select associative dimension to convert or
[LInear/Horizontal/Vertical/Aligned/ANgular/Radial/Diameter/Form]
<Aligned>:_Linear
Specify first constraint point or [Object] <Object>: Enter
Select object: *Select the vertical line on the left.*
Specify dimension line location: *Move the cursor to left side and place the vertical dimension at the required location. Enter* ***20*** *in the text box and press ENTER.*

5. Similarly, apply dimensions to the horizontal and vertical lines.

6. Choose the **Radial** button from the **Dimensional** panel of the **Parametric** tab; you are prompted to select an arc or a circle.

7. Select the arc and enter **30** as the value of radius.

8. Choose the **Diameter** button from the **Dimensional** panel of the **Parametric** tab; you are prompted to select the circle.

9. Select the circle and enter **30** as diameter.

10. Choose the **Parameters Manager** button from the **Manage** panel of the **Ribbon**; the **Parameters Manager** palette will be displayed with the names of dimensional constraints and their values, as shown in Figure 11-35.

Figure 11-35 The *Parameters Manager* palette after applying all dimensions

11. Double-click on **40** in the **Expression** column of the **Parameters Manager** palette and enter **d1/3.75**.

12. Resize the palette. Double-click in the **Description** column corresponding to the value **40** and enter **Hole location**.

13. Similarly, add the expression and description to all dimensions, except the vertical dimension 20, as given below.

Value to be selected from the Expression column	Expression to be added	Description to be added
35	d1/30*7	Slot Location
25	d1/6	Slot Width
5	d1/30	Slot Depth
30 (Diameter value)	d1/5	Hole Diameter
30 (Radius value)	d1/5	Radius

Note
*If the **Type** and **Description** columns are not added by default, right-click on the title of a column and choose the appropriate options.*

14. The **Parameters Manager** palette, after you have changed all expressions is shown in Figure 11-36.

Name	Expression	Value	Type	Description
⊟ **Dimen...**				
d1	150	150.0000	Horizontal	
d2	20	20.0000	Vertical	
d3	d1/3.75	40.0000	Vertical	Hole location
d4	d1/30*7	35.0000	Horizontal	Slot Location
d5	d1/6	25.0000	Horizontal	Slot Width
d6	d1/30	5.0000	Vertical	Slot Depth
dia1	d1/5	30.0000	Diameter	Hole Diamter
rad1	d1/5	30.0000	Radial	Radius

8 of 8 parameters displayed

*Figure 11-36 The **Parameters Manager** palette after adding all expressions*

For the value 20, you need to create a user-defined expression as follows:

15. Choose the **Create a new user parameter** button in the **Parameters Manager** palette; the **User variables** node will be added with the default name, user1 selected.

16. Change the name user1 to **d150**, the corresponding expression to **20,** and description to **If d1 = 150**.

17. Similarly, add three more user variables and their corresponding expressions and descriptions as given below.

User variable to be added	Expression to be added	Description to be added
d75	10	If d1 = 75
d300	40	If d1 = 300
d100	13.3	If d1 = 100

Note
*You can also write a suitable expression for the dimension 20 in the **Dimensional Constraints** node. For the purpose of example, the expression for the dimension 20 is given as a user variable.*

18. If the dimensional constraints are displayed, choose the **Show Dynamic Constraints** button from the **Dimensional** panel of the **Ribbon** to hide the dimensional constraints.

19. Now, change the dimensional constraints **d1** and **d2**, as specified below. Observe the change in the sketch in the drawing area and the dimension value in the **Value** column of the **Parameters Manager** palette.

$$d1 = 75, \qquad d2 = d75$$
$$d1 = 300, \qquad d2 = d300$$
$$d1 = 100, \qquad d2 = d100$$
$$d1 = 150, \qquad d2 = d150$$

Self-Evaluation Test

Answer the following questions and then compare them to those given at the end of this chapter:

1. AutoCAD provides you _____ types of geometric constraints.

2. The _____ constraints are applied between two entities in a sketch and they control the shape of the drawing.

3. Whenever you apply a geometric constraint between two entities, the entity selected first will be affected. (T/F)

4. You cannot apply the coincident constraint to all keypoints at a time. (T/F)

5. It is always recommended to first apply dimensional constraints so that the shape of the sketched entity does not change on applying geometric constraints. (T/F)

6. You can add equations while applying dimensional constraints. (T/F)

7. Which of the following buttons should be chosen in the **Dimensional** panel of the **Ribbon** to apply both horizontal and linear dimensional constraints?

   (a) **Aligned**   (b) **Linear**
   (c) **Lateral**   (d) None of the above

8. In which of the following states will the sketch exist, if only few constraints are defined?

   (a) Under-defined   (b) Fully-defined
   (c) Over-defined   (d) All of the above

9. The _____ constraint forces the selected lines to lie on the same infinite line.

10. Invoke the _____ palette to add equations to dimensional constraints.

Review Questions

Answer the following questions:

1. The _____ makes you to change or modify the shape and size of an entity of a sketch at any stage of the design process.

2. The _____ constraint is used to apply a curvature continuity between a spline and an entity connected together.

3. A _____ constraint is displayed in brackets and is driven by a dimensional or annotational constraint.

4. The dimensions that you apply using the tools in the _____ panel do not change the size of an entity.

5. The symmetric constraint forces two selected lines, arcs, circles, or splines to remain equidistant from the centerline. (T/F)

6. Using the fix constraint, you can only fix the points. (T/F)

7. You can also enter arithmetic symbols directly in the text box that is displayed on placing a dimensional constraint. (T/F)

8. If you hide the display of a constraint, the constraint applied gets deleted. (T/F)

9. How many geometric constraints can be applied automatically?

   (a) Six   (b) Ten
   (c) Eight   (d) Twelve

10. Which of the following constraints forces two lines or polylines to have equal length and the selected circles, arcs, or an arc and circle to have equal radii?

   (a) Concentric   (b) Equal
   (c) Collinear   (d) Same

Exercise 1

Draw the sketch shown in Figures 11-35 and 11-36. Then, add geometric and dimensional constraints to it. **

Figure 11-35 *Sketch for Exercise 1*

Figure 11-36 *Sketch for Exercise 1*

Answers to Self-Evaluation Test

1. twelve, **2.** Geometric, **3.** F, **4.** F, **5.** F, **6.** T, **7. Linear**, **8.** Under-defined, **9.** collinear, **10. Parameters Manager**

Chapter *12*

Model Space Viewports, Paper Space Viewports, and Layouts

Learning Objectives

After completing this chapter, you will be able to:

- *Understand the concepts of model space and paper space.*
- *Create tiled viewports in the model space using various commands.*
- *Create floating viewports in layouts using various commands.*
- *Shift from paper space to model space using the **MSPACE** command.*
- *Shift from model space to paper space using the **PSPACE** command.*
- *Control the visibility of viewport layers using the **VPLAYER** command.*
- *Set linetype scaling in paper space using the **PSLTSCALE** command.*
- *Control the display of annotative objects in viewports.*

MODEL SPACE AND PAPER SPACE/LAYOUTS

For ease in designing, AutoCAD provides two different types of environments, model space, and paper space. The paper space is also called layout. The model space is basically used for designing or drafting work. This is the default environment that is active when you start using AutoCAD. Almost the entire design is created in the model space. The other environment is the paper space. This environment is used for plotting drawings or generating drawing views for the solid models. By default, the paper space provides you with two layouts. A layout can be considered a sheet of paper on which you can place the design created in the model space and then print it. You can also assign different plotting parameters to these layouts for plotting. Almost all the commands of the model space also work in the layouts. Note that you cannot select the drawing objects created in model space when they are displayed in the viewports in layouts. However, you can snap on to the different points of the drawing objects such as the endpoints, midpoints, center points, and so on, using the **OSNAP** options.

You can shift from one environment to the other by choosing the **Model** tab or the **Layout1/ Layout2** tabs at the bottom of the drawing area. If these tabs are not available by default, right-click on the Status Bar and choose the **Paper/Model** option; the **Model** and **Layout** buttons will be available on the Status Bar. Next, right-click on any of these buttons and choose **Display Layout and Model Tabs** option from the shortcut menu; the **Model**, **Layout 1**, and **Layout 2** tabs will be displayed at the bottom of the drawing area. You can also invoke the model/layout tabs using the **Quick View Layouts** and **Quick View Drawings** buttons available at the bottom right of the Status Bar. The **Quick View Layouts** button enables a quick display of all the layouts and the model of the current drawing in a series of thumbnails. You can select a thumbnail to view the corresponding layout. Each thumbnail contains two buttons, **Plot** and **Publish**. Using these buttons, you can plot or publish the selected layout. On choosing the **Quick View Layouts** button, a control panel will be displayed along with a series of thumbnails of the existing model and layout tabs. Choose the **New Layout** button in this control panel to create a new layout. On doing so, a new thumbnail is attached to the end of series of thumbnails. The **Publish** button of the control panel enables you to publish all the layouts. You can pin the control panel for frequent use and close when not required. The shortcut menu displayed on right-clicking any thumbnail is the same as that of the **Model** and **Layout** tabs.

The **Quick View Drawings** button is similar to **Quick View Layouts**. On choosing this button, all the opened drawings with their model space and layouts are displayed. All the opened drawings are displayed in larger thumbnails whereas their layouts are displayed in smaller thumbnails. However, they are enlarged when you move the cursor over them. It allows you to quickly switch between the opened drawings and its layouts. The opened drawing thumbnails contain two buttons; **Save** and **Close**. These buttons enable you to save and close the opened drawings without actually opening them. The control panel displayed on choosing the **Quick View Drawings** button allows you to create a new drawing or open an existing one. In this case also, you can pin the control panel for frequent use, and close it, if no longer required.

You can also shift from one environment to the other using the **TILEMODE** system variable. The default value of this variable is **1**. If the value of this system variable is set to **0**, you will be shifted to the layouts and if its value is set to **1**, you will be shifted to model space. The viewports created in the model space are called tiled viewports and viewports in layouts are called floating viewports.

MODEL SPACE VIEWPORTS (TILED VIEWPORTS)

Ribbon: View > Viewports > New   **Menu Bar:** View > Viewports > New Viewports
Toolbar: Viewports > Display Viewports Dialog or Layouts > Display Viewports Dialog
Command: VPORTS

A viewport in the model space is defined as a rectangular area of the drawing window in which you can create the design. When you start AutoCAD, only one viewport is displayed in the model space. You can create multiple non-overlapping viewports in the model space that display different views of the same object, see Figure 12-1. Each of the viewports will become the individual drawing area. This is generally used while creating solid models. You can view the same solid model from different positions by creating the tiled viewports and defining the distinct coordinate system configuration for each viewport. You can also use the **PAN** or **ZOOM** command to display different portions or different levels of the detail of the drawing in each viewport. The tiled viewports can be created using the **VPORTS** command.

Figure 12-1 *Screen display with multiple tiled viewports*

Creating Tiled Viewports

As mentioned earlier, the display screen in the model space can be divided into multiple non-overlapping tiled viewports. All these viewports are created only in the rectangular shape.

This number depends on the equipment and the operating system on which AutoCAD is running. Each tiled viewport contains a view of the drawing. The tiled viewports touch each other at the edges without overlapping. While using tiled viewports, you are not allowed to edit, rearrange, or turn individual viewports on or off. These viewports are created using the **VPORTS** command when the system variable **TILEMODE** is set to 1 or the **Model** tab is active. When you choose the **Named** or **New** button from the **Viewports** panel in the **View** tab of the **Ribbon** (Figure 12-2), the **Viewports** dialog box is displayed. You can use this dialog box to create new viewport configurations and save them. The options under both the tabs of the **Viewports** dialog box are discussed next.

Figure 12-2 The ***New*** *button in the* ***Viewports*** *panel*

New Viewports Tab

The **New Viewports** tab of the **Viewports** dialog box (Figure 12-3) provides the options related to standard viewport configurations. You can also save a user-defined configuration using this tab. The name for the new viewport configuration can be specified in the **New name** edit box. If you do not enter a name in this edit box, the viewport configuration you create is not saved and, therefore, cannot be used later. A list of standard viewport configurations is listed in the **Standard viewports** list box. This list also contains the ***Active Model Configuration***, which is the current viewport configuration. From the **Standard viewports** list, you can select and apply any one of the listed standard viewport configurations. A preview image of the selected configuration is displayed in the **Preview** window. The **Apply to** drop-down list has the **Display** and **Current viewport** options. Selecting the **Display** option

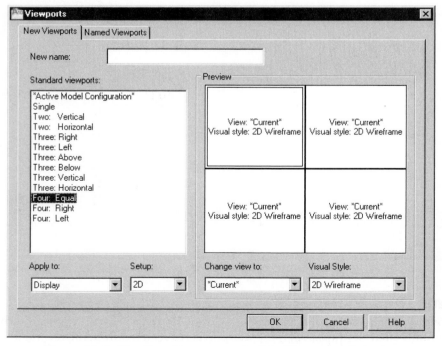

Figure 12-3 The ***New Viewports*** *tab of the* ***Viewports*** *dialog box*

applies the selected viewport configuration to the entire display and selecting the **Current viewport** option applies the selected viewport configuration to only the current viewport. With this option, you can create more viewports inside the existing viewports. The changes will be applied to the current viewport and the new viewports will be created inside the current viewport.

From the **Setup** drop-down list, you can select **2D** or **3D**. When you select the **2D** option, it creates the new viewport configuration with the current view of the drawing in all the viewports initially. Using the **3D** option applies the standard orthogonal and isometric views to the viewports. For example, if a configuration has four viewports, they are assigned the **Top**, **Front**, **Right**, and **South East Isometric** views, respectively. You can also modify these standard orthogonal and isometric views by selecting from the **Change view to** drop-down list and replacing the existing view in the selected viewport. For example, you can select the viewport that is assigned the **Top** view and then choose **Bottom** view from the **Change view to** drop-down list to replace it. The preview image in the **Preview** window reflects the changes you make. If you use the **2D** option, you can select a named viewport configuration to replace the selected one. From the **Visual Style** drop-down list, you can select the visualization mode of your model. For example, you can select a viewport and then choose **3D Hidden** from the **Visual Style** drop-down list. Now, the model will be displayed with the hidden edges invisible in this viewport. Choose **OK** to exit the dialog box and apply the created or selected configuration to the current display in the drawing. When you save a new viewport configuration, it saves the information about the number and position of viewports, the viewing direction and zoom factor, and the grid, snap, coordinate system, and UCS icon settings.

Named Viewports Tab

The **Named Viewports** tab of the **Viewports** dialog box (Figure 12-4) displays the name of the current viewport next to **Current name**. The names of all the saved viewport configurations in a drawing are displayed in the **Named viewports** list box. You can select any one of the named viewport configurations and apply it to the current display. A preview image of the selected configuration is displayed in the **Preview** window. Choose **OK** to exit the dialog box and apply the selected viewport configuration to the current display. In the **Named viewports** list box, you can select a name and right-click to display a shortcut menu. Choosing **Delete** deletes the selected viewport configuration and choosing **Rename** allows you to rename the selected viewport configuration.

MAKING A VIEWPORT CURRENT

The viewport you are working in is called the current viewport. You can display several model space viewports on the screen, but you can work in only one of them at a time. You can switch from one viewport to another even when you are in the middle of a command. For example, you can specify the start point of the line in one viewport and the endpoint of the line in the other viewport. The current viewport is indicated by a border that is heavy compared to the borders of the other viewports. Also, the graphics cursor appears as a drawing cursor (screen crosshairs) only when it is within the current viewport. Outside the current viewport this cursor appears as an arrow cursor. You can enter points and select objects only from the

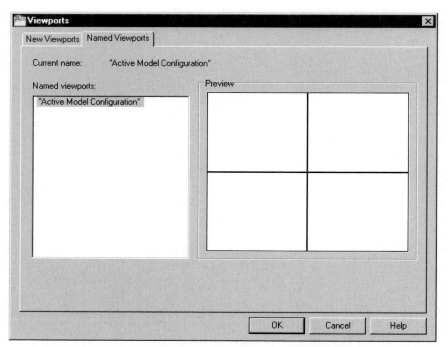

*Figure 12-4 The **Named Viewports** tab of the **Viewports** dialog box*

current viewport. To make a viewport current, you can select it with the pointing device. Another method of making a viewport current is by assigning its identification number to the **CVPORT** system variable. The identification numbers of the named viewport configurations are not listed in the display.

JOINING TWO ADJACENT VIEWPORTS

AutoCAD provides you with an option of joining two adjacent viewports. However, remember that the viewports you wish to join should result in a rectangular-shaped viewport only. As mentioned earlier, the viewports in the model space can only be in rectangular shape. Therefore, you will not be able to join two viewports, in case they do not result in a rectangular shape. The viewports can be joined by choosing **View > Viewports > Join Viewports** from the **Viewports** panel. On choosing this button, you will be prompted to select the dominant viewport. A dominant viewport is one whose display will be retained after joining. After selecting the dominant viewport, you will be prompted to select the viewport to be joined. Figure 12-5 shows the viewport configuration before joining and Figure 12-6 shows the viewport configuration after joining.

Tip
*You can also use the Command line to create, save, restore, delete, or join the viewport configurations. This is done using the **-VPORTS** command.*

Figure 12-5 *Selecting the dominant viewport and the viewport to be joined*

Figure 12-6 *Viewports after joining*

PAPER SPACE VIEWPORTS (FLOATING VIEWPORTS)

As mentioned earlier, the viewports created in the layouts are called floating viewports. This is because unlike in model space, the viewports in the layouts can be overlapping and of any shape. In layouts, there is no restriction of the shape of the viewports. You can even convert a closed object into a viewport in the layouts. Figure 12-7 shows a layout with floating viewports. The method of creating floating viewports is discussed next.

Figure 12-7 Screen display with multiple floating viewports

Creating Floating Viewports

Ribbon: View > Viewports > Named   **Menu Bar:** View > Viewports > New Viewports
Toolbar: Viewports > Display Viewports Dialog   **Command:** VPORTS

This command is used for creating the floating viewports in layouts. However, when you invoke this command in the layouts, the dialog box displayed is slightly modified. Instead of the **Apply to** drop-down list in the **New Viewports** tab, the **Viewport Spacing** spinner is displayed, see Figure 12-8. This spinner is used to set the spacing between the adjacent viewports. The rest of the options in both the **New Viewports** and the **Named Viewports** tabs of the **Viewports** dialog box are the same as those discussed under the tiled viewports. When you select a viewport configuration and choose **OK**, you will be prompted to specify the first and the second corner of a box that will act as a reference for placing the viewports. You will also be given an option of **Fit**. This option fits the configuration of viewports such that they fit exactly in the current display.

Tip
You can also use the **+VPORTS** *command to display the* **Viewports** *dialog box. When you invoke this command, you will be prompted to specify the* **Tab Index***. Enter* **0** *to display the* **New Viewports** *tab and enter* **1** *to display the* **Named Viewports** *tab.*

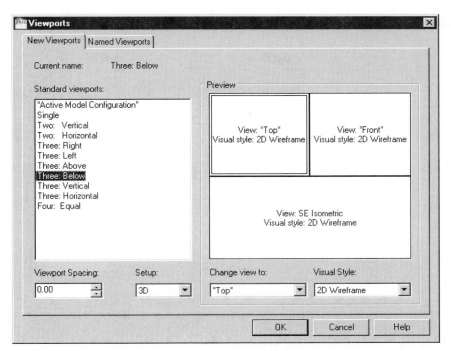

Figure 12-8 The New Viewports tab of the Viewports dialog box displayed in layouts

Creating Polygonal Viewports

Ribbon: View > Viewports > Create Polygonal   **Toolbar:** Viewports > Polygonal Viewport
Menu Bar: View > Viewports > New Viewports   **Command:** VPORTS

As mentioned earlier, you can create floating viewports of any closed shape. To create a polygonal viewport, choose **View > Viewports > Create Polygonal** from the **Ribbon**. The viewports thus created can even be self-intersecting in shape. The prompt sequence that will follow when you choose this button is given next.

> Specify start point: *Specify the start point of the viewport.*
> Specify next point or [Arc/Length/Undo]: *Specify the next point or select an option.*
> Specify next point or [Arc/Close/Length/Undo]: *Specify the next point or select an option.*

The various options in the prompt sequence are discussed next.

Arc

This option is used to switch to the arc mode for creating the viewports. When you invoke this option, the options for creating the arcs will be displayed. You can switch back to the line mode by choosing the **Line** option.

Length

This option is used to specify the length of the next line of the viewport. The line will be drawn in the direction of the last drawn line segment. In case the last drawn segment was an arc, the line will be drawn tangent to it.

Undo

This option is used to undo the last drawn segment of the polygonal viewport.

Close

This option is used to close the polygon and create the viewport. The last entity that will be used to close the polygon will depend upon whether you were in the arc mode or in the line mode. If you were in the line mode, the last entity will be a line. If you were in the arc mode, the last entity will be an arc.

Figure 12-9 shows the polygonal viewport created using the combination of lines and arcs.

Figure 12-9 A polygonal viewport

Converting an Existing Closed Object into a Viewport

Ribbon:	View > Viewports > Create Polygonal > Create from Object	
Toolbar:	Viewports > Convert Object to Viewport	Command: VPORTS

One of the major enhancements in the recent releases of AutoCAD is the option that allows you to convert an existing closed object into a viewport. However, remember that the object selected should be a single entity. The objects that can be converted into a viewport include polygons drawn using the **POLYGON** command, rectangles drawn using the **RECTANG** command, polylines (last segment closed using the **Close** option), circles, ellipses, closed splines, or regions. To convert any of these objects into a viewport, choose **View > Viewports > Create Polygonal > Create from Object** from the **Ribbon**. You will be prompted to select the object that will be converted into a viewport. Figure 12-10 shows a viewport created using a polygon of nine sides and Figure 12-11 shows a viewport created using a closed spline.

Note

*When you shift to the layouts, the **Page Setup** dialog box is displayed for printing and a rectangular viewport is created that fits the drawing area. If you want, you can retain or delete this viewport using the **ERASE** command.*

TEMPORARY MODEL SPACE

Sometimes, when you create a floating viewport in the layout, the drawing is not displayed completely inside it, see Figures 12-10 and 12-11. In such cases, you need to zoom or pan the drawings to fit them in the viewport. But when you invoke any of the **ZOOM** or the **PAN** commands in the layouts, the display of the entire layout is modified instead of the display

Figure 12-10 *A viewport created using a polygon of nine sides*

Figure 12-11 *A viewport created using a closed spline*

Chapter 12

inside of the viewport. Now, to change the display of the viewports, you will have to switch to the temporary model space. The temporary model space is defined as a state when the model space is activated in the layouts. The temporary model space is exactly similar to the actual model space and you can make any kind of modifications in the drawing from temporary model space. Therefore, the main reason for invoking the temporary model space is that you can modify the display of the drawing. The temporary model space can be invoked by choosing the **Paper** button from the status bar. You can also switch to the temporary model space by double-clicking inside the viewports. You will see that the model space UCS icon is automatically displayed when you switch to the temporary model space. Also, the extents of the viewport become the extents of the drawing. You can use the **ZOOM** and **PAN** commands to fit the model inside the viewport. The temporary model space can also be invoked using the **MSPACE** command.

Once you have modified the display of the drawing in the temporary model space, you have to switch back to the paper space. This is done by choosing the **Model** button from the status bar. You can also switch back to the paper space by double-clicking anywhere in the layout outside the viewport, or by using the **PSPACE** command.

Example 1   *Mechanical*

In this example, you will draw the object shown in Figure 12-12 and then create a floating viewport of the shape shown in Figure 12-13 to display the object in the layouts. The dimensions of the viewport are in the paper space. Do not dimension the object.

1. Start a new drawing and then draw the object shown in Figure 12-12.

2. Choose the **Layout1** tab to switch to the layout.

3. A rectangular viewport will be automatically created in this layout. Choose the **Erase**

Figure 12-12 *Model for Example 1*

Figure 12-13 *Shape for the floating viewport*

button from the **Modify** panel in the **Home** tab of the **Ribbon**. You will be prompted to select the object. Enter **L** in this prompt to delete the last object, that is, the viewport, in this case.

4. Choose the **Polygon** button from the **Draw** panel in the **Home** tab of the **Ribbon** and create the required hexagon.

5. Choose **View > Viewports > Create Polygonal > Create from Object** from the **Ribbon**; you are prompted to select an object. Select the hexagon and it will be converted into a viewport.

6. You will see that the complete object is not displayed inside the viewport. Therefore, you have to modify its display. Double-click inside the viewport to switch to the temporary model space.

7. The viewport border will become thick, suggesting that you have switched to the temporary model space. Now, using the **ZOOM** and the **PAN** commands, fit the drawing inside the viewport.

8. Choose the **Model** button from the status bar to switch back to the paper space. The drawing is now displayed fully inside the viewport, see Figure 12-14.

Figure 12-14 *Displaying the drawing inside the polygonal viewport*

EDITING VIEWPORTS

You can perform various editing operations on the viewports. For example, you can control the visibility of the objects in the viewports, lock their display, clip the existing viewports using an object, and so on. All these editing operations are discussed next.

Controlling the Display of Objects in Viewports

The display of the objects in the viewports can be turned on or off. If the display is turned off, the objects will not be displayed in the viewport. However, the object in the model space is not affected by this editing operation. To control the display of the objects, select the viewport entity in the layout and right-click to display the shortcut menu. In this menu, choose **Display Viewport Objects > No**. If there is only one viewport, you will be prompted to confirm whether you really want to turn off all active viewports. Enter **Y** to turn off the display. However, if there are more than one viewports, the visibility of the selected viewports will be automatically turned off when you choose **Display Viewport Objects > No** from the shortcut menu. Similarly, you can again turn on the display of the objects by choosing **Display Viewport Objects > Yes** from the shortcut menu. This shortcut menu is displayed on selecting the viewport and right-clicking. This editing operation can also be done using the **OFF** option of the **MVIEW** or the **-VPORTS** command.

Locking the Display of Objects in Viewports

To avoid accidental modification in the display of objects in the viewports, you can lock their display. If the display of a viewport is locked, the commands such as **ZOOM** and **PAN** do not work in it. Also, you cannot modify the view in the locked viewport. For example, if the display of a viewport is locked, you cannot zoom or pan the display or change the view in that viewport even if you switch to the temporary model space. To lock the display of the viewports, select it and right-click on it to display the shortcut menu. In this menu, choose **Display Locked > Yes**. Now, the display of this viewport will not be modified. However, you can draw objects or delete objects in this viewport by switching to the temporary model space. Similarly, you can unlock the display of the objects in the viewports by choosing **Display Locked > No** from the shortcut menu. You can also lock or unlock the display of the viewports using the **Lock** option of the **MVIEW** command or the **-VPORTS** command.

Controlling the Display of Hidden Lines in Viewports

While working with three-dimensional solid or surface models, there are a number of occasions where you have to plot the solid models such that the hidden lines are not displayed. Plotting solid models in the model space (Tilemode=1) can be easily done by selecting the **Hidden** option from the **Shade plot** drop-down list. This drop-down list is available in the **Shaded viewport options** area in **Plot** dialog box. If this area is not available by default, you need to choose the **More Options** button in the dialog box. This option is not available in layouts. In this case, you will have to control the display of the hidden lines in the viewports. To control the display of the hidden lines, select the viewport and right-click to display the shortcut menu. In this menu, choose **Shade plot > Hidden**. Although the hidden lines will be displayed in the viewports, now they will not be displayed in the printouts. The display of the hidden lines can also be controlled using the **Shadeplot** option of the **MVIEW** command or the **-VPORTS** command.

Note
*You can also use the other options in the **Shade plot** drop-down list. These options are explained in detail in Chapter 14 (Plotting Drawings).*

Tip

Apart from the previously mentioned editing operations, you can also move, copy, rotate, stretch, scale, or trim the viewports using the respective commands. You can also use the grips to edit the viewports.

Clipping Existing Viewports

Ribbon:	View > Viewports > Clip	**Command:** VPCLIP	
Toolbar:	Viewports > Clip existing viewport	**Menu Bar**: Modify > Clip > Viewport	

You can modify the shape of the existing viewport by clipping it using an object or by defining the clipping boundary. The viewports can be clipped using the **VPCLIP** command. This command can also be invoked by choosing **Clip** from the **Viewports** panel in the **View** tab of the **Ribbon**. The prompt sequence that will follow when you invoke this command is given next.

Select viewport to clip: *Select the viewport to be clipped.*
Select clipping object or [Polygonal/Delete] <Polygonal>: *Select an object for clipping the viewport or specify an option.*

Select clipping object Option

This option is used to clip the viewport using a selected closed loop. The objects that can be used for clipping the viewports include circles, ellipses, closed polylines, closed splines, and regions. As soon as you select the clipping object, the original viewport will be deleted and the selected object will be converted into a viewport. The portion of the display that was common to both the original viewport and the object selected will be displayed. You can, however, change the display of the viewport using the **ZOOM** command or the **PAN** command in the viewports. Figure 12-15 shows a viewport and an object that will be used to clip the viewport and Figure 12-16 shows the new viewport created after clipping.

Figure 12-15 Selecting the object for clipping the viewport

Figure 12-16 New viewport created after clipping

Polygonal

This option is used to create a polygonal boundary for clipping the viewports. When you invoke this option, the options for creating a polygonal boundary will be displayed. You can draw a polygonal boundary for clipping the viewport using these options.

Delete

This option is used to delete the new clipping boundary created using an object or using the **Polygonal** option. The original viewport is restored when you invoke this option, which will be available only if the viewport has been clipped at least once. If the viewport is clipped more than once, you can restore only the last viewport clipping boundary.

Maximizing Viewports

Status Bar:	Maximize Viewport

While working with floating viewports, you may need to invoke the temporary model space to modify the drawing. One of the options is that you double-click inside the viewport to invoke the temporary model space and make the changes in the drawing. But in this case, the shape and size of the floating viewport will control the area of the temporary model space. If the viewport is polygonal and small in size, you may have to zoom and pan the drawing a number of times. To avoid this, AutoCAD allows you to maximize a viewport on the screen. This provides you all the space in the drawing area to make the changes in the drawing.

To maximize a floating viewport, choose the **Maximize Viewport** button from the Status bar. The viewport is automatically maximized in the drawing area and the **Maximize Viewport** button is replaced by the **Minimize Viewport** button. If there are more than one floating viewports, two arrows will be displayed on either side of the **Minimize Viewport** button. These arrows can be used to switch to the display in the other floating viewports. After making the changes in the drawing, choose the **Minimize Viewport** button to restore the original display of the layout. When you do so, the view and the magnification in all the viewports is the same as that before maximizing them. Also, the visibility of the layers remains the same as that before maximizing the viewport.

MANIPULATING THE VISIBILITY OF VIEWPORT LAYERS

Command:	VPLAYER

You can control the visibility of layers inside a floating viewport with the **VPLAYER** or **LAYER** command. The **VPLAYER** command is used to control the visibility of layers in individual floating viewports. For example, you can use the **VPLAYER** command to freeze a layer in the selected viewport. The contents of this layer will not be displayed in the selected viewport, although in the other viewports, they will be displayed. This command can be used from either temporary model space or paper space. The only restriction is that **TILEMODE** be set to 0 (Off); that is, you can use this command only in the **Layout** tab.

Command: **VPLAYER** Enter
Enter an option [?/Color/Ltype/LWeight/PStyle/Freeze/Thaw/Reset/Newfrz/Vpvisdflt]:

? Option

You can use this option to obtain a listing of the frozen layers in the selected viewport. When you enter ?, you will be prompted to select the viewport. The AutoCAD text window will be displayed showing all the layers that are frozen in the current layer. If you invoke this option, when you are in temporary model space, AutoCAD will temporarily shift you to the paper space to let you select the viewport.

Color Option

The **Color** option is used to assign a color to all the objects in the layer (or layers) in the viewports. You can either use True colors or you can specify a Color book, if you have previously installed any. The changes can be applied on the current, selected, or all the viewports.

Ltype Option

You can use this option to specify any linetype to a layer(or layers) in the current, selected, or all the viewports.

LWeight Option

The **LWeight** option is used to specify a line width to a particular layer or layers. You can specify the width between 0.0mm to 2.11 mm. Then you are prompted to specify the name(s) of the layer(s) to assign the line width. In this case also, you can apply the changes to the current, selected, or all the viewports.

PStyle Option

The **PStyle** option is used to specify the plot style. On selecting this option, you will be prompted to specify the name of the plot style to be applied. On entering **?** AutoCAD text window will be displayed with the list of existing plot style. If you enter **Normal**, you will be prompted to specify the viewports. This option is explained in detail in later chapters.

Freeze Option

The **Freeze** option is used to freeze a layer (or layers) in one or more viewports. When you select this option, you will be prompted to specify the name(s) of the layer(s) you want to freeze. If you want to specify more than one layer, the layer names must be separated by commas. You can also select an object whose layer you want to freeze. Once you have specified the name of the layer(s), AutoCAD prompts you to select the viewport(s), to freeze the specified layer(s). You can select one or all the viewports. The layers will be frozen after you exit this command.

Thaw Option

With this option, you can thaw the layers that have been frozen in the viewports using the

VPLAYER command or the **LAYER** command. Layers that have been frozen, thawed, turned on, or turned off globally are not affected by **VPLAYER Thaw**. For example, if a layer has been frozen, the objects on it are not regenerated on any viewport, even if **VPLAYER Thaw** is used to thaw that layer in any viewport. If you want to thaw more than one layer, they must be separated by commas. You can thaw the specified layers in the current, selected, or all the viewports.

Reset Option

With the **Reset** option, you can set the visibility of the layer(s) in the specified viewports to their current default setting. The visibility defaults of a layer can be set by using the **Vpvisdflt** option of the **VPLAYER** command. When you invoke the **Reset** option, you will be prompted to specify the names of the layers to be reset. You can reset the layers in the current viewport, in selected viewports, or in all the viewports.

Newfrz (New Freeze) Option

With this option, you can create new layers that are frozen in all the viewports. This option is used mainly where you need a layer that is visible only in one viewport. This can be accomplished by creating the layer with the **Newfrz** option and then thawing it in the viewport where you want to make the layer visible. On invoking this option, you will be prompted to specify the name(s) of the new layer(s) that will be frozen in all the viewports. To specify more than one layer, separate the layer names with commas. After you specify the name(s) of the layer(s), AutoCAD creates frozen layers in all viewports. Also, the default visibility setting of the new layer(s) is set to Frozen; therefore, if you create any new viewports, the layers created with **VPLAYER Newfrz** are also frozen in them.

Vpvisdflt (Viewport Visibility Default) Option

With this option, you can set a default for the visibility of the layer(s) in the subsequent new viewports. When a new viewport is created, the frozen/thawed status of any layer depends on the **Vpvisdflt** setting for that particular layer. When you invoke this option, you will be prompted to specify the names of the layer(s) whose visibility is to be changed. After specifying the name(s), you will be prompted to specify whether the layers should be frozen or thawed in the new viewports.

CONTROLLING THE LAYERS IN VIEWPORTS USING THE LAYER PROPERTIES MANAGER DIALOG BOX

You can use the **Layer Properties Manager** dialog box to control the layer display properties in viewports. When you open the **Layer Properties Manager** dialog box from a floating viewport, some extra properties are added to it, see Figure 12-17. These properties are used to override the global property of the layer in a particular viewport, while retaining the global layer properties in other floating viewports and the model space. These properties are discussed next.

Chapter 12

Figure 12-17 *Controlling the visibility of layers in the viewports using the **Layer Properties Manager** dialog box*

VP Freeze

When the **TILEMODE** option is turned off, you can freeze or thaw the selected layers in the current floating viewport by selecting the **VP Freeze** option. You can freeze a layer in the current floating viewport if it is thawed in the model space but you cannot thaw a layer in the current viewport if it is frozen in the model space. The frozen layers still remain visible in other viewports.

VP Color

The swatch under the **VP Color** column is used to change the display color of the objects in the selected layer in the current floating viewport. In all other floating viewports and model space, the color of the objects remains unaffected.

VP Linetype

The field under the **VP Linetype** column is used to change the linetype defined for the selected layer in the active floating viewport. In all other floating viewports and model space, the linetype remains unaffected.

VP Lineweight

The field under the **VP Lineweight** column is used to override the lineweight defined for the selected layer in the active floating viewport. In all other floating viewports and the model space, the lineweight remains unaffected.

VP Plot Style

The field under the **VP Plot Style** column is used to override the plot style settings assigned to the selected layer in the active floating viewport. The plot style for all other floating viewports and model space remains unaffected. You cannot override the plot styles for the color dependent plot styles. The override for the plot style also does not affect the plot if the visual style is set to **Realistic** or **Conceptual**.

 Note
*By default, all the overrides that you have defined for the current active viewport will get highlighted in light blue color in the **Layer Properties Manager** dialog box. Also, the icon of the selected layer under the **Status** column will change to inform you that some of the properties of the selected layer have been overridden by the new one.*

Removing Viewport Overrides

To remove the viewport overrides defined for the active viewports, open the **Layer Properties Manager** dialog box in the active viewport and right-click on the layer property from which you want to remove the viewport override; the shortcut menu will be displayed. Next, choose the **Remove Viewport Overrides for** option to display the cascading menu. You can make a choice to remove the override for a single property / all properties of the selected layer / all layers in the current viewport / all viewports, see Figure 12-18. You can also remove the layer property overrides for all the layers of the selected viewports from the paper space. To do so, make the paper space current; select the viewports boundaries of the viewports for which you want to remove the override and choose the **Remove Viewport Overrides for All Layers** option from the shortcut menu.

*Figure 12-18 Suboptions in the **Remove Viewport Override for** option of the **Layer** shortcut menu*

 Note
*For more information about the **Layer Properties Manager** dialog box, see Chapter 4.*

PAPER SPACE LINETYPE SCALING (PSLTSCALE SYSTEM VARIABLE)

By default, the linetype scaling is controlled by the **LTSCALE** system variable. Therefore, the display size of the dashes depends on the **LTSCALE** factor, on the drawing limits, or on the drawing units. If you have different viewports with different zoom (XP) factors, the size of the dashes will be different for these viewports. Figure 12-19 shows three viewports with different sizes and different zoom (XP) factors. You will notice that the dash length is different in each of these three viewports..

Generally, it is desirable to have identical line spacing in all viewports. This can be achieved with the **PSLTSCALE** system variable. By default, **PSLTSCALE** is set to 0. In this case, the size of the dashes depends on the **LTSCALE** system variable and on the zoom (XP) factor of the viewport where the objects have been drawn. If you set **PSLTSCALE** to 1 and **TILEMODE** to 0, the size of the dashes for objects in the model space are scaled to match the **LTSCALE**

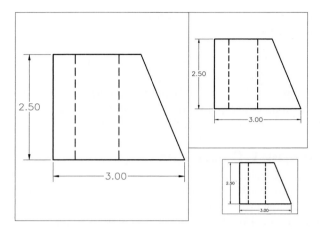

Figure 12-19 *Varying sizes of the dashed lines with*
PSLTSCALE *= 0*

of objects in the paper space viewport, regardless of their zoom scale. In other words, if **PSLTSCALE** is set to 1, even if the viewports are zoomed to different sizes, the length of the dashes will be identical in all viewports. Figure 12-20 shows three viewports with different sizes. Notice that the dash length is identical in all the three viewports.

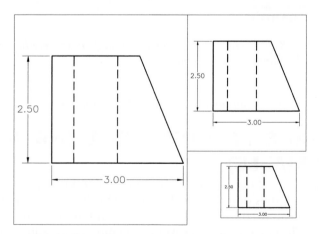

Figure 12-20 *Similar sizes of the dashed lines with*
PSLTSCALE *= 1*

Tip
*You can control the scale factor for displaying the objects in the viewports using the **Viewport Scale Control** drop-down list. This drop-down list is available in the **Viewports** toolbar.*

INSERTING LAYOUTS

Status Bar: Quick View Layouts > New Layout   **Command:**   LAYOUT
Insert: Insert > Layouts > New Layout   **Toolbar:**   Layouts > New Layout

 The **LAYOUT** command is used to create a new layout. It also allows you to rename, copy, save, and delete existing layouts. A drawing designed in the **Model** tab can be composed for plotting in the **Layout** tab. The prompt sequence is as follows.

Enter layout option [Copy/Delete/New/Template/Rename/SAveas/Set/?]<Set>:

The options in the prompt sequence are discussed next.

New Option

This option is used to create a new layout. On choosing this option, you will be prompted to specify the name of the new layout. A new tab with the new layout name will appear in the drawing. Alternatively, you can right-click on the **Model** or the **Layout** tab and choose **New layout** from the shortcut menu to add a new layout. You can also, choose the **Quick View Layouts** button in the Status Bar and then choose **New Layout** button available below the thumbnails to create a new layout. The new layout tab will be added at the end of the existing layout tabs with the default name **Layout N**, where **N** is a natural number starting from one and acquires a value that has not been used in the layout names of the current drawing.

Copy Option

This option copies a layout. When you invoke this option, you will be prompted to specify the layout that has to be copied. Upon specifying the layout, you will be prompted to specify the name of the new layout. If you do not enter a name, the name of the copied layout is assumed with an incremental number in the brackets next to it. For example, Layout 1 is copied as Layout1 (2). The name of the new layout appears as a new tab next to the copied layout tab. Alternatively, you can right-click on the **Model** or **Layout** tab or on the **Quick View Layouts** tile and then choose **Move or**

Figure 12-21 The Move or Copy dialog box

Copy from the shortcut menu to move or copy the selected layout; the **Move or Copy** dialog box will be displayed, see Figure 12-21. Select the layout from the **Before layout** area to move it above the previous layout and then choose the **OK** button. For example, if you want to move the Layout 2 before Layout 1, then select Layout 2 from the **Move or Copy** dialog box; the Layout 2 will be moved before Layout 1. To create a copy of the selected layout select the **Create a copy** check box.

Tip
You can also drag and drop layouts to move them. To create copies of the selected layouts, press the CTRL key as you drag and drop the layouts.

Delete Option

This option deletes an existing layout. On invoking this option, you will be prompted to specify the name of the layout to be deleted. The current layout is the default layout for deleting. Remember that the **Model** tab cannot be deleted. You can also right-click on the **Model** or the **Layout** tab or on the **Quick View Layouts** tiles and choose **Delete** from the shortcut menu; the AutoCAD alert window will be displayed. Choose the **OK** button to delete the selected layout.

Template Option

This option creates a new template based on the existing layout template in the *.dwg*, *.dwt*, or *.dxf* files. This option invokes the **Select Template From File** dialog box, see Figure 12-22. You can also right-click on the **Model** or the **Layout** tab or on the **Quick View Layouts** tiles and choose the **From template** option from the shortcut menu to invoke the **Select Template From File** dialog box.

*Figure 12-22 The **Select Template From File** dialog box*

The layout and geometry from the specified template or drawing file is inserted into the current drawing. After the *dwt*, *dwg*, or *dxf* file is selected, the **Insert Layout(s)** dialog box is displayed, as shown in Figure 12-23.

Choosing the **Layout from Template** button from the **Layouts** toolbar creates a layout using an existing template or drawing file.

Note

If you insert a template that has a title block, it will be inserted as a block and all the text in the title block will be inserted as attributes. You will learn more about blocks and attributes in later chapters.

*Figure 12-23 The **Insert Layout(s)** dialog box*

Rename Option

This option allows you to rename a layout. On choosing this option, you will be prompted to specify the name of the layout to be renamed. On specifying the name, you will be prompted to specify the new name of the layout. The layout names have to be unique and can contain up to 255 characters, out of which only 32 are displayed in the tab. The characters in the name are not case-sensitive. You can also right-click on the **Model** or the **Layout** tab or on the **Quick View Layouts** tiles and choose **Rename** from the shortcut menu to rename the selected layout.

 Tip
*You can also double-click on the **Layout** tab to rename it.*

SAveas Option

This option is used to save a layout in the drawing template file. On choosing this option, you will be prompted to specify the layout that has to be saved. If you specify the name of the layout to be saved, the **Create Drawing File** dialog box will be displayed. In this dialog box, you can enter the name of the template in the **File name** edit box. The layout templates can be saved in the *.dwg*, *.dwt*, or the *.dxf* format.

Set Option

This option sets a layout as the current layout. When you invoke this option, you will be prompted to specify the name of the layout that has to be made current.

? Option

This option lists all the layouts available in the current drawing. The list is displayed in the Command line. You can open the AutoCAD Text Window to view the list by pressing the F2 key.

IMPORTING LAYOUTS TO SHEET SETS

One of the recent enhancements in AutoCAD is that it enables you to add a layout from the drawing to the current sheet set. To add a layout to the current sheet set, right-click on the

Model or the **Layout** tab and choose the **Import Layout as Sheet** option from the shortcut menu; the **Import Layout as Sheet** dialog box will be invoked. This dialog box displays all the layouts available in the selected drawing. Select the check box on the left of the drawing file name under the **Drawing Name** column and choose the **Import Checked** button to import the selected layouts to the current sheet set.

Note
A layout can be imported to only one sheet set. If a layout already belongs to a sheet set, you have to create a copy of the drawing to import its layout to another sheet set.

Tip
You can also drag and drop the layout from a drawing to add it to the sheet set. But, make sure that you save the drawing file before you drag and drop the layout.

INSERTING A LAYOUT USING THE WIZARD

Command:	LAYOUTWIZARD

This command displays the **Layout Wizard** that guides you step-by-step through the process of creating a new layout.

DEFINING PAGE SETTINGS

Ribbon: Output > Plot > Page Setup Manager	**Command:** PAGESETUP
Toolbar: Layouts > Page Setup Manager	

This command allows you to specify the layout and plot device settings for each new layout. You can also right-click on the **Model** or the current **Layout** tab and choose **Page Setup Manager** from the shortcut menu to invoke this command. When you invoke this command, the **Page Setup Manager** dialog box is displayed, which will be discussed in Chapter 14.

Example 2	*Mechanical*

In this example, you will create a drawing in the model space and then use the paper space to plot the drawing. The drawing to be used is shown in Figure 12-24.

1. Increase the limits to 75, 75 and then draw the sketch shown in Figure 12-24.

2. Choose the **Layout1** tab; AutoCAD displays **Layout1** on the screen with the default viewport. Delete this viewport. Right-click on **Layout1** tab and choose **Page Setup Manager** to display the **Page Setup Manager** dialog box. **Layout1** is automatically selected in the **Current page setup** list box.

3. Choose the **Modify** button to display the **Page Setup - Layout1** dialog box. Select the printer or plotter from the **Name** drop-down list in the **Printer/plotter** area. In this example, **HP Lasejet4000** is used. From the drop-down list in the **Paper size** area, select

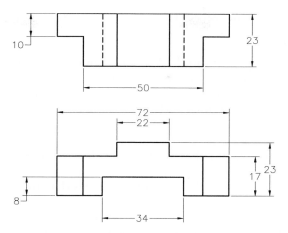

Figure 12-24 Drawing for Example 2

the paper size that is supported by your plotting device. In this example, the paper size is **A4 (210 x 297mm)**. Choose the **OK** button to accept the settings and exit the dialog box. Close the **Page Setup Manager** dialog box.

4. Choose **View > Viewports > New** from the **Ribbon**; the **Viewports** dialog box is displayed. Select the **Single** option from the **Standard viewports** list box. The prompt sequence is as follows:

    Specify corner of viewport or
    [ON/OFF/Fit/Shadeplot/Lock/Object/Polygonal/Restore/2/3/4] <Fit>: *Specify the first corner of the viewport near the bottom left corner of the drawing window.*
    Specify opposite corner: **@297,210**
    Regenerating model.

5. Use the **ZOOM** command in the paper space to zoom to the extents of the viewport.

6. Double-click in the viewport to switch to the temporary model space and use the **ZOOM** command to zoom the drawing to 2XP. In this example, it is assumed that the scale factor is 2:1; therefore, the zoom factor is 2XP.

7. Create the dimension style with the text height of 1.5 and arrowhead height of 1.25. Define all the other parameters based on the text and arrowhead heights and then select the **Annotative** check box and **Scale dimensions to layout** radio button from the **Scale for dimension features** area of the **Fit** tab in the **New Dimension Style** dialog box.

8. Using the new dimension style, dimension the drawing. Make sure that you do not change the scale factor. You can use the **PAN** command to adjust the display.

9. Double-click in the paper space to switch back to the paper space. Choose the **Plot**

button from the **Plot** panel of the **Output** tab in the **Ribbon** to display the **Plot** dialog box.

10. Choose the **Window** option from the **What to plot** drop-down list in the **Plot area**. The dialog box will be temporarily closed and you will be prompted to specify the first and the second corner of the window. Define a window close to the boundary of the viewport such that the viewport is not included in it.

11. As soon as you define both the corners of the window, the **Plot** dialog box will be redisplayed on the screen. Select **1:1** from the **Scale** drop-down list of the **Plot scale** area.

12. Select the **Center the plot** check box from the **Plot offset (origin set to printable area)** area.

13. Choose the **Preview** button to display the plot preview. You can make any adjustments, if required, by redefining the window.

14. After you are satisfied with the preview, right-click and choose **Plot** from the shortcut menu. The drawing will be printed with the scale of 2:1. This means that two plotted units will be equal to one actual unit. Save this drawing with the name *Example2.dwg*.

WORKING WITH THE MVSETUP COMMAND

The **MVSETUP** command is a very versatile command and can be used both on the **Model** tab as well as on the **Layout** tab. This means that this command can be used when the **Tilemode** is set to **1** or when it is set to **0**. The uses of this command in both the drawing environments are discussed next.

Using the MVSETUP Command on the Model Tab

On the **Model** tab, this command is used to set the units, scale factor for the drawing, and the size of the paper. On invoking this command, you will be prompted to specify whether or not you want to enable the paper space. Enter **NO** at this prompt to use this command on the **Model** tab. The prompt sequence that will follow when you enter **NO** is given next.

Enter units type [Scientific/Decimal/Engineering/Architectural/Metric]: *Specify the unit. In this case,* ***Architectural*** *is selected.*

Architectural Scales
=====================

(480) 1/40"=1'
(240) 1/20"=1'
(192) 1/16"=1'
(96) 1/8"=1'
(48) 1/4"=1'
(24) 1/2"=1'

(16) 3/4"=1'
(12) 1"=1'
(4)   3"=1'
(2)   6"=1'
(1)   FULL

Enter the scale factor: *Specify the scale factor for displaying the drawing.*
Enter the paper width: *Specify the width of the paper.*
Enter the paper height: *Specify the height of the paper.*

A box of the specified width and height will be drawn and the drawing will be displayed inside the box. The display of the drawing will depend upon the scale factor you have specified.

Using the MVSETUP Command on the Layout Tab

The way this command works on the **Layout** tab is entirely different from that on the **Model** tab. On the **Layout** tab it is used to insert a title block, create an array of viewports, align the objects in the viewports, and so on. The prompt sequence that will follow when you invoke this command on the **Layout** tab is given next.

[Align/Create/Scale viewports/Options/Title block/Undo]: *Select an option.*

Align Option

This option is used to align the objects in one of the viewports with that of another viewport. You can specify horizontal, vertical, angled, or rotated alignment. After you have selected the option you will be prompted to select the base point. This base point is the point which will be used as the reference point for moving the objects in the specified viewport. Once you have specified the base point, you will be prompted to specify the point in the viewport that will be panned. This point will be aligned with the base point and thus the objects in this viewport will be moved.

Create Option

This option is used to create the viewports in the layouts. On invoking this option, you will be prompted to specify whether you want to create a viewport, delete the objects in it, or undo the changes of this option. The default option is **Create**. If this option is selected, the AutoCAD Text Window is displayed. This window provides you with four options. These options are discussed next.

0:	**None**	1:	**Single**
2:	**Std. Engineering**	3:	**Array of Viewports**

If you enter **0**, the viewports will not be created and the previous prompt will be displayed again. If you enter **1**, you will be prompted to specify two points for the bounding box of the viewport. A new viewport will be created inside the specified bounding box. If you enter **2**, you will be prompted to specify two corners of the bounding box. Inside the specified bounding box, four standard engineering viewports will be created. The first viewport will display the model from the top view, the second will display the model from the SE isometric view, the third will display the front view of the model (*ZX* plane of the current UCS), and the fourth will display the right-side view of the model (*YZ* plane of the current UCS). This option is generally used while working with 3D models. You will also be prompted to specify the distances between the viewports in the X direction and in the Y direction.

Note
The details about the UCS and different views will be discussed in Chapters 24 (The User Coordinate System) and 25 (Getting Started with 3D), respectively.

If you enter **3**, you will be allowed to create an array of viewports along the *X* axis and the *Y* axis. You will be prompted to define a bounding box, in which the viewports will be created. After you have defined the bounding box, you will be prompted to specify the number of viewports in the X and Y directions. Then you will be prompted to specify the distances between the viewports in the X and Y directions. Figure 12-25 shows an array of viewports in the layout with the distance between the viewports in the *X* axis and *Y* axis directions set to zero.

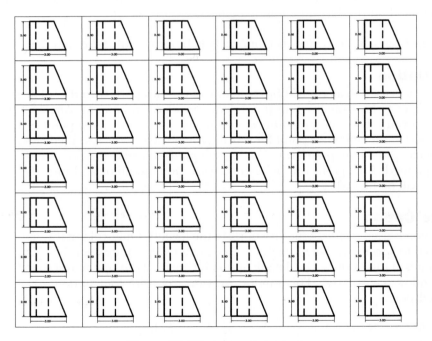

Figure 12-25 *Array of viewports*

Scale viewports Option

This option is used to modify the scale factors of the viewports. This scale factor is defined by the ratio of the paper space units to model space units. You can select one or more than one viewports to modify the scale factors. If you select more than one viewport, you will be prompted to specify whether you want to define a uniform scale factor for the viewports or define individual scale factors. You will be first prompted to specify the number of the paper space units in the viewport and then the model space units in it. The display in the viewports will be automatically modified once you have defined both the values.

Options Option

This option is used to set the parameters for inserting the title block. The parameters that can be set include layers, limits, units, and external reference options. Using the **Layers** option, you can predefine the layer into which the title block will be inserted. Using the **Limits** option, you can specify whether or not the limits should be reset to the extents of the title block after it is inserted. Using the **Units** option, you can set the units in which the size of the title block will be translated. Using the **Xref** option, you can set whether the title block, after inserting, will become the entity of the current drawing or remain as an Xref object.

Title block Option

This option is used to insert the title block of the desired size in the current layout, delete the entities from the current layout, or reset the origin of the current layout. The default option is that for inserting a title block. If you select this option, the **AutoCAD Text Window** will be invoked, displaying the different sizes and formats of the title blocks that can be inserted. The various options that will be displayed are given next.

0:	None	1:	ISO A4 Size(mm)
2:	ISO A3 Size(mm)	3:	ISO A2 Size(mm)
4:	ISO A1 Size(mm)	5:	ISO A0 Size(mm)
6:	ANSI-V Size(in)	7:	ANSI-A Size(in)
8:	ANSI-B Size(in)	9:	ANSI-C Size(in)
10:	ANSI-D Size(in)	11:	ANSI-E Size(in)
12:	Arch/Engineering (24 x 36in)		
13:	Generic D size Sheet (24 x 36in)		

You can enter the number corresponding to the desired title block and it will be inserted in the current layout. The properties will depend on the parameters set using the **Options** option. You can also add or delete the title blocks in this default list. Figure 12-26 shows a layout with an A4 size title block and a viewport created in it.

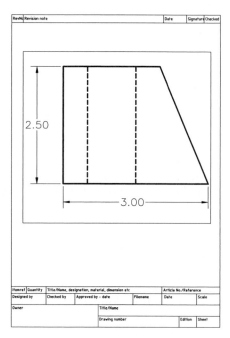

Figure 12-26 *Layout with the title block*

CONVERTING THE DISTANCE BETWEEN MODEL SPACE AND PAPER SPACE

Toolbar: Text > Convert distance between spaces   **Command:** SPACETRANS

 While working with drawings in layouts, you may need to find a distance value that is equivalent to a specific distance in the model space. For example, you may need to write text whose height should be equal to a similar text written in the model space. To convert these distances between the model space and layouts, AutoCAD provides the **SPACETRANS** command. Note that this command will not work in the model space. This command will work only in the layouts or in the temporary model space invoked from the layouts. When you invoke this command in the paper space, AutoCAD prompts you to specify the model space distance. Enter the original distance value that was measured in the model space. AutoCAD displays the paper space equivalent to the specified distance.

Similarly, when you invoke this command in the temporary model space, AutoCAD prompts you to specify the paper space distance. Enter the distance value measured in the paper space. AutoCAD displays the model space equivalent to the specified distance.

CONTROLLING THE DISPLAY OF ANNOTATIVE OBJECTS IN VIEWPORTS

One of the recent enhancements in AutoCAD is the increased control over the display of annotative objects in viewports and model space. Some new buttons have been added to the status bar to control the annotation scale in viewports and model space separately. When you activate a floating viewport, a new **VP Scale** button is added to the status bar. The **VP Scale** button lists the same set of scales as the **Annotation Scale**. You can set an annotation display scale either from the **VP Scale** button or the **Annotation Scale** button, and the other scales will be updated accordingly. The viewport will zoom to a appropriate scale so that the annotation objects can be displayed at the specified scale.

Also, when you activate a floating viewport, the **Lock/Unlock Viewport** button is added to the status bar. With the help of this button, you can toggle between the lock and unlock states of the viewport. The **VP Scale** and **Annotation Scale** buttons are not accessible when the viewport is locked. If the viewport is unlocked and you zoom the drawing instead of specifying the viewport scale, the annotation scale will not be changed and the current scale representation will remain intact and visible. But, the viewport scale will be changed to display the actual scale of the viewport.

The example given next will explain various concepts related to creating and controlling the display of annotations in viewports.

 Note
When the paper space is active, the annotation scale is always 1:1 and it cannot be modified.

Example 3   *Mechanical*

In this example, you will draw the object, as shown in Figure 12-27, and create the floating viewports, as shown in Figure 12-28. The left viewport has a scale of 3/8"=1'-0" and the right viewport has a scale of 1-1/2"=1'-0". All the annotations that are created in the model space should be displayed with a text height of 0.08" on the sheet even if the same annotation appears in multiple viewports.

Figure 12-27 *Model for Example 3*

Figure 12-28 *Displaying the drawing in two viewports*

1. Start a new file in the **2D Drafting & Annotation** workspace, and draw the object, as shown in Figure 12-27.

2. Set the **Annotation Scale** to 3/8"=1'-0" from the Status Bar.

3. Create an annotative text style with the **Paper Text Height** equal to **0.08**.

4. Create an annotative dimension style with the values given next and draw the dimensions refer to Figure 12-27.

    Arrow size = **0.07**
    Text height = **0.08**

5. Create an annotative multileader style with the values given next and draw the multileaders, refer to Figure 12-27.

    Landing distance = **0.075**
    Arrowhead size = **0.07**
    Text height = **0.08**

6. Change the annotation scale to 1-1/2"=1'-0", select all the dimensions from the right of the object and choose the **Add Current Scale** button from the **Annotation Scaling** panel of the **Annotate** tab in the **Ribbon**; the annotation scale of 1-1/2"=1'-0" is added to the selected dimensions. Notice that the selected dimensions and multileaders appear smaller than the other annotations because they reflect the current annotation scale of 1-1/2"=1'-0".

7. Turn the **Annotation Visibility** button off; now you can see the annotations that support a scale of 1-1/2"=1'-0". This helps you to find out the dimensions to which the scale of 1-1/2"=1'-0" has been assigned.

8. Select the annotations from the right of the object and adjust their locations with the help of grips, such that the placement of dimensions remains similar to the one shown in Figure 12-27. Notice that when you select the annotation object, its different scale representations are displayed with faded dashed lines.

9. From the Status Bar, set the annotation scale to 3/8"=1'-0" and then change it back to 1-1/2"=1'-0". And now you will notice that the same annotation objects not only change the size, but also change the location.

10. Choose the **Layout1** tab to switch to the layouts

11. A rectangular viewport is automatically created in this layout. Choose the **Erase** button from the **Modify** panel of the **Home** tab in the **Ribbon**; you are prompted to select the object. Enter **L** in this prompt to delete the last object, which in this case is the viewport.

12. Draw two rectangles side-by-side with dimensions **3'X3.5'** and **2.25'X3.5'** respectively.

13. Choose **View > Viewports > Create Polygonal > Create from Object** from the **Ribbon**. You are prompted to select the object. Select one of the rectangles and it is converted into a viewport. Similarly, convert the second rectangle to viewport.

14. Activate the viewport to the left and set its annotation scale to 3/8"=1'-0" from the Status Bar.

15. Similarly, activate the viewport to the right and set its annotation scale to 1-1/2"=1'-0". Pan the view so that it looks similar to Figure 12-28.

Self-Evaluation Test

Answer the following questions and then compare them those given at the end of this chapter:

1. Viewports in the model space can be of any shape. (T/F)

2. Viewports in the model space can overlap each other. (T/F)

3. Two different tiled viewports can be joined. (T/F)

4. You cannot insert any additional layout in the current drawing. (T/F)

5. The _____ command is used to create the tiled viewports.

6. The viewports in layouts are called _____ viewports.

7. The two working environments provided by AutoCAD are _____ and _____.

8. The _____ command is used to insert a title block in the current layout.

9. When you join two adjacent viewports, the resultant viewport is _____ in shape.

10. The default viewport that is created in the new layout is _____ in shape.

Review Questions

Answer the following questions:

1. Only the viewports created in layouts can be polygonal in shape. (T/F)

2. You cannot lock the display of a floating viewport. (T/F)

3. An existing closed loop can be converted into a viewport in the model space. (T/F)

Chapter 12

4. You can create an array of viewports using the **MVSETUP** command in the model space. (T/F)

5. Which of the following commands can be used to control the display of objects in viewports?

    (a) **MVIEW**   (b) **DVIEW**
    (c) **LAYOUT**   (d) **MSPACE**

6. Which of the following commands can be used to switch to the temporary model space?

    (a) **MVIEW**   (b) **DVIEW**
    (c) **LAYOUT**   (d) **MSPACE**

7. Which of the following options of the **MVIEW** command in the paper space can be used to hide the hidden lines of solid models in printing?

    (a) **Hide**   (b) **Shadeplot**
    (c) **Create**   (d) **None**

8. Which of the following commands can be used for clipping an existing floating viewport?

    (a) **MVIEW**   (b) **DVIEW**
    (c) **VPCLIP**   (d) **MSPACE**

9. Which of the following options of the **VPLAYER** command is used to create a layer that will be frozen in all the viewports?

    (a) **Freeze**   (b) **Thaw**
    (c) **Newfrz**   (d) **Reset**

10. The _____ option of the **MVSETUP** command can be used to set the parameters related to the inserting of the title blocks.

11. You can work only in the _____ tiled viewport.

12. The _____ command can be used to set similar linetype scale for all the viewports.

13. The _____ command is used to switch back to paper space from the temporary model space.

14. The _____ dialog box is used to save a viewport configuration in the model space.

15. Layers that have been frozen, thawed, switched on, or switched off globally are not affected by the _____ command.

Exercise

Exercise 1   *Mechanical*

In this exercise, you will perform the following operations:

a. In the model space, make the drawing of the shaft shown in Figure 12-29.
b. Create three tiled viewports in the model space and then display the drawing in all the three tiled viewports.
c. Create a new layout with the name **Title Block** and insert a title block of ANSI A size in this layout.
d. Create two viewports, one for the drawing and one for the detail "A". See Figure 12-29 for the approximate size and location. The dimensions in detail "A" viewport must not show in the other viewport. Also, adjust the LTSCALE factor for hidden and center lines.
e. Plot the drawing.

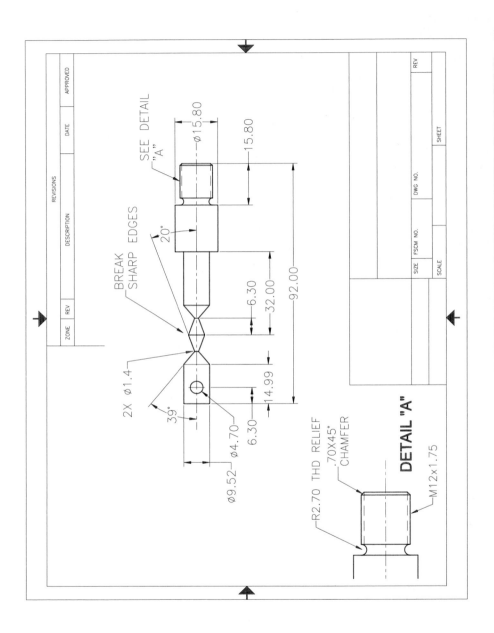

Figure 12-29 Drawing for Exercise 1

Answers to Self-Evaluation Test

1. F, **2. F**, **3.** T, **4.** F, **5. VPORTS**, **6.** floating, **7.** model space and paper space/layouts, **8. MVSETUP**, **9.** rectangular, **10.** rectangular.

Chapter 13

Template Drawings

- *Create template drawings.*
- *Load template drawings using dialog boxes and the Command line.*
- *Do an initial drawing setup.*
- *Customize drawings with layers and dimensioning specifications.*
- *Customize drawings with layouts, viewports, and paper space.*

CREATING TEMPLATE DRAWINGS

One way to customize AutoCAD is to create template drawings that contain initial drawing setup information and if desired, visible objects and text. When the user starts a new drawing, the settings associated with the template drawing are automatically loaded. If you start a new drawing from the scratch, AutoCAD loads default setup values. For example, the default limits are (0.0,0.0), (12.0,9.0) and the default layer is 0 with white color and a continuous linetype. Generally, these default parameters need to be reset before generating a drawing on the computer using AutoCAD. A considerable amount of time is required to set up the layers, colors, linetypes, lineweights, limits, snaps, units, text height, dimensioning variables, and other parameters. Sometimes, border lines and a title block may also be needed.

In production drawings, most of the drawing setup values remain the same. For example, the company title block, border, layers, linetypes, dimension variables, text height, LTSCALE, and other drawing setup values do not change. You will save considerable time if you save these values and reload them when starting a new drawing. You can do this by creating template drawings that contain the initial drawing setup information configured according to the company specifications. They can also contain a border, title block, tolerance table, block definitions, floating viewports in the paper space, and perhaps some notes and instructions that are common to all drawings.

STANDARD TEMPLATE DRAWINGS

AutoCAD comes with standard template drawings like Acad.dwt, Acadiso.dwt, Ansi a (portrait) -color dependent plot styles.dwt, Din a1 -named plot styles.dwt, and so on. The ansi, din, and iso template drawings are based on the drawing standards developed by ANSI (American National Standards Institute), DIN (German), and ISO (International Organization for Standardization). When you start a new drawing with **STARTUP** system variable set to 1, the **Create New Drawing** dialog box will be displayed. To load the template drawing, choose the **Use a Template** button; the list of standard template drawings is displayed. From this list, you can select any template drawing according to your requirements. If you want to start a drawing with the default settings, select the **Start from Scratch** button in the **Create New Drawing** dialog box. The following are some of the system variables, with the default values that are assigned to the new drawing.

System Variable Name	Default Value
CHAMFERA	0.0000
CHAMFERB	0.0000
COLOR	Bylayer
DIMALT	Off
DIMALTD	2
DIMALTF	25.4
DIMPOST	None
DIMASO	On
DIMASZ	0.18
FILLETRAD	0.0000
GRID	0.5000

GRIDMODE	0
ISOPLANE	Left
LIMMIN	0.0000,0.0000
LIMMAX	12.0000,9.0000
LTSCALE	1.0
MIRRTEXT	0 (Text not mirrored like other objects)
TILEMODE	1 (On)

Example 1

Create a drawing template using the **Advanced Setup** wizard of the **Use a Wizard** button with the following specifications and save it with the name *proto1.dwt*.

Units	Engineering with precision 0'-0.00"
Angle	Decimal degrees with precision 0
Angle Direction	Counterclockwise
Area	144'x96'

Step 1

Set the value of the **STARTUP** system variable to 1 to show the **Startup** dialog box. Choose the **New** button from the **Quick Access** toolbar to display the **Create New Drawing** dialog box. Choose the **Use a Wizard** button and choose the **Advanced Setup** option, as shown in Figure 13-1. Choose **OK**. The **Units** page of the **Advanced Setup** dialog box is displayed.

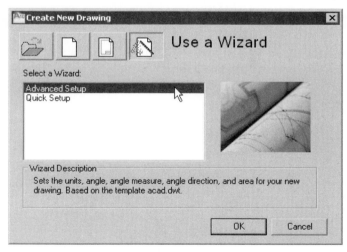

*Figure 13-1 The **Advanced Setup** wizard option of the **Create New Drawing** dialog box*

Step 2

Select the **Engineering** radio button. Select **0'-0.00"** precision from the **Precision** drop-down list, refer to Figure 13-2 and then, choose the **Next** button; the **Angle** page of the **Advanced Setup** dialog box is displayed.

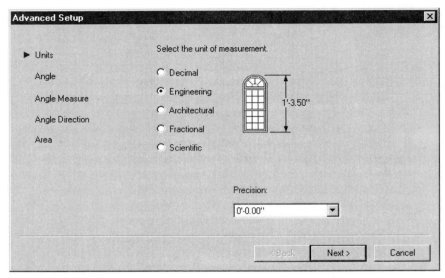

Figure 13-2 *The* **Units** *page of the* **Advanced Setup** *dialog box*

Step 3

In the **Angle** page, select the **Decimal Degrees** radio button and select **0** from the **Precision** drop-down list, as shown in Figure 13-3. Choose the **Next** button; the **Angle Measure** page of the **Advanced Setup** dialog box is displayed.

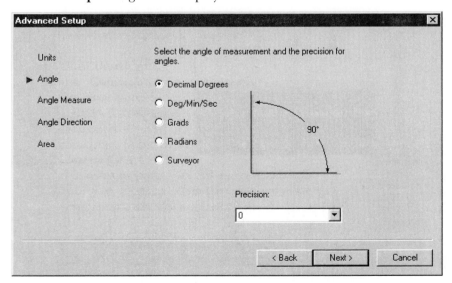

Figure 13-3 *The* **Angle** *page of the* **Advanced Setup** *dialog box*

Step 4

In the **Angle Measure** page, select the **East** radio button. Choose the **Next** button to display the **Angle Direction** page.

Step 5

Select the **Counter-Clockwise** radio button and then choose the **Next** button. The **Area** page is displayed. Specify the area as 144' and 96' by entering the value of the width and length as **144'** and **96'** in the **Width** and **Length** edit boxes and then choose the **Finish** button. Use the **All** option of the **ZOOM** command to display new limits on the screen.

Step 6

Now, save the drawing as *proto1.dwt* using the **Save** button from the **Quick Access** toolbar. You need to select **AutoCAD Drawing Template (*dwt)** from the **Files of type** drop-down list in the dialog box. Choose the **Save** button; the **Template Options** dialog box will be displayed on the screen, as shown in Figure 13-4. Enter the description about the template in the **Description** edit box and choose the **OK** button. Now, the drawing will be saved as *proto1.dwt* on the default drive.

*Figure 13-4 The **Template Options** dialog box*

 Note
*To customize only units and area, you can use the **Quick Setup** option in the **Create New Drawing** dialog box.*

Example 2

Create a drawing template using the following specifications. The template should be saved with the name *proto2.dwt*.

Limits	18.0,12.0
Snap	0.25
Grid	0.50
Text height	0.125
Units	3 digits to the right of decimal point
	Decimal degrees
	2 digits to the right of decimal point
	0 angle along positive X axis (east)
	Angle positive if measured counterclockwise

Step 1
Starting a new drawing

Start AutoCAD and choose **Start from Scratch** from the **Create New Drawing** dialog box. From the **Default Settings** area, select the **Imperial (feet and inches)** radio button, as shown in Figure 13-5. Choose **OK** to open a new file.

Chapter 13

*Figure 13-5 The **Start from Scratch** button of the **Create New Drawing** dialog box*

Step 2
Setting limits, snap, grid, and text size

The **LIMITS** command can be invoked by entering **LIMITS** at the Command prompt.

> Command: **LIMITS**
> Reset Model space limits:
> Specify lower left corner or [ON/OFF] <0.0000,0.0000>: [Enter]
> Specify upper right corner <12.0000,9.0000>: **18,12**

After setting limits, the next step is to expand the drawing display area. Use the **ZOOM** command with the **All** option to display new limits on the screen.

Now, right-click on the **Snap Mode** or **Grid Display** button in the Status Bar to display a shortcut menu. Choose the **Settings** option from the shortcut menu to display the **Drafting Settings** dialog box. You can also choose the **Osnap Settings** button from the **Object Snap** toolbar to display the **Drafting Settings** dialog box. Choose the **Snap and Grid** tab. Enter **0.25** and **0.25** in the **Snap X spacing** and **Snap Y spacing** edit boxes, respectively. Enter **0.5** and **0.5** in the **Grid X spacing** and **Grid Y spacing** edit boxes, respectively. Then choose **OK**.

Note
*You can also use **SNAP** and **GRID** commands to set these values.*

The size of the text can be changed by entering **TEXTSIZE** at the Command prompt.

> Command: **TEXTSIZE**
> Enter new value for TEXTSIZE <0.2000>: **0.125**

Step 3
Setting units

Choose **Drawing Utilities > Units** from the **Application Menu** or enter **UNITS** at the Command prompt to invoke the **Drawing Units** dialog box, as shown in Figure 13-6. In the **Length** area, select **0.000** from the **Precision** drop-down list. In the **Angle** area, select **Decimal Degrees** from the **Type** drop-down list and **0.00** from the **Precision** drop-down list. Also make sure the **Clockwise** check box in the **Angle** area is not selected.

Figure 13-6 The **Drawing Units** *dialog box*

Choose the **Direction** button from the **Drawing Units** dialog box to display the **Direction Control** dialog box (Figure 13-7) and then select the **East** radio button. Exit both the dialog boxes.

Figure 13-7 The **Direction Control** *dialog box*

Chapter 13

Step 4

Now, save the drawing as *proto2.dwt* using the **Save** button from the **Quick Access** toolbar. You must select **AutoCAD Drawing Template (*dwt)** from the **Files of type** drop-down list in the dialog box. Choose the **Save** button; the **Template Options** dialog box will be displayed on the screen. Enter the description about the template in the **Description** edit box and choose the **OK** button from the **Template Options** dialog box. The drawing will be saved as *proto2.dwt* on the default drive. You can also save this drawing to some other location using the **Save Drawing As** dialog box.

LOADING A TEMPLATE DRAWING

You can use the template drawing to start a new drawing file. To use the preset values of the template drawing, start AutoCAD or select the **New** button from the **Quick Access** toolbar. The dialog box that appears will depend on whether you have set the **STARTUP** system variable to **1** or **0**. If you have set this value as **1**, the **Create New Drawing** dialog box will appear. Choose the **Use a Template** option. All templates that are saved in the default **Template** directory will be shown in the **Select a Template** list box, see Figure 13-8. If you have saved the template in any other file, choose the **Browse** button. The **Select a template file** dialog box will be displayed. You can use this dialog box to browse the directory in which the template file is saved.

Figure 13-8 *Default templates in the **Create New Drawing** dialog box*

If you have set the **STARTUP** system variable to 0, the **Select template** dialog box appears when you choose the **New** button. This dialog box also displays the default **Template** folder and all template files saved in it, see Figure 13-9. You can use this dialog box to select the template file you that want to open.

Figure 13-9 *The **Select template** dialog box that appears when the option to show the **Startup** dialog box is not selected*

Using any of the previously mentioned dialog boxes, select the *proto1.dwt* template drawing. AutoCAD will start a new drawing that will have the same setup as that of the template drawing *proto1.dwt*.

You can have several template drawings, each with a different setup. For example, **PROTOB** for a 18" by 12" drawing, **PROTOC** for a 24" by 18" drawing. Each template drawing can be created according to user-defined specifications. You can then load any of these template drawings, as discussed previously.

CUSTOMIZING DRAWINGS WITH LAYERS AND DIMENSIONING SPECIFICATIONS

Most production drawings need multiple layers for different groups of objects. In addition to layers, it is a good practice to assign different colors to different layers to control the line width at the time of plotting. You can generate a template drawing that contains the desired number of layers with linetypes and colors according to your company specifications. You can then use this template drawing to make a new drawing. The next example illustrates the procedure used for customizing a drawing with layers, linetypes, and colors.

Example 3

Create a template drawing *proto3.dwt* that has a border and the company's title block, as shown in Figure 13-10.

Figure 13-10 *The template drawing for Example 3*

This template drawing will have the following initial drawing setup:

Limits	48.0,36.0
Text height	0.25
Border line lineweight	0.012"
Ltscale	4.0

DIMENSIONS
Overall dimension scale factor 4.0
Dimension text above the extension line
Dimension text aligned with dimension line

LAYERS

Layer Name	Line Type	Color
0	Continuous	White
OBJ	Continuous	Red
CEN	Center	Yellow
HID	Hidden	Blue
DIM	Continuous	Green
BOR	Continuous	Magenta

Step 1
Setting limits, text size, polyline width, polyline, and linetype scaling

Start a new drawing with default parameters by selecting the **Start from Scratch** option in the **Create New Drawings** dialog box. In the new drawing file, use the AutoCAD commands to set up the values as given for this example. Also, draw a border and a title block, as shown in Figure 13-10. In this figure, the hidden lines indicate drawing limits. The border lines are 1.0 units inside the drawing limits. For border lines, increase the lineweight to a value of 0.012".

Use the following procedure to produce the prototype drawing for Example 3.

1. Invoke the **LIMITS** command by choosing **Application Menu > Format > Drawing Limits** or by entering **LIMITS** at the Command prompt. The following is the prompt sequence

   Command: **LIMITS**
   Reset Model space limits:
   Specify lower left corner or [ON/OFF] <0.0000,0.0000>: [Enter]
   Specify upper right corner <12.0000,9.0000>: **48,36**

2. Increase the drawing display area by choosing the **All** option of the **ZOOM** command.

3. Enter **TEXTSIZE** at the Command prompt to change the text size.

   Command: **TEXTSIZE**
   Enter new value for TEXTSIZE <0.2000>: **0.25**

4. Next, draw the border using the **RECTANG** command. The prompt sequence to draw the rectangle is:

   Command: **RECTANG**
   Specify first corner point or [Chamfer/Elevation/Fillet/Thickness/Width]: **1.0,1.0**
   Specify other corner point or [Area/Dimensions/Rotation]: **47.0,35.0**

5. Now, select the rectangle and change its lineweight to **0.012"**. Make sure that the **Show/Hide Lineweight** button is chosen in the Status Bar.

6. Enter **LTSCALE** at the Command prompt to change the linetype scale.

   Command: **LTSCALE**
   Enter new linetype scale factor<Current>: **4.0**

Step 2
Setting dimensioning parameters

You can use the **Dimension Style Manager** dialog box to set the dimension variables. Choose the inclined arrow displayed on the **Dimensions** panel title bar in the **Annotate** tab of the **Ribbon**; the **Dimension Style Manager** dialog box will be displayed, as shown in Figure 13-11.

*Figure 13-11   The **Dimension Style Manager** dialog box*

You can also invoke this dialog box by entering **DIMSTYLE** at the Command prompt. Choose the **New** button from the **Dimension Style Manager** dialog box; the **Create New Dimension Style** dialog box will be displayed. Specify the new style name as **MYDIM1** in the **New Style Name** edit box, as shown in the Figure 13-12 and then choose the **Continue** button. The **New Dimension Style:MYDIM1** dialog box is displayed.

*Figure 13-12   The **Create New Dimension Style** dialog box*

Overall dimension scale factor

To specify a dimension scale factor, choose the **Fit** tab of the **New Dimension Style: MYDIM1** dialog box. Set the value in the **Use overall scale of** as 4 in the **Scale for dimension features** area, as shown in Figure 13-13.

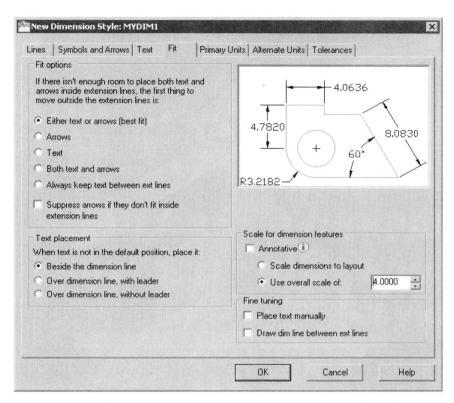

Figure 13-13 *The* **Fit** *tab of the* **New Dimension Style: MYDIM1** *dialog box*

Dimension text over the dimension line
In the **Fit** tab of the **New Dimension Style: MYDIM1** dialog box. Select the **Over dimension line, with leader** radio button from the **Text placement** area.

Dimension text aligned with the dimension line
In the **Text Alignment** area of the **Text** tab of the **New Dimension Style: MYDIM1** dialog box, select the **Aligned with the dimension line** radio button and then choose **OK** button.

Setting the new dimension style to current
A new dimension style with the name **MYDIM1** is shown in the **Styles** area of the **Dimension Style Manager** dialog box. Select this dimension style and then choose the **Set Current** button to make it the current dimension style. Choose the **Close** button to exit this dialog box.

Step 3
Setting layers
Choose **Home > Layers > Layer Properties** from the **Ribbon** or choose **Format > Layer** from the menu bar to invoke the **Layer Properties Manager** dialog box. You can also invoke the **Layer Properties Manager** dialog box from the **Layers** toolbar. Choose the **New Layer**

button in the **Layer Properties Manager** dialog box and rename Layer1 as OBJ. Choose the color swatch of the OBJ layer to display the **Select Color** dialog box. Select the **Red** color and choose **OK**; the red color is assigned to the OBJ layer. Again, choose the **New Layer** button in the **Layer Properties Manager** dialog box and rename the Layer1 as CEN. Choose the linetype swatch to display the **Select Linetype** dialog box.

If different linetypes are not already loaded, choose the **Load** button to display the **Load or Reload Linetypes** dialog box. Select the **CENTER** linetype from the **Available Linetypes** area and choose **OK**; the **Select Linetype** dialog box will reappear. Select the **CENTER** linetype from the **Loaded linetypes** area and choose **OK**. Choose the color swatch to display the **Select Color** dialog box. Select the **Yellow** color and choose **OK**; the color yellow and linetype center will be assigned to the layer CEN.

Similarly, different linetypes and different colors can be set for different layers mentioned in the statement of this example, as shown in Figure 13-14.

Figure 13-14 *Partial display of the* **Layer Properties Manager** *dialog box*

Step 4
Adding title block
Next, add the title block and the text, as shown in Figure 13-10. After completing the drawing, save it as *proto3.dwt*. You have created a template drawing (PROTO3) that contains all the information given in Example 3.

CUSTOMIZING A DRAWING WITH LAYOUT

The Layout (paper space) provides a convenient way to plot multiple views of a three-dimensional (3D) drawing or multiple views of a regular two-dimensional (2D) drawing. It takes quite some time to set up the viewports in the model space with different vpoints and scale factors. You can create prototype drawings that contain predefined viewport settings, with vpoint and the other desired information. If you create a new drawing or insert a drawing, the views are automatically generated. The following example illustrates the procedure for generating a prototype drawing with paper space and model space viewports:

Example 4

Create a drawing template, as shown in Figure 13-15, with four views in Layout3 (Paper space) that display front, top, side, and 3D views of the object. The plot size is 10.5 by 8 inches. The plot scale is 0.5 or 1/2" = 1". The paper space viewports should have the following vpoint settings:

Viewports	Vpoint	View
Top right	1,-1,1	3D view
Top left	0,0,1	Top view
Lower right	1,0,0	Right side view
Lower left	0,-1,0	Front view

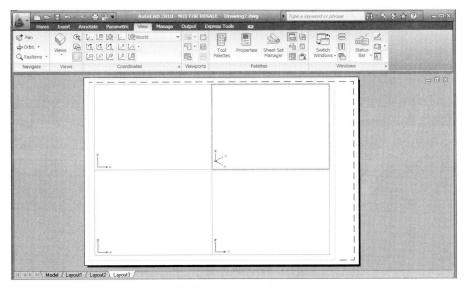

Figure 13-15 *Paper space with four viewports*

Start AutoCAD and create a new drawing. Use the following commands and options to set up various parameters:

Step 1
First, you need to create a new layout using the **LAYOUT** command. You can also choose **New Layout** button from the **Quick View Layouts** panel available on choosing the **Quick View Layouts** button on the Status Bar. A new layout is automatically created with the default name. Alternatively, you can also use the **LAYOUT** command to create a new layout. The Command prompt that is displayed is given next.

Command: **LAYOUT**
Enter layout option [Copy/Delete/New/Template/Rename/SAveas/Set/?] <set>: **N**
Enter new Layout name <Layout3>: Enter

Step 2

Next, choose the new layout (**Layout3**) tab. The new layout (**Layout3**) will be displayed with the default viewport. Use the **ERASE** command to erase this viewport. Invoke the **Page Setup Manager** dialog box from the shortcut menu that will by displayed when you right click on the **Quick View Layouts** button and choose the **Modify** button to modify the default page setup. In the **Page Setup - Layout3** dialog box, select the required printer from the **Printer/plotter** area. In this example, HP LaserJet 4000 has been used.

Step 3

From the **Paper size** area, select the paper size that is supported by the selected plotting device. In this example, the paper size is A4. Choose the **OK** button to accept the settings and return to the **Page Setup Manager** dialog box. Choose **Close** in this dialog box to close the **Page Setup Manager** dialog box.

Step 4

Next, you need to set up a layer with the name VIEW for viewports and assign it green color. Invoke the **Layer Properties Manager** dialog box. Choose the **New Layer** button and name the Layer1 as VIEW. Choose the color swatch of the VIEW layer to display the **Select Color** dialog box. Select the color **Green** and choose the **OK** button. This color will be assigned to **View** layer. Also, make the VIEW layer current and then exit the **Layer Properties Manager** dialog box.

Step 5

To create four viewports, use the **MVIEW** command. You can directly choose **View > Viewports > 4Viewports** from the menu bar or enter **MVIEW** command at the Command prompt. The following is the prompt sequence when you invoke the **MVIEW** command.

> Command: **MVIEW**
> Specify corner of viewport or
> [ON/OFF/Fit/Shadeplot/Lock/Object/Polygonal/Restore/LAyer/2/3/4] <Fit>: **4**
> Specify first corner or [Fit] <Fit>: **0.25,0.25**
> Specify opposite corner: **10.25,7.75**

Step 6

Choose the **PAPER** button in the Status Bar to activate the model space or enter **MSPACE** at the Command prompt.

> Command: **MSPACE** (or **MS**)

Make the lower left viewport active by clicking on it. Next, you need to change the viewpoints of different paper space viewports using the **VPOINT** command. To invoke this command, choose **View > 3D Views > Viewpoint** from the menu bar or enter **VPOINT** at the Command prompt. The viewpoint values for different viewports are given in the statement of Example 4. To set the view point for the lower left viewport the Command prompt sequence is given next.

Command: **VPOINT**
Current view direction: VIEWDIR= 0.0000,0.0000,1.0000
Specify a view point or [Rotate] <display compass and tripod>: **0,-1,0**

Similarly, use the **VPOINT** command to set the viewpoint of the other viewports.

Make the top left viewport active by selecting a point in the viewport and then use the **ZOOM** command to specify the paper space scale factor to 0.5. The **ZOOM** command can be invoked by choosing **View > Zoom > Scale** from the **Application Menu** or by entering **ZOOM** at the Command prompt. The Command prompt sequence is given next.

Command: **ZOOM**
Specify corner of window, enter a scale factor (nX or nXP), or
[All/Center/Dynamic/Extents/Previous/Scale/Window/Object] <real time>: **0.5XP**

Now, make the next viewport active and specify the zoom scale factor. Do the same for the remaining viewports.

Step 7
Use the **MODEL** button in the Status Bar to change to paper space and then set a new layer PBORDER with yellow color. Make the PBORDER layer current, draw a border, and if needed, a title block using the **PLINE** command. You can also change to paper space by entering **PSPACE** at the Command prompt.

The **PLINE** command can be invoked by choosing the **Polyline** button from the **Draw** toolbar or choosing **Home > Draw > Polyline** from the **Ribbon**. The **PLINE** command can also be invoked by entering **PLINE** at the Command prompt.

Command: **PLINE**
Specify start point: **0,0**
Current line-width is 0.0000
Specify next point or [Arc/Halfwidth/Length/Undo/Width]: **0,8.0**
Specify next point or [Arc/Close/Halfwidth/Length/Undo/Width]: **10.5,8.0**
Specify next point or [Arc/Close/Halfwidth/Length/Undo/Width]: **10.5,0**
Specify next point or [Arc/Close/Halfwidth/Length/Undo/Width]: **C**

Step 8
The last step is to choose the **Model** tab (or change the **TILEMODE** to 1), if not already active and save the prototype drawing as template. To test the layout that you just created, make the 3D drawing, as shown in Figure 13-18 or make any 3D object. Switch to **Layout 3** tab; you will find four different views of the object (Figure 13-16). If the object views do not appear in the viewports, use the **PAN** commands to position the views in the viewports. You can freeze the VIEW layer so that the viewports do not appear on the drawing. You can plot this drawing from the Layout3 with a plot scale factor of 1:1 and the size of the plot will be exactly as specified.

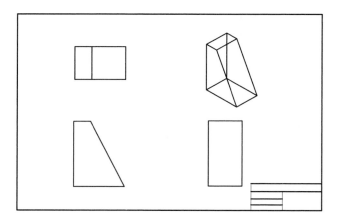

Figure 13-16 *Four views of a 3D object in paper space*

CUSTOMIZING DRAWINGS WITH VIEWPORTS

In certain applications, you may need multiple model space viewport configurations to display different views of an object. This involves setting up the desired viewports and then changing the viewpoint for different viewports. You can create a prototype drawing that contains a required number of viewports and the viewpoint information. If you insert a 3D object in one of the viewports of the prototype drawing, you will automatically get different views of the object without setting viewports or viewpoints. The following example illustrates the procedure for creating a prototype drawing with a standard number (four) of viewports and viewpoints.

Example 5

Create a prototype drawing with four viewports, as shown in Figure 13-17. The viewports should have the following viewpoints:

Viewports	Vpoint	View
Top right	1,-1,1	3D view
Top left	0,0,1	Top view
Lower right	1,0,0	Right side view
Lower left	0,-1,0	Front view

Step 1
Start AutoCAD and create a new drawing from scratch.

Step 2
Setting viewports
Viewports and corresponding viewpoints can be set with the **VPORTS** command. You can also choose the **Display Viewports Dialog** button from the **Viewports** toolbar or choose **View > Viewports > New** from the **Ribbon** to display the **Viewports** dialog box, as shown

Figure 13-17 *Viewports with different viewpoints*

in Figure 13-18. Select **Four: Equal** from the **Standard viewports** area. In the **Preview** area four equal viewports are displayed. Select **3D** from the **Setup** drop-down list. The four viewports with the different viewpoints will be displayed in the **Preview** area as Top, Front, Right and SE Isometric, respectively. **Top** represents the viewpoints as (0,0,1), **Front** represents the

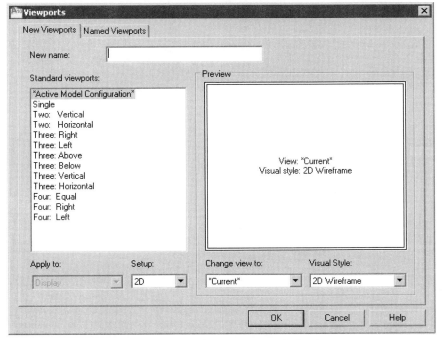

Figure 13-18 *The **Viewports** dialog box*

viewpoints as (0,-1,0), **Right** represents the viewpoints as (1,0,0) and **SE Isometric**, represents the viewpoints as (1,-1,1) respectively. Choose the **OK** button. Save the drawing as *proto5.dwt*.

Viewports and viewpoints can also be set by entering **-VPORTS** and **VPOINT** at the Command prompt, respectively.

Step 3
Start a new drawing and draw the 3D tapered block, as shown in Figure 13-19.

Step 4
Again, start a new drawing, TEST, using the prototype drawing *proto5.dwt*. Make the top right viewport current and then insert or create a drawing, as shown in Figure 13-19. Four different views will be automatically displayed on the screen, as shown in Figure 13-20.

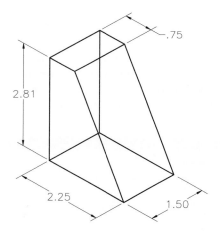

Figure 13-19 *3D tapered block*

Note
The method for creating 3D models is discussed in Chapter 25, Getting Started with 3D.

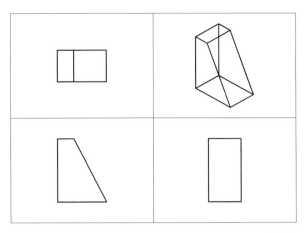

Figure 13-20 *Different views of a 3D tapered block*

CUSTOMIZING DRAWINGS ACCORDING TO PLOT SIZE AND DRAWING SCALE

For controlling the plot area, it is recommended to use layouts. You can make the drawing of any size, use the layout to specify the sheet size, and then draw the border and title block. However, you can also plot a drawing in the model space and set up the system variables so that the plotted drawing is to your specifications. You can generate a template drawing according to plot size and scale. For example, if the scale is 1/16" = 1' and the drawing is to be plotted on a 36" by 24" area, you can calculate drawing parameters like limits, **DIMSCALE**, and **LTSCALE** and save them in a template drawing. This will save considerable time in the initial drawing setup and provide uniformity in the drawings. The next example explains the procedure involved in customizing a drawing according to a certain plot size and scale.

 Note
You can also use the paper space to specify the paper size and scale.

Example 6

Create a drawing template (**PROTO6**) with the following specifications:

Plotted sheet size	36" by 24" (Figure 13-20)
Scale	1/8" = 1.0'
Snap	3'
Grid	6'
Text height	1/4" on plotted drawing
Linetype scale	Calculate
Dimscale factor	Calculate
Units	Architectural

	Precision, 16-denominator of the smallest fraction
	Angle in degrees/minutes/seconds
	Precision, 0d00'
	Direction control, base angle, east
	Angle positive, if measured counterclockwise
Border	Border should be 1" inside the edges of the plotted drawing sheet, using PLINE 1/32" wide when plotted (Figure 13-21)

Figure 13-21 *Border of the template drawing*

Step 1
Calculating limits, text height, linetype scale, dimension scale, and polyline width

In this example, you need to calculate some values before you set the parameters. The limits of the drawing depend on the plotted size of the drawing and its scale. Similarly, **LTSCALE** and **DIMSCALE** depend on the plot scale of the drawing. The following calculations explain the procedure for finding the values of limits, ltscale, dimscale, and text height.

Limits

Given:
Sheet size   36" x 24"
Scale   1/8" = 1'
        or 1" = 8'

Calculate:
X Limit
Y Limit
Since sheet size is 36" x 24" and scale is 1/8"=1'
Therefore,   X Limit = 36 x 8' = 288'
         Y Limit = 24 x 8' = 192'

Text height

Given:
Text height when plotted = 1/4"
Scale   1/8" = 1'

Calculate:
Text height
Since scale is 1/8" = 1'
          or 1/8" = 12"
            or   1" = 96"
Therefore, scale factor = 96
        Text height = 1/4" x 96
                    = 24" = 2'

Linetype scale and dimension scale

Known:
Since scale is 1/8" = 1'
          or 1/8" = 12"
            or   1" = 96"

Calculate:
LTSCALE and DIMSCALE
Since scale factor = 96
Therefore, LTSCALE = Scale factor = 96
Similarly, DIMSCALE = 96
(All dimension variables, like DIMTXT and DIMASZ, will be multiplied by 96.)

Polyline Width

Given:
Scale is 1/8" = 1'

Calculate:
PLINE width
Since scale is 1/8" = 1'
          or   1" = 8'
          or   1" = 96"

Therefore,
PLINE width = 1/32 x 96
             = 3"

After calculating the parameters, use the following AutoCAD commands to set up the drawing and save the drawing as *proto6.dwt*.

Chapter 13

Step 2
Setting units
Start a new drawing and enter **UNITS** at the Command prompt to display the **Drawing Units** dialog box. Select **Architectural** from the **Type** drop-down list in the **Length** area. Select **0'-01/16"** from the **Precision** drop-down list. Make sure the **Clockwise** check box in the **Angle** area is not selected. Select **Deg/Min/Sec** from the **Type** drop-down list and select **0d00'** from the **Precision** drop-down list in the **Angle** area. Now, choose the **Direction** button to display the **Directional Control** dialog box. Select the **East** radio button, if it is not selected, in the **Base Angle** area and then choose **OK**.

Step 3
Setting limits, snap and grid, textsize, linetype scale, dimension scale, dimension style and pline
To set limits, enter **LIMITS** at the Command prompt.

>   Command: **LIMITS**
>   Specify lower left corner or [ON/OFF] <0'-0",0'-0">:**0,0**
>   Specify upper right corner <1'-0",0'-9">: **288',192'**

Invoke the **All** option of the **ZOOM** command to increase the drawing display area.

Right-click on the **Snap Mode** or **Grid Display** button in the Status Bar to invoke the shortcut menu. In the shortcut menu, choose **Settings** to display the **Drafting Settings** dialog box. You can also choose the **Osnap Settings** button from the **Object Snap** toolbar. In the dialog box, choose the **Snap and Grid** tab. Enter **3'** and **3'** in the **Snap X spacing** and **Snap Y spacing** edit boxes, respectively. Enter **6'** and **6'** in the **Grid X spacing** and **Grid Y spacing** edit boxes, respectively. Then choose **OK**.

You can also set these values by entering **SNAP** and **GRID** at the Command prompt.

The size of the text can be changed by entering **TEXTSIZE** at the Command prompt.

>   Command: **TEXTSIZE**
>   Enter new value for TEXTSIZE <current>: **2'**

To set the **LTSCALE**, choose the **Format > Linetype** from the menu bar or enter **LINETYPE** at the Command prompt to invoke the **Linetype Manager** dialog box. Choose the **Show details** button. Specify the **Global scale factor** as **96** in the **Global scale factor** edit box.

You can also change the scale of the linetype by entering **LTSCALE** at the Command prompt.

Choose the inclined arrow displayed on the **Dimensions** panel title bar from the **Annotate** tab of the **Ribbon** to invoke the **Dimension Style Manager** dialog box, as shown in Figure 13-11. Choose the **New** button from the **Dimension Style Manager** dialog box to invoke the **Create New Dimension Style** dialog box. Specify the new style name as **MYDIM2** in the **New Style Name** edit box and then choose the **Continue** button. The **New Dimension**

Style: MYDIM2 dialog box will be displayed. Choose the **Fit** tab and set the value in the **Use overall scale of** spinner to **96** in the **Scale for dimension features** area. Now, choose the **OK** button to again display the **Dimension Style Manager** dialog box. Choose **Close** to exit the dialog box.

You can invoke the **PLINE** command by choosing **Home > Draw > Polyline** button from the **Ribbon** or entering **PLINE** at the Command prompt.

> Command: **PLINE**
> Specify start point: **8',8'**
> Current line-width is **0'-0"**
> Specify next point or [Arc/Halfwidth/Length/Undo/Width]: **W**
> Specify starting width<0'-00">: **3**
> Specify ending width<0'-3">: **ENTER**
> Specify next point or [Arc/Halfwidth/Length/Undo/Width]: **280,0**
> Specify next point or [Arc/Close/Halfwidth/Length/Undo/Width]: **0,184**
> Specify next point or [Arc/Close/Halfwidth/Length/Undo/Width]: **-280,0**
> Specify next point or [Arc/Close/Halfwidth/Length/Undo/Width]: **C**

Now, save the drawing as *proto6.dwt*.

Self-Evaluation Test

Answer the following questions and then compare them to those given at the end of this chapter:

1. The template drawings are saved in _____ format.

2. To use a template file, select the _____ option in the **Create New Drawing** dialog box.

3. To start a drawing with the default setup, select the _____ option in the **Create New Drawing** dialog box.

4. If the plot size is 36" x 24", and the scale is 1/2" = 1', then X Limit = _____ and Y Limit = _____.

5. You can use AutoCAD's _____ command to set up a viewport in the paper space.

Review Questions

Answer the following questions:

1. The default value of **DIMSCALE** is _____.

2. The default value of **DIMTXT** is _____.

3. The default value of **SNAP** is _____.

4. Architectural units can be selected by using the _____ command or the _____ command.

5. Name three standard template drawings that come with AutoCAD software _____ , _____ , and _____.

6. If the plot size is 24" x 18", and the scale is 1:20, the X Limit = _____ and Y Limit = _____.

7. If the plot size is 200 x 150 and limits are (0.00,0.00) and (600.00,450.00), the **LTSCALE** factor = _____.

8. _____ provides a convenient way to plot multiple views of a 3D drawing or multiple views of a regular 2D drawing.

9. You can use the _____ command to change to paper space.

10. You can use AutoCAD's _____ command to change to model space.

11. The values that can be assigned to **TILEMODE** are _____ and _____.

12. In the model space, if you want to reduce the display size by half, the scale factor you enter in the **ZOOM > Scale** option is _____.

Exercises

Exercise 1   *General*

Create a drawing template (*protoe1.dwt*) with the following specifications:

Units	Architectural with precision 0'-0 1/16
Angle	Decimal Degrees with precision 0.
Base angle	East.
Angle direction	Counterclockwise.
Limits	48' x 36'

Exercise 2   *General*

Create a drawing template (*protoe2.dwt*) with the following specifications:

Limits	36.0,24.0
Snap	0.5

Grid	1.0
Text height	0.25
Units	Decimal
	Precision 0.00
	Decimal degrees
	Precision 0
	Base angle, East
	Angle positive if measured counterclockwise

Exercise 3   *General*

Create a drawing template (*protoe3.dwt*) with the following specifications:

Limits	48.0,36.0
Text height	0.25
PLINE width	0.03
LTSCALE	4.0
DIMSCALE	4.0
Plot size	10.5 x 8

LAYERS

Layer Name	Line Type	Color
0	Continuous	White
OBJECT	Continuous	Green
CENTER	Center	Magenta
HIDDEN	Hidden	Blue
DIM	Continuous	Red
BORDER	Continuous	Cyan

Exercise 4   *General*

Create a prototype drawing with the following specifications (the name of the drawing is *protoe4.dwt*):

Limits	36.0,24,0
Border	35.0,23.0
Grid	1.0
Snap	0.5
Text height	0.15
Units	Decimal (up to 2 places)
LTSCALE	1
Current layer	Object

LAYERS

Layer Name	Linetype	Color
0	Continuous	White
Object	Continuous	Red
Hidden	Hidden	Yellow
Center	Center	Green
Dim	Continuous	Blue
Border	Continuous	Magenta
Notes	Continuous	White

This prototype drawing should have a border line and a title block as shown in Figure 13-22.

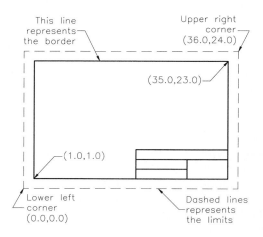

Figure 13-22 *Prototype drawing*

Exercise 5   *General*

Create a template drawing shown in Figure 13-23 with the following specifications and save it with the name *protoe5.dwt*:

Plotted sheet size	36" x 24" (Figure 13-22)
Scale	1/2" = 1.0'
Text height	1/4" on plotted drawing
LTSCALE	24
DIMSCALE	24
Units	Architectural
	32-denominator of smallest fraction to display
	Angle in degrees/minutes/seconds
	Precision 0d00"00"
	Angle positive if measured counterclockwise
Border	Border is 1-1/2" inside the edges of the plotted drawing sheet, using the PLINE 1/32" wide when plotted.

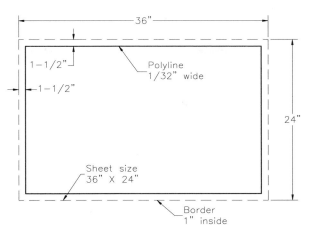

Figure 13-23 *Drawing for Exercise 5*

Exercise 6   *General*

Create a prototype drawing, as shown in Figure 13-24 with the following specifications (the name of the drawing is *protoe6.dwt*):

Plotted sheet size   24" x 18" (Figure 13-23)
Scale   1/2"=1.0'
Border   The border is 1" inside the edges of the plotted
                       drawing sheet, using the PLINE 0.05" wide when
                       plotted (Figure 13-24)
Dimension text over the dimension line
Dimensions aligned with the dimension line
Calculate overall dimension scale factor
Enable the display of alternate units
Dimensions to be associative.

Chapter 13

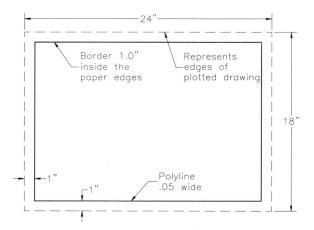

Figure 13-24 *Prototype drawing*

Exercise 7   *Mechanical*

Create a template with the necessary specifications to draw the Articulated Rod shown in Figure 13-25. Also, use this template to draw the views of the Articulated Rod.

Figure 13-25 *Views and dimensions of the Articulated Rod*

Answers to Self-Evaluation Test
1. *.dwt*, 2. **Use a Template**, 3. **Start from Scratch**, 4. 72",48", 5. **MVIEW**.

Chapter *14*

Plotting Drawings

Learning Objectives

After completing this chapter, you will be able to:
- *Set plotter specifications and plot drawings.*
- *Configure plotters and then edit plotter configuration files.*
- *Create, use, and modify plot styles and plot style tables.*
- *Plot sheets in a sheet set.*

PLOTTING DRAWINGS IN AutoCAD

After you have completed a drawing, you can store it on the computer storage device such as the hard drive or diskette. However, to get its hard copy, you should plot the drawing on a sheet of paper using a plotter or printer. A hard copy is a handy reference for professionals working on site. With pen plotters, you can obtain a high-resolution drawing. Basic plotting has already been discussed in Chapter 2, Getting Started with AutoCAD. You can plot drawings in the **Model** tab or any of the layout tabs. A drawing has a **Model** and two layout tabs (**Layout1**, **Layout2**) by default. Each of these tabs has its own settings and can be used to create different plots. You can also create new layout tabs using the **LAYOUT** command. This is discussed in Chapter 11.

PLOTTING DRAWINGS USING THE PLOT DIALOG BOX

Quick Access Toolbar: Plot	**Toolbar:** Standard > Plot	**Command:** PLOT
Application Menu: Print > Plot	**Ribbon:** Output > Plot > Plot	

The **PLOT** command is used to plot a drawing. When you invoke this command, the **Plot** dialog box is displayed. This dialog box can be invoked by right-clicking on the **Model** tab or any of the **Layout** tabs and choosing the **Plot** option from the shortcut menu. Figure 14-1 shows the expanded form of the **Plot** dialog box.

Figure 14-1 The expanded form of the **Plot** dialog box

In this dialog box, some values were set when AutoCAD was first configured. You can examine these values and if they conform to your requirements, you can start plotting directly. Otherwise, you can alter these values to define plot specifications by using the options in the **Plot** dialog box. The available plot options are discussed next.

Page setup Area

The **Name** drop-down list in this area displays all the saved and named page setups. A page setup contains the settings required to plot a drawing on a sheet of paper to create a layout. It consists of all the settings related to the plotting of a drawing such as the scale, pen settings, and so on, and also includes the plot devices being used. These settings can be saved as a named page setup, which can be later selected from this drop-down list and used for plotting a drawing. If you select **Previous plot** from the drop-down list, the settings used for the last drawing plotted are applied to the current drawing. You can also import the existing page setup from the other files by selecting the **Import** option from the drop-down list. The page setup can be imported from any drawing file(*.dwg*), from a template file(*.dwt*), or from a drawing interchange format file(*.dxf*). You can choose the base for the page setup on a named

page setup, or you can add a new named page setup by choosing the **Add** button, which is located next to the drop-down list. When you choose this button, AutoCAD displays the **Add Page Setup** dialog box, as shown in Figure 14-2. Enter the name of the new page setup in this dialog box and choose **OK**. All the settings that you configure in the current **Plot** dialog box will be saved under this page setup.

*Figure 14-2 The **Add Page Setup** dialog box*

Tip
To create a new page setup based on the existing one, select the existing page setup from the ***Name*** *drop-down list and make modifications in it. Next, choose the* ***Add*** *button; the modified page setup will be saved.*

Printer/plotter Area

This area displays all information about the configured printers and plotters currently selected from the **Name** drop-down list. It displays the plotter driver and the printer port being used. It also displays the physical location and some description text about the selected plotter or printer. All the currently configured plotters are displayed in the **Name** drop-down list.

Note
To add plotters and printers to the ***Name*** *drop-down list, choose* ***Plot > Plotter Manager*** *from the* ***Output*** *tab of the* ***Ribbon***. *Double-click on the* ***Add-A-Plotter Wizard*** *icon in this window to display the* ***Add Plotter*** *wizard. You can use this wizard to add a plotter to the list of configured plotters. Once the plotter is added, the plotter configuration file (PC3) for the plotter will be created. This file consists of all settings needed by the specific plotter to plot the drawing. (The* ***Plotters*** *window will be discussed later in this chapter in the section Adding Plotters).*

Chapter 14

Properties

To check information about a configured printer or plotter, choose the **Properties** button. When you choose this button, the **Plotter Configuration Editor** dialog box will be displayed. This dialog box lists all the details of the selected plotter under three tabs: **General**, **Ports**, and **Device and Document Settings**. The **Plotter Configuration Editor** will be discussed later in the "Editing Plotter Configuration" section of this chapter.

Plot to file

If you select this check box, AutoCAD plots the output to a file rather than to the plotter. Depending on the plotter selected, the file can be plotted in the *.dwf*, *.plt*, *.jpg*, or *.png* format. On selecting this check box and choosing **OK** from the **Plot** dialog box, the **Browse for Plot File** dialog box will be displayed. Specify the file name and its location in this dialog box.

Partial Preview Window

The window displayed below the **Properties** button is called the **Partial Preview** window. The preview in this window dynamically changes as you modify the parameters in the **Plot** dialog box. The outer rectangle in this window is the paper you selected. It also shows the size of the paper. The inner hatched rectangle is the section of the paper that is used by the image. If the image extends beyond the paper, a red border is displayed around the paper.

Paper size Area

The drop-down list in this area displays all standard paper sizes for the selected plotting device. You can select any size from the list to make it current. If **None** has been selected currently from the **Name** drop-down list in the **Printer/plotter** area, AutoCAD will display the list of all standard paper sizes.

Number of copies Area

You can use the spinner available in this area to specify the number of copies that you want to plot. If multiple layouts and copies are selected and some of the layouts are set for plotting to a file or AutoSpool, they will produce a single plot. AutoSpool allows you to send a file for plotting while you are working on another program.

Plot area Area

Using the **What to plot** drop-down list provided in this area, you can specify the portion of the drawing to be plotted. You can also control the way the plotting will be carried out. The options in the **What to plot** drop-down list are described next.

Display

If you select this option, the portion of the drawing that is currently being displayed on the screen is plotted.

Extents

This option resembles the **Extents** option of the **ZOOM** command and prints the drawing to the extents of the objects. If you add more objects to the drawing, they are also included in the plot and the extents of the drawing are recalculated. If you reduce the drawing extents by erasing, moving, or scaling the objects, the extents of the drawing are again recalculated. You can use the **Extents** option of the **ZOOM** command to determine which objects shall be plotted. If you use the **Extents** option when the perspective view is on and the position of the camera is not outside the drawing extents, the following message is displayed: **Plot of perspective view has been scaled to fit available area**.

Limits

This option is available only if you plot from the **Model** tab. Selecting this option plots the complete area defined within the drawing limits.

Note
*To be able to clearly view the differences between the three previously listed plotting options, it may be a good idea to make sure that the default scale options have been selected. Alternatively, select the **Fit to paper** check box from the **Plot scale** area of the dialog box if you are in the **Model** tab, and select **1:1** if you are working in any one of the layout tabs.*

Window

On selecting this option, you need to specify the section of the drawing to be plotted by defining a window. To define a window, select the **Window** option from the **What to plot** drop-down list. The **Plot** dialog box will be temporarily closed and you will be prompted to specify two points on the screen that define a window; the area within this window will be plotted. Once you have defined the window, the **Plot** dialog box is redisplayed on the screen. You will notice that the **Window** button is displayed on the right of the **What to plot** drop-down list now. To reselect the area to be plotted, choose the **Window** button. If the **Model** tab is chosen, you will notice that the previously selected area is displayed in white and the remaining area is displayed in gray. After selecting the area to be plotted, you can choose the **OK** button in the dialog box to plot the drawing.

Note
*Sometimes, while using the **Window** option, the area you have selected may appear clipped off. This may happen because the objects are too close to the window you have defined on the screen. You need to redefine the window in this situation. Such errors can be avoided by using the **Preview** option discussed later.*

View

Selecting the **View** option enables you to plot a view that was created with the **VIEW** command. The view must be defined in the current drawing. If no view has been created, the **View** option is not displayed. When you select this option, a drop-down list is displayed in this area. You can select a view for plotting from this drop-down list and then choose **OK** in the **Plot** dialog box. While using the **View** option, the specifications of the plot will depend on the specifications of the named view.

Layout

This option is available only when you are plotting from the layout. This option prints the entire content of the drawing that lies inside the printable area of the paper selected from the drop-down list in the **Paper size** area.

Plot offset (origin set to printable area) Area

This area allows you to specify an offset of the plotting area from the lower left corner of the paper. The lower left corner of a specified plot area is positioned at the lower left margin of the paper by default. If you select the **Center the plot** check box, AutoCAD automatically centers the plot on the paper by calculating the X and Y offset values. You can specify an offset from the origin by entering positive or negative values in the **X** and **Y** edit boxes. For example, if you want the drawing to be plotted 4 units to the right and 4 units above the origin point, enter **4** in both the **X** and **Y** edit boxes. Depending on the units you have specified in the **Paper size** area of the dialog box, the offset values are either in inches, in millimeters, or in pixels.

Plot scale Area

This area controls the drawing scale of the plot area. The **Scale** drop-down list has thirty-three architectural and decimal scales by default apart from **Custom** option. The default scale setting is **1:1** when you are plotting a layout. However, if you are plotting in a **Model** tab, the **Fit to paper** check box is selected. The **Fit to paper** option allows you to automatically fit the entire drawing on the paper. It is useful when you have to print a large drawing using a printer that uses a smaller size paper.

Whenever you select a standard scale from the drop-down list, the scale is displayed in the edit boxes as a ratio of the plotted units to the drawing units. You can also change the scale factor manually in these edit boxes. When you do so, the **Scale** edit box displays **Custom**. For example, for an architectural drawing, which is to be plotted at the scale 1/4"=1'-0", you can enter either 1/4"=1'-0" or 1=48 in the edit boxes.

Note

*The **PSLTSCALE** system variable controls the paper space linetype scaling and has a default value of 1. This implies that irrespective of the zoom scale of the viewports, the linetype scale of the objects in the viewports remains the same. If you want the linetype scale of the objects in different viewports with different magnification factors to appear different, you should set the value of the **PSLTSCALE** variable to 0. This has been discussed in detail in Chapter 11 (Model Space Viewports, Paper Space Viewports, and Layouts).*

The **Scale lineweights** check box is available only if you are plotting in a layout tab. This option is not available in the **Model** tab. If you select the **Scale lineweights** check box, you can scale lineweights in proportion to the plot scale. Lineweights generally specify the linewidth of the printable objects and are plotted with the original lineweight size, regardless of the plot scale.

Note

*You can save your custom scales that you use during plotting or in layouts in the **Edit Scale List** dialog box. To invoke the **Edit Scale List** dialog box, choose **Annotation Scaling > Scale List** from the **Annotate** tab of the **Ribbon**. The **Scale List** area lists the default scales in AutoCAD. Choose the **Add** button to invoke the **Add Scale** dialog box. Enter a name for the custom scale in the **Name appearing in the scale list** edit box. Next, enter the scale in the **Scale Properties** area and choose the **OK** button; the name specified for the custom scale will be listed in the **Scale List** area. In this way, you can keep a track of the custom scales used.*

Plot style table (pen assignments) Area

This area will be available when you expand the **Plot** dialog box. This area allows you to view and select a plot style table, edit the current plot style table, or create a new plot style table. A plot style table is a collection of plot styles. A plot style is a group of pen settings that are assigned to an object or layer and that determine the color, linetype, thickness, line ending, line joining and the fill style of drawing objects when they are plotted. It is a named file that allows you to control the pen settings for a plotted drawing.

You can select the required plot style from the drop-down list in this area. Whenever you select a plot style, AutoCAD displays the **Question** message box asking you to specify whether the selected plot style should be assigned to all layouts or not. If you choose **Yes**, the selected plot style will be used to plot from all layouts. You can also select **None** from the drop-down list to plot a drawing without using any plot styles. You can assign different plot style tables to a drawing and plot the same drawing differently each time. The use of plot styles will be discussed later in this chapter.

You can also select **New** from the drop-down list to create a new plot style. When you select this option, a wizard will be started that will guide you through the process of creating a new plot style.

Note

You will learn more about creating plot styles later in this chapter.

Edit

You can edit a plot style table you have selected from the **Plot style table** drop-down list by choosing the **Edit** button. This button is not available when you have selected **None** from the drop-down list. When you choose the **Edit** button, AutoCAD displays the **Plot Style Table Editor**, where you can edit the selected plot style table. This dialog box has three tabs: **General**, **Table View**, and **Form View**. The **Plot Style Table Editor** will be discussed later in the "Using Plot Styles" section of this chapter.

Shaded viewport options Area

The options in this area are used to print a shaded or a rendered image. These options are discussed next.

Chapter 14

Shade plot

This drop-down list is used to select a technique that will be used for plotting the drawings. This drop-down list will be available only while plotting from the model space. If you select **As displayed** from this drop-down list, the drawing will be plotted as it is displayed on the screen. If the drawing is hidden, shaded, or rendered, it will be printed as it is. The hidden geometry consists of objects that lie behind the facing geometry and displays the object as it would be seen in reality. If you select the **Wireframe** option, the model will be printed in wireframe mode displaying all hidden geometries, even if it is shaded in the drawing. Selecting the **Hidden** option will plot the drawing with the hidden lines of the object suppressed. On selecting the **3D Hidden**, **3D Wireframe**, **Conceptual**, or **Realistic** option, the drawing will be plotted in **3D Hidden**, **3D Wireframe**, **Conceptual**, or **Realistic** visual style respectively, even if the model is displayed in some other visual styles. Similarly, selecting the **Rendered** option will plot the rendered image of the drawing. The other options in the drop-down list are used to specify the quality for plotting the shaded and rendered model. But, if you are plotting a drawing from the paper space, the display settings of the viewport will be used for plotting it. To set the display settings of the viewport, select the viewport and right-click; a shortcut menu will be displayed. Choose the **Shade plot** option from the shortcut menu and then select the desired shading option.

Quality

This drop-down list is used to select printing quality in terms of dots per inch (dpi) for the printed drawing. The **Draft** option prints the drawing with 0 dpi, which results in the wireframe printout. The **Preview** option prints the drawing at 50 dpi, the **Normal** option prints the drawing at 100 dpi, the **Presentation** option prints the drawing at 200 dpi, the **Maximum** option prints the drawing at the selected plotting device's maximum dpi. You can also specify a custom dpi by selecting the **Custom** option from this drop-down list. The custom value of dpi can be specified in the **DPI** drop-down list, which is enabled below the **Quality** drop-down list when on selecting the **Custom** option.

Note

*Selecting the **Draft** option from the **Quality** drop-down list prints the drawing in wireframe even if you select **Rendered** from the **Shade plot** drop-down list. Also, to plot the drawings in layouts with the hidden lines suppressed, you need to use the **Shade plot** option discussed in Chapter 11, Model Space Viewports, Paper Space Viewports, and Layouts. You will learn more about wireframe, hidden, shaded, and rendered models in later chapters.*

Plot options Area

This area displays six options that can be selected as per the plot requirements. They are described next.

Plot in background

Select this check box to perform the printing and plotting operation in the background. While plotting in the background, you can return immediately to your drawing without waiting for the printing or plotting operation to finish. By default, the background plotting is turned off for plotting and it is turned on for publishing. The **BACKGROUNDPLOT** system variable is used to control the default settings for background plotting.

Plot object lineweights

This check box is not available if the **Plot with plot styles** check box is selected. To activate this option, clear the **Plot with plot styles** check box. Select this check box, if you need to plot the drawing with specified lineweights. To plot the drawing without specified lineweights, clear this check box.

Plot with plot styles

When you select the **Plot with plot styles** check box, AutoCAD plots using the plot styles applied to the objects in the drawing and defined in the plot style table. The different property characteristics associated with the different style definitions are stored in the plot style tables and can be easily attached to the geometry. This setting replaces the pen mapping used in earlier versions of AutoCAD.

Plot paperspace last

This check box is not available when you are in the **Model** tab because no paper space objects are present in the **Model** tab. This option is available when you are working in the layout tab. By selecting the **Plot paperspace last** check box, you can plot the model space geometry before paper space objects. Usually the paper space geometry is plotted before the model space geometry. This option is also useful when there are multiple tabs selected for plotting and you want to plot the model space geometry before the layout tabs.

Hide paperspace objects

This check box is also available in the **Layout** tab and is used to specify whether or not the objects drawn in layouts will be hidden while plotting. If this check box is selected, the objects created in layouts will be hidden.

Plot stamp on

This check box is selected to turn the plot stamp on. The plot stamp is a user-defined information that will be displayed on the sheet while plotting. You can set the plot stamp information when you select this check box. When you select this check box, the **Plot Stamp Settings** button is displayed on the right of this check box. You can choose this button to display the **Plot Stamp** dialog box to set the parameters for the plot stamp.

Save changes to layout

This check box is selected to save the changes made using the **Plot** dialog box and apply them to the layout selected to be plotted.

Drawing orientation Area

This area provides options that help you specify the orientation of the drawing on the paper for the plotters that support landscape or portrait orientation. You can change the drawing orientation by selecting the **Portrait** or **Landscape** radio button, with or without selecting the **Plot upside-down** check box. The paper icon displayed on the right side of this area indicates the media orientation of the selected paper and the letter icon (A) on it indicates the orientation of the drawing on the page. The **Landscape** radio button is selected by default for AutoCAD drawings and orients the length of the paper along the *X* axis, that is horizontally. If we assume this orientation to be at a rotation angle of 0-degree, when selecting the **Portrait** radio button, the plot is oriented with the width along the *X* axis, which is equivalent to the plot being rotated through a rotation angle of 90-degree. Similarly, if you select both the **Landscape** radio button and the **Plot upside-down** check box at the same time, the plot gets rotated through a rotation angle of 180-degree and if you select both the **Portrait** radio button and the **Plot upside-down** check box at the same time, the plot gets rotated through a rotation angle of 270-degree. The AutoCAD screen conforms to the landscape orientation by default.

Preview

When you choose the **Preview** button, AutoCAD displays the drawing on the screen just as it would be plotted on the paper. Once the regeneration is performed, the dialog boxes on the screen disappear, and an outline of the paper size is shown. In the plot preview (Figure 14-3), the cursor is replaced by the **Zoom Realtime** icon. This icon can be used to zoom in and out interactively by pressing and moving the left mouse button. You can right-click to display a shortcut menu and then choose **Exit** to exit the preview or press the ENTER or ESC key to return to the dialog box. You can also choose **Plot** to plot the drawing right away or choose the other zooming options available.

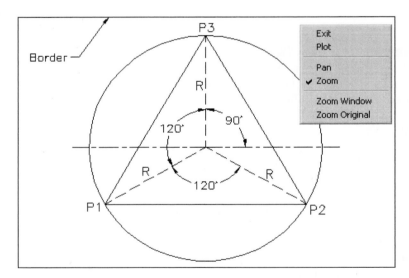

Figure 14-3 *Full plot preview with the shortcut menu*

Tip
*You can also choose **Plot Preview** from the **Quick Access Toolbar** to preview the plot, and at the same time, bypass the **Plot** dialog box.*

After finishing all the settings and other parameters, if you choose the **OK** button in the **Plot** dialog box, AutoCAD starts plotting the drawing in the file or plotters as specified. Also, the

Plot Job Progress dialog box (Figure 14-4), where you can view the actual progress of plotting will be displayed.

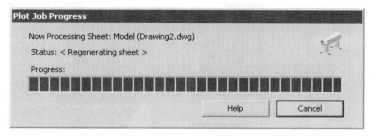

Figure 14-4 The **Plot Job Progress** *dialog box*

ADDING PLOTTERS

Application Menu:	Print > Manage Plotters	**Command:** PLOTTERMANAGER
Ribbon:	Output > Plot > Plotter Manager	

In AutoCAD, the plotters can be added using the **PLOTTERMANAGER** command. This command is discussed next.

The PLOTTERMANAGER Command

When you invoke the **PLOTTERMANAGER** command, AutoCAD will display the **Plotters** window, see Figure 14-5.

The **Plotters** window is basically a Windows Explorer window. It displays all the configured plotters and the **Add-A-Plotter Wizard** icon. You can right-click on any one of the icons belonging to the plotters that have already been configured to display a shortcut menu. You can choose **Delete** from the shortcut menu to remove a plotter from the list of available plotters in the **Name** drop-down list in the **Plot or Page Setup** dialog box. You can also choose **Rename** from the shortcut menu to rename the plotter configuration file or choose **Properties** to view the properties of the configured device.

Add-A-Plotter Wizard

If you double-click on the **Add-A-Plotter Wizard** icon in the **Plotters** window, AutoCAD guides you to configure a nonsystem plotter for plotting your drawing files. AutoCAD stores all the information of a configured plotter in configured plot (PC3) files. The PC3 files are stored in the *Documents and Settings\<owner>\Application Data\Autodesk\AutoCAD 2010\R18\enu\Plotters* folder by default. The steps for configuring a new plotter using the **Add-A-Plotter Wizard** are as follows:

1. Open the **Plotters** window by choosing the **Plotter Manager** button from the **Plot** panel in the **Ribbon** and double-click on the **Add-A-Plotter Wizard** icon.
2. In the **Add Plotter - Introduction Page**, read the introduction carefully, and then choose the **Next** button to advance to the **Add Plotter - Begin** page.

*Figure 14-5 The **Plotters** window*

3. On the **Add Plotter - Begin** page, the **My Computer** radio button is selected by default.
    Choose the **Next** button; the **Add Plotter - Plotter Model** page is displayed.

4. On this page, select a manufacturer and model of your nonsystem plotter from the
    Manufacturers and **Models** list boxes, respectively. Now, choose the **Next** button. The
    Add Plotter - Import Pcp or Pc2 page is displayed.

    If your plotter is not present in the list of available plotters, and you have a driver disk for
    your plotter, choose the **Have Disk** button to locate the *.hif* file from the driver disk, and
    install the driver supplied with your plotter.

5. In the **Add Plotter - Import Pcp or Pc2** page, if you want to import configuring
    information from a PCP or a PC2 file created with a previous version of AutoCAD, you
    can choose the **Import File** button and select the file. Otherwise, simply choose the **Next**
    button to advance to the next page.

6. On the **Add Plotter - Ports** page, select the port from the list to be used when plotting,
    and choose **Next**.

7. On the **Add Plotter - Plotter Name** page, you can specify the name of the currently configured plotter or the default name will be entered automatically. Choose **Next**.

8. When you reach the **Add Plotter - Finish** page, you can choose the **Finish** button to exit the **Add-A-Plotter Wizard**.

You can also choose the **Edit Plotter Configuration** button to display the **Plotter Configuration Editor** dialog box where you can edit the current plotter's configuration. Also, in this page you can choose the **Calibrate Plotter** button to display the **Calibrate Plotter** wizard. This wizard allows you to calibrate your plotter by setting up a test measurement. After test plotting, it compares the plot measurements with the actual measurements and computes a correction factor.

Once you have chosen **Finish** to exit the wizard, a PC3 file for the newly configured plotter will be displayed in the **Plotters** window. This PC3 file contains all the settings needed by the plotter to plot. Also, the newly configured plotter name is added to the **Name** drop-down list in the **Plotter configuration** area of the **Plot Device** tab in the **Plot** dialog box. You can now use the plotter for plotting.

EDITING THE PLOTTER CONFIGURATION

You can modify the properties of a selected plot device by using the **Plotter Configuration Editor** dialog box. This dialog box can be invoked in several ways. As discussed earlier, when using the **Plot** or **Page Setup** dialog box, you can choose the **Properties** button in the **Printer/plotter** area to display the **Plotter Configuration Editor** dialog box, see Figure 14-6.

You can modify the default settings of a plotter while configuring it, by choosing the **Edit Plotter Configuration** button on the **Add Plotter - Finish** page of the **Add Plotter** wizard. You can also select the PC3 file for editing in the **Plotters** window using Windows Explorer (by default, PC3 files are stored in the *\Documents and Settings\<owner>\Application Data\Autodesk\AutoCAD 2010\R18\enu\Plotters* folder) and double-click on the file or right-click on the file and choose **Open** from the shortcut menu. The three tabs in the **Plotter Configuration Editor** dialog box are discussed next.

General Tab

This tab contains basic information about the configured plotter or the PC3 file. You can make changes only in the **Description** area. The rest of the information in the tab is read only. This tab contains information on the configured plotter file name, plotter driver type, HDI driver file version number, name of the system printer (if any), and the location and name of the PMP file (if any calibration file is attached to the PC3 file).

Ports Tab

This tab contains information about the communication between the plotting device and your computer. You can choose between a serial (local), parallel (local), or network port.

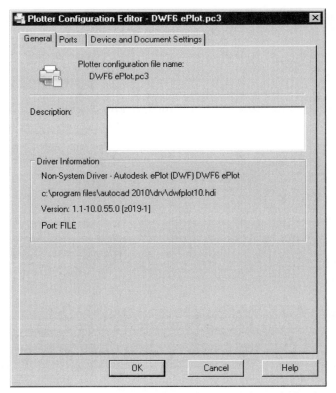

Figure 14-6 *The **Plotter Configuration Editor** dialog box*

The default settings for parallel and serial ports are **LPT1** and **COM1**, respectively. You can also change the port name, if your device is connected to a different port. You can also select the **Plot to File** radio button, if you want to save the plot as a file. You can select **AutoSpool**, if you want plotting to occur automatically, while you continue to work on another application.

Device and Document Settings Tab

This tab contains the plotting options specific to the selected plotter and are displayed as a tree view in the window. For example, if you configure a nonsystem plotter, you have the option to modify the pen characteristics. You can select any plotter properties from the tree view displayed in the window to change the values as required. Whenever you select an icon from the tree view in the window, the corresponding information is displayed in an area below. For example, if you select **PMP File Name <None>** in the tree view of the window, a **PMP file** area is displayed below. This area contains the current settings and options to modify it. Information that is displayed within brackets (<>) can be modified. By default, **Custom properties** is displayed as highlighted in the window because it contains the properties that are modified commonly. The **Access Custom Dialog** area is displayed below the window. Choosing the **Custom Properties** button in this area displays a dialog box specific to the selected plotter. This dialog box has several properties of the selected plotter grouped and displayed under various areas and can be modified here.

Once you have made the desired changes, choose **OK** to exit the dialog box and then choose **Save As** in the **Plotter Configuration Editor** dialog box to save the changes you just made to the PC3 file. You can also import old plot configuration files (PCP or PC2) from the previous releases of AutoCAD using the **Import** button and save some of the information from these settings as a new PC3 file. When you choose the **Import** button, the **Plotting Components** dialog box is displayed. This dialog box displays what to use while importing AutoCAD 14 information into an AutoCAD 2010 drawing. For example, it tells you that to import PCP or PC2 file setting into a drawing in AutoCAD 2010, you should use the **Import PCP or PC2 Plot Settings** Wizard. Choose **OK** to exit this dialog box; the **Import** dialog box is displayed. Here you can select the file to be imported and then choose the **Import** button.

Tip

It is better to create a new PC3 file for a plotter and keep the original file as it is so that you encounter no error while using the specific printer later. The PC3 files determine the proper function of a plotter and any modifications may lead to errors.

IMPORTING PCP/PC2 CONFIGURATION FILES

If you want to import a PCP or PC2 configuration file or plot settings created by previous releases of AutoCAD into the **Model** tab or the current layout for the drawing, you can use the **PCINWIZARD** command to display the **Import PCP or PC2 Plot Settings** Wizard. All information from a PCP or PC2 file regarding plot area, rotation, plot offset, plot optimization, plot to file, paper size, plot scale, and pen mapping can be imported. Read the **Introduction** page of the wizard that is displayed carefully and then choose the **Next** button. The **Browse File name** page is displayed. Here, you can either enter the name of the PCP or PC2 file directly in the **PC2 or PCP file name** edit box or then choose the **Browse** button to display the **Import** dialog box, where you can select the file to be imported. After you specify the file for importing, choose **Import** to return to the wizard. Choose the **Next** button to display the **Finish** page. After importing the files, you can modify the rest of the plot settings for the current layout.

SETTING PLOT PARAMETERS

Before starting with the drawing, you can set various plotting parameters in the **Model** tab or in the Layouts. The plot parameters that can be set include the plotter to be used, for example, plot style table, the paper size, units, and so on. All these parameters can be set using the **PAGESETUP** command discussed next.

Working with Page Setups

Toolbar: Layouts > Page Setup Manager	**Application Menu:** Print > Page Setup
Ribbon: Output > Plot > Page Setup Manager	**Command:** PAGESETUP

As discussed earlier, a page setup contains the settings required to plot a drawing. Each layout, as well as the **Model** tab, can have a unique page setup attached to it. You can use the **PAGESETUP** command to create named page setups that can be used later. A page setup consists of specifications for the layout page, plotting device, paper size, and settings for the layouts to be plotted. The **PAGESETUP** command can also be invoked from the shortcut menu by right-clicking on the current **Model** or **Layout** tab and choosing **Page Setup Manager**. Remember that the **Page Setup Manager** option will be available in the shortcut menu only for the current **Model** or **Layout** tab.

When you invoke the **PAGESETUP** command, AutoCAD displays the **Page Setup Manager** dialog box. The names of the tabs displayed in the **Current page setup** list box of the **Page setups** area depend on the tab from which you invoke this dialog box. For example, if you invoke this dialog box from the **Model** tab, it displays only **Model** in this list box. However, if you invoke this dialog box from the **Layout** tab, it displays the list of all the layouts that are invoked at least once. Figure 14-7 shows the **Page Setup Manager** dialog box invoked from the **Layout** tab. In this case, both **Layout1** and **Layout2** were activated at least once.

You can use this dialog box to create a new page setup, modify the existing page setup, or import a page setup from an existing file.

*Figure 14-7 The **Page Setup Manager** dialog box displayed on invoking from the **Layout1** tab*

Creating a New Page Setup

To create a new page setup, choose the **New** button from the **Page Setup Manager** dialog box. The **New Page Setup** dialog box will be displayed, as shown in Figure 14-8. Enter the name of the new page setup in the **New page setup name** text box. The existing page setups with which you can start are shown in the **Start with** area. You can select any of the page setups listed in this area and choose **OK** to proceed.

Figure 14-8 *The **New Page Setup** dialog box*

When you choose **OK**, the **Page Setup** dialog box will be displayed. This dialog box is similar to the **Plot** dialog box, see Figure 14-9.

Figure 14-9 *The **Page Setup** dialog box*

Modifying a Page Setup

To modify a page setup, select the page setup from the **Current page setup** list box and choose the **Modify** button. The **Page Setup** dialog box will be displayed. Modify the parameters in this dialog box and exit it.

Note
*If you select the **Display when creating a new layout** check box, the **Page Setup Manager** will be displayed whenever you invoke a layout for the first time.*

Importing a Page Setup

Command:	PSETUPIN

AutoCAD allows you to import a user-defined page setup from an existing drawing and use it in the current drawing or base the current page setup for the drawing on it. This option is available by choosing the **Import** button from the **Page Setup Manager** dialog box. It is also possible to bypass this dialog box and directly import a page setup from an existing drawing into a new drawing layout by using the **PSETUPIN** command. This command facilitates importing a saved and named page setup from a drawing into a new drawing. The settings of the named page setup can be applied to layouts in the new drawing. When you choose the **Import** button from the **Page Setup Manager** dialog box or invoke the **PSETUPIN** command, the **Select Page Setup From File** dialog box is displayed, as shown in Figure 14-10. You can use this dialog box to locate a *.dwg*, *.dwt*, or *.dwf* file whose page setups have to be imported.

*Figure 14-10 The **Select Page Setup From File** dialog box*

After you select the file, AutoCAD displays the **Import Page Setups** dialog box, as shown in Figure 14-11. You can also enter **-PSETUPIN** at the Command prompt to display prompts at the command line.

*Figure 14-11 The **Import Page Setups** dialog box*

Note

*If a page setup with the same name already exists in the current file, the **AutoCAD Alert** box is displayed and you will be informed that "A page setup with the same name already exists in the current file, do you want to redefine it?" If you choose **Yes** in this dialog box, the current page setup will be redefined.*

USING PLOT STYLES

The plot styles can change the complete look of a plotted drawing. You can use this feature to override a drawing object's color, linetype, and lineweight. For example, if an object is drawn on a layer that is assigned the red color and no plot style is assigned to it, the object will be plotted as red. However, if you have assigned a plot style to the object with the color blue, the object will be plotted as blue, irrespective of the layer color it was drawn on. Similarly, you can change the Linetype, Lineweight, end, join, and fill styles of the drawing, and also change the output effects such as dithering, grayscales, pen assignments, and screening. Basically, you can use **Plot Styles** effectively to plot the same drawing in various ways.

Every object and layer in the drawing has a plot style property. The plot style characteristics are defined in the plot style tables attached to the **Model** tab, layouts, and viewports within the layouts. You can attach and detach different plot style tables to get different looks for your plots. Generally, there are two plot style modes. They are **Color-Dependent** and **Named**. The **Color-dependent** plot styles are based on object color and there are **255** color-dependent plot styles. It is possible to assign each color in the plot style a value for the different plotting properties and these settings are then saved in a color-dependent plot style table file that has a *.ctb* extension. Similarly, **Named** plot styles are independent of the object color and you can assign any plot style to any object regardless of that object's color. These settings are saved in a named plot style table file that has *.stb* extension. Every drawing in AutoCAD 2010 is in either of the plot style modes.

Adding a Plot Style

Application Menu:	Print > Manage Plot Styles	**Command:**	STYLESMANAGER

All plot styles are saved in the *\Application Data\Autodesk\AutoCAD 2010\R18\enu\Plot Styles* folder. If you enter **STYLESMANAGER** at the Command prompt, AutoCAD displays the **Plot Styles** window, see Figure 14-12. This window displays icons for all available plot styles, in addition to the **Add-A-Plot Style Table Wizard** icon. You can double-click on any of the plot style icons to display the **Plot Style Table Editor** dialog box and edit the selected plot style. When you double-click on the **Add-A-Plot Style Table Wizard** icon, the **Add Plot Style Table** wizard is displayed and you can use it to create a new plot style.

Add-A-Plot Style Table Wizard

If you want to add a new plot style table to your drawing, double-click on the **Add-A-Plot Style Table Wizard** in the **Plot Styles** window to display the **Add Plot Style Table** wizard. The following steps are required for creating a new plot style table using the wizard.

1. Read the introduction page carefully and choose the **Next** button.

2. In the **Begin** page, select the **Start from scratch** radio button and choose **Next**. Selecting this option creates a new plot style table. Therefore, the **Browse File** page is not available.

    In addition to the **Start from scratch** option, this page has three more options. They are **Use an existing plot style table**, **Use My R14 Plotter Configuration (CFG)**, and **Use a PCP or PC2 file**. When you use the **Use an existing plot style table** option, an existing plot style table is used as a base for the new plot style table you are creating. In such a situation, the **Table Type** page of the wizard is not available and is not displayed because the table type will be based on the existing plot style table you are using to create a new one. With the **Use My R14 Plotter Configuration (CFG)** option, the pen assignments from the Release 14 *acad.cfg* file are used as a base for the new table you are creating. If you are using the **Use a PCP or PC2** option, the pen assignments saved earlier in a Release 14 PCP or PC2 file are used to create the new plot style.

Figure 14-12 The **Plot Styles** window

3. The **Pick Plot Style Table** page allows you to select the **Color-Dependent Plot Style Table** or the **Named Plot Style Table** according to your requirement. Select the **Color-Dependent Plot Style Table** radio button and then choose the **Next** button.

4. Since you have selected the **Start from scratch** radio button in the **Begin** page, the **Browse File** page is not available and the **File name** page is displayed. However, if you had selected any of the other three options on the **Begin** page, the **Browse File** page would have been displayed. You can select an existing file from the drop-down list available in this page or choose the **Browse** button to display the **Select File** dialog box. You can then browse and select a file from a specific folder and choose **Select** to return to the wizard. You can also enter the name of the existing plot style table on which you want to base the new plot style table, directly in the edit box. After you have specified the file name, choose **Next** to display the **File name** page of the wizard. In the **File name** page, enter a file name for the new plot style table and choose **Next**. The **Finish** page is displayed.

 Note
*If you are using the pen assignments from the Release 14 acad.cfg file to define the new plot style table, you also have to specify the printer or plotter to use from the drop-down list available in the **Browse File** page of the wizard.*

5. The **Finish** page gives you the option of choosing the **Plot Style Table Editor** button to display the **Plot Style Table Editor** and then edit the plot style table you have created. If you select the **Use this plot style table for new and pre-AutoCAD 2010 drawings** check

Chapter 14

box in this page of the wizard, the plot style table that you have created will become the default plot style table for all the drawings you create. This check box is available only if the plot style mode you have selected in the wizard is the same as the default plot style mode specified by you in the **Default plot style behavior for new drawings** area of the **Plot Style Table Settings** dialog box. This dialog box can be invoked by choosing the **Plot Style Table Settings** button from the **Plot and Publish** tab of the **Options** dialog box. Choose **Finish** in the **Finish** page of the wizard to exit the wizard. A new plot style table gets added to the **Plot Styles** window and can be used for plotting.

Note

*You can also choose **Add Plot Style Table/ Add Color-Dependent Plot Style Table** from the **Tools > Wizards** menu to display wizards that are similar to the **Add Plot Style Table** wizard except that in the **Begin** page, the **Use an existing plot style table** option is not available. Also the **Table Type** page is not available and the **Finish** page has an additional option to use the new plot style table for the current drawing.*

Plot Style Table Editor

When you double-click on any of the plot style table icons in the **Plot Styles** window, the **Plot Style Table Editor** is displayed, where you can edit the particular plot style table. You can also choose the **Plot Style Table Editor** button in the **Finish** page to display the **Plot Style Table Editor**. Alternatively, expand the **Plot** dialog box and choose the button adjacent to the drop-down list in the **Plot style table (pen assignments)** area to display the **Plot Style Table Editor** dialog box.

The **Plot Style Table Editor** has three tabs: **General**, **Table View**, and **Form View**. You can edit all properties of an existing plot style table using the different tabs. The description of the tabs is as follows.

General Tab

This tab provides information about the file name, location of the file, version, and scale factor. All information except the description is read only. You can enter a description about the plot style table in the **Description** text box here. If you select the **Apply global scale factor to non-ISO linetypes** check box, all the non-ISO linetypes in a drawing are scaled by the scale factor specified in the **Scale factor** edit box below the check box. If this check box is cleared, the **Scale factor** edit box is not available.

Table View Tab

This tab displays all plot styles along with their properties in a tabular form and they can be edited individually in the table, see Figure 14-13.

In case of a named plot style table, you can edit existing styles or add new styles by choosing the **Add Style** button. On doing so, a new column with the default style name **Style1** will be added in the table. You can change the style name if you want. To edit various properties in the table, select the particular value to be modified; the corresponding drop-down list will be displayed. You can select a value from this drop-down list. This manner of editing is

*Figure 14-13 The **Plot Style Table Editor** with the **Table View** tab chosen*

similar to the one you use when editing the properties of objects using the **Properties** palette. The **Normal** plot style table is not available, and therefore, it cannot be edited. This plot style is assigned to layers by default.

You can select a particular plot style by clicking on the gray bar above the column and the entire column gets highlighted. You can select several plot styles by pressing the SHIFT key and selecting more plot styles. All the plot styles that are selected are highlighted. If you choose the **Delete Style** button now, the selected plot styles are removed from the table.

Choosing the **Edit Lineweights** button displays the **Edit Lineweights** dialog box. You can select the units for specifying the lineweights in the **Units for Listing** area of the dialog box, and can use either **Millimeters (mm)** or **Inches (in)**. You can also edit the value of a lineweight by selecting it in the **Lineweights** list box and choosing the **Edit Lineweight** button. After you have edited the lineweights, choose the **Sort Lineweights** button to rearrange the lineweight values in the list box. Choose **OK** to exit the **Edit Lineweights** dialog box and return to **Plot Style Table Editor**.

With a color-dependent plot style table, the **Table View** tab displays all 255 plot styles, one for each color, and the properties can be edited in the table in the same way as discussed for

Chapter 14

named plot styles. The only difference is that you cannot add a new plot style or delete an existing one, and therefore, the **Delete Style** and **Add Style** buttons are not available. The properties that can be defined in a plot style are discussed next.

Name. This field displays the name of the color in the case of the color-dependent plot styles and the name of the style in the case of the named plot styles.

Description. Here you can enter the description about the plot style.

Color. The color you assign to the plot style overrides the color of the object in the drawing. The default value is Use object color.

Enable dithering. Dithering is described as a mixing of various colored dots to produce a new color. You can enable or disable this property by selecting from the drop-down list that is available in this field. Dithering is enabled by default and is independent of the color selected.

Convert to grayscale. If you are using a plotter that supports grayscaling, selecting **On** from the drop-down list in this field applies a grayscale to the objects color. By default, Off is selected and the object's color is used.

Use assigned pen #. This property is applied only to pen plotters. The pens range from 1 to 32. The default value is Automatic, which implies that the pen used will be based on the plotter configuration.

Virtual pen #. Nonpen plotters can behave like pen plotters using virtual pens. The value of this property lies between 1 and 255. The default value is Automatic. This implies that AutoCAD will assign a virtual pen automatically from the AutoCAD Color Index (ACI).

Screening. This property of a plot style indicates the amount of ink used while plotting. The value ranges between 0 and 100. The default value is 100, which creates the plot in its full intensity. Similarly, a value of 0 produces white color. This may be useful when plotting on a colored background.

Linetype. Like the property of color, the linetype assigned to a plotstyle overrides the object linetype on plotting. The default value is Use object linetype.

Adaptive adjustment. This property is applied by default and implies that the linetype scale of a linetype is applied such that on plotting, the linetype pattern will be completed.

Lineweight. The lineweight value assigned here overrides the value of the object lineweight on plotting. The default value is Use object lineweight.

Line End Style. This determines the manner in which a plotted line ends. The effect of this property is more noticeable when the thickness of the line is substantial. The line can end in a **Butt**, **Square**, **Round**, or **Diamond** shape. The default value is Use object end style.

Line Join style. You can select the manner in which two lines join in a plotted drawing. The available options are **Miter**, **Bevel**, **Round**, and **Diamond**. The default value is Use object join style.

Fill Style. The fill style assigned to a plot style overrides the objects fill style, when plotted. The available options are **Solid**, **Checkerboard**, **Crosshatch**, **Diamonds**, **Horizontal Bars**, **Slant Left**, **Slant Right**, **Square Dots**, and **Vertical Bars**.

Form View Tab

This tab displays all properties in one form, see Figure 14-14. Also, it displays all available plot styles in the **Plot styles** list box. You can select any style from the list box and then edit its properties in the **Properties** area.

*Figure 14-14 The **Plot Style Table Editor** with the **Form View** tab chosen*

If you want to add a new plot style to the named plot style table, choose the **Add Style** button when the **Form View** tab is chosen; the **Add Plot Style** dialog box with the default plot style name **Style 1** will be displayed. You can change the name of a plot style in the **Plot Style** edit box, see Figure 14-15. Now, when you choose the **OK** button, the new style will be added in the **Plot Styles** list box and you can select it for editing. If you want to delete a style, select it and then choose the **Delete Style** button. With color-dependent plot styles, the **Form View** tab also does not provide options that allow you to add or delete plot styles.

After editing, choose the **Save & Close** button to save and return to the **Plot Styles** window. You can also choose the **Save As** button to display the **Save As** dialog box and save a plot style table with another name. To remove the changes you made to a plot style table, choose the **Cancel** button.

Applying Plot Styles

Figure 14-15   *The **Add Plot Style** dialog box*

The **Model**, or any of the layout tabs, can be assigned a plot style table. The plot style table can be either named or color-dependent, as discussed earlier. These plot style modes for a new drawing can be determined from the **Default plot style behavior for new drawings** area of the **Plot Style Table Settings** dialog box. This dialog box can be invoked by choosing the **Plot Style Table Settings** button from the **Plot and Publish** tab of the **Options** dialog box. The **Use color dependent plot styles** option is selected by default and the drawings are assigned a color-dependent plot style. If you select the **Use named plot styles** radio button, the new drawings (not the current drawing) will be assigned a named plot style. The **PSTYLEPOLICY** system variable also controls the default plot style modes of the new drawings. A value of 0 implies a named plot style mode and a value of 1 implies a color-dependent plot style mode.

You can select a plot style table that you want to use as a default for drawings from the **Default plot style behavior for new drawings** area of the **Plot Style Table Settings** dialog box. If the **None** option is selected, the drawing will be plotted with the object properties as displayed on the screen. When you select the **Use named plot styles** radio button from the **Default plot style behavior for new drawings** area of the **Plot Style Table Settings** dialog box, the **Default plot style for layer 0** and **Default plot style for objects** drop-down lists will be available. You can select the default plot styles that you want to assign to Layer 0 and to the objects in a drawing from these drop-down lists, respectively. If you choose the **Add or Edit Plot Style Tables** button from the **Plot Style Table Settings** dialog box, the **Plot Styles** window will be displayed. Here, you can double-click on any plot style table icon available and edit it using the **Plot Style Table Editor** that is displayed.

To change the plot style table for a current layout, you have to invoke the **Page Setup** or **Plot** dialog box and then select a plot style table from the **Plot styles** drop-down list in the **Plot style table (pen assignments)** area of the dialog box. A color-dependent plot style table can be selected and applied to a tab only if the default plot style mode already has been set to color dependent. Similarly, to apply a named plot style table to a tab, select the **Use named plot styles** radio button from the **Plot Style Table Settings** dialog box.

A color-dependent plot style cannot be applied to objects or layers and therefore the **Plot Style Control** drop-down list is not available in the **Properties** toolbar when a drawing has a color-dependent plot style mode. The plot styles also appear grayed out in the **Layer Properties Manager** dialog box and cannot be selected and changed. However, named plot styles can be applied to objects and layers. A plot style applied to an object overrides the plot style applied to the layer on which the object is drawn. To apply a plot style to a layer, invoke the **Layer Properties Manager** dialog box where all the layers in the selected tab are displayed. Select a layer to which you want to apply a plot style and select the default plot style (Normal) currently applied to the layer; the **Select Plot Style** dialog box will be displayed, see Figure 14-16. The **Plot styles** list box in this dialog box displays all the plot styles present in the plot style table attached to the current tab. You can select another plot style table to be attached to the current tab from the **Active plot style table** drop-down list. You will notice that all the plot styles in the selected plot style table are displayed in the list box now. You can also choose the **Editor** button to display the **Plot Style Table Editor** to edit plot style tables, as discussed earlier. The **Select Plot Style** dialog box also displays the name of the original plot style assigned to the object adjacent to **Original**. Also, the new plot style to be assigned to the selected object is displayed next to **New**.

Figure 14-16 The Select Plot Style dialog box

You can apply a named plot style to an object using the **Plot Style Control** drop-down list in the **Properties** toolbar or the **Properties** palette. The process is the same as that applied for layers, colors, linetypes, and lineweights. This named plot style is applied to an object irrespective of the tab on which it is drawn. If the plot style assigned to an object is present in the plot style table of the tab in which it is present, the object is plotted with the specified plot style. However, if the plot style assigned to the object is not present in the plot style table assigned to the tab on which it is drawn, the object will be plotted with the properties that are displayed on the screen. The default plot style assigned to an object is **Normal** and the default plot style assigned to a layer is **ByLayer**.

Setting the Current Plot Style

You can use the **PLOTSTYLE** command to set the current plot style of new objects or of the selected object. When you enter **PLOTSTYLE** at the Command prompt, and if no object has been selected in the drawing, AutoCAD displays the **Current Plot Style** dialog box, see Figure 14-17. However, if any object selection is there in the drawing, then AutoCAD displays the **Select Plot Style** dialog box (Figure 14-16), which has been discussed earlier. You can select a plot style from the list box and choose **OK** to assign it to the selected objects in the drawing. All the plot styles present in the current plot style table that are assigned to the current tab are displayed in the list box in the **Current Plot Style** dialog box. You can select any one of these plot styles and choose **OK**. Now, when you create new objects, they will have the plot style that you had set current in the **Current Plot Style** dialog box. The parameters of this dialog box are described next.

Chapter 14

Figure 14-17 The **Current Plot Style** *dialog box*

Current plot style

The name of the current plot style is displayed adjacent to this label.

Plot style list box

This list box lists all the available plot styles that can be assigned to an object, including the default plot style, **Normal**.

Active plot style table

This drop-down list displays the names of all the available plot style tables. The current plot style table attached to the current layout or viewport is displayed in the edit box.

Editor

If you choose the **Editor** button, adjacent to the drop-down list, AutoCAD displays the **Plot Style Table Editor** to edit the selected plot style table.

Attached to

The tab to which the selected plot style table is attached, such as **Model** or any one of the layout tabs, is displayed next to this label.

Note
*As soon as the drawing is plotted, the **Plot/Publish Details Report Available** button is displayed in the Status Bar tray. Choose this button to see the plot and publish details. This option has already been discussed in Chapter 1.*

Exercise 1   *Mechanical*

Create the drawing shown in Figure 14-18. Create a named plot style table *My Named Table.stb* with three plot styles: Style 1, Style 2, and Style 3, in addition to the Normal plot style. The Normal plot style is used for plotting the object lines. These three styles have the following specifications:

Figure 14-18 Drawing for Exercise 1

Style 1. This style has a value of Screening = 50. The dimensions, dimension lines, and the text in the drawing must be plotted with this style.

Style 2. This style has a value of Lineweight = 0.800. The border and title block must be plotted with this style.

Style 3. This style has a linetype of Medium Dash. The centerlines must be plotted with this plot style.

PLOTTING SHEETS IN A SHEET SET

Using the **Sheet Set Manager**, you can easily plot all the sheets available in a sheet set. However, before plotting the sheets in a sheet set, you need to make sure that you have selected the required printer in the page setup of all the sheets in the sheet set. This is because the printer set in the page setup of the sheet will be automatically selected to plot the sheet.

To print the sheets after setting the page setup, right-click on the name of the sheet set in the **Sheet Set Manager** and choose **Publish > Publish to Plotter** from the shortcut menu. If the value of the **BACKGROUNDPLOT** system variable is set to **2**, which is the default value, all the sheets will be automatically plotted in the background and you can continue working on the drawings.

Note
You will learn more about publishing in Chapter 28.

Self-Evaluation Test

Answer the following questions and then compare them to those given at the end of this chapter:

1. All settings about a plotter are saved in the *.PC3* file. (T/F)

2. Different objects in the same drawing can be plotted in different colors with different linetypes and line widths. (T/F)

3. You can partially or fully preview a drawing before plotting. (T/F)

4. The **PSLTSCALE** system variable controls the paper space linetype scaling and has a default value of 1. (T/F)

5. The size of a plot can be specified by selecting any paper size from the _____ drop-down list in the **Plot** dialog box.

6. If you want to store a plot in a file and not have it printed directly on a plotter, select the _____ check box in the **Printer/plotter** area.

7. The scale for a plot can be specified in the _____ edit box in the **Plot** dialog box.

8. If you select the _____ option from the **What to plot** drop-down list, the portion of the drawing that is in the current display is plotted.

9. Before you plot sheets in the sheet set, it is important that you set the _____ for the individual sheets in their page setups.

10. You can set the quality of a plot using the _____ drop-down list in the **Shaded viewport options** area.

Review Questions

Answer the following questions:

1. The **Page Setup Manager** dialog box is displayed when you invoke the **PAGESETUP** command. (T/F)

2. By selecting the **View** option from the **What to plot** drop-down list in the **Plot** area, you can plot a view that was created with the **VIEW** command in the current drawing. (T/F)

3. If you do not want the hidden lines of a 3D object created in the **Model** tab, you can select the **Hidden** option from the **Shade plot** drop-down list in the **Shaded viewport options** area. (T/F)

4. The orientation of a drawing can be changed using the **Plot** dialog box. (T/F)

5. Which of the following check boxes is available in the **Plot options** area of the **Plot** dialog box, while plotting in a **Layout** tab?

    (a) **Plot paperspace last**   (b) **Hide objects**
    (c) **Plot with plot styles**   (d) **None**

6. Which of the following commands when invoked display the **Plotters** window?

    (a) **PLOTTER**   (b) **PLOTTERMANAGER**
    (c) **PLOTSTYLE**   (d) **None**

7. With which command is it possible to bypass the **Plot/Page Setup** dialog box and directly import a page setup from an existing drawing into a new drawing layout?

    (a) **PSETUPIN**   (b) **PLOTTERMANAGER**
    (c) **PLOTSTYLE**   (d) **None**

8. Which of the following commands is used to create a new plot style?

    (a) **STYLE**   (b) **STYLESMANAGER**
    (c) **PLOTSTYLE**   (d) **None**

9. Which of the following commands can be used to import a PCP file or PC2 files?

    (a) **PCINWIZARD**   (b) **PCIN**
    (c) **PLOTSTYLE**   (d) **None**

10. You can view a plot on the specified paper size before actually plotting it by choosing the _____ button from the **Plot** dialog box.

11. In the preview window of the **Printer/plotter** area, the _____ rectangle represents the section of the paper that is used by the image.

12. You can modify the properties of a selected plotting device by using the _____.

13. The plot style modes available are _____ and _____.

14. The **Plot Style Table Editor** has three tabs: _____ , _____ , and _____.

Exercises

Exercise 2   *Mechanical*

Create the drawing shown in Figure 14-19 and plot it according to the following specifications. Also, create and use a plot style table with the specified plot styles.

Figure 14-19 *Drawing for Exercise 2*

1. The drawing is to be plotted on 10 X 8 inch paper.

2. The object lines must be plotted with a plot style Style 1. Style 1 must have a value of lineweight = 0.800 mm.

3. The dimension lines must be plotted with plot style Style 2. Style 2 must have a value of screening = 50.

4. The centerlines must be plotted with plot style Style 3. Style 3 must have a linetype of Medium Dash and screening = 50.

5. The border and title block must be plotted with plot style Style 4. The value of the lineweight should be =0.25 mm.

Exercise 3   *Mechanical*

Create the drawing shown in Figure 14-20 and plot the drawing according to your specifications.

Figure 14-20 *Drawing for Exercise 3*

Problem-Solving Exercise 1   *Mechanical*

Make the drawing shown in Figure 14-21 and plot it according to your specifications.

Figure 14-21 Drawing for Problem-Solving Exercise 1

Answers to Self-Evaluation Test
1. T, **2**. T, **3**. F, **4**. T, **5. Paper size**, **6. Plot to file**, **7. Custom**, **8. Display**, **9.** Paper size **10. Quality**

Chapter 15

Hatching Drawings

Learning Objectives

After completing this chapter, you will be able to:
- *Use the **HATCH** command to hatch an area using various patterns.*
- *Use the boundary hatch with Predefined, User defined, and Custom hatch patterns as options.*
- *Specify pattern properties.*
- *Preview and apply hatching.*
- *Use advanced hatching options and Ray Casting.*
- *Create annotative hatch.*
- *Edit associative hatch and hatch boundary using the **HATCHEDIT** command, **Properties** palette, and Grips.*
- *Hatch inserted blocks.*
- *Align hatch lines in adjacent hatch areas.*
- *Hatch by using the **HATCH** command at the Command prompt.*

HATCHING

In many drawings, such as sections of solids, the sectioned area needs to be filled with some pattern. Different filling patterns make it possible to distinguish between different parts or components of an object. Also, the material of which an object is made, can be indicated by the filling pattern. You can also use these filling patterns in graphics for rendering architectural elevations of buildings, or indicating the different levels in terrain and contour maps. Filling objects with a pattern is known as hatching (Figure 15-1). This hatching process can be accomplished by using the **HATCH** command, the **Ribbon**, or the **Tool Palettes** window.

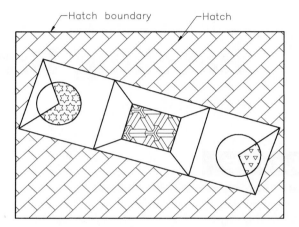

Figure 15-1   *Illustration of hatching*

Before using the **HATCH** command, you need to understand some terms that are used while hatching. The following subsection describes some of the terms.

Hatch Patterns

AutoCAD supports a variety of hatch patterns (Figure 15-2). Every hatch pattern is composed of one or more hatch lines or a solid fill. The lines are placed at specified angles and spacing. You can change the angle and the spacing between the hatch lines. These lines may be broken into dots and dashes, or may be continuous, as required. The hatch pattern is trimmed or repeated, as required, to fill exactly the specified area. The lines comprising the hatch are drawn in the current drawing plane. The basic mechanism behind hatching is that the line objects of the pattern you have specified are generated and incorporated in the desired area in the drawing. Although a hatch can contain many lines, AutoCAD normally groups them together into an internally generated object and treats them as such for all practical purposes. For example, if you want to perform an editing operation, such as erasing the hatch, all you need to do is select any point on the hatch and press ENTER; the entire pattern gets deleted. If you want to break a pattern into individual lines to edit them, you can use AutoCAD's **EXPLODE** command.

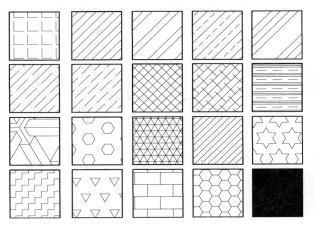

Figure 15-2 *Some hatch patterns*

Hatch Boundary

Hatching can be used on parts of a drawing enclosed by a boundary. This boundary may be lines, circles, arcs, polylines, 3D faces, or other objects, and at least part of each bounding object must be displayed within the active viewport.

HATCHING DRAWINGS USING THE HATCH AND GRADIENT DIALOG BOX

Ribbon:	Draw > Hatch	Toolbar:	Draw > Hatch
Command:	HATCH or H		

The **HATCH** command allows you to hatch a region enclosed within a boundary (closed area) by selecting a point inside the boundary or by selecting the objects to be hatched. This command automatically designates a boundary and ignores other objects (whole or partial) that may not be a part of this boundary. When you invoke the **HATCH** command, the **Hatch and Gradient** dialog box will be displayed, as shown in Figure 15-3. You can also enter **HATCH** at the pointer input to invoke the **Hatch and Gradient** dialog box. The default appearance of the **Hatch and Gradient** dialog box is a little different from the one shown in Figure 15-3. This is because options like **Islands**, **Boundary retention** are not visible by default. To make these options visible choose the **More Options** button available at the lower right corner of the **Hatch and Gradient** dialog box. You can perform the hatching operation using this dialog box.

Figure 15-3 The **Hatch and Gradient** *dialog box*

Example 1   *General*

In this example, you will hatch a circle using the default hatch settings. Later, in the chapter you will learn how to change the settings to get the desired hatch pattern.

1. Invoke the **Hatch and Gradient** dialog box by choosing the **Hatch** button from the **Draw** panel in the **Ribbon**.

2. Choose the **Add: Pick points** button from the **Boundaries** area of the **Hatch and Gradient** dialog box; the dialog box disappears.

3. Select a point inside the circle (P1) (Figure 15-4) and press ENTER; the **Hatch and Gradient** dialog box is displayed again. You can now preview the hatching by choosing the **Preview** button. Alternatively, after specifying an internal point, right-click to display a shortcut menu and choose the **Preview** option.

4. After checking the preview, press ESC to complete the point specification; the dialog box reappears.

5. Choose **OK** to apply the hatch to the selected object, Figure 15-5. Note that if you right-click while checking the preview, the hatch will be applied.

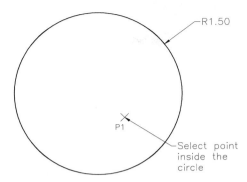

R1.50

P1

Select point
inside the
circle

Hatch pattern
created inside
the circle

Figure 15-4 *Specifying a point to hatch the circle*   **Figure 15-5** *Drawing after hatching*

Tip
The **FILLMODE** *system variable is set to On by default (value of the variable is 1) and hence the hatch patterns are displayed. In case of large hatching areas, you can set the* **FILLMODE** *to Off (value of the variable is 0), so that the hatch pattern is not displayed and the regeneration time is saved.*

You can also hatch a selected face of a solid model. However, to hatch a face of the solid model, you need to align the UCS at that face. You will learn more about this in later chapters.

HATCH AND GRADIENT DIALOG BOX OPTIONS

The **Hatch and Gradient** dialog box has several options that let you control the various aspects of hatching, such as pattern type, scale, angle, boundary parameters, associativity, and so on. The following is a description of these options. The options have been grouped by the tab name in which they are found in the **Hatch and Gradient** dialog box. The **Hatch and Gradient** dialog box has two tabs: **Hatch** and **Gradient**. The description of these tabs is as follows.

Hatch Tab

This tab provides the options that define the appearance of the hatching patterns that can be applied (Figure 15-3). For the quick hatching of an object, this tab can be used easily.

Type and pattern Area

This area contains the options that allows you to choose the type of pattern and preview the chosen pattern type.

Type. The **Type** drop-down list (Figure 15-6) displays the types of patterns that can be used for hatching drawing objects. This list lets you choose the type of hatch pattern you want. The three types of hatch patterns available are **Predefined**, **User defined**, and **Custom**. The predefined type of patterns come with AutoCAD and are stored in the

Figure 15-6 *The* **Type** *drop-down list displaying hatch pattern types*

acad.pat and *acadiso.pat* files. The predefined type of hatch pattern is the default type. When you select **User defined** from the **Type** drop-down list, a pattern of lines that is based on the current linetype is created. You can set the angle of the lines in the pattern and also the spacing between them. When you select **Custom** from the **Type** drop-down list, you can select a pattern that has been defined in a custom PAT file.

Note

*When using the patterns defined in the custom PAT files, make sure that their search path has been added in the **Files** tab of the **Options** dialog box.*

Pattern. The **Pattern** drop-down list displays the names of all predefined patterns available in AutoCAD. You can select any pattern from this list and the selected pattern is displayed in the **Swatch** box. The selected pattern is stored in the **HPNAME** system variable. ANSI31 is the default pattern in the **HPNAME** system variable. The **Pattern** drop-down list is available only if you have selected **Predefined** from the **Type** drop-down list.

To select a particular hatch pattern, you can also choose the [**...**] button that displays the **Hatch Pattern Palette** dialog box (Figure 15-7). This dialog box shows the images and names of the available hatch patterns. This dialog box is also displayed by clicking on the pattern displayed in the **Swatch** box. The **Hatch Pattern Palette** dialog box has four tabs. They are described next.

*Figure 15-7 The **Hatch Pattern Palette** dialog box (**ANSI** tab)*

ANSI. This tab displays all ANSI (American National Standards Institute)-defined hatch patterns that come with AutoCAD. ANSI31 is the default pattern.

ISO. This tab displays all the ISO (International Standard Organization)-defined hatch patterns that come with AutoCAD.

Other Predefined. This tab displays all predefined hatch patterns other than ANSI or ISO-defined patterns.

Custom. This tab displays all custom patterns described in any custom *PAT* file that is added in the search path of AutoCAD 2010.

Tip
*Using the **Hatch Pattern Palette** dialog box for selecting a hatch pattern is more convenient since you can view all pattern names and the corresponding patterns simultaneously.*

Swatch. This box displays the selected pattern. When you click in the **Swatch** box, the **Hatch Pattern Palette** dialog box is displayed.

Custom Pattern. This drop-down list is available only if you have selected **Custom** from the **Type** drop-down list. It displays all the available custom hatch patterns defined in the custom PAT files that have been added to the AutoCAD search path. The recently used custom hatch patterns in a current drawing session appear in the list. The selected pattern is stored in the **HPNAME** system variable. You can also choose the **[...]** button; the **Hatch Pattern Palette** dialog box (**Custom** tab) appears. This tab displays the custom pattern paths.

Angle and scale Area
The options in the **Angle and scale** area are discussed next.

Angle. The **Angle** drop-down list (Figure 15-8) displays the angles (at an increment of 15-degree) by which you can rotate the hatch pattern with respect to the *X* axis of the current UCS. You can select any angle from this drop-down list or you can also enter an angle of rotation of your choice in the **Angle** edit box. The angle value is stored in the **HPANG** system variable. The angle of hatch lines of a particular hatch pattern is governed by the values specified in the hatch definition. For example, in the **ANSI31** hatch pattern definition, the specified angle of hatch lines is 45-degree. If you select an angle of 0, the angle of hatch lines will be 45-degree. If you enter an angle of 45-degree, the angle of the hatch lines will be 90-degree.

Scale. The **Scale** drop-down list (Figure 15-8) displays scale factors by which you can expand or contract the selected hatch pattern. The scale factors that are displayed range between 0.25 to 2 at increments of 0.25. You can also enter the scale factor of your choice in the **Scale** edit box. The **Scale** drop-down list is not available when you are using user-defined

Figure 15-8 The Angle and Scale drop-down lists

hatch patterns. The scale value is stored in the **HPSCALE** system variable. The value 1 does not mean that the distance between the hatch lines is 1 unit. The distance between the hatch lines and other parameters of a hatch pattern is governed by the values specified in the hatch definition. For example, in the ANSI31 hatch pattern definition, the specified distance between the hatch lines is 0.125. If you select a scale factor of 1, the distance between the lines will be 0.125. If you enter a scale factor of 0.5, the distance between the hatch lines will be 0.5 X 0.125 = 0.0625.

Relative to Paper Space. If this check box is selected, then AutoCAD will automatically scale the hatched pattern relative to the paper space units. This option can be used to display the hatch pattern at a scale that is appropriate for your layout. This option is available only in a layout.

ISO pen width. The **ISO pen width** drop-down list is available only for ISO hatch patterns. You can select the desired pen width value from the **ISO pen width** drop-down list. The value selected specifies the ISO-related pattern scaling.

User defined Hatch Patterns

To define a simple pattern, you can select **User defined** from the **Type** drop-down list. You will notice that the **Angle** and **Spacing** edit boxes and the **Double** check box are available in the dialog box. The **Scale** drop-down list is not available.

Angle

Just as with predefined patterns, you can specify in the **Angle** edit box an angle with respect to the *X* axis of the current UCS through which the hatch pattern is rotated. This value is stored in the **HPANG** system variable.

Double

This check box is available only for user-defined patterns. When you select this check box, AutoCAD doubles the original pattern by drawing a second set of lines at right angles to the original lines in the hatch pattern. For example, if you have a parallel set of lines as a user-defined pattern and if you select the **Double** check box, the resulting pattern has two sets of lines intersecting at 90-degree. You can notice the effect of selecting the **Double** check box in Figure 15-9. If the **Double** check box is selected, the **HPDOUBLE** system variable is set to 1.

Spacing

The **Spacing** edit box is available only when you have selected **User defined** from the **Type** drop-down list. The value you enter in the **Spacing** edit box sets the spacing between the lines in a user-defined hatch pattern. This spacing is stored in the **HPSPACE** system variable. The default spacing value is 1.0000.

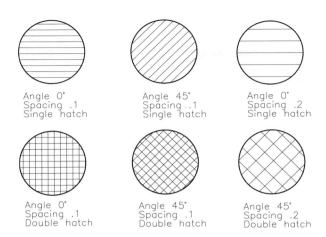

Figure 15-9 *Specifying angle and spacing for user-defined hatch patterns*

Custom Hatch Patterns

When you select **Custom** from the **Type** drop-down list in the **Hatch and Gradient** dialog box, you will notice that the **Custom pattern** drop-down list is available. The **Angle** and **Scale** drop-down lists are also available.

Custom pattern

In AutoCAD, hatch patterns are normally stored in the file named *acad.pat*. You can select a hatch pattern from an individual file by selecting **Custom** from the **Type** drop-down list. The **Custom pattern** drop-down list is available now. You can choose the [**...**] button to display the **Custom** tab of the **Hatch Pattern Palette** dialog box. Here, the list box on the left side of the dialog box displays the paths of the custom hatch patterns. When you select a path, the corresponding pattern is displayed in a preview window on the left. Choose **OK** to return to the **Hatch and Gradient** dialog box. The corresponding pattern is displayed in the **Swatch** window. You can also select a pattern from the **Custom pattern** drop-down list. You can also enter the name of the custom pattern in the field, if you know the name of the hatch pattern. The name is held in the **HPNAME** system variable. If AutoCAD does not locate the entered pattern in the *acad.pat* file, it searches for it in a file with the same name as the pattern. You can specify an angle and scale for the custom pattern using the **Angle** and **Scale** drop-down lists, as in the case of predefined patterns.

Tip
When you are hatching large areas, it is better to use a larger scale factor for the hatch pattern that will help you save time regenerating and plotting. Also, while using a larger scale factor for the hatch pattern, the drawing appears to be neater and not too cluttered.

Exercise 1	Mechanical

In this exercise, you will hatch the given drawing using the hatch pattern named **STEEL**. Set the scale and the angle to match the drawing shown in Figure 15-10.

Chapter 15

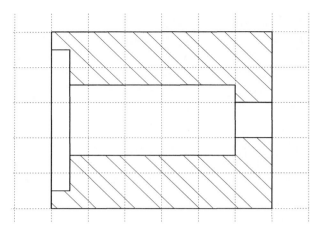

Figure 15-10 Drawing for Exercise 1

Selecting Hatch Boundary

The boundary selection options in the **Hatch and Gradient** dialog box allow you to define the hatch boundary. This is done by selecting a point inside a closed area to be hatched or by selecting the objects. There are other options also available here for removing islands, viewing the selection, using properties from an existing hatched object, and controlling associativity of the hatch pattern.

Boundaries Area

The options under **Boundaries** area of **Hatch and Gradient** dialog box are discussed next.

Add: Pick points. When you choose this button, AutoCAD automatically constructs a boundary from the objects that form a closed area. This is the easiest method to specify a region for hatching. When you choose the **Add: Pick points** button, the dialog box disappears temporarily to allow you to select a point within the closed area to be hatched. Select a point inside the object and a boundary is defined around the selected point. The following prompts appear when you choose the **Add: Pick points** button in the **Hatch and Gradient** dialog box.

> Pick internal point or [Select objects/remove Boundaries]: *Select a point inside the object to hatch.*
> Selecting everything visible...
> Analyzing the selected data...
> Analyzing internal islands...
> Pick internal point or [Select objects/remove Boundaries]: *Select another internal point or press ENTER to end selection.*

You can also select objects or remove boundaries from the existing selection to define the hatch boundary. To select objects enter **S** at the **Pick internal point or [Select objects/ remove Boundaries]** prompt. Now, you can select the objects to define hatch boundary. The prompt sequence is given next.

Pick internal point or [Select objects/remove Boundaries]: **S** Enter
Select objects or [picK internal point/remove Boundaries]: *Select the object to hatch.*
Select objects or [picK internal point/remove Boundaries]: *Select another object to hatch or press ENTER to end the selection.*

Enter **K** at the Select objects or [picK internal point/remove Boundaries] prompt to specify the hatch boundary by picking the internal point or enter **B** to remove the boundaries from the existing selection. The prompt sequence when you enter **B** is given next.

Select objects or [picK internal point/remove Boundaries]: **B** Enter
Select objects or [Add boundaries]: *Select objects to clear from selection*
Select objects or [Add boundaries/Undo]: *Select objects to clear from selection or press ENTER to end the selection.*

To add the boundaries, enter **A** at the **Select objects or [Add boundaries]** prompt. As you enter **A** at the specified prompt, you will be routed back to the **Pick internal point or [Select objects/remove Boundaries]** prompt. To undo the last selection of boundaries, enter **U** at the **Select objects or [Add boundaries/Undo]** prompt.

Once the internal points have been selected, press ENTER or choose the **Enter** option from the shortcut menu displayed on right-clicking in the drawing area; the dialog box reappears. Choose **OK** to apply the hatch pattern to the region surrounding the selected internal point. If you want to hatch an object and leave another object within it, without hatching, select a point inside the object you want to hatch. By default, when using the **HATCH** command, selecting a point creates multiple boundaries; a boundary of the internal object is also created, and the region between these two boundaries is selected for hatching. In Figure 15-11, the **Add: Pick points** button has been used to hatch the square but not the triangle simply by selecting the point between the square and the triangle (using the default settings of the dialog box).

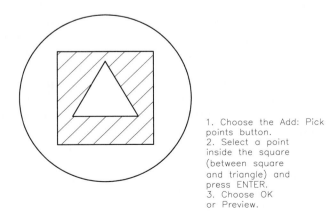

1. Choose the Add: Pick
points button.
2. Select a point
inside the square
(between square
and triangle) and
press ENTER.
3. Choose OK
or Preview.

Figure 15-11 Defining multiple hatch boundaries by selecting a point

Chapter 15

Tip

*Suppose, you have selected a point within an area for hatching and later you realize that you have selected the wrong area, in such a case, you can enter **U** or **Undo** at the Command prompt to undo the last selection. You can also undo a hatch pattern that you have already applied to an area by entering **Undo** at the Command prompt.*

Boundary Definition Error. Sometimes, while selecting the boundary to be hatched, AutoCAD displays an error. This error may occur due to various reasons. AutoCAD displays different types of Boundary Definition Error message boxes, depending on the kind of error occurred while selecting the boundary. For example, if you pick a point inside any boundary that is not closed, a message box will be displayed, as shown in Figure 15-12, informing that the hatch boundary has gaps. Choose the **Close** button and then create a closed area as boundary. You can also specify the gap tolerance value in the **Tolerance** edit box of the **Hatch** tab in the **Hatch and Gradient** dialog box so that hatch will be created if the gap is within the permissible limit. If you select the same boundary twice, the message that the boundary duplicates an existing boundary will be displayed.

*Figure 15-12 The **Boundary Not Closed** message box*

Add: Select objects. This option in the **Hatch and Gradient** dialog box lets you select objects that form the boundary for hatching. It is useful when you have to hatch an object and disregard objects that lie inside it or intersect with it. When you select this option, AutoCAD will prompt you to select objects. You can select the objects individually or use the other object selection methods. In Figure 15-13, the **Select Objects** button is used to select the triangle. It uses the triangle as the hatch boundary, and everything inside it is hatched. The text inside the triangle, in this case, also gets hatched. To avoid hatching of such internal objects, select them at the next **Select objects** prompt. In Figure 15-14, the text is selected to exclude it from hatching. When using the **Pick points** option, the text automatically gets selected to be excluded from the hatching.

Tip

If you have an area to be hatched that has many objects intersecting with it, it is easier to select the entire object to be hatched rather than choosing the internal points within each of the smaller regions created by the intersection.

You can also turn off layers that contain text or lines that make it difficult for you to select hatch boundaries.

1. Choose the Add: Select objects button.
2. Select the triangle for hatching and press ENTER.
3. Choose Preview or OK.

1. Choose the Add: Select objects button.
2. Select the triangle for hatching and press ENTER.
3. Select the text.
4. Choose Preview or OK.

Figure 15-13 *Using the* ***Select Objects*** *button to specify the hatching boundary*

Figure 15-14 *Using the* ***Select Objects*** *button to exclude the text from hatching*

Remove boundaries. This option is used to remove boundaries and islands from the hatching area. Boundaries inside another boundary are known as islands. If you choose the **Add: Pick points** button to hatch an area, the inside boundaries, which are known as islands, will not be hatched by default. But, if you want to hatch the islands, you can choose the **Remove boundaries** button in the **Hatch and Gradient** dialog box. This applies, for example, if you have a rectangle with circles within, as shown in Figure 15-15. You can use the **Pick points** option to select a point inside the rectangle; AutoCAD will select both the circles and the rectangle. To remove the circles (islands), you can use the **Remove boundaries** option. When you select this option, AutoCAD will prompt you to select the islands to be removed. Select the islands to be removed and press ENTER to return to the dialog box. Choose **OK** to apply the hatch. Similarly, you can remove boundaries.

Hatching without removing the islands

Using the Remove boundaries option to remove the islands

Figure 15-15 *Using the* ***Remove Boundary*** *option to remove islands from the hatch area*

Tip
It may be a good idea to use the ***Select Objects*** *option to select the object containing islands, if you want to remove the islands from the hatching area.*

Recreate boundary. This option is required while editing a hatch and is used to recreate boundaries around the existing hatch pattern. On choosing this button, the dialog box is temporarily closed and you are prompted to specify whether you want to recreate boundaries as a region or as a polyline. You can also associate the hatch to the new boundary.

View Selections. This option lets you view the selected boundaries before you apply the hatch pattern. This option is not available when you have not selected any points or boundaries. When you choose the **View Selections** button, the dialog box closes temporarily and the selected boundaries in the drawing are displayed as highlighted.

Options Area

The options in this area allow you to specify the draw order and some commonly used properties of the hatch pattern.

Annotative

Select this check box to create an annotative hatch. An annotative hatch is defined relative to the paper space size. The scale of the annotative hatch objects changes in the viewport or layout according to the annotation scale assigned to the hatch objects and the annotation scale specified for that particular layout or viewport.

Associative

This check box is selected by default, which implies that when you modify the boundary of the hatch object, the hatch patterns will be automatically updated to fill up the new area. One of the major advantages with the associative hatch feature is that you can edit the hatch pattern or edit the geometry that is hatched without having to modify the associated pattern or boundary separately. After editing, AutoCAD automatically regenerates the hatch and the hatch geometry to reflect the changes. The hatch pattern can be edited by using the **HATCHEDIT** command or by choosing **Hatch Edit** from the shortcut menu whereas the hatch geometry can be edited by using grips or AutoCAD editing commands.

Create separate hatches

While hatching multiple closed areas that are not nested together, selecting this check box ensures that a separate hatch is created for each closed area. If this check box is cleared, a single hatch is created for all the selected closed areas.

Draw order

The drop-down list is used to assign a draw order to the hatch. If you want to send the hatch behind all the entities, select the **Send to back** option. Similarly, if you want to place the hatch in front of all the entities, select the **Bring to front** option. If you want to place the hatch behind the hatch boundary, select **Send behind boundary**. Similarly, if you want to place the hatch in front of the boundary, select **Bring in front of boundary**. You can also select the **Do not assign** option, if you do not want to assign the draw order to the hatch.

Inherit Properties

The **Inherit Properties** button is available below the **Options** area and is used to hatch the specified boundaries using the properties of an existing hatch. When you invoke the **HATCH** command and choose the **Inherit Properties** button, the following command sequence appears. Note that this prompt sequence does not appear when you specify the area to hatch first and then choose the **Inherit Properties** button.

> Select hatch object: *Select the object from which you have to inherit the hatch pattern.*
> Inherited Properties: Name <current>, Scale <current>, Angle <current>
> Pick internal point or [Select objects/remove Boundaries]: *Pick a point inside the object to hatch.*
> Selecting everything visible...
> Analyzing the selected data...
> Analyzing internal islands...
> Pick internal point or [Select objects/remove Boundaries]: Enter

When you press ENTER, the **Hatch and Gradient** dialog box reappears with the name of the inherited hatch pattern in the **Pattern** edit box. The inherited hatch pattern is also displayed in the **Swatch** box. The selected pattern now becomes the current hatch pattern. If you then want to adjust the hatch properties, such as the angle or scale of the pattern, you can do so.

Preview

This button is used to preview the hatching before it is actually applied. This option is available only after you have selected the area to be hatched. When you choose the **Preview** button, the dialog box is temporarily closed and the object selected for hatching is temporarily filled with the specified hatch pattern. You can also preview the drawing by right-clicking in the drawing area, after selecting the boundary and choosing **Preview** from the shortcut menu. After previewing the drawing, pick a point on the screen or press ESC to redisplay the **Hatch and Gradient** dialog box so that you can make modifications in the hatching operation. You can also press ENTER or right-click after previewing the drawing to accept the hatching and apply it to the selected boundary. Note that because you accept the current hatch settings by pressing ENTER or by right-clicking; the **HATCH** command is completed and the **Hatch and Gradient** dialog box is not displayed.

Exercise 2   *Mechanical*

In this exercise, you will hatch the front section view of the drawing in Figure 15-16 using the hatch pattern for brass. Two views, top, and front are shown. In the top view, the cutting plane line indicates how the section is cut and the front view shows the full section view of the object. The section lines must be drawn only where the material is actually cut.

Chapter 15

Figure 15-16 *Drawing for Exercise 2*

Hatch origin Area

Pattern alignment is an important feature of hatching, as on many occasions, you need to hatch adjacent areas with similar or sometimes identical hatch patterns while keeping the adjacent hatch patterns properly aligned. Proper alignment of hatch patterns is taken care of automatically by generating all lines of every hatch pattern from the same reference point.

The reference point is normally at the origin point (0,0). Figure 15-17 shows two adjacent hatch areas. The area on the right is hatched using the pattern ANSI32 at an angle of 0-degree and the area on the left is hatched using the same pattern at an angle of 90-degree. When you hatch these areas, the hatch lines may not be aligned, as shown in Figure 15-17(a). The options in the **Hatch origin** area allow you to specify the origin of hatch so that they get properly aligned, as shown in Figure 15-17(b). The options in the **Hatch origin** area are discussed next

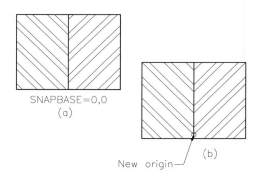

Figure 15-17 *Aligning hatch patterns using the SNAPBASE variable*

Use current origin

By default, the **Use current origin** radio button is selected. This implies that the origin of the hatch pattern to be created is the origin of the current drawing.

Specified origin

When you select the **Specified origin** radio button, other options available in **Hatch Origin** area are highlighted. Choose the **Click to set new origin** button and specify the origin point for the hatch pattern in the drawing area. Select the **Default to boundary extents** check box;

the drop-down list is now available and you can select the options available in the drop-down list to specify the origin of hatch pattern. If you select the **Bottom left** option from the drop-down list, the origin of the hatch pattern will be at the bottom left corner of the boundary. The preview corresponding to the option chosen is displayed in the preview area at the right of the drop-down list. Selecting the **Store as default origin** check box stores the origin just selected as the default origin for all the hatch patterns to be created now onwards.

Note
*The reference point for hatching can also be changed using the **SNAPBASE** system variable.*

Exercise 3   *Mechanical*

In this exercise, you will hatch the given drawing using the hatch pattern ANSI31. Align the hatch lines as shown in the drawing (Figure 15-18).

Figure 15-18 *Drawing for Exercise 3*

MORE OPTIONS

When you choose the **More Options** button available on the right of the **Help** button in the **Hatch and Gradient** dialog box, it expands and provides the following options.

Islands Area

The options in this area are used to specify the island detection method of creating the hatch. These options are discussed next.

Island detection

The options available under the **Islands** area are available only if the **Island detection** check box is selected. When the check box is selected, the method of island detection is the Flood type. When the check box is cleared the island detection is the Ray Casting type. The two methods of island detection are discussed next.

Flood. This method of defining the hatch boundary includes the islands as the boundary objects. This is the default method of island detection.

Ray Casting. This method casts rays in all directions or a particular direction from the point specified to the nearest object and then traces the boundary in a counterclockwise direction, thus excluding islands as boundary objects. Ray casting is discussed later in detail.

Island display style

These options are used to select the style of the island detection during hatching (Figure 15-19). There are three styles available in this area, **Normal**, **Outer**, and **Ignore**. To select a particular style, you can select the corresponding radio button. The effect of using the particular style is displayed in the form of an illustration in the image tile placed above the particular radio button and is also shown in Figure 15-20. You can also choose the image tiles to select a particular island detection style. The three styles available in this option are discussed next.

Figure 15-19 Island detection styles

Figure 15-20 Using hatching styles

Note
The selection of an island detection style carries meaning only if the objects to be hatched are nested (that is, one or more selected boundaries is within another boundary).

Normal. The **Normal** style is selected by default. This style hatches inward starting at the outermost boundary. If it encounters an internal boundary, it turns off the hatching. An internal boundary causes the hatching to turn off until another boundary is encountered. In this manner, alternate areas of the selected object are hatched, starting with the outermost area. Thus, areas separated from the outside of the hatched area by an odd number of boundaries are hatched, while those separated by an even number of boundaries are not.

Outer. This particular option also lets you hatch inward from an outermost boundary, but the hatching is turned off if an internal boundary is encountered. Unlike the previous case, it does not turn the hatching on again. The hatching process, in this case, starts from both ends of each hatch line; only the outermost level of the structure is hatched, hence, the name **Outer**.

Ignore. In this option, all areas bounded by the outermost boundary are hatched. The option ignores any hatch boundaries that are within the outer boundary. All islands are completely ignored and everything within the selected boundary is hatched.

Tip

*It is also possible to set the pattern and the island detection style at the same time by using the **HPNAME** system variable. AutoCAD stores the Normal style code by adding **N** to the pattern name. Similarly, **O** is added for the Outer style, and **I** for the Ignore style to the value of the **HPNAME** system variable. For example, if you want to apply the BOX pattern using the outer style of island detection, you should enter the **HPNAME** value to the **BOX, O**. Now, when you apply the hatch pattern, the BOX pattern is applied using the outer style of hatching.*

Note

*The **Normal**, **Outer**, and **Ignore** options are also available in the shortcut menu when you select an internal point in the drawing object and right-click in the drawing area. The shortcut menu will be discussed later.*

Boundary retention Area

When you select an internal point in a region to be hatched, a boundary is created around it. By default, these boundaries are removed as soon as the hatch pattern is applied. The **Boundary retention** area provides options that allow you to specify whether these boundaries are to be retained as objects or not. You can also specify the type of object it can be saved as.

Retain boundaries check box. The **Retain boundaries** check box is cleared by default, which implies that the hatch boundaries are not saved. If you select this check box, you can retain the defined boundary. When a hatching is successful and you want to keep the boundary as a polyline or region so that you can use it again, you can select the **Retain boundaries** check box. When you select this check box, the **Object type** drop-down list is available.

Object type drop-down list. The **Object type** drop-down list is available when you select the **Retain boundaries** check box. From this list, you can select the type of object AutoCAD will create when you define a boundary for hatching. It has two options: **Polyline** and **Region**. **Polyline** is selected by default and the boundary created around the hatch area is a polyline. Similarly, when you select **Region**, the boundary of the hatch area is a region. Regions are two-dimensional areas that can be created from closed shapes or loops.

Boundary set Area

When hatching takes place normally, AutoCAD evaluates the entire drawing visible on the screen to determine the boundary. For larger drawings, this may take a lot of time and therefore, slow down the process of hatching. The **Boundary set** area provides options that allow you to specify what is considered when hatching. You can either select an option from the drop-down list or you can create a new boundary set. The boundary set comprises the objects that the **HATCH** command uses when constructing the boundary. The default boundary set is **Current viewport**, which comprises everything that is visible in the current viewport. Boundary hatching is made faster by specifying a boundary set because, in this case, AutoCAD does not have to examine everything on screen.

This option allows you to define a boundary area so that only a specific portion of the drawing is considered for hatching. You use this option to create a new boundary set. When you choose the **New** button, the dialog box is temporarily cleared from the screen to allow the selection of objects to be included in the new boundary set. While constructing the boundary set, AutoCAD uses only those objects that you select and that are hatchable. If a boundary set already exists, it is replaced by the new one. If you do not select any hatchable objects, no new boundary set is created. AutoCAD retains the current set, if there is any. Once you have selected the objects to form the boundary set, press ENTER and the dialog box reappears on the screen. You will notice that **Existing set** gets added to the drop-down list. When you invoke the **HATCH** command and you have not formed a boundary set, there is only one option, **Current viewport**, available in the drop-down list. The benefit of creating a selection set is that when you select a point or select the objects to define the hatch boundary, AutoCAD will search only for the objects that are in the selection set. By confining the search to the objects in the selection set, the hatching process is faster. If you select an object that is not a part of the selection set, AutoCAD ignores it. When a boundary set is formed, it becomes the default for hatching until you exit the **HATCH** command or select the **Current viewport** option from the drop-down list.

Tip
To improve the hatching speed in large drawings, you should zoom into the area to be hatched, so that defining the boundary to be hatched is easier. Also, since AutoCAD does not have to search the entire drawing to find hatch boundaries, the hatch process gets faster.

Gap tolerance Area

The **Gap tolerance** edit box is used to set the value up to which the open area will be considered closed when selected for hatching, using the **Add: Pick points** method. The default value of the gap tolerance is 0. As a result, the open area will not be selected for hatching. You can set the value of the gap tolerance using the **Gap Tolerance** edit box. If the gap in the open area is less than the value specified in this edit box, the area will be considered closed and will be selected for hatching.

Inherit options Area

The options available in this area are used to set the hatch origin while working with the **Inherit Properties** tool. By default, the **Use current origin** radio button is selected because of which, the current origin will be taken as the hatch origin. If you select the **Use source hatch origin** radio button, the hatch pattern will accept the origin of the source hatch as the origin of the current hatch pattern.

The HATCH Shortcut Menu

After invoking the **HATCH** command and selecting a point that lies in the interior of the object to be hatched, or the object to be hatched itself, right-click to display the shortcut menu (Figure 15-21). This shortcut menu displays the options that are also available through the **Hatch and Gradient** dialog box. If you choose **Preview**, you can preview the hatching before applying it. If you choose **Enter**, the dialog box reappears. Choosing **Clear All** removes all the previous selections made. You can also select another island detection mode and also change from one method of hatching to another. Choose the **Remove Boundaries** option to remove the boundaries from the selected hatching area. You can specify the hatch origin using the options available in the **Hatch Origin** cascading menu.

Figure 15-21 *The HATCH shortcut menu*

Gradient Tab

The **Gradient** tab of the **Hatch and Gradient** dialog box is used to specify the appearance of the gradient fill that can be applied to the selected objects. Figure 15-22 shows the **Gradient** tab and the related options in the expanded **Hatch and Gradient** dialog box.

Figure 15-22 *The **Gradient** tab of the **Hatch and Gradient** dialog box*

Color Area

The **Color** area provides the options to specify the colors for the gradient fill. The options in this area are discussed next.

Chapter 15

One color. The **One color** radio button is used to specify the gradient fill comprising different shades and tints of a single color. By default, this radio button is selected in the dialog box. You can select the desired color by choosing the **Color Swatch** button available below the **One Color** radio button. You can also change the shade and tint of the selected color from the Shade and Tint slider available on the right of the **Color Swatch**.

Two color. The **Two color** radio button is used to specify the gradient fill comprising of two colors. When you select this radio button, AutoCAD displays the color swatch for color 1 and color 2. Note that in this case, the **Shade and Tint** slider is not available.

Color Swatch

The **Color Swatch** is used to specify the color of the gradient fill. You can choose the [...] button available on the right side of the **Color Swatch** to display the **Select Color** dialog box. You can select the desired color using this dialog box.

Shade and Tint Slider

The **Shade and Tint** Slider is used to specify the shade and tint of the color selected for the gradient fill.

Gradient Patterns

You can specify the desired display pattern of the gradient fill by selecting any one of the nine fixed patterns available below the **Color** area.

Tip
*You can specify the gradient pattern by entering its value in the **GFNAME** system variable. The default value of this variable is 1. As a result, the first gradient pattern is selected.*

Orientation Area

You can specify the angle and symmetricity of the fill.

Centered. The **Centered** radio button is used to specify a symmetrical gradient fill. If this check box is not selected, then the gradient fill is moved up and to the left. When you clear this check box, the gradient patterns are automatically updated.

Angle. The **Angle** drop-down list displays the angles (at an increment of 15-degree) by that you can rotate the gradient fill with respect to the X axis of the current UCS. You can select any angle from this drop-down list or you can also enter an angle of your choice in the **Angle** edit box. The angle value is stored in the **GFANG** system variable.

Tip
Using the shortcut menu is more convenient and it saves time.

CREATING ANNOTATIVE HATCH

You can create annotative hatch having similar annotative properties like text and dimensions. The annotative hatch are defined relative to the viewport scaling, you only have to specify the hatch scale according to its display on the sheet. The display size of the hatch in the model space will be controlled by the current annotation scale multiplied by the paper space height. You can also control the annotative properties of the hatch pattern as discussed in Chapter 7.

To create an annotative hatch, choose the **Hatch** button from the **Draw** panel in the **Home** tab of the **Ribbon**; the **Hatch and Gradient** dialog box will be displayed. Choose the **Add: Select objects** button, select the objects that you want to hatch and press the ENTER key; the **Hatch and Gradient** dialog box will be displayed again. From the **Options** area in this dialog box, select the **Annotative** check box and choose the **OK** button. You can also convert the existing non-annotative hatch into the annotative hatch. To do so, select the non-annotative hatch in the drawing and select **Yes** from the **Annotative** drop-down list in the **Pattern** list of the **Properties** palette.

Note
*The annotative hatch pattern will only be displayed in the drawing area for the annotative scale that is set current while creating the hatch. To display that hatch pattern at any other annotative scale, you have to add that scale to the selected hatch using the **OBJECTSCALE** command.*

HATCHING THE DRAWING USING THE TOOL PALETTES

| **Ribbon:** | View > Palette > Tool Palettes | **Command:** | TOOLPALETTES |
| **Toolbar:** | Standard Annotation > Tool Palettes Window | | |

You can use the **Tool Palettes** window shown in Figure 15-23 to insert predefined hatch patterns and blocks in the drawings. By default, AutoCAD displays **Tool Palettes** as a window on the right of the drawing area. A number of tabs are available in this window, such as **Command Tool Samples**, **Hatches and Fills**, **Civil**, **Structural**, **Electrical**, and so on. In this chapter, you will learn to insert **Imperial Hatches**, **ISO Hatches** and **Gradient Samples** in the **Hatches and Fills** tab of the **Tool Palettes** window. The **Imperial Hatches** list provides the options to insert the hatch patterns that are created using the Imperial units. The **ISO Hatches** list provides the options to insert the hatch patterns that are created using the Metric units. You will notice that the hatch patterns provided in both these lists are similar. The basic difference between the two lists is their scale factor. The **Gradient Samples** list provides option to add the color gradients to objects as hatch.

Tip
*The **Tool Palettes** window can be turned on and off by pressing the CTRL+3 keys.*

AutoCAD provides two methods to insert the predefined hatch patterns from the **Tool Palettes** window: **Drag and Drop** method and **Select and Place** method. Both these methods are discussed next.

Drag and Drop Method

To insert the predefined hatch pattern from the **Tool Palettes** using this method, move the cursor over the desired predefined pattern in the **Tool Palettes**. You will notice that as you move the cursor over the hatch pattern, the hatch icon gets converted into a 3D icon. Also, a tooltip is displayed that shows the name of the hatch pattern. Press and hold the left mouse button and drag the cursor within the area to be hatched. Release the left mouse button, and you will notice that the selected predefined hatch pattern is added to the drawing.

Select and Place Method

You can also add the predefined hatch patterns to the drawings using the select and place method. To add the hatch pattern, move the cursor over the desired pattern in the **Tool Palettes**; the pattern icon will be changed to a 3D icon. Next, click the left mouse button; the selected hatch pattern will be attached to the cursor and you will be prompted to specify the insertion point of the hatch pattern. Now, move the cursor within the area to be hatched and click the left mouse button; the selected hatch pattern will be inserted at the specified location.

Modifying the Properties of the Predefined Patterns Available in the Tool Palettes

To modify the properties of the predefined hatch patterns, move the cursor over the hatch pattern in the **Tool Palettes** and right-click on it to display the shortcut menu. Using the options available in this shortcut menu, you can cut or copy the desired hatch pattern available in one tab of **Tool Palettes** and paste it on the other tab. You can also delete and rename the selected hatch pattern using the **Delete** and **Rename** options, respectively. To update and change the image displayed on the selected hatch pattern, choose the **Update tool image** or **Specify image** option. To modify the properties of the selected hatch pattern, choose **Properties** from the shortcut menu, as shown in Figure 15-24. The **Tool Properties** dialog box will be displayed, as shown in Figure 15-25.

Figure 15-23 The **Hatch** *tab of* **Tool Palettes**

Figure 15-24 *Choosing* **Properties** *from the shortcut menu*

In the **Tool Properties** dialog box, the name of the selected hatch pattern is displayed in the **Name** edit box. You can also rename the hatch pattern by entering a new name in the **Name** edit box. The **Image** area available on the left of the **Name** edit box displays the image of the selected hatch pattern. You can change the display image by choosing the **Specify image** option from the shortcut menu that will be displayed when you right-click on the image. If you enter a description of the hatch pattern in the **Description** text box, it is stored with the hatch definition in the **Tool Palettes**. Now, when you move the cursor over the hatch pattern in **Tool Palettes** and pause for a second, the description of the hatch pattern appears along with its name in the tooltip. The **Tool Properties** dialog box displays the properties of the selected hatch pattern under the following categories.

Figure 15-25 The **Tool Properties** dialog box

Pattern

In this category, you can change the pattern type, pattern name, angle, and scale of the selected pattern. You can modify the spacing of the **User defined** patterns in the **Spacing** text box. Also, the ISO pen width of the ISO patterns can be redefined in the **ISO pen width** text box. The **Double** drop-down list is available only for the **User-defined** hatch patterns. You can select Yes or No from the **Double** drop-down list to determine the hatch pattern to be doubled at right angles to the original pattern or not. When you choose the [...] button in the **Type** property field, AutoCAD displays the **Hatch Pattern Type** dialog box. You can select the type of hatch pattern from the **Pattern Type** drop-down list. If the **Predefined** pattern type is selected, you can specify the pattern name by either selecting it from the **Pattern** drop-down list or from the **Hatch Pattern Palette** dialog box displayed on choosing the **Pattern** button. Similarly, if you select the **Custom** pattern type, you can enter the name of the custom pattern in the **Custom Pattern** edit box. If you select the **User defined** pattern type, both the **Pattern** drop-down list and the **Custom Pattern** edit box are not available.

General

In this category, you can specify the general properties of the hatch pattern such as color, layer, linetype, plot style, and line weight for the selected hatch pattern. The properties of a particular field can be modified from the drop-down list available on selecting that field. Choose **OK** to apply the changes and close the dialog box.

Chapter 15

HATCHING AROUND TEXT, DIMENSIONS, AND ATTRIBUTES

When you select a point within a boundary to be hatched and if it contains text, dimensions, and attributes, the hatch lines do not pass through the text, dimensions, and attributes present in the object being hatched by default. AutoCAD places an imaginary box around these objects that does not allow the hatch lines to pass through it. Remember that if you are using the select objects option to select objects to hatch, you must select the text/attribute/ shape along with the object in which it is placed when defining the hatch boundary. If multiple line text is to be written, the **MTEXT** command is used. You can also select both the boundary and the text when using the window selection

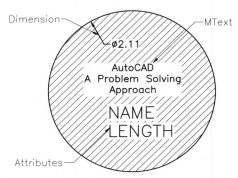

Figure 15-26 Hatching around dimension, multiline text, and attributes

method. Figure 15-26 shows you how hatching takes place around multiline text, attributes, and dimensions.

EDITING HATCH PATTERNS
Using the HATCHEDIT Command

| **Ribbon:** | Home > Modify > Edit Hatch | **Tool Palette:** | Modify > Edit Hatch |
| **Toolbar:** | Modify II > Edit Hatch | **Command:** | HATCHEDIT |

The **HATCHEDIT** command (Figure 15-27) is used to edit the hatch pattern. When you invoke this command and select the hatch for editing, the **Hatch Edit** dialog box (Figure 15-28) is displayed on the screen.

The **Hatch Edit** dialog box can also be invoked from the shortcut menu or by double-clicking on the hatch pattern. After selecting a hatched object, right-click in the drawing area, and choose **Hatch Edit** from the shortcut menu.

*Figure 15-27 Choosing **Edit Hatch** from the **Modify** panel*

AutoCAD will display the **Hatch Edit** dialog box, where the hatch pattern can be changed or modified accordingly. This dialog box is the same as the **Hatch and Gradient** dialog box, except that only the options that control the hatch pattern are available. The available options work in the same way as they do in the **Hatch and Gradient** dialog box.

The **Hatch Edit** dialog box has two tabs, just like the **Hatch** and **Gradient** dialog box. They are, **Hatch** and **Gradient**. In the **Hatch** tab, you can redefine the type of hatch pattern by selecting another type from the **Type** drop-down list. If you are using the **Predefined** pattern,

Figure 15-28 *The* **Hatch** *tab of the* **Hatch Edit** *dialog box*

you can select a new hatch pattern name from the **Pattern** drop-down list. You can also change the scale or angle by entering new values in the **Scale** or **Angle** edit box. When using the **User defined** pattern, you can redefine the spacing between the lines in the hatch pattern by entering a new value in the **Spacing** edit box. If you are using the **Custom** pattern type, you can select another pattern from the **Custom pattern** drop-down list. You can also redefine the island detection style by selecting either of the **Normal**, **Outer**, or **Ignore** styles from the **Islands** area of the dialog box. You can also convert a non-annotative hatch into an annotative hatch and vice-versa. If you want to copy the properties from an existing hatch pattern, choose the **Inherit Properties** button, and then select the hatch whose properties you want to be inherited. You can also change the draw order of the hatch patter using the options available in the **Draw order** area. If the **Associative** radio button in the **Options** area of the dialog box is selected, the pattern is associative. This implies that whenever you modify the hatch boundary, the hatch pattern is automatically updated. The appearance of the gradient fill can be specified using the **Gradient** tab of the dialog box. You can specify the gradient fill comprising different shades and tints of a single color or double colors by selecting the **One color** and the **Two color** radio buttons, respectively. You can also specify the color of the gradient fill or the shade and tint of a single color using the color swatch and the **Shade and Tint** Slider. If the Centered button is selected, AutoCAD will apply a symmetrical gradient fill. Also, AutoCAD provides you with an option to select the desired display pattern of the gradient fill by selecting any one of the nine fixed patterns for the gradient fill.

Note
If a hatch pattern is associative, the hatch boundary can be edited using grips and editing commands and the associated pattern is modified accordingly. This is discussed later.

In Figure 15-29, the object is hatched using the **ANSI31** hatch pattern. With the **HATCHEDIT** command, you can edit the hatch using the **Hatch Edit** dialog box to change the hatch to the one shown in Figure 15-30. You can also edit an existing hatch through the command line by entering **-HATCHEDIT** at the Command prompt.

Figure 15-29 ANSI31 hatch pattern

*Figure 15-30 The modified hatch pattern using the **HATCHEDIT** command*

IGNORING HATCH PATTERN ENTITIES WHILE SNAPPING

In the older releases of AutoCAD, the hatch pattern elements were even considered, while trying to snap to a point close to them. This causes a problem in snapping to the exact point. Now AutoCAD allows you to set the option that ensures that the hatch pattern elements are not considered while snapping. This option is set using the **Ignore hatch objects** check box in the **Object Snap Options** area of the **Drafting** tab of the **Options** dialog box. This check box is selected by default.

Using the Properties Command

Quick Access Toolbar: Properties (*Customize to Add*)	**Command:** PROPERTIES
Ribbon: View > Palettes > Properties	

You can also use the **PROPERTIES** command to edit the hatch pattern. When you select a hatch pattern for editing, and invoke the **PROPERTIES** command, AutoCAD displays the **Properties** palette for the hatch, see Figure 15-31. You can also invoke the **Properties** palette by selecting the hatch pattern and right-clicking to display a shortcut menu. Choose **Properties** from the shortcut menu; the **Properties** palette is displayed. This palette displays the properties of the selected pattern under the following categories.

General

In this category, you can change the general pattern properties like color, layer, linetype, linetype scale, lineweight, and so on. When you click on the field corresponding to a particular property, a drop-down list is displayed from where you can select an option or value. Whenever a drop-down list does not get displayed, you can enter a value in the field.

Pattern

In this category, you can change the pattern type, pattern name, annotative property, annotation scale, angle, scale, spacing, associativity, and island detection style properties of the pattern. In the case of ISO patterns, you can redefine the ISO pen width of the pattern modify the spacing of the **User defined** patterns. You can also determine whether you want to have a **User defined** pattern as double or not. When you choose the [**...**] button in the **Type** property field, AutoCAD displays the **Hatch Pattern Type** dialog box, see Figure 15-32. Here, you can select the type of hatch pattern from the **Pattern Type** drop-down list. If you have selected a **Predefined** pattern type, you can select a pattern name

Figure 15-31 The Properties palette (Hatch)

from the **Pattern** drop-down list or choose the **Pattern** button to display the **Hatch Pattern Palette** dialog box. You can then select a pattern here in one of the tabs and then choose **OK** to exit the dialog box. Now, choose **OK** in the **Hatch Pattern Type** dialog box to return to the **Properties** palette. You will notice that the pattern name you have selected is displayed in the **Pattern** name field. Similarly, if you select the **Custom** pattern type, you can enter the name of the custom pattern in the **Custom Pattern** edit box. If you select a **User defined** pattern type, both the **Pattern Type** drop-down list and the **Custom Pattern** edit box will not be available. When you click in the **Pattern name** property field, the [**...**] button is displayed. When you choose the [**...**] button, the **Hatch Pattern Palette** dialog box is displayed with the ANSI31 pattern selected. Select the **Yes** or **No** option from the **Annotative** drop-down list to convert the existing hatch into annotative or non-annotative respectively. When you click on the **Annotation scale** property field, the [**...**] button will be displayed. On choosing the [**...**] button, the **Annotation Object Scale** dialog box will be invoked. You can add other annotation scales to the selected hatch with the help of

Figure 15-32 The Hatch Pattern Type dialog box

this dialog box. The values of the angle, scale, and spacing properties can be entered in the corresponding fields. A value for the ISO pen width can be selected from the **ISO pen width** drop-down list. For the user-defined hatch patterns, you can select **Yes** or **No** from the **Double** drop-down list to specify if you want the hatch pattern to double at right angle to the original pattern or not. You can also select **Yes** or **No** from the **Associative** drop-down list to determine if a pattern is associative or not. Similarly, you can select an island detection style from the **Island detection style** drop-down list.

Tip
*It is more convenient to choose the [...] button in the **Pattern Type** edit box to display the **Hatch Pattern Type** dialog box and enter both the type and name of the pattern at the same time.*

Geometry
In this category, the elevation of the hatch pattern can be changed. A new value of elevation for the hatch pattern can be entered in the **Elevation** field. This category also displays the area and the cumulative area of the hatch.

EDITING HATCH BOUNDARY
Using Grips
One of the ways you can edit the hatch boundary is by using grips. You can select the hatch pattern or the hatch boundaries. If you select the hatch pattern, the hatch highlights and a grip is displayed at the centroid of the hatch. A centroid for a region that is coplanar with the *XY* plane is the center of that particular area. However, if you select an object that defines the hatch boundary, the object grips are displayed at the defining points of the selected object. Once you change the boundary definition, and if the hatch pattern is associative, AutoCAD will reevaluate the hatch boundary and then hatch the new area. When you edit the hatch boundary, make sure that there are no open spaces in it. AutoCAD will not create a hatch if the outer boundary is not closed. Figure 15-33 shows the result of moving the circle and text, and shortening the bottom edge of the hatch boundary of the object shown in Figure 15-30.

The objects were edited by using grips and since the pattern is associative, when modifications are made to the boundary object, the pattern automatically fills up the new area. If you select the hatch pattern, AutoCAD will display the hatch object grip at the centroid of the hatch. You can select the grip to make it hot and then edit the hatch object. You can stretch, move, scale, mirror, or rotate the hatch pattern. Once an editing operation takes place on the hatch pattern, it may be moved, rotated, or scaled; then, the associativity is lost and AutoCAD displays the message: "**Hatch boundary associativity removed - no more scales may be added.**" in the prompt window.

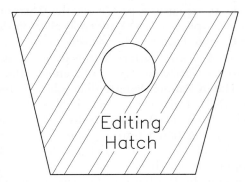

Figure 15-33 The result of using grips to edit the hatch boundary of the object shown in Figure 15-30

The **Retain boundaries** check box in the **Hatch and Gradient** dialog box is cleared when a hatch is created by default. Hence, all objects selected or found by the selection or point selection processes during hatch creation are associated with the hatch as boundary objects.

Editing any of them may affect the hatch. If, however, the **Retain boundaries** check box was selected when the hatch was created, only the retained polyline boundary created by **HATCH** is associated with the hatch as its boundary object. Editing it will affect the hatch, and editing the objects selected or found by selection or point selection processes during the hatch created will have no effect on the hatch. You can, of course, select and simultaneously edit the objects selected or found by the selection or point selection processes, as well as the retained polyline boundary created by the **HATCH** command.

Trimming the Hatch Patterns

One of the recent additions in editing hatches is that you are now allowed to trim the hatch patterns using a cutting edge. For example, refer to Figure 15-34. This figure shows a drawing before trimming the hatch. In this drawing, the outer loop was selected as the object to be hatched. This is the reason the space between the two vertical lines on the right is also hatched. Figure 15-35 shows the same drawing after trimming the hatch using the vertical lines as the cutting edge. You will notice that even after trimming some of the portions of the hatch, it is a single entity.

Figure 15-34 *Before trimming the hatch*

Figure 15-35 *After trimming the hatch using the vertical lines as the cutting edge*

Using AutoCAD Editing Commands

When you use the editing commands such as **MOVE**, **ROTATE**, **SCALE**, and **STRETCH**, associativity is maintained, provided all objects that define the boundary are selected for editing. If any of the boundary-defining objects is missing, the associativity will be lost and AutoCAD will display the message **Hatch boundary associativity removed - no more scales may be added.** When you rotate or scale an associative hatch, the new rotation angle and the scale factor are saved with the hatch object data. This data is then used to update the hatch. When using the **ARRAY**, **COPY**, and **MIRROR** commands, you can array, copy, or mirror just the hatch boundary, without the hatch pattern. Similarly, you can erase just the hatch boundary, using the **ERASE** command and associativity is removed. If you explode an associative hatch pattern, the associativity between the hatch pattern and the defining boundary is removed. Also, when the hatch object is exploded, each line in the hatch pattern becomes a separate object.

When the original boundary objects are being edited, the associated hatch gets updated, but new boundary objects do not have any effect. For example, when you have a square island within a circular boundary and then you erase the square, the hatch pattern is updated to fill up the entire circle. But, once the island is removed, another island cannot be added, since it was not calculated as a part of the original set of boundary objects.

HATCHING BLOCKS AND XREF DRAWINGS

When you apply hatching to inserted blocks and xref drawings, their internal structure is treated as if the block or xref drawing were composed of independent objects. This means that if you have inserted a block that consists of a circle within a rectangle and you want the internal circle to be hatched, all you have to do is invoke the **Hatch and Gradient** dialog box and then choose the **Add: Pick points** button to select a point within the circle to generate the desired hatch, see Figure 15-36. However, if you choose the **Select objects** option from the shortcut menu that is displayed on right-clicking after choosing the **Add: Pick points** button, you will be prompted to select an object. Select an object; entire block will be selected and a hatch will be created, as shown in Figure 15-36.

Hatch created using the Add: Pick points button

Hatch created using the Select Objects option

Figure 15-36 Hatching inserted blocks

When you xref a drawing, you can hatch any part of the drawing that is visible. Also, if you hatch an xref drawing and then use the **XCLIP** command to clip it, the hatch pattern is not clipped, although the hatch boundary associativity is removed. Similarly, when you detach the xref drawing, the hatch pattern and its boundaries are not detached, although the hatch boundary associativity is removed.

CREATING A BOUNDARY USING CLOSED LOOPS

Ribbon: Home > Draw > Boundary	**Command:** BOUNDARY

The **BOUNDARY** command is used for creating a polyline or region around a selected point within a closed area, in a manner similar to the one used for defining a hatch boundary. When this command is invoked, AutoCAD displays the **Boundary Creation** dialog box, as shown in Figure 15-37.

The options in the **Boundary Creation** dialog box are similar to those in the **Hatch and Gradient** dialog box discussed earlier. Although, only the options related to boundary selections are available, the **Pick Points** button in the **Boundary Creation** dialog box is used to create a boundary by selecting a point inside the object, whose boundary you want to create.

When you choose the **Pick Points** button, the dialog box is cleared from the screen temporarily and you are prompted to select the internal point. Select a point that lies within the boundary of the object. Once you select an internal point, a polyline or a region will be formed around the boundary (Figure 15-38). To end this process, press ENTER or right-click at the **Pick internal point** prompt. Whether the boundary created is a polyline or a region is determined by the option you have selected from the **Object type** drop-down list in the **Boundary retention** area. **Polyline** is the default option. You can edit the boundary that has been created with the editing commands. The boundary is selected by using the **Last** object selection option.

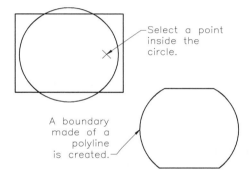

Figure 15-37 *The **Boundary Creation** dialog box*

Figure 15-38 *Using the **BOUNDARY** command to create a polyline boundary*

Similar to the **HATCH** command, you can also define a boundary set here by choosing the **New** button in the **Boundary set** area of the dialog box. The dialog box will temporarily close to allow you to select objects to be used to create a boundary. The default boundary set is **current viewport**, which means everything visible in the current viewport consists of the boundary set. As discussed earlier, the boundary set option is especially useful when you are working with large and complex drawings, where examining everything on the screen becomes a time-consuming process.

The **Boundary Creation** dialog box also provides you the option of determining the island detection method. The **Island detection** check box is available below the **Pick Points** button and is selected by default. As a result, the island detection method is Flood **t**ype and islands are included as boundary objects. If you clear this check box, the Ray casting method of island detection is enabled, where rays shall be cast from the selected internal point in all directions (by default, using the dialog box). The nearest closed object it encounters is used for boundary creation. If the selected point does not lie within the encountered boundary, a **Boundary Definition Error** dialog box is displayed and you have to select another internal point.

Note
*The **HPBOUND** system variable controls the type of boundary object created using the **BOUNDARY** command. If the value is 1, the object created a polyline and if the value is 0, the object created is a region.*

Exercise 4   *Mechanical*

See Figure 15-39 and then create the drawing shown in Figure 15-40 using the **BOUNDARY** command to create a hatch boundary. Copy the boundary from the drawing shown in Figure 15-40, and then hatch.

Figure 15-39 Drawing for Exercise 4

Figure 15-40 Final drawing for Exercise 4

OTHER FEATURES OF HATCHING

1.   In AutoCAD, the hatch patterns are separate objects. The objects store the information about the hatch pattern boundary with reference to the geometry that defines the pattern for each hatch object.

2.   When you save a drawing, the hatch patterns are stored with the drawing. Therefore, the hatch patterns can be updated even if the hatch pattern file that contains the definitions of hatch patterns is not available.

3.   If the system variable **FILLMODE** is 0 (Off), the hatch patterns are not displayed. To see the effect of a changed **FILLMODE** value, you must use the **REGEN** command to regenerate the drawing after the value has been changed.

4.   You can edit the boundary of a hatch pattern even when the hatch pattern is not visible (**FILLMODE**=0).

5.   The hatch patterns created in the earlier releases are automatically converted into AutoCAD 2010 hatch objects when the hatch pattern is edited.

6.   When you save an AutoCAD 2010 drawing in Release 12 format (DXF), the hatch objects are automatically converted to Release 12 hatch blocks.

7. You can use the **CONVERT** command to change the pre-Release 14 hatch patterns to AutoCAD 2010 objects. You can also use this command to change the pre-Release 14 polylines to AutoCAD 2010 optimized format to save memory and disk space.

Self-Evaluation Test

Answer the following questions and then compare them to those given at the end of this chapter:

1. Different filling patterns make it possible to distinguish between different parts or components of an object. (T/F)

2. If AutoCAD does not locate the entered pattern in the *acad.pat* file, it searches for it in a file with the same name as the pattern. (T/F)

3. When a boundary set is formed, it does not become the default boundary set for hatching. (T/F)

4. You can edit the hatch boundary and the associated hatch pattern by using grips. (T/F)

5. If you need to hatch a drawing that has a gap, then you need to specify the gap value in the _____ edit box in the **Hatch and Gradient** dialog box.

6. One of the ways to specify a hatch pattern from the group of stored hatch patterns is by selecting one from the _____ drop-down list.

7. The value you enter in the _____ edit box lets you rotate the hatch pattern with respect to the *X* axis of the current UCS.

8. The _____ button in the **Hatch and Gradient** dialog box lets you select objects that form the boundary for hatching.

9. You can use the _____ option after you have selected the area to be hatched to see the hatching before it is actually applied.

10. The selection of a hatch style carries meaning only if the objects to be hatched are _____.

Review Questions

Answer the following questions:

1. The **HATCH** command does not allow you to hatch a region enclosed within a boundary (closed area) by selecting a point inside the boundary. (T/F)

2. A Boundary Definition Error message box is displayed if AutoCAD finds that the selected point is not inside the boundary or that the boundary is not closed. (T/F)

3. The **Tolerance** edit box is used to set the value up to which the open area will be considered closed when selected for hatching using the **Add: Pick points** method. (T/F)

4. When you use editing commands such as **MOVE**, **SCALE**, **STRETCH**, and **ROTATE**, associativity is lost, even if all the boundary objects are selected. (T/F)

5. The hatching procedure in AutoCAD does not work on inserted blocks. (T/F)

6. Patterns drawn using the **HATCH** command are associative. (T/F)

7. Which of the following system variables has to be set to 1, if the **Double** hatch box is selected?

   (a) **HPSPACE**   (b) **HPDOUBLE**
   (c) **HPANG**   (d) **HPSCALE**

8. For which of the following hatch patterns, will the **ISO pen width** drop-down list be available?

   (a) **ANSI**   (b) **Predefined**
   (c) **Custom**   (d) **ISO**

9. Which of the following buttons should be chosen, if you want to have the same hatching pattern, style, and properties as that of the existing hatch?

   (a) **Inherit Properties**   (b) **Add: Select objects**
   (c) **Add: Pick points**   (d) **Remove Islands**

10. Which of the following variables can you use to align the hatches in adjacent hatch areas?

    (a) **SNAPBASE**   (b) **FILLMODE**
    (c) **SNAPANG**   (d) **HPANG**

11. In which of the following system variables, will the specified hatch spacing value be stored?

    (a) **HPDOUBLE**   (b) **HPANG**
    (c) **HPSCALE**   (d) **HPSPACE**

12. One of the advantages of using the _____ command is that you don't have to select each object comprising the boundary of the area you want to hatch, as in the case of the _____ command.

13. To select a custom hatch pattern, first select **Custom** from the **Type** drop-down list and then select the name of a previously stored hatch pattern from the _____ drop-down list.

14. There are three hatching styles from which you can choose: _____, _____, and _____.

15. If you select the _____ style, all areas bounded by the outermost boundary are hatched, ignoring any hatch boundaries that lie within the outer boundary.

Exercises

Exercise 5   *Mechanical*

Hatch the drawings in Figures 15-41 and 15-42 using the hatch pattern to match.

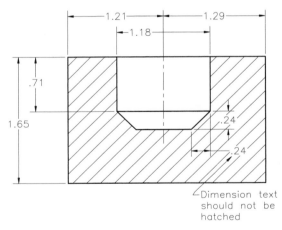

Figure 15-41 *Drawing for Exercise 5*

Figure 15-42 *Drawing for Exercise 5*

Exercise 6   *General*

Hatch the drawing in Figure 15-43 using the hatch pattern **ANSI31**. Use the **SNAPBASE** variable to align the hatch lines, as shown in the drawing.

Figure 15-43 *Drawing for Exercise 6*

Exercise 7   *Mechanical*

Figure 15-44 shows the top and front views of an object. It also shows the cutting plane line. Based on the cutting plane line, hatch the front views in section. Use the hatch pattern of your choice.

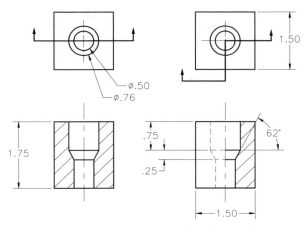

Figure 15-44 Drawing for Exercise 7

Exercise 8   *Mechanical*

Figure 15-45 shows the top and front views of an object. Hatch the front view in full section. Use the hatch pattern of your choice.

Figure 15-45 Drawing for Exercise 8

Exercise 9   *Mechanical*

Figure 15-46 shows the front and side views of an object. Hatch the side view in half section. Use the hatch pattern of your choice.

Chapter 15

Figure 15-46 *Drawing for Exercise 9*

Exercise 10   *Mechanical*

Figure 15-47 shows the front view with broken section and top views of an object. Hatch the front view as shown. Use the hatch pattern of your choice.

Figure 15-47 *Drawing for Exercise 10*

Exercise 11   *Mechanical*

Figure 15-48 shows the front, top, and side views of an object. Hatch the side view in section using the hatch pattern of your choice.

Figure 15-48 Drawing for Exercise 11

Problem-Solving Exercise 1   *Mechanical*

Figure 15-49 shows an object with front and side views with aligned section. Hatch the side view using the hatch pattern of your choice.

Chapter 15

Figure 15-49 *Drawing for Problem-Solving Exercise 1*

Problem-Solving Exercise 2   *Mechanical*

Figure 15-50 shows the front view, side view, and the detail "A" of an object. Hatch the side view and draw the detail drawing as shown.

Figure 15-50 *Drawing for Problem-Solving Exercise 2*

Problem-Solving Exercise 3   *Mechanical*

Create the drawing shown in Figure 15-51. Assume the missing dimensions.

Figure 15-51 *Drawing for Problem-Solving Exercise 3*

Problem-Solving Exercise 4   *Mechanical*

Create the drawing shown in Figure 15-52. Assume the missing dimensions.

Figure 15-52 *Drawing for Problem-Solving Exercise 4*

Chapter 15

Problem-Solving Exercise 5

Mechanical

Draw a detail drawing whose top, side, and section views are given in Figure 15-53. Then, hatch the section view. **

Figure 15-53 *Views and dimensions of the drawing for Problem-Solving Exercise 5*

Answers to Self-Evaluation Test

1. T, **2.** T, **3.** F, **4.** T, **5. Tolerance**, **6. Pattern**, **7. Angle**, **8. Select objects**, **9. Preview**, **10.** nested

Chapter 16

Working with Blocks

Learning Objectives

After completing this chapter, you will be able to:

• *Create blocks using the **Block Definition** dialog box.*
• *Insert blocks using the **Insert** dialog box and the **MINSERT** commands.*
• *Understand the concept of dynamic blocks and **Block Editor**.*
• *Add parameters and assign action to them to create dynamic blocks.*
• *Create drawing files using the **Write Block** dialog box.*
• *Use the **DesignCenter** to locate, preview, copy, or insert blocks and existing drawings.*
• *Use the **Tool Palettes** window to insert blocks.*
• *Perform editing operations on blocks in-place using the **REFEDIT** command.*
• *Split a block into individual objects using the **EXPLODE** and **XPLODE** commands.*
• *Rename blocks and delete unused blocks.*

THE CONCEPT OF BLOCKS

The ability to store parts of a drawing, or the entire drawing, such that they need not be redrawn when required in the same drawing or another drawing is a great benefit to the user. These parts of a drawing, entire drawings, or symbols (also known as **blocks**) can be placed (inserted) in a drawing at the location of your choice, with the desired orientation, and scale factor. The block is given a name (block name). The block is referenced (inserted) by its name. All the objects within a block are treated as a single object. You can **MOVE**, **ERASE**, or **LIST** the block as a single object, that is, you can select the entire block simply by selecting anywhere on it. As for the edit and inquiry commands, the internal structure of a block is immaterial, since a block is treated as a primitive object, like a polygon. If a block definition is changed, all references to the block in the drawing are updated to incorporate the changes.

A block can be created using the **BLOCK** command. You can also save objects in a drawing, or an entire drawing as a drawing file using the **WBLOCK** command. The main difference between the two is that a wblock can be inserted in any other drawing, but a block can be inserted only in the drawing file, in which it was created.

One of the recent enhancements in AutoCAD is the annotative blocks that can be used as symbol to annotate your drawing. The annotative blocks are defined in terms of paper space height. The annotative blocks are displayed in floating viewports and their respective size in model space is calculated by multiplying the current annotation scale set for those spaces and the paper space height of the block.

Another feature of AutoCAD is that instead of inserting a symbol as a block (which results in adding the content of the referenced drawing to the drawing in which it is inserted), you can reference the other drawings (Xref) in the current file. This means that the contents of the referenced drawing are not added to the current drawing file, although they become part of that drawing on the screen. This is explained in detail in Chapter 16.

Advantages of Using Blocks

Blocks offer the following advantages:

1. Drawings often have some repetitive features. Instead of drawing the same feature again and again, you can create its block and insert it wherever required. This helps you to reduce drawing time and better organize your work.
2. Blocks can be drawn and stored for future use. You can thus create a custom library of objects required for different applications. For example, if your drawings are concerned with gears, you could create their blocks and then integrate them with custom menus. In this manner, you could create an application environment of your own.
3. The size of a drawing file increases as you add objects to it. AutoCAD keeps track of information about the size and position of each object in the drawing, for example, the points, scale factors, annotation scale, radii, and so on. If you combine several objects into a single object by forming a block with the **BLOCK** command, there will be a single scale factor, annotation scale, rotation angle, position, and so on, for all objects in the block, thereby saving storage space. Each object, repeated in the multiple block insertions,

needs to be defined only once in the block definition. Ten insertions of a gear, made of forty-three lines and forty-one arcs, require only ninety-four objects (10+43+41), while ten individually drawn gears require 840 objects [10 X (43+41)].

4. If the specification for an object changes, the drawing needs to be modified. This is a very tedious task, if you need to detect each location where the change is to be made and edit it individually. But, if this object has been defined as a block, you can redefine it. Also, where ever that object appears, it is revised automatically. This has been dealt with in Chapter 17, Defining Block Attributes.

5. Attributes (textual information) can be included in blocks. Different attribute values can be changed in each insertion of a block.

6. You can create symbols and then store them as blocks with the **BLOCK** command. Later on, with the **INSERT** command, you can insert the blocks in the drawing, in which they were defined. There is no limit to the number of times you can insert a block in a drawing.

7. You can store symbols as drawing files using the **WBLOCK** command and later insert them into any other drawing, using the **INSERT** command.

8. Blocks can have varying X, Y, and Z scale factors and rotation angles from one insertion to another (Figure 16-1).

9. You can insert multiple copies of a block about a specified path, using the **DIVIDE** or **MEASURE** command.

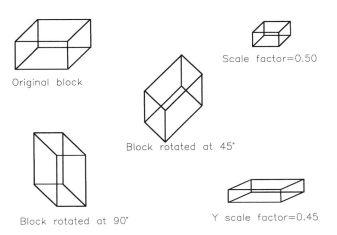

Figure 16-1 Blocks with different specifications

FORMATION OF BLOCKS
Drawing Objects for Blocks

The first step in creating blocks is to draw the object(s) to be converted into a block. You can consider any symbol, shape, or view that may be used more than once for conversion into a block (Figure 16-2). Even a drawing, that is to be used more than once, can be inserted as a block.

ELECTRICAL SYMBOLS

ARCHITECTURAL SYMBOLS

MECHANICAL SYMBOLS

Figure 16-2 *Some common drafting symbols created to be stored in a library*

Tip
Examine the sketch of the drawing you are about to begin and carefully look for any shapes, symbols, and components that are to be used more than once. They can then be drawn once and stored as blocks.

Checking Layers

You should be particularly careful about the layers on which the objects to be converted into a block are drawn. The objects must be drawn on layer 0, if you want the block to inherit the linetype and color of the layer on which it is inserted. If you want the block to retain the linetype and color of the layer on which it was drawn, draw the objects defining the block on that layer. For example, if you want the block to have the linetype and color of the layer OBJ, draw the objects comprising the block on layer OBJ. Now, even if you insert the block on any other layer, the linetype and color of the block will be that of the layer OBJ.

If the objects in a block are drawn with the color, linetype, and lineweight as **ByBlock**, this block when inserted, assumes the properties of the current layer. To set the color to ByBlock, select **ByBlock** from the **Object Color** drop-down list of the **Properties** Palette or from the **Properties** panel in the **Home** tab of the **Ribbon**. To set the linetype to ByBlock, select **ByBlock** from the **Selects a linetype** drop-down list in the **Properties** panel of the **Home** tab in the **Ribbon**. Similarly, you can specify the lineweight as **ByBlock**.

Note
*As discussed earlier, before creating a block, you should check the layer on which the objects will be drawn because a layer determines the color and linetype of the block. To change the property of a layer associated with an object, you need to invoke the **Properties** palette by choosing **View > Palettes > Properties** from the **Ribbon**, from **Standard** toolbar, by pressing CTRL+1, or from the shortcut menu. You can also change the properties by entering **Properties** option at the **CHANGE** command or the **CHPROP** command.*

CONVERTING ENTITIES INTO A BLOCK

Ribbon:	Insert > Block > Create or Home > Block > Create
Toolbar:	Draw > Make Block **Command:** BLOCK

You can convert the entities in the drawing window into a block, by using the **BLOCK** command or by choosing **Create** button from the **Block** panel in the **Insert** tab or the **Home** tab of the **Ribbon**, as shown in Figure 16-3. Alternatively, you can do so by choosing the **Make** option from the **Draw** toolbar. When you invoke this command, the **Block Definition** dialog box is displayed, as shown in Figure 16-4. You can use the **Block Definition** dialog box to save any object or objects as a block.

Figure 16-3 Choosing the Create button from the Block panel

Figure 16-4 The Block Definition dialog box

In the **Block Definition** dialog box, you can enter the name of the block you want to create in the **Name** edit box. All the block names present in the current drawing are displayed in the **Name** drop-down list in an alphabetical and numerical order. This way you can verify whether a block you have defined has been saved. By default, the block name can have up to 255 characters. Also, it can contain letters, digits, blank spaces as well as any special characters including the $ (dollar sign), - (hyphen), and _ (underscore), provided they are not being used for any other purpose by AutoCAD or Windows.

Note

*The block name is controlled by the **EXTNAMES** system variable and its default value is 1. If the value is set to 0, the block name will be only thirty-one characters long and will not include spaces or any other special characters, apart from a $ (dollar sign), - (hyphen), or _ (underscore).*

If a block already exists, with the block name that you have specified in the **Name** edit box, and you choose **OK** in the **Block Definition** dialog box, the **Block-Redefine Block** message box will be displayed informing you that the block with this name is already defined. It will also prompt you to specify if you want to redefine the block. In this dialog box, you can either redefine the existing block by choosing the **Redefine** button or you can exit it by choosing the **No** button. You can then use another name for the block in the dialog box.

After you have specified a block name, you are required to specify the insertion base point, which will be used as a reference point to insert the block. Usually, either the center or the lower left corner of the block is defined as the insertion base point. Later on, when you insert the block, you will notice that it appears at the insertion point, and you can insert it with reference to this point. The point you specify as the insertion base point is taken as the origin point of the block's coordinate system. You can specify the insertion point by choosing the **Pick point** button in the **Base point** area of the dialog box. The dialog box is temporarily removed, and you can select a base point on the screen. Alternatively, you can also enter the coordinates in the **X**, **Y**, and **Z** edit boxes. Once the insertion base point is specified, the dialog box reappears and the *X, Y,* and *Z* coordinates of the insertion base point are displayed in the **X**, **Y**, and **Z** edit boxes respectively. If no insertion base point is selected, AutoCAD assumes the insertion point coordinates to be 0,0,0, which are the default coordinates.

Tip

In drawings where the insertion point of a block is important, for example, in the case of inserting the block of a door symbol in architectural plans, it may be a good idea to first specify the requisite scale factors and rotation angles and then the insertion point. This way, you are able to preview the block symbol as it is to be inserted.

After specifying the insertion base point, you are required to select the objects that will constitute the block. Until the objects are selected, AutoCAD displays a warning: **No objects selected**, at the bottom of the **Objects** area. Choose the **Select objects** button. The dialog box is temporarily removed from the screen. You can select the objects on the screen using any selection method. After completing the selection process, right-click or press the ENTER key to return to the dialog box. The number of objects selected is displayed at the bottom of the **Objects** area of the dialog box.

The **Objects** area of the **Block Definition** dialog box also has a **QuickSelect** button. On choosing this button, the **Quick Select** dialog box is displayed, which allows you to define a selection set based on the properties of objects. **Quick Select** is used in cases where the drawings are very large and complex. The **Quick Select** dialog box has been discussed earlier in Chapter 6, *Editing Sketched Objects-II*.

In the **Objects** area, if you select the **Retain** radio button, the selected objects that form the block, are retained on the screen as individual objects. If you select the **Convert to block** radio button, the selected objects are converted into a block and will not be displayed on the screen after the block has been defined. Rather, the created block will be displayed in place of the original objects. If you select the **Delete** radio button, the selected objects will be deleted from the drawing, after the block has been defined.

Tip
*When you select the **Delete** radio button from the **Objects** area while defining a block, the selected objects are removed from the drawing. To restore them, you can use the **OOPS** command. Using **U** at the Command prompt or choosing the **Undo** button from the **Standard** toolbar, will undo the block definition from the drawing.*

If you select the **Annotative** check box in the **Behavior** area, the block created will become annotative. The block will now acquire annotative properties like text, dimension, hatch, and so on. On selecting the **Annotative** check box, the **Match block orientation to layout** check box will get highlighted. If you select the **Match block orientation to layout** check box, the orientation of the block in paper space viewport will always remain aligned to the orientation of the layout, while inserting the block. If you select the **Scale uniformly** check box, the block can only be scaled uniformly in all directions, while inserting. In such cases, if you create an annotative block, you can scale it only uniformly, while inserting. The **Scale uniformly** check box will be selected by default and remains unavailable for modification when you select the **Annotative** check box. If you select the **Allow exploding** check box, the block can be exploded into separate entities using the **EXPLODE** command, whenever required.

In the **Settings** Area, the **Block unit** drop-down list displays the units that will be used when inserting the current block. For example, if the block is created with inches as the units and you select **Feet** from the **Block unit** drop-down list, the block will be scaled to feet. You can also invoke the **Insert Hyperlink** dialog box by choosing the **Hyperlink** button in the **Settings** Area. This dialog box allows you to link any specific files, websites, or named views with the current drawing. You can also link the drawing to a default email address or you can select one from the recently used email addresses.

The **Open in block editor** check box is available at the bottom of **Block Definition** dialog box. If this check box is selected, the **Block Editor** will be opened as soon as you close the **Block Definition** dialog box. Also, the entities of the block will be displayed in the authoring area of the block editor. You will learn more about the **Block Editor** in the next section.

After setting all parameters in the **Block Definition** dialog box, choose the **OK** button to complete defining the block.

Tip
It is a good idea to create a unit block, especially for drawings where the same block is to be inserted with various scale factors. A unit block can be defined as a block that is created within a unit square area (1 unit x 1 unit). This way, every time you specify a scale factor, the block simply gets enlarged or reduced by the said amount directly, since the multiplication factor is 1.

Note
*You can also create blocks using the command line by entering **-BLOCK** at the Command prompt.*

Exercise 1

General

Draw a circle of radius 1unit and then draw multiple circles inside it, as shown in Figure 16-5. Next, convert them into a block and name it as CIRCLE. Refer to the following figure for details.

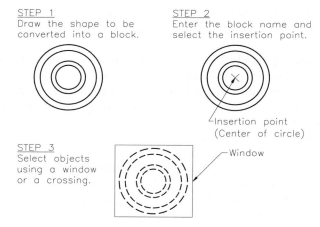

Figure 16-5 *Using the **BLOCK** command*

INSERTING BLOCKS

Ribbon:	Insert > Block > Insert or Home > Block > Insert
Toolbar:	Insert > Insert Block or Draw > Insert Block **Command:** INSERT

The blocks created in the current drawing are inserted using the **INSERT** command (Figure 16-6). An inserted block is called a block reference. You should determine the layer and location to insert the block, and also the angle by which you want the block to be rotated prior to its insertion. If the layer on which you want to insert the block is not current, you can use the **Layer** drop-down list in the **Layers** panel of the **Home** tab in the **Ribbon** or choose the **Set Current** button in the **Layer Properties Manager** palette to make it current. When you invoke the **INSERT** command, the **Insert** dialog box is displayed, as shown in Figure 16-7. You can specify different parameters of the block or external file to be inserted in the **Insert** dialog box.

Figure 16-6 *Choosing the Insert button*

Name

This drop-down list is used to specify the name of the block to be inserted. Select the name

*Figure 16-7 The **Insert** dialog box*

from this drop-down list. You can also enter a name for it. All blocks, created in the current drawing, are available in the **Name** drop-down list.

Note
*The last block name inserted becomes the default name for the next insertion and is displayed in the **Name** drop-down list.*

The **Browse** button is used to insert external files. When you choose this button, the **Select Drawing File** dialog box is displayed, as shown in Figure 16-8.

This dialog box is similar to the standard **Select File** dialog box, which has been discussed earlier in Chapter 1. You can select a drawing file from the files listed in the current directory. You can also change the directory by selecting the desired directory from the **Look in** drop-down list. Once you select the drawing file, choose the **Open** button; the drawing file name is displayed next to the **Name** drop-down list of the **Insert** dialog box. The **Path** option displays the path of the external file selected to be inserted. Now, if you want to change the block name, just change the name in the **Name** drop-down list. In this manner, the drawing can be inserted with a different name. Changing the original drawing does not affect the inserted drawing.

Note
*If the name you have specified in the **Name** edit box does not exist as a block in the current drawing, AutoCAD will search the drives and directories on the path (specified in the **Options** dialog box) for a drawing of the same name. If the block is found, it will be inserted.*

*Also, suppose you have inserted a block in a drawing and then you want to insert a drawing with the same name as the block, AutoCAD will display a message saying that XX is already defined as a block and asks what do you want to do. If you choose **Redefine**, the block in the drawing will get replaced by the drawing with the same name, that is, the block gets redefined.*

Figure 16-8 The **Select Drawing File** *dialog box*

Tip
*You can create a block in the current drawing from an existing drawing file. This saves the time, by avoiding redrawing the object as a block. Locate and select an existing drawing file using the **Browse** button; choose the **Open** button after selecting the existing drawing. Next, choose the **OK** button; the **Insert** dialog box will be closed, prompting you to specify the insertion point. Now, instead of specifying the insertion point, press the ESC key; the selected file will be converted into a block, but will not be inserted into the drawing.*

Insertion point Area

When a block is inserted, its coordinate system is aligned parallel to the current UCS. In the **Insertion point** area, you can specify the X, Y, and Z coordinate locations of the block insertion point in the **X**, **Y**, and **Z** edit boxes, respectively. If you select the **Specify On-screen** check box, the **X**, **Y**, and **Z** edit boxes will not be available. You can specify the insertion point on the screen. By default, the **Specify On-screen** check box is selected, and hence, you can specify the insertion point on the screen.

Scale Area

In this area, you can specify the X, Y, and Z scale factors of the block to be inserted in the **X**, **Y**, and **Z** edit boxes, respectively. By selecting the **Specify On-screen** check box, you can specify the scale of the block at which it has to be inserted on the screen. The **Specify On-screen** check box is cleared by default and the block is inserted with the scale factors of 1 along the three axes. Also, if the **Uniform Scale** check box is selected, the X scale factor value is assumed for the Y and Z scale factors also. This means that if this check box is selected, you need to specify only the X scale factor. All dimensions in the block are multiplied by the same scale factors that you specify. These scale factors allow you to stretch or compress a block along the *X* and *Y* axes, respectively. You can also insert 3D objects into a drawing by specifying the third scale factor (since 3D objects have three dimensions), the Z scale factor. Figure 16-9 shows variations of the X and Y scale factors for the block CIRCLE, during insertion.

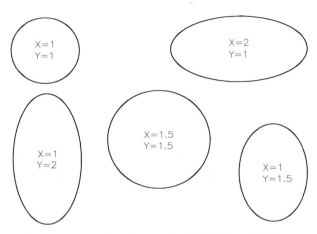

Figure 16-9 *Block inserted with different scale factors*

Tip
By specifying negative scale factors, you can insert a mirror image of a block along a particular axis. A negative scale factor for both the X and Y axes is the same as rotating the block reference through 180-degree, since it mirrors the block reference in the opposite quadrant of the coordinate system. The effect of the negative scale factor on the block (DOOR) can be marked by a change in position of the insertion point marked with an (X), as shown in Figure 16-10.

Also, specifying a scale factor of less than 1 inserts the block reference smaller than the original size. A scale factor greater than 1 inserts the block reference larger than its original size.

Rotation Area
You can enter the angle of rotation for the block to be inserted in the **Angle** edit box. The insertion point is taken as the location about which the rotation takes place. Selecting the **Specify On-screen** check box enables you to specify the angle of rotation on the screen. This check box is cleared by default and the block is inserted at an angle of zero-degree.

Block Unit Area
This area displays the information related to the block unit specified in the **Block Definition** dialog box and the scale factor used for the unit to insert the block. For example, if you have defined a block with the block unit as feet and you insert the same block in a drawing whose

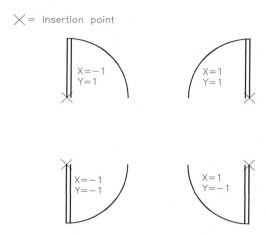

Figure 16-10 Block inserted using negative scale factors

unit is inches, the **Block Unit** area will display **Unit** as feet and **Factor** as 12. The information displayed in this area is read-only.

 Note
*You can control the units for the insertion of the block through the **INSUNITS** system variable. The **Factor** information displayed in the **Block Unit** area will change accordingly.*

Explode Check Box

By selecting this check box, the block is inserted as a collection of individual objects. The function of the **Explode** check box is identical to that of the **EXPLODE** command. Once a block is exploded, the X, Y, and Z scale factors become identical. Therefore, you are provided access to only one scale factor edit box (**X** edit box), the **Y** and **Z** edit boxes are not available, and the **Uniform Scale** check box also gets selected. The X scale factor is assigned to the Y and Z scale factors too. You must enter a positive scale factor value.

 Note
*If you select the **Explode** check box before insertion, the block reference will be exploded and the objects will be inserted with their original properties such as layer, color, and linetype.*

Once you have entered the relevant information in the dialog box, choose the **OK** button. The **Insert** dialog box is removed from the screen. If you have selected the **Specify On-Screen** check boxes, you can specify the insertion point, scale, and angle of rotation with a pointing device. Whenever the insertion point is to be specified on the screen (by default), you can specify the scale factors and rotation angle values. These values will override the values specified in the dialog box using the command line. Before specifying the insertion point on the screen, you can right-click to display the shortcut menu, which has all the options available through the command line. You can choose the options for insertion from the dynamic preview. On being prompted to specify the insertion point, press the down arrow key to see the options in the dynamic preview. However, if you have already specified an insertion point in the dialog box and have selected the **Specify On-screen** check boxes in the **Scale** or **Rotation** areas of the **Insert** dialog box, you are allowed to specify only the scale factors or the rotation angle at the Command line. The following Command prompt will be displayed when all three **Specify On-screen** check boxes are selected:

Specify insertion point or [Basepoint/Scale/X/Y/Z/Rotate]: *You can specify an insertion point on the screen or select an option.*
Enter X scale factor, specify opposite corner, or [Corner/XYZ] <1>: *Specify scale factor in X- axis or select an option or ENTER to accept the default scale factor of 1.*
Enter Y scale factor <use X scale factor>: *Specify scale factor in Y- axis or ENTER to use the X scale factor.*
Specify rotation angle <0>: *Specify the angle of rotation of the inserted block.*

Tip
Using the Command line allows you to override scale factors and rotation angles already specified in the dialog box.

The Command line options are discussed next.

Basepoint

If you enter **B** at the **Specify insertion point** prompt, or select the option from the dynamic preview, you are again prompted to specify the base point. Specify the base point, by clicking the cursor at the desired location. This point will now act as the insertion point. All prompts are also displayed at the cursor input so there is no need to always take a look on the command prompt. The prompt sequence for this option is given next.

Specify insertion point or
[Basepoint/Scale/X/Y/Z/Rotate]: B `Enter`
Specify base point: *Specify base point*
Specify insertion point or [Basepoint/Scale/X/Y/Z/Rotate]: *Specify insertion point*

Scale

On entering **S** at the **Specify insertion point** prompt, you are asked to enter the scale factor. After entering it, the block assumes the specified scale factor, and AutoCAD lets you drag the block until you locate the insertion point on the screen. The *X, Y,* and *Z* axes are uniformly scaled by the specified scale factor. The prompt sequence for this option is given next.
Specify insertion point or [Basepoint/Scale/X/Y/Z/Rotate]: **S**
Specify scale factor for XYZ axes<1>: *Enter a value to preset general scale factor.*
Specify insertion point or [Basepoint/Scale/X/Y/Z/Rotate]: *Specify insertion point*

X, Y, Z

With X, Y, or Z as the response to the **Specify insertion point** prompt, you can specify the X, Y, or Z scale factors before specifying the insertion point. These options are only available when the **Uniform Scale** check box is cleared from the **Insert** dialog box. The prompt sequence when you use the **X** option is given next.

Specify insertion point or [Basepoint/Scale/X/Y/Z/Rotate]: **X**
Specify X scale factor<1>: *Enter a value for the X scale factor.*
Specify insertion point or [Basepoint/Scale/X/Y/Z/Rotate]: *Select the insertion point.*
Specify rotation angle <0>: *Specify angle of rotation.*

Rotate

On entering **R** at the **Specify insertion point** prompt, you are asked to specify the rotation angle. You can enter the angle of rotation or specify it by specifying two points on the screen. Dragging is resumed only when you specify the angle. Also, the block assumes the specified rotation angle. The prompt sequence for this option is given next.

Specify insertion point or [Basepoint/Scale/X/Y/Z/Rotate]: **R**
Specify rotation angle<0>: *Enter a value for the rotation angle.*
Specify insertion point or [Basepoint/Scale/X/Y/Z/Rotate]: *Select the insertion point.*

The XYZ and Corner Options

When you specify the insertion point and scale factors on the screen, the following prompt appears after you have selected an insertion point.

Enter X scale factor, specify opposite corner or [Corner/XYZ]<1>: *Enter X scale factor, specify opposite corner or select an option.*

Here, the **XYZ** option can be used to enter a 3D block reference and the successive prompts allow you to specify the X, Y, and Z scale factors individually.

With the **Corner** option, you can specify both X and Y scale factors at the same time. When you invoke this option, and **DRAGMODE** is turned off, and you are prompted to specify the other corner (the first corner of the box is the insertion point). You can also enter a coordinate value, instead of moving the cursor. The length and breadth of the box are taken as the X and Y scale factors for the block. For example, if the X and Y dimensions (length and width) of the box are the same as that of the block, the block will be drawn without any change. The points selected should be above and to the right of the insertion block. Mirror images will be produced if points selected are below or to the left of the insertion point. The prompt sequence is given next.

Enter X scale factor, specify opposite corner or [Corner/XYZ]<1>: **C**
Specify opposite corner: *Select a point as the corner.*

The **Corner** option also allows you to use a dynamic scaling technique when **DRAGMODE** is turned to **Auto (default)**. You can also move the cursor at the **Enter X scale factor** prompt to change the block size dynamically and, when the size meets your requirements, select a point.

Note

*1. You should try to avoid using the **Corner** option, if you want to have identical X and Y scale factors because it is difficult to select a point whose X distance equals its Y distance, unless the **SNAP** mode is on. If you use the **Corner** option, it is better to specify the X and Y scale factors explicitly or select the corner by entering coordinates.*

*2. You can also insert blocks using the command window by entering the **-INSERT** command or simply **-I** .*

Exercise 2   *General*

Create a block with the name SQUARE with the insertion base point at 1,2. Insert this block in the drawing. The X scale factor is 2 units, the Y scale factor is 2 units, and the angle of rotation is 35-degree. It is assumed that the block SQUARE is already defined in the current drawing.

Exercise 3   *General*

a. Insert the block CIRCLE created in Exercise 1. Use different X and Y scale factors to get different shapes after inserting this block.
b. Insert the block CIRCLE created in Exercise 1. Use the **Corner** option to specify the scale factor.

Exercise 4   *General*

a. Construct a triangle and form a block of it. Name the block as TRIANGLE. Now, set the Y scale factor of the inserted block as 2.
b. Insert the block TRIANGLE with a rotation angle of 45-degree. After defining the insertion point, enter the X and Y scale factor of 2.

CREATING AND INSERTING ANNOTATIVE BLOCKS

You can create an annotative block by selecting the **Annotative** check box from the **Behavior** area of the **Block Definition** dialog box. The annotative block acquires the annotative properties like text, dimension, hatch, and so on. To convert an existing non-annotative block into an annotative one, choose the **Create** button from the **Block** panel of the **Insert tab** in the **Ribbon**. Select the block to be converted to annotative from the **Name** drop-down list of the **Block Definition** dialog box. Select the **Annotative** check box from the **Behavior** area and then choose the **OK** button. AutoCAD will display a message box stating that the selected block is already defined as a block. If you want to redefine the block, choose the **Redefine** button; the specified block will be converted to annotative.

The annotative blocks are indicated by an annotative symbol attached to the block in the preview displayed in the **Insert** dialog box. While inserting the annotative block in the AutoCAD session for the first time, the **Select Annotation Scale** dialog box will be displayed on the screen, see Figure 16-11. Select the annotation scale from the drop-down list and then choose the **OK** button. The block will be inserted at the specified annotation scale. The Annotative blocks are inserted at a scale value decided by the multiplication of the current annotation scale and the block scale specified in the **Scale** area of the **Insert** dialog box.

Note
*While inserting the annotative blocks, the settings specified for the **INSUNITS** system variable are ignored. Also, the **Block Unit** area of the **Insert** dialog box will display no change in the factor value.*

*Figure 16-11 The **Select Annotation Scale** dialog box*

Example 1   *Electronics*

In this example, you will draw the object shown in Figure 16-12, and convert it into an annotative block, named NOR Gate. Next, you will insert the NOR Gate block into the drawing at the annotation scales of 1:1, 1:2, and 1:8 and notice the changes in the size of the annotative blocks inserted in the drawing.

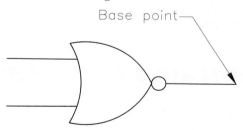

Figure 16-12 Drawing of block for Example 1

1. Start a new file in the **2D Drafting & Annotation** workspace and draw the object, as shown in Figure 16-12.

2. Invoke the **Block Definition** dialog box by choosing **Insert > Block > Create** from the **Ribbon**.

3. Enter **NOR Gate** as the name of the block in the **Name** edit box. Next, choose the **Select objects** button from the **Objects** area; the **Block Definition** dialog box will disappear from the screen. Select the object drawn and press the ENTER key; the **Block Definition** dialog box will appear. Next, select the **Delete** radio button from the **Objects** area.

4. Choose the **Pick point** button from the **Base point** area; the **Block Definition** dialog box will disappear from the screen. Specify the base point, as shown in Figure 16-12.

5. Select the **Annotative** check box from the **Behavior** area. Next, choose the **OK** button from the **Block Definition** dialog box; the selected objects disappear from the screen and an annotative block is defined with the name **NOR Gate**.

6. Now, you need to change the annotation scale of the drawing to 1:8. To do so, click on the down-arrow on the right of the **Annotation Scale** button in the Status Bar, and choose the scale 1:8 from the flyout displayed.

7. To insert the block into the drawing at the annotation scale of 1:8, choose **Insert > Block > Insert** from the **Ribbon**; the **Insert** dialog box is displayed.

8. Select the **NOR Gate** block from the **Name** drop-down list. Select the **Specify On-screen** check box from the **Insertion point** area and choose the **OK** button.

9. Specify the insertion point for the block by clicking on the screen; the block is inserted into the drawing at an annotation scale of 1:8.

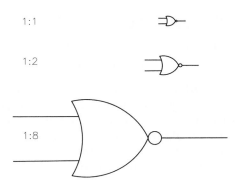

10. Similarly, set the annotation scale to **1:2** in the Status Bar and insert the block. Then set the annotation scale to **1:1** and insert the block.

11. Notice the difference in the size of the inserted blocks, see Figure 16-13. This automated variation in size occurs due to annotative blocks. You will notice that the **Annotation Visibility** button on the right of the **Annotation Scale** button is on.

Figure 16-13 *Annotative blocks inserted at different annotation scales*

12. Now, set the current annotation scale to **1:8**. Choose **Annotate > Annotation Scaling > Add/Delete Scales** from the **Ribbon**. Select the block displayed at the annotation scale of 1:8 and press ENTER; the **Annotation Object Scale** dialog box is displayed. Choose the **Add** button; the **Add Scales to Object** dialog box is displayed. Press and hold the CTRL key and select the scale of 1:2 and 1:1 from the **Scale List** area. Next, choose the **OK** button.

13. The selected annotation scales get associated with the selected block. In this way, you can associate different annotation scales to a single annotative block.

14. Select the block to which you had added the annotation scales; the preview of the block with all annotation scales associated to the block is displayed, as shown in Figure 16-14. The bigger block with the current annotation (1:8) scale is displayed with dark dotted lines and the smaller blocks with other annotation scales (1:1, 1:2) are displayed with the faded dashed lines.

15. To move the block with the current annotation scale (1:8) to the other locations, select the blue grip displayed at the insertion point and move the block to the desired location. This way you can place block with different annotation scales to any desired location.

Figure 16-14 *Different annotation scales associated to a single block*

Block Editor

Ribbon:	Insert > Block > Block Editor or Home > Block > Edit		
Toolbar:	Standard > Block Editor	**Command:**	BEDIT

The **Block Editor** is an important feature. This application is used to edit existing blocks or create new blocks. To invoke the **Block Editor**, choose the **Block Editor** button from the **Block** panel of the **Insert** tab in the **Ribbon**, as shown in Figure 16-15, or enter **BE** (shortcut for the **BEDIT** command) at the pointer input. You can also double-click on the existing block to invoke this command. On doing so, the **Edit Block Definition** dialog box is displayed, as shown in Figure 16-16.

Figure 16-15 *Choosing the* **Block Editor** *button from the* **Block** *panel in the* **Ribbon**

Figure 16-16 *The* **Edit Block Definition** *dialog box*

If you want to create a new block, enter its name in the **Block to create or edit** text box and choose **OK**; the **Block Editor** will be invoked where you can draw the entities in the new block. Similarly, if you want to edit a block, select it from the list box provided in this dialog box; its preview will be displayed in the **Preview** area. Also, its related description, if any, will be displayed in the **Description** area. Choose the **OK** button; the **Block Editor** is invoked.

The default appearance of the **Block Editor** is shown in Figure 16-17. The drawing area of the **Block Editor** has a dull background and is known as the authoring area. You can edit existing entities or add new ones to the block in the authoring area. In addition to the authoring area, the **Block Editor** tab and **Block Authoring Palettes** are provided in the **Block Editor** that contain tools to create dynamic blocks. This will be discussed in detail, later in this chapter. Now, you can edit a block using any tool, as you did in the drawing. Choose the **Save Block** button from the **Open/Save** panel of the **Block Editor** tab when you are finished with the editing, save the changes and then choose the **Close Block Editor** button from the **Close** panel in the **Block Editor** tab of the **Ribbon** to return to the drawing again. Like this, you can edit the block at any time of your design process.

Figure 16-17 *Appearance of the drawing area in the* **Block Editor**

DYNAMIC BLOCKS

The concept of dynamic blocks is very new in AutoCAD. It provides you the flexibility of modifying the geometry of the inserted blocks dynamically or using the **Properties** window. You can create dynamic blocks by adding **Parameters** and **Actions** to existing blocks using the **Action Parameters** panel in the **Block Editor** tab (see Figure 16-18) or the **Block Authoring Palettes**.

Figure 16-18 *The* **Block Editor** *tab*

Block Editor Tab

The **Block Editor** tab, shown in Figure 16-18, provides the tools to create and modify dynamic blocks. Additionally, you can create visibility states for dynamic blocks using the **Visibility** panel of the **Block Editor** tab. All these tools are discussed next.

Edit Block

 This button is chosen to invoke the **Edit Block Definition** dialog box. Note that if you have modified the current block using the tools in the **Block Editor**, you will be prompted to specify whether you want to save the changes in the current block. On doing so, the **Edit Block Definition** dialog box will be displayed. You can select another block to edit or enter the name of the new block that you want to create. Next, proceed to the **Block Editor** to edit the block or create a new one.

Save Block

This button is chosen to save the changes that you have made to the block in the **Block Editor**.

Save Block As

This button is chosen to save a block with a different name. Expand the **Open/Save** panel and choose this button; the **Save Block As** dialog box will be displayed, as shown in Figure 16-19. Enter the name with which you want to save the current block in the **Block Name** edit box and then choose the **OK** button; the block will be saved with the specified name and it will become the current block in the **Block Editor**.

*Figure 16-19 The **Save Block As** dialog box*

Test Block*

On choosing this button, the **Test Block Window** will be displayed. If you have made any changes to the block in the **Block Editor**, you can check them in this window without saving the block. After testing the block, choose **Close > Close Test Block Window** to close the **Test Block Window**.

Authoring Palettes

The **Authoring Palettes** button is chosen from the **Manage** panel to toggle on/off the display of the **Block Authoring Palettes** in the authoring area. The **Block Authoring Palettes** contain the tools for creating dynamic blocks.

Action Parameters

Parameters are predefined rules that are applied to blocks to enable them to react dynamically based on the action or the set of actions assigned to them. To add parameters to the existing blocks, enter **BPARAMETER** at the Command prompt or choose **Action**

Chapter 16

Parameters > Point from the **Ribbon** and then choose one of the buttons from the flyout displayed, as shown in Figure 16-20. If you enter **BPARAMETER** at the Command prompt when the **Dynamic Input** button is chosen, all parameters will be listed in the dynamic preview. These parameters are discussed later in this chapter.

Action

Choose **Action Parameters > Move** from the **Ribbon**; a flyout will be displayed. You can choose a button from this flyout to assign actions to the parameters added to the dynamic blocks. Depending on the type of parameter selected, the possible actions are displayed in the dynamic preview. The types of actions in the **Block Editor** are discussed later in this chapter.

Attribute Definition

Choose **Action Parameters > Attribute Definition** from the **Block Editor** tab of the **Ribbon** to invoke the **Attribute Definition** dialog box to assign attributes to the block. Attributes are discussed in Chapter 17.

Update Fields

Choose **Data > Update Fields** from the **Insert** tab of the **Ribbon** to regenerate the dynamic block such that the text size of the parameters and actions are updated.

Close

When you choose this button, the **Block Editor** is closed and you will be returned to the drawing environment. If this button is chosen before saving the changes in the current block, the **Block - Changes Not Saved** message box will be displayed, as shown in Figure 16-21, and you will be prompted to specify whether or not you want to save changes in the current block.

Figure 16-20 *Options displayed on choosing the* **Point** *button*

Figure 16-21 The **Block - Changes Not Saved** *message box*

Figure   16-22   The **Parameters** *tab of the* **Block Authoring Palettes**

Adding Parameters and Assigning Actions to Dynamic Blocks (Block Authoring Palettes)

The **Block Authoring Palettes**, shown in Figure 16-22, contain the tools to add parameters and actions to the dynamic blocks. You can also invoke these tools from the flyout that is displayed on choosing the down arrow beside the **Point** button in the **Action Parameters** panel of the **Block Editor** tab of the **Ribbon**, refer to Figure 16-20. On doing so, different types of parameters and actions will be displayed at the pointer input or the Command prompt. The command sequence will be the same in both the cases. The types of parameters and actions that can be added to dynamic blocks are discussed next.

Point Parameter

The **Point** parameter is used to assign the **Stretch** or **Move** action to a dynamic block. A grip point will be displayed at the location where you place the **Point** parameter. This grip point will be used as the base point when you assign an action to the **Point** parameter.

To assign a **Point** parameter, invoke the **Point** tool from the **Action Parameters** tab of the **Block Authoring Palettes**. The following prompt sequence will be displayed when you invoke this parameter from the **Action Parameters** tab:

> Specify parameter location or [Name/Label/Chain/Description/Palette]: *Select a keypoint on the block to add the **Point** parameter.*
> Specify label location: *Specify a point to place the label.*
> Enter number of grips [0/1] <1>: *Enter 0 or 1* Enter

Tip
You can double-click on the label to find out the actions that are assigned with that parameter. For example, if you double-click on the point parameter, the move and stretch actions will be available to be performed.

The options available at the Command prompt are discussed next.

Name. Names of a parameter is displayed in the **Properties** palette when you select the **Point** parameter in the **Block Editor**. The default name for the point parameter is Point. However, you can change the name of the parameter using this option. The prompt sequence to modify the name of a parameter is given next.

> Specify parameter location or [Name/Label/Chain/Description/Palette]: **N** Enter
> Enter parameter name <Point>: *Enter new name for the parameter.*
> Specify parameter location or [Name/Label/Chain/Description/Palette]: *Specify a location for the parameter.*

Label. Labels of a parameter are displayed in the authoring area, as shown in Figure 16-23, and in the **Properties** palette in the **Block Editor**. The default label of a point parameter is Position. Using the **Label** option, you can modify the label of the point parameter. The prompt sequence for modifying the label is as follows:

Specify parameter location or [Name/Label/ Chain/Description/Palette]: **L** `Enter`
Enter position property label <Position>: *Enter the new label for the **Point** parameter*
Specify parameter location or [Name/Label/ Chain/Description/Palette]: *Specify a location for the parameter*

Chain. This option is used when the current point parameter is assigned an action, that is also associated with other parameters. In this case, if the **Chain** option is set to **Yes** and the other parameters are edited, the point parameter is also edited with them. If this option is set to **No**, the point parameter is not edited, when the other parameters are edited.

*Figure 16-23 Dynamic block with a **Point Parameter***

Description. This option allows you to enter the description about the parameter. This description is displayed in the **Properties** palette in the **Block Editor**.

Palette. This option is used to specify whether or not the base X and base Y coordinate values of the keypoint where the parameter label is placed will be displayed in the **Properties** palette when you select the block outside the **Block Editor**.

Assigning the Move Action to the Point Parameter

To assign the move action to the **Point** parameter, double-click on the label and select the **Move** option from the dynamic preview or from the Command line. You can also choose the **Move** button from the **Action Parameters** panel of the **Ribbon**. Alternatively, you can choose the **Move** tool from the **Actions** tab of the **Block Authoring Palettes**. The prompt sequence that will follow when you invoke it from the **Block Authoring Palettes** is given next.

Select parameter: *Select the point Parameter.*
Specify selection set for action
Select objects: *Select objects on which action will be applied.*
Select objects: `Enter`

You need to specify the distance multiplier and angle offset properties to the actions using the **Properties** palette. The **Move** action will not work until these options are set.

Distance Multiplier. This option is used during the dynamic modification of the blocks, using grips of the parameters. Using this option, you can specify the distance by which the selected entities of the dynamic block will move with respect to the one unit movement of the cursor, while dragging the cursor using the grips. To set the distance multiplier, select the action parameter and invoke the **Parameters** palette. Now, enter the required multiplier value in the **Distance multiplier** field in the **Overrides** rollout of the **Properties** palette. By default, the value of the distance multiplier factor is 1. As a result, the entities will move by the same distance as moved by the cursor. If you set this value to 0.5, the entities will move half the distance moved by the cursor, as shown in Figure 16-24.

Figure 16-24 *Results of different values of the distance multiplier*

Angle Offset. This option is used to specify the angular offset for the selected entities of the dynamic block. The angle offset works when the entities are dragged using the parameter grip. To set the angle offset, select the action parameter and invoke the **Parameters** palette. Now, enter the required value in the **Angle offset** field in the **Overrides** rollout of the **Properties** palette. By setting this value, you can ensure that the entities move at an angle from the trace line. For example, if you set the value of the angle offset to 30, the entities will be moved at an angle of 30-degree with respect to the trace line, which is displayed while dragging the entities using the parameter grips. Figure 16-25 shows the manipulation of entities in the drawing with the default angle offset of 0-degree and an angle offset of 30-degree.

Figure 16-25 *Result of the angle offset during the dynamic manipulation*

Assigning the Stretch Action to the Point Parameter

You can invoke the **Stretch** action by double-clicking on the parameter or by choosing it from the **Block Authoring Palettes**. On invoking this action and selecting the parameter, you are prompted to specify the stretch frame. The entities lying inside the stretch frame will be moved during the dynamic manipulation. The entities that are crossed by the stretch frame will be stretched. The following prompt sequence will be displayed on invoking this action:

Select parameter: *Select the parameter to associate the action.*
Specify first corner of stretch frame or [CPolygon]: *Specify the first corner of stretch frame or enter CP to define a crossing polygon for specifying the stretch frame.*
Specify opposite corner: *Specify opposite corner of stretch frame.*
Specify objects to stretch
Select objects: *Select objects for stretching.*
Select objects: Enter
Specify action location or [Multiplier/Offset]: *Specify a point for action location or specify the distance multiplier or angular offset.*

Figure 16-26 shows the stretch frame and the entities selected for the stretch action and Figure 16-27 shows the preview of the dynamic manipulation.

Figure 16-26 *Selecting entities for the **Stretch** action*

Figure 16-27 *Dynamically stretching the entities of a block*

Linear Parameter

The linear parameter is displayed as the distance between two keypoints and is similar to a linear or an aligned dimension. This parameter can be associated with Array, Move, Stretch, or Scale actions. Figure 16-28 shows a block with a linear parameter added to it. To add this parameter, choose the **Linear** tool from the **Block Authoring Palettes**; the following prompt sequence will be displayed:

Specify start point or [Name/Label/Chain/Description/Base/Palette/Value set]: *Specify start point for the parameter.*

Figure 16-28 *The **Linear** parameter added to a block*

> Specify endpoint: *Specify endpoint for the Parameter.*
> Specify label location: *Specify a point where the label will be placed*

You can specify the base of the parameter and the distance within which the entities are to be moved at the Command prompt when you are prompted to specify the start point. Some of the options that will be available when you invoke the **Linear Parameter** tool have already been discussed in the **Point** parameter. The remaining options are discussed next.

Base. This option lets you to specify the base of the parameter. This base point will be stationary while editing the endpoint of the block. You can specify the option of keeping the start point or the midpoint stationary while using the **Base** option.

Value set. This option allows you to specify a set of values for parameters such that the manipulation will be in accordance with these specified values. The values for the parameter can be specified by providing a list or by specifying the increment and range. If there is a specified value set for a parameter, small vertical grey lines are displayed at those values in the **Block Editor**. These vertical grey lines are also displayed in the drawing when you select the parameter grip to manipulate the block.

After adding the **Linear** parameter, you can perform the array, move, scale, and stretch actions. Assign the **Move** or **Stretch** action as discussed earlier. To test the action, choose the **Test Block** button from the **Open/Save** panel and then select the block; the linear parameter grip will be displayed. Left-click on the grip to select it and specify the new location; the selected entities will act according to the action performed.

Assigning the Scale Action to the Linear Parameter
Invoke the **Scale** action tool from the **Actions** tab of **Block Authoring Palettes**, or double-click on the linear parameter and select the **Scale** option. The prompt sequence that follows when you assign the **Scale** action is as follows:

> Select parameter: *Select the linear parameter to associate the action.*
> Specify selection set for action
> Select objects: *Select the objects for scaling.*
> Select objects: Enter

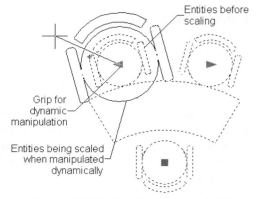

Save the changes and close the **Block Editor**. Then select the block; the **Linear** parameter grips will be displayed. Click on a grip to select it; the **Specify point location or [Base Point/Undo/Exit]** prompt will be displayed. Specify the new base point, if needed. Moving the cursor in the second or third quadrant scales the entities by a factor more than 1, as shown in Figure 16-29. If you move the cursor in the first or fourth quadrant, the entities will be scaled by a factor less than 1.

Figure 16-29 Entities being scaled by dynamic manipulation

You can scale the blocks by using a property called **Base type**. To specify the **Base type** property, invoke the **Block Editor** and display the **Properties** palette for the **Scale** action. Next, specify the base type in the **Overrides** rollout. There are two types of base type, dependent and independent. The base type can be dependent on the original base point of the parameter to which the scale action is assigned. You can also make the base type independent so that the entities can be scaled based on the specified base point and the second point.

Assigning the Array Action to the Linear Parameter

By adding array action to the entities in the block, you can create the array dynamically by using grips. Invoke the **Array** action tool from **Block Authoring Palettes** or double-click on the linear parameter and select the **Array** option. You will be prompted to select the objects to create the array. Next, you will be prompted to enter the distance between the columns. The number of instances in the array will depend on the distance by which you move the cursor after selecting the parameter grip. Figure 16-30 shows the preview of an array being created dynamically using the **Array** action.

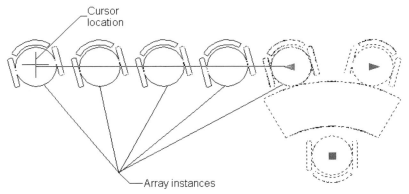

Figure 16-30 *Creating an array dynamically using the* ***Array*** *action*

Polar Parameter

The **Polar** parameter is defined by the distance between the two keypoints and an angle. The **Move**, **Scale**, **Stretch**, **Polar Stretch**, or **Array** actions can be associated with the **Polar** parameter. To add a polar parameter, choose **Point > Polar** from the **Action Parameters** panel of the **Block Editor**. The following prompt sequence will be displayed:

> Specify base point or [Name/Label/Chain/Description/Palette/Value set]: *Specify start point for the parameter.*
> Specify endpoint: *Specify endpoint for the Parameter.*
> Specify label location: *Specify a point where Label will be placed.*
> Enter number of grips [0/1/2] <2>: *Specify the number of grips to be displayed*

Note that the working of the **Value set** option is different in case of the **Polar** parameter. You can invoke the **Properties** palette for the **Polar** parameter and change the value sets in the **Value Set** parameters, or specify them when you are prompted to specify the base point. The prompt sequence for this option is given next.

Specify base point or [Name/Label/Chain/Description/Palette/Value set]: **V** Enter

Enter distance value set type [None/List/Increment] <None>: **L** Enter

Enter list of distance values (separated by commas): *Specify list of values for the **Distance*** Enter

Enter angle value set type [None/List/Increment] <None>: **L** Enter

Enter list of angle values (separated by commas): *Specify list of values for the **Angle** parameter.*

Specify base point or [Name/Label/Chain/Description/Palette/Value set]:

Similarly, you can use the **Increment** option to specify the increment and range for the distance and angle parameters.

Assigning the Polar Stretch Action to the Polar Parameter

By specifying the polar stretch action, you can stretch the entities and at the same time rotate them. As you move the cursor, the entities specified for rotation will rotate by the same angle that has been moved by the cursor. Remember that the polar stretch action can only be applied to the polar parameter. To add this action, choose **Polar Stretch** action from the **Block Authoring Palettes** or double-click on the parameter and select the polar stretch option. The following is the prompt sequence:

Select parameter: *Select the parameter to associate the action.*

Specify parameter point to associate with action or enter [sTart point/Second point] <Second>: *Specify the point that will be associated with the action.*

Specify first corner of stretch frame or [CPolygon]: *Specify the first corner of stretch frame.*

Specify opposite corner: *Specify opposite corner of stretch frame.*

Specify objects to stretch

Select objects: *Select the object to stretch, refer to Figure 16-31.*

Select objects: Enter

Specify objects to rotate only

Select objects: *Select the object that needs to be rotated only, refer to Figure 16-31.*

Select objects: Enter

Figure 16-31 *Selections for adding the polar stretch action*

Set the **Distance multiplier** and **Angle offset** in the **Properties** palette, as discussed earlier. Next, save the changes and close the **Block Editor**. Select the block; the polar parameter grips will be displayed. Click on the grip to select it; you will be prompted to specify the point location. Specify the new point; some of the entities will stretch, whereas the others will rotate, as shown in Figure 16-32.

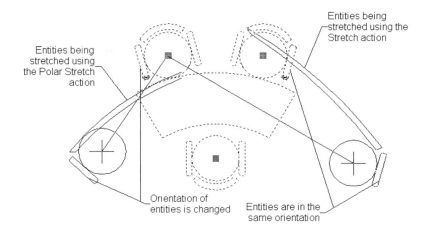

Figure 16-32 *Difference between the stretch action and the polar stretch action*

Tip
*While you assign different actions and parameters to the block entities in the **Block Editor**, you no need to exit the block editor and check the actions. Choose the **Test Block** button in the **Open/Save** panel in the **Block Editor**; a new test window will be displayed. Here you can check the changes applied to the block. To exit this new window, choose **Close > Close Test Block Window** from the **Close** panel.*

XY Parameter

The XY parameter is used to define the X and Y distances between a specified base point and an endpoint in the dynamic block. You can assign the array, move, scale, and stretch actions to the XY parameter. To add this parameter, choose the **XY** parameters tool from **Block Authoring Palettes**. The prompt sequence that will be displayed is given next.

Specify base point or [Name/Label/Chain/ Description/Palette/Value set]: *Specify the base point from where the X and Y distance will be measured.*
Specify endpoint: *Specify endpoint up to which the distances will be measured.*

Figure 16-33 shows the XY parameter. Double-click on the parameter to display the dynamic preview. The dynamic preview lists the actions that can be assigned to this parameter.

Figure 16-33 *XY parameter assigned to a point and the actions that can be assigned to this parameter*

Rotation Parameter

The rotation parameter is used to define an angular variation for the selected entities of the dynamic block. You can assign the rotate action to this parameter, which ensures that the selected entities of the dynamic block can be rotated around a predefined base point. To add this parameter, choose the **Rotation** parameter from **Block Authoring Palettes**. The prompt sequence that will be displayed is as follows:

> Specify base point or [Name/Label/Chain/Description/Palette/Value set]: *Specify base point around which the selected entities of the dynamic block will be rotated.*
> Specify radius of parameter: *Specify the radius, which defines the imaginary circle along the circumference of which the entities will be rotated.*
> Specify default rotation angle or [Base angle] <0>: *Specify default rotation angle at which the grip point will be placed for rotating the entities dynamically.*
> Specify label location: *Specify a point where the label will be placed.*

When you specify the radius of the parameter, you will be prompted to specify the default rotation angle or base angle. The base angle defines the datum for measuring the rotation angle. By default, this value is 0-degree, which means that the angles will be measured from the positive X-axis. You can also modify the base angle to any other value, such that the angles are measured from the new base angle. For example, if you modify the base angle value to 30-degree, the default rotation angle will be measured from an axis defined at 30-degree from the X-axis, as shown in Figure 16-34.

Assigning the Rotate Action to the Rotation Parameter

The **Rotate** action can only be assigned to the **Rotation** parameter. This action allows you to rotate the selected components of the dynamic block using grips, as shown in Figure 16-35.

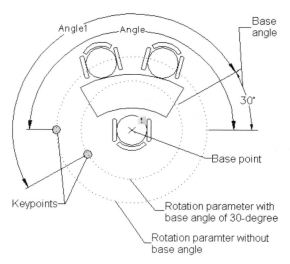

Figure 16-34 The **Rotation** parameters with and without the base angle

Figure 16-35 The **Dynamic** manipulation of entities after adding the **Rotate** action

To add the **Rotate** action, double-click on the **Rotation** parameter. The following prompt sequence will be displayed:

> Specify selection set for action
> Select objects: *Select the entities to be rotated.*
> Select objects: `Enter`

Save the changes and close the **Block Editor**. Select the block; the **Rotation** parameter grips will be displayed on the block. Click on a grip to select it; you will be prompted to specify the rotation angle. Specify the angle; the entities will rotate as specified.

Alignment Parameter

The **Alignment** parameter is used to align the entire dynamic block with the other entities in the drawing. The alignment is defined using the base point of alignment, at which the dynamic block grip is created. The block is rotated and moved while aligning. Remember that because the entire block is aligned using this parameter, you do not need to assign any action to this parameter. When you invoke this parameter, the following prompt sequence is displayed:

> Specify base point of alignment or [Name]: *Specify base point to align the dynamic block.*
> Alignment type = Perpendicular
> Specify alignment direction or alignment type [Type] <Type>: *Specify the alignment direction by defining another point in the drawing window.*

Remember that the dynamic block will be aligned at the angle that is defined while defining the alignment direction in the **Specify alignment direction or alignment type [Type] <Type>** prompt.

Type. Enter **T** at the **Specify alignment direction or alignment type [Type] <Type>** prompt to specify the type of alignment. The types of alignment are discussed next.

Perpendicular. This alignment type ensures that the dynamic block is aligned perpendicular to the entities in the drawing.

Tangent. This alignment type ensures that the dynamic block is aligned tangent to the entities in the drawing.

After adding the alignment parameter, exit the block editor and click on the block; the dynamic block grip is displayed at the base point of the alignment parameter. Select the grip and then move the dynamic block close to an existing entity in the drawing. The dynamic block aligns itself to the entity and the alignment angle depends on the alignment direction that you defined.

Flip Parameter

The **Flip** parameter is used to add a flip action to the dynamic block. The flip action reverses the orientation of the dynamic block. Thus in simple terms, the flip parameter is used to create a mirrored image of the dynamic block and remove the original block. The mirror line is defined by the reflection line, which you specify while defining the flip parameter, see Figure 16-36. Remember that depending on the location of the reflection axis, the location of the flipped entities also changes. When you add the flip action, an arrow is displayed at the base point. You can click on this arrow to flip the orientation of the selected entities of the dynamic block. The following prompt sequence will be displayed when you invoke the **Flip** parameter tool from the **Block Authoring** palette:

Specify base point of reflection line or [Name/Label/Description/Palette]: *Specify the base point from where the reflection line will start.*
Specify endpoint of reflection line: *Specify the endpoint of the reflection line, thus defining its orientation.*
Specify label location: *Specify location for label of the parameter.*

Assigning the Flip Action to the Flip Parameter

As mentioned earlier, you can assign the **Flip** action to the **Flip** parameter so as to reverse the orientation of the selected entities of the block. To assign the **Flip** action, double-click on the **Flip** parameter; you will be prompted to select the objects to be flipped. Select the objects and press ENTER; the flip action will be applied. Figure 16-37 shows the orientation and location of the selected entities before and after flipping.

Figure 16-36 *Defining the **Flip** parameter*

Figure 16-37 *Dynamically flipping the entities of the dynamic block*

Visibility Parameter

The visibility parameter is used to create visibility states of a dynamic block. Using the visibility states, you can control the display of entities in the dynamic block and specify whether they would be visible or hidden. When you invoke this tool, you will be prompted to specify the location of the parameter. As soon as the visibility parameter is added, the following options will be available in the **Visibility** panel of the **Block Editor**.

Visibility Mode. This button toggles the value of the **Visibility Mode** in between 0 and 1. The default value is set to 0 and so the entities that have been made invisible will not be shown in the authoring area. If the value of this mode is set to 1, the entities that you make invisible will also be displayed in the authoring area, but in gray color.

Make Invisible. This button is chosen to make some of the entities invisible in the current visibility state. When you choose this button, you are prompted to select the entities to be hidden.

Make Visible. This button is chosen to make the entities visible that were made invisible in the current visibility state using the **Make Invisible** button. When you choose this button, all the entities that were hidden are displayed in the authoring area, even if the **Visibility Mode** is set to 0. Select the entities that you want to make visible; the selected entities will be made visible after you exit the **Make Visible** tool.

Visibility States. This button is used to create an additional visibility state. On choosing this button, the **Visibility States** dialog box will be displayed, as shown in Figure 16-38. By default, VisibilityState0 is available and listed in the **Visibility states** area. To create a new visibility state, choose the **New** button; the **New Visibility State** dialog box will be displayed, as shown in Figure 16-39. The options of this dialog box are discussed next.

*Figure 16-38 The **Visibility States** dialog box*

*Figure 16-39 The **New Visibility State** dialog box*

Visibility state name. Specify the name of the new visibility state in this edit box.

Visibility options for new states. The options in this area are used to control the visibility of the new visibility state. By default, the **Leave visibility of existing objects unchanged in new state** radio button is selected, implying that the visibility of entities in the new state will be similar to the visibility of the objects in the current state. If you select the **Hide all existing objects in new state** radio button, all entities will be hidden in the new visibility state. If you select the **Show all existing objects in new state** radio button, all objects will be visible in the new visibility state.

The current visibility state is displayed in the drop-down list in the **Visibility** panel of the **Block Editor** tab. To make any other visibility state as the current visibility state, select it in the drop-down list. To rename an existing visibility state, select its name from the **Visibility states** list box available in the **Visibility states** dialog box and choose the **Rename** button.

After defining the visibility states, exit the **Block Editor**. Now, you can hide or show the entities as per the requirements of the drawing. When you select the dynamic block for which the visibility state has been defined, the lookup grip will be displayed, which resembles an arrow head. If you click on the lookup grip, a shortcut menu will be displayed, listing all available visibility states. In the list, the current visibility state is displayed with a check mark on its left. In Figure 16-40, **Visibility State0** is the current visibility state. Figure 16-41 shows the drawing in which the **Visibility State1** has been made as the current visibility state and therefore, the table is hidden in it.

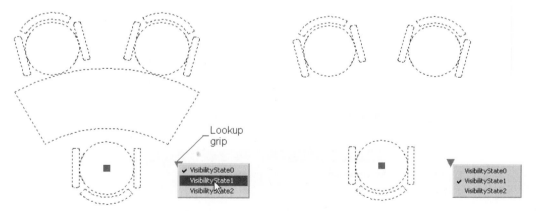

Figure 16-40 *Shortcut menu that is displayed when you click on the lookup grip*

Figure 16-41 *Changing the visibility state using the shortcut menu*

Lookup Parameter

The **Lookup** parameter is added to assign a lookup action to the dynamic block. Using the lookup action, you can control the values of the parameters assigned to the dynamic block. To add this parameter, choose its button from the **Block Authoring Palettes** and specify the location of the parameter.

Assigning the Lookup Action to the Lookup Parameter

To assign the lookup action to the lookup parameter, double-click on it; you will be prompted to specify the action location. On specifying it, the **Property Lookup Table** dialog box will be displayed, as shown in Figure 16-42.

Figure 16-42 The **Property Lookup Table** *dialog box*

To add a lookup property, choose the **Add Properties** button; the **Add Parameter Properties** dialog box will be displayed, as shown in Figure 16-43. This dialog box displays the parameters assigned to the current dynamic block. Select single or multiple properties and choose **OK**; all properties will be added as rows in the **Input Properties** area of the **Properties Lookup Table** dialog box. Save the changes and close the **Block Editor**. Next, select the block; the lookup grips will be displayed. Select a lookup grip; a shortcut menu listing all property values will be displayed. Select the required values and perform the action.

Base Point Parameter

The **Base Point** parameter is used to create a base point for the block, which will be used as the insertion base point while inserting this block. Note that the blocks that are inserted are also modified, depending on the new base point that you specify using this parameter. To add this parameter, choose the **Base Point** button from the **Block Authoring Palettes** and select the point that you want to use as the base point.

*Figure 16-43 The **Add Parameter Properties** dialog box*

ADDING PARAMETER AND ACTION SIMULTANEOUSLY USING PARAMETER SETS

In the previous section, you learned how to add a parameter and then assign an action to it. AutoCAD also allows you to add parameters and actions simultaneously using the **Parameters Sets** tab of the **Block Authoring Palettes**, see Figure 16-44. Various parameter-action combinations that can be defined for a dynamic block are available in this tab as individual sets. However, note that the entities that will be affected by the specified action are not automatically defined using the parameter sets. You need to double-click on the action to select the objects. For example, instead of first applying the **Point** parameter and then adding the move action to it, you can directly apply the **Point Move** set. After you specify the parameter and the action, double-click on **Move** in the authoring area and select the entities of the dynamic block that will be moved using this action.

*Figure 16-44 The **Parameter Sets** tab of the **Block Authoring Palettes***

Tip

*To delete any action applied to a parameter, right-click on the action bar displayed on the parameter. Next, choose the **Delete** option from the shortcut menu displayed. You can also create a new selection set or modify the existing selection set by choosing the **Action Selection Set** option. To rename an action, choose the **Rename Action** option from the shortcut menu.*

INSERTING BLOCKS USING THE DESIGNCENTER

You can use the **DesignCenter** to locate, preview, copy, or insert blocks or existing drawings into the current drawing. Choose **View > Palettes > DesignCenter** from the **Ribbon** to display the **DesignCenter** window. By default, the tree pane is displayed. Choose the **Tree View Toggle** button to display the tree pane on the left side, if it is not already displayed. Expand **My Computer** to display the *C:/Program Files/AutoCAD 2010/Sample/DesignCenter* folder, by clicking on the plus (+) signs on the left of the respective folders. Click on the plus sign adjacent to the folder to display its contents. Select a drawing file you wish to use to insert blocks from and then click on the plus sign adjacent to the drawing again. All the icons depicting the components such as blocks, dimension styles, layers, linetypes, text styles, and so on, in the selected drawing are displayed. Select **Blocks** by clicking on it in the **tree pane**; the blocks in the drawing are displayed in the palette. Select the block you wish to insert and drag and drop it into the current drawing. Later, you can move it in the drawing to the desired location. You can also right-click on the block name to display the shortcut menu, as shown in Figure 16-45.

Choose **Insert Block** from the shortcut menu; the **Insert** dialog box is displayed. Here, you can specify the insertion point, scale, and rotation angle in the respective edit boxes. On selecting the **Specify On-screen** check box in the **Insert** dialog box, you will be allowed to

Figure 16-45 *Using the **DesignCenter** to insert blocks*

specify these parameters on the screen. The selected block is inserted into the current drawing. You can also choose **Copy** from the shortcut menu; the selected block will be copied to the clipboard. Now, you can right-click in the drawing area to display a shortcut menu and choose **Paste**; the preview of the selected block will be attached to the cursor and as you move the cursor, the block also moves with it. You can now select a point to paste the block. The advantage of this option over drag and drop is that you can select an insertion point at the time of pasting. In the case of a drag and drop operation, you need to move the inserted block to a specific point after it has been dropped on the screen.

Note
*For a detailed explanation of the **DesignCenter**, see Chapter 6, Editing Sketched Objects-II.*

USING TOOL PALETTES TO INSERT BLOCKS

You can use the **Tool Palettes** window, as shown in Figure 16-46, to insert predefined blocks in the current drawing. The **Tool Palettes** window has many tabs. You can view the complete list of tabs by clicking on stacks of the tabs available at the end of the list of tabs. In this chapter, you will learn how to insert blocks using the tabs of the **Tool Palettes**.

Inserting Blocks in the Drawing

AutoCAD provides two methods to insert blocks from the **Tool Palettes**; **Drag and Drop** method and **Select and Place** method. Both these methods of inserting blocks using the **Tool Palettes** are discussed next.

Drag and Drop Method

To insert blocks from the **Tool Palettes** in the drawing, using this method, move the cursor over the desired predefined block in the **Tool Palettes**. You will notice that as you move the cursor over the block, the block icon gets converted into a 3D icon. Also, a tooltip is displayed that shows the name and description of the block. Press and hold the left mouse button and drag the cursor to the drawing area. Release the left mouse button, and you will notice that the selected block is inserted in the drawing. You may need to modify the drawing display area to view the block.

*Figure 16-46 The **Tool Palettes** window*

Remember that when a block is inserted from the **Tool Palettes** using the drag and drop method, you are not prompted to specify its rotation angle or scale. The blocks are automatically inserted with their default scale factor and rotation angle.

Select and Place Method

You can also insert the desired block in the drawings using the select and place method. To insert the block using this method, move the cursor over the desired block in the **Tool Palettes**;

the block icon is changed to a 3D icon. Press the left mouse button; the selected block is attached to the cursor and the **Specify insertion point or [Basepoint/Scale/X/Y/Z/Rotate]** prompt is displayed. Modify any parameter using the prompt sequence and then move the cursor to the required location in the drawing area. Click on the left mouse button; the selected block is inserted at the specified location.

Modifying Properties of the Blocks in the Tool Palettes

To modify the properties of a block, move the cursor over it in the **Tool Palettes** and right-click to display the shortcut menu. Using the options in this shortcut menu, you can cut or copy the desired block available in one tab of **Tool Palettes** and paste it on the other tab. You can also delete and rename the selected block using the **Delete** and **Rename** options, respectively. To modify the properties of the block, choose **Properties** from the shortcut menu. The **Tool Properties** dialog box is displayed, as shown in Figure 16-47.

In the **Tool Properties** dialog box, the name of the selected block is displayed in the **Name** edit box. You can rename the block by entering a new name in the **Name** edit box. The **Image** area on the left of the **Name** edit box displays the image of the selected block. If you enter a description of the block in the **Description** text box, it is stored with the block definition in the **Tool Palettes**. Now, when you move the cursor over the block in

*Figure 16-47 The **Tool Properties** dialog box*

Tool Palettes and pause for a second, the description along with its name appears in the tooltip. The **Tool Properties** dialog box displays the properties of the selected block under the following categories.

Insert

In this category, you can specify the insertion properties of the selected block such as its name, original location of the block file, scale, and rotation angle. The **Name** edit box specifies the name of the block. The **Source file** edit box displays the location of the file, in which the selected block is created. When you choose the [**...**] button in the **Source file** edit box, AutoCAD displays the location of the file in the **Select Linked Drawing** dialog box. The **Scale** edit box is used to specify the scale factor of the block. The block will be inserted in the drawing according to the scale factor specified in this edit box. You can enter the angle of rotation in the **Rotation** edit box. The **Explode** edit box is used to specify whether the block will be exploded while inserting or will be inserted as a single entity.

General

In this category, you can specify the general properties of the block such as the **Color**, **Layer**, **Linetype**, **Plot style**, and **Lineweight** for the selected block. The properties of a particular field can be modified from the drop-down list available on selecting that field.

Custom

In this category, you can specify the custom properties of the block, such as the dimensions and related parameters.

ADDING BLOCKS IN TOOL PALETTES

By default, the **Tool Palettes** window displays the predefined blocks in AutoCAD. You can also add the desired block and the drawing file to the **Tool Palettes** window. This is done using the **DesignCenter**. AutoCAD provides two methods for adding blocks from the **DesignCenter** to the **Tool Palettes**; **Drag and Drop** method and **Shortcut menu**. These two methods are discussed next.

Drag and Drop Method

To add blocks from the **DesignCenter** in the **Tool Palettes**, move the cursor over the desired block in the **DesignCenter**. Press and hold the left mouse button on the block and drag the cursor to the **Tool Palettes** window. You will notice that a box with a + sign is attached to the cursor and a black line appears on the **Tool Palettes** window, as shown in Figure 16-48. If you move the cursor up and down in the **Tool Palettes**, the black line also moves between the two consecutive blocks. This line is used to define the position of the block to be inserted in the **Tool Palettes** window. Release the left mouse button and you will notice that the selected block is added to the location specified by the black line in the **Tool Palettes**.

*Figure 16-48 Specifying the location of the block to be inserted in the **Tool Palettes** window*

Shortcut Menu

You can also add the desired block from the **DesignCenter** to the **Tool Palettes** using the shortcut menu. To add the block, move the cursor over the desired block in the **DesignCenter** and right-click on it to display a shortcut menu. Choose **Create Tool Palette** from it. You will notice that a new tab with the name **New Palette** is added to the **Tool Palettes**. And, the block is added in the new tab of the **Tool Palettes**. Also, a text box appears that displays the current name of the tab. You can change its name by entering a new one in this text box.

You can also add a number of blocks in a drawing to the **Tool Palettes** using the following two methods:

Select the **Blocks** folder of any drawing in the **DesignCenter** and right-click in the **Tree View** area of the **DesignCenter**; a shortcut menu will be displayed. Choose **Create Tool Palette** from the shortcut menu; a new tab will be added to the **Tool Palettes** with the same name as that of the drawing file selected in the **DesignCenter**. This new tab contains all blocks which were in the folder that you selected from the **DesignCenter**.

You can also add all the blocks in a drawing by right-clicking on the drawing in the **tree view** of the **DesignCenter**; a shortcut menu is displayed. Choose **Create Tool Palette** from it; you will notice that a new tab is added to the **Tool Palettes**, which contains all the blocks that were available in the selected drawing. You will also notice that the new tab has the same name as that of the selected drawing.

MODIFYING EXISTING BLOCKS IN THE TOOL PALETTES

If you modify an existing block that was added to the **Tool Palettes** and then insert it using the **Tool Palettes** in the same or a new drawing, you will notice that the modified block is inserted and not the original block. However, if you insert the modified block from the **Tool Palettes** in the drawing in which the original block was already inserted, AutoCAD inserts the original block and not the modified one. This is because the file already has a block of the same name in its memory.

To insert the modified block, you first need to delete the original block from the current drawing, using the **ERASE** command. Next, you need to delete the block from the memory of the current drawing. The unused block can be deleted from the memory of the current drawing, using the **PURGE** command. To invoke this command, enter **PURGE** at the command window. The **Purge** dialog box is displayed. Choose the (+) sign located on the left of **Blocks** in the tree view available in the **Items not used in drawing** area. You will notice that a list of blocks in the drawing is shown. Select the original block to be deleted from the memory of the current drawing and then choose the **Purge** button. The **Confirm Purge** dialog box is displayed, which confirms the purging of the selected item. Choose **Yes** in it and then choose the **Close** button to exit the **Purge** dialog box. Next, when you insert the block using the **Tool Palettes**, the modified block is inserted in the drawing.

 Note
*You will learn more about the **PURGE** command in Chapter20, Grouping and Advanced Editing of Sketched Objects. Deleting unused blocks using the command line is, however, discussed later in this chapter.*

*If you create a block with the name that is defined in the **Tool Palettes**, the block in the **Tool Palettes** is redefined in the current drawing. However, when you open a new drawing and insert the block using the **Tool Palettes**, the original block will be inserted.*

LAYERS, COLORS, LINETYPES, AND LINEWEIGHTS FOR BLOCKS

A block possesses the properties of the layer on which it is drawn. The block may be composed of objects drawn on several different layers, with different colors, linetypes, and lineweights. All this information is preserved in the block. At the time of insertion, each object in the block is drawn on its original layer with the original linetype, lineweight, and color, irrespective of the current drawing layer, object color, object linetype, and object lineweights. You may want all instances of a block to have identical layers, linetype properties, lineweight, and color. This can be achieved by allocating all the properties explicitly to the objects forming the block. On the other hand, if you want the linetype and color of each instance of a block to be set according to the linetype and color of the layer on which it is inserted, draw all the objects forming the block on layer 0 and set the color, lineweight, and linetype to **By Layer**. Objects with a **By Layer** color, linetype, and lineweight can have their colors, linetypes, and lineweights changed after insertion by changing the layer settings. If you want the linetype, lineweight, and color of each instance of a block to be set according to the current explicit linetype, lineweight, and color at the time of insertion, set the color, lineweight, and linetype of its objects to **By Block**. You can use the **PROPERTIES**, **CHPROP**, or **CHANGE** command to change some of the characteristics associated with a block (such as layer).

 Note
The block is inserted on the current layer, but the objects comprising the block are drawn on the layers on which they were drawn when the block was being defined.

For example, assume block B1 includes a square and a triangle that were originally drawn on layer X and layer Y, respectively. Let the color assigned to the layer X be red and to layer Y be green. Also, let the linetype assigned to layer X be continuous and for layer Y be hidden. Now, if we insert B1 on layer L1 with color yellow and linetype dot, block B1 will be on layer L1, but the square will be drawn on layer X with color red and linetype continuous. The triangle will be drawn on layer Y with the color green and the linetype hidden.

The **By Layer** option instructs AutoCAD to assign objects within the block the color and linetype of the layers on which they were created. There are three exceptions:

1. If objects are drawn on a special layer (layer 0), they are inserted on the current layer. These objects assume the characteristics of the current layer (the layer on which the block is inserted) at the time of insertion, and can be modified after insertion by changing that layer's settings.

2. Objects created with the special color **By Block** are generated with the color that is current at the time of insertion of the block. This color may be explicit or **By Layer**. You are thus allowed to construct blocks that assume the current object color.

3. Objects created with the special linetype **By Block** are generated with the linetype that is prevalent at the time the block is inserted. Blocks are thus constructed with the current object linetype, which may be **By Layer** or explicit.

Note

If a block is created on a layer that is frozen at the time of insertion, it is not shown on the screen.

Tip

*If you provide drawing files to others for their use, using only **BYLAYER** settings provide the greatest compatibility with varying office standards for layer/color/linetype/lineweight. This is because they can be changed more easily after insertion.*

NESTING OF BLOCKS

The concept of having one block within another block is known as the **nesting of blocks**. For example, you can insert several blocks by selecting them, and then, with the **BLOCK** command, create another block. Similarly, if you use the **INSERT** command to insert a drawing, containing several blocks, into the current drawing, it creates a block containing nested blocks in the current drawing. There is no limit to the degree of nesting. The only limitation in nesting of blocks is that blocks that reference themselves cannot be inserted. The nested blocks must have different block names. Nesting of blocks affects layers, colors, and linetypes. The general rule is given next.

If an inner block has objects on layer 0, or objects with linetype or color **By Block**, these objects may be said to behave like fluids. They "float up" through the nested block structure until they find an outer block with fixed color, layer, or linetype. These objects then assume the characteristics of the fixed layer. If a fixed layer is not found in the outer blocks, then the objects with color or linetype **By Block** are formed; that is, they assume the color white and the linetype CONTINUOUS.

To clarify the concept of nested blocks, consider the following example.

Example 2   *General*

1. Draw a rectangle on layer 0, and form its block, named X.
2. Change the current layer to OBJ, set its color to red, and linetype to hidden.
3. Draw a circle on OBJ layer.
4. Insert the block X in the OBJ layer.
5. Combine the circle with the block X (rectangle) to form a block Y.
6. Now, insert block Y in any layer (say, layer CEN) with the color green and linetype continuous.

You will notice that block Y is generated in red and its linetype is hidden. However, the block X, which is nested in block Y and created on layer 0, will not be generated in the color (green) and linetype (continuous) of the layer CEN. This is because, the object (rectangle) on layer 0 floated up through the nested block structure and assumed the color and linetype of the first outer block (Y) with a fixed color (red), layer (OBJ), and linetype (hidden). If both the blocks (X and Y) were on layer 0, the objects in the block Y would assume the color and linetype of the layer on which the block was inserted.

Example 3   *General*

1. Change the color of layer 0 to red.
2. Draw a circle with color **By Block** and then form its block, B1. It appears white because its color is set to **By Block** (Figure 16-49).
3. Set the color to **By Layer** and draw a square. The color of the square is red.
4. Insert block B1. Notice that the block B1 (circle) assumes red color.
5. Create another block B2 consisting of the Block B1 (circle) and square.
6. Create a layer L1 with green color and hidden linetype. Make it current. Insert block B2 in layer L1.
7. Explode block B2. Notice the change.
8. Explode block B1, circle. You will notice that the circle changes to white because it was drawn with the color set to **By Block** .

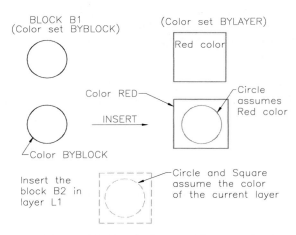

Figure 16-49 *Blocks versus layers and colors*

Example 4   *General*

Part A

1. Draw a unit square on layer 0 and make it a block named B1.
2. Draw a circle of radius 0.5 and change it into a block named B2.
3. Insert block B1 into the drawing with an X scale factor of 3 and a Y scale factor of 4.
4. Now, insert the block B2 in the drawing and position it at the top of B1.
5. Make a block of the entire drawing and name it Plate.
6. Insert the block **Plate** in the current layer.
7. Create a new layer with different colors and linetypes and insert blocks B1, B2, and Plate.

Keep in mind the layers on which the individual blocks and the inserted block were made.

Part B
Try nesting the blocks drawn on different layers and with different linetypes.

Part C
Change the layers and colors of the different blocks you have drawn so far.

INSERTING MULTIPLE BLOCKS

Command:	MINSERT

The **MINSERT** (multiple insert) command is used for the multiple insertion of a block. This command comprises features of the **INSERT** and **ARRAY** commands. **MINSERT** is similar to the **-INSERT** command because it is used to insert blocks at the command level. The difference between these two commands is that the **MINSERT** command inserts more than one copy of the block in a rectangular fashion, similar to the command **ARRAY.** Also, blocks inserted by **MINSERT** cannot be exploded. Preceding the block name with an asterisk does not explode it on insertion. With the **MINSERT** command, only one block reference is created. But, in addition to the standard features of a block definition (insert point, X/Y scaling, rotation angle, and so on), this block has repeated row and column counts. In this manner, using this command saves time and disk space. The prompt sequence is very similar to that of the **-INSERT** and **-ARRAY** commands. Consider an example in which you have to array a block named BENCH, with the following specifications: number of rows = 4, number of columns = 3, unit cell distance between rows = 0.50, and unit cell distance between columns = 2.0. The prompt sequence for creating the arrangement of benches in an auditorium (Figure 16-50) is given next.

Command: **MINSERT**
Enter block name or [?] <current>: **BENCH**
Units: Inches   Conversion: 1.0000
Specify insertion point or [Basepoint/Scale/Rotate]: *Press ENTER or specify a base point and press ENTER*
Specify scale factor <1>: *Press ENTER or specify a value and then press ENTER*
Specify rotation angle <0>: *Press ENTER or specify a value and then press ENTER*
Enter number of rows (---)<1>: **4**
Enter number of columns (|||) <1>: **3**
Enter distance between rows or specify unit cell (---): **0.50**
Specify distance between columns (|||): **2.0**

Note
*You cannot use an annotative block for the **MINSERT** command.*

All options of the **MINSERT** command are similar to the ones discussed earlier in the **-INSERT** command. The **?** option of the **Enter block name or [?]:** prompt lists the names of the blocks. When you specify the insertion point, AutoCAD prompts you to specify the scale

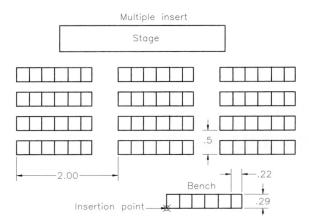

Figure 16-50 *Arrangement of benches in an auditorium*

factor. Once you specify the scale factor, you will be prompted to specify the rotation angle. If you need to insert the block at an angle, specify the angle and press ENTER. Next, you need to specify the number of rows, columns, distance between the rows, and distance between the columns in succession. The array of blocks will be created.

The prompts pertaining to the number of rows and columns require a positive nonzero integer. For the number of rows exceeding 1, you are prompted to specify the distance between the rows. Similarly, if the number of columns exceeds 1, you need to specify the distance between columns. The distance between rows and columns can be a positive or negative value. While specifying the distance between the rows and columns, you can also specify a unit cell, by selecting two points on the screen that form a box whose width and height represent the distance between columns and rows respectively. While inserting a block, if you specify the block rotation, each copy of the inserted block is rotated through that angle. If you specify the rotation angle in the **MINSERT** command, the whole **MINSERT** block (array) is rotated by the specified rotation angle. If both the block and the **MINSERT** block (array) are rotated through the same angle, the block does not seem rotated in the **MINSERT** pattern. This is the same as an array with the rotation angle equal to 0-degree. If you want to rotate individual blocks within the array, you should first generate a rotated block and then **MINSERT** it.

Note
*In the array of blocks generated with the use of the **MINSERT** command, you can alter the number of rows or columns or the spacing between them. This is done using the **Misc** rollout of the **Properties** palette.*

*When you right-click while specifying the insertion point at the **Specify insertion point** prompt, a shortcut menu is displayed where all the command line options are available.*

Exercise 5

General

1. Create a triangle with each side equal to 3 units.
2. Generate a block of the triangle.
3. Use the **MINSERT** command to insert the block to create a 3 by 3 array. The distance between the rows is 1 unit, and the distance between columns is 2 units.
4. Again, use the **MINSERT** command to insert the block to create a 3 by 3 array. The distance between the rows is 2 units and between the columns is 3 units. The array is rotated through 15-degree.

CREATING DRAWING FILES USING THE WRITE BLOCK DIALOG BOX

Command:	WBLOCK

The blocks or symbols created by the **BLOCK** command can be used only in the drawing, in which they were created. This is a shortcoming, because you may need to use a particular block in different drawings. The **WBLOCK** command is used to export symbols by writing them to new drawing files that can then be inserted in any drawing. With the **WBLOCK** command, you can create a drawing file (*.dwg* extension) of the specified blocks, selected objects in the current drawing, or the entire drawing. All the used named objects (linetypes, layers, styles, and system variables) of the current drawing are inherited by the new drawing created with the **WBLOCK** command. This block can then be inserted in any drawing.

*Figure 16-51 The **Write Block** dialog box*

When you invoke the **WBLOCK** command, the **Write Block** dialog box is displayed, as shown in Figure 16-51. This dialog box converts the blocks into drawing files and also saves objects as drawing files. You can also save the entire current drawing as a new drawing file.

The **Write Block** dialog box has two main areas; **Source** and **Destination**. The **Source** area allows you to select objects and blocks, specify insertion base points and convert them into drawing files. In this area of the dialog box, different default settings are displayed, depending upon the selection you make. By default, the **Objects** radio button is selected. In the **Destination** area, the **File name and path** edit box displays *new block.dwg* as the new file name and its location. The **Block** drop-down list is also not available. Now, you can select objects in a drawing and save them as a wblock, and can enter a name and a path for the file.

Sometimes the current drawing consists of blocks. To save a block as a wblock, you can select the **Block** radio button. When the **Block** radio button is selected, the **Block** drop-down list is available. The **Block** drop-down list displays all the block names in the current drawing and you can select a block name to convert it into a wblock. The **Base point** and **Objects** areas are not available, since the insertion points and objects have already been saved with the block definition. Also, you will notice that in the **Destination** area, by default, the **File name and path** edit box displays the name of the selected block. This means that you can keep the name of the wblock the same as the selected block or you can change it. Selecting the **Entire drawing** radio button, selects the current drawing as a block and saves it as a new file. When you use this option, the **Base point** and **Objects** areas are not available.

The **Base point** area allows you to specify the base point of a wblock, which is used as an insertion point. You can either enter values in the **X**, **Y**, and **Z** edit boxes or choose the **Pick point** button to select it on the screen. The default value is 0, 0, 0. The **Objects** area allows you to select objects to save as a file. You can use the **Select Objects** button to select objects or use the **QuickSelect** button to set parameters in the **Quick Select** dialog box to select objects in the current drawing. The number of objects selected is displayed at the bottom of the **Objects** area. If the **Retain** radio button is selected in the **Objects** area, the selected objects in the current drawing are kept as such, after they have been saved as a new file. If the **Convert to block** radio button is selected, the selected objects in the current drawing will be converted into a block with the same name as the new file, after being saved as a new file. Selecting the **Delete from drawing** radio button deletes the selected objects from the current drawing after they have been saved as a file.

Note

*Both the **Base point** and **Objects** areas are available in the **Write Block** dialog box only when the **Objects** radio button is selected in the **Source** area of the dialog box.*

The **Destination** area sets the file name, location, and units of the new file, in which the selected objects are saved. In the **File name and path** edit box, you can specify the file name and the path of the block or the selected objects. You can choose the [...] button to display the **Browse for Drawing File** dialog box, where you can specify the path where the new file will be saved. From the **Insert Units** drop-down list, you can select the units the new file will use when inserted as a block. The settings for units are stored in the **INSUNITS** system variable and the default option Inches has a value of 1. On specifying the required information in the dialog box, choose **OK**. The objects or the block is saved as a new file in the path specified by you. A **WBLOCK Preview** window with the new file contents is displayed. This preview image is stored and displayed in the DesignCenter, when using it to insert drawings and blocks.

Note

Whenever a drawing is inserted into the current drawing, it acts as a single object. It cannot be edited unless exploded.

Tip

*Using the **Entire drawing** radio button in the **Write Block** dialog box to save the current drawing as a new drawing is a good way to reduce the drawing file size. This is because all unused blocks, layers, linetypes, text styles, dimension styles, multiline styles, shapes, and so* *on are removed from the drawing. The new drawing does not contain any information that is no longer needed. This **Entire drawing** option is faster than the **-PURGE** command (which also removes unused named objects from a drawing file) and can be used whenever you have completed a drawing and want to save it. The **-PURGE** command has been discussed later in this chapter.*

DEFINING THE INSERTION BASE POINT

Ribbon:	Insert > Block > Set Base Point or Home > Block > Set Base Point
Command:	BASE

The **BASE** command lets you set the insertion base point for a drawing, just as you set the base insertion point using the **BLOCK** command. This base point is defined so that when you insert the drawing into some other drawing, the specified base point is placed on the insertion point. By default, the base point is at the origin (0,0,0). When a drawing is inserted on a current layer, it does not inherit the color, linetype, or thickness properties of the current layer. On invoking the **BASE** command, the prompt sequence is given next.

Enter base point <0.0000,0.0000,0.0000>: *Specify a base point or press ENTER to accept the default value.*

Tip
*You can insert a drawing that you want to refer to for checking dimensions or certain features into the current drawing for reference. Later on, you can use the **UNDO** command to delete the inserted drawing. This way you can save stationery and time spent in printing.*

Note
The inserted drawings become a part of the current drawing and increase the file size. Also, when the current drawing is modified, the inserted drawing does not get updated. Hence, it is better to xref a drawing into the current drawing instead. External references are discussed in Chapter 18, Understanding External References.

Exercise 6 *General*

1. Create a drawing file named CHAIR using the **WBLOCK** command. Make a listing of your *.dwg* files and make sure that *CHAIR.dwg* is listed. Quit the drawing editor.
2. Begin a new drawing and insert the drawing file into it. Save the drawing.

EDITING BLOCKS

Ribbon:	Insert > Reference > Edit Reference	**Command:**	REFEDIT
Toolbar:	Refedit > Edit reference in place		

You can edit blocks by breaking them into parts and then making modifications and redefining them or by editing them in place. Both these methods have been discussed here.

Note

*The blocks inserted using the **MINSERT** command can be refedited. However, the non-uniformly scaled blocks cannot be refedited.*

Editing Blocks in Place

You can edit blocks in the current drawing by using the **REFEDIT** command, referred to as the in-place reference editing. This feature of AutoCAD allows you to make minor changes to blocks, wblocks, or drawings that have been inserted in the current drawing without breaking them up into component parts or opening the original drawing and redefining them. This command saves valuable time of going back and forth between drawings and redefining blocks.

Note

*You cannot use the **REFEDIT** command on blocks that have been inserted using the **MINSERT** command. This command is discussed in reference to xrefs in Chapter 18, Understanding External References.*

You can invoke the **REFEDIT** command by choosing the **Edit Reference** button from the **Reference** panel in the **Insert** tab of the **Ribbon**, as shown in Figure 16-52, or by entering **REFEDIT** at the Command prompt. On doing so, the **Reference Edit** dialog box will be displayed. Alternatively, select the block, right-click on any block reference and then choose the **Edit Block In-Place** option from the shortcut menu displayed. When you invoke the **REFEDIT** command, the following prompt is displayed.

*Figure 16-52 The **Edit Reference** button in the **Reference** panel of the **Insert** tab*

> Select reference: *Select the block to edit.*

Once you have selected the block reference to be edited, the **Reference Edit** dialog box is displayed. The **Reference Edit** dialog box has two tabs; **Identify Reference** and **Settings**. The description of these tabs is as follows.

Identify Reference Tab

The **Identify Reference** tab, as shown in Figure 16-53, provides information for identifying the reference edit, selecting, and editing the references.

Reference name. The **Reference name** list box displays the name of the selected reference. A plus sign (+) next to the name implies that the selected block reference contains nested references. When you click on the plus sign, the nested reference is displayed in a tree view, as shown in Figure 16-54. If a block does not contain any nested blocks, this plus sign is not displayed next to the reference name in the list box. In the figure, the block Valve-2 contains a nested block Valve. An image of the selected block is displayed in the **Preview** box. If you want to see a preview of the nested block, select its name in the list box; the corresponding preview image is displayed in the **Preview** area.

Figure 16-53 *The **Identify Reference** tab of the **Reference Edit** dialog box*

Path. It displays the location of the selected reference. Note that if the selected reference is a block, no path is displayed.

The **Automatically select all nested objects** radio button is selected by default and allows you to automatically include all the selected objects in the reference editing session. The selected objects also include the nested objects.

The **Prompt to select nested objects** radio button, when selected, allows you to individually select the nested objects in the reference editing session. If this radio button is selected and you choose the **OK** button, the **Reference Edit** dialog box is closed. You are prompted to select the nested objects in the reference that you want to edit.

Settings Tab

The **Settings** tab, as shown in Figure 16-55, provides options for editing references. The **Create unique layer, style, and block names** check box is selected by default and allows you to control the layer and symbol names of the objects that are extracted from the selected reference. If you clear this check box, the names of the layers remain the same as in the reference drawing. Similarly, the **Display attribute definitions for editing** check box, when selected, allows you to edit attributes and attribute definitions associated with the block references. The modified attribute definitions are effective for future insertions, and no changes are made in the current insertions. Attributes are explained in detail in the next chapter. The **Lock objects not in working set** check box, when selected, allows you to lock all the objects that are not in the working set.

*Figure 16-54 The **Reference name** list box displaying the nested references*

*Figure 16-55 The **Settings** tab of the **Reference Edit** dialog box*

After selecting the required block to edit, choose **OK** to exit the dialog box. If you have selected the **Prompt to select nested objects** radio button, the following prompt sequence will be displayed:

> Select nested objects: *Select objects within the block that you want to modify.*
> n entities added
> Select nested objects: Enter
> n items selected
> Use **REFCLOSE** or the **Refedit** toolbar to end reference editing session.

Once you have selected the objects to be edited and pressed ENTER, the total number of selected objects will be displayed. Also, you will notice that the **Edit Reference** panel has been added to the **Ribbon** and the objects, except the selected ones have faded out. You can control the fading intensity of the objects by using the options from the **Fade Control** area of the **Display** tab in the **Options** dialog box, see Figure 16-56. In this area, you can use the slider bar to increase or decrease the fading intensity or you can enter a value in the edit box. You can also control the fading intensity of other objects at the time of editing by using the **XFADECTL** variable. By default, the fading intensity value is set at 50.

The objects that have been selected for editing are referred to as the working set. You can use any drawing and editing command to modify this working set. Sometimes, you may need to add an external object from the current drawing to the block. To do so, choose the **Add to Working Set** button in the **Edit Reference** panel, see Figure 16-57. Next, select the objects to be added to the working set and press ENTER; these objects will then get deleted from the current drawing and added to the current working set. Similarly, to remove objects from a working set, choose the **Remove from Working set** button from the **Edit Reference** panel. You are prompted to select the objects from the working set that you want to remove. On completing the selection, press ENTER. The objects that have been removed from the working set appear faded and get added back to the current drawing. You can also use the **REFSET** command to add or remove objects from a working set. But, this command can be used only when a reference has already been selected using the **REFEDIT** command.

Figure 16-56 The **Fade Control** area of the **Display** tab in the **Options** dialog box

Figure 16-57 The **Edit Reference** Panel

After adding or removing objects from the existing set, choose the **Save Changes** button in the **Edit Reference** panel to save the changes; the **AutoCAD** information box will be displayed informing that all reference edits will be saved. This information box also prompts you to specify whether you want to continue saving or cancel the command. Choose **OK** to save the modifications or choose **Cancel** to cancel the command. If you choose **OK**, the modifications are applied to the current block as well as to all future insertions.

Note

The Automatic Save feature of AutoCAD is disabled during reference editing.

To exit reference editing without saving any changes in the block reference, choose the **Discard Changes** button from the **Edit Reference** panel; the **AutoCAD** information box will be displayed, as shown in Figure 16-58, informing that all changes that have been made in the object will be discarded. To continue, choose **OK**, and to get back to reference editing, choose **Cancel**.

Figure 16-58 AutoCAD information box displayed on choosing the **Discard Changes** or **Save Changes** button from the **Edit Reference** panel

Note

*You can also use the **-REFEDIT** command to display prompts on the Command line.*

Exploding Blocks Using the XPLODE Command

Command:	XPLODE

With the **XPLODE** command, you can explode a block or blocks into component objects and simultaneously control their properties such as layer, linetype, color, and lineweight. The scale factor of the object to be exploded should be equal. Note that if the scale factor of the objects to be exploded is not equal, you need to change the value of the **EXPLMODE** system variable to 1. Note that, if the **Allow exploding** check box in the **Behavior** area of the **Block Definition** dialog box was cleared while creating the block, you will not be able to explode the block. The command prompts for this command are as follows.

> Command: **XPLODE**
> Select objects to XPlode
> Select objects: *Using any object selection method, select objects, and then press ENTER.*

On pressing ENTER, AutoCAD reports the total number of objects selected and also the number of objects that cannot be exploded. If you select multiple objects to explode, AutoCAD further prompts you to specify whether the changes in the properties of the component objects should be made individually or globally. The prompt is given next.

> XPlode Individually/<Globally>: *Enter i, g, or press ENTER to accept the default option.*

If you enter **i** at the above prompt, AutoCAD will modify each object individually, one at a time. The next prompt is given below.

> Enter an option [All/Color/LAyer/LType/LWeight/Inherit from parent block/Explode] <Explode>: *Select an option.*

The options available at the Command line are discussed next.

All

This option sets all the properties such as color, layer, linetype, and lineweight of the selected objects, after exploding them. AutoCAD prompts you to enter new color, linetype, lineweight, and layer name for the exploded component objects.

Color

This option sets the color of the exploded objects. The prompt is given next.

> New color [Truecolor/COlorbook]<BYLAYER>: *Enter a color option or press ENTER.*

When you enter **BYLAYER**, the component objects take on the color of the exploded object's layer and when you enter **BYBLOCK**, they take on the color of the exploded object.

Layer

This option sets the layer of the exploded objects. The default option is inheriting the current layer. The command prompt is as follows.

> Enter new layer name for exploded objects <current>: *Enter an existing layer name or press ENTER.*

LType

This option sets the linetype of the components of the exploded object. The command prompt is given next.

> Enter new linetype name for exploded objects <ByLayer>: *Enter a linetype name or press ENTER to accept the default options.*

LWeight

This option sets the lineweight of the components of the exploded object. The command prompt is given next.

> Enter new lineweight: *Enter a lineweight or press ENTER to accept the default.*

Inherit from parent block

This option sets the properties of the component objects to that of the exploded parent object, provided the component objects are drawn on layer 0 and the color, lineweight, and linetype are **BYBLOCK**.

Explode

This option explodes the selected object exactly as in the **EXPLODE** command.

Selecting the **Globally** option, applies changes to all the selected objects at the same time. The options are similar to the ones discussed in the **Individually** option.

RENAMING BLOCKS

Command:	RENAME

Blocks can be renamed using the **RENAME** command. To rename a block enter **RENAME** at the Command prompt and press ENTER; the **Rename** dialog box will be displayed, see Figure 16-59. This dialog box allows you to modify the name of an existing block. In the **Rename** dialog box, the **Named Objects** list box displays the categories of object types that can be renamed, such as blocks, layers, dimension styles, linetypes, Multileader styles, material, table styles and text styles, UCSs, views, and viewports. You can rename all of these except layer 0 and continuous linetype. When you select **Blocks** from the **Named Objects** list, the **Items** list box displays all the block names in the current drawing. When you select a block name to rename from the **Items** list box, it is displayed in the **Old Name** edit box. Enter the new name to be assigned to the block in the **Rename To** edit box. Choosing the **Rename To** button applies the change in name to the old name. Choose **OK** to exit the dialog box. For example, to rename a block named Bracket to **Valve-3**, select **Bracket** from the **Items** list box; it is displayed in the **Old Name** edit box. Enter **Valve-3** in the **Rename To** edit box and choose the **Rename To** button; the **Valve-3** block appears in the **Items** list box. Now, choose **OK** to exit the dialog box.

*Figure 16-59 The **Rename** dialog box*

Note
The layer 0 and Continuous linetype cannot be renamed and therefore do not appear in the ***Items*** *list box, when* ***Layers*** *and* ***Linetypes*** *are selected in the* ***Named Objects*** *list box.*

*To rename a block enter **-RENAME or -REN** at the Command window.*

DELETING UNUSED BLOCKS

Sometimes, after completing a drawing, you may notice that the drawing contains several named objects, such as dimstyles, textstyles, layers, blocks, and so on that are not being used. Since these unused named objects unnecessarily occupy disk space, you may want to remove them. Unused blocks can be deleted with the **-PURGE** command. For example, to delete an unused block named Drawing2, the prompt sequence is given next.

Command: **-PURGE**
Enter type of unused objects to purge
[Blocks/Dimstyles/LAyers/LTypes/MAterial/MUltileaderstyles/Plotstyles/SHapes/ textSTyles/Mlinestyles/Tablestyles/Visualstyles/Regapps/Zero-length geometry/Empty Text Objects/All]: **B**
Enter name(s) to purge <*>: **Drawing2**
Verify each name to be purged? [Yes/No] <Y>: Enter
Purge block "Drawing2"? <N>: **Y**

If there are no objects to be removed, AutoCAD displays a message that there are no unreferenced objects to be purged.

Note

*The unused blocks can also be deleted using the **PURGE** command discussed in Chapter 20 (Object Grouping and Editing Commands).*

*Preceding the name of the wblock with an (*) asterisk when entering the name of the wblock while using the **-WBLOCK** command has the same effect as using the **-PURGE** command. Also, you can select the **Entire drawing** radio button in the **Write Block** dialog box when creating a block using the **WBLOCK** command to get the same effect. But, the **WBLOCK** command is faster and deletes the unused named objects automatically, while the **-PURGE** command allows you to select the type of named objects you want to delete, and it also gives you an option to verify the objects before the deletion occurs.*

APPLYING CONSTRAINTS TO BLOCKS*

While creating assembly drawings, you may need some parts that have same geometry but different sizes. AutoCAD 2010 allows you to create the drawings of these parts by using geometric and dynamic constraints, if you could control these constraints in a block table. Note that you need to apply the geometric and dynamic constraints to blocks in the **Block Editor**.

To create the parts having same geometry but different sizes, draw the sketch of the parts and then convert them into a block. Next, invoke the **Block Editor** to edit the block. In the **Block Editor**, apply the necessary geometric and dimensional constraints and make the sketch a fully-defined sketch. Now, choose the **Block Table** button from the **Dimensional** panel of the **Block Editor** tab; you will be prompted to specify the location of the parameter. Specify a suitable location; you will prompted to specify the number of grips. If you enter **0**, no grip will be displayed in the drawing area. It is recommended to have atleast one grip. After specifying the number of grips, the **Block Properties Table** will be displayed, as shown in Figure 16-60.

In the **Block Properties Table**, choose the **Adds properties which appear as columns in the table** button to add all or some of the dimensional constraints added to the block. On choosing this button, the **Add Parameter Properties** dialog box will be displayed, as shown in Figure 16-61. Next, select the parameter properties to be added to the **Block Properties Table** by pressing and holding the CTRL key and then choose **OK**; the selected properties will be listed in the **Block Properties Table**. If you need to add any new parameter to the table, choose the **Creates a new user parameter and adds it to the table** button from the **Block Properties Table**; the **New Parameter** dialog box will be displayed, as shown in Figure 16-62. Specify the parameters and choose the **OK** button; the new parameter will be added to the **Block Properties Table**. You can also check the errors in the table. To do so, choose the **Audits the block property table for errors** button from the **Block Properties Table**. After specifying all parameters, choose the **OK** button from the **Block Properties Table**. Next, save the changes and exit the **Block Editor**. Now, if you select the block, the lookup grip will be displayed. Click on the grip; different values will be displayed, as shown in Figure 16-63. Next, choose the required value; the block will change accordingly. Note that if you have selected the **Block properties must match a row in the table** check box in the **Block Properties Table**, then the lookup grip will not be displayed and you cannot change the block sizes dynamically.

Figure 16-60 *The **Block Properties Table***

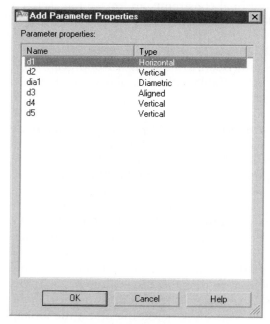

Figure 16-61 *The **Add Parameter
Properties** dialog box*

Figure 16-62 *The **New Parameter** dialog box*

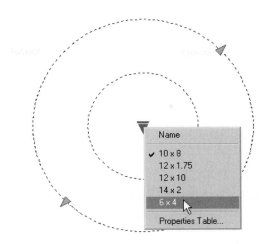

Figure 16-63 *Shortcut menu displayed on selecting the lookup grip*

Example 5

In this example, you will create the sketch of a bolt, as shown in Figure 16-64. Then, you will convert it to block and apply constraints such that you can create bolts of diameter 6, 10, 12.5, 15, 18, 24, and 25.

Figure 16-64 *Sketch for Example 5*

Note
The bolt shown in Figure 16-64 does not have any standard dimension or annotation. It is given for the purpose of example. However, you can follow the procedure given next to design a standard bolt.

1. Start a new file with the *acad.dwt* template in the **2D Drafting & Annotation** workspace. Draw the sketch assuming that the value of d=10 and satisfying other relations shown in Figure 16-64. Make sure that the endpoints are snapped so that it becomes a closed sketch, as shown in Figure 16-65.

2. Choose the **Create** button from the **Block** panel in the **Home** tab of the **Ribbon** and convert the sketch into a block. Name the block as **Bolt**.

3. Choose the **Edit** button from the **Block** panel in the **Home** tab; the **Edit Block Definition** dialog box is displayed. Select **Bolt** and invoke the **Block Editor** to edit the block.

4. Draw a vertical line passing through the midpoint of the bottom horizontal line.

5. Choose the **Construction Geometry** button from the **Manage** panel of the **Block Editor** tab, select the vertical line drawn in the previous step, and press ENTER; you will be prompted to enter an option.

6. Enter **C** and press ENTER; the vertical line is converted to a construction line, as shown in Figure 16-66. Note that you can also create a centerline in the drawing mode but for the purpose of example, this method is given.

Figure 16-65 The closed sketch of the bolt   *Figure 16-66 Block after converting the vertical line to a construction line*

7. Choose the **AutoConstrain** button from the **Geometric** panel and select all objects to apply all possible constraints. Next, press ENTER.

8. Choose the **Equal** button from the **Geometric** panel and apply the equal constraints between fillets, chamfers, vertical lines tangential to fillets, horizontal lines tangential to fillets, and vertical lines at the top.

9. Choose the **Symmetric** button from the **Geometric** panel and apply the symmetric constraints between all entities on either side of the construction line.

10. Hide all geometric constraints by choosing the **Hide All** button from the **Geometric** panel of the **Block Editor**.

11. Apply the dimensional constraints by using the tools in the **Dimensional** panel of the **Block Editor**. Figure 16-67 shows the sketch with all dimensional constraints applied. Also, make sure that you snap the end points of the vertical line while applying constraints to the base diameter and to the length of the bolt. While applying the dimensional constraints in **Block Editor**, you are prompted to specify the number of grips as soon as you specify the value. For this tutorial, enter **0** as the number of grip.

12. Choose the **Parameters Manager** button from the **Manage** panel; the **Parameters Manager** palette is displayed.

Next, you need to rename the constraints in the **Name** column of the **Properties Manager** palette, as shown in Figure 16-68.

Figure 16-67 *Sketch with all dimensional constraints*

Figure 16-68 *New name of the constraints*

13. To rename a constraint, select it in the **Name** column and press the F2 key or double-click on the respective field; an edit box will be displayed. Change the name of the constraint.

14. Apply a new expression to each renamed constraint, as shown in Figure 16-64. The **Parameters Manager** palette after renaming the constraints and applying appropriate expressions is shown in Figure 16-69.

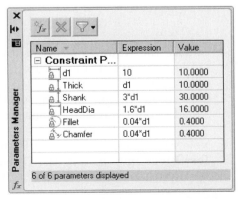

Name	Expression	Value
⊟ **Constraint P...**		
d1	10	10.0000
Thick	d1	10.0000
Shank	3*d1	30.0000
HeadDia	1.6*d1	16.0000
Fillet	0.04*d1	0.4000
Chamfer	0.04*d1	0.4000

6 of 6 parameters displayed

Figure 16-69 *The Parameters Manager*

15. Choose the **Block Table** button from the **Dimensional** panel of the **Block Editor**; you are prompted to specify the parameter location.

16. Select the midpoint of the horizontal line at the bottom; you are prompted to specify the number of grips.

17. Enter **1** and press ENTER; the **Block Properties Table** is displayed.

18. Choose the **Creates a new user parameter and adds it to the table** button; the **New Parameter** dialog box is displayed.

19. Enter **Size** in the **Name** edit box. Next, select the **String** option from the **Type** drop-down list and choose the **OK** button; a new column named **Size** is added to the **Block Properties Table**.

20. Choose the **Add properties which appear as columns in the table** button from the **Block Properties Table**; the **Add Parameter Properties** dialog box is displayed.

21. Select the **d1** parameter and choose the **OK** button; the selected parameter is added to the **Block Properties Table**.

22. Clear the **Block properties must match a row in the table** check box, if not cleared already.

23. Enter the following values in the **Size** and **d1** columns:

Size	d1
M6	6
M10	10
M12.5	12.5
M15	15
M18	18
M24	24
M25	25

24. Choose the **Audits the block property table for errors** button; a message box is displayed stating that no errors have been found.

25. Close the message box and choose **OK** in the **Block Properties Table**; the **Size** parameter is added to the **Parameters Manager** palette under the **User Parameters** node.

26. Now, you need to test the block by invoking the **Test Block Window**. To do so, choose the **Test Block** button from the **Open/Save** panel of the **Block Editor**; the **Test Block Window** is opened. You can view the name of the new window at the title bar.

27. Select the block; the block is displayed in dashed lines and two grips are also displayed, one at the insertion point of the block and other at the midpoint of the horizontal line at the bottom. The triangular grip at the midpoint is known as lookup grip.

28. Click on the lookup grip; a shortcut menu is displayed.

29. Choose different sizes and check whether the block resizes without any error.

30. After checking the block, choose **Close > Close Test Block Window** from the **Ribbon**; the **Block Editor** is displayed.

31. Choose the **Close Block Editor** button to close the block editor; a message box is displayed.

32. Choose the **Save changes and exit the Block Editor** option from the message box.

33. Select the block and click on the lookup grip; different possible sizes are displayed. Select the required size.

34. Save the drawing and exit.

Note
In this example, the block has been created without creating a fully-defined sketch. But it is recommended to create a fully-defined sketch while working with dynamic blocks.

Self-Evaluation Test

Answer the following questions and then compare them to those given at the end of this chapter:

1. Individual objects in a block cannot be erased using the **Block Editor**. (T/F)

2. A block can be mirrored by providing a scale factor of -1 for X. (T/F)

3. Blocks created by the **BLOCK** command can be used in any drawing. (T/F)

4. An existing block cannot be redefined. (T/F)

5. The _____ command lets you create a drawing file (.*dwg* extension) of a block defined in the current drawing.

6. The _____ command can be used to change the name of a block.

7. The _____ command is used to place a previously created block in a drawing.

8. You can delete the unreferenced blocks using the _____ command.

9. You can use the _____ to locate, preview, copy, and insert blocks or existing drawings into the current drawing.

10. The _____ command is used for in-place reference editing.

Review Questions

Answer the following questions:

1. An entire drawing can be converted into a block. (T/F)

2. The objects in a block possess the properties of the layer on which they are drawn, such as color and linetype. (T/F)

3. If the objects forming a block were drawn on layer 0 with color and linetype **BYLAYER**, then at the time of the insertion, each object that makes up a block is drawn on the current layer with the current linetype and color. (T/F)

4. Objects created with the special color **BYBLOCK** are generated with the color that is current at the time the block was inserted. (T/F)

5. In the array generated with the **MINSERT** command, there is no way to change the number of rows or columns or the spacing between them, after insertion. The whole **MINSERT** pattern is considered as one object that cannot be exploded. (T/F)

6. Which one of the following actions cannot be assigned to the **Point** parameter?

    (a) **Move**   (b) **Array**
    (c) **Stretch**   (d) **None**

7. Which command should you use to get back the objects that consist of the block and have been removed from the drawing?

    (a) **OOPS**   (b) **BLIPS**
    (c) **BLOCK**   (d) **UNDO**

8. By what amount is a block rotated, if the value of both the X and Y scale factors is -1?

    (a) 90   (b) 180
    (c) 270   (d) 360

9. When you insert a drawing into a current drawing, how many of the blocks belonging to the inserted drawing are brought into the current drawing?

    (a) One   (b) None
    (c) All   (d) Two

Chapter 16

10. What command is used to create a rectangular array of a block?

    (a) **ARRAY**   (b) **INSERT**
    (c) **MINSERT**   (d) **3DARRAY**

11. The **Entire drawing** option of the **WBLOCK** command has the same effect as the **PURGE** command. The only difference is that with the **PURGE** command,
_____ .

12. The _____ tab of the **Block Authoring Palettes** allows you to insert a parameter set.

13. The automatic save feature is _____ during in-place reference editing.

14. You cannot use the **REFEDIT** command on blocks inserted using the _____ command.

15. The Layer 0 and the Continuous linetype _____ be renamed using the **RENAME** command.

Exercises

Exercise 7   *Mechanical*

Draw part (a) of Figure 16-70 and define it as a block named A. Then, using the block insert command, insert the block in the plate, as shown.

Figure 16-70 *Drawing for Exercise 7*

Exercise 8   *Piping*

Draw the diagrams in Figure 16-71 using blocks.
a. Create a block for the valve, Figure 16-71(a).
b. Use a thick polyline for the flow lines.

Figure 16-71 *Drawing for Exercise 8*

Exercise 9   *General*

Draw part (a) of Figure 16-72 and define it as a block named B. Then, using the relevant
insertion method, generate the pattern as shown. Note that the pattern is rotated at 30-degree.

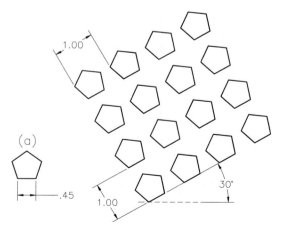

Figure 16-72 *Drawing for Exercise 9*

Exercise 10   *General*

Draw part (a) of Figure 16-73 and define it as a block named Chair. The dimensions of the chair can be referred from the Problem Solving Exercise 3 of Chapter 5. Then, using the block insert command, insert the chair around the table as shown.

Figure 16-73 Drawing for Exercise 10

Exercise 11   *General*

Create the sketch of a bolt shown, as shown in Figure 16-74. Then, convert it to a block and apply constraints such that you can create bolts of diameter 6, 10, 12.5, 15, 18, 24, and 25.

Figure 16-74 Drawing for Exercise 11

Exercise 12   *Mechanical*

Create the drawing shown in Figure 16-75 and use the block B1 to create the outer features.**

Figure 16-75 *Dimensions for the drawing*

AUTOCAD PART II

Author's Web sites

For Faculty: Please contact the author at stickoo@calumet.purdue.edu to access the Web site that contains the following:

1. PowerPoint presentations, program listings, and drawings used in this textbook.
2. Syllabus, chapter objectives and hints, and questions with answers for every chapter.

For Students: You can download drawings, exercises, tutorials, programs, and free chapters by accessing the author's Web site www.cadcim.com/students/students.htm or use the link www.cengage.com/cad/autodeskpress. On this Web page, choose the Online Companions option and click on the AutoCAD 2010: A Problem-Solving Approach link.

Chapter 17

Defining Block Attributes

Learning Objectives

After completing this chapter, you will be able to:
- *Understand the meaning of attributes and define them with a block.*
- *Create annotative block attributes.*
- *Edit attribute tag names.*
- *Insert blocks with attributes and assign values to attributes.*
- *Extract attribute values from inserted blocks.*
- *Control attribute visibility.*
- *Perform global and individual editing of attributes.*
- *Insert a text file in a drawing to create Bill Of Materials.*

UNDERSTANDING ATTRIBUTES

AutoCAD has provided a facility that allows the user to attach information to blocks. This information can then be retrieved and processed by other programs for various purposes. For example, you can use this information to create a bill of material for a project, find the total number of computers in a building, or determine the location of each block in a drawing. Attributes can also be used to create blocks (such as title blocks) with prompted or preformatted text, to control text placement. The information associated with a block is known as attribute value or attribute. AutoCAD references the attributes with a block through tag names.

Before assigning attributes to a block, you must create an attribute definition by using the **ATTDEF** command. The attribute definition describes the characteristics of the attribute. You can define several attribute definitions (tags) and include them in the block definition. Each time you insert the block, AutoCAD will prompt you to enter the value of the attribute. The attribute value automatically replaces the attribute tag name. The information (attribute values) assigned to a block can be extracted and written to a file by using AutoCAD's **EATTEXT** command. This file can then be inserted in the drawing as a table or processed by other programs to analyze the data. The attribute values can be edited by using the **EATTEDIT** command. The display of attributes can be controlled with the **ATTDISP** command.

DEFINING ATTRIBUTES

Ribbon: Insert > Attributes > Define Attributes	**Command:** ATTDEF

When you invoke the **ATTDEF** command, the **Attribute Definition** dialog box is displayed, see Figure 17-1. The block attributes can be defined through this dialog box. When creating an attribute definition, you must define the mode, attributes, insertion point, and text information for each attribute. All this information can be entered in the dialog box. The following is the description of each area of the **Attribute Definition** dialog box.

Mode Area

The **Mode** area of the **Attribute Definition** dialog box has six check boxes: **Invisible**, **Constant**, **Verify**, **Preset**, **Lock position**, and **Multiple lines**. These determine the display and edit features of the block attributes. For example, if you select the **Invisible** check box, the attribute becomes invisible which means, it is not displayed on the screen. Similarly, if the **Constant** check box is selected, the attribute becomes constant. This means that its value is predefined and cannot be changed. These options are described next.

Invisible

This option lets you create an attribute that is not visible on the screen, by default. Select this check box if you want the attribute to be invisible.

Tip
*The **Invisible** mode is especially useful when you do not want the attribute values to be displayed on the screen to avoid cluttering the drawing. Also, if the attributes are invisible, it takes less time to regenerate the drawing.*

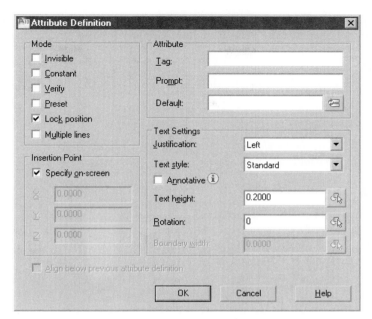

Figure 17-1 *The **Attribute Definition** dialog box*

You can make the invisible attribute visible by using the **ATTDISP** command discussed later in this chapter, in the section "Controlling Attribute Visibility".

Constant

This option lets you create an attribute that has a fixed value and cannot be changed after block insertion. When you select this mode, the **Prompt** edit box and the **Verify** and **Preset** check boxes are disabled. Since the value is constant, there is no need to be prompted for new values. This check box is cleared by default and you can use different attribute values for the blocks.

Verify

This option allows you to verify the attribute value you have entered when inserting a block by asking you twice for the data. If the value is incorrect, you can correct it by entering the new value. If this check box is cleared, you are not prompted for verification of the attribute values.

Preset

This option allows you to create an attribute that is automatically set to the default value. The attribute values are not requested when you insert a block and the default values are used. But unlike a constant attribute, the preset attribute value can be edited later.

Lock position

This check box is selected by default and it is used to lock the position of the attributes with respect to the block object. In case of non-dynamic blocks, if the attributes are created with the **Lock position** check box cleared, then a separate grip will be displayed on that attribute. This grip can be used to modify the location of the attribute. When the dynamic blocks are created with the **Lock position** check box cleared, the separate grip does not get displayed on the attribute. Besides this, the attribute cannot be manipulated by the parameters and action.

Multiple lines

Select this check box to specify that the default value of the attribute may be a single-line or multiple line text.

Note

*With all the check boxes cleared in the **Mode** area, all the prompts will be displayed at the command line and the values will be visible on the screen. This is also referred to as the **Normal** mode.*

Attribute Area

The **Attribute** area (Figure 17-2) in the **Attribute Definition** dialog box has three edit boxes: **Tag**, **Prompt**, and **Default**, where you can enter values. You can enter up to 256 characters in these edit boxes. If the first character to be entered in any one of these edit boxes is a space, you should start with a backslash (\). But if the first character is a backslash (\), you should start the value to be entered with two backslashes (\\). The three edit boxes have been described next.

*Figure 17-2   The **Attribute** area of the **Attribute Definition** dialog box*

Tag

This is like a label that is used to identify an attribute. For example, the tag name COMPUTER can be used to identify an item. Here you can enter the tag names as uppercase, lowercase, or both, but all lowercase letters are automatically converted into uppercase when displayed. The tag name cannot be null. Also, it must not contain any blank spaces.

Tip

It is advisable to specify a tag name that reflects the contents of the item being tagged. For example, the tag name COMP or COMPUTER is an appropriate name for labeling computers.

Prompt

The text that you enter in the **Prompt** edit box is used as a prompt when you insert a block that contains the defined attribute. For example, if COMPUTER is the tag, you can enter WHAT IS THE MEMORY? or ENTER MEMORY: in the **Prompt** edit box.AutoCAD will then prompt you with this same statement when you insert the block with which the attribute is defined. If you have selected the **Constant** check box in the **Mode** area, the **Prompt** edit box is not available because no prompt is required if the attribute is constant. If you do not enter anything in the **Prompt** edit box, the entry made in the **Tag** edit box is used as the prompt.

Default

The entry in the **Default** edit box defines the default value of the specified attribute. If you do not enter a value, the attribute takes the value entered in the **Tag** edit box. The entry of a value is optional.

Insert Field

 This button is used to insert a field as the value of the attribute. When you choose this button, the **Field** dialog box is displayed that can be used to insert the required field.

Multiline Editor

 This button is displayed in place of the **Insert Field** button when the **Multiple lines** check box is selected from the **Mode** area. Choose this button to invoke the **Text Formatting** window on the screen. You can enter the text that you want to assign as the default value of the attribute in the text window.

Note

*The **Text Formatting** toolbar displayed here has only some commonly used options. The full **Text Formatting** toolbar can be displayed by changing the value of the **ATTIPE** system variable to 1.*

Insertion Point Area

The **Insertion Point** area of the **Attribute Definition** dialog box (Figure 17-3) lets you define the insertion point of the block attribute text. You can define the insertion point by entering the values in the **X**, **Y**, and **Z** edit boxes or by specifying it on the screen. To specify the insertion point on the screen, select the **Specify on-Screen** check box. Now, set the parameters in all the other fields and areas and then choose **OK**. You will be prompted to select the start point of the attribute.

Figure 17-3 The Insertion Point area

Just below the **Insertion Point** area of the dialog box is a check box labeled **Align below previous attribute definition**. This check box is not available when you use the **Attribute Definition** dialog box for the first time. After you have defined an attribute and when you press ENTER to display the **Attribute Definition** dialog box again, this check box is available. You can select this check box to place the subsequent attribute text just below the previously defined attribute automatically. When you select this check box, the **Insertion Point** area and the **Text Settings** areas of the dialog box are not available and AutoCAD assumes previously defined values for text such as text height, text style, text justification, and text rotation. The text is automatically placed on the following line.

Text Settings Area

The **Text Settings** area of the **Attribute Definition** dialog box (Figure 17-4) lets you control the justification, text style, annotation property, height, rotation, and boundary width of the attribute text. To set the text justification, select a justification type from the **Justification** drop-down list. The default option is **Left** for the single line text and **Top left** for the multiline text. Similarly, you can use the **Text style** drop-down list to select a text style. All the text styles defined in the current drawing are displayed in the **Text style** drop-down list. The default text style is **Standard**. To write annotative attributes, select the **Annotative** text style from the **Text style** drop-down list. Select the **Annotative** check box to assign the annotative property to the attribute text. The annotative attributes can be attached to both the annotative or non-annotative blocks. Choose the information **(i)** button on the side of the **Annotative** check box to collect more information about the annotative object. You can specify the text height and text rotation in the **Text height** and **Rotation** edit boxes, respectively. You can also define the text height by choosing the **Text height** button. When you choose this button, AutoCAD temporarily exits the **Attribute Definition** dialog box and lets you enter the height value by selecting points on the screen or from the

*Figure 17-4 The **Text Settings** area of the **Attribute Definition** dialog box*

command line. Once the height on the screen is defined, the **Attribute Definition** dialog box will reappear and the defined text height will be displayed in the edit box. Similarly, you can define the text rotation by choosing the **Rotation** button and then selecting points on the screen or by entering the rotation angle at the command line. The **Boundary width** edit box is used to specify the maximum width for the multiline text. If the width of the text line is more than this value, the text will automatically go to the second line. This edit box will be available only when you select the **Multiple lines** check box from the **Mode** area.

Note
The text style must be defined before it can be used to specify the text.

*If you select a style that has the predefined height, AutoCAD automatically disables the **Text height** edit box.*

*If you have selected the **Align** option from the **Justification** drop-down list, the **Insertion Point** area and the **Text height** and **Rotation** edit boxes in the **Text Settings** area are disabled.*

*If you have selected the **Fit** option from the **Justification** drop-down list, the **Insertion Point** area and the **Rotation** edit box are disabled. To specify the location of the attribute, choose **OK** from the dialog box. AutoCAD prompts you to specify the first and second endpoints of the text baseline. The text will be fit between the two endpoints that you specify.*

After you complete the settings in the **Attribute Definition** dialog box and choose **OK**, the attribute tag text is inserted in the drawing at the specified insertion point. Now, you can use the **BLOCK** or **WBLOCK** commands to select all the objects and attributes to define a block.

Note
*You can use the **-ATTDEF** command to display the dynamic prompt in the drawing area. The options in the **Attribute Definition** dialog box are available through the dynamic prompt too. See Figure 17-5.*

*Figure 17-5 Dynamic prompt for the **-ATTDEF** command*

Example 1   *General*

In this example, you will define the following attributes for a computer and then create a block using the **BLOCK** command. The name of the block is COMP.

Mode	Tag name	Prompt	Default value
Constant	ITEM		Computer
Preset, Verify	MAKE	Enter make:	CAD-CIM
Verify	PROCESSOR	Enter processor type:	Unknown
Verify	HD	Enter Hard-Drive size:	40 GB
Invisible, Verify	RAM	Enter RAM:	256 MB

1. Draw the computer, as shown in Figure 17-6. Assume the dimensions, or measure the dimensions of the computer you are using for AutoCAD.

2. Choose **Insert > Attributes > Define Attributes** from the **Ribbon**; the **Attribute Definition** dialog box is displayed.

3. Define the first attribute as shown in the preceding table. Select **Constant** check box in the **Mode** area because the mode of the first attribute is constant. In the **Tag** edit box, enter the tag name, **ITEM**. Similarly, enter **COMPUTER** in the **Default** edit box. Note that the **Prompt** edit box is not available because the mode is constant.

4. In the **Insertion Point** area, select the **Specify on-screen** check box to define the text insertion point.

5. In the **Text Settings** area, specify the justification, style, annotative property, height, and rotation of the text.

6. Choose the **OK** button once you have entered information in the **Attribute Definition** dialog box. Select a point below the insertion base point (P1) of the computer to place the text.

7. Press ENTER to invoke the **Attribute Definition** dialog box again. Enter the mode and attribute information for the second attribute as shown in the table at the beginning of Example 1. You need not define the insertion point and text options again. Select the **Align below previous attribute definition** check box that is located just below the **Insertion Point** area. You will notice that when you select this check box, the **Insertion Point** and **Text Settings** area are not available. Now, choose the **OK** button. AutoCAD places the attribute text just below the previous attribute text.

8. Similarly, define the remaining attributes also (Figure 17-7).

9. Now, use the **BLOCK** command to create a block. Make sure that the **Retain** radio button is selected in the **Objects** area. The name of the block is **COMP**, and the insertion point of the block is P1, midpoint of the base. When you select objects for the block, make sure you also select attributes.

ITEM

MAKE

PROCESSOR

HD

RAM

Figure 17-6 *Drawing for Example 1*   *Figure 17-7* *Attributes defined below the*
                                                                  computer

10. Insert the block created. You will notice that the order of prompts is the same as the order of attributes selection.

EDITING ATTRIBUTE DEFINITION

Menu Bar: Modify > Object > Text > Edit   **Command:** DDEDIT

You can edit the text and attribute definitions, before you define the block, using the **DDEDIT** command. After invoking this command, AutoCAD will prompt you to **Select an annotation object or [Undo]**. If you select an attribute created using the **Attribute Definition** dialog box, the **Edit Attribute Definition** dialog box is displayed (Figure 17-8). You can also invoke the **Edit Attribute Definition** dialog box by double-clicking on the attribute definition.

You can enter the new values in the respective edit boxes. Once you have entered the changed values, choose the **OK** button in the dialog box. After you exit the dialog box, AutoCAD will continue to prompt you to select another text or attribute object (Attribute tag). If you have finished editing and do not want to select another attribute object to edit, press ENTER to exit the command.

Figure 17-8 *The* **Edit Attribute Definition** *dialog box*

Using the Properties Palette

The **PROPERTIES** command has been already discussed in Chapter 4, Working with Drawing Aids. It can also be used to edit defined attributes. Select the attribute to be modified and right-click to display a shortcut menu. Choose **Properties** from the shortcut menu; the **Properties** palette will be displayed, see Figure 17-9. You will notice that **Attribute Definition** is displayed in the text box located at the top of the palette. Also, you will find that all properties of the selected attribute are displayed under four headings. They are **General**, **3D Visualization**, **Text**, and **Misc**. You can change these values in their corresponding fields. For example, you can modify the color, layer, linetype, thickness, linetype scale, and so on, of the selected attribute under the **General** head. Similarly, you can modify the tag name, prompt, and value of the selected attribute in the **Tag**, **Prompt**, and **Value** fields under the **Text** heading. Under the **Text** head, you can also modify the text style, justification, annotation scale, text height, rotation angle, width factor, and make the angle values of the selected attribute oblique.

*Figure 17-9 The **Properties** palette to modify attributes*

You can also determine if you want the attribute text to appear upside down or backwards or not under the **Misc** heading. Here, you can also modify the attribute modes, which have been already defined. When you select a particular mode, for example **Invisible**, a drop-down list is available in the corresponding field. This list displays two options, **Yes** and **No**. If you select **Yes**, the attribute is made invisible and if you select **No**, the attribute is made visible. To select a particular mode, you should choose **Yes** from their corresponding drop-down lists.

Note
Remember that if you change the justification of the first attribute below which you have aligned the remaining attributes, the remaining attributes will not be updated automatically. You will have to change justifications of the remaining attributes individually to align them below the first attribute.

If you change the justification of the attributes from align to some other, the size of the text will also be changed.

*You can also use the **CHANGE** command to edit the attribute objects using the Command line.*

INSERTING BLOCKS WITH ATTRIBUTES

The value of the attributes can be specified during block insertion, either at the command line or in the **Edit Attributes** dialog box. When you use the **INSERT** command or the **-INSERT** command to insert a block in a drawing (discussed earlier in Chapter 16, Working with

Blocks), and after you have specified the insertion point, scale factors, and rotation angle, the **Edit Attributes** dialog box (Figure 17-10) is displayed, if the system variable **ATTDIA** is set to **1**.

*Figure 17-10   The **Edit Attributes** dialog box*

The default value for **ATTDIA** is **0**, which disables the dialog box. The prompts and their default values, which you had specified with the attribute definition, are then displayed on the command line after you have specified the insertion point, scale, and rotation angle for the block to be inserted.

In the **Edit Attributes** dialog box, the prompts that were entered at the time of attribute definition in the dialog box are displayed with their default values in the corresponding fields. If an attribute has been defined with the **Constant** mode, it is not displayed in the dialog box because a constant attribute value cannot be edited. You can enter the attribute values in the fields located next to the attribute prompt. If no new values are specified, the default values are displayed. Eight attribute values are displayed at a time in the dialog box. If there are more attributes, they can be accessed by using the **Next** or **Previous** buttons. The block name is displayed at the top of the dialog box. After entering the new attribute values, choose the **OK** button. AutoCAD will place these attribute values at the specified location.

Tip
*It is more convenient to use the **Edit Attributes** dialog box because you can view all the attribute values at a glance and can correct them before placing. Therefore, it is a good idea to set the **ATTDIA** value to 1, before you insert a block with attributes.*

*Attributes can also be defined from the command line by setting the system variable **ATTDIA** to 0 (default value). Now, when you use the **INSERT** command, AutoCAD does not display the **Edit Attributes** dialog box. Instead, AutoCAD prompts you to enter the attribute values for various attributes that have been defined in the block at the pointer input.*

Example 2   *General*

In this example, you will insert the block (COMP) that was defined in Example 1. The following is the list of the attribute values for computers.

ITEM	MAKE	PROCESSOR	HD	RAM
Computer	Gateway	486-60	150 MB	16 MB
Computer	Zenith	486-30	100 MB	32 MB
Computer	IBM	386-30	80 MB	8 MB
Computer	Del	586-60	450 MB	64 MB
Computer	CAD-CIM	Pentium-90	100 Min	32 MB
Computer	CAD-CIM	Unknown	600 MB	Standard

1. Draw the floor plan drawing, as shown in Figure 17-11 (assume the dimensions).

2. Set the system variable **ATTDIA** to 1. Use the **INSERT** command to insert the blocks. When you invoke the **INSERT** command, the **Insert** dialog box is displayed. Enter **COMP** in the **Name** edit box and choose **OK** to exit the dialog box. Select an insertion point on screen to insert the block. After you specify the insertion point, the **Edit Attributes** dialog box is displayed, where you can change the attribute values, if you need to, in the different edit boxes.

3. Repeat the **INSERT** command to insert other blocks, and define their attribute values, as shown in Figure 17-12.

4. Save the drawing for further use.

MANAGING ATTRIBUTES

Ribbon:	Insert > Attributes > Manage	**Command:** BATTMAN
Menu Bar:	Modify > Object > Attribute > Block Attribute Manager	

The **BATTMAN** command allows you to manage attribute definitions for blocks in the current drawing through the **Block Attribute Manager** dialog box. This dialog box is shown in Figure 17-13. By default, any attribute changes you make are applied to all existing block references in the current drawing. Immediate changes in the default value of the attribute in the existing drawing can only be seen upon regeneration, if the attribute that is edited is in the **Constant** mode. If the mode is other than **Constant**, the changes in the attribute can only be seen for new block insertions.

When you invoke this command, AutoCAD displays the number of blocks that exist in the current drawing below the list box in the dialog box. The block can be in the model space or in the layout. The names of the existing blocks in the current drawing are displayed in the **Block** drop-down list. Attributes of the selected block are displayed in the attribute list. By default, the **Tag**, **Prompt**, **Default**, **Modes**, and **Annotative** attribute properties are displayed in the attribute list. You can edit the attribute definitions in blocks, remove attributes from blocks, and change the order in which you were prompted for attribute values while inserting the block.

Figure 17-11 *Floor plan drawing for Example 2*

Figure 17-12 *The floor plan after inserting blocks and defining their attributes*

Figure 17-13 *The **Block Attribute Manager** dialog box*

Select block

This button allows you to select a block from the drawing area whose attributes you want to edit. When you choose the **Select block** button, the dialog box closes temporarily until you select block from the drawing or cancel by pressing ESC. Once you have selected the block, the dialog box is redisplayed. The name of the block you have selected is displayed in the **Block** drop-down list. If you modify the attributes of a block and then select a new block before saving the attribute changes you made, you will be prompted to save the changes before selecting another block through an alert message box (Figure 17-14).

*Figure 17-14 The **Block Attribute Manager Alert** message box*

Block Drop-down List

This drop-down list in the **Block Attribute Manager** dialog box allows you to choose the block whose attributes you want to modify. It lists all the blocks in the current drawing that have attributes.

Sync

This button of the **Block Attribute Manager** dialog box allows you to update all instances of the selected block with the attribute properties currently defined. This does not affect any of the values assigned to attributes in each block when attributes were defined. On choosing this button, you can update attributes in all block references in the current drawing with the changes you made to the block definition. For example, you may have used the **Block Attribute Manager** dialog box to modify attribute properties of several block definitions in your drawing. But you do not want it to automatically update the existing block references when you made the changes. Now, when you are satisfied with the attribute changes you made, you can apply those changes to all blocks in the current drawing by using the **Sync** button.

You can also use the **ATTSYNC** command to update attribute properties in block references to match their block definition. This command can also be invoked by choosing the **Synchronize** button from the **Attributes** panel in the **Blocks & References** tab of the **Ribbon** or by choosing the **Synchronize Attributes** button from the **Modify II** toolbar.

Note
*Changing the attribute properties of an existing block does not affect the values assigned to the block. For example, in a block containing an attribute whose **Tag** is Cost and default value is 200. The default value 200 remains unaffected, if the **Tag** is changed to UnitCost.*

Move Up

This button moves the selected attribute tag earlier in the prompt sequence. Now, when you insert the block, the attribute you have just moved up will appear earlier in the prompt sequence. The **Move Up** button is not available when a constant attribute is selected and if the attribute is at the topmost position in the attribute list.

Move Down

This button moves the selected attribute tag later in the prompt sequence. The **Move Down** button is not available when a constant attribute is selected and when the attribute is at the bottom position in the attribute list.

Tip
*When you define a block, the order in which you select the attributes determines the order in which you are prompted for the attribute information when you insert the block. You can use the **Block Attribute Manager** to change the order of prompts that request attribute values.*

Edit

This button of the **Block Attribute Manager** allows you to modify the attribute properties. To edit the attribute definition in blocks, select the attribute and then choose **Edit** from the dialog box; the **Edit Attribute** dialog box will be displayed. In the dialog box, **Active Block** displays the name of the block that you have selected and whose properties have to be edited.

Note
*You can also double-click on the attribute or choose **Edit Attribute** > **Single** from the **Attributes** panel in the **Insert** tab of the **Ribbon** and then select the attribute to display the **Enhanced Attribute Editor** dialog box.*

The options in the **Edit Attribute** dialog box are discussed next.

Attribute Tab

The options under the **Attribute** tab of the **Edit Attribute** dialog box are used to modify the mode and data of attributes (see Figure 17-15). The options under the **Mode** area can be selected for the block. The **Data** area of the dialog box has three edit boxes: **Tag**, **Prompt**, and **Default**, where you can enter the data to be changed.

Text Options Tab

The options under this tab (Figure 17-16) are used to set the properties that define the way an attribute's text is displayed in the drawing. The **Text Options** are discussed next.

Text Style. This drop-down list specifies the text style for attribute text. Default values for this text style are assigned to the text properties displayed in this dialog box.

Justification. This drop-down list specifies how attribute text is justified.

Figure 17-15 The **Attribute** *tab of the* **Edit Attribute** *dialog box*

Figure 17-16 The **Text Options** *tab of the* **Edit Attribute** *dialog box*

Height. This edit box specifies the height of the attribute text. If the height is defined in the current text style, this edit box is not available.

Rotation. This edit box specifies the rotation angle of the attribute text.

Backwards. This check box specifies whether or not the text is displayed backwards. Select this box to display the text backwards.

Upside down. This check box specifies whether or not the text is displayed upside down. Select this box display the text upside down.

Width Factor. This edit box sets the character spacing for attribute text. Entering a value less than 1.0 condenses the text and a value greater than 1.0 expands it.

Oblique Angle. In this edit box, the angle at which the attribute text is slanted away from its vertical axis is specified.

Annotative. This check box is used to convert the non-annotative attribute to the annotative attribute and vice-versa.

Boundary Width. This edit box sets the width of the boundary that will be displayed while writing the multiline text for the attributes' default value. The value specified here is the relative scale value. If you specify a value less than 1, it will decrease the boundary width. Whereas, if you specify a value more than 1, it will increase the boundary width. Accordingly, the paragraph width for the multiline text will increase or decrease. You can modify the value of the boundary width only when you select the **Multiple lines** check box from the **Mode** area of the **Attribute Definition** dialog box.

Note
*If you change the above-mentioned properties of the **Constant** attribute, you may need to regenerate the drawing to view the effects.*

Properties

The options under this tab of the **Edit Attribute** dialog box (Figure 17-17) are used to define the layer that the attribute is on and the color, lineweight, and linetype for the attribute's line. Also, if the drawing uses plot styles, you can assign a plot style to the attribute. The options of the **Properties** tab are discussed next.

*Figure 17-17 The **Properties** tab of the **Edit Attribute** dialog box*

Layer. This drop-down list specifies the layer that the attribute is on. The layer on which the attribute was defined is displayed here in this drop-down list. This cannot be changed using the dialog box.

Linetype. This drop-down list specifies the linetype of attribute text.

Color. This drop-down list specifies the attribute's text color.

Lineweight. This drop-down list specifies the lineweight of attribute text. The changes you make to this option do not come into effect if the **LWDISPLAY** system variable is off.

Plot Style. This drop-down list specifies the plot style of the attribute. If the current drawing uses color-dependent plot styles, the **Plot style** list in this dialog box is not available.

Auto preview changes

This check box controls whether or not the drawing area is immediately updated to display any visible attribute changes you make. If the **Auto preview changes** check box is selected, changes are immediately visible. If this check box is cleared, changes are not immediately visible.

Tip
*The buttons such as **Move Up**, **Move Down**, **Edit**, **Remove**, and **Settings** can also be invoked by the shortcut menu, which appears by right-clicking on the attributes in the attribute list in the **Block Attribute Manager** dialog box.*

Settings

Choosing this button of the **Block Attribute Manager** opens the **Block Attribute Settings** dialog box (Figure 17-18), where you can customize how attribute information is listed in the **Block Attribute Manager**.

Display in list

You can specify the attribute properties you want to be displayed in the list by selecting the check boxes available in this area of the **Block Attribute Settings** dialog box. This area shows the check boxes of all the properties that are displayed in the attribute list. Only the check boxes of the properties selected are displayed in the attribute list. These properties can be viewed in the attribute list by scrolling it. The **Tag** check box is always selected by default.

*Figure 17-18 The **Block Attribute Settings** dialog box*

This area of the dialog box has two buttons, **Select All** and **Clear All**, which select and clear all the check boxes, respectively.

Emphasize duplicate tags

This check box turns duplicate tag emphasis on and off. If this check box is selected, duplicate attribute tags are displayed in red type in the attribute list. For example, if a block has PRICE as a tag more than once, the attribute PRICE will be displayed in red in the attribute list. If this check box is cleared, duplicate tags are not emphasized in the attribute list.

Chapter 17

Apply changes to existing references

This check box of the dialog box specifies whether or not to update all existing instances of the block whose attributes you are modifying. Therefore, if this check box is selected, all instances of the block with the new attribute definitions are updated. If this check box is cleared, only new insertions of the block with the new attribute definitions are displayed.

Use **Sync** in the **Block Attribute Manager** to apply changes immediately to existing block instances. This temporarily overrides the **Apply changes to existing references** option.

Tip
*If constant attributes or nested attributed blocks are affected by your changes, use the **REGEN** command to update the display of those blocks in the drawing area.*

Remove

This button removes the selected attribute from the block definition. If the **Apply changes to existing references** check box is selected in the **Block Attribute Settings** dialog box before you choose the **Remove** button, the attribute is removed from all instances of the block in the current drawing. If this check box is cleared before choosing the **Remove** button, the attribute is removed only from the attribute list and the blocks that will be inserted henceforth. You can remove attributes from block definitions and from all existing block references in the current drawing. Attributes removed from existing block references do not disappear in the drawing area until you regenerate the drawing using the **REGEN** command. You cannot remove all attributes from the block, at least one attribute should remain. However, if you want to remove all the attributes, you have to redefine the block.

Note that the **Remove** button is not available for blocks with only one attribute.

EXTRACTING ATTRIBUTES

Ribbon:	Insert > Linking & Extraction > Extract Data	Menu Bar:	Data Extraction
Command:	EATTEXT or DATAEXTRACTION		

The **EATTEXT** command allows you to extract the block attribute information and property information such as drawing summary from single or multiple drawings. The settings that were used to extract data are stored in a *data extraction (.dxe)* file format. When this command is invoked, the **Data Extraction** wizard is displayed on the screen (Figure 17-19). This wizard includes the following pages:

Begin

The **Begin** page, as shown in Figure 17-19, allows you to save the settings that are meant to be used for extracting data to a new *.dxe* file or to modify an existing *.dxe* file. You can also use the saved *.dxe* or *.blk* file as the template of settings for extracting data.

This wizard page has two radio buttons: **Create a new data extraction** and **Edit an existing data extraction**. If you select the **Use previous extraction as a template [.dxe or .blk]** check box, you will be allowed to select an existing template file. To select a template file, choose

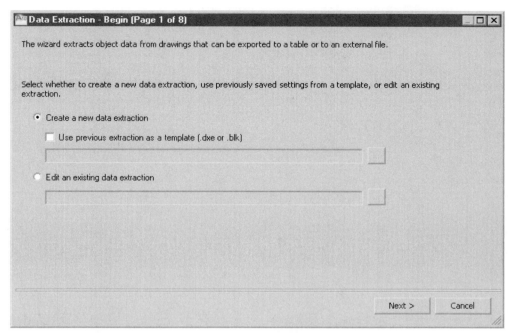

Figure 17-19 *The* **Begin** *page of the* **Data Extraction** *wizard*

the **Browse[...]** button available on the right of the template name edit box. A *.blk* template file is a text file that allows you to specify the attribute and property values that you want to extract and the information about the drawing that you want to retrieve. The file can be created by using any text editor such as Notepad or WordPad. While using a *.blk* template file, you will be prompted to save the data extraction to a *.dxe* file to extract the information. Select the **Edit an existing data extraction** check box to modify an existing *.dxe* file. Now, you will be allowed to select the existing data file. To select a template file, choose the **Browse[...]** button available on the right of the template name edit box.

Choose the **Next** button from the **Begin** page of the **Data Extraction** wizard; the **Save Data Extraction As** dialog box will be invoked. Next, specify the desired name and location in this dialog box to save the *.dxe* file and choose the **Save** button; the **Define Data Source** page will be displayed, as shown in Figure 17-20.

Define Data Source

The **Define Data Source** page, as shown in Figure 17-20 allows you to select the object either from a single drawing or from multiple drawings. The **Data source** area of this dialog box has two radio buttons, namely **Drawings/Sheet set** and **Select objects in the current drawing**.

Drawings/Sheet set

This radio button enables you to specify the drawings and folder from which you can extract the data. If this radio button is selected, the **Add Folder** and **Add Drawings** buttons will be

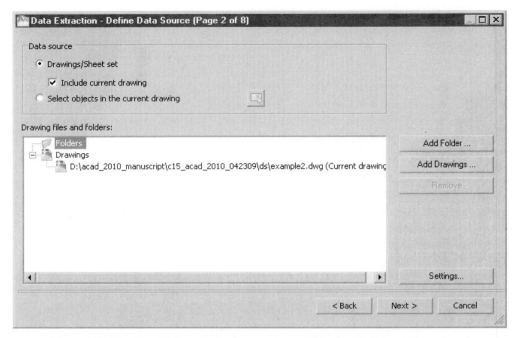

*Figure 17-20 The **Define Data Source** page of the **Data Extraction** wizard*

enabled. The name of the drawings selected will be displayed with full path in the **Drawing files and folders** area.

Choose the **Add Folder** button; the **Add Folder Options** dialog box will be invoked, see Figure 17-21. All the drawings within the added folder will be considered for extracting the data. To add a folder, choose the **Browse[...]** button available on the right of the **Folder** display box. Select the **Automatically include new drawings added in this folder to the data extraction** check box in the **Options** area to automatically consider the new drawing added to this folder for data extraction.

*Figure 17-21 The **Add Folder Options** dialog box*

Select the **Include subfolders** check box to include the drawings within the subfolders of the folder selected to be added. Select the **Utilize wild-card characters to select drawings** check box to include only the files matching the specified wild-card characters for data extraction. By default, *.dwg is used as the wild-card character and that is why, all the files with extension *.dwg*(all the drawing files) are considered for the data extraction. For example, if you want AutoCAD to consider all the drawing files having "computer" in their name, enter the wild-card character ***computer*.dwg**. Next, specify the desired settings and choose the **OK** button.

Similarly, you can add a single drawing for the data extraction by choosing the **Add Drawings** button of the **Define Data Source** page of the **Data Extraction** wizard.

Select the **Include current drawing** check box in the **Define Data Source** page to add the current drawing for the data extraction including the folders and drawings.

Select objects in the current drawing

This radio button enables you to add the selected object from the current drawing for the data extraction. If you select this radio button, the **Select objects in the current drawing** button will be enabled. Choose this button to select the objects from the drawing.

Drawing files and folders

This area of the dialog box displays the tree view of all the drawing and folder names with their locations that are selected for the data extraction. All the drawing and folders are displayed under the **Folders** and **Drawings** heads respectively. Only the checked drawings under the folder are included for the data extraction. You can also remove the selected drawing or folder from the list of drawing to be used for data extraction. To remove the file or folder from the list of files and folders to be used for the data extraction, select them from the **Drawing files and folders** area and choose the **Remove** button.

Settings

Choose the **Settings** button from the **Define Data Source** page to invoke the **Data Extraction - Additional Settings** dialog box, as shown in Figure 17-22. This dialog box has two areas: **Extraction settings** and **Extract from**. By default, the **Extract objects from blocks** and **Extract objects from xrefs** check boxes are selected from the **Extraction settings** area. The first two check boxes allow you to specify whether you want to extract the block attribute information from the external reference drawings or the nested blocks, or both. If you select the **Include xrefs in block counts** check box, then the xrefs will automatically be included while counting the blocks. The two radio buttons

*Figure 17-22 The **Data Extraction - Additional Settings** dialog box*

available in the **Extract from** area allow you to control the settings for counting the blocks. If you select the **Objects in model space** radio button, only the blocks currently inserted into the model space will be counted. If you select the **All objects in drawing** radio button, all the blocks from the current drawing will be counted. Choose the **OK** button to exit the **Data Extraction - Additional Settings** dialog box. Choose the **Next** button from the **Define Data Source** page; the **Select Objects** page of the **Data Extraction** wizard will be displayed.

Select Objects

The **Select Objects** page, as shown in Figure 17-23, allows you to filter out the objects to be included for data extraction. The objects that are checked in the **Objects** list box will only be considered for data extraction. The objects that are present in the current drawing are checked by default. Apart from the check box column, three more columns namely, **Object**, **Display Name**, and **Type** are displayed in the **Objects** list box. These columns display the information about the objects that are available to be selected and their type based on blocks and non-blocks.

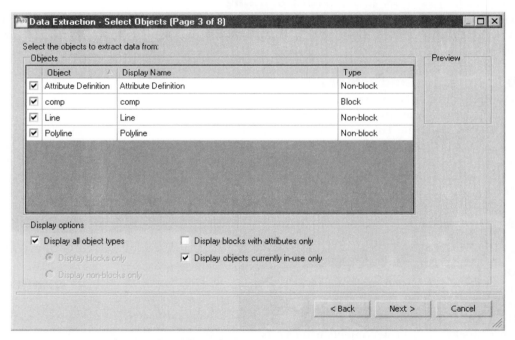

*Figure 17-23 The **Select Objects** page of the **Data Extraction** wizard*

The **Display options** area is used to control the objects to be displayed in the **Objects** list box. You can sort out the display from the blocks and non-blocks, blocks having attributes, and objects used in the current drawing. After specifying the objects to be displayed, choose the **Next** button from the **Select Objects** page to move to the **Select Properties** page.

Tip
*You can reverse the display order of the entities by clicking on the respective column head in the **Objects** list box. The direction of the arrow displayed at the column head of the selected column represents whether the data is arranged in the ascending or the descending order.*

Select Properties

This page, as shown in Figure 17-24, allows you to select the properties of the objects to be included for data extraction. Only the properties that are checked in the **Properties** list box are considered for the data extraction. Apart from the check box column, three more columns namely, **Property**, **Display Name**, and **Category** are displayed in the **Properties** list box. These columns display the information about the available properties and their category.

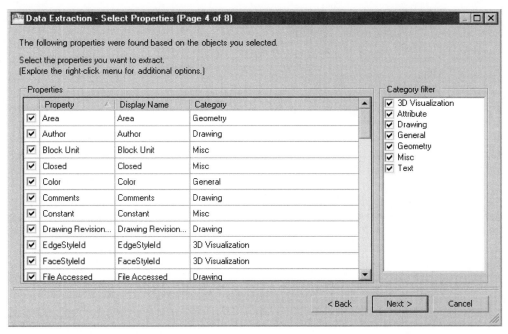

*Figure 17-24 The **Select Properties** page of the **Data Extraction** wizard*

The **Category filter** area is used to control the properties further to be displayed in the **Properties** list box. To do so, select the categories of the properties to be included in the **Properties** list box. For example, if you select the **Attribute** check box in the **Category filter** area, only the information about the attributes defined for the drawing will be extracted and displayed in the **Properties** list box. All the properties and categories displayed here are similar to the one displayed in the **Properties** palette.

You can also edit the names below the **Display Name** column. Select the row for which you want to change the display name and double-click on the **Display Name** column or right-click to display the shortcut menu. Next, choose the **Edit Display Name** option from the shortcut menu and enter a new name. After specifying the properties, choose the **Next** button from the **Select Properties** page to display the **Refine Data** page.

Refine Data

This page, as shown in Figure 17-25, allows you to format the data extracted from the drawing.

The properties selected from the **Select Properties** page of the **Data Extraction** wizard are displayed as the column head on the **Refine Data** page and the extracted data is displayed under these column heads of the table.

*Figure 17-25 The **Refine Data** page of the **Data Extraction** wizard*

The **Combine identical rows** check box is used to merge the rows having the same data. The sum of the merged rows having the same data will be displayed below the **Count** column head. The **Show count column** check box is used to display the **Count** column head on the extracted data table. The **Show name column** is used to display the **Name** column head on the extracted data table. All the check boxes explained above are selected by default.

You can also add columns from an excel sheet to the data extracted from the drawing. Choose the **Link External Data** button; the **Link External Data** dialog box will be displayed, see Figure 17-26. The options in the **Link External Data** dialog box are discussed next.

Data link Area

This area is used to specify the excel sheet to be attached with the drawing. The **Choose the source of the external data** drop-down list displays all the external data links that are attached to the current drawing. Select the external data link from which you want to include the columns. To attach a new excel file to the drawing, choose the **Launch Data Link Manager** button; the **Data Link Manager** dialog box will be displayed on the screen. The other options in the **Data Link Manager** dialog box have been discussed earlier in Chapter 7 (*Creating Text and Tables*). The rest of the options in the **Link External Data** dialog box will be available as soon as you specify the external data to be linked with the drawing.

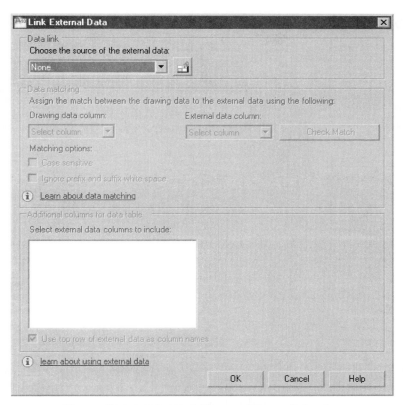

Figure 17-26 *The* **Link External Data** *dialog box*

Data matching Area

The options in this area are used to establish a relationship between the data extracted from the drawing and the attached external excel sheet. There must be at least one cell with the matching data in the selected column of the drawing data and the external data column respectively. This matching data in the external and internal data acts as a key cell to specify the proper insertion of the external data into the extracted data.

The **Drawing data column** drop-down list displays all the property column heads of the data extracted by the drawing. Select the column head from this drop-down list that is to be used for matching. The **External data column** displays the column names from the external data. Select the column name from where you want to insert data into the extracted data. Select the **Case sensitive** check box to check the case of the alphabets while matching the data of the selected columns from the extracted and external data. Select the **Ignore prefix and suffix white space** check box to ignore the blank spaces in the front and back of the entries in the columns that are used for matching.

Specify preferences for matching the data and choose the **Check Match** button to compare the data of the specified columns from the extracted and external data. If the data matching is successful, the **Valid Key** information box will be displayed, see Figure 17-27. Otherwise, the **Check Key** information box will be displayed on the screen suggesting the possible reasons of conflict between the two datum, see Figure 17-28. Unless there is a proper matching between the two datum, the merging of the external data into the extracted data will not take place.

Figure 17-27 *The **Valid Key** information box*

Figure 17-28 *The **Check Key** information box*

Additional columns for data table Area

This area displays all the columns of the external data. Select the check box on the left of each column entry to include that column into the data extraction. The **Use top row of external data as column names** check box will be used to specify the first row of the selected columns as the column head for the extracted data. Otherwise, the column heads are displayed as **Column 1**, **Column 2**, and so on. By default, this check box is selected.

Specify the preferences for the external data to be included in the current data extraction in the **Link External Data** dialog box and choose the **OK** button. The specified columns of the external data will be added to the extracted data and displayed in the **Refine Data** page of the **Data Extraction** wizard. The drawing data column used for matching and the external data column included in the extracted data will be represented with an external data link symbol. This symbol will be displayed on the side of the column head in the **Refine Data** page, as shown in Figure 17-29.

Figure 17-29 *The external data link symbol*

You can also change the display order of rows in the extracted data. To do so, choose the **Sort Columns Options** button from the **Refine Data** page; the **Sort Columns** dialog box will be displayed, see Figure 17-30. You can define the sort order for a single column or a group of columns at a time with this dialog box. The options in the **Sort Column** dialog box are discussed next.

The **Define sort order** list box has two columns with the **Column** and **Order** column heads. Click on the drop-down arrow below the **Column** head to display all the column head entries for the extracted data. Select the column for which you want to change the display order of

*Figure 17-30 The **Sort Columns** dialog box*

the rows. Choose the **Ascending** or **Descending** option from the **Order** drop-down list to specify the display order. Choose the **Add** button to add more rows in the **Define sort order** list box so as to specify the display order for other columns of the extracted data. The **Remove** button enables you to remove the selected sort order from the **Define sort order** list box. Choose the **Move Up** or **Move Down** button to move the selected row in the sort order, one level up or one level down respectively in the **Define sort order** list box.

Choose the **Full Preview** button; the preview will be displayed in the **Full Preview** window. A partial view of the **Full Preview** window is shown in Figure 17-31.

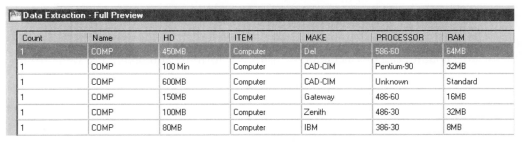

*Figure 17-31 The **Full Preview** window*

When you right-click on the table displayed on the **Refine Data** page, a shortcut menu will be displayed, see Figure 17-32. The options in the shortcut menu are discussed next.

Sort Descending/Sort Ascending

These options sort the data of the selected column in the descending/ ascending order.

Sort Columns Options

This option is used to invoke the **Sort Columns** dialog box, which has been discussed earlier.

Rename Column

This option is used to rename the column head of the selected
column.

Hide Column

This option is used to hide the selected column from the table
of the extracted data. The selected column is only removed
from the view, it is not deleted.

Show Hidden Columns

This option is used to display the hidden columns. This option
becomes available only when some columns are hidden from
the view. You can choose to display all the hidden columns or
only the selected one.

Set Column Data Format

This option is used to invoke the **Set Cell Format** dialog box.
You can set the data format for the selected row through this
dialog box.

Insert Formula Column

*Figure 17-32   The shortcut
menu for the* ***Refine Data*** *page*

This option is used to insert a formula column in the extracted
data. Choose the **Insert Formula Column** option to invoke
the **Insert Formula Column** dialog box, see Figure 17-33. You can define and insert a formula
column using this dialog box. The entries in the **Formula** column are governed by the
mathematical formula defined by using one or more columns. The entry in a particular cell
of the formula column can be calculated by using the formula and taking the value from the
same row under the column that has been used to define the formula. Enter the column head
name to be displayed on the formula column in the **Column name** edit box. The **Formula**
area is used to define the formula. To select the column to define the formula, double-click
on the column in the **Columns** list box, or drag and drop the selected column into the
Formula area. To use an operator such as +,-,*,and / in the formula, choose the desired
operator button from the **Formula** area. You can also make entry in the **Formula** area through
the keyboard. Choose the **Validate** button to check the validity of the formula. If the
defined formula is valid, AutoCAD will display the **Valid Equation** message window, as shown
in Figure 17-34, and if the formula defined is invalid, the **Invalid Equation** message window
will be displayed, as shown in Figure 17-35. The **Formula** column is always inserted to the
right of the selected column and is indicated by a formula symbol on the side of the column
head name, see Figure 17-36.

Edit Formula Column

This option is used to edit the pre-defined formula. Choose this option to invoke the **Edit
Formula Column** dialog box.

Figure 17-33 *The **Insert Formula Column*** *dialog box*

Figure 17-34 *The **Valid Equation** message window*

Figure 17-35 *The **Invalid Equation** message window*

Figure 17-36 *The **Formula** and the **Filter** symbols*

Remove Formula Column

This option is used to delete the formula column from the extracted data. The **Edit Formula Column** and the **Remove Formula Column** options are available in the shortcut menu only when you right-click on the formula column.

Combine Record Mode

This option is used to control the display of the numeric data into the selected columns by displaying them either as separate values or collapsing the identical property rows into one and displaying their total sum. If any formula column is added and you apply the **Combine Record Mode** option to the formula column, then in case of the **Separate Values** sub-option, the formula will be applied after combining the identical rows. But in case of **Sum Values** sub-option, the formula will be applied to the individual rows and then the identical rows will be combined together and the resultant value will be displayed as the sum of those individual columns. This option will be available only when the selected columns have numerical data and the **Combined identical rows** check box is selected.

Show Count Column/Show Name Column

These options are used to display the **Count** column and the **Name** column, respectively. These options are selected by default.

Insert Totals Footer

This option is used to add a footer row under the selected column. This row can display the sum, maximum or minimum value in the selected column, and the average of all values above the footer row in that column. This option is only available for the columns that have numerical data.

Remove Totals Footer

This option is used to remove the totals footer from the selected row.

Filter Options

This option is used to decide the filter criterion for the selected column. Choose this option to invoke the **Filter Column** dialog box, see Figure 17-37. Only the entries that fulfill the specified filter criterion will be displayed on the table. The column on which the filter is applied is represented by a **Filter** symbol on the side of the column name, refer to Figure 17-36.

Figure 17-37 The **Filter Column** *dialog box*

Reset Filter/Reset All Filters

These options are used to remove the applied filters from the selected columns or from all the columns of the table at a time respectively.

Copy to Clipboard

This option is used to copy the selected table to the clipboard.

After formatting the data from the extracted files, choose the **Next** button to display the **Choose Output** page.

Choose Output

The **Choose Output** page, as shown in Figure 17-38, allows you to specify the format of the output file and the location where to save it. Select the **Insert data extraction table into drawing** check box to insert the extracted data into the drawing as a table. You will be prompted to specify the insertion point of the table when the wizard is finished. Select the **Output data to external file (.xls .csv .mdb .txt)** check box to save the extracted data into a separate file

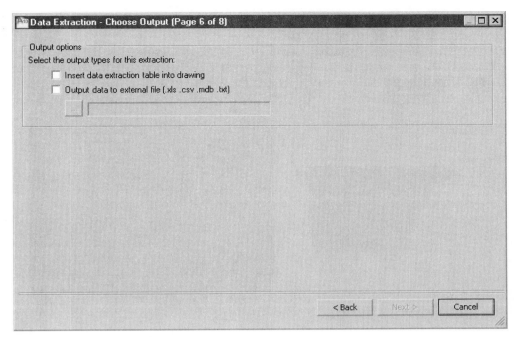

*Figure 17-38 The **Choose Output** page of the **Data Extraction** wizard*

at the end of the wizard. Select the **Browse [...]** button to specify the location to save this file. Choose the **Next** button from the **Choose Output** page to move to the **Table Style** page.

Table Style

The **Table Style** page, as shown in Figure 17-39, will be available only if the **Insert data extraction table into drawing** check box is selected from the **Choose Output** page. The various options in the **Table Style** page are discussed next.

Table style Area

You can select the table style from the **Select the table style to use for the inserted table** drop-down list. By default, only the **Standard** table style is available. To create a new table style, choose the **Table Style** button on the right of the **Select the table style to use for the inserted table** drop down list; the **Table style** dialog box will be displayed. You can use this dialog box to create a new table style or modify an existing table style to be used for the extracted attributes.

Formatting and structure Area

The **Use table in table style for label rows** radio button will be available only when the table style that you have specified contains a table template. Select this radio button to insert the extracted data into the pre-defined tables with the pre-defined header and footers. Select the **Manually setup table** radio button to enter the title, header, and data cell styles manually. Enter the title for the table in the **Enter a title for your table** text box. Select the title style

*Figure 17-39 The **Table Style** page of the **Data Extraction** wizard*

defined in the specified table styles in the **Title cell style** drop-down list. Similarly, you can specify the header and data cell style in their respective drop-down lists. The **Use property names as additional column headers** check box is selected by default. This option enables you to include the column headers into the table and use the displayed name as the header row. The preview of the table style is displayed on the right of the **Table Style** page. After specifying the **Table Style** formatting and structure, choose the **Next** button to move to the **Finish** page.

Finish

The **Finish** page, as shown in Figure 17-40, completes the extraction of data. Choose the **Finish** button and specify the insertion point in the drawing area to insert the extracted data. Note that this is possible only if the **Insert data extraction table into drawing** check box is selected on the **Choose Output** page of the **Data Extraction** wizard.

The ATTEXT Command for Attribute Extraction

Command: ATTEXT

The **ATTEXT** command allows you to use the **Attribute Extraction** dialog box (Figure 17-41) for extracting attributes. The information about the **File Format**, **Template File**, and **Output File** must be entered in the dialog box to extract the defined attribute. Also, you must select the blocks whose attribute values you want to extract. If you do not specify a particular block, all the blocks in the drawing are used.

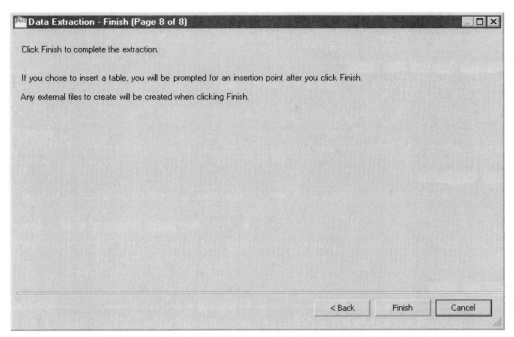

*Figure 17-40 The **Finish** page of the **Data Extraction** wizard*

File Format Area

This area of the dialog box lets you select the file format. You can select either of the three radio buttons available in this area. They are **Comma Delimited File (CDF)**, **Space Delimited File (SDF)**, and **DXF Format Extract File (DXX)**. The format selection is determined by the application that you plan to use to process data. Both CDF and SDF formats can be used with the database software. All of these formats are printable.

*Figure 17-41 The **Attribute Extraction** dialog box*

Comma Delimited File (CDF). When you select this radio button, the extracted attribute information is displayed in a CDF format. Here, each character field is enclosed in single quotes, and the records are separated by a delimiter (comma by default). A CDF file is a text file with the extension *.txt*. Of the three formats available, this is the most cumbersome.

Space Delimited File (SDF). In SDF format, the records are of fixed width as specified in the template file. The records are separated by spaces and the character fields are not enclosed in single quotes. The SDF file is a text file with the extension *.txt*. This file format is the most convenient and easy to use.

DXF Format Extract File (DXX). If you select this file format, you will notice that the **Template File** button and edit box in the **Attribute Extraction** dialog box are not available. This is because extraction in this file format does not require any template. The file created by this option contains only block references, attribute values, and end-of-sequence objects. This is the most complex of the three file formats available and is related to programming. The extension of these files is *.dxx*.

Select Objects

Select the blocks with attributes whose attribute information you want to extract. You can use any object selection method. Once you have selected the blocks that you want to use for attribute extraction, right-click or press ENTER. The **Attribute Extraction** dialog box is redisplayed and the number of objects you have selected is displayed adjacent to **Number found**.

Template File

When you choose the **Template File** button, the **Template File** dialog box is displayed, where you are allowed to select a template file that has been defined already. After you have selected a template file, choose the **Open** button to return to the **Attribute Extraction** dialog box. The name of the selected file is displayed in the **Template File** edit box. The template file is saved with the extension of the file as *.txt*. The following are the fields that you can specify in a template file (the comments given on the right are for explanation only; they must not be entered with the field description):

BL:LEVEL	Nwww000	(Block nesting level)
BL:NAME	Cwww000	(Block name)
BL:X	Nwwwddd	(X coordinate of block insertion point)
BL:Y	Nwwwddd	(Y coordinate of block insertion point)
BL:Z	Nwwwddd	(Z coordinate of block insertion point)
BL:NUMBER	Nwww000	(Block counter)
BL:HANDLE	Cwww000	(Block's handle)
BL:LAYER	Cwww000	(Block insertion layer name)
BL:ORIENT	Nwwwddd	(Block rotation angle)
BL:XSCALE	Nwwwddd	(X scale factor of block)
BL:YSCALE	Nwwwddd	(Y scale factor of block)
BL:ZSCALE	Nwwwddd	(Z scale factor of block)
BL:XEXTRUDE	Nwwwddd	(X component of block's extrusion direction)
BL:YEXTRUDE	Nwwwddd	(Y component of block's extrusion direction)
BL:ZEXTRUDE	Nwwwddd	(Z component of block's extrusion direction)
Attribute tag		(The tag name of the block attribute)

The extract file may contain several fields. For example, the first field might be the item name and the second field might be the price of the item. Each line in the template file specifies one field in the extract file. Any line in a template file consists of the name of the field, the width of the field in characters, and its numerical precision (if applicable). For example, consider the case given next.

```
ITEM        N015002
BL:NAME     C015000
```

Where   **BL:NAME** ------ Field name
        Blankspaces --- Blank spaces (must not include the tab character)
        C ------------------ Designates a character field
        N ------------------ Designates a numerical field
        015 -------------- Width of field in characters
        002 -------------- Numerical precision

BL:NAME
or **ITEM**   Indicates the field names; can be of any length.

C   Designates a character field; that is, the field contains characters or
                        it starts with characters. If the file contains numbers or starts with
                        numbers, then C will be replaced by N. For example, **N015002**.

015   Designates a field that is fifteen characters long.

002   Designates the numerical precision. In this example, the numerical
                        precision is 2, or two places following the decimal. The decimal point
                        and the two digits following the decimal are **included in the field
                        width**. In the next example, (000), the numerical precision, is not
                        applicable because the field does not have any numerical value (the
                        field contains letters only).

After creating a template file, when you choose the **Template File** button, the **Template File** dialog box is displayed, where you can browse and select a template file (Figure 17-42).

Note
You can put any number of spaces between the field name and the character C or N (ITEM N015002). However, you must not use the tab characters. Any alignment in the fields must be done by inserting spaces after the field name.

In the template file, a field name must not appear more than once. The template file name and the output file name must be different.

The template file must contain at least one field with an attribute tag name because the tag names determine which attribute values are to be extracted and from which blocks. If several blocks have different block names but the same attribute tag, AutoCAD will extract attribute values from all the selected blocks. For example, if there are two blocks in the drawing with the attribute tag PRICE, then when you extract the attribute values, AutoCAD will extract the value from both blocks (if both blocks were selected). To extract the value of an attribute, the

*Figure 17-42 The **Template File** dialog box to select a template file*

tag name must match the field name specified in the template file. AutoCAD automatically converts the tag names and the field names to uppercase letters before making a comparison.

Output File

When you choose the **Output File** button, the **Output File** dialog box is displayed. You can select an existing file here, if you want the extracted or output file to be saved as an existing file. You can enter a name in the **File name** edit box in the **Output File** dialog box (Figure 17-43) and then choose the **Save** button, if you want to save the output file as a new file. By default, the output file has the same name as the drawing name. For example, a drawing named *Drawing1.dwg* will have an output file by the name *Drawing1.txt* by default. Once a name for the output file is specified, it is displayed in the **Output File** edit box in the **Attribute Extraction** dialog box. You can also enter the file name in this edit box. As discussed earlier, AutoCAD appends *.txt* file extension for CDF or SDF files and *.dxx* file extension for DXF files.

Note
*You can also use the **-ATTEXT** command to extract attributes using the command line. Here, you are prompted to specify a file format for the extract information and you can also select specific blocks to extract their attribute information. You can specify a template file using the **Select Template File** dialog box and an extract file using the **Create extract file** dialog box.*

*Figure 17-43 The **Output File** dialog box to save an output file*

Example 3   *General*

In this example, you will extract the attribute values defined in Example 2. Extract the values of NAME, COUNT, MAKE, PROCESSOR, HD, RAM, and ITEM. These attribute values must be saved in a *.txt* format named TEST and arranged, as shown in the table given next.

COMP	1	Del	586-60	450 MB	64 MB	Computer
COMP	1	CAD-CIM	Pentium-90	100 Min	32 MB	Computer
COMP	1	CAD-CIM	Unknown	600 MB	Standard	Computer
COMP	1	Gateway	486-60	150 MB	16 MB	Computer
COMP	1	Zenith	486-30	100 MB	32 MB	Computer
COMP	1	IBM	386-30	80 MB	8 MB	Computer

1. Open the drawing that you saved in Example 2.

2. Choose the **Link & Extraction > Extract Data** from the **Insert** tab of the **Ribbon**; the **Data Extraction** wizard is displayed. Select the **Create a new data extraction** radio button from the **Begin** page and choose the **Next** button.

3. Enter **TEST** in the **File name** edit box of the **Save Data Extraction As** dialog box and choose the **Save** button.

4. Select the **Drawings/Sheet set** radio button and make sure that the **Include current drawing** check box is selected on the **Define Data Source** page. Choose the **Next** button.

5. Clear all check boxes in the **Objects** column except the **COMP** object in the **Object** column and choose the **Next** button.

6. Clear all check boxes except the **Attribute** check box from the **Category filter** area of the **Select Properties** page of the **Data Extraction** wizard. Also, make sure that the check box on the left of the **HD**, **ITEM**, **MAKE**, **PROCESSOR**, and **RAM** property under the **Property** column are selected. Choose the **Next** button.

7. Arrange the display order of the columns according to the table displayed above, by dragging and dropping the column heads to the desired position in the **Refine Data** page of the **Data Extraction** wizard. Make sure that the **Show count column** and **Show name column** check box are selected. Choose the **Full Preview** button and check the data displayed in the **Data Extraction - Full Preview** window and match it with the table displayed above. Close the window and choose the **Next** button from the **Refine Data** page of the **Data Extraction** wizard.

8. Select the **Output data to external file (.xls .csv .mdb .txt)** check box and choose the **Browse [...]** button. Specify the location where you want to save the text file, select **Text/ Template/Extract File (*.txt)** option from the **Files of type** drop-down list, enter **TEST.txt** in the **File name** edit box, and choose the **Save** button. Choose the **Next** button from the **Choose Output** page.

9. Choose the **Finish** button from the **Finish** page of the **Data Extraction** wizard.

CONTROLLING ATTRIBUTE VISIBILITY

Ribbon: Attributes > Retain Display/Display All/Hide All	**Command:** ATTDISP

The **ATTDISP** command allows you to change the visibility of all attribute values. Normally, the attributes are visible unless they are defined invisible by using the **Invisible** mode. The invisible attributes are not displayed, but they are a part of the block definition. You can select any one of the options of the **ATTDISP** command to turn the display of the attributes completely on or off. If the attributes are created in the **Invisible** mode and you have selected the **Normal** option from the **ATTDISP** command, the attributes continue to be invisible. You can also select the options of the **ATTDISP** command from the **Attributes** panel in the **Insert** tab of **Ribbon**. The options that can be chosen are **Retain Display/ Display All/Hide All**. The prompt sequence for the **ATTDISP** command is given next.

Command: **ATTDISP**
Enter attribute visibility setting [Normal/ON/OFF] <Current>: *Specify an option and press ENTER.*

When you choose **ON**, all attribute values will be displayed, including the attributes that are defined with the **Invisible** mode. You can also do it by choosing **Insert > Attributes > Display All** from the **Ribbon.** If you choose **OFF**, all attribute values will become invisible. You can also do it by choosing **Insert > Attributes > Hide All** from the **Ribbon.** Similarly, if you choose **Normal**, AutoCAD will display the attribute values the way they are defined, that is, the attributes that were defined invisible will stay invisible and the attributes that were defined visible will be visible. You can also do this by choosing **Insert > Attributes > Retain Display** from the **Ribbon.**

In Example 2, the RAM attribute was defined with the **Invisible** mode. Therefore, the RAM values are not displayed with the block. If you want to make the RAM attribute values visible (Figure 17-44), choose **Attributes > Display All** from the **Ribbon.**

Figure 17-44 Using -ATTEDIT to change the attribute values

Tip
*After you have defined the attribute values and saved them with the block definition, it may be a good idea to use the **OFF** option of the **ATTDISP** command. By doing this, the drawing is simplified and also the regeneration time is reduced.*

EDITING BLOCK ATTRIBUTES

Ribbon:	Attributes > Edit Attribute > Single	**Command:**	EATTEDIT
Toolbar:	Modify II > Single		

 The block attributes can be edited using the **EATTEDIT** command.

Editing Attributes Using the Enhanced Attribute Editor

The **EATTEDIT** command allows you to edit the block attribute values through the **Enhanced Attribute Editor** dialog box (Figure 17-45). When you invoke this command, AutoCAD prompts you to select the block whose values you want to edit. After selecting the block, the **Enhanced Attribute Editor** dialog box is displayed. This dialog box allows you to change the

Attributes, Text Options, and **Properties** of the block attribute. This dialog box can also be invoked by double-clicking on the block or the attribute in the current drawing. The **Enhanced Attribute Editor** dialog box displays all the attributes of the selected block and their properties. You can change the properties of the attribute you want. If an attribute has been defined with the **Constant** mode, it is not displayed in the dialog box because a constant attribute value cannot be edited. The dialog box displays the following tabs.

Attribute Tab

The options under this tab are used to change the value of the attribute. Also, here the **Mode** of the attribute is not displayed. The other attributes of the block such as **Tag**, **Prompt**, and **Value** are displayed, as shown in Figure 17-45.

*Figure 17-45 The **Attribute** tab of the **Enhanced Attribute Editor** dialog box*

Text Options Tab

Under this tab, the options related to the text of attributes can be modified. This includes **Text Style, Justification, Height, Rotation, Backwards, Upside down, Width Factor, Oblique Angle, Annotative**, and **Boundary width**, as shown in Figure 17-46.

Properties Tab

This tab can be used to edit the properties of attributes such as **Layer, Linetype, Color, Lineweight**, and **Plot style**. Choose the **Properties** tab of **Enhanced Attribute Editor** to display the dialog box, as shown in Figure 17-47.

Editing Attributes Using the Edit Attributes Dialog Box

The **ATTEDIT** command is used to edit the block attribute values through the **Edit Attributes** dialog box. Invoke this command; AutoCAD prompts you to select the block whose values are to be edited. After selecting the block, the **Edit Attributes** dialog box (Figure 17-48) is displayed.

The dialog box is similar to the **Enter Attributes** dialog box and shows the prompts and the

Figure 17-46 *The* **Text Options** *tab of the* **Enhanced Attribute Editor** *dialog box*

Figure 17-47 *The* **Properties** *tab of the* **Enhanced Attribute Editor** *dialog box*

attribute values of the selected block. If an attribute has been defined with the **Constant** mode, it is not displayed in the dialog box because a constant attribute value cannot be edited. To make any changes, select the existing value and enter a new value in the corresponding edit box. After you have made the modifications, choose the **OK** button. The attribute values are updated in the selected block.

If a selected block has no attributes, AutoCAD will display the alert message **That block has no editable attributes**. Similarly, if the selected object is not a block, AutoCAD again displays the alert message **That object is not a block**.

Note
You cannot use the **ATTEDIT** *command to perform global editing of attribute values, or to modify position, height, or style of the attribute value. The global editing can be performed using the* **-ATTEDIT** *command, which is discussed in the next section.*

Figure 17-48 *The **Edit Attributes** dialog box*

Example 4   *General*

In this example, you will use the **EATTEDIT** command to change the attribute of the first computer (150 MB to 2.1 GB), which is located in Room-1.

1. Open the drawing that was created in Example 2. The drawing has six blocks with attributes. The name of the block is **COMP**, and it has five defined attributes, one of them is invisible. Zoom in so that the first computer is displayed on the screen (Figure 17-49).

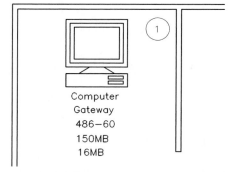

2. Invoke the **EATTEDIT** command. AutoCAD will prompt you to select a block. Select the computer located in Room-1.

Figure 17-49 *Zoomed view of the first computer*

AutoCAD will display the **Enhanced Attribute Editor** dialog box (refer to Figure 17-45) that shows the attribute prompts and the attribute values.

3. Edit the values, choose **Apply,** and then the **OK** button in the dialog box. When you exit the dialog box, the attribute values are updated.

Tip

*You can also edit attributes with the **FIND** command. This command can be invoked by entering **FIND** at the pointer input. When you invoke this command, the **Find and Replace** dialog box is displayed.*

*When you know that a set of attributes in a drawing may have to be changed in the future, you can make a group out of them by using the **GROUP** command. Then, you can choose the **Select objects** button in the **Find and Replace** dialog box and enter **G** at the **Select objects** prompt and enter the name of the group. All the objects in the group will get selected.*

Global Editing of Attributes

Ribbon: Insert > Attributes > Edit Attribute > Multiple   **Command:**   -ATTEDIT

The **-ATTEDIT** command allows you to edit the attribute values independently of the blocks that contain the attribute reference. For example, if there are two blocks, COMPUTER and TABLE, with the attribute value PRICE, you can globally edit this value (PRICE) independently of the block that references these values. You can also edit the attribute values one at a time. For example, you can edit the attribute value (PRICE) of the block TABLE without affecting the value of the other block, COMPUTER. When you enter the **-ATTEDIT** command, AutoCAD displays the following prompt:

Command: **-ATTEDIT**
Edit attributes one at a time? [Yes/No]<Y>: **N**
Performing global editing of attribute values

If you enter **N** at this prompt, it means that you want to do the global editing of the attributes. However, you can restrict the editing of attributes by block names, tag names, attribute values, and visibility of attributes on the screen.

Editing Visible Attributes Only

After you select global editing, AutoCAD will display the following prompt:

Edit only attributes visible on screen? [Yes/No] <Y>: **Y**

If you enter **Y** at this prompt, AutoCAD will edit only those attributes that are visible and displayed on the screen. The attributes might not have been defined with the **Invisible** mode, but if they are not displayed on the screen, they are not visible for editing. For example, if you zoom in, some of the attributes may not be displayed on the screen. Since the attributes are not displayed on the screen, they are invisible and cannot be selected for editing.

Editing All Attributes

If you enter **N** at the previously mentioned prompt, AutoCAD flips from graphics to text screen and displays the message stating: '**Drawing must be regenerated afterwards**'.

Now, AutoCAD will edit all attributes even if they are not visible or displayed on the screen. Also, changes that you make in the attribute values are not reflected immediately. Instead, the attribute values are updated and the drawing is regenerated after you are done with the command.

Editing Specific Blocks

Although you have selected global editing, you can confine the editing of attributes to specific blocks by entering the block name at the prompt. For example,

Enter block name specification <*>: **COMP**

When you enter the name of the block, AutoCAD will edit the attributes that have the given block (COMP) reference. You can also use the wild-card characters to specify the block names. If you want to edit attributes in all blocks that have attributes defined, press ENTER.

Editing Attributes with Specific Attribute Tag Names

Like blocks, you can confine attribute editing to those attribute values that have the specified tag name. For example, if you want to edit the attribute values that have the tag name MAKE, enter the tag name at the following AutoCAD prompt:

Enter attribute tag specification <*>: **MAKE**

When you specify the tag name, AutoCAD will not edit attributes that have a different tag name, even if the values being edited are the same. You can also use the wild-card characters to specify the tag names. If you want to edit attributes with any tag name, press ENTER.

Editing Attributes with a Specific Attribute Value

Like blocks and attribute tag names, you can confine attribute editing to a specified attribute value. For example, if you want to edit the attribute values that have the value 100 MB, enter the value at the following AutoCAD prompt:

Enter attribute value specification <*>: **100MB**

When you specify the attribute value, AutoCAD will not edit attributes that have a different value, even if the tag name and block specification are the same. You can also use the wild-card characters to specify the attribute value. If you want to edit attributes with any value, press ENTER.

Sometimes the value of an attribute is null, and these values are not visible. If you want to select the null values for editing, make sure you have not restricted the global editing to visible attributes. To edit the null attributes, enter \ at the following prompt:

Enter attribute value specification <*>: \

After you enter this information, AutoCAD will prompt you to select the attributes. You can select the attributes by selecting individual attributes or by using one of the object selection options (Window, Crossing, or individually).

Select Attributes: *Select the attribute values parallel to the current UCS only.*

After you select the attributes, AutoCAD will prompt you to enter the string you want to change and the new string. A string is a sequence of consecutive characters. It could also be a portion of the text. AutoCAD will retrieve the attribute information, edit it, and then update the attribute values.

> Enter string to change: *Enter the value which is to be modified.*
> Enter new string: *Enter the new value.*

The following is the complete command prompt sequence of the **-ATTEDIT** command. It is assumed that the editing is global and for visible attributes only.

> Command: **-ATTEDIT**
> Edit attributes one at a time? [Yes/No] <Y>: N
> Performing global editing of attribute values.
> Edit only attributes visible on screen? [Yes/No] <Y>: N
> Drawing must be regenerated afterwards.
> Enter block name specification <*>: Enter
> Enter attribute tag specification <*>: Enter
> Enter attribute value specification <*>: Enter
> n attributes selected.
> Enter string to change: *Enter the value to be modified.*
> Enter new string: *Enter the new value.*

Note
AutoCAD regenerates the drawing at the end of the command automatically unless the system variable **REGENAUTO** *is off, which controls automatic regeneration of the drawing.*

If you select an attribute defined with the **Constant** *mode using the* **-ATTEDIT** *command, the prompt displays 0 found, since attributes with* **Constant** *mode are not editable.*

Example 5   *General*

In this example, you will use the drawing from Example 2 to edit the attribute values that are highlighted in the following table. The tag names are given at the top of the table (ITEM, MAKE, PROCESSOR, HD, RAM). The RAM values are invisible in the drawing.

	ITEM	MAKE	PROCESSOR	HD	RAM
COMP	Computer	Gateway	486-60	150 MB	**16 MB**
COMP	Computer	Zenith	486-30	100 MB	**32 MB**
COMP	Computer	IBM	386-30	80 MB	**8 MB**
COMP	Computer	Del	586-60	450 MB	**64 MB**
COMP	Computer	**CAD-CIM**	Pentium-90	100 Min	**32 MB**
COMP	Computer	**CAD-CIM**	**Unknown**	600 MB	Standard

Make the following changes in the highlighted attribute values shown in the above table.

1.   Change Unknown to Pentium.

2. Change CAD-CIM to Compaq.

3. Change MB to Meg for all attribute values that have the tag name RAM. (No changes should be made to the values that have the tag name HD.)

The following is the prompt sequence to change the attribute value from **Unknown** to **Pentium**.

1. Enter the **-ATTEDIT** command at the Command prompt. At the next prompt, enter **N** and press ENTER.

    Command: **-ATTEDIT**
    Edit attributes one at a time? [Yes/No] <Y>: **N**
    Performing global editing of attribute values.

2. We want to edit only those attributes that are visible on the screen, so press ENTER at the following prompt:

    Edit only attributes visible on screen? [Yes/No] <Y>: 〔Enter〕

3. As shown in the table, the attributes belong to a single block, COMP. In a drawing, there could be more blocks. To confine the attribute editing to the COMP block only, enter the name of the block (COMP) at the next prompt.

    Enter block name specification <*>: **COMP**

4. At the next two prompts, enter the attribute tag name and the attribute value specification. When you enter these two values, only those attributes that have the specified tag name and attribute value will be edited.

    Enter attribute tag specification<*>: **Processor**
    Enter attribute value specification<*>: **Unknown**

5. Next, AutoCAD will prompt you to select attributes. Use any object selection option to select all the blocks. AutoCAD will search for the attributes that satisfy the given criteria (attributes belong to the block COMP, the attributes have the tag name Processor, and the attribute value is Unknown). Once AutoCAD locates such attributes, they will be highlighted.

6. At the next two prompts, enter the string you want to change, and then enter the new string.

    Enter string to change: **Unknown**
    Enter new string: **Pentium**

7. The following is the Command prompt sequence to change the make of the computers from **CAD-CIM** to **Compaq**.

Command: **-ATTEDIT**
Edit attributes one at a time? [Yes/No] <Y>: **N**
Performing global editing of attribute values.
Edit only attributes visible on screen? [Yes/No] <Y>: [Enter]
Enter block name specification <*>: **COMP**
Enter attribute tag specification <*>: **MAKE**
Enter attribute value specification <*>: [Enter]
Select Attributes: *Use any selection method to select the attributes.*
n attributes selected.
Select Attributes: [Enter]
Enter string to change: **CAD-CIM**
Enter new string: **Compaq**

8. The following is the Command prompt sequence to change **MB** to **Meg**.

Command: **-ATTEDIT**
Edit attributes one at a time? [Yes/No] <Y>: **N**
Performing global editing of attribute values.
At the next prompt, you must enter **N** because the attributes you want to edit (tag name, RAM) are not visible on the screen.

Edit only attributes visible on screen? [Yes/No] <Y>: **N**
Drawing must be regenerated afterwards.
Enter block name specification <*>: **COMP**
At the next prompt, about the tag specification, you must specify the tag name because the text string MB also appears in the hard drive size (tag name, HD). If you do not enter the tag name, AutoCAD will change all MB attribute values to Meg.

Enter attribute tag specification <*>: **RAM**
Enter attribute value specification <*>:
n Attributes selected
Enter string to change: **MB**
Enter new string: **Meg**

9. Choose **Attributes > Display All** from the **Ribbon** to display invisible attributes on the screen.

Note
*You can also use the **-ATTEDIT** command to edit the attribute values individually. Attribute value, position, height, angle, style, layer, and color can be changed using this command.*

Example 6   *General*

In this example, you will use the drawing of Example 5 to edit the attributes individually (Figure 17-50). Make the following changes in the attribute values:

Figure 17-50 Using -ATTEDIT to change the attribute values individually

a. Change the attribute value 100 Min to 100 MB.

b. Change the height of all attributes with the tag name RAM to 0.075 units.

1. Load the drawing that you had saved in Example 5.

2. Double-click on the **100Min** attribute value in the drawing. The **Enhanced Attribute Editor** dialog box is displayed.

3. In the **Value** edit box, change **100 Min** to **100MB**. Choose **Apply** and then choose **OK**.

4. To change the height of attribute text, double-click on the value of **RAM** in the drawing; the **Enhanced Attribute Editor** dialog box is displayed.

5. In the dialog box, choose the **Text Options** tab and change height to **0.075** in the **Height** edit box.

6. Choose **Apply** and then choose **OK**.

7. Repeat Steps 5 and 6 to change the height of the other attribute values.

Tip
*When you are defining attributes and have certain attributes whose values are not known at that time, you should enter AAAA or something similar. Later on you can replace such text with the values that you have obtained, using the **ATTEDIT** or **-ATTEDIT** commands. This is easier than adding an attribute later after the block has already been defined.*

REDEFINING A BLOCK WITH ATTRIBUTES

If a block with attributes has been inserted a number of times in a drawing and you wish to now edit the block geometry, or add or delete attributes from them, it may take a lot of time. You can use the **ATTREDEF** command to redefine an existing block with attributes, and once redefined, all the copies of the block with attributes get updated automatically. But before using this command, you have to explode a copy of the block containing the attributes and make the changes or you may use the new geometry. If you do not explode a block and try to use this command, an error message is displayed: **New block has no attributes**. The prompt sequence is as follows:

Command: **ATTREDEF**
Enter name of the block you wish to redefine: *Enter the name of the block to be redefined and press ENTER.*
Select objects for new Block...
Select objects: *Select the new block geometry with attributes and press ENTER.*
Specify insertion base point of the new Block: *Specify an insertion base point.*

After you select an insertion base point, the block is redefined and all the copies of the block in the drawing are updated. If you do not include an existing attribute when selecting objects for the new block, it will not be a part of the new block definition.

Note
*The **ATTREDEF** command removes all the formatting changes incorporated into the property by the **ATTEDIT** or **EATTEDIT** command.*

In-place Editing of Blocks with Attributes

Ribbon:	Insert > Reference > Edit Reference	**Command:**	REFEDIT
Toolbar:	Refedit > Edit Reference In-Place		

The **REFEDIT** command has already been discussed earlier in Chapter 16, Working with Blocks. This command can also be used for in-place editing of blocks with attributes. After you invoke the **REFEDIT** command, select a block with attributes that you want to edit. The **Reference Edit** dialog box is displayed with the name of the block in the **Reference name** list box. To be able to edit attributes, you should select the **Display attribute definitions for editing** check box in the **Settings** tab of the dialog box and then choose **OK** to exit. When you select this check box, it displays the attributes and makes them available for editing. At the next prompt, **Select nested objects**, you can select any part of the block with attributes that need to be edited and press ENTER. If you want to edit some other objects that are part of the block, you can select them too. All the selected objects become a part of the working set. You will notice that when you select the block with attributes, the tags replace the assigned attribute values and objects in the drawing that were not selected appear faded.

You can now edit the block geometry or edit only the attributes. Enter the **DDEDIT** command at the Command prompt and select the attribute to be edited at the next prompt; the **Edit**

Attribute Definition dialog box is displayed. Make changes in the values displayed in the **Tag**, **Prompt**, and **Default** edit boxes and choose **OK** to exit the dialog box. You can then select another attribute to edit and so on until you have made all the modifications in the attributes you intended and then press ENTER to exit the command. Now, choose the **Save Changes** button in the **Edit Reference** panel of the **Ribbon**. An AutoCAD message box is displayed prompting you whether you want to save the changes made or cancel the command and discard the changes. Choose **OK** to save the completed changes and exit the **REFEDIT** command.

Once the block has been redefined, the blocks that already exist in the drawing are not updated, unless a change is made to a default value of an attribute that has been defined with the **Constant** mode. The modifications made using the **REFEDIT** command can be seen in the blocks inserted later.

Note
*You can also use the **-REFEDIT** command to change block attributes. The prompts are displayed at the pointer and command line. All the options available through the **REFEDIT** command are available here also. The **DDEDIT** command is used to make modifications in the attribute tags. Then, you can either choose the **Save Changes** button in the **Edit Reference** panel of the **Ribbon** or enter the **REFCLOSE** command to save and exit the command.*

INSERTING TEXT FILES IN THE DRAWING

Ribbon:	Annotate > Text > Multiline Text	Command:	MTEXT
Menu Bar:	Draw > Text > Multiline Text		

After you have extracted attribute information into a file, you may want to insert this text into a drawing. You can insert this text file in a drawing by using the **Import Text** option that is displayed when you right-click in the text window of the **In-Place Text Editor**. The **In-Place Text Editor** is displayed using the **MTEXT** command.

After you have invoked the **MTEXT** command, AutoCAD prompts you to enter the insertion point and other corner of the paragraph text box, within which the text file will be placed. After you specify these points, the **In-Place Text Editor** is displayed. To insert the text file *TEST.txt* (created in Example 3), right-click in the text window of the **In-Place Text Editor** to display the shortcut menu. In this shortcut menu, choose **Import Text**. AutoCAD displays the **Select File** dialog box.

In the **Select File** dialog box, you can select the text file **TEST** from the list box and then choose the **Open** button. The imported text is displayed in the text window of the **In-Place Text Editor** (Figure 17-51). Note that only the ASCII files are properly interpreted. Now, click on the screen, outside the text editor, to get the imported text in the selected area on the screen (Figure 17-52).

You can also use the **MText** command to change the text style, height, direction, width, rotation, line spacing, and attachment.

Figure 17-51 The **In-Place Text Editor** *displaying the imported text*

Block Name	Count	MAKE	PROCESSOR	HD	RAM	ITEM
COMP	1	Gateway	486−60	150MB	16MB	Computer
COMP	1	Zenith	486−30	100MB	32MB	Computer
COMP	1	IBM	386−30	80MB	8MB	Computer
COMP	1	Del	586−60	450MB	64MB	Computer
COMP	1	CAD−CIM	Pentium−90	100Min	32MB	Computer
COMP	1	CAD−CIM	Unknown	600MB	Standard	Computer

Figure 17-52 The imported text file on the screen

Self-Evaluation Test

Answer the following questions and then compare them to those given at the end of this chapter:

1. Like the **Constant** attribute, the **Preset** attribute cannot be edited. (T/F)

2. In case of tag names, any lowercase letter is automatically converted to uppercase. (T/F)

3. You can use the **EATTEDIT** command to modify the justification, height, or style of the attribute value. (T/F)

4. If you select the **ON** option of the **ATTDISP** command, even the attributes defined with the **Invisible** mode are displayed. (T/F)

5. The entry in the **Value** edit box of the **Attribute Definition** dialog box defines the _____ of the specified attribute.

6. If you have selected the **Align** option from the **Justification** drop-down list in the **Text Options** tab of the **Enhanced Attribute Editor** dialog box, the **Height** and **Rotation** edit boxes will be _____.

7. You can use the _____ and _____ commands or the **Properties** palette to edit text or attribute definitions.

8. The default value of the **ATTDIA** variable is _____, which disables the dialog box.

9. In the _____ file, the records are not separated by a comma and the character fields are not enclosed in single quotes.

10. The _____ command is used to redefine an existing block with attributes in a drawing, and once redefined, all copies of the block in the drawing are updated.

Review Questions

Answer the following questions:

1. If you do not enter anything in the **Prompt** edit box, the entry made in the **Tag** edit box will be used as prompt. (T/F)

2. You can also use the **Find and Replace** dialog box to modify attribute values. (T/F)

3. When using the **REFEDIT** command for in-place editing of attributes, the existing blocks in a drawing do not get updated. Only the subsequent insertions display the modified block. (T/F)

4. You can use ? and * in the string value. When these characters are used in string values, AutoCAD does not interpret them as wild-card characters. (T/F)

5. The template file name and the output file name can be the same. (T/F)

6. Not selecting any of the check boxes in the **Mode** area of the **Attribute Definition** dialog box displays all the prompts at the command line and the values will be visible on the screen. This is also referred to as which mode?

    (a) Formal   (b) Normal
    (c) Abnormal   (d) None of the above

7. Which of the following system variables when set to 0 will suppress the display of prompts for the new values?

    (a) **ATTDIA**   (b) **ATTREQ**
    (c) **ATTMODE**   (d) **ATTDEF**

8. Selecting the **OFF** option of which command turns off the visibility of all attribute values?

    (a) **ATTDISP**   (b) **ATTEXT**
    (c) **ATTEDIT**   (d) **ATTDEF**

9. AutoCAD regenerates the drawing at the end of the **-ATTEDIT** command, unless which of the following is turned off?

    (a) **AUTOSNAP**   (b) **REGENAUTO**
    (c) **ATTDIA**   (d) **ATTMODE**

10. If you need a leading blank space in the string to be changed, the character string must start with which one of the following characters?

(a) space ( )   (b) backlash (\)
(c) asterisk (*)   (d) colon (:)

11. You can insert the text file in the drawing by choosing the _____ button in the **In-Place Text Editor** dialog box displayed when using the **MTEXT** command.

12. The function of the **Preset** option is _____.

13. If you select the **Constant** check box in the **Mode** area of the **Attribute Definition** dialog box, the **Prompt** edit box will _____.

14. You should select the _____ check box in the **Attribute Definition** dialog box to automatically place the subsequent attribute text just below the previously defined attribute.

15. The attribute extract file is saved as a _____ file.

Exercises

Exercise 1   *Electronics*

In this exercise, you will define the following attributes for a resistor and then create a block using the **BLOCK** command. The name of the block is **RESIS**.

Mode	Tag name	Prompt	Default value
Verify	RNAME	Enter name	RX
Verify	RVALUE	Enter resistance	XX
Verify, Invisible	RPRICE	Enter price	00

1. Draw the resistor, as shown in Figure 17-53.

2. Enter **ATTDEF** at the AutoCAD Command prompt to invoke the **Attribute Definition** dialog box.

3. Define the attributes, as shown in the preceding table, and position the attribute text as shown in Figure 17-53.

4. Use the **BLOCK** command to create a block. The name of the block is **RESIS**, and the insertion point of the block is at the left end of the resistor. When you select the objects for the block, make sure you also select the attributes.

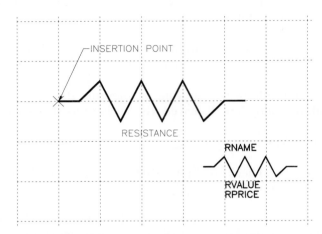

Figure 17-53 *Drawing of a resistor for Exercise 1*

Exercise 2   *Electronics*

In this exercise, you will use the **INSERT** command to insert the block that was defined in Exercise 1 (RESIS). The following is the list of the attribute values for the resistances in the electric circuit.

RNAME	RVALUE	RPRICE
R1	35	.32
R2	27	.25
R3	52	.40
R4	8	.21
RX	10	.21

1. Draw the electric circuit diagram, as shown in Figure 17-54 (assume the dimensions).

2. Set the system variable **ATTDIA** to 1. Use the **INSERT** command to insert the blocks, and define the attribute values in the **Attribute Definition** dialog box.

3. Repeat the **INSERT** command to insert other blocks, and define their attribute values as given in the table. Save the drawing as *attexr2.dwg* (Figure 17-55).

Figure 17-54 *Electric circuit diagram without resistors for Exercise 2*

Figure 17-55 *Electric circuit diagram with resistors for Exercise 2*

Exercise 3 *Electronics*

In this example, you will extract the attribute values that were defined in Exercise 2. Extract the values of RNAME, RVALUE, and RPRICE. These attribute values must be saved in a Tab Delimited File format named **RESISLST** and arranged, as shown in the following table.

RESIS	1	R1	35	.32
RESIS	1	R2	27	.25
RESIS	1	R3	52	.40
RESIS	1	R4	8	.21
RESIS	1	RX	10	.21

1. Load the drawing **ATTEXR2** that you saved in Exercise 2.

2. Use the **EATTEXT** command to invoke the **Data Extraction** wizard, and select the **Select Objects** radio button. Now, choose **Select Objects** button and select the objects in the current drawing, whose attributes have to be extracted. Now, choose **Next** in the wizard.

3. On the **Additional Settings** page, clear both the check boxes of Xrefs and Nested blocks and choose **OK** and **Next**.

4. On the **Use Template** page, select the **No Template** radio button and choose **Next**.

5. On the **Select Attribute** page, under the Block Name heading, select the check box named **RESIS**. In the attribute list, clear all the check boxes and select **RNAME, RVALUE**, and **RPRICE** check boxes. Choose **Next**.

6. **The View output** page shows the result of your query. Here, choose the view you want the data to be displayed in the exported file. Select the Next button.

7. Ignore the **Save template** page and choose **Next**.

8. On the **Export** page, enter the name of the output file, **RESISLST**, and save it in **Tab Delimited File** format. Choose **Finish**. Use Notepad to list the output file, **RESISLST**. The output file will be similar to the file shown at the beginning of Exercise 3.

Exercise 4   *Electronics*

In this exercise, you will use the **EATTEDIT** command to change the attributes of the resistances that are highlighted in the following table. You will also extract the attribute values and insert the text file in the drawing.

1. Load the drawing **ATTEXR2** that was created in Exercise 2. The drawing has five resistances with attributes. The name of the block is RESIS, and it has three defined attributes, one of them invisible.

2. Use the **EATTEDIT** command to edit the values that are highlighted in the following table.

RESIS	R1	**40**	.32
RESIS	R2	**29**	.25
RESIS	R3	52	**.45**
RESIS	R4	8	**.25**
RESIS	**R5**	10	.21

3. Extract the attribute values and write the values to a text file.

4. Use the **MTEXT** command to insert the text file in the drawing.

Exercise 5   *Electronics*

Use the information given in Exercise 3 to extract the attribute values, and write the data to the output file. The data in the output file should be Comma Delimited CDF. Use the **ATTEXT** and **-ATTEXT** commands to extract the attribute values.

Exercise 6   *Electronics*

In this exercise, you will create the blocks with the required attributes. Next, draw the circuit diagram shown in Figure 17-56 and then extract the attributes to create a Bill of Materials.

Figure 17-56 Drawing of the circuit diagram for Exercise 6

Exercise 7   *Electronics*

In this exercise, you will create the blocks with the required attributes. Next, draw the circuit diagram shown in Figure 17-57 and then extract the attributes to create a Bill Of Materials.

Figure 17-57 *Drawing for Exercise 7*

Answers to Self-Evaluation Test
1. F, **2.** T, **3.** T, **4.** T, **5.** default value, **6.** disabled, **7.** **EATTEDIT**, **CHANGE**, **8.** 0, **9.** Space Delimited, **10.** **ATTREDEF**

Chapter *18*

Understanding External References

Learning Objectives

After completing this chapter, you will be able to:
* *Understand External References and their applications.*
* *Understand dependent symbols.*
* *Use the **XREF** command and its options.*
* *Use the **Attach**, **Unload**, **Reload**, **Detach**, and **Bind** options.*
* *Edit the path of an xref.*
* *Understand the difference between the **Overlay** and **Attachment** options.*
* *Understand the use of the **ATTACH** command.*
* *Work with Underlays*
* *Use the **DesignCenter** to attach a drawing as an xref.*
* *Use the **XBIND** command to add dependent symbols.*
* *Use the **CLIP** command to clip xref drawings.*
* *Understand demand loading.*
* *Use the **REFEDIT** command for In-place editing.*

EXTERNAL REFERENCES

The external reference feature allows you to reference an external drawing without making that drawing a permanent part of the existing drawing. For example, assume that we have an assembly drawing ASSEM1 that consists of two parts, SHAFT and BEARING. The SHAFT and BEARING are separate drawings created by two CAD operators or provided by two different vendors. We want to create an assembly drawing from these two parts. One way to create an assembly drawing is to insert these two drawings as blocks by using the **INSERT** command. Now assume that the design of BEARING has changed due to customer or product requirements. To update the assembly drawing, you have to make sure that you insert the BEARING drawing after the changes have been made. If you forget to update the assembly drawing, then the assembly drawing will not reflect the changes made in the piece part drawing. In a production environment, this could have serious consequences.

You can solve this problem by using the **external reference** facility, which lets you link the piece part drawings with the assembly drawing. If the xref drawings (piece part) get updated, the changes are automatically reflected in the assembly drawing. This way, the assembly drawing stays updated, no matter when the changes were made in the piece part drawings. There is no limit to the number of drawings that you can reference. You can also have **nested references**. For example, the piece part drawing BEARING could be referenced in the SHAFT drawing, then the SHAFT drawing could be referenced in the assembly drawing ASSEM1. When you open or plot the assembly drawing, AutoCAD automatically loads the referenced drawing SHAFT and the nested drawing BEARING. When using External References, several people working on the same project can reference the same drawing and all the changes made are displayed everywhere the particular drawing is being used.

If you use the **INSERT** command to insert the piece parts, the piece parts become a permanent part of the drawing, and therefore, the drawing has a certain size. However, if you use the external reference feature to link the drawings, the piece part drawings are not saved with the assembly drawing. AutoCAD only saves the reference information with the assembly drawing; therefore, the size of the drawing is minimized. Like blocks, the xref drawings can be scaled, rotated, or positioned at any desired location, but they cannot be exploded. You can also use only a part of the Xref by making clipped boundary of Xrefs.

Tip
External referenced drawings are useful for creating parts or subassemblies and then putting them together in one drawing to create the main assembly. You can also use it for laying out the contents of a drawing with multiple views before plotting.

DEPENDENT SYMBOLS

When you use the **INSERT** command to insert a drawing, the information about the named objects is lost, if names are duplicated. However. if they are unique, they are imported. The named objects are entries such as blocks, text styles, and layers. For example, if the assembly drawing has a layer Hidden of green color and HIDDEN linetype, and the piece part Bearing has also a layer Hidden of blue color and HIDDEN2 linetype, the values set in the assembly drawing will override the values of the inserted drawing when the Bearing drawing is inserted in the assembly drawing (Figure 18-1). As a result, in the assembly drawing, the layer Hidden will retain green color and HIDDEN linetype, ignoring the layer settings of the inserted drawing. Only those layers that have the same names are affected. The remaining layers that have different layer names are added to the current drawing.

Figure 18-1 *Partial view of the layer settings of the current drawing in the* ***Layer Properties Manager*** *dialog box*

In the xref drawings, the information about the named objects is not lost because AutoCAD will create additional named objects such as specified layer settings, as shown in Figure 18-2. For xref drawings, these named objects become dependent symbols (features such as layers, linetypes, object color, text style, and so on).

Figure 18-2 *Xref creates additional layers*

The layer Hidden of the xref drawing (c26d1) is appended with the name of the xref drawing Bearing, and the two are separated by the vertical bar symbol (|). The names of these layers appear in light gray color in the **Layer Properties Manager** or **Layer** drop-down list of the **Layers** panel in the **Home** tab of the **Ribbon**. These layers can neither be selected nor be made current. The layer name Border changes to c26d1|Border. Similarly, CEN is renamed as c26d1|CEN and OBJ is renamed as C26d1|OBJ (Figure 18-2). The information added to the current drawing is not permanent. It is added only when the xref drawing is loaded. If you detach the xref drawing, the dependent symbols are automatically erased from the current drawing.

When you xref a drawing, AutoCAD does not let you reference the symbols directly. For example, you cannot make the dependent layer, c26d1|Border, as current layer. Therefore, you cannot add any objects to that layer. However, you can change the color, linetype, lineweight, plotstyle, or visibility (on/off, freeze/thaw) of the layer in the current drawing. If the **Retain changes to Xref layers** check box in the **External References (Xrefs)** area of the **Open and Save** tab of the **Options** dialog box is cleared, which also implies that the system variable **VISRETAIN** is set to 0, the settings are retained only for the current drawing session. This means that when you save and exit the drawing, the changes are discarded and the layer settings return to their default status. If this check box is selected (default), which also implies that the **VISRETAIN** variable is set to 1, layer settings such as color, linetype, on/off, and freeze/thaw are retained, and they are saved with the drawing and used when you open the drawing the next time. Whenever you open or plot a drawing, AutoCAD reloads each Xref in the drawing and as a result, the latest updated version of the drawing is loaded automatically.

Note
*You cannot make the xref-dependent layers current in a drawing. Only when the xref drawing is bounded to the current drawing using the **XBIND** command, you can make the xref-dependent layers a permanent part of the current drawing and use them. The **XBIND** command will be discussed later in this chapter.*

MANAGING EXTERNAL REFERENCES IN A DRAWING

Ribbon:	View > Palettes > External References Palette	
Menu Bar:	Insert > External References	**Command:** XREF

When you choose the **External References** button from the **Palettes** panel (Figure 18-3), AutoCAD displays the **External References** palette (Figure 18-4). The **External References** palette displays the status of each Xref in the current drawing and the relation between the various Xrefs.

*Figure 18-3 The **Palettes** panel*

It allows you to attach a new xref, detach, unload, load an existing one, change an attachment to an overlay, or an overlay to an attachment. You can also open a reference drawing for editing, from this palette. Additionally, it allows you to edit an xref's path and bind the xref definition to the drawing.

Apart from the methods mentioned in the command box, you can also invoke the **External References** palette by selecting an Xref in the current drawing and then right-clicking in the drawing area to display a shortcut menu. Next, choose **External References** from the shortcut menu; the **External References** palette is displayed.

The upper right corner of the palette has two buttons: **List View** and **Tree View**.

List View

Choosing the **List View** button displays the xrefs present in the drawing in alphabetical order. This is the default view. The list view displays information about xrefs in the current drawing under the following headings:

Reference Name

This column lists the name of all existing references in the current drawing.

Status

This column lists the current status of each xref in the drawing. It lists whether an xref is loaded, unloaded, unreferenced, not found, orphaned, unresolved, or marked to be reloaded. A loaded xref implies that the xref is attached to the current drawing. You can then unload it and then reload it using the options in the dialog box (this will be discussed later). An xref selected to be unloaded or reloaded displays **Unload** and **Reload**, respectively, under the **Status** column. If the xref has nested references that cannot be found, the status is **Unreferenced**, and if the parent of the nested reference gets unloaded, or cannot be found, its status is described as **Orphaned**. An unreferenced xref will not be displayed. If the xref is not found in the search paths defined, its status is **Not Found**. A missing xref or one that cannot be found is **Unresolved**.

Figure 18-4 The External References palette

Size

The file size of each xref is listed here.

Type

This column lists whether the xref is an attachment or overlay.

Date

This column lists the date on which the xref drawing was last saved.

Saved Path

This column lists the path of the xref, that is, the route taken to locate the particular referenced drawing.

Choosing any of these headings, sorts and lists the Xrefs in the current drawing according to that particular title. For example, choosing **Reference Name**, sorts and lists the xrefs as per name. The column widths can be increased or decreased as per requirements. When you place your cursor at the edge of a column title button, the cursor changes to a horizontal resizing cursor. Now, press the pick button of your mouse and drag the column edge to increase or decrease its width. After you increase the column widths, it is possible that the width of the columns extend beyond the list box width. In such a case, a horizontal scroll bar appears at the bottom of the list box. You can use the scroll bar to view the columns that extend beyond the width of the list box.

Chapter 18

Choosing the **Tree View** button, displays the xrefs in the drawing in a hierarchical tree view. It displays information on nested xrefs and their relationship with one another. Xrefs are indicated by an icon of a paper with a paper clip. This icon appears faded when the xref has been unloaded, and if there is a missing xref, a question mark appears. Similarly, an upward arrow indicates that the xref was reloaded and an arrow pointing downward indicates that the xref is unloaded. You can also choose **List View** and **Tree View** by pressing the F3 and F4 keys, respectively.

Attaching an Xref Drawing (Attach Option)

The **Attach** button is available at the upper left corner of the **External References** palette. If you choose the down arrow on this button, a flyout will be displayed. Select the **Attach DWG** option from the flyout to attach an xref drawing to the current drawing. This option can also be invoked by right-clicking on the **File References** area. The following examples illustrate the process of attaching an xref to the current drawing. In this example, it is assumed that there are two drawings, SHAFT and BEARING. SHAFT is the current drawing that is loaded on the screen (Figure 18-5) and the BEARING drawing is saved on the disk. If you want to xref the BEARING drawing in the SHAFT drawing, you need to follow the steps given below.

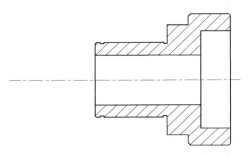

Figure 18-5 Current drawing - SHAFT

1.   The first step is to make sure that the SHAFT drawing is on the screen (draw the shaft drawing with assumed dimensions).

 Tip
One of the drawings need not be on the screen. You could attach both drawings, BEARING and SHAFT, to an existing drawing, even if it is a blank drawing.

2.   Invoke the **XREF** command to display the **External References** palette. In this palette, choose the **Attach DWG...** button; the **Select Reference File** dialog box will be displayed. Select the drawing that you want to attach (BEARING) and then choose the **Open** button. The **Attach External Reference** dialog box will be displayed on the screen (Figure 18-6). In the **Attach External Reference** dialog box (Figure 18-6), the name of the file that you have selected to be attached to the current drawing as an xref is displayed in the **Name** edit box. You can also specify a name of the file to attach from the **Name** drop-down list.

 Note
*AutoCAD also searches for the xref file in the paths defined in the **Project Files Search Path** folder in the **Files** tab of the **Options** dialog box. This folder does not have any path defined in it and displays **Empty** when the tree view is expanded. To define a search path, highlight **Empty** and choose the **Add** button in the dialog box. You can enter a project name here, if you want. Now, expand the tree view, select **Empty** again and choose the **Browse** button. The*

*Browse for Folder window is displayed. Select the folder that is to be searched for the file and choose **OK**. Then, choose **Apply** and **OK** in the **Options** dialog box to exit it.*

Figure 18-6 *The **Attach External Reference** dialog box*

In the **Reference Type** area, select the **Attachment** radio button if it is not already selected (default option). The **Overlay** option is discussed later in this chapter. The **Path type** drop-down list is used to specify whether you want to attach a drawing with full path, relative path, or no path. If you select the **Full path** option, the precise location of the xreffed drawing is saved. If you select the **Relative path** option, the position of the xreffed drawing with reference to the host drawing is saved. If you select the **No path** option, AutoCAD will search for the xreffed drawing in only that folder in which the host drawing is saved. You can either specify the insertion point, scale factors, and rotation angle in the respective **X**, **Y**, **Z**, and **Angle** edit boxes or select the **Specify On-screen** check boxes to use the pointing device to specify them on the screen. By default, the X, Y, and Z scale factors are 1 and the rotation angle is 0. The **Block Unit** area provides information regarding the units of the inserted block. The **Unit** edit box displays the unit of the block. The **Factor** edit box displays the scale factor, depending on the unit of the block and that of the current drawing. If you choose the **Show Details** button, the path of the file is displayed adjacent to **Found in**. Also, the saved path of the file is displayed adjacent to **Saved path**. Choose the **OK** button from the **Attach External Reference** dialog box to accept the default values and exit the dialog box. Specify the insertion point; the drawing is attached to the current drawing, but it appears faded. After attaching the BEARING drawing as an xref, save the current drawing with the file name SHAFT (Figure 18-7). You can control the fading intensity of the attached objects by using the options from the **Fade Control** area of the **Display** tab in the **Options** dialog box. In this area, you can either use the slider bar to increase or decrease the fading intensity or you can enter a value in the edit box. You can also use the **XFADECTL** variable to control the fading intensity of other objects while editing them. By default, the fading intensity value is set to 50.

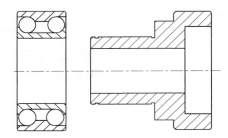

Figure 18-7 *Attaching the BEARING drawing as an xref*

Note
*You can also use the **-XREF** command to attach a drawing from the pointer input. Except the **Open** option, all the options available in the **External References** palette are available through the dynamic preview too.*

3. Open the drawing BEARING and make the changes shown in Figure 18-8 (draw polylines on the sides). Now, save the drawing with the file name BEARING.

4. Reload the drawing BEARING on the screen. You will notice that the xref drawing BEARING is automatically updated (Figure 18-9). This is the most useful feature of the **XREF** command. You could also have inserted the BEARING drawing as a block, but if you had updated the BEARING drawing, the drawing in which it was inserted would not have been updated automatically.

When you attach an xref drawing, AutoCAD remembers the name of the attached drawing. If you xref the drawing again, AutoCAD displays a message to that effect as given next.

Xref "BEARING" has already been defined.
Using existing definition.
Specify insertion point or [Scale/X/Y/Z/Rotate/PScale/PX/PY/PZ/PRotate]: *Specify the location to place another copy of the xref.*

Note
*If the xref drawing you that want to attach is currently being edited, AutoCAD will attach the drawing that was last saved through the **SAVE**, **WBLOCK**, or **QUIT** command.*

Points to Remember about Xref

1. When you enter the name of the xref drawing, AutoCAD checks for block names and xref names. If a block exists with the same name as the name of the xref drawing in the current drawing, the **XREF** command is terminated and an error message is displayed.

2. When you xref a drawing, the objects that are in the model space are attached. Any objects that are in the paper space are not attached to the current drawing.

3. The layer 0, DEFPOINTS, and the linetype CONTINUOUS are treated differently. The current drawing layers 0, DEFPOINTS, and linetype CONTINUOUS will override the layers and linetypes of the xref drawing. For example, if the layer 0 of the current drawing

Figure 18-8 *The modified xref drawing BEARING*

Figure 18-9 *After loading the SHAFT drawing, BEARING is automatically updated*

is white and the layer 0 of the xref drawing is red, the white color will override the red.

4. The xref drawings can be nested. For example, if the BEARING drawing contains the reference INRACE and you xref the BEARING drawing to the current drawing, the INRACE drawing is automatically attached to the current drawing. If you detach the BEARING drawing, the INRACE drawing gets detached automatically.

5. You can rename an xref under the **File References** column name in the list box of the **External References** palette by highlighting the xref and then clicking on it again. You can now enter a new name. An AutoCAD warning is displayed: **Caution! "XXXX" is an externally referenced block. Renaming it will also rename its dependent symbols**.

6. When you xref a drawing, AutoCAD stores the name and path of the drawing by default. If the name of the xref drawing or the path where the drawing was originally stored has changed or you cannot find it in the path specified in the **Options** dialog box, AutoCAD cannot load the drawing, plot it, or use the **Reload** option of the **XREF** command.

Detaching an Xref Drawing (Detach Option)

The **Detach** option can be used to detach or remove the xref drawings. If there are any nested xref drawings defined with the xref drawings, they are also detached. Once a drawing is detached, it is erased from the screen. To detach an xref drawing, select the file name in the **External References** palette list box to highlight it and then right-click on it to select the **Detach** option. After selecting **Detach** option, the xref is completely removed from the current drawing.

You can also use the **-XREF** command to detach the xref drawings. When AutoCAD prompts for an xref file name to be detached, you can enter the name of one xref drawing or the name of several drawings separated by commas. You can also enter * (asterisk), in which case, all referenced drawings, including the nested drawings, will be detached.

Chapter 18

Updating an Xref Drawing (Reload Option)

When you load a drawing, AutoCAD automatically loads the referenced drawings. The **Reload** option of the **XREF** command lets you update the xref drawings and nested xref drawings at any time. You do not need to exit the drawing editor and then reload the drawing. To reload the xref drawings, invoke the **External References** palette, select the xreffed drawing in the list box, and then right-click on it to choose the **Reload** option from the shortcut menu. AutoCAD will scan for the referenced drawings and the nested xref drawings and load the most recently saved version of the drawing.

You can reload all the attached xref drawings at one time by selecting **Reload All References** from the drop-down list of the **Refresh** button on the upper left corner of the **External References** palette.

The **Reload** option is generally used when the xref drawings are currently being edited and you want to load the updated drawings. The xref drawings are updated based on what is saved on the disk. Therefore, before reloading an xref drawing, you should make sure that the xref drawings that are being edited have been saved. If AutoCAD encounters an error while loading the referenced drawings, the **XREF** command is terminated, and the entire reload operation is canceled.

You can also reload the xref drawings by using the **-XREF** command. When you enter the **-XREF** command, AutoCAD will prompt you to enter an option. Enter **R** for the **Reload** option. Next, you are prompted to enter the name of the xref drawing. You can enter the name of one xref drawing or the names of several drawings separated by commas. If you enter * (asterisk), AutoCAD will reload all xref and nested xref drawings.

Unloading an Xref Drawing (Unload Option)

The **Unload** option allows you to temporarily remove the definition of an xref drawing from the current drawing. However, AutoCAD retains the pointer to the xref drawings. When you unload the xref drawings, the drawings are not displayed on the screen. You can reload the xref drawings by using the **Reload** option.

Tip
It is recommended that you unload the referenced drawings if they are not being used. After unloading the xref drawings, the drawings load much faster and need less memory.

Adding an Xref Drawing (Bind Option)

The **Bind** option lets you convert the xref drawings to blocks in the current drawing. The bound drawings, including the nested xref drawings (that are no longer xrefs), become a permanent part of the current drawing. The bound drawing cannot be detached or reloaded. You can use this option when you want to send a copy of your drawing to a customer for review. Because the xref drawings are a part of the existing drawing, you do not need to include the xref drawings or the path information. You can also use this option to safeguard the master drawing from accidental editing of the piece parts. To bind the xref drawings, select the file names in the **External References** palette and then right-click and choose the

Bind option from the shortcut menu; the **Bind Xrefs** dialog box (Figure 18-10) is displayed. AutoCAD provides two methods to bind the xref drawing in the **Bind Type** area of the dialog box. These methods are discussed next.

Bind

When you use the **Bind** radio button, AutoCAD

*Figure 18-10 The **Bind Xrefs** dialog box*

binds the selected xref definition to the current drawing. All the xrefs are converted to blocks and the named objects are renamed. For example, if you xref the drawing BEARING with a layer named OBJECT, a new layer BEARING|OBJECT is created in the current drawing. When you bind this drawing, the xref dependent layer BEARING|OBJECT will become a locally defined layer BEARING0OBJECT (Figure 18-11). If the BEARING0OBJECT layer already exits, AutoCAD will automatically increment the number, and the layer name becomes BEARING1OBJECT.

Insert

When you use the **Insert** option, AutoCAD inserts the xref drawing. The xrefs get converted into blocks. For example, if you xref the drawing SHAFT with a layer named OBJECT, a new layer SHAFT|OBJECT is created in the current drawing. If you use the **Insert** option to bind the xref drawing, the layer name SHAFT|OBJECT is renamed as OBJECT (Figure 18-11). If the object layer already exists, then the values set in the current drawing override the values of the inserted drawing.

*Figure 18-11 The **Layer Properties Manager** dialog box*

Note

It is possible to bind only an individual or several xref-dependent named objects instead of the entire drawing into the current drawing. This will be discussed later in this chapter.

*You can also use the **-XREF** command to bind the xref drawings. In such cases, prompts are displayed on the command line.*

Editing an XREF's Path

By default, AutoCAD saves the path of the referenced drawing and displays it in the **Saved Path** column in the **External References** palette. As mentioned earlier, when AutoCAD loads the drawing containing a referenced file, and if it is not able to find the file at the location specified in the **Saved Path** column of the **External References** palette, it searches for the file in the current directory, and in the **Support File Search Path** locations specified in the **Files** tab of the **Options** dialog box. If a file with the same name is found here, it is loaded. Now, when you invoke the **External References** palette, you will notice that when you select an xref name in the list box to highlight it, the path displayed in the **Saved Path** column for the xref file is different from the one displayed in the **Found At** edit box. To update the path of the xref file, choose the **Save Path** button. The new path is saved and displayed in the **Saved Path** column.

If AutoCAD cannot locate the specified file even in the directories specified in the **Files** tab of the **Options** dialog box, it will display an error message saying that it cannot find the specified file. The path of the file is displayed as a marker text in the current drawing. Now, when you invoke the **External References** palette, the status of the drawing is shown as **Not Found**. To specify a new path for the xref file, select the xref file name in the list box of the dialog box and choose the **Browse** button. The **Select new path** dialog box is displayed where you can locate the drawing to be used as xref. Once you have found the file, choose the **Open** button to return to the **External References** palette. The new path is displayed in the **Saved Path** column and the **Found At** edit box. The specified xref file is reloaded and replaces the marker text in the drawing, when you choose the **OK** button in the **External References** palette. If you remember the new location of the xref file, you can also enter it in the **Found At** edit box. For example, if a drawing, which was originally in the C:\CAD\Proj1 subdirectory, has been moved to A:\Parts directory, the path must be edited so that AutoCAD is able to load the xref drawing.

THE OVERLAY OPTION

As discussed earlier, when you attach an xref to a drawing, the **Attach External References** dialog box is displayed. The **Reference Type** area of this dialog box has two radio buttons, **Attachment** and **Overlay**. You can select any of these radio buttons to xref a drawing. The **Attachment** radio button is selected by default. The advantage of selecting the **Overlay** radio button is that you can access the desired drawing instead of the drawing along with its xreffed attachments. For example, consider three people working on three different drawings that are a part of the same project. The first designer is working on the layout of walls of a room, the second designer is working on the furniture layout of the room, and the third on the electrical layout of that room. The names of the drawings are WALLS, FURNITURE, and ELECTRICAL, respectively. Assume that the designer working on the walls layout selects the **Attachment** radio button to xref the FURNITURE drawing so that he or she can check the furniture layout according to the wall structure. After insertion, the WALLS drawing will comprise the wall structure (current drawing) along with the furniture layout (xreffed drawing). Now, if the designer working on the electrical layout xrefs the WALLS drawing to check the location of electrical fittings with respect to the walls, he/she will get the drawing that has the furniture layout as well as the wall layout. This is because the FURNITURE drawing was xreffed in the WALLS drawing by selecting the **Attachment** radio button.

In the above example, the designer working on the ELECTRICAL drawing may not require the FURNITURE drawing. This is because at this stage, he/she is more interested in checking the electrical fittings with respect to the wall structure. So the furniture layout that is xreffed with the wall structure needs to be avoided. This can be done by selecting the **Overlay** radio button while X-referencing the FURNITURE drawing in the WALLS drawing. This means that the designer working on the wall structure needs to xref the furniture layout by selecting the **Overlay** radio button. Now, if the wall structure is xreffed in some other drawing, the furniture layout will not appear.

One of the problems on with the **Attachment** option is that you cannot have a circular reference. For example, assume you are designing the plant layout of a manufacturing unit. One person is working on the floor plan (see Figure 18-12), and the second person is working on the furniture layout in the offices (Figure 18-13). The names of the drawings are FLOORPLN and OFFICES, respectively. The person working on the office layout uses the **Attachment** option to insert the FLOORPLN drawing so that he or she has the latest floor plan drawing. The person who is working on the floor plan wants to reference the OFFICES drawing.

Figure 18-12 *Drawing files hierarchy*

Now, if the **Attachment** option is used to reference the drawings, AutoCAD displays an error message. This is because by attaching the OFFICES drawing, a circular reference is created. The AutoCAD message displayed is "**Circular references detected. Continue?**" (Figure 18-14). If you choose the **No** button, the **XREF** command is canceled and no drawing is referenced. But, if you choose the **Yes** button, the following message is displayed **Breaking circular reference from "offices" to "current drawing"** and the particular file you wanted to reference is referenced.

Figure 18-13 *Sample plant layout including OFFICES drawing*

However, to overcome this problem of circular reference, you can use the **Overlay** option to overlay the OFFICES drawing. This is a very useful option because the **Overlay** option lets different operators avoid circular reference and share the drawing data without affecting

Chapter 18

the drawing. Overlaying allows you to view a referenced drawing without having to attach it to the current drawing. This option can be invoked by selecting the **Overlay** radio button in the **Attach External References** dialog box, which is displayed after you have selected a drawing to reference. Also, when a drawing that has a nested overlay is overlaid, the nested overlay is not visible in the current drawing. This is another difference between attaching an xref and overlaying an xref to a drawing. This feature is especially useful when you want to reference a drawing that another user who is referencing your drawing does not need. Although the attachment will reference the nested reference too, the overlay option ignores nested references.

Figure 18-14 The AutoCAD Alert message box

Example 1   *Architectural*

In this example, you will use the **Attachment** and **Overlay** options to attach and reference the drawings. Two drawings, PLAN and PLANFORG are given. The PLAN drawing (Figure 18-15) consists of the floor plan layout, and the PLANFORG drawing (Figure 18-16) has the details of the forging section only. The CAD operator who is working on the PLANFORG drawing wants to xref the PLAN drawing for reference. Also, the CAD operator working on the PLAN drawing should be able to xref the PLANFORG drawing to complete the project. The following steps illustrate how to accomplish the defined task without creating a circular reference.

How circular reference is caused?

1. Load the drawing PLANFORG, invoke the **XREF** command, and choose the **Attach DWG** button from the **External References** palette. Select the PLAN drawing from the list box of the **Select Reference File** dialog box and choose the **Open** button; the **Attach External Reference** dialog box is displayed. In this dialog box, the name of the PLAN drawing is displayed in the **Name** edit box, and the **Attachment** radio button is selected by default in the **Reference Type** area. Choose **OK** to exit the dialog box and specify an insertion point on the screen. Now, the drawing consists of the PLANFORG and PLAN. Save the drawing.

2. Open the drawing file PLAN. Next, invoke the **XREF** command and attach the PLANFORG drawing to the PLAN drawing using the same steps as described in Step 1. AutoCAD will display the message that the circular reference has been detected and will ask you if you want to continue. If you choose to continue by choosing **Yes** in the AutoCAD message box, the circular reference is broken and you are allowed to reference the specific drawing.

    The possible solution for the operator working on the PLANFORG drawing is to detach the PLAN drawing. This way, the PLANFORG drawing does not contain any reference to the PLAN drawing and would not cause any circular reference. The other solution is to use the **Overlay** option, as follows.

Figure 18-15 *The PLAN drawing*   **Figure 18-16** *The PLANFORG drawing*

How to prevent circular reference?

3. Open the PLANFORG drawing (Figure 18-17) and select the **Overlay** radio button in the **Attach External References** dialog box, which is displayed after you have selected the PLAN drawing to reference. The PLAN drawing is overlaid on the PLANFORG drawing (Figure 18-18).

Figure 18-17 *The PLANFORG drawing*   **Figure 18-18** *The PLANFORG drawing after overlaying the PLAN drawing*

4. Open the drawing file PLAN (Figure 18-19), and select the **Attachment** radio button in the **Attach External Reference** dialog box, which is displayed when you have selected the PLANFORG drawing in the **Select Reference File** dialog box to attach as an xref to the PLAN drawing. You will notice that only the PLANFORG drawing is attached (Figure 18-20). The drawing that was overlaid in the PLANFORG drawing (PLAN) does not appear in the current drawing. This way, the CAD operator working on the PLANFORG drawing can overlay the PLAN drawing, and the CAD operator working on the PLAN drawing can attach the PLANFORG drawing, without causing a circular reference.

Chapter 18

Figure 18-19 *The PLAN drawing*

Figure 18-20 *The PLAN drawing after attaching the PLANFORG drawing*

WORKING WITH THE ATTACH COMMAND*

| **Ribbon:** | Insert > Reference > Attach | **Command:** | ATTACH |

You can use the **ATTACH** command to attach a DWG, DGN, DWF, PDF, or image file without invoking the **External References** palette. Alternatively, choose **Insert > Reference > Attach** from the **Ribbon** (Figure 18-21). When you invoke this command, AutoCAD displays the **Select Reference File** dialog box. To attach a *.dwg, .dgn, .dwf, .pdf,* or image file, specify it in the **Files of type** drop-down list in the **Select Reference** dialog box; the corresponding files will be listed in the dialog box. This command makes attaching a drawing file easier, since most of the xref operations involve simply attaching a drawing file. Select the drawing file to be attached and choose the **Open** button; the **Attach External Reference** dialog box is displayed. Select the **Attachment** radio button under the **Reference Type** area. You can specify the insertion point, scale, and rotation angle on screen or in the respective edit boxes.

Figure 18-21 *The **Reference** panel*

Tip

*When you attach or reference a drawing that has a drawing order created by using the **DRAWORDER** command, the drawing order is not maintained in the xref. To maintain the drawing order, first open the xref drawing and specify the drawing order in it. Now, use the **WBLOCK** command to convert it into a drawing file and the **XATTACH** command to attach the newly created drawing file to the current drawing. In this way the drawing order will be maintained.*

Note

*To select a .dgn, .dwf, .pdf, or image file, specify the file type in the **Files of type** drop-down list in the **Select Reference File** dialog box; the files will be attached as an underlay. Underlays are discussed in the next section.*

Note

*AutoCAD maintains a log file (.xlg) for xref drawings if the **XREFCTL** system variable is set to 1. This file lists information about the date and time of loading and other xref operations to be completed. This .xlg file is saved in the current drawing with the same name as the current drawing and is updated each time the drawing is loaded or any xref operations are carried out.*

WORKING WITH UNDERLAYS*

You can attach a DWF, DGN, or PDF file as an underlay to the current drawing file. The underlay files are not a part of original drawing files. Therefore, if you add a file as an underlay, it does not increase the file size of the current drawing. The procedure to add a file as an underlay is similar to attaching a drawing file using the **ATTACH** command. After you select the file to attach, the **Attach <XXXX> Underlay** dialog box will be displayed, where <XXXX> is the file name. Figure 18-22 shows the **Attach PDF Underlay** dialog box that is displayed on selecting a pdf file from the **Select Reference File** dialog box.

*Figure 18-22 The **Attach PDF Underlay** dialog box*

If the selected *.pdf* file has multiple pages, all pages of the pdf file are listed in the **Select one or more pages from the PDF file** area. Select the pages to be attached from this area. If you have selected multiple pages as well as the **Specify on-screen** check box from the **Insertion point** area, then you are prompted to specify different insertion points for different pages.

The files attached as an underlay behave like blocks. The general modify commands like move, copy, rotate, mirror, and so on can be applied on them. However, you cannot bind a file that is attached as an underlay or modify the attached file in the current drawing file.

Editing an Underlay

As discussed earlier, you cannot edit a file that is attached as an underlay. However, you can control the appearance of the underlay by adjusting the contrast, fade, and monochrome display. To do so, select the attached object; a contextual tab will be displayed. Figure 18-23 shows the **DGN Underlay** tab that is displayed on selecting a DGN file.

*Figure 18-23 The **DGN Underlay** tab*

You can use the options in this contextual tab to adjust the fade, contrast, and monochrome display of the underlay. You can create a new clipping boundary similar to that of an attached drawing. By default, the **Enable Snap** button is chosen in the **Options** panel. Therefore, you can snap entities in the underlaid object. If you deselect this button, you cannot snap entities in the underlaid objects. You can use the **Show Underlay** button from the **Options** panel to display/hide the underlaid objects. Both **Enable Snap** and **Show Underlay** are toggle buttons. If the attached underlay has layers, you can hide/display the selected layers. To do so, choose the **Edit Layers** button from the contextual tab; the **Underlay Layers** dialog box will be displayed, as shown in Figure 18-24. If multiple drawings are attached as an underlay, all file names will be listed in the **Reference Name** drop-down list. Select the file from the **Reference Name** drop-down list for which you need to hide the layers; the corresponding layers will be listed in the list box. Click on the bulb icon of the layer to be hidden. Alternatively, select the layer from the list box and right-click; a shortcut menu will be displayed. Choose the **Layer(s) Off** option to hide the selected layer. To hide multiple layers, press the CTRL key and select the layers to be hidden. Next, choose **Layer(s) Off**. Then, press ESC to exit the editing mode; the contextual tab will disappear.

*Figure 18-24 The **Underlay Layers** dialog box*

If you have attached a file as an underlay, you can invoke the **Underlay Layers** dialog box without invoking the contextual tab. To do so, choose **Insert > Reference > Underlay Layers** from the **Ribbon**. Similarly, you can switch on/off the OSNAP settings for the drawing that are underlaid by choosing the **Snap to Underlays ON / Snap to Underlays OFF** button respectively, from the **Reference** panel in the **Insert** tab of the **Ribbon**.

Note

*If there are more underlays of different file types and you have set different osnap setting for each file type, then the **Underlay Osnaps Vary** button will be displayed in the **Reference** panel of the **Insert** tab. However, if you choose the **Snap to Underlays ON / Snap to Underlays OFF** button, the corresponding osnap setting will be applied to all underlays.*

OPENING AN XREFFED OBJECT IN A SEPARATE WINDOW

If you are in the host drawing and you want to open a selected xreffed object in a separate window without using the **Select File** dialog box, you can use the **XOPEN** command. When you invoke this command, you are prompted to select the xref. Select the xreffed object that you want to open in a separate window. When you select the desired xref, AutoCAD opens the DWG file of the xreffed object in a separate window. You can now make the desired changes in the DWG file of the xreffed object. Save the changes and then close the drawing. When you open the host drawing, you will notice that the **Communication Center** displays a message that the external reference file has changed and the xreffed drawing needs to be reloaded. Reload the xreffed drawing using the **External References** palette and you will notice that the host drawing is updated.

USING THE DESIGNCENTER TO ATTACH A DRAWING AS AN XREF

The **DesignCenter** can also be used to attach an xref to a drawing. To do so, choose **Insert > Content > DesignCenter** from the **Ribbon** to display the **DesignCenter**. In the **DesignCenter**, choose the **Tree View Toggle** button, if it is not chosen already, to display the tree pane. Expand the **Tree view** and double-click on the folder whose contents you want to view. The contents of the selected folder are displayed in the palette. From the thumbnails of drawings displayed on the right-side of the palette, right-click on the drawing to be attached as an xref; a shortcut menu is displayed, as shown in Figure 18-25. Choose **Attach as Xref**; the **Attach External Reference** dialog box is displayed. Alternatively, you can also use the right mouse button to drag and drop the drawing into the current drawing; a shortcut menu is displayed again. Choose **Attach as Xref** and the **Attach External Reference** dialog box is displayed.

In this dialog box, the **Name** edit box displays the name of the selected file to be inserted as an xref. Select the **Attachment** radio button in the **Reference type** area, if not already selected. Specify the **Insertion point**, **Scale**, and **Rotation** in the respective edit boxes or select the **Specify On-screen** check boxes to specify this information on the screen. Choose **OK** to exit the dialog box; the selected object is attached as xref to the current drawing.

*Figure 18-25 The **DesignCenter***

ADDING XREF DEPENDENT NAMED OBJECTS

Menu Bar:	Modify > Object > External Reference > Bind	**Command:**	XBIND

You can use the **XBIND** command to add the selected named objects such as blocks, dimension styles, layers, line types, and text styles of the xref drawing to the current drawing. The following example describes how to use the **XBIND** command (Figure 18-26).

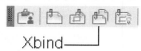

*Figure 18-26 The **Reference** toolbar*

1. Load the drawing Bearing that was created earlier when the **Attach** option of the **XREF** command was discussed. Make sure the drawing has the following layer setup. Otherwise, create the following layers, using the **Layer Properties Manager** dialog box.

Layer Name	Color	Linetype
0	White	Continuous
Object	Red	Continuous
Hidden	Blue	Hidden2
Center	White	Center2
Hatch	Green	Continuous

2. Draw a circle and use the **BLOCK** command to create a block of it. The name of the block is SIDE. Save the drawing as BEARING.

3. Start a new drawing with the following layer setup:

Layer Name	Color	Linetype
0	White	Continuous
Object	Red	Continuous
Hidden	Green	Hidden

4. Use the **XATTACH** command or the **Attach** option of the **XREF** command and attach the Bearing drawing to the current drawing. When you xref the drawing, the layers will be added to the current drawing, as discussed earlier in this chapter.

5. Now, invoke the **XBIND** command. The **Xbind** dialog box is displayed on the screen. This dialog box has two areas with list boxes. They are **Xrefs** and **Definitions to Bind**. If you want to bind the blocks defined in the xref drawing Bearing, first click on the plus sign adjacent to the xref Bearing. Icons for named objects in the drawing are displayed in a tree view (Figure 18-27). Click on the plus sign next to the Block icon. AutoCAD lists the blocks defined in the xref drawing (Bearing). Select the block Bearing|SIDE and then choose the **Add** button. The block name will be added to the **Definitions to Bind** area list box. Choose **OK** to exit the dialog box. AutoCAD will bind the block with the current drawing and a message at the command line is displayed: **1 Block(s) bound**. The name of the block will change to Bearing0SIDE. You can invoke the **Block Definition** dialog box using the **BLOCK** command and check the **Name** drop-down list to see if the block with the name Bearing0SIDE has been added to the drawing. If you want to insert the block, you must enter the new block name (Bearing0SIDE). You can also rename the block to another name that is easier to use.

*Figure 18-27 The **Xbind** dialog box*

If the block contains a reference to another xref drawing, AutoCAD binds that xref drawing and all its dependent symbols to the current drawing also. Once you **XBIND** the dependent symbols, AutoCAD does not delete them, not even when the xref drawing is detached. For example, the block Bearing0SIDE will not be deleted even when you detach the xref drawing or end the drawing session.

Chapter 18

You can also use the **-XBIND** command to bind the selected dependent symbols of the xref drawing, using the command line.

6. Similarly, you can bind the dependent symbols, Bearing|STANDARD (textstyle), Bearing|Hidden, and Bearing|Object layers of the xref drawing. Click on the plus signs adjacent to the respective icons to display the contents and then select the layer or textstyle you want to bind and choose the **Add** button. The selected named objects are displayed in the **Definitions to Bind** area list box. If you have selected a named object that you do not want to bind, select it in the **Definitions to Bind** area list box and choose the **Remove** button. Once you have finished selecting the named objects that you want to bind to the current drawing, choose **OK**. A message indicating the number of named objects that are bound to the current drawing is displayed at the command line.

Once bound, the layer names will change to Bearing0Hidden and Bearing0Object. If the layer name Bearing0Hidden was already there, the layer will be named Bearing1Hidden. These two layers become a permanent part of the current drawing. Even if the xref drawing is detached or the current drawing session is closed, the layers are not discarded.

CLIPPING EXTERNAL REFERENCES*

Ribbon:	Insert > Reference > Clip	Command:	CLIP

The **CLIP** command (Figure 18-28) is used to trim an xref after it has been attached to a drawing to display only portion of the drawing (Figure 18-29). After you have attached an xref to a drawing, you can trim it to display only a portion of the drawing by using the **CLIP** command. However, clipping xref does not, any way, modify the referenced drawing. On invoking the **CLIP** command, you are prompted to select an object to clip. Select a DWG, DGN, IMAGE, or PDF; the respective clipping options will be displayed. You can also invoke the **CLIP** command by selecting an xref and then right-clicking in the drawing area to display a shortcut menu. Next, choose the corresponding clip command from it. For example, to clip an attached .*dwg* file, select the attached .*dwg* file, right-click, and choose the **CLIP Xref** option; the following prompt sequence will be displayed.

*Figure 18-28 The **Reference** panel*

Command: **-XCLIP** *1 found*
Select objects: *Select the xref that needs to be clipped.*
Select objects: *Press ENTER*
Enter clipping option
[ON/OFF/Clipdepth/Delete/generate

*Figure 18-29 Using the **XCLIP** command to clip*

Polyline/New boundary] <New>: *Press ENTER to specify a new clip boundary or enter an option.*
[Select polyline/Polygonal/Rectangular/Invert clip] <Rectangular>: *Press ENTER to select the **Rectangular** option, and then specify two corners of the rectangular boundary.*

You can also right-click in the drawing area at the **Enter clipping option** prompt to display a shortcut menu. This shortcut menu displays all the options available at the command line and can be selected here. All the clipping options available are discussed next.

New boundary

The clipping boundary can be specified by using the **Rectangular**, **Polygonal**, or **Select polyline** options. The **Rectangular** option generates a rectangular boundary and the **Polygonal** option allows you to specify a boundary of any shape. You can draw the boundary using the **PLINE** or the **POLYGON** commands and then, use the **Select polyline** option to select this polyline as the clipping boundary. The **Invert clip** option inverts the direction of the clipping. For example, if the objects outside the clipping boundary are hidden, by default, the **Invert clip** option will hide the objects inside the clipping boundary and display the objects outside the boundary. After selecting the **Invert clip** option, you need to redefine the clipping boundary using any one of the options discussed above. The **Invert clip** option can only be used with the 2D Wireframe visual style. If a boundary already exists, AutoCAD will ask you if the old boundary should be deleted. You can enter **YES** if you want to delete the old boundary and define a new one. If you enter **NO**, the old boundary is retained and the **XCLIP** command ends.

ON/OFF

The **ON/OFF** option allows you to specify whether to display the clipped portion or not. When the clipping boundary is off, you can see the complete xref drawing and when it is on, the drawing that is within the clipping polygon is displayed.

Delete

The **Delete** option completely deletes the predefined clipping boundary and the entire xref gets displayed. The **ERASE** command cannot be used to delete the clipping boundary.

generate Polyline

If you have a clipped boundary, select this option to create a polyline coinciding with the boundary of the clipped xref drawing. You can edit the polyline boundary without affecting the clipped drawing. For example, if you stretch the boundary, it does not affect the xref drawing. The edited boundary can be used later to specify a new clipping boundary.

 Note
*Polyline generated using the **generate Polyline** option of the **XCLIP** command should not be confused with the clipping boundary. Although both coincide with each other, you can select and edit only the polyline. When you select the clipping boundary, the xref object is also selected.*

Clipdepth

This option allows you to define the front and back clipping planes for 3D xref objects or blocks. The objects between the front and back planes will be displayed (Figures 18-30 and 18-31). You will be able to understand this option better after you have dealt with 3D drawings later in the book. In the figures that follow, the clipping boundary is defined aligned to the front face of the object, and therefore, the clipping planes will get defined parallel to it. The xref must contain a clip boundary before specifying a clip depth. If you use the **Clipdepth** option, it prompts you to specify a front clip point and a back clip point. Specifying these points, creates a clipping frame passing through them and parallel to the clipping boundary. If the front clipping plane is specified behind the back clipping plane, AutoCAD displays an error message and the clipdepth is also not applied. The **Distance** option for specifying the front or back clip points creates a clipping plane at a specified distance from the clipping boundary parallel to it. The **Remove** option removes both the front and back clipping planes and the entire object is visible.

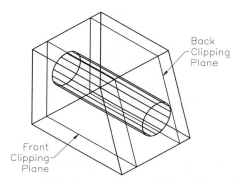

Figure 18-30 *3D xref object before clipping*

Figure 18-31 *3D xref object after clipping*

Similarly, you can clip the other underlaid xrefs. The options available for clipping the underlaid Xrefs are **ON**, **OFF**, **Delete**, and **New Boundary**. These options are the same as discussed above.

DISPLAYING CLIPPING FRAME*

You can use the **FRAME** system variable to turn on or off the display of clipping boundary of all clipped Xrefs. When the value of **FRAME** system variable is 0 (default), the clipping boundary will not be displayed. When it is 1, the clipping boundary will be displayed and plotted. When it is 2, the clipping boundary will be displayed, but not plotted. To

Figure 18-26 *The Reference toolbar*

control the display of frames of individual attached Xrefs, you need to invoke the individual system variables like **XCLIPFRAME**, **DGNFRAME**, **PDFFRAME**, and **IMAGE FRAME**. Alternatively, choose the appropriate button from the **Reference** panel in the **Insert** tab of the **Ribbon** (Figure 18-32). If there are multiple underlays of different file types and you have set different frame settings for each file type, then the **Frames Vary** button will be displayed in the **Reference** panel of the **Insert** tab.

DEMAND LOADING

The demand loading feature loads only that part of the referenced drawing that is required in the existing drawing. For example, demand loading provides a mechanism by which objects on frozen layers are not loaded. Also, only the clipped portion of the referenced drawing can be loaded. This makes the xref operation more efficient since less disk space is used, especially when the drawing is reopened. Demand loading is enabled by default. You can modify its settings in the **External References (Xrefs)** area of the **Open and Save** tab of the **Options** dialog box (Figure 18-33). You can select any of the three settings available in the **Demand load Xrefs** drop-down list in this dialog box. They are **Disabled**, **Enabled**, and **Enabled with copy**. These options correspond to a value of **0**, **1**, and **2** of the **XLOADCTL** system variable, respectively, and are discussed next.

Setting	Value of XLOADCTL	Features
Disabled	0	1. Turns off demand loading. 2. Loads entire xref drawing file. 3. The file is available on the server and other users can edit the xref drawing.
Enabled	1	1. Turns on demand loading. 2. The referenced file is kept open. 3. Makes the referenced file read-only for other users.
Enabled with copy	2	1. Turns on demand loading with the copy option. 2. A copy of a referenced drawing is opened. 3. Other users can access and edit the original referenced drawing file.

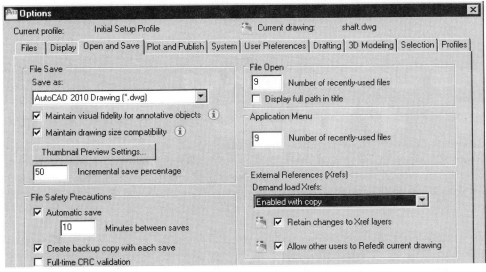

Figure 18-33 Setting **XLOADCTL** using the **Options** dialog box

You can also set the value of **XLOADCTL** at the command line. When you are using the **Enabled with copy** option of demand loading, the temporary copies of the xref are saved in the AutoCAD temporary files directory (defined in the **Temporary External Reference File Location** folder in the **Files** tab of the **Options** dialog box) or in a user-specified directory. The **XLOADPATH** system variable creates a temporary path to store demand loaded Xrefs.

Spatial and Layer Indexes

As mentioned earlier, the demand loading improves performance when the drawing contains referenced files. To make it work effectively and to take full advantage of demand loading, you must store a drawing with Layer and Spatial indexes. The layer index maintains a list of objects in different layers and the spatial index contains lists of objects based on their location in 3D space.

Layer and spatial indexes are created using the **Save Drawing As** dialog box. Choose the **Tools** button available on the upper right corner of the dialog box to display a shortcut menu. Choose the **Options** option from the shortcut menu to display the **Saveas Options** dialog box. Choose the **DWG Options** tab if it is not already chosen and select the type of index you want to save the file with from the **Index type** drop-down list. **None** is the default option, that is, no indexes are created. The other options available are: **Layer**, **Spatial**, and **Layer & Spatial**. Once you have selected the type of index to create, choose the **OK** button to exit the dialog box and choose the **Save** button in the **Save Drawing As** dialog box to save the drawing with the indexes. The **INDEXCTL** variable also controls the creation of layer index and spatial index and its value can be set using the command line. The following are the settings of the **INDEXCTL** system variable, and they correspond to the **Index type** options available in the **DWG Options** tab of the **Save Options** dialog box.

Setting	Features	Index Type option
0	No index created	None
1	Layer index created	Layer
2	Spatial index created	Spatial
3	Layer and spatial index created	Layer and Spatial

EDITING REFERENCES IN-PLACE*

Ribbon:	Insert > Reference > Edit In-Place	**Command:**	REFEDIT
Toolbar:	Refedit > Edit Reference In-Place		

Often you have to make minor changes in the xref drawing if you want to save yourself from the trouble of going back and forth between drawings. In this situation, you can use in-place reference editing to select the xref, modify it, and then save it after modifications. This feature of AutoCAD has already been discussed in detail with reference to blocks in Chapter 16, Working with Blocks. The following steps explain the referencing editing process.

1. To edit in-place, select an external referenced object; the **External Reference** tab will be added to the **Ribbon**. Choose the **Edit Reference In-place** option from the **External Reference** tab; the **Reference Edit** dialog box will be displayed. After selecting the external

referenced object, you can also invoke the shortcut menu and choose the **Edit Xref In-place** option from it to invoke the **Reference Edit** dialog box.

2. In the **Reference name** list box of the **Reference Edit** dialog box, the selected reference and its nested references are listed. Select the specific xref you wish to edit from the list.

3. In the **Settings** tab of the dialog box, select both the **Create unique layer, style, and block names** and the **Lock objects not in working set** check boxes. The **Display attribute definitions for editing** check box is disabled by default. In the **Identify Reference** tab of the dialog box, select the **Prompt to select nested objects** radio button available below **Path**. Choose **OK** to return to the current drawing. At the **Select nested objects:** prompt, select the objects you want to modify and press ENTER. The **Edit Reference** panel is added to **Ribbon** (Figure 18-34) and the objects that have not been selected for modifications become faded. The fading is controlled by the **XFADECTL** system variable or by using the **Xref display** slider bar in the **Display** tab of the **Options** dialog box. You can add or remove objects to the working set by choosing the respective buttons in the **External Reference** panel in the **Edit Reference** tab of the **Ribbon** or by using the **REFSET** command. Once you have defined a working set, all the standard AutoCAD commands can be used to modify them.

*Figure 18-34 The **Edit Reference** panel*

4. When you choose the **Save Changes** button in the **Edit Reference** panel or use the **REFCLOSE** command, the modifications made are saved in the drawing used as a reference as well as in the current drawing. Choosing the **Discard Changes** button in the **Edit Reference** panel does not save the changes made. After you have saved the changes to the xref, AutoCAD displays a message: **All reference edits will be saved**. Choose **OK** to return to the drawing area. The current drawing is redisplayed.

Note

*In the **External References (Xrefs)** area of the **Open and Save** tab of the **Options** dialog box, if you select the **Allow other users to Refedit current drawing** check box, the current drawing can be edited in-place by other users even when it is open and is being referenced by another file. This option is selected by default and can also be controlled by the **XEDIT** system variable. The default value of **XEDIT** is 1 and can be changed using the command line.*

Self-Evaluation Test

Answer the following questions and then compare them to those given at the end of this chapter:

1. If an assembly drawing has been created by inserting a drawing, the drawing will be updated automatically, if a change is made in the drawing that was inserted. (T/F)

2. The external reference facility helps you keep a drawing updated no matter when the changes were made in the piece part drawings. (T/F)

3. Objects can be added to a dependent layer. (T/F)

4. While using the **Attachment** option during referencing a drawing, the drawing will reference the nested references too, while the **Overlay** option ignores the nested references. (T/F)

5. The _____ are entries such as blocks, layers, and text styles.

6. If you use the **INSERT** command to insert a drawing, the information about the named objects is lost if the names are _____ and if the names are _____, the drawing will be imported.

7. The _____ button of the **External References** palette is used to attach an xref drawing to the current drawing.

8. The _____ feature loads only that part of the referenced drawing that is required in the existing drawing.

9. The _____ option can be used to overcome the problem of circular reference.

10. If the **Retain changes to Xref layers** check box is _____ in the **External References (Xrefs)** area of the **Open and Save** tab of the **Options** dialog box, the layer settings such as color, linetype, on/off, and freeze/thaw are retained. The settings are saved with the drawing and are used when you xref the drawing the next time.

Review Questions

Answer the following questions:

1. If the xref drawings get updated, changes are not automatically reflected in the assembly drawing when you open an assembly drawing. (T/F)

2. There is a limit to the number of drawings you can reference. (T/F)

3. It is not possible to have nested references. (T/F)

4. Like blocks, the xref drawings can be scaled, rotated, or positioned at any desired location. (T/F)

5. You can change the color, linetype, or visibility (on/off, freeze/thaw) of a dependent layer. (T/F)

6. Which of the following features lets you reference an external drawing without making this drawing a permanent part of the existing drawing?

    (a) demand loading   (b) external reference
    (c) external clipping   (d) insert drawing

7. If an xref has nested references that cannot be found, which of the following will be displayed under the status heading of the **List View** button in the **External References** palette?

    (a) **Orphaned**   (b) **Not found**
    (c) **Unreferenced**   (d) **Unresolved**

8. Which of the following commands can be used from the command line to attach a drawing?

    (a) **-XBIND**   (b) **-XREF**
    (c) **XCLIP**   (d) None of the above

9. Which of the following system variables, when set to 1, will allow AutoCAD to maintain a log file (*.xlg*) for xref drawings?

    (a) **XLOADCTL**   (b) **XLOADPATH**
    (c) **XREFCTL**   (d) **INDEXCTL**

10. Which of the following system variables, when set to 0, will not allow the clipping boundary to be displayed?

    (a) **FRAME**   (b) **XLOADCTL**
    (c) **INDEXCTL**   (d) **XREFCTL**

11. In the _____ drawings, the information regarding dependent symbols is not lost.

12. What is the function of the **FRAME** system variable? Explain.

13. AutoCAD maintains a log file for xref drawings, if the _____ variable is set to 1.

14. It is possible to edit xrefs in-place using the _____ command.

15. You can use the _____ command to add the selected dependent symbols from the xref drawing to the current drawing.

Chapter 18

Exercise

Exercise 1   *Mechanical*

In this exercise, you will start a new drawing and xref the drawings Part-1 and Part-2. You will also edit one of the piece parts to correct the size and use the **XBIND** command to bind some of the dependent symbols to the current drawing. The following are detailed instructions for completing this exercise:

1. Start a new drawing, Part-1, and set up the following layers:

Layer Name	Color	Linetype
0	White	Continuous
Object	Red	Continuous
Hidden	Blue	Hidden2
Center	White	Center2
Dim-Part1	Green	Continuous

2. Draw Part-1 with dimensions, as shown in Figure 18-35. Save the drawing as Part-1.

3. Start a new drawing, Part-2, and set up the following layers:

Layer Name	Color	Linetype
0	White	Continuous
Object	Red	Continuous
Hidden	Blue	Hidden
Center	White	Center
Dim-Part2	Green	Continuous
Hatch	Magenta	Continuous

4. Draw Part-2 with dimensions as shown in the Figure 18-36. Save the drawing as Part-2.

5. Start a new drawing, ASSEM1, and set up the following layers.

Layer Name	Color	Linetype
0	White	Continuous
Object	Blue	Continuous
Hidden	Yellow	Hidden

6. Xref the two drawings Part-1 and Part-2 so that the centers of the two drilled holes coincide. Notice the overlap as shown in Figure 18-37. Save the assembly drawing as ASSEM1.

7. Open the drawing Part-1 and correct the mistake so that there is no overlap. You can do it by editing the line (1.175 dimension) so that the dimension is 1.160.

Figure 18-35 *Drawing of Part-1*

Figure 18-36 *Drawing of Part-2*

Figure 18-37 *Assembly drawing after attaching Part-1 and Part-2*

Chapter 18

8. Open the assembly drawing ASSEM1 and notice the change in the overlap. The assembly drawing gets updated automatically.

9. Study the layers and notice how AutoCAD renames the layers and the linetypes assigned to each layer. Check to see if you can make the layers, belonging to Part-1 or Part-2, current.

10. Use the **XBIND** command to bind the Object and Hidden layers that belong to the drawing Part-1. Check again to see if you can make one of these layers current.

11. Use the **Detach** option to detach the xref drawing Part-1. Study the layer again, and notice that the layers that were edited with the **XBIND** command have not been erased. Other layers belonging to Part-1 are erased.

12. Use the **Bind** option of the **XREF** command to bind the xref drawing Part-2 with the assembly drawing ASSEM1. Open the xref drawing Part-1 and add a border or make any changes in the drawing. Now, open the assembly drawing ASSEM1 and check to see if the drawing is updated.

Answers to Self-Evaluation Test
1. F, **2.** T, **3.** F, **4.** T, **5.** named objects, **6.** duplicated, unique, **7. Attach**, **8.** demand loading, **9. Overlay**, **10.** selected

Chapter *19*

Working with Advanced Drawing Options

Learning Objectives

After completing this chapter, you will be able to:

- *Define multiline style and specify properties of multilines using the **MLSTYLE** command.*
- *Draw various types of multilines using the **MLINE** command.*
- *Edit multilines using the **MLEDIT** command.*
- *Draw **NURBS** splines using the **SPLINE** command.*
- *Edit **NURBS** splines using the **SPLINEDIT** command.*

UNDERSTANDING THE USE OF MULTILINES

The **Multiline** feature allows you to draw composite lines that consist of multiple parallel lines. Multilines consist of parallel lines, which are called elements. You can draw these lines with the **MLINE** command. Before drawing multilines, you need to set the multiline styles. This can be accomplished using the **MLSTYLE** command. Also, editing of the multilines is made possible by the **MLEDIT** command.

DEFINING THE MULTILINE STYLE

Menu Bar: Format > Multiline Style	**Command:** MLSTYLE

You can create the style of multilines using the **MLSTYLE** command. You can specify the number of elements in the multiline and the properties of each element. The style also controls the start and end caps, start and end lines, and the color of multilines and fill. When you invoke this command, AutoCAD displays the **Multiline Style** dialog box (Figure 19-1). In this dialog box, you can set the spacing between the parallel lines, linetype pattern, colors, solid fill, and capping arrangements. By default, the multiline style (STANDARD) has two lines that are offset at 0.5 and -0.5.

 Note
You cannot modify the style of a multiline that is used in the current drawing. You first need to delete the multiline from all the places where it is used and then you can modify its style.

*Figure 19-1 The **Multiline Style** dialog box*

Styles List Box

The **Styles** list box displays the names of the styles. By default, the **STANDARD** style is available in this list box. If several styles have been defined, all of them will be listed in this list box.

Set Current

If several styles have been defined, select a style from the **Styles** list box and choose the **Set Current** button to make the style as the current style. The current multiline style is saved in the **CMLSTYLE** system variable.

New

The **New** button is chosen to create a new multiline style. On choosing this button, the **Create New Multiline Style** dialog box is displayed, as shown in Figure 19-2. The **New Style**

Name edit box lets you enter the name of the new multiline style you want to define. The **Start With** drop-down list is available only if some styles have already been defined. You can select among the existing styles as a reference to create the new style. The new style will start with the properties associated with the style selected from the **Start With** drop-down list. As soon as you specify the name of the new style, the **Continue** button is highlighted. To create a new style, enter the name of the new style in the **New Style Name** edit box and then choose the **Continue** button; the **New Multiline Style** dialog box will be displayed, as shown in Figure 19-3.

*Figure 19-2 The **Create New Multiline Style** dialog box*

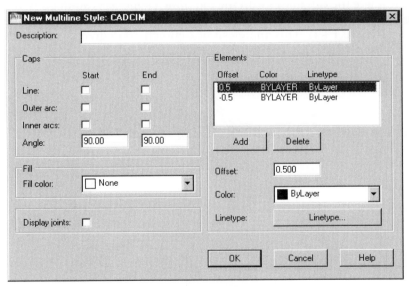

*Figure 19-3 The **New Multiline Style** dialog box*

The Title bar of the **New Multiline Style** dialog box displays the name of the new style. By default, some properties in the dialog box are defined at some preset values. These values correspond to the selection made in the **Start With** drop-down list.

Description

This edit box allows you to enter the description of the new multiline style. The description, including spaces, can be up to 255 characters long. This is an optional edit box and you can leave it blank if you want.

Caps Area

The options under this area are used to set the type and location of the start and end caps of the multilines. These options are discussed next.

Chapter 19

Line. Selecting the corresponding **Start** and **End** check boxes of this option caps the multiline with a line at the start point and endpoint (Figure 19-4).

Outer arc. Selecting the corresponding **Start** and **End** check boxes of this option draws a semicircular arc between the endpoints of the outermost lines (Figure 19-4).

Inner arcs. This option draws a semicircular arc between the two innermost lines at the start point and endpoint of a multiline. The arc is drawn between the even-numbered inner lines (Figure 19-4). For example, if there are two inner lines, an arc will be drawn at their ends. On the other hand, if there are three inner lines, the two outer lines are capped with the arc and the middle line is not capped.

Angle. By default, the end caps of a multiline are drawn tangentially to the multiline elements, that is, the angle the cap makes with the multiline elements is 90-degree. You can control the cap angle at the start and end of a multiline by entering angles in the corresponding **Angle** edit boxes (Figure 19-4). This angle can be between 10-degree and 170-degree.

Figure 19-4 Drawing multilines with different end cap specifications

Fill Area
The options under this area are used to fill the multiline with a specified color. By default, **None** is selected from the **Fill color** drop-down list. As a result, there is no fill effect in the multiline. To specify a fill color, select it from this drop-down list. Selecting the **Select Color** option displays the **Select Color** dialog box, using which you can select additional colors.

Display joints
If you select the **Display joints** check box, AutoCAD will draw a line across all elements of the multiline at the vertex points, where the multiline changes direction. The joints of the multiline are also called miter joints. If you draw only one multiline segment, no line is drawn at the miter joint.

Elements Area

The options in this area modify the element properties of the multiline style. The element properties include the number of lines, color, and linetype of the lines comprising the multiline. The following options are available for setting the properties of the individual lines (elements) that constitute the multiline.

Offset, Color, and Linetype

This list box displays the offset, color, and linetype of each line that constitutes the current multiline style. The lines are always listed in the descending order, based on the offset distance. For example, a line with a 0.5 offset will be listed first and a line with a 0.25 offset will be listed next. You can select an element to be modified from the list box; the selected element is highlighted.

Add

This button allows you to add new lines to the current line style. The maximum number of lines that you can add is 16. When you choose the **Add** button, AutoCAD inserts a line with the offset distance of 0.0, **Color** = BYLAYER, and **Linetype** = Bylayer. The new element gets added between two existing elements. After the line is added, you can change its offset distance, color, or linetype. The offset distance can be entered in the **Offset** edit box. Similarly, you can select the options in the **Color** drop-down list to specify a color for the selected element of the multiline. Also, you can choose the **Linetype** button to display the **Select Linetype** dialog box. You can use this dialog box and its options to load and select the linetype you want to apply to the selected element of a multiline.

Delete

This button allows you to delete a selected element from the multiline style. The highlighted element is removed from the **Elements** list box.

Offset

This edit box allows you to change the offset distance of the selected line in the **Elements** list box. The offset distance is defined with respect to the origin (0,0). The offset distance can be a positive or a negative value, which enables you to center the lines.

Color

This drop-down list allows you to assign a color to the selected line. When you choose the **Select Color** option from the drop-down list, AutoCAD displays the **Select Color** dialog box. You can select a color from the dialog box or enter a color number or name in the **Color** edit box.

Linetype

This button allows you to assign a linetype to the selected line. When you choose this button, AutoCAD displays the **Select Linetype** dialog box. After selecting a linetype in this dialog box, choose **OK** to exit. You can also choose the **Load** button in the **Select Linetype** dialog box to display the **Load or Reload Linetypes** dialog box, where you can select a linetype you want to use. The selected linetype is displayed in the **Loaded Linetype** listbox of the **Select**

Linetype dialog box. You can now select the specific linetype and apply it to a selected element of a multiline.

After you have added the elements to a multiline and set their properties, choose **OK** button to exit the **New Multiline Style** dialog box. An image of the current multiline style is displayed in the **Preview of** area in the **Multiline Style** dialog box. The new style is now listed in the **Style** list box.

Load

Choose this button from the **Multiline Style** dialog box to load a multiline style, that was saved earlier, from an external multiline library file (*acad.mln*). When you choose this button, AutoCAD displays the **Load Multiline Styles** dialog box, see Figure 19-5. This dialog box displays all the multiline styles available in the current *mln* directory. If the style is saved in another directory or file, choose the **File** button to display the **Load Multiline Style From File** dialog box and then select the multiline file that you want to load. Once the file is loaded, all styles saved under it are displayed in the list box. You can select the style and then choose **OK** to make it current.

Figure 19-5 The Load Multiline Styles dialog box

Save

This button lets you save a multiline style to an external file (*.mln*). When you choose this button, AutoCAD displays the **Save Multiline Style** dialog box. This dialog box lists the names of the available predefined multiline style files (*.mln*). From the file listing, select or enter the name of the file where you want to save the current multiline style and choose the **Save** button. By default, the multiline styles get saved in the *acad.mln* file in the *C:\Documents and Settings\User\Application Data\Autodesk\AutoCAD 2010\R18\enu\Support* folder.

Rename

This button allows you to rename the current multiline style. To do so, select the multiline style to be renamed and choose the **Rename** button; the text will be displayed in an edit box. Rename the text and click once on the list box. Note you cannot rename the **STANDARD** multiline style.

Preview of Area

The **Multiline Style** dialog box also displays the current multiline configuration in the **Preview of** area. The panel will display the color, linetype, caps and relative spacing of the lines. As you select another multiline style, the image in the preview area also gets updated accordingly. The name of the multiline style whose preview is being shown is displayed above the **Preview of** area.

DRAWING MULTILINES

Menu Bar: Draw > Multilines	**Command:** MLINE or ML

The **MLINE** command is used for drawing multilines. The following is the prompt sequence for the **MLINE** command:

> Current settings: Justification = Top, Scale = 1.00, Style = CURRENT
> Specify start point or [Justification/Scale/STyle]: *Select a start point or enter an option.*
> Specify next point: *Select the second point.*
> Specify next point or [Undo]: *Select the next point or enter **U** for undo.*
> Specify next point or [Close/Undo]: *Select the next point, enter **U**, or enter **C** for close. You can also right-click to display the shortcut menu and choose **Enter**, **Cancel**, **Recent Input**, **Dynamic Input Close**, **Undo**, **Snap overrides**, **Pan**, **Zoom**, **SteeringWheels**, or **Quick calc**.*

When you invoke the **MLINE** command, it always displays the current status of the multiline justification, scale, and style name. Remember that all line segments created in a single **MLINE** command are a single entity. The options provided under this command are discussed next.

Justification Option

The justification determines how a multiline is drawn between the specified points. Three justifications are available for the **MLINE** command. They are **Top**, **Zero**, and **Bottom**.

Top

This justification produces a multiline, in which the top line coincides with the points you selected on the screen. Since the line offsets in a multiline are arranged in descending order, the line with the largest positive offset will coincide with the selected points. This is the default justification option, see Figure 19-6.

Zero

This option will produce a multiline, in which the zero offset position of the multiline coincides with the selected points. Multilines will be centered if the positive and negative offsets are equal, see Figure 19-6.

Bottom

This option will produce a multiline, in which the bottom line (the line with the least offset distance) coincides with the selected point when the line is drawn from the left to right, see Figure 19-6.

Scale Option

The **Scale** option allows you to change the scale of the multiline. For example, if the scale factor is 0.5, the distance between the lines (offset distance) will be reduced to half, as shown in Figure 19-7. Therefore, the width of the multiline will be half of what was defined in the multiline style. A negative scale factor will reverse the order of the offset lines. The multilines

Chapter 19

will now be drawn such that the line with the maximum negative offset distance is at the top and the line with the maximum positive offset distance is at the bottom. For example, if you enter a scale factor of -0.5, the order in which the lines are drawn will be reversed, and the offset distances will be reduced by half. Here it is assumed that the lines are drawn from left to right. If the lines are drawn from right to left, the offsets are reversed. Also, if the scale factor is 0, AutoCAD forces the multiline into a single line. However, the line will still possess the properties of a multiline. This scale factor does not affect the linetype scale factor (**LTSCALE**).

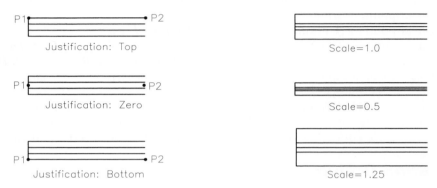

Figure 19-6 *Justifications in multilines*   *Figure 19-7* *Different multiline scales*

STyle Option

The **STyle** option allows you to change the current multiline style. The style must be defined before using the **STyle** option to change the style.

Note
All options explained above are available at the dynamic prompt. To access the options through dynamic prompt, press the down arrow key from the keyboard whenever dynamic prompting is available at the cursor. Now, select the option using the cursor.

EDITING MULTILINES USING GRIPS

Menu Bar: Modify > Object > Multiline	**Command:** MLEDIT

Multilines can be easily edited using grips. When you select a multiline, the grips appear at the endpoints, based on the justification used when drawing multilines. For example, if the multilines are top-justified, the grips will be displayed at the endpoint of the first (top) line segment. Similarly, for zero and bottom-justified multilines, the grips are displayed on the centerline and bottom line, respectively.

Note
*Multilines do not support certain editing commands such as **BREAK**, **CHAMFER**, and **FILLET**. However, commands such as **COPY**, **MOVE**, **ROTATE**, **MIRROR**, **STRETCH**, **SCALE**, and **EXPLODE** as well as most object snap modes can be used with multilines.*

Tip
*You must use the **MLEDIT** command to edit multilines. The **MLEDIT** command has several options that make it easier to edit these lines.*

EDITING MULTILINES USING THE DIALOG BOX

When you invoke the **MLEDIT** command, AutoCAD displays the **Multilines Edit Tools** dialog box, see Figure 19-8. This dialog box contains five basic editing tools. To edit a multiline, first select the editing operation you want to perform by clicking on the image tile. Once you have selected the editing option, AutoCAD will prompt you to select the first and second multiline. After editing one set of multilines, you will be prompted again to select the first multiline or undo the last editing. Press ENTER to exit this command after you have finished editing. The following is the list of options for editing multilines.

Cross Intersection
Closed Cross (CC), Open Cross (OC), Merged Cross (MC)

Tee Intersection
Closed Tee (CT), Open Tee (OT), Merged Tee (MT)

Corner Joint (CJ)
Adding and Deleting Vertices
Add Vertex (AV), Delete Vertex (DV)

Cutting and Welding Multilines
Cut Single (CS), Cut All (CA), Weld All (WA)

Cross Intersection (CC/ OC/MC)

With the **MLEDIT** command options, you can create three types of cross intersections: closed, open, and merged. You must be careful about the order, in which you select the multilines for editing.

*Figure 19-8 The **Multilines Edit Tools** dialog box*

The order in which the multilines are selected determines the edited shape of a multiline (Figure 19-9). The cross intersection can belong to a self-intersecting multiline or two different multilines.

When you invoke the **MLEDIT** command, the **Multilines Edit Tools** dialog box is displayed. Choose the **Closed Cross** image tile; the dialog box is removed from the screen and the following prompts appear in the command line:

Select first mline: *Select the first multiline.*
Select second mline: *Select the second multiline.*

Figure 19-9 *Three types of cross intersections*

Select first mline or [Undo]: *Select another multiline, press ENTER to end the command, or enter **U** to undo the last operation.*

If you enter **U**, AutoCAD will undo the last editing operation and prompt you to select the first multiline. However, if you do not enter **U** and select another multiline, AutoCAD will prompt you to select the second multiline.

Tee Intersection (CT/OT/MT)

With the **MLEDIT** command options, you can create three types of tee-shaped intersections: closed, open, and merged. As with the cross intersection, you must be careful how you select the objects because the order in which you select them determines the edited shape of a multiline (Figure 19-10). The prompt sequence for the tee intersection is the same as that for the cross intersection.

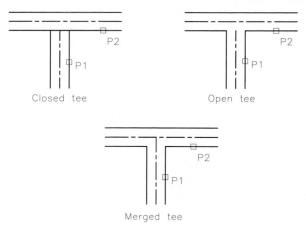

Figure 19-10 *Three types of tee intersections*

When you invoke the **MLEDIT** command, the **Multilines Edit Tools** dialog box is displayed. Choose the **Closed Tee** image tile; the dialog box is removed from the screen and the following prompts appear in the command line.

Select first mline: *Select the first multiline.*
Select second mline: *Select the intersecting multiline.*
Select first mline or [Undo]: *Select another multiline, enter **U** to undo the last operation, or press ENTER to end the command.*

Corner Joint (CJ)

The **Corner Joint** option creates a corner joint between the two selected multilines. The multilines must be two separate objects (multilines) or a self-intersecting multiline (Figure 19-11). When you specify the two multilines, AutoCAD trims or extends the first multiline to intersect with the second one. The prompt sequence that follows when you invoke this option is given next.

Select first mline: *Select the multiline to trim or extend.*
Select second mline: *Select the intersecting multiline.*
Select first mline or [Undo]: *Select another multiline, enter **U** to undo the last operation, or press ENTER to end the command.*

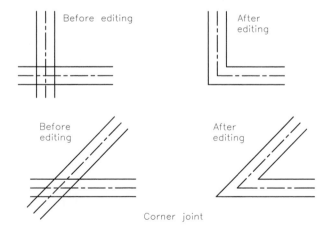

Figure 19-11 Editing corner joints

Adding and Deleting Vertices (AV/DV)

You can use the **MLEDIT** command to add or delete the vertices of a multiline (Figure 19-12). When you select a multiline for adding a vertex, AutoCAD inserts a vertex point at the point where you clicked on the object while selecting it. Later, you can move the vertex by using grips. Similarly, you can also use the **MLEDIT** command to delete vertices by selecting the object whose vertex point you want to delete. AutoCAD removes the vertex that is nearest to the point where you click to select the multiline segment.

Chapter 19

When you invoke the **MLEDIT** command, the **Multilines Edit Tools** dialog box is displayed. Choose the **Add Vertex** image tile; the dialog box is removed from the screen and the following prompts appear in the command line.

Select mline: *Select the multiline for adding vertex.*
Select mline or [Undo]: *Select another multiline for modifications; enter* **U** *to undo the last operation, or press ENTER to exit the command.*

A same prompt sequence appears at the command line when you use the **Delete Vertex** option.

Cutting and Welding Multilines (CS/CA/WA)

You can also use the **MLEDIT** command to cut or weld multilines. When you cut a multiline, it does not create two separate multilines. They continue to be a part of the same object (multiline). You can cut away an area from any of the elements of a multiline or from the entire multiline. Also, the points selected for cutting the multiline do not have to be on the same element of the multiline (Figure 19-13). You can select the two points that define the cut on different elements of the multiline. When cutting a single element of the multiline, if you select the two cut points on different elements, the element you selected first gets cut by the distance defined by the two points.

In the **Multilines Edit Tools** dialog box, choose the **Cut Single** image tile; the dialog box is removed from the screen, and the following prompts appear in the command line.

Select mline: *Select the multiline. (The point where you select the multiline specifies the first cut point.)*
Select second point: *Select the second cut point.*
Select mline or [Undo]: *Select another multiline to cut, enter* **U** *to undo the last operation, or press ENTER to exit the command.*

The **Weld** option reconnects the multilines that have been cut by using the **Cut Single** or **Cut All** option of the **MLEDIT** command (Figure 19-13). You cannot weld or join separate multilines.

Figure 19-12 Adding and deleting vertices   *Figure 19-13 Cutting and welding multilines*

In the **Multilines Edit Tools** dialog box, choose the **Weld All** image tile; the dialog box is removed from the screen, and the following prompts appear in the command line.

Select mline: *Select the multiline.*
Select second point: *Select a point on the other side of the cut.*
Select mline or [Undo]: *Select another multiline to modify, enter U to undo the last operation, or press ENTER to exit the command.*

Note
*The following system variables store values for various aspects of multilines. The **CMLJUST** variable stores the justification of the current multiline (0-Top, 1-Middle, 2-Bottom). The **CMLSCALE** variable stores the scale of the current multiline (Default scale = 1.0000) and the **CMLSTYLE** variable stores the name of the current multiline style (Default style = STANDARD).*

Tip
*You can also edit multilines by entering **-MLEDIT** command at the pointer input. All options are displayed at the dynamic prompt, see Figure 19-14.*

Figure 19-14 *Dynamic prompt for the **-MLEDIT** command*

Example 1   *Architectural*

In the following example, you will create a multiline style that represents a wood-frame wall system. The wall system consists of ½" wallboard, 4 X ½" wood stud, and ½" wallboard.

1. Enter **MLSTYLE** at the Command prompt to display the **Multiline Style** dialog box. Also, the current style, **STANDARD**, will be displayed in the **Styles** list box.

2. Choose the **New** button to invoke the **Create New Multiline Style** dialog box. Enter **MYSTYLE** in the **New Style Name** edit box and choose **Continue** to display the **New Multiline Style: MYSTYLE** dialog box.

3. Enter **Wood-frame Wall System** in the **Description** edit box.

4. Select the **0.5** line definition in the **Elements** list box. In the **Offset** edit box, replace **0.500** with **1.00**. This redefines the first line as being 1.00" above the centerline of the wall.

5. Similarly, select the **-0.5** line definition in the **Elements** list box. In the **Offset** edit box, replace **-0.500** with **-1.00**. This redefines the second line as being 1.00" below the centerline of the wall.

Chapter 19

6. Now, choose the **Add** button; a new line is added with the offset value of 0 and is highlighted in the **Elements** list box.

7. In the **Offset** edit box, replace **0.000** with **1.50**.

8. Select **Yellow** from the **Color** drop-down list.

9. Repeat steps 6 through 8, but this time use the value **-1.50** in step 7 to add another line to the current multiline style. The color of this line should be red.

10. Choose the **OK** button and return to the **Multiline Style** dialog box. The new multiline style will be displayed in the **Preview of** area.

11. Choose the **Set Current** button to make **MYSTYLE** current.

12. Choose the **OK** button from the **Multiline Style** dialog box to return to the drawing area. To test the new multiline style, use the **MLINE** command and draw a series of lines (Figure 19-15).

1/2" Wallboard
4 x 1/2" Wood Stud
1/2" Wallboard

Figure 19-15 Multiline style created for Example 1

CREATING REVISION CLOUDS

Ribbon:	Home > Draw > Revision Cloud or Annotate > MarkUp > Revision Cloud		
Tool Palettes:	Draw > Revision Cloud	Toolbar:	Draw > Revision Cloud
Menu Bar:	Draw > Revision Cloud	Command:	REVCLOUD

The **REVCLOUD** command is used to create a cloud shaped polyline. Figure 19-16 shows the use of the **REVCLOUD** command. This command can be used to highlight the details of a drawing. The prompt sequence that will follow when you choose **Home > Draw > Revision Cloud** from the **Ribbon** is as follows:

Minimum arc length: 0.5000   Maximum arc length: 0.5000   Style: Normal
Specify start point or [Arc length/Object/Style] <Object>: *Specify the start point of the revision cloud.*
Guide crosshairs along cloud path...

As you move the cursor, different arcs of the cloud with varied radii are drawn. When the start point and endpoints meet, the revision cloud is completed and you get a message.

Revision cloud finished.

You can define the length of the arcs to be drawn using the **Arc length** option. Also, all the arcs drawn are of constant length. You can convert a closed loop into a revision cloud using the **Object** option. Note that the selected closed loop should be a single entity such as an ellipse, a circle, a rectangle, a polyline, and so on. The **Style** option is used to define the arc style for the revision cloud. The default style is **Normal** that creates a revision cloud similar to the one shown in Figure 19-16. You can change the style to **Calligraphy**, and this creates a revision cloud similar to the one shown in Figure 19-17.

Figure 19-16 Creating revclouds

Figure 19-17 Calligraphy revclouds

Note
*The **REVCLOUD** stores the last used arc length in the system registry. This value is multiplied by **DIMSCALE** to provide consistency when the program is used with drawings that have different scale factors.*

CREATING WIPEOUTS

Ribbon:	Home > Draw > Wipeout or Annotate > Markup > Wipeout
Command:	WIPEOUT

The **WIPEOUT** command creates a polygonal area to cover the existing objects with the current background color. The area defined by this command is governed by the wipeout frame. The frame can be turned on and off for editing and plotting the drawings, respectively.

This command can be used to add notes and details to the drawing. The prompt sequence that will follow when you choose **Home > Draw > Wipeout** from the **Ribbon** is as follows.

Specify first point or [Frames/Polyline] <Polyline>: *Specify the start point of the wipeout.*
Specify next point: *Specify the next point of the wipeout.*
Specify next point or [Undo]: *Specify the next point of the wipeout.*
Specify next point or [Close/Undo]: *Specify the next point of the wipeout.*

Figure 19-18 shows a drawing before creating wipeout and Figure 19-19 shows a drawing after creating wipeout.

Figure 19-18 Drawing before creating a wipeout   *Figure 19-19* Drawing after creating a wipeout

If you do not want the frame of the wipeout to be displayed, enter **F** at the **Specify first point or [Frames/Polyline] <Polyline>** prompt and turn the frame off. The display of frames of all the existing wipeouts will be turned off. Also, the display of frames of all the new wipeouts will be turned off.

CREATING NURBS

| Ribbon: | Home > Draw > Spline | **Menu Bar:** Draw > Spline |
| **Tool Palettes:** | Draw > Spline | **Toolbar:** Draw > Spline | **Command:** SPLINE |

The NURBS is an acronym for **NonUniform Rational Bezier-Spline**. These splines are considered true splines. In AutoCAD, you can create NURBS using the **SPLINE** command. The spline created with the **SPLINE** command (Figure 19-20) is different from the spline created using the **PLINE** command. The nonuniform aspect of the spline enables the spline to have sharp corners because the spacing between the spline elements that constitute a spline can be irregular. Rational means that irregular geometry such as arcs, circles, and ellipses can be combined with free-form curves. The Bezier-spline (B-spline) is the core that enables accurate fitting of curves to input data with Bezier's curve-fitting

Figure 19-20 Invoking the **SPLINE** command from the **Draw** panel

interface. Not only are spline curves more accurate compared to smooth polyline curves, but they also use less disk space. The following is the prompt sequence for creating the spline shown in Figure 19-21.

Specify first point or [Object]: *Select point P1.*
Specify next point: *Select point P2.*
Specify next point or [Close/Fit tolerance] <start tangent>: *Select point P3.*
Specify next point or [Close/Fit tolerance] <start tangent>: *Select point P4.*
Specify next point or [Close/Fit tolerance] <start tangent>: *Select point P5.*
Specify next point or [Close/Fit tolerance] <start tangent>: *Select point P6.*
Specify next point or [Close/Fit tolerance] <start tangent>: *Press ENTER to end the process of point specification.*
Specify start tangent: *Press ENTER to accept the default start tangent.*
Specify end tangent: *Select point P7.*

You can specify the start and end tangents to change the direction in which the spline curve starts and ends at the **Specify start tangent:** and **Specify end tangent:** prompts. The **SPLINE** command options are discussed next.

Options for Creating Splines

The options under this command for creating the splines are as follows.

Object

This option allows you to change a 2D or 3D spline fitted polyline into a NURBS. The original splined polyline is deleted if the system variable **DELOBJ** is set to 1, which is the default value of the variable. You can change a polyline into a splined polyline using the **Spline** option of the **PEDIT** command.

Close

This option allows you to create closed NURBS. When you use this option, AutoCAD will automatically join the endpoint of the spline with the start point, and you will be prompted to define the start tangent only.

Fit Tolerance

This option allows you to control the fit of the spline between specified points. By default, this value is zero and so the spline passes through the points through which it is created. Using this option, you can specify the tolerance values that will govern the spline creation, see Figure 19-22. The splines will be offset from the specified point by a distance equal to the tolerance value. The smaller the value of the tolerance, the closer the spline will be to the specified points.

Start and End Tangents

This allows you to control the tangency of the spline at the start point and endpoint of the spline. If you press ENTER at these prompts, AutoCAD will use the default value. By default, the tangency is determined by the slope of the spline at the specified point.

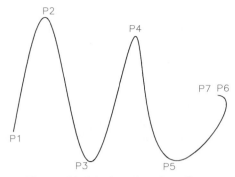

Figure 19-21 *Creating the spline*

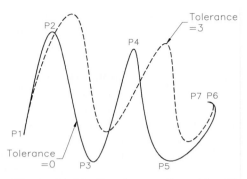

Figure 19-22 *Creating a spline with a **Fit Tolerance** of 0 and 3*

EDITING SPLINES

Ribbon:	Home > Modify > Edit spline	Menu Bar: Modify > Object > Spline
Tool Palettes:	Modify > Edit spline	**Toolbar:** Modify II > Edit Spline
Command:	SPLINEDIT	

The NURBS can be edited using the **SPLINEDIT** command (Figure 19-23). With this command, you can fit data in the selected spline, close or open the spline, move vertex points, and refine or reverse a spline. Apart from the ways mentioned in the preceding command box, you can also invoke the **SPLINEDIT** command by choosing **Spline** from the shortcut menu that is displayed when you select a spline and right-click. The prompt sequence that will follow when you choose **Home > Modify > Edit Spline** button in the **Ribbon** is given next.

Figure 19-23 *Choosing the **Edit Spline** button from the **Modify** panel*

Select spline: *Select the spline that is to be edited if not selected already, using the above mentioned shortcut menu.*
Enter an option [Fit data/Close/Move vertex/Refine/rEverse/convert to Polyline/Undo]: *Select any one of the options.*

Options for Editing Splines

The options under this command for editing the splines are described next.

Fit Data

When you draw a spline, the spline is fit to the specified points (data points). The **Fit data** option allows you to edit these points. You can add, delete, or move the data points. These data points or control points are also referred to as fit points. For example, if you want to redefine the start and end tangents of a spline, select the **Fit Data** option, then select the **Tangents** option. The prompt sequence that will follow when you invoke this option is given next.

Select spline: *Select the spline that is to be edited.*
Enter an option [Fit data/Close/Move vertex/Refine/rEverse/convert to Polyline/Undo]:
F
Enter a fit data option
[Add/Close/Delete/Move/Purge/Tangents/toLerance/eXit] <eXit>:**T**

The start and end tangent points can be selected or their coordinates can be entered. The options available within the **Fit data** option are as follows:

Add. You can use this option to add new fit points to the spline. When you invoke this option, you will be prompted to specify the control point. This control point should be one of the existing control points on the spline. After selecting the existing control point, you will be prompted to specify the location of the new control points. The fit point you select and the next fit point appear as selected grips. You can now add a fit point between these two selected fit points, as shown in Figure 19-24. If you select the start point or endpoints of the spline, only those points are highlighted. When you select the start point of the spline, you are prompted to specify whether you want to add the new fit point before or after the start point of the spline. AutoCAD will continue prompting for the location of new control points until you press ENTER at the **Specify new point <exit>** prompt.

Close/Open. This option allows you to close an open spline or open a closed spline, see Figure 19-25. If the spline is open, the **Close** option is available and if the spline is closed, the **Open** option is displayed.

Delete. This option allows you to delete a selected fit point from the spline. You can continue deleting fit points from a spline until only two fit points are left in the spline.

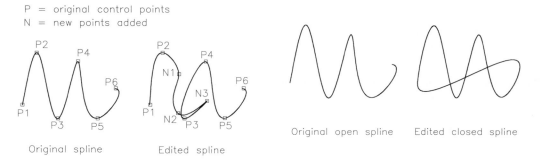

Figure 19-24 *The original spline and the edited spline after adding fit points*

Figure 19-25 *An open spline and a closed spline*

Move. You can move the fit points by using this option. When you select this option, the start point of the spline is highlighted and the following prompt sequence appears:

Specify new location or [Next/Previous/Select point/eXit] <N>: *Select a new location for the start point using the mouse pick button or enter any one of these options.*

Chapter 19

You can enter **N** if you want to select the next point, **P** if you want to select a previous point, or **S** if you want to select any other point. If you enter **X** at the preceding prompt, you can exit the command. Figure 19-26 shows the movement of data points in a spline. Remember that this option is used to move only the data points on the spline and not the control points on the Bezier control frame.

Purge. This option allows you to remove fit point data from a spline. This reduces the file size, which is useful when a drawing, for example, a landscape contour map, contains large number of splines. Purging simplifies the spline definition and the drawing file size. But, once the fit point data has been removed from a spline, editing a spline gets difficult. Also, the **Fit data** option is not displayed when you again use the **SPLINEDIT** command on the edited spline.

Tangents. This option allows you to modify the tangents of the start and endpoints of the selected spline, see Figure 19-27. When you select this option, you will be prompted to specify the start and end tangents for the spline. You can specify the start and end tangents or use the systems default tangents.

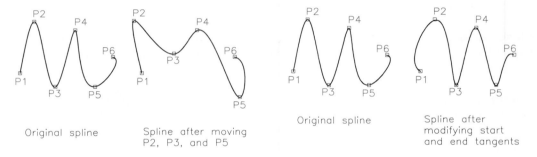

Figure 19-26 The original spline and the edited spline after moving the data points

Figure 19-27 The original spline and the spline after modifying the start and end tangents

toLerance. This option allows you to modify the fit tolerance values of a selected spline. As discussed while creating the splines, different tolerance values produce different splines. A smaller tolerance value creates a spline that passes very closely through the definition points of a spline. When you invoke this option, you will be prompted to specify the tolerance value for the spline.

eXit. This option allows you to exit the **Fit data** option of editing splines.

Close/Open
This option allows you to close an open spline or open a closed spline. When you select the **Close** option, AutoCAD lets you open, move the vertex, or refine or reverse the spline.

Move vertex

When you draw a spline, it is associated with the Bezier control frame. The **Move vertex** option allows you to move the vertices of the control frame. To display this frame with the spline, set the value of the **SPLFRAME** system variable to 1. The default value of this system variable is 0 and the frames are not displayed with the spline. The **Move vertex** option is similar to the **Move** option of the **Fit data** option of editing the splines. The prompt sequence that follows when you invoke this option is given next.

> Enter an option [Fit data/Close/Move vertex/Refine/rEverse/convert to Polyline/Undo]: **M**
> Specify new location or [Next/Previous/Select point/eXit] <N>:

Refine

This option allows you to refine a spline by adding more control points in it, elevating the order, or adding weight to vertex points. The prompt sequence that will follow when you invoke this option is given next.

> Enter an option [Fit data/Close/Move vertex/Refine/rEverse/convert to Polyline/Undo]: **R**
> Enter a refine option [Add control point/Elevate order/Weight/eXit] <eXit>:

Add control point. This option allows you to add more control points on the spline. When you invoke this option, you will be prompted to specify the location of the new point on the spline. You can directly specify the location of the new point using the left mouse button. A new control point will be added at the specified location.

Elevate order. This option allows you to increase the order of a spline curve. An order of the curve can be defined as the highest power of the algebraic expression that defines the spline plus 1. For example, the order of a cubic spline will be $3 + 1 = 4$. Using the **Elevate order** option, you can increase this order for a selected spline. This results in more control points on the curve and a greater possibility of controlling the spline. The value of the spline order varies from 4 to 26. You can only increase the order and not decrease it. For example, if a spline order is 18, you can elevate its order to any value greater than 18, but not less than 18.

Weight. You can also add weight to any of the vertices of the spline by using this option. When weight is added to a particular vertex, the spline gets pulled more towards it. Similarly, a lower value of weight of a particular spline vertex will result in the spline getting pulled less towards that particular vertex. In other words, adding weight to a particular point will force the selected point to maintain its tangency with the point. The more weight added to the point, the more is the distance through which the spline will remain tangent to the point. By default, the spline gets pulled equally towards the vertices of the spline. The default value of weight provided to each control point is 1.0 and can have only positive values. Once you have added the weight to a point, you can proceed to the next point. You can also directly select the point to which the weight has to be added. The prompt sequence for using this option is given next.

Enter a refine option [Add control point/Elevate order/Weight/eXit] <eXit>: **W**
Enter new weight (current = 1.0000) or [Next/Previous/Select point/eXit] <N>:

rEverse

This option allows you to reverse the spline direction. This implies that when you apply this option to a spline, its start point becomes its endpoint and vice versa.

convert to Polyline

This option allows you to convert a spline to polyline. On specifying this option, you will be prompted to specify a precision value. Specify a value between 0 and 99, and then press ENTER; the spline will be converted to polyline. The precision value determines the accuracy of the resulting polyline.

Undo

This option will undo the previous editing operation applied to a spline within the current session of the **SPLINEDIT** command. You can continue to use this option till you reach the spline as it was when you started to edit it.

Self-Evaluation Test

Answer the following questions and then compare them to those given at the end of this chapter:

1. The name of the multiline style that you enter in the **New Style Name** edit box of the **Create New Multiline style** dialog box can have up to thirty-one characters and they contain spaces. (T/F)

2. The **Zero** option will produce a multiline so that the zero offset position of the multiline coincides with the selected points. (T/F)

3. Multilines cannot be edited using grips. (T/F)

4. When using the **Top** justification option to draw a multiline, the line with the largest positive offset coincides with the points selected on the screen. (T/F)

5. When a multiline is cut, two separate multilines are created. (T/F)

6. The maximum number of lines you can add to a multiline style is up to _____.

7. When using the **Inner arcs** option for multilines, the capping arc is drawn between the _____ numbered inner lines.

8. The **Multilines Edit tools** dialog box contains _____ basic editing tools.

9. The _____ option of the **Fit data** option of the **SPLINEDIT** command reduces the file size by removing the fit data of the selected splines.

10. The _____ system variable stores the justification of the current multiline.

Review Questions

Answer the following questions:

1. By default, the *.mln* files are saved in the *AutoCAD2010/Support* folder. (T/F)

2. Once the **Purge** option of the **Fit Data** option of the **SPLINEDIT** command is used on a spline, it becomes difficult to edit it and the **Fit Data** option of the **SPLINEDIT** command is no longer available for the particular spline. (T/F)

3. You can break or trim a multiline. (T/F)

4. You cannot change the lineweights of the lines that constitute a multiline. (T/F)

5. You cannot weld or join separate multilines. (T/F)

6. In which file are the multiline styles saved by default?

    (a) *acad.mln*   (b) *acad.pat*
    (c) *acad.mpr*   (d) *acad.plt*

7. Which of the following is the default start and end angle for multilines?

    (a) 90   (b) 180
    (c) 45   (d) 0

8. AutoCAD forces the multiline into a single line if the scale factor of the multiline is

    (a) 0   (b) 1
    (c) 2   (d) 5

9. Which of the following commands does not work on multilines?

    (a) **MIRROR**   (b) **FILLET**
    (c) **STRETCH**   (d) **EXPLODE**

10. Which option can be used to reverse the direction of spline creation while editing the splines?

    (a) **Fit data**   (b) **Refine**
    (c) **rEverse**   (d) **Weight**

Chapter 19

11. The current multiline style is saved in the _____ system variable and the default value is **STANDARD**.

12. Three justifications are available for the **MLINE** command. They are _____, _____, and _____.

13. Using the **MLEDIT** command options, you can create three types of cross intersections. They are _____, _____, and _____.

14. If the Fit tolerance value of a spline is _____, the spline passes exactly through the fit points of the spline.

15. To display the Bezier control frame with the spline, set the value of the _____ system variable to 1.

Exercises

Exercise 1   *Mechanical*

Create the drawing shown in Figure 19-28. Assume the missing dimensions.

Figure 19-28 Drawing for Exercise 1

Exercise 2   *Architectural*

Create the drawing shown in Figure 19-29. Use the **MULTILINE** command to draw the walls. The wall thickness is 12 inches. Assume the missing dimensions.

Figure 19-29 Drawing for Exercise 2

Problem-Solving Exercise 1   *General*

Create the drawing shown in Figure 19-30. Some of the reference dimensions are given in the drawing. Assume the missing dimensions so that the drawing looks similar to the one given below.

Figure 19-30 Drawing for Problem-Solving Exercise 1

Problem-Solving Exercise 2

General

Create the drawing shown in Figure 19-31. Some of the reference dimensions are given in the drawing. Assume the missing dimensions so that the drawing looks similar to the one given below.

Figure 19-31 Drawing for Problem-Solving Exercise 2

Problem-Solving Exercise 3   *General*

Create the drawing shown in Figure 19-32. Some of the reference dimensions are given in the drawing. Assume the missing dimensions so that the drawing looks similar to the one given below.

Figure 19-32 Drawing for Problem-Solving Exercise 3

Problem-Solving Exercise 4   *Architecture*

Create the drawing shown in Figure 19-33. Some of the reference dimensions are given in the drawing. Assume the missing dimensions so that the drawing looks similar to the one given below.

Figure 19-33 Drawing for Problem-Solving Exercise 4

Problem-Solving Exercise 5   *Architecture*

Create the drawing shown in Figure 19-34. Some of the reference dimensions are given in the drawing. Assume the missing dimensions so that the drawing looks similar to the one given below.

Figure 19-34 Drawing for Problem-Solving Exercise 5

Problem-Solving Exercise 6

Architecture

Create the drawing shown in Figure 19-35. Some of the reference dimensions are given in the drawing. Assume the missing dimensions so that the drawing looks similar to the one given below.

Figure 19-35 Drawing for Problem-Solving Exercise 6

Problem-Solving Exercise 7   *Architecture*

Create the drawing shown in Figure 19-36. Some of the reference dimensions are given in the drawing. Assume the missing dimensions so that the drawing looks similar to the one given below.

Figure 19-36 *Drawing for Problem-Solving Exercise 7*

Problem-Solving Exercise 8   *Architecture*

Create the drawing shown in Figure 19-37. Some of the reference dimensions are given in the drawing. Assume the missing dimensions so that the drawing looks similar to the one given below.

Figure 19-37 *Drawing for Problem-Solving Exercise 8*

Problem-Solving Exercise 9   *Architecture*

Create the drawing shown in Figure 19-38. Some of the reference dimensions are given in the drawing. Assume the missing dimensions so that the drawing looks similar to the one given below. **

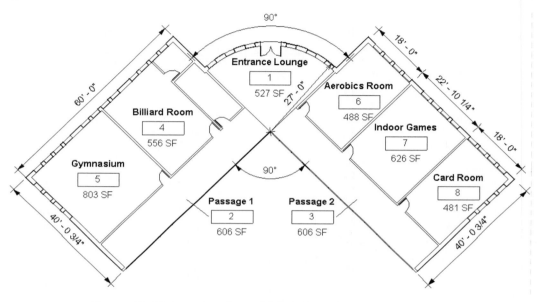

Figure 19-38 *Drawing for Problem-Solving Exercise 9*

Chapter 20

Grouping and Advanced Editing of Sketched Objects

Learning Objectives

After completing this chapter, you will be able to:

- *Group sketched objects.*
- *Select and cycle through defined groups.*
- *Change properties and location of sketched objects.*
- *Perform editing operations on polylines.*
- *Explode compound objects and undo previous commands.*
- *Rename named objects and remove unused named objects.*

GROUPING SKETCHED OBJECTS USING THE OBJECT GROUPING DIALOG BOX

Command: GROUP or G

You can use the **Object Grouping** dialog box (Figure 20-1) to group AutoCAD objects and assign a name to the group. Once you have created groups, you can select the objects by the group name. The individual characteristics of an object are not affected by forming groups. Groups are simply a mechanism that enables you to select all the objects in a group together for editing. It makes the object selection process easier and faster. Objects can be members of several groups. Also, a larger group can contain another smaller group. This may be referred to as nested groups.

*Figure 20-1 The **Object Grouping** dialog box*

Although an object belongs to a group, you can still select an object as if it did not belong to any group. Groups can be selected by entering the group name or by selecting an object that belongs to the group. You can also highlight the objects in a group or sequentially highlight the groups to which an object belongs. The **Object Grouping** dialog box can be displayed by invoking the **GROUP** command.

Group Name List Box

The **Group Name** list box in the **Object Grouping** dialog box displays the names of the existing groups. The list box also displays whether a group is selectable or not, under the **Selectable** column. A selectable group is one in which all members of the group are selected on selecting a single member. Members that are in frozen or locked layers are not selected. You can also make a group non-selectable; this allows you to select the objects in the group individually.

Group Identification Area

The options under the **Group Identification** area are as follows:

Group Name

The **Group Name** edit box displays the name of the selected group. You can also use the **Group Name** edit box to enter the name of a new group. You can enter any name, but it is recommended that you use a name that reflects the type of objects in the group. For example, the group name WALLS can include all lines that form walls. Group names can be up to thirty-one characters long and can include letters, numbers, and special characters ($, _, and -). Group names cannot have spaces.

Description

The **Description** edit box displays the description of the selected group. It can be used to enter the description of a group. The description text can be up to sixty-four characters long, including spaces.

Find Name

The **Find Name** button is used to find the group name or names associated with an object. When you select this button, the dialog box temporarily disappears from the screen and AutoCAD prompts you to select an object of the group. Once you select the object, the **Group Member List** dialog box (Figure 20-2) will be displayed, showing the group name(s) to which the selected object belongs.

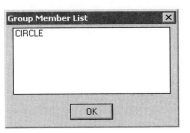

*Figure 20-2 The **Group Member List** dialog box*

Highlight

The **Highlight** button is used to highlight the objects in the selected group name. This button is available only when you select a group from the **Group Name** list box. When you choose the **Highlight** button, the objects that are members of the selected group are highlighted in the drawing and the **Object Grouping** message box (Figure 20-3) is displayed. Choose the **Continue** button or press ENTER to return to the **Object Grouping** dialog box.

*Figure 20-3 The **Object Grouping** message box*

Include Unnamed

The **Include Unnamed** check box is used to display the names of the unnamed groups in the **Object Grouping** dialog box. The unnamed groups are created when you select the **Unnamed** check box in the **Create Group** area or when you copy the objects that belong to a group. AutoCAD automatically groups and assigns a name to the copied objects. The format of the name is An (for example, *A1, *A2, *A3). If you select the **Include Unnamed** check box, the unnamed group names (*A1, *A2, *A3, ...) will also be displayed in the **Group Name** list box of the dialog box. The unnamed groups can also be created by not assigning a name to the group (see **Unnamed** in the next section, "**Create Group** Area").

Create Group Area

The options under this area are discussed next.

New

The **New** button is used to create a new group. Enter the name of the group in the **Group Name** edit box and then choose this button. Once you choose the **New** button, the **Object Grouping** dialog box will be temporarily closed and you will be prompted to select the objects to be included in the group. After selecting the objects, the dialog box will be redisplayed and a new group will be created from the selected objects. The new group name is now displayed in the **Group Name** list box of the dialog box. The group names in the **Group Name** list box are listed alphabetically.

Selectable

The **Selectable** check box allows you to define a group as selectable. A selectable group has a property that if you select any one object in the group, the entire group will be selected. If you clear this check box while defining a group, the new group created will not be selectable. Therefore, you need to select all members of the group individually as the entire group will not be selected upon selecting the individual entities of the group.

Even if a group is defined as selectable, and the **PICKSTYLE** system variable is set to 0, you will not be able to select the entire group by selecting one of its members. You will be able to individually select members of the group. You can also turn the group selectability on or off by selecting or clearing the **Object grouping** check box in the **Selection modes** area of the **Selection** tab of the **Options** dialog box or by using the SHIFT+CTRL+A toggle keys.

Unnamed

When you create a group, you can assign it a name or leave it unnamed. If you select the **Unnamed** check box, AutoCAD will automatically assign a group name to the selected objects and these unnamed groups will be listed in the **Group Name** list box. To view the name of the unnamed group the **Include Unnamed** check box should be selected. The format of the name is A*n* (*A1, *A2, *A3 ...), where *n* is a number incremented with each new unnamed group.

Change Group Area

All the options under this area, other than the **Re-order** button, will be available only when you highlight a group name by selecting it in the **Group Name** list box of the dialog box. The options in this area modify the properties of the existing group. The options are as follows.

Remove

The **Remove** button is used to remove objects from the selected group. Once you select the group name from the **Group Name** list box and then choose the **Remove** button, the dialog box will be temporarily closed and AutoCAD will display the following prompts at the Command line.

Select objects to remove from group...
Remove objects: *(Select the objects that you want to remove from the selected group.)*

When you select objects to be removed on the screen, you will notice that all the objects belonging to the selected group are highlighted. Now, as soon as you select the objects to be removed from the group, they no longer appear highlighted. If you remove all objects from the group, the group name will still exist, unless you use the **Explode** option to remove the group definition (discussed later in this section). If you remove objects and then add them later in the same drawing session, the objects retain their order in the group.

Add

The **Add** button is used to add objects to the selected group. When you select this option, AutoCAD will prompt you to select the objects you want to add to the selected group. The prompt sequence is similar to the **Remove** option. Similar to using the Remove option, all objects belonging to the selected group are highlighted and as soon as you select an object to be added to the group, it also gets highlighted.

Rename

The **Rename** button is used to rename the selected group. To rename a group, first select the group name from the **Group Name** list box, enter the new name in the **Group Name** edit box, and then choose the **Rename** button. AutoCAD will rename the specified group. You can use this option to rename unnamed groups that were discussed earlier.

Re-Order

The **Re-Order** button allows you to change the order of the objects in the selected group. The objects are numbered in the order in which you select them when selecting objects for the group. Sometimes, when creating a tool path, you may want to change the order of these objects to get a continuous tool motion. You can do this using the **Re-Order** button. This option is also useful in some batch operations where one object needs to be on top of another object for display reasons. When you choose the **Re-Order** button, AutoCAD displays the **Order Group** dialog box (Figure 20-4) where you can change the order of the group members. The following example explains the use of the **Re-Order** button using the **Order Group** dialog box.

Example 1   *General*

Figure 20-5 shows four objects that are in group G1. Determine the order of these objects and then reorder them in a clockwise direction.

1. Invoke the **Object Grouping** dialog box, and make the group (G1) current by selecting it in the **Group Name** list box.

2. Choose the **Re-Order** button in the dialog box; the **Order Group** dialog box is displayed. Select group G1 in the **Group Name** list box, if not already selected, to highlight it. All the group names present in the current drawing are displayed in this list box. The

*Figure 20-4 The **Order Group** dialog box*

Description area displays the corresponding description text that was entered in the **Description** edit box of the **Object Grouping** dialog box when defining the particular group.

3. Use the **Highlight** button to highlight the current order of individual objects in the selected group. Figure 20-5(a) shows the order of objects in group G1. When you choose the **Highlight** button, the **Object Grouping** dialog box (Figure 20-6) is displayed. You can use the **Next** and **Previous** buttons to cycle through the group objects. To create a tool path so that the tool moves in the clockwise direction, you need to reorder the objects, as shown in Figure 20-5(b).

Figure 20-5 Changing the order of group objects

4. To get a clockwise tool path, you must switch object numbers 1 and 2. You can do this by entering the necessary information in the **Order Group** dialog box. Enter **1** in the **Remove from position {0-3}** edit box and **2** in the **Enter new position number for the object {0-3}** edit box. Enter **1** in the **Number of objects (1-4)** edit box, since there is only one object to be replaced.

*Figure 20-6 The **Object Grouping** dialog box*

5. After entering the information, choose the **Re-Order** button to define the new order of the objects. You can confirm the change by choosing the **Highlight** button again and cycling through the objects.

Description

The **Description** button lets you change the description of the selected group. To change the description of a group, first select the group name from the **Group Name** list box, enter the new description in the **Description** edit box, and then choose the **Description** button. AutoCAD will update the description of the specified group.

Explode

The **Explode** option deletes group definition of the selected group. The objects that were in the group become regular objects without a group reference. If you had made copies of the group that was exploded in a drawing, they shall remain as unnamed groups in the drawing. You can select the **Include Unnamed** check box in the **Group Identification** area of the **Object Grouping** dialog box to view the names of the unnamed groups in the **Group Name** list box. You can then select them in the list box and explode them or rename them as required.

Selectable

The **Selectable** button changes the selectable status of the selected group. To change the selectable status of the group, first select the group name from the **Group Name** list box, and then choose the **Selectable** button. This button acts like a toggle key between the options of selectable and not selectable. If the selectable status for a group is displayed as "**Yes**", you can choose the **Selectable** button to change it to "**No**". Selectable implies that the entire group gets selected when one object in the group is selected. If the selectable status is "**No**", you cannot select the entire group by selecting one object in the group.

SELECTING GROUPS

You can select a group by name by entering **G** at the **Select objects** prompt. For example, if you have to move a particular group, choose the **Move** button from the **Modify** panel in the **Home** tab of the **Ribbon** and then enter the following prompt sequence:

> Select objects: **G**
> Enter group name: *Enter the group name.*
> n found
> Select objects: Enter

Instead of entering **G** at the **Select Objects:** prompt, if you select any member of a selectable group, all the group members get selected. Make sure that the **PICKSTYLE** system variable is set to 1 or 3. Also, the group selection can be turned on or off by pressing SHIFT+CTRL+A. If the group selection is off and you want to erase a group from a drawing, choose the **Erase** button from the **Modify** panel. The prompt sequence for erasing the group is given next.

> Select objects: *Press SHIFT+CTRL+A*
> <Groups on> *Select an object that belongs to a group. (If the group has been defined as selectable, all objects belonging to that group will be selected.)*

If the group has not been defined as selectable, you cannot select all objects in the group, even if you turn the group selection on by pressing the SHIFT+CTRL+A keys. This setting can also be changed in the **Selection** tab of the **Options** dialog box. You can select or clear the **Object Grouping** check box in the **Selection Modes** area of the dialog box to turn group selectability on or off.

Tip

The combination of SHIFT+CTRL+A is used as a toggle to turn the group selection on or off.

Note

*You can also use the Command line form of the **GROUP** command by entering **-GROUP** at the Command prompt.*

CHANGING PROPERTIES OF AN OBJECT

AutoCAD provides you different options for changing the properties of an object. These options are discussed next.

Using the Properties Palette

Ribbon:	View > Palettes > Properties	Toolbar:	Standard > Properties
Quick Access:	Properties (*Customize to Add*)	Command:	PROPERTIES or PR

The categories displayed in the **Properties** palette depend on the type of object selected. The **General** category displays the general properties of objects, such as color, layer, linetype, linetype scale, plot style, lineweight, hyperlink, and thickness. The procedure to change the general properties of objects using the **Properties** palette has already been discussed earlier in Chapter 4, Working with Drawing Aids. In this chapter, the **Geometry** category containing properties that control the geometry of an object will be discussed. Depending on the type of object selected, the **Geometry** category will contain a set of different properties. If you have selected many types of objects in a drawing, the edit box at the top of the **Properties** palette display **All, General,** and **3D Visualization** categories. If you select a type of object from the drop-down list, the corresponding categories of the object properties will be displayed in the palette. You can also invoke the **Properties** palette from the shortcut menu displayed on selecting the object.

Selecting a Line

If you have selected a line in the drawing, the **Properties** palette appears similar to what is shown in Figure 20-7, where you can change the properties of the selected object.

Start X/Start Y/Start Z. The **Start X/Start Y/Start Z** fields in the **Geometry** category display the start point coordinates of the selected line. You can enter new values in these fields or use the **Pick Point** button that appears in the field you click in. When you choose the **Pick Point**

button, a rubber-band line gets attached to the cursor on its start point and the endpoint of the line is fixed. You can now move the cursor and specify a new start point for the line.

End X/End Y/End Z. Similar to the start point, you can also modify the location of the endpoint of a line by entering new coordinate values in the **End X**, **End Y**, **End Z** fields. You can also choose the **Pick Point** button in the respective fields and select a new location for the endpoint of the selected line. This **Pick Point** button appears in the fields when you click on them.

Note
*The **Delta X**, **Delta Y**, **Delta Z**, **Length**, and **Angle** fields are not available for change but the values in these fields get updated as the start and endpoints of the line are changed.*

Selecting a Circle

When you select a circle, the **Geometry** category displays the categories as shown in Figure 20-8. You can modify the values in these fields. The effects of modifying the values are discussed next.

Center X/Center Y/Center Z. Here, under the **Center X/Center Y/Center Z** fields of the Geometry category, you can specify a new location of the center point of the circle by entering new coordinate values in the **Center X**, **Center Y**, and **Center Z** fields. When you click in any of these fields, the **Pick Point** button is displayed. You can choose this button to locate a new location of the center point of the circle. The circle will be drawn at the new location with the same radius as before. It is as if the circle has moved from the old location to a new one.

Radius/Diameter. You can enter a new value for the radius or diameter of the circle in the **Radius** or **Diameter** fields respectively. The circle's radius or diameter is modified as per the new values you have entered. As you modify the values of the radius or the diameter, the values in the **Circumference** and **Area** fields get modified accordingly.

*Figure 20-7 The **Properties** palette*

*Figure 20-8 The **Geometry** category of the **Properties** palette for a circle*

Note
*You can also enter new values of the circumference or area in the **Circumference** and **Area** fields, respectively, if you want to create a circle with a given value of circumference or area. You will notice that the radius and diameter values are modified accordingly. The **Normal X**, **Normal Y**, and **Normal Z** fields are not available for modifications.*

Chapter 20

Selecting the Multiline Text

If you select multiline text written in a drawing and invoke the **Properties** palette, apart from the **General** category, three more categories of properties are displayed: **Text, 3D Visualization** and **Geometry**. The properties under these two categories (Figure 20-9) and how they can be modified are discussed next.

Figure 20-9 *The **Geometry** and **Text** categories of the **Properties** palette for text*

Contents. The current contents of the text are displayed in the **Contents** field under the **Text** category of the **Properties** palette. To modify the contents, click in the field and the [**...**] button is displayed at the right-corner of the field. You can then choose the [**...**] button to display the **In-Place Text Editor** where you can make the necessary modifications using the options in the dialog box. Once you have made the changes, choose **OK** to exit the text editing and return to the **Properties** palette. The changes you made are reflected in the value of the **Contents** field of the **Properties** palette and in the drawing.

Style. When you click in the **Style** field in the **Properties** palette, a drop-down list of the defined text styles is displayed. By default, only the Standard style is displayed. If you had defined more text styles in the drawing, they can be selected from the drop-down list and applied to the current text in the drawing.

Annotative. You can convert the non-annotative text to annotative and vice-versa by selecting the **Yes** and **No** options respectively from the **Annotative** drop-down list.

Justify. All the text justification options are available in the **Justify** drop-down list. This drop-down list is available when you click in the **Justify** field. By default, the Top left justification is applied to the text. You can select a text justification from the drop-down list to be applied to the selected text in the drawing.

Direction. This drop-down list displays the possible directions for the text. This option is generally available for the multiline text objects.

Text height. You can enter a new value for the height of the selected text in the drawing. The default text height is 0.2. But, you can make the text smaller or larger in the drawing by simply entering a new value in the **Text height** field of the **Properties** palette.

Rotation. This field displays the current angle of the selected text and is **0**-degree, by default. You can change the angle of rotation of the text by entering an angle by which you want to rotate the selected text in the **Rotation** field and then pressing ENTER. The effect is immediately visible in the drawing.

Line space factor. The effect of changing the value of this field is more evident when you have selected a paragraph or multiline text. This value controls the spacing between the lines in a paragraph or multiline text. A scale factor of 1 is the default value. You can increase or decrease the spacing between the lines of the selected paragraph text by entering a new scale factor of your choice and pressing ENTER. The result is immediately visible in the drawing.

Line space distance. This field is used to modify the distance between the spacing of each line of the multiline text. Click in this field and enter the new value in it.

Line space style. This field when selected displays two line spacing options in the corresponding drop-down list. They are **Exactly** and **At least**. **At least** is the default option.

Background mask. This field specifies whether or not any background mask is assigned to the multiline text.

Defined width/Defined height. These fields display the defined width/height of the text window in the **In-Place Text Editor**. To change the value, click in the field and enter the new value in it.

Columns. Columns are of three types, None, Static, or Dynamic. To edit the column settings, choose the **[...]** button; the **Column Settings** dialog box will be displayed. You can change the column settings through this dialog box.

Position X/Position Y/Position Z. You can change the location of the insertion point of the selected text by entering new *X/Y/Z* coordinates for the text in these fields. You can also use the **Pick Points** button that is displayed in these fields when you click in them. When you choose the **Pick Points** button, a rubber-band line is attached between the cursor and the selected text and you can move the cursor and specify a new location for the text on the screen. The values of the three fields get updated automatically.

Selecting the Annotative Block Reference

When you select a block in the drawing and invoke the **Properties** palette, apart from the **General** category, three more categories containing the properties of the block are available. They are **Geometry**, **3D Visualization**, and **Misc**. The properties under these categories are discussed next.

Position X/Position Y/Position Z. Here, under the **Position X/Position Y/Position Z** fields of the **Geometry** category you can enter new *X*, *Y*, and *Z* coordinates for the selected block. The block reference shall move to the new location as you specify the coordinates. You can also click in any one of the fields to display the **Pick Points** button and choose it to specify a new location for the block on the screen.

Scale X/Scale Y/Scale Z. You can specify new *X*, *Y*, and *Z* scale factors for the selected block in the **Scale X**, **Scale Y**, and **Scale Z** fields, respectively. The current scale factors for the block are displayed in the **Scale X**, **Scale Y**, and **Scale Z** fields and as you change them, the modifications are reflected in the drawing.

Chapter 20

Name. This field under the **Misc** category displays the name of the selected block reference and is not available for modifications.

Rotation. This field specifies the current rotation angle of the selected block. You can specify a new angle of rotation for the selected block in this field. After specifying an angle, when you press ENTER, the selected block in the drawing is rotated through the specified angle.

Annotative. This field specifies whether the block is annotative or non-annotative. This information is read-only.

Annotative scale. This field displays the current annotation scale of the drawing. You can add more annotation scales to the drawing in the **Annotation Object Scale** dialog box that will be displayed when you choose the **[...]** button from the **Annotative scale** field.

Match orientation to layout. This field displays whether or not the orientation of the text used in the block is forced to align with the orientation of the floating viewports. This information is read-only.

Block Unit. This field specifies the UNIT of the block.

Unit factor. This field specifies the conversion factor between the block unit and the drawing unit.

Selecting the Annotative Attribute

When you select an attribute before converting it into a block, apart from the **General** category, the other four categories under which the properties of an attribute are displayed are **Text**, **General**, **3D Visualization**, and **Misc**. How to use the **Properties** palette to edit an attribute has already been discussed in Chapter 17, Defining Block Attributes.

Tag. You can modify the text of the tag by entering a new value in the **Tag** field under the **Text** category. After you have modified the tag text and pressed ENTER, the old attribute text in the drawing is replaced by the new text.

Annotative. This field specifies whether the attribute is annotative or non-annotative. Select the appropriate option from the **Annotative** drop-down list to change the annotative property of the attributes.

Annotative scale. This field displays the current annotation scale of the drawing. You can add more annotation scales to the drawing in the **Annotation Object Scale** dialog box that will be displayed when you choose the **[...]** button from the **Annotative scale** field.

Prompt. You can also change the text of the prompt and enter a new prompt value in the **Prompt** field.

Value. This field displays the default value of the selected attribute. You can change this default value by entering a new value in the **Value** field.

Style. When you click in this field, the **Style** drop-down list is available. You can select a style for the attribute text from this drop-down list. Only the text styles that already have been defined are available in this drop-down list and can be selected.

Justify. You can select an attribute text justification from the **Justification** drop-down list.

Paper text height. This field displays the height of the annotative text that is to be maintained on the paper.

Model text height. This field displays the height of the text that is to be displayed in the model space. The **Model text height** value depends on the **Paper text height** value and the **Annotative scale**.

Height. You can change the height of the selected non-annotative attribute text by entering a new value in this field.

Rotation. You can rotate the selected attribute text by specifying an angle in the **Rotation** field.

Width factor. You can modify the width factor of the attribute text by entering a new value in the **Width factor** field of the **Properties** palette.

Obliquing. The attribute text can also be made slanting by entering an obliquing angle in the **Obliquing** field of the **Properties** palette.

Direction. This field displays the flow direction of the attribute text. You can select the desired flow direction from the **Horizontal**, **Vertical**, or **By style** options (defined in the Textstyle) in the **Direction** drop-down list.

Boundary Width. This field displays the width of the text window in the **In-Place Text Editor.** The value displayed here is the relative value; specifying a value less than **1** will decrease the boundary width and a value more than **1** will increase the boundary width. The value in the **Boundary Width** is only available when the **Multiple lines** check box is selected from the **Mode** area of the **Attribute Definition** dialog box. Alternatively, you can select the **Yes** option from the **Multiple lines** drop-down list in the **Misc** area of the **Properties** Palette.

Text alignment X/Text alignment Y/Text alignment Z. The **Text alignment X/Text alignment Y/Text alignment Z** fields of the **Text** category display the X, Y, and Z coordinates of the alignment point of the selected attribute. Note that these fields will be available only if you modify the properties such as the text alignment of the attributes, while defining them. If these fields are available and you click on them, the **Pick Point** button appears in the field. You can use this button to specify the new location of the attribute.

Note

*If you modify the text alignment of the first attribute and define the remaining attributes below the previous, the remaining attributes will also have the **Text Alignment** fields active.*

Misc. As discussed earlier, under the **Misc** category, you can redefine the attribute modes that have been defined at the time of the attribute definition. You can also make the attribute text appear upside-down or backwards by selecting **Yes** from the **Upside down** drop-down list and the **Backward** drop-down list, respectively.

Note

*Choose the **QuickCalc** button from the respective fields to open the Quick Calculator. Using the Quick Calculator, you can enter the numerical values as mathematical equations. These equations may include scientific calculations. You can also perform units conversion using the Quick Calculator.*

Exercise 1   *General*

Draw a hexagon on layer OBJ in red color. Let the linetype be hidden. Now, use the **PROPERTIES** command to change the layer of the hexagon to some other existing layer, the color to yellow, and the linetype to continuous. Also, in the **Properties** palette, under the **Geometry** category, specify a vertex in the **Vertex** field or select one using the arrow (Next or Previous) buttons and then relocate it. The values of the **Vertex X** and **Vertex Y** fields change as the coordinate values of the vertices change. Also, notice the change in the value in the **Area** field. You can also assign a start and end lineweight to each of the hexagon segments between the specified vertices. Use the **LIST** command to verify that the changes have taken place.

EXPLODING COMPOUND OBJECTS

Ribbon:	Modify > Explode	Toolbar:	Modify > Explode
Menu Bar:	Modify > Explode	Command:	EXPLODE

The **EXPLODE** command is used to split compound objects such as blocks, polylines, regions, polyface meshes, polygon meshes, multilines, 3D solids, 3D meshes, bodies, or dimensions into the basic objects that make them up (Figure 20-10). For example, if you explode a polyline or a 3D polyline, the result will be ordinary lines or arcs (tangent specification and width are not considered). When a 3D polygon mesh is exploded, the result is 3D faces. Polyface meshes are turned into 3D faces, points, and lines. Upon exploding 3D solids, the planar surfaces of the 3D solid turn into regions, and nonplanar surfaces turn into bodies. Multilines are changed to lines. Regions turn into lines, ellipses, splines, or arcs. 2D polylines lose their width and tangent specifications and 3D polylines explode into lines. When a body is exploded, it changes into single-surface bodies, curves, or regions. When a leader is exploded, the components are lines, splines, solids, block inserts, text or multiline text, tolerance objects, and so on. Multiline text explodes into a single line text. This command is especially useful when you have inserted an entire drawing and you need to alter a small detail. After you invoke the **EXPLODE** command, you are prompted to select the objects you want to explode. After selecting the objects, press ENTER or right-click to explode the selected objects and then end the command.

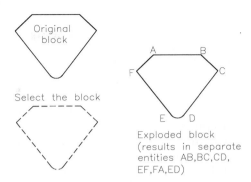

Figure 20-10 Use of the EXPLODE command

When a block or dimension is exploded, there is no visible change in the drawing. The drawing remains the same except that the color and linetype may have changed because of floating layers, colors, or linetypes. The exploded block is turned into a group of objects that can be modified separately. To check whether the explosion of the block has taken place, select any object that was a part of the block. If the block has been exploded, only that particular object will be highlighted. With the **EXPLODE** command, only one nesting level is exploded at a time. Hence, if there is a nested block or a polyline in a block and you explode it, the inner block or the polyline will not be exploded. Attribute values are deleted when a block is exploded, and the attribute definitions are redisplayed.

Note
*Remember that the blocks inserted using the **MINSERT** command and x-refs cannot be exploded. This command was discussed in relation to the blocks in Chapter 16.*

Tip
*If you want to insert a block in the form of its separate components while using the **INSERT** command, select the **Explode** check box in the **Insert** dialog box. Also, if you are using the **-INSERT** command, type * in front of the block name at the **Enter block name or [?]:** prompt, the block will be inserted in the drawing as separate component object, and not as an entire block.*

EDITING POLYLINES

A polyline can assume various characteristics such as width, linetype, joined polyline, and closed polyline. You can edit polylines, polygons, or rectangles to attain the desired characteristics using the **PEDIT** command. In this section, we will be discussing how to edit simple 2D polylines. The following are the editing operations that can be performed on an existing polyline using the **PEDIT** command. These operations are discussed in detail later in this chapter.

1.   A polyline of varying widths can be converted to a polyline of uniform width.

Chapter 20

2. An open polyline can be closed and a closed one can be opened.

3. You can remove bends and curved segments between two vertices to make a straight polyline.

4. A polyline can be split up into two and individual polylines or polyarcs connected to one another can be joined into a single polyline.

5. If you select an entity that is not a polyline to edit using the **PEDIT** command, you are prompted to specify whether you want the entity to be converted into a polyline. You can avoid this prompt by setting the value of the **PEDITACCEPT** variable to 1.

6. Multiple polylines can be edited.

7. The appearance of a polyline can be changed by moving and adding vertices.

8. Curves of arcs and B-spline curves can be fitted to all vertices in a polyline, with the specification of the tangent of each vertex being optional.

9. The linetype generation at the vertices of a polyline can be controlled.

10. Multiple polylines can be joined together to form a single polyline.

Editing Single Polyline

Ribbon:	Home > Modify > Edit Polyline	Toolbar:	Modify II > Edit Polyline
Menu Bar:	Modify > Object > Polyline	Command:	PEDIT

Apart from the methods displayed in the **PEDIT** command box, as shown, you can also invoke the **PEDIT** command by choosing **Polyline Edit** from the shortcut menu that is displayed when you select a polyline and right-click. You can use the **PEDIT** command to edit any type of polyline. When you invoke this command, the following prompt sequence is displayed:

Select polyline or [Multiple]:

If the selected entity is not a polyline, the following message will be displayed at the Command prompt.

Object selected is not a polyline.
Do you want to turn it into one? <Y>:

If you want to turn the object into a polyline, respond by entering Y and pressing ENTER or by simply pressing ENTER. To let the object be as it is, enter N. AutoCAD will then prompt you to select another polyline or object to edit. As mentioned earlier, you can avoid this prompt by setting the value of the **PEDITACCEPT** variable to **1**. The subsequent prompts and editing options depend on the type of polyline that has been selected. AutoCAD provides you the option of either selecting a single polyline or multiple polylines. In this case, a single 2D polyline is selected, the next prompt displayed is given next.

Enter an option [Close/Join/Width/Edit vertex/Fit/Spline/Decurve/Ltype gen/Reverse/Undo]: *Enter an option or press ENTER to end command.*

Note
Depending on the type of polyline selected, the **PEDIT** *command options change. In this chapter, the* **2D polyline** *options have been discussed. The* **3D polylines** *and* **3D Polygon Mesh** *editing options will be discussed in Chapter 25, Getting Started with 3D.*

The options available in the **Single** polyline selection method are discussed next.

Close (C) Option

This option is used to close an open polyline. **Close** creates the segment that connects the last segment of the polyline to the first. You will get this option only if the polyline is not closed. Figure 20-11 illustrates this option.

Open (O) Option

If an open polyline is closed, the **Close** option is replaced by the **Open** option. Entering **O**, for open, removes the closing segment, see Figure 20-11.

Join (J) Option

This option appends lines, polylines, or arcs whose endpoints meet a selected polyline at any of its endpoints and adds (joins) them to it (Figure 20-12). This option can be used only if a polyline is open. After this option has been selected, AutoCAD asks you to select objects. Once you have chosen the objects to be joined to the original polyline, AutoCAD examines them to determine whether any of them has an endpoint in common with the current polyline, and joins such an object with the original polyline. The search is then repeated using new endpoints. They will not join if the endpoint of the object does not exactly meet the polyline. The line touching a polyline at its endpoint to form a T will not be joined. If two lines meet a polyline in a Y shape, only one of them will be selected, and this selection is unpredictable. To verify which lines have been added to the polyline, use the **LIST** command or select a part of the object. All the segments that are joined to the polyline will be highlighted.

Width (W) Option

The **W** option allows you to define a new, unvarying width for all segments of a polyline (Figure 20-13). It changes the width of a polyline with a constant or a varying width. The desired new width can be specified either by entering the width at the keyboard or by specifying the width as the distance between the two specified points. Once the width has been specified, the polyline assumes it. Here, you will change the width of the given polyline in figure from 0.02 to 0.05 (Figure 20-14). After you invoke the **PEDIT** command, the next prompts are as follows:

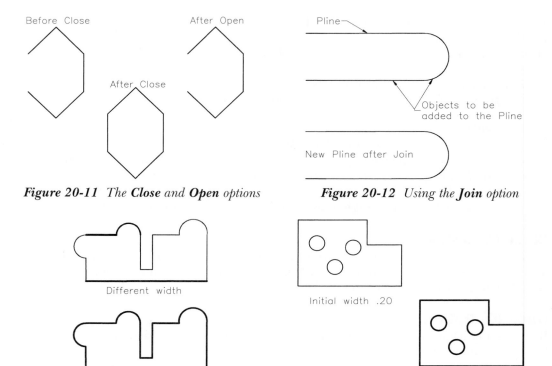

Figure 20-11 *The **Close** and **Open** options*   **Figure 20-12** *Using the **Join** option*

Figure 20-13 *Making the width of a polyline uniform*   **Figure 20-14** *Entering a new width for all segments*

Select polyline or [Multiple]: *Select a polyline.*
Enter an option [Close/Join/Width/Edit vertex/Fit/Spline/Decurve/Ltype gen/Reverse/Undo]: **W**
Specify new width for all segments: **0.05**

Note
*Circles drawn using the **CIRCLE** command cannot be changed to polylines. However, polycircles can be drawn using the **Pline > Arc** option (by drawing two semicircular polyarcs) or by using the **DONUT** command (you can modify the thickness using the **PEDIT > Width** option).*

Edit vertex (E) Option

The **Edit vertex** option lets you select a vertex of a polyline and perform different editing operations on the vertex and the segments following it. A polyline segment has two vertices. The first one is at the start point of the polyline segment; the other one is at the endpoint of the segment. When you invoke this option, an X marker appears on the screen at the first vertex of the selected polyline. If a tangent direction has been specified for this particular vertex, an arrow is generated in that direction. After this option has been selected, the next prompt appears with a list of options for this prompt. The prompt sequence, after you invoke the **PEDIT** command, is given next.

Select polyline or [Multiple]: *Select the polyline to be edited.*
Enter an option [Close/Join/Width/Edit vertex/Fit/Spline/Decurve/Ltype gen/Reverse/ Undo]: **E**
Enter a vertex editing option
[Next/Previous/Break/Insert/Move/Regen/Straighten/Tangent/Width/eXit]<N>: *Enter an editing option or press ENTER to accept default.*

All the options for the **Edit vertex** option are discussed next.

Next and Previous Options. These options move the X marker to the next or the previous vertex of a polyline. The default value in the **Edit vertex** option is one of these two options. The option that is selected as default is the one you chose last. In this manner, the **Next** and **Previous** options help you to move the X marker to any vertex of the polyline by selecting one of these two options, and then pressing ENTER to reach the desired vertex. These options cycle back and forth between the first and last vertices, but cannot move past the first or last vertices, even if the polyline is closed (Figure 20-15).

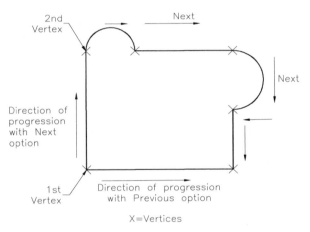

*Figure 20-15 The **Next** and **Previous** options*

The prompt sequence for using this option, after you invoke the **PEDIT** command, is given next.

Select polyline or [Multiple]: *Select the polyline to be edited.*
Enter an option [Close/Join/Width/Edit vertex/Fit/Spline/Decurve/Ltype gen/Reverse/ Undo]: **E**
Enter a vertex editing option
[Next/Previous/Break/Insert/Move/Regen/Straighten/Tangent/Width/eXit] <N>: *Enter N or P to move to the next or previous vertices respectively.*

Break Option. With the **Break** option, you can remove a portion of the polyline, as shown in Figure 20-16 or break it at a single point. The breaking of the polyline can be specified between any two vertices. By specifying two different vertices, all the polyline segments and vertices between the specified vertices are erased. If one of the selected vertices is at the endpoint of the polyline, the **Break** option will erase all the segments between the first

vertex and the endpoint of the polyline. The
exception to this is that AutoCAD does not erase
the entire polyline, if you specify the first vertex at
the start point (first vertex) of the polyline and the
second vertex at the endpoint (last vertex) of the
polyline. If both vertices are at the endpoint of the
polyline, or only one vertex is specified and its
location is at the endpoint of the polyline, no
change is made to the polyline. The last two
selections of vertices are treated as invalid by
AutoCAD, which acknowledges this by displaying
the message ***Invalid***.

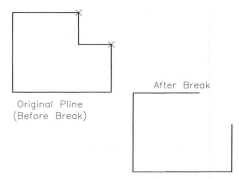

Figure 20-16 *Use of the **Break** option*

To use the **Break** option, first you need to move the marker to the first vertex where you want
the split to start. The placement of the marker can be achieved with the help of the **Next** and
Previous options. Once you have selected the first vertex to be used in the **Break** operation,
invoke the **Break** option by entering **B** at the Command prompt. AutoCAD takes the vertex
where the marker (X) is placed as the first point of the breakup. The next prompt asks you to
specify the position of the next vertex for the breakup. You can enter **GO** at this prompt if
you want to split the polyline at one vertex only. Otherwise, use the **Next** or **Previous** option
to specify the position of the next vertex and then enter GO. The polyline segments between
the two selected vertices is erased. The prompt sequence after you have invoked the
PEDIT command is given next.

> Select polyline or [Multiple]: *Select the polyline to be edited.*
> Enter an option [Close/Join/Width/Edit vertex/Fit/Spline/Decurve/Ltype gen/Reverse/
> Undo]: **E**
> Enter a vertex editing option
> [Next/Previous/Break/Insert/Move/Regen/Straighten/Tangent/Width/eXit]<N>: *Enter N
> or P to locate the first vertex for the **Break** option.*
> Enter a vertex editing option
> [Next/Previous/Break/Insert/Move/Regen/Straighten/Tangent/Width/eXit]<N>: **B**

Once you invoke the **Break** option, AutoCAD treats the vertex where the marker (X) is
displayed as the first point for splitting the polyline. The next prompt issued is given next.

> Enter an option [Next/Previous/Go/eXit] <N>: *Enter **G** if you want to split the polyline at one
> vertex only or move the X marker using the **Next** or **Previous** option to specify the position of the
> next vertex for breakup.*

After you have specified the next position of the X marker using the **Next** and **Previous**
options, entering **Go** deletes the polyline segment between the two markers specified. Now,
exit the **Enter a vertex editing option** prompt using the **eXit** option.

Insert Option. The **Insert** option allows you to define a new vertex and add it to the polyline
(Figure 20-17). You can invoke this option by entering I for Insert. You should invoke this option
only after you have moved the marker (X) to the vertex after which the new vertex is to be

added. The new vertex is inserted immediately after the vertex with the X mark. After you have invoked the **PEDIT** command, the prompt sequence for using the **Insert** option is as follows:

Select polyline or [Multiple]: *Select the polyline to be edited.*
Enter an option [Close/Join/Width/Edit vertex/Fit/Spline/Decurve/Ltype gen/Reverse/ Undo]: **E**
Enter a vertex editing option
[Next/Previous/Break/Insert/Move/Regen/Straighten/Tangent/Width/eXit]<N>: *Move the marker to the vertex after which the new vertex is to be inserted.*
Enter a vertex editing option
[Next/Previous/Break/Insert/Move/Regen/Straighten/Tangent/Width/eXit]<N>: **I**
Specify location for new vertex: *Move the cursor and select to specify the location of the new vertex or enter the coordinates of the new location.*

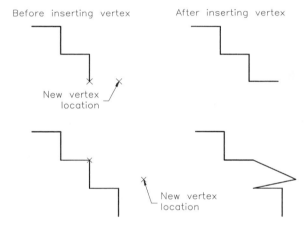

*Figure 20-17 Using the **Insert** option to define new vertex points*

Move option. This option lets you move the X-marked vertex to a new position (Figure 20-18). Before invoking the **Move** option, you must move the X marker to the vertex you want to relocate by selecting the **Next** or **Previous** option. The prompt sequence for relocating a vertex after you invoke the **PEDIT** command is given next.

Select polyline or [Multiple]: *Select the polyline to be edited.*
Enter an option [Close/Join/Width/Edit vertex/Fit/Spline/Decurve/Ltype gen/Reverse/ Undo]: **E**
Enter a vertex editing option
[Next/Previous/Break/Insert/Move/Regen/Straighten/Tangent/Width/eXit]<N>: *Enter **N** or **P** to move the X marker to the vertex you want to relocate.*
Enter a vertex editing option
[Next/Previous/Break/Insert/Move/Regen/Straighten/Tangent/Width/eXit]<N>: **M**
Specify new location for marked vertex: *Specify the new location for the selected vertex by using the pick button of your pointing device or by entering its coordinate values.*
Enter a vertex editing option
[Next/Previous/Break/Insert/Move/Regen/Straighten/Tangent/Width/eXit]<N>: *Enter an option or enter **X** to exit.*

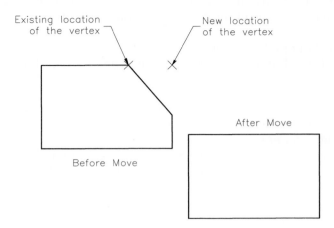

Figure 20-18 *The Move option*

Regen Option. The **Regen** option regenerates the polyline to display the effects of edits you have made, without having to exit the vertex mode editing. It is used most often with the **Width** option.

Straighten Option. The **Straighten** option can be used to straighten polyline segments or arcs between specified vertices (Figure 20-19). It deletes the arcs, line segments, or vertices between the two specified vertices and substitutes them with one polyline segment.

The prompt sequence to use this option after you have invoked the **PEDIT** command is given next.

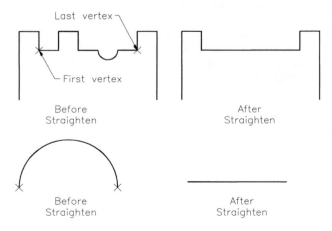

Figure 20-19 *Using the **Straighten** option to straighten polylines*

Select polyline or [Multiple]: *Select the polyline to be edited.*
Enter an option [Close/Join/Width/Edit vertex/Fit/Spline/Decurve/Ltype gen/Reverse/
Undo]: **E**
Enter a vertex editing option
[Next/Previous/Break/Insert/Move/Regen/Straighten/Tangent/Width/eXit]<N>: *Move
the marker to the desired vertex from where you want to start applying the* **Straighten** *option with
the* **Next** *or* **Previous** *option.*
Enter a vertex editing option
[Next/Previous/Break/Insert/Move/Regen/Straighten/Tangent/Width/eXit]<N>: **S**
Enter an option [Next/Previous/Go/eXit] <N>: *Move the marker to the next desired vertex,
until you reach the vertex you want to straighten.*
Enter an option [Next/Previous/Go/eXit] <N>: **G**

The polyline segments between two marker locations are replaced by a single straight line
segment. If you specify a single vertex, the segment following the specified vertex is
straightened, if it is curved.

Tangent option. The **Tangent** option is used to associate a tangent direction to the current
vertex (marked by X). The tangent direction is used in the curve fitting or the **Fit** option of
the **PEDIT** command. This option is discussed in detail in the subsequent section on curve
fitting. The prompt issued on using the **Tangent** option is given next.

Specify direction of vertex tangent: *Specify a point or enter an angle.*

You can specify the direction of the vertex tangent by entering an angle at the previous
prompt or by selecting a point to express the direction with respect to the current vertex. You
can then move the marker to another vertex using the **Next** or **Previous** option and change
its direction of tangent or enter X to exit the **Enter a vertex editing option** prompt.

Width option. The **Width** option lets you change the starting and ending widths of a polyline
segment that follows the current vertex (Figure 20-20). By default, the ending width is equal
to the starting width and hence, you can get a polyline segment of uniform width by pressing
ENTER at the **Specify ending width for next segment <starting width>** prompt. You can
also specify different starting and ending widths to get a varying-width polyline. The prompt
sequence is given next.

Enter an option [Close/Join/Width/Edit vertex/Fit/Spline/Decurve/Ltype gen/Reverse/
Undo]: **E**
Enter a vertex editing option
[Next/Previous/Break/Insert/Move/Regen/Straighten/Tangent/Width/eXit]<N>: *Move the
marker to the starting vertex of the segment whose width is to be altered, using the* **Next** *or* **Previous**
option.
Enter a vertex editing option
[Next/Previous/Break/Insert/Move/Regen/Straighten/Tangent/Width/eXit]<N>: **W**
Specify starting width for next segment <current>: *Enter the revised starting width.*
Specify ending width for next segment <starting width>: *Enter the revised ending width or
press ENTER to accept the default option of keeping the ending width of the segment equal to the
starting width.*

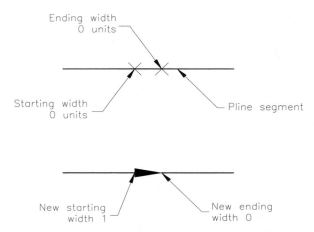

Figure 20-20 *Using the* **Width** *option to change the width of a polyline*

If no difference is noticed in the appearance of the polyline, you may need to use the **Regen** option. The segment with the revised widths is redrawn after invoking the **Regen** option or when you exit the vertex mode editing.

eXit Option. This option lets you exit the vertex mode editing and return to the **PEDIT** prompt.

Fit Option

The **Fit** or **Fit curve** option generates a curve that passes through all the corners (vertices) of the polyline, using the tangent directions of the vertices (Figure 20-21).

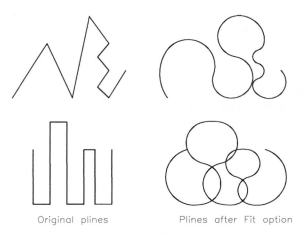

Original plines   Plines after Fit option

Figure 20-21 *The* **Fit** *option*

The curve is composed of a series of arcs passing through the corners (vertices) of the polyline. This option is used when you draw a polyline with sharp corners and you need to convert it into a series of smooth curves. An example of this is a graph. In a graph, we need to show a curve by joining a series of plotted points. The process involved is called curve fitting; therefore, this option is known as **Fit**. The vertices of the polyline are also known as the control points.

The closer these control points are, the smoother is the curve. Therefore, if the **Fit** option does not give optimum results, insert more vertices into the polyline or edit the tangent directions of vertices and then use the **Fit** option on the polyline. Before using this option, you may give each vertex a tangent direction.

The curve is then constructed, according to the tangent directions you have specified. The following prompt sequence illustrates the **Fit** option after you invoke the **PEDIT** command.

> Select polyline or [Multiple]: *Select the polyline to be edited.*
> Enter an option [Close/Join/Width/Edit vertex/Fit/Spline/Decurve/Ltype gen/Reverse/Undo]: **F**

If the tangent directions need to be edited, use the **Edit vertex** option of the **PEDIT** command. Move the X marker to each of the vertices that need to be changed. Now, you can invoke the **Tangent** option, and either enter the tangent direction in degrees or select points. The chosen direction is expressed by an arrow placed at the vertex. The prompt sequence, after you invoke the **PEDIT** command, is given next.

> Select polyline or [Multiple]: *Select the polyline to be edited.*
> Enter an option [Close/Join/Width/Edit vertex/Fit/Spline/Decurve/Ltype gen/Reverse/Undo]: **E**
> Enter a vertex editing option
> [Next/Previous/Break/Insert/Move/Regen/Straighten/Tangent/Width/eXit]<N>: **T**
> Specify direction of vertex tangent: *Specify a direction in + or - degrees, or select a point in the desired direction. Press ENTER.*

Once you specify the tangent direction, use the **eXit** option to return to the previous prompt and use its **Fit** option.

Spline Option

The **Spline** option (Figure 20-22) also smoothens the corners of a straight segment polyline, as does the **Fit** option, but the curve passes through only the first and the last control points (vertices), except in the case of a closed polyline. The spline curve is stretched toward the other control points (vertices) but does not pass through them, as in the case of the **Fit** option. The greater the number of control points, the greater the force with which the curve is stretched toward them. The prompt sequence after you invoke the **PEDIT** command is as follows:

> Select polyline or [Multiple]: *Select the polyline.*
> Enter an option [Close/Join/Width/Edit vertex/Fit/Spline/Decurve/Ltype gen/Reverse/Undo]: **S**

The generated curve is a B-spline curve. The **frame** is the original polyline without any curves in it. If the original polyline has arc segments, these segments are straightened when the spline's frame is formed. A frame that has width produces a spline curve that tapers smoothly from the width of the first vertex to that of the last. Any other width specification between the first width specification and the last is neglected. When a spline is formed from a polyline, the frame is displayed as a polyline with a zero width and a continuous linetype.

Figure 20-22 *The **Spline** option*

Also, AutoCAD saves its frame so that it may be restored to its original form. Tangent specifications on control point vertices do not affect spline construction.

By default, the spline frames are not shown on the screen, but you may want them displayed for reference. In this case, the system variable **SPLFRAME** needs to be manipulated. The default value for this variable is zero. If you want to see the spline frame as well, set it to 1.

Now, whenever the **Spline** option is used on a polyline, the frame will also be displayed. Most editing commands such as **MOVE**, **ERASE**, **COPY**, **MIRROR**, **ROTATE**, and **SCALE** work similarly for both the Fit curves and Spline curves. They work on both the curve and its frame, whether the frame is visible or not. The **EXTEND** command changes the frame by adding a vertex at the point the last segment intersects with the boundary. If you use any of the previously mentioned commands and then use the **Decurve** option to decurve the spline curve and later use the **Spline** option again, the same spline curve is generated. The **BREAK**, **TRIM**, and **EXPLODE** commands delete the frame. The **DIVIDE**, **MEASURE**, **FILLET**, **CHAMFER**, **AREA**, and **HATCH** commands recognize only the spline curve and do not consider the frame. The **STRETCH** command first stretches the frame and then fits the spline curve to it.

When you use the **Join** option of the **PEDIT** command, the spline curve is decurved and the original spline information is lost. The **Next** and **Previous** options of the **Edit vertex** option of the **PEDIT** command move the marker only to the points on the frame, whether visible or not. The **Break** option discards the spline curve. The **Insert**, **Move**, **Straighten**, and **Width** options refit the spline curve. Object Snaps consider the spline curve and not the frame; therefore, if you wish to snap to the frame control points, restore the original frame.

There are two types of spline curves:
Quadratic B-spline
Cubic B-spline

Both of them pass through the first and the last control points, which is characteristic of the spline curve. Cubic curves are very smooth. A cubic curve passes through the first and last control points, and the curve is closer to the other control points. Quadratic curves are not as smooth as the cubic ones, but they are smoother than the curves produced by the **Fit curve** option. A quadratic curve passes through the first and last control points, and the rest of the curve is tangent to the polyline segments between the remaining control points (Figure 20-23).

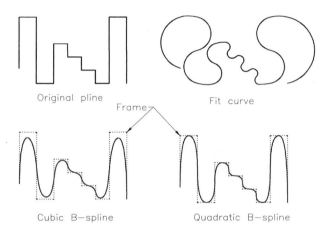

Figure 20-23 *Comparison of fit curve, quadratic B-spline, and cubic B-spline*

Generation of different types of spline curves. If you want to edit a polyline into a B-spline curve, you are first required to enter a relevant value in the **SPLINETYPE** system variable. A value of 5 produces the quadratic curve, whereas a value of 6 produces a cubic curve. The default value is 6, which implies that when we use the **Spline** option of the **PEDIT** command, a cubic curve is produced by default. You can change the value of the **SPLINETYPE** variable using the command line.

You can also set the value of the variable by selecting the **Polyvars** option in the **PEDIT** screen menu. The screen menu can be invoked using the **Display screen menu** option available on the **Display** tab of the **Options** dialog box. On doing so, the **Set Spline Fit Variables** dialog box (Figure 20-24) will be displayed. You can select any type of curve from the available list. The **SPLINETYPE** variable is automatically set to the value of the type you have chosen from the dialog box. In case you are editing 3D polygon meshes, the smoothness of the meshes is controlled by the **SURFTYPE** variable. The value of the variable can be set on the command line. You can select the type of polygon mesh you want when you are using the **Smooth Surface** option of the **PEDIT** command from the **Set Spline Fit Variables** dialog box.

Chapter 20

Figure 20-24 The **Set Spline Fit Variables** *dialog box*

SPLINESEGS. The system variable **SPLINESEGS** governs the number of line segments used to construct the spline curves, and therefore you can use this variable to control the smoothness of the curve. The default value for this variable is 8. With this value, a reasonably smooth curve that does not need much regeneration time is generated. The greater the value of this variable, the smoother the curve, but greater the regeneration time, the more space occupied by the drawing file.

Figure 20-25 shows cubic curves with different values for the **SPLINESEGS** variable.

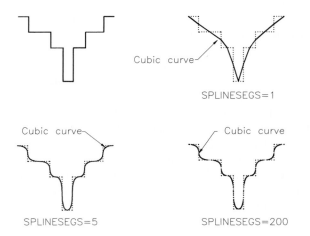

Figure 20-25 Using the **SPLINESEGS** variable

Decurve Option

The **Decurve** option straightens the curves generated after using the **Fit** or **Spline** option on a polyline. They return to their original shape (Figure 20-26). The polyline segments are straightened using the **Decurve** option. The vertices inserted after using the **Fit** or **Spline** option are also removed. Information entered for the tangent reference is retained for use in future fit curve operations. You can also use this command to straighten out any curve drawn with the help of the **Arc** option of the **PLINE** command. Enter **D** at the **Enter an option [Close/Join/Width/Edit vertex/Fit/Spline/Decurve/Ltype gen/Reverse/Undo]** prompt to invoke the **Decurve** option of the **PEDIT** command.

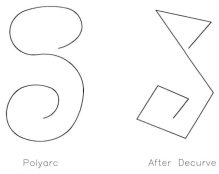

Polyarc   After Decurve

Figure 20-26 *The **Decurve** option*

Note
*If you use edit commands such as the **BREAK** or **TRIM** on spline curves, the **Decurve** option cannot be used.*

Ltype gen Option

You can use this option to control the linetype pattern generation for linetypes other than Continuous with respect to the vertices of the polyline. This option has two modes ON and OFF. If turned off, the break in the noncontinuous linetypes will be avoided at the vertices of the polyline and a continuous segment will be displayed at the vertices (Figure 20-27). If turned on, this option generates the linetype in a continuous pattern such that the gaps may be displayed at the vertices. This option is not applicable to polylines with tapered segments. The command prompt displayed when you select a polyline and invoke the **Ltype gen** option of the **PEDIT** command is given next.

> Enter polyline linetype generation option [ON/OFF] <Off>: *Enter ON or press ENTER to accept the default value.*

The linetype generation for new polylines can also be controlled with the help of the **PLINEGEN** system variable, which acts as a toggle. The default value is 0 (off). Entering the value **1** turns on the **Ltype gen** option.

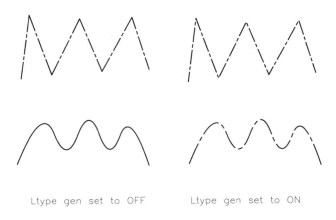

*Figure 20-27 Using the **Ltype gen** option*

Reverse Option

The **Reverse** option is used to change the start point of the polyline. On entering **R** and pressing ENTER at the **Enter an option [Close/Join/Width/Edit vertex/Fit/Spline/Decurve/ Ltype gen/Reverse/Undo]** Command prompt, the start point of the polyline will be changed. This can be confirmed by using the **Edit vertex** option of the **PEDIT** command before and after using the **Reverse** option.

Undo (U) Option

The **Undo** option negates the effect of the most recent **PEDIT** operation. You can go back as far as you need to in the current **PEDIT** session by using the **Undo** option repeatedly until you get the desired screen. If you started editing by converting an object into a polyline, and you want to change the polyline back to the object from which it was created, the **Undo** option of the **PEDIT** command will not work. In this case, you will have to exit to the Command prompt and use the **UNDO** command to undo the operation.

Editing Multiple Polylines

Selecting the **Multiple** option of the **PEDIT** command allows you to select more than one polyline for editing. You can select the polylines using any of the objects selection techniques. After the objects to be edited are selected, press ENTER or right-click to proceed with the command. The prompt sequence that will follow is given next.

Enter an option [Close/Open/Join/Width/Fit/Spline/Decurve/Ltype gen/Reverse/Undo]:

Join

This option is used to join more than one polylines that may or may not be in contact with each other. Even the polylines that are not coincident can be joined using this option. The polylines are joined using a specified distance value called the **Fuzz distance**. After you select the **Join** option, the sequence of prompts is as follows.

Join Type = Extend
Enter fuzz distance or [Jointype] <current>: *Enter a distance or **J** for changing the **Jointype**.*

The prompt sequence to follow when you enter **J** at the above prompt is given next.

Enter join type [Extend/Add/Both] <Extend>:

Extend. This option extends or trims the selected polylines at the endpoints that are nearest to each other, see Figure 20-28. Keep in mind that the segments of the selected polylines that are nearest to each other should not be parallel. This means that the segments that are nearest to each other should intersect at some point when extended.

Add. This option joins the selected polylines by drawing a straight line between the nearest endpoints, see Figure 20-29. The fuzz distance should be greater than the actual distance between the two endpoints to be joined.

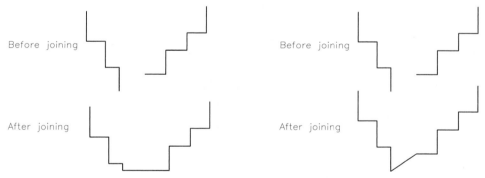

Figure 20-28 *Joining two polylines using the* ***Extend*** *option*

Figure 20-29 *Joining two polylines using the* ***Add*** *option*

Both. This option joins the endpoints by extending or trimming if possible; otherwise, it adds a line segment between the two endpoints.

Exercise 2   *General*

a.   Draw a line from point (0,0) to point (6,6). Convert the line into a polyline with a starting width 0.30 and a ending width 0.00.

b.   Draw different polylines of varying width segments and then use the **Join** option to join all the segments into one polyline. After joining the segments, make their width uniform, with the **Width** option.

Exercise 3   *General*

a.   Draw a staircase-shaped polyline and use the different options of the **PEDIT** command to generate a fit curve, a quadratic B-spline, and a cubic B-spline, and then convert the curves back to the original polyline.

Chapter 20

b. Draw square-wave-shaped polylines, and use different options of the **Edit vertex** option to navigate around the polyline, split the polyline into two at the third vertex, insert more vertices at different locations in the original polyline, and convert the square-wave-shaped polyline into a straight-line polyline.

UNDOING PREVIOUS COMMANDS

Command:	UNDO

The **LINE** and **PLINE** commands have the **Undo** option, which can be used to undo (nullify) the changes made within these commands. The **UNDO** command is used to undo a previous command or to undo more than one command at a time. This command can be invoked by entering **UNDO** at the Command prompt. **Undo** is also available in the **Quick Access** toolbar, **Edit** menu, and the **Standard** toolbar. The **UNDO** button in the **Quick Access** toolbar can only undo the previous command and only one command at a time. If you right-click in the drawing area; a shortcut menu will be displayed. Choose **Undo**. It is equivalent to **UNDO > 1** or to enter **U** at the Command prompt. The **U** command can undo only one operation at a time. You can use **U** as many times as you want until AutoCAD displays the message: **Nothing to undo**. When an operation cannot be undone, a message is displayed but no action takes place. External commands such as **PLOT** cannot be undone.

Undoing Previous Commands using the Quick Access Toolbar

Click the arrow on the **Undo** button in the **Quick Access** toolbar; the **Undo** list box will be displayed. All commands used recently will be listed in the **Undo** list box. Select the commands that you want to undo by placing the cursor in the list box and moving downward. The Commands selected are highlighted in blue color, as shown in Figure 20-30. The Status Bar below the **Undo** list box displays the number of commands selected. Click the left mouse button to confirm the selections.

*Figure 20-30 The **Undo** list box*

Undoing Previous Commands using the Command Line

To **Undo** previous commands using the Command Line, enter **UNDO** at the Commands line (if you are working in dynamic mode, the command will be entered at pointer input). The prompt sequence for the **UNDO** command is given next.

Command: **UNDO**
Current settings: Auto = On, Control = All, Combine = Yes, Layer = Yes
Enter the number of operations to undo or [Auto/Control/BEgin/End/Mark/Back] <1>:
Enter a positive number, an option, or press ENTER to undo a single operation.

The various options of this command are discussed next.

Number (N) Option

This is the default option. This number represents the number of previous command sequences to be deleted. For example, if the number entered is 1, the previous command is undone; if the number entered is 4, the previous four commands are undone, and so on. This is identical to invoking the **U** command four times, except that only one regeneration takes place. AutoCAD lets you know which commands were undone by displaying a message after you press ENTER. The prompt sequence is given next.

Command: **UNDO**
Enter the number of operations to undo or [Auto/Control/BEgin/End/Mark/Back]<1>:3
PLINE LINE CIRCLE

Control (C) Option

This option lets you determine how many of the options are active in the **UNDO** command. You can disable the options you do not need. With this option, you can even disable the **UNDO** command. To access this option, type C. You will get the following prompt.

Enter an UNDO control option [ALL/None/One/Combine/Layer] <All>: *Enter an option or press ENTER to accept the default.*

ALL

The **All** option activates all the features (options) of the **UNDO** command.

None (N)

This option turns off **UNDO** and the **U** command. If you have used the **BEgin** and **End** options or **Mark** and **Back** options to create **UNDO** information, all of that information is undone. The prompt sequence for invoking the **Control** option is as follows.

Command: **UNDO**
Enter the number of operations to undo [Auto/Control/BEgin/End/Mark/Back] <1>: **C**
Enter an UNDO control option [ALL/None/One/Combine/Layer] <All>: **N**

If you try to use the **U** command now, while the **UNDO** command has been disabled, AutoCAD gives you the following message.

Command: **U**
U command disabled. Use **UNDO** command to turn it on.

The prompt sequence for the **UNDO** command after issuing the **None** option is given next.

Command: **UNDO**
Enter an UNDO control option [All/None/One/Combine/Layer] <All>: *Enter O or press ENTER to accept the default.*

To enable the **UNDO** options again, you must enter the **All** or **One** (one mode) option.

One (O)

This option restrains the **UNDO** command to a single operation. All **UNDO** information saved earlier during editing is scrapped. The prompt sequence is as follows.

Enter the number of operations to undo or [Auto/Control/BEgin/End/Mark/Back] <1>: **C**
Enter an UNDO control option [All/None/One/Combine/Layer] <All>: **O**

If you then enter the **UNDO** command, you will get the prompt that is given next.

Command: **UNDO**
Enter an option [Control] <1>: *Press ENTER to undo the last operation or enter **C** to select another control option.*

In response to this last prompt, you can now either press ENTER to undo only the previous command, or go into the **Control** options by entering C. AutoCAD acknowledges undoing the previous command with messages such as given next.

Command: **UNDO**
Enter an option [Control] <1>: [Enter]
CIRCLE
Everything has been undone

Tip
*AutoCAD stores all information about all the **UNDO** operations taking place in a drawing session. When you use the **None** control option of the **UNDO** command, this information is removed and valuable disk space is made available. In case of limited disk space availability, it may be a good idea to use the **One** suboption of the **UNDO Control** command. This allows you to use both the **U** and **UNDO** commands and at the same time makes the disk space available.*

Combine
This option allows you to combine all the consecutive **Zoom** and **Pan** operations as a single operation, while using the **UNDO** and **REDO** commands.

Layer
This option allows you to combine layer property actions as a single action.

Auto (A) Option
Enter A to invoke this option. The following prompt is displayed.

Enter UNDO Auto mode [ON/OFF] <On>: *Select ON or OFF or press ENTER to accept the default setting.*

This option is on by default and every menu item is a single operation. When you use the **Begin** and **End** options to group a series of commands as a single operation, entering **U** at the Command prompt removes the objects individually and not as a group, if the **Auto** option is on. But if **Auto** is off, entering **U** at the Command prompt will remove the entire group of objects you had grouped using the **Begin** and **End** options in one single operation.

Also, if you have put a marker in between the start and end operations and if **Auto** is On, the group of commands till the marker, is undone altogether on using the **Back** option. This group of commands is considered to be a single group and is undone as a single command, although, entering **U** at the Command prompt removes the objects individually until the marker is encountered. But, if the **Auto** option is off, entering **U** at the Command prompt, or using the **Back** option of the **UNDO** command, undoes the entire group of operations irrespective of the marker being there.

BEgin (BE) and End (E) Options

A group of commands is treated as one command for the **U** and **UNDO** commands (provided **Auto** option is off) by embedding the commands between the **BEgin** and **End** options of the **UNDO** command. If you anticipate the removal of a group of successive commands later in a drawing, you can use this option, since all of the commands after the **BEgin** option and before the **End** option are treated as a single command by the **U** command (if the **Auto** option is Off). For example, the following sequence illustrates the possibility of removal of two commands.

> Command: **UNDO**
> Enter the number of operations to undo or [Auto/Control/BEgin/End/Mark/Back] <1>: **BE**
> Command: **CIRCLE**
> Specify center point for Circle or [3P/2P/Ttr (tan tan radius)]: *Specify the center.*
> Specify radius of circle or [Diameter] <current>: *Specify the radius of the circle.*
> Command: **PLINE**
> Specify start point: *Select the first point.*
> Specify next point or [Arc/Halfwidth/Length/Undo/Width]: *Select the next point.*
> Specify next point or [Arc/Close/Halfwidth/Length/Undo/Width]: *Press ENTER.*
> Command: **UNDO**
> Enter the number of operations to UNDO or [Auto/Control/BEgin/End/Mark/Back]: **E**
> Command: **U**

To start the next group once you have finished specifying the current group, use the **End** option to end this group. Another method is to enter the **BEgin** option to start the next group while the current group is active. This is equivalent to issuing the **End** option followed by the **BEgin** option. The group is complete only when the **End** option is invoked to match a **BEgin** option. If **U** or the **UNDO** command is issued after the **BEgin** option has been invoked and before the **End** option has been issued, only one command is undone at a time until it reaches the juncture where the **BEgin** option has been entered. If you want to undo the commands issued before the **BEgin** option was invoked, you must enter the **End** option so that the group is complete. This is demonstrated in the following example:

Example 2   *General*

Enter the following commands in the sequence given below, and notice the changes that take place on screen.

CIRCLE
POLYGON
UNDO BEgin (Make sure that the Auto is OFF)
PLINE
TRACE
U
DONUT
UNDO End
TEXT
U
U
U
U

The first **U** command will undo the **TRACE** command. If you repeat the **U** command, the **PLINE** command will be undone. Any further invoking of the **U** command will not undo any previously drawn object (**POLYGON** and **CIRCLE**, in this case), because after the **PLINE** is undone, you have an **UNDO BEgin**. Only after you enter **UNDO End**, can you undo the **POLYGON** and the **CIRCLE**. In the example, the second **U** command will undo the **TEXT** command, the third **U** command will undo the **DONUT** and **PLINE** commands (these are enclosed in the group), the fourth **U** command will undo the **POLYGON** command, and the fifth **U** command will undo the **CIRCLE** command. Whenever the commands in a group are undone, the name of each command or operation is not displayed as it is undone. Only the name, **GROUP**, is displayed.

Tip
*You can use the **BEgin** and **End** options only when the **UNDO Control** is set to **All** and the **Auto** option is off.*

Mark (M) and Back Options

The **Mark** option installs a marker in the Undo file. The **Back** option lets you undo all the operations until the mark. In other words, the **Back** option returns the drawing to the point where the previous mark was inserted. For example, if you have completed a portion of your drawing and do not want anything up to this point to be deleted, you can insert a marker and then proceed. Then, even if you use the **UNDO Back** option, it will work only until the marker. You can insert multiple markers, and with the help of the **Back** option, you can return to the successive mark points. The following prompt sequence is used to introduce a mark point:

Command: **UNDO**
Enter the number of operations to undo or [Auto/Control/BEgin/End/Mark/Back]: **M**

The prompt sequence for using the **Back** option is given next.

Command: **UNDO**
Enter the number of operations to undo or [Auto/Control/BEgin/End/Mark/Back]: **B**

Once all the marks have been exhausted with the successive **Back** options, any further invoking of the **Back** option displays the message: **This will undo everything. OK? <Y>**. If you enter Y (Yes) at this prompt, all operations carried out since you entered the current drawing session will be undone. If you enter N (No) at this prompt, the **Back** option will be disregarded.

You cannot undo certain commands and system variables, for example, **DIST**, **LIST**, **DELAY**, **NEW**, **OPEN**, **QUIT**, **AREA**, **HELP**, **PLOT**, **QSAVE** and **SAVE**, among many more. Actually, these commands have no effect that can be undone. Commands that change operating modes (**GRID**, **UNITS**, **SNAP**, **ORTHO**) can be undone, though the effect may not be apparent at first. This is the reason why AutoCAD displays the command names as they are undone.

REVERSING THE UNDO OPERATION

Quick Access Toolbar: Redo   **Toolbar:** Standard > Redo   **Command:** REDO

If you right-click in the drawing area, a shortcut menu is displayed. Choose **Redo** to invoke the **REDO** command. The **REDO** command brings back the process you removed previously using the **U** and **UNDO** commands. This command undoes the last **UNDO** command, but it must be entered immediately after the **UNDO** command and the objects previously undone reappear on the screen. Click on the **Redo** button in the **Quick Access** toolbar. The **Redo** list box

*Figure 20-31 The **Redo** list box*

(Figure 20-31) is displayed; it lists the commands that have been recently undone. Select the commands that you want to **Redo** and click the left mouse button to confirm the selection.

Note
*Consecutive **Zoom** and **Pan** operations can be undone or redone by using the **Undo** or **Redo** command only once so that the drawing returns to the initial zoom state. To disable this feature, clear the **Combine zoom and pan commands** check box in the **Undo/Redo** area of the **User Preferences** tab of the **Options** dialog box.*

RENAMING NAMED OBJECTS

Menu Bar: Format > Rename   **Command:**   RENAME
Ribbon: Tools > Drawing Utilities > Rename

You can edit the names of the named objects such as blocks, dimension styles, layers, linetypes, styles, UCS, views, and viewports using the **Rename** dialog box. You can select the named object from the list in the **Named Objects** area of the dialog box. The corresponding names of all the objects of the specified type that can be renamed are displayed in the **Items** area.

Chapter 20

For example, if you want to rename the layer named HID to HIDDEN, the process will be as follows:

1. Enter **RENAME** at the Command prompt and then press ENTER; the **Rename** dialog box will be displayed, as shown in Figure 20-32.

Figure 20-32 The **Rename** dialog box

2. Select **Layers** in the **Named Objects** list box. All the layer names in the current drawing that can be renamed are displayed in the **Items** list box.

3. Select **HID** in the **Items** list box to highlight it. HID will be displayed in the **Old Name** edit box.

4. Enter **HIDDEN** in the **Rename To** edit box, and choose the **OK** button.

Now, the layer named **HID** will be renamed to **HIDDEN**. You can view the change in the **Layer** drop-down list in the **Layers** panel of the **Home** tab of the **Ribbon** or invoke the **Layer Properties Manager** dialog box to notice this change. You can rename blocks, dimension styles, linetypes, styles, UCS, views, and viewports in the same way.

Note
*The **RENAME** command cannot be used to rename drawing files created using the **WBLOCK** command. You can change the name of a drawing file by using the **Rename** option of your Operating System.*

*The Layer **0** and the **Continuous** linetype cannot be renamed and, therefore, do not appear in the **Items** list box of the **Rename** dialog box.*

REMOVING UNUSED NAMED OBJECTS

Application Menu:	File > Drawing Utilities > Purge	**Command:** PURGE

This is another editing operation used for deletion and it was discussed earlier in relation to blocks. You can delete unused named objects such as blocks, layers, dimension styles, linetypes, text styles, shapes, and so on with the help of the **PURGE** command. When you create a new drawing or open an existing one, AutoCAD records the named objects in that drawing and notes other drawings that reference the named objects. Usually only a few of the named objects in the drawing (such as layers, linetypes, and blocks) are used. For example, when you create a new drawing, the prototype drawing settings may contain various text styles, blocks, and layers which you do not want to use. Also, you may want to delete particular unused named objects such as unused blocks in an existing drawing. Deleting inactive named objects is important and useful because doing so reduces the space occupied by the drawing. With the **PURGE** command, you can select the named objects you want to delete. You can use this command at any time in the drawing session. When you invoke the **PURGE** command, the **Purge** dialog box is displayed (Figure 20-33).

View items you can purge

When this radio button is selected, AutoCAD displays the tree view of all the named objects that are in the current drawing and those can be purged. When this radio button is selected, the **Items not used in drawing** area, and the two check boxes, **Confirm each item to be purged** and **Purge nested items**, are displayed in the dialog box. These are discussed as follows:

Items not used in drawing

This area lists all the named objects that can be purged. You can list the items of any object type by choosing the plus (+) sign or by double-clicking on the named object in the tree view. Select them to be purged. You can select more than one item by holding down the SHIFT or the CTRL key and selecting the items.

Confirm each item to be purged

This check box is selected to confirm before purging the selected item. The **Confirm Purge** dialog box is displayed when items are selected to be purged and after the **Purge** or **Purge All** button is chosen. You can confirm or skip the items to be purged in this dialog box.

Purge nested items

This check box, when selected, removes all the nested items not in use. This removes the nested items only when you choose the **Purge All** button or select blocks.

Purge zero length geometry and empty text objects

While drawing a line, arc, or polyline if the same point is selected as the start and the end points by mistake, then a zero length geometry will be created. Also, if there is only empty space created by mistake while creating a text, then the **Purge zero-length geometry and empty text objects** check box will be available. You can purge the zero length geometry or the empty spaces by selecting this check box .

Chapter 20

Figure 20-33 *The* **Purge** *dialog box with items that can be purged*

View items you cannot purge

This radio button is selected to display the tree view of the items that you cannot purge. These are the items that are used in the current drawing or are the default items that cannot be removed (Figure 20-34). Choosing this radio button displays the tree view in the **Items currently used in drawing** area of the dialog box. When you select an object in this tree view, information about why you cannot purge this object is displayed below the tree view.

Note
The **Entire Drawing** *option of the* **Write Block** *dialog box, which is invoked on using the* **WBLOCK** *command or the* **-WBLOCK** *option, has the same effect as the* **PURGE All** *command. The only difference is that in this case the unused named objects are removed automatically without any verification, though this method is much faster as compared to the previous one.*

Standard objects created by AutoCAD (such as layer 0, STANDARD text style, Dimension style, and linetype CONTINUOUS) cannot be removed by the **PURGE** *command, even if these objects are not used.*

*Figure 20-34 The **Purge** dialog box with items that cannot be removed*

SETTING SELECTION MODES USING THE OPTIONS DIALOG BOX

Application Menu: Options	**Command:** OPTIONS

When you select a number of objects, the selected objects form a selection set. Selection of the objects is controlled in the **Options** dialog box (Figure 20-35) that is invoked by choosing the **Options** button from the **Application Menu**. Six selection modes are provided in the **Selection** tab of this dialog box. You can select any one of these modes or a combination of various modes.

Noun/verb selection

By selecting the **Noun/verb selection** check box, you can select the objects (noun) first and then specify the operation (verb) (command) to be performed on the selection set. This mode is active by default. For example, if you want to move some objects when the mode **Noun/verb selection** is enabled, first select the objects to be moved, and then invoke the

Chapter 20

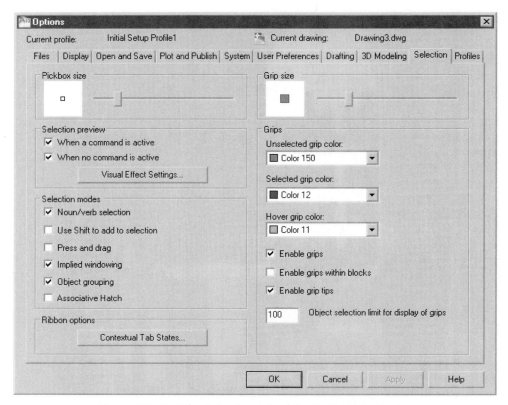

Figure 20-35 *The* ***Selection*** *tab of the* ***Options*** *dialog box*

MOVE command. The objects selected are highlighted automatically when the **MOVE** command is invoked, and AutoCAD does not issue any **Select objects** prompt. The following are some of the commands that can be used on the selected objects when the **Noun/verb selection** mode is active.

ARRAY	**BLOCK**	**CHANGE**	**CHPROP**	**COPY**
DVIEW	**EXPLODE**	**ERASE**	**LIST**	**MIRROR**
MOVE	**Properties**	**ROTATE**	**SCALE**	**WBLOCK**

The following are some of the commands that are not affected by the **Noun/verb selection** mode. You are required to specify the objects (noun) on which an operation (command/verb) is to be performed, after specifying the command (verb).

BREAK	**CHAMFER**	**DIVIDE**	**EDGESURF**	**EXTEND**
FILLET	**MEASURE**	**OFFSET**	**REVSURF**	**RULESURF**
TABSURF	**TRIM**			

When the **Noun/Verb selection** mode is active, the **PICKFIRST** system variable is set to 1 (On). In other words, you can also activate the **Noun/verb selection** mode by setting the **PICKFIRST** system variable to 1 (On).

Use Shift to add to selection Option

The next option in the **Selection Modes** area of the **Selection** tab of the **Options** dialog box is **Use Shift to add to selection**. Selecting this option establishes the additive selection mode, which is the normal method of most Windows programs. In this mode, you have to hold down the SHIFT key when you want to add objects to the selection set. For example, suppose X, Y, and Z are three objects on the screen and you have selected the **Use Shift to add to selection** check box in the **Selection Modes** area of the **Options** dialog box. Select object X. It is highlighted and put in the selection set. After selecting X, and while selecting object Y, if you do not hold down the SHIFT key, only object Y is highlighted and it replaces object X in the selection set. On the other hand, if you hold down the SHIFT key while selecting Y, it is added to the selection set (which contains X), and the resulting selection set contains both X and Y. Also, both X and Y are highlighted. To summarize the concept, objects are added to the selection set only when the SHIFT key is held down while objects are selected. Objects can be discarded from the selection set by reselecting these objects while the SHIFT key is held down. If you want to clear an entire selection set quickly, draw a blank selection window anywhere in a blank drawing area. You can also right-click to display a shortcut menu and choose **Deselect All**. All selected objects in the selection set are discarded from it.

When the **Use Shift to add to selection** mode is active, the **PICKADD** system variable is set to 0 (Off). In other words, you can activate the **Use Shift to add to selection** mode by setting the **PICKADD** system variable to off.

Press and drag

This selection mode is used to govern the way you can define a selection window or a crossing window. When you select this option, you can create the window by pressing the pick button to select one corner of the window and continuing to hold down the pick button and dragging the cursor to define the other diagonal point of the window. When you have the window you want, release the pick button. If the **Press and drag** mode is not active, you have to select twice to specify the two diagonal corners of the window to be defined. This mode is not active by default. This implies that to define a selection window or a crossing window, you have to select twice to define their two opposite corner points.

When the **Press and drag** mode is active, the **PICKDRAG** system variable is set to 1 (On). In other words, you can activate the **Press and drag** mode by setting **PICKDRAG** to On.

Implied windowing

By selecting this option, you can automatically create a Window or Crossing selection when the **Select objects** prompt is issued. The selection window or crossing window, in this case, is created in the following manner: At the **Select objects** prompt, select a point in the empty space on the screen. This becomes the first corner point of the selection window. After this, AutoCAD asks you to specify the other corner point of the selection window. If the first corner point is to the right of the second corner point, a Crossing selection is defined; if the first corner point is to the left of the second corner point, a Window selection is defined. The **Implied windowing** check box is selected by default.

Chapter 20

If this option is not active, or if you need to select the first corner in a crowded area where selecting would select an object, you need to specify Window or Crossing at the **Select objects** prompt, depending on your requirement.

When the **Implied windowing** mode is active, the **PICKAUTO** system variable is set to 1 (On). In other words, you can activate the **Implied windowing** mode by setting the value of the **PICKAUTO** system variable to 1 (On).

Object grouping

This turns the automatic group selection on and off. When this option is on and you select a member of a group, the whole group is selected. You can also activate this option by setting the value of the **PICKSTYLE** system variable to 1. (Groups were discussed earlier in this chapter.)

Tip
*You can clear the **Object grouping** check box in the **Selection Modes** area of the **Selection** tab of the **Options** dialog box to be able to select the objects of a group individually for editing without having to explode the group.*

Associative Hatch

If the **Associative Hatch** check box is selected in the **Selection modes** area of the **Selection** tab of the **Options** dialog box, the boundary object is also selected when an associative hatch is selected. You can also select this option by setting the value of the **PICKSTYLE** system variable to 2 or 3. It is recommended that you select the **Associative Hatch** check box for most drawings.

Pickbox Size

The **Pickbox** slider bar controls the size of the pickbox. The size ranges from 0 to 20. The default size is 3. You can also use the **PICKBOX** system variable.

Self-Evaluation Test

Answer the following questions and then compare them to those given at the end of this chapter:

1. Even if a group is defined as selectable, if the **PICKSTYLE** system variable is set to 0, you will not be able to select the entire group by selecting one of its members. (T/F)

2. The **PROPERTIES** command is used to change the properties associated with an object. (T/F)

3. The **Properties** palette can be used to modify the geometry of objects apart from their general properties. (T/F)

4. The **RENAME** command can be used to change the name of a drawing file created by the **WBLOCK** command. (T/F)

5. A group of commands is treated as one command for the **U** and **UNDO** commands by embedding them between the _____ and _____ options of the **UNDO** command.

6. If the **Press and Drag** mode is not active, you have to select _____ to specify the two diagonal points of a selection window.

7. The _____ option of the **PEDIT** command's main prompt can be used to change the starting and ending widths of a polyline separately to a desired value.

8. The _____ option of the **PEDIT** command allows you to select more than one polyline.

9. You can move past the first and last vertices in a closed polyline by using either the _____ option or the _____ option.

10. While using the _____ command, if you select a line or an arc, AutoCAD provides you with the option of converting them into a polyline first.

Review Questions

Answer the following questions:

1. When you use the **GROUP** command to form object groups, AutoCAD lets you sequentially highlight the groups of which the selected object is a member. (T/F)

2. The standard objects created by AutoCAD (such as layer 0, STANDARD text style, and linetype CONTINUOUS) cannot be removed by the **PURGE** command. (T/F)

3. After exploding an object, the object remains identical except that the color and linetype may change because of floating layers, colors, or linetypes. (T/F)

4. The **PURGE** command has the same effect as that of the **WBLOCK Entire drawing** or the **-WBLOCK** method. The only difference is that with the **PURGE** command, deletion takes place automatically. (T/F)

5. If you have made a copy of a group without naming the copy, its name is displayed in which of the following notations in the **Group Name** list box?

   (a) $A1   (b) @A1
   (c) %A1   (d) *A1

6. Which of the following options deletes the selected group from a drawing?

   (a) **Remove**   (b) **Reorder**
   (c) **Explode**   (d) **Rename**

7. Which of the following system variables controls the smoothness of a curve?

   (a) **SPLINETYPE**   (b) **SPLINESEGS**
   (c) **PLINEGEN**   (d) **PICKFIRST**

8. Which option of the **PEDIT** command can straighten a curve drawn with the help of **PLINE** command?

   (a) **Join**   (b) **Close**
   (c) **Decurve**   (d) None of the above

9. Which of the following properties cannot be changed using the **CHANGE** command?

   (a) **Lineweight**   (b) **Color**
   (c) **Plotstyle**   (d) **Thickness**

10. The _____ command can be used to change the name of a block.

11. If the _____ option of the **UNDO** command is off, any group of commands grouped together using the **Begin** and **End** options of the **UNDO** command are undone together.

12. Circles drawn using the **CIRCLE** command can be changed to _____.

13. The _____ option in the **Group** command lets you change the order of the objects in the selected group.

14. A group can be selected by entering _____ at the AutoCAD **Select objects**: prompt.

15. The _____ option of the **UNDO** command disables the **UNDO** and **U** commands entirely.

Exercises

Exercise 4   *Graphics*

Draw part (a) in Figure 20-36 and then using the **Properties** and relevant **PEDIT** command options, convert it into parts (b), (c), and (d). The linetype used in part (d) is HIDDEN.

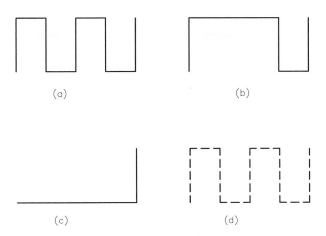

Figure 20-36 *Drawing for Exercise 4*

Exercise 5   *Mechanical*

Draw the sketch shown in Figure 20-37 using the **LINE** command. Change the object to a polyline with a width of 0.05.

Figure 20-37 *Drawing for Exercise 5*

Exercise 6   *Graphics*

Draw part (a) in Figure 20-38 and then using the relevant **PEDIT** command options, convert it into drawing (b).

Chapter 20

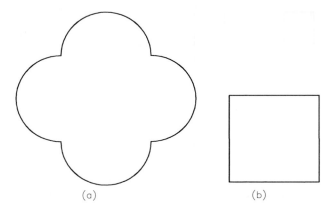

Figure 20-38 *Drawing for Exercise 6*

Exercise 7   *Graphics*

Draw part (a) in Figure 20-39 and then using the relevant **PEDIT** options, convert it into
drawings (b), (c), and (d). Identify the types of curves.

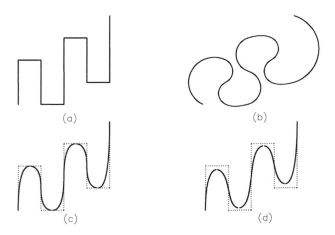

Figure 20-39 *Drawing for Exercise 7*

Exercise 8   *Graphics*

Draw part (a) in Figure 20-40 and then using the relevant **PEDIT** options, convert it into
drawing (b).

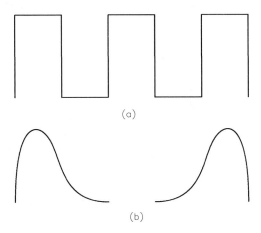

(a)

(b)

Figure 20-40 *Drawing for Exercise 8*

Exercise 9 *Graphics*

Draw part (a) in Figure 20-41 and then using the relevant **PEDIT** options, convert it into part (b) of the drawing.

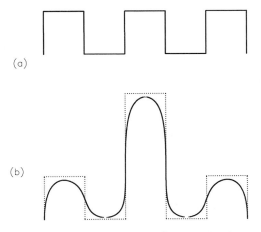

(a)

(b)

Figure 20-41 *Drawing for Exercise 9*

Exercise 10 *Mechanical*

Draw the illustration shown at the top of Figure 20-42; then use the required commands to obtain the illustration shown at the bottom. You can also use grips to obtain this illustration. Assume the missing dimensions.

Chapter 20

Figure 20-42 *Drawing for Exercise 10*

Exercise 11   *Mechanical*

Create the drawing shown in Figure 20-43. Use the splined polylines to draw the break lines.
Dimension the drawing as shown.

Figure 20-43 *Drawing for Exercise 11*

Chapter *21*

Working with Data Exchange and Object Linking and Embedding

Learning Objectives

After completing this chapter, you will be able to:

- *Import and export .dxf files using the* **SAVEAS** *and* **OPEN** *commands.*
- *Convert scanned drawings into the drawing editor using the* **DXB** *file format.*
- *Attach the raster images to the current drawing.*
- *Manage raster images.*
- *Edit raster images.*
- *Understand the embedding and linking functions of the* **OLE** *feature of Windows.*

UNDERSTANDING THE CONCEPT OF DATA EXCHANGE IN AutoCAD

Different companies have developed different software for applications such as CAD, desktop publishing, and rendering. This non-standardization of software has led to the development of various data exchange formats that enable transfer (translation) of data from one data processing software to another. This chapter will cover various data exchange formats provided in AutoCAD. AutoCAD uses the *.dwg* format to store drawing files. This format is not recognized by most other CAD software such as Intergraph, CADKEY, and Microstation. To eliminate this problem so that files created in AutoCAD can be transferred to other CAD software for further use, AutoCAD provides various data exchange formats such as DXF (data interchange file) and DXB (binary drawing interchange).

CREATING DATA INTERCHANGE (DXF) FILES

The DXF file format generates a text file in ASCII code from the original drawing. This allows any computer system to manipulate (read/write) data in a DXF file. Usually, the DXF format is used for CAD packages based on microcomputers. For example, packages like SmartCAM use DXF files. Some desktop publishing packages, such as PageMaker and Ventura Publisher, also use DXF files.

Creating a Data Interchange File

The **SAVE** or **SAVEAS** commands are used to create an ASCII file with a *.dxf* extension from an AutoCAD drawing file. Once you invoke any of these commands, the **Save Drawing As** dialog box is displayed. You can also use the **-WBLOCK** command. The **Create Drawing File** dialog box is displayed, where you can enter the name of the file in the **File name** edit box and select **DXF [*dxf]** from the **File of type** drop-down list. By default, the DXF file to be created assumes the name of the drawing file from which it will be created. However, you can specify any file name for the DXF file by typing it in the **File name** edit box. Select the extension as **DXF [*dxf]**. This can be observed in the **Files of type** drop-down list where you can select the output file format.

In the **Save Drawing As** dialog box, choose the **Tools** button to display the flyout. From the flyout, choose **Options** to display the **Saveas Options** dialog box, see Figure 21-1. In this dialog box, choose the **DXF Options** tab and enter the degree of accuracy for the numeric values. The default value for the degree of accuracy is sixteen decimal places. You can enter a value between 0 and 16 decimal places.

In this dialog box, you can also select the **Select objects** check box, which allows you to specify objects you want to include in the

Figure 21-1 The Saveas Options dialog box

DXF file. In this case, the definitions of the named objects such as block definitions, text styles, and so on are not exported. Selecting the **Save thumbnail preview image** check box, saves a preview image with the file that can be previewed in the **Preview** window of the **Save Drawing As** dialog box.

Select the **ASCII** radio button. Choose **OK** to return to the **Save Drawing As** dialog box. Choose the **Save** button here. Now an ASCII file with a *.dxf* extension has been created, and this file can be accessed by other CAD systems. This file contains data on the objects specified. By default, DXF files are created in the ASCII format. However, you can also create binary format files by selecting the **BINARY** radio button in the **Saveas Options** dialog box. Binary DXF files are more efficient and occupy only 75 percent of the ASCII DXF file. You can access a binary format file more quickly than a ASCII format file.

Information in a DXF File

The DXF file contains data of the objects which are specified using the **Select objects** check box in the **Saveas Options** dialog box. You can change the data in this file to your requirement. To examine the data in this file, load the ASCII file in the word processing software. A DXF file is composed of the following parts.

Header

In this part of the drawing database, all variables in the drawing and their values are displayed.

Classes

This section deals with the internal database.

Tables

All named objects such as linetypes, layers, blocks, text styles, dimension styles, and views are listed in this part.

Blocks

The objects that define blocks and their respective values are displayed in this part.

Entities

All entities in the drawing, such as lines, circles, and so on, are listed in this part.

Objects

Objects in the drawing are listed in this part.

Converting DXF Files into Drawing Files

You can import a DXF file into a new AutoCAD drawing file with the **OPEN** command. After you invoke the **OPEN** command, the **Select File** dialog box is displayed. From the **Files of type** drop-down list, select **DXF [*.dxf]**. In the **File name** edit box, enter the name of the file you want to import into AutoCAD or select the file from the list. Choose the **Open** button.

Once this is done, the specified DXF file is converted into a standard DWG file, regeneration is carried out, and the file is inserted into the new drawing. Now, you can perform different operations on this file just as with the other drawing files.

Tip
A data interchange file (DXF) can also be inserted in the current drawing using the INSERT command.

OTHER DATA EXCHANGE FORMATS

The other formats that can be used to exchange data from one data processing format to the other are discussed next.

DXB File Format

Menu Bar: Insert > Drawing Exchange Binary	**Command:** DXBIN

AutoCAD offers another file format for data interchange, DXB. This format is much more compressed than the binary DXF format and is used when you want to translate a large amount of data from one CAD system to another. For example, when you scan a drawing, a DXB file is created. This file has a huge amount of data in it. To import a DXB file, enter **DXBIN** command at the Command prompt; the **Select DXB File** dialog box will be displayed. In the **File name** edit box, enter the name of the file (in DXB format) you want to import into AutoCAD. Choose the **Open** button. Once this is done, the specified DXB file will be converted into a standard DWG file and is inserted into the current drawing.

Note
Before importing a DXB file, you must create a new drawing file. No editing or drawing setup (limits, units, and so on) can be performed in this file. This is because if you import a DXB file into an old drawing, the settings (definitions) of layers, blocks, and so on, of the file being imported are overruled by the settings of the file into which you are importing the DXB file.

Creating and Using an ACIS File

Application Menu: Export > Other Formats	**Command:** EXPORT, ACISOUT

Trimmed Nurbs surfaces, regions, and solids can be exported to an ACIS file with an ASCII format. To do this, choose **Export > Other Formats** from the **Application Menu**. The **Export Data** dialog box is displayed. Enter a file name in the **File name** edit box. From the **Files of type** drop-down list, select **ACIS [*.sat]** and then choose the **Save** button. You will be prompted to select the solids, regions, or ACIS bodies that you want to add to the file. Select the objects you wish to save as an ACIS file and press ENTER. AutoCAD appends the file extension *.sat* to the file name. You can also use the **ACISOUT** command to perform this function. It prompts you to select objects before the **Create ACIS File** dialog box is displayed on the screen. Keep in mind that only solids, regions, or ACIS bodies can be selected to add to the *.sat* file.

To import an ACIS file, choose **Import > Import** from **Insert** tab of the **Ribbon**; the **Import File** dialog box will be displayed. In the **Files of type** drop-down list, select the *.sat* file format. Next, select a file to be imported and choose the **Open** button; the file will be inserted. Alternatively, enter **ACISIN** at the command prompt and import the file using the **Select ACIS File** dialog box.

Importing 3D Studio Files

Menu Bar:	Insert > 3D Studio	**Command:**	IMPORT, 3DSIN

You can import 3D geometry, views, lights, and objects with surface characteristics created in 3D Studio and saved in *.3ds* format. The **3DSIN** command displays the **3D Studio File Import** dialog box. Select the file you wish to import and choose the **Open** button. You can also invoke this command by choosing **3D Studio** from the **Insert** menu.

Creating and Using a Windows Metafile

Application Menu:	Export > Other Formats	**Command:** EXPORT, WMFOUT

The Windows Metafile format (WMF) file contains screen vector and raster graphics format. In the **Export Data** dialog box or the **Create WMF File** dialog box, enter the file name. Select **Metafile [*.wmf]** from the **Files of type** drop-down list. The extension *.wmf* is appended to the file name. Next, save the settings and then select the objects you want to save in this file format.

The **WMFIN** command displays the **Import WMF** dialog box. Window metafiles are imported as blocks in AutoCAD. Select the *.wmf* file you want to import and choose the **Open** button. Specify an insertion point, rotation angle, and scale factor. Specify the scaling by entering a scale factor using the **corner** option to specify an imaginary box whose dimensions correspond to the scale factor, or by entering **xyz** to specify 3D scale factors. You can also import a *.wmf* file by using the **Import File** dialog box that is displayed on choosing **Import > Import** from the **Insert** tab of the **Ribbon**.

Creating and Using a V8 DGN File

Application Menu:	Export > DGN	**sCommand:** EXPORT, DGNEXPORT

The **V8 DGN** (*.DGN*) file format is used by the Microstation software. A **.dwg* file created in AutoCAD can be exported to Microstation using this format. To do so, choose **Export > DGN** from the **Application Menu**; the **Export DGN File** dialog box will be displayed. Enter a file name in the **File name** edit box. From the **Files of type** drop-down list, select **V8 DGN [*.dgn]** and then choose the **Save** button; the **Export DGN Settings** dialog box will be displayed. In this dialog box, you can specify how the externally referenced files in the current drawing are taken care of when converting the current drawing into the DGN file. You can also specify the seed files (standard template files containing the settings for the units to be used for conversion, working units, resolution of the drawing, and so on) to be used for conversion in this dialog box.

To import a DGN file, choose **Open > DGN** from the **Application Menu**; the **Import DGN File** dialog box will be displayed. Alternatively, you can use the **DGNIMPORT** command to import a DGN file. From the **Files of type** drop-down list, select the **MicroStation DGN [*.dgn]** option. Next, select or enter the name of the design file and choose the **Open** button; the **Import DGN Settings** dialog box will be invoked. You can exchange the data of the drawing files and design files using the **Export/Import DGN Settings** dialog box. For example, you can translate DGN elements to DWG objects, Levels to Layers, Line Styles to Linetypes, Cells to Blocks, Sheet Model to DWG Layouts, etc. You can save and reuse, modify, rename or delete the mapping translations using the **DGNMAPPING** command.

Creating a BMP File

Application Menu:	Export > Other Formats	Command:	EXPORT, BMPOUT

This is used to create bitmap images of the objects in your drawing. In the **Export Data** dialog box, enter the name of the file, select **Bitmap [*.bmp]** from the **Files of type** drop-down list and then choose **Save**. Select the objects you wish to save as bitmap and press ENTER. Entering **BMPOUT** displays the **Create Raster File** dialog box. Enter the file name and choose **Save**. Select the objects to be saved as bitmap. The file extension *.bmp* is appended to the file name.

RASTER IMAGES

A raster image consists of small square-shaped dots known as pixels. In a colored image, the color is determined by the color of pixels. The raster images can be moved, copied, or clipped, and used as a cutting edge with the **TRIM** command. They can also be modified by using grips. You can also control the image contrast, transparency, and quality of the image. AutoCAD stores images in a special temporary image swap file whose default location is the Windows **Temp** directory. You can change the location of this file by modifying it under **Temporary Drawing File Location** in the **Files** tab of the **Options** dialog box.

The images can be 8-bit gray, 8-bit color, 24-bit color, or bitonal. When the image transparency is set to On, the image file formats with transparent pixels is recognized by AutoCAD and transparency is allowed. The transparent images can be in color or grayscale. AutoCAD supports the formats given next.

Image Type	File Extension	Description
BMP	*.bmp, .dib, .rle*	Windows and OS/2 Bitmap Format
CALS-I	*.gp4, .mil, .rst*	Mil-R-Raster I
FLIC	*.flc, .fli*	Flic Autodesk Animator Animation
GEOSPOT	*.bil*	GeoSPOT (BIL files must be accompanied with HDR and PAL files with connection data in the same directory.)
IG4	*.ig4*	Image Systems group 4

IGS	*.igs*	Image Systems Grayscale
JFIF or JPEG	*.jpg, jpeg*	Joint Photographics Expert group
PCX	*.pcx*	Picture PC Paintbrush Picture
PICT	*.pct*	Picture Macintosh Picture
PNG	*.png*	Portable Network Graphic
RLC	*.rlc*	Run-length Compressed
TARGA	*.tga*	True Vision Raster based Data format
TIFF/LZW	*.tif, .tiff*	Tagged image file format

When you store images as Tiled Images, that is, in the Tagged Image File Format (TIFF), you can edit or modify any portion of the image; only the modified portion is regenerated, thus saving time. Tiled images load much faster as compared to the untiled images.

Attaching Raster Images

Ribbon: Insert > Reference > Attach   **Tool Palettes:** Draw > Attach Image
Toolbar: Reference > Attach Image or Insert > Attach Image   **Command:** IMAGEATTACH

Attaching Raster images do not make them part of the drawing. To attach an image, choose **Insert > Reference > Attach** from the **Ribbon**; the **Select Reference File** dialog box will be displayed (Figure 21-2). You can also invoke this dialog box by using the **IMAGEATTACH** command. A preview of the selected image is displayed in the **Preview** area of the dialog box. Select the image and choose the **Open** button.

AutoCAD displays the **Attach Image** dialog box (Figure 21-3). The name of the selected image is displayed in the **Name** drop-down list. You can select another file by using the **Browse** button. The **Name** drop-down list displays the names of all the images in the current drawing. Select the **Specify on-screen** check boxes to specify the **Insertion point**, the **Scale**, and **Rotation** Angle on the screen. Alternately, you can clear these check boxes and enter values in the respective edit boxes. Choosing the **Show Details** button expands the **Image** dialog box and provides the image information such as the Horizontal and Vertical resolution values, current AutoCAD unit, and Image size in pixels and units. Choose **OK** to return to the drawing screen.

You can also attach and scale an image from the Internet. In the **Select Reference File** dialog box, choose the **Search the Web** button. Once you have located the image file you wish to attach to the current drawing, enter the URL address in the **Look in** drop-down list and the image file name in the **Name or URL** edit box. Choose the **Open** button twice. Specify the insertion point, scale, and rotation angle in the **Attach Image** dialog box. You can also right-click on the image you wish to attach to the current drawing and choose **Properties** from the shortcut menu to display the **Properties** window. From here, you can select, cut, and paste the URL address in the **File name** edit box of the **Select Reference File** dialog box.

Figure 21-2 *The **Select Reference File** dialog box*

Figure 21-3 *The **Attach Image** dialog box*

Managing Raster Images

Ribbon: View > Palette > External References Palette   **Command:** IMAGE
Toolbar: Reference > External References   **Menu Bar:** Insert > External References

 Choose **View > Palette > External References Palette** from the **Ribbon**; the **External References** palette will be displayed (Figure 21-4). If the image has not been inserted earlier, then right-click in the **File References** area and choose **Attach Image** from the shortcut menu. You can view the image information either as a list or as a tree view by choosing the respective buttons located at the right corner of the **File References** head of the **External References** palette. The **List View** displays the names of all the images in the drawing, its loading status, size, date last modified on, and its search path. The **Tree View** displays the images in a hierarchy, which shows its nesting levels within blocks and Xrefs. The **Tree View** does not display the status, size, or any other information about the image file. You can rename an image file in this dialog box.

Figure 21-4 The External References palette

Preview

To display the **Preview** area, choose the **Preview** button located at the right corner of the **Details** head. In this area, you can see the preview of the image. This area displays the preview of the image file attached to the current drawing and file details of the attached image file. To get the file details, choose the **Details** button located adjacent to the **Preview** button. This lists the information about the image file such as file name, saved path, and so on, as shown in Figure 21-5. Once you have referenced the current drawing to an image file, you can replace it with another image file by using the **Found At** edit box. To do so, choose the **Browse [...]** button in the corner of the edit box to display the **Select Image File** dialog box. Browse and select the image file and then choose the **Open** button. The new path will now be displayed in the **Found At** edit box. Also, the location of the image file will be automatically updated in the **Saved Path** edit box.

> **Tip**
> *You can also use the Command line for managing the image files using the -IMAGE command. Except the Open option, all the other options of the External References palette will be available in the Command line.*

When you select the attached image and right-click on it, the shortcut menu will be displayed. The options in this shortcut menu are discussed next.

Open

This option enables you to open the attached image in its default viewing software from its current location in the AutoCAD file.

Attach

This option allows you to attach an image file with the current drawing. You can also attach an image file by choosing the **Attach Image** option from the flyout on the upper left corner of the **External References** palette; the **Select Reference File** dialog box will be displayed. Select the file that you want to attach to the drawing and the selected image is displayed in the **Preview** box. Choose the **Open** button; the **Image** dialog box is displayed. The name of the file and its path and extension are displayed in the **Attach Image** dialog box. You can use the **Browse** button to select another file. Choose the **OK** button in the **Attach Image** dialog box and specify a point to insert the image. Next, AutoCAD will prompt you to enter the scale factor. You can enter the scale factor or specify a point on the screen.

Details			
Reference Na...	Tower_iso1
Status	Loaded
Size	521 KB
Type	JPG
Date	2/10/2005 8:33:00 AM
Saved Path	D:\acad_2010_manuscript\c19_aca
Found At	D:\acad_2010_manuscript\c19_aca
Color System	RGB
Color Depth	32
Pixel Width	612
Pixel Height	2295
Resolution	300.00 per inch
Default Size	2.04 x 7.65 inches

*Figure 21-5   The **Details** area of **External References** palette*

Detach

This option detaches the selected and all associated raster images from the current drawing. Once the image is detached, no information about the image is retained in the drawing.

Reload

Reload simply reloads an image. The changes made to the image since the last insert will be loaded on the screen.

Unload

The **Unload** option unloads an image. An image is unloaded when it is not needed in the current drawing session. When you unload an image, AutoCAD retains information about the location and size of the image in the database of the current file. Unloading a file does not unlink the file from the drawing and the image frame is displayed. If you reload the image, the image will appear at the same point and in the same size as the image was before unloading. Unloading the raster images enhances AutoCAD performance. Also, the unloaded images are not plotted. If multiple images are to be loaded and the memory is insufficient, AutoCAD automatically unloads them.

EDITING RASTER IMAGE FILES
Clipping Raster Images

Ribbon:	Insert > Reference > Clip	**Toolbar:**	Reference > Clip Image
Menu Bar:	Modify > Clip > Image	**Command:**	IMAGECLIP

 This command is used to clip the boundaries of images and provide a desired shape to them. To invoke this command, select the image to be clipped; the **Image** tab will be added to the **Ribbon**. Choose the **Create Clipping Boundary** button; you will be prompted to specify the boundary. Specify a rectangular or polygonal boundary; the image will be clipped at the specified boundary.

You can also invoke this command by selecting an image to be clipped and then right-clicking to display the shortcut menu. Choose **Image > Clip** to invoke the **IMAGECLIP** command. When you invoke the command, AutoCAD will prompt you to select the image to be clipped. Select the raster image by selecting the boundary edge of the image. The image boundary must be visible to select it. The clipping boundary must be specified in the same plane as the image or a plane that is parallel to it. The following prompt sequence is displayed when you choose **Insert > Reference > Clip** from the **Ribbon**.

> Select image to clip: *Select the image.*
> Enter image clipping option [ON/OFF/Delete/New boundary] <New>:

New boundary
This option is used to define a boundary for clipping the image. When you invoke this option, you will be prompted to specify whether or not you want to delete the old boundary, if a boundary already exists. If you enter **No** at the **Delete old boundary? [No/Yes] <Yes>** prompt, this command will end. You can enter **Yes** to specify the new clipping boundary. The boundary can be defined in a rectangular shape or a polygonal shape.

ON
This option is used to turn the image clipping on. The image will be clipped using the last boundary.

OFF
This option is used to turn the image clipping off. If you have a clipped image, you can redisplay the entire image.

Delete
This option is used to delete the existing clipping boundary.

One such clipping is shown in Figures 21-6 and 21-7. Figure 21-6 shows the actual image and Figure 21-7 shows the clipped image where the clipping boundary becomes the midpoint of all the four sides of the image.

Tip
*You can improve the performance by turning on or off the image highlighting that is visible when you select an image. This can be done by selecting or clearing the **Highlight raster image frame only** check box in the **Display** tab of the **Options** dialog box. You can also use the **IMAGEHLT** system variable. By default, **IMAGEHLT** is set to 0, that is, only the raster image frame will be highlighted.*

Figure 21-6 Image before clipping

Figure 21-7 Image after clipping

Adjusting Raster Images

Ribbon:	Insert > Reference > Adjust	**Modify:** Insert > Image > Adjust
Toolbar:	Reference > Adjust Image	**Command:** IMAGEADJUST/ADJUST

The **IMAGEADJUST** command allows you to adjust the brightness, contrast, and fading of the raster image. In AutoCAD 2010, if you select an attached image, the **Image** tab will be added to the **Ribbon**. Adjust the brightness, contrast, and fade using the corresponding option in the **Adjust** panel of the **Image** tab.

Alternatively, choose **Adjust Image** from the **Reference** toolbar to invoke the **Image Adjust** dialog box (Figure 21-8), which you can use to adjust the image. As you adjust the brightness, contrast, or fade, the changes are dynamically made in the image. If you choose the **Reset** button, the values are returned to the default values (Brightness= 50, Contrast= 50, and Fade= 0). You can also choose **Properties** in the shortcut menu to display the **Properties** palette. In this palette, you can enter new values of **Brightness**, **Contrast**, and **Fade**. If you choose **Insert > Reference > Adjust** from the **Ribbon**, you need set to the values in the command window.

Figure 21-8 The **Image Adjust** dialog box

Modifying the Image Quality

Menu Bar:	Modify > Object > Image > Quality	**Command:** IMAGEQUALITY
Toolbar:	Reference > Image Quality	

The **IMAGEQUALITY** command allows you to control the quality of the image that affects the display performance. A high-quality image takes a longer time to display. When you change the quality, the display changes immediately without causing a **REGEN**. The images are always plotted using a high-quality display. Although draft quality images appear more grainy, they are displayed more quickly.

Modifying the Transparency of an Image

Toolbar:	Reference > Image Transparency	**Command:** TRANSPARENCY
Menu Bar:	Modify > Object > Image > Transparency	

When you attach an image with transparent background to a drawing, the transparent background turns opaque. The **TRANSPARENCY** command is used to control the transparency of the image that is created in a transparent background. To control the transparency of the attached image, select it; the **Image** tab will be added to the **Ribbon**. Choose the **Transparency** button from the **Ribbon** to turn on the transparency of the image; the background of the image will become transparent. Figure 21-9 shows two images one over the other with the transparency of the top image turned off. Figure 21-10 shows the figure with the transparency of the top image turned on. You can also right-click on an image and choose **Properties** from the shortcut menu displayed. In the **Properties** palette, select an option from the **Transparency** drop-down list displayed under the **Misc** list.

Figure 21-9 *Image before turning the transparency on*

Figure 21-10 *Image after turning the transparency on*

Note
You have to choose images in such a way that they can give the effect of transparency. Dull and opaque images will not give the effect of transparency.

Controlling the Display of Image Frames

Menu Bar:	Modify > Object > Image > Frame	**Command:** IMAGEFRAME
Toolbar:	Reference > Image Frame	

The **IMAGEFRAME** command is used to turn the image boundary on or off. If the image boundary is off, the image cannot be selected with the pointing device, and therefore, cannot be accidentally moved or modified.

CHANGING THE DISPLAY ORDER

Ribbon:	Home > Modify > Draw Order	**Command:** DRAWORDER
Toolbar:	Draw Order	

The **Draw Order** toolbar is used to change the display order of images and other objects. This toolbar provides you with four buttons to modify the display order of a selected entity.

If you choose the **Bring to Front** button, the selected object will be moved to the front of all the entities. If you choose the **Send to Back** button, the selected object will be moved behind all the entities. If you choose the **Bring Above Objects** button, the selected object will be moved above the referenced objects. If you choose the **Send Under Objects** button, the selected object will be moved under the referenced objects.

Other Editing Operations

You can also perform other editing operations such as copy, move, and stretch to edit the raster image. Remember that you can also use the image as the trimming edge for trimming objects. However, you cannot trim an image. You can insert the raster image several times or make multiple copies of it. Each copy can have a different clipping boundary. You can also edit the image using grips. You can use the **Properties** window to change the image layer, boundary linetype, and color, and perform other editing commands such as changing the scale, rotation, width, and height by entering new values in the respective edit boxes. You can also display an image or turn off the display by selecting options from the **Show Image** drop-down list in the **Misc** field of the **Properties** window. If you select **Yes**, the image is displayed and if you select **No**, the display of the image is turned off. You can turn off the display of the image when you do not need it, thus improving the performance.

Scaling Raster Images

The scale of the inserted image is determined by the actual size of the image and the unit of measurement (inches, feet, and so on). For example, if the image is 1" X 1.26" and you insert this image with a scale factor of 1, its size on the screen will be 1 AutoCAD unit by 1.26 AutoCAD units. If the scale factor is 5, the image will be five times larger. The image that you want to insert must contain the resolution information (DPI). If the image does not contain this information, AutoCAD treats the width of the image as one unit.

WORKING WITH POSTSCRIPT FILES

Menu Bar:	File > Export	**Application Menu:**	Export > Other Format
Command:	EXPORT, PSOUT		

PostScript is a page description language developed by Adobe Systems. It is mostly used in DTP (desktop publishing) applications. AutoCAD allows you to work with PostScript files. You can create and export PostScript files from AutoCAD, so that these files can be used for DTP applications. PostScript images have a higher resolution than raster images. The extension for these files is *.eps* (Encapsulated PostScript).

Creating the PostScript Files

As just mentioned, any AutoCAD drawing file can be converted into a PostScript file. This can be accomplished with the **PSOUT** command. When you invoke this command, the **Create PostScript File** dialog box is displayed, as shown in Figure 21-11. You can also use the **EXPORT** command to display the **Export Data** dialog box. The options are similar in both the dialog boxes.

*Figure 21-11 The **Create PostScript File** dialog box*

In the **File name** edit box, enter the name of the PostScript (EPS) file you want to create. When you use the **EXPORT** command, the **Export Data** dialog box is displayed. Select **Encapsulated PS (*.eps)** from the **Files of type** drop-down list. Next, choose the **Save** button to accept the default setting and create the PostScript file. You can also choose **Options** from the **Tools** flyout of the **Create PostScript File** dialog box to change the settings through the

PostScript Out Options dialog box (Figure 21-12) and then save the file. The **PostScript Out Options** dialog box provides the following options.

*Figure 21-12 The **PostScript Out Options** dialog box*

Prolog Section Name

In this edit box, you can assign a name for a prolog section to be read from the *acad.psf* file.

What to plot Area

The **What to plot** area of the dialog box has the following options.

Display. If you specify this option when you are in the model space, the display in the current viewport is saved in the specified EPS file. Similarly, if you are in the layouts, the current view is saved in the specified EPS file.

Extents. If you use this option, the PostScript file created will contain the section of the AutoCAD drawing that currently holds objects. In this way, this option is similar to the **ZOOM Extents** option. If you add objects to the drawing, they are also included in the PostScript file to be created because the extents of the drawing are also altered. If you reduce the drawing extents by erasing, moving, or scaling objects, then you must use the **ZOOM Extents** or **ZOOM All** option. Only then does the **Extents** option of the **PSOUT** command understand the extents of the drawing to be exported. If you are in the model space, the PostScript file is created in relation to the model space extents; if you are in the paper space, the PostScript file is created in relation to the paper space extents.

Limits. With this option, you can export the whole area specified by the drawing limits. If the current view is not the plan view [viewpoint (0,0,1)], the **Limits** option exports the area just as the **Extents** option would.

View. Any view created with the **VIEW** command can be exported with this option. This option will be available only when you have created a view using the **VIEW** command. Choose the **View** button to display the **View Name** dialog box from where you can select the view.

Window. In this option, you need to specify the area to be exported with the help of a window. When this radio button is selected, the **Window** button is also available. Choose the **Window** button to display the **Window Selection** dialog box where you can select the **Pick** button and then specify the two corners of the window on the screen. You can also enter the coordinates of the two corners in the **Window Selection** dialog box.

Preview Area

The **Preview** area of the dialog box has two types of formats to preview images: **EPSI** and **TIFF**. If you want a preview image with no format, select the **None** radio button. If you select **TIFF** or **EPSI**, you are required to enter the pixel resolution of the screen preview in the **Pixels** area. You can select a preview image size of 128 X 128, 256 X 256, or 512 X 512.

Size Units Area

In this area, you can set the paper size units to **Inches** or **MM** by selecting their corresponding radio buttons.

Scale Area

In this area, you can set an explicit scale by specifying how many drawing units are to be output per unit. You can select the **Fit to paper** check box so that the view to be exported is made as large as possible for the specified paper size.

Paper Size Area

You can select a size from the list or enter a new size in the **Width** and **Height** edit boxes to specify a paper size for the exported PostScript image.

OBJECT LINKING AND EMBEDDING (OLE)

With Windows, it is possible to work with different Windows-based applications by transferring information between them. You can edit and modify the information in the original Windows application, and then update this information in other applications. This is made possible by creating links between different applications and then updating those links, which in turn update or modify the information in the corresponding applications. This linking is a function of the OLE feature of Microsoft Windows. The OLE feature can also join together separate pieces of information from different applications into a single document. AutoCAD and other Windows-based applications such as Microsoft Word, Notepad, and Windows WordPad support the Windows OLE feature.

For the OLE feature, you should have a source document where the actual object is created in the form of a drawing or a document. This document is created in an application called a **server** application. AutoCAD for Windows and Paintbrush can be used as server applications. Now this source document is to be linked to (or embedded in) the **compound** (destination) document, which is created in a different application, known as the **container** application.

AutoCAD for Windows, Microsoft Word, and Windows WordPad can be used as container applications.

Clipboard

The transfer of a drawing from one Windows application to another is performed by copying the drawing or the document from the server application to the Clipboard. The drawing or document is then pasted in the container application from the Clipboard. Hence, a Clipboard is used as a medium for storing the documents while transferring them from one Windows application to another. The drawing or the document on the Clipboard stays there until you copy a new drawing, which overwrites the previous one, or until you exit Windows. You can save the information present on the Clipboard with the *.clp* extension.

Object Embedding

You can use the embedding function of the OLE feature when you want to ensure that there is no effect on the source document even if the destination document has been changed through the server application. Once a document is embedded, it has no connection with the source. Although editing is always done in the server application, the source document remains unchanged. Embedding can be accomplished by means of the following steps. In this example, AutoCAD for Windows is the server application and MS Word is the container application.

1. Create a drawing in AutoCAD (the server application).

2. Open MS Word (container application).

3. It is preferable to arrange both the container and server windows so that both are visible, as shown in Figure 21-13.

4. In AutoCAD, copy the drawing using the CTRL+C key; the **COPYCLIP** command is invoked. This command can be used to embed the drawings. You can also invoke this command from the **Edit** menu (Choose **Copy**), by entering **COPYCLIP** at the command line. The next prompt, **Select objects**, allows you to select the entities you want to transfer. You can either select the full drawing by entering **ALL** or select some of the entities by selecting them. You can use any of the object selection methods for selecting the objects. With this command, the selected objects are automatically copied to the Windows Clipboard.

5. After the objects are copied to the Clipboard, make the MS Word window active. To get the drawing from the Clipboard to the MS Word application (client), you need to paste it in the Word application. Choose **Edit > Paste** from the menu bar in MS Word (Figure 21-14). You can also use **Edit > Paste Special** from the menu bar, which will display the **Paste Special** dialog box (Figure 21-15). In this dialog box, select the **Paste** radio button (default) for embedding, and then choose **OK**. The drawing is now embedded in the MS Word window.

Figure 21-13 *AutoCAD graphics screen with the MS Word window*

Figure 21-14 *Pasting a drawing to the MS Word application*

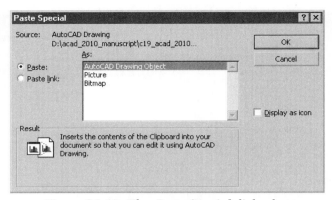

Figure 21-15 *The* ***Paste Special*** *dialog box*

6. Your drawing is now displayed in the MS Word window, but it may not be displayed at the proper position. You can get the drawing in the current viewport by moving the scroll bar up or down in the MS Word window. You can also save your embedded drawing by choosing the **Save** button. It displays the **Save As** dialog box where you can enter a file name. You can now exit AutoCAD.

7. You can now edit your embedded drawing. Editing is performed in the server application, which in this case is AutoCAD. You can get the embedded drawing into the server application (AutoCAD) directly from the container application (MS Word) by double-clicking on the drawing in MS Word. The other method is by choosing **Edit > AutoCAD Drawing Object > Edit** from the menu bar after selecting the drawing. (This menu item has replaced **Object**, which was present before pasting the drawing.)

8. Now you are in AutoCAD, with your embedded drawing displayed on the screen, but as a temporary file with a file name such as [Drawing in Document]. Here, you can edit the drawing by changing the color and linetype or by adding and deleting the text, entities, and so on. In this example, the cup and plate have been hatched.

9. After you have finished modifying your drawing, choose **File > Update Microsoft Word** from the menu bar in the server (AutoCAD) see Figure 21-16. To display the menu bar, click on the down arrow in the **Quick Access Toolbar** and select **Show Menu Bar**. AutoCAD automatically updates the drawing in MS Word (container application). Now, you can exit this temporary file in AutoCAD.

Note
Do not zoom or pan the drawing in the temporary file. If you do so, this will be included in the updating and the new file will display only that portion of the drawing that lies inside the original area.

10. This completes the embedding function, so you can exit the container application. While exiting, a message box that asks whether or not to save changes in MS Word is displayed.

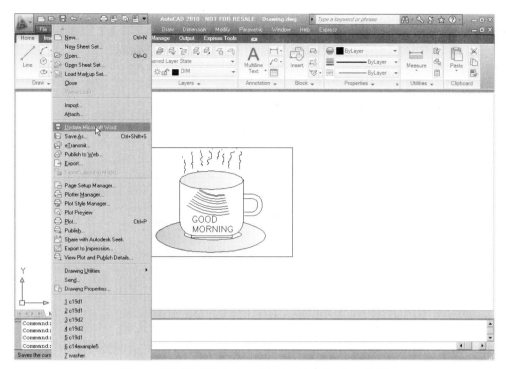

Figure 21-16 *Updating the drawing in AutoCAD*

Linking Objects

The linking function of OLE is similar to the embedding function. The only difference is that here a link is created between the source document and the destination document. If you edit the source, you can simply update the link, which automatically updates the client. This allows you to place the same document in a number of applications, and if you make a change in the source document, the clients will also change by simply updating the corresponding links. Consider AutoCAD for Windows to be the server application and MS Word to be the container application. Linking can be performed by means discussed next.

1. Open a drawing in the server application (AutoCAD). If you have created a new drawing, then you must save it before you can link it with the container application. Then, open MS Word (container application).

2. It is preferable to arrange both the container and the server windows so that both are visible.

3. In AutoCAD, choose **Copy Link** from the **Edit** menu from the menu bar to invoke the **COPYLINK** command. This command can be used in AutoCAD for linking the drawing. This command copies all objects that are displayed in the current viewport directly to the Clipboard. Here, you cannot select the objects for linking. If you want only a portion of the drawing to be linked, you can zoom into that view so that it is displayed in the current viewport prior to invoking the **COPYLINK** command. This command also creates a new view of the drawing having a name OLE1.

4. Make the MS Word window active. To get the drawing from the Clipboard to the MS Word (container) application, choose **Paste Special** from the **Edit** menu. This will display the **Paste Special** dialog box. In this dialog box, select the **Paste link** radio button for linking. Choose **OK**. Note that the **Paste link** radio button in the **Paste Special** dialog box will not be available if you have not saved the drawing.

5. The drawing is now displayed in the MS Word window and is linked to the original drawing. You can also save your linked drawing by choosing **Save** from the **Standard** toolbar. It displays a **Save As** dialog box where you can enter a file name.

6. You can now edit your original drawing. Editing can be performed in the server application, which in this case, is AutoCAD for Windows. You can edit the drawing by changing the color and linetype or by adding and deleting text, entities, and so on. Then save your drawing in AutoCAD by using the **QSAVE** command. You can now exit AutoCAD.

7. You will notice that the drawing is automatically updated, and the changes made in the source drawing are present in the destination drawing also. This automatic updating is dependent on the selection of the **Automatic** radio button (default) in the **Links** dialog box (Figure 21-17). The **Links** dialog box can be invoked by choosing **Links** from the **Edit** menu of the Word application. For updating manually, you can select the **Manual** radio button in the dialog box. In the manual case, after making changes in the source document and saving it, you need to invoke the **Links** dialog box and then choose the **Update Now** button. This will update the drawing in the container application and display the updated drawing in the MS Word document.

Note
The dialog box may be different in your case as it depends on the version of MS Word you are currently using.

8. Exit the container application after saving the updated file.

*Figure 21-17 The **Links** dialog box*

Linking Information into AutoCAD

Similarly, you can also embed and link information from a server application into an AutoCAD drawing. You can also drag selected OLE objects from another application into AutoCAD, provided this application supports Microsoft ActiveX and the application is running and visible on the screen. Dragging and dropping is like cutting and pasting. If you press the CTRL key while you drag the object, it is copied to AutoCAD. Dragging and dropping an OLE object into AutoCAD embeds it into AutoCAD.

Linking Objects into AutoCAD

Start any server application such as MS Word and open a document in it. Select the information you wish to use in AutoCAD with your pointing device and choose the **Copy** button to copy this data to the Clipboard. Open the AutoCAD drawing you wish to link this data to. Choose **Paste Special** from the **Edit** menu or use the **PASTESPEC** command. The **Paste Special** dialog box will be displayed (Figure 21-18). In the **As** list box, select the data format you wish to use. For example, for a MS Word document, select **Microsoft Office Word Document**. Picture format uses a Metafile format. Select the **Paste Link** radio button to paste the contents of the Clipboard to the current drawing. If you select the **Paste** radio button, the data is embedded and not linked. Choose **OK** to exit the dialog box. Specify the insertion point; the data is displayed in the drawing.

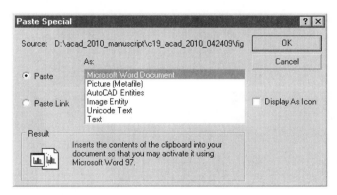

Figure 21-18 The **Paste Special** dialog box

You can also use the **INSERTOBJ** command by entering **INSERTOBJ** at the Command prompt, or by choosing **OLE Object** from the **Data** tab of the **Insert** panel of the **Ribbon**, or by choosing the **OLE Object** button in the **Insert** toolbar. This command links an entire file to a drawing from within AutoCAD. Using this command displays the **Insert Object** dialog box, as shown in Figure 21-19.

Select the **Create from File** radio button. Also, select the **Link** check box. Choosing the **Browse** button, displays the **Browse** dialog box. Select a file you want to link from the list box or enter a name in the **File name** edit box and choose the **Open** button. The path of the file is displayed in the **File** edit box of the **Insert Object** dialog box. If you select the **Display As Icon** check box, an icon is displayed in the dialog box as well as in the drawing. Choose **OK** to exit the dialog box; the selected file is linked to the AutoCAD drawing.

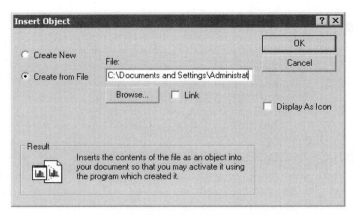

Figure 21-19 The **Insert Object** *dialog box*

AutoCAD updates the links automatically by default, whenever the server document changes, but you can use the **OLELINKS** command to display the **Links** dialog box (Figure 21-20) where you can change these settings. This dialog box can also be displayed by choosing **OLE Links** from the **Edit** menu. In the **Links** dialog box, select the link you want to update and then choose the **Update Now** button. Then choose the **Close** button. If the server file location changes or if it is renamed, you can choose the **Change Source** button in the **Links** dialog box to display the **Change Source** dialog box. In this dialog box, locate the server file name and location and choose the **Open** button. You can also choose the **Break Link** button in the **Links** dialog box to disconnect the inserted information from the server application. This is done when the linking between the two is no longer required.

Figure 21-20 The **Links** *dialog box*

Embedding Objects into AutoCAD

Open the server application and select the data you want to embed into the AutoCAD drawing. Copy this data to the Clipboard by choosing the **Copy** button in the toolbar or **Copy** from the

Edit menu. Open the AutoCAD drawing and choose **Paste** from the AutoCAD **Edit** menu. You can use the **PASTECLIP** command also. Specify the insertion point; the selected information is embedded into the AutoCAD drawing.

You can also create and embed an object into an AutoCAD drawing starting from AutoCAD itself using the **INSERTOBJ** command. Alternatively, choose **Insert > Data > OLE Object** from the **Ribbon** to invoke the **Insert Object** dialog box (Figure 21-21). In this dialog box, select the **Create New** radio button and select the application you wish to use from the **Object Type** list box. Choose **OK**. The selected application opens. Now, you can create the information you wish to insert in the AutoCAD drawing and save it before closing the application. You can edit information embedded in the AutoCAD drawing by double-clicking on the inserted OLE object; the server application will open. The modifications made in the server application get automatically reflected in the AutoCAD drawing.

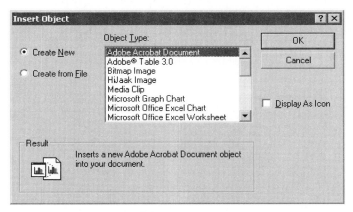

Figure 21-21 *The* **Insert Object** *dialog box*

Working with OLE Objects

Select an OLE object and right-click to display a shortcut menu; choose **Properties** from the shortcut menu; the **Properties** window is displayed. You can also select the OLE object and use the **OLESCALE** command. Specify a new height and a new width in the **Height** and **Width** edit boxes in the **Geometry** area. Enter a value in percentage of the current values in the **Height** and **Width** edit boxes. Here, if you select the **Lock Aspect Ratio** check box, whenever you change either the height or the width under **Scale** or **Size**, the respective width or height change automatically to maintain the aspect ratio. If you want to change only the height or only the width, clear this check box. Choose **OK** to apply the changes.

Choosing the **Reset** button restores the selected OLE objects to their original size, that is, the size they were when inserted. If the AutoCAD drawing contains an OLE object with the text with different fonts and you wish to select and modify specific text, you can select a particular font and point size from the drop-down lists under the **Text size** area and enter the value in drawing units in the box after the equal (=) sign. For example, if you wish to select a text in the **Times New Roman** font, of point size **10** and modify it to size **0.5** drawing units, select **Times New Roman** and **10** point size from the drop-down lists and enter **0.5** in the text box after the equal (=) sign. The text that was in **Times New Roman** and with a point size 10 will change to 0.5 drawing units in height.

The pointing device can also be used to modify and scale an OLE object in the AutoCAD drawing. Selecting the object displays the object frame and the move cursor. The move cursor allows you to select and drag the object to a new location. The middle handles allows you to select the frame and stretch it. It does not scale objects proportionately. The corner handles scale the object proportionately.

Select an OLE object and right-click to display the shortcut menu. Choosing **Cut** removes the object from the drawing and pastes it on the Clipboard, **Copy** places a copy of the selected object on the clipboard and **Clear** removes the object from the drawing and does not place it on the clipboard. Choosing **Object** displays the **Convert**, **Open**, and **Edit** options. Choosing **Convert** displays the **Convert** dialog box where you can convert objects from one type to another and **Edit** opens the object in the Server application where you can edit it and update it in the current drawing. **Undo** cancels the last action. **Bring to Front** and **Send to Back** options place the OLE objects in the front of or back of the AutoCAD objects. The **Selectable** option turns the selection of the OLE object on or off. If the **Selectable** option is on, the object frame is visible and the object is selected.

If you want to change the layer of an OLE object, select the object and right-click to display the shortcut menu; choose **Cut**. The selected object is placed on the clipboard. Change the current layer to the one you want to change the OLE object's layer to using the **Layer Properties button** dialog box. Now, choose **Paste** from the **Edit** menu to paste the contents of the clipboard in the AutoCAD drawing. The OLE object is pasted in the new layer, in its original size.

The **OLEHIDE** system variable controls the display of OLE objects in AutoCAD. The default value is 0, that is, all the OLE objects are visible. The different values and their effects are as follows:

0   All OLE objects are visible
1   OLE objects are visible in paper space only
2   OLE objects are visible in model space only
3   No OLE objects are visible

The **OLEHIDE** system variable affects both screen display and printing.

Self-Evaluation Test

Answer the following questions and then compare them to those given at the end of this chapter:

1.   The DXF file format generates a text file in ASCII code. (T/F)

2. You can directly open a DXF file in AutoCAD using the **OPEN** command. (T/F)

3. An image attached to AutoCAD cannot be detached. (T/F)

4. The file in the container application is automatically modified only when a link is maintained between the server application and the container application. (T/F)

5.   You can import a DXB file into an AutoCAD drawing file by using the _____ command.

6. The _____ system variable controls the display of OLE objects in AutoCAD.

7. If you import a DXB file into an old drawing, the settings of layers, blocks, and so on of the file being imported are _____ by the settings of the file into which you are importing the DXB file.

8. The _____ is used as a medium for storing documents while transferring them from one Windows application to the other.

9. You can get an embedded drawing into the server application directly from the container application by _____ on the drawing.

10. The **COPYLINK** command copies a drawing in the _____ to Clipboard.

Review Questions

Answer the following questions:

1. In AutoCAD, you can clip the boundary of an image and change them to desired shape. (T/F)

2. The scanned files can be imported into the current session of AutoCAD using the **DXBIN** command. (T/F)

3. You can export an AutoCAD drawing into 3D Studio MAX. (T/F)

4. Raster images can be attached using the **IMAGEATTACH** command as well as the **IMAGE** command. (T/F)

5. Which of the following commands can be used to open a DXF format file?

    (a) **OPEN**   (b) **NEW**
    (c) **START**   (d) **None**

6. Which of the following commands is used to detach an attached image file?

    (a) **IMAGE**   (b) **IMAGEATTACH**
    (c) **IMAGECLIP**   (d) **IMAGEADJUST**

7. Which of the following commands is used to modify the brightness of an image file?

    (a) **IMAGE**   (b) **IMAGEATTACH**
    (c) **IMAGECLIP**   (d) **IMAGEADJUST**

8. Which of the following commands is used to modify the frames by retaining only the desired portion of the images?

    (a) **IMAGE**   (b) **IMAGEATTACH**
    (c) **IMAGECLIP**   (d) **IMAGEADJUST**

9. The _____ command is used to create an ASCII format file with the *.dxf* extension from AutoCAD drawing files.

10. Binary DXF files are _____ efficient and occupy only 75 percent of the ASCII DXF file.

11. File access for files in binary format is _____ than for the same file in ASCII format.

12. In a _____ file, information is stored in the form of a dot pattern on the screen. This bit pattern is also known as _____.

13. When you select an image file to be attached, the _____ dialog box is displayed where you can define the insertion point, scale factor, and rotation angle for the image.

14. The clipping boundary for the raster images can be of _____ or the _____ shape.

Exercise

Exercise 1   *General*

In this exercise, you will create a cup and a plate, refer to Figure 21-22 for dimensions. Assume the missing dimensions. Below this cup and plate, enter the following text in MS Word: **These objects are drawn in AutoCAD 2010.** Then using OLE, paste the text in the current drawing.

Figure 21-22 Dimensions for Exercise 1

Answers to Self-Evaluation Test
1. T, **2.** T, **3.** F, **4.** T, **5.** **DXBIN**, **6.** **OLEHIDE**, **7.** overruled, **8.** Clipboard, **9.** double-clicking, **10.** current display

Chapter 22

Technical Drawing with AutoCAD

Learning Objectives

After completing this chapter, you will be able to:

- *Understand the concepts of multiview drawings.*
- *Understand X, Y, and Z axes, XY, YZ and XZ planes, and parallel planes.*
- *Draw orthographic projections and position views.*
- *Dimension a drawing.*
- *Understand basic dimensioning rules.*
- *Draw sectional views using different types of sections.*
- *Hatch sectioned surfaces.*
- *Understand how to use and draw auxiliary views.*
- *Create assembly and detail drawings.*

MULTIVIEW DRAWINGS

When designers design a product, they visualize its shape in their minds. To represent that shape on paper or to communicate the idea to people, they must draw a picture of the product or its orthographic views. Pictorial drawings, such as isometric drawings, convey the shape of the object, but it is difficult to show all of its features and dimensions in an isometric drawing. Therefore, in industry, multiview drawings are the accepted standards for representing products. Multiview drawings are also known as orthographic projection drawings. To draw different views of an object, it is very important to visualize the shape of the product. The same is true when you look at different views of an object to determine its shape. To facilitate visualizing the shapes, you must picture the object in 3D space with reference to the X, Y, and Z axes. These reference axes can then be used to project the image into different planes. This process of visualizing objects with reference to different axes is, to some extent, natural in human beings. You might have noticed that sometimes, when looking at objects that are at an angle, people tilt their heads. This is a natural reaction, an effort to position the object with respect to an imaginary reference frame (X, Y, and Z axes).

UNDERSTANDING X, Y, AND Z AXES

To understand the X, Y, and Z axes, imagine a flat sheet of paper on the table. The horizontal edge represents the positive X axis, and the other edge along the width of the sheet, represents the positive Y axis. The point where these two axes intersect is the origin. Now, if you draw a line perpendicular to the sheet passing through the origin, the line defines the positive Z axis (Figure 22-1). If you project the X, Y, and Z axes in the opposite direction beyond the origin, you will get the negative X, Y, and Z axes (Figure 22-2).

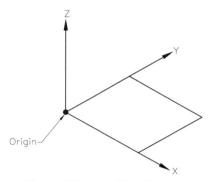

Figure 22-1 *X, Y, and Z axes*

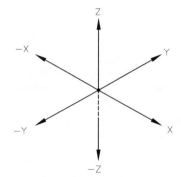

Figure 22-2 *Positive and negative axes*

The space between the X and Y axes is called the XY plane. Similarly, the space between the Y and Z axes is called the YZ plane, and the space between the X and Z axes is called the XZ plane (Figure 22-3). A plane parallel to these planes is called a parallel plane (Figure 22-4).

ORTHOGRAPHIC PROJECTIONS

The first step in drawing an orthographic projection is to position the object along the

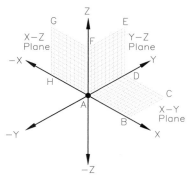

Figure 22-3 *XY, YZ, and XZ planes*

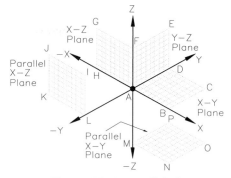

Figure 22-4 *Parallel planes*

imaginary *X*, *Y*, and *Z* axes. For example, if you want to draw orthographic projections of the step block shown in Figure 22-5, position the block so that the far left corner coincides with the origin, and then align it with the *X*, *Y*, and *Z* axes.

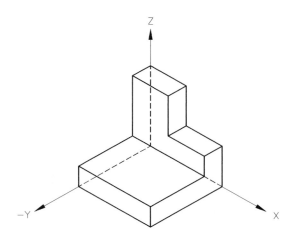

Figure 22-5 *Aligning the object with the X, Y, and Z axes*

Now you can look at the object from different directions. Looking at the object along the negative *Y* axis and toward the origin is called the front view. Similarly, looking at it from the positive X direction is called the right side view. To get the top view, you can look at the object from the positive *Z* axis. See Figure 22-6.

To draw the front, side, and top views, project the points onto the parallel planes. For example, to draw the front view of the step block, imagine a plane parallel to the *XZ* plane located at a certain distance in front of the object. Now, project the points from the object onto the parallel plane (Figure 22-7), and join them to complete the front view.

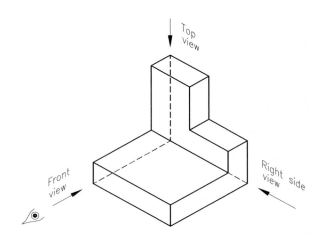

Figure 22-6 *Viewing directions for Front, Side, and Top views*

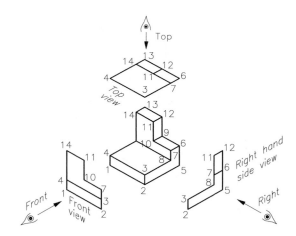

Figure 22-7 *Projecting points onto parallel planes*

Repeat the same process for the side and top views. To represent these views on paper, position them, as shown in Figure 22-8.

Another way of visualizing different views is to imagine the object enclosed in a glass box (Figure 22-9).

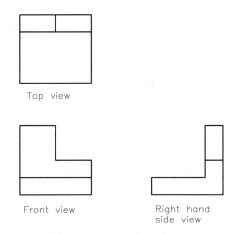

Figure 22-8 *Representing views on paper*

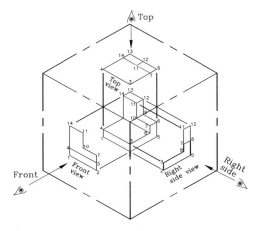

Figure 22-9 *Objects inside a glass box*

Now, look at the object along the negative *Y* axis and draw the front view on the front glass panel. Repeat the process by looking along the positive *X* and *Z* axes, and draw the views on the right side and the top panel of the box (Figure 22-10).

To represent the front, side, and top views on paper, open the side and the top panels of the glass box (Figure 22-11). The front panel is assumed to be stationary.

Chapter 22

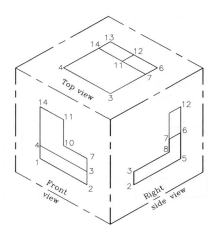

Figure 22-10 *Front, top, and side views*

Figure 22-11 *Opening the side and the top panels*

After opening the panels through 90-degree, the orthographic views will appear, as shown in Figure 22-12.

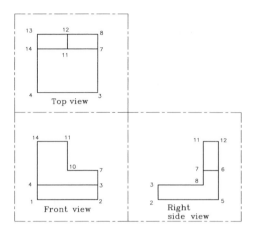

Figure 22-12 *Views after opening the box*

POSITIONING ORTHOGRAPHIC VIEWS

Orthographic views must be positioned, as shown in Figure 22-13.

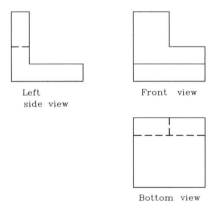

Figure 22-13 *Positioning orthographic views*

The right side view must be positioned directly on the right side of the front view. Similarly, the top view must be directly above the front view. If the object requires additional views, they must be positioned, as shown in Figure 22-14.

The different views of the step block are shown in Figure 22-15.

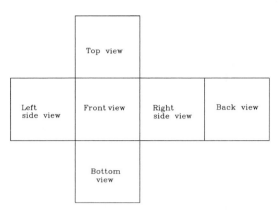

Figure 22-14 *Standard placement of orthographic views*

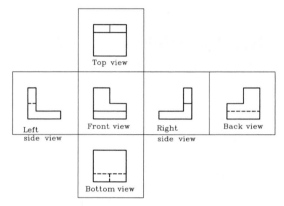

Figure 22-15 *Different views of the step block*

Example 1   *Mechanical*

In this example, you will draw the required orthographic views of the object shown in Figure 22-16. Assume the dimensions of the drawing.

Drawing the orthographic views of an object involves the following steps:

1. Look at the object, and determine the number of views required to show all of its features. For example, the object in Figure 22-16 will require three views only (front, side, and top).

2. Based on the shape of the object, select the side that you want to show as the front view. Generally, the front view is the one that shows the maximum number of features or

Figure 22-16 *Step block with hole and slot*

that gives a better idea about the shape of the object. Sometimes, the front view is determined by how the part will be positioned in an assembly.

3. Picture the object in your mind, and align it along the imaginary *X, Y,* and *Z* axes (Figure 22-17).

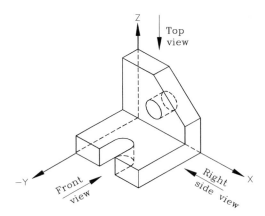

Figure 22-17 *Align the object with the imaginary X, Y, and Z axes*

4. Look at the object along the negative *Y* axis, and project the image on the imaginary *XZ* parallel plane, (Figure 22-18).

5. Draw the front view of the object. If there are any hidden features, they must be drawn with hidden lines. The holes and slots must be shown with centerlines.

6. To draw the right side view, look at the object along the positive *X* axis and project the image onto the imaginary *YZ* parallel plane.

Chapter 22

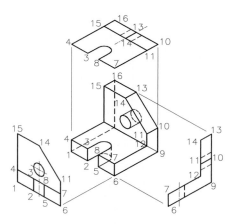

Figure 22-18 *Project the image onto the parallel planes*

7. Draw the right side view of the object. If there are any hidden features, they must be drawn with hidden lines. The holes and slots, when shown in the side view, must have one centerline.

8. Similarly, draw the top view to complete the drawing. Figure 22-19 shows different views of the given object.

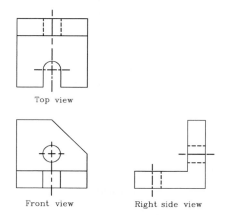

Figure 22-19 *Front, side, and top views*

Exercises 1 through 4　　　　　　　　　　　　　*Mechanical*

Draw the required orthographic views of the following objects in Figure 22-20 through Figure 22-23. The distance between the dotted lines is 0.5 units.

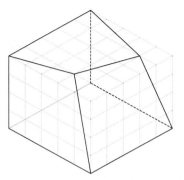

Figure 22-20 *Drawing for Exercise 1 (the object is shown as a surfaced wireframe model)*

Figure 22-21 *Drawing for Exercise 2*

Figure 22-22 *Drawing for Exercise 3*

Figure 22-23 *Drawing for Exercise 4*

DIMENSIONING

Dimensioning is one of the most important elements in a drawing. When you dimension, you not only give the size of a part, you give a series of instructions to a machinist, an engineer, or an architect. The way the part is positioned in a machine, the sequence of machining operations, and the location of different features of the part depend on how you dimension it. For example, the number of decimal places in a dimension (2.000) determines the type of machine, which will be used to do that machining operation. The machining cost of such an operation is significantly higher than a dimension that has only one digit after the decimal (2.0). If you are using a computer numerical control (CNC) machine, locating a feature may not be a problem, but the number of pieces you can machine without changing the tool depends on the tolerance assigned to a dimension. A closer tolerance (+.0001, -.0005) will definitely increase the tooling cost and ultimately the cost of the product. Similarly, if a part is to be forged or cast, the radius of the edges and the tolerance you provide to these dimensions determine the cost of the product, the number of defective parts, and the number of parts you get from the die.

While dimensioning, you must consider the manufacturing process involved in making a part and the relationships that exist among different parts in an assembly. If you are not familiar with any operation, get help. You must not assume things when dimensioning or making a piece part drawing. The success of a product, to a large extent, depends on the way you dimension a part. Therefore, never underestimate the importance of dimensioning in a drawing.

Dimensioning Components

A dimension consists of the following components (Figure 22-24):

Extension line
Arrows or tick marks
Dimension line
Leader lines
Dimension text

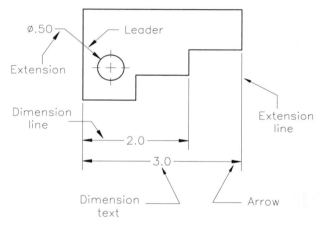

Figure 22-24 *Dimensioning components*

The extension lines are drawn to extend the points that are dimensioned. The length of the extension lines is determined by the number of dimensions and the placement of the dimension lines. These lines are generally drawn perpendicular to the surface. The dimension lines are drawn between the extension lines, at a specified distance from the object lines. The dimension text is a numerical value that represents the distance between the points. The dimension text can also consist of a variable (A, B, X, Y, Z12,...), in which case the value assigned to it is defined in a separate table. The dimension text can be centered around or on the top of the dimension line. Arrows or tick marks are drawn at the end of the dimension line to indicate the start and end of the dimension. Leader lines are used when dimensioning a circle, arc, or any nonlinear element of a drawing. They are also used to attach a note to a feature or to give the part numbers in an assembly drawing.

Basic Dimensioning Rules

1. You should make the dimensions in a separate layer/layers. This makes it easy to edit or control the display of dimensions (freeze, thaw, lock, and unlock). Also, the dimension layer/layers should be assigned a unique color so that at the time of plotting, you can assign the desired pen to plot the dimensions. This helps to control the line width and the contrast of dimensions at the time of plotting.

2. The distance of the first dimension line should be at least 0.375 units (10 units for metric drawing) from the object line. In CAD drawing, this distance may be 0.75 to 1.0 units (19 to 25 units for metric drawings). Once you decide on the spacing, it should be maintained throughout the drawing.

3. The distance between the first dimension line and the second dimension line must be at least 0.25 units. In CAD drawings, this distance may be 0.25 to 0.5 units (6 to 12 units for metric drawings). If there are more dimension lines (parallel dimensions), the distances between them must be the same (0.25 to 0.5 units). Once you decide on the spacing (0.25 to 0.5), the same spacing should be maintained throughout the drawing. An appropriate snap setting is useful for maintaining this spacing. If you are using baseline dimensioning, you can use AutoCAD's **DIMDLI** variable to set the spacing. You must present the dimensions so that they are not crowded, especially when there is not much space (Figure 22-25).

Figure 22-25 *Arrow size, text height, and spacing between dimension lines*

4. For parallel dimension lines, the dimension text can be staggered if there is not enough room between the dimension lines to place the dimension text. You can use the AutoCAD Object Grips feature or the **DIMTEDIT** command to stagger the dimension text (Figure 22-26).

Note
You can change the grid, snap or UCS origin, and snap increment to make it easier to place the dimensions. You can also add the following lines to the AutoCAD menu file or toolbar buttons.

SNAP;R;\0;SNAP;0.25;GRID;0.25
SNAP;R;0,0;0;SNAP;0.25;GRID;0.5

The first line sets the snap to 0.25 units and allows the user to define the new origin of snap and grid display. The second line sets the grid to 0.5 and snap to 0.25 units. It also sets the origin for grid and snap to (0,0).

5. All dimensions should be given outside the view. However, the dimensions can be shown inside the view if they can be easily understood there and cause no confusion with the other dimensions or details (Figure 22-27).

Figure 22-26 Staggered dimensions

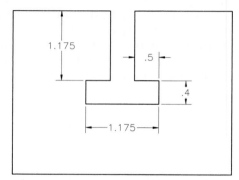

Figure 22-27 Dimensions inside the view

6. Dimension lines should not cross the extension lines (Figure 22-28). You can accomplish this by giving the smallest dimension first and then the next largest dimension (Figure 22-29).

Figure 22-28 Dimension lines crossing each other

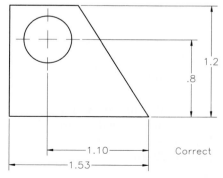

Figure 22-29 Dimensions placed correctly

7. If you decide to have the dimension text aligned with the dimension line, then the entire dimension text in the drawing must be aligned (Figure 22-30). Similarly, if you decide to have the dimension text horizontal or above the dimension line, then to maintain uniformity in the drawing, the entire dimension text must be horizontal (Figure 22-31) or above the dimension line (Figure 22-32).

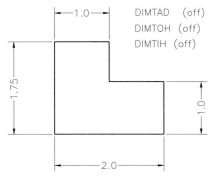

Figure 22-30 *Dimension text aligned with the dimension line*

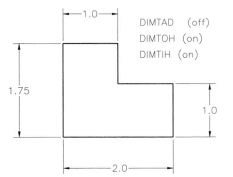

Figure 22-31 *Dimension text horizontal*

Figure 22-32 *Dimension text above the dimension line*

8. If you have a series of continuous dimensions, they should be placed in a continuous line (Figure 22-33). Sometimes you may not be able to give the dimensions in a continuous line even after adjusting the dimension variables. In that case, give the dimensions that are parallel (Figure 22-34).

Chapter 22

Figure 22-33 *Continuous Dimensions*

Figure 22-34 *Parallel dimensions*

9. You should not dimension the hidden lines. The dimension should be given where the feature is visible (Figure 22-35). However, in some complicated drawings, you might be justified to dimension a detail with a hidden line.

10. The dimensions must be given where the feature that you are dimensioning is obvious and shows the contour of the feature (Figure 22-36).

Figure 22-35 *Do not dimension with hidden lines*

Figure 22-36 *Dimensions should be given where they are obvious*

11. The dimensions must not be repeated; this makes it difficult to update a dimension (Figure 22-37).

Figure 22-37 *Dimension must not be repeated*

12. The dimensions must be given depending on how the part will be machined, and also on the relationship that exists between the different features of the part (Figure 22-38).

Figure 22-38 *When dimensioning, consider the machining processes involved*

13. When a dimension is not required but you want to give it for reference, it must be a reference dimension. The reference dimension must be enclosed with parentheses (Figure 22-39).

14. If you give continuous (incremental) dimensions for dimensioning various features of a part, the overall dimension must be omitted or given as a reference dimension (Figure 22-40). Similarly, if you give the overall dimension, one of the continuous (incremental) dimensions must be omitted or given as a reference dimension. Otherwise,

Figure 22-39 *Reference dimension*

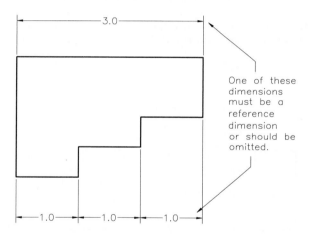

Figure 22-40 *Referencing or omitting a dimension*

there will be a conflict in tolerances. For example, the total positive tolerance on the three incremental dimensions shown in Figure 22-40 is 0.06. Therefore, the maximum size based on the incremental dimensions is (1 + 0.02) + (1 + 0.02) + (1 + 0.02) = 3.06. Also, the positive tolerance on the overall 3.0 dimension is 0.02. Based on this dimension, the overall size, of the part must not exceed 3.02. This causes a conflict in tolerances: with incremental dimensions, the total tolerance is 0.06, whereas with the overall dimension, the total tolerance is only 0.02.

15. If the dimension of a feature appears in a section view, you must not hatch the dimension text (Figure 22-41). You can accomplish it by selecting the dimension object when defining the hatch boundary. You can also accomplish this by drawing a rectangle around the dimension text and then hatching the area after excluding the rectangle from the hatch boundary. (You can also use the **EXPLODE** command to explode the dimension and

then exclude the dimension text from hatching. This is not recommended because by exploding a dimension, the associativity of the selected dimension is lost.)

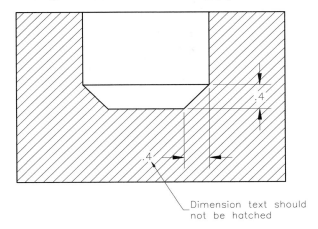

Figure 22-41 *Dimension text should not be hatched*

16. When dimensioning a circle, the diameter should be preceded by the diameter symbol (Figure 22-42). AutoCAD automatically puts the diameter symbol in front of the diameter value. However, if you override the default diameter value, you can use %%c followed by the value of the diameter (%%c1.25) to put the diameter symbol in front of the diameter dimension.

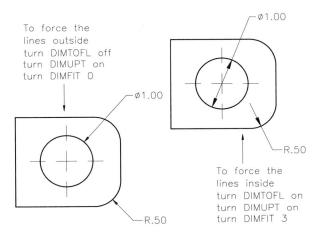

Figure 22-42 *Diameter should be preceded by the diameter symbol*

17. The circle must be dimensioned as a diameter, never as a radius. The dimension of an arc must be preceded by the abbreviation R (R1.25), and the center of the arc should be indicated by drawing a small cross. You can use the AutoCAD **DIMCEN** variable to control

the size of the cross. If the value of this variable is 0, AutoCAD does not draw the cross in the center when you dimension an arc or a circle. You can also use the **DIMCENTER** command or the **CENTER** option of the **DIM** command to draw a cross at the center of the arc or circle.

18. When dimensioning an arc or a circle, the dimension line (leader) must be radial. Also, you should place the dimension text horizontally (Figure 22-43).

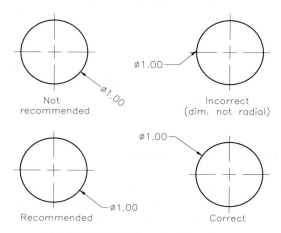

Figure 22-43 *Dimensioning a circle*

19. A chamfer can be dimensioned by specifying the chamfer angle and the distance of the chamfer from the edge. It can also be dimensioned by specifying the distances, as shown in Figure 22-44.

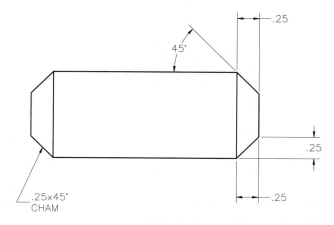

Figure 22-44 *Different ways of specifying a chamfer*

20. A dimension that is not in scale should be indicated by drawing a straight line under the dimension text (Figure 22-45). This line can be drawn by using the **DDEDIT** command. If you invoke this command, AutoCAD will prompt you to select an annotative object. Select the object; the **In-Place Text Editor** will be displayed and a new **Multiline Text** tab will be added to the **Ribbon**. Select the text (< >), and then choose **Multiline Text > Formatting > U** from the **Ribbon** to underline the text.

21. A bolt circle should be dimensioned by specifying the diameter of the bolt circle, the diameter of the holes, and the number of holes in the bolt circle (Figure 22-46).

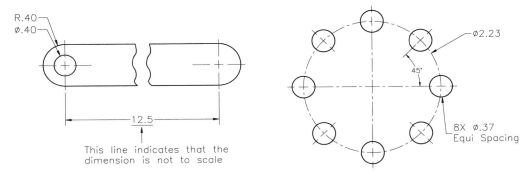

Figure 22-45 *Specifying dimensions that are not in scale*

Figure 22-46 *Dimensioning a bolt circle*

Exercises 5 through 10   *Mechanical*

Draw the required orthographic views of the following objects, and then give the dimensions (refer to Figures 22-47 through 22-52). The distance between the grid lines is 0.5 units.

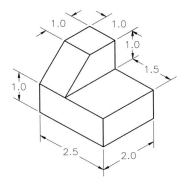

Figure 22-47 *Drawing for Exercise 5*

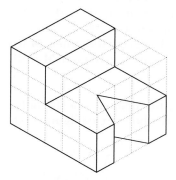

Figure 22-48 *Drawing for Exercise 6*

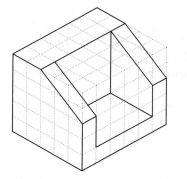

Figure 22-49 *Drawing for Exercise 7*

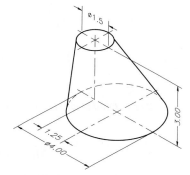

Figure 22-50 *Drawing for Exercise 8*

Figure 22-51 *Drawing for Exercise 9*

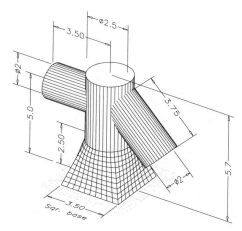

Figure 22-52 *Drawing for Exercise 10 (assume the missing dimensions)*

SECTIONAL VIEWS

In the principal orthographic views, the hidden features are generally shown by hidden lines. In some objects, the hidden lines may not be sufficient to represent the actual shape of the hidden feature. In such situations, sectional views can be used to show the features of the object that are not visible from outside. The location of the section and the direction of sight depend on the shape of the object and the features that need to be shown. Several ways to cut a section in the object are discussed next.

Full Section

Figure 22-53 shows an object that has a drilled hole, counterbore, and taper. In the orthographic views, these features will be shown by hidden lines (Figure 22-54).

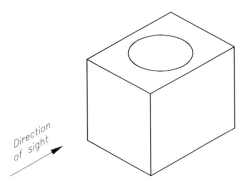

Figure 22-53 *Rectangular object with hole*

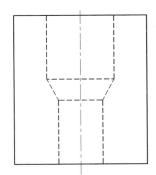

Figure 22-54 *Front view without section*

To better represent the hidden features, the object must be cut so that the hidden features are visible. In the full section, the object is cut along its entire length. To get a better idea of a full section, imagine that the object is cut into two halves along the centerline, as shown in Figure 22-55. Now remove the left half and look at the right half in the direction that is perpendicular to the sectioned surface. The view you get after cutting the section is called a full section view (Figure 22-56).

Figure 22-55 *One-half of the object removed*

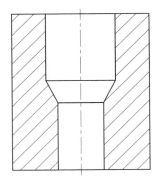

Figure 22-56 *Full section view*

Chapter 22

In this section view, the features that would be hidden in a normal orthographic view are visible. Also, the part of the object where the material is actually cut is indicated by section lines. If the material is not cut, the section lines are not drawn. For example, if there is a hole, no material is cut when the part is sectioned, and so the section lines must not be drawn through that area of the section view.

Half Section

If the object is symmetrical, it is not necessary to draw a full section view. For example, in Figure 22-57 the object is symmetrical with respect to the centerline of the hole, so a full section is not required. Also, in some objects it may help to understand and visualize the shape of the hidden details better to draw the view in half section. In half section, one-quarter of the object is cut, as shown in Figure 22-58. To draw a view in half section, imagine one-quarter of the object removed, and then look in the direction that is perpendicular to the sectioned surface.

Figure 22-57 *Front view in half section*

Figure 22-58 *One-quarter of the object removed*

You can also show the front view with a solid line in the middle, as shown in Figure 22-59. Sometimes the hidden lines, representing the remaining part of the hidden feature, are not drawn, as in Figure 22-60.

Figure 22-59 *Front view in half section*

Figure 22-60 *Front view in half section with the hidden features not drawn*

Broken Section

In the broken section, only a small portion of the object is cut to expose the features that need to be drawn in the section. The broken section is designated by drawing a thick zigzag line in the section view (Figure 22-61).

Revolved Section

The revolved section is used to show the true shape of the object at the point where the section is cut. The revolved section is used when it is not possible to show the features

Figure 22-61 *Front view with broken section*

clearly in any principal view. For example, for the object in Figure 22-62, it is not possible to show the actual shape of the middle section in the front, side, or top view. Therefore, a revolved section is required to show the shape of the middle section.

The revolved section involves cutting an imaginary section through the object and then looking at the sectioned surface in a direction that is perpendicular to it. To represent the shape, the view is revolved through 90-degree and drawn in the plane of the paper, as shown Figure 22-62. Depending on the shape of the object, and for clarity, it is recommended to provide a break in the object so that its lines do not interfere with the revolved section.

Figure 22-62 *Front view with the revolved section*

Removed Section

The removed section is similar to the revolved section, except that it is shown outside the object. The removed section is recommended when there is not enough space in the view to show it or if the scale of the section is different from the parent object. The removed section can be shown by drawing a line through the object at the point where the revolved section is desired and then drawing the shape of the section, as shown in Figure 22-63.

The other way of showing a removed section is to draw a cutting plane line through the object where you want to cut the section. The arrows should point in the direction, in which you are looking at the sectioned surface. The section can then be drawn at a convenient place in the drawing. The removed section must be labeled, as shown in Figure 22-64. If the scale has been changed, it must be mentioned with the view description.

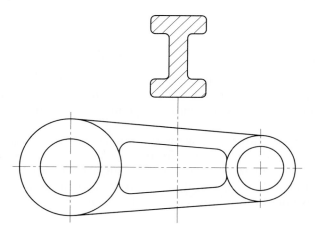

Figure 22-63 *Front view with the removed section*

Figure 22-64 *Front view with the removed section drawn at a convenient place*

Offset Section

The offset section is used when the features of the object that you want to section are not in one plane. The offset section is designated by drawing a cutting plane line that is offset through the center of the features that need to be shown in the section (Figure 22-65). The arrows indicate the direction in which the section is viewed.

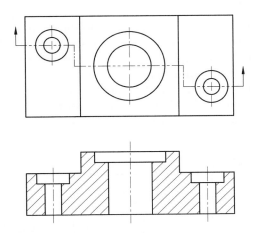

Figure 22-65 *Front view with offset section*

Aligned Section

In some objects, cutting a straight section might cause confusion in visualizing the shape of the section. Therefore, the aligned section is used to represent the shape along the cutting plane (Figure 22-66). Such sections are widely used in circular objects that have spokes, ribs, or holes.

Figure 22-66 *Side view in section (aligned section)*

Cutting Plane Lines

Cutting plane lines are thicker than object lines (Figure 22-67). You can use the **PLINE** command to draw the polylines of the desired width, generally 0.005 to 0.01. However, for drawings that need to be plotted, you should assign a unique color to the cutting plane lines and then assign that color to the slot of the plotter that carries a pen of the required tip width. (For details, see Chapter 14, Plotting Drawings.)

In the industry, generally three types of lines are used to show the cutting plane for sectioning. The first line consists of a series of dashes 0.25 units long. The second type consists of a series of long dashes separated by two short dashes (Figure 22-68). The length of the long dash can vary from 0.75 to 1.5 units, and the short dashes are about 0.12 units long. The space between the dashes should be about 0.03 units. Third, sometimes the cutting plane lines might clutter the drawing or cause confusion with other lines in the drawing. To avoid this problem, you can show the cutting plane by drawing a short line at the end of the section (Figure 22-68 and Figure 22-69). The line should be about 0.5 units long.

Figure 22-67 *Cutting plane line*

Note

*In AutoCAD, you can define a new linetype that you can use to draw the cutting plane lines. Refer to the chapter on Creating Linetypes and Hatch Patterns that is provided for free download, for more information on defining linetypes. Add the following lines to the **ACLT.LIN** file, and then load the linetypes before assigning it to an object or a layer.*

*CPLANE1,___ ___ ___ ___
A,0.25,-0.03
*CPLANE2,___ ___ ___ ___
A,1.0,-0.03,0.12,-0.03,0.12,-0.03

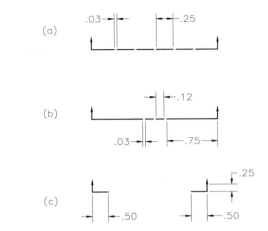

Figure 22-68 *Cutting plane lines*

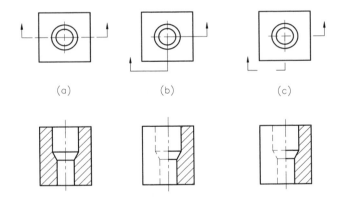

Figure 22-69 *Application of cutting plane lines*

Spacing for Hatch Lines

The spacing between the hatch (section) lines is determined by the space that is being hatched (Figure 22-70). If the hatch area is small, the spacing between the hatch lines should be smaller compared to a large hatch area.

In AutoCAD, you can control the spacing between the hatch lines by specifying the scale factor at the time of hatching. If the scale factor is 1, the spacing between the hatch lines is the same as defined in the hatch pattern file for that particular hatch. For example, in the following hatch pattern definition, the distance between the lines is 0.125.

*ANSI31, ANSI Iron, Brick, Stone masonry
45, 0, 0, 0, .125

When the hatch scale factor is 1, the line spacing will be 0.125; if the scale factor is 2, the spacing between the lines will be 0.125 x 2 = 0.25.

Direction of Hatch Lines

The angle for the hatch lines should be 45-degree. However, if there are two or more hatch areas next to one another representing different parts, the hatch angle must be changed so that the hatched areas look different (Figure 22-71).

Figure 22-70 Hatch line spacing

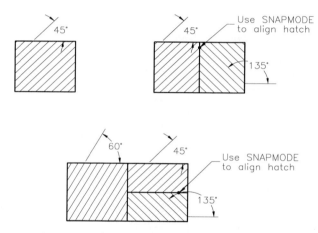

Figure 22-71 Hatch angle for adjacent parts

Also, if the hatch lines fall parallel to any edge of the hatch area, the hatch angle should be changed so that the lines are not parallel or perpendicular to any object line (Figure 22-72).

Points to Remember

1. Some parts, such as bolts, nuts, shafts, ball bearings, fasteners, ribs, spokes, keys, and other similar items, if sectioned, should not be shown in the section. In case of ribs, if the cutting plane passes through the center plane of the feature, they should not be section-lined. If the cutting

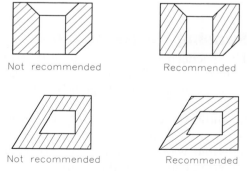

Figure 22-72 Hatch angle

plane passes crosswise through ribs they must be section-lined.

2. Hidden details should not be shown in the section view unless the hidden lines represent an important detail or help the viewer to understand the shape of the object.

3. The section lines (hatch lines) must be thinner than the object lines. You can accomplish this by assigning a unique color to the hatch lines and then assigning the color to that slot on the plotter that carries a pen with a thinner tip.

4. The section lines must be drawn on a separate layer for display and editing purposes.

Exercises 11 and 12   *Mechanical*

In the following drawings (Figures 22-73 and 22-74), the views have been drawn without a section. Draw these views in the section as indicated by the cutting plane lines in each object.

Figure 22-73 Drawing for Exercise 11: draw the front view in section

Figure 22-74 Drawing for Exercise 12: draw the left side view in section

Exercises 13 and 14   *Mechanical*

Draw the required orthographic views of the following objects shown in Figures 22-75 and 22-76. Show the front view in the section when the object is cut so that the cutting plane passes through the holes. Also, draw the cutting plane lines in the top view.

Figure 22-75 *Drawing for Exercise 13: draw the front view in full section*

Figure 22-76 *Drawing for Exercise 14: draw the front view with offset section*

Exercise 15   *Mechanical*

Draw the required orthographic views of the objects in Figures 22-77 and 22-78 with the front view in the section. Also, draw the cutting plane lines in the top view to show the cutting plane. The material thickness is 0.25 units. (The object has been drawn as a surfaced 3D wiremesh model.)

Figure 22-77 *Draw the front view in half section*

Figure 22-78 *Draw the front view in half section*

AUXILIARY VIEWS

As discussed earlier, most objects generally require three principal views (front view, side view, and top view) to show all features of the object. Round objects may require just two views. Some objects have inclined surfaces. It may not be possible to show the actual shape of the inclined surface in one of the principal views. To get the true view of the inclined surface, you must look at the surface in a direction that is perpendicular to the inclined surface. Then you can project the points onto the imaginary auxiliary plane that is parallel to the inclined surface. The view you get after projecting the points is called the auxiliary view, as shown in Figures 22-79 and 22-80.

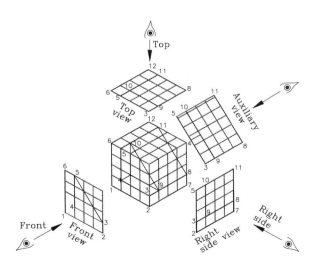

Figure 22-79 *Project points onto the auxiliary plane*

The auxiliary view in Figure 22-80 shows all the features of the object as seen from the auxiliary view direction. For example, the bottom left edge is shown as a hidden line. Similarly, the lower right and upper left edges are shown as continuous lines. Although these lines are technically correct, the purpose of the auxiliary view is to show the features of the inclined surface. Therefore, in the auxiliary plane, you should draw only those features that are on the inclined face, as shown in Figure 22-81. Other details that will help you to understand the shape of the object may also be included in the auxiliary view. The remaining lines should be ignored because they tend to cause confusion in visualizing the shape of the object.

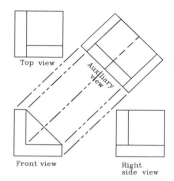

Figure 22-80 *Auxiliary, front, side, and top views*

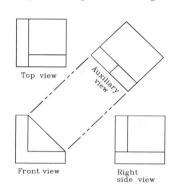

Figure 22-81 *Auxiliary, front, side, and top views*

How to Draw Auxiliary Views

The following example illustrates how to use AutoCAD to generate an auxiliary view.

Example 2   *Mechanical*

Draw the required views of the hollow triangular block with a hole in the inclined face, as shown in Figure 22-82. (The block has been drawn as a solid model.)

The following steps are involved in drawing different views of this object.

1. Draw the required orthographic views: the front view, side view, and the top view as shown in Figure 22-83(a). The circles on the inclined surface appear like ellipses in the front and top views. These ellipses may not be shown in the orthographic views because they tend to clutter them [Figure 22-83(b)].

2. Determine the angle of the inclined surface. In this example, the angle is 45-degree. Use

Figure 22-82 *Hollow triangular block*

Figure 22-83(a) *Front, side, and top views*   ***Figure 22-83(b)*** *The ellipses may not be shown in the orthographic views*

the **SNAP** command to rotate the snap by 45-degree (Figure 22-84) or rotate the UCS by 45-degree around the *Z* axis.

Command: **SNAP** [Enter]
Specify snap spacing or [ON/OFF/Aspect/Rotate/Style/Type] <0.5000>: **R** [Enter]
Specify base point <0.0000,0.0000>: *Select P1.*
Specify rotation angle <0>: **45** [Enter]

Using the rotate option, the snap will be rotated by 45-degree and the grid lines will also be displayed at 45-degree (if the **GRID** is on). Also, one of the grid lines will pass through point P1 because it was defined as the base point.

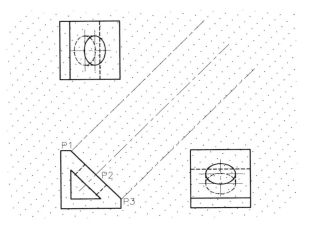

Figure 22-84 *Grid lines at 45-degree*

3. Turn **ORTHO** on, and project points P1, P2, and P3 from the front view onto the auxiliary plane. Now you can complete the auxiliary view and give the dimensions. The projection lines can be erased after the auxiliary view is drawn (Figure 22-85).

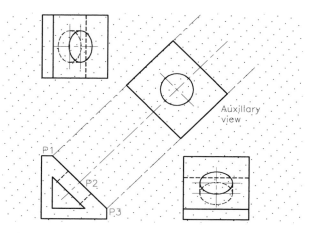

Figure 22-85 Auxiliary, front, side, and top views

Exercise 16   *General*

Draw the required orthographic and auxiliary views for the object in Figure 22-86. The object is drawn as a surfaced 3D wiremesh model.

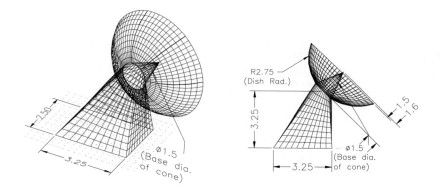

Figure 22-86 Drawing for Exercise 16

Exercises 17 and 18   *General*

Draw the required orthographic and auxiliary views for the following objects shown in Figure 22-87 and Figure 22-88. The objects are drawn as 3D solid models and are shown at different angles (viewpoints are different).

Figure 22-87 *Drawing for Exercise 17*

Figure 22-88 *Drawing for Exercise 18*

DETAIL DRAWING, ASSEMBLY DRAWING, AND BILL OF MATERIALS

Detail Drawing

A detail drawing is a drawing of an individual component that is a part of the assembled product. Detail drawings are also called **piece part drawings**. Each detail drawing must be drawn and dimensioned to completely describe the size and shape of the part. It should also contain information that might be needed in manufacturing the part. The finished surfaces should be indicated by using symbols or notes and all the necessary operations on the drawing. The material of which the part is made and the number of parts that are required for the production of the assembled product must be given in the title block. Detail drawings should also contain part numbers. This information is used in the bill of materials and the assembly drawing. The part numbers make it easier to locate the drawing of a part. You should make a detail drawing of each part, regardless of the part's size, on a separate drawing. When required, these detail drawings can be inserted in the assembly drawing by using the **XREF** command.

Assembly Drawing

The assembly drawing is used to show the parts and their relative positions in an assembled product or a machine unit. The assembly drawing should be drawn so that all parts can be shown in one drawing. This is generally called the main view. The main view may be drawn in full section so that the assembly drawing shows nearly all the individual parts and their locations. Additional views should be drawn only when some of the parts cannot be seen in the main view. The hidden lines, as far as possible, should be omitted from the assembly drawing because they clutter it and might cause confusion. However, a hidden line may be drawn if it helps to understand the product. Only assembly dimensions should be shown in the assembly drawing. Each part should be identified on the assembly drawing by the number used in the detail drawing and in the bill of materials. The part numbers should be given as shown in Figure 22-89. It consists of a text string for the detail number, a circle (balloon), a leader line, and an arrow or dot. The text should be made at least 0.2 inches (5 mm) high and enclosed in a 0.4 inch (10 mm) circle (balloon). The center of the circle must be located not less than 0.8 inches (20 mm) from the nearest line on the drawing. Also, the leader line should be radial with respect to the circle (balloon). The assembly drawing may also contain an exploded isometric or isometric view of the assembled unit.

Bill of Materials

A bill of materials is a list of parts placed on an assembly drawing just above the title block (Figure 22-89). The bill of materials contains the part number, part description, material, quantity required, and drawing numbers of the detail drawings (Figure 22-90). If the bill of materials is placed above the title block, the parts should be listed in ascending order so that the first part is at the bottom of the table (Figure 22-91). The bill of materials may also be placed at the top of the drawing. In that case, the parts must be listed in descending order with the first part at the top of the table. This structure allows room for any additional items that may be added to the list.

Chapter 22

Figure 22-89 *Assembly drawing with title block, bill of materials, and surfaced 3D wireframe model*

Figure 22-90 *Detail drawing (piece part drawing) of Part Number 2*

Figure 22-91 *Detail drawings of Part Numbers 1 and 3*

Self-Evaluation Test

Answer the following questions and then compare them to those given at the end of this chapter:

1. A line perpendicular to the *X* and *Y* axes defines the _____ axis.

2. The front view shows the maximum number of features or gives a better idea about the shape of the object. (T/F)

3. The number of decimal places in a dimension (for example, 2.000) determines the type of machine that will be used to do that machining operation. (T/F)

4. The dimension layer/layers should be assigned a unique color so that at the time of plotting you can assign a desired pen to plot the dimensions. (T/F)

5. All dimensions should be placed inside the view. (T/F)

6. The dimensions can be shown inside the view if they can be easily understood there and cause no confusion with the other object lines. (T/F)

7. The circle must be dimensioned as a _____, never as a radius.

8. If you give continuous (incremental) dimensioning for dimensioning various features of a part, the overall dimension must be omitted or given as a(n) _____ dimension.

9. If the object is symmetrical, it is not necessary to draw a full section view. (T/F)

10. The removed section is similar to the _____ section, except that it is shown outside the object.

11. In AutoCAD, you can control the spacing between hatch lines by specifying the _____ at the time of hatching.

12. When dimensioning, you must consider the manufacturing process involved in making a part and the relationship that exists between different parts in an assembly. (T/F)

13. The distance between the first dimension line and the second dimension line must be _____ to _____ units.

14. The reference dimension must be enclosed in _____.

15. A dimension must be given where the feature is visible. However, in some complicated drawings, you might be justified to dimension a detail with a hidden line. (T/F)

Review Questions

Answer the following questions:

1. Multiview drawings are also known as _____ drawings.

2. The space between the X and Y axes is called the XY plane. (T/F)

3. A plane that is parallel to the XY plane is called a parallel plane. (T/F)

4. If you look at the object along the negative Y axis and toward the origin, you will get the side view. (T/F)

5. The top view must be directly below the front view. (T/F)

6. Before drawing orthographic views, you must look at the object and determine the number of views required to show all features of the object. (T/F)

7. By dimensioning, you not only give the size of a part, but also give a series of instructions to a machinist, an engineer, or an architect. (T/F)

8. What are the components of a dimension? _____
    _____ .

9. Why should you make the dimensions in a separate layer/layers?

10. The distance of the first dimension line should be _____ to _____ units from
    the object line.

11. You can change grid, snap or UCS origin, and snap increments to make it easier to place
    the dimensions. (T/F)

12. For parallel dimension lines, the dimension text can be _____ if there is not enough
    room between the dimension lines to place the dimension text.

13. You should not dimension with hidden lines. (T/F)

14. A dimension that is not to scale should be indicated by drawing a straight line under the
    dimension text. (T/F)

15. Dimensions must be given where the feature that you are dimensioning is obvious and
    shows the contour of the feature. (T/F)

16. When dimensioning a circle, the diameter should be preceded by the _____.

17. When dimensioning an arc or a circle, the dimension line (leader) must be _____.

18. In radial dimensioning, you should place the dimension text vertically. (T/F)

19. If the dimension of a feature appears in a section view, you must hatch the dimension
    text. (T/F)

20. Dimensions must not be repeated; this makes it difficult to update dimensions, and they
    might get confusing. (T/F)

21. A bolt circle should be dimensioned by specifying the _____ of the bolt circle,
    the _____ of the holes, and the _____ of the holes in the bolt circle.

22. In the _____ section, the object is cut along the entire length of the object.

23. The part of the object where the material is actually cut is indicated by drawing _____
    lines.

24. The _____ section is used to show the true shape of the object at the point where
    the section is cut.

25. The _____ section is used when the features of the object that you want to section are not in one plane.

26. Cutting plane lines are thinner than object lines. You can use the AutoCAD **PLINE** command to draw the polylines of the desired width. (T/F)

27. The spacing between the hatch (section) lines is determined by the space being hatched. (T/F)

28. The angle for the hatch lines should be 45-degree. However, if there are two or more hatch areas next to one another, representing different parts, the hatch angle must be changed so that the hatched areas look different. (T/F)

29. Some parts, such as bolts, nuts, shafts, ball bearings, fasteners, ribs, spokes, keys, and other similar items that do not show any important feature, if sectioned, must be shown in section. (T/F)

30. Section lines (hatch lines) must be thicker than object lines. You can accomplish this by assigning a unique color to the hatch lines and then assigning the color to that slot on the plotter carrying a pen with a thicker tip. (T/F)

31. The section lines must be drawn on a separate layer for _____ and _____ purposes.

32. In a broken section, only a _____ of the object is cut to _____ the features that need to be drawn in section.

33. The assembly drawing is used to show the _____ and their _____ in an assembled product or a machine unit.

34. Why shouldn't hidden lines be shown in an assembly drawing? _____ _____.

Exercises

Exercises 19 through 24   *Mechanical*

Draw the required orthographic views of the following objects (the isometric view of each object is given in Figures 22-92 through 22-97). The dimensions can be determined by counting the number of grid lines. The distance between the isometric grid lines is assumed to be 0.5 units. Also dimension the drawings.

Figure 22-92 *Drawing for Exercise 19*

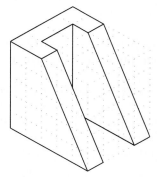

Figure 22-93 *Drawing for Exercise 20*

Figure 22-94 *Drawing for Exercise 21*

Figure 22-95 *Drawing for Exercise 22*

Figure 22-96 *Drawing for Exercise 23*

Figure 22-97 *Drawing for Exercise 24*

Answers to Self-Evaluation Test
1. Z, **2.** T, **3.** T, **4.** T, **5.** F, **6.** T, **7.** diameter, **8.** reference, **9.** T, **10.** revolved, **11.** scale factor, **12.** T, **13.** 0.25 to 0.5, **14.** parentheses, **15.** T

Chapter 23

Isometric Drawings

Learning Objectives

After completing this chapter, you will be able to:
- *Understand isometric drawings, isometric axes, and isometric planes.*
- *Set isometric grid and snap.*
- *Draw isometric circles in different isoplanes.*
- *Dimension isometric objects.*
- *Write text with isometric styles.*

ISOMETRIC DRAWINGS

Isometric drawings are generally used to help visualize the shape of an object. For example, if you are given the orthographic views of an object, as shown in Figure 23-1, it takes time to put information together to visualize the shape. However, if an isometric drawing is given, as shown in Figure 23-2, it is much easier to conceive the shape of the object. Thus, isometric drawings are widely used in industry to help in understanding products and their features.

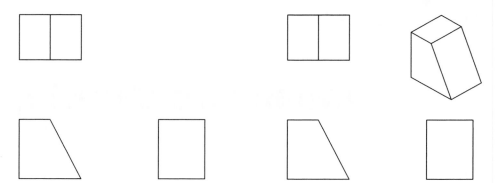

Figure 23-1 *Orthographic views of an object*

Figure 23-2 *Orthographic views with an isometric drawing*

An isometric drawing should not be confused with a three-dimensional (3D) drawing. An isometric drawing is just a two-dimensional (2D) representation of a 3D drawing on a 2D plane. A 3D drawing is the 3D model of an object on the *X*, *Y*, and *Z* axes. In other words, an isometric drawing is a 3D drawing on a 2D plane, whereas a 3D drawing is a true 3D model of the object. The model can be rotated and viewed from any direction. A 3D model can be a wireframe model, surface model, or solid model.

ISOMETRIC PROJECTIONS

The word isometric means "equal measure" because the angles between the three principal axes of an isometric drawing are 120-degree each (Figure 23-3). An isometric view is obtained by rotating the object 45-degrees around the imaginary vertical axis, and then tilting the object forward through a 35°16' angle. If you project the points and edges on the frontal plane, the projected length of the edges will be approximately 81 percent (isometric length/actual length = 9/11), which is shorter than the actual length of the edges. However, isometric drawings are always drawn to a full scale because their purpose is to help the user visualize the shape of the object. Isometric drawings are not meant to describe the actual size of the object. The

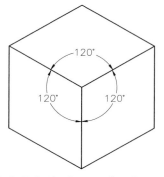

Figure 23-3 *Principal axes of an isometric drawing*

actual dimensions, tolerances, and feature symbols must be shown in the orthographic views. Also, you should avoid showing any hidden lines in the isometric drawings, unless they show an important feature of the object or help in understanding its shape.

ISOMETRIC AXES AND PLANES

Isometric drawings have three axes: right horizontal axis (P0,P1), vertical axis (P0,P2), and left horizontal axis (P0,P3). The two horizontal axes are inclined at 30-degree to the horizontal, or X axis (X1,X2). The vertical axis is at 90-degrees, as shown in Figure 23-4.

When you draw an isometric drawing, the horizontal object lines are drawn along or parallel to the horizontal axis. Similarly, the vertical lines are drawn along or parallel to the vertical axis. For example, to make an isometric drawing of a rectangular block, the vertical edges of the block are drawn parallel to the vertical axis. The horizontal edges on the right side of the block are drawn parallel to the right horizontal axis (P0,P1), and the horizontal edges on the left side of the block are drawn parallel to the left horizontal axis (P0,P3). It is important to remember that the angles do not appear true in isometric drawings. Therefore, the edges or surfaces that are at an angle are drawn by locating their endpoints. The lines that are parallel to the isometric axes are called isometric lines. The lines that are not parallel to the isometric axes are called nonisometric lines.

Similarly, the planes can be isometric planes or nonisometric planes.

Isometric drawings have three principal planes, isoplane right, isoplane top, and isoplane left, as shown in Figure 23-5. The isoplane right (P0,P4,P10,P6) is defined by the vertical axis and the right horizontal axis. The isoplane top (P6,P10,P9,P7) is defined by the right and left horizontal axes. Similarly, the isoplane left (P0,P6,P7,P8) is defined by the vertical axis and the left horizontal axis.

Figure 23-4 Isometric axes

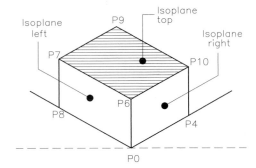

Figure 23-5 Isometric planes

SETTING THE ISOMETRIC GRID AND SNAP

You can use the **SNAP** command to set the isometric grid and snap. The isometric grid lines are displayed at 30-degree to the horizontal axis. Also, the distance between the grid lines is determined by the vertical spacing, which can be specified by using the **GRID** or **SNAP** command. The grid lines coincide with three isometric axes, which make it easier to create isometric drawings. The following command sequence and Figure 23-6 illustrates the use of the **SNAP** command to set the isometric grid and snap of 0.5 units.

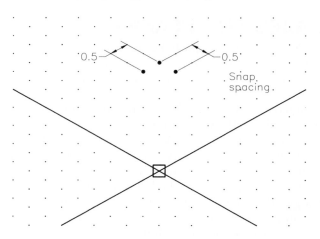

Figure 23-6 *Setting the isometric grid and snap*

Command: **SNAP**
Specify snap spacing or [ON/OFF/Aspect/Style/Type] <0.5000>: **S**
Enter snap grid style [Standard/Isometric] <S>: **I**
Specify vertical spacing <0.5000>: *Enter a new snap distance.*

Note
*When you use the **SNAP** command to set the isometric grid, the grid lines may not be displayed. To display the grid lines, turn the grid on using the **GRID** command or press F7.*

You cannot set the aspect ratio for the isometric grid. Therefore, the spacing between the isometric grid lines will be the same.

You can also set the isometric grid and snap by using the **Drafting Settings** dialog box shown in Figure 23-7, which can be invoked by entering **DSETTINGS** at the Command prompt.

You can also invoke this dialog box by right-clicking on **Snap**, **Grid**, **Polar,** or **Osnap** button available in the Status Bar and choosing **Settings** from the shortcut menu.

The isometric snap and grid can be turned on/off by choosing the **Grid On (F7)** check box located in the **Snap and Grid** tab of the **Drafting Settings** dialog box. The **Snap and Grid** tab also contains the radio buttons to set the snap type and style. To display the grid on the screen, make sure the grid is turned on.

When you set the isometric grid, the display of crosshairs also changes. The crosshairs are

Figure 23-7   *The **Drafting Settings** dialog box*

displayed at an isometric angle, and their orientation depends on current isoplane. You can toggle among isoplane right, isoplane left, and isoplane top by pressing the CTRL and E keys (CTRL+E) simultaneously or using the function key, F5. You can also toggle among different isoplanes by using the **Drafting Settings** dialog box or by entering the **ISOPLANE** command at the Command prompt:

    Command line: **ISOPLANE**
    Enter isometric plane setting [Left/Top/Right] <Top>: **T**
    Current Isoplane: **Top**

The Ortho mode is often useful when drawing in Isometric mode. In Isometric mode, Ortho aligns with the axes of the current isoplane.

Example 1   *Mechanical*

In this example, you will create the isometric drawing shown in Figure 23-8.

1.   Use the **SNAP** command to set the isometric grid and snap. The snap value is 0.5 units.

    Command: **SNAP**
    Specify snap spacing or [ON/OFF/Aspect/Style/Type] <0.5000>: **S**

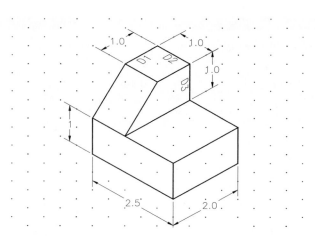

Figure 23-8 Isometric drawing for Example 1

Enter snap grid style [Standard/Isometric] <S>: **I**
Specify vertical spacing <0.5000>: **0.5**   (*or press ENTER.*)

2. Change the isoplane to isoplane left by pressing the F5 key. Enter the **LINE** command and draw lines between points P1, P2, P3, P4, and P1, as shown in Figure 23-9.

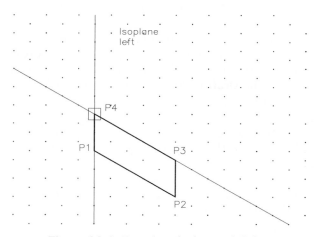

Figure 23-9 Drawing the bottom left face

Tip
*You can change the limits and increase the size of the crosshairs using the **Crosshair size** slider bar in the **Display** tab of the **Options** dialog box.*

3. Change the isoplane to isoplane right by pressing the F5 key. Invoke the **LINE** command and draw the lines, as shown in Figure 23-10.

4. Change the isoplane to isoplane top by pressing the F5 key. Invoke the **LINE** command and draw the lines, as shown in Figure 23-11.

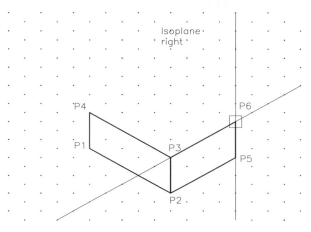

Figure 23-10 Drawing the bottom right face

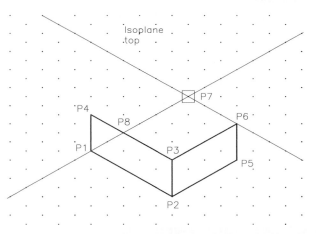

Figure 23-11 Drawing the top face

5. Similarly, draw the remaining lines, as shown in Figure 23-12.

6. The front left end of the object is tapered at an angle. In isometric drawings, oblique surfaces (surfaces at an angle to the isometric axis) cannot be drawn like other lines. You must first locate the endpoints of the lines that define the oblique surface and then draw the lines between those points. To complete the drawing, shown Figure 23-8, draw a line from P10 to P8 and from P11 to P4, see Figure 23-13.

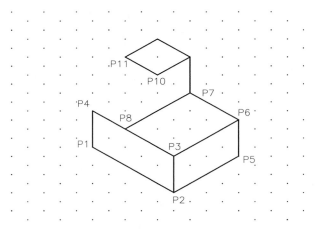

Figure 23-12 *Drawing the remaining lines*

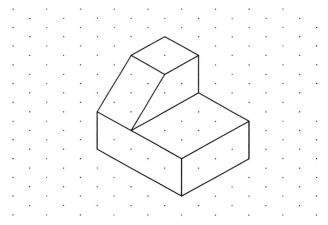

Figure 23-13 *Isometric drawing with the tapered face*

DRAWING ISOMETRIC CIRCLES

The isometric circles are drawn by using the **ELLIPSE** command and then selecting the **Isocircle** option. You must have the isometric snap on for the **ELLIPSE** command to display the **Isocircle** option. If the isometric snap is not on, you cannot draw an isometric circle. Before entering the radius or diameter of the isometric circle, you must make sure that you are in the required isoplane. For example, to draw a circle in the right isoplane, you must toggle through the isoplanes until the required isoplane (right isoplane) is displayed. You can also set the required isoplane current before entering the **ELLIPSE** command. The crosshairs and the shape of the isometric circle will automatically change as you toggle through different isoplanes. As you enter the radius or diameter of the circle, AutoCAD draws the isometric circle in the selected plane, see Figure 23-14.

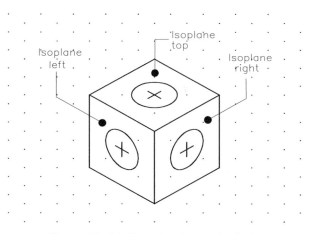

Figure 23-14 *Drawing isometric circles*

The prompt sequence to draw an isometric circle is:

Command: **ELLIPSE**
Specify axis endpoint of ellipse or [Arc/Center/Isocircle]: **I**
Specify center of isocircle: *Select a point.*
Specify radius of isocircle or [Diameter]: *Enter circle radius.*

Creating Fillets in the Isometric Drawings

To create fillets in isometric drawings, you first need to create an isometric circle and then trim its unwanted portion. Remember that there is no other method to directly create an isometric fillet.

DIMENSIONING ISOMETRIC OBJECTS

Isometric dimensioning involves two steps: (1) dimensioning the drawing using the standard dimensioning commands; (2) editing the dimensions to change them to oblique dimensions.

The following example illustrates the process involved in dimensioning an isometric drawing.

Example 2   *Mechanical*

In this example, you will dimension the isometric drawing created in Example 1.

1. Dimension the drawing of Example 1, as shown in Figure 23-15. You can use the aligned or linear dimensions to dimension the drawing. Remember that when you select the points, you must use the **Intersection** or **Endpoint** object snap to snap the endpoints of the object you are dimensioning. AutoCAD automatically leaves a gap between the object line and the extension line, as specified by the **DIMGAP** variable.

Chapter 23

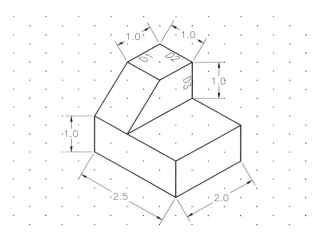

Figure 23-15 *The dimensioned isometric drawing before obliquing*

2. The next step is to edit the dimensions. You can choose **Annotate > Dimensions > Oblique** from the **Ribbon**. After selecting the dimension that you want to edit, you are prompted to enter the obliquing angle. The obliquing angle is determined by the angle that the extension line of the isometric dimension makes with the positive *X* axis. The following prompt sequence is displayed when you invoke this option from the **Ribbon**:

    Select object: *Select the dimension (D1).*
    Select object: *Press ENTER.*
    Enter obliquing angle (Press ENTER for none): **150**

For example, the extension line of the dimension labeled D1 makes a 150-degree angle with the positive *X* axis [Figure 23-16(a)], therefore, the oblique angle is 150-degree. Similarly, the extension lines of the dimension labeled D2 make a 30-degree angle with the positive *X* axis [Figure 23-16(b)], therefore, the oblique angle is 30-degree. After you edit all dimensions, the drawing should appear as shown in Figure 23-17.

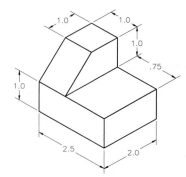

Figure 23-16 *Determining the oblique angle*   **Figure 23-17** *Object with isometric dimensions*

ISOMETRIC TEXT

You cannot use regular text when placing the text in an isometric drawing because the text in an isometric drawing is obliqued at a positive or negative 30-degree. Therefore, you must create two text styles with oblique angles of 30-degree and negative 30-degree. You can use the **-STYLE** command or the **Text Style** dialog box to create a new text style. Figure 23-18 shows the **Text Style** dialog box with a new text style, **ISOTEXT1** and its corresponding values. (For more details, refer to "Creating Text Styles" in Chapter 7.)

Figure 23-18 The **Text Style** *dialog box*

Similarly, you can create another text style, **ISOTEXT2**, with a negative 30-degree oblique angle. When you place the text in an isometric drawing, you must also specify the rotation angle for the text. The text style and the text rotation angle depend on the placement of the text in the isometric drawing, as shown in Figure 23-19.

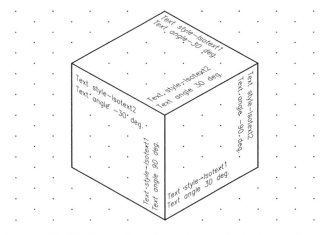

Figure 23-19 Text style and rotation angle for isometric text

Self-Evaluation Test

Answer the following questions and then compare them to those given at the end of this chapter:

1. The word isometric means "_____" because the three angles between the three principal axes of an isometric drawing are _____ degrees each.

2. The ratio of isometric length to the actual length in an isometric drawing is approximately _____.

3. The angle between the right isometric horizontal axis and the X axis is _____ degrees.

4. Isometric drawings have three principal planes: isoplane right, isoplane top, and _____.

5. To toggle among isoplane right, isoplane left, and isoplane top, you can use _____ key combination or _____ function key.

6. You can only use the aligned dimension option to dimension an isometric drawing. (T/F)

7. The isometric snap must be turned on to display the **Isocircle** option with the **ELLIPSE** command. (T/F)

8. When placing text in an isometric drawing, you need to specify the rotation angles.(T/F)

9. The lines that are not parallel to the isometric axes are called _____.

10. You should avoid showing any hidden lines in isometric drawings. (T/F)

Review Questions

Answer the following questions:

1. Isometric drawings are generally used to help in _____ the shape of an object.

2. An isometric view is obtained by rotating the object _____ degree around the imaginary vertical axis, and then tilting the object forward through a _____ angle.

3. If you project the points and edges onto the frontal plane, the projected length of the edges will be approximately _____ percent shorter than their actual length.

4. You should avoid showing the hidden lines in the isometric drawings, unless they help in understanding its shape. (T/F)

5. Isometric drawings have three axes: right horizontal axis, vertical axis, and _____.

6. The angles do not appear true in isometric drawings. (T/F)

7. The lines parallel to the isometric axis are called _____.

8. You can use _____ command to set the isometric grid and snap.

9. Isometric grid lines are displayed at _____ degrees to the horizontal axis.

10. It is possible to set the aspect ratio for the isometric grid. (T/F)

11. You can also set the isometric grid and snap by using the **Drafting Settings** dialog box, which can be invoked by entering _____ at the Command prompt.

12. Isometric circles are drawn by using the **ELLIPSE** command and then selecting the _____ option.

13. You can draw an isometric circle without turning the isometric snap on. (T/F)

14. Only aligned dimensions can be edited to change them to oblique dimensions. (T/F)

15. To place the text in an isometric drawing, you must create two text styles with oblique angles of _____ degrees and negative _____ degrees.

Exercises

Exercises 1 through 6   *General*

Draw the following isometric drawings (Figures 23-20 through 23-25). The dimensions can be determined by counting the number of grid lines. The distance between the isometric grid lines is assumed to be 0.5 units. Dimension the odd-numbered drawings.

Figure 23-20 Drawing for Exercise 1

Figure 23-21 Drawing for Exercise 2

Figure 23-22 *Drawing for Exercise 3*

Figure 23-23 *Drawing for Exercise 4*

Figure 23-24 *Drawing for Exercise 5*

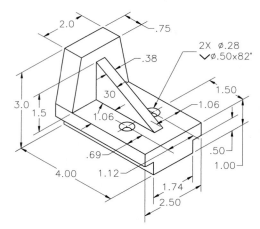

Figure 23-25 *Drawing for Exercise 6*

Answers to Self-Evaluation Test

1. equal measure, 120, **2.** 9/11, **3.** 30, **4.** isoplane left, **5.** CTRL+E, or F5, **6.** F, **7.** T, **8**. T, **9.** nonisometric lines, **10**. T

Chapter 24

The User Coordinate System

Learning Objectives

After completing this chapter, you will be able to:
- *Understand the concept of World Coordinate System (WCS).*
- *Understand the concept of User Coordinate System (UCS).*
- *Control the display of UCS icon using the **UCSICON** command.*
- *Change the current UCS icon type using the **UCSICON** command.*
- *Use the **UCS** command.*
- *Understand different options of changing UCS using the **UCS** command.*
- *Change UCS using the **Dynamic UCS** button.*
- *Manage UCS through a dialog box using the **UCSMAN** command.*
- *Understand different system variables related to the UCS and the UCS icon.*

THE WORLD COORDINATE SYSTEM (WCS)

As discussed earlier, when you start a new AutoCAD drawing, by default, the world coordinate system (WCS) is established. The objects you have drawn until now use the WCS. In the WCS, the *X*, *Y*, and *Z* coordinates of any point are measured with respect to the fixed origin (0,0,0). By default this origin is located at the lower left corner of the screen. This coordinate system is fixed and cannot be moved. The WCS is generally used in 2D drawings, wireframe models, and surface models. However, it is not possible to create a solid model keeping the origin and the orientation of the *X*, *Y*, and *Z* axes at the same place. The reason for this is that in case you want to create a feature on the top face of an existing model, you can do it easily by shifting the working plane using the **Elevation** option of the **ELEV** command, see Figure 24-1. But, using the **ELEV** command it is not possible to create a feature on the faces other than the top and the bottom faces of an existing model, as shown in Figure 24-2.

Figure 24-1 *Creating a new feature on the top face*

Figure 24-2 *Creating a new feature on the side face*

This problem can be solved using a concept called the user coordinate system (UCS). Using the UCS, you can relocate and reorient the origin and *X*, *Y*, and *Z* axes and establish your own coordinate system, depending on your requirement. The UCS is mostly used in 3D drawings, where you may need to specify points that vary from each other along the *X*, *Y*, and *Z* axes. It is also useful for relocating the origin or rotating the *X* and *Y* axes in 2D work, such as ordinate dimensioning, drawing auxiliary views, or controlling the hatch alignment. The UCS and its icon can be modified using the **UCSICON** and the **UCS** commands.

CONTROLLING THE VISIBILITY OF THE UCS ICON

Ribbon:	View > Coordinates > Show UCS Icon / Show UCS Icon at Origin /Hide UCS Icon		
Menu Bar:	View > Display > UCS Icon	**Command:**	UCSICON

This command is used to control the visibility and location of the UCS icon, which is a geometric representation of the directions of the current *X*, *Y*, and *Z* axes. AutoCAD displays different UCS icons in the model space and paper space, as shown in Figures 24-3 and 24-4. By default, the UCS icon is displayed near the bottom left corner of the drawing area. You can change the location and visibility of this icon using the **UCSICON** command. The prompt sequence for the **UCSICON** command is given next.

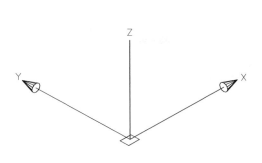

Figure 24-3 *Model space UCS icon*   **Figure 24-4** *Paper space UCS icon*

Enter an option [ON/OFF/All/Noorigin/ORigin/Properties] <ON>: *You can specify any option or press ENTER to accept the default option.*

ON

This option is used to display the UCS icon on the screen. You can also display the UCS icon by choosing **View > Coordinates > Show UCS Icon** from the **Ribbon**.

OFF

This option is used to make the UCS icon invisible from the screen. When you choose this option, the UCS icon will no longer be displayed on the screen. You can again turn on the display using the **On** option of the **UCSICON** command. Alternatively, choose **View > Coordinates > Hide UCS Icon** from the **Ribbon** to make the UCS icon invisible.

All

This option is used to apply changes to the UCS icon in all the active viewports. If this option is not used, the changes will be applied only to the current viewport.

Noorigin

This option is used to display the UCS icon at the lower left corner of the viewport, irrespective of the actual location of the origin of the current UCS. Alternatively, choose **View > Coordinates > Show UCS Icon at Origin** from the **Ribbon**.

ORigin

This option is used to place the UCS icon at the origin of the current UCS.

Properties

When you invoke this command, the **UCS Icon** dialog box will be displayed, as shown in Figure 24-5. You can also display this dialog box by choosing **View > Coordinates > UCS Icon, Properties** from the **Ribbon**. The options in this dialog box are discussed next.

Chapter 24

UCS icon style Area

2D. If this radio button is selected, the 2D UCS icon will be displayed on the screen instead of the 3D UCS icon, see Figure 24-6.

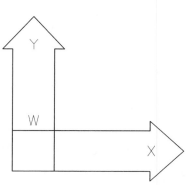

Figure 24-5 The **UCS Icon** *dialog box*

Figure 24-6 *2D UCS icon at the World position*

3D. If this radio button is selected, the 3D UCS icon will be displayed on the screen. This is the default option. AutoCAD displays the 3D UCS icon by default.

Cone. If this check box is cleared, the cones at the end of the X and Y axes of the 3D UCS icon will not be displayed. Instead, the arrows will be displayed. This option is not available if you select the **2D** radio button.

Line width. This drop-down list provides the width values that can be assigned to the 3D UCS icon. The default value for the line width is **1**. This drop-down list will not be available, if the **2D** radio button is selected.

UCS icon size Area

The slider bar provided in this area is used to increase the size of the UCS icon. You can also enter the value of the size of the UCS icon directly in the edit box. The value can vary between 5 and 95. The default value of the size of the UCS icon is 12.

UCS icon color Area

The **UCS icon color** area in this dialog box has two drop-down lists, **Model space icon color** and **Layout tab icon color**. The options of these drop-down lists are used to change the color of the UCS icon. By default, the color in the Model space is black and in the Layout is blue. You can assign any color to the UCS icon. By default, there are seven colors in these drop-down lists. However, you can also select a color from the **Select Color** dialog box which will be displayed after you select **Select Color** from the **Model space icon color** or the **Layout tab icon color** drop-down list.

DEFINING THE NEW UCS

Ribbon:	View > Coordinates > UCS	Toolbar:	UCS
Menu Bar:	Tools > New UCS	Command:	UCS

The **UCS** command is used to set a new coordinate system by shifting the working plane (*XY* plane) to the desired location. For certain views of the drawing, it is better to have the origin of measurements at some other point on or relative to your drawing objects. This makes locating the features and dimensioning the objects easier. The change in the UCS can be viewed by the change in the position and orientation of the UCS icon, which is placed by default at the lower left corner of the drawing window. The

Figure 24-7 Choosing the UCS button

origin and orientation of a coordinate system can be redefined by using the **UCS** command. Alternatively, choose **Coordinates > UCS** from the **View** tab of the **Ribbon**, as shown in Figure 24-7. The prompt sequence is as follows:

> Command: **UCS**
> Current ucs name: *WORLD*
> Specify origin of UCS or [Face/NAmed/OBject/Previous/View/World/X/Y/Z/ZAxis]
> <World>: *Select an option.*

If the **UCSFOLLOW** system variable is set to 0, any change in the UCS will not affect the drawing view.

If you choose the default option of the **UCS** command, you can establish a new coordinate system by specifying a new origin point, a point on the positive side of the new *X* axis, and a point on the positive side of the new *Y* axis. The direction of the *Z* axis is determined by applying the right-hand rule, about which you will learn in the next chapter. This option changes the orientation of the UCS to any angled surface. The prompt sequence that will follow, is given next.

> Specify origin of UCS or [Face/NAmed/OBject/Previous/View/World/X/Y/Z/ZAxis]
> <World>: *Specify the origin point of the new UCS.*
> Specify point on X-axis or <accept>: *Specify a point on the positive portion of the X axis*
> Specify point on the XY plane or <accept>: *Specify a point on the positive portion of the Y axis to define the orientation of the UCS completely.*

Note
*Depending upon the type of UCS required, various commands can be invoked by choosing different buttons from the **Coordinates** panel in the **View** tab of the **Ribbon**.*

A null response to the point on the *X* or *Y* axis prompt will lead to a coordinate system, in which the *X* or *Y* axis of the new UCS is parallel to that of the previous UCS.

W (World) Option

 With this option, you can set the current UCS back to the WCS, which is the default position. When the UCS is placed at the world position, a small rectangle is displayed at the point where all the three axes meet in the UCS icon, see Figure 24-8. If the UCS is moved from its default position, this rectangle is no longer displayed, indicating that the UCS is not at the world position, as shown in Figure 24-9.

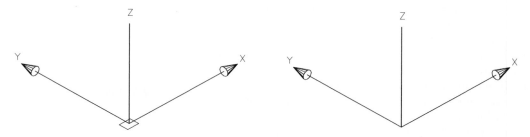

Figure 24-8 UCS at the world position   *Figure 24-9 UCS not at the world position*

 Tip
In case you have selected the 2D UCS instead of the 3D UCS icon, the W will not be displayed, if the UCS is not at the world position.

F (Face) Option

This option aligns the new UCS with the selected face of the solid object. The prompt sequence that will follow when you choose this button is given next.

Select face of solid object: *Select the face to align the UCS.*
Enter an option [Next/Xflip/Yflip] <accept>: *Select an option or accept the selected face to align.*

The **Next** option locates the new UCS on the next adjacent face or the back face of the selected edge. **Xflip** rotates the new UCS by 180-degree about the *X* axis and **Yflip** rotates it about the *Y* axis. Pressing ENTER at the **Enter an option [Next/Xflip/Yflip] <accept>** accepts the location of the new UCS as specified.

OB (OBject) Option

With the **OB** (**OBject**) option of the **UCS** command, you can establish a new coordinate system by pointing to any object in an AutoCAD drawing. However, the objects that cannot be used as an object for defining a UCS are a 3D polyline, 3D mesh, viewport object, or xline. The positive *Z* axis of the new UCS is in the same direction as the positive *Z* axis of the object selected. If the *X* and *Z* axes are given, the new *Y* axis is determined by the right-hand rule. The prompt sequence that will follow when you choose this button is given next.

Select object to align UCS: *Select the object to align the UCS.*

In Figure 24-10, the UCS is relocated using the **OBject** option and is aligned to the circle. The origin and the *X* axis of the new UCS are determined by the following rules.

Arc. When you select an arc, its center becomes the origin for the new UCS. The *X* axis passes through the endpoint of the arc that is closest to the point selected on the object.

Circle/Cylinder/Ellipse. The center of the circle becomes the origin for the new UCS, and the *X* axis passes through the point selected on the object (Figure 24-11).

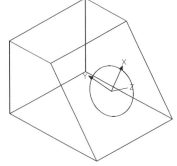

Figure 24-10 *Relocating the UCS using a circle*   **Figure 24-11** *UCS at a new location*

Line/Mline/Ray/Leader. The new UCS origin is the endpoint of the line nearest to the point selected on the line. The *X* axis is defined so that the line lies on the *XY* plane of the new UCS. Therefore, in the new UCS, the *Y* coordinate of the second endpoint of the line is 0.

Tip
The linear edges of solid models or regions are considered as individual lines when selected to align a UCS.

Spline. The origin of the new UCS is the endpoint of the spline that is nearest to the point selected on the spline. An imaginary line will be drawn between the two endpoints of the spline and the *X* axis will be aligned along this imaginary line.

Trace. The origin of the new UCS is the "**start point**" of the trace. The new *X* axis lies along the direction of the selected trace.

Dimension. The middle point of the dimension text becomes the new origin. The *X* axis direction is identical to the direction of the *X* axis of the UCS that existed when the dimension was drawn.

Point. The position of the point is the origin of the new UCS. The directions of the *X*, *Y*, and *Z* axes will be same as those of the previous UCS.

Solid. The origin of the new UCS is the first point of the solid. The X axis of the new UCS lies along the line between the first and second points of the solid.

2D Polyline. The start point of the polyline or polyarc is treated as the new UCS origin. The X axis extends from the start point to the next vertex.

3D Face. The first point of the 3D face determines the new UCS origin. The X axis is determined from the first two points, and the positive side of the Y axis is determined from the first and fourth points. The Z axis is determined by applying the right-hand rule.

Shape/Text/Insert/Attribute/Attribute Definition. The insertion point of the object becomes the new UCS origin. The new X axis is defined by the rotation of the object around its positive Z axis. Therefore, the object you select will have a rotation angle of zero in the new UCS.

Tip
Except for 3D faces, the XY plane of the new UCS will be parallel to the XY plane existing when the object was drawn; however, X and Y axes may be rotated.

V (View) Option

 The **V** (**View**) option of the **UCS** command allows you define a new UCS whose XY plane is parallel to the current viewing plane. The current viewing plane, in this case is the screen of the monitor. Therefore, a new UCS is defined that is parallel to the screen of the monitor. The origin of the UCS defined in this option remains unaltered. This option is used mostly to view a drawing from an oblique viewing direction or for writing the text for the objects on the screen. As soon as you choose this button, a new UCS is defined parallel to the screen of the monitor.

X/Y/Z Options

With these options, you can rotate the current UCS around a desired axis. You can specify the angle by entering the angle value at the required prompt or by selecting two points on the screen with the help of a pointing device. You can specify a positive or a negative angle. The new angle is taken relative to the X axis of the existing UCS. The **UCSAXISANG** system variable stores the default angle by which the UCS is rotated around the specified axis, by using the **X**/ **Y**/ **Z** options of the **New** option of the **UCS** command. The right-hand thumb rule is used to determine the positive direction of rotation of the selected axis.

X Option

In Figure 24-12, the UCS is relocated using the **X** option by specifying an angle about the X axis. The first model shows the UCS setting before the UCS was relocated and the second model shows the relocated UCS.

The prompt sequence that will follow when you choose this button is given next.

Specify rotation angle about X axis <90>: *Specify the rotation angle.*

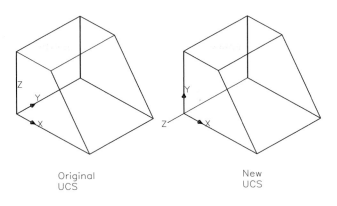

Figure 24-12 *Rotating the UCS about the X axis*

Y Option

In Figure 24-13, the UCS is relocated using the **Y** option by specifying an angle about the Y axis. The first model shows the UCS setting before the UCS was relocated and the second model shows the relocated UCS. The prompt sequence that will follow when you choose this button is given next.

Specify rotation angle about Y axis <90>: *Specify the angle.*

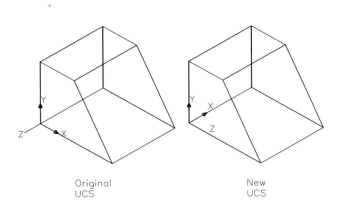

Figure 24-13 *Rotating the UCS about the Y axis*

Z Option

In Figure 24-14, the UCS is relocated using the **Z** option by specifying an angle about the Z axis. The first model shows the UCS setting before the UCS was relocated and the second model shows the relocated UCS. The prompt sequence that will follow when you choose this button is given next.

Specify rotation angle about Z axis <90>: *Specify the angle.*

Chapter 24

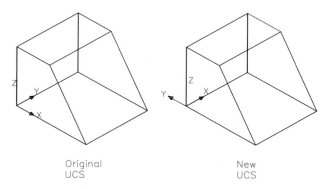

Figure 24-14 The UCS rotated about the Z axis

Tip
*The rotation about the Z axis can be used as an alternative to using the **SNAP** command to rotate the snap and grid.*

ZA (ZAxis) Option

This option is used to change the coordinate system by selecting the origin point of the *XY* plane and a point on the positive *Z* axis. After you specify a point on the *Z* axis, AutoCAD determines the *X* and *Y* axes of the new coordinate system accordingly. The prompt sequence that will follow when you choose **View > Coordinates > Z Axis Vector** from the **Ribbon** is given next.

> Specify new origin point or [Object] <0,0,0>: *Specify the origin point, as shown in Figure 24-15.*
> Specify point on positive portion of Z-axis <default>: **@ 0,-1,0**

Now, the front face of the model will become the new work plane (Figure 24-16) and all the new objects will be oriented accordingly. If you give a null response to the **Specify point on the positive portion of Z axis <current>** prompt, the *Z* axis of the new coordinate system will be parallel to (in the same direction as) the *Z* axis of the previous coordinate system.

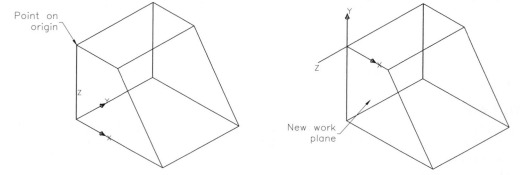

Figure 24-15 Specifying a point on the origin   *Figure 24-16* Relocating the UCS using the **ZA** option

Null responses to the origin point and the point on the positive *Z* axis establish a new coordinate system, in which the direction of the *Z* axis is identical to that of the previous coordinate system; however, the *X* and *Y* axes may be rotated around the *Z* axis. The positive *Z* axis direction is also known as the extrusion direction.

When the **Object** option is selected at the **Specify new origin point or [Object] <0,0,0>** prompt, then you will be prompted to select an object. Select an open object and the UCS will be placed at the end point nearest to the point of the object selection, with its Z-axis along the tangent direction at that end point and pointing away from the object.

P (Previous) Option

The **P** (**Previous**) option restores the current UCS settings to the previous UCS settings. The last ten UCS settings are saved by AutoCAD. You can go back to the previous ten UCS settings in the current space using the **Previous** option. If the **TILEMODE** is off, the last ten coordinate systems in paper space and in model space are saved. When you choose this button, the previous UCS settings are automatically restored.

NA (NAmed) Option

This option is used to name and save the current UCS settings, restore a previously saved UCS setting, view a previously saved UCS list, and delete the saved UCS from the list. The prompt sequence for this option is given next.

    Command: **UCS**
    Current ucs name: *WORLD*
    Specify origin of UCS or [Face/NAmed/OBject/Previous/View/World/X/Y/Z/ZAxis]
    <World>: **NA**
    Enter an option [Restore/Save/Delete/?]: *Enter an option.*

Depending on your requirement, you can select any one of the following options:

R (Restore) Option

The **Restore** in the **UCS** command is used to restore a previously saved named UCS. Once it is restored, it becomes the current UCS. However, the viewing direction of the saved UCS is not restored. You can also restore a named UCS by selecting it from the **UCS** dialog box that appears when you invoke **View > Coordinates > Named** from the **Ribbon**. As this option does not have a button to restore the Named UCS, you need to choose this option using the **UCS** command at the Command prompt. The prompt sequence that will follow when you invoke the **UCS** command is given next.

    Enter an option [Restore/Save/Delete/?]: *R*
    Enter name of UCS to restore or [?]: *Specify the name of UCS or ENTER to list all the UCS*

You can specify the name of the UCS to be restored or list the UCSs that can be restored by pressing ENTER at the previous prompt. The prompt sequence that will be followed is given next.

Chapter 24

Enter UCS name(s) to list <*>: *Specify the name of the UCS to list or give a null response to list all available UCSs.*

If you give a null response at the previously mentioned prompt, then the AutoCAD text window will be opened listing all the available UCSs.

S (Save) Option

With this option, you can name and save the current UCS settings. The following points should be kept in mind, while naming the UCS.

1. The name can be up to 255 characters long.
2. The name can contain letters, digits, blank spaces and the special characters $ (dollar), - (hyphen), and _ (underscore).

As this option does not have a button, you can invoke this option using the command prompt. The prompt sequence that will follow when you invoke the **UCS** command is given next.

Current ucs name: *WORLD*
Specify origin of UCS or [Face/NAmed/OBject/Previous/View/World/X/Y/Z/ZAxis]: **NA**
Enter an option [Restore/Save/Delete/?]: S
Enter name to save current UCS or [?]: *Specify a name to the UCS or press ENTER to list.*

Enter a valid name for the UCS at this prompt. AutoCAD saves it as a UCS. You can also list the previously saved UCSs by entering **?** at this prompt. The next prompt sequence is:

Enter UCS name(s) to list <*>:

Enter the name of the UCS to list or give a null response to list all the available UCSs.

D (Delete) Option

The **D** (**Delete**) option is used to delete the selected UCS from the list of saved coordinate systems. This option also does not have a button. As this option does not have a button, you can invoke this option using the Command prompt. The prompt sequence that will follow when you invoke the **UCS** command is given next.

Current ucs name: *WORLD*
Specify origin of UCS or [Face/NAmed/OBject/Previous/View/World/X/Y/Z/ZAxis]: **NA**
Enter an option [Restore/Save/Delete/?]: **D**
Enter UCS name(s) to delete <none>: *Specify the name of the UCS to delete.*

The UCS name you enter at this prompt is deleted. You can delete more than one UCS by entering the UCS names separated with commas or by using wild cards.

? Option

By invoking this option, you can list the name of the specified UCS. This option gives you the name, origin, and *X*, *Y*, and *Z* axes of all the coordinate systems relative to the existing UCS. If the current UCS has no name, it is listed as WORLD or UNNAMED. The choice between these two names depends on whether the current UCS is the same as the WCS. The prompt sequence that will be displayed when you invoke this option using the Command prompt is given next.

> Current ucs name: *WORLD*
> Specify origin of UCS or [Face/NAmed/OBject/Previous/View/World/X/Y/Z/ZAxis]: **NA**
> Enter an option [Restore/Save/Delete/?]: **?**
> Enter UCS name(s) to list <*>:

A (Apply) Option

The **Apply** option applies the current UCS settings to a specified viewport or to all the active viewports in a drawing session. If the **UCSVP** system variable is set to 1, each viewport saves its UCS settings. The **Apply** option is not displayed in the Command prompt as an option for the **UCS** command. Therefore, you have to choose the **Apply** button from the **UCS** toolbar. The prompt sequence that will follow when you choose this button is given next.

> Pick viewport to apply current UCS or [All]<current>: *Pick inside the viewport or enter **A** to apply the current UCS settings to all viewports.*

O (Origin) Option

This option is used to define a new UCS by changing the origin of the current UCS, without changing the directions of the X,Y, Z axes, see Figures 24-17 and 24-18. The new point defined will now be the origin point (0,0,0) for all the coordinate entries from this point on, until the origin is changed again. You can specify the coordinates of the new origin or click on the screen using the pointing device. Alternatively, choose **View > Coordinates > Origin** from the **Ribbon** to specify the coordinates of the new origin. The prompt sequence that will follow when you choose this button is given next.

> Specify new origin point <0,0,0>: *Specify the origin point, as shown in Figure 24-17.*

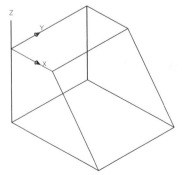

Figure 24-17 *Defining a new origin for the UCS*   **Figure 24-18** *Relocating UCS to the new origin*

Tip
If you do not provide a Z coordinate for the origin, the current Z coordinate value will be assigned to the UCS.

3 (3-Point) Option

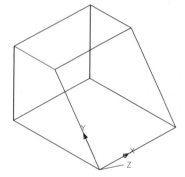

The **3-Point** option is the default option of the **UCS** command which changes the orientation of the UCS to any angled surface. With this option, you can establish a new coordinate system by specifying a new origin point, a point on the positive side of the new X axis, and a point on the positive side of the new Y axis (Figure 24-19). The direction of the Z axis is determined by applying the right-hand rule, about which you will learn in the next chapter. The prompt sequence that will follow when you choose this button is given next.

Specify new origin point <0,0,0>: *Specify the origin point of the new UCS.*
Specify point on X-axis <or accept>: *Specify a point on the positive portion of the X axis.*
Specify point on XY plane <or accept>: *Specify a point on the positive portion of the Y axis.*

A null response to the **Specify new origin point <0,0,0>** prompt will lead to a coordinate system, in which the origin of the new UCS is identical to that of the previous UCS. Similarly, null responses to the point on the X or Y axis prompt will lead to a coordinate system, in which the X or Y axis of the new UCS is parallel to that of the previous UCS. In Figure 24-20, the UCS has been relocated by specifying three points (the origin point, a point on the positive portion of the X axis, and a point on the positive portion of the Y axis).

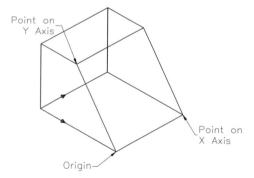

Figure 24-19 Relocating the UCS using the 3-Point option

Figure 24-20 UCS at a new position

Example 1

In this example, you will draw a tapered rectangular block. After drawing it, you will align the UCS to the inclined face of the block using the **3-Point** option of the **UCS** command. Then, you will draw a circle on the inclined face. The dimensions for the block and the circle are given in Figure 24-21.

This example can be divided into the following steps:

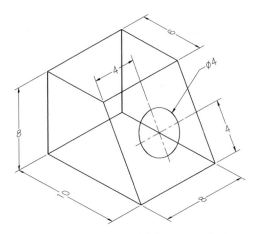

Figure 24-21 *Model for Example 1*

- Draw edges on the bottom face of the tapered block.
- Draw edges on the top face of the tapered block.
- Draw the remaining edges of the block.
- Align the UCS to the inclined face.
- Draw the circle on the inclined face.

These steps are discussed next.

1. Start a new drawing and choose the **Line** button. The prompt sequence is given next.

   Specify first point: **1,1**
   Specify next point or [Undo]: **@10,0**
   Specify next point or [Undo]: **@0,8**
   Specify next point or [Close/Undo]: **@-10,0**
   Specify next point or [Close/Undo]: C

2. As the top face of the model is at a distance of 8 units from the bottom face, in the positive Z direction, therefore, you will have to define a UCS at a distance of 8 units to create the top face. This can be done by choosing **View > Coordinates > Origin** from the **Ribbon**. The prompt sequence is as follows:

   Specify new origin point <0,0,0>: **0,0,8**

3. Choose the **Line** button from the **Draw** panel in the **Home** tab of the **Ribbon**. The prompt sequence is as follows:

   Specify first point: **1,1**
   Specify next point or [Undo]: **@6,0**
   Specify next point or [Undo]: **@0,8**

Specify next point or [Close/Undo]: **@-6,0**
Specify next point or [Close/Undo]: **C**

4. When you open a new drawing, by default you view the model from the top view. Therefore, when viewing from the top, the three edges of the top side overlap with the corresponding bottom edges. Hence, you will not be able to see all edges. However, you can change the viewpoint to clearly view the model in 3D. Choose **View > Views > SE Isometric** from the **Ribbon** to proceed to the SE isometric view. Now, you can see the 3D view of the objects, see Figure 24-22.

5. Join the remaining edges of the model using the **Line** button. The model, after joining all the edges, should look similar to the one shown in Figure 24-23.

     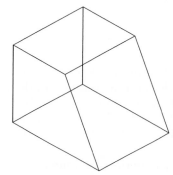

Figure 24-22 *3D view of the objects*   *Figure 24-23* *Model after joining all the edges*

6. Now, you need to draw the circle on the inclined face. First, you need to make the inclined face as the current working plane. Therefore, choose **View > Coordinates > 3-Point** from the **Ribbon**. The prompt sequence is as follows.

Specify new origin point <0,0,0>: *Select point P1, as shown in Figure 24-24.*
Specify point on positive portion of X-axis <default>: *Select point P2, refer to Figure 24-24.*
Specify point on positive-Y portion of the UCS XY plane <default>: *Select point P3, refer to Figure 24-24.*

7. Choose **Home > Draw > Circle** from the **Ribbon** and draw a circle on the inclined face.

8. The final model should look similar to the one shown in Figure 24-25.

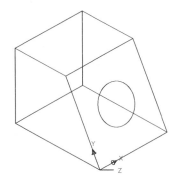

Figure 24-24 *Model after aligning the UCS on the inclined face*

Figure 24-25 *Final model for Example 1*

MANAGING THE UCS THROUGH THE DIALOG BOX

The **UCSMAN** command displays the **UCS** dialog box (Figure 24-26). This dialog box can be used to restore the saved and orthographic UCSs, specify UCS icon settings, and rename UCSs. This dialog box has three tabs: **Named UCSs**, **Orthographic UCSs**, and **Settings**.

Named UCSs Tab

Ribbon:	View > Coordinates > Named	**Toolbar:**	UCS
Menu Bar:	Tools > Named UCS	**Command:**	UCSMAN

The list of all the coordinate systems defined (saved) on your system is displayed in the list box of the **View** tab. The first entry in this list is always **World**, which means WCS. The next entry is **Previous**. If you have defined any other coordinate systems in the current editing session, it will also be displayed in the list box. Selecting the **Previous** entry and then choosing the **OK** button repeatedly allows you to go backward through the coordinate systems defined in the current editing session. **Unnamed** is the next entry in the list, if you have not named the current coordinate system. If there are a number of viewports and unnamed settings, only the current viewport UCS name is displayed in the list. The current coordinate system is indicated by a small pointer icon to the left of the coordinate system name. The current UCS name is also displayed next to **Current UCS** on top of the **UCS** dialog box. To make other coordinate system current, select its name in the UCS Names list, choose the **Set Current** button, and choose **OK**. Alternatively, select the required UCS from the **Named UCS Combo Control** drop-down list available in the coordinate panel of the **View** tab. To delete a coordinate system, select its name, and then right-click to display a shortcut menu. The options are **Set Current**, **Rename**, **Delete**, and **Details**. Choose the **Delete** button to delete the selected UCS name. To rename a coordinate system, select its name and then right-click to display a shortcut menu and choose **Rename**. Now, enter the new name. You can also double-click on the name you need to modify, and then change the name.

Figure 24-26 *The **Named UCSs** tab of the **UCS** dialog box*

Tip
*All the changes and updating of the UCS information in the drawing are carried out only after you choose the **OK** button.*

If you want to check the current coordinate system's origin and *X*, *Y*, and *Z* axis values, select a UCS from the list and then choose the **Details** button. The **UCS Details** dialog box (Figure 24-27) containing that information is displayed. Alternatively, right click on a specific UCS name in the list box and choose the **Details** option from the shortcut menu.

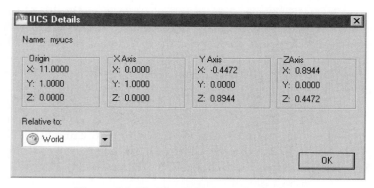

Figure 24-27 *The **UCS Details** dialog box*

Setting the UCS to Preset Orthographic UCSs Using the Orthographic UCSs Tab

Different orthographic views are listed in the **Orthographic UCSs** tab, as shown in Figure 24-28. You can select any one of the six orthographic views of the WCS or the user defined coordinate system using the **UCS** dialog box. To define an orthographic coordinate system relative to a UCS, select it from the **Relative to** drop-down list, select the orthographic view from the list of preset orthographic views, and then choose the **Set Current** button; the UCS icon is placed at the corresponding view. You can also right-click on the specific orthographic UCS name in the list and choose **Set Current** from the shortcut menu. The name of the current UCS is displayed above the list of orthographic view. If the current settings have not been saved and named, the **Current UCS** name displayed is **Unnamed**.

Figure 24-28 *The **Orthographic UCSs** tab of the **UCS** dialog box*

The **Depth** field in the list box of this dialog box displays the distance between the *XY* plane of the selected orthographic UCS setting and a parallel plane passing through the origin of the UCS base setting. The system variable **UCSBASE** stores the name of the UCS, which is considered the base setting; that is, it defines the origin and orientation. You can enter or modify values of **Depth** by double-clicking on the depth value of the selected UCS in the list box. This displays the **Orthographic UCS depth** dialog box (Figure 24-29), where you can enter new depth values. You can also right-click on a specific orthographic UCS in the list box to display a shortcut menu. Choose **Depth** to display the **Orthographic UCS depth** dialog box. Enter a value in the edit box. If you need to pick a position from the drawing area, choose the **Select new origin** button available next to the edit box; the dialog box will disappear allowing you to specify a new origin or depth on the screen.

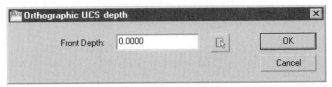

Figure 24-29 *The **Orthographic UCS depth** dialog box*

Chapter 24

Choosing the **Details** button displays the **UCS Details** dialog box with the origin and the *X*, *Y*, *Z* coordinate values of the selected UCS. You can also choose **Details** from the shortcut menu that is displayed on right-clicking on a UCS in the list box. The shortcut menu also has a **Reset** option. Choosing **Reset** from the shortcut menu restores the selected UCS to its default settings. Also, the depth value, if assigned, becomes zero.

Settings Tab

The **Settings** tab of the **UCS** dialog box (Figure 24-30) displays and modifies the UCS and UCS icon settings of a specified viewport. If you choose the inclined arrow on the Coordinates panel, the **UCS** dialog box will be displayed with the **Settings** tab chosen. The options in this dialog box are discussed next.

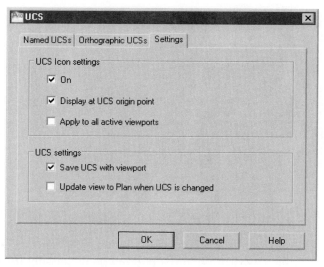

*Figure 24-30   The **Settings** tab of the **UCS** dialog box*

UCS Icon settings Area

Selecting the **On** check box displays the UCS icon in the current viewport. This process is similar to using the **UCSICON** command and enables you to set the display of the UCS icon to on or off. If you select the **Display at UCS origin point** check box, the UCS icon is displayed at the origin point of the coordinate system in use in the current viewport. If the origin point is not visible in the current viewport, or if the check box is cleared, the UCS icon is displayed in the lower left corner of the viewport. Selecting the **Apply to all active viewports** check box applies the current UCS icon settings to all the active viewports in the current drawing.

UCS settings Area

This area of the **Settings** tab in the **UCS** dialog box specifies the UCS settings for the current viewport. Selecting the **Save UCS with viewport** check box saves the UCS with the viewport. This setting is stored in the **UCSVP** system variable. If you clear this check box, the **UCSVP** variable will be set to 0 and the UCS of the viewport will reflect the UCS of the current viewport. When you select the **Update view to Plan when UCS is changed** check box, the plan view is restored after the UCS in the viewport is changed. When the selected UCS is restored, the plan view is also restored. The value is stored in the **UCSFOLLOW** system variable.

SYSTEM VARIABLES

The coordinate value of the origin of the current UCS is held in the **UCSORG** system variable. The X and Y axis directions of the current UCS are held in the **UCSXDIR** and **UCSYDIR** system variables, respectively. The name of the current UCS is held in the **UCSNAME** variable. All these variables are read-only. If the current UCS is identical to the WCS, the **WORLDUCS** system variable is set to 1; otherwise, it holds the value 0. The current UCS icon setting can be examined and manipulated with the help of the **UCSICON** system variable. This variable holds the UCS icon setting of the current viewport. If more than one viewport is active, each one can have a different value for the **UCSICON** variable. If you are in paper space, the **UCSICON** variable will contain the setting for the UCS icon of the paper space. The **UCSFOLLOW** system variable controls the automatic display of a plan view when you switch from one UCS to another. If **UCSFOLLOW** is set to 1 and you switch from one UCS to another, a plan view is automatically displayed. The **UCSAXISANG** variable stores the default angle value for the X, Y, or Z axis around which the UCS is rotated using the **X, Y, Z** options of the **New** option of the **UCS** command. The **UCSBASE** variable stores the name of the UCS that acts as the base; that is, it defines the origin and orientation of the orthographic UCS setting. The **UCSVP** variable decides whether the UCS settings are stored with the viewport or not.

Self-Evaluation Test

Answer the following questions and then compare them to those given at the end of this chapter:

1. The World Coordinate System can be moved from its position. (T/F)

2. The **View** option of the **UCS** command allows you to define a new UCS whose *XY* plane is parallel to the current viewing plane. (T/F)

3. An ellipse cannot be used as an object for defining a new UCS. (T/F)

4. If you do not specify the Z coordinate of the new point while moving the UCS, the previous Z coordinates will be taken for the new value. (T/F)

5. By default, the _____ is established as the coordinate system in the AutoCAD environment.

6. The _____ coordinate system can be moved and rotated to any desired location.

7. Once a saved UCS is restored, it becomes the _____ UCS.

8. The _____ command controls the display of the UCS icon.

9. You can change the 3D UCS icon to a 2D UCS icon using the _____ option of the
    UCSICON command.

10. The _____ point of the 3D face defines the origin of the new UCS.

Review Questions

Answer the following questions:

1. The **All** option of the **Apply** button in the **UCS** toolbar applies the changes to the UCS
    icon in all the active viewports. (T/F)

2. The line width of the UCS icon can be changed. (T/F)

3. The size of the UCS icon can vary between 5 and 100. (T/F)

4. The name of the current UCS is stored in the **UCSNAME** variable. (T/F)

5. Which option is used to restore the previous UCS settings?

    (a) **ZAxis**   (b) **Restore**
    (c) **Previous**   (d) **Save**

6. Which option is used to list the names of all the saved UCSs?

    (a) **ZAxis**   (b) **Restore**
    (c) **?**   (d) **Save**

7. Which system variable is used to control the automatic display of the plan view when you
    switch to the new UCS?

    (a) **UCSORG**   (b) **UCSFOLLOW**
    (c) **UCS**   (d) **UCSBASE**

8. Which system variable stores the coordinates of the origin of the current UCS?

    (a) **UCSICON**   (b) **UCSORG**
    (c) **UCSVP**   (d) **UCSFOLLOW**

9. Which command is used to manage the UCS using a dialog box?

    (a) **UCS**   (b) **UCSMAN**
    (c) **UCSICON**   (d) **UCSBASE**

10. The _____ variable is used to control the X axis direction of the current UCS.

11. The _____ direction is known as the extrusion direction.

12. The lineweight of the UCS icon can be changed using the _____ option of the **UCSICON** command.

13. The _____ option of the **UCS** command is used to change the origin of the current UCS.

14. The _____ variable is used to control the *Y* axis direction of the current UCS.

15. The _____ variable is used to store the default angle value by which the UCS will be rotated using the **X**, **Y**, and **Z** options of the **New** option of the **UCS** command.

Exercises

Exercises 1 Through 4   *Mechanical*

Draw the objects shown in Figures 24-31 through 24-34. Assume the dimensions. Use the **UCS** command to align the ucs icon and then draw the objects.

Figure 24-31 *Drawing for Exercise 1*

Figure 24-32 *Drawing for Exercise 2*

Figure 24-33 *Drawing for Exercise 3*

Figure 24-34 *Drawing for Exercise 4*

Chapter 24

Exercise 5   *General*

Create the drawings shown in Figure 24-35. Assume its missing dimensions. Use the **UCS** command to align the ucs icon and then draw the objects. The left side of the transition makes a 90-degree angle with the bottom.

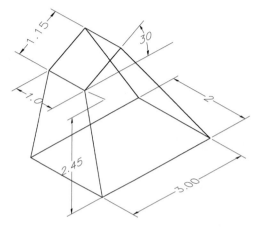

Figure 24-35 *Model for Exercise 5*

Exercise 6   *General*

Create the drawing shown in Figure 24-36. Assume its missing dimensions. Use the **UCS** command to position the ucs icon and then draw the objects. The center of the top polygon is offset at 0.75 units from the center of the bottom polygon.

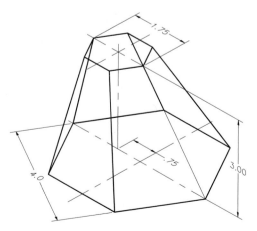

Figure 24-36 *Model for Exercise 6*

Exercises 7 Through 9 *General*

In this exercise, you will draw the model shown in Figures 24-37 through 24-39. Assume the missing dimensions.

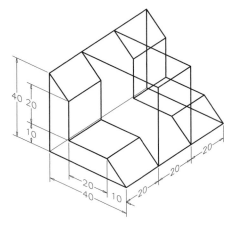

Figure 24-37 Model for Exercise 7

Figure 24-38 Model for Exercise 8

Figure 24-39 Model for Exercise 9

Exercises 10   *General*

In this exercise, you will draw the model shown in Figure 24-40. Assume the missing
dimensions. **

Figure 24-40 *Model for Exercise 10*

Chapter 25

Getting Started with 3D

Learning Objectives

After completing this chapter, you will be able to:

- *Start 3D environment in AutoCAD for creating models.*
- *Understand the use of three-dimensional (3D) drawing.*
- *Understand various types of 3D models.*
- *Understand conventions in AutoCAD.*
- *Use the **DDVPOINT** command to view objects in 3D space.*
- *Use the **VPOINT** command to view objects in 3D space.*
- *Use the **ViewCube** to view objects easily in 3D space.*
- *Understand various types of 3D coordinate systems.*
- *Set the thickness and elevation for objects.*
- *Dynamically view 3D objects using the **3DORBIT** command.*
- *Use the **DVIEW** command for viewing the model.*
- *Use the **SteeringWheel** for viewing the model.*

STARTING THREE DIMENSIONAL (3D) MODELING IN AUTOCAD

In AutoCAD, you can start the 3D Modeling in a separate workspace. All commands and options required to create the 3D design are displayed in this workspace, by default. To start a new file in the 3D workspace, invoke the **Select template** dialog box and select the *acad3D.dwt* template. This template file supports the 3D environment. The other template files that support the 3D workspace are *acadiso3D.dwt*, *acad-Named Plot Styles3D.dwt*, and *acadISO-Named Plot Styles3D.dwt*. You can switch between the **2D Drafting & Annotation** workspace and the **3D Modeling** workspace using the **Workspace Switching** button available at the bottom-right corner of the Status Bar. Figure 25-1 shows the **3D Modeling** workspace of AutoCAD.

Figure 25-1 *The **3D Modeling** workspace of AutoCAD*

USE OF THREE-DIMENSIONAL DRAWING

The real world, and literally whatever you see around in it, is three-dimensional (3D). Each and every design idea you think of is 3D. However, until some time back, it was not possible to materialize these ideas in the design industry due to the lack of the third dimension. This is because the drawings were made on the drawing sheets, which are two dimensional objects with only *X* and *Y* coordinates. Therefore, it was not possible to draw the 3D objects. As prototyping is a very long and costly affair, therefore, the designers had to suppress their ideas and convert the 3D designs into 2D drawings.

The use of computers and the CADD Technology has brought a significant improvement in materializing the design ideas and creating the engineering drawings. You can create a proper 3D object in the computer using the CADD Technology. This technology allows you to create models with the third coordinate called the *Z* coordinate, in addition to the *X* and *Y* coordinates. Apart from materializing the design ideas, the 3D models have the following advantages.

1. **To generate the drawing views**. Once you have created the 3D model, its 2D drawing views can be automatically generated.

2. **To provide realistic effects**. You can provide realistic effects to the 3D models by assigning a material and providing light effects to them. For example, an architectural drawing can be made more realistic and presentable by assigning the material to the walls and interiors and adding lights to it. You can also create its bitmap image and use it in presentations.

3. **To create the assemblies and check them for interference**. You can assemble various 3D models and create the assemblies. Once the components are assembled, you can check them for interference to reduce the errors and material loss during manufacturing. You can also generate the 2D drawing views of the assemblies.

4. **To create an animation of the assemblies**. You can animate the assemblies and view the animation to provide the clear display of the mating parts.

5. **To apply Boolean operations**. You can apply Boolean operations to the 3D models.

6. **To calculate the mass properties**. You can calculate the mass properties of the 3D models.

7. **Cut Sections**. You can cut sections through the solid models to view the shape at that cross-section.

TYPES OF 3D MODELS

In AutoCAD, depending on their characteristics, the 3D models are divided into the following three categories.

Wireframe Models

These models are created using simple AutoCAD entities such as lines, polylines, rectangles, polygons, or some other entities such as 3D faces, and so on. To understand the wireframe models, consider a 3D model made up of match sticks or wires. These models consist of only the edges and hence you can see through them. You cannot apply the Boolean operations on these models and cannot calculate their mass properties. Wireframe models are generally used in the bodybuilding of the vehicles. Figure 25-2 shows a wireframe model.

Surface Models

These models are made up of one or more surfaces. They have a negligible wall thickness and are hollow inside. To understand these models, consider a wireframe model with a cloth wrapped around it. You cannot see through it. These models are used in the plastic molding industry, shoe manufacturing, utensils manufacturing, and so on.

You can directly create a surface or mesh. Alternatively, you can create wireframe model and then convert it into a mesh or surface model. Remember that you cannot perform Boolean operations in the surface models. Figure 25-3 shows a surface model created using a single surface or mesh and Figure 25-4 shows a surface model created using a combination of surfaces.

Figure 25-2 A wireframe model   *Figure 25-3 A surface model created using a single surface*

Solid Models

These are the solid filled models having mass properties. To understand a solid model, consider a model made up of metal or wood. You can perform the Boolean operations on these models, cut a hole through them, or even cut them into slices (Figure 25-5).

Figure 25-4 A semi finished surface model created using more than one surface

Figure 25-5 A solid model

CONVENTIONS IN AutoCAD

It is important for you to know the conventions in AutoCAD before you proceed with 3D. AutoCAD follows these three conventions:

1. By default, any drawing you create in AutoCAD will be in the world XY plane.

2. The right-hand thumb rule is followed in AutoCAD to identify the X, Y, and Z axis directions. This rule states that if you keep the thumb, index finger, and the middle finger of the right hand mutually perpendicular to each other, as shown in Figure 25-6(a), then the thumb of the right hand will represent the direction of the positive X axis, the index finger will represent the direction of the positive Y axis, and the middle finger of the right hand will represent the direction of the positive Z axis, see Figure 25-6(b).

3. The right-hand thumb rule is followed in AutoCAD to determine the direction of rotation or revolution in the 3D space. The right-hand thumb rule states that if the thumb of the right hand points in the direction of the axis of rotation, then the direction of the curled fingers will define the positive direction of rotation, see Figure 25-6(c).

This rule will be used during the rotation of 3D models or the UCS and also, during the creation of revolved surfaces and revolved solids.

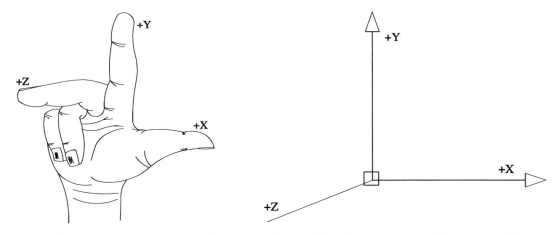

Figure 25-6(a) *The orientation of fingers as per the right-hand thumb rule*

Figure 25-6(b) *Orientation of the X, Y, and Z axes*

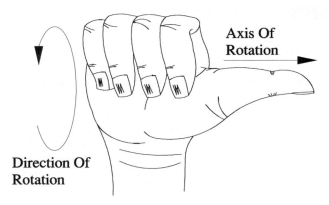

Figure 25-6(c) *Right-hand thumb rule showing the axis and direction of rotation*

CHANGING THE VIEWPOINT TO VIEW 3D MODELS

Until now, you have been drawing only the 2D entities in the *XY* plane and in the Plan view. But in the Plan view, it is very difficult to find out whether the object displayed is a 2D entity or a 3D model. For example, see Figure 25-7. In this view, it appears as though the cuboid displayed is a rectangle. The reason for this is that, by default, you view the objects in the Plan view from the direction of the positive *Z* axis. This confusing situation can be avoided by viewing the object from a direction that also displays the *Z* axis. In order to do that, you need to change the viewpoint so that the object is displayed in the space with all three axes, as shown in Figure 25-8. In this view, you can also see the *Z* axis of the model along with the *X* and *Y* axes. It will now be clear that the original object is not a rectangle but a 3D model. The viewpoint can be changed by using the **DDVPOINT** command, the **VPOINT** command, or by using the options in the **View** panel of the **Ribbon**.

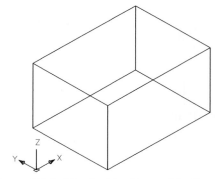

Figure 25-7 *Cuboid looking like a rectangle in the Plan view*

Figure 25-8 *Viewing the 3D model after changing the viewpoint*

Tip
Remember that changing the viewpoint does not affect the dimensions or the location of the model. When you change the viewpoint, the model is not moved from its original location. Changing the viewpoint only changes the direction from which you view the model.

Changing the Viewpoint Using the Viewpoint Presets Dialog Box

Menu Bar: View > 3D Views > Viewpoint Presets   **Command:** DDVPOINT

The **DDVPOINT** command is used to invoke the **Viewpoint Presets** dialog box for setting the viewpoint to view the 3D models. The dialog box sets the viewpoint with respect to the angle from the X axis and the angle from the XY plane. Figure 25-9 shows different view direction parameters.

Viewpoint Presets Dialog Box Options
The options in the **Viewpoint Presets** dialog box (Figure 25-10) are discussed next.

Absolute to WCS. This radio button is selected to set the viewpoint with respect to the world coordinate system (WCS).

Relative to UCS. This radio button is selected to set the viewpoint with respect to the current user coordinate system (UCS).

Note
The WCS and UCS have already been discussed in Chapter 24.

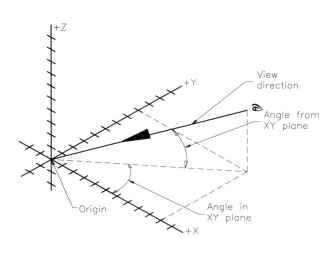

Figure 25-9 *View direction parameters*

Figure 25-10 *The **Viewpoint Presets** dialog box*

Chapter 25

From: X Axis. This edit box is used to specify the angle in the *XY* plane from the *X* axis. You can directly enter the required angle in this edit box or select the desired angle from the image tile (Figure 25-11). There are two arms in this image tile and by default, they are placed one over the other. The gray arm displays the current viewing angle and the black one displays the new viewing angle. On selecting a new angle from the image tile, it is automatically displayed in the **X Axis** edit box. This value can vary between 0 and 359.9.

Figure 25-11 The image tile for selecting the angle in the XY plane

From: XY Plane. This edit box is used to specify the angle from the *XY* plane. You can directly enter the desired angle in it or select the angle from the image tile, as shown in Figure 25-12. This value can vary from -90 to +90.

Set to Plan View. This button is chosen to set the viewpoint to the Plan view of the WCS or the UCS. If the **Absolute to WCS** radio button is selected, the viewpoint is set to the Plan view of the WCS. If the **Relative to UCS** radio button is selected, the viewpoint is set to the Plan view of the current UCS. Figures 25-13 and 25-14 show a 3D model from different viewing angles.

Figure 25-12 The image tile for selecting the angle from the XY plane

Figure 25-13 Viewing the 3D model with the angle in the XY plane as 225 and angle from the XY plane as 30

Figure 25-14 Viewing the 3D model with the angle in the XY plane as 145 and angle from the XY plane as 25

Tip
If you set a negative angle from the XY plane, the Z axis will be displayed as a dotted line, suggesting a negative Z direction, see Figures 25-15 and 25-16.

In case of confusion in identifying the direction of the X, Y, and Z axes, use the right-hand thumb rule.

Figure 25-15 *Viewing the 3D model with the angle in the XY plane as 315 and angle from the XY plane as -25*

Figure 25-16 *Viewing the 3D model with the angle in the XY plane as 225 and angle from the XY plane as -25*

Changing the Viewpoint Using the Command Line

Menu Bar:	View > 3D Views > Viewpoint	Command:	VPOINT

The **VPOINT** command is also used to set the viewpoint for viewing the 3D models. Using this command, the user can specify a point in the 3D space that will act as the viewpoint. The prompt sequence that will follow when you invoke this command is given next.

> Current view direction: VIEWDIR=0.0000,0.0000,1.0000
> Specify a view point or [Rotate] <display compass and tripod>:

Specifying a Viewpoint

This is the default option and is used to set a viewpoint by specifying its location of (viewer) using the *X*, *Y*, and *Z* coordinates of that particular point. AutoCAD follows a convention of the sides of the 3D model for specifying the viewpoint. The convention states that if the UCS is at the World position (default position), then

1. The side at the positive *X* axis direction will be taken as the right side of the model.
2. The side at the negative *X* axis direction will be taken as the left side of the model.
3. The side at the negative *Y* axis direction will be taken as the front side of the model.
4. The side at the positive *Y* axis direction will be taken as the back side of the model.
5. The side at the positive *Z* axis direction will be taken as the top side of the model.
6. The side at the negative *Z* axis direction will be taken as the bottom side of the model.

Some standard viewpoint coordinates and the view they display are given next.

Value	View	Value	View	Value	View
1,0,0	Right side	-1,0,0	Left side	0,1,0	Back
0,-1,0	Front	0,0,1	Top view	0,0,-1	Bottom view
1,1,1	Right, Back, Top	-1,1,1	Left, Back, Top	1,-1,1	Right, Front, Top
-1,-1,1	Left, Front, Top	1,1,-1	Right, Back, Bottom	-1,1,-1	Left, Back, Bottom
1,-1,-1	Right, Front, Bottom	-1,-1,-1	Left, Front, Bottom		

Chapter 25

You can also enter the values in decimal, but the resultant views will not be the standard views. Figures 25-17 through 25-20 show the 3D model from different viewpoints.

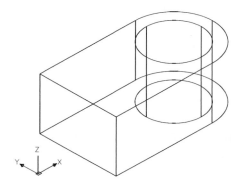

Figure 25-17 *Viewing the model from the viewpoint -1,-1,1*

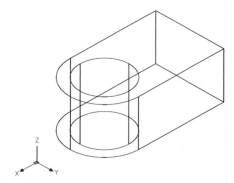

Figure 25-18 *Viewing the model from the viewpoint 1,1,1*

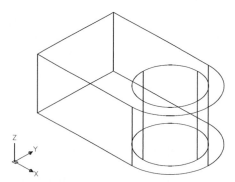

Figure 25-19 *Viewing the model from the viewpoint 1,-1,1*

Figure 25-20 *Viewing the model from the viewpoint -1,1,1*

Compass and Tripod

When you press ENTER at the **Specify a viewpoint or [Rotate] <display compass and tripod>** prompt, a compass and an axis tripod are displayed, as shown in Figure 25-21. You can directly set the viewpoint using this compass. The crosshairs appear on the compass. You can select any point on this compass to specify the viewpoint. The compass consists of two circles, a smaller circle and a bigger circle. Both these circles are divided into four quadrants: first, second, third, and fourth. The resultant view will depend upon the quadrant and the circle, on which you select the point. In the first quadrant, both the X and Y axes are positive; in the second quadrant, the X axis is negative and the Y axis is positive; in the third quadrant, both the X and Y axes are negative; and in the fourth quadrant, the X axis is positive and the Y axis is negative. Now, if you select a point inside the inner circle, it will be in the positive Z axis direction. If you select the point outside the inner circle and inside the outer circle, it will be in the negative Z axis direction. Therefore, if the previously mentioned statements are added, the following is concluded.

Z

Y

X

Axis Tripod

Crosshairs

Compass

Figure 25-21 *The axis tripod and the compass*

1. If you select a point inside the smaller circle in the first quadrant, the resultant view will be the Right, Back, and Top view. If you select a point outside the smaller circle and inside the bigger circle in this quadrant, the resultant view will be the Right, Back, and Bottom view (Figure 25-22).

2. If you select a point inside the smaller circle in the second quadrant, the resultant view will be the Left, Back, and Top view. If you select a point outside the smaller circle and inside the bigger circle in this quadrant, the resultant view will be the Left, Back, and Bottom view (Figure 25-22).

3. If you select a point inside the smaller circle in the third quadrant, the resultant view will be the Left, Front, and Top view. If you select a point outside the smaller circle and inside the bigger circle in this quadrant, the resultant view will be the Left, Front, and Bottom view (Figure 25-22).

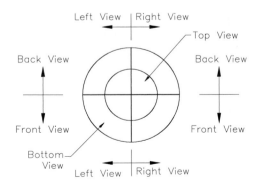

Figure 25-22 *The directions for the compass*

4. If you select a point inside the smaller circle in the fourth quadrant, the resultant view will be the Right, Front, and Top view. If you select a point outside the smaller circle and inside the bigger circle in this quadrant, the resultant view will be the Right, Front, and Bottom view (Figure 25-22).

Rotate

This option is similar to setting the viewpoint in the **Viewpoint Presets** dialog box, which is displayed by invoking the **DDVPOINT** command. When you select this option, you will be prompted to specify the angle in the *XY* plane and the angle from the *XY* plane. The angle in the *XY* plane will be taken from the *X* axis.

Changing the Viewpoint Using the Toolbar or the Ribbon

Apart from the **Viewpoint Presets** dialog box and the **VPOINT** command, you can also set the viewpoint either by using the **View** toolbar or by using the **Views** panel in the **View** tab of the **Ribbon**.

The **View** drop-down list (Figure 25-23) in the **View** panel consists of six preset orthographic views: **Top**, **Bottom**, **Left**, **Right**, **Front**, and **Back** and four preset isometric views: **SW Isometric**, **SE Isometric**, **NE Isometric**, and **NWIsometric**. This drop-down list also allows you to retrieve the previously set views or any previously created named views. All the above mentioned controls are also available in the **View Manager** dialog box that is invoked on selecting the **View Manager** option from the **View** drop-down list.

Choose the **New** button from the **View Manager** dialog box to create a new view; the **New View / Shot Properties** dialog box will be displayed, as shown in Figure 25-24. Enter a name in the **View name** edit box to save the view. You can save different types of views into different categories using the options in this dialog box.

Figure 25-23 The View drop-down list

The different types of views available in the **View type** drop-down list are **Still**, **Cinematic**, and **Recorded Walk**. But only **Still** is used for creating views and the other two are used for creating shots. The **View Properties** tab of this dialog box has three main areas: **Boundary**, **Settings** and **Background**. The **Boundary** area allows you to save the current display or you can select an area of display by choosing the **Define window** radio button. In the **Settings** area, you can control the position of UCS to save the view, select the section of the view, and control the display of the model. The drop-down list in the **Background** area is used to select the type of background to save the view. The preview of the selected background is displayed in the preview area next to the drop-down list. You need to select the **Save sun properties with view** check box to save the view with the sun properties.

Tip

*When you select any of the preset orthographic views from the **View** drop-down list, the UCS is also aligned to that view.*

Note

*The **View** panel in the **Home** tab is available only when the 3D Modeling workspace is activated.*

Changing the Viewpoint Using the ViewCube

You can change the view of the models easily and quickly using the ViewCube. The ViewCube

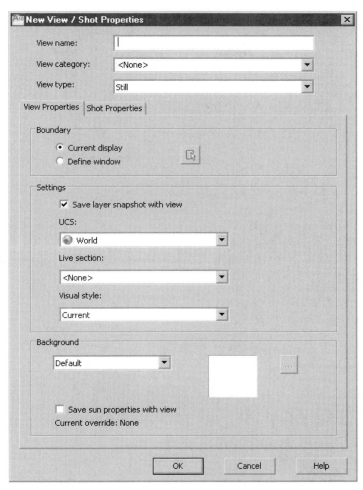

Figure 25-24 *The* ***New View / Shot Properties*** *dialog box*

is displayed when the **3D Modeling** workspace is enabled and it allows you to switch between the standard and isometric views, roll the current view, or return to the Home view of the model. A ViewCube consists of Cube, Home, Compass, and WCS menu. When you move the cursor over the ViewCube, it becomes active, and the area at the pointer tip gets highlighted. The highlighted area can be a face, a corner, or an edge of the ViewCube. These highlighted areas are called hotspots. There are 6 faces, 12 edges, and 8 corners on a cube see Figure 25-25. So, you can get 26 views using a ViewCube. You can select the required hotspot to restore a view. You can also go back to the Home view by clicking on the Home icon of the ViewCube. You can set any existing view as the Home view by choosing the **Set Current View as Home** option from the shortcut menu that is displayed on right-clicking on the ViewCube. By default, a model is displayed in the perspective view. To display the model in parallel projection, right-click on the ViewCube and choose **Parallel** from the shortcut menu.

Figure 25-25 *The ViewCube*

The compass on the ViewCube indicates the geographic location of the model. The **N**, **E**, **S**, and **W** alphabets on the compass indicate the North, East, South, and West directions of the 3D model. You can pick and drag the compass ring to rotate the current view in the same plane. You can choose between the **UCS** and **WCS** from the WCS menu and even create a new UCS from the menu. When you right-click on the ViewCube, a shortcut menu will be displayed. Choose the **ViewCube Settings** option from it; the **ViewCube Settings** dialog box will be displayed. (See Figure 25-26). This dialog box allows you to adjust the display of the compass ring and UCS menu, size, appearance, and location of the ViewCube.

 Note
*You can use the **View Cube** toggle button available in the **Views** panel of the **View** tab of the **Ribbon** to hide or unhide the ViewCube.*

The ViewCube is displayed only when you use the 3D templates such as acadISO -Named Plot Styles3D.dwt, acad3D.dwt, acadiso3D.dwt, or acad -Named Plot Styles3D.dwt templates.

Figure 25-26 *The **ViewCube Settings** dialog box*

3D COORDINATE SYSTEMS

Similar to the 2D coordinate systems, there are two main types of coordinate systems. They are as follows:

Absolute Coordinate System

This type of coordinate system is similar to the 2D absolute coordinate system, in which the coordinates of the point are calculated from the origin (0,0). The only difference is that here, along with the X and Y coordinates, the Z coordinate is also included. For example, to draw a line in 3D space starting from the origin to a point say 10,6,6, the procedure to be followed is given next.

> Command: **LINE**
> Specify first point: **0,0,0**
> Specify next point or [Undo]: **10,6,6**
> Specify next point or [Undo]: Enter

Figure 25-27 shows the model drawn using the absolute coordinate systems.

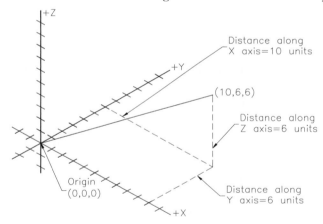

Figure 25-27 *Drawing a line from origin to 10,6,6*

Example 1

In this example, you will draw the 3D wireframe model shown in Figure 25-28.

1. Start a new file with the *acad3d.dwt* template in the **3D Modeling** workspace and make sure that the **Dynamic Input** button is not chosen. Choose **Home > Draw > Line** from the **Ribbon**. The prompt sequence is given next.

> Specify first point: **2,2,0**
> Specify next point or [Undo]: **8,2,0**
> Specify next point or [Undo]: **8,6,0**
> Specify next point or [Close/Undo]: **2,6,0**
> Specify next point or [Close/Undo]: **C**

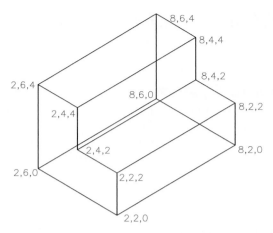

Figure 25-28 *3D wireframe model for Example 1*

2. Enter **DDVPOINT** at the Command prompt to display the **Viewpoint Presets** dialog box.

3. Set the value of the **X Axis** edit box to **225**.

4. Set the value of the **XY Plane** edit box to **38**, as shown in Figure 25-29. Choose **OK**.

5. Choose **Home > Draw > Line** from the **Ribbon**. The prompt sequence is as given next.

Specify first point: **2,2,0**
Specify next point or [Undo]: **2,2,2**
Specify next point or [Undo]: **2,4,2**
Specify next point or [Close/Undo]: **8,4,2**
Specify next point or [Close/Undo]: **8,2,2**
Specify next point or [Close/Undo]: **2,2,2**
Specify next point or [Close/Undo]: Enter

6. Choose **Home > Draw > Line** from the **Ribbon** or press the ENTER key to repeat the last command. The prompt sequence is given next.

Specify first point: **2,4,2**
Specify next point or [Undo]: **2,4,4**
Specify next point or [Undo]: **2,6,4**
Specify next point or [Close/Undo]: **2,6,0**
Specify next point or [Close/Undo]: Enter

Figure 25-29 *The **Viewpoint Presets** dialog box*

7. Enter **DDVPOINT** at the Command prompt to display the **Viewpoint Presets** dialog box.

8. Set the value of the **X Axis** edit box to **335**.

9. Set the value of the **XY Plane** edit box to **35**. Choose **OK**.

10. Choose **Home > Draw > Line** from the **Ribbon**. The prompt sequence is given next.

    Specify first point: **8,2,0**
    Specify next point or [Undo]: **8,2,2**
    Specify next point or [Undo]: Enter

11. Choose **Home > Draw > Line** from the **Ribbon** or press the ENTER key to repeat the last command. The prompt sequence is given next.

    Specify first point: **8,4,2**
    Specify next point or [Undo]: **8,4,4**
    Specify next point or [Undo]: **8,6,4**
    Specify next point or [Close/Undo]: **8,6,0**
    Specify next point or [Close/Undo]: Enter

12. Choose **Home > Draw > Line** from the **Ribbon** or press the ENTER key to repeat the last command. The prompt sequence is given next.

    Specify first point: **8,6,4**
    Specify next point or [Undo]: **2,6,4**
    Specify next point or [Undo]: Enter

13. Choose **Home > Draw > Line** from the **Ribbon** or press the ENTER key to repeat the last command. The prompt sequence is given next.
    Specify first point: **8,4,4**
    Specify next point or [Undo]: **2,4,4**
    Specify next point or [Undo]: Enter

14. Enter **VPOINT** at the Command prompt. The prompt sequence is given next.

    Current view direction: VIEWDIR=0.7424,-0.3462,0.5736
    Specify a view point or [Rotate] <display compass and tripod>: **-1,-1,1**

15. The final model should look similar to the one shown in Figure 25-30. You can also get a similar view of this model using the ViewCube. The view shown above is **Southwest Isometric**. Using the compass on the ViewCube, you can get the Southwest view of the solid model. The Southwest view of the model has three hotspots, bottom, edge, and top. You need to select the top hotspot to get the view similar to the above. Refer to Figure 25-31 for the Southwest hotspot. However, you can check different hotspots and view the model in different angles using the ViewCube.

Chapter 25

Figure 25-30 *Final 3D model for Example 1*   **Figure 25-31** *ViewCube showing Southwest Isometric*

Relative Coordinate System

There are the following three types of relative coordinate systems in 3D:

Relative Rectangular Coordinate System

This coordinate system is similar to the relative rectangular coordinate system of 2D, except that in 3D you also have to enter the Z coordinate along with the X and Y coordinates. The syntax of the relative rectangular system for the 3D is **@X coordinate, Y coordinate, Z coordinate**.

Relative Cylindrical Coordinate System

In this coordinate system, you can locate the point by specifying its distance from the reference point, the angle in the *XY* plane, and its distance from the *XY* plane. The syntax of the relative cylindrical coordinate system is **@Distance from the reference point in the XY plane < Angle in the XY plane from the X axis, Distance from the XY plane along the Z axis**. Figure 25-32 shows the components of the relative cylindrical coordinate system.

Relative Spherical Coordinate System

In this coordinate system, you can locate the point by specifying its distance from the reference point, the angle in the *XY* plane, and the angle from the *XY* plane. The syntax of the relative spherical coordinate system is **@Length of the line from the reference point < Angle in the XY plane from the X axis < Angle from the XY plane**. Figure 25-33 shows the components of the relative spherical coordinate system.

Tip
The major difference between the relative cylindrical and relative spherical coordinate system is that in the relative cylindrical coordinate system, the specified distance is the distance from the reference point in the XY plane. On the other hand, in the relative spherical coordinate system, the specified distance is the total length of the line in the 3D space.

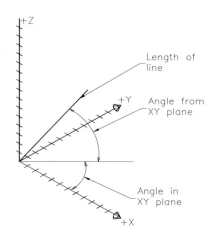

Figure 25-32 *Various components of the relative cylindrical coordinate system*

Figure 25-33 *Various components of the relative spherical coordinate system*

Example 2

In this example, you will draw the 3D wireframe model shown in Figure 25-34. Its dimensions are given in Figures 25-35 and 25-36.

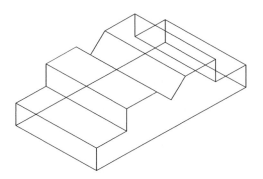

Figure 25-34 *Wireframe model for Example 2*

1. Start a new file in the *acad3d.dwt* template and choose **Views > View > SW Isometric** from the **View** tab of the **Ribbon**. Alternatively, select the Southwest top corner hotspot on the **View Cube** (refer to the compass of the ViewCube to get the Southwest view of the model).

2. As the start point of the model is not given, you can start the model from any point. Choose **Home > Draw > Line** from the **Ribbon** to invoke the **LINE** command. The prompt sequence is given next.

   Specify first point: **4,2,0**
   Specify next point or [Undo]: **@0,0,2**

Chapter 25

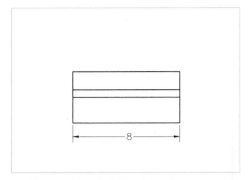

Figure 25-35 *Front view of the model*   **Figure 25-36** *Side view of the model*

Specify next point or [Undo]: **@3,0,0**
Specify next point or [Close/Undo]: **@0,0,2**
Specify next point or [Close/Undo]: **@3,0,0**
Specify next point or [Close/Undo]: **@2<0<315**
Specify next point or [Close/Undo]: **@2<0<45**
Specify next point or [Close/Undo]: **@3,0,0**
Specify next point or [Close/Undo]: **@0,0,-2**
Specify next point or [Close/Undo]: **@3,0,0**
Specify next point or [Close/Undo]: **@0,0,-2**
Specify next point or [Close/Undo]: **C**

3. Choose **Home > Draw > Line** from the **Ribbon**.. The prompt sequence is as follows:

Specify first point: **4,2,0**
Specify next point or [Undo]: **@0,8,0**
Specify next point or [Undo]: **@0,0,2**
Specify next point or [Close/Undo]: **@3,0,0**
Specify next point or [Close/Undo]: **@0,0,2**
Specify next point or [Close/Undo]: **@3,0,0**
Specify next point or [Close/Undo]: **@2<0<315**
Specify next point or [Close/Undo]: **@2<0<45**
Specify next point or [Close/Undo]: **@3,0,0**
Specify next point or [Close/Undo]: **@0,0,-2**
Specify next point or [Close/Undo]: **@3,0,0**
Specify next point or [Close/Undo]: **@0,0,-2**
Specify next point or [Close/Undo]: **@0,-8,0**
Specify next point or [Close/Undo]: Enter

4. Complete the model by joining the remaining edges using the **LINE** command.

5. The final 3D model should look similar to the one shown in Figure 25-37.

Figure 25-37 *Final 3D wireframe model for Example 2*

Exercise 1   *General*

In this exercise, you will draw the 3D wireframe model shown in Figure 25-38. Its dimensions are given in the figure itself. You can start the model from any point.

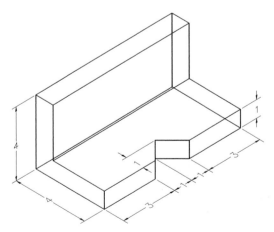

Figure 25-38 *Wireframe model for Exercise 1*

TRIM, EXTEND, AND FILLET COMMANDS IN 3D

As discussed in earlier chapters, you can use the options of these commands in 3D space. You can also use the **PROJMODE** and **EDGEMODE** variables. The **PROJMODE** variable sets the projection mode for trimming and extending. A value of 0 implies **True 3D mode**, that is, no projection. In this case, the objects must intersect in 3D space to be trimmed or extended. It is similar to using the **None** option of the **Project** option of the **EXTEND** and **TRIM** commands. A value of 1 projects onto the *XY* plane of the current UCS and is similar to using the **UCS** option of the **Project** option of the **TRIM** or **EXTEND** commands. A value of 2 projects onto the current view plane and is like using the **View** option of the **Project** option of the **TRIM** or **EXTEND** commands. The value of the **EDGEMODE** system variable controls the cutting or trimming boundaries. A value of 0 considers the selected edge with no extension. You can also use the **No extend** option of the **Edge** option of the **TRIM** or **EXTEND** commands. A value of 1 considers an imaginary extension of the selected edge. This is similar to using the **Extend** option of the **Edge** option of the **TRIM** or **EXTEND** commands.

You can fillet coplanar objects whose extrusion directions are not parallel to the *Z* axis of the current UCS, by using the **FILLET** command. The fillet arc exists on the same plane and has the same extrusion direction as the coplanar objects. If the coplanar objects to be filleted exist in opposite directions, the fillet arc will be on the same plane but will be inclined toward the positive *Z* axis.

SETTING THICKNESS AND ELEVATION FOR THE NEW OBJECTS

You can create the objects with a preset elevation or thickness using the **ELEV** command. This command is discussed next.

ELEV Command

Command:	ELEV

This is a transparent command and is used to set the elevation and thickness for new objects. The following prompt sequence is displayed, when you invoke this command.

> Specify new default elevation <0.0000>: *Enter the new elevation value.*
> Specify new default thickness <0.0000>: *Enter the new thickness value.*

Elevation

This option specifies the elevation value for new objects. Setting the elevation is nothing but moving the working plane from its default position. By default, the working plane is on the world *XY* plane. You can move this working plane using the **Elevation** option of the **ELEV** command. However, remember that the working plane can be moved only along the *Z* axis (Figure 25-39). The default elevation value is 0.0 and you can set any positive or negative value. All objects that will be drawn hereafter will be with the specified elevation. The Elevation option sets the elevation only for the new objects, and the existing objects are not modified using this option.

Thickness

This option specifies the thickness values for the new objects. It can be considered as another method of creating the surface models. Specifying the thickness creates extruded surface models. The thickness is always taken along the *Z* axis direction (Figure 25-40). The 3D faces will be automatically applied on the vertical faces of the objects drawn with a thickness. The **Thickness** option sets the thickness only for the new objects and the existing objects are not modified.

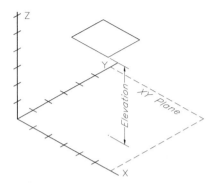

Figure 25-39 *Object drawn with elevation*

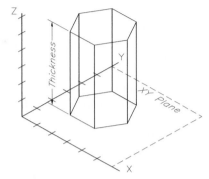

Figure 25-40 *Object drawn with thickness*

Note
*You can suppress the hidden edges in the model by entering **HIDE** at the Command prompt. The **HIDE** command is discussed in the next section.*

Tip
The elevation value will be reset to 0.0 when you change the coordinate system to WCS.

*The rectangles drawn using the **RECTANG** command do not consider the thickness value set using the **ELEV** command. To draw a rectangle with a thickness, use the **Thickness** option of the **RECTANG** command (Figure 25-41). To draw an ellipse with a thickness, first set the value of the **PELLIPSE** system variable to **1**.*

To write a text with a thickness, first write a simple text and then change its thickness using any of the commands that modify the properties (Figure 25-42).

Figure 25-41 *A point, line, ellipse, circle, and a polygon drawn with a thickness of 5 units*

Figure 25-42 *Text with and without thickness*

SUPPRESSING THE HIDDEN EDGES

Menu Bar: View > Hide   **Toolbar:** Render > Hide   **Command:** HIDE

Whenever you create a surface or a solid model, the edges that lie at the back (also called the hidden edges) are also visible. As a result, the model appears like a wireframe model. You need to manually suppress the hidden lines in the 3D model using the **HIDE** command. When you invoke this command, all objects on the screen are regenerated. Also, the 3D models are displayed with the hidden edges suppressed. If the value of the **DISPSILH** system variable is set to **1**, the model is displayed only with the silhouette edges. The internal edges that have facets will not be displayed. The hidden lines are again included in the drawing when the next regeneration takes place. Figure 25-43 shows surface models displaying the hidden edges. Figure 25-44 shows surface models without the hidden edges.

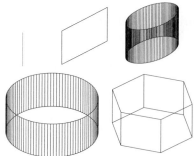

Figure 25-43 *3D models displaying all edges*

Figure 25-44 *Models without the hidden edges*

Tip
The **HIDE** command considers Circles, Solids, Traces, Polylines with width, Regions, and 3D Faces as opaque surfaces that will hide objects.

CREATING 3D POLYLINES

Ribbon: Home > Draw > 3d Polylines   **Menu Bar:** Draw > 3D Polyline
Command: 3DPOLY

The **3DPOLY** command is used to draw straight polylines in a 2D plane or 3D space. This command is similar to the **POLYLINE** command except that you can draw the polylines in a plane other than the *XY* plane also. However, this command does not provide the **Width** or the **Arc** option, and therefore you cannot create a 3D polyline with a width or an arc. The **Close** and the **Undo** options of this command are similar to those of the **POLYLINE** command.

Tip
*You can fit a spline curve about the 3D polyline using the **PEDIT** command. Use the **Spline curve** option to fit the spline curve and the **Decurve** option to remove it.*

CONVERTING WIREFRAME MODELS INTO SURFACE MODELS

You can convert a wireframe model into a surface model using the **3DFACE** command and the **PFACE** command. These commands are discussed next.

Creating 3D Faces

Menu Bar: Draw > Modeling > Meshes > 3D Face	**Command:** 3DFACE

This command is used to create 3D faces in space. You can create three-sided or four-sided faces using this command. You can specify the same or a different Z coordinate value for each point of the face. On invoking this command, you will be prompted to specify the first, second, third, and the fourth point of the 3D face. Once you have specified the first four points, it will again prompt you to specify the third and fourth point. In this case, it will take the previous third and fourth point as the first and second point, respectively. This process will continue until you press ENTER in the **Specify third point or [Invisible] <exit>** prompt. Keep in mind that the points must be specified in the natural clockwise or the counterclockwise direction to create a proper 3D face. Figure 25-45 shows a simple 3D wireframe model.

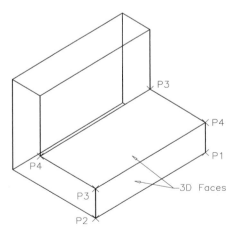

Figure 25-45 *3D faces applied to the wireframe model*

However, in case of complex wireframe models (Figure 25-46), it is not easy to apply 3D faces. This is because when you apply a 3D face, it generates edges about all the points that you specify. These edges will be displayed on the wireframe model, as shown in Figure 25-46. You can avoid these unwanted visible edges by applying the 3D faces with invisible edges using the **Invisible** option. It is very important to note here that the edge that will be created after you enter **I** at the **Specify next point or [Invisible] <exit>** prompt will not be the invisible edge. The edge that will be created after this edge will be the invisible edge. Therefore, to make the P3, P4 edge invisible, specify P1 as the first point and P2 as the second point. Before specifying P3 as the third point, enter **I** at the **Specify third point or [Invisible] <exit>** prompt. Similarly, follow the same procedure for creating the other invisible edges (Figure 25-47).

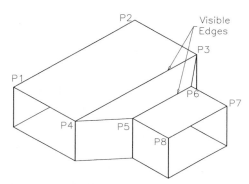

Figure 25-46 *Wireframe model with the edges*

Figure 25-47 *Wireframe model without the edges*

Creating Polyface Meshes

Command: PFACE

The **PFACE** command is similar to the **3DFACE** command. This command also allows you to create a mesh of any desired surface shape by specifying the coordinates of the vertices and assigning the vertices to the faces in the mesh. The difference between them is that on using the **3DFACE** command, you do not select the vertices that join another face twice. With the **PFACE** command, you need to select all the vertices of a face, even if they are coincident with the vertices of another face. In this way, you can avoid generating unrelated 3D faces that have coincident vertices. Also, with this command, there is no restriction on the number of faces and vertices the mesh can have.

Command: **PFACE**
Specify location for vertex 1: *Specify the location of the first vertex.*
Specify location for vertex 2 or <define faces>: *Specify the location of the second vertex.*
Specify location for vertex 3 or <define faces>: *Specify the location of the third vertex.*
Specify location for vertex 4 or <define faces>: *Specify the location of the fourth vertex.*
Specify location for vertex 5 or <define faces>: *Specify the location of the fifth vertex.*
Specify location for vertex 6 or <define faces>: *Specify the location of the sixth vertex.*
Specify location for vertex 7 or <define faces>: *Specify the location of the seventh vertex.*
Specify location for vertex 8 or <define faces>: *Specify the location of the eighth vertex.*
Specify location for vertex 9 or <define faces>: Enter

After defining the locations of all the vertices, press ENTER and assign vertices to the first face.

> Face 1, vertex 1:
> Enter a vertex number or [Color/Layer]: **1**
> Face 1, vertex 2:
> Enter a vertex number or [Color/Layer] <next face>: **2**
> Face 1, vertex 3:
> Enter a vertex number or [Color/Layer] <next face>: **3**
> Face 1, vertex 4:
> Enter a vertex number or [Color/Layer] <next face>: **4**

Once you have assigned the vertices to the first face, give a null response at the next prompt.

> Face 1, vertex 5:
> Enter a vertex number or [Color/Layer] <next face>:

You can also change the color and layer of the first face with the **PFACE** command. For example, to change the color of the first face to red, following is the prompt sequence.

> Face 1, vertex 5:
> Enter a vertex number or [Color/Layer] <next face>: **C**
> New Color [Truecolor/COlorbook] <BYLAYER>: **RED**

Now, the first face will be red.

After assigning the vertices, give a null response at the next prompt. AutoCAD provides you with the following prompt sequence, in which you need to assign vertices to the second face.

> Face 2, vertex 1:
> Enter a vertex number or [Color/Layer]: **1**
> Face 2, vertex 2:
> Enter a vertex number or [Color/Layer]: **4**
> Face 2, vertex 3:
> Enter a vertex number or [Color/Layer]: **5**
> Face 2, vertex 4:
> Enter a vertex number or [Color/Layer]: **4**

Once you have assigned the vertices to the second face, give a null response at the next prompt.

> Face 2, vertex 5:
> Enter a vertex number or [Color/Layer]:
> Face 3, vertex 1:
> Enter a vertex number or [Color/Layer]:

Similarly, assign vertices to all the faces (Figure 25-48). To make an edge invisible, specify a negative number for its first vertex. The display of the invisible edges of 3D solid surfaces is governed by the **SPLFRAME** system variable. If **SPLFRAME** is set to 0, invisible edges are not displayed. If this variable is set to a number other than 0, all invisible edges are displayed.

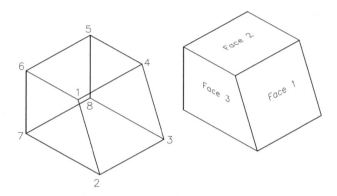

Figure 25-48 *Using the **PFACE** command*

Controlling the Visibility of the 3D Face Edges

Command: EDGE

This command is used to control the visibility of the selected edges of the 3D faces. You can hide the edges of the 3D faces or display them using this command (Figures 25-49 and 25-50). On invoking this command, you will be prompted to specify the 3D face edge to toggle its visibility or display. To hide an edge, select it. To display the edges, enter **D** at the **Specify edge of 3dface to toggle visibility or [Display]** prompt. You can display all the invisible 3D face edges or the selected edges using this option.

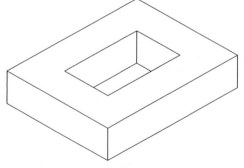

Figure 25-49 *Model with the visible 3D face edges*   *Figure 25-50* *Model after hiding the edges*

Exercise 2   *General*

In this exercise, you will apply the 3D faces to the 3D wireframe model created in Example 2.

DRAWING STANDARD SURFACE PRIMITIVES

Command: 3D

Apart from creating the surface using the previously mentioned commands, you can also create some standard 3D surface primitives like box, cone, dish dome, pyramid, torus, sphere, mesh, and wedge. These standard surface primitives can be created using the **3D** command. When you enter this command at the Command line, it will be executed using the Command line. You have to select the required surface using the Command line. While creating 3D meshes, you will not get a dynamic preview of the operations you are performing.

Creating a Box-Shaped Surface Mesh

This option is used to create a 3D polygon mesh in the shape of a box, see Figure 25-51. On selecting this option, you will be prompted to specify the corner of the box. Once you specify it, you will be prompted to specify the length, width, and the height of the box. To draw a cubical surface mesh, enter **C** at the **Specify width of box or [Cube]** prompt. The specified length will be taken as the length of the cubical surface mesh. Note that the length of the box will be taken along the *X* axis of the current UCS, the width along the *Y* axis, and the height along the *Z* axis. After specifying the length, width, and height, you will be prompted to specify the rotation angle of the box. You can also specify the rotation angle using the **Reference** option. The base point of the rotation in this case will be the corner of the box that you had specified.

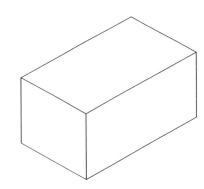

Figure 25-51 *3D box-shaped surface mesh*

Note
*You can also get all options of the Command prompt at the pointer input when the **Dynamic Input** button is active in the Status Bar.*

Creating a Cone-Shaped Surface Mesh

You can create a surface mesh model of a cone (Figure 25-52) by using any one of the methods for creating surface mesh models. If you use the **3D** command, the prompt sequence will be as follows:

Command: **3D**
Enter an option
[Box/Cone/DIsh/DOme/Mesh/Pyramid/Sphere/Torus/Wedge]: **CONE**

You can draw a truncated cone or a pointed cone as explained next.

Pointed Cone

Three parameters are required to draw a pointed cone:

1.   Center point of the base of the cone.
2.   Diameter/radius of the base of the cone.
3.   Height of the cone.

Choose the **Cone** option from the **3D** command. The prompt sequence for drawing a pointed cone is given next.

> Command: **AI_CONE**
> Specify center point for base of cone: *Specify the center of base.*
> Specify radius for base of cone or [Diameter]: *Specify the radius of the base.*

Figure 25-52   3D cone-shaped surface mesh

The next prompt asks you for the diameter/radius of the top. The default value is zero (0), which results in a pointed cone.

> Specify radius for top of cone or [Diameter] <0>: Enter
> Specify height of cone: *Specify the height of the cone.*
> Enter number of segments for surface of cone <16>: *Specify the number of segments for the cone.*

Note
The value of segments for the surface of cone can vary from 2 to 32767.

Truncated Cone

A truncated cone is a cone whose top side (pointed side) has been truncated (Figure 25-53). To draw a truncated cone, you require the following four parameters:

1.   Center point of the base of the cone.
2.   Diameter/radius of the base of the cone.
3.   Diameter/radius of the top of the cone.
4.   Height of the cone.

The prompt sequence for drawing a truncated cone is given next.

Figure 25-53   Truncated 3D cone-shaped surface mesh

> Command: **AI_CONE**
> Specify center point for base of cone: *Specify the center of base.*
> Specify radius for base of cone or [Diameter]: *Specify the radius of the base.*

Specify radius for top of cone or [Diameter] <0>: *Specify the radius of the top.*
Specify height of cone: *Specify the height of the cone.*
Enter number of segments for surface of cone <16>: *Specify the number of segments for the cone.*

Tip
To draw a cylindrical surface mesh, draw a truncated cone with an identical base and top diameters or radii.

Creating a Surface Mesh Dish

A dish is a hemisphere with the convex surface facing toward the bottom. You can create a surface mesh model of a dish (Figure 25-54) by using any one of the methods for drawing surface mesh models.

Command: **3D**
Enter an option
[Box/Cone/DIsh/DOme/Mesh/Pyramid/Sphere/Torus/Wedge]: **DI**
Specify center point of dish: *Specify the location of the center of the dish.*
Specify radius of dish or [Diameter]: *Specify the radius of the dish or enter D for the diameter.*
Enter number of longitudinal segments for surface of dish <16>: *Specify the number of longitudinal (vertical) segments for the dish.*
Enter number of latitudinal segments for surface of dish <8>: *Specify the number of latitudinal (horizontal) segments for the dish.*

Creating a Surface Mesh Dome

A dome is a hemisphere with the convex surface facing up (toward the top). You can create a surface mesh model of a dome (Figure 25-55) by using any one of the methods for drawing surface mesh models.

Command: **3D**
Enter an option
[Box/Cone/DIsh/DOme/Mesh/Pyramid/Sphere/Torus/Wedge]: **DO**

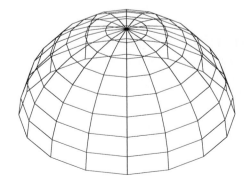

Figure 25-54 3D surface dish   *Figure 25-55* 3D surface dome

Specify center point of dome: *Specify the location of the center of the dome.*
Specify radius of dome or [Diameter]: *Specify the radius of the dome.*
Enter number of longitudinal segments for surface of dome <16>: *Specify the number of longitudinal (vertical) segments for the dome.*
Enter number of latitudinal segments for surface of dome <8>: *Specify the number of latitudinal (horizontal) segments for the dome.*

Creating a Planar Mesh

You can create a planar mesh (Figure 25-56) using any one of the methods for drawing surface mesh objects. The type of mesh obtained is also known as planar mesh because it is generated in a single plane. The prompt sequence is given next.

Command: **3D**
Enter an option
[Box/Cone/DIsh/DOme/Mesh/Pyramid/
Sphere/Torus/Wedge]: **M**

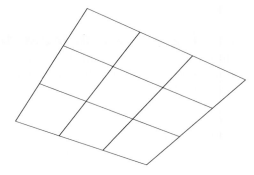

Figure 25-56 3D Planar surface mesh

In the next four prompts, AutoCAD asks you to specify the four corner points of the mesh.

Specify first corner point of mesh: *Specify the first corner point.*
Specify second corner point of mesh: *Specify the second corner point.*
Specify third corner point of mesh: *Specify the third corner point.*
Specify fourth corner point of mesh: *Specify the fourth corner point.*

Once you have specified the four corner points, AutoCAD prompts you to specify the number of rows (M) and the number of columns (N) of vertices. The number of vertices in both directions should be in the range of 2 to 256; that is, M and N should lie in the range of 2 to 256.

Enter mesh size in the M direction: *Enter the number of rows in the mesh.*
Enter mesh size in the N direction: *Enter the number of columns in the mesh.*

Creating a Pyramid Surface Mesh

You can create a surface mesh model of a pyramid by using any one of the methods for drawing surface meshes. Choose the **Pyramid** option from the **3D** command. Alternatively, you can enter **AI_PYRAMID** at the Command prompt.

Command: **AI_PYRAMID**
Specify first corner point for base of pyramid: *Specify the location of the first base point.*
Specify second corner point for base of pyramid: *Specify the location of the second base point.*
Specify third corner point for base of pyramid: *Specify the location of the third base point.*

Once you have located the three base points on the screen, the next prompt issued is **Specify fourth corner point for base of pyramid or [Tetrahedron]**. At this prompt, you are given the choice of specifying the fourth base point of the pyramid or constructing a tetrahedron. Actually, a pyramid is a polyhedron with any polygonal base (from 3 to n base sides) and all other sides as triangles that all meet at the top. Therefore, a tetrahedron or an Egyptian-style pyramid is a special case of pyramids, which can never have a flattened top in true geometrical terms. A more correct description of the flattened top case would be that AutoCAD enables you to create a truncated pyramid tetrahedron with a flattened top. The **Ridge** option simply contradicts the definition of a pyramid; it may be created with the **Pyramid** command, but it is not really a true pyramid.

Specify fourth corner point for base of pyramid or [Tetrahedron]: *Specify the location of the fourth base point.*

The next prompt issued is given below.

Specify apex point of pyramid or [Ridge/Top]:

At this prompt, you can select any one of the following options.

Pyramid using Apex Point Option

The **Apex point** is the default option(Figure 25-57). With this option, you can specify the location of the apex of the pyramid. You must enter the X, Y, Z coordinates to specify the location of the apex or use the .X, .Y, .Z filters. If you simply select a point on the screen, the X and Y coordinates of the apex will assume the X and Y coordinates of the point selected, and the elevation, or the Z coordinate, will be the same as the Z coordinate of the base points. Hence, the apex of the pyramid will have the same elevation as the base of the pyramid. Only if you are in the 3D space, for example, your viewpoint is (1,-1,1), you can specify the apex by selecting a point on the screen.

Pyramid using Ridge Option

This option (Figure 25-58) can be invoked by entering **R** at the **Specify apex point of pyramid or [Ridge/Top]** prompt. When you enter this option, the most recently drawn line

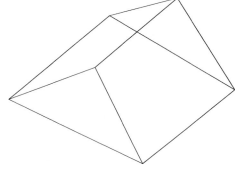

Figure 25-57 *Surfaced pyramid with an apex*   *Figure 25-58* *Surfaced pyramid with a ridge*

of the base is highlighted. This indicates that the first point on the ridge will be perpendicular to the highlighted line. The next prompts are given below.

Specify first ridge end point of pyramid: *Specify a point inside the highlighted line.*
Specify second ridge end point of pyramid: *Specify a point inside the second highlighted line.*

Pyramid using Top Option

You can invoke this option (Figure 25-59) by entering **T** at the **Specify apex point of pyramid or [Ridge/Top]** prompt. When you invoke this option, a rubber-band line is attached to the first corner of the base. The rubber-band line stretches from the corner point to the crosshairs. This allows you to select the first top point with respect to the first corner, the second top point with respect to the second corner, the third top point with respect to the third corner, and the fourth top point with respect to the fourth corner. You should use filters to locate the points. If you specify the same location (coordinates) for all four top points, the result is a regular pyramid with its apex at the location where all the four top points are located.

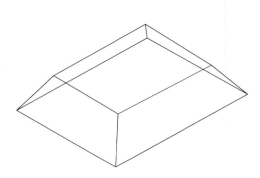

Figure 25-59 *Surfaced pyramid with a top*

Specify first corner point for top of pyramid: *Specify the location of the first top point.*
Specify second corner point for top of pyramid: *Specify the location of the second top point.*
Specify third corner point for top of pyramid: *Specify the location of the third top point.*
Specify fourth corner point for top of pyramid: *Specify the location of the fourth top point.*

Instead of entering the fourth base point for the pyramid, you can enter **T** for Tetrahedron at the **Specify fourth corner point for base of pyramid or [Tetrahedron]** prompt to draw a tetrahedron. You will construct two types of tetrahedrons. The first has four triangular sides and an apex, and the second has a flat top. You will first draw the tetrahedron with an apex and then a flat-topped tetrahedron.

Tetrahedron using Apex Option

As mentioned before, with this option you can draw a tetrahedron that has four triangular sides and an apex (Figure 25-60). Choose the **Pyramid** option from the **3D** command. The following is the prompt sequence to draw a tetrahedron with an apex.

Command: **AI_PYRAMID**
Specify first corner point for base of pyramid: *Specify the location of the first base point.*
Specify second corner point for base of pyramid: *Specify the location of the second base point.*
Specify third corner point for base of pyramid: *Specify the location of the third base point.*
Specify fourth corner point for base of pyramid or [Tetrahedron]: **T**
Specify apex point of tetrahedron or [Top]: *Specify the apex point.*

Tetrahedron using Top Option

With this option, you can draw a tetrahedron that has a flat top (Figure 25-61). This option can be invoked by entering **T** at the **Specify apex point of tetrahedron or [Top]** prompt.

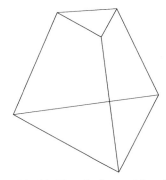

Figure 25-60 *Tetrahedron with an apex*   *Figure 25-61* *Tetrahedron with a flat top*

The next prompts displayed are given next.

> Specify apex point of tetrahedron or [Top]: **T**
> Specify first corner point for top of tetrahedron: *Specify the location of the first top point.*
> Specify second corner point for top of tetrahedron: *Specify the location of the second top point.*
> Specify third corner point for top of tetrahedron: *Specify the location of the third top point.*

Tip
*Specify the apex point and all the top points with the **.X**, **.Y**, **.Z** filters, or enter the exact coordinate values.*

Creating a Spherical Surface Mesh

The spherical surface mesh (Figure 25-62) has its central axis parallel to the Z axis, and the latitudinal lines are parallel to the *XY* plane of the current UCS. You can also append AI_ as a prefix to SPHERE and use it as a command. The prompt sequence is given next.

> Command: **3D**
> Enter an option
> [Box/Cone/DIsh/DOme/Mesh/Pyramid/Sphere/Torus/Wedge]: **S**
> Specify center point of sphere: *Specify the location of center point.*
> Specify radius of sphere or [Diameter]: *Specify the diameter/radius of the sphere.*
> Enter number of longitudinal segments for surface of sphere <16>: *Specify the number of longitudinal (vertical) segments for the sphere.*
> Enter number of latitudinal segments for surface of sphere <16>: *Specify the number of latitudinal (horizontal) segments for the sphere.*

Creating a Surface Mesh Torus

A torus is like a thick, hollow ring. You can create a surface mesh model of a torus (Figure 25-63) by using any one of the methods for drawing surface mesh models.

> Command: **3D**
> Enter an option
> [Box/Cone/DIsh/DOme/Mesh/Pyramid/Sphere/Torus/Wedge]: **T**
> Specify center point of torus: *Specify the center of the torus.*
> Specify radius of torus or [Diameter]: *Specify the radius of the torus.*
> Specify radius of tube or [Diameter]: *Specify the diameter of the tube.*
> Enter number of segments around tube circumference <16>: *Specify the number of segments for the tube.*
> Enter number of segments around torus circumference <16>: *Specify the number of segments for the torus.*

The *XY* plane of the current UCS is parallel to the torus. If the center point of the torus has a zero *Z* coordinate value (the center point of the torus lies in the present construction plane), then the *XY* plane bisects it.

Figure 25-62 Spherical surface mesh   *Figure 25-63* Surfaced torus

Creating a Surface Mesh Wedge

You can draw a surface mesh model of a right-angled wedge (Figure 25-64) by using any one of the methods for drawing the surface mesh models. For example, you can initiate the creation of a 3D surface mesh model of a wedge by entering the **3D** command at the Command prompt and then entering WEDGE at the next prompt. In the following example, you will draw a surface mesh model of a wedge using the **3D** command. Let the length of the wedge be 5 units, width 3 units, height 1 unit, and rotation 0-degree.

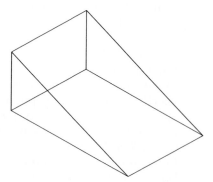

Figure 25-64 3D wedge-shaped surface mesh

Command: **3D**
Enter an option
[Box/Cone/DIsh/DOme/Mesh/Pyramid/Sphere/Torus/Wedge]: **W**
Specify corner point of wedge: *Specify the corner point of the wedge.*
Specify length of wedge: **5**
Specify width of wedge: **3**
Specify height of wedge: **1**
Specify rotation angle of wedge about the Z axis: *Specify rotation around the Z axis.*

The sloped side of the wedge is always opposite the first corner specified at the **Specify corner point of wedge:** prompt. The base of the wedge is always parallel to the construction plane (*XY* plane) of the current UCS. The height is measured parallel to the Z axis.

Tip
*To draw a wedge, in which the triangular sides are not parallel, use the **Ridge** option of **AI_PYRAMID** command.*

CREATING PLANAR SURFACES

| **Ribbon:** | Home > Modeling > Planar Surface | **Command:** | PLANESURF |
| **Toolbar:** | Modeling > Planar Surface | | |

 This command is used to generate 2D surfaces on the working plane by specifying two diagonally opposite points. These two points specify the rectangular shape area to be covered by this planar surface, see Figure 25-65. The prompt sequence for this command is discussed next.

Command: **PLANESURF**
Specify first corner or [Object] <Object>: *Specify the first corner point of the planar surface.*
Specify other corner: *Specify the diagonally opposite point of the planar surface.*

You can also convert an existing object to a surface by entering **O** at the **Specify first corner or [Object] <Object>** prompt. While selecting the object, you can directly select a closed boundary or a number of individual objects that result in a closed boundary, see Figure 25-66. The number of lines displayed in the surface is controlled by the **SURFU** and **SURFV** system variables. The **SURFU** variable controls the number of lines to be displayed in the M direction and the **SURFV** variable controls the number of lines to be displayed in the N direction.

THE 3DMESH COMMAND

| **Ribbon:** | Home > 3D Modeling > 3D Mesh | **Command:** | 3DMESH |
| **Menu Bar:** | Draw > Modeling > Meshes > 3D Mesh | | |

 This command is used to create a free-form polygon mesh using a matrix of M X N size. When you invoke this command, you will be prompted to specify the size of the mesh in the M direction and the N direction, where M X N is the total number of

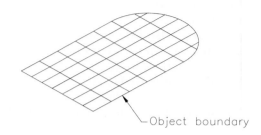

Figure 25-65 *Planar surface created by specifying corner points*

Figure 25-66 *Planar surface created by selecting an object*

vertices in the mesh. Once you have specified the size, you need to specify the coordinates of each and every vertex individually. For example, if the size of the mesh in the M and N directions is 12 X 12, then you will have to specify the coordinates of all the 144 vertices. You can specify any 2D or 3D coordinates for the vertices. Figure 25-67 shows a 4 X 4 freeform mesh. This command is mostly used for programming. For a general drawing, use the **Mesh** option of the **3D** command. The 3D polygon mesh is always open in the M and N directions. You can edit this mesh or the mesh created using the **3D** command or the **PEDIT** command.

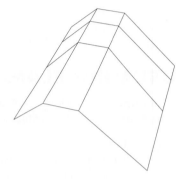

Figure 25-67 *3D mesh*

EDITING THE SURFACE MESH

The surface meshes can be edited using the **PEDIT** command, which is discussed next.

The PEDIT Command

Ribbon:	Home > Modify > Edit Polyline	**Menu Bar:**	Modify > Object > Polyline
Toolbar:	Modify II > Edit Polyline	**Command:**	PEDIT

You can select the **Polyline Edit** option from the shortcut menu displayed on selecting the surface mesh and right-clicking on it. The behavior of the **PEDIT** command differs, depending on whether the object selected is a line, polyline, or surface mesh. The prompt sequence that will be followed when you choose this button is given next.

Select polyline or [Multiple]: *Select the surface mesh.*
Enter an option [Edit vertex/Smooth surface/Desmooth/Mclose/Nclose/Undo]: *Select an option.*

Edit Vertex

This option is used for individual editing of the vertices of the mesh in the M direction or in the N direction. When you invoke this option, a cross will appear on the first point of the mesh and the following sub-options will be provided.

Next. This sub-option is used to move the cross to the next vertex.

Previous. This sub-option is used to move the cross to the previous vertex.

Left. This sub-option is used to move the cross to the previous vertex in the N direction of the mesh.

Right. This sub-option is used to move the cross to the next vertex in the N direction of the mesh.

Up. This sub-option is used to move the cross to the next vertex in the M direction of the mesh.

Down. This sub-option is used to move the cross to the previous vertex in the M direction of the mesh.

Move. This sub-option is used to redefine the location of the current vertex. You can define the new location using any of the coordinate systems or you can directly select the location on the screen, see Figures 25-68 and 25-69.

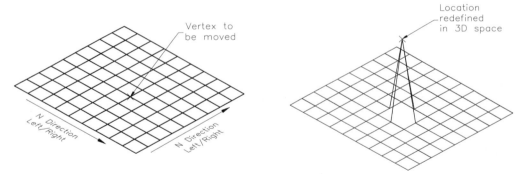

Figure 25-68 *Vertex before moving*   *Figure 25-69* *Vertex after moving*

REgen. This sub-option is used to regenerate the 3D mesh.

eXit. This sub-option is used to exit the **Edit Vertex** option.

Smooth surface

This option is used to smoothen the surface of the 3D mesh by fitting a smooth surface, as shown in Figures 25-70 and 25-71. The smoothness of the surface will depend upon the **SURFU** and **SURFV** system variables. The value of these system variables can be set between 2 and 200.

The type of the smooth surface to be fitted is controlled by the **SURFTYPE** system variable. If you set the value of the **SURFTYPE** system variable to **5**, it will fit a Quadratic B-spline

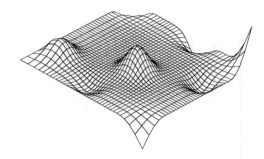

Figure 25-70 *3D mesh before fitting the smooth surface*

Figure 25-71 *3D mesh after fitting the smooth surface*

surface. If you set the value to **6**, it will fit a Cubic B-spline surface. If you set the value to **7**, it will fit a Bezier surface.

Desmooth

This option is used to remove the smooth surface that was fitted on the 3D mesh using the **Smooth surface** option and restore to the original mesh.

Mclose

This option is used to close the 3D mesh in the M direction, as shown in Figures 25-72 and Figure 25-73.

Nclose

This option is used to close the 3D mesh in the N direction, as shown in Figures 25-72 and 25-73.

Mopen

This option will be available if the 3D mesh is closed in the M direction. It is used to open the 3D mesh in the M direction (Figures 25-72 and 25-73).

Nopen

This option will be available if the 3D mesh is closed in the N direction. It is used to open the 3D mesh in the N direction (Figures 25-72 and 25-73).

Undo

This option is used to undo all the operations to the start of the **PEDIT** command.

eXit

This option is used to exit the **PEDIT** command.

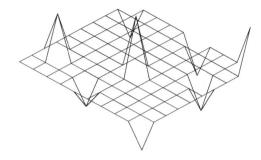

Figure 25-72 3D mesh open in the M and N
directions

Figure 25-73 3D mesh closed in the M and N
directions

DYNAMIC VIEWING OF 3D OBJECTS

Status Bar:	SteeringWheels	**Ribbon:**	View > Navigate > SteeringWheels
Command:	NAVSWHEEL	**Menu Bar:**	View > SteeringWheels

The advanced options for the dynamic viewing of solid models are one of the major enhancements in the recent releases of AutoCAD. These options make the entire process of the solid modeling very interesting. You can shade the solid models and dynamically rotate them on the screen to view them from different angles. You can also define clipping planes and then rotate the solid models such that whenever them pass through the cutting planes, they are sectioned just to be viewed. The original model can be restored as soon as you exit these commands. The options and commands used to perform all these functions are discussed next.

Using the SteeringWheels

The SteeringWheels enables you to easily view and navigate through your 3D models. It is an ideal tool for navigating through your models. It includes not only the common navigating tools such as **Zoom**, **Pan**, and **Orbit** but also an option to move the user specified center of the model to the center of the display window. Using this tool, you can also rewind and get back your previous views. It is a time saving tool, since it combines many common navigation tools into a single tool.

The SteeringWheel is divided into different sections known as Wedges. Each wedge on a wheel represents a command. You can activate a particular navigation option by clicking on the respective wedges and then modify the view of the model by dragging the cursor, refer to Figure 25-74. The navigation tools available in the SteeringWheels are discussed next.

Figure 25-74 The SteeringWheel

Chapter 25

ZOOM

This wedge is used in the same way as you use camera's zoom lens. It makes the object appear closer or far away, without changing the position of the camera. To invoke the **ZOOM** tool, move the cursor to the **ZOOM** wedge and drag. A magnifier is attached to the cursor that allows you to zoom in or out with reference to a base point. However, you can change the base point or pivot point to zoom in or zoom out of the model by choosing the **CENTER** wedge. You can even enable the incremental zoom of 25 percent on a single click from the **Zoom** area of the **SteeringWheels Settings** dialog box. To invoke this dialog box, right-click on the SteeringWheels; a shortcut menu will be displayed. Choose the **SteeringWheels Settings** option from the shortcut menu; the **SteeringWheels Settings** dialog box will be displayed. You can even choose the **Fit to Window** option from the shortcut menu of the SteeringWheels to view the entire drawing in the viewport.

REWIND

This wedge enables you to quickly restore the previous views. To invoke this tool, move the cursor to the **REWIND** wedge and click; a series of frames from the previous view orientation are displayed. You can now quickly pick the required view and control the thumbnail previews generated on changing various views using the **Rewind thumbnail** area of the **SteeringWheels Settings** dialog box.

PAN

This wedge allows you to drag the view to a new location in the drawing window. To invoke this tool, move the cursor to the **PAN** wedge and drag. A plus arrow is attached to the cursor, enabling you to drag the view in any direction. You can choose the **Fit to Window** option from the shortcut menu of the SteeringWheels to view the entire drawing in the viewport.

ORBIT

This wedge allows you to visually maneuver around the 3D objects to obtain different views. When you move your cursor to the **ORBIT** wedge and drag, a circular arrow is attached to the cursor. Now, you can rotate the model with respect to a base point. Similar to the **ZOOM** option, in this case also, you can specify the required center point of the view and rotate the model using the **CENTER** wedge option. You can also set the viewpoint upside down using the options in the **SteeringWheels Settings** dialog box. Moreover, you can choose the **Fit to Window** option to view the full drawing view.

CENTER

This wedge helps you specify the pivot or center point of the view to orbit or zoom a model. When you move your cursor to this wedge and drag, a pivot ball is attached to the cursor. You can now adjust your center or target point by placing the cursor at the required point. However, you can restore the center of the viewport by choosing the **Restore Original Center** option from the shortcut menu of the SteeringWheels. You can also restore the home view of the model by choosing the **Go Home** option from the shortcut menu.

LOOK

This tool works similar to turning the camera on a tripod without changing the distance between the camera and the target. This means the source remains the same, but the target changes. To invoke this tool, move the cursor to the **LOOK** wedge; an arc is attached to the cursor which enables you to change your view direction. You can also invert the vertical direction of the **Look** tool by selecting the **Invert vertical axis for Look tool** check box from the **SteeringWheels Settings** dialog box. You can choose the **Fit to Window** option to view the entire drawing in the viewport.

WALK

This tool moves the camera around the model as if you are walking away or towards the focus direction of the camera. In the walk 3D navigation, the camera moves parallel to the XY plane and gives the effect as if you are walking. To invoke this tool, move the cursor to the **WALK** wedge. An arrow is attached to the cursor showing the direction of the camera. You can drag the cursor to move around the model and also control the walk speed using the options in the **Walk Tool** area of the **SteeringWheels Settings** dialog box. Using this dialog box, you can also constraint the walk angle to the ground plane.

UP/DOWN

This option slides the view along the Y-axis of the screen (which is like being in an elevator). When you move your cursor to the **UP/DOWN** wedge, a vertical slider is attached to the cursor. You can slide from up to down in the slider to view the model from up to down. Using the options in the shortcut menu of the SteeringWheels, you can go back to the original view, restore the original center, or fit the entire drawing in the viewport.

Modes of SteeringWheels

There are four working modes of SteeringWheels, see Figure 25-75. These working modes are discussed next.

Figure 25-75 Different modes of the SteeringWheels

2D Navigation Wheel

You can invoke the SteeringWheel that has only the 2D navigation tools. To do so, choose **View > Navigate > SteeringWheels > 2D** from the **Ribbon**; the 2D SteeringWheel will be displayed with **PAN**, **ZOOM**, and **REWIND** tools.

View Object Wheel

This mode contains only the viewing tools such as **CENTER**, **ZOOM**, **REWIND**, and **ORBIT**. You can invoke this mode by choosing **Basic Wheels > View Object Wheel** from the shortcut menu of the SteeringWheels or by choosing **View > Navigate > SteeringWheels > Basic View Object** from the **Ribbon**.

Tour Building Wheel

This mode is ideal for navigating inside the building. It includes **FORWARD**, **REWIND**, **LOOK**, and **UP/DOWN** tools. You can invoke this mode by choosing **Basic Wheels > Tour Building Wheel** from the shortcut menu of the SteeringWheels or by choosing **View > Navigate > SteeringWheels > Basic Tour Building** from the **Ribbon**.

Full Navigation Wheel

This mode contains both the view and the navigating tools. It enables you to select a variety of tools from a single wheel. You can invoke this tool by choosing the **Full Navigation Wheel** option from the shortcut menu of the SteeringWheels or by choosing **View > Navigate > SteeringWheels > Full Navigation** from the **Ribbon**.

The SteeringWheels are of two types: Big Wheels and Mini Wheels. Both the wheels have the same functioning modes. The Big Wheels mode is displayed by default, whereas the Mini Wheels mode is invoked by choosing the **Mini View Object Wheel**, **Mini Tour Building Wheel**, or **Mini Full Navigation Wheel** option from the SteeringWheels shortcut menu or by choosing appropriate buttons from the flyout that is displayed on choosing **View > Navigate > SteeringWheels** from the **Ribbon**. You can customize the size, appearance, and behavior of the SteeringWheel using the **SteeringWheels Settings** dialog box, refer to Figure 25-76. You can control the size and opacity of the Big Wheels and Mini Wheels with the help of a slider available next to the wheels. Also, using this dialog box you can control the tooltips and messages that are displayed when you move the cursor over any wedges.

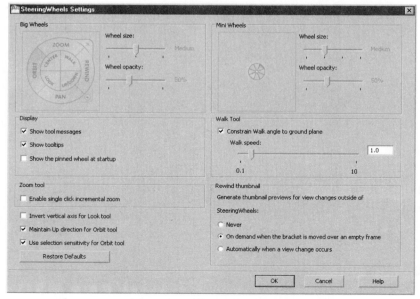

Figure 25-76 *The **SteeringWheels Settings** dialog box*

Dynamically Rotating the View of the Model

Ribbon: View > Navigate > Orbit   **Menu Bar:** View > Orbit > Constrained Orbit
Toolbar: 3D Navigation > Constrained Orbit   **Command:** 3DORBIT

 The **3DORBIT** command allows you to visually maneuver around 3D objects to obtain different views. This is one of the most important commands for the advanced 3D viewing options. All other advanced 3D viewing options can be invoked inside this command or using the **3D Navigation** toolbar, see Figure 25-77. This command activates a 3D Orbit view in the current viewport. You can click and drag your pointing device to view the 3D object from different angles. In 3D orbit viewing, the target is considered stationary and the camera location is considered to be moving around it. The cursor looks like a sphere encircled by two arc shaped arrows. This is known as **Orbit mode** cursor, and clicking and dragging the pointing device allows you to rotate the view freely. You can move the **Orbit mode** cursor horizontally, vertically, and diagonally. If you drag the pointing device horizontally, the camera will move parallel to the XY plane of the WCS. If you move your pointing device vertically, then the camera will move along the Z axis. The **3DORBIT** command is a transparent command and can be invoked inside any other command. You can select individual objects or the entire drawing to view before you invoke the **3DORBIT** command.

*Figure 25-77 The **3D Navigation** toolbar*

Tip
Press and hold the SHIFT key and the middle mouse button to temporarily enter the constrained orbit mode.

You can invoke the other advanced viewing options using the shortcut menu that is displayed upon right-clicking in the drawing window when you are inside the **3DORBIT** command. The shortcut menu that is displayed is shown in Figure 25-78. The options of this shortcut menu are discussed next.

Exit
This option is used to exit the **3DORBIT** command. You can also exit this command by pressing ESC.

Current Mode
This option displays the currently active navigation mode in which you obtained this shortcut menu.

*Figure 25-78 The shortcut menu displayed when the **3DORBIT** command is active*

Other Navigation Modes

This option enables you to choose other navigation modes of the **3D Navigation** toolbar. In this way, you can toggle between the navigation modes without exiting the command. The suboptions provided by the **Other Navigation Modes** are as follows:

Free Orbit. This suboption can be invoked by entering the **3DFORBIT** command or by choosing **View > Navigate > Orbit > Free Orbit** from the **Ribbon**. This command activates a 3D orbit view in the current viewport. When you invoke the **3DFORBIT** command, an arcball appears. This arcball is a circle with four smaller circles placed such that they divide the bigger circle into quadrants. The UCS icon is replaced by a shaded 3D UCS icon. In the 3D orbit viewing, the target is considered stationary and the camera location is considered to be moving around it. The target is the center of the arcball. The **3DFORBIT** command is a transparent command and can be invoked inside any other command.

The cursor icon changes as you move the cursor over the 3D Orbit view. The different icons indicate the different directions in which the view is being rotated. They are as follows.

Orbit mode. When you move the cursor within the arcball, the cursor looks like a sphere encircled by two lines. This icon is the **Orbit mode** cursor, and clicking and dragging the pointing device allows you to rotate the view freely as if the cursor was grabbing a sphere surrounding the objects and moving it around the target point. You can move the **Orbit mode** cursor horizontally, vertically, and diagonally.

Roll mode. When you move the cursor outside the arcball, it changes to look like a sphere encircled by a circular arrow. When you click and drag, the view is rotated around an axis that extends through the center of the arcball, perpendicular to the screen.

Orbit Left-Right. When you move the cursor to the small circles placed on the left and right of the arcball, it changes into a sphere surrounded by a horizontal ellipse. When you click and drag, the view is rotated around the *Y* axis of the screen.

Orbit Up-down. When you move the cursor to the circles placed in the top or bottom of the arcball, it changes into a sphere surrounded by a vertical ellipse. Clicking and dragging the cursor rotates the view around the horizontal axis or the *X* axis of the screen, and passes it through the middle of the arcball.

Figure 25-79 shows a model being rotated using the **3DFORBIT** command.

Figure 25-79 *3D orbit view, arcball with circular graphics, and the 3D shaded UCS icon*

Tip
*Press and hold the SHIFT+CTRL keys and the middle button of the mouse to temporarily enter the **Free Orbit** navigation mode.*

*If you are in the **3D Orbit** navigation mode, press and hold the SHIFT key to temporarily enter the **Free Orbit** navigation mode.*

Continuous Orbit. This suboption can be invoked by entering the **3DCORBIT** command. It allows you to set the objects you select in a 3D view into continuous motion in a free-form orbit. You can also invoke this command by choosing **View > Navigate > Orbit > Continuous Orbit** from the **Ribbon**. While using this suboption, the cursor icon changes to the continuous orbit cursor. Clicking in the drawing area and dragging the pointing device in any direction starts to move the object(s) in the direction you have dragged it. When you release the pointing device button, that is, stop clicking and dragging, the object continues to move in its orbit in the direction you specified. The speed at which the cursor is moved determines the speed at which the objects spin. You can click and drag again to specify another direction for rotation in the continuous orbit. While using this suboption, you can right-click in the drawing area to display the shortcut menu and choose the other suboptions without exiting the command. Choosing **Pan**, **Zoom**, **Orbit**, **Adjust Clipping Plane** ends the continuous orbit.

Adjust Distance. This suboption can be invoked by entering the **3DDISTANCE** command or by choosing **other Navigation Modes > Adjust Distance** from the shortcut menu. Alternatively, choose **View > Camera > Adjust Distance > 3D Adjust Distance** from the menu bar. This suboption is similar to taking the camera closer to or farther away from the target object. Therefore, it makes the object appear closer or farther away. You can press the left button and drag the mouse to adjust the distance between the target and the camera. This option is used to reduce the distance between the camera and the object when the object rotates by a large angle in the **3DORBIT** command.

Swivel Camera. This suboption can be invoked by entering the **3DSWIVEL** command or by choosing **View > Camera > Swivel** from the menu bar. It works similar to turning the camera on a tripod without changing the distance between the camera and the target. You can press the left button and drag the mouse to swivel the camera.

Walk. This suboption can be invoked by entering the **3DWALK** command or by choosing **Render > Animations > Walk** from the **Ribbon** after customizing. It is similar to the **Walk** option in the SteeringWheel. This suboption is used to move the camera around the model with the help of a keyboard, as if you are walking away or toward the focus direction of the camera. In walk 3D navigation, the camera moves parallel to the XY plane and this seems as if you are walking. Choose the **Walk** button; the **POSITION LOCATOR** window, showing the top view of the model with the location of the camera indicated by a red spot, will be displayed; the **POSITION LOCATOR** window gives a dynamic preview of the location and movement of the camera with respect to the object, see Figure 25-80.

Chapter 25

Figure 25-80 *The **POSITION***
LOCATOR *window*

The following keys are used to control the location of the camera with respect to the object.

Motion Type	Key
Move the camera in the forward direction	UP ARROW or W key
Move the camera in the backward direction	DOWN ARROW or S key
Move the camera in the left direction	LEFT ARROW or A key
Move the camera in the right direction	RIGHT ARROW or D key
Change the focus direction of the camera	Press and hold the left mouse button in the drawing area and drag the mouse in the desired direction
Change the orientation of the XY plane	Press and hold the middle mouse button in the drawing area and drag the mouse in the desired direction
Change the distance of the camera from the object	Move the cursor to the green triangular area in the **POSITION LOCATOR** window and pan to the desired location
Change the coverage area of the camera	Move the cursor onto the camera in the **POSITION LOCATOR** window and pan the camera

Fly. This suboption can be invoked by entering the **3DFLY** command or by choosing **Render > Animations > Walk > Fly** from the **Ribbon** after customizing. This suboption is used to move the camera through the model with the help of a keyboard as if you are flying away or toward the focus direction (Figure 25-80) of the camera. In the fly 3D navigation, the camera can even move at an inclination with the XY plane and this seems as if you are flying around the object. The **POSITION LOCATOR** window will be displayed on the screen in **3DFLY** command also. The position of the camera with respect to the object can be controlled by using the same procedure as discussed in the **3DWALK** command.

Note
To control the settings related to walk and fly, invoke the command and right-click. Choose the ***Walk and Fly Settings*** *option; the* ***Walk and Fly Settings*** *dialog box will be displayed. You can specify the settings in this dialog box.*

The ***Walk*** *and* ***Fly*** *3D navigation modes can be activated only in the perspective projection. Activating the* ***Walk*** *or* ***Fly*** *3D navigation mode in the parallel projection will display the* ***Warning*** *window, as shown in Figure 25-81.*

Figure 25-81 The ***Warning*** *window for activating the* ***Walk*** *and* ***Fly*** *3D navigation modes in the parallel projection*

 Tip
You can toggle between the 3D navigation modes from the shortcut menu or by entering the number displayed in front of each option in the shortcut menu.

Zoom. This suboption can be invoked by entering the **3DZOOM** command. This command can also be invoked by choosing **View > Zoom** from the menu bar. It is similar to the **ZOOM** option in the SteeringWheel and functions like the camera's zoom lens. You can invoke the other options of the Zoom by entering **3DZOOM** at the Command prompt. Alternatively, the options can also be displayed by right-clicking in the drawing area when you are inside this command.

Pan. The **Pan** suboption is invoked by entering the **3DPAN** command. This command can also be invoked by choosing **View > Pan** from the menu bar. It is similar to the **PAN** option in the SteeringWheel. This command also starts the 3D interactive viewing. You can right-click in the drawing area, when you are inside this command, to display a shortcut menu that displays all the 3D Orbit options.

Enable Orbit Auto Target
This option, if chosen, ensures that the target point is focused on the object while rotating

the view. If this option is cleared, the target point is set at the center of the viewport, in which the **3DORBIT** command is invoked.

Animation Settings

This option, if chosen, displays the **Animation Settings** dialog box, see Figure 25-82. In this dialog box, you can change the settings for visual style, resolution, number of frames to be captured in one second, and the format in which you want to save the file while creating the animation of 3D navigation. Creating animations will be discussed in the next chapter.

Zoom Window/Extents/Previous

These options are used to zoom the solid model using a window, zoom the objects to the extents, or retrieve the previous zoomed views of the model.

Parallel

This option is chosen to display the selected model using the parallel projection method. It is the default method when a new AutoCAD file is started with 2D templates. In this method, no parallel lines in the model converge at any point.

Perspective

This option is chosen to use the one point perspective method to display the model. It is the default method when a new AutoCAD file is started with 3D templates. In this method, all the parallel lines in the model converge at a single point, thus providing the realistic view of the model when viewed with the naked eye, see Figure 25-83.

Figure 25-82 *The **Animation Settings** dialog box*

Figure 25-83 *Viewing the model using the perspective projection*

Reset View

This option, when chosen automatically, restores the view that was current when you started rotating the solid model in the 3D orbit. Note that you will not exit the **3DORBIT** command if you choose this option.

Preset Views

This option is used to make current any of the six standard orthogonal views or the four isometric views.

Named Views

If you have created some views using the **VIEW** command, then the existing named views will be displayed as the suboptions under the **Named Views** option in the **3D Orbit** shortcut menu. Use this option to activate the saved view.

Visual Styles

This option is invoked by the **SHADEMODE** command and the suboptions provided under this option are used to specify the mode of shading of the model while it is being rotated in the 3D orbit. The suboptions used to shade the solid model are discussed next.

3DHidden. This suboption is used to display the solid model after hiding the hidden lines. You will not be able to see through the solid model.

3DWireframe. This suboption is used to display the solid model as a wireframe model. You can see through the solid model while the model is being rotated in the 3D orbit view.

Conceptual. This suboption is used to display different faces of the model with a combination of two different colors. The color on the faces of the model keeps changing as you change the view of the model. The edges between the faces are smooth and at the same time, the tangent edges are hidden. The **Conceptual** visual style gives a better visualization of different faces on the model, see Figure 25-84.

Realistic. This suboption applies the color on all faces of the solid model. The color of the shade will depend on the shade of the solid model. The edges between the faces are smooth and the tangent edges are made visible (Figure 25-85). The material that you attach to the object is also displayed in this visual style.

Figure 25-84 Shading using the **Conceptual** shade

Figure 25-85 Shading using the **Realistic** shade

Visual Aids

The **3DORBIT** command provides you with three visual aids for the ease of visualizing the solid model in the 3D orbit view. These visual aids are **Compass**, **Grid**, and **UCS Icon**. A check mark is displayed in front of the options you have selected. You can select none, one, two, or all three of the visual aids. Choosing **Compass** displays a sphere drawn within the arcball with three lines. These lines indicate the X, Y, and Z axis directions. Choosing the

Grid displays a grid at the current *XY* plane. You can specify the height by using the **ELEVATION** system variable. The **GRID** system variable controls the display options of the **Grid** before using the **3DORBIT** command. The size of the grid will depend upon the limits of the drawing. Choosing the **UCS Icon** displays the UCS icon. If this option is not selected, then AutoCAD turns off the display of the UCS icon. While using any of the 3D Orbit options, interactive 3D viewing is on and a shaded 3D view UCS icon is displayed. The *X* axis is red, the *Y* axis is green, and the *Z* axis is blue or cyan. Figure 25-86 shows a realistic shaded solid model displaying the compass and grid, as well as the UCS icon.

Figure 25-86   Realistic shaded solid model showing compass, grid, and UCS icon

Clipping the View of the Model Dynamically

Command: 3DCLIP

3DCLIP command is used to adjust the clipping planes for sectioning the selected solid model. Note that the actual solid model is not modified by this command. This command is used only for the purpose of viewing. The original solid model will be restored when you exit this command. With this option, you can adjust the location of the clipping planes. When you choose this option, the **Adjust Clipping Planes** window will be displayed, as shown in Figure 25-87. In this window, the object appears rotated at 90-degree to the top of the current view in the window. This facilitates the display of cutting planes. Setting the location of the front and back clipping planes in this window is reflected in the result in the current view. The various options of the **3DCLIP** command are displayed in the **Adjust Clipping Planes** toolbar in the window. There are seven buttons provided in this window. The functions of these buttons are discussed next.

 Adjust Front Clipping. This button is chosen to locate the front clipping plane, which is defined by the line located in the lower end of the **Adjust Clipping Planes** window. You can see the result in the 3D Orbit view if the front clipping is on.

 Adjust Back Clipping. This button is chosen to locate the back clipping plane, which is defined by the line located in the upper end of the **Adjust Clipping Planes** window. You can see the result in the 3D Orbit view if the back clipping is on.

Create Slice. Choosing this button causes both the front and back clipping planes to move together. The slice is created between the two clipping planes and is displayed in the current 3D view. You can first adjust the front and the back clipping planes individually by choosing the respective options. Next, by choosing the **Create Slice** button, activate both the clipping planes simultaneously and display the result in the current view.

Pan. Displays the pan cursor that you can use to pan the object in **Adjust Clipping Planes** window. Hold down the left mouse button and drag it in any direction. The pan cursor stays active until you click another button.

 Zoom. Displays a magnifying-glass cursor that you can use to enlarge or reduce the clipping plane. To enlarge the image in the **Adjust Clipping Planes** window.

 Front Clipping On/Off. Turns on or off the sectioning of the solid model using the front clipping plane, see Figure 25-88. You can toggle between front clipping on and off by choosing the **Front Clip On/Off** button in the **3D Navigation** toolbar.

 Back Clipping On/Off. Turns on or off the sectioning of the solid model using the back clipping plane. You can toggle between back clipping on and off by choosing the **Back Clip On/Off** button from the **3D Navigation** toolbar.

*Figure 25-87 The **Adjust Clipping Planes** window*

Figure 25-88 The interior of a house displayed by setting the front clipping on and then hiding the hidden lines

Self-Evaluation Test

Answer the following questions and then compare them to those given at the end of this chapter:

1. The **DDVPOINT** command is used to change the viewpoint to view the solid model. (T/F)

2. The wireframe models can be converted into surface models as well as solid models. (T/F)

3. The right-hand thumb rule is followed in AutoCAD to identify the *X*, *Y*, and *Z* axes direction. (T/F)

4. The right-hand thumb rule is followed in AutoCAD to find the direction of rotation in the 3D space. (T/F)

5. The _____ and _____ commands can be used to change the viewpoint for viewing the models in the 3D space.

6. Changing the viewpoint moves the 3D model from its default position. (T/F)

7. Using the **DDVPOINT** command, you can set the viewpoint with respect to _____ and _____ .

8. The various types of 3D coordinate systems are _____ .

9. Setting the clipping options in the **3DCLIP** command also sets them in the **DVIEW** command. (T/F)

10. The _____ command is used to draw standard 3D surface primitives.

Review Questions

Answer the following questions:

1. The 3D model can be continuously rotated in the 3D orbit. (T/F)

2. The **ELEV** command is a transparent command. (T/F)

3. You can directly write a text with thickness. (T/F)

4. Which command is used to set the shading mode for the 3D objects?

    (a) **SHADEMODE**   (b) **SHADE**
    (c) **MODESHADE**   (d) **None**

5. Which command is used to create a 3D polyline?

    (a) **POLYLINE**   (b) **3DPOLY**
    (c) **3DPOLYLINE**   (d) **POLY3D**

6. Which command is used to set the elevation and thickness for new objects?

    (a) **ELEVATION**   (b) **THICKNESS**
    (c) **ELEV**   (d) **THICK**

7. Which command is used to suppress the hidden edges in the 3D model?

    (a) **SUPPRESS**   (b) **HIDE**
    (c) **3DHIDE**   (d) **EDGE**

8. In which view are you by default when you open a new drawing?

    (a) SE Isometric View   (b) SW Isometric View
    (c) Plan View   (d) Bottom View

9   The _____ command creates profile images of 3D solids by displaying only the edges and silhouettes of the curved surfaces of the solids in the current view.

10. To draw an ellipse with thickness, you need to set the value of the _____ system variable to _____.

Exercises

Exercises 3 and 4   *General*

In these exercises, you will create the models shown in Figures 25-89 and 25-90. Assume the missing dimensions.

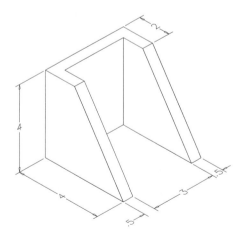

Figure 25-89 *Model for Exercise 3*

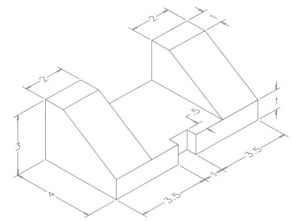

Figure 25-90 *Model for Exercise 4*

Exercise 5

Mechanical

In this exercise, you will create the model shown in Figure 25-91. The dimensions for the model are given in Figure 25-92. **

Figure 25-91 Solid model of the Plummer Block Casting

Figure 25-92 Dimensions of the Plummer Block Casting

Answers to Self-Evaluation Test

1. T, **2**. F, **3**. F, **4**. F, **5. DDVPOINT, VPOINT**, **6**. F, **7**. angle in the *XY* plane from the *X* axis, angle from the *XY* plane, **8**. absolute, relative rectangular, relative cylindrical, and relative spherical, **9**. T, **10. 3D**

Chapter 26

Creating Solid Models

Learning Objectives

After completing this chapter, you will be able to:

- *Understand solid modeling.*
- *Create standard solid primitives and polysolids.*
- *Understand regions and create them.*
- *Understand the use of Boolean operations.*
- *Modify the visual styles of solids.*
- *Define new UCS using the DUCS, ViewCube, and **Ribbon**.*
- *Use the **EXTRUDE** and **REVOLVE** commands to create solid models.*
- *Use the **SWEEP** and **LOFT** commands to create complex solid models.*
- *Use the **PRESSPULL** command.*

WHAT IS SOLID MODELING?

Solid modeling is the process of building objects that have all attributes of an actual solid object. For example, if you draw a wireframe or a surface model of a bushing, it is sufficient to define the shape and size of the object. However, in engineering, the shape and size alone are not enough to describe an object. For engineering analysis, you need more information such as volume, mass, moment of inertia, and material properties (density, Young's modulus, Poisson's ratio, thermal conductivity, and so on). When you know these physical attributes of an object, it can be subjected to various tests to make sure that it performs as required by the product specifications. It eliminates the need for building expensive prototypes and makes the product development cycle shorter. Solid models also make it easy to visualize the objects because you always think of and see the objects as solids. With computers getting faster and software getting more sophisticated and affordable, solid modeling has become the core of the manufacturing process. AutoCAD solid modeling is based on the ACIS solid modeler, which is a part of the core technology.

CREATING PREDEFINED SOLID PRIMITIVES

The Solid primitives form the basic building blocks for a complex solid. ACIS has seven predefined solid primitives that can be used to construct a solid model such as Box, Wedge, Cone, Cylinder, Pyramid, Sphere, and Torus. The number of lines in a solid model representation is controlled by the value assigned to the **ISOLINES** variable. These lines are called tessellation lines. The number of lines determines the number of computations needed to generate a solid. If the value is high, it will take significantly more time to generate a solid on the screen. Therefore, the value you assign to the **ISOLINES** variable should be realistic. When you enter commands for creating solid primitives, AutoCAD Solids will prompt you to enter information about the part geometry. The height of the primitive is always along the positive *Z* axis and perpendicular to the construction plane. Similar to surface meshes, solids are also displayed as wireframe models unless you hide, render, or shade them. The **FACETRES** system variable controls the smoothness in the shaded and rendered objects and its value can go up to 10.

Creating a Solid Box

Ribbon:	Home > Modeling > Box	**Menu Bar:**	Draw > Modeling > Box
Toolbar:	Modeling > Box	**Command:**	BOX

You can use the **BOX** command to create a solid rectangular box or a cube. Start a new file with the *acad3D.dwt* template. In 3D drawing templates, you can dynamically preview the operations that you perform. The methods to create a solid box are discussed next.

Creating the Box Dynamically

To create the box dynamically, choose **Home > Modeling > Box** from the **Ribbon**. Specify the first corner point of the box and then specify the diagonally opposite corner point for defining the base of the box, as shown in Figure 26-1. Now, specify the height of box by moving the cursor away from the base. After getting the desired height, click in the drawing window to create the box or specify the height of the box in the dynamic input box, see Figure 26-1.

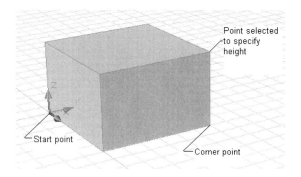

Figure 26-1 *Box created dynamically*

Two Corner Option

This is the default option using which you can create a solid box by defining the first corner of the box and then its other corner (Figure 26-2). Note that the length of the box will always be taken along the X axis, the width along the Y axis, and the height along the Z axis. Therefore, in this case, when you specify the other corner, the value along the X axis will be taken as the length of the box and the value along the Y axis will be taken as the width of the box. Next, you will be prompted to specify the height of the box. When you choose this button, the following prompt sequence will be displayed.

Specify first corner or [Center] <0,0,0>: **2,2,0**
Specify other corner or [Cube/Length]: **@5,4,0**   *(Length = 5, Width = 4)*
Specify height or [2Point]: **3**

Center-Length Option

The center of the box is the point where the center of gravity of the box lies. This option is used to create a box by specifying the center of the box as well as the length, width, and height of the box (Figure 26-3). The following prompt sequence is displayed when you choose the **Box** button.

Specify first corner or [Center] <0,0,0>: **C**
Specify center <0,0,0>: **4,4**
Specify corner or [Cube/Length]: **L**
Specify length: **8**
Specify width: **6**
Specify height or [2point]: **3**

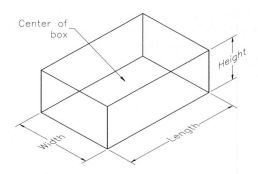

Figure 26-2 *Creating the box using the* **Two Corner** *option*

Figure 26-3 *Creating the box using the* **Center-Length** *option*

Tip
*The **Corner-Length** option is similar to the **Center-Length** option, except that in the **Corner-Length** option, you will define the first corner of the box and then the length, width, and height of the box. The **2point** option is used to specify the height of the box by specifying 2 points using the pointing device.*

Corner-Cube Option

This option is used to create a cube starting from a specified corner as shown in Figure 26-4. You are creating a cube, and so you will be prompted to enter the length of the cube, that will also act as the width and height. The prompt sequence for creating the cube using the **Corner-Cube** option is given next.

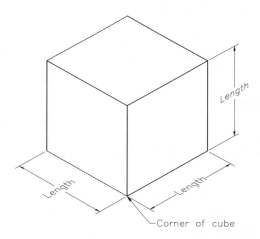

Figure 26-4 *Creating the cube using the* **Corner-Cube** *option*

Specify first corner or [Center] <0,0,0>: **2,2**
Specify other corner or [Cube/Length]: **C**
Specify length: **5**

Tip
*The **Center-Cube** option is similar to the **Corner-Cube** option, except that in the **Center-Cube** option, you will have to define the center and the length of the cube.*

Creating a Solid Cone

Ribbon:	Home > Modeling > Box > Cone	**Toolbar:**	Modeling > Cone
Menu Bar:	Draw > Modeling > Cone	**Command:** CONE	

The **CONE** command creates a solid cone with an elliptical or circular base. This command provides you with the option of defining the cone height or the location of the cone apex. Defining the location of the apex will also define the height of the cone and the orientation of the cone base from the *XY* plane. The methods for creating a solid cone are discussed next.

Creating the Cone Dynamically

Start a new file with the *acad3D.dwt* template file. In the 3D drawing templates, you can dynamically preview the operations that you perform. To do so, choose **Home > Modeling > Box > Cone** from the **Ribbon**. Specify the center point and the radius of the base, as shown in Figure 26-5. Next, specify the height of cone by moving the cursor away from the base. After getting the desired height, click in the drawing window to create the cone. You can also enter all parameters defining the cone in the dynamic input box. Figure 26-5 shows a cone created dynamically.

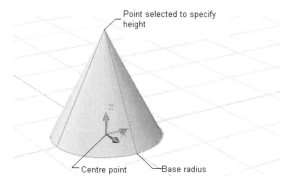

Figure 26-5 *Cone created dynamically*

Circular Cone

This method is used to create a cone with a circular base (Figure 26-6). The options for defining the circular base are **Center Radius**, **Center Diameter**, **3 Point**, **2 Point**, and **Tangent Tangent Radius**. The prompt sequence that will be displayed when you choose the **Cone** button is given next.

Specify center point for base or [3P/2P/Ttr/
Elliptical]: *Specify the center of the base.*
Specify base radius or [Diameter]
<default>: *Specify the radius or Enter **D** to
specify the diameter of the cone.*
Specify height or [2Point/Axis
endpoint/Top radius]<default>: *Specify
the height of the cone or enter an option or
press the ENTER key to accept the default
value.*

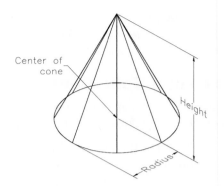

Figure 26-6 *A circular cone*

To specify the height using the **2Point**
option, enter **2P** at the **Specify height or [2
Point/Axis endpoint/Top radius]** prompt. You
need to specify two points. The distance between these two specified points will be considered
as the height of the cone. The prompt sequence that will follow is given next.

Specify first point: *Specify the first point.*
Specify second point: *Specify the second point on the screen.*

To specify the height using the **Axis endpoint** option, choose **Axis endpoint** from the shortcut
menu, at the **Specify height or [2Point/Axis endpoint/Top radius]** prompt. Using this option,
you can specify the endpoint of the axis, whose starting point is the center of the base. Apart
from fixing the height of the cone, this option also specifies the orientation of the cone in 3D
space. This means you can also create an inclined cylinder, see Figure 26-7. The prompt
sequence is as follows:

Specify axis endpoint: *Specify the location of the apex or axis endpoint of the right circular
cone in 3D space.*

If you want to create a frustum of a cone, choose the **Top radius** option from the shortcut
menu, at the **Specify height or [2Point/Axis endpoint/Top radius]** prompt. After specifying
the base radius, specify the radius of the top circle, see Figure 26-8. The prompt sequence is
as follows:

Specify top radius <default>: *Specify the radius for the top of the cone frustum.*
Specify height or [2Point/Axis endpoint] <default>: *Specify the height of the frustum or
choose an option.*

Elliptical Cone

Using this method, you can create a cone that has an elliptical base. To create an elliptical
cone, choose the **Elliptical** option from the shortcut menu, at the **Specify center point for
base of cone or [3P/2P/Ttr/Elliptical] <0,0,0>** prompt. You will be prompted to create the
base ellipse for the cone. You can create it using any of the methods for drawing the ellipse.
However, you cannot specify the rotation, in this case, for defining the other axis. You need
to specify the length of the other axis. Once you have entered all these values, you will be

Figure 26-7 *Creating a cone using the **Axis endpoint** option*

Figure 26-8 *Creating a cone using the **Top radius** option*

prompted to specify the cone height or choose from the **2point**, **Axis endpoint**, and **Top radius**. The procedure to create an elliptical cone using these options is similar to the procedure used to create a circular cone. Figure 26-9 shows an elliptical cone.

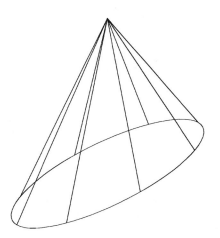

Figure 26-9 *An elliptical cone*

Creating a Solid Cylinder

Ribbon:	Home > Modeling > Box > Cylinder	**Toolbar:**	Modeling > Cylinder
Menu Bar:	Draw > Modeling > Cylinder	**Command:**	CYLINDER

You can use the **CYLINDER** command to create a solid cylinder. Similar to the **CONE** command, this command provides you with two options for creating the cylinder: **circular cylinder** and **elliptical cylinder**. This command also allows you to define the height of the cylinder or choose from the **2Point** or **Axis endpoint** options. The methods to create a solid cylinder are discussed next.

Creating the Cylinder Dynamically

Start a new file with the *acad3D.dwt* template file. In this template, you can dynamically preview the operations that you perform. To create a cylinder dynamically, choose **Home > Modeling > Box > Cylinder** from the **Ribbon**. Specify the center point of the base and then specify the radius of the base, as shown in Figure 26-10. Now, specify the height of the cylinder by moving the cursor away from the base. After getting the desired height, click in the drawing window to create the cylinder. You can also enter all parameters that define the cylinder in the dynamic input box. Figure 26-10 shows a cylinder created dynamically.

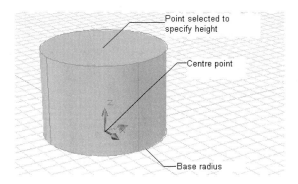

Figure 26-10 *Cylinder created dynamically*

Circular Cylinder

This option is used to create a cylinder with a circular base (Figure 26-11). This circular base can be defined by using the **Center Radius**, **Center Diameter**, **3Point**, **2Point**, or **Tangent Tangent Radius** options. The prompt sequence that follows when you choose the **Cylinder** button is given next.

> Specify center point of base or [3P/2P/Ttr/Elliptical]: *Specify the location of the center point or choose one of the options.*
> Specify base radius or [Diameter]<default>: *Specify the radius or choose the option for specifying the diameter.*
> Specify height of cylinder or [2Point/Axis endpoint]<default>: *Specify the height of the cylinder or choose an option.*

To specify the height using the **2Point** option, choose the **2Point** option from the shortcut menu, at the **Specify height** of cylinder **or [2Point/Axis endpoint]** prompt. You need to specify two points. The distance between these two specified points will be considered as the height of the cylinder. The prompt sequence is given next.

Specify first point: *Specify the first point.*
Specify second point: *Specify the second point on the screen.*

To specify the height using the **Axis endpoint** option, choose the **Axis endpoint** option from the shortcut menu, at the **Specify height or [2 Point/Axis endpoint]** prompt. Using this option, you can specify the endpoint of the axis, or in other words, the location of the center of the other end. Apart from fixing the height of the cylinder, this option also specifies the orientation of the cylinder in 3D space. This means you can also create an inclined cylinder, see Figure 26-12. The prompt sequence is given next.

Specify axis endpoint: *Specify the location of the center of the other end.*

You can also create an inclined cylinder by specifying the center of the other end, see Figure 26-12. This is done by entering **C** at the **Specify height or [2 Point/Axis endpoint]** prompt. The prompt sequence is given next.

Specify axis endpoint: *Specify the location of the center of the other end.*

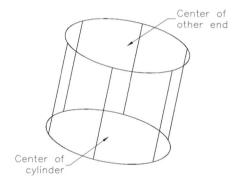

Figure 26-11 *Creating a circular cylinder*   ***Figure 26-12*** *Specifying the center of the other end*

Elliptical Cylinder

This option is used to create a cylinder with an elliptical base. The elliptical cylinder can be created by choosing the **Elliptical** option from the shortcut menu, at the **Specify center point of base or [3P/2P/Ttr/Elliptical]** prompt. You will be prompted to create the ellipse for the base. You can create the ellipse for the base using any of the methods for creating the ellipse. You can specify the height of the cylinder or the axis endpoint of the other end of the cylinder, see Figures 26-13 and 26-14.

Figure 26-13 *Creating an elliptical cylinder*   **Figure 26-14** *Specifying the center of the other end*

Creating a Solid Sphere

Ribbon:	Home > Modeling > Box > Sphere	**Toolbar:**	Modeling > Sphere
Menu Bar:	Draw > Modeling > Sphere	**Command:**	SPHERE

The **SPHERE** command is used to create a solid sphere, as shown in Figure 26-15. Start a new file with the *acad3D.dwt* template file. In this template, you can dynamically preview the operations that you perform. To create a sphere, choose **Home > Modeling > Box > Sphere** from the **Ribbon**; you will be prompted to specify the center of the sphere. On specifying the center, you can create the sphere by defining its radius or diameter. Instead of specifying the center of the sphere, you can also specify its circumference by choosing any one of the **3P/ 2P/Ttr** options. The prompt sequence, when you choose the **Sphere** button, is given next.

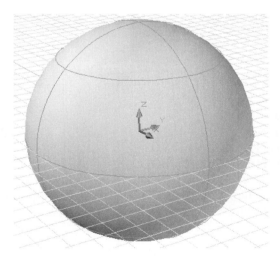

Figure 26-15 *Solid sphere created*

Specify center point or [3P/2P/Ttr]: *Specify the location of the center of the sphere or choose an option.*
Specify radius of sphere or [Diameter]<default>: *Specify the radius or choose an option.*

If you choose the **3P** option, you need to specify three points in the drawing window through which the circumference of the sphere will pass. The radius or diameter of the sphere is not required in this case because the three points itself fix the size of the sphere. The prompt sequence, when you choose the **3P** option, is given next.

Specify first point: *Specify the first through point.*
Specify second point: *Specify the second through point.*
Specify third point: *Specify the third through point of the sphere.*

If you choose the **2P** option, you need to specify two points in the drawing area through which the circumference of the sphere will pass. These two points are diametrically opposite points, so you need not to specify the radius or diameter of the sphere. The prompt sequence, when you choose the **2P** option, is given next.

Specify first end point of diameter: *Specify the first point.*
Specify second end point of diameter: *Specify the second diametrically opposite point.*

If you choose the **Ttr** option, you need to specify two objects to which the sphere will be tangent. These objects can be circles, arcs or lines. You will then need to specify the radius or the diameter of the sphere. The value of the radius fixes the size and location of the sphere while maintaining the tangent constraint. The prompt sequence, when you choose the **Ttr** option, is given next.

Specify point on object for first tangent: *Select a tangent point on an object.*
Specify point on object for second tangent: *Select a tangent point on an object.*
Specify radius <default>: *Specify a radius or press ENTER to specify the default radius value.*

Creating a Solid Torus

| **Ribbon:** | Home > Modeling > Box > Torus | **Toolbar:** | Modeling > Torus |
| **Menu Bar:** | Draw > Modeling > Torus | **Command:** | TORUS |

You can use the **TORUS** command to create a torus that is like the shape of tyre-tube, as shown in Figure 26-16. Start a new file with the *acad3D.dwt* template file. In the 3D drawing templates, you can dynamically preview the operations that you perform. To create a torus, choose **Home > Modeling > Box > Torus** from the **Ribbon**; you will be prompted to enter the diameter or the radius of the torus and the diameter or the radius of tube (Figure 26-17). The radius of torus is the distance from the center of the torus to the center-line of the tube. This radius can have a positive or a negative value. If the value is

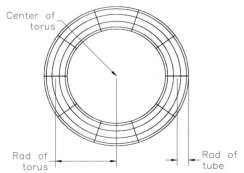

Figure 26-16 *Torus created dynamically*

Figure 26-17 *Parameters associated with a torus*

negative, the torus has a rugby-ball like shape (Figure 26-18). A torus can be self-intersecting. If both the radii of the tube and the torus are positive and the radius of the tube is greater than the radius of the torus, the resulting solid looks like an apple (Figure 26-19). Instead of specifying the radius or diameter of the torus, you can also specify the points through which the circumference of the torus will pass by choosing any one of **3P**, **2P**, or **Ttr** options. These options work in the same way as discussed in the **SPHERE** command. The torus is centered on the construction plane. The top half of the torus is above the construction plane and the other half is below it. The prompt sequence, when you choose the **Torus** button, is given next.

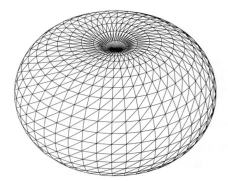

Figure 26-18 *Torus with a negative radius value*

Figure 26-19 *Torus with radius of tube more than the radius of torus*

Specify center point or [3P/2P/Ttr]: *Specify the location of the center of the torus or choose an option.*
Specify radius of torus or [Diameter]: *Specify the radius of the torus or Enter* **D***.*
Specify radius of tube or [2Point/Diameter]: *Specify the radius of the tube or choose an option.*

Creating a Solid Wedge

| **Ribbon:** | Home > Modeling > Box > Wedge | **Toolbar:** | Modeling > Wedge |
| **Menu Bar:** | Draw > Modeling > Wedge | **Command:** | WEDGE |

This command is used to create a solid wedge and is similar to the **BOX** command. This means that this command provides you with the options of creating the wedge that is similar to those of the **BOX** command.

Creating a Pyramid

| **Ribbon:** | Home > Modeling > Box > Pyramid | **Toolbar:** Modeling > Pyramid |
| **Menu Bar:** | Draw >Modeling > Pyramid | **Command:** PYRAMID |

The **Pyramid** option allows you to create a solid pyramid where all the faces, other than the base, are triangular and converge at a point called apex (Figure 26-20). The base of a pyramid can be any polygon, but is typically a square. To create a pyramid, choose **Home > Modeling > Box > Pyramid** from the **Ribbon**. The prompt sequence that will follow is given next.

Command: _pyramid 4 sides Circumscribed
Specify center point of base or [Edge/Sides]: *Specify the center point.*
Specify base radius or [Inscribed] <4>: *Specify the base radius.*
Specify height or [2Point/Axis endpoint/Top radius] <10>: *Specify the height to create the pyramid,* refer to Figure 26-20.

The methods for creating a solid pyramid are discussed below.

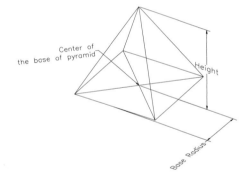

Figure 26-20 *Dynamically creating the Pyramid*

On invoking the **PYRAMID** command, you can specify the center point of the base polygon or the length of the edges, or the number of sides of the polygon. If you select the **Edge** option in the command window, you will be prompted to pick two points from the drawing area that determine the length and orientation of the edge of the base polygon. If you select the **Sides** option, you will be prompted to specify the number of sides of the base polygon. Note that in AutoCAD, the number of sides of a pyramid can vary from 3 to 32.

Next, you need to specify the base radius of the circle in which the base polygon is created. The base polygon is either circumscribed around a circle or inscribed within a circle. By default, the base of the pyramid is circumscribed. After specifying the base radius, you need to specify the height of the pyramid or select an option.

To specify the height using the **2Point** option, you need to specify any two points in the drawing area. The distance between the two specified points will be considered as the height of the pyramid.

To specify the height using the **Axis endpoint** option, you need to specify the endpoint of the pyramid. The pointed tip at the top of the pyramid called apex. This option defines the length and orientation of the Pyramid in 3D Space.

You can also create the frustum of a pyramid by selecting the **Top radius** option. Here, you are prompted to specify the radius of the top of the pyramid frustum. Then, you need to specify the height of the frustum either dynamically or by selecting the **2Point** or the **Axis endpoint** option.

Creating a Polysolid

| **Ribbon:** | Home > Modeling > Polysolid | **Toolbar**: | Modeling > Polysolid |
| **Menu Bar:** | Draw > Modeling > Polysolid | **Command:** | POLYSOLID or PSOLID |

The **POLYSOLID** command is similar to the **POLYLINE** command with the difference that this command creates a solid with a rectangular cross-section of a specified width and height, see Figure 26-21. This command can also convert existing lines, 2D polylines, arcs, and circles to a polysolid feature. The prompt sequence, when you will invoke the **POLYSOLID** command, is given next.

Command: **POLYSOLID**
Height = current, Width = current, Justification = current
Specify start point or [Object/Height/Width/Justify]<Object>: *Specify the start point for the profile of the polysolid. Else, press the ENTER key to specify an object to convert to a polysolid or enter an option.*

The options available in this command are discussed next.

Next Point of Polysolid

When you specify the start point for the profile of the solid, this option is displayed. This option is used to specify the next point of the current polysolid segment. If additional polysolid segments are added to the first polysolid, AutoCAD automatically makes the endpoint of the previous polysolid, the start point of the next polysolid segment. The prompt sequence is given next.

Command: **POLYSOLID**
Height = current, Width = current, Justification = current
Specify start point or [Object/Height/Width/Justify] <Object>: *Specify the start point of the polysolid.*
Specify next point or [Arc/Undo]: *Specify the endpoint of the first polysolid segment.*

Specify next point or [Arc/Close/Undo]: *Specify the endpoint of the second polysolid segment or press the ENTER key to exit the command.*

Arc. This option is used to switch from drawing linear polysolid segments to drawing curved polysolid segments. The prompt sequence that will follow is given next.

Specify next point or [Arc/Close/Undo]: A [Enter]
Specify end point of the arc or [Close/Direction/Line/Second point/Undo]: *Specify the endpoint of the arc or choose an option to create the arc.*

The **Close** option closes the polysolid by creating an arc shaped trajectory from the most recent endpoint to the initial start point and exits the **POLYSOLID** command. Usually, the arc drawn with the **POLYSOLID** command, is tangent to the previous polysolid segment. This means that the starting direction of the arc is the ending direction of the previous segment. The **Direction** option allows you to specify the tangent direction of your choice for the arc segment to be drawn. You can specify the direction by specifying a point. The **Line** option switches the command to the **Line** mode. The **Second point** option selects the second point of the arc in the **3P** arc option. The **Undo** option reverses the changes made in the previously drawn polysolid.

Close. This option is available when at least two segments of the polysolid are drawn. It closes the polysolid segment by creating a linear polysolid from the recent endpoint to the initial point.

Undo. This option erases the most recently drawn polysolid segment. You can use this option repeatedly until you reach the start point of the first polysolid segment.

Object

This option converts an existing 2D object into a polysolid. The 2D objects that you convert into a polysolid include lines, 2D polylines, arcs, and circles. In Figure 26-21, the object which is converted into a polysolid is created by the 2D polyline. The prompt sequence for this option is given next.

Command: **POLYSOLID**
Height = current, Width = current, Justification = current
Specify start point or [Object/Height/Width/Justify]<Object>: [Enter]
Select object: *Select an object to convert it into a polysolid.*

Figure 26-21 *Object converted to a polysolid using the **Object** option*

Height

This option is used to specify the height of the rectangular cross-section of the polysolid. The value that you specify will be set as the default value for this command.

The prompt sequence for this option is given next.

> Command: **POLYSOLID**
> Height = current, Width = current, Justification = current
> Specify start point [Object/Height/Width/Justify]<Object>: **H** [Enter]
> Specify height <current>: *Specify the value of the height or press the ENTER key to accept the current value.*
> Specify start point [Object/Height/Width/Justify]<Object>: *Specify the start point of the polysolid or choose an option.*

Width

This option is used to specify the width of the rectangular cross-section of a polysolid. The prompt sequence for this option is given next.

> Command: **POLYSOLID**
> Height = current, Width = current, Justification = current
> Specify start point [Object/Height/Width/Justify]<Object>: **W** [Enter]
> Specify width <current>: *Specify the width or press the ENTER key to accept the current value.*
> Specify start point [Object/Height/Width/Justify]<Object>: *Specify the start point of the polysolid or choose an option.*

Justify

The justification option determines the position of the created polysolid with respect to the starting point. The three justifications that are available for a **POLYSOLID** command are **Left**, **Center**, and **Right**.

Left. This justification produces a polysolid with its left extent fixed at the start point and the width added to the right of the start point, when you view the polysolid from the direction of its creation, see Figure 26-22.

Center. This justification produces a polysolid with its center fixed at the start point and the width added equally to both the sides of the start point when you view the polysolid from the direction of its creation, see Figure 26-23.

Right. This justification produces a polysolid with its right extent fixed at the start point and the width added to the left of the start point, when you view the polysolid from the direction of its creation, see Figure 26-24.

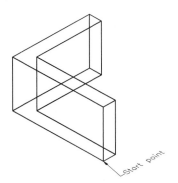

Figure 26-22 *Polysolid with the **Left** justification*

Figure 26-23 *Polysolid with the **Center** justification*

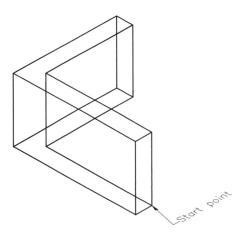

Figure 26-24 *Polysolid with the **Right** justification*

Creating a Helix

Ribbon:	Home > Draw > Helix	Toolbar:	Modeling > Helix
Menu Bar:	Draw > Helix	Command:	HELIX

The **HELIX** command is used to create a 3D helix. For generating a helical curve, you need to first specify the center for the base of the helix. Next, you need to enter the base radius, radius at the top of the helix, and height of the helix. A helix of converging or diverging shape can also be generated by specifying different values for the base and top radius. You can also control the number of turns in a helix and specify whether the twist will be in the clockwise direction or counter clockwise direction. Figure 26-25 shows a helix with the parameters to be defined.

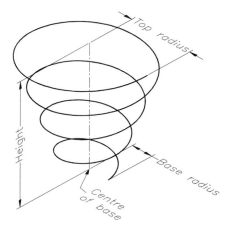

Figure 26-25 *Helix with associated parameters*

The prompt sequence for the **HELIX** command is given next.

> Command: **HELIX**
> Number of turns = current   Twist = CCW
> Specify center point of base: *Specify a point for the center of the helix base.*
> Specify base radius or [Diameter] <1.0000>: *Specify the value for the base radius, enter* ***D*** *to specify the diameter, or press the ENTER key to select the default value for the radius.*
> Specify top radius or [Diameter] <current>: *Specify the value for the top radius of the helix, enter* ***D*** *to specify the diameter, or press the ENTER key to select the current value of the base radius.*
> Specify helix height or [Axis endpoint/Turns/turn Height/tWist] <default>: *Specify the height of the helix, press the ENTER key to select the default value or choose an option.*

The options available in this command are discussed next.

Axis endpoint

To specify the height using the **Axis endpoint** option, choose the **Axis endpoint** option from the shortcut menu, at the **Specify helix height or [Axis endpoint/Turns/turn Height/tWist]** prompt. Using this option, you can specify the endpoint of the axis, whose starting point is assumed to be at the center of the base of the helix. Apart from fixing the height of the helix, this option also specifies the orientation of the helix in 3D space or in other words, you can create an inclined helix also, see Figure 26-26.

Turns

To specify the number of turns in the helix, choose the **Turns** option from the shortcut menu, at the **Specify helix height or [Axis endpoint/Turns/turn Height/tWist]** prompt. Using this option, you can change the number of turns in the helical curve. The default number of turns for the helix command is three.

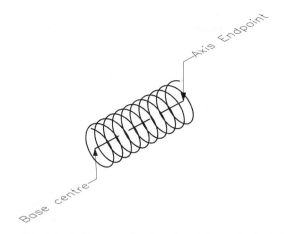

Figure 26-26 *Helix created using the **Axis endpoint** option*

 Note
In a helical curve, the number of turns cannot be more than 500.

turn Height

To specify the pitch of the helix, choose the **turn Height** option from the shortcut menu, at the **Specify helix height or [Axis endpoint/Turns/turn Height/tWist]** prompt. Using this option, instead of specifying the whole height of the helix, you can specify the height of one complete single turn of the helix. The helical curve will be generated with the total height equal to the turn height multiplied by the number of turns.

tWist

To specify the starting direction of the helical curve, choose the **tWist** option from the shortcut menu, at the **Specify helix height or [Axis endpoint/Turns/turn Height/tWist]** prompt. By using this option, you can specify whether the curves will be generated in the clockwise or the counterclockwise direction. Figure 26-27 shows a helix with counterclockwise turns and Figure 26-28 shows a helix with clockwise turns. The default value for the starting direction of the helical curve is counterclockwise.

Figure 26-27 *Helix created using the **CCW** option*   *Figure 26-28* *Helix created using the **CW** option*

Tip
*You can use a generated helix as a trajectory for the **SWEEP** command, which will be discussed later in this chapter.*

MODIFYING THE VISUAL STYLES OF SOLIDS

Ribbon: View > 3D Palettes > Visual Styles		**Command:**	VISUALSTYLES
Menu Bar: Tools > Palettes > Visual Styles			

AutoCAD provides you with various predefined modes of visual styles. These visual styles are grouped together in the **Visual Styles Manager**, see Figure 26-29. To invoke the **Visual Styles Manager**, choose **View > 3D Palettes > Visual Styles** from the **Ribbon**. Alternatively, click on the arrow on the bottom right of the **Visual Styles** panel in the **Render** tab of the **Ribbon**. There are five visual styles available in the **Available Visual Styles in Drawing** rollout of the **Visual Styles Manager**. To apply a visual style, you need to drag and drop it from the rollout to the drawing area. Alternatively, choose **Realistic** from the **Visual Styles** panel of the **Render** tab of the **Ribbon**; a flyout with various visual styles will be displayed. Choose the required visual style; the model will be updated to the new visual style automatically. You can modify the properties of the selected visual style. To do so, invoke the **Visual Styles Manager**; the properties of the selected visual style will be displayed below the **Available Visual Styles in Drawing** rollout. You can modify a property depending upon your requirement. You can also create a user-defined visual style based on the current visual style applied to the model. To do so, choose the **Create New Visual Style** button from the tool strip at the bottom of the

*Figure 26-29 The **Visual Styles Manager** palette*

Available Visual Styles in Drawing rollout; the **Create New Visual Style** dialog box will be displayed. Enter a new name and description for the newly created visual style. Next, choose **OK**; a new visual style will be created. Set the properties of the new visual style by modifying the default properties. Alternatively, set the properties of a visual style and then select **Save as a New Visual Style** from the **Visual Styles** flyout; you will be prompted to enter a name to save the current drawing. Enter a name and press ENTER; the new visual style will be saved.

Available Visual Styles in Drawing
The options available in this rollout are discussed next.

2D/3D Wireframe
On applying this visual style, all hidden lines will be displayed along with the visible lines in

the model. Sometimes, it becomes difficult to recognize the visible lines and hidden lines, if you set this visual style for complex models.

3D Hidden

On applying this visual style, the hidden lines in the model will not be displayed, and only the edges of the faces that are visible in the current viewport will be displayed.

Conceptual

In this visual style, the model will be displayed as shaded and the edges of the visible faces of the model will also be displayed.

Realistic

This visual style is the same as **Conceptual**, but with a more realistic appearance. Moreover, if materials are applied to a model, this visual style will display the model along with the materials applied.

2D Wireframe options

The options in this area are discussed next.

Contour Lines

The lines that are displayed on the curved surface of a solid are known as contour lines. The default value for the number of contours to be displayed on each curved surface is 4 and it is controlled by system variable **ISOLINES**. Figure 26-30 shows a cylinder with four contour lines and Figure 26-31 shows a cylinder with eight contour lines.

Figure 26-30 Cylinder with 4 contour lines   *Figure 26-31* Cylinder with 8 contour lines

Draw true silhouettes

This option is used to specify whether or not you want to display the true silhouette edges of the model when you hide it using the **HIDE** command. If you select **Yes** from this drop-down list, the silhouette edges will be displayed when you hide the model, see Figure 26-32. However, when you select **No** from this drop-down list, the silhouette edges will not be displayed when you hide the model, see Figure 26-33.

Figure 26-32 Model with the true silhouette edges on

Figure 26-33 Model with the true silhouette edges off

2D Hide - Obscured Lines

The options in this area are used to set the linetype and color of the hidden lines (also called obscured lines).

Linetype

This drop-down list is used to set the linetype for the obscured lines. You can select the desired linetype from this drop-down list. The linetype of the hidden lines can also be modified using the **OBSCUREDLTYPE** system variable. The default value of this variable is 0. As a result, the hidden lines are suppressed when you invoke the **HIDE** command. You can set any value between 0 and 11 for this system variable. The following table gives the details of the different values of this system variable and the corresponding linetypes that will be assigned to the hidden lines.

Value	Linetype	Sample
0	Off	None
1	Solid	————————————
2	Dashed	— — — — — — — — —
3	Dotted	··················
4	Short Dash	— — — — — —
5	Medium Dash	— — — — — —
6	Long Dash	—— —— —— ——
7	Double Short Dash	— — —

8	**Double Medium Dash**	—— —— ——
9	**Double Long Dash**	—— —— ——
10	**Medium Long Dash**	—— — — — —
11	**Sparse Dot**	▪ ▪ ▪ ▪ ▪ ▪

Figure 26-34 shows a model with a hidden linetype changed to dashed lines and Figure 26-35 shows the same model with hidden lines changed to dotted lines.

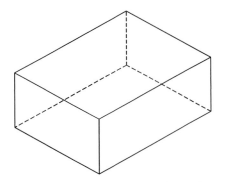

Figure 26-34 *Model with hidden lines changed to dashed lines*

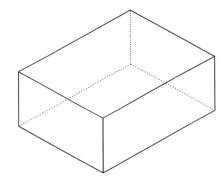

Figure 26-35 *Model with hidden lines changed to dotted lines*

Color

The **Color** drop-down list is used to define the color for the obscured lines. If you define a separate color, the hidden lines will be displayed with that color, when you invoke the **HIDE** command. You can select the required color from this drop-down list. This can also be done using the **OBSCUREDCOLOR** system variable. Using this variable, you can define the color to be assigned to the hidden lines. The default value is 257. This value corresponds to the **ByEntity** color. You can enter the number of any color at the sequence that will follow when you enter this system variable. For example, if you set the value of the **OBSCUREDLTYPE** variable to **2**, and that of the **OBSCUREDCOLOR** to **1**, the hidden lines will appear in red dashed lines, when you invoke the **HIDE** command.

Note
*The linetype and the color for the hidden lines defined using the previously mentioned variables are valid only when you invoke the **HIDE** command. They do not work, if the model is regenerated.*

2D - Intersection Edges

The options in this area are used to display a 3D curve at the intersection of two surfaces. Note that the 3D curve is displayed only after invoking the **HIDE** command. These options are discussed next.

Visible

If you select **Yes** from the **Visible** drop-down list, a 3D curve will define the intersecting portion of the 3D surfaces or solid models.

Color

The **Color** drop-down list is used to specify the color of the 3D curve that is displayed at the intersection of the 3D surfaces or solid models.

2D Hide - Miscellaneous

This head in the **Visual Styles Manager** is used to set the percentage for halo gap.

Halo gap %

The **Halo gap %** area is used to specify, in terms of percentage of unit length, the distance by which the edges that are hidden by a surface or solid model will be shortened, see Figures 26-36 and 26-37.

Note

The obscured linetype in Figures 26-36 and 26-37 is changed to dashed.

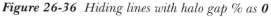

Figure 26-36 *Hiding lines with halo gap % as **0***   *Figure 26-37* *Hiding lines with halo gap % as 40*

Similarly, you can specify the settings for faces, edges, and environment required for other visual styles. These settings are discussed next.

Controlling the Display of Edges

You can control the display of edges by assigning different colors and linetypes to the edges or by giving them some special effects such as Overhang and Jitter. Both the effects give a hand-drawn effect with the only difference being that the sketches drawn in **Overhang** extend beyond the model, whereas in **Jitter**, the sketches are neatly drawn, as shown in Figure 26-38. However, you can adjust the amount of these effects on the model using the sliders that are available in the **Edge Effects** panel of the **Render** tab of the **Ribbon**. Note that you need to choose the required buttons from this panel to activate the sliders. You can now slide or directly enter a required value in the edit box available next to the slider. The model keeps on updating automatically as you move the slider. You can use either one or both the effects on the model.

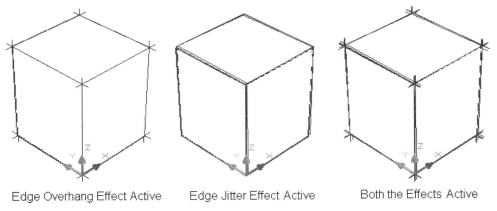

Edge Overhang Effect Active   Edge Jitter Effect Active   Both the Effects Active

*Figure 26-38 The **Overhang** and **Jitter** effects of the edges*

You can also control the type of edge to be displayed in the viewport from the **Isolines** flyout in the **Edge Effects** panel. The options in the flyout are **No Edges**, **Isolines**, and **Facet Edges**, see Figure 26-39. By default, the **Isolines** option is chosen enabling you the display of the contour lines on the surface of the solid. If you do not want the edges to be displayed, choose **No Edges** from the flyout. The third option is **Facet Edges**. The Facet Edges are the edges between the planar surfaces. On choosing this option, the **Obscured Edges** and the **Intersection Edges** toggle buttons, available in the extended panel of **Edge Effects**, get activated. These are used to highlight the obscured or hidden edges and the intersection edges of the models. When you choose these buttons, the respective drop-down lists get activated, enabling you to assign colors to the hidden and intersection edges. These settings are available in the **Edge Settings** head of the **Visual Styles Manager**.

No Edges   Isolines   Facet Edges

Figure 26-39 Different types of edge display

You can also adjust the width of the Silhouette edges by choosing the **Silhouette Edges** button from the **Edge Effects** panel. The width can be specified using the slider next to this button and its value ranges between 1 to 25. You can also set the color of the silhouette edge from the **Edge Color** drop-down list available in the **Edge Effects** panel of the **Ribbon**.

Controlling the Display of Faces

AutoCAD enables you to control the display of the faces by giving color effects and shading to the solids or surfaces. The shading of models can be controlled using the flyout available in the extended **Visual Styles** panel of the **Render** tab in the **Ribbon**. The options available in this flyout are **No Face Style**, **Realistic Face Style**, and **Warm-Cool Face Style,** refer to Figure 26-40. By default, the **Realistic Face Style** is chosen and therefore gives a real world effect to the model. The **Warm-Cool Face Style** enhances the display by using light colors, instead of dark colors. If the **No Face Style** is chosen then no style is applied to the model and only the edges are displayed. These settings can also be performed in the **Face Settings** head in the **Visual Styles Manager**.

No Face Style   Realistic   Warm-Cool
                       Face Style   Face Style

Figure 26-40 *Different face styles*

You can also control the appearance of the faces in the solids or surfaces using the lighting flyout available in the extended **Visual Styles** panel in the **Render** tab of the **Ribbon**. The options available in this flyout are **Facet Lighting**, **Smooth Lighting**, and **Smoothest Lighting**. In the **Facet Lighting** option, the color is applied to each face, thereby making the object appear flat. In the **Smooth Lighting** option, the color applied is the gradient between the two vertices of the face. This option is chosen by default. In the **Smoothest Lighting** option, the color is calculated for per pixel lighting using the **Manual Performance Tuning** dialog box. This dialog box is invoked by choosing the **Manual Tune** button from the **Adaptive Degradation and Performance Tuning** dialog box. This dialog box is in turn invoked by choosing **Render> Visual Styles > Performance Tuner** in the **Ribbon.**

Different color modifiers can be set for the faces by choosing an option in the **Normal** flyout in the **Visual Styles** panel of the **Render** tab of the **Ribbon**. The options in this flyout are **Normal**, **Monochrome**, **Tint**, and **Desaturate**. No face color modifier will be applied, if the **Normal** option is chosen. If you choose the **Monochrome** option, all the faces will be shaded and have a single color. Choose the **Tint** option to use the same color to shade all faces by changing the hue and saturation values of the color. Choose the **Desaturate** option to soften the color by reducing the saturation by 30%, refer to Figure 26-41.

Figure 26-41 *Different face modes*

You can also control the transparency and shininess for the solids and surfaces. To set the transparency, choose the **X-Ray Effect** button; the **Opacity** slider gets active. Specify the desired transparency level. Figure 26-42 shows the object with and without the transparency set. To set the shininess on the object, choose the **Highlight Intensity** button available on the **Face Settings** head of the **Visual Styles Manager**; **the Highlight Intensity** edit box will be available. Now, you set the shininess in the **Highlight Intensity** edit box. Note that you cannot set this value when material is applied on it.

Figure 26-42 *Different face transparencies*

It is important to note that putting too many effects may slow down your system. Therefore, you need to comprise between efficiency and effects. The **Performance Tuner** helps in doing so by turning off some effects automatically to maintain the performance of the drawing. You can also control the order in which the features gets turned off by choosing **Render > Visual Styles > Performance Tuner** from the **Ribbon**.

Controlling Backgrounds and Shadows

You can control the display of background using the **Backgrounds** drop-down list in the **Environment Settings** head in the **Visual Styles Manager**. The background can be a single solid color, gradient fill, an image, or the sun and the sky as the background. This is explained in detail in Chapter 30 (Rendering and Animating Designs).

Shadows are also as very important as backgrounds to provide a real effect to the rendering. Shadows are generated when light is applied. The lights can be user-defined lights or the sunlight. Choose **Render > Lights > No Shadows** from the **Ribbon**; a flyout is displayed. The options available in this flyout are **No Shadows**, **Ground Shadows**, and **Full Shadows**. The **Ground Shadows** are the shadows that objects cast on the ground. The **Full Shadows** are the shadows casted by one object on other objects. The shadows appear darker when they overlap. Note that the full shadows cannot be displayed when the **Enhanced 3D Performance** is off. To do so, invoke the **Adaptive Degradation and Performance Tuning** dialog box and choose the **Manual Tune** button; the **Manual Performance Tuning** dialog box is displayed. Select the **Enable hardware acceleration** check box, and select the **Enhanced 3D Performance** to turn off the effect.

Using shadows can slow down the system. So, you need to turn off the shadows while working, and turn them on when you need the shadows. You can set the shadow display from the **Properties** palette. The options under the **Shadow** area of the **Properties** palette are **Casts and Receives Shadows**, **Casts shadows**, **Receives shadows** and **Ignore shadows**. The shadows will be explained in detail in later chapters.

CREATING COMPLEX SOLID MODELS

Until now you have learned how to create simple solid models using the standard solid primitives. However, the real-time designs are not just the simple solid primitives, but complex solid models. These complex solid models can be created by modifying the standard solid primitives with the Boolean operations or directly by creating complex solid models by extruding or revolving the regions. All these options of creating complex solid models are discussed next.

Creating Regions

Ribbon:	Home >Draw > Region	Toolbar:	Draw > Region
Menu Bar:	Draw > Region	Command:	REGION

The **REGION** command is used to create regions from the selected loops or closed entities. Regions are the 2D entities with properties of 3D solids. You can apply the Boolean operation on the regions and you can also calculate their mass properties. Bear in mind that the 2D entity you want to convert into a region should be a closed loop. Once you have created regions, the original object is deleted automatically. However, if the value of the **DELOBJ** system variable is set to **0**, the original object is retained. The valid selection set for creating the regions are closed polylines, lines, arcs, splines, circles, or ellipses. The current color, layer, linetype, and lineweight will be applied to the regions.

CREATING COMPLEX SOLID MODELS BY APPLYING BOOLEAN OPERATIONS

You can create complex solid models by applying the Boolean operations on the standard solid primitives. The various Boolean operations that can be performed are union, subtract, intersect, and interfere. The commands used to apply these Boolean operations are discussed next.

Combining Solid Models

Ribbon:	Home > Solid Editing > Union	**Toolbar:**	Modeling > Union
Menu Bar:	Modify > Solid Editing > Union	**Command:**	UNION

 The **UNION** command is used to apply the union Boolean operations on the selected set of solids or regions. You can create a composite solid or region by combining them using this command. You can combine any number of solids or regions. When you invoke this command, you will be asked to select the solids or regions to be added. Figure 26-43 shows two solid models before union and Figure 26-44 shows the composite solid created after union.

 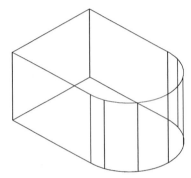

Figure 26-43 Solid models before union   *Figure 26-44 Composite solid created after union*

Subtracting One Solid From the Other

Ribbon:	Home > Solid Editing > Subtract	**Toolbar:**	Modeling > Subtract
Menu Bar:	Modify > Solid Editing > Subtract	**Command:**	SUBTRACT

 This command is used to create a composite solid by removing the material common to the selected set of solids or regions. On invoking this command, you will be prompted to select the set of solids or regions to subtract from. Once you have selected it, you will be prompted to select the solids or regions to subtract. The material common to the first selection set and the second selection set is removed from the first selection set. The resultant object will be a single composite solid, see Figures 26-45 and 26-46.

> **Note**
> *If you subtract a single solid from two solids, the two solids join after the subtract operation.*

Intersecting Solid Models

Ribbon:	Home > Solid Editing > Intersect	**Command:**	INTERSECT
Menu Bar:	Modify > Solid Editing > Intersect	**Toolbar:**	Modeling > intersect

The **INTERSECT** command is used to create a composite solid or region by retaining

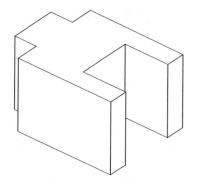

Figure 26-45 *Solid models before subtracting*   *Figure 26-46* *Composite solid created after subtracting*

the material common to the selected set of solids or regions. When you invoke this command, you will be asked to select the solids or regions to intersect. The material common to all the selected solids or regions will be retained to create a new composite solid (Figures 26-47 and 26-48).

Figure 26-47 *Solid models before intersecting*   *Figure 26-48* *Solid model created after intersecting*

Checking Interference in Solids

Ribbon: Home > Solid Editing > Interference Checking	**Command:** INTERFERE
Menu Bar: Modify > 3d Operations> Interference Checking	

The **INTERFERE** command is used to create a composite solid model by retaining the material common to the selected sets of solids. This command is generally used for analyzing interference between the mating parts of an assembly. On invoking this command, you will be prompted to select the first set of objects. Once you have selected them, you will be prompted to select the second set of objects. Select the second set of objects and press ENTER; the **Interference Checking** dialog box will be displayed. Also, the interference will be highlighted in the drawing area. The **Interfering Objects** area in this dialog box displays the number of objects selected in both the first and second sets and the number of interferences found between them. Choose the **Previous** or the **Next** button in the **Highlight** area to view the previous or next interferences. If the **Zoom to Pair** check box is selected in this area, the interference is zoomed while cycling through the interference objects. You can also use the navigation tools in the **Interference Checking** dialog box to view the interfering area clearly. Clear the **Delete interference objects created on close** check box, if you need to retain the interference objects after closing the dialog box. After closing the dialog box, you can move the interfering objects using the **Move** tool. Figure 26-49 shows two mating components of an assembly with interference between them and Figure 26-50 shows the interference solid created and moved out.

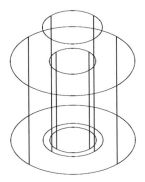

Figure 26-49 *Two mating components with interference*

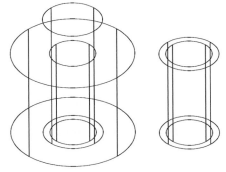

Figure 26-50 *Interference solid created using the* **INTERFERE** *command*

You can also specify the visual style and color for the interfering objects. To do so, enter **INTERFERE** at the Command prompt and press ENTER; the **Select first set of Objects or [Nested selection/Settings]** prompt will be displayed. Enter **S** at the Command prompt and press ENTER; the **Interference Settings** dialog box will be displayed. In this dialog box, specify the required parameters.

Example 1

In this example, you will create the solid model shown in Figure 26-51.

1. Increase the limits to 100,100. Zoom to the limits of the drawing.

2. Choose **View > Views > Views > SW Isometric** from the **Ribbon**.

3. Enter **UCSICON** at the Command prompt. The following prompt sequence is displayed:

    Enter an option [ON/OFF/All/Noorigin/ORigin/Properties] <ON>: N

4. Invoke the orthomode, if it is not on.

5. Set the visual style as **3D Wireframe** in the **Visual Styles** drop-down list available in the **View** panel in the **Home** tab of the **Ribbon**.

Figure 26-51 *Solid model for Example 1*

6. Choose **Home > Modeling > Box** from the **Ribbon**; the prompt sequence that follows is given next.

    Specify first corner or [Center] <0,0,0>: **10,10**
    Specify other corner or [Cube/Length]: **L**
    Specify length: **80**
    Specify width: **42**
    Specify height: **48**

7. Again, choose **Home > Modeling > Box** from the **Ribbon**; the following command sequence is displayed:

    Specify first corner or [Center] <0,0,0>: *Specify a point on the screen.*
    Specify other corner or [Cube/Length]: **L**
    Specify length: **42**
    Specify width: **15**
    Specify height: **30**

8. Move the new box inside the old box using the midpoints of both the boxes. Copy the new box to the other side of the box, refer to Figure 26-52.

9. Choose **Home > Solid Editing > Subtract** from the **Ribbon**; the following sequence will be displayed at the prompt:

    Select solids and regions to subtract from.
    Select objects: *Select the bigger box.*
    Select objects: Enter
    Select solids and regions to subtract.
    Select objects: *Select one of the smaller boxes.*

Select objects: *Select other smaller boxes.*
Select objects: [Enter]

10. Change the visual style to **3D Hidden.** The model, after subtracting the boxes, should look similar to the one shown in Figure 26-53.

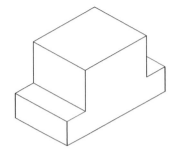

Figure 26-52 *Boxes moved inside the bigger box using midpoints*

Figure 26-53 *Model after subtracting the smaller boxes*

11. Again, change the visual style to **3D Wireframe** and choose **Home > Modeling > Box** from the **Ribbon**. The following command sequence will be displayed:

    Specify first corner or [Center] <0,0,0>: *Specify a point on the screen.*
    Specify other corner or [Cube/Length]: **L**
    Specify length: **20**
    Specify width: **80**
    Specify height: **29**

12. Invoke the **Move** command and select the box created in the previous step.

13. Specify the midpoint of the lower edge as the base point; the **Specify second point or <use first point as displacement>** prompt is displayed.

14. Enter **From** and press ENTER; you will be prompted to specify the base point.

15. Specify the midpoint of the lower edge of the solid created by **Subtract** and move the cursor vertically upward.

16. Enter **9** at the Command prompt and press ENTER; the box is moved, as shown in Figure 26-54.

17. Choose **Home > Solid Editing > Subtract** from the **Ribbon**. The following command sequence will be displayed:

    Select solids and regions to subtract from.

Select objects: *Select the model.*
Select objects: [Enter]
Select solids and regions to subtract.
Select objects: *Select the box.*
Select objects: [Enter]

The model, after changing the visual style to **3D Hidden**, should look similar to the one shown in Figure 26-55.

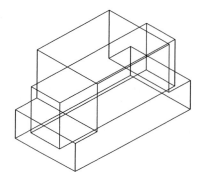

Figure 26-54 Model before subtraction

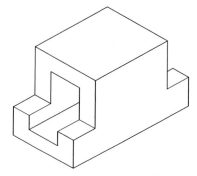

Figure 26-55 Model after subtraction

18. Choose **Home > Modeling > Box > Cylinder** from the **Ribbon**. The following sequence will be displayed at the prompt:

    Specify center point of base or [3P / 2P / Ttr/ Elliptical]: *Press SHIFT and right-click. Choose the **Mid Between 2 Points** option. Select the midpoint of the two longer edge at the top of the model.*
    Specify base radius or [Diameter]: **7**
    Specify height or [2Point / Axis endpoint]: **-10** *(The negative value of the height creates the cylinder extruded in the negative Z direction, see Figure 26-56)*

19. Subtract this cylinder from the existing model, refer to Figure 26-57.

Figure 26-56 Model before subtraction

Figure 26-57 Model after subtraction

20. Create a new UCS on the front face of the model.

21. Choose **Home > Modeling > Cylinder** from the **Ribbon**. The following command sequence will be displayed:

Specify center point of base or [3P / 2P / Ttr/ Elliptical]: *Use the **From** option to specify the center point of the cylinder at a distance of **27** units from the midpoint of the longer edge of the base.*
Specify base radius or [Diameter]: **15**
Specify height or [2Point / Axis endpoint]: **-42**

22. Subtract this cylinder from the model. The final model should look similar to the one shown in Figure 26-58.

Figure 26-58 Final solid model for Example 1

Exercise 1   *Mechanical*

In this exercise, you will create the solid model shown in Figure 26-59. Save this drawing with the name \Ch-26\Exercise1.dwg.

Figure 26-59 Solid model for Exercise 1

DYNAMIC UCS

This option temporarily aligns the XY plane of the current UCS with the selected face of an existing solid. Choose the **Allow / Disallow Dynamic UCS** toggle button from the Status Bar to turn it on. Now, invoke a command and move the cursor near the face on which you want to create the new sketch or solid; the face will be highlighted in dashed lines. Click the left mouse button on the desired face; the XY plane of the UCS will automatically be aligned with the selected face. Also, the origin of the UCS will be placed at the location where you clicked on the face. Now, you can create the desired sketch or solid at the selected face. After the completion of the feature creation, the UCS will again move automatically to its original position.

For example, if you want to sketch a circle on the slant face of the model shown in Figure 26-60, choose **Home > Draw > Circle** from the **Ribbon** and then choose the **Dynamic UCS** button from the Status Bar to turn it on. Next, move the cursor near the slant face of the model; the slant face of the model will be highlighted. Now, left-click on that face at the location where you want to place the UCS. The UCS will automatically align to that face, see Figure 26-61. Now, you can create the circle of the required dimension on the slant face.

If you want a better visualization of how the X, Y, and Z axes of the UCS will be oriented, right-click on the **DUCS** button and choose the **Display crosshair labels** option from the shortcut menu. This will enable you to see the X, Y and Z labels attached with the axes and they will keep on changing as you move the cursor to different faces.

Note
The option of DUCS works only when any command is active.

Figure 26-60 Model with UCS at origin   *Figure 26-61 UCS dynamically moved on the slant face of the model*

Tip
To toggle between the on and off states of the DUCS button, use the F6 function key. To temporarily deactivate the DUCS button, press and hold SHIFT+Z keys.

DEFINING THE NEW UCS USING THE ViewCube AND THE RIBBON

In the earlier chapters, you learned to create new UCS using the **UCS** command. You can also define a new UCS, which is parallel to the current viewing plane using the ViewCube and the **Ribbon**. To define a new UCS whose XY plane is parallel to the current viewing plane, click on a hotspot in the ViewCube; the view corresponding to the hotspot will become normal to screen. Next, choose **WCS > New UCS** from the **WCS** flyout in the ViewCube; you will be prompted to specify the origin. Enter **V** at the Command prompt and press ENTER; a new UCS will be created at the current viewing plane. Now, if you draw a sketch, it will be in the new UCS. To save this new UCS, choose **Coordinates > Named** from the **View** panel of the **Ribbon**; the **UCS** dialog box will be displayed with the **Unnamed** option. Rename it and choose **OK**; the new name of the UCS will be saved and also displayed in the **WCS** flyout of the ViewCube.

To define a new UCS using the **Ribbon**, choose **Views > Views** from the **View** panel of the **Ribbon**; the list of predefined orthographic views will be listed. Select an orthographic view; the model will orient to that view. Also, you will notice that the ViewCube is also oriented according to the selected view, but **Top** is displayed as the view and **Unnamed** is displayed at the **WCS** flyout. Now, if you draw a sketch, it will be in the new UCS. However, if there are different faces parallel to the new UCS, you can draw sketches on those faces.

CREATING EXTRUDED SOLIDS

Ribbon:	Home > Modeling > Extrude	**Command:**	EXTRUDE
Menu Bar:	Draw > Modeling > Extrude	**Toolbar:**	Modeling > Extrude

Sometimes, the shape of the solid model is such that it cannot be created by just applying the Boolean operations on the standard solid primitives. For such cases, AutoCAD provides the **EXTRUDE**, **REVOLVE**, **SWEEP**, **LOFT**, and **PRESSPULL** commands. Using these commands, you can create solid models of any complex shape.

The **EXTRUDE** command is used to create a complex solid model by extruding a closed 2D entity or a region along the Z axis direction or in any specified direction or about a specified path. Remember that for extrusion, the original entity must be a closed loop or a region. The solid model can be created by extruding along the Z axis direction or in any specified direction or about a specified path. All these options of creating the extruded solid are discussed next.

Extruding along the Z Axis

This is the default option and is used to create a solid model by extruding a closed 2D entity or a region along the Z axis direction. Figure 26-62 shows a region to be converted into an extruded solid and Figure 26-63 shows the solid created upon extruding the region. The prompt sequence that will be displayed when you choose the **Extrude** button is given next.

Current wire frame density: ISOLINES=4
Select objects to extrude: *Select the region.*
Select objects to extrude: Enter
Specify height of extrusion or [Direction / Path / Taper angle]: *Specify the height.*

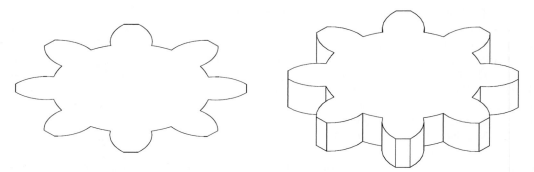

Figure 26-62 Base object for extruding   *Figure 26-63* Solid created upon extruding

You can also specify the taper angle for the extruded solid by selecting the **Taper angle** option **at the Specify height of extrusion or [Direction/Path/Taper angle]** prompt. The positive value of the taper angle will taper in from the base object and the negative value will taper out of the base object (Figure 26-64).

Extruding along a Direction

This option is used to create a solid by extruding a closed 2D entity or a region in any desired direction by specifying two points. You can create an inclined extruded object by selecting a start point and an endpoint. The distance between the start point and endpoint acts as the height of the extrusion. Figure 26-65 shows the result of extruding a base object using the **Direction** option.

Figure 26-64 Results of various taper angles   *Figure 26-65* Extrusion with the **Direction** option

Extruding along a Path

This option is used to create a solid by extruding a closed 2D entity or a region about a specified path. Bear in mind that the path of extrusion should be normal to the plane of the base object. If the path consists of more than one entity, all of them should be first joined using the **PEDIT** command so that the path remains a single entity. This option is generally used for creating complex pipelines and also by architects and interior designers for creating beadings. The path used for extrusion can be a closed entity or an open entity and the valid entities that can be used as path are lines, circles, ellipses, polygons, arcs, polylines, or splines. You cannot specify the taper angle when you use a path. Figure 26-66 shows a base object and a path about which it has to be extruded and Figure 26-67 shows the solid created upon extruding the base entity about the specified path.

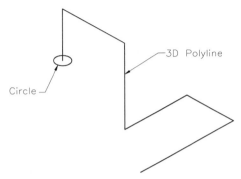

Figure 26-66 *The base object and the path for extrusion*

Figure 26-67 *Solid model created upon extruding the base entity about the specified path*

CREATING REVOLVED SOLIDS

| **Ribbon:** | Home > Modeling > Extrude > Revolve | **Command:** REVOLVE |
| **Menu Bar:** | Draw > Modeling > Revolve | **Toolbar:** Modify > Revolve |

This command is used to create a complex solid by revolving closed 2D entities or regions about a specified revolution axis. The entire 2D entity should be on one side of the revolution axis. Self-intersecting and crossed entities cannot be revolved using this command. Remember that the direction of revolution is determined using the right-hand thumb rule. The axis of rotation can be defined by specifying two points, using the *X* or the *Y* axis of the current UCS, or using an existing object. The following prompt sequence will be issued when you choose this button:

Current wire frame density: ISOLINES=4
Select objects: *Select the object or region to revolve.*
Select objects: Enter
Specify axis start point or
define axis by [Object/X/Y/Z] <Object>:

Specifying the Start Point for the Axis of Revolution

This option is used to define the axis of revolution using two points: the start point and the endpoint of the axis of revolution. The positive direction of the axis will be from the start point to the endpoint and the direction of revolution will be defined using the right-hand thumb rule. Before revolving, make sure the complete 2D entity is on one side of the axis of revolution.

Object

This option is used to create a revolved solid by revolving a selected 2D entity or a region about a specified object. The valid entities that can be used as an object for defining the axis of revolution can be a line or a single segment of a polyline. If the polyline selected as the object consists of more than one entity, then AutoCAD draws an imaginary line from the start point of the first segment to the endpoint of the last segment. This imaginary line is then taken as the object for revolution.

X

This option uses the positive direction of the *X* axis of the current UCS for revolving the selected entity. If the selected entity is not completely on one side of the *X* axis, it will give you an error message that it cannot revolve the object.

Y

This option uses the positive direction of the *Y* axis of the current UCS as the axis of revolution for creating the revolved solid.

Z

This option uses the positive direction of the *Z* axis of the current UCS as the axis of revolution for creating the revolved solid.

After setting the revolution axis option, you will be prompted to specify the angle of revolution. The default value is 360-degree. You can enter the required value at this prompt. Figure 26-68 shows the entity to be revolved and the revolution axis along with the revolved solid. Figure 26-69 shows the same solid revolved through an angle of 270-degree.

Figure 26-68 Creating a revolved solid

Figure 26-69 Solid revolved to an angle

CREATING SWEPT SOLIDS

Ribbon:	Home > Modeling > Extrude > Sweep	**Toolbar:** Modeling > Sweep
Menu Bar:	Draw > Modeling > Sweep	**Command:** SWEEP

 This tool is used to sweep an open or a closed profile along a path. To create this solid, you need an object and a path. The object is a cross-section for the sweep feature and the path is the course taken by the object while creating the swept solid. Figure 26-70 shows the object to be swept and the path to be followed and Figure 26-71 shows the resulting sweep feature.

The path specified for the profile and the object for the cross-section of the swept solid can be open or closed. If the object for the cross-section is open or closed, but is not forming a single region, the generated feature will result in a surface sweep. If the object for the section is closed and it forms a single region, the generated feature will be a solid. Figure 26-72 shows the object and the closed path and Figure 26-73 shows the resulting sweep feature.

Note
*The path selected for the **SWEEP** command should not form a self-intersecting loop.*

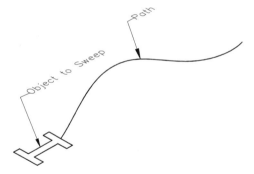

Figure 26-70 Object to be swept and path

Figure 26-71 Resulting sweep feature

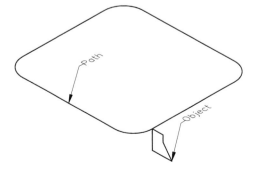

Figure 26-72 Object to be swept with a closed path

Figure 26-73 Resulting sweep feature

You can select more than one section at a time to be swept along a path, but all these sections should be drawn on the same plane. The prompt window that will be followed is discussed next.

> Command: **SWEEP**
> Current wire frame density: ISOLINES=4
> Select objects to sweep: *Select the object to be swept.*
> Select objects to sweep: *Select more objects to be swept or press ENTER.*
> Select sweep path or [Alignment/Base point/Scale/Twist]: *Select the path to be followed by the object or enter an option.*

Depending on your requirement, you can invoke the following options at the **Select sweep path or [Alignment/Base point/Scale/Twist]** prompt:

Alignment

This option allows you to specify whether the object will be oriented normal to the curve at the start point. By default, the sweep feature is created with a section normal to the path at the start point. To avoid this, choose **No** from the shortcut menu at the **Align sweep object perpendicular to path before sweep [Yes/No]** prompt. Figure 26-74 shows the object and the path for the sweep command. Figure 26-75 shows the resulting sweep feature with the aligned option **Yes** and Figure 26-76 shows the resulting sweep feature with the aligned option **No**.

Base Point

This option is used to specify the point on the object that will be attached to the path while sweeping it. You can specify the location of the cross-section while sweeping it when prompted to specify

Figure 26-74 Object and path for sweep command

Figure 26-75 Swept solid with the aligned option **Yes**

Figure 26-76 Swept solid with the aligned option **No**

the base point. Figure 26-77 shows the path and the cross-section with base points P1 and P2. Figure 26-78 shows the resulting sweep feature with P1 as the base point and Figure 26-79 shows the resulting sweep feature with P2 as the base point.

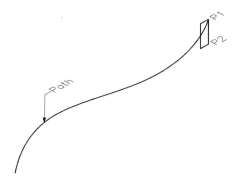

Figure 26-77 *Path with object and P1, P2 as base points*

Figure 26-78 *Sweep with P1 as the base point*

Figure 26-79 *Sweep with P2 as the base point*

Scale

This option scales the object uniformly from the start of the path to the end of the path while sweeping it. Note that the size of the start section will remain the same and end section will be scaled by the specified value. Also, the transition from the beginning to the end is smooth. Figure 26-80 shows a pentagon swept along a path with a scale factor of 0.25.

Twist

You can rotate the object about a path uniformly, from the starting to the end with a specified angle. The value of the twist should be less than 360 degrees. Figure 26-81 shows a sweep feature with 300 degrees of twist.

Figure 26-80 *Sweep with the **Scale** option*

Figure 26-81 *Sweep with the **Twist** option*

Note

A helix can also be used as the path for the sweep command. Figure 26-82 shows the object and a helical path generated using the HELIX command and Figure 26-83 shows the resulting swept solid.

Figure 26-82 *Helical path and object to be swept*

Figure 26-83 *Sweep feature with the helical path*

CREATING LOFTED SOLIDS

Ribbon: Home > Modeling > Extrude > Loft	**Command:** LOFT	
Menu Bar: Draw > Modeling > Loft	**Toolbar:** Modeling > Loft	

The **LOFT** command is used to create a feature by blending two or more similar or dissimilar cross-sections together to get a free form shape. These similar or dissimilar cross-sections may or may not be parallel to each other. You can select an open as well as closed sections to create a loft. However, note that if you select an open section first, all the sections should be open. Similarly, if you select a closed section, all the sections should be closed. The open section results in a surface and the closed section results in a solid. To create a loft feature, choose **Home > Modeling > Extrude >Loft** from the **Ribbon**. Next, select at least two sections in a sequence in which you want to blend them. The prompt sequence that will be followed after choosing the **Loft** button is given next.

Command: **LOFT**
Select cross-sections in lofting order: *Select the first section from where the blending will start.*
Select cross-sections in lofting order: *Select the next section up to which the blending of material will propagate, starting from the first one.*
Select cross-sections in lofting order: *Select the next section or press ENTER to finish the selection of sections.*
Enter an option [Guides/Path/Cross-sections only] <Cross-sections only>: *Invoke any one of the three options.*

Depending on requirement, you can invoke any one of the following three options.

Guide

AutoCAD enables you to select guide curves between the sections of the loft feature. These guide curves define the shape of the material addition between the selected sections for the loft feature. Guide curves also specify the points of different sections to be joined while adding material between them. This reduces the possibility of an unnecessary twist in the resulting feature. You can select as many guide curves as you require. But these guide curves should pass through or should touch each section that you selected for loft feature. Also the guides should start from the first section and terminate at the last section. Figure 26-84 shows the sections to be lofted and the guide curves and Figure 26-85 shows the resulting model.

Figure 26-84 Section and guide curves for the loft feature

Figure 26-85 Resulting loft feature by using the guide curve

Path

The path option is used to create a loft feature by blending more than one section along the direction specified by the path curve. Note that you can select only one path curve. The selected path curve may or may not touch all the sections, but it should pass through all the sketching planes of the sections. Figure 26-86 shows the sections and the path along which the blending of material will propagate and Figure 26-87 shows the resulting loft feature.

Figure 26-86 Sections and path to be used for the **Loft** option

Figure 26-87 Resulting loft feature after using the **Path** option

Cross sections only

When you select the **Cross sections only** option, the **Loft Settings** dialog box will be displayed, as shown in Figure 26-88. This dialog box is used to control the shape of the material addition between the sections without specifying any guide curve or path. Various options in the **Loft Settings** dialog box are discussed next.

Ruled

If you select the **Ruled** radio button, a straight blend will be created by connecting different sections with straight lines. The loft feature created by this option will have sharp edges at the sections. Figure 26-89 shows the loft feature when the **Ruled** radio button is selected.

*Figure 26-88 The **Loft Settings** dialog box*

Smooth Fit

If you select the **Smooth Fit** radio button, a smooth loft will be created between the sections. The loft feature created by this option will have smooth edges at the intermediate section. Figure 26-90 shows the loft feature when the **Smooth Fit** radio button is selected.

*Figure 26-89 Loft solid with the **Ruled** radio button selected*

*Figure 26-90 Loft solid with the **Smooth Fit** radio button selected*

Normal to

This option controls the shape of the loft between the sections. When you select this radio button, the drop-down list will be enabled. The following options are available in this drop-down list.

Start cross section. In this option, the normal of the lofted feature, is normal to the start section. The loft feature starts normal to the start section and follows a smooth polyline as it approaches the next section. Figure 26-91 shows the section selected to create the loft and Figure 26-92 shows the loft created by specifying the **Start cross section** option.

End cross section. In this option, the normal to the blending material is normal to the end section. At the second last section, the blending material starts along a spline and while reaching up to end section, it becomes normal to it, as shown in Figure 26-93.

Start and End cross sections. In this option, the normal to the blending material is normal to both the start and end section.

All cross sections. In this option, the normal to the blending material always remains normal to all cross-sections, as shown in Figure 26-94.

Figure 26-91 *Figure showing start and end sections*

Figure 26-92 *Loft solid with blending material normal to the start section*

Figure 26-93 *Loft solid with blending material normal to the end section*

Figure 26-94 *Loft solid with blending material normal to all sections*

Draft angles

In this option, you can define the angle at the start section and the end section of the loft. Remember that you cannot define an angle for the intermediate sketches. You can also specify the distance up to which the blending material will follow this draft angle before bending toward the intermediate sections. The various options for specifying the draft angle are discussed next.

Start angle. This edit box is used to specify the draft angle at the start section of the loft.

Figure 26-95 shows the two sections to be blended and Figure 26-96 shows the loft with the start and end angles as 0-degree. Figures 26-97 and 26-98 shows the loft feature with the start angle as 90 and 180-degree, respectively.

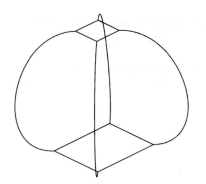

Figure 26-95 *Start and end sections of the loft feature*

Figure 26-96 *Loft feature with 0-degree start and end angles*

Figure 26-97 *Loft feature with 90-degree start and end angles*

Figure 26-98 *Loft feature with 180-degree start angle and 90-degree end angle*

Start magnitude. The value specified in this region signifies the distance up to which the blending material will follow the start angle before bending toward the next section. Figure 26-99 shows a loft feature with 180-degree as the start and end angles, and 15 as the start magnitude. Figure 26-100 shows a loft feature with 180-degree as the start and the end angles and 2 as the start magnitude. Notice the difference in the extent up to which the blended material follows the start angle in the two figures.

End angles. This edit box is used to specify the draft angle at the end section of the loft. Figure 26-95 shows two sections to be blended and Figure 26-96 shows the loft with the start and end angles as 0-degree. Figure 26-97 and Figure 26-98 show the loft feature with the end angle as 90 and 180-degree, respectively.

End magnitude. The value specified in this region signifies the distance up to which the blending material will follow the end angle before approaching the last section. Figure 26-99 shows a loft feature with 180-degree as the start and end angles, and 2 as the end magnitude. Figure 26-100 shows a loft feature with 180-degree as the start and the end angles, and 10 as the end magnitude. Notice the difference in the extent up to which the blended material follows the end angle in the two figures.

Figure 26-99 *Loft feature with 180-degree as the start and end angles, 15 as the start magnitude, and 2 as the end magnitude*

Figure 26-100 *Loft feature with 180-degree as the start and end angles, 2 as the start magnitude, and 10 as the end magnitude*

Close surface or solid

The **Close surface or solid** check box is selected to close the loft feature by joining the end section with the start section. Figure 26-101 shows a loft feature created with this check box cleared and Figure 26-102 shows the same sections lofted by this check box selected.

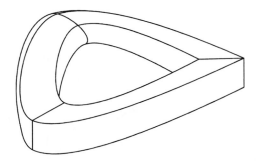

Figure 26-101 *Loft solid with the **Close surface or solid** check box cleared*

Figure 26-102 *Loft solid with the **Close surface or solid** check box selected*

Note

*To visualize the effect of the **Close surface or solid** option, the loft feature should contain at least three sections. Otherwise, the extra blending material will be added at the same location where the loft feature would have been created without selecting this check box, thereby resulting in no change.*

Preview changes

If this check box is selected, you can dynamically visualize the effect of changing various options on the screen. AutoCAD regenerates the model for any change in the options.

Tip
You can also use a point as a section for the loft feature, but it should always be the start section or the end section of the loft feature.

CREATING PRESSPULL SOLIDS

Ribbon:	Home > Modeling> Press/Pull	**Toolbar:** Modeling > Press/Pull
Command:	PRESSPULL		

The **PRESSPULL** command is used to remove or add material of any desired shape from a model. As the name presspull indicates, if you press a closed boundary inside an existing model, it will remove the material of the shape of the closed boundary. However, if you pull a closed boundary outside an existing model, it will add material to the model. To perform this operation, choose **Home > Presspull** from the **Ribbon**. Next, click inside a closed boundary that you want to press or pull and then move the mouse in the desired direction. You can specify the desired height by entering a value or by clicking the mouse. Figure 26-103 shows the solid model with the closed area to be presspulled. Figure 26-104 shows the resulting model by pressing the closed region and Figure 26-105 shows the resulting model by pulling the closed area.

Figure 26-103 Model with closed area to be presspulled

If the sketch that you select to press or pull has nested loops, as shown in Figure 26-106, the inner closed loops will be deleted from the outer loop, as shown in Figure 26-107. The closed area you select for the presspull should consist of lines, polylines, 3D face, edges, regions, and faces of a 3D solid. But all these entities should form a coplanar closed area.

Figure 26-104 *Model with the closed area pressed*   **Figure 26-105** *Model with closed area pulled*

Figure 26-106 *Area to be presspulled with closed*   **Figure 26-107** *Resulting model after the*
boundary inside it   *presspull operation is performed*

Tip
*To perform the presspull operation without choosing the **Presspull** button, press and hold the
CTRL+ALT keys and then click inside the closed area to be pressed or pulled.*

Example 2

In this example, you will create the solid model shown in Figure 26-108.

1. Start a new file using the *acad3D.dwt* template.

2. Increase the limits and then zoom to the limits of the drawing.

3. Switch on the ortho mode. Using lines and arc, create the base of the model with the
    dimensions shown in Figure 26-109. You can also use the **PLINE** command to create this
    sketch.

4. Choose **Home > Draw > Region** from the **Ribbon**. The prompt sequence is given next.

Figure 26-108 *Solid model for Example 2*

Select objects: *Select the complete base.*
Select objects: Enter
1 loop extracted.
1 Region created.

Note
*If you have created the sketch using the **PLINE** command, you do not need to convert it into a region.*

5. Choose **Home > Modeling > Extrude** from the **Ribbon**. The following sequence will be displayed at the prompt:

    Select objects to extrude: *Select the region.*
    Select objects to extrude: Enter
    Specify height of extrusion or [Direction/ Path/ Taper angle]: **16**

6. Change the viewpoint to the SE Isometric view.

7. Choose **Home > Modeling > Box > Wedge** from the **Ribbon**. The following sequence will be displayed at the prompt:

    Specify first corner or [Center] <0,0,0>: *Pick a point on the screen.*
    Specify other corner or [Cube/Length]: **L**
    Specify length: **40**
    Specify width: **12**
    Specify height or [2Point]: **28**

8. Move this wedge using the endpoint so that it is properly aligned with the base, see Figure 26-110.

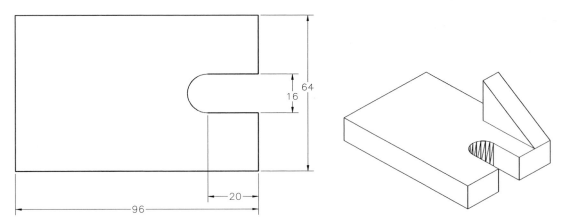

Figure 26-109 *Base for the solid model*

Figure 26-110 *Wedge aligned with the base*

9. Choose the **X** button from the **Coordinates** panel in the **View** tab of the **Ribbon**. The following sequence will be displayed at the prompt:

    Specify rotation angle about X axis <90>: Enter

10. Draw the object shown in Figure 26-111 and then convert it into a region.

11. Choose **Home > Extrude** from **Ribbon**. The prompt sequence is as follows:

    Current wire frame density: ISOLINES=4
    Select objects to extrude: *Select the region.*
    Select objects to extrude: Enter
    Specify height of extrusion or [Direction/ Path/ Taper angle]: **42**

    The model after extruding the region is shown in Figure 26-112.

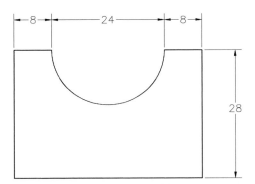

Figure 26-111 *The region to be extruded*

Figure 26-112 *Solid aligned with the base*

12. Choose the **Y** button from the **Coordinates** panel in the **View** tab of the **Ribbon**. The
prompt sequence is as follows:

Specify rotation angle about Y axis <90>: Enter

13. Draw the object shown in Figure 26-113 and then convert it into a region.

14. Extrude and then move it so that it is properly aligned with the base, see Figure 26-114.

15. Choose the **Union** button from the **Solid Editing** panel in the **Home** tab of the **Ribbon**.
The prompt sequence is as follows:

Select objects: *Select all objects.*
Select objects: Enter

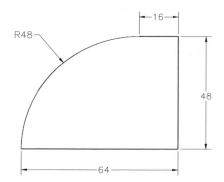

Figure 26-113 *The region before extrusion*

Figure 26-114 *Next solid aligned with the base*

The final solid model should look similar to the one shown in Figure 26-115.

Figure 26-115 *Solid model for Example 2*

Example 3

In this example, you will create the solid model shown in Figure 26-116.

Figure 26-116 *Solid model for Example 3*

1. Start a new file using the *acad3D.dwt* template.

2. Choose **Views > Views > Left** from the **View** panel; the UCS is aligned to the left view. Next, draw the path and the cross-section of the sweep feature, as shown in Figure 26-117.

3. Choose **Home > Draw > Region** from the **Ribbon**. The prompt sequence is as follows:

    Select object: *Select both the circles.*
    Select objects: Enter
    2 loops extracted.
    2 Regions created.

4. Choose **Home > Solid Editing > Subtract** from the **Ribbon**. The following sequence will be displayed at the prompt:

    Select objects: *Select the region of diameter 29.*
    Select objects: Enter
    Select solids and regions to subtract..
    Select objects: *Select the region of diameter 21.*
    Select objects: Enter

5. Choose **Home > Modeling > Extrude > Sweep** from the **Ribbon.** The following sequence will be displayed at the prompt:

Select objects to sweep: *Select the region created in Step 4.*
Select objects to sweep: [Enter]
Select sweep path or [Alignment/Base point/Scale/Twist]: *Select the arc created in Step 2.*

6. Choose **View > Views > Views > SE isometric** from the **Ribbon** and change the **UCS** to **World** using the ViewCube. Also, choose **Home > View > Visual Styles > Realistic** from the **Ribbon**. The resulting model should look similar to the one shown in Figure 26-118.

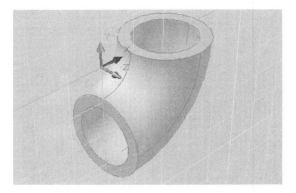

Figure 26-117 *Sketch for base pipe feature*   *Figure 26-118* *Resulting sweep feature*

7. Choose **Top** from the ViewCube and draw the sketch with the dimensions shown in Figure 26-119. The model, after drawing the sketch and turning off the dimension layer, is shown in Figure 26-120.

Figure 26-119 *Sketch for the area to presspull*   *Figure 26-120* *Location of the sketch*

8. Choose **Home > Modeling > Press/Pull** from the **Ribbon** and then click inside the rectangular area at the location shown as Area A in Figure 26-121. Next, move the mouse in the upward direction, enter the value **5** at the Command prompt, and press ENTER.

9. Again, choose the **Press/Pull** button and pull the area shown as Area B in Figure 26-121

by the same value as specified in Step 8. The resulting model should be similar to the one shown in Figure 26-122.

Figure 26-121 *Two areas to presspull*

Figure 26-122 *Resulting model after Step 10*

10. Similarly, create the flange for the other face of the pipe. The resulting model should be similar to the one shown in Figure 26-123.

11. Choose **Home > Solid Editing > Union** from the **Ribbon**. The following sequence will be displayed at the prompt:

Select objects: *Select all the objects.*
Select objects: Enter

The final solid model should look similar to the one shown in Figure 26-124.

Figure 26-123 *Resulting model after Step 12*

Figure 26-124 *Final model after Union*

Self-Evaluation Test

Answer the following questions and then compare them to those given at the end of this chapter:

1. The **CONE** command can be used to create a solid cone with a circular and an elliptical base. (T/F)

2. Open entities can be converted into regions. (T/F)

3. An ellipse can be converted into a polysolid. (T/F)

4. You can twist a cross-section while sweeping it. (T/F)

5. The _____ command can be used to add as well as remove material from a model.

6. In AutoCAD, you can create a pyramid of maximum 42 sides. (T/F)

7. The _____ command is used to generate a helical curve.

8. The number of paths for a loft feature cannot be more than _____ .

9. Atleast _____ cross-sections are required to create a loft feature.

10. You can extrude the selected region about a path using the _____ option of the **EXTRUDE** command.

Review Questions

Answer the following questions:

1. An open entity can be revolved. (T/F)

2. You cannot apply Boolean operations on regions. (T/F)

3. The entity to be revolved should lie completely on one side of the revolution axis. (T/F)

4. You cannot apply the **PRESSPULL** command on an open object. (T/F)

5. There cannot be more than one guide for the loft feature. (T/F)

6. A curve generated by the **HELIX** command can be used as a path for the swept solids. (T/F)

7. Which value of the taper angle will taper the extruded model from the base?

    (a) Positive   (b) Negative
    (c) Zero   (d) None

8. Which command is used to create a cube?

    (a) **BOX**   (b) **CUBOID**
    (c) **POLYGON**   (d) **CYLINDER**

9. Which command is used to check the interference between the selected solid models?

    (a) **INTERFERE**   (b) **INTERSECT**
    (c) **INTERFERENCE**   (d) **CHECK**

10. The **PRESSPULL** command is used to

    (a) Add material   (b) Remove material
    (c) Both add and remove material   (d) Neither add nor remove material

11. The _____ direction is known as the extrusion direction.

12. The _____ command is used to create an apple-like structure.

13. The _____ command is used to create a revolved solid.

14. Guides for the **LOFT** command should always _____ the cross-section.

15. The _____ option of the **REVOLVE** command is used to select a 2D entity as the revolution axis.

Exercises

Exercise 2

In this exercise, you will create the solid model shown in Figure 26-125. Assume the missing dimensions.

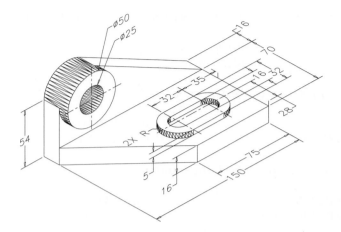

Figure 26-125 *Model for Exercise 2*

Exercise 3

In this exercise, you will create the solid model shown in Figure 26-126. The dimensions for the model are given in the same figure. Assume the missing dimensions.

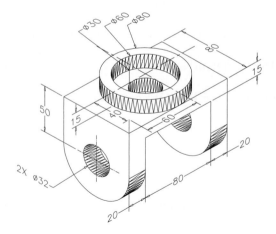

Figure 26-126 *Solid model for Exercise 3*

Exercise 4

In this exercise, you will create the solid model shown in Figure 26-127. The dimensions for the model are shown in Figures 26-128 and 26-129.

Figure 26-127 *Model for Exercise 4*

Figure 26-128 *Top view of the model*

Figure 26-129 *Front view of the model*

Problem-Solving Exercise 1

In this exercise, you will create the solid model with dimensions shown in Figure 26-130 and Figure 26-131.

Figure 26-130 *Solid model for Problem-Solving Exercise 1*

Figure 26-131 *Viewports for Problem-Solving Exercise 1*

Answers to Self-Evaluation Test

1. T, **2.** F, **3.** F, **4.** T, **5. PRESSPULL**, **6. F**, **7. HELIX**, **8. one**, **9.** two, **10. Path**

Chapter 27

Modifying 3D Objects

Learning Objectives

After completing this chapter, you will be able to:
• *Use the* **FILLET** *and the* **CHAMFER** *commands for solid models.*
• *Rotate and mirror solid models in 3D space.*
• *Create an array in 3D space.*
• *Align the solid models using the* **ALIGN** *command.*
• *Extract the edges of a solid model.*
• *Slice the solid models and create cross-sections.*
• *Convert objects to surfaces and solids.*
• *Convert surfaces to solids.*

FILLETING SOLID MODELS

Ribbon:	Home > Modify > Fillet	**Menu Bar:**	Modify > Fillet
Toolbar:	Modify > Fillet	**Command:**	FILLET

As mentioned earlier, the **FILLET** command is used to round the edges or corners of the models. This is generally done to reduce the stress concentration area in the model. The behavior of this command is different while working with 2D entities from the behavior while working with solid models. Therefore, it is very important for the user to first understand the use of this command to fillet the edges of the solid models. Figure 27-1 shows two lines that are selected to be filleted. Now, as these lines are nothing but 2D entities, when you select these two lines to fillet, the result will be as shown in Figure 27-2.

Figure 27-1 *Lines before filleting*   **Figure 27-2** *Lines after filleting*

This shows that if actually there was a vertical edge at the corner of the two lines shown in Figure 27-1, then it would have been filleted. However, in 3D models, you directly have the vertical edges and therefore, you have to select the vertical edge to be filleted. Figure 27-3 shows a solid model. To fillet this model, you just need to select the vertical edges, see Figure 27-4.

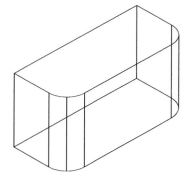

Edges selected to fillet

Figure 27-3 *Selecting the edges to be filleted*   **Figure 27-4** *Model after filleting the edges*

The prompt sequence that will be followed when you choose the **Fillet** button is given next.

Current settings: Mode = TRIM, Radius = 0.0000
Select first object or [Undo/Polyline/Radius/Trim/Multiple]: *Select the edges to fillet. You will be allowed to select only one edge at this moment.*
Enter fillet radius: *Enter the required fillet radius.*
Select an edge or [Chain/Radius]: *Select the other edges to fillet or enter an option.*

Undo

This option is used to revert to the previous fillet, when you are working in multiple modes of **FILLET** command.

Chain

This option is used to select the other tangential edges on the model. For example, you can select all tangential edges on the top or the bottom face of the solid model shown in Figure 27-5 for filleting in just a single attempt using the **Chain** option. Figure 27-6 shows the resulting fillet.

Radius

This option is used to redefine the fillet radius.

Figure 27-5 *Selecting the edges using the* **Chain** *option*

Figure 27-6 *The fillet created using the* **Chain** *option*

CHAMFERING SOLID MODELS

Menu Bar:	Modify > Chamfer	Toolbar:	Modify > Chamfer
Command:	CHAMFER	Ribbon:	Home > Modify > Fillet > Chamfer

The **CHAMFER** command is used to bevel the edges of the solid models. This command is also used to reduce the area of the stress concentration in the solid models. The working of this command is also different while working with solid models. The following prompt sequence is displayed when you choose the **Chamfer** button.

(TRIM mode) Current chamfer Dist1 = current, Dist2 = current
Select first line or [Undo/Polyline/Distance/Angle/Trim/mEthod/Multiple]: *Select the edge to chamfer. One of the faces associated with the edge will be selected and highlighted.*

Base surface selection...
Enter surface selection option [Next/OK (current)] <OK>: *Give a null response if you want to make this face as the base surface. Otherwise enter N at this prompt.*
Specify base surface chamfer distance <default value>: *Specify the distance.*
Specify other surface chamfer distance <default value>: *Specify the distance.*
Select an edge or [Loop]: *Select the edge to fillet.*

Figures 27-7 and 27-8 show the solid model before and after chamfering, respectively.

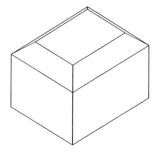

Figure 27-7 *Solid model before chamfering*   *Figure 27-8* *Solid model after chamfering*

Undo

This option reverses the previous chamfer operation performed using the current command.

Loop

This option is used to select all the edges composed of a loop on the selected face of the solid model. To use this option, you will have to select any of the edges that comprises the loop at the **Select an edge or [Loop]** prompt.

ROTATING SOLID MODELS IN 3D SPACE

Command: ROTATE3D

The **ROTATE** command is used to rotate the selected solid model in the 3D space about a specified axis. Once again the right-hand thumb rule will be used to determine the direction of rotation of the solid model in 3D space. The prompt sequence that will follow when you choose this command is given next.

Current positive angle: ANGDIR=counterclockwise ANGBASE=0
Select objects: *Select the solid model.*
Select objects: Enter
Specify first point on axis or define axis by
[Object/Last/View/Xaxis/Yaxis/Zaxis/2Points]: *Specify a point on the axis of rotation or select a option.*

2points Option

This is the default option for rotating solid models. This option allows you to rotate the solid model about an axis specified by two points. The direction of the axis will be from the first point to the second point. Using this direction of the axis, you can calculate the direction of rotation of the solid model by applying the right-hand thumb rule. The prompt sequence that will follow when you invoke this option is given next.

[Object/Last/View/Xaxis/Yaxis/Zaxis/2points]:2P [Enter]
Specify first point on axis: *Specify the first point of the rotation axis. See Figure 27-9.*
Specify second point on axis: *Specify the second point of the rotation axis. See Figure 27-9.*
Specify rotation angle or [Reference]: *Specify the angle of rotation.*

Object Option

This option is used to rotate the solid model in the 3D space using a 2D entity. The 2D entities that can be used are lines, circles, arcs, or 2D polyline segments. If the selected entity is a line or a straight polyline segment, then it will be directly taken as the rotation axis. However, if the selected entity is an arc or a circle, then an imaginary axis will be drawn starting from the center and normal to the plane, in which the arc or circle is drawn. The object will then be rotated about this imaginary axis. The prompt sequence that will follow when you invoke this option is given next.

[Object/Last/View/Xaxis/Yaxis/Zaxis/2points]: O [Enter]
Select a line, circle, arc, or 2D-polyline segment: *Select the 2D entity, as shown in Figure 27-10.*
Specify rotation angle or [Reference]: *Specify the angle of rotation.*

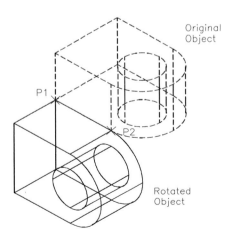

Figure 27-9 *Rotating the solid model using the* **2points** *option*

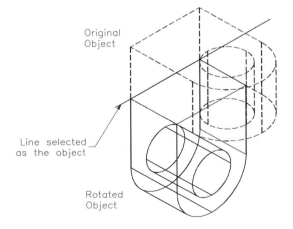

Figure 27-10 *Rotating the solid model using the* **Object** *option*

Last Option

This option uses the same axis that was last selected to rotate the solid model.

View Option

This option is used to rotate the solid model about the viewing plane. In this case, the viewing plane is the screen of the computer. This option draws an imaginary axis starting from the specified point and continues normal to the viewing plane. The model is then rotated about this axis. The prompt sequence that will follow when you invoke this option is given next.

Specify first point on axis or define axis by
[Object/Last/View/Xaxis/Yaxis/Zaxis/2points]: **V**
Specify a point on the view direction axis <0,0,0>: *Specify the point on the view plane, as shown in Figure 27-11.*
Specify rotation angle or [Reference]: *Specify the angle of rotation.*

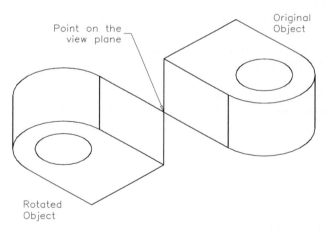

Figure 27-11 *Rotating the solid model using the **View** option through an angle of 180-degree*

Xaxis Option

This option is used to rotate the solid model about the positive *X* axis of the current UCS. On invoking this option, you will be prompted to select a point on the *X* axis. The prompt sequence that will follow when you invoke this command is given next.

Specify first point on axis or define axis by
[Object/Last/View/Xaxis/Yaxis/Zaxis/2points]: **X**
Specify a point on the X axis <0,0,0>: *Specify the point on the X axis.*
Specify rotation angle or [Reference]: *Specify the angle of rotation.*

Yaxis Option

This option is used to rotate the solid model about the positive *Y* axis of the current UCS. When you invoke this option, you will be prompted to select a point on the *Y* axis. The prompt sequence that will follow when you invoke this command is given next.

> Specify first point on axis or define axis by
> [Object/Last/View/Xaxis/Yaxis/Zaxis/2points]: **Y**
> Specify a point on the Y axis <0,0,0>: *Specify the point on the Y axis.*
> Specify rotation angle or [Reference]: *Specify the angle of rotation.*

Zaxis Option

This option is used to rotate the solid model about the positive *Z* axis of the current UCS. When you invoke this option, you will be prompted to select a point on the *Z* axis. The prompt sequence that will follow when you invoke this command is given next.

> Specify first point on axis or define axis by
> [Object/Last/View/Xaxis/Yaxis/Zaxis/2points]: **Z**
> Specify a point on the Z axis <0,0,0>: *Specify the point on the Z axis.*
> Specify rotation angle or [Reference]: *Specify the angle of rotation.*

ROTATING SOLID MODELS ABOUT AN AXIS

Ribbon:	Home > Modify > 3D Rotate	**Toolbar:** Modeling > 3D Rotate	
Menu Bar:	Modify > 3D Operations > 3D Rotate	**Command:** 3DROTATE	

 This command is used to dynamically rotate an object or subobjects such as edges and faces with respect to a specified base point and about a selected axis. To perform this operation, choose **Home > Modify > 3D Rotate** from the **Ribbon** and select the object or subobject to be rotated. After you select the object, the Rotate Grip tool will be displayed attached to the cursor, see Figure 27-12. Specify the base point by placing the base point of the rotate grip tool at the desired location. Now, specify the axis about which you want to rotate the object. To do this, move the cursor close to the axis handles (Figure 27-12). The axis handle closest to the cursor will change to golden color and an axis passing through its center will be displayed, as shown in Figure 27-13. Select the desired axis handle.

The prompt sequence that is displayed when you invoke this tool is as follows:

> Select Object: *Select an object or subobject to rotate*
> Select Object: *Select more objects or press the ENTER key*
> Specify base point: *Select a point, which will remain fixed or about which the object will rotate*
> Pick a rotation axis: *Select the axis by clicking on the axis handle about which the object will rotate.*
> Specify angle start point or type an angle: *Select the angle measurement start location by using the mouse or enter the rotation angle to complete the step.*
> Specify angle end point: *specify the angle measurement end location by using the mouse.*

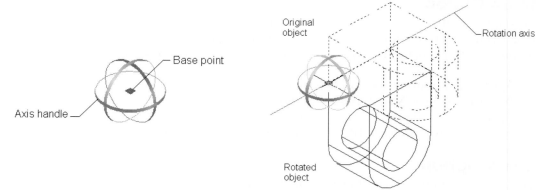

Figure 27-12 *The Rotate Grip tool*   **Figure 27-13** *Model rotated through 90-degree about X axis*

MIRRORING SOLID MODELS IN 3D SPACE

Ribbon:	Home > Modify > 3D Mirror	**Command:**	MIRROR3D
Menu Bar:	Modify > 3D Operations > 3D Mirror		

The **MIRROR3D** command is used to mirror the solid models about a specified plane in the space. The prompt sequence that will follow when you choose **Home > Modify > 3D Mirror** from the **Ribbon** is given next.

Select objects: *Select the solid model to be mirrored.*
Select objects: ⏎
Specify first point of mirror plane (3 points) or
[Object/Last/Zaxis/View/XY/YZ/ZX/3points] <3points>: *Specify the first point on the mirror plane or select an option.*

3points Option

This is the default option for mirroring the solid models. As discussed earlier, a line can be defined by the two points from which the line passes. Similarly, a plane can be defined by the three points through which it passes. This option allows you to specify the three points from which the mirroring plane passes. The prompt sequence that follows when you invoke this command is given next.

Specify first point of mirror plane (3 points) or
[Object/Last/Zaxis/View/XY/YZ/ZX/3points] <3points>: ⏎
Specify first point on mirror plane: *Specify the first point on the plane. See Figure 27-14.*
Specify second point on mirror plane: *Specify the second point on the plane. See Figure 27-14.*
Specify third point on mirror plane: *Specify the third point on the plane. See Figure 27-14.*
Delete source objects? [Yes/No] <N>: ⏎

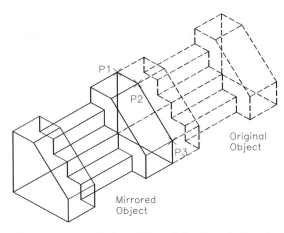

*Figure 27-14 Mirroring the solid model using the **3points** option*

Object

This option is used to mirror the solid model using a 2D entity. The 2D entities that can be used to mirror the solids are circles, arcs, and 2D polyline segments. The prompt sequence that follows when you invoke this option is given next.

Specify first point of mirror plane (3 points) or
[Object/Last/Zaxis/View/XY/YZ/ZX/3points] <3points>: **O**
Select a circle, arc, or 2D-polyline segment: *Select the 2D entity, as shown in Figure 27-15.*
Delete source objects? [Yes/No] <N>: Enter

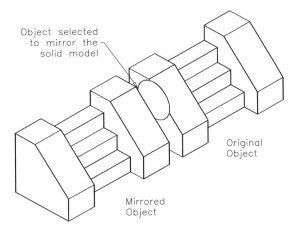

*Figure 27-15 Mirroring the solid model using the **Object** option*

Zaxis

This option allows you to define a mirroring plane using two points. The first point is the point on the mirroring plane and the second point is a point on the positive direction of the Z axis of that plane. The prompt sequence that follows when you invoke this option is given next.

Specify first point of mirror plane (3 points) or
[Object/Last/Zaxis/View/XY/YZ/ZX/3points] <3points>: **Z**
Specify point on mirror plane: *Specify the point on the plane, as shown in Figure 27-16.*
Specify point on Z-axis (normal) of mirror plane: *Specify the point on the Z direction of the plane, as shown in Figure 27-16.*
Delete source objects? [Yes/No] <N>: Enter

View Option

This is one of the interesting options provided for mirroring solid models. This option is used to mirror the selected solid model about the viewing plane. The viewing plane, in this case, is the screen of the monitor (Figure 27-17). The prompt sequence that will follow when you invoke this command is given next.

Specify first point of mirror plane (3 points) or
[Object/Last/Zaxis/View/XY/YZ/ZX/3points] <3points>: **V**
Specify point on view plane <0,0,0>: *Specify the point on the view plane.*
Delete source objects? [Yes/No] <N>: Enter

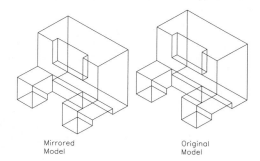

Figure 27-16 *Mirroring the solid model using the* ***Zaxis*** *option*

Figure 27-17 *A solid model mirrored using the* ***View*** *option and then the mirrored model is isolated using the* ***MOVE*** *command*

Tip

As the model is mirrored about the view plane, the new model will be placed over the original model when you view it from the current viewpoint. The best option to view the mirrored model is to hide the hidden lines using the ***HIDE*** *command. You can also move the new model away from the last model using the* ***MOVE*** *command or change the viewpoint for viewing both the models simultaneously.*

XY/YZ/ZX

These options are used to mirror the solid model about the *XY*, *YZ*, or *ZX* planes of the current UCS.

Tip

*All these planes will be considered with reference to the current orientation of the UCS and not with its world position. This means that when you select the **XY** option to mirror the solid model, then the XY plane of the current UCS will be considered to mirror the model and not the XY plane of the world UCS.*

Example 1

In this example, you will create the solid model shown in Figure 27-18. Assume the missing dimensions.

Figure 27-18 Solid model for Example 1

1. Start a new 3D template drawing file in **3D Modeling** workspace. Choose **Views > Views > Top** from the **View** panel of the **Ribbon** and increase the limit to 100,100.

2. Create the base of the model and then change the view to Southeast isometric view using the ViewCube, see Figure 27-19.

3. Choose **Views > Views > Front** from the **View** panel of the **Ribbon**.

4. Draw the sketch for the next object using the **LINE, ARC**, and **CIRCLE** commands.

5. Create the object by using the **REGION, SUBTRACT**, and **EXTRUDE** commands. The resulting object is shown in Figure 27-20.

6. Choose **Home > Modify > 3D Mirror** from the **Ribbon**. The prompt sequence that will follow is given next.

Figure 27-19 *Base of the model*

Figure 27-20 *Model after creating the next object and moving it*

Select objects: *Select the last object.*
Select objects: Enter
Specify first point of mirror plane (3 points) or
[Object/Last/Zaxis/View/XY/YZ/ZX/3points] <3points>: **Z**
Specify point on mirror plane: *Select P1, as shown in Figure 27-21.*
Specify point on Z-axis (normal) of mirror plane: *Select P2, as shown in Figure 27-21.*
Delete source objects? [Yes/No] <N>: Enter

7. The object after mirroring and hiding the hidden lines should look similar to that shown in Figure 27-22.

P1
(Center Point)

P2
(Endpoint)

Figure 27-21 *Selecting the points to be mirrored*

Figure 27-22 *Model after mirroring*

8. Choose **Views > Views > Left** from the **View** panel of the **Ribbon**.

9. Draw the sketch for the next object using the **LINE**, **ARC**, and **CIRCLE** commands.

10. Create the object by using the **REGION**, **SUBTRACT**, and **EXTRUDE** commands. The resulting object is shown in Figure 27-23.

11. Choose **Home > Solid Editing > Union** from the **Ribbon** and then unite all objects, see Figure 27-24.

Figure 27-23 *Model after creating the next object and moving it*

Figure 27-24 *The model after union*

12. Choose **Home > Modify > Fillet** from the **Ribbon**. The following prompt sequence is displayed:

Current settings: Mode = TRIM, Radius = 0.5000
Select first object or [Undo/Polyline/Radius/Trim/Multiple]: *Select first edge, as shown in Figure 27-25.*
Enter fillet radius <0.5000>: **2**
Select an edge or [Chain/Radius]: *Select the second edge, as shown in Figure 27-25.*
Select an edge or [Chain/Radius]: *Select the third edge, as shown in Figure 27-25.*
Select an edge or [Chain/Radius]: Enter
3 edge(s) selected for fillet.

13. The final model for Example 3 should look similar to the one shown in Figure 27-26.

Figure 27-25 *Selecting the edges for filleting*

Figure 27-26 *The final model for Example 3*

MOVING MODELS IN 3D SPACE

Ribbon:	Home > Modify > 3D Move	**Toolbar:**	Modeling > 3D Move
Menu Bar:	Modify > 3D Operations > 3D Move	**Command:**	3DMOVE

Sometimes, you need to move objects such as solid models, edges, or faces in 3D space. In such cases, you can use the **3DMOVE** command to translate the model at the desired location. You can move the selected object or subobject at an orientation, parallel to a plane or along an axial direction. To perform this operation, choose **Home > Modify > 3D Move** from the **Ribbon**. Next, select the object or subobject to be moved; the **Move** grip tool will be displayed attached to the cursor, see Figure 27-27. Specify the base point by placing the **Move** grip tool at the desired location; the dynamic preview of the movement of the object with respect to the base

*Figure 27-27 The **Move** grip tool*

point will be displayed. To move the object in 3D space with respect to the selected base point, specify the distance and angle. To move the object along a particular plane, move the cursor near the desired plane in the **Move** grip tool; the plane, including the corresponding axis handle, will turn golden in color. Click at that location; the object will now move only parallel to the selected plane. To move the object along a particular direction, move the cursor near the desired axis handle; it will turn golden in color and a construction line of the color of the selected axis will be visible. Click at that location; the object will now move only along the selected direction.

The prompt sequence that will be followed when you choose the **3D Move** button is given next.

> Select objects: *Select objects or subobjects to be moved.*
> Select objects: Enter
> Specify base point or [Displacement] <Displacement>: *Select the base point with respect to which you want to move the object or select a particular plane/axis from the axis handle.*
> Specify second point or <use first point as displacement>: *Specify the distance by which you want to move the object.*

If you press the ENTER key at the **Specify second point or <use first point as displacement>** prompt, AutoCAD interprets the first point as the relative value of the displacement in the X axis, Y axis, and Z axis directions. This value will be added in the X and Y axes coordinates and the object will be automatically moved to the resultant location. For example, model a cylinder with its center at (3,3,3) and then select the center point of the cylinder as the base point. Now **Specify second point or <use first point as displacement>** prompt, press the

ENTER key. You will notice that the cylinder is moved such that its center is now placed at (6,6,6). This is because 3 units (initial coordinates) are added along the X, Y, and Z directions.

CREATING ARRAYS IN 3D SPACE

Ribbon:	Home > Modify > 3D Array	**Toolbar:**	Modeling > 3D Array
Menu Bar:	Modify > 3D Operations > 3D Array	**Command:**	3DARRAY

As mentioned earlier, the arrays are defined as the method of creating multiple copies of the selected object in a rectangular or a polar fashion. The 3D arrays can also be created similar to the 2D arrays. The only difference is that in 3D array, another factor called the Z axis is also taken into consideration. There are two types of 3D arrays. Both of these types are discussed next.

3D Rectangular Array

This is the method of arranging the solid model along the edges of a box. In this type of array you need to specify three parameters. They are the rows (along the Y axis), the columns (along the X axis), and the levels (along the Z axis). You will also have to specify the distances between the rows, columns, and the levels. The 3D rectangular array can be easily understood by taking an example shown in Figure 27-28. This figure shows two floors of a building. Initially, only one chair is placed on the ground floor. Now, if you create a 2D rectangular array, the chairs will be arranged only on the ground floor. However, when you create the 3D rectangular array, then the chairs will be arranged on the first floor (along the Z axis) as well as the ground floor, see Figure 27-29. In this example, the number of rows is three, columns is four, and levels is two.

Figure 27-28 *The model before creating the rectangular array*

Figure 27-29 *The model after creating the array in 3D space*

The prompt sequence that will follow when you choose this command from the **Modify** panel is given next.

> Select objects: *Select the object to array.*
> Select objects: Enter
> Enter the type of array [Rectangular/Polar] <R>: Enter
> Enter the number of rows (---) <1>: *Specify the number of rows along the X axis.*

Enter the number of columns (|||) <1>: *Specify the number of columns along the Y axis.*
Enter the number of levels (...) <1>: *Specify the number of levels along the Z axis.*
Specify the distance between rows (---): *Specify the distance between the rows.*
Specify the distance between columns (|||): *Specify the distance between the columns.*
Specify the distance between levels (...): *Specify the distance between the levels.*

3D Polar Array

The 3D polar arrays are similar to the 2D polar arrays. The only difference is that in 3D, you will have to specify an axis about which the solid models will be arranged. For example, consider the solid models shown in Figure 27-30. This figure shows a circular plate that has four holes. Now, to place the bolts in this plate, you can use the 3D polar array as shown in Figure 27-31. The axis for an array is defined using the centers at the top and the bottom faces of the circular plate. The prompt sequence that will follow is given next.

Select objects: *Select the object to array.*
Select objects: Enter
Enter the type of array [Rectangular/Polar] <R>: **P**
Enter the number of items in the array: *Specify the number of items.*
Specify the angle to fill (+=ccw, -=cw) <360>: Enter
Rotate arrayed objects? [Yes/No] <Y>: Enter
Specify center point of array: *Specify the first point on the axis.*
Specify second point on axis of rotation: *Specify the second point on the axis.*

Figure 27-30 Model before creating array   *Figure 27-31 Model after creating array*

ALIGNING SOLID MODELS

| **Ribbon:** | Home > Modify > Align | **Command:** ALIGN |
| **Menu Bar:** | Modify > 3D Operations > Align | |

The **ALIGN** command is a very versatile and highly effective command. It is extensively used in solid modeling. As the name suggests, this command is used to align the selected solid model with another solid model. In addition to this, it can also be used to translate, rotate, and scale the selected solid model. This command uses pairs of source and destination points to align the solid model. The source point is a point with which you want to align the object.

The destination point is a point on the destination object at which you want to place the source object. You can specify one, two, or three pairs of points to align the objects. However, the working of this command will be different for all the three cases. All three cases are discussed next.

Aligning the Objects Using One Pair of Points

When you align the object using just one pair of points, the working of this command will be similar to the **MOVE** command. This means that the source object will be as it is moved from its original location and will be placed on the destination object. Here, the source point will work as the first point of displacement and the destination point will work as the second point of displacement. A reference line will be drawn between the source and the destination points. This line will disappear once you exit the command. The prompt sequence is as follows:

> Select objects: *Select the object to align.*
> Select objects: `Enter`
> Specify first source point: *Select S1, as shown in Figure 27-32.*
> Specify first destination point: *Select D1, as shown in Figure 27-32.*
> Specify second source point: `Enter`

Figure 27-33 shows the models after aligning.

Figure 27-32 Models before aligning   *Figure 27-33 Models after aligning*

Aligning the Objects Using Two Pairs of Points

The second case is to align the objects using two pairs of source and destination points. This method of aligning the objects forces the selected object to translate and rotate once. You are also provided with an option of scaling the source object to align with the destination object. The prompt sequence that follows is given next.

> Select objects: *Select the object to align.*
> Select objects: `Enter`
> Specify first source point: *Select S1, as shown in Figure 27-34.*
> Specify first destination point: *Select D1, as shown in Figure 27-34.*
> Specify second source point: *Select S2, as shown in Figure 27-34.*
> Specify second destination point: *Select D2, as shown in Figure 27-34.*

Specify third source point or <continue>: [Enter]
Scale objects based on alignment points? [Yes/No] <N>: **Y**   *(You can also enter N at this prompt if you do not want to scale the object).*

Figure 27-35 shows the objects after aligning and scaling.

Figure 27-34 *Objects before aligning*

Figure 27-35 *Objects after aligning and scaling*

Aligning the Objects Using Three Pairs of Points

The third case is to align the objects using three pairs of points. This option forces the selected object to rotate twice and then translate, see Figures 27-36 and 27-37. The first pair of source and destination points is used to specify the base point of alignment, the second pair of source and destination points is used to specify the first rotation angle, and the third pair of source and destination points is used to specify the second rotation angle. In this case, you will not be allowed to scale the object. The prompt sequence that follows is given next.

Select objects: *Select the object to align.*
Select objects: [Enter]
Specify first source point: *Select S1, as shown in Figure 27-36.*
Specify first destination point: *Select D1, as shown in Figure 27-36.*
Specify second source point: *Select S2, as shown in Figure 27-36.*
Specify second destination point: *Select D2, as shown in Figure 27-36.*
Specify third source point or <continue>: *Select S3, as shown in Figure 27-36.*
Specify third destination point: *Select D3, as shown in Figure 27-36.*

Figure 27-36 *Objects before aligning*

Figure 27-37 *Objects after aligning and rotating*

ALIGNING SOLIDS BY DEFINING AN ALIGNMENT PLANE

Ribbon:	Home > Modify > 3D Align	**Toolbar:**	Modeling > 3D Align
Menu Bar:	Modify > 3D Operations > 3D Align	**Command:**	3DALIGN

The **3DALIGN** command is used to align the selected solids or the copy of those solids with another solid object. To do so, define a plane for the source and destination objects that are to be aligned to each other. In this command, you are prompted to specify the first, second, and third points continuously on the source to define a plane for the source. Then, you will be prompted to specify the first, second, and third points on the destination object to define a plane for the destination. As a result, these planes will be automatically aligned to each other. While specifying the points on the destination object, you get a dynamic preview for the placement of the source object. The following is the significance of each point that you specify on the source and destination objects.

First point on source. This is the base point on the source object that later coincides with the first point of the destination object.

Second point on source. This point defines the direction of the X axis on the new XY plane that will be created on the source object.

Third point on source. This point is specified on the new XY plane to define the direction of the Y axis. This completely defines the new XY plane of the source. After specifying this point, the source object is rotated to make this new XY plane parallel to the XY plane of the current UCS.

First point on destination. This is the base point of the destination. The base point of the source will coincide with this point.

Second point on destination. This point defines the direction of the X axis on the new XY plane of the destination object. The X axis of the source will get aligned to the X axis of the destination object.

Third point on destination. This point is specified on the new XY plane to define the direction of the Y axis. This completely defines the new XY plane of the destination.

The prompt sequence that will be followed when you choose the **3D Align** button from the **Modeling** toolbar is given next.

> Select objects: *Select the objects to align.*
> Select objects: Enter
> Specify source plane and orientation ...
> Specify base point or [Copy]: *Select S1, as shown in Figure 27-38, or Enter **C** to align the copy of the selected object and leave the original one as it is.*

Specify second point or [Continue] <C>: *Select S2, as shown in Figure 27-38.*
Specify third point or [Continue] <C>: *Select S3, as shown in Figure 27-38.*
Specify destination plane and orientation ...
Specify first destination point: *Select D1, as shown in Figure 27-38.*
Specify second destination point or [eXit] <X>: *Select D2, as shown in Figure 27-38.*
Specify third destination point or [eXit] <X>: *Select D3, as shown in Figure 27-38.*

Figure 27-39 shows the model after aligning.

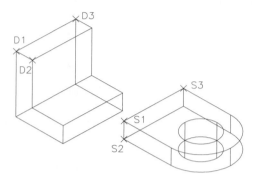

Figure 27-38 *Objects before aligning*

Figure 27-39 *Objects after aligning*

If you press the ENTER key at the **Specify second point or [Continue] <C>** prompt, then AutoCAD assumes the X and Y axes of the new plane to be parallel and in the same direction of the XY plane of the current UCS. Also, you will be prompted to select the first destination point.

If you press the ENTER key at the **Specify second destination point or [eXit] <X>** prompt, then AutoCAD assumes the X and Y axes of the new plane for the destination to be parallel and in the same direction as the XY plane of the current UCS. Also, the XY plane of the source object will get aligned to this plane.

EXTRACTING EDGES OF A SOLID MODEL

Ribbon:	Home > Solid Editing > Extract Edges	Command:	XEDGES
Menu Bar:	Modify > 3D Operations > Extract Edges		

This command is used to automatically generate the wireframe from a previously created solid model. The wireframe will be generated at the same location and with the same orientation where the parent model was located. The prompt sequence that will be followed when you choose the **XEDGES** command is given next.

Command: **XEDGES**
Select objects: *Select object from which you want to extract the edges.*
Select objects: *Select more objects or* Enter

Figure 27-40 shows a model from which the edges are to be extracted and Figure 27-41 shows the wireframe extracted from that model after moving it to another location.

Figure 27-40 *Model to extract edges*

Figure 27-41 *Wireframe obtained by edge extraction*

Note
*The **XEDGES** command can only be applied to regions, surfaces, and solids.*

Tip
To extract edges from a particular edge or face of the model, press and hold the CTRL key during the selection of the objects.

CONVERTING OBJECTS TO SURFACES

Ribbon: Home > Solid Editing > Convert to Surface	**Command:** CONVTOSURFACE
Menu Bar: Modify > 3D Operations > Convert to Surface	

This command is used to convert objects into surfaces. The objects that can be converted to surfaces include arcs and lines with nonzero thickness, open zero width polyline having a nonzero thickness, 2D solids, donuts, boundary, regions, and 3D faces created on a plane.

To convert objects to surfaces, choose **Home > Solid Editing > Convert to Surface** from the **Ribbon** and then select the objects to be converted into a surface. Figure 27-42 shows a boundary to be converted into a surface and Figure 27-43 shows the resulting surface.

Tip
*You can convert 3D Solids with curved faces to surfaces by using the **EXPLODE** command. The curved faces of the solid will be converted into surfaces and planar faces will be converted into regions.*

Figure 27-42 *The boundary to be converted into a surface*

Figure 27-43 *The resulting surface*

CONVERTING OBJECTS TO SOLIDS

| **Ribbon:** | Home > Solid Editing > Convert to Solid | **Command:** | CONVTOSOLID |
| **Menu Bar:** | Modify > 3D Operation > Convert to Solid | | |

This command is used to convert objects with some thickness into solid models. The solid generated by this option is similar to an extruded solid having height equal to the thickness of the object. The objects that can be converted into solids are open polylines with nonzero uniform width and thickness, closed polylines with thickness, and circles with thickness.

To convert objects into solids, choose **Home > Solid Editing > Convert to Solid** from the **Ribbon** and then select the objects to be converted into solids. Figure 27-44 shows a closed zero width polyline having a nonzero thickness and Figure 27-45 shows the resulting solid.

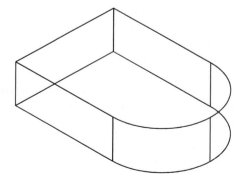

Figure 27-44 *Polyline with thickness*

Figure 27-45 *Resulting solid*

CONVERTING SURFACES TO SOLIDS

| **Ribbon:** | Home > Solid Editing > Thicken | **Command:** | THICKEN |
| **Menu Bar:** | Modify > 3D Operation > Thicken | | |

This command is used to convert any surface into a solid by adding material of a specified thickness to the surface. The material addition takes place in the direction normal to the surface. To convert surfaces into solids, choose **Home > Solid Editing > Thicken** from the **Ribbon**. Next, select the surface to thicken and then specify the thickness of the material to be added to the surface. Figure 27-46 shows the surface to be thickened and Figure 27-47 shows the resulting solid after thickening.

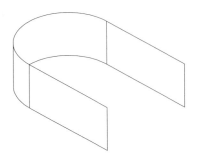

Figure 27-46 Surface to be thickened

Figure 27-47 Resulting solid after thickening

Note
*You cannot thicken 3D primitive surfaces generated by the **3D** command.*

SLICING SOLID MODELS

Ribbon:	Home > Solid Editing > Slice	Command:	SLICE
Menu Bar:	Modify > 3D Operation > Slice		

 As the name suggests, the **SLICE** command is used to slice the selected solid with the help of a specified plane. You will be given an option to select the portion of the sliced solid that has to be retained. You can also retain both the portions of the sliced solids.

The prompt sequence that follows is given next.

Select objects to slice: *Select the object to slice.*
Select objects to slice: Enter
Specify start point of slicing plane or [planar Object/Surface/Zaxis/View/XY/YZ/ZX/3points] <3points>: *Specify a point on the slicing plane or select an option.*

3points

This option is used to slice a solid using a plane defined by three points, see Figures 27-48 and 27-49. The prompt sequence that follows is given next.

Select objects to slice: *Select the object to be sliced.*
Select objects to slice: [Enter]
Specify start point of slicing plane or [planar Object/Surface/Zaxis/View/XY/YZ/ZX/3points] <3points>: [Enter]
Specify first point on plane: *Specify the point P1 on the slicing plane, as shown in Figure 27-48.*
Specify second point on plane: *Specify the point P2 on the slicing plane, as shown in Figure 27-48.*
Specify third point on plane: *Specify the point P3 on the slicing plane, as shown in Figure 27-48.*
Specify a point on desired side of the plane or [keep Both sides]: *Select the portion of the solid to retain or enter B to retain both the portions of the sliced solid.*

The model, after slicing, is shown in Figure 27-49.

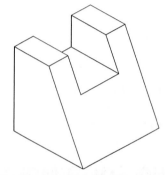

Figure 27-48 *Defining the slicing plane*   **Figure 27-49** *Model after slicing*

Object

This option is used to slice the solid model using an object. The objects that can be used to slice the solid model include arcs, circles, ellipses, 2D polylines, and splines. The prompt sequence that will follow is given next.

Select objects to slice: *Select the object to slice.*
Select objects to slice: [Enter]
Specify start point of slicing plane or [planar Object/ Surface/Zaxis/View/XY/YZ/ZX/ 3points] <3points>: **O**
Select a circle, ellipse, arc, 2D-spline, 2D-polyline to define the slicing plane: *Select the object to slice the solid.*
Specify a point on desired side or [keep Both sides]: *Specify the portion of the solid to retain.*

Surface

This option is used to slice the solid model using a surface. Note that the ruled, tabulated, revolved, or edge surfaces cannot be used for slicing the solids. Figure 27-50 shows the object to be sliced and the slicing surface, and Figure 27-51 shows the model after slicing. The prompt sequence that will follow is given next.

Select objects to slice: *Select the object to slice.*
Select objects to slice: [Enter]
Specify start point of slicing plane or [planar Object/Surface/Zaxis/View/XY/YZ/ZX/3points] <3points>: **S**
Select a surface: *Select the surface to slice the solid.*
Select solid to keep or [keep Both sides] <Both>: *Specify the portion of the solid to retain.*

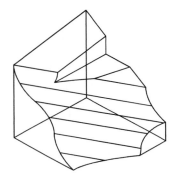

Figure 27-50 *Model with slicing surface*   ***Figure 27-51*** *Model after slicing*

Zaxis

This option is used to slice the solid using a plane defined by two points. The first point is the point on the section plane and the second point is the point in the direction of the Z axis of the plane. The prompt sequence that follows is given next.

Select objects to slice: *Select the object to slice.*
Select objects to slice: [Enter]
Specify start point of slicing plane or [planar Object/Surface/Zaxis/View/XY/YZ/ZX/3points] <3points>: **Z**
Specify a point on the section plane: *Specify the point on the section plane.*
Specify a point on the Z-axis (normal) of the plane: *Specify the point in the direction of the Z axis of the section plane.*
Specify a point on desired side or [keep Both sides]: *Select the portion to retain.*

View

This option is used to slice the selected solid about the viewing plane, see Figures 27-52 and 27-53. The viewing plane in this case will be the screen of the monitor. You will be prompted to specify a point on the solid model through which the viewing plane will pass. The prompt sequence that will follow is given next.

Select objects to slice: *Select the object to slice.*
Select objects to slice: [Enter]
Specify start point of slicing plane or [planar Object/Surface/Zaxis/View/XY/YZ/ZX/3points] <3points>: **V**

Specify a point on the current view plane <0,0,0>: *Specify the point P1, as shown in Figure 27-52.*
Specify a point on desired side or [keep Both sides] <Both>: *Specify the portion of the solid to retain.*

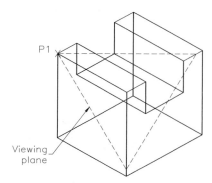

Figure 27-52 *Defining the slicing plane*

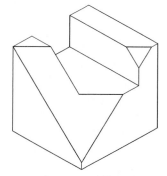

Figure 27-53 *Model after slicing*

XY, YZ, ZX

These options are used to slice the selected solid about the *XY, YZ,* or the *ZX* plane respectively. When you invoke this option, you will be prompted to select the point on the plane. The prompt sequence that will follow is given next.

Select objects to slice: *Select the object to slice.*
Select objects to slice: [Enter]
Specify start point of slicing plane or [planar Object/ Surface/Zaxis/View/XY/YZ/ZX/ 3points] <3points>: *Select the XY, YZ, or the ZX plane.*
Specify a point on the XY-plane <0,0,0>: *Specify the point on the slicing plane.*
Specify a point on desired side or [keep Both sides] <Both>: *Specify the portion to retain.*

Creating the Cross-Sections of the Solids

Command:	SECTION

The **SECTION** command is very similar to the **SLICE** command. The only difference is that this command does not chop the solid. Instead, it creates a cross-section along the selected section plane. The cross-section thus created is a region. Note that the regions are created in the current layer and not in the layer in which the sectioned solid is stored. The prompt sequence that follows when you enter the **SECTION** command is given next.

Select objects: *Select the object to section.*
Select objects: [Enter]
Specify first point on section plane by [Object/Zaxis/View/XY/YZ/ZX/3points] <3points>:

3points

This option is used to define the section plane using three points.

Object

This option is used to specify the section plane using a planar object. The objects that can be used to create the sections are arcs, circles, ellipses, splines, or 2D polylines.

View

This option uses the current viewing plane to define the section plane. You will be prompted to specify the point on the current view plane. It will then automatically create a cross-section parallel to the current viewing plane and passing through the specified point.

XY, YZ, ZX

These options are used to define the section planes that are parallel to the *XY*, *YZ*, or the *ZX* planes, respectively. You will be prompted to specify the point through which the selected plane will pass, see Figures 27-54 and 27-55.

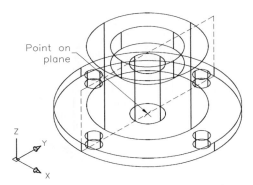

Figure 27-54 *Creating the cross-section along the YZ plane*

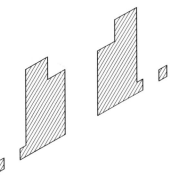

Figure 27-55 *Cross-section created, isolated, and hatched for clarity*

Tip
The number of regions created as cross-sections will be equal to the number of solids selected to create the cross-section.

*You can also hatch the cross-section created using the **SECTION** command. Select the entire region as the object for hatching. You may have to define a new UCS based on the section to hatch it.*

Example 2

In this example, you will draw the solid model shown in Figure 27-56. The dimensions for the model are shown in Figures 27-57 and 27-58. After creating it, slice it as shown in Figure 27-59.

Figure 27-56 *Model for Example 2*

Figure 27-57 *Top view of the model*

Figure 27-58 *Front view of the model*

Figure 27-59 *The sliced solid*

1. Start a new 3D template drawing file and then draw the sketch for the base of the model in the top view, as shown in Figure 27-60. Choose **Home > Draw > Region** from the **Ribbon** to convert the model into a region.

2. Choose **Home > Modeling > Extrude** from the **Ribbon** and extrude the region to a distance of 1.5.

3. Choose **Views > Views > SW Isometric** from the **Ribbon**. The model should look similar to the one given in Figure 27-61.

4. Rotate the UCS about the X axis through an angle of 90-degree. Draw the sketch for the next feature and convert it into a region, see Figure 27-62.

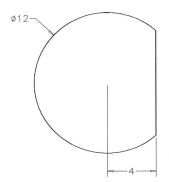

Figure 27-60 *Sketch for the base of the model*

Figure 27-61 *Base of the model*

5. Choose **Home > Modeling > Extrude > Revolve** from the **Ribbon** and then revolve the sketch to an angle of 360-degree. Move it using the center point of the lower face to the center of the lower face of the base of the model. Choose **Home > Solid Editing > Union** from the **Ribbon** and then unite both the objects, see Figure 27-63.

Figure 27-62 *Sketch for the next feature*

Figure 27-63 *Model after union*

6. Create a new cylinder of diameter 1 unit and height same as that of the base of the model. The center of the base of the cylinder should lie at a distance of 1.5 units from the quadrant of the lower face of the base of the model.

7. Choose **Home > Modify > 3D Array** from the **Ribbon**. The prompt sequence that follows is given next:

    Initializing... 3DARRAY loaded.
    Select objects: *Select the cylinder.*
    Select objects: Enter
    Enter the type of array [Rectangular/Polar] <R>: **P**
    Enter the number of items in the array: **3**

Specify the angle to fill (+=ccw, -=cw) <360>: ⌶Enter⌶
Rotate arrayed objects? [Yes/No] <Y>: ⌶Enter⌶
Specify center point of array: *Select the center of the base of the model.*
Specify second point on axis of rotation: *Select the center of the top face of the model.*

8. Subtract all three cylinders from the model by choosing **Home > Solid Editing > Subtract** from the **Ribbon**.

9. Create the central hole to be subtracted from the cylinder of diameter 4 units and height 6 units. The model should now look similar to the one shown in Figure 27-64.

10. Fillet the edges with radius 0.25 using the **Fillet** tool.

11. Relocate the UCS at the world position.

12. Choose **Home > Solid Editing > Slice** from the **Ribbon** to cut the solid model to half. The prompt sequence is as follows:

Select objects: *Select the model.*
Select objects: ⌶Enter⌶
Specify first point on slicing plane by [Object/Zaxis/View/XY/YZ/ZX/3points] <3points>: **ZX**
Specify a point on the ZX-plane <0,0,0>: *Select the center of the top face of the model.*
Specify a point on desired side of the plane or [keep Both sides]: *Specify a point on the back side of the model to retain it.*

13. The final model for Example 2 should look similar to the one shown in Figure 27-65.

Figure 27-64 *Model after creating the central hole fillet*

Figure 27-65 *Final model after slicing*

Self-Evaluation Test

Answer the following questions and then compare them to those given at the end of this chapter:

1. The movement of a model is dynamically visible in the **3DMOVE** command. (T/F)

2. The edges of the solid model can be filleted using the **3DFILLET** command. (T/F)

3. You can select an ellipse as an object to rotate the solid model using the **ROTATE3D** command. (T/F)

4. The **CONVTOSURFACE** command can convert an open zero width polyline having a nonzero thickness to a surface. (T/F)

5. The **SECTION** command creates a cross-section and also slices the model. (T/F)

6. The _____ rule is used to determine the direction of rotation of the solid model in 3D space.

7. The _____ command is used to extract edges from a solid model.

8. The _____ command is used to move, rotate, and scale the solid model in a single attempt.

9. The **SECTION** command creates a _____ along the plane of a section.

10. The _____ command is used to align solids by defining an alignment plane.

Review Questions

Answer the following questions:

1. A surface can be used to slice a model using the **SLICE** command. (T/F)

2. The **CONVTOSOLID** command can convert objects with a zero thickness into solid. (T/F)

3. You can also extract an individual edge of a solid model using the **XEDGES** command. (T/F)

4. You can select all tangential edges of the selected solid model for filleting using the **Chain** option of the **FILLET** command. (T/F)

5. Regions can be converted to surfaces by the **CONVTOSURFACE** command. (T/F)

Chapter 27

6. Which option of the **ROTATE3D** command is used to select a 2D entity as an object for rotating the solid models?

    (a) **2D**   (b) **Last**
    (c) **Object**   (d) **Entity**

7. Which command is used to convert a surface to a solid?

    (a) **CONVTOSOLID**   (b) **CONVQUILT**
    (c) **CONVTOSURFACE**   (d) **THICKEN**

8. Which object can be converted into a solid using the **CONVTOSOLID** command?

    (a) Rectangle   (b) Circle
    (c) Ellipse   (d) Polygon

9. Which option of the **MIRROR3D** command is used to mirror the object about the view plane?

    (a) **Object**   (b) **Last**
    (c) **View**   (d) **Zaxis**

10. What is the least number of points that you have to define on the source object to align it using the **3DALIGN** command?

    (a) One   (b) Two
    (c) Three   (d) Four

11. The _____ command gives a dynamic preview while aligning the solid model.

12. The _____ option of the **ROTATE3D** command is used to select the same axis that was last selected to rotate the solid model.

13. In the **3DARRAY** command, the levels are arranged along the _____ axis.

14. The object having some _____ can be converted into a solid using the **CONVTOSOLID** command.

15. The _____ option is used to slice a model with a sketched entity.

Exercises

Exercise 1

In this exercise, you will create the solid model shown in Figure 27-66. The dimensions for the model are shown in Figures 27-67 and 27-68. After creating it, slice it to get the model shown in Figure 27-69.

Figure 27-66 *Model for Exercise 1*

Figure 27-67 *Top view of the model*

Figure 27-68 *Front view of the model*

Figure 27-69 *Model after slicing*

Exercise 2

In this exercise, you will create the solid model shown in Figure 27-70. The dimensions are given in the drawing. The fillet radius is 5mm.

Figure 27-70 *Solid model for Exercise 2*

Exercise 3

In this exercise, you will create the solid model shown in Figure 27-71. The dimensions for the model are given in the same figure. The fillet radius is 0.13 units.

Figure 27-71 *Solid model for Exercise 3*

Exercise 4

In this exercise, you will create the solid model shown in Figure 27-72. The dimensions for the model are given in the same figure. Assume the missing dimensions.

Figure 27-72 *Solid model for Exercise 4*

Problem-Solving Exercise 1   *Mechanical*

For the Gear Puller shown in Figure 27-73, draw the following:

Piece part drawings are shown in Figures 27-74 through 27-78, assembly drawing with the Bill of Materials is shown in Figure 27-79, and the exploded view of the solid model is shown in Figure 27-80.

Figure 27-73 *Assembled solid model of the Gear Puller*

P1: Yoke

Figure 27-74 *Front and side views of the Yoke*

P3: Screw

Figure 27-75 *Front view of the Screw*

P2: Arm

Figure 27-76 *Front and top views of the Arm*

Figure 27-77 *Views of the Spring Pin and Handle*

Figure 27-78 *Views of the Spread Nut and Knob*

P.N.	Qty	Material	Description
1	1	1020 STL	Yoke
2	3	1020 STL	Arm
3	1	1020 STL	Screw
4	1	1020 STL	Spread nut
5	1	1020 STL	Handle
6	2	1020 STL	Knob
7	3	STL	Spring pin

Figure 27-79 *Assembled view and BOM of the Gear Puller*

Exploded View of
Solid Model

Figure 27-80 *Exploded solid model of the Gear Puller*

Problem-Solving Exercise 2   *Mechanical*

Create the model shown in Figure 27-81. The dimensions of the model are given in Figure 27-82. **

Figure 27-81 *Solid model for Problem-Solving Exercise 2*

Figure 27-82 *Dimensions for the solid model*

Problem-Solving Exercise 3 *Mechanical*

Create the model shown in Figure 27-83. The dimensions of the model are given in Figure 27-84. **

Figure 27-83 *Solid model for Problem-Solving Exercise 3*

Section A—A

Figure 27-84 *Dimensions for the solid model*

Problem-Solving Exercise 4   *Mechanical*

Create the model shown in Figure 27-85. The sectioned view and dimensions of the model are given in Figure 27-86. **

Figure 27-85   *Solid model for Problem-Solving Exercise 4*

Figure 27-86   *Sectioned view and dimensions for the solid model*

Problem-Solving Exercise 5 *Mechanical*

Create the model shown in Figure 27-87. The sectioned view and dimensions of the model are given in Figure 27-88. Note that the hidden lines have been removed for clarity. **

Figure 27-87 Solid model for Problem-Solving Exercise 5

Figure 27-88 Sectioned view and dimensions for the solid model

Problem-Solving Exercise 6   *Mechanical*

Create the model shown in Figure 27-89. The dimensions of the model are given in Figure 27-90. **

Figure 27-89 *Solid model for Problem-Solving Exercise 6*

Figure 27-90 *Dimensions for the solid model*

Problem-Solving Exercise 7

Mechanical

Create the model shown in Figure 27-91. The dimensions of the model are given in Figure 27-92. **

Figure 27-91 Solid model for Problem-Solving Exercise 7

Figure 27-92 Dimensions for the solid model

Problem-Solving Exercise 8   *Mechanical*

Create the model shown in Figure 27-93. The dimensions of the model are given in Figure 27-94. **

Figure 27-93 *Solid model*

Figure 27-94 *Dimensions of the model for Problem-Solving Exercise 8*

Answers to Self-Evaluation Test

1. T, **2.** F, **3.** F, **4.** T, **5.** F, **6.** right-hand thumb, **7. XEDGES**, **8. ALIGN**, **9.** cross-section, **10. 3DALIGN**

Chapter 28

Editing 3D Objects

Learning Objectives

After completing this chapter, you will be able to:

- *Edit solid models using the **SOLIDEDIT** command.*
- *Generate sections of the model.*
- *Generate drawing views of the solid model using the **SOLVIEW** command.*
- *Generate profiles and section in the drawing views using the **SOLDRAW** command.*
- *Create profile images of the solid model using the **SOLPROF** command.*
- *Calculate mass properties of the solid models using the **MASSPROP** command.*
- *Understand the use of **Action Recorder**.*
- *Use the **ShowMotion** tool for presentation purpose.*

EDITING SOLID MODELS

Ribbon:	Home > Solid Editing	**Menu Bar:**	Modify > Solid Editing
Toolbar:	Solid Editing	**Command:**	SOLIDEDIT

One of the major enhancements in the recent releases of AutoCAD is editing the solid models. This has made solid modeling in AutoCAD very user-friendly. In AutoCAD you can edit the solid models using the **SOLIDEDIT** command. You can also select the options for editing the solid models from the **Home > Solid Editing** panel in the **Ribbon**. The prompt sequence that will follow when you invoke this command is given next.

> Solids editing automatic checking: SOLIDCHECK=1
> Enter a solids editing option [Face/Edge/Body/Undo/eXit] <eXit>:

Face Option

This option allows you to edit the faces of solid models. The suboptions provided under this option are **Extrude, Move, Rotate, Offset, Taper, Delete, Copy, coLor** and **mAterial**. The next prompt is given.

> Enter a face editing option
> [Extrude/Move/Rotate/Offset/Taper/Delete/Copy/coLor/mAterial/Undo/eXit]<eXit>:
> *Enter an option.*

Extrude

This option allows you to extrude selected faces of a solid model to a specific height or along a selected path (Figure 28-1). You can directly invoke this option by choosing the **Extrude faces** button from the **Solid Editing** toolbar or by choosing **Solid Editing > Extrude Faces** from the **Ribbon**. The prompt sequence that will follow when you choose this button is given next.

> Select faces or [Undo/Remove]: *Select faces to extrude or enter an option.*
> Select faces or [Undo/Remove/ALL]: *Select another face, an option, or press ENTER.*

The **Undo** option cancels the selection of the most recent face you have selected. The **Remove** option allows you to remove a previously selected face from the selection set for extrusion. The **ALL** option selects all the faces of the specified solid. After you have selected a face for extrusion, the next prompt is given below:

> Specify height of extrusion or [Path]: *Enter a height value or enter **P** to select a path.*

The **Path** option allows you to select a path for extrusion based on a specified line or curve. At the **Select extrusion path** prompt, select the line, circle, arc, ellipse, elliptical arc, polyline, or spline to be specified as the extrusion path. This path should not lie on the same plane as the selected face and should not have areas of high curvature.

A positive value for the height of extrusion extrudes the selected face outward, and a negative value extrudes the selected face inward. On specifying the height of extrusion, the next prompt is given below.

Specify angle of taper for extrusion <0>: *Specify a value between -90 and +90-degree or press ENTER to accept the default value of 0.*

A positive angle value tapers the selected face inward and a negative value tapers the selected face outward (Figure 28-1).

Move

 This option allows you to move the selected faces from one location to another without changing the orientation of the solid (Figure 28-2). For example, you can move holes from one location to another without actually modifying the solid model. You can invoke this option by choosing **Home > Solid Editing > Extrude Faces > Move Faces** from the **Ribbon**. The prompt sequence that will follow when you choose this button is given next.

Select faces or [Undo/Remove]: *Select one or more faces or enter an option.*
Select faces or [Undo/Remove/ALL]: *Select one or more faces, enter an option, or press ENTER.*
Specify a base point or displacement: *Specify a base point.*
Specify a second point of displacement: *Specify a point or press ENTER.*

The face moves to the specified location.

Figure 28-1 Using the **Extrude Faces** option to extrude a solid face with a positive taper angle

Figure 28-2 Using the **Move Faces** option to move the hole from the original position

 Tip
If the feature you want to move consists of more than one face, then you must select all faces to move the entire feature.

Offset

The **Offset** option allows you to offset the selected faces of a solid model uniformly through a specified distance on the model. Offsetting takes place in the direction of the positive side of the face. For example, you can offset holes to a larger or smaller size in a 3D solid through a specified distance. In Figure 28-3, the hole on the solid has been offset to a bigger size. You can invoke this option by choosing **Home > Solid Editing > Extrude Faces > Offset Faces** from the **Ribbon**. The following prompt sequence is displayed when you choose this button:

Select faces or [Undo/Remove]: *Select one or more faces or enter an option.*
Select faces or [Undo/Remove/ALL]: *Select one or more faces, enter an option, or press ENTER.*
Specify the offset distance: *Specify the offset distance.*

Rotate

This option allows you to rotate selected faces of a solid model through a specified angle. For example, you can rotate cuts or slots around a base point through an absolute or relative angle (Figure 28-4). The direction, in which the rotation takes place is determined by the right-hand thumb rule. You can invoke this option by choosing **Home > Solid Editing > Extrude Faces > Rotate Faces** from the **Ribbon.**

The prompt sequence that will follow when you choose this button is given below.

Select faces or [Undo/Remove]: *Select face(s) or enter an option.*
Select faces or [Undo/Remove/ALL]: *Select face(s), or enter an option, or press ENTER.*
Specify an axis point or [Axis by object/View/Xaxis/Yaxis/Zaxis] <2points>: *Select a point, enter an option, or press ENTER.*

If you press ENTER to use the **2points** option, you will be prompted to select the two points that will define the axis of rotation. The prompt sequence is given next.

Specify the first point on the rotation axis: *Specify the first point.*
Specify the second point on the rotation axis: *Specify the second point.*
Specify a rotation angle or [Reference]: *Specify the angle of rotation or enter **R** to use the reference option.*

Before offsetting   After offsetting   Before rotating   After rotating
the faces   the faces   the faces   the faces

Figure 28-3 *Using the **Offset Faces** option to offset a solid face with a positive taper angle*

Figure 28-4 *Using the **Rotate Faces** option to rotate the faces from the original position*

The **Axis by object** option allows you to select an object to define the axis of rotation. The axis of rotation is aligned with the selected object; for example, if you select a circle, the axis of rotation is aligned along the axis that passes through the center of the circle and is normal to the plane of the circle. The selected objects can also be a line, ellipse, Arc, 2D, 3D, or LW polyline or even a spline. The prompt sequence is given next.

Select a curve to be used for the axis: *Select the object to be used to define the axis of rotation.*
Specify a rotation angle or [Reference]: *Specify rotation angle or use the reference angle option.*

The **Xaxis**, **Yaxis**, **Zaxis** options align the axis of rotation with the axis that passes through the specified point. The Command prompt is:

Specify the origin of rotation <0,0,0>: *Select a point through which the axis of rotation will pass.*
Specify a rotation angle or [Reference]: *Specify rotation angle or use the reference angle option.*

Taper

 This option allows you to taper selected face(s) through a specified angle (Figure 28-5). The tapering takes place depending on the selection sequence of the base point and the second point along the selected face. It also depends on whether the taper angle is positive or negative. A positive value tapers the face inward and a negative value tapers the face outward. You can enter any value between -90-degree and +90-degree. The default value of 0-degree extrudes the face along the axis that is perpendicular to the selected face. If you have selected more than one face, they will all be tapered through the same angle. You can invoke this option by choosing **Home > Solid Editing > Extrude Faces > Taper faces** from the **Ribbon**. The prompt sequence that will follow when you choose this button is given below.

Select faces or [Undo/Remove/ALL]: *Select one or more faces, enter an option, or press ENTER.*
Specify the base point: *Specify a base point.*
Specify another point along the axis of tapering: *Specify a point.*
Specify the taper angle: *Specify an angle between –90 and +90-degree.*

Delete

 This option removes the selected face(s) from the specific 3D solid. It also removes chamfers and fillets (Figure 28-6). The prompt sequence is as follows:

Select faces or [Undo/Remove]: *Select the face to remove.*
Select faces or [Undo/Remove/ALL]: *Select one or more faces, enter an option, or press ENTER.*
Solid validation started.
Solid validation completed.

Copy

The **Copy** option allows you to copy a face as a body or region (Figure 28-7). The prompt sequence that follows when you choose this button is as follows:

Select faces or [Undo/Remove]: *Select the faces to copy.*

Select faces or [Undo/Remove/ALL]: *Select one or more faces, enter an option, or press ENTER.*
Specify a base point or displacement: *Specify a base point.*
Specify a second point of displacement: *Specify the second point.*

Figure 28-5 *Tapering the faces using the **Taper Faces** option*

Figure 28-6 *Deleting the faces using the **Delete Faces** option*

Figure 28-7 *Copying the faces using the **Copy Faces** option*

coLor

 This option is used to change the color of the selected face. The prompt sequence that will follow when you choose this button is given next.

Select faces or [Undo/Remove/ALL]: *Select one or more faces, enter an option, or press ENTER.*

When you press ENTER at the previous prompt, the **Select Color** dialog box will be displayed. You can assign the color to the selected face using this dialog box.

mAterial

This option is used to apply material to the selected face. The prompt sequence that will be followed when you invoke this option is given next.

> Select faces or [Undo/Remove]: *Select one or more faces to apply the material.*
> Select faces or [Undo/Remove/ALL]: *Select one or more faces, enter an option, or* [Enter]
> Enter new Material name <ByLayer>: *Enter the name of the material or* [Enter]

Undo

This option allows you to cancel the changes made to the faces of the solid model during the **SOLIDEDIT** command. This option will be available only when you invoke the **SOLIDEDIT** command using the command line.

eXit

This option allows you to exit the face editing operations in the **SOLIDEDIT** command. This option will be available only when you invoke this command using the command line.

Edge Option

With the **Edge** option, you can modify properties of the edges of a solid. You can copy and change the color of edges. This option provides the **Copy**, **coLor**, **Undo**, and **eXit** suboptions. These options are discussed next.

Copy

This option allows you to copy individual edges of a solid model. The edges are copied as lines, arcs, circles, ellipses, or splines. You can also invoke this option by choosing **Home > Solid Editing > Extract Edges > Copy edges** from the **Ribbon**. The prompt sequence that follows when you choose this button is given next.

> Select edges or [Undo/Remove]: *Select the edges to be copied.*
> Select edges or [Undo/Remove]: [Enter]
> Specify a base point or displacement: *Specify the first point of displacement.*
> Specify a second point of displacement: *Specify the second point of displacement.*

COLor

You can use this option to change the color of selected edge of a 3D solid. This option can also be invoked by choosing **Home > Solid Editing > Extract Edges > Color Edges** from the **Ribbon**. The prompt sequence that follows when you choose this button is given next.

> Select edges or [Undo/Remove]: *Select one or more edges or enter an option.*
> Select edges or [Undo/Remove]: [Enter]

After you select edges and press ENTER, the **Select Color** dialog box is displayed. You can assign the color to the selected edges using this dialog box. You can also assign colors by entering their names or numbers. Individual edges can be assigned individual colors.

The **Undo** option cancels the modifications made during the **Edge** editing of the **SOLIDEDIT** command, and **eXit** option allows you to exit the **Edge** editing of this command.

Body Option

This option is used to edit the entire body of the solid model. The **Body** option provides the **Imprint**, **seParate solids**, **Shell**, **cLean**, **Check**, **Undo**, and **eXit** suboptions. These suboptions are discussed next.

Imprint

This option allows you to imprint an object on the 3D solid object. Remember that the object to be imprinted should intersect one or more faces of the solid object. The objects that can be imprinted are arcs, circles, lines, 2D and 3D polylines, ellipses, splines, regions, bodies, and 3D solids. To invoke this option, choose **Home > Solid Editing > Extract Edges > Imprint** from the **Ribbon**. The prompt sequence that follows when you choose this button is given next.

> Select a 3D solid: *Select an object.*
> Select an object to imprint: *Select an object to be imprinted.*
> Delete the source object [Yes/No] <N>: *Select an option or press ENTER.*
> Select an object to imprint: *Select a new object or press ENTER.*

SeParate solids

This option allows you to separate 3D solids with disjointed volumes into separate 3D solids. The solids with disjointed volumes are created by the union of two solids that are not in contact with each other. Keep in mind that the composite solids created using the Boolean operations that share the same volume cannot be separated using this option. To invoke this option, choose **Home > Solid Editing > Separate** from the **Ribbon**. The prompt sequence that will follow when you choose this button is given next.

> Select a 3D solid: *Select a solid object with multiple lumps.*

Shell

This is one of the most extensively used options of the **SOLIDEDIT** command. This option allows you to create a shell of a specified 3D solid. The shelling is defined as a process of scooping out the material from the solid model in such a manner that the walls with some thickness are left. One 3D object can have only one shell. You can exclude certain faces before the scooping of the material takes place. This means the selected face will not be included during the shelling, see Figure 28-8. To invoke this option, choose **Home > Solid Editing > Separate > Shell** from the **Ribbon**. The prompt sequence is as follows.

> Select a 3D solid: *Select an object.*
> Remove faces or [Undo/Add]: *Select one or more faces or enter an option.*
> Remove faces or [Undo/Add/ALL]: *Select a face, enter an option, or press ENTER.*
> Enter the shell offset distance: *Specify the wall thickness.*

Before
shelling

After
shelling

Faces selected
to remove

Figure 28-8 Creating a shell in the solid model

Tip
If you have specified a positive distance, the shell is created on the inside perimeter of the solid, and in case it is negative, the shell is created on the outside perimeter of the 3D solid.

Clean

This option allows you to remove shared edges or vertices sharing the same surface or curve definition on either side of the edge or vertex. It removes all redundant edges and vertices, imprinted as well as unused geometry. It also merges faces that share the same surface. To invoke this option, choose **Home > Solid Editing > Separate > Clean** from the **Ribbon**. The prompt sequence is as follows.

Select a 3D solid: *Select the solid model to clean.*

Check

This option can be used to check if a solid object created by you is a valid 3D solid object. This option is independent of the settings of the **SOLIDCHECK** variable. The **SOLIDCHECK** variable turns on or off the solid validation for the current drawing session while using the **SOLIDEDIT** options. Solid validation is turned on when the value of the **SOLIDCHECK** variable is set to 1, which is the default value. If the solid object is a valid 3D solid object, you can use solid editing options to modify it without AutoCAD displaying ACIS error messages. Solid editing can take place only on valid 3D solids. To invoke this option, choose **Home > Solid Editing > Separate > Check** from the **Ribbon**. The prompt sequence is as follows:

Select a 3D solid: *Select the solid you want to check.*
This object is a valid ShapeManager solid.

The **Undo** option cancels the editing modifications made and the **eXit** option allows you to exit the body editing options.

Undo Option

This option allows you to cancel or undo the editing action taken.

Exit Option

This option allows you to exit the **SOLIDEDIT** command.

Example 1   *Mechanical*

In this example, you will create the solid model shown in Figure 28-9. After creating the model, modify it, as shown in Figure 28-10, using the options in the **Solid Editing** panel.

Figure 28-9 *Model for Example 1*   **Figure 28-10** *Model after modification*

1. Create a box and two cylinders at appropriate position, refer to Figure 28-9.

2. Combine all objects to make them a single solid model.

3. Choose **Home > Solid Editing > Separate > Shell** from the **Ribbon**. The prompt sequence is as follows:

   Select a 3D solid: *Select the solid.*
   Remove faces or [Undo/Add/ALL]: *Click on the top face of the bigger cylinder.*
   Remove faces or [Undo/Add/ALL]: *Click on the front face of the smaller cylinder.*
   Remove faces or [Undo/Add/ALL]: *Click on the left face of the box.*
   Remove faces or [Undo/Add/ALL]: Enter
   Enter the shell offset distance: **0.25**

   Solid validation started.
   Solid validation completed.
   Enter a body editing option
   [Imprint/seParate solids/Shell/cLean/Check/Undo/eXit] <eXit>: *Press ESC.*

4. Choose **Home > Solid Editing >Extrude Faces > Taper Faces** from the **Ribbon**. The prompt sequence is given next.

Solids editing automatic checking: SOLIDCHECK=1
Enter a solids editing option [Face/Edge/Body/Undo/eXit] <eXit>: _face
Enter a face editing option
[Extrude/Move/Rotate/Offset/Taper/Delete/Copy/coLor/Undo/eXit] <eXit>: _taper
Select faces or [Undo/Remove]: *Click on the cylindrical faces of the smaller cylinder.*
Remove faces or [Undo/Add/ALL]: Enter
Specify the base point: *Select the center of the front face of the smaller cylinder.*
Specify another point along the axis of tapering: *Select the center at the back face of the smaller cylinder.*
Specify the taper angle: **-10**
Solid validation started.
Solid validation completed.
Enter a face editing option
[Extrude/Move/Rotate/Offset/Taper/Delete/Copy/coLor/Undo/eXit] <eXit>: *Press ESC.*

5. The final model should look similar to the one shown in Figure 28-10.

GENERATING A SECTION BY DEFINING A SECTION PLANE

Ribbon:	Home > Section > Section Plane	**Command:**	SECTIONPLANE
Menu Bar:	Draw > Modeling > Section Plane		

This command is used to create a sectioned object by defining a sectioning plane. By default, the section plane cuts all the 3D objects, surfaces, and regions lying in the line of this plane even if the section plane does not actually pass through them. Various methods of defining the section plane are discussed next.

Defining a Section Plane by Selecting a Face

In this option, the section plane will be aligned to the plane of the selected face of the 3D solid. Choose **Home > Section > Section Plane** from the **Ribbon** and select the desired face on the solid object; a section plane will be displayed aligned with the selected face, as shown in Figure 28-11. The prompt sequence is given next.

Command: **SECTIONPLANE**
Select face or any point to locate section line or [Draw section/Orthographic]: *Select a face on the 3D solid where you want to align the section plane.*

Now, you can move the section plane along its normal direction to place it at the intended location. To do so, select a section line visible on that section plane and then select the triangular arrow grip displayed at the midpoint of the section line, see Figure 28-12. While moving the section plane, you will get a dynamic preview of the remaining section of the model. Figure 28-13 shows the remaining section of the solid model after moving the section plane, refer to Figure 28-11, to the center of the model. You can also flip the direction of the sectioned portion of the model. To do this, click on the section line and then click on the

Figure 28-11 Model with a section plane

Figure 28-12 Various grips visible on the section line

direction grip, see Figure 28-14. To rotate the section plane, right-click on the section line and then choose the **Rotate** option from the shortcut menu. Next, select the base point about which you want to rotate the section plane and specify the rotation angle. Figure 28-15 shows the remaining section of the model after rotating the section plane by -90 degree about the midpoint of the section line.

Figure 28-13 Remaining section of the model after moving the section plane

Figure 28-14 Resulting model after flipping the direction of the section plane

Instead of displaying only the section plane, you can show the section boundary with the section plane. The section boundary displays a two dimensional rectangular sketch. Only the object lying within this boundary will be displayed as the remaining section. To create a section boundary, click on the menu grip and choose the **Section Boundary** option. You can also show the section volume with the section boundary and the section plane. The section volume displays a three dimensional box and only the object lying within this boundary will be displayed as the remaining section. To create a section volume, choose the **Section Volume** option from the menu grip. Figure 28-16 shows a model with a section plane, section boundary, and section volume. The size of the section boundary and the section plane can also be edited with the help of grips.

Generating the section by a section plane does not affect the original model. The original model is hidden and only the remaining section of the model is displayed. If you delete the section line from the drawing, the model will be displayed without the section.

Figure 28-15 *Resulting model after rotating the section plane by -90 degree*

Figure 28-16 *Model after choosing the **Section Volume** option from the menu grip*

AutoCAD also allows you to display both the sectioned and unsectioned portion of 3D solid. Select the section line and right-click to display the shortcut menu. Choose the **Activate live sectioning** option from the shortcut menu, if it is not chosen by default. You can also choose **Section > Live Section** from the **Home** tab of the **Ribbon**. Again, select the section line, invoke the shortcut menu and choose the **Show cut-away geometry** option from it. Now, a translucent and red colored sectioned model will be displayed.

You can add jogs to the section plane. A jog provides a 90-degree bend to the section plane for a certain distance and after that the section plane again continues in the original direction. To create a jog, select the section line and choose **Add jog to section** from the shortcut menu; you will be prompted to specify a point on the section line where you want to add the jog. After adding the jog, you can edit its length and location with the help of grips. You can add multiple jogs to the section plane. You can also choose **Section > Add Jog** from the **Home** tab of the **Ribbon** to add the jog.

Note
The procedure to modify the location and appearance of a section plane, displaying the sectioned portion of the model, and adding jogs to the section plane by grips remains the same in all the other options of defining the section plane that are discussed next.

Defining a Section Plane by Specifying Two Points

In this option, the section plane is generated such that it passes through the two points that you specify in the drawing area. If you specify the first point on the 3D object, it generates a section plane at the selected face, but if you specify the first point at any location in the drawing area other than the 3D object, this point acts as a base point about which the section plane can rotate. Specifying the second point will fix the through point of the section plane. A section plane will be generated passing through these two points, see Figure 28-17. The prompt sequence for this option is given next.

Command: **SECTIONPLANE**
Select face or any point to locate section line or [Draw section/Orthographic]: *Specify a point on the screen but not on the 3D object.*

Specify through point: *Specify the other point through which the section line will pass*.

Once you generate the section plane, you will not get a preview of the sectioned object. This is because the **Activate live sectioning** option is not chosen by default when you generate a section plane by specifying two points. To activate the live sectioning, select the section line and choose **Activate live sectioning** from the shortcut menu. Figure 28-17 displays a model sectioned by specifying two points with the live sectioning turned on. You can also activate the live sectioning by entering **LIVESECTION** at the Command prompt. The prompt sequence for this command is given next.

Command: **LIVESECTION**
Select section object: *Select the section line on the section plane*.

Defining a Section Plane by Drawing the Section Line

Using this option, you can draw a section line and generate the section plane that passes along it. You can section the 3D object in many planes by picking points through which the section plane will pass. Choose **Home > Section > Section Plane** from the **Ribbon** and invoke the **Draw Section** option. Now, specify the point from which the section line will start and then specify the points through which it will pass. After specifying all the points, press the ENTER key and then specify the side of the object to retain after sectioning. By default, the section plane in this case will be displayed as the section boundary with the live sectioning deactivated. Figure 28-18 shows the model sectioned by drawing a section line with the live sectioning activated. The prompt sequence for generating the section plane by drawing the section line is given next.

Figure 28-17 *Model sectioned by specifying two points*

Figure 28-18 *Model sectioned by drawing the section line*

Command: **SECTIONPLANE**
Select face or any point to locate section line or [Draw section/Orthographic]: **D** Enter
Specify start point: *Specify the start point of the section line*.
Specify next point: *Specify the next point through which the section line will pass*.
Specify next point or ENTER to complete: *Specify next point of the section line or* Enter
Specify point in direction of section view: *Specify a point on the desired side of the section line to specify the portion of the solid to retain*.

Note

A section line should be drawn only on one sketching plane. Otherwise, you will not be able to visualize the sectioned model after activating the live sectioning. Also, the drawn section line should not be self-intersecting.

Defining a Section Plane by Selecting a Preset Orthographic Plane

In this option, a section plane will be generated parallel to preset orthographic planes such as front, top, right so on, and passing through an imaginary center of a box. This box is calculated by enveloping all the solids, surfaces, and regions present in the drawing. Choose **Home > Section > Section Plane** from the **Ribbon** and invoke the **Orthographic** option from the shortcut menu; you will be prompted to select the predefined orthographic plane with which the generated section plane will be aligned. On selecting this, you get a sectioned solid whose material is chopped from the side you selected as the preset orthographic plane. In this option, you will instantaneously get the sectioned object because live sectioning is activated automatically. The prompt sequence for defining the section plane by selecting a preset orthographic plane is given next.

Command: **SECTIONPLANE**
Select face or any point to locate section line or [Draw section/Orthographic]: **O** Enter
Align section to: [Front/bAck/Top/Bottom/Left/Right] <Bottom>: *Specify one of the preset orthographic planes.*

Generating 2D and 3D Sections

As mentioned earlier, sectioning a model by generating a section plane does not affect the original model. The original model is hidden and only the remaining section of the sectioned model is displayed. On deleting the section plane, the sectioned object will get deleted. If you want to retain the generated section as a 2D or 3D section with the original model, you can use the **Generate 2D/3D section** option.

To generate a 2D or a 3D section, select the section line and choose the **Generate 2D/3D section** option from the shortcut menu; the **Generate Section/Elevation** dialog box will be displayed. Choose the **Show Detail** button to expand this dialog box, as shown in Figure 28-19. The options in this dialog box are discussed next.

Section Plane Area

If there are multiple section planes in the model, choose the **Select section plane** button from this area and select the section plane for which you need to create 2D/3D sections.

2D/3D Area

This area allows you to specify the type of section you want to generate. The **2D Section / Elevation** option generates a two dimensional sketch of the sectioned object on the XY plane of the current UCS. The **3D section** generates a 3D sketch of the sectioned object on a plane parallel to the section plane. Figure 28-20 shows the model with its 2D section and Figure 28-21 shows the model with its 3D section.

Figure 28-19 *The expanded **Generate Section/Elevation** dialog box*

Figure 28-20 *Model with 2D section*

Figure 28-21 *Model with 3D section*

Source Geometry Area

This area allows you to specify the objects of the drawing that you want to include while creating the 2D or 3D section. The **Include all objects** radio button, if selected, includes all the 3D objects, surfaces, and regions in the drawing. It also includes the external reference objects and blocks. The **Select objects to include** option allows you to select the objects to be included in the section generation. If you select the **Select objects to include** radio button and then choose the **Select objects** button, the **Generate Section/Elevation** dialog box will disappear allowing to select the solids. After selecting the solids, surfaces, or regions, press the ENTER key; the **Generate Section/Elevation** dialog box will reappear, displaying the number of selected objects below the **Select objects** button.

Destination Area

This area gives the option to specify how you want to use the generated section. The **Insert as new block** option inserts the generated section in the same drawing file as a block. The **Replace existing block** option inserts the generated section by replacing an existing block. Choose the **Select block** button to select the block to be replaced by the newly generated section block. The **Export to a file** option allows you to create a new drawing file with the generated section. To save it as a new file, choose the **Browse [...]** button to specify the location where you want to save the new drawing file.

Section Settings

This button is used to invoke the **Section Settings** dialog box, see Figure 28-22. You can also invoke this dialog box by clicking on the inclined arrow in the **Section** panel of the **Ribbon**. If you select the **2D Section/ Elevation** option in the 2D/3D area, the **Section Settings** dialog box will be displayed with the **2D section / elevation block creation settings** radio button selected by default. Similarly, if you select the **3D Section/ Elevation** option in the 2D/3D area, the **Section Settings** dialog box will be displayed with the **3D section block creation settings** option selected by default. The **Section Settings** dialog box allows you to modify the properties of the sections generated. These properties depend on the radio button selected.

Intersection Boundary. This option modifies the display of the lines that are created to outline the boundary of the intersected solid with the section plane.

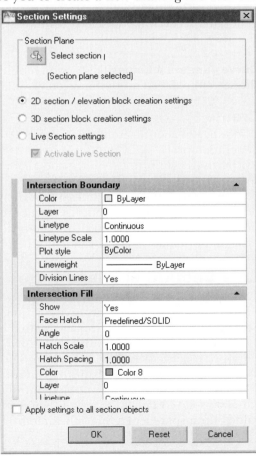

*Figure 28-22 The **Section Settings** dialog box*

Intersection Fill. This option modifies the display of the filled portion representing the presence of material at the intersection plane. The **Show** drop-down list allows you to select whether you want to display the section filled. Selecting the **Select hatch pattern type** option from the **Face Hatch** drop-down list, displays the **Hatch Pattern Type** dialog box, see Figure 28-23. Choose the **Pattern** button from this dialog box to select the predefined standard hatch patterns to be displayed as the intersection fill.

Figure 28-23 *The* **Hatch Pattern Type** *dialog box*

Background Lines. This option modifies the display of the hidden edges of the model. In case of a 2D section, you can choose to display the background lines.

Cut-away Geometry. This option modifies the display of the cut-away solid part in the generated section, if you select **Yes** in the **Show** drop-down list of the **Cut-away Geometry** category.

Curve Tangency Lines. This option modifies the display of curved lines that are tangent to the section plane. This option is only available when you generate a 2D section.

Create

Choosing this button from the **Generate Section/Elevation** dialog box finishes the section generation process. If you insert the generated section as a new block, you will be prompted to specify the insertion point, scale factor, and the rotation angle for the new generated section to be inserted as a block. If you replace an existing block, the generated section will be inserted with the same values of insertion point, scale factor, and rotation angle as for the block selected to be replaced. If you choose to export the generated section to a new file, a new file will be created and the section will be created at the same location with the same scale and orientation as the model in the previous file.

Live Section settings

This option is used to control the dynamic display of a sectioned solid in the drawing. Select the section line and choose **Live Section settings** from the shortcut menu; the **Section Settings** dialog box will be displayed with the **Live Section settings** option selected by default. The options available to control the display of the live sectioning of the model are similar to the options discussed in the **Section Settings** dialog box. The **Face Transparency** and **Edge Transparency** options in the **Cut-away Geometry** area are only available for the live section settings. These two options control the transparency of the edge and the face of the cut-away solid for a better view of the section and the remaining portion of the model. The value of transparency signifies the amount of light that passes through the cut-away part.

GENERATING DRAWING VIEWS OF THE SOLID MODEL

As mentioned earlier, one of the major advantages of working with the solid models is that the drawing views of the solid model are automatically generated once you have the solid model with you. All you have to do is to specify the type of drawing view required. This is done using the **SOLVIEW**, **SOLDRAW**, and **SOLPROF** commands. All these commands are discussed next.

SOLVIEW Command

| Ribbon: | Home > Modeling > Solid View | Command: | SOLVIEW |
| Menu Bar: | Draw > Modeling > Setup > View | | |

The **SOLVIEW** command is an externally defined ARX application (*solids.arx*). This command is used to create the documentation of the solid models in the form of the drawing views. This command guides you through the process of creating orthographic or auxiliary views. The **SOLVIEW** command creates floating viewports using orthographic projections to layout sectional view and multiview drawings of solids, while you are in the layout tab.

For the ease of documentation, this command automatically creates layers for visible lines, hidden lines, and section hatching for each view. **SOLVIEW** also creates layers for dimensioning that are visible in individual viewports. You can use these layers to draw dimensions in individual viewports. The following is the naming convention for layers.

Layer name	Object type
View name-VIS	Visible lines
View name-HID	Hidden lines
View name-DIM	Dimensions
View name-HAT	Hatch patterns (for sections)

The view name is the user-defined layer name given to the view when created.

Note
The information stored on these layers is deleted and updated when you run **SOLDRAW**. *Therefore, do not place any permanent drawing information on these layers.*

The prompt sequence is as follows:

Command: **SOLVIEW**
Enter an option [Ucs/Ortho/Auxiliary/Section]: *Select an option to generate the drawing view.*

Ucs Option

This option allows you to create a profile view relative to a UCS. This is the first view that has to be generated. The reason for this is that the other options require existing viewports. The UCS can either be the current UCS or any other named UCS. The view generated using this option will be parallel to the *XY* plane of the current UCS. The next prompt sequence is:

> Enter an option [Named/World/?/Current] <Current>: *Enter an option or press ENTER to accept the current UCS.*

The **Named** option uses a named UCS to create the profile view. The prompt sequence is as follows:

> Enter name of UCS to restore: *Enter the name of the UCS you want to use.*
> Enter view scale <1.0>: *Enter a positive scale factor.*

This view scale value you enter is equivalent to zooming into your viewport by a factor relative to the paper space. The default ratio is 1:1. The next prompt sequence is as follows:

> Specify view center: *Specify the location of the view.*
> Specify view center <specify viewport>: `Enter`

This point is based on the Model space extents. The next prompt sequence is as follows.

> Specify the first corner of viewport: *Specify a point.*
> Specify opposite corner of viewport: *Specify a point.*
> Enter view name: *Enter a name for the view.*

The **World** option uses the *XY* plane of the WCS to create a profile view. The prompt sequences are the same as the **Named** option.

The **?** option lists all the named UCSs. The prompt sequence is given next.

> Enter UCS names to list<*>: *Enter a name or press ENTER to list all the named UCSs.*

The **Current** option uses the *XY* plane of the current UCS to generate the profile view. This is also the default option.

Ortho Option

An orthographic view is generated by projecting the lines on to a plane normal to an existing view. The **Ortho** option generates an orthographic view from the existing view. Keep in mind that the orthographic views created in AutoCAD are folded views and not the unfolded views. AutoCAD prompts you to specify the side of the viewport from where you want to project the view. The resultant view will depend on the side of the viewport selected. The prompt sequence is given next.

Specify side of viewport to project: *Select the edge of a viewport.*
Specify view center: *Specify the location of the orthographic view.*
Specify view center <specify viewport>: ⌨Enter
Specify first corner of viewport: *Select the first point.*
Specify opposite corner of viewport: *Select the second point.*
Enter view name: *Specify the view name.*

Auxiliary Option

This option creates an auxiliary view from an existing view. An auxiliary view is the one that is generated by projecting the lines from an existing view on to a plane that is inclined at a certain angle. AutoCAD prompts you to define the inclined plane used for the auxiliary projection. The prompt sequence is as follows.

Specify first point of inclined plane: *Specify the first point on the inclined plane.*
Specify second point of inclined plane: *Specify the second point on the inclined plane.*
Specify side to view from: *Specify a point.*
Specify view center: *Specify the location of the view.*
Specify view center <specify viewport>: ⌨Enter
Specify first corner of viewport: *Specify the first corner of the viewport.*
Specify opposite corner of viewport: *Specify the second corner of the viewport.*
Enter view name: *Enter the view name.*

Section Option

This option is used to create the section views of the solid model. The type of the section view depends on the alignment of the section plane and the side from which you view the model. Initially, the crosshatching is not displayed in the section view. To display the crosshatching, invoke the **SOLDRAW** command. AutoCAD prompts you to define the cutting plane, and the prompt sequence is as follows.

Specify first point of cutting plane: *Specify the first point on the cutting plane.*
Specify second point of cutting plane: *Specify the second point on the cutting plane.*

Once you have defined the cutting plane, you can specify the side to view from and the view scale, relative to paper space. The prompt sequence is given next.

Specify side to view from: *Specify the point in the direction from which the section is viewed.*
Enter view scale <current>: *Enter a positive number.*
Specify view center: *Specify the location for the view.*
Specify view center <specify viewport>: ⌨Enter
Specify first corner of viewport: *Specify the first corner of the viewport.*
Specify opposite corner of viewport: *Specify the second corner of the viewport.*
Enter view name: *Enter the view name.*

Chapter 28

SOLDRAW Command

Ribbon :	Home > Modeling > Solid Drawing	Command:	SOLDRAW
Menu Bar:	Draw > Modeling > Setup > Drawing		

The **SOLDRAW** command generates profiles and sections in the viewports that were created with the **SOLVIEW** command. The basic function of this command is to clean up the viewports created by the **SOLVIEW** command. It creates the visible and hidden lines that represent the silhouettes and edges of solids in a viewport, and then projects them in a plane that is perpendicular to the viewing direction. When you use the **SOLDRAW** command with the sectional view, a temporary copy of the solid is created. This temporary copy is then chopped at the section plane defined by you. Then **SOLDRAW** creates the visible half as the section and discards the copy of the solid model. The crosshatching is automatically created using the current hatch pattern (HPNAME), hatch angle (HPANG), and hatch scale (HPSCALE). In this case, the layer **View name- HID** is frozen. If there are any existing profiles and sections in the selected viewports, they are deleted and replaced with new ones. The prompt sequence is as follows.

Select viewports to draw...
Select objects: *Select viewports to be drawn.*

SOLPROF Command

Ribbon :	Home > Modeling > Solid Profile	Command:	SOLPROF
Menu Bar:	Draw > Modeling > Setup > Profile		

The **SOLPROF** command creates profile images of 3D solids by displaying only the edges and silhouettes of the curved surfaces of the solids in the current view. Note that before you invoke this command, you must be in the temporary model space of the layout. This means that the value of the **TILEMODE** variable should be set to 0, but you should switch to the model space using the **MSPACE** command. The prompt sequence that follows when you choose the **Solid Profile** button is given next.

Select objects: *Select the object in the viewport.*
Select objects: Enter
Display hidden profile lines on separate layer? [Yes/No] <Y>: *Specify an option.*
Project profile lines onto a plane? [Yes/No] <Y>: *Specify an option.*
Delete tangential edges? [Yes/No] <Y>: *Specify an option.*
One solid selected.

FLATSHOT Command

Ribbon:	Home > Section > Flatshot	Command:	FLATSHOT

The **FLATSHOT** command creates a 2D sketch of a model by projecting the model in the current view on the XY plane of the current UCS. The created sketch is inserted in the

drawing as a block. A flatshot can be created in
any preset orthographic view or created view in
the parallel projection. If you create a flatshot in
the perspective view, a 3D wireframe of the whole
model will be created by projecting the model in
the current view on the XY plane of the current
UCS. Choose **Home > Section > Flatshot**
from the **Ribbon**; the **Flatshot** dialog box will be
displayed, see Figure 28-24. The options in this
dialog box are discussed next.

Destination Area

The options in this area are similar to those
discussed in the **Generating 2D and 3D Sections**
heading.

Foreground lines Area

The options in this area control the color and line
type of the edges in the front at the time of the
flatshot creation.

Obscured lines Area

The options in this area control the display, color,
and line type of the hidden edges at the time of
the flatshot creation.

Create

Choosing this button finishes the flatshot
creation process.

*Figure 28-24 The **Flatshot** dialog box*

Include tangential edges

Select this check box to include the tangential edges of the curved surface into the flatshot.

Note
The 3D objects that are sectioned by the section plane will be flatshot as if they are not sectioned.

Tip
*All 3D objects in the current drawing are included in the flatshot creation. If you do not want
to include some objects in the flatshot creation, transfer them to a separate layer and freeze the
layer before creating the flatshot.*

Example 2   *Mechanical*

In this example, you will use the **SOLVIEW** and **SOLDRAW** commands to generate the top view and the sectioned front view of the solid model shown in Figure 28-25. Assume the missing dimensions for the model.

Figure 28-25 *Solid model to generate the drawing views*

1. Create the solid model by assuming the dimensions.

2. Switch to **Layout 1** and delete the default viewport.

3. Choose **Home > Modeling > Solid View** from the **Ribbon**. The prompt sequence is as follows:

   Enter an option [Ucs/Ortho/Auxiliary/Section]: **U**
   Enter an option [Named/World/?/Current] <Current>: **W**
   Enter view scale <1.0000>: ⏎
   Specify view center: *Specify the view location close to the top left corner of the drawing window.*
   Specify view center <specify viewport>: ⏎
   Specify first corner of viewport: *Specify the point of the viewport, see Figure 28-26.*
   Specify opposite corner of viewport: *Specify the second point, see Figure 28-26.*
   Enter view name: **Top**
   Enter an option [Ucs/Ortho/Auxiliary/Section]: ⏎

4. Double-click in the viewport to switch to the temporary model space. Zoom the drawing to the extents using the **Extents** option of the **ZOOM** command. Now, switch back to the paper space using the **PSPACE** command. The layout after generating the top view should look similar to the one shown in Figure 28-26.

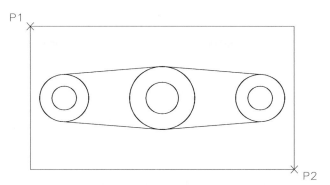

Figure 28-26 *Generating the top view*

5.   Again, choose **Home > Modeling > Solid View** from the **Ribbon.** The prompt sequence is as follows:

Enter an option [Ucs/Ortho/Auxiliary/Section]: **S**
Specify first point of cutting plane: *Select the left quadrant of the left cylindrical portion in the top view.*
Specify second point of cutting plane: *Select the right quadrant of the right cylindrical portion in the top view.*
Specify side to view from: *Select a point below the top view in the viewport.*
Enter view scale <current>: Enter
Specify view center: *Specify the location of the view below the top view.*
Specify view center <specify viewport>: Enter
Specify first corner of viewport: *Select first corner of the viewport.*
Specify opposite corner of viewport: *Select the second corner of the viewport.*
Enter view name: **Front**
Enter an option [Ucs/Ortho/Auxiliary/Section]: Enter

6.   Choose **Home > Modeling > Solid Drawing** from the **Ribbon** to invoke the **SOLDRAW** command. The prompt sequence is as follows:

Select viewports to draw..
Select objects: *Select the first viewport.*
Select objects: *Select the second viewport.*
Select objects: Enter

7.   The drawing views should look similar to those shown in Figure 28-27.

Tip
*You can change the scale and the type of hatch pattern using the **Properties** palette. You can also set different linetypes and colors to the different lines in the views by selecting their respective layers using the **LAYER** command.*

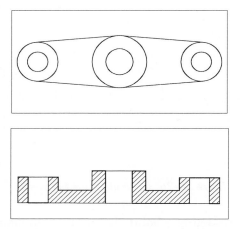

Figure 28-27 *Layout after generating the drawing views*

CALCULATING THE MASS PROPERTIES OF THE SOLID MODELS

Menu Bar:	Tools > Inquiry > Region/Mass Properties	Command:	MASSPROP
Toolbar:	Inquiry > Region/Mass Properties		

The mass properties for the solids or the regions can be automatically calculated using the **MASSPROP** command. The various mass properties are displayed in the AutoCAD Text Window. You can also write these properties to a file. This file will be in the *.mpr* format. The properties thus calculated can be used in analyzing the solid models. Note that the mass properties are calculated by taking the density of the material of the solid model as one unit.

It is very important for you to first understand the various properties that will be calculated using this command. The properties are:

Mass. This property provides the measure of mass of a solid. This property will not be available for the regions.

Area. This property provides the measure of area of the region. This property will be available only for the regions.

Volume. This property tells you the measure of the space occupied by the solid. Since AutoCAD assigns a density of 1 to solids, the mass and volume of a solid are equal. This property will be available only for the solids.

Perimeter. This property, providing the measure of the perimeter of the region, will be available only for the regions.

Bounding Box. If the solid were to be enclosed in a 3D box, the coordinates of the diagonally opposite corners would be provided by this property. Similarly, for regions, the X and the Y coordinates of the bounding box are displayed.

Centroid. This property provides the coordinates of the center of mass for the selected solid. The density of a solid is assumed to be unvarying.

Moments of Inertia. This property provides the mass moments of inertia of a solid about the three axes. The equation used to calculate this value is given next.

$$\text{mass\_moments\_of\_inertia} = \text{object\_mass} * (\text{radius of axis})^2$$

The radius of axis is nothing but the radius of gyration. The values obtained are used to calculate the force required to rotate an object about the three axes.

Products of Inertia. The value obtained with this property helps you to determine the force resulting in the motion of the object. The equation used to calculate this value is:

$$\text{product \_of\_inertia YX,XZ} = \text{mass} * \text{dist centroid\_to\_YZ} * \text{dist centroid\_to\_XZ}$$

Radii of Gyration. The equation used to calculate this value is given next.

$$\text{gyration\_radii} = (\text{moments\_of\_inertia/body\_mass})^{1/2}$$

Principal Moments and X-Y-Z Directions about Centroid. This property provides you with the highest, lowest, and middle value for the moment of inertia about an axis passing through the centroid of the object.

The prompt sequence that follows when you invoke the **MASSPROP** command is given next.

Select objects: *Select the solid model.*
Select objects: Enter

---------------- SOLIDS ----------------		
Mass:	67278.5917	
Volume:	67278.5917	
Bounding box:	X: 1.4012 -- 151.4012	
	Y: 118.6406	-- 158.6406
	Z: 0.0000	-- 25.0000
Centroid:	X: 76.4012	
	Y: 138.6406	
	Z: 9.2026	
Moments of inertia:	X: 1308864825.5181	

	Y: 507244127.3062
	Z: 1799135898.4724
Products of inertia:	XY: 712635884.7490
	YZ: 85837337.1138
	ZX: 47302745.3493
Radii of gyration:	X: 139.4790
	Y: 86.8301
	Z: 163.5285

Principal moments and X-Y-Z directions about centroid:
            I: 9991363.2928 along [1.0000 0.0000 0.0000]
            J: 108831209.8607 along [0.0000 1.0000 0.0000]
            K: 113244795.9294 along [0.0000 0.0000 1.0000]
Press ENTER to continue: Enter

Write analysis to a file? [Yes/No] <N>: *Specify an option.*

If you enter **Y** at the last prompt, the **Create Mass and Area Properties File** dialog box will be displayed. This dialog box is used to specify the name of the *.mpr* file. The mass properties will then be written in this file.

RECORDING YOUR DRAWING STEPS USING ACTION RECORDER

Ribbon:	Manage > Action Recorder > Record	**Command:**	ACTRECORD
Menu Bar:	Tools > Action Recorder > Record		

The **Action Recorder** helps you to record the drawing steps quickly and easily. These steps are called actions. These actions are then saved to a named macro for repeated use. To record an action, choose **Manage > Action Recorder > Record** from the **Ribbon** or enter the **ACTRECORD** command at the Command prompt. On invoking this command, a red circle appears near the cursor indicating that the actions are recorded. Now, if you perform any action, it will be recorded. Also, the **Action Tree** will be expanded and the actions you perform will be updated simultaneously, as shown in Figure 28-28. If you need any message to be displayed during playback, you can do so by choosing **Manage > Action Recorder > Insert Message** from the **Ribbon**; the **Insert User Message** dialog box will be displayed. Enter the message and choose the **OK** button; the message will be added in the **Action Tree**. This message will be displayed during playback. Similarly, if you want the user to input any values while playback, you can choose **Action Recorder > Request Input** and record the action.

After recording the actions, choose the **Stop** button from the **Ribbon** to stop recording; the **Action Macro** dialog box will be displayed, as shown in Figure 28-29. Specify the name for the action macro in the **Action Macro Command Name** edit box. If you want to give any description of the action, enter it in the **Description** edit box. The check boxes in the **Restore pre-playback view** area are used to invoke the playback message box when you play the action macro. If you select the **Check for inconsistencies when playback begins** check box

Figure 28-28 The *Action Recorder*
panel recording the actions performed

Figure 28-29 The *Action Macro* dialog box

and if there is inconsistency in the workspace, template, or any other, while you playback the action macro, then a warning box will be displayed. Remember that if you invoke any dialog boxes such as **Hatch and Gradient** or **OSnap**, and change any settings during recording, then those settings will not be recorded. So, during the playback, the corresponding dialog box will be displayed again and you need to set the parameters again in it.

The **Folder Path** display box in the **Action Macro** dialog box displays the default path where the action macro is saved. The **ACTRECPATH** command is used to change the path where the action macro is stored, by default. On entering this command, the following prompt sequence will be displayed:

Command: **ACTRECPATH**
Enter new value for ACTRECPATH, or . for none <"C:\Documents and
Settings\User\Application Data\Autodesk\A...">: *Enter the new location within the quotes (" ")*

To playback the recorded action macro, select the action macro from the **Available Action Macro** drop-down list in the **Action Recorder** panel and choose the **Play** button; the action macro will start playing. You can use an existing action macro in a new drawing. To do so, open a new drawing file in the same template as used for the action macro. Next, choose the **Tools** tab from the **Ribbon** and choose **Play** from the **Action Recorder** panel of the **Ribbon**.

USING SHOWMOTION FOR PRESENTATION PURPOSES

Status Bar:	ShowMotion	**Menu Bar:**	View > ShowMotion
Command:	NAVSMOTION		

The **ShowMotion** tool is used to create animated views of the current drawing. This can be used to quickly step through different named views in your drawing, enabling you to play the animated views for presentations purposes or to review your drawings. This command can be invoked by choosing the **ShowMotion** button that is available on the Status Bar. Alternatively, you can choose **View > ShowMotion** from the menu bar or enter **NAVSMOTION** at the Command prompt.

On invoking this command, a control panel for **ShowMotion** with five buttons is displayed above the **Status Bar**. The **Play all** button is used to play all the saved animations, which include shots and view categories, in a sequence. While the animations are being played, the **Play all** button automatically gets converted to the **Pause** button, enabling you to pause the playback and vice-versa. The **Stop** button stops playing the animation. The **Turn on Looping** button of the **ShowMotion** allows the animation to continue playing until you choose the **Stop** button or press the ESC key. A new animated view or a shot is created by choosing the **New Shot** button. On choosing this button, the **New View / Shot Properties** dialog box will be displayed, refer to Figure 28-30. The options in this dialog box are discussed next.

View name. You can specify the name for the shot in the **View name** edit box. These names will be displayed below the respective shot thumbnails.

View category. You can enter the name of the category in this edit box. You can also group a sequence of shots of the same view categories. The different view categories are displayed as a series of thumbnails at the bottom of the drawing area.

View type. You can take three different types of shots using this drop-down list. These are **Still, Cinematic** and **Recorded Walk**. You can give transition effects between two shots, and the length or duration of the effects can be specified in the **Transition** area. Select an option in the **Transition type** drop-down list to specify the transition effects between two shots. You can see the preview of any view type in the preview box available in the **Motion** area of the **Shot Properties** tab of the dialog box.

If you select the **Still** option from the **View type** drop-down list, the position of camera will be fixed and you can set the duration of the shot in the **Motion** area of the dialog box.

If you select the **Cinematic** option from the **View type** drop-down list, you can move the camera position and set the duration of the recording. Select the type of movement in the **Movement type** drop-down list and specify the duration and distance in the respective spinners. On selecting the **Crane up** or **Crane down** option in the **Movement type** drop-down list, the **Shift left** check box becomes available. Consequently, you can specify the amount by which the camera shifts to left or right during the movement. On selecting the **Always look at camera pivot point** check box, the view is always focused at the camera.

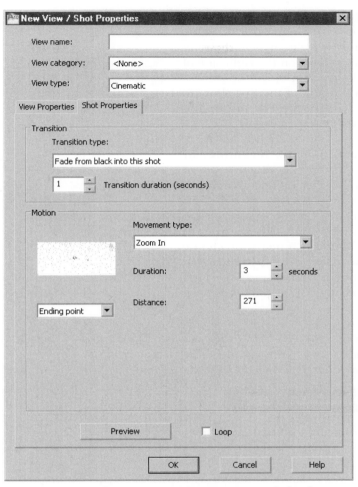

Figure 28-30 *The* *New View/Shot Properties* *dialog box*

If you select the **Recorded Walk** option from the **View type** drop-down list, you can create an animated view. To do so, choose the **Start Recording** button available in the **Motion** area of the **Shot Properties** tab. The dialog box disappears and a navigation tool is attached to the cursor. Next, press and hold the left mouse button and move the cursor. Once you release the left mouse button, the dialog box reappears. Choose the **OK** button to exit the dialog box.

Playing the Animation

Each **View Category** and its Shots are displayed as thumbnails at the bottom of the drawing area. Each **View Category** thumbnail contains its corresponding name at its bottom. The series of shots under each category are displayed as smaller thumbnails above the **View category** thumbnails, as shown in Figure 28-31. However, the thumbnails are enlarged when the cursor moves over them. Each thumbnail contains two buttons, **Play** and **Go**. Choosing the **Play** button on the shot thumbnail plays that particular animated shot whereas choosing the **Play** button on the **View Category** thumbnail plays the sequence of shots under that particular category. On choosing the **Go** button available on the right side of the thumbnails, you can move the camera to the camera position of the first shot. To play all shots, choose the **Play all** button from the control panel.

You can rename, delete, recapture, or update a shot using the shortcut menu displayed on right-clicking on a thumbnail. You can also re-order the sequence of shots by moving them to the left or right using the options in the shortcut menu, refer to Figure 28-31. To create a new shot, choose the **New View / Shot** option from the shortcut menu. To modify the shots, choose the **Properties** option from the shortcut menu; the **View / Shot Properties** dialog box is displayed. Now, you can change the required properties in this dialog box.

*Figure 28-31   The Control Panel and the **View Category** thumbnails*

Note
*On invoking the **NEWSHOT** command in the **Layout** tab, the **Recorded Walk** option will not be available in the **View type** drop-down list. Only the **Cut to Shot** option will be available in the **Transition type** drop-down list of the **New View /Shot Properties** dialog box.*

Self Evaluation Test

Answer the following questions and then compare them to those given at the end of this chapter:

1. The solid models can be edited using the **EDITSOLID** command. (T/F)

2. You can create a shell in the solid model using the **Shell** option of the **SOLIDEDIT** command. (T/F)

3. A section plane cuts all 3D objects, surfaces, and regions only if the section plane actually passes through them. (T/F)

4. The **ACTRECORD** command is used to create animated views of the current drawing. (T/F)

5. The _____ command is used to generate the drawing views of the solid model.

6. The mass properties of the selected solid model can be written into a file of _____ format.

7. The _____ command creates a 2D sketch of a model by projecting the model in the current view on the XY plane of the current UCS.

8. The _____ option of the **DVIEW** command is used to set the perspective projection.

9. Choose the _____ option from the **Solid Editing** panel of the **Ribbon** to separate 3D solids from the 3D solids with disjointed volumes.

10. The impression on the solid models can be removed using the _____ option of the **SOLIDEDIT** command.

Review Questions

Answer the following questions:

1. You can edit the wireframe models by using the **SOLIDEDIT** command. (T/F)

2. You can move a selected face of the solid model from one location to the other. (T/F)

3. You can assign different colors to the different faces of the solid model. (T/F)

4. After adding the jog, you can edit its length and location with the help of grips. (T/F)

5. The sketch created using the **FLATSHOT** command is inserted in the drawing as a block. (T/F)

6. Which of the following commands is used to generate the profiles and sections in the viewports created using the **SOLVIEW** command?

    (a) **SOLDRAW**   (b) **SOLPROF**
    (c) **SOLEDIT**   (d) **None**

7. Which of the following options is used to specify a section plane?

    (a) by Specifying Two Points   (b) by Selecting a Preset Orthographic Plane
    (c) by Drawing the Section Line   (d) All of the above

8. Which option of the **SOLIDEDIT** command is used to edit the faces of the solid model?

    (a) **Body**   (b) **Face**
    (c) **Solid**   (d) **None**

9. Which option of the **SOLVIEW** command is used to generate the drawing view relative the current UCS?

    (a) **New**   (b) **Section**
    (c) **UCS**   (d) **Ortho**

10. Which of the following commands is used to calculate mass properties of the solid model?

   (a) **PROPERTIES**   (b) **MASSPROP**
   (c) **CALCULATE**   (d) **MASSPROPERTIES**

Exercises

Exercise 1   *Mechanical*

In this exercise, you will create the solid model shown in Figure 28-32. The dimensions for the solid model are shown in Figures 28-33 and 28-34. After creating the model, generate its top and sectioned front view. Also, set the front clipping, and then shade the model using the Realistic shading and dynamically rotate it using the SteeringWheel.

Figure 28-32 Solid model for Exercise 1

Figure 28-33 Dimensions for Exercise 1

Figure 28-34 Dimensions for Exercise 1

Exercise 2   *Architectural*

In this exercise, you will create the solid model shown in Figure 28-35. After creating the model, shade and rotate it in the 3D space using the **3DORBIT** command. Assume the missing dimensions.

Figure 28-35 *Solid model for Exercise 2*

Problem-Solving Exercise 1   *General*

Draw the Piece part drawings shown in Figure 28-36. Then assemble them and create the Bill of Materials, as shown in Figure 28-37. The solid model of the assembled Tool Organizer is shown in Figure 28-38.

Chapter 28

Figure 28-36 *Piece part drawing of the Tool Organizer*

Part	Qty.	Description		
1	1	Lower Shelf	.75 × 7.25 × 48.00	
2	1	Upper Shelf	.75 × 3.75 × 48.00	
3	1	Lower Rail	.75 × 8.75 × 48.00	
4	1	Left Side Support	.75 × 2.75 × 32.50	
5	1	Right Side Support	.75 × 2.75 × 32.50	
6	1	Right Stile	.75 × 2.75 × 22.75	
7	1	Upper Rail	.75 × 2.75 × 48.00	
8	1	Left Stile	.75 × 2.75 × 22.75	
9	1	Tempered Pegboard	.25 × 24.00 × 48.00	
10	4	3 1/2 – No. 8 FH Wood Screw – Steel		
11	20	2 1/4 – No. 8 FH Wood Screw – Steel		
12	16	3/4 – No. 6 RH Wood Screw – Steel		

Use wood screws to assemble the parts

Enter university/college name

Tool Organizer
Assembly Drawing

Name:

Chapter 28

Figure 28-37 *Assembly drawing and BOM of the Tool Organizer*

Front view

Back view

Enter university/college name

Tool Organizer
Solid Model

Name:

Figure 28-38 Solid model of the Tool Organizer

Problem-Solving Exercise 2   *Mechanical*

Draw the Piece part drawings shown in Figure 28-39. Then assemble them and create the Bill of Materials, as shown in Figure 28-40. The solid model of the assembled Work Bench is shown in Figure 28-41.

Figure 28-39 Piece part drawings of the Work Bench

Chapter 28

PART NO.	QTY	DESCRIPTION
11	3	1/8" DIA SCREWS 4" LONG.
10	50	1/8"DIA SCREWS 2.5" LONG.
9	4	3/8"DIA HEXAGON BOLTS 4.5" LONG
8	2	2"*6"*39-1/4" Pine Board
7	4	2"*6"*31-1/4" Pine Board
6	2	2"*6" *60-1/2" Pine Board
5	5	2"*6" *19" Pine Board
4	1	2"*6" *42" Pine Board
3	2	2"*6" *27-1/2" Pine Board
2	3	2"*6"*6' Pine Board
1	2	2"*6"*6' Pine Board

WORK BENCH

ASSEMBLY DRAWING

Figure 28-40 *Assembly drawing and BOM of the Work Bench*

Figure 28-41 *Solid model of the Work Bench*

WORK BENCH

Solid Model

Chapter 28

Problem-Solving Exercise 3   *Mechanical*

Create the model shown in Figure 28-42. Also, create the section view of the model as shown in Figure 28-43. The dimensions of the model are given in Figure 28-44. **

Figure 28-42 Solid model

Figure 28-43 Section view of the model

Figure 28-44 Dimensions of the model for Problem-Solving Exercise 3

Problem-Solving Exercise 4

Mechanical

Create various components of the Pipe Vice assembly and then assemble them by moving, rotating, and aligning. The Pipe Vice assembly is shown in Figure 28-45. For your reference, the exploded view of the assembly is shown in Figure 28-46. The dimensions of the individual components are shown in Figures 28-47 through 28-52.

Figure 28-45 *The Pipe Vice assembly*

Figure 28-46 *Exploded view of the Pipe Vice assembly*

Figure 28-47 *Dimensions of the Screw, Handle, and the Handle Stop*

Chapter 28

Figure 28-48 *Side view of the Base*

Figure 28-49 *Front view of the Base*

Figure 28-50 *Isometric view of the Base*

Figure 28-51 *Sectioned top view of the Base*

Figure 28-52 *Dimensions of the Movable Jaw*

Answers to Self-Evaluation Test

1. F, **2**. T, **3**. F, **4**. F, **5. SOLVIEW**, **6**. *.mpr*, **7. FLATSHOT**, **8. Distance**, **9. Separate**, **10. Clean**

Chapter 29

Mesh Modeling

Learning Objectives

After completing this chapter, you will be able to:

- *Create mesh models.*
- *Create standard mesh primitives.*
- *Create mesh surfaces.*
- *Smoothen, refine, and crease mesh objects.*
- *Split and extrude mesh faces.*

INTRODUCTION

A mesh model is similar to a solid model without mass and volume properties. The mesh modeling tools are available in the **3D Modeling** workspace. You can create mesh primitives as you create solid primitives and then, create freeform designs from these mesh primitives. The mesh models are easily editable and can be smoothened to give a rounded effect. Meshes can also be refined at required areas for precision. You can perform editing operations such as grip editing, gizmo editing, extrusion on the mesh model, and so on to create freeform designs. Some of the important features of mesh modeling are that you can deform a mesh model using gizmos, add creases to sharpen the mesh model, split mesh faces to segment the meshes as required, and extrude mesh faces. You will learn all these techniques later in this chapter.

CREATING MESH PRIMITIVES*

Mesh primitives are standard shapes that help you create complex models. These standard shapes include box, cone, cylinder, pyramid, sphere, wedge, and torus. You can create mesh primitives as you create solid models. By default, mesh primitives are not smooth. However, you can make these primitives smooth by adding smoothness to them. Note that as you smoothen the mesh, it makes the model complex and the system performance low. Therefore, you must perform the editing operations first and then smoothen the mesh to reduce the complexity of the model. The procedure for creating mesh primitives is given next.

Creating a Mesh Box

Ribbon:	Mesh Modeling > Primitives > Mesh Box	Command:	MESH

 You can create a 3D mesh box similar to a solid or surface box. To do so, enter **MESH** at the Command prompt; the following command sequence will be displayed.

Command: **MESH**
Current smoothness level is set to: 0
Enter an option [Box/Cone/CYlinder/Pyramid/Sphere/Wedge/Torus/SEttings] <Box>: **B** `Enter`
Specify first corner or [Center]: *Specify the first corner.*
Specify other corner or [Cube/Length]: *Specify the diagonally opposite corner.*
Specify height or [2Point] <1>: *Specify the height of the mesh box.*

After you specify the height of the mesh box, it will be created in the drawing area. Figure 29-1 shows the mesh box created. You can create a mesh box in five ways using the above prompt. These ways are listed below:

1. Dynamically.
2. By first specifying the two corners of the rectangular base and then, the height of the box.
3. By first specifying the center and corner of the box and then, the height of the box.
4. By specifying a corner and the length of the edge of cube.
5. By specifying the length, width, and height of the box.

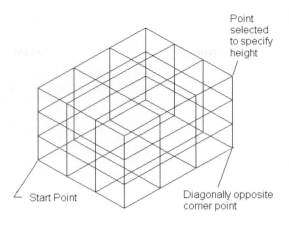

Figure 29-1 *The mesh box created*

You can also create the mesh box using the **Ribbon**. To do so, choose **Mesh Modeling > Primitives > Mesh Box** from the **Ribbon** and follow the command sequence.

The number of faces/divisions in a mesh box depends upon the values specified in the **Mesh Primitive Options** dialog box. This dialog box can be invoked by clicking on the inclined arrow available in the **Primitives** panel of the **Mesh Modeling** tab of the **Ribbon**, see Figure 29-2. Specify the required values in the **Tessellation Divisions** rollout of this dialog box. Next, invoke the **MESH** command to create a new mesh model with the newly specified values. You can also change the appearance of the mesh by changing the level of smoothness of the model. Various areas of the **Mesh Primitive Options** dialog box are discussed next.

The Mesh Area

The **Mesh** area displays a list of mesh primitives and its corresponding parameters. You can select the required primitive from the **Mesh Primitive** list to control its appearance. The **Tessellation Divisions** rollout in this area displays the parameters of the selected primitive. You can edit the corresponding parameters to set the number of divisions on it. For example, if you choose **Cylinder** from the **Mesh Primitive** node in the **Mesh** area, the **Axis**, **Height,** and **Base** parameters will be available in the **Tessellation Divisions** rollout. Now, you can edit the number of faces along the axis, height, and base. Note that to create a mesh primitive with desired settings, first you need to set the values of the parameters and then create the mesh primitive.

The Preview Area

The **Preview** area displays the preview of the selected meshed primitive. The buttons in the **Preview** area are used to pan, zoom, or rotate the mesh primitive. To perform any of these operations, choose the respective button from the **Preview** area and then click in the preview screen. If you want the preview to be automatically updated according to the settings specified in the **Tessellation Divisions** rollout, select the **Auto-update** check box in the **Preview Update** area. Otherwise, choose the **Update** button to update the preview manually.

*Figure 29-2 The **Mesh Primitive Options** dialog box*

The **Preview's smoothness level** drop-down list in the **Preview** area is used to specify the level of smoothness of the mesh primitive. By default, Level 0 is set in the drop-down list, which means, the mesh primitives created will not be smooth by default. Figures 29-3 to 29-5 show a mesh box with various levels of smoothness. You can set the level of smoothness even after the primitive has been created. To do so, select the required primitive; the **Quick Properties** panel will be displayed. Select the level of smoothness from the **Smoothness** drop-down list in the **Quick Properties** panel; the mesh primitive will be smoothened.

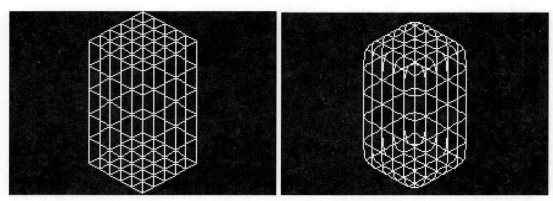

Figure 29-3 *Mesh box with smoothness level=0*   **Figure 29-4** *Mesh box with smoothness level=2*

Like mesh box, the **MESH** command can also be used to create pointed or frustum mesh cones with circular or elliptical base, mesh cylinders with circular or elliptical bases, mesh pyramids, mesh spheres, mesh wedges, and mesh torus. You can set the corresponding parameters of the mesh primitive in the **Mesh Primitive Options** dialog box. The procedure to create primitives is the same as that of solids.

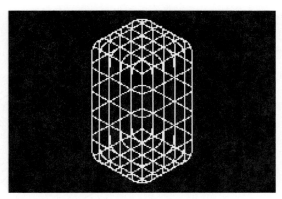

Figure 29-5 Mesh box with smoothness level=4

CREATING SURFACE MESHES

Apart from creating basic mesh primitives, you can create various types of surface meshes. Surface meshes are created by filling the space between objects consisting of straight or curved edges. There are various types of surface meshes such as revolved surface, edge surface, ruled surface, and tabulated surface mesh. These surface meshes as well as the commands used to create them are discussed next.

Creating Revolved Surface Meshes

Ribbon: Mesh Modeling > Primitives > Revolved Surface	Command: REVSURF

 In AutoCAD, you can create a revolved surface by using the **REVSURF** command or by choosing **Mesh Modeling > Primitives > Revolved Surface** from the **Ribbon**. A revolved surface is formed by revolving a path curve about a specified axis of rotation. You can select an open or a closed entity as the path curve. The direction of the revolution depends upon the side from which you select the revolution axis. Since it is not possible to define the clockwise and counterclockwise directions in 3D space, therefore, AutoCAD uses the right-hand thumb rule (refer to Chapter 25 "Getting Started with 3D") to determine the direction of revolution. To create a revolved surface mesh, enter **REVSURF** at the Command prompt; the following command sequence will be displayed:

Command: **REVSURF**
Current wire frame density: SURFTAB1=6 SURFTAB2=6
Select object to revolve: *Select the object to revolve around the axis.*
Select object that defines the axis of revolution: *Select the axis of revolution.*
Specify start angle <0>: Specify second point: *Specify the start angle to start the mesh at an offset from the profile curve.*
Specify included angle (+=ccw, -=cw) <360>: *Specify the included angle that indicates the distance through which the profile curve will be swept.*

Figure 29-6 shows a path curve and a revolution axis and Figure 29-7 shows the resultant revolved surface. You can specify the start angle and the included angle for the revolved surface, see Figures 29-8 and 29-9.

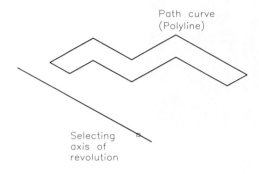

Figure 29-6 Selecting the object to revolve and the axis of revolution

Figure 29-7 The resultant revolved surface

Figure 29-8 Revolved surface with the start angle as 0 and included angle as 270

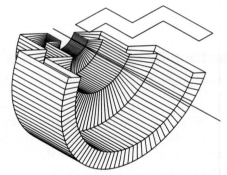

Figure 29-9 Revolved surface with the start angle as 25 and included angle as 180

The smoothness of the revolved surface and the total number of lines that AutoCAD window will draw depend upon the value of the **SURFTAB1** system variable. AutoCAD draws tabulated lines in the direction of revolution. The number of tabulated lines is defined by the value of the **SURFTAB2** system variable (Figure 29-10).

Creating Edge Surface Meshes

Ribbon: Mesh Modeling > Primitives > Edge Surface   **Command:** EDGESURF

The **EDGESURF** command is used to create a 3D polygon mesh from the four adjoining edges that form a topologically closed loop. Alternatively, you can choose **Mesh Modeling > Primitives > Edge Surface** from the **Ribbon** to create a 3D polygon mesh. The entities that you select as edges can be lines, arcs, splines, open 2D polylines, or open 3D polylines. While creating an Edge Surface mesh, you can select the

Figure 29-10 *Different values of the **SURFTAB1** and **SURFTAB2** system variables*

edges of the closed loop in any sequence. The **SURFTAB1** and **SURFTAB2** system variables control the appearance of the mesh and can be used to set the wireframe density. The edge that is selected first will define the M-direction (**SURFTAB1** direction). The other two edges that start from the endpoints of the first edge will define the N-direction (**SURFTAB2** direction), see Figures 29-11 and 29-12.

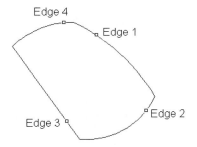

Figure 29-11 *Edges to be selected*

Figure 29-12 *An Edge Surface mesh created*

Creating Ruled Surface Meshes

Ribbon: Mesh Modeling > Primitives > Ruled Surface   **Command:** RULESURF

You can create a ruled surface between two defining entities by using the **RULESURF** command or by choosing **Mesh Modeling > Primitives > Ruled Surface** from the **Ribbon**. The defining entities can be lines, circles, ellipses, arcs, polylines, curves, or splines. You can create a ruled surface mesh between two closed entities, two open entities, a point and a closed entity, and a point and an open entity. However, you cannot create a ruled surface between a closed and an open entity. Also, two points do not define a surface and therefore, a ruled surface cannot be created between them.

If the two defining entities are closed, the selection point does not make any difference in the resultant surface. In case of circles, the start point of the surface is determined by the combination of direction of the *X* axis and the current value of the **SNAPANG** system variable. In case of closed polylines, the ruled surface starts from the last vertex of the polyline and proceeds backward along the polyline segment.

If the two defining entities are open, the selection points do make a difference in the resultant surface. If you select open entities at the same end, it creates a straight polygon mesh. If you select open entities at opposite ends, it creates a self-intersecting polygon mesh, refer to Figures 29-13 and 29-14.

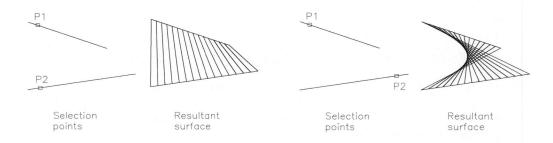

Figure 29-13 *Selecting objects at same ends*   *Figure 29-14* *Selecting objects at opposite ends*

The smoothness of the ruled surface can be increased using the **SURFTAB1** system variable, see Figures 29-15 and 29-16. The default value of this system variable is 6, but it can accept any integer between 2 and 32766 as its value. Note that this value must be increased before creating the surface.

Figure 29-15 *The ruled surface created when the value of the SURFTAB1 system variable is 6*

Figure 29-16 *The ruled surface created when the value of the SURFTAB1 system variable is 50*

Creating Tabulated Surface Meshes

Ribbon: Mesh Modeling > Primitives > Tabulated Surface	**Command:** TABSURF

 You can create a tabulated surface mesh by using the **TABSURF** command or by choosing **Mesh Modeling > Primitives > Tabulated Surface** from the **Ribbon**. The tabulated surface mesh is formed when a profile is swept along the direction defined by the direction vector. The profile curves can be lines, circles, arcs, splines, 2D polylines, 3D polylines, or ellipses. The length of the tabulated surface will be equal to the length of the direction vector, which inturn will be determined by the point selected for the direction vector, see Figures 29-17 and 29-18.

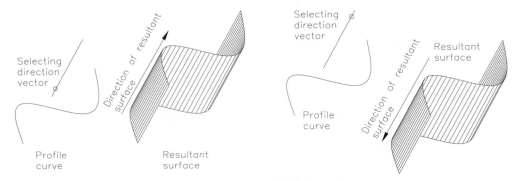

Figure 29-17 *Direction of the tabulated surface*   **Figure 29-18** *Direction of the tabulated surface*

If there are more than one direction vectors, all intermediate points will be ignored and an imaginary vector will be drawn between the first and last points on the polyline. The imaginary vector will be considered as the direction vector and the tabulated surface will be created along it, as shown in Figure 29-19.

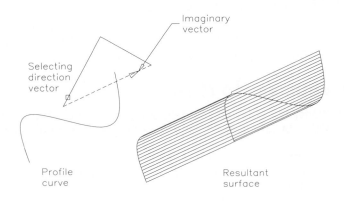

Figure 29-19 *Tabulated surface created along the imaginary vector*

Tip
You cannot select a curved entity as the direction vector for defining the direction of the tabulated surface.

You can increase the smoothness of the tabulated surface by increasing the value of the **SURFTAB1** *system variable.*

Note
*Other than surface meshes, the **REVSURF**, **EDGESURF**, **RULESURF**, and **TABSURF** commands can be also used to create surface models.*

Exercise 1   *Architectural*

In this exercise, you will create the table shown in Figure 29-20 by using various types of surfaces. Assume the dimensions for the table.

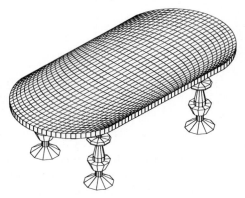

Figure 29-20 Model for Exercise 1

MODIFYING MESH OBJECTS*

Mesh models can be modified to create freeform designs. This can be done by changing their smoothing levels, refining specific areas, adding creases to sharpen the edges, and so on. Various types of modifications that can be applied on meshes are discussed next.

Adding Smoothness to Meshes

| **Ribbon:** Mesh Modeling > Mesh > Smooth Object | **Command:** MESHSMOOTH |
| **Menu Bar:** Modify > Mesh Editing > Smooth More/Less | **Toolbar:** Smooth Mesh |

Adding smoothness to a mesh results in roundness of the sharp edges of the meshed model. To change the level of smoothness of the mesh, select it; the **Quick Properties** panel will be displayed. In this panel, select the required level of smoothness from the **Smoothness** drop-down list; the selected smoothness level will be applied to the mesh. Note that **Level 0** indicates minimum smoothness, whereas **Level 4** indicates maximum smoothness. You can switch between different levels of smoothness with the help of the **Quick Properties** panel. Remember that the higher level of smoothness makes the mesh complex and thus, decreases the system performance. Figure 29-21 shows a mesh with different levels of smoothness applied to it. It is recommended to first perform the required editing of the mesh and then apply smoothness to it.

Note
*You can also be used to set the level of smoothness of the mesh from the **Properties** palette.*

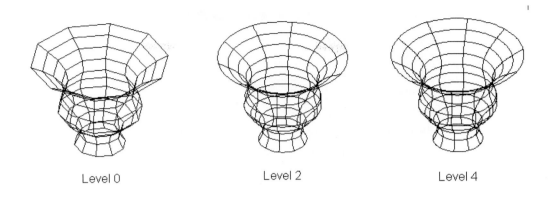

Level 0   Level 2   Level 4

Figure 29-21 Mesh with different levels of smoothness

If you need to apply smoothness to objects other than the mesh objects, first convert the objects into mesh models. The objects that can be converted into mesh models include 3D solids, 3D surfaces, 3D faces, polygon meshes, regions, and closed polylines. To convert an object into a mesh object, choose **Mesh Modeling > Mesh > Smooth Object** from the **Ribbon**; you will be prompted to select the object to convert. Select an object and press ENTER; the 3D object will be converted into a mesh object. Now, you can perform the editing operations such as smoothening, refining, splitting, and so on on it. Note that the smoothening operation works best for primitive solids. But, if you select objects other than the primitive solids, a message box will be displayed, as shown in Figure 29-22. Choose **Create mesh** from the message box; the selected objects will be converted into mesh models with the default settings in the **Mesh Tessellation Options** dialog box.

*Figure 29-22 The **Smooth Mesh - Non-primitive Objects Selected** message box*

However, you can modify these settings as per your requirement by invoking the **Mesh Tessellation Options** dialog box. To invoke this dialog box, click on the arrow in the **Mesh** panel of the **Mesh Modeling** tab of the **Ribbon**. Alternatively, enter **MESHOPTIONS** at the Command prompt; the **Mesh Tessellation Options** dialog box will be displayed, as shown in Figure 29-23.

Chapter 29

*Figure 29-23 The **Mesh Tessellation Options** dialog box*

The options in this dialog box are discussed next.

Select objects to tessellate

This button enables you to select objects to tessellate while the **MESHOPTIONS** command is active. On choosing this button, the **Mesh Tessellation Options** dialog box will disappear allowing you to select objects. Select the required objects from the drawing area and press ENTER; the dialog box will reappear and the number of objects selected will be displayed below the **Select objects to tessellate** button. Now, you can specify the required parameters in the **Mesh Tessellation Options** dialog box.

Mesh Type and Tolerance Area

The options in this area are discussed next.

Mesh type

This drop-down list is used to specify the mesh type that you want to create. This drop-down list has three options: **Smooth Mesh Optimized**, **Mostly Quads**, and **Triangle**. By default, the **Smooth Mesh Optimized** option is selected and therefore, the shape of the mesh face will be the optimized mesh shape that can adapt to the shape of the mesh object. If you select the **Mostly Quads** option, the shape of the mesh face will be quadrilateral to its maximum possible, and if you select the **Triangle** option, the shape of the mesh face will be triangular to its maximum possible.

Mesh distance from original faces

This edit box is not enabled if the **Smooth Mesh Optimized** option is selected in the **Mesh Type** drop-down list. This edit box helps you specify how closely the converted mesh object will adhere to the original shape of solid or surface. Smaller the values specified in this edit box, lesser the deviation from the original faces, and therefore, greater the number of faces in the mesh created. This means small value results in more accurate mesh objects but with slower performance of system.

Maximum angle between new faces

The **Maximum angle between new faces** edit box is not enabled if the **Smooth Mesh Optimized** option is selected in the **Mesh Type** drop-down list. This edit box allows you to specify the maximum angle between the surface normal and continuous mesh faces. Larger the value specified in this edit box, more dense will be the mesh at higher curvatures and less dense at flat areas. This setting is used to refine the appearance of the curved features like holes or fillets.

Maximum aspect ratio for new faces

This edit box is not enabled in case the **Smooth Mesh Optimized** option is selected. This edit box allows you to set the maximum aspect ratio, which is, height/width ratio of new faces. This option is mostly used to avoid long and thin cylindrical faces. Note that smaller the value specified in this edit box, better will be the shape of the mesh.

Maximum edge length for new faces

This edit box allows you to set the maximum length of the edges of the mesh objects that are created by converting solids and surfaces. Smaller the value specified in this edit box, better will be the shape/appearance of the mesh.

Meshing Primitive Solids

The options in this area are discussed next.

Use optimized representation for 3D primitive solids

Select this check box to apply the mesh settings specified in the **Mesh Primitive Options** dialog box and clear this check box to apply the settings specified in the **Mesh Tessellation Options** dialog box.

Mesh Primitives

The **Mesh Primitives** button will be enabled only when the **Use optimized representation for 3D primitive solids** check box is selected. On choosing the **Mesh Primitives** button, the **Mesh Primitive Options** dialog box will be displayed, where you can specify the required settings for the mesh object.

Smooth Mesh After Tessellation

This area will be active only when the **Smooth Mesh Optimized** option is selected in the **Mesh type** drop-down list. This is because the mesh objects converted on selecting any of the other two options from the drop-down list do not have any smoothness.

Chapter 29

Apply smoothness after tessellation
Select this check box to apply smoothness to newly created mesh objects.

Smoothness level
This spinner will be available only when the **Apply smoothness after tessellation** check box is selected and you can set the level of smoothness of the mesh object in it.

Preview
This button is available when the objects required to be converted into meshes are selected. You can choose this button to preview the effect of the current settings in the drawing area. To go back to the dialog box, press ESC, or press ENTER to accept the mesh tessellation options.

Note

*To increase the smoothness of mesh by one level, choose the **Smooth More** button from the **Mesh** panel; the mesh will become smoother (rounded). Similarly, if you choose the **Smooth Less** button, the smoothness of the mesh will be decreased by one level.*

Refining the Meshes

Ribbon: Mesh Modeling > Mesh > Refine Mesh	**Command:** MESHREFINE
Menu Bar: Modify > Mesh Editing > Refine Mesh	**Toolbar:** Smooth Mesh

Refining the mesh means increasing the number of divisions in the mesh of an object, resulting in a dense mesh. You can refine the complete mesh object or a particular face of an object. Refining is mostly used for finer and detailed models, in which modification of small or refined divisions of the mesh is required. But remember that, denser the mesh, slower will be performance of the system. To refine the mesh, choose **Mesh Modeling > Mesh > Refine Mesh** from the **Ribbon**; you will be prompted to select the mesh object or faces of the mesh object to refine. Select the entity and press ENTER; the selected entity will get refined. Note that the **Refine Mesh** option works only when the smoothness option of the selected mesh object is set to **Level 1** or higher. Also note that, as you refine the object, its level of smoothness becomes zero. To refine it again, you need to set the smoothness level to **1** or higher, and then follow the same procedure of refining. Figure 29-24 shows the mesh object before and after refinement.

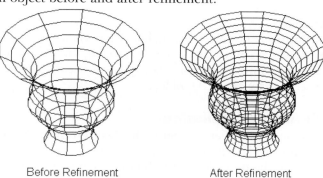

Before Refinement   After Refinement

Figure 29-24 Object before and after refinement

To refine the faces of the selected mesh object, first you need to set the selection criteria to faces. To do so, right-click anywhere on the screen, and then choose **Subobject Selection Filter > Face** from the menu displayed. Alternatively, choose **Mesh Modeling > Subobject > No Filter > Face** from the **Ribbon**. After the selection criteria is set, invoke the **Refine Mesh** tool. Next, you need to select a face of the mesh object. To do so, press CTRL and click on the required mesh face, and then press ENTER. The specified face will be subdivided into four faces, as shown in Figure 29-25.

Face selected to refine   Face after refinement

Note that refining an object resets the smoothness level to 0, whereas refining face/s does not affect the smoothness level.

Figure 29-25 *Face before and after refinement*

Adding Crease to Meshes

Ribbon: Mesh Modeling > Mesh > Add Crease	**Command:** MESHCREASE
Menu Bar: Modify > Mesh Editing > Crease	**Toolbar:** Smooth Mesh

You can also modify a mesh object by adding creases to it. Creases are added to an object in order to sharpen the mesh sub-objects, thereby, deforming the mesh object. You can sharpen the faces, edges, or vertices of a mesh object. To add crease to the selected mesh object, choose **Mesh Modeling > Mesh > Add Crease** from the **Ribbon**; you will be prompted to select the required mesh sub-objects. Select the mesh sub-objects and press ENTER; the **Specify crease value [Always] <Always>:** prompt will be displayed. By default, the crease is set to **Always**, which means the crease will retain its sharpness even if you smoothen or refine the mesh object. You can also specify a crease value at which the crease will start losing its sharpness. The crease value set to **-1** is the same as the crease value set to **Always**. If you specify **0** as the crease value, the existing crease will be removed. On sharpening a sub-object, the edges adjoining the selected sub-object are updated to the modified model. Figures 29-26 through 29-31 show the effect of applying crease on various sub-objects.

Note
*To make specific selection of entities, right-click in the drawing area and choose the desired sub-option from the **Subobject Selection Filter** option displayed in the shortcut menu.*

The crease will not be visible when the level of smoothness is 0. So, increase the smoothness level to make the crease visible.

You can remove the crease added to an object. To do so, choose the **Remove Crease** button from the **Ribbon**. Next, select the crease that you want to remove and press ENTER; the selected crease will be removed from the object and the corresponding faces will be updated. Alternatively, to remove the crease from the object, enter **MESHUNCREASE** at the Command prompt.

Chapter 29

Figure 29-26 *Faces selected to be creased*

Figure 29-27 *Faces after adding crease*

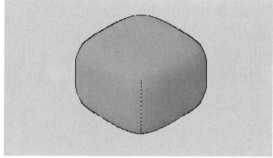

Figure 29-28 *Edge selected to be creased*

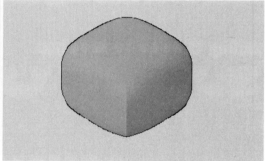

Figure 29-29 *Edge after adding crease*

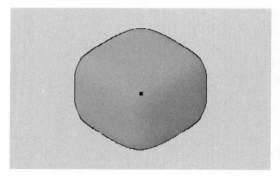

Figure 29-30 *Vertex selected to be creased*

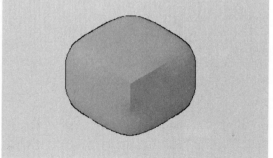

Figure 29-31 *Vertex after adding crease*

EDITING MESH FACES*

You can edit the faces of a mesh object either by splitting or extruding its faces. In this process, the extruded part and the splitted part merge with the mesh object and act as a single entity. The splitting and extruding operations are useful for detailed and advanced modeling. To edit the faces of a mesh object, you can use the editing tools in the **Mesh Edit** panel of the **Mesh Modeling** tab of the **Ribbon**. These tools are discussed next.

Splitting the Mesh Faces

Ribbon: Mesh Modeling > Mesh Edit > Split Mesh Face   **Command:** MESHSPLIT
Menu Bar: Modify > Mesh Editing > Split Mesh

Split
Mesh Face
In the process of Refining, the entire mesh model is split into various divisions, whereas in Splitting, a particular location of split is defined. As a result, splitting avoids larger deformations for smaller modifications.

You can edit a mesh object by splitting it using the **Split Mesh Face** tool. This tool is used to divide the selected mesh face into two parts. To split the mesh, choose **Mesh Modeling > Mesh Edit > Split Mesh Face** from the **Ribbon**. Alternatively, enter **MESHSPLIT** at the Command prompt; you will be prompted to select the face of mesh to split. Select the face to split; a knife symbol will be displayed with the cursor and you will be prompted to specify the cut in the mesh. Now, specify the start and end points of the split line; the mesh will split along the line joining the specified points, and the adjoining faces will separate accordingly. Figure 29-32 shows the points to be selected for splitting the face and Figure 29-33 shows the resulting mesh model after splitting.

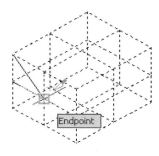

Figure 29-32 *Points selected for splitting the mesh*

Tip
In case of detailed modeling, it is advisable to first split the mesh faces and then refine the mesh model.

Extruding the Mesh Faces

Ribbon: Mesh Modeling > Mesh Edit > Extrude Face   **Command:** EXTRUDE
Menu Bar: Draw > Modeling > Extrude   **Toolbar:** Modeling

Extrude
Face
Extruding the mesh models is similar to extruding the 3D models. The only difference between the two processes is that in 3D modeling, the extruded part is created as a separate entity, whereas in mesh modeling, the extruded part merges with the mesh object and acts as a single entity. To extrude the mesh faces, choose **Mesh Modeling > Mesh Edit > Extrude Face** from the **Ribbon**. Alternatively, enter **EXTRUDE** at the Command prompt; the command sequence that will be displayed is given next.

Command: **EXTRUDE**
Current wire frame density: ISOLINES=8
Select objects to extrude: *Select one or more faces to extrude*
Specify height of extrusion or [Direction/Path/Taper angle] <100>: *Specify the height of extrusion.*

You can extrude the mesh faces dynamically, by specifying height, by specifying the direction of extrusion, by specifying the path of extrusion, or by specifying an angle of taper for extrusion.

Chapter 29

All these options work in the same way as in the 3D modeling. Figure 29-34 shows the mesh box with extruded split faces. The splitting and extruding operations on mesh models result in new designs.

Note
To select the faces that you want to extrude, click on the required faces of the mesh object keeping the CTRL key pressed. Continue till the selection is completed. To remove a face from the selection, press SHIFT + CTRL and click on the face to be deselected.

Figure 29-33 *Mesh model after splitting*   ***Figure 29-34*** *Mesh model with extruded split faces*

The following example illustrates the use of the Mesh Edit tools:

Example 1   *General*

In this example, you will apply various mesh edit tools on a mesh box with dimensions = 100 X 100 X 20.

Creating a Mesh Box
To create a mesh box, first you need to specify the parameters of the tessellation divisions on the mesh box.

1.   Start a new file with the *acad3D.dwt* template in the **3D Modeling** workspace.

2.   Invoke the **Mesh Primitive Options** dialog box by clicking on the inclined arrow displayed at the lower-right corner of the **Primitives** panel in the **Mesh Modeling** tab of the **Ribbon**.

3.   Select the **Box** option from the **Mesh Primitives** list box and set the corresponding tessellation divisions to **4**. Choose the **OK** button from the **Mesh Primitive Options** dialog box to accept the settings and close the dialog box.

4.   Change the visual style to **3D Wireframe**.

5. Choose **Mesh Modeling > Primitives > Mesh Box** from the **Ribbon**; you are prompted to specify the start point.

6. Specify the start point of the mesh box by clicking at the desired location in the drawing area; the **Specify other corner or [Cube/Length]** prompt is displayed.

7. Enter **L** at the above prompt and specify the following dimensions to create a mesh box, as shown in Figure 29-35.

    Length: **100**, Width: **100**, and Height: **20**

Figure 29-35 The mesh box created

Splitting the Selected Faces

Now, you need to split the faces of the mesh box.

1. Choose **Mesh Modeling > Mesh Edit > Split Mesh Face** from the **Ribbon**; you are prompted to select the required face to split.

2. Select the inner face on top of the box; you are prompted to specify the starting point of the split line.

3. Specify the start and end points of the cutting line, as shown in Figure 29-36; the selected face gets split into two faces diagonally.

4. Similarly, split the other faces, as shown in Figure 29-37.

Figure 29-36 Face selected to split

Figure 29-37 The model after splitting

Extruding the Selected Faces

Now, you need to extrude the specified split faces.

1. Press CTRL and select the faces, as shown in Figure 29-38.

2. Choose **Mesh Modeling > Mesh Edit > Extrude Face** from the **Ribbon**; you are prompted to specify the height of extrusion.

3. Enter **60** in the Dynamic Input box as the value of extrusion; the selected faces are extruded.

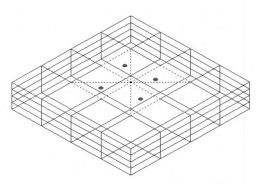

Figure 29-38 *Faces selected to be extruded*

The final extruded model is shown in Figure 29-39.

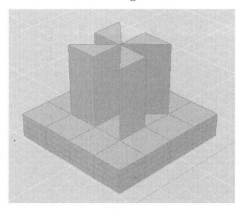

Figure 29-39 *The final model*

CONVERTING MESH OBJECTS*

You have already learned that 3D solids and surfaces can be converted into mesh objects using the **MESHSMOOTH** command so that they can be smoothed, refined, and creased. Now, you will learn how to convert mesh objects into 3D solids or surfaces so that you can perform the solid modeling techniques such as union, subtraction, intersection, on them. Also, you can specify the settings such that the mesh objects converted into 3D solids or surfaces are smooth or faceted, or have merged faces.

Converting Mesh Objects into Solids

Ribbon: Mesh Modeling > Convert Mesh > Convert to Solid   **Command:** CONVTOSOLID
Menu Bar: Modify > 3D Operations > Convert to Solid

To convert a mesh object into solid, choose **Mesh Modeling > Convert Mesh > Convert to Solid** from the **Ribbon**. Alternatively, enter **CONVTOSOLID** at the Command prompt and then press ENTER; you will be prompted to select a mesh object. Select the required mesh object and press ENTER; the selected mesh object will be converted into a solid with the default settings. To convert a mesh object into solid with the user-defined settings, you need to specify the new settings before the conversion. You can also specify whether the converted solid will be smooth or faceted, or have merged faces.

Note
Other than the meshes, the closed polygons and circles with thickness, as well as 3D surfaces can also be converted into solids. However, the objects like exploded solids, separate and continuous edges forming a surface, separate and continuous surfaces forming volumes, meshes with gaps between faces, and meshes with intersecting boundaries cannot be converted into solids.

To specify pre-defined settings for the solids that will be created from meshes, choose **Mesh Modeling > Convert Mesh > Smooth, optimized** from the **Ribbon**; a flyout with pre-defined settings will be displayed. These pre-defined settings are discussed next.

Smooth, optimized
On choosing this option, all tessellations of the mesh in the same plane will be merged into a single face in the solid. Also, the edges of the solid model will be rounded. The Figure 29-40 shows a mesh box and Figure 29-41 shows the solid created from the mesh box on choosing the **Smooth, optimized** option.

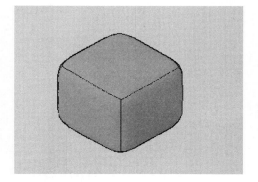

Figure 29-40 A mesh box

Figure 29-41 The solid created on choosing the *Smooth, optimized* option

Smooth, not optimized
On choosing this option, the edges of the mesh will be rounded when it is converted to solid, but the resulting faces will not be merged, and will remain same as in the mesh object, as shown in Figure 29-42.

Chapter 29

Faceted, optimized

On choosing this option, all tessellations of the mesh in the same plane will be merged into a single face in the solid. Also, the edges in the resulting solid will not be rounded, but creased, as shown in Figure 29-43.

*Figure 29-42 The solid created on choosing the **Smooth, not optimized** option*

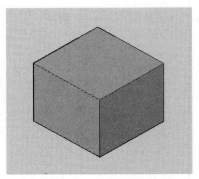

*Figure 29-43 The solid created on choosing the **Faceted, optimized** option*

Faceted, not optimized

On choosing this option, the faces of the converted solid will not be merged but the edges will be rounded, as shown in Figure 29-44.

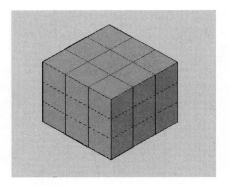

*Figure 29-44 The solid created on choosing the **Faceted, not optimized** option*

Converting Mesh Objects into Surfaces

Ribbon: Mesh Modeling > Convert Mesh > Convert to Surface **Command:** CONVTOSOLID
Menu Bar: Modify > 3D Operations > Convert to Surface

 Like converting a mesh object into solid, you can also convert a mesh object into surface. To do so, choose **Mesh Modeling > Convert Mesh > Convert to Surface** from the **Ribbon**; you will be prompted to select a mesh object. Select the required mesh object; the mesh object will be converted into a surface with default settings. You can change the default settings of the surface created in the same way as you did with solids. Figure 29-45 shows a mesh object and Figures 29-46 through 29-49 show the surfaces created on applying different settings on this object.

Figure 29-45 *A mesh object*

Figure 29-46 *The surface created on choosing the **Smooth, optimized** option*

Figure 29-47 *The surface created on choosing the **Smooth, not optimized** option*

Figure 29-48 *The surface created on choosing the **Faceted, optimized** option*

Figure 29-49 *The surface created on choosing the **Faceted, not optimized** option*

Note
Other than meshes, regions, lines and arcs with some thickness, planar 3D faces, and 2D solids can also be converted into surfaces. Refer to Chapter 25.

Tip
*You can create section views of the meshed models as you did for solid models in Chapter 26. The tools for creating section views are available in the **Section** panel of the **Home** tab or the **Mesh Modeling** tab of the **Ribbon**.*

WORKING WITH GIZMOS

Gizmo is a tool that helps you move, rotate, or scale 3D objects or their sub-objects along a specified axis or plane. When you select objects or sub-objects in any 3D visual style, a gizmo is displayed. The gizmo are of three types, move, rotate, and scale. To set the gizmo that you want to be displayed by default, choose the required gizmo from the **Subobject** panel in the **Mesh Modeling** tab of the **Ribbon**. Also, if you do not want any gizmo to be displayed, specify it in the **Subobject** panel. Various types of gizmos are discussed next.

Move Gizmo

When **Move Gizmo** is chosen from the **Subobject** panel in the **Mesh Modeling** tab of the **Ribbon**, the move gizmo is displayed on the first selected object or sub-object, see Figure 29-50. The move gizmo tool has three axis handles, three planes, and a base point. Using this gizmo, you can move the selected faces or edges of objects along any axis or plane, and thus, create freeform designs as per your requirement.

To move the selected object along any particular axis, click on the axis handle of the gizmo and drag along the axis; the selected object will move accordingly. To move the selected object anywhere along a plane, click on the rectangle formed in between the axes handles; the object will move along the specified plane. To relocate the base point of the gizmo, right-click on the gizmo, and then choose **Relocate Gizmo** from the shortcut menu displayed. You can also switch between various gizmos using this shortcut menu.

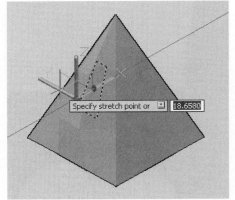

Figure 29-50 The move gizmo on the mesh object

Rotate Gizmo

The rotate gizmo is displayed on the mesh object when **Rotate Gizmo** is chosen from the **Subobject** panel in the **Mesh Modeling** tab of the **Ribbon**. The rotate gizmo is displayed as a sphere with three axes and a base point on it. Figure 29-51 shows the scale gizmo on the mesh model. This tool is used to rotate the selected objects about a particular axis. By default, the gizmo is displayed at the center of the selected object. As you move the cursor over the rotate gizmo, the nearest axis of the gizmo will be highlighted. Click when the required axis gets highlighted and specify the rotation angle. You can relocate the gizmo as well as switch between various gizmos using the shortcut menu that is displayed on right-clicking on the gizmo.

Scale Gizmo

The scale gizmo is displayed when **Scale Gizmo** is chosen from the **Subobject** panel in the **Mesh Modeling** tab of the **Ribbon**. This tool is used to scale the selected objects uniformly along an axis or a plane. As you move the cursor toward the gizmo, the axis of scale will be displayed. Click and drag the axis to scale the object in that direction. To scale the object uniformly, click on the triangle that is formed between the axes. Figure 29-52 shows the scale gizmo on the mesh model.

Therefore, in this section you learned that all the three gizmos can be used to create freeform designs from primitive meshes. The next example illustrates the use of these gizmos.

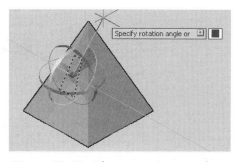

Figure 29-51 *The rotate gizmo on the mesh object*

Figure 29-52 *The scale gizmo on the mesh object*

Note
*You can choose the **No Gizmo** option from the **Subobject** panel in the **Mesh Modeling** tab of the **Ribbon**, if you do not want any gizmo to be displayed on selecting the objects in the 3D visual style.*

Example 2   *General*

In this example, first you will create an elliptical mesh cylinder with the following dimensions: major axis = 55, minor axis = 15, and height= 100. Then, you will modify it using various gizmos.

Creating an Elliptical Cylinder

First, you need to start a new drawing file, specify the settings for the number of faces in the cylinder and then create the cylinder.

1. Start a new file with the *acad3D.dwt* template in the **3D Modeling** workspace.

2. Choose the inclined down-arrow displayed at the lower right corner of the **Primitives** panel of the **Mesh Modeling** tab in the **Ribbon**; the **Mesh Primitive Options** dialog box is displayed.

3. In this dialog box, select **Cylinder** from the **Mesh Primitives** area and set the following data in the **Tessellation Divisions** rollout.

    Axis: **10**
    Height: **5**
    Base: **3**

4. Choose **OK** to accept the settings specified in the rollout and exit the dialog box.

5. Switch on the **Ortho Mode** and invoke the **MESH** command to create the mesh cylinder; the command sequence that will follow is given next.

    Command: **MESH**
    Current smoothness level is set to: 0
    Enter an option [Box/Cone/CYlinder/Pyramid/Sphere/Wedge/Torus/SEttings] <Box>: *Enter **CY** to create the cylinder.*
    Specify center point of base or [3P/2P/Ttr/Elliptical]: *Enter **E** to create elliptical base.*
    Specify endpoint of first axis or [Center]: *Enter **C** to specify the center of base.*
    Specify center point: *Specify the center point for the base of the cylinder.*
    Specify distance to first axis: *Drag the cursor along X axis and enter **55** in the Dynamic Input box to specify a point on the first axis.*
    Specify endpoint of second axis: *Drag the cursor along Y axis and enter **15** in the Dynamic Input box to specify a point on the second axis.*
    Specify height or [2Point/Axis endpoint]: *Enter **100** to specify the height of the cylinder.*

6. Change the view to **SW Isometric** and the visual style to **3D Wireframe**. Figure 29-53 shows the meshed model.

Moving the Edges
To move the edges, you first need to filter the selection to Edges and then move them.

1. To filter the selection, choose **Mesh Modeling > Subobject > No Filter > Edge** from the **Ribbon**.

2. Change the visual style to **Realistic**.

3. Press CTRL and select the required edge, as shown in Figure 29-54; the move gizmo is displayed by default.

Note
*If the move gizmo is not displayed by default, choose the **Move Gizmo** option in the **Subobject** panel of the **Ribbon**.*

4. Click on the X handle of the move gizmo such that the base of the cylinder is elongated; the preview of the moved edge is displayed. Next, click on the screen at the required distance (here, it is **20**). Similarly, move the edge on the other side to the same distance.

Figure 29-53 The mesh model created   *Figure 29-54* Edge selected to be moved

5. Now, you need to smoothen the mesh model. To do so, select it; the **Quick Properties** panel is displayed (if it does not appear, choose the **Quick Properties** button from the Status Bar). In this panel, select **Level 1** from the **Smoothness** drop-down list.

Rotating the Edges

Now, you need to rotate the edges. Make sure the **Edge** option is chosen from the filter flyout.

1. To invoke the rotate gizmo, choose **Mesh Modeling > Subobject > Move Gizmo > Rotate Gizmo** from the **Ribbon**.

2. Select the required edges from the mesh model, as shown in Figure 29-55; the rotate gizmo is displayed on the first selected edge.

3. Click on the red handle of the rotate gizmo such that the adjoining faces are rotated inward; the preview of rotation is displayed. Click when the desired rotation is achieved (here, it is **60** degrees).

Similarly, select the other edges below the selected edges in the same face and rotate the face. The final model, after rotating the edges, is shown in Figure 29-56.

Moving the Inner Faces

Now, you first need to change the selection filter from edges to faces and then, move faces.

1. To select faces, choose **Face** from the filter flyout in the **Ribbon**.

2. Choose **Move Gizmo** from the **Subobjects** panel of the **Ribbon**.

3. Select the faces in the inner circular region at the top of the mesh model; the move gizmo is displayed on the selected faces.

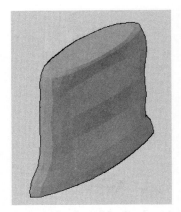

Figure 29-55 *Edges selected to be rotated*   **Figure 29-56** *Final model after rotating the edges*

4. Click on the Z handle of the gizmo and drag it; the selected faces elongate. Click when the desired height is displayed.

Scaling the Entire Model
Now, you can elongate the model using the scale gizmo.

1. Choose **Scale Gizmo** from the gizmo flyout.

2. Select the model; the scale gizmo is displayed, as shown in Figure 29-57.

3. Click on the Z handle; the model gets elongated in the Z-direction. Click in the drawing area at the required distance. The final model is created, as shown in Figure 29-58.

Figure 29-57 *The scale gizmo displayed*   **Figure 29-58** *The final model*

Self-Evaluation Test

Answer the following questions and then compare them to those given at the end of this chapter:

1. A mesh model is similar to a solid model without mass and volume properties. (T/F)

2. You can specify values for the number of faces in the primitive mesh in the _____ dialog box.

3. The _____ command is used to create mesh primitives.

4. The _____ command is used to create a mesh by adjoining four continuous edges or curves.

5. Which of the following objects can be converted into mesh objects?

    (a) 3D objects   (b) Closed Polygons
    (c) Regions   (d) All the above

6. Which of the following options is used to convert 3D objects into meshed objects?

    (a) **Refine Mesh**   (b) **Smooth More**
    (c) **Smooth Object**   (d) None of these

7. You can invoke the _____ tool to sharpen faces, edges, or vertices of mesh objects.

8. Which of the following options is used to specify the default settings for the solid created from the meshed object?

    (a) **Smooth, optimized**   (b) **Smooth, not optimized**
    (c) **Faceted, optimized**   (d) All the above

9. The mesh editing tools are available in the _____ panel of the **Mesh Modeling** tab of the **Ribbon**.

10. To retain the sharpness of the crease even when you smoothen or refine the mesh, you need to set the crease to _____.

Review Questions

Answer the following questions:

1. The _____ and _____ commands are used to set the wireframe density of primitives.

2. In smoothening, the complexity of a model is decreased. (T/F)

3. You can set the level of smoothness of mesh objects using the **Properties** palette. (T/F)

4. To create mesh cones with frustum, you need to choose the _____ option from the **MESH CONE** command.

5. The defining curves to create ruled surfaces can be lines, arcs, polylines, curves, and splines. (T/F)

6. The **Mesh Tessellation Options** dialog box will be displayed on entering _____ at the Command prompt.

7. The **Refine Mesh** tool works only when the smoothening of the mesh object is set to **Level 1** or higher. (T/F)

8. Refining an object resets the smoothening level to 0, whereas refining a face does not reset the smoothing level. (T/F)

9. To remove creases from the model, you need to enter _____ at the Command prompt.

10. Which of the following can be used to relocate a gizmo?

    (a) Mouse   (b) Shortcut menu
    (c) **Ribbon**   (d) All the above

Exercises

Exercise 2   *General*

Create the 3D model of a globe shown in Figure 29-59 arbitrarily, using the mesh modeling tools.

Figure 29-59 *Model of a globe*

Exercise 3   *General*

Create the drawing shown in Figure 29-60. First, create the base unit (transition) and then the dish and cone.

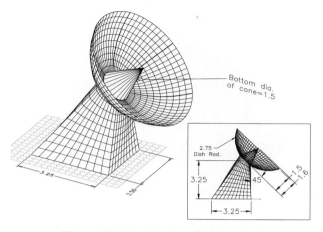

Figure 29-60 *Drawing for Exercise 3*

Exercise 4   *General*

Create the drawing shown in Figure 29-61. First, create one of the segments of the base and then use the **ARRAY** command to complete it.

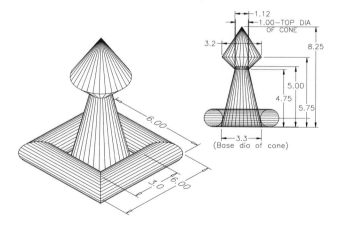

Figure 29-61 *Drawing for Exercise 4*

Chapter 29

Exercise 5   *General*

In this exercise, you will create the model of a web camera, as shown in Figure 29-62. The reference dimensions for the model are given in Figure 29-63. Assume the missing dimensions.**

Figure 29-62 *Model of a web camera*

Figure 29-63 *Dimensions for the model*

Answers to Self-Evaluation Test
1. T, 2. **Mesh Primitive Options**, 3. **MESH**, 4. **EDGESURF**, 5. d, 6. c, 7. **Add Crease**, 8. d, 9. **Mesh Edit**, 10. **Always**

Chapter 30

Rendering and Animating Designs

Learning Objectives

After completing this chapter, you will be able to:
* *Understand rendering and the need to render objects.*
* *Attach materials to models and modify materials.*
* *Define and modify new materials.*
* *Insert and modify light sources.*
* *Select rendering types and render objects.*
* *Define and modify rendering settings.*
* *Replay and print renderings.*
* *Define and apply background and fog.*
* *Configure, load, and unload AutoCAD Render.*
* *Create animations from the **Ribbon**.*
* *Create animations by defining the motion path.*

UNDERSTANDING THE CONCEPT OF RENDERING

A rendered image makes it easier to visualize the shape and size of a 3D object as compared to a wireframe image or a shaded image. A rendered object also makes it easier to express your design ideas to other people. For example, to make a presentation of your project or design, you do not need to build a prototype. You can use a rendered image to explain your design more clearly because you have a complete control over the shape, size, color, and surface material and lighting of the rendered image. Additionally, any required changes can be incorporated into the object, and it can be rendered to check or demonstrate the effect of these changes. Thus, rendering is a very effective tool for communicating ideas and demonstrating the shape of an object.

Generally, the process of giving a realistic effect to the models (rendering) can be divided into the following four steps:

1.   Selecting the material to be assigned to the design.

2.   Attaching material to the design.

3.   Applying additional effects to the model by assigning lights at different locations.

4.   Using the **RENDER** command to render the objects.

SELECTING AND ATTACHING MATERIALS

AutoCAD provides you with a number of materials that can be attached to the design. Once you have selected the material, you can assign it directly to the object. You can also assign the material to individual faces, blocks, and layers. The commands that are used to select and attach materials are discussed next.

Selecting a Material for a Design

Tool Palettes: Materials or Materials Library   **Command**: MATLIB

To get a realistic rendered image of an object, it is necessary to assign materials to it. Therefore, after creating the design, you need to select the material to be assigned to the objects in it. By default, a global material is assigned to all objects. The color of this material depends on the color of the object. To assign other materials, you need to select them from the **Tool Palettes - Materials** window and add them to the current drawing. The **Tool Palettes - Materials** will not be displayed by default. To invoke it, choose **View > Palettes > Tool Palettes** from the **Ribbon**. Next, right-click on the title bar of the palette and choose **Materials** from the shortcut menu. The **Tool Palettes - Materials** window will be displayed, as shown in Figure 30-1. All the available materials will be displayed as groups in the **Tool Palettes - Materials** window. Similarly, you can also use the **Tool Palettes - Materials Library**, if you have installed the materials library with AutoCAD. By default, these material libraries are installed in the **Tool Palettes File Locations** path specified in the **Files** tab of the **Options** dialog box.

To add a material from the tool palette to the current drawing, right-click on the desired material in the **Tool Palettes - Materials** or **Tool Palettes - Materials Library** and choose the **Add to Current Drawing** option from the shortcut menu. If the selected material already exists in the current drawing, the **material - Already Exists** window will be displayed on the screen, see Figure 30-2. From this window, you can choose to create a copy of the same material, overwrite the existing material with the same name, or cancel the process of the material inclusion in the current drawing.

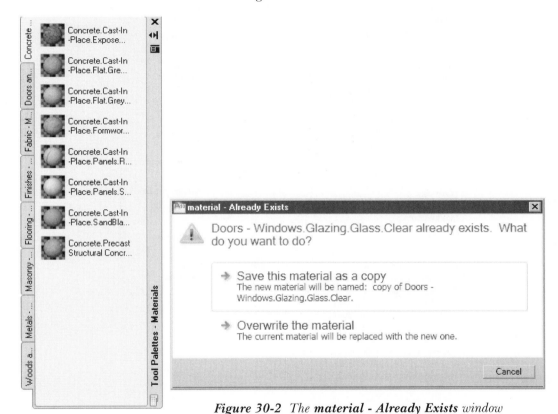

*Figure 30-2 The **material - Already Exists** window*

*Figure 30-1 The **Tool Palette - Materials** window*

Tip
*To get more options of the inbuilt materials of AutoCAD, right-click on the title bar of the **Tool Palettes** and choose **All Palettes** from the shortcut menu; the **Tool Palettes - All Palettes** window will be displayed. Click at the location shown in Figure 30-3 and you will get the options for materials in AutoCAD.*

Chapter 30

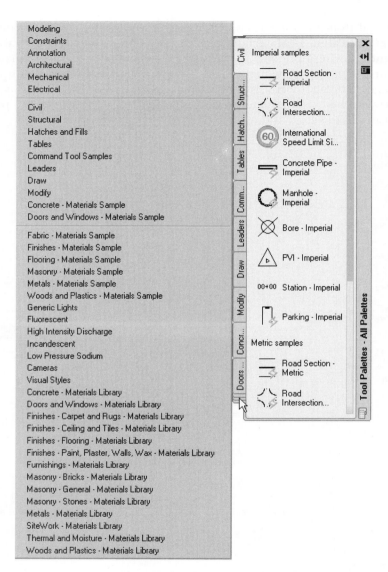

*Figure 30-3 The **Tool Palette - All Palettes** window*

Attaching Selected Materials to Objects

Ribbon:	View > 3D Palettes > Materials	Command:	MAT or MATERIALS
Menu Bar:	View > Render > Materials	Toolbar:	Render > Materials...

Once you have selected the materials from the **Tool Palettes - Materials** window, you need to attach them to the objects in the current drawing using the **MAT** command. This command is also used to create a new material or modify the existing material. To invoke this command, choose **View > 3D Palettes > Materials** from the **Ribbon**. When you choose this button, the **Materials** palette will be displayed, as shown in Figure 30-4. You can also invoke this palette by clicking the inclined arrow on the **Materials** panel in the **Render** tab of the **Ribbon**. The options in **Materials** palette are given next.

Available Materials in Drawing Area

The list box in this area displays all materials that are imported from the **Tool Palettes-Materials** window into the current drawing. In addition to the imported materials, the default global material is also displayed in this area. The selected material is displayed with a yellow boundary around it. The options in this area are given next.

Toggle Display Mode. If you choose this button at the upper right corner of the **Available Materials in Drawing** area, you can toggle between the display of the material swatches as small or large ones.

 Swatch Geometry. This button is used to change the display of the geometry in the material swatch. You can choose the display geometry as a box, cylinder, or sphere from the **Swatch Geometry** flyout. The selected geometry type affects only the selected material swatch in the **Available Materials in Drawing** area.

 Checkered Underlay On. This button is used to display a check board as the background in the selected material swatch. By default, the display of this background is on. To switch it off, choose this button. The display of the checkered underlay helps you to visualize the extent of the opacity of the material.

*Figure 30-4 The **Materials** palette*

 Preview Swatch Lighting Model. This button allows you to preview the Lighting of the Model. On choosing this button, a flyout is displayed from where you can select either the source light or the back light.

Create New Material. This button allows you to create your own material. To create any new material, choose the **Create New Material** button; the **Create New Material** dialog box will be displayed, see Figure 30-5. Specify a name of the material you are creating and give some description about it. Choose the **OK** button to accept the material. Now, this material will be added in the **Available Materials in Drawing** area. You can define and edit the properties of this new material in the **Material Editor** area, which will be discussed later in this chapter.

Purge from Drawing. This button is chosen to delete all the unused material from the current drawing. If the **Available Materials in Drawing** area displays any material that is not assigned to an object in the current drawing, select the unused material and then choose the **Purge from Drawing** button; the selected material will be removed from the current drawing. Note that you cannot purge the global material from the current drawing.

Indicate Materials in Use. This button is used to mark the material swatch that had been applied to some object in the current drawing. Once you choose this button, the used materials will always be displayed with the AutoCAD drawing file symbol at the lower right corner, as shown in Figure 30-6.

Figure 30-5 *The **Create New Material** dialog box*

Figure 30-6 *Symbol indicating the material used in the current drawing*

Apply Material to Objects. Choose this button to attach the selected material to the object in the current drawing. You have to first select the material from the materials swatch and then choose this button to attach it. The cursor will change to the paintbrush cursor and you will be prompted to select the object from the drawing window to which the material has to be attached. If the visual style is set to **Realistic**, the material will be displayed instantaneously on the objects that you select. Once you have selected all the objects to which the current material is attached, press the ENTER key. You can also apply the material to the objects by selecting the material swatch and then choosing the **Apply Material** option from the shortcut menu. You can also apply the materials to an object by pressing the left mouse button and dragging and dropping the selected material swatch from the **Available Materials in Drawing** area to the object.

Tip
*You can attach materials to objects directly from the **Tool Palettes - Materials** palette. Choose the material from **Tool Palettes** and then select the objects on which you want to apply the material. You can also select the material and drag it to the desired object or choose **Apply Material to Objects** from the shortcut menu in the **Tool Palettes - Materials** palette. The applied materials will automatically be added to the **Materials** palette.*

Remove Materials from Selected Objects. This button detaches the material from the selected objects. When you choose this button, the cursor will change to the paintbrush cursor and you will be prompted to select the object from which you want to detach the selected material. Select the objects; the material will be removed. By default, the global material will again be attached to the selected objects from which you detach the material.

Note
*The remaining options of the **Materials** palette will be discussed in the **Creating New Materials** and **Modifying Materials** sections of this chapter.*

Attaching Material by Layers

Ribbon: Render > Materials > Attach By Layer	**Command:** MATERIALATTACH

This button is chosen to assign a particular material to a layer. This means that all the objects that are placed on a layer will be rendered using the specified material. When you choose this button, the **Material Attachment Options** dialog box will be displayed, as shown in Figure 30-7. This dialog box displays all the materials in the current drawing in the **Material Name** column, the layers in the **Layer** column, and the material attached to this layer in the **Material** column. To assign a material to a layer, select it from the **Material Name** column and drag it to the desired layer in the **Layer** column. All the objects placed in the selected layer will now be rendered using the specified material. After attaching the material to a layer, the **Detach** button is automatically added on the right of the **Material** column. You can detach the material from the selected layer by choosing this **Detach** button.

*Figure 30-7 The **Material Attachment Options** dialog box*

Chapter 30

UNDERSTANDING ELEMENTARY RENDERING

Ribbon:	Render > Render > Render	**Menu Bar:**	View > Render > Render
Toolbar:	Render > Render	**Command:**	RENDER (RR)

In this section, you will learn about elementary or basic rendering. When you choose this button, the **Render** window will be displayed on the screen and the rendering process will start displaying the rendered image in the **Render** window, see Figure 30-8.

Figure 30-8 The **Render** window

Rendering can be performed on a model in three ways. These are as follows:

The first type is the basic rendering. In this, you can render the object without actually displaying the effect of the material. A global material is attached to the model and rendered. The color of the rendered image will depend on the color of the object. Basic rendering does not give a realistic effect to the model, see Figure 30-9.

In the second type of rendering, you can attach a material to the model and then render it. Figure 30-10 shows a model to which a material has been applied and then rendered.

Figure 30-9 *Model rendered without material*   **Figure 30-10** *Model rendered with material*

The third type of rendering is photorealistic. In this type of rendering, additional parameters such as the mirror effect, reflection, and refraction can be defined for the material, see Figure 30-11.

 Note

The remaining options of **Rendering** *will be discussed in the* **Understanding Advanced Rendering** *section of this chapter.*

Figure 30-11 *Model rendered by adding effects such as mirror*

Example 1   *Architectural*

In this example, you will draw the staircase shown in Figure 30-12. You can assume arbitrary dimensions. Attach the materials to the objects and then render the model. The materials that you have to assign to the model are given next.

> Stairs
> Finishes.Flooring.Wood.Parquet.1
> Balusters and Handrail
> Finishes.Painting.Paint.White

1. Create the model shown in Figure 30-12 assuming arbitrary dimensions.

2. Invoke **Tool Palettes - All Palettes**, right-click on the title bar of the **Tool Palette** and choose **Materials Library**; the **Tool Palettes - Material Library** is displayed.

Figure 30-12 *Model for Example 1*

3. Right-click on **Finishes.Flooring.Wood.Parquet.1** in the **Finishes-Flooring** tab and choose the **Apply Material To Objects** option from the shortcut menu; the cursor changes to the paintbrush cursor. Select the stairs and then press the ENTER key.

4. As explained in Step 3, select **Finishes.Painting.Paint.White** in the **Finishes-Paint, Plaster, Walls, Wax** tab and apply it to the balusters and handrails.

 Note
*In the **Materials** window, you will notice that all three materials selected in Step 3 and Step 4 are automatically attached to the current drawing and displayed in the **Available Materials in Drawing** field.*

5. Choose the **Render** button from the **Render** toolbar to invoke the **Render** window. The rendering process starts and the rendered image is displayed in the **Render** window. The rendered image should look similar to the one shown in Figure 30-13.

Figure 30-13 Rendered image of staircase

ADDING LIGHTS TO THE DESIGN

Ribbon: Render > Lights	**Toolbar:** Lights or Render > New Point Light
Command: LIGHT	**Menu Bar:** View > Render > Light

Lights are vital to rendering a realistic image of an object. Without proper lighting, the rendered image may not show the features the way you expect. The lights can be added to the design using the **LIGHT** command.

AutoCAD supports two types of lighting arrangements: Standard workflow and photometric workflow. The standard lighting workflow is the lighting arrangement that was followed in the pre-2008 releases of AutoCAD, whereas the photometric lighting workflow was introduced in AutoCAD 2008 and now, it is the default lighting workflow in AutoCAD. The photometric workflow provides more accurate control over lighting through the photometric properties. The photometric workflow supports two types of lighting units: International (for example lux), and American (for example foot-candles). You can select the lighting workflow and its unit from the **Units for specifying the intensity of lighting** drop-down list in the **Lighting** area of the **Drawing Units** dialog box. To invoke this dialog box, choose **Drawing Utilities > Units** from the **Application Menu**. Alternatively, expand the **Lights** panel in the **Render** tab of the **Ribbon** and specify the units. You can also change the lighting workflow by setting the **LIGHTINGUNITS** system variable. Setting the **LIGHTINGUNITS** system variable to 0 represents the standard (generic) lighting whereas, setting the **LIGHTINGUNITS** system variable to 1 and 2 represents the photometric lighting in American and International SI units respectively.

Note
*The Standard workflow is enabled if **Generic lighting units** is selected as a unit in the **Lights** panel. Similarly, the Photometric workflow is enabled if **American lighting units** or **International lighting units** is selected as a unit in the **Lights** panel.*

All types of light sources supported by AutoCAD are discussed next.

Default Light

You can visualize the default light as the natural light source that equally illuminates all surfaces of the objects, see Figure 30-14. The default light consists of two distance lights that do not have a source, and therefore, no location or direction. It is the default light that is automatically applied to the design. However, you can modify the brightness, contrast, and mid tone settings of the environment by using the respective button from the **Lights** panel in the **Render** tab of the **Ribbon**, see Figure 30-15.

Figure 30-14 Default light emitting constant illumination

*Figure 30-15 The expanded **Lights** panel of the **Ribbon***

The slider bar in front of the **Brightness** button and the **Contrast** button can be used to increase or decrease the brightness and contrast of the default light. By default, the color of the default light is white. Figures 30-16 and 30-17 show the same model rendered using different values of brightness and contrast of the default light.

Figure 30-16 Model rendered with less ambient light brightness and contrast

Figure 30-17 Model rendered with more ambient light brightness and contrast

When you choose any New Light button to create user-defined lights, the **Lighting - Viewport Lighting Mode** message will be displayed on the screen, see Figure 30-18. If you want to render the model by using the user-defined lights only, then choose the **Turn off the default lighting (recommended)** button. But if you want to render the model using both the default and user defined lights, then choose the **Keep the default lighting turned on** button. However, you will notice that on keeping the default lights turned on, the effects of the user-defined lights are not displayed properly, so you need to turn off the default lighting manually. If you invoke the **LIGHT** command again after turning off the default light, the **Viewport Lighting Mode** message will not be displayed on the screen.

*Figure 30-18 The **Lighting - Viewport Lighting Mode** message box*

Point Light

Ribbon: Render > Lights > Create Light > Point **Command:** POINTLIGHT
Toolbar: Lights > New Point Light **Menu Bar:** View > Render > Light > New Point Light

A point light source emits light in all directions and the intensity of the emitted light is uniform. You can visualize an electric bulb as a point light source. In AutoCAD render, if you select to add a point light, you can set the options of casting the shadow of the objects in the design. Figure 30-19 shows a point light source that radiates light uniformly in all directions.

As this is not the default light source, therefore, you will have to add it manually. This light source can be added by choosing **Render > Lights > Create Light > Point** from the **Ribbon.** The prompt sequence for the **POINTLIGHT** command is given next.

Figure 30-19 Point light source emitting light uniformly in all directions

   Command: **POINTLIGHT**

Specify source location <0,0,0>: *Specify the location where you want to place the point light.*
Enter an option to change [Name/Intensity factor/Status/Photometry/shadoW/
Attenuation/filterColor/eXit] <eXit>: *Specify an option or press* [Enter]

The following options can be selected from this prompt:

Name

This option is used to specify the name of the point light. AutoCAD, by default, specifies a
name to the newly created light. You can change the name by using this option.

Intensity factor

This option is used to specify the intensity of the point light. The prompt sequence that will
be followed is given next.

Enter an option to change [Name/Intensity factor/Status/Photometry/shadoW/
Attenuation/filterColor/eXit] <eXit>: **I** [Enter]
Enter intensity (0.00 - max float) <1.0000>: *Enter a value for the intensity of the light or
press the* [Enter] *key to accept the default value.*

Status

This option is used to turn the point light on and off.

Photometry

This property will be available only when you work in
the photometric environment that means if the
American lighting units or **International lighting units**
is selected as a unit in the **Lights** panel. The
Photometry option is used to control the intensity, unit
of intensity, color, and type of point lights, so that you
can define the lights accurately according to the real
world requirement. Instead of specifying the
photometric properties of the point light in the command

*Figure 30-20 The **Photometric
properties** rollout of the **Properties**
palette in the point light*

itself, you can also specify or modify all the photometric properties later in the **Properties**
palette of the point light, see Figure 30-20. The options in the **Photometric properties** rollout
are discussed next.

Lamp intensity. This property is used to specify the intensity, flux, or illuminance of the
lamp. By default, the intensity of the light is specified in Candela. To change the properties,
select the **Lamp intensity** edit box and choose the swatch displayed on the right; the **Lamp
Intensity** dialog box will be displayed, see Figure 30-21. Select the **Intensity (Candela)** radio
button to specify the intensity of the point light in Candela. The **Flux (Lumen)** radio button
is used to specify the flux of light in Lumen. The flux is the rate of total energy leaving the
light source. The spinner on the right of the above mentioned radio button is used to change
the value for both the intensity and flux depending on the radio button selected. The
Illuminance (Lux) radio button is used to specify the amount of energy received per unit

area by an object. In this case, you have to specify the distance at which the object is placed from the light source because the photometric lights are divergent by nature and their intensities will reduce with the increase of distance between the light source and the object. You can also increase or decrease the resulting intensity, flux, or illuminance of the light source by specifying a multiplying factor in the **Intensity factor** spinner of the **Resulting intensity** area. The resulting intensity will be the product of intensity input and the intensity factor.

Resulting intensity. The **Resulting intensity** property displays the resulting final intensity of the light source. This information is read-only.

*Figure 30-21 The **Lamp Intensity** dialog box*

Lamp color. This property is used to specify the lamp color of the light source. To specify or modify the lamp color, select the **Lamp color** edit box and choose the swatch displayed on the right; the **Lamp Color** dialog box will be displayed, see Figure 30-22. Select the **Standard colors** radio button to select the standard predefined colored lights listed in the **Standard colors** drop-down list, see Figure 30-23. The preview of the selected light will be displayed in the color swatch on the right of the **Standard colors** drop-down list. An ideal black body emits a particular color at a particular temperature. To specify the lamp color with respect to temperature, select the **Kelvin colors** radio button. You can specify the temperature in Kelvin in the spinner just below the **Kelvin colors** radio button. The **Filter color** drop-down list in the **Resulting color** area is used to mix some colors with the lamp color. The final resulting color of the lamp will be a mixture of the lamp color and filter color. The preview of the resulting lamp color will be displayed in the color swatch in the **Resulting color** area.

*Figure 30-22 The **Lamp Color** dialog box*

*Figure 30-23 The **Standard colors** drop-down list*

Resulting color. The **Resulting color** property displays the resulting final color of the light source. This information is read-only.

shadoW

The sub-options provided under this option are used to control the display of the shadows of the objects after rendering. The sub-options are given next.

Off. This option is used to turn off the display of the shadow in rendering.

Sharp. This option is used to display the shadow of the objects with sharp edges.

soFtmapped. This option is used to display the shadow with blurred edges. This gives a more realistic effect to the rendered image. If this sub-option is selected, you will have to specify the map size and softness of the shadow. The map size is calculated in terms of pixels. The greater the number of pixels, the better is the shadow. However, if the map size is large, the time taken to render the model will also increase. The shadow softness is used to specify the softness for the shadow. If the value of this option is increased, the shadow will be blurred. Figure 30-24 shows a model with a shadow map size 128 and softness 5, and Figure 30-25 shows a model with a shadow map size 1024 and softness 2.

Figure 30-24 *Shadow map size =128 and softness = 5*

Figure 30-25 *Shadow map size =1024 and softness = 2*

Tip
You can view the best shadow effect by keeping the value of the shadow map size 512 or 1024 and the shadow softness between 3 and 4.

softsAmpled. This option is used to specify the shape of the point light lamp. The shadow of the object will be displayed depending upon the shape of the point light lamp. The prompt sequence for this option is given next.

Enter an option to change [Name/Intensity factor/Status/Photometry/shadoW/
Attenuation/filterColor/eXit] <eXit>: **W** [Enter]
Enter [Off/Sharp/soFtmapped/softsAmpled] <Sharp>: **A** [Enter]
Enter an option to change [Shape/sAmples/Visible/eXit]<eXit>: **S** [Enter]
Enter shape [Linear/ Disk/ Rect/ Sphere/ Cylinder] <Sphere>: *Enter an option to specify the shape of the slit from which the point light will pass through.*

After specifying the shape of the slit, you will be prompted to specify the dimensions of the shape. For example, the radius of the sphere or the length and width of a rectangle. The prompt sequence for the **Cylinder** option is given next.

> Enter shape [Linear/Disk/Rect/Sphere/Cylinder] <Sphere>: **C** Enter
> Enter radius <0.0000>: *Specify the radius of the cylindrical slit.*
> Enter Length <0.0000>: *Specify the height of the cylindrical slit.*
> Enter an option to change [Shape/sAmples/Visible/eXit]<eXit>: *Specify an option or press* Enter

The **sAmples** option in the above prompt is used to specify the size of the shadow samples for light. The default value of the shadow sample is 16. The **Visible** option is used to specify whether to display the shape of the lamp in rendering or not. Figure 30-26 displays the rendered image of a cylindrical shape point light without the visibility of the point light shape. Figure 30-27 displays the same rendered image with the point light shape visible.

Figure 30-26 *Rendered image with the shape of cylindrical point light not visible*

Figure 30-27 *Rendered image with the shape of point light visible*

Note
The shape of the point light is also displayed in the drawing area close to the point light. When you move the cursor over the point light, the shape of the point light is highlighted in brown color.

Attenuation

The light intensity is defined as the amount of light falling per unit of area. The intensity of light is inversely proportional to the distance between the light source and the object. Therefore, the intensity decreases as the distance increases. This phenomenon is called Attenuation. In AutoCAD, it occurs only with spotlights and a point light.

In Figure 30-28, light is emitted by a point source. Assume the amount of light incident on Area1 is I. Therefore, the intensity of light on Area1 = I/Area. As light travels farther from the source, it covers a larger area. The amount of light falling on Area2 is the same as on Area1 but the area is larger. Therefore, the intensity of light for Area2 is smaller (Intensity of light for Area2 = I/Area). Area1 will be brighter than Area2 because of a higher light intensity.

The prompt sequence that will be followed after selecting this option is given next.

> Enter an option to change [attenuation Type/Use limits/ attenuation start Limit/attenuation End limit/eXit] <eXit>: *Select a sub-option or press the* [Enter] *key.*

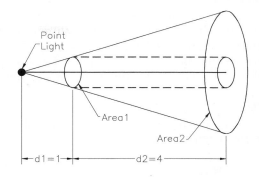

Figure 30-28 *Figure explaining the concept of attenuation*

The sub-options that you can select at the above prompt are:

attenuation Type. AutoCAD render provides you with the following three types of attenuation.

None. If you select the **None** sub-option for the light falloff, the brightness of the objects will be independent of the distance. This means that the objects that are far away from the point light source will be as bright as those close it.

Inverse linear. In this sub-option, the light falling on the object (brightness) is inversely proportional to the distance of the object from the light source (Brightness = 1/Distance). As the distance increases, the brightness decreases. For example, assume that the intensity of the light source is I and the object is located at a distance of 2 units from the light source. Now, the brightness or intensity = I/2. If the distance is 8 units, the intensity (light falling on the object per unit area) = I/8. The brightness is a linear function of the distance of the object from the light source.

inverse Squared. In this sub-option, the light falling on the object (brightness) is inversely proportional to the square of the distance between the object and the light source (Brightness = 1/Distance^2). For example, assume that the intensity of the light source is I and the object is located at a distance of 2 units from the light source. Now, the brightness or intensity = I/(2)^2 = I/4. If the distance is 8 units, the intensity (light falling on the object per unit area) = I/(8)^2 = I/64.

Use limits. This option is used to specify whether you want to use the sub-options of **attenuation start Limit and attenuation End limit**. The effect of these limits can only be visualized in rendering, if the sub-option of **Use limits** is turned on.

attenuation start Limit. This option is used to specify the distance from the centre of the light to where the light will start.

attenuation End limit. This option is used to specify the distance from the center of the light at which the light will finish. The objects beyond this limit will not be illuminated by this point light.

Chapter 30

filterColor

This option is used to specify the color for the point light. You can define a custom color by using the RGB combination. You can select a color by choosing the **Index color** sub-option. Select the sub-option **Hsl** to define the color by Hue, Saturation, and Luminance. You can also specify the color of the point light from the color book by choosing the **colorBook** sub-option. Next, enter the name of the color book from which you want to choose the color.

eXit

This option is used to finish the process of defining the point light properties and to exit the **POINTLIGHT** command.

Note
The lightings in the photometric workflow always follow the inverse squared attenuation type, and the attenuation properties cannot be modified later on for photometric lights.

Tip
*Instead of specifying all the above-mentioned properties of the point light in the command itself, you can also specify or modify all the properties later on from the **Properties** palette of the point light.*

Example 2   *Architectural*

In this example, you will apply the point light at the lintel of the window and at the lintel of the door of the model, as shown in Figures 30-29 and 30-30. Choose your own materials to make the model look realistic and then render it.

Figure 30-29 *Model for Example 2*

Figure 30-30 *Model after rendering*

1. Create the model shown in Figure 30-29.

2. Choose materials from **Tool Palettes - Materials Library** and attach them to the model as described in Example 1. (Recommended: **Masonry - Stones > Masonry.Stone.Limestone.Ashlar.Coursed** for Walls, **Finishes - Ceiling and Tiles > Finishes.Tiling.Ceramic.Mosaic.Blue** for flooring, and **Finishes - Flooring > Finishes.Flooring.VCT.Diamonds** for roof.)

3. Choose **Render > Lights > Create Light > Point** from the **Ribbon** to invoke the **POINTLIGHT** command.

4. Choose the **Turn off the default lighting (recommended)** button from the **Lighting - Viewport Lighting Mode** message box. (This message box will be displayed when you create the light for the first time).

5. The prompt sequence for the **POINTLIGHT** command is given next.

    Command: **POINTLIGHT**
    Specify source location <0,0,0>: *Specify the location of the point light at the lintel of the window.*
    Enter an option to change [Name/Intensity factor/Status/Photometry/shadoW/Attenuation/filterColor/eXit] <eXit>: **N** Enter
    Enter light name <Pointlight>: **1** Enter
    Enter an option to change [Name/Intensity factor/Status/Photometry/shadoW/Attenuation/filterColor/eXit] <eXit>: **C** Enter
    Enter true color (R,G,B) or enter an option [Index color/Hsl/colorBook]<255,255,255>: **I** Enter
    Enter color name or number (1-255): **Yellow** Enter
    Enter an option to change [Name/Intensity factor/Status/Photometry/shadoW/Attenuation/filterColor/eXit] <eXit>: **I** Enter
    Enter intensity (0.00 - max float) <1.0000>: *Specify the intensity based on the size of the objects in the drawing and press the* Enter *key*
    Enter an option to change [Name/Intensity factor/Status/Photometry/shadoW/Attenuation/filterColor/eXit] <eXit>: **S** Enter
    Enter status [oN/oFf] <On>: Enter
    Enter an option to change [Name/Intensity factor/Status/Photometry/shadoW/Attenuation/filterColor/eXit] <eXit>: **P** Enter
    Enter a photometric option to change [Intensity/Color/eXit] <I>: **X** Enter
    Enter an option to change [Name/Intensity factor/Status/Photometry/shadoW/Attenuation/filterColor/eXit] <eXit>: **W** Enter
    Enter [Off/Sharp/soFtmapped/softsAmpled] <Sharp>: **S** Enter
    Enter an option to change [Name/Intensity factor/Status/Photometry/shadoW/Attenuation/filterColor/eXit] <eXit>: Enter

6. Similarly, you can set the second light at the lintel of the door. Set the color of the light to green.

7. Choose the **Render** button in the **Render** panel under the **Render** tab of the **Ribbon** to invoke the **Render** window. The rendering process will start and display the rendered image in the **Render** window. The rendered image should be similar to the one shown in Figure 30-31.

Tip
You can increase the smoothness of the rendered curved object by increasing the value of the ***FACETRES*** *system variable. The default value of this variable is 0.5. Set this value close to **6**.*

Figure 30-31 *Rendered model of a house*

Spotlight

Ribbon: Render > Lights > Create Light > Spot	**Toolbar:** Lights > New Spotlight
Menu Bar: View > Render > Light > New Spotlight	**Command:** SPOTLIGHT

A spotlight emits light in the defined direction with a cone-shaped light beam. This light has a focused beam that starts from a point and is targeted at another point. This type of light is generally used to illuminate a specific point, such as illuminating a person on a stage. The phenomenon of **attenuation** also applies to spotlights. This light source can be added by choosing **Render > Lights > Create Light > Spot** from the **Ribbon**. When you choose this button, you will be prompted to enter two values. As already mentioned, this type of light source has a start point and a target point. Therefore, you will first be prompted to specify the light location and then the light target. The prompt sequence for the **SPOTLIGHT** command is given next.

Command: **SPOTLIGHT**
Specify source location <0,0,0>: *Specify the point where you want to place the spotlight.*
Specify target location <0,0,-10>: *Specify the location on the object where you want to focus the spotlight.*
Enter an option to change [Name/Intensity factor/Status/Photometry/Hotspot/Falloff/shadoW/Attenuation/filterColor/eXit] <eXit>: *Specify an option or press* Enter

Most of the options in the **SPOTLIGHT** command are similar to those in the **POINTLIGHT** command. The options that are different are discussed next.

Hotspot/Falloff

As mentioned earlier, the spotlight has a focused beam of light that is targeted at a particular point. Therefore, there are two cones that comprise the spotlight.

The hotspot is the same as the cone that carries the highest intensity light beam. In this cone, the light beam is the most focused and is defined in terms of an angle, see Figure 30-32. The prompt sequence for the **Hotspot** option is given next.

> Enter an option to change [Name/Intensityfactor/Status/Photometry/Hotspot/Falloff/
> shadoW/Attenuation/filterColor/eXit]<eXit>: **H** Enter
> Enter hotspot angle (0.00-160.00) <45.0000>: *Specify the interior angle of the cone in which the intensity of light will be high or press the* Enter *key to accept the default value.*

The other cone is called a falloff and it specifies the full cone of light. It is the area around the hotspot where the intensity of the beam of light is not very high, as shown in Figures 30-32(a) and 30-32(b). It is defined in terms of an angle. The value of the hotspot and the falloff can vary from 0 to 160. The prompt sequence for the **Falloff** option is given next.

> Enter an option to change [Name/Intensity factor/Status/Photometry/Hotspot/Falloff/
> shadoW/Attenuation/filterColor/eXit]<eXit>: **F** Enter
> Enter falloff angle (0.00-160.00) <current>: *Specify the interior angle of the cone in which the intensity of the light will be high or press the* Enter *key to accept the default value.*

Figure 30-32(a) *The hotspot and falloff cones*

Figure 30-32(b) *The hotspot and falloff after rendering*

Tip
The value of hotspot should always be less than the falloff. If you set the value of the hotspot more than the value of the falloff, then the value of the falloff will automatically change to one degree greater than the value of the hotspot.

Distant Light

A distant light source emits a uniform parallel beam of light in a single direction only (Figure 30-33). The intensity of the light beam does not decrease with the distance. It remains constant. For example, the sunrays can be assumed to be a distant light source because the light rays are parallel. When you use a distant light source in a drawing, the location of the light source does not matter; only the direction is critical. Distant light is mostly used to light objects or a backdrop uniformly and for getting the effect of the sunlight.

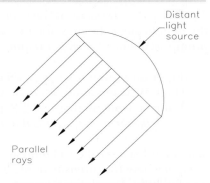

Figure 30-33 Distant light source

The distant lights are inaccurate and their use in photometric workflow is not preferred. Therefore, the **Distant** button in the **Lights** panel under the **Render** tab of the **Ribbon** is not available for the use in the photometric workflow. However, to use the distant light in photometric workflow, choose **Lights > Create Light > Distant** from the **Ribbon** or use the command window; the **Lighting - Viewport Lighting Mode** message box will be displayed on the screen. Choose the **Turn off the default lighting (recommended)** button so that the effect of the distant lights can be seen. On doing so, the **Lighting - Photometric Distant Lights** message box is displayed, see Figure 30-34. Choose the **Allow distant lights** button from the **Lighting - Photometric Distant Lights** message box to invoke the **DISTANTLIGHT** command. The prompt sequence for the **DISTANTLIGHT** command is given next.

Command: **DISTANTLIGHT**
Specify light direction FROM <0,0,0> or [Vector]: *Specify the start point to represent the viewing direction of the distant light or type* **V** *and press the ENTER key.*
Specify light direction TO <1,1,1>: *Specify the endpoint to represent the viewing direction of the distant light.*
Enter an option to change [Name/Intensity factor/Status/Photometry/shadoW/filterColor/eXit] <eXit>: *Specify an option.*

If you select the **Vector** option at the **Specify light direction FROM <0,0,0> or [Vector]** prompt, then you will have to specify the vector direction in which the distant light will focus. The options in the **DISTANTLIGHT** command are similar to those in the **POINTLIGHT** command.

Figure 30-34 The **Lighting - Photometric Distant Lights** *message box*

Sun Light

Ribbon: Render > Sun & Location (Inclined Arrow)	**Toolbar:** Light > Sun Properties
Menu Bar: View > Render > Light > Sun Properties	**Command:** SUNPROPERTIES

This light is similar to the distant light. The rays of the sun are parallel and have the same intensity at any distance. The angle of the rays with the object depend on the location, date, and time that you specify. You can also change the intensity and color of the sunlight. Unlike distant light, sunlight can only be one in number. In the photometric workflow, you can also control the illumination effect when the sky comes between the sunlight and the object. This helps in adding more realistic lighting caused by the interaction between the sun and the sky.

To use sunlight, choose the inclined arrow on the **Sun & Location** panel of the **Render** tab of the **Ribbon**; the **Sun Properties** window will be displayed, as shown in Figure 30-35. The options in this window are discussed next.

General

In this area, you can switch on or off the sunlight and change its intensity and color. You can also specify whether the object will cast a shadow due to the sunlight.

Figure 30-35 The **Sun Properties** window

Sky Properties

These properties are only available while working in the photometric workflow. You can use the sky in between the model and the sun for the illumination of the model. The options

Chapter 30

available in the **Status** edit box are **Sky Off**, **Sky Background**, and **Sky Background and Illumination**. These options are also available in a flyout in the **Sun & Location** panel of the **Render** tab in the **Ribbon**. Figure 30-36 displays a model rendered by selecting the **Sky Off** option, Figure 30-37 displays a model rendered by selecting the **Sky Background** option, and Figure 30-38 displays the model rendered by selecting the **Sky Background and Illumination** option. The **Intensity Factor** edit box is used to magnify the brightness of the light reaching the object after passing through the sky. Figure 30-38 displays the rendered model by choosing the **Sky Background and Illumination** option and intensity factor set to 1 and Figure 30-39 displays the same model rendered by selecting the **Sky Background and Illumination** option and intensity factor set to **0.2**. The **Haze** edit box determines the amount of scattering in the atmosphere and the haze value can vary from 0 to 15. Figure 30-38 displays a rendered model with haze value set to **0** and Figure 30-40 displays the same model rendered with haze value set to **7**.

Horizon. The options in this area are used to control the location and appearance of the ground. Figure 30-41 displays a rendered model created by selecting the **Sun & Sky Background** and setting that view as current. The ground color for the rendered image is set to cyan.

Figure 30-36 Rendered model with the **Sky Off** option selected

Figure 30-37 Rendered model with the **Sky Background** option selected

Figure 30-38 Rendered model with the **Sky Background and Illumination** option selected and intensity and haze value set to **1** and **0**, respectively

Figure 30-39 Rendered model with the **Sky Background and Illumination** option selected and intensity value set to **0.2**

*Figure 30-40 Rendered model with the **Sky Background and Illumination** option selected and intensity and haze value set to **1** and **7**, respectively*

*Figure 30-41 Rendered model with the view saved as **Sun & Sky Background** and ground color set to **Cyan***

Advanced. The properties under the **Advanced** rollout are used to produce more artistic effects in rendering such as the color of the night sky, whether to use the Aerial perspective view on or off, and so on.

Sun Disk appearance. The properties under this rollout are used to control the appearance of the sun disk such as its size, intensity, and so on, when a view is saved with a **Sun & Sky Background**.

Note

*You can also specify or modify the **Sky Properties** from the **Adjust Sun & Sky Background** dialog box that will be displayed when you choose the **Sky Properties** button displayed at the upper right corner of the **Sky Properties** title bar of the **Sun Properties** window, see Figure 30-42.*

*Figure 30-42 The **Adjust Sun & Sky Background** dialog box*

Sun Angle Calculator

In this area, you can set the date and time in the **Date** and **Time** edit boxes, respectively. You can also switch on or off the effect of day light settings in the rendering. Apart from this, you can also view some read only information in this field such as the azimuth and altitude of the sun light direction. You can also view the default light source vector as the coordinates in the X, Y, and Z list boxes in the **Source Vector** area.

Rendered Shadow Details

This area gives you the option to control the shadow cast by the object in rendering. All the options in this field are similar to the **POINTLIGHT** command.

Geographic Location

This area displays information about the geographic location of the place where you want to view the light effect. You can edit this information by choosing the **launch Geographic Location** button at the upper right corner of the **Geographic Location** area. Alternatively, choose **Render > Sun & Location > Set Location** from the **Ribbon**; the **Geographic Location-Define Geographic Location** window will be displayed with various options to define the location. To define the location, you can import the exact location data like longitude, latitude, and altitude from the .kml or .kmz files. Alternatively, you can pick your desired location from Google Earth, if it is already installed on your system or you can directly enter the location values. On choosing the **Enter the location values** button from the **Geographic Location- Define Geographic Location** window, the **Geographic Location** dialog box will be displayed, as shown in Figure 30-43. The options in this dialog box are given next.

Figure 30-43 The **Geographic Location** dialog box

You can set the latitude and longitude in the **Latitude** and **Longitude** edit boxes, respectively, and the directions as north, south, east, and west in the **North** and **West** drop-down lists in the dialog box. By default, the *Y* axis is assumed to be in the north direction in AutoCAD. You can also change this default direction in the **Angle** edit box.

You can select various locations by choosing the **Use Map** button available in the **Latitude & Longitude** area of the **Geographic Location** dialog box; the **Location Picker** dialog box will be displayed. You can select different locations from the map in this dialog box; a red colored cursor is displayed on the selected locations. You can select the required map from the **Region** drop-down list. Also, all the nearest cities and the time zone, in which the selected city is located, will be listed in the **Nearest City** and **Time Zone** drop-down lists, respectively. Select the city and the time zone from these drop-down lists. You can also select a location in the World map. On doing so, the corresponding city and its latitude, longitude, and time zone are displayed in the respective drop-down lists. Choose the **OK** button; a **Geographic Location - Time Zone Updated** message box is displayed prompting you to either accept the updated time or return to the previous time. On doing so, you can go back to the **Sun Angle Calculator** area in the **Sun Properties** palette and set the date and time. Figures 30-44 through 30-49 show different positions of the sun at different times and locations.

Figure 30-44 City: Chicago, Time: 11:00, Date: 1st May

Figure 30-45 City: Chicago, Time: 16:00, Date: 1st May

Figure 30-46 City: Tokyo, Time: 11:00, Date: 1st May

Figure 30-47 City: Tokyo, Time: 16:00, Date: 1st May

Figure 30-48 *City: New Delhi, Time: 11:00, Date: 1st May*

Figure 30-49 *City: New Delhi, Time: 16:00, Date: 1st May*

You can also switch the sun light on and off, set the sky background illumination, change the date and time, open and close the **Sun Properties** palette and the **Geographic Location - Define Geographic Location** windows directly from the **Render** tab of the **Ribbon**, see Figure 30-50.

Figure 30-50 *The **Render** tab of the **Ribbon***

 Note
*1. If the **Sun Status** button is chosen in the **Sun & Location** panel, sunlight is turned on.*
2. Except for the geographic location, the rest of the settings for the sun and the sky are saved for a particular viewport, and not for the entire drawing.

Example 3   *Architectural*

In this example, you will create the model shown in Figures 30-51. Choose materials of your choice. Set the sun light using the geographic location and render it, as shown in Figure 30-52. The geographic location and other parameters are given next.

City: New York, Date: 20 June, 2009 Time: 04:00 pm

1.   Create the model, as shown in Figure 30-51.

2.   Select the materials and apply them to the model, as explained in Example 1.

3.   Choose the inclined arrow on the **Sun & Location** panel of the **Render** tab of the **Ribbon** to invoke the **Sun Properties** window.

Figure 30-51 Model for Example 3 before rendering

Figure 30-52 Model for Example 3 after rendering

4. In this window, select the **Sky Background and Illumination** option from the **Status** drop-down list and set the **Intensity Factor** value to **2.0** under the **Sky Properties** rollout.

5. Set a color of your choice for the ground in the **Ground color** edit box under the **Horizon** rollout.

6. Choose **Render > Sun & Location > Set Location** from the **Ribbon**; the **Geographic Location - Define Geographic Location** window is displayed. Choose the **Enter the location values** button from the window; the **Geographic Location** dialog box is displayed. Choose the **Use Map** button in this dialog box; the **Location Picker** dialog box is displayed. Select **North America** from the **Region** drop-down list. Next, select **New York, NY** from the **Nearest City** drop-down list. You can also select a city by clicking in the map of **North America** displayed in the **Location Picker** dialog box. Choose **OK** to exit the dialog box; the **Geographic Location - Time Zone Updated** message box is displayed prompting you to update the time zone or return to the previous time zone. Choose **Accept updated time zone** from the message box; the updated locations are displayed in the **Geographic Location** dialog box. Choose the **OK** button to exit the dialog box.

7. Choose **Render > Sun & Location > Sun Status** from the **Ribbon**; the **Lighting - Viewport Lighting Mode** message box is displayed. Choose the **Turn off the default lighting (recommended)** option.

8. Set the date and time as **6/20/2009** and **4:00 pm** in the **Date** and **Time** edit boxes, respectively in the **Sun Angle Calculator** rollout of the **Sun Properties** palette.

9. Choose the **Render** button from the **Output** panel in the **Ribbon** to invoke the **Render** window. The rendering process will start and display the rendered image in the **Render** window.

Chapter 30

CONVERTING PRE-AutoCAD 2010 LIGHTS TO AutoCAD 2010 FORMAT

For converting a model created in AutoCAD 2009 or earlier version of AutoCAD to AutoCAD 2010 format such that it supports both the standard workflow and the photometric workflow, set the value of **LIGHTINGUNITS** system variable to 1 or 2. The models and lights created prior to AutoCAD 2008 can be converted to the lights of AutoCAD 2010 by the **CONVERTOLDLIGHTS** command. The prompt sequence is given next.

> Command: **CONVERTOLDLIGHTS**
> Loading materials...done.
> 2 light(s) converted

In the prompt sequence, only 2 lights have been converted for the current drawing. Therefore, the prompt sequence displays **2 light(s) converted**. The prompt sequence will display the number of lights converted and also the lights that are not converted. Note that after converting the lights, their effect is not always the same as in their original versions. Therefore, you have to modify some parameters such as intensity, color, and so on.

MODIFYING LIGHTS

As mentioned earlier, the lights and the lighting effects are important in rendering in order to create a realistic representation of an object. The sides of an object that face the light must appear brighter and the sides that are on the other side of the object must be darker. This smooth gradation of light produces a realistic image of the object. If the light intensity is uniform on the entire surface, the rendered object will not look realistic. For example, if you use the **SHADE** command to shade an object, it does not look realistic because the displayed model lacks any kind of gradation of light. Any number of lights can be added to the drawing. The color, location, and direction of all the lights can be specified individually. As mentioned earlier, you can specify an attenuation for the point lights and the spotlights. AutoCAD also allows you to change the color, position, and intensity of any light source. The point light and the spotlight can be interchanged. But you cannot change a distant light into a point light or a spotlight. All lights, except the default light and the sun light, attached to the current drawing are displayed in the light list. Choose the inclined arrow on the **Lights** panel of the **Render** tab of the **Ribbon**; the **Lights in Model** palette will be displayed, as shown in Figure 30-53. From this list box, select the light you want to modify, right-click, and then choose the **Properties** option from the shortcut menu displayed. Depending on the type of light selected, the **Properties** palette will be displayed. For example, if you select a distant light for modifications, the **Properties** palette with the distant light properties will be displayed. One such **Properties** palette is shown in Figure 30-54. All options in this window are similar to those under the **POINTLIGHT** command.

You can also modify the lights dynamically with the help of grips. This method provides a dynamic preview of the modifications made in lights. The icons of the point light and the spot light, displaying their location in the drawing, can also be switched on or off by choosing **Render > Lights > Light glyph display** from the **Ribbon**.

Figure 30-53 *The **Lights in Model** palette*

Figure 30-54 *The **Properties** palette for distant light*

CREATING NEW MATERIALS

AutoCAD enables you to create new materials according to your requirements. This is done using the **Materials** window. To create a new material, choose the **Create New Material** button, see Figure 30-55; the **Create New Material** dialog box will be displayed. Specify a name for the material you are creating and give its description. Choose the **OK** button to accept the material. This material will be added in the **Available Materials in Drawing** area. You can define its properties in the **Material Editor** area. The options in this field are given next.

You can define the properties of the material based on four types of materials: **Realistic**, **Realistic Metal**, **Advanced**, and **Advanced Metal**. To use any of these as base material, select the required option from the **Type** drop-down list. Next, you can also select a standard material template from the **Template** drop-down list and modify the desired properties according to your requirement. The templates are only available for **Realistic** and **Realistic Metal** types.

Three colors can be defined for an object: **Diffuse**, **Ambient**, and **Specular**. Diffuse color is the actual color of the object. The ambient color is applied on the faces illuminated by the default light, and the specular color is used to specify a shiny portion of an object illuminated by lights. You can apply only a diffused color to the **Realistic** and **Realistic Metal** templates. For **Advanced Metal**, you can define an ambient and a diffuse color. For the **Advanced** template, you can define all three colors, see Figure 30-56. Click on the color

Figure 30-55 *The Create New Material button of the Materials window*

Figure 30-56 *The color options for the Advanced template*

button to define the color or select the **By Object** check box to assign the color of the object to the material. To define the same color for all the three types, click on the lock icon.

There are some other properties that add extra effects to the material. These are: **Shininess, Opacity, Reflection, Refraction index, Translucency, Self-Illumination**, and **Luminance**. Shininess defines the reflective quality of a surface. If a surface is more shiny, it will reflect more light and the color in the affected portion by the light will be lighter. Opacity controls the transparency of the material. The material with no opacity is transparent. Reflection controls the reflectivity of the material. A hundred percent reflective material will display the surrounding objects on the material surface. Refraction index defines the extent by which the view of an object will distort if it is placed behind a translucent object. For a refractive index of 1, the view of the object will not distort at all and as this value goes on increasing, the extent of distortion will also go on increasing. The maximum value of refractive index is 3. Translucency controls the amount of light transmitted and scattered within the object. At 0 translucency, all the light will be transmitted through the material without scattering. As the value of translucency goes on increasing, the amount of light scattered within the material will also go on increasing. Self-illumination makes the material itself a source of light. As its value goes on increasing, it appears that the light is coming from the material itself. To change these properties, use the slider bar in front of these options. Luminance controls the amount of light reflected from a surface. It is a measure of how bright or dark a surface will appear. While defining luminance, the **Self-Illumination** option is not available. Enter the value of luminance in the edit box in terms of candela/meter^2. The **Two sided material** check box is used to apply material on both the positive and negative faces of the material. If this check box is not selected, the material will be applied only to the face in the direction of positive normal while rendering.

The **Diffuse map** option is used to add some texture to the color of the material. Nine types of diffuse maps can be added to the diffused color of the object: **Texture Map**, **Checker**, **Gradient Ramp**, **Marble**, **Noise**, **Speckle**, **Tiles**, **Waves**, and **Wood**. The diffused color of the object is replaced by the map color. Select the **Diffuse map** check box to activate the addition of the map to the material.

If you select the **Texture map** option from the drop-down list, the **Select Image** button will be displayed below it, see Figure 30-57. Choose this button; the **Select Image File** dialog box will be invoked. Select the texture image from the location where you stored it and choose the **Open** button in the **Select Image File** dialog box. Now, more buttons will be added in the **Diffuse map** area, see Figure 30-58. Use the slider bar to decide the extent of the effect of the texture map on the material. At 0 value, the effect of the texture map will be neutralized and only the material will be displayed. Also, as this value increases, the effect of the texture map becomes more dominating on the material and at the value of 100, only the texture map will be displayed on the object. To edit the image, choose the **Click for Texture Map settings** button. All the rollouts in the **Materials** window will be replaced by the **Texture Map**, **Scaling & Tiling**, and **Offset & Preview** rollouts in the **Materials** window. The **Texture Map** rollout has a slider bar that controls the extent of the effect of the texture map on the material and a button displaying the name of the currently mapped image. To return from the material editing options to the basic material setting options of the **Materials** window, choose the **Home to Material Settings** button and to move one level up in the nested mapping, use the **Up One Level to Parent Map** button, see Figure 30-59. In case of texture mapping, the function of both the keys will be the same only because you cannot perform nested mapping for texture maps. You can perform nested mapping for **Checker**, **Gradient Ramp**, **Noise**, **Tiles**, and **Waves** map types.

Figure 30-57 The **Diffuse map** area before applying the texture map image

Figure 30-58 The **Diffuse map** area after applying the texture map image

Chapter 30

Home to Material —   Up one level to
Setting   Parent Map

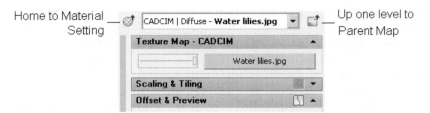

*Figure 30-59 Partial view of the **Materials** window*

Note
*You can control the display of materials and textures from the flyout in the **Materials** panel of the **Render** tab in the **Ribbon**. You can select either the **Materials/Textures Off** or **Materials On/Textures Off**, or **Materials/Textures On** option from the flyout. The **Scaling & Tiling** and **Material Offset & Preview** rollouts will be discussed in the later part of this chapter.*

Choose the **Delete map information from material** button to delete the mapped texture image from the material, refer to Figure 30-58; the **Materials - Delete Nested Maps** warning box is displayed to confirm the deletion of the diffuse map and all the nested maps. By default, the settings for a particular map type are synchronized. For example, if you are changing the setting for the texture map type in the **Diffuse map** area, then the texture map settings, if used in bump, reflection, and opacity map will also change simultaneously. This synchronization is represented by the connected synchronization button symbol displayed on the right of the **Delete map information from material** button, see Figure 30-58. Choose this button to toggle the synchronization between the diffused, bump, reflection, and opacity maps; the **Material - Desynchronize Map Setting** message box is displayed to confirm the desynchronization. The unsynchronized map setting status will be represented by a disconnected faded button displayed at the same location. The **Preview map channel procedural results** button is used to invoke the **Diffuse Map Preview** window for the selected map. The changes made in the map property settings will be displayed dynamically in this window.

If you select the **Marble** option from the drop-down list, the material will be mapped with the image of the marble. To edit the image of the marble structure, choose the **Click for Marble settings** button on the right of the **Map type** drop-down list, see Figure 30-60. The **Marble** rollout will get added in the **Materials** window, see Figure 30-61. Click on the **Stone color** swatch to choose the color of the base stone and the **Vein color** swatch to change the color of the veins displayed on the base stone. Choose the **Swaps the Colors** button to interchange the stone and vein color. The **Vein spacing** and **Vein width** edit boxes are used to control the relative spacing between the veins and the relative size of the vein width, respectively.

Similarly, if you select the other options from the **Map type** drop-down list, the material will be mapped with the selected map type. To edit the attached map type, choose the respective settings button; the respective map settings rollout will get added in the **Materials** window. Using this rollout, you can edit the attached map type according to your requirement. For example, if you select the **Wood** option from the **Map type** drop-down list, the material will

Figure 30-60 *The **Diffuse map** area with the **Marble** option selected from the **Map type** drop-down list*

Figure 30-61 *The **Marble** settings rollout*

be mapped with the image of the wood. To edit the image of the wood structure displayed on the material, choose the **Click for Wood settings** button; the **Wood** rollout will get added in the **Materials** window, see Figure 30-62. Click on the respective color buttons to change the color of the wood base and the pattern. Choose the **Swaps the Colors** button to interchange the wood base and pattern colors. The **Radial noise** and the **Axial noise** options are used to decide the relative occurrence of patches of the wood in radial and axial directions, respectively. The **Grain thickness** option decides the relative size of the grain patches in the material. All the values specified in this edit box are relative to the size of the object on which it will be applied.

Figure 30-62 *The **Wood** settings rollout*

The **Reflection map** area is used to define the environment reflecting qualities of a material. The **Reflection map** area is only available for the **Advanced** and **Advanced Metal** templates. Select the **Reflection map** check box to apply reflection map settings on the object. An image file or some predefined textures can be reflected from the object as if the surrounding of the material is composed of this image file. If you select the **Texture map** option from the **Map type** drop-down list, the **Select Image** button will be displayed below it. Choose this button to select an image file to be reflected from the object. Similarly, you can also select some predefined reflection maps from the **Map type** drop-down list such as **Checker**, **Gradient Ramp**, **Marble**, **Noise**, and so on. After selecting the map type, more buttons will be added to the **Reflection map** area. Use the slider bar to decide the extent of the effect of the texture map on the material. At 0 value, the material will not reflect at all and as this value increases, the reflectivity of the material also increases, and at 100 value the material becomes a perfect mirror. The button next to the slider bar displays the name of the texture file on it. Choose this button to change the attached texture file. The icons next to this button are used to delete, synchronize, and preview the attached texture file on the object.

The **Opacity map** area is used to define an imaginary transparency or opacity effect on the material. Select the **Opacity map** check box to apply reflection map settings on the object. At 0 value, the material appears to be totally transparent and as the value goes on increasing, the object appears to be more and more opaque. The options in the **Opacity map** area are similar to the one discussed above in the **Reflection map** area.

Chapter 30

The **Bump map** area is used to add a realistic bumpy or irregular effect to the surface of the material. Select the **Bump map** check box to apply reflection map settings on the object. The attached map type in this area will appear to be engraved on the surface of the material, with the light colors being projected outward or inward. The dark colors will be left as such. Use the slider bar to adjust the extent of the bump. The value of the bump ranges from -1000 to 1000. For negative values, the brighter colors will appear to be projected inwards and for positive values the brighter colors will appear to be projected outwards.

When the light falls on a colored object in the drawing, it bounces back to the nearby objects, and some effect of the color of the original object is displayed on the nearby objects. This phenomenon is known as indirect illumination. The **Advanced Lighting Override** rollout is used to control the factors affecting the rendering of the objects due to indirect illumination, see Figure 30-63. This option is only available for the **Realistic** and **Realistic Metal** templates. The options in the **Advanced Lighting Override** rollout are discussed next.

Figure 30-63 *The* ***Advanced Lighting Override*** *rollout of the* ***Materials*** *window*

The **Color Bleed Scale** option is used to control the saturation of the colors reflected from the object on which it is falling. The **Indirect Bump Scale** option is used to control the bump mapping in the portions illuminated by the indirect light. The **Reflectance Scale** option is used to control the amount of diffused light energy that the material reflects. The **Transmittance Scale** option is used to control the amount of light energy allowed to pass through the material.

Tip
*AutoCAD stores its bitmap images in C:\Documents and Settings\All Users\Application Data\Autodesk\AutoCAD 2010\R18\enu\Textures folder. These bitmap images are in the *.jpg format.*

Note
You can create a material similar to the original AutoCAD material. This is done by selecting the original material from the ***Available Materials in Drawing*** *rollout and then choosing the* ***Create New Material*** *button.*

The other options in the ***Materials*** *window, the concept of mapping materials on an object, adjusting and modifying the maps will be discussed in later topics of this chapter.*

CONVERTING PRE-AutoCAD 2010 MATERIALS TO AutoCAD 2010 FORMAT

The **3DCONVERSIONMODE** system variable is used to control the settings for the automatic conversion of the materials created in the previous releases of AutoCAD. If the variable is set to 0, no conversion of material will take place. If it is set to a value of 1, the conversion will take place automatically. If the value is set to 2, then the user will be prompted when materials need to be converted and will have the option to convert or not to convert.

You can convert the materials created in the previous releases of AutoCAD manually by the **CONVERTOLDMATERIALS** command. The prompt sequence is given next.

Command: **CONVERTOLDMATERIALS**
Loading materials...done.
4 material(s) created.
7 object(s) updated.

The prompt sequence indicates that the 4 materials applied to seven objects have been converted to the AutoCAD 2010 format. This prompt sequence also indicates the materials which are not converted to AutoCAD 2010 format, if there are any.

MODIFYING MATERIAL

Ribbon:	View > 3D Palettes > Advanced Render Settings	**Command:**	RPREF
Menu Bar:	View > Render > Advanced Render Settings		
Toolbar:	Render > Advanced Render Settings		

You can also modify AutoCAD materials using the **Materials** window. To modify a material, select it from the **Available Materials in Drawing** area in the **Materials** window and then modify the material in the **Material Editor** area. The display of options in the **Material Editor** area depends on the type of material selected to be modified.

UNDERSTANDING ADVANCED RENDERING

As mentioned earlier, rendering allows you to control the appearance of the objects in the design. This is done by defining the surface material and the reflective quality of the surface and by adding lights to get the desired effects. The basic rendering has already been discussed earlier in this chapter. You will now learn about the options of the **Advanced Render Settings** window, shown in Figure 30-64.

Rendering Quality

The drop-down list at the top of the **Advanced Render Settings** window contains five standard qualities of rendering as presets. These are **Draft**, **Low**, **Medium**, **High**, and **Presentation**. The lowest quality of rendering is done in the **Draft** preset. The quality of rendering goes on increasing with the options and it is the best in the **Presentation** preset. However, you can also create your own render presets by selecting the **Manage Render Preset** option from the drop-down list; the **Render Presets Manager** dialog box is displayed, which will be explained in detail later in this chapter.

You can also render the model using the **Advanced Render Settings** window by choosing the **Render** button displayed at its upper right corner. The various rollouts of the **Advanced Render Settings** window are given next.

General Rollout

The **General** rollout of the **Advanced Render Settings** window has the following areas:

Render Context Area

The **Determines if file is written** button on the right of the **Render Context** area is used to specify if you want to save the rendered image as a file. The other options in the **Render Context** area are given next.

Procedure. This field is used to specify the portion of the model to be included in the rendering. When you click on this field, it turns into a drop-down list. The **View** option renders the current view of the model in the present window without opening the render window. The **Crop** option allows you to specify an area of the drawing to be included in the rendering. When you choose the **Render** button, you will be prompted to specify an area on the screen before rendering. The **Crop** option is available only if you select the **Viewport** option in the **Destination** drop-down list. This option can also be invoked by the **RENDERCROP** command. The **Selected** option is used to render only the selected objects.

Figure 30-64 *The Advanced Render Settings window*

Destination. This field is used to specify where to render and show the model. The **Window** option opens a separate **Render** window to render and display the rendered image. The **Viewport** option renders the model in the current viewport only and displays the rendered image in the drawing itself.

Output file name. This edit box is used to specify the file name, file type, and location to save the rendered image. This option will be available only when you choose the **Save File** button on the right of the **Render Context** area.

Output size. This field is used to select the standard output resolution of the rendered image file. However you can also specify your desired value. To do so, click on this field and choose the **Specify Output Size** option; the **Output Size** dialog box will be displayed, see Figure 30-65.

Figure 30-65 *The **Output Size** dialog box*

Exposure Type. This field is used to control the tone settings. The **Automatic** option changes the tone operator according to the current viewport tone operator settings. The **Logarithmic** option is used to apply the log exposure control.

Physical Scale. This used to specify the physical scale.

Materials Area

This area is used to specify how the materials will be included in the rendering. The options of this area are given next.

Apply materials. This option specifies whether to render the model with or without the material attached to it by the user. If you render the model by selecting the **off** option from the **Apply materials** rollout, the global material will be attached to the model while rendering.

Texture filtering. This option is used to specify how the texture maps are being filtered.

Force 2-sided. This option is used to specify whether to render one side or both the sides of a surface. If this option is on, both sides of a face will be rendered.

Sampling Area

The images on the monitor are displayed as discrete pixels and so the slant and curved edges are displayed as distorted lines. Sampling is used to remove this distortion. The higher the value of Minimum Sample and Maximum Sample, the smoother will be the resulting edges in the rendering. But the rendering time increases considerably with the increase in the number of the samples. Figure 30-66 shows a model rendered with Min. sample=1 and Max. sample=16, and Figure 30-67 shows a model with Min. sample=1/64 and Max. sample=1/4. Compare the display of the slant and curved edges in the two figures.

Figure 30-66 Model rendered keeping the Min. sample=1 and Max. sample=16

Figure 30-67 Model rendered keeping the Min. sample=1/64 and Max. sample=1/4

Note

*Some of the options in the **Advanced Render Settings** window such as rendering a cropped region, selecting render presets, specifying the sampling limits, saving the rendered image to a file, and specifying the resolution size of the rendered file can be directly changed from the **Render** panel in the **Render** tab of the **Ribbon**, see Figure 30-68.*

Chapter 30

Figure 30-68 *The* **Render** *panel of the* **Render** *tab of the* **Ribbon**

Shadows Area

This area is used to control the display of shadows in the rendered image. The **Bulb** button on the right of the **Shadows** area is used to specify whether to display shadows in rendering. The other options of the **Shadows** area will be highlighted only if this button is on. The options in this area are given next.

Mode. This field is used to specify the order in which the shadow is generated in the **Rendering** window. There are three modes to specify the order of the shadow generation: **Simple**, **Sorted**, and **Segment**. The **Simple** mode is the default mode of shadow generation.

Shadow map. This field is used to specify whether the shadow in rendering will be cast as a mapped image or as ray traced. If you switch on the option of **Shadow map**, the shadows generated will not be sharp. If you switch off the **Shadow map**, then the shadow will be generated as ray-traced. The shadows generated by this option are more accurate and sharp as compared to shadow map.

Sampling Multiplier. This field is used to specify the multiplying factor to control the overall shadow sampling for rendering.

Ray Tracing Rollout

This area controls the reflection and refraction caused by the model while ray-tracing. The **Bulb** button on the right of the **Ray Tracing** area is used to specify whether ray-tracing will be performed while calculating the reflection and refraction. All the other options of this area will be available only if this button is on. The other options in this area are discussed next.

Max reflections

This option controls the total number of times the ray will be reflected by the model, after which the ray-tracing will stop.

Max refractions

This option controls the total number of times the ray will be refracted by the model, after which the ray-tracing will stop.

Max depth

This option controls the total number of times, a ray reflects and refracts, after which the ray-tracing will stop. This is irrespective of whether the ray has attained the limits set using the **Max reflections** and **Min refractions** options.

Indirect Illumination Rollout

The **Indirect Illumination** rollout of the **Advanced Render Settings** window has the following areas:

Global Illumination Area

This area controls the effect of colors in the indirectly illuminated portions. For example, if a white object has dark blue objects surrounding it, then this white object will appear light blue in color. To calculate such indirect illumination effects, the global illumination uses a photon map. The quality and accuracy of the illumination is controlled by the **Photons/sample** and **Radius** options.

Photons/sample. The brightness of the indirect illumination depends on the number of photons. The larger the number of photons, the brighter will be the indirect illumination. But in this case, the sharpness of the indirect illumination will decrease.

Use radius. If this option is turned on, the photon radius is decided by the value specified in the **Radius** option. If this option is turned off, each photon radius becomes equal to 1/10 of the radius of the full scene.

Radius. Increasing the radius of the photon increases the amount of smoothing. This results into more natural indirect illumination.

Max depth/Max reflections/Max refractions. These options are similar to those explained in the **Ray Tracing** area. Here, these options are applicable for the photons used for indirect illumination.

Final Gather Area

Final gathering is used to improve global illumination. It increases the number of rays used to calculate global illumination. Because global illumination is calculated from a photon map, dark corners in the lighting may occur. By activating the final gathering, you can increase the number of rays used to calculate global illumination, which reduces these defects.

Mode. This field is used to turn the final gathering on, off, or set it to automatic mode so that the final gathering can be turned on, off, or function automatically, depending on the status of the skylight.

Rays. This option is used to specify the number of rays required for indirect illumination.

Radius mode. This option is used to select the radius mode for the final gathering calculation. If the radius mode is turned on, the radius specified in the **Max radius** option in world units is used for the final gathering. If the radius mode is turned off, the maximum radius is the default value of 10 percent of the maximum model radius, in world units. If the **View** option is selected in the **Radius mode** edit box, then the maximum radius will be considered in terms of pixels instead of world units.

Max radius. This option controls the maximum radius within which the final gathering will be calculated. Reducing this value increases the quality of rendering.

Use min. This option controls whether the minimum radius is to be included in the process of the final gathering.

Min radius. This option controls the minimum radius for the final gathering. Increasing this value increases the quality of rendering.

Light Properties Area

This area is used to define the properties of the lights used for indirect illumination. However, the properties defined here affect all the lights present in the drawing. The options of this area are given next.

Photons/light. This option controls the number of photons emitted per light for calculating the global illumination. Increasing this number, increases the quality of the rendered image.

Energy multiplier. This option multiplies the intensity of the indirect illumination with the factor specified in the **Energy multiplier** field.

Diagnostic Rollout

The **Diagnostic** rollout of the **Advanced Render Settings** window has the following areas:

Visual Area

The options in this area are used to control the visibility of visual aids such as grids, photon density, and so on during rendering. The options in this area are given next.

Grid. This option renders the model such that its coordinate spaces are displayed on the model. The coordinates can be displayed in three different ways: **Object**, **World**, and **Camera**. In the **Object** option, the local coordinate spaces of each object are displayed on it. In the **World** option, the world coordinate space is displayed and it remains the same for all the objects. In the **Camera** option, the camera coordinate spaces are displayed on the object. Camera coordinates consist of rectangular coordinates superimposed on the model from the viewing direction, see Figure 30-69.

Grid size. This option is used to specify the size of the grid to be displayed on the rendered model.

Photon. This option displays the distribution of photons on the rendered model. The global illumination should be on to display the photon distribution. The **Density** option displays

Figure 30-69 *Rendered model with camera coordinates*

the distribution of the photon density. The location of the highest photon density will be displayed in red and the location of the minimum photon density will be displayed in light color, see Figure 30-70. The **Irradiance** option displays the photon based on its irradiance.

The maximum irradiance will be displayed in red color and the minimum irradiance will be displayed in light color.

Samples. By default, this option is turned off. Figure 30-71 shows the rendered model with the **Samples** option turned on.

Figure 30-70 *Rendered model displaying the photon density*

Figure 30-71 *Rendered model with the Sample option turned on*

BSP. This option renders a visualization of the parameters defined for the tree. The **Depth** option shows the depth of the tree with top faces in bright red and deep faces in increasingly light colors. The **Size** option shows the size of leaves in the tree with differently sized leaves indicated by different colors.

Processing Rollout

The **Processing** rollout of the **Advanced Render Settings** window has the following options:

Tile size

This option is used to specify the size of the tile that is displayed on the rendering screen while carrying out the rendering processes. The rendering time is inversely proportional to the size of the tile.

Tile order

This option is used to specify the order of appearance of the next tile on the render screen. Choose the tile order depending on the orders in which you want the rendered image to appear on the screen. There are six orders in which the next tile appears on the screen: **Hilbert**, **Spiral**, **Left to right**, **Right to left**, **Top to bottom**, and **Bottom to top**. **Hilbert** is the default tile order.

Memory limit

This option controls the memory allocated to the rendering processes.

Managing Render Presets

The **Render Presets Manager** dialog box, shown in Figure 30-72, is used to save any changes made to the render settings, such that they can be used later on as render presets. You can also select any standard preset render setting to get a preview of the settings and make them the current settings, to be used for the rendering. But you cannot change the settings of the standard render preset from this dialog box. To open this dialog box, select the **Manage**

Figure 30-72 The **Render Presets Manager** dialog box

Render Presets option from the Render Presets drop-down list at the top of the Advanced Render Settings window. You can also invoke this dialog box using the RENDERPRESETS command. The options available in the Render Presets Manager dialog box are given next.

Render Presets List

This list displays all render presets saved in the current drawing and the render setting that is set as the current for rendering. The render presets are displayed in two categories, Standard and Custom. The Standard Render Presets displays the standard render presets provided in AutoCAD. The Custom Render Presets displays the user-defined render presets.

Properties Panel

This panel displays the settings of the render properties. You can also edit these properties for the custom presets, but the standard presets cannot be edited. The properties displayed here are similar to those explained in Advanced Render Settings.

Set Current

This button is used to set current the selected render preset so that the render uses this preset.

Create Copy

This button is used to create a copy of the selected render preset or a new render preset, based on the selected render preset. Choose this button; the Copy Render Preset dialog box will be displayed on the screen, see Figure 30-73. Enter a name and description about the

Figure 30-73 The Copy Render Preset dialog box

render preset you are creating and choose the OK button. Now, change the render properties of the newly created render preset in the Properties panel.

Delete

This button is used to delete the selected custom render preset. You cannot delete the standard render preset.

Thumbnail Viewer

This area displays the preview of the selected render settings if an image file with these render settings is selected in the Thumbnail Image option of the Preset Info area of the Properties panel.

CONTROLLING THE RENDERING ENVIRONMENT

You can enhance the rendering by adding effects such as the surrounding environment of the model or by adding colors or images to the background of the model. The commands to do so are discussed next.

Chapter 30

Rendering with a Background

You can render a model by changing its background. You can set a single color, combination of two colors, combination of three colors, an image, or sun and sky as the background of the model. To render the model with a background, you need to save a named view of the model with a background by using the **VIEW** command, as explained in the **View Manager** dialog box (refer to Chapter-6). Next, set the named view with a background as the current view and then render the model with the desired render settings. Figure 30-74 shows a rendered model with its named view saved with a

Figure 30-74 The rendered model with a gradient background

gradient background, Figure 30-75 shows a rendered model with its named view saved with an image as a background, and Figure 30-76 shows a rendered model with its named view saved with the sun and sky as a background.

Figure 30-75 The rendered model with Image as the background

Figure 30-76 The rendered with the Sun & Sky as the background

Rendering with Fog

Ribbon:	Render > Render > Environment	**Toolbar:** Render > Render Environment
Menu Bar:	View > Render > Render Environment	**Command:** FOG

You can use this option to add a misty effect to an object when it is rendered. You can also assign a color to the fog. When you choose **Render > Render > Environment** from the **Ribbon**, the **Render Environment** dialog box will be displayed, as shown in Figure 30-77. This dialog box can also be invoked using the **FOG** command.

Click on the **Enable Fog** field and select **On** from the drop-down list to enable the fog. Enabling the fog does not affect the other settings of the rendering. Click on the color edit box to specify the color of the fog using various options specified in this drop-down list. Click on the **Fog Background** field and select **On** from the drop-down list to enable the **Fog and**

Background options at the same time. The **Near Distance** and **Far Distance** options define where the fog begins and ends. The values specified in the **Near Distance** and **Far Distance** edit boxes are the percentages of the distances between the camera and the back clipping plane. The **Near Fog Percentage** and **Far Fog Percentage** edit boxes describe the percentage of the fog at the near and far distances. Figure 30-78 shows a rendering displaying the fog effect.

Figure 30-77 The **Render Environment** *dialog box*   *Figure 30-78* A design rendered with the fog

Adjusting the Lighting Exposure to Rendered Image

Ribbon: Render > Render > Adjust Exposure
Command: RENDEREXPOSURE

 You can control the global lighting effect of a model dynamically even after it has been rendered. This saves a lot of time needed to render the drawing each time when you make some change in the render settings. The settings changed at this stage will only affect the most recent rendered output file. To change the settings after rendering, choose the **Adjust Exposure** button from the **Render** panel in the **Render** tab of the **Ribbon**; the **Adjust Rendered Exposure** dialog box will be displayed, with the thumbnail preview of the most recent rendered file, see Figure 30-79. You can use this dialog box to control the brightness, contrast, mid tones, and the exposure of exterior objects illuminated by the sun. You can also specify whether to include the background for exposure control or not.

Figure 30-79 The **Adjust Rendered Exposure** *dialog box*

Chapter 30

MAPPING MATERIALS ON OBJECTS

Ribbon: Render > Materials > Material Mapping **Command:** MATERIALMAP or SETUV
Menu Bar: View > Render > Mapping **Toolbar:** Mapping or Render > Mapping

In AutoCAD, mapping is defined as the method of adjusting the coordinates and the bitmap of the pattern of the material attached to the solid model. Mapping is required when the pattern of the material is not properly displayed on the object after rendering. Therefore, to display the material properly, you need to adjust the coordinates and the bitmap of the material.

Figure 30-80 The Mapping flyout

To perform mapping on a model, choose **Render > Materials > Material Mapping** from the **Ribbon**; a flyout with different types of mapping will be displayed, see Figure 30-80. The options in the mapping flyout are also available in the form of buttons in the **Mapping** toolbar. Choose the mapping type that you want to use on the object. The options for the mapping type are: Planar, Box, Spherical, and Cylindrical. A material mapper grip tool will be displayed on the object. Use the grip tools on the object to adjust the display of the image on the model or face of the model. You can move or rotate the mapping about an axis by selecting the axis and right-clicking. Figure 30-81 shows a model with the **Planar** mapping and Figure 30-82 shows a model with the **Cylindrical** mapping.

*Figure 30-81 Model with the **Planar** mapping*

*Figure 30-82 Model with the **Cylindrical** mapping*

The prompt sequence for the **MATERIALMAP** command is given next.

> Command: **MATERIALMAP**
> Select an option [Box/Planar/Spherical/Cylindrical/copY mapping to/Reset mapping]<Box>: *Specify an option for the material mapping or press* Enter.

You can invoke the following options depending on your requirements.

Box

This option maps the image on a box in a fashion that the image is repeated on each side of the object. The prompt sequence for this option is given next.

Select an option [Box/Planar/Spherical/Cylindrical/copY mapping to/Reset mapping]<Box>: **B** Enter
Select faces or objects: *Select a face or object on which the image will be mapped.*
Select faces or objects: *Select more faces or objects on which the image will be mapped or press the* Enter *key.*
Accept the mapping or [Move/Rotate/reseT/sWitch mapping mode]: *Accept the mapping as it is or select an option to modify it.*

The sub-options that you can select at this prompt are given next.

Move

Selecting this sub-option displays the move grip tool on the object. You can move and adjust the mapped image along any selected axis handle.

Rotate

Selecting this sub-option displays the rotate grip tool on the object. You can rotate and adjust the mapped image along any selected axis handle.

reseT

Selecting this sub-option resets all the adjustments made in the mapping coordinate to the default one.

sWitch mapping mode

Selecting this sub-option redisplays the main prompt sequence on the screen to change the type of mapping.

Planar

This option maps the image on the object as it is being projected on that object. The projected image will be scaled automatically according to the size of the object. The prompt sequence for this option is given next.

Select an option [Box/Planar/Spherical/Cylindrical/copY mapping to/Reset mapping]<Box>: **P** Enter
Select faces or objects: *Select a face or object on which the image will be mapped.*
Select faces or objects: *Select more faces or objects on which the image will be mapped or press the* Enter *key.*
Accept the mapping or [Move/Rotate/reseT/sWitch mapping mode]: *Accept the mapping as it is or select an option to modify it.*

The sub-options available at this prompt are similar to the ones discussed in the **Box** option.

Spherical

This option maps the image by warping the image horizontally and vertically. The top edge and the bottom edge are compressed to a point at the north and south pole of the object, respectively. The prompt sequence for this option is similar to that discussed in the **Box** option.

Cylindrical

This option is used to map the image by wrapping the image on a cylinder. The height of the image is scaled along the cylinder axis. The prompt sequence for this option is similar to that discussed in the **Box** option.

copY mapping to

You can also acquire the mapping from an existing object and copy it on the selected object. The prompt sequence for this option is given next

> Select an option [Box/Planar/Spherical/Cylindrical/copY mapping to/Reset mapping]<Box>: **Y** Enter
> Select faces or objects: *Select the source object to copy the mapping information.*
> Select the objects to copy the mapping to: *Select the object to apply the copied mapping information.*
> Select faces or objects: *Select more objects or press* Enter

You can also perform this operation by choosing the **Copy Mapping Coordinates** button from the expanded options of the **Materials** panel in the **Render** tab of the **Ribbon**.

Reset mapping

Selecting this option resets all the adjustments made in the mapping coordinates of the selected object to the default one produced by AutoCAD.

You can also perform this operation by choosing the **Reset Mapping Coordinates** button from the expanded options **of the Materials** panel in the **Render** tab of the **Ribbon**.

Adjusting Bitmap Placement on a Model

You can control the scale, tiling, and offset properties of a mapped image through the **Material Scaling & Tiling** rollout and **Material Offset & Preview** rollout of the **Materials** window.

The **Material Scaling & Tiling** rollout is available only for the **Texture Map**, **Checker**, **Gradient Ramp**, and **Tiles** map types, see Figure 30-83. In the **Scale**

Figure 30-83 The Material Scaling & Tiling rollout of the Materials window

units drop-down list, you can specify the units to be used in scaling. Select the **None** option to use a fixed scale. Select the **Fit to Gizmo** option to fit the image to the face or object. You can also specify the method to be followed to create a pattern of the image. The options available in the **U Tile** and **V Tile** are **Tile**, **Mirror**, and **None**. You can also set the **U Tile** and **V Tile** value that controls the coordinates of the material on a swatch in the **Material Scaling & Tiling** rollout.

The **Material Offset & Preview** rollout allows you to change the scale factor of the bitmap image of the material, see Figure 30-84. You can also specify the offset values for the bitmap image of the material using this dialog box.

Figures 30-85 and 30-86 show the models rendered with different U and V scale factors.

*Figure 30-84 The **Material Offset & Preview** rollout of the **Materials** window*

Figure 30-85 Model rendered with U Tile = 8, V Tile = 2

Figure 30-86 Model rendered with U Tile = 4, V Tile = 1

RENDERING A MODEL WITH DIFFERENT RENDER SETTINGS

The rendering depends on the view that is current and the lights that are defined in the drawing. Sometimes, the current view or the lighting setup may not be enough to show all features of an object. You might need different views with a certain light configuration to

Chapter 30

show different features of the object. But when you change the view or define the lights for a rendering, the previous setup is lost. To bypass this, you can use the rendering information in the history pane of the **Render** window. In the lower portion of the **Render** window, the history pane is located in which you can get the recent history of rendered images of the model in the drawing, see Figure 30-87. From this history pane, you can regain the previous rendering settings and can also save the rendered images. The data stored in the history pane includes: file name of the rendering, image size, view name (if no named view is used, the view is stored as current view), render time, and the name of the render preset used for the rendering.

Output File Name	Output Size	View	Render time	Render preset
c26exm4-Temp033	1024 x 768	Current	00:03:02	Presentation
c26exm4-Temp032	320 x 240	Current	00:00:28	High
c26exm4-Temp031	640 x 480	Current	00:00:12	Medium
c26exm4-Temp029	640 x 480	Current	00:00:44	Medium

*Figure 30-87   The history pane in the **Render** window*

Right-clicking on a history entry displays a shortcut menu that contains the following options:

Render Again

This option restarts the renderer for the selected history entry with the same settings.

Save

This option displays the **Render Output File** dialog box where you can specify the name and format of the image file. Next, choose the **Save** button; the **BMP Image Options** dialog box will be displayed on the screen. You can choose the color quality of the image file, see Figure 30-88. Select the desired radio button and then choose the **OK** button to save the image file.

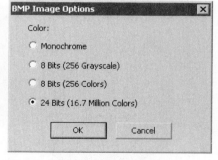

*Figure 30-88   The **BMP Image Options** dialog box*

Save Copy

This option saves the image to a new location without affecting the location stored in the entry.

Make Render Settings Current

This option loads the render settings associated with the selected history entry, if multiple history entries are present with different render presets.

Remove From The List

This option removes the entry from the history pane.

Delete Output File

This option removes the rendered image from the **Image** pane, but the entry remains in the history pane.

Note
Any history entry that is not saved with the drawing only exists for the duration of the current drawing session. Once you close the file or exit the program, the unsaved history entries are lost.

Example 4

In this example, you will open the model of Example 2 of this chapter and then create two rendering settings. Then, you will render the model with these settings and save the rendered image in a file. The first setting should display the effect of light 1 and the second setting should display the effect of light 2.

1. Open the drawing of Example 2.

2. Choose **Render > Lights > Light glyph Display** in the **Ribbon**.

3. Select light 2 and choose the **Properties** option from the shortcut menu; the **Properties** window is displayed. In this window, select the **Off** option from the **On/Off Status** drop-down list in the **General** rollout.

4. Choose the **Render** button from the **Render** panel in the **Render** tab of the **Ribbon** to perform rendering. The rendered model is displayed in the **Render** window with the effect of light 1 only, as shown in Figure 30-89.

5. Next, switch off light 1 and switch on light 2, as explained in the previous steps and render the model. The rendered model is displayed in the **Render** window with the effect of light 2 only, as shown in Figure 30-90.

6. The history pane of the **Render** window displays both the render entries. Select one entry and choose the **Save** option from the shortcut menu. Now, specify the location, name, and format in which you want to save the image file in the **Render Output File** dialog box and choose the **Save** button.

7. Specify the Color Mode and choose the **OK** button in the **BMP Image Options** dialog box.

8. Similarly, save the second history entry, as explained in Step 6.

Note
*In future, to render the model with the same rendering settings, select the history entry with that setting and choose the **Make Render Settings Current** option from the shortcut menu.*

Figure 30-89 Rendered model with the effect of light 1

Figure 30-90 Rendered model with the effect of light 2

OBTAINING RENDERING INFORMATION

The right side area of the **Render** window is known as the statistics pane. This portion displays the details about the rendering and render settings in effect when the image was created. The information stated in the statistics pane is derived from the settings made in the **Render Presets Manager** dialog box and the information that is generated at the time of the rendering.

SAVING A RENDERING

A rendered image can be saved by rendering to a file or by rendering to the screen and then saving it. Redisplaying a saved rendering image requires very less time compared to the time involved in rendering. Various methods of saving the rendered image are discussed next.

Saving the Rendered Image to a File

You can save a rendered image directly to a file. One of the advantages of saving a rendered image to a file is that you can redisplay the rendered image in less time as compared to rendering the image again. Another advantage of saving a rendered image is that when you render a file in the viewport, the resolution of the image on rendering depends on the resolution of your current display. Whereas, if you render an image to a file, you can specify a higher resolution. Later, you can display this rendered image that has a higher resolution. The rendered images can be saved in different formats, such as BMP, PCX, TGA, TIF, JPEG, and PNG. The following steps explain the procedure of saving a rendering to a file.

1. Open the **Advanced Render Settings** window. Choose the **Determines if file is written** button on the right of the **Render Context** area.

2. Click on the **Output file name** field and choose the browse button; the **Render Output File** dialog box is displayed.

3. Specify the file name and file type. Then, choose the **Save** button.

4. Specify the Color Mode and choose **OK** in the **BMP Image Options** dialog box.

5. Choose the **Render** button at the top of the **Advanced Render Settings** window. The rendered image is saved at the specified location.

Saving the Viewport Rendering

In the **Advanced Render Settings** window, if you have selected the **Viewport** option in the **Destination** field, then the model will be rendered in the viewport. A rendered image in the viewport can be saved by using the **SAVEIMG** command. This command can be invoked by choosing **Tools > Display Image > Save** from the menu bar. The **Render Output File** dialog box will be displayed when you invoke this command. This dialog box is used to specify the file name and type. The valid output file formats are BMP, PCX, TGA, TIF, JPEG and PNG.

The steps given next explain the procedure of saving a viewport rendering.

1. Choose **Tools > Display Image > Save** from the menu bar to invoke the **Render Output File** dialog box.

2. Specify the file name and file type. Then, choose the **Save** button.

3. Specify the Color Mode and choose the **OK** button in the **BMP Image Options** dialog box.

Chapter 30

4.   Choose the **Render** button from the **Render** panel in the **Render** tab of the **Ribbon**. The rendered image is saved at the specified location.

Saving the Rendered Image from Render Window

A rendered image in the **Render** window can be saved using the **Save** option from the **File** menu of the **Render** window. In this case, the rendered image can be saved in any of these formats: BMP, PCX, TGA, TIF, JPEG, and PNG. The following steps explain the procedure for saving a render-window image:

1.   Choose the **Render** button from the **Render** > **Render**; the rendered image will be displayed in the **Render** window.

2.   Choose the **Save** option from the shortcut menu available by right-clicking on the temporary image. Alternatively, use the **File** menu of the **Render** window to save the rendered image; the **Render Output File** dialog box is displayed.

3.   Specify the file name and file type. Then, choose the **Save** button.

4.   Specify the Color Mode and choose the **OK** button in the **BMP Image Options** dialog box.

PLOTTING RENDERED IMAGES

Plotting the rendered design was one of the biggest problems in the earlier releases of AutoCAD. You had to go through a number of steps to plot a rendered drawing. However, from AutoCAD 2007, plotting of rendered designs became extremely easy. To plot a rendered design, use the following steps:

1.   Open the design to be plotted and then invoke the **Plot** dialog box. Remember that you do not need to first render the design and then invoke the **Plot** dialog box. You can also render the design while plotting. However, the render settings should be specified in advance.

2.   Select the printer that supports rendered printing from the **Printer/plotter** area and select the other options.

3.   Select the **Rendered** option from the **Shade plot** drop-down list in the **Shaded viewport options** area. If this area is not available by default, then choose **More Options** from the bottom right corner of the **Plot** dialog box.

4.   Set the other options in this dialog box and then plot the design. The design will be first rendered and then plotted in the rendered form.

   Note
*If the **Rendered** option is used for a highly complex set of objects, the hardcopy output might contain only the viewport border.*

UNLOADING AutoCAD RENDER

When you select any AutoCAD **RENDER** command, AutoCAD Render is loaded automatically. If you do not need AutoCAD Render, you can unload it by entering the **ARX** at the Command prompt. You can reload AutoCAD Render by invoking the **RENDER** command or any other command associated with rendering (such as **FOG**, **LIGHT**, and so on).

> Command: **ARX**
> Enter an option [?/Load/Unload/Commands/Options]: **U**
> Enter ARX/DBX file name to unload: **ACRENDER**
> ACRENDER successfully unloaded.

CREATING ANIMATIONS

One of the latest enhancements in AutoCAD is the motion path animation. You can create a walk through or fly through animation. You can record and play back these presentations, whenever required. This helps you communicate your design intent dynamically and with more impact. The tools to create an animation are not displayed in the **Ribbon** by default. To display them, choose the **Render** tab and right-click on any one of the buttons; a shortcut menu will be displayed. Choose **Panels > Animations** from the shortcut menu; the **Animations** panel will be added to the **Ribbon**. The methods of creating animations are discussed next.

Creating Animation of 3D Navigations

You can create, record, playback, and save animations of any 3D navigation modes, such as **Walk** and **Fly** navigation modes. Before creating animations, you should specify the animation settings. To do so, choose the **Walk** or **Fly** button and right-click. Choose **Animation Settings** from the shortcut menu; the **Animation Settings** dialog box will be displayed, see Figure 30-91. In this dialog box, you can specify the visual style in which you want to view the model in animation, resolution of the frames, number of frames to be captured in one second, and the format (AVI, MPG, or WMV) in which you want to save the file while creating the animation of the 3D navigation.

*Figure 30-91 The **Animation Settings** dialog box*

Recording Animation

To create an animation, start any 3D navigation command. The **Animation Record** button in the **Animations** panel in the **Tools** tab of the **Ribbon** will be enabled. Choose this button to start recording, see Figure 30-92. Navigate through the drawing model and all navigations will get recorded. You can also change the navigation mode while recording. To do so, choose the **Animation Pause** button from the **Animations** panel of the **Ribbon**. Next, right-click in the drawing area and choose an option from the **Other Navigation Modes**. Choose the

*Figure 30-92 The **Animations** panel in the **Render** tab of the **Ribbon***

Animation Record button again and the recording will start with the changed 3D navigation mode without any break in recording. If you are in the **Walk** or **Fly** navigation mode, you can also adjust the settings from the **POSITION LOCATOR** window, in the same way as explained above.

Playing Animation

To preview a recorded animation, choose the **Animation Play** button from the **Animations** panel in the **Render** tab of the **Ribbon**; the recording will pause and the **Animation Preview** window will be displayed, see Figure 30-93. The **Animation Preview** window, displays the animation. If you want to preview the animation in a different visual style, you can select another style from the drop-down list. The options in the visual style drop-down list are **Current**, **3D Hidden**, **3D Wireframe**, **Conceptual**, and **Realistic**. After playing the animation, you can also resume the recording by choosing the **Animation Record** button displayed in red color in the **Animation Preview** window.

*Figure 30-93 The **Animation Preview** window*

Saving Animation

If you are satisfied with the animation playback, choose the **Save** button in the **Animation Preview** dialog box or the **Animations** panel of the **Ribbon**; the **Save As** dialog box will be displayed on the screen. Select the file location, file name, type of file and choose the **Save** button. In **Files of type** drop-down list of the **Save As** dialog box, you will get only the file type selected in the **Animation Settings** dialog box. To change the file type, choose the **Animation Settings** button from the **Save As** dialog box; the **Animation Settings** dialog box will be displayed. Select the desired file type from the **Format** drop-down list of the **Animation Settings** dialog box.

Creating Animation by Defining the Path of Camera Movement

Ribbon: Render > Animations > Animation Motion Path	**Command:** ANIPATH
Menu Bar: View > Motion Path Animations	

In this method of creating an animation, you can control the camera motion by linking the camera movement and its target to a point or a path. If you want the camera or target to remain still while creating animation, link the camera or target to a point. However, if you want the camera or target to move along a path while creating animation, link the camera or target to a path. To create a motion path animation, choose **Render > Animations > Animation Motion Path** in the **Ribbon**; the **Motion Path Animation** dialog box will be displayed on the screen, see Figure 30-94. Using this dialog box, specify the settings for the camera motion path and create the animation file. The options of **Motion Path Animation** dialog box are discussed next.

Figure 30-94 The Motion Path Animation dialog box

Camera Area

The options in this area are used to specify the camera movement. Select the **Point** or **Path** radio button, based on your requirement. Select the **Point** radio button, if you want the camera to remain stationary at the specified point in the drawing throughout the animation. Select the **Path** radio button, if you want the camera to move along the specified path in the drawing. Next, choose the **Pick Point** / **Select Path** button available next to the radio buttons to select the point/path in the drawing. Specify the point/path in the drawing; the **Point Name** or **Path Name** dialog box will be displayed, depending on whether the **Point** or **Path** radio button is selected as the camera link, respectively, see Figure 30-95. Specify a name for the

Chapter 30

point/path and choose the **OK** button. You can also select a previously created point/path from the drop-down list that displays a list of named points or paths.

Figure 30-95 shows the Path Name dialog box with Name field containing "Path1".

Target Area
The options in this area are used to specify the movement of the target with the movement of

Figure 30-95 *The **Path Name** dialog box*

the camera. Select the **Point** or **Path** radio button, based on your requirement. To specify a target point, choose the **Pick Point** button option and specify a point in the drawing. Enter a name for the point and choose the **OK** button in the **Point Name** dialog box. To specify a target path, click the **Select Path** button and specify a path in the drawing. Enter a name for the path and choose the **OK** button in the **Path Name** dialog box. To specify an existing target point or path, select it from the drop-down list.

Note
You cannot link both the camera and the target to a point.

To create a path, you can link a camera to a line, arc, elliptical arc, circle, polyline, 3D polyline, or spline.

Tip
You must create the path object before you create the motion path animation. The path that you create will not be visible in the animation.

Animation settings Area
The options in this area are used to control the output quality and display of the animation file. The options in this area are discussed next.

Frame rate (FPS). The frames per second is the speed at which the animation will be played. The value of the FPS can vary from 1 to 60. The default value for the FPS is 30.

Number of frames. This option is used to specify the total number of frames to be created in the animation. This value divided by the FPS determines the duration of the animation file.

Duration (seconds). This option specifies the duration of the animation in seconds. Any change in the duration automatically resets the **Number of frames** value.

Visual style. This option displays a drop-down list of the visual styles and the render presets that can be applied to an animation file.

Note
You cannot create an animation of 3D navigation with a rendered model, but creating an animation by motion path is also possible with a rendered model.

Format. This drop-down list is used to specify the file format for the animation. You can save an animation to an **AVI**, **MOV**, **MPG**, and **WMV** file format.

Note

*The **MOV** format is available only if QuickTime Player is installed. The **WMV** format is available only if Microsoft Windows Media Player 9 or later is installed. Otherwise, the **AVI** format is the default selection.*

Resolution. This drop-down list is used to specify the resolution of the resulting animation in terms of the width and height. The default value is 320 x 240.

Corner deceleration. This check box is used to specify the speed of the camera movement at the corners. If this check box is selected, the camera will move at a slower rate as it turns at a corner.

Reverse. This check box, if selected, will reverse the direction of the animation display.

When previewing show camera preview. This check box, if selected, will display the **Animation Preview** dialog box when you choose the **Preview** button so that you can preview the animation before you save it.

Preview. This button is used to get a preview of the animation in the **Animation Preview** dialog box, before saving it.

After you have finished adjusting the points, paths, and settings, choose **Preview** to view the animation, Next, choose the **OK** button to save the animation. The **Save As** dialog box will be displayed. Specify the location, file name, and file type of the animation to save and choose the **OK** button to save the animation.

Example 5

In this example, you will create and animate the model shown in Figure 30-96 and save this animation to a file.

1. Create the model shown in Figure 30-96.

2. Select the materials of your choice and attach them to the model, as described in Example 1. The rendered image of the model is shown in Figure 30-97.

3. Change the view of the model to the top view and create the path to attach the camera, see Figure 30-98.

4. Move the path created in the previous step along the Z axis by a distance equal to half the height of the model.

5. Invoke the **Motion Path Animation** dialog box(Figure 30-99) from the **Animations** panel in the **Render** tab of the **Ribbon**.

Figure 30-96 *Model for Example 5*

Figure 30-97 *Rendered model for Example 5*   *Figure 30-98* *Top view of the model displaying the path to attach the camera*

6. Select the **Path** radio button and choose the **Select Path** button in the **Camera** area of the **Motion Path Animation** dialog box. Next, select the path created in the previous step; the **Path Name** dialog box is displayed. Enter the name of the path as **Path1**.

7. Select the **Point** radio button and choose the **Select Point** button in the **Target** area of the **Motion Path Animation** dialog box. Next, specify a point in the middle of the model, which will act as the target of the camera; the **Point Name** dialog box is displayed. Enter the name as **Point1**.

8. Change the values of the **Animation settings** area to the values shown in Figure 30-99.

Figure 30-99 *The **Motion Path Animation** dialog box for Example 5*

9. Choose the **Preview** button to take a preview of the animation in the **Animation Preview** dialog box. Next, close the **Animation Preview** dialog box.

10. Choose the **OK** button to save the animation; the **Save As** dialog box is displayed. Specify the location, file name, and file type of the animation to save and then choose the **Save** button from the **Save As** dialog box.

Self-Evaluation Test

Answer the following questions and then compare them to those given at the end of this chapter:

1. A rendered image makes it easier to visualize the shape and size of a 3D object. (T/F)

2. You can unload AutoCAD Render by invoking the **ARX** command. (T/F)

3. The falloff occurs with a distant source of light. (T/F)

4. You can change the color of the ambient light. (T/F)

5. You cannot take the printout of a model without rendering it. (T/F)

6. The _____ command is used to attach material to a layer in a drawing.

7. The _____ source emits a focused beam of light in the defined direction.

Chapter 30

8. The _____ and _____ commands are used to adjust the coordinates and the bitmap of the pattern of the material attached to the solid model.

9. The _____ and _____ commands are used to apply a fog effect to the rendering.

10. A _____ light source emits light in all directions, and the intensity of the emitted light is uniform.

Review Questions

Answer the following questions:

1. You can also import materials to the current drawing directly using the **RMAT** command. (T/F)

2. You can assign materials to the object using the AutoCAD Color Index. (T/F)

3. The phenomenon of attenuation occurs in case of direction lights. (T/F)

4. You can link both the camera and the target to a point. (T/F)

5. Which light does not have a source, and therefore, no location or direction?

    (a) **Default**   (b) **Point**
    (c) **Spot**   (d) **Distant**

6. The intensity of light decreases as the distance increases. This phenomenon is called:

    (a) **Attenuation**   (b) **Frequency**
    (c) **Light Effect**   (d) **None**

7. Which of the following lights allows you to define the geographic location?

    (a) **Sun**   (b) **Point**
    (c) **Spot**   (d) **Distant**

8. Which of the following commands can be used to define the background of the object while it is being rendered?

    (a) **FOG**   (b) **ARX**
    (c) **BACKGROUND**   (d) **VIEW**

9. Which of the following commands is used to create a motion path animation?

    (a) **ANIMAP**   (b) **ANITRAC**
    (c) **ANIPATH**   (d) **ANIMATION**

10. In the _____ source the brightness of the incident light on an object is inversely proportional to the distance of the object from the light source.

11. The _____ command allows you to set the rendering preferences for the rendering.

12. A spotlight source emits a _____ beam of light in one direction only.

13. Redisplaying a saved rendered image takes _____ time as compared to the time involved in rendering.

14. The _____ command is used to render only the selected portion of a drawing.

15. Attenuation is defined as _____.

Exercises

Exercise 1   *General*

Create the 3D drawings, as shown in Figure 30-100. Next, render the drawing after applying materials and inserting lights at appropriate locations to get a realistic model. Assume the missing dimensions.

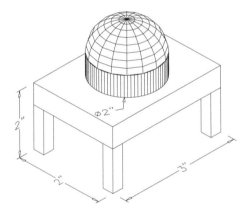

Figure 30-100 *Drawing for Exercise 1*

Exercise 2   *Architectural*

Create the model (Figure 30-101) and apply materials of your choice. Apply the Point light and the Spotlight at the sill of the window and at bottom of the door. Also, attach a background image to it and then render the model. Assume the missing dimensions.

Chapter 30

Figure 30-101 *Model for Exercise 2*

Exercise 3   *General*

In this exercise, you will display the effect of a fog on the model shown in Figure 30-102.
Choose your own materials.

Figure 30-102 *Model for Exercise 3*

Tip
*The display of fog is based on the trial and error method. Use different combinations of
options available in the **Fog** dialog box. Use white color as the background to clearly
visualize the fog.*

Exercise 4 *General*

Render the Tool Organizer of Chapter 28. Assign a dark red wood material to the peg board and white oak to the remaining members, Figure 30-103.

Figure 30-103 *Model for Exercise 4*

Exercise 5 *General*

Render the Work Bench of Chapter 28. Assign white oak to all members. Adjust the mapping coordinates so that the rendered image is seen as shown in Figure 30-104.

Figure 30-104 *Model for Exercise 5*

Exercise 6   *Mechanical*

Create the solid model shown in Figure 30-105. Also, assign brass material to the solid and render it.

Figure 30-105 *Views and dimensions of the model for Exercise 6*

Answers to Self-Evaluation Test
1. T, 2. T, 3. F, 4. F, 5. F, 6. MATERIALATTACH, 7. spotlight, 8. MATERIALMAP, SETUV,
9. RENDERENVIRONMENT, FOG, 10. distant

Index

Symbols

2D Navigation Wheel 25-43
2D Wireframe 26-20
3D Coordinate Systems 25-15
3D Hidden 26-21
3D Wireframe 26-20
3DARRAY command 27-15
3DFACE command 25-25
3DFLY command 25-49
3DMESH command 25-37
3DMOVE command 27-14
3DORBIT command 25-45
3DROTATE command 27-7
3DWALK command 25-47

A

Absolute Coordinate System 2-7, 25-15
Add Parameter Properties dialog box 16-59
Add-A-Plot Style Table Wizard 14-20
Add-A-Plotter Wizard 14-11
Adding a Plot Style 14-20
Adding a Subset to a Sheet Set 1-44
Adding an Xref Drawing 18-10
Adding Blocks in Tool Palettes 16-40
Adding Crease to Meshes 29-15
Adding Equations 7-45
Adding Geometric Tolerance 8-48
Adding leaders to the existing Multileader 8-41
Adding Lights to the Design 30-10
Adding Parameter and Action
    Simultaneously 16-36
Adding Plotters 14-11
Adding Sheets to a Sheet Set or a Subset 1-45
Adding Smoothness to Mesh 29-10
Adding Xref Dependent Named Objects 18-20
Additional Help Resources 1-52
Additional Resources 1-51

Adjusting Bitmap Placement on a Model 30-50
Adjusting the Lighting Exposure to Rendered
    Image 30-47
Adjusting the Raster Images 21-12
Advanced Render Settings Window 30-39
Advanced Setup dialog box 13-5
Advanced Setup wizard 13-3
Aligned Dimensions 8-14
Aligning Multileaders 8-42
Aligning Solid Models 27-16
Allow/Disallow Dynamic UCS 1-8
Alternate Dimension Units 10-23
Alternate Units 8-5
Angular Dimensions 8-20
Animation Preview dialog box 30-59
Animation Settings dialog box 25-51, 30-56
ANIPATH command 30-59
Annotation Scale 1-9, 7-2
Annotative Dimensions 8-8, 8-56
Annotative Hatch 15-23
Annotative Multileaders 8-56
Annotative Objects 7-2
Annotative Tolerances 8-56
Application Menu 1-13
Application Status Bar 1-6
Applying Plot Styles 14-26
ARC command 3-2
Arc Length Dimensions 8-15
Archiving a Sheet Set 1-45
AREA command 6-24
Ariel View 6-47
Arraying Sketched Objects 5-33
Arrowheads 8-3
Assembly Drawing 22-38
Assigning Annotative Property and Annotation
    Scale 7-2
Assigning Color to Layer 4-9
Assigning Linetype to Layer 4-8
Assigning Lineweight to Layer 4-10

Assigning Multiple Annotation Scales to
    Annotative 7-4
Assigning Plot Style to Layer 4-10
Associative Dimensions 8-6
Attach PDF Underlay dialog box 18-16
Attaching an Xref 18-6
Attaching Material by Layers 30-7
Attaching Raster Images 21-7
Attaching Selected Materials to Objects 30-4
ATTDEF command 17-2
ATTDISP command 17-38
ATTEXT Command for Attribute
    Extraction 17-32
Attribute Definition dialog box 17-3
Attribute Visibility 17-38
AutoCAD dialog boxes 1-16
AutoCAD Screen Components 1-5
Automatic Timed Save and Back up files 1-29
Automatically Add Scale 1-9
AutoSnap 4-35
AutoStack Properties dialog box 7-25
AutoTracking 4-58
Auxiliary Views 22-33
Average 7-44

B

Background Mask 7-14
Backgrounds and Shadows 26-27
Baseline Dimensions 8-17
BATTMAN command 17-11
Bill of Materials 22-38
Bind Xrefs dialog box 18-10
Block Attribute Manager dialog box 17-13
Block Definition dialog box 16-5
Block Editor 16-18
Block Editor toolbar 16-18
Block Properties Table 16-59
Blocks 16-3
BMP Image Options dialog box 30-53
BOUNDARY command 15-33
BOX 5-6
Breaking Sketched Objects 5-40
Broken Section 22-25

C

CAD Standards 1-11
Calculating the Mass Properties of the Solid
    Model 28-26
Cell 7-44
Cell Border Properties dialog box 7-42
Cell Styles 7-46
Center 25-42
Center Marks and Centerlines 8-26
CHAMFER command 27-3
Chamfering Sketches 5-21
Chamfering Solid Models 27-3
Changing Linetype Scale Using the
    PROPERTIES Command 4-28
Changing Properties of an Object 20-8
Changing Properties Using Grips 6-14
Changing Properties Using the Properties Palette
    6-14
Changing the Display Order 21-14
Changing the Point Size 3-35
Changing the Point Type 3-34
Changing the Viewpoint to View the 3D Models
    25-6
Changing the Viewpoint Using the Toolbar or
    the Ribbon 25-12
Changing the viewpoint using the view cube
    25-12
Changing the Viewpoint Using the Viewpoint
    Presets 25-7
Check 28-9
Checking Interference in Solids 26-30
Checking Spelling 7-51
CIRCLE command 2-19
Clean 28-9
Clean Screen 1-10
Clipping Existing Viewports 12-14
Clipping External References 18-22
Clipping Raster Images 21-10
Closing a Drawing 1-31
Coincident Constraint 11-4
Collecting Multiple Notes to Single Landing
    8-45
Collinear Constraint 11-5
Color edges 28-7
Color faces 28-6
Color Options for the Advanced
    Template 30-33

Column Settings dialog box 7-20
Columns 7-19
Combining Geometric Characteristics 8-50
Combining Solid Models 26-29
Command Window 1-5
Composite Position Tolerancing 8-51
Concentric Constraint 11-7
Concept of Blocks 16-2
Conceptual visual style 26-21
CONE command 26-5
Constraint Settings dialog box 11-9
Constraints to Blocks 16-57
Continued Dimensions 8-18
Controlling Attribute Visibility 17-38
Controlling Dimension Text Format 10-11
Controlling the Backgrounds and
    Shadows 26-27
Controlling the Display of Annotative Objects
    in Viewport 12-30
Controlling the Display of Faces 26-26
Controlling the Display of Layers 4-6
Controlling the Display of the Annotative
    Objects 7-5
Controlling the Display of the Hidden Lines in
    Viewport 12-13
Controlling the Display of the Objects in
    Viewport 12-13
Controlling the Layer Settings 4-16
Controlling the Layers in Viewports 12-17
Controlling the Rendering Environment 30-45
Controlling the Visibility of the 3D Face Edges
    25-28
Controlling the Visibility of the UCS Icon 24-2
Conventions in AutoCAD 25-5
Convert Distance Between Model Space and
    Paper Space 12-30
Converting a Dimension into a True Associative
    Dimension 8-33
Converting a Mesh Object into Solid 29-21
Converting an Existing Closed Object into a
    Viewport 12-10
Converting Entities into a Block 16-5
Converting Mesh Objects into Surfaces 29-23
Converting Objects to Solids 27-22
Converting Objects to Surfaces 27-21
Converting Surfaces to Solids 27-22

Converting Wireframe Models into Surface
    Models 25-25
CONVERTOLDLIGHTS command 30-30
CONVERTOLDMATERIALS command 30-37
Coordinate Systems 2-6
COPY command 5-9
Copy edges 28-7
Copy faces 28-5
COPYBASE command 5-10
Copying Objects using the base point 5-10
Copying Sketched Objects 5-9
Count 7-44
CPolygon 5-4
Create New Dimension Style dialog box 13-12
Create New Drawing dialog box 1-18
Create New File dialog box 13-7
Create Sheet Set dialog box 1-41
Creating 3D Faces 25-25
Creating 3D Polylines 25-24
Creating a Boundary Using Closed
    Loops 15-33
Creating a Helix 26-17
Creating a New Page Setup 14-16
Creating a New Table Style 7-36
Creating a New Workspace 1-48
Creating a Polysolid 26-14
Creating a Pyramid 26-13
Creating a Selection Set 5-2
Creating a Sheet Set 1-40
Creating a Solid Box 26-2
Creating a Solid Cone 26-5
Creating a Solid Cylinder 26-7
Creating a Solid Sphere 26-10
Creating a Solid Torus 26-11
Creating a Solid Wedge 26-13
Creating Aligned Dimensions 8-14
Creating and Inserting Annotative Blocks 16-15
Creating and Restoring Dimension Styles 10-2
Creating Angular Dimensions 8-20
Creating Animation by Defining the Path of
    Camera 30-59
Creating Animations 30-57
Creating Annotative Dimensions, Tolerances,
    Leader 8-56
Creating Annotative Hatch 15-23
Creating Annotative text 7-51
Creating Arc Length Dimensions 8-15

Creating Arrays in the 3D Space 27-15
Creating Baseline Dimensions 8-17
Creating Continued Dimensions 8-18
Creating Data Interchange (DXF) Files 21-2
Creating Diameter Dimensions 8-23
Creating Dimension Breaks 8-29
Creating Extruded Solids 26-37
Creating Fillets in the Isometric Drawings 23-9
Creating Fillets Using the Polyline Option 5-19
Creating Floating Viewports 12-8
Creating Jogged Dimensions 8-24
Creating Linear Dimensions 8-10
Creating Lofted Solids 26-44
Creating Multiline Text 7-11
Creating New Layers 4-4
Creating NURBS 19-16
Creating Ordinate Dimensions 8-26
Creating Planar Surfaces 25-37
Creating Polyface Meshes 25-26
Creating Polygonal Viewports 12-9
Creating Predefined Solid Primitives 26-2
Creating Presspull solids 26-50
Creating Radius Dimensions 8-25
Creating Regions 26-28
Creating Revision Clouds 19-14
Creating Revolved Solids 26-39
Creating Rotated Dimensions 8-16
Creating Swept Solids 26-41
Creating Template Drawings 13-2
Creating Text Styles 7-48
Creating the Cross-Sections of the Solids 27-26
Creating Tiled Viewports 12-3
Creating Title Sheet Table in a Sheet Set 7-54
Creating Views 6-43
Creating Wipeouts 19-15
Creation of Backup Files 1-30
Custom Hatch Patterns 15-9
Customer Involvement Program 1-52
Customizing a Drawing with Layout 13-14
Customizing Annotation Scale 7-3
Customizing Drawings According to Drawing
      Scale 13-21
Customizing Drawings According to Plot
      Size 13-21
Customizing Drawings with Dimensioning
      Specifications 13-9

Customizing Drawings with Layers 13-9
Customizing Drawings with Layers and
      Dimensioning 13-9
Customizing Drawings with Viewports 13-18
Cycling Through Groups 20-8
Cycling Through Snaps 4-48

D

Data Exchange 21-2
Data Exchange Formats 21-4
Data Link 7-32
DATAEXTRACTION command 17-18
DDEDIT command 7-29, 17-8
DDPTYPE command 3-34
DDVPOINT command 25-7
Default Light 30-11
Define Data Source 17-19
Defining a Section Plane by Drawing the
      Section Lines 28-14
Defining a Section Plane by Selecting a
      Face 28-11
Defining a Section Plane by Selecting a Preset
      Orthographic Views 28-15
Defining a Section Plane by Specifying Two
      Points 28-13
Defining Attributes 17-2
Defining Page Settings 12-24
Defining the New UCS 24-5
Definition Points 8-7
Delete faces 28-5
Deleting Unused Blocks 16-56
Demand Loading 18-25
Dependent Symbols 18-2
DesignCenter 4-28
Detaching an Xref 18-9
Detail Drawing 22-37
Determining Text Height 7-50
DGN Underlay tab 18-19
Diameter Dimensions 8-23
DIMBREAK command 8-29
DIMEDIT command 9-5
Dimension Line 8-3
Dimension Style Families 10-29
Dimension Style Manager dialog
      box 10-2, 13-13
Dimension Style Overrides 10-32

Dimension Text 8-3
Dimension Text and Arrowheads 10-16
Dimension Text Format 10-11
Dimension toolbar 8-8
Dimensional Constraints 11-17
Dimensioning 22-11
Dimensioning a Number of Objects
    Together 8-9
Dimensioning Components 22-12
Dimensioning Isometric Objects 23-9
Dimensioning Rules 22-13
DIMINSPECT command 8-31
DIMJOGLINE command 8-25
Direct Distance Entry 2-13
Direction Control dialog box 13-7
Display Commands 2-24
Display of Annotative Objects in
    Viewports 12-30
Display of Edges 26-24
Display of Faces 26-26
Display of Image Frames 21-14
Displaying Clipping Frame 18-24
DIST command 6-27
Distant Light 30-22
Dividing Sketched Objects 5-43
DONUT command 3-31
Download from source 7-46
Drafting Settings dialog box 4-28
Draw toolbar 1-14
Drawing Arcs 3-2
Drawing Area 1-5
Drawing Circles 2-19
Drawing Coordinates 1-6
Drawing Ellipses 3-17
Drawing Elliptical Arcs 3-20
Drawing Infinite Lines 3-36
Drawing Isometric Circles 23-8
Drawing Leaders 8-34
Drawing Lines in AutoCAD 2-2
Drawing Multileaders 8-39
Drawing Multilines 19-7
Drawing Polylines 3-26
Drawing Properties dialog box 6-33
Drawing Rectangles 3-11
Drawing Standard Surface Primitives 25-29
Drawing Status Bar 1-9
Drawing Status Information 6-30

Drawing Straight Lines Using the ORTHO
    Mode 4-33
Drawing Units dialog box 2-27, 13-7
Drawing XLINE 3-36
Dynamic Blocks 16-19
Dynamic Input 1-8
Dynamic Input Mode 1-36
Dynamic UCS 26-35
Dynamic Viewing of 3D Objects 25-41
Dynamically Rotating the View of the Model
    25-45

E

Edge Surface 29-7
Edit Attribute dialog box 17-15
Edit Block Definition dialog box 16-18
Edit Reference panel 18-27
Editing an XREF's Path 18-11
Editing Attribute Definition 17-8
Editing Attributes Using the Edit Attributes
    Dialog box 17-40
Editing Block Attributes 17-39
Editing Blocks 16-49
Editing Dimension Text 9-6
Editing Dimensions by Stretching 9-2
Editing Dimensions by Trimming and
    Extending 9-3
Editing Dimensions using the Properties Palette
    9-8
Editing Dimensions with Grips 9-8
Editing Hatch Boundary Using Grips 15-31
Editing Hatch Patterns 15-27
Editing Multilines Using Grips 19-8
Editing Multiple Polylines 20-30
Editing Plotter Configuration 14-13
Editing Polylines 20-15
Editing Raster Image Files 21-10
Editing References in Place 18-26
Editing Sketches 5-8
Editing Solid Models 28-2
Editing Splines 19-18
Editing Text 7-28
Editing Text Using the Properties Palette 7-29
Editing Text Using the Properties Palette 7-29
Editing the Surface Mesh 25-38
Editing Viewports 12-12

Editing with Grips 6-2
Editor Settings 7-23
Elementary Rendering 30-8
Enable Orbit Auto Target 25-49
Enable Pointer Input 1-36
Enhanced Attribute Editor dialog box 17-40
Entering Special Characters 7-11
Equal Constraint 11-8
Exploding Compound Objects 20-14
Extending Sketched Objects 5-27
Extension Lines 8-4
External References 18-2
External References dialog box 21-9
External References Palette 18-5
Externally Referenced Dimension Styles 10-34
Extracting Attributes 17-18
Extracting Edges of Solid Model 27-20
EXTRUDE command 26-37
Extrude faces 28-2
Extruding the Mesh Faces 29-17

F

F (Face) Option 24-6
Field 7-43
FILLET command 27-2
Filleting Objects with a Different UCS 5-20
Filleting Sketches 5-17
Filleting Solid Models 27-2
Fillets Using the Multiple Option 5-20
Find and Replace 7-22
Finding and Replacing Text 7-54
Fix Constraint 11-5
FLATSHOT Command 28-22
Flipping Dimension Arrow 9-5
FOG command 30-46
Formatting the Tolerances 10-25
Formula 7-43
Freeze Option 12-16
Full Section 22-23
Fully Defined Sketch 11-1, 11-19
Fully-defined Sketch 11-19
Function and Control Keys 4-52
Fundamental Dimensioning Terms 8-3

G

Generating a Section by Defining a Section
    Plane 28-11
Generating Center Marks and Centerlines 8-26
Generating the Drawing Views of the Solid
    Model 28-19
Geographic Location dialog box 30-27
Geometric Characteristics and Symbols 8-47
Geometric Dimensioning and Tolerancing 8-46
Geometric Tolerance 8-48
Global and Current Linetype Scaling 4-26
Global Editing of Attributes 17-43
Grid Display 1-7
Grip Settings 6-3
GROUP command 20-2

H

Half Section 22-24
Hatch and Gradient Dialog Box Options 15-5
Hatch Boundary 15-3
Hatch Edit dialog box 15-29
Hatch Patterns 15-2
HATCHEDIT command 15-27
Hatching 15-2
Hatching Around Text, Dimensions, and
    Attributes 15-26
Hatching Blocks and Xref Drawings 15-33
Hatching Drawings Using the Hatch and
    Gradient Dialog box 15-3
Hatching the Drawing Using the Tool Palettes
    15-23
HELIX command 26-17
Help Menu 1-49
History Pan 30-53
Horizontal Constraint 11-2
Hotspot/Falloff 30-21

I

ID command 6-28
Identifying the Location of a Point on the Screen
    6-28
Ignoring Hatch Pattern Entities While Snapping
    15-29
IMAGEATTACH command 21-7
Importing a Page Setup 14-18

Importing Layouts to Sheet Sets 12-23
Importing PCP/PC2 Configuration Files 14-15
Imprint 28-8
In-place Editing of Blocks with
     Attributes 17-49
InfoCenter Bar 1-52
Initial Setup wizard 1-3
Insert a Block in a Table Cell dialog box 7-43
Insert Formula Column dialog box 17-28
Insert Table dialog box 7-31
Inserting Blocks 16-8
Inserting Blocks Using the DesignCenter 16-37
Inserting Blocks with Attributes 17-9
Inserting Multiple Blocks 16-45
Inserting Table in the Drawing 7-30
Inserting Text Files in the Drawing 17-50
Insertion Base Point 16-49
INTERFERE command 26-30
INTERSECT command 26-29
Invoking Commands in AutoCAD 1-11
Isolating and Unisolating Layers 4-15
Isometric Axes and Planes 23-3
Isometric Circles 23-8
Isometric Drawings 23-2
Isometric Grip and Snap 23-3
Isometric Projections 23-2
Isometric Text 23-11

J

Jogged Dimensions 8-24
Joining Sketched Objects 5-44
Joining Two Adjacent Viewports 12-6
Justification of the Text 7-30

L

Layer Filters Properties dialog box 4-12
Layer Properties Manager dialog
     box 4-4, 13-14
Layer States 4-13
Layers, Colors, Linetypes, and Lineweights of
     Blocks 16-42
Layers panel 4-3
LAYOUT command 12-21
LAYOUTWIZARD command 12-24
Leader 8-4
Leader Settings dialog box 8-34

Lengthening Sketched Objects 5-31
LIGHT command 30-10
Light panel of the Ribbon 30-28
LIGHTINGUNITS system variable 30-10
Limits 2-32, 8-5
LINE command 2-2
Line Spacing 7-17
Linear Dimensions 8-10
LINETYPE command 4-21
Linetype Control list box 4-23
Linetype Manager dialog box 4-20
Lineweight dialog box 4-11
Lineweight Settings dialog box 4-22
Link Cell 7-46
Link External Data dialog box 17-24
Links dialog box 21-22
Listing Information 6-29
Loading a Template Drawing 13-8
Loading Hyperlinks 6-13
Lock or Unlock a Layer 4-7
Locking the Display in Viewports 12-13
LOFT command 26-44
Loft Settings dialog box 26-47
Look tool 25-42
LTSCALE Factor for Plotting 4-27
LWEIGHT command 4-22

M

Maintaining Equal spacing between
     Dimensions 8-28
Make a Layer Plottable or Nonplottable 4-7
Making a Layer Current 4-5
Making a Viewport Current 12-5
Manage Cell Content 7-45
Manage Cell Styles dialog box 7-40
Manage Xrefs 1-11
Managing Attributes 17-11
Managing External References 18-4
Managing Raster Images 21-9
Managing Render Presets 30-44
Managing the Display of Columns 4-11
Managing the UCS Through the dialog
     box 24-17
Manipulating the Visibility of Viewport Layers
     12-15
Mapping flyout 30-49

Mapping Materials on Objects 30-48
Marble dialog box 30-34
MASSPROP command 28-26
Match Cells 7-46
Matching Properties of Sketched Objects 6-15
MATCHPROP command 6-15
Material Attachment Options dialog box 30-6
Material Condition Modifier 8-50
MATERIALATTACH command 30-7
Materials - Tool Palettes 30-3
Maximizing Viewports 12-15
Measuring Sketched Objects 5-42
Measuring the Distance Between Two
        Points 6-27
Menu Bar 1-14
Mesh Primitive Options dialog box 29-5
Mesh Primitives 29-2
Mirroring Sketched Objects 5-38
Mirroring Solid Models in 3D Space 27-8
MLEADERALIGN command 8-42
MLEADERCOLLECT command 8-45
MLEADEREDIT command 8-41
MLSTYLE command 19-2
Model Space and Paper Space
        Dimensioning 9-11
Model Space and Paper Space/Layouts 12-2
Model Space Viewports (Tiled Viewports) 12-3
Modes of SteeringWheel 25-43
Modify Multileader Style dialog box 10-39
Modifying Tables 7-40
Modifying a Page Setup 14-18
Modifying AutoCAD Settings Using the
        Options dialog box 2-41
Modifying Existing Blocks in the Tool Palettes
        16-41
Modifying lights 30-30
Modifying Material 30-37
Modifying Tables 7-40
Modifying the Justification of the Text 7-30
Modifying the Properties of the Predefined
        Pattern 15-25
Modifying the Scale of the Text 7-35
Modifying the Workspace Settings 1-48
Motion Path Animation dialog box 30-58
MOVE command 5-8
Move faces 28-3
Move Gizmo 29-24

Moving Models in 3D Space 27-14
Moving Sketched Objects 5-8
MTEXT command 7-11
Multileader Style Manager dialog box 10-35
Multileaders 8-38
Multiline Edit Tools dialog box 19-9
Multiline Style dialog box 19-2
Multiline Text 7-11
Multiple Copies 5-9
Multiview Drawings 22-2

N

NA (NAmed) Option 24-11
Nesting of Blocks 16-43
New Dimension Style dialog box 10-4
New Features Workshop 1-51
New Multiline Style dialog box 19-3
New Parameter dialog box 16-59
New Table Style 7-36
New Table Style dialog box 7-37
New UCS 24-5
New View dialog box 6-44
New View/Shot Properties dialog box 6-45
New VP Freeze 4-6
Newfrz (New Freeze) Option 12-17

O

OB (OBject) Option 24-6
Object Grouping dialog box 20-2
Object Linking and Embedding (OLE) 21-17
Object Properties 4-23
Object Selection Methods 2-17
Object Snap 1-7, 4-34
Object Snap toolbar 4-34
Object Snap Tracking 1-7
Obtaining Rendering Information 30-54
OFFSET command 5-11
Offset faces 28-4
Offseting Sketched Objects 5-11
Open a Drawing 1-19
Opening an Existing Drawing Using the Select
        File 1-32
Opening an Xreffed Object in a Separate
        Window 18-19
Options dialog box (Files tab) 2-43
Orbit 25-42

Ordinate Dimensions 8-26
Ortho Mode 1-7, 4-33
Orthographic Projections 22-2
Other Features of Hatching 15-35
Other Navigation Modes 25-46
Over-defined Sketch 11-19
Overlay Option 18-12

P

P (Previous) Option 24-11
Page Setup Manager dialog box 14-16
PAGESETUP command 12-24
Pagesetup command 14-15
Pan 1-8, 25-42
PAN command 6-42
Panning Drawings 6-42
Paper Space Linetype Scaling (PSLTSCALE
        System Variable 12-19
Paper space viewports (Floating
        Viewports) 12-7
Paragraph 7-15
Paragraph dialog box 7-17
Parallel Constraint 11-6
Parameters Manager palette 11-27
Parameters tab of the Block Authoring Palettes
        16-21
Paste Special dialog box 21-23
PASTEBLOCK command 5-11
PASTEORIG command 5-11
Pasting contents using the original coordinates
        5-11
PEDIT command 20-16, 25-38
Perpendicular Constraint 11-6
Photometry 30-13
Placing Views on a Sheet of a Sheet Set 6-47
PLANESURF command 25-37
Playing the Animation 28-31
PLINE command 3-26
PLOT command 14-2
Plot dialog box 14-2
Plot Style Table Editor 14-22, 14-25
Plot/Publish Details Report Available 1-11
PLOTTERMANAGER command 14-11
Plotting Drawings 2-36
Plotting Drawings in AutoCAD 14-2
Plotting Drawings Using the Plot Dialog Box
        14-2

Plotting Rendered Images 30-56
Plotting Sheets in a Sheet Set 14-29
POINT command 3-34
Point Light 30-12
Point Type 3-34
Point/Path Name dialog box 30-59
Polar Tracking 1-7, 4-51
POLYGON command 3-23
POLYSOLID command 26-14
Position Locator Window 25-49
Positioning Orthographic Views 22-7
PostScript Files 21-15
PRESSPULL command 26-50
Primary Dimension Units 10-19
Projected Tolerance Zone 8-53
PROPERTIES command 4-24, 20-8
Properties Palette 4-24
Properties panel 4-23
PSETUPIN command 14-18

Q

QLEADER command 8-34
QSAVE command 1-26
QSELECT command 6-16
Quick Properties 1-8
Quick Select dialog box 6-17
Quick View Drawings 1-8, 12-2
Quick View Layouts 1-8, 12-2
Quitting AutoCAD 1-35

R

Radius Dimensions 8-25
Raster Images 21-6
RAY command 3-39
Realistic Visual Style 26-21
Realtime Zooming 6-34
Reconciling New Layers 4-15
Recording your drawing steps Using Action
        Recorder 28-28
Redefining a Block With Attributes 17-49
REDRAW command 6-32
Refedit toolbar 16-52
Reference Edit dialog box 16-50
Reference toolbar 18-4
Refining the Mesh 29-14
REGEN command 6-33

REGION command 26-28
Relative Coordinate System 2-9, 25-18
Relative Cylindrical Coordinate System 25-18
Relative Rectangular Coordinate System 25-18
Relative Rectangular Coordinates 2-9
Relative Spherical Coordinate System 25-18
Reload Linetypes dialog box 4-9
Removing Leaders from the Multileader 8-41
Removing the Dimension Associativity 8-33
Removing Unused Named Objects 20-39
Removing Viewport Overrides 12-19
Renaming Blocks 16-55
Renaming Named Objects 20-37
RENDER command 30-8
Render Environment dialog box 30-46
Render Panel of the Ribbon 30-41
Render Presets Manager dialog box 30-45
Render Window 30-8
Rendered Shadow Details 30-26
Rendering a Model with Different Render
      Settings 30-51
Rendering Quality 30-37
Rendering with Fog 30-46
Resaving All the Sheets in a Sheet Set 1-46
Reset Option 12-17
REVCLOUD command 19-14
Reversing the Undo Operation 20-37
REVOLVE command 26-39
REVSURF command 29-5
REWIND 25-42
Ribbon 1-12
RMAT command 30-4
Rotate faces 28-4, 28-5
Rotate Gizmo 29-25
Rotated Dimensions 8-16
Rotating Sketched Objects 5-13
Rotating solid Models about an axis 27-7
Rotating Solid Models in 3D Space 27-4
RULESURF command 29-7
Running Object Snap Mode 4-46

S

Save Block As dialog box 16-21
Save Drawing As dialog box 1-26
Saving a Rendering 30-55
Saving the Rendered Image from Render
      Window 30-56

Saving the Rendered Image to a File 30-55
Saving the Viewport Rendering 30-55
Saving Your Work 1-26
Scale Gizmo 29-25
Scale of the Text 7-29
Scaling Sketched Objects 5-16
SECTION command 27-26
Sectional Views 22-23
SECTIONPLANE command 28-11
Select Linetype dialog box 4-9
Select Plot Style dialog box 4-10
Select Template dialog box 1-18
Selecting a Material for a Design 30-2
Selecting Dimensioning Commands 8-8
Selecting Groups 20-7
Selecting Hatch Boundary 15-10
Selection tab of the Options dialog box 6-4
Selective Display of Layers 4-11
Separate solids 28-8
Setting a Table Style Current 7-40
Setting Limits 2-32
Setting Selection Modes Using the Options
      dialog box 20-41
Setting Snap 4-31
Setting the Current Plot Style 14-27
Setting the Isometric Grid and Snap 23-3
Setting the Plot Parameters 14-15
Setting the UCS to Preset Orthographic UCSs
      24-18
Setting Thickness and Elevation for the New
      Object 25-22
Setting Units 2-26
Sheet Set Manager 1-42
Shell 28-8
Shortcut Menu 1-15
Show/Hide Lineweight 1-8
ShowMotion 1-9
Single Line Text 7-6
Sky Properties 30-23
Sky Properties Palette 30-23
Slicing Solid Models 27-23
Smooth Constraint 11-8
Snap Mode 1-7
SOLDRAW command 28-22
SOLIDEDIT command 28-2
SOLPROF command 28-22
SOLVIEW command 28-19

SPACETRANS command 12-30
Special Characters 7-11
SPHERE command 26-10
SPLINE command 19-16
SPLINEDIT command 19-18
Splitting the Mesh Faces 29-17
Spotlight 30-20
Stack Properties 7-24
Standard Template Drawings 13-2
Starting a New Drawing 1-18
Starting AutoCAD 1-2
Starting Three Dimensional (3D) Modeling
    environment 25-2
Status Bar Menu 1-10
SteeringWheel 1-9
Stretching Sketched Objects 5-30
STYLESMANAGER command 14-20
Subset Properties dialog box 1-44
Substituting Fonts 7-47
SUBTRACT command 26-30
Sum 7-44
Sun Angle Calculator 30-25
Sun Light 30-23
Suppressing the Hidden Edges 25-24
SWEEP command 26-42
Symmetric Constraint 11-7
System Variables 24-21

T

TABLE command 7-31
Table in the Drawing 7-30
TABSURF command 29-9
Tangent Constraint 11-7
Taper faces 28-5
Template Drawings 13-2
Temporary Model Space 12-10
Test Block 16-20
TEXT command 3-39
Text Height 7-50
Text Mirroring 5-44
Text Quality and Text Fill 7-53
Text Styles 7-48
Thaw Option 12-16
Thickness And Elevation 25-22
Time-Related Information 6-30
Title Sheet Table in a Sheet Set 7-54

Tolerances 8-5, 10-25
Tool Palettes 1-13
Tool Palettes - All Palettes 30-4
Toolbar 1-15
Toolbar/Window Positions Unlocked 1-9
TORUS command 26-11
Transparency of an Image 21-13
Trim, Extend, and Fillet Commands in 3D
    25-21
Trimming Sketched Objects 5-24
Trimming the Hatch Patterns 15-32
True Associative Dimensions 8-33
Turn a Layer On or Off 4-6
Types of 3D models 25-3
Types of Grips 6-2

U

UCS command 24-5
UCS dialog box 24-19
UCSICON command 24-2
Under-defined Sketch 11-19
Underlay Layers dialog box 18-19
Underlays 18-17
Understanding Advanced Rendering 30-37
Understanding Elementary Rendering 30-8
Understanding the Concept of Rendering 30-2
Understanding the Concept of Sheet Sets 1-39
Understanding the Use of Multilines 19-2
Understanding X, Y, and Z Axes 22-2
Undoing Previous Commands 20-32
UNION command 26-29
Unisolating Layers 4-15
Unloading an Xref 18-10
Unloading AutoCAD Render 30-57
Unreconciled New Layers 1-11
Up/Down 25-43
Updating an Xref 18-10
Updating Dimensions 9-7
Use of Three-dimensional drawing 25-2
User defined Hatch Patterns 15-8
Using AutoCAD Editing Commands 15-32
Using AutoTracking 4-50
Using Feature Control Frames with Leaders
    8-52
Using Show Motion for Presentation Purposes
    28-30

Using the Drawing Recovery Manager to Recover File 1-30
Using the MVSETUP Command on the Model Tab 12-26
Using the PROPERTIES Command 15-29
Using the SteeringWheels 25-41
Using Tool Palettes to Insert Blocks 16-38

V

V (View) Option 24-8
V8 DGN File 21-5
Validate Digital Signatures 1-11
Vertical Constraint 11-3
View Manager dialog box 6-43, 6-45
View toolbar 6-42
Viewpoint Presets dialog box 25-7
Viewports dialog box 13-19
Visibility States dialog box 16-32
Visual Styles Manager palette 26-21
VP Color 12-18
VP Freeze 12-18
VP Linetype 12-18
VP Lineweight 12-18
VP Plot Style 12-18
VPCLIP command 12-14
VPLAYER command 12-15
VPOINT command 25-9
VPORTS command 12-8, 12-9, 12-10
Vpvisdflt (Viewport Visibility Default) Option 12-17

W

W (World) Option 24-6
Walk tool 25-43
What is Solid Modeling 26-2
WIPEOUT command 19-15
Working with Layers 4-3
Working with Object Snaps 4-34
Working with OLE Objects 21-25
Working with Page Setups 14-15
Working with the DESIGNCENTER 4-28
Working with the MVSETUP Command 12-26
World Coordinate System (WCS) 24-2
Write Block dialog box 16-46
Writing a Single Line Text 3-39, 7-6

X

X/Y/Z Options 24-8
XATTACH command 18-16
Xbind dialog box 18-21
XCLIP command 18-22
XEDGES command 27-20
XPLODE command 16-53
XREF 18-4

Z

ZA (ZAxis) Option 24-10
Zoom 1-9, 25-42
ZOOM command 6-33
Zooming Drawings 6-33